MW00988944

AC/DC

PRINCIPLES AND APPLICATIONS

Second Edition

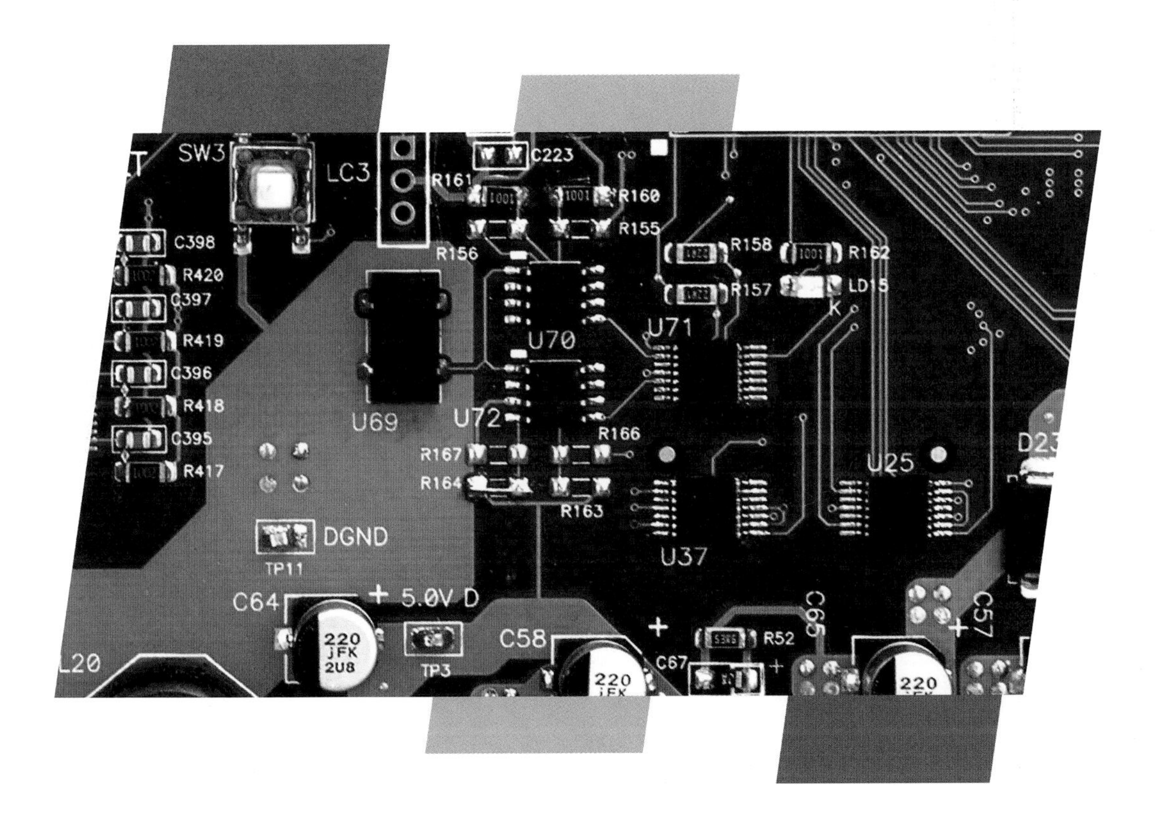

AMERICAN TECHNICAL PUBLISHERS
Orland Park, Illinois

Paul T. Shultz

American Technical Publishers Editorial Staff

Editor in Chief:
Jonathan F. Gosse
Vice President—Production:
Peter A. Zurlis
Assistant Production Manager:
Nicole D. Bigos
Digital Media Coordinator:
Adam T. Schuldt
Art Supervisor:
Sarah E. Kaducak
Supervising Copy Editor:
Catherine A. Mini
Technical Editor:
James T. Gresens
Copy Editor:
Talia J. Lambarki
Cover Design:
Bethany J. Fisher
Illustration/Layout:
Bethany J. Fisher
Nicholas W. Basham
Thomas E. Zabinski
Digital Resources:
Robert E. Stickley
Cory S. Butler

2 3 4 5 6 7 8 9 – 16 – 9 8 7 6 5 4

Printed in the United States of America

ISBN 978-0-8269-1357-9
eISBN 978-0-8269-9501-8

 This book is printed on recycled paper.

Dedication

In loving memory of my father, George Patrick Shultz

Preface

This book started with a manuscript written by my father, George Patrick Shultz, a seasoned educator and experienced author. The intended product was a two-volume series covering both AC and DC topics. After completing a 900-page DC theory manuscript, it became clear that a book that included both AC and DC topics would have to be edited significantly to be a size that would be most useful in the field. However, my father found it extremely difficult to edit his own work as he labored over the merit of each word, picture, and equation.

To expedite the process, he asked me to edit the manuscript while continuing his work. But as with many projects, writing, editing, and production are slow, time-consuming processes. Unfortunately, my father passed away without seeing his efforts in final form. Before he died he asked me to complete the book. The manuscript was completed with the content restructured to incorporate pertinent AC topics and the latest technology.

My father spent his entire adult life educating tradesmen, high school students, and those with special needs. His dedication to education enabled many people to live productive, successful lives. I believe my father would have enjoyed this book, and I hope that it will continue his legacy of helping people learn.

Paul T. Shultz

Acknowledgements

The author and publisher are grateful to the following companies, organizations, and individuals for providing photographs, information, and technical assistance:

ABB Inc., Drives & Power Electronics
ASTM International
Bacharach Inc.
Baldor Electric Co.
British Standards Institute (BSI)
Cummins Power Generation
Fluke Corporation
GE Motors & Industrial Systems
GE Thermometrics
Gould Inc.
Henkel Corporation
Honeywell
Ideal Industries, Inc.
Industrial Scientific Corporation
International Manufacturing Services, Inc.
March Manufacturing, Inc.
Megger Group Limited
Milwaukee Electric Tool Corporation
MTE Corporation
NREL
Omron Electronics, Inc.
Panduit Corp.
Siemens
Snorkel
Square D/Schneider Electric
Texas Instruments
Verband Deutscher Elektrotechniker Testing and Certification Institute (VDE)
U.S. Air Force
U.S. Navy

Contents

Introduction

AC/DC Principles and Applications is a comprehensive textbook that discusses electrical and electronic theory and its application to alternating- and direct-current circuits. AC and DC series and parallel circuits are covered individually to distinguish differences between the various types of circuits. Each chapter includes electrical, electronic, and mathematical concepts and applications that address the skill-set needs of technicians and equipment designers and tie key concepts and applications to real-world systems and equipment.

This new edition includes electrical theory, builds on circuit fundamentals, and reinforces comprehension through examples of electronics applications. Special emphasis is placed on the development of AC and DC circuit design skills.

The author of the book, Paul T. Shultz, has over 25 years of broad electrical engineering experience and currently serves as Director of Engineering for Chesapeake Sciences Corporation. In his current position, he directs development efforts for both commercial and military end users using his background in electrical and optical high-resolution data acquisition sensor systems, network-based telemetry systems, optical mass-storage systems, and multiprocessor digital-processing systems. Paul has received many awards and has written and presented numerous technical papers. His extensive hands-on knowledge and expertise are reflected throughout this textbook.

Features

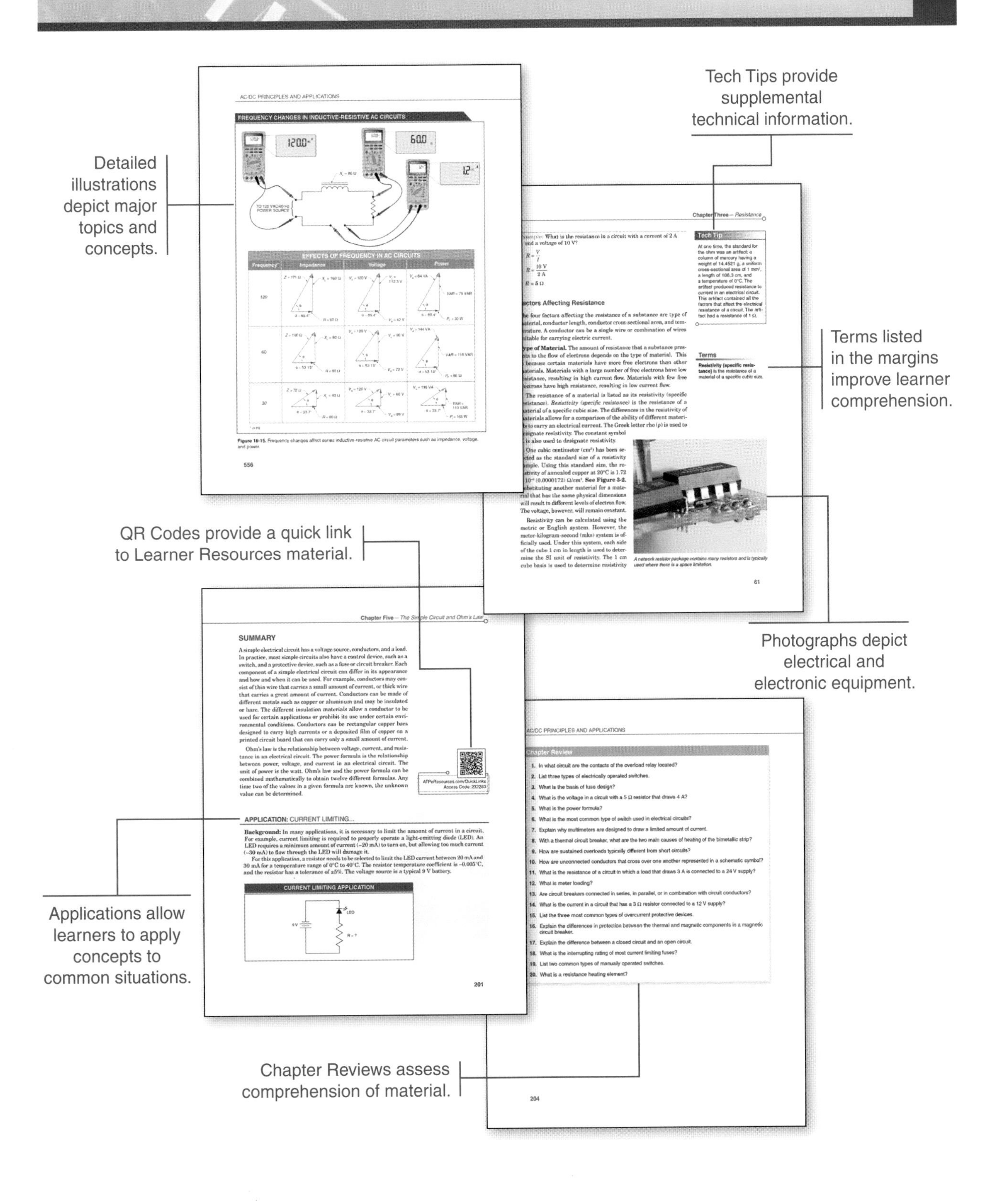
Detailed illustrations depict major topics and concepts.
Tech Tips provide supplemental technical information.
Terms listed in the margins improve learner comprehension.
QR Codes provide a quick link to Learner Resources material.
Photographs depict electrical and electronic equipment.
Applications allow learners to apply concepts to common situations.
Chapter Reviews assess comprehension of material.

Learner Resources

AC/DC Principles and Applications also includes access to Learner Resources that reinforce textbook content and enhance learning. These online resources can be accessed using either of the following methods:

- Key ATPeResources.com/QuickLinks into a web browser and enter QuickLinks™ access code 232263.
- Use a Quick Response (QR) reader app to scan the QR Code with a mobile device.

The Learner Resources include the following:

- Quick Quizzes™ that provide interactive questions for each chapter, with embedded links to textbook references and the Illustrated Glossary
- An Illustrated Glossary that serves as a helpful reference to commonly used terms, with selected terms linked to textbook illustrations
- Circuit Fundamentals, which provides a link to Multisim 9 Student Demo software and circuit activities and calculations
- Flash Cards that provide a self-study/review of common electrical and electronic terms and their definitions
- A Media Library that consists of videos and animations that reinforce textbook content
- ATPeResources.com, which provides access to additional online resources that support continued learning

To obtain information on other related training material including the eTextbook for this title, visit the American Technical Publishers website at www.atplearning.com.

The Publisher

1

Basic Concepts of Electricity

OBJECTIVES

- Explain how the basic concepts of matter and its subatomic structure apply to electricity.
- Define and understand electricity.
- Explain the laws of electric charges and how they apply to electrification.
- Describe the basic units of measure and how they are used with electricity.
- List and describe derived units of measure.
- Explain scientific notation.

The discovery of electricity was one of humanity's most significant discoveries. The science of electricity was vital to many technological innovations that allowed the conveniences we experience today. The importance of electricity is quickly realized when electrical power is interrupted.

Knowledge of fundamental concepts is needed to begin the study of electricity. Because the study of electricity is a science, it necessary to measure its characteristics using standards and units that are recognized across scientific fields. The units of measure used in describing electricity are derived from seven basic units and units derived from these basic units. Many electrical quantities are either very small or very large. To express these extreme values, scientific notation is used.

To understand the basic concepts of electricity, the building blocks of matter and its subatomic structure must be understood. Subatomic particles are used to explain the nature of electricity.

Terms

Matter is anything that has mass and occupies space. All objects consist of matter.

A **solid** is matter that has a definite volume and shape.

A **liquid** is matter that has a definite volume but not a definite shape.

A **gas** is matter that has no definite volume or definite shape.

MATTER

Matter is anything that has mass and occupies space. All objects consist of matter. Matter can exist in a solid, liquid, or gaseous state. **See Figure 1-1.** A *solid* is matter that has a definite volume and shape. A *liquid* is matter that has a definite volume but not a definite shape. A *gas* is matter that has no definite volume or definite shape.

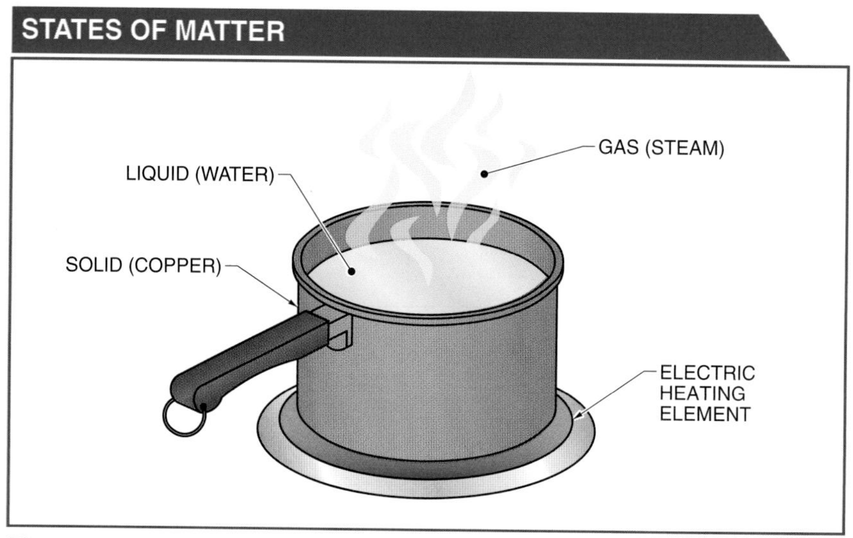

Figure 1-1. Matter can exist in a solid, liquid, or gaseous state.

Terms

A **conductor** is a material that has low electrical resistance and permits electrons to move through it easily.

Metal, wood, and plastic are examples of matter in a solid state. Water, milk, and gasoline are examples of matter in a liquid state. Hydrogen, oxygen, and carbon dioxide are examples of matter in a gaseous state. Some matter, such as water, can exist in the solid, liquid, or gaseous state. The state that water exists in is changed by the addition or removal of heat.

All matter has some electrical properties. The electrical behavior of matter varies according to the physical makeup of the matter. Some matter, such as copper, allows electricity to easily move through it and can act as a conductor. A *conductor* is a material that has low electrical resistance and permits electrons to move through it easily. Most metals are good conductors. Copper is the most commonly used conductor. Silver is a better conductor than copper, but is too expensive for most applications. Aluminum is not as good a conductor as copper, but costs less and is lighter. This enables its use in high-voltage applications such as overhead power lines.

Conductors are available as wire, cord, or cable and may be bare, insulated, or covered. Bare conductors contain no covering or insulation. Insulated conductors are the most common conductors used in an electrical system. Insulated conductors may be solid or stranded. Covered conductors have a covering that are rated for continuous operation at 167°F to 194°F (75°C to 90°C). The *National Electrical Code® (NEC®)* is a standard on practices for the design and installation of electrical products. Coverings for conductors are specified in Article 310 of the National Electrical Code® (NEC®). **See Figure 1-2.**

Terms

The **National Electrical Code® (NEC®)** is a standard on practices for the design and installation of electrical products.

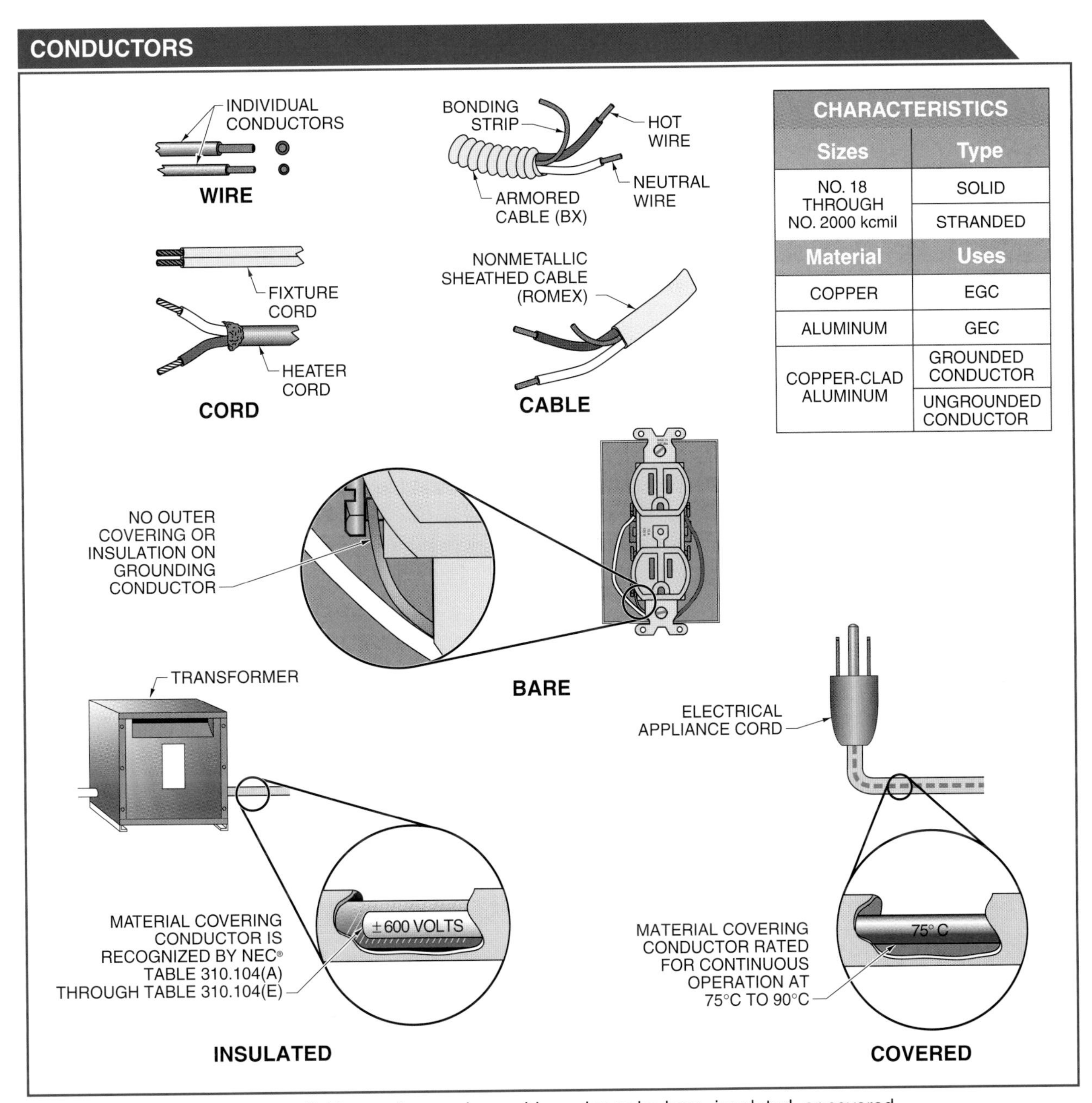

Figure 1-2. Conductors are available as wire, cord, or cable and may be bare, insulated, or covered.

A bare copper conductor is typically used as a grounding wire to dissipate unwanted electrical charges to earth or ground.

Factors that affect the resistance of conductors are the conductor cross-sectional area, length, material, and temperature. A conductor with a large cross-sectional area has less resistance than a conductor with a small cross-sectional area. The longer the conductor, the greater the resistance. Short conductors have less resistance than long conductors of the same cross-sectional area. Copper is a better conductor than aluminum, and may carry more current for a given size. Per the NEC®, all conductors must be copper unless otherwise specified. Temperature also affects resistance in a conductor. The higher the conductor temperature, the greater the resistance.

Terms

An **insulator** is a material that has a very high resistance to the flow of electrons.

A **semiconductor** is a material that exhibits electrical conductivity between that of a conductor (high conductivity) and that of an insulator (low conductivity).

Some matter, such as rubber, plastic, air, glass, and paper, does not allow electricity to move through it easily and can act as an insulator. An *insulator* is a material that has a very high resistance to the flow of electrons. Materials used for insulation must have a very high resistance. The resistance of insulation decreases when degraded by moisture and/or damaged by overheating.

All electrical conductors must be protected against possible contact with other conductors, metal parts, and people. Conductor insulation protects the conductor from damage and isolates the electrical power contained within the conductor. However, not all energized parts of an electrical circuit are protected by insulation. When energized parts of an electrical circuit are exposed, such as where conductors are terminated at fuse or circuit breaker panels, the distance (air gap) is used as the insulator. The greater the distance between energized electrical conductors or parts, the greater the resistance. The higher the voltage, the greater the air gap that is required to create a high enough resistance to prevent undesired electron flow, such as with deadly arc flashes.

Conductor insulation is classified by temperature ratings as 140°F (60°C), 167°F (75°C), and 194°F (90°C). **See Figure 1-3.** The insulation required depends on the application. Common types of residential wiring categorized by the NEC® are underground feeder (UF) and nonmetallic-sheathed cable (NM and NMC). In high-voltage applications, conductor insulation properties must be greater than in low-voltage applications.

A *semiconductor* is a material that exhibits electrical conductivity between that of a conductor (high conductivity) and that of an insulator (low conductivity). Semiconductors include carbon, germanium, and silicon. Semiconductor materials do not conduct electricity easily and are not good insulators. The properties of different matter must be understood when designing electrical components and circuits, working with or near electrical equipment, and troubleshooting electrical circuits.

CONDUCTOR INSULATION TEMPERATURE RATING

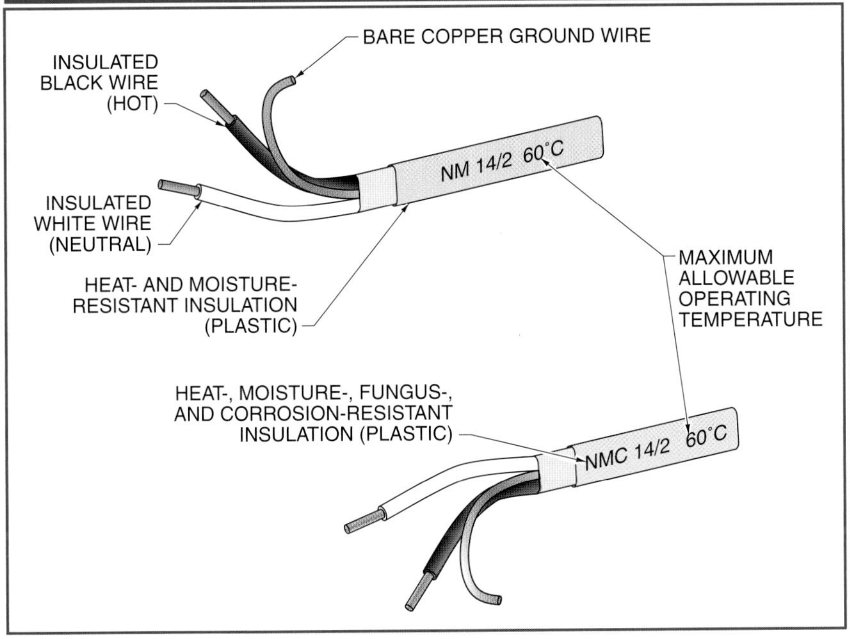

Figure 1-3. Conductor insulation is classified by temperature ratings marked directly on the insulation.

Atoms

An *atom* is the smallest particle that an element can be reduced to and still maintain the properties of that element. The three principle parts of an atom are the electron, the neutron, and the proton.

The *electron* is the negatively charged particle of an atom. The *neutron* is the neutral particle with a mass approximately the same as a proton and exists in the nucleus of an atom. The *proton* is the positively charged particle in the nucleus of an atom. Every atom has a definite number of protons, neutrons, and electrons. **See Figure 1-4.** The proton of an atom has a positive (+) charge, the electron has a negative (–) charge, and a neutron has no charge. Protons and neutrons combine to form the nucleus of an atom. Since the neutron has no charge, the nucleus has a positive charge because of the protons. The number of protons in an atom determines the atom's weight (mass) and its atomic number. For example, a hydrogen (H) atom has the fewest protons of all atoms, has the least amount of mass, and is assigned the atomic number of one. Of conductive materials, gold (Ag) has the most protons and the greatest atomic mass, and is assigned an atomic number of 79, while aluminum (Al) has an atomic number of 13, copper (Cu) an atomic number of 29, and silver (Ag) an atomic number of 47.

Terms

An **atom** is the smallest particle that an element can be reduced to and still maintain the properties of that element.

The **electron** is the negatively charged particle of an atom.

The **neutron** is the neutral particle with a mass approximately the same as a proton and exists in the nucleus of an atom.

The **proton** is the positively charged particle in the nucleus of an atom.

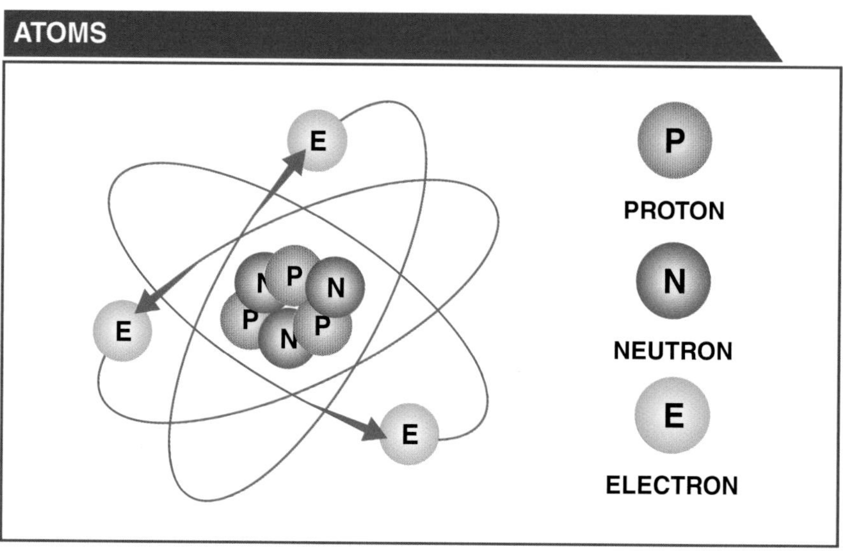

Figure 1-4. All atoms have a definite number of protons, neutrons, and electrons.

A positively charged atom or ion is produced when there are fewer electrons than protons. A negatively charged atom or ion is produced when there are more electrons than protons. The electrons orbit the nucleus of an atom, completing billions of trips around the nucleus each millionth of a second. In all atoms, the electrons are arranged in shells at various distances from the nucleus according to the amount of energy they have. The total number of shells an atom has varies from one to seven. The shells are numbered innermost to outermost, 1, 2, 3, 4, 5, 6, and 7 or lettered, K, L, M, N, O, P, and Q. **See Figure 1-5.**

Conductor terminals are typically composed of copper or aluminum because these materials have good electrical properties.

Each shell can hold only a specific number of electrons. The innermost shell can hold two electrons, the second shell can hold eight electrons, the third shell can hold 18 electrons, the fourth shell can hold 32 electrons, the fifth shell can hold 50 electrons, the sixth shell can hold 72 electrons, and the seventh shell can hold 98 electrons. The shells are filled starting with the inner shells and working outward, so that when the inner shell is filled with as many electrons as it can hold, the next shell is started. Most materials used in electrical/electronic applications contain four shells or less.

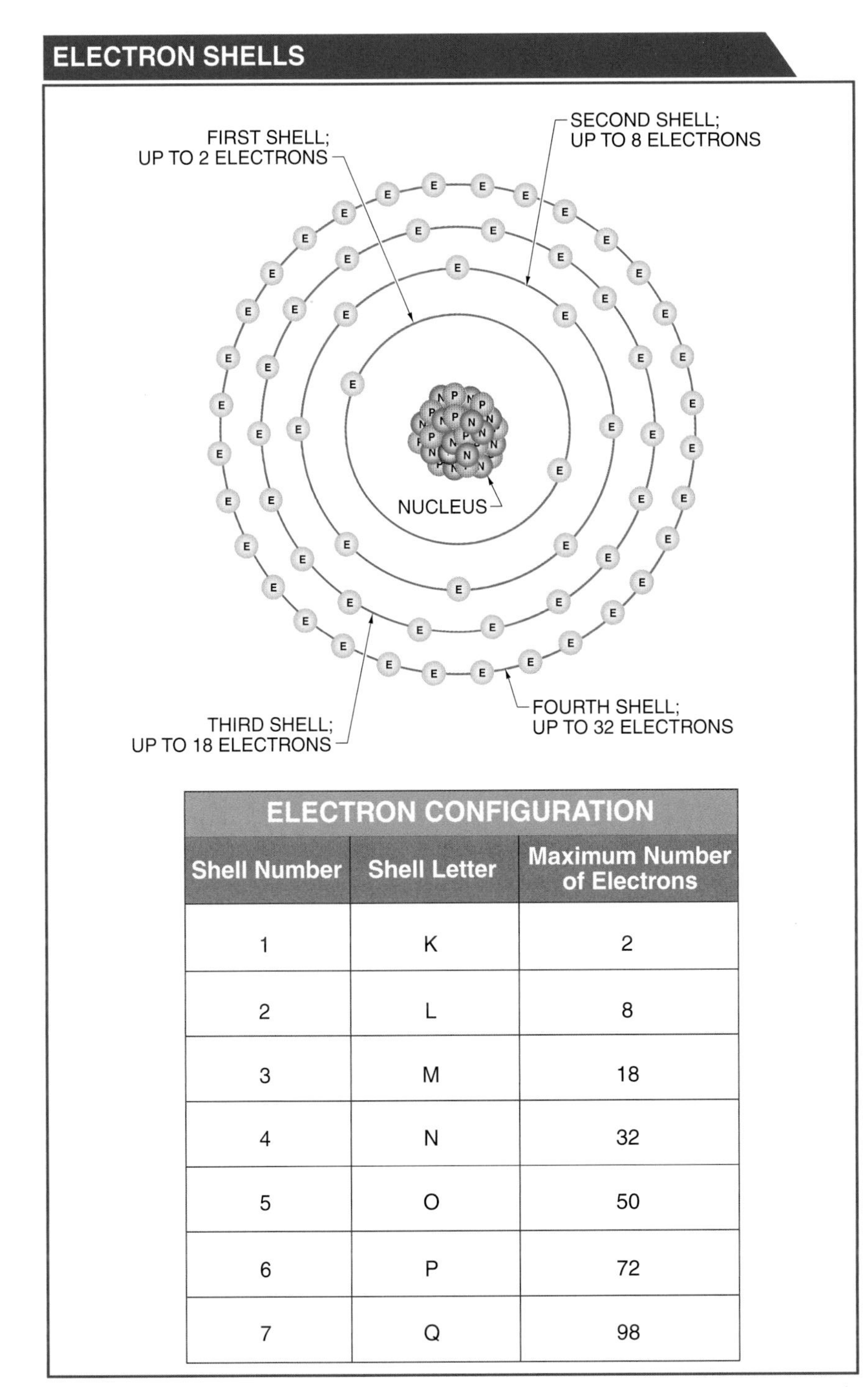

ELECTRON CONFIGURATION		
Shell Number	Shell Letter	Maximum Number of Electrons
1	K	2
2	L	8
3	M	18
4	N	32
5	O	50
6	P	72
7	Q	98

Figure 1-5. The total number of shells an atom has varies from one to seven and are numbered 1, 2, 3, 4, 5, 6, and 7, or lettered K, L, M, N, O, P, and Q.

Terms

An **element** is a substance that cannot be chemically broken down and contains atoms of only one variety.

Elements

An *element* is a substance that cannot be chemically broken down and contains atoms of only one variety. Elements are the basic materials that make up all matter. All solids, liquids, and gases are made up of elements. Some matter may be made up of only one element, but most matter is made up of more than one element. For example, a pure copper kettle is made up of the copper (Cu) element and the water inside the kettle is made up of hydrogen (H) and oxygen (O) elements. Oxygen is one of the most plentiful elements on earth. **See Figure 1-6.**

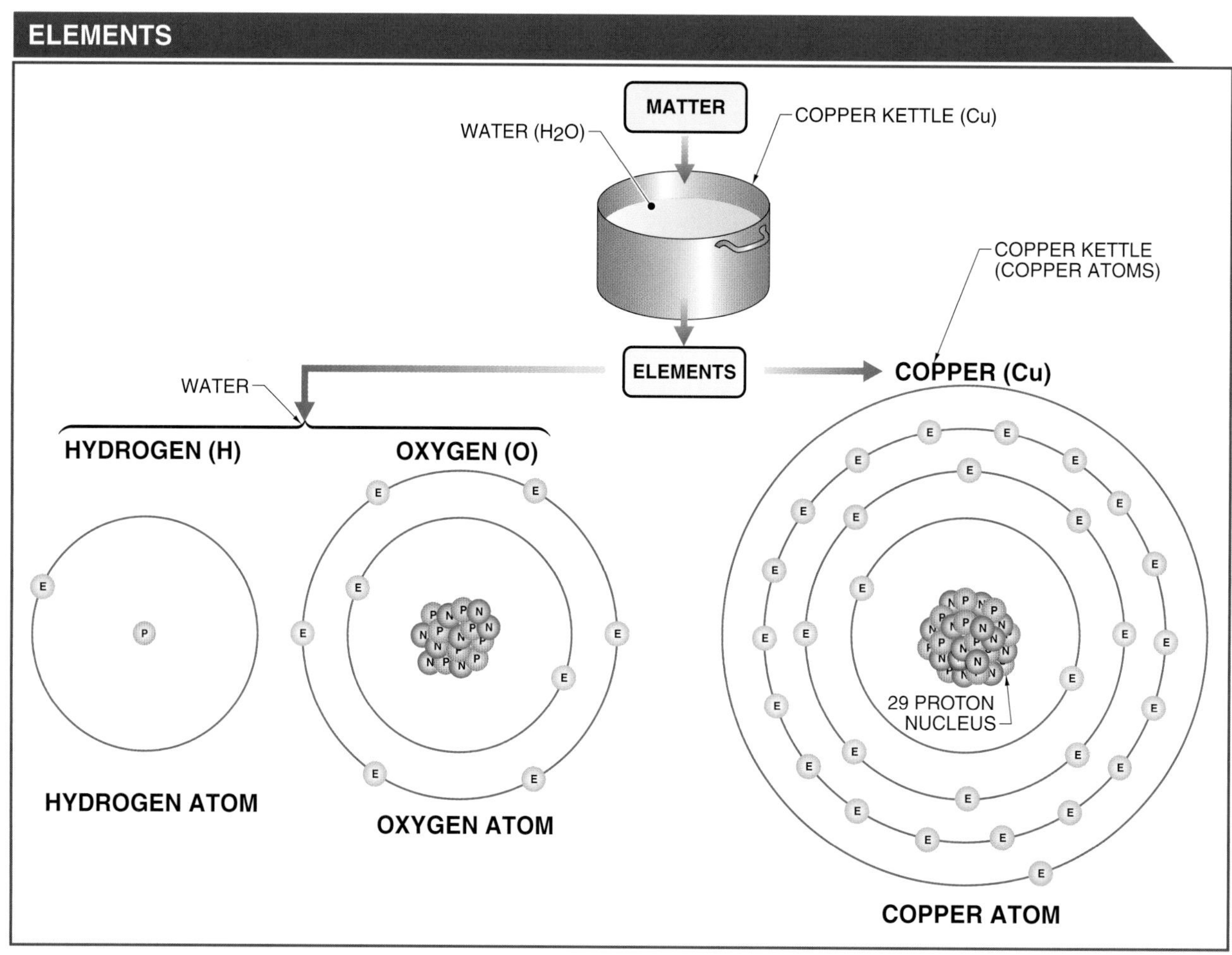

Figure 1-6. All matter is made up of one or more elements.

There are 118 elements, 98 of which are natural. The others are artificial (manufactured). The number of elements increases as more elements are discovered and developed. Each element has a symbol that is an abbreviation for the element. The abbreviations for the elements are internationally recognized and commonly used as shown in the Periodic Table of Elements. **See Appendix.** Element symbols are used to determine material composition and to write formulas.

Solid elements such as iron (Fe), copper (Cu), aluminum (Al), silver (Ag), and gold (Au) are used in electrical applications because they are good conductors. Liquid such as water (H_2O) and oils, and gaseous elements, such as oxygen (O), helium (He), sodium (Na), mercury-vapor (Hg), argon (Ar), and neon (Ne), are used in specific applications. For example, neon lights are lamps that contain neon, helium, or argon gas under low pressure. Depending on which gas is used and the color of the gas tube, many different-colored light outputs are possible. Neon is used to produce red and orange light. Argon is used to produce blue and green light. Helium is used to produce gold and white light. **See Figure 1-7.**

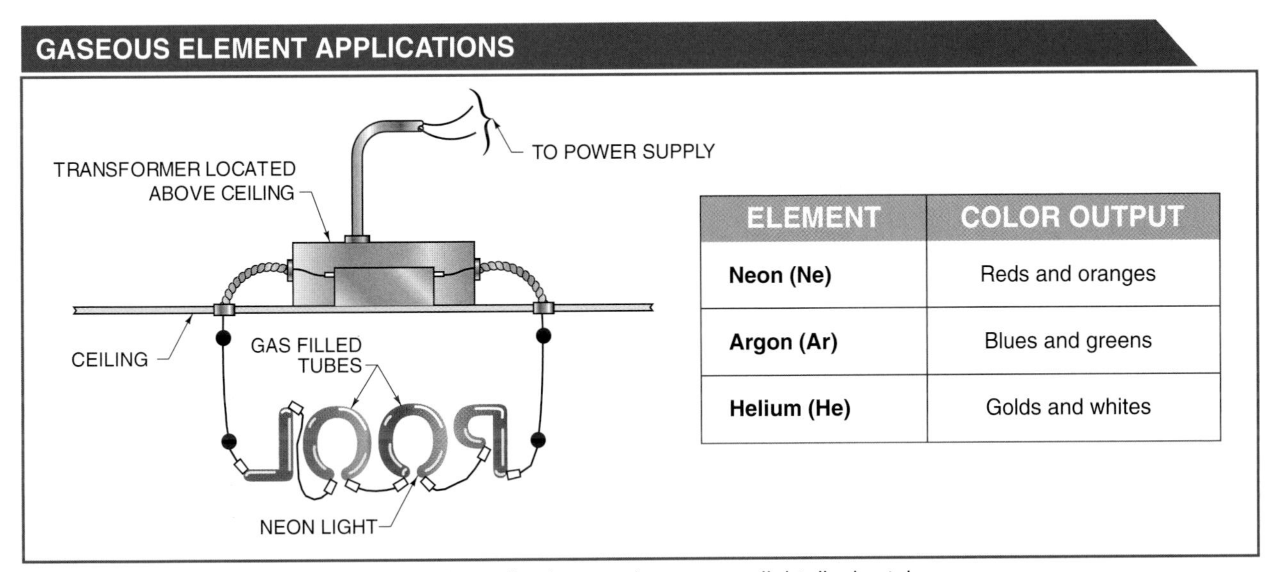

ELEMENT	COLOR OUTPUT
Neon (Ne)	Reds and oranges
Argon (Ar)	Blues and greens
Helium (He)	Golds and whites

Figure 1-7. Gaseous elements are used in applications such as a neon light display tube.

The ability to distinguish between different element symbols is required because of their use on equipment. For example, conductors are commonly made of copper, aluminum, and copper-clad aluminum (Cu-clad). Copper and aluminum are the most commonly used electrical material. Copper-clad aluminum is a conductor that has copper bonded to an aluminum core. Some electrical equipment cannot be used with an aluminum conductor. This equipment is marked as "CU" and "Cu-clad." Verifying that a conductor matches the equipment rating ensures a safe and reliable electrical installation.

In addition to referring to element symbols when installing equipment, element symbols are also referred to when specifying equipment such as switchgear, motors, transformers, cabling, junction boxes, and conduit. Electrical contacts are used to switch (start and stop) the flow of electricity in devices such as switches and relays. The element the contacts are composed of determines how effectively the contacts operate.

Relay contacts are available in several different materials and configurations. Material such as silver (Ag), silver-cadmium (AgCd), gold-plated silver (AgAu), and tungsten (W) are typical elements used to make electrical contacts. The material used is determined by the application. If the wrong type of material is used, equipment is subject to decreased operating life and malfunctions. The proper material must be specified when ordering contacts for a given application. Most contacts include silver, because silver has the highest electrical conductive property of all materials. However, because pure silver is soft, it sticks, welds (from electrical arcing), and is subject to sulfidation when used for the incorrect application. *Sulfidation* is the formation of film on the contact surface. Sulfidation increases the resistance to the flow of electricity through the contacts.

Omron Electronics, Inc.

Relay contacts are available in several different materials and configurations, the designations of which are marked on the relay contact case.

Terms

Sulfidation is the formation of film on the contact surface.

Silver is alloyed with other metals to create contacts that provide the optimum performance for the requirements of different electrical circuit applications. For example, silver is alloyed with cadmium (Cd) to produce a silver-cadmium contact. Silver-cadmium contacts have good electrical characteristics and low resistance. This helps the contact resist arcing but not sulfidation. Silver-cadmium contacts are used in circuits that switch high currents at more than 12 V because the high current burns off the sulfidation.

Some electrical contacts are used in applications that require frequent switching, others are used in applications that require very little switching. When contacts are seldom switched, sulfidation can occur and damage the silver contacts. To prevent this problem, gold-plated silver contacts are used in applications that require infrequent switching. However, gold-plated contacts are not used in high-current applications because the gold burns off quickly at high currents. Tungsten contacts are used in high-current applications because tungsten has a high melting temperature and is less affected by arcing. Tungsten contacts are also used when high repetitive switching is required. Because tungsten is wear-resistant, it is also used in many nonelectrical applications such as tool steel alloys.

Compounds

A *compound* is a combination of the atoms of two or more elements. A *molecule* is the smallest particle that a compound can be reduced to and still possesses the chemical properties of the original compound. If a compound is divided further, only atoms of elements remain. For example, a drop of water contains billions of water molecules. If a drop of water is divided until only a single water molecule remains, the final drop of water still possesses all the chemical properties of water. However, if one molecule of water is divided, only two hydrogen atoms and one oxygen atom remain. **See Figure 1-8.**

Terms

A **compound** is a combination of the atoms of two or more elements.

A **molecule** is the smallest particle that a compound can be reduced to and still possesses the chemical properties of the original compound.

COMPOUNDS

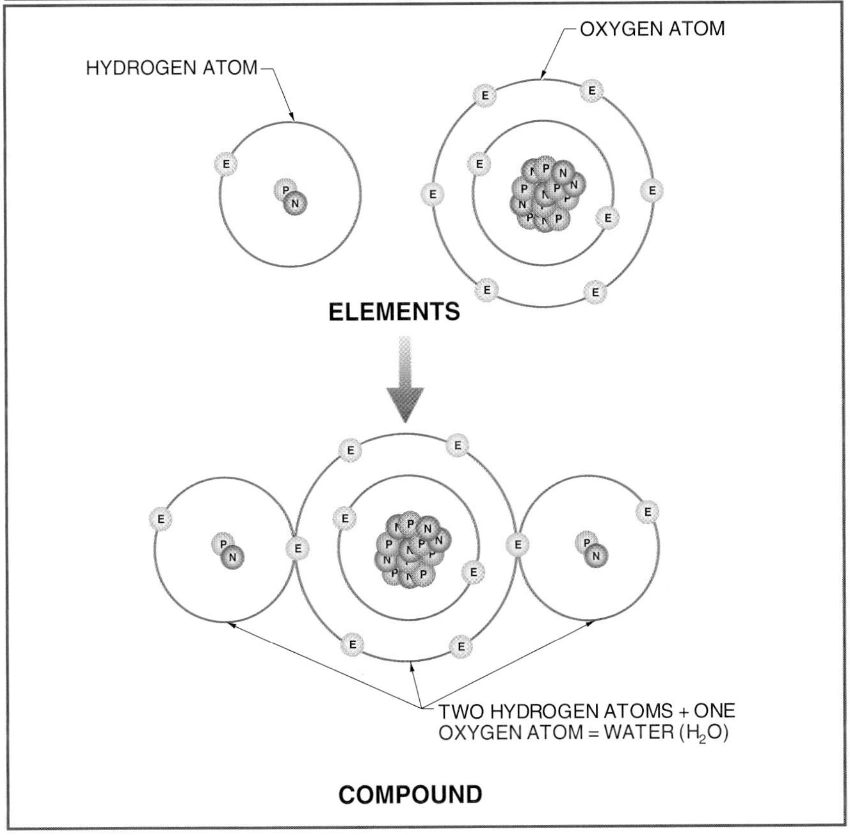

Figure 1-8. A compound is a combination of the atoms of two or more elements.

Every compound has a definite composition described by a chemical formula. For example, water (H_2O) is an example of a simple compound composed of hydrogen and oxygen elements. Water consists of two atoms of hydrogen linked to one atom of oxygen. The chemical formula for water is written H_2O because there are exactly twice as many hydrogen atoms as oxygen atoms in any sample of water.

While there are only 118 elements, these elements can be combined in many different configurations to form millions of compounds. Elements can even be combined in different proportions to produce different compounds. For example, the elements of carbon and hydrogen can combine to form methane (CH_4) or propane (C_3H_8).

When some elements are combined, the properties of the compound may differ considerably from that of the individual elements. For example, the element sodium (Na) is a soft metal that reacts violently with water, and the element chlorine is a poisonous gas. Common table salt (NaCl) is formed when the elements sodium and chlorine are combined with oxygen. Elements such as carbon (C), silicon (Si), and germanium (Ge) are combined with other elements to produce specific compounds that have unique electrical characteristics. For example, carbon is used to make electrical resistors. Germanium and silicon are used to make electronic components such as diodes and transistors.

The electrical properties of compounds are important in water treatment, gas and oil refinement, food processing, chemical and pharmaceutical manufacturing, medical applications, and research. In such areas, product testing is accomplished by passing low-voltage electricity through the compounds and taking measurements that provide information about the product being tested. Also, electricians performing tasks in hazardous or confined spaces use gas detectors. A gas detector can display information such as the amount of carbon monoxide (CO) in the air in parts per million (ppm). **See Figure 1-9.**

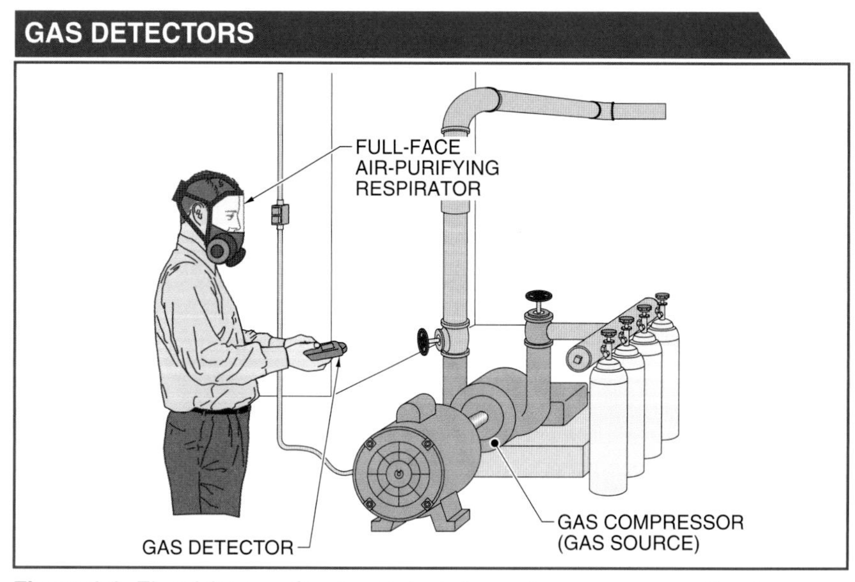

Figure 1-9. Electricians performing tasks in hazardous areas or confined spaces of a facility are required to use gas detectors.

Valence Electrons

Valence electrons are electrons in the outermost shell of an atom. Valence electrons determine the conductive or insulative properties of a material. Most elements do not have a completed outer shell with the maximum allowable number of valence electrons. Conductors normally have only one or two valence electrons in their outer shell. **See Figure 1-10.** Insulators normally have numerous electrons in their outer shell, which is either almost or completely filled with electrons. Semiconductors fall between the low resistance offered by a conductor and the high resistance offered by an insulator. Semiconductors are made from materials that have four valence electrons.

Terms

Valence electrons are electrons in the outermost shell of an atom.

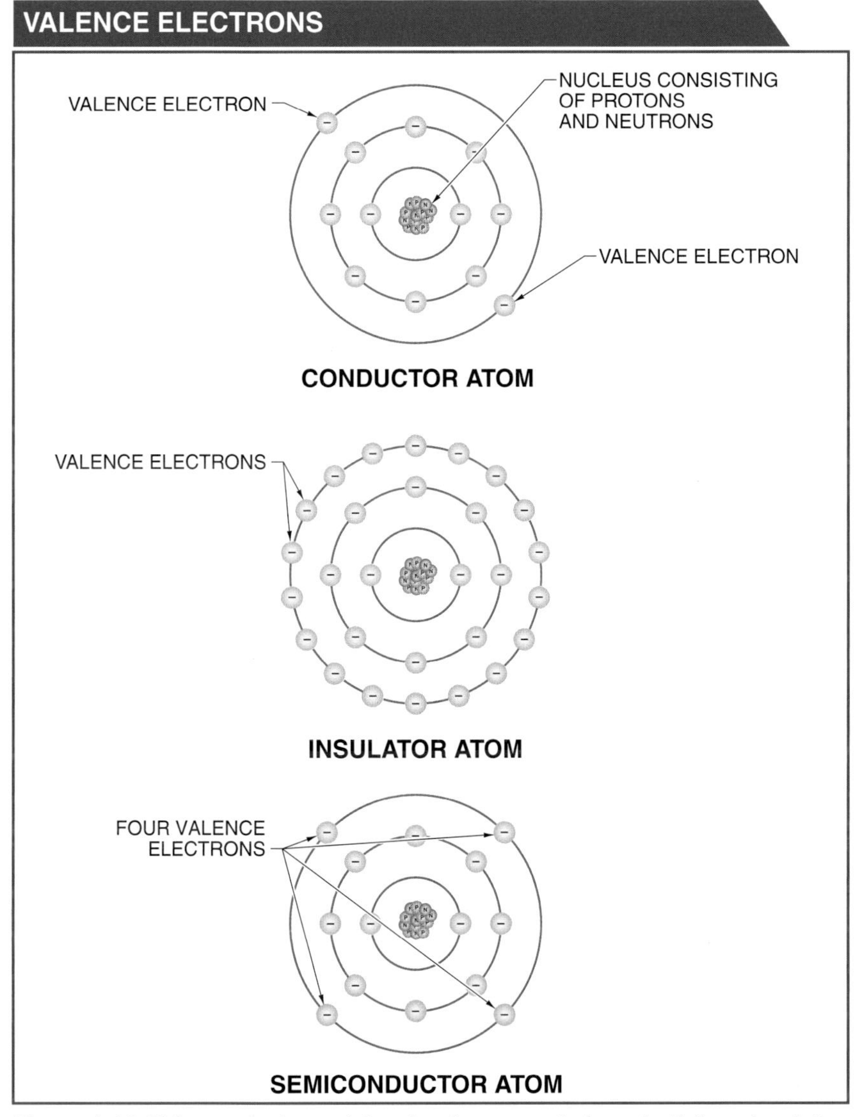

Figure 1-10. Valence electrons determine the amount of conductivity or insulating characteristics of a given material.

In a conductor atom, an outside force can be applied to force the atom to lose or gain valence electrons. Electrons can be forced to move by chemical reaction, friction, pressure, heat, light, or magnetism. These forces can occur naturally, as with lightning and static electricity, or they can be produced, as in a battery or generator.

ELECTRICITY

Electricity is the movement of electrons from atom to atom. *Current* is the flow of electrons in a conductor. Electrons enter a conductor, which provides a path for the current to flow. Electron movement in a conductor operates similarly to the movement of marbles in a tube. For example, if a tube is filled with a single line of marbles, as one marble is forced into one end of the tube, another marble is forced out of the opposite end of the tube. The same principle is true for electrons in a conductor. Electrons are present throughout the length of the conductor. If an electron enters one end of the conductor, another electron is forced out of the opposite end of the conductor. **See Figure 1-11.** This movement of electrons is also referred to as static electricity.

Terms

Electricity is the movement of electrons from atom to atom.

Current is the flow of electrons in a conductor.

Static electricity is an electrical charge at rest.

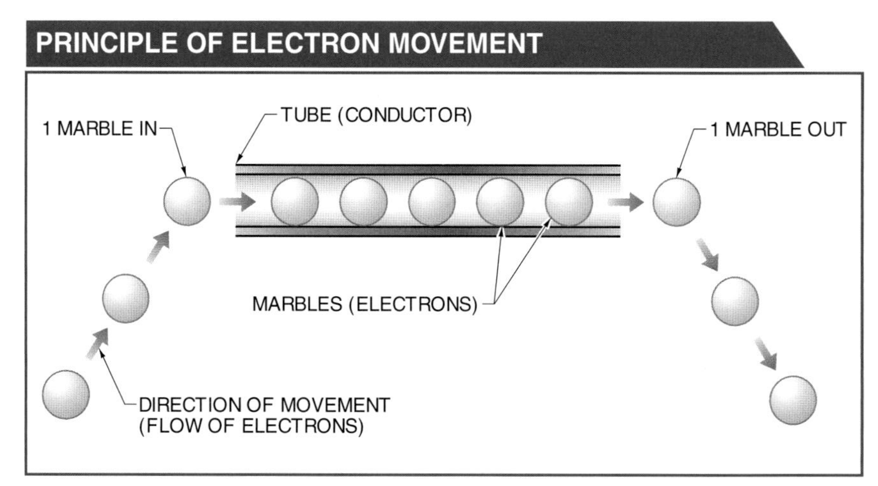

Figure 1-11. Electron movement in a conductor operates similarly to the movement of marbles in a tube.

Static electricity is an electrical charge at rest. The electrical shock felt when making contact with a nonenergized object is an example of the transfer of a static charge by contact. Lightning is an example of the transfer of a static charge by spark. Static electricity has some limited practical uses, such as in electrostatic spray-painting and electrostatic air filters.

Generated electricity is electricity produced either by pressure (piezoelectricity), light (photocell), heat (thermocouple), chemical action (battery), or magnetism (generator). Generated electricity is used in circuits that are specifically designed to carry electrical charges through a controlled path in order to operate specific loads.

Of the basic forms of energy (electrical, chemical, nuclear, light, heat, and mechanical), electrical energy is the most widely used form of energy for transmission. Electrical energy is used to provide energy in the form of power for electrical applications such as lighting, heating, cooling, cooking, communication, and transportation.

Tech Tip

In 600 BC, the Greeks discovered a substance, amber (fossilized sap), that when rubbed against wool, caused other substances to be attracted to it.

LAW OF ELECTRIC CHARGES

An atom is normally electrically neutral because there are as many electrons (–) in it as there are protons (+). However, an atom can lose or gain a few electrons. The gaining or losing of electrons produces an electric charge in the atom. All charged particles exert forces on one another, even if they are not in physical contact. This exerted force is due to the electric field that surrounds the charged particles. The electric field that surrounds the charged particles is known as an electrostatic field. The electrostatic field radiates out perpendicularly in all directions from the positive charge and radiates perpendicularly from all directions toward the negative charge.

The *law of electric charges* is a law that states that opposite charges attract, and like charges repel. Therefore, a positively charged particle attracts a negatively charged particle and repels another positively charged particle. A negatively charged particle attracts a positively charged particle and repels another negatively charged particle. **See Figure 1-12.**

Terms

Generated electricity is electricity produced either by pressure (piezoelectricity), light (photocell), heat (thermocouple), chemical action (battery), or magnetism (generator).

The **law of electric charges** is a law that states that opposite charges attract, and like charges repel.

Semiconductors

The basic material used in most semiconductor devices is either germanium or silicon. In their natural states, germanium and silicon are pure crystals. These pure crystals do not have enough free electrons to support a significant current flow. To prepare these crystals for use as semiconductor devices, their structures must be altered to permit significant current flow. Doping is the addition of impurities to the crystal structure of a semiconductor. In doping, some of the atoms in the crystal are replaced with atoms of other elements. The addition of new atoms in the crystal structure creates N type material and P type material. Together, N-type and P-type semiconductors are the basic building blocks of most solid-state electronic devices.

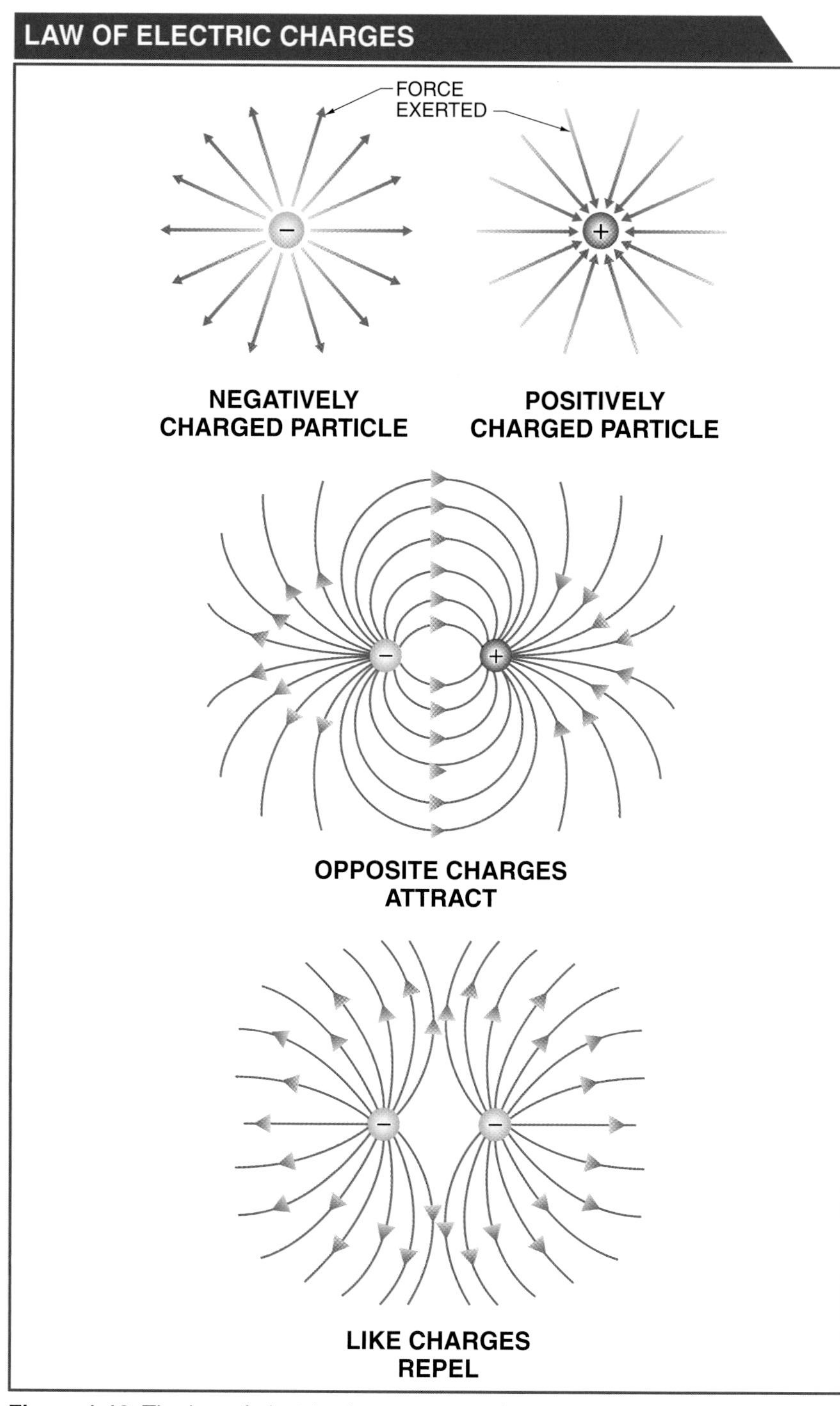

Figure 1-12. The law of electric charges states that opposite charges attract and like charges repel.

Terms

Electrification is the process of charging an object. The three types of electrification are contact electrification, conduction electrification, and induction electrification.

Electrification

Electrification is the process of charging an object. The three types of electrification are contact electrification, conduction electrification, and induction electrification. With each type of electrification, electrons are removed from one body and transferred to another. One body becomes negatively charged while the other becomes positively charged. Contact electrification occurs any time two objects with different levels of charge make physical contact. For example, an electric switch uses contact electrification. One side of the switch is connected to a source charge, and the other side is connected to a load. When the switch is closed, electrons flow through the switch contacts to equalize the charge across the switch. **See Figure 1-13.**

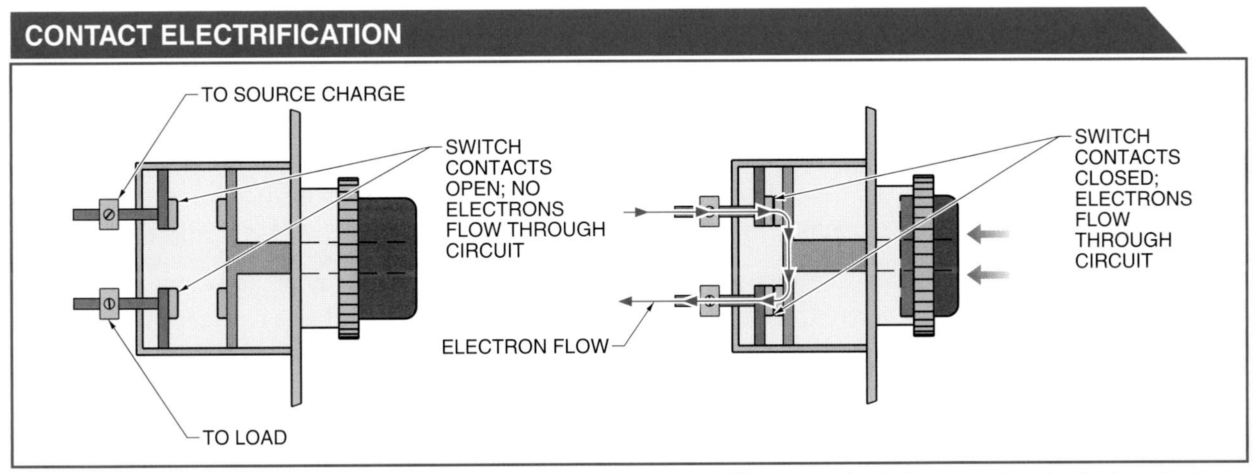

Figure 1-13. Contact electrification occurs any time two objects with different levels of charge make physical contact.

Conduction electrification occurs when a conductor, such as a copper wire, connects two objects with different charges. Electrons move from the negative object to the positive object through the conductor. Conduction electrification is the most common type of electrification and is used to transmit and distribute electrical energy (power) from power generating plants to end users.

Induction electrification occurs when a negatively charged object is brought close to a neutral object. Electrons are repelled from the negatively charged object into the neutral object. The electrons on the neutral object flow to ground if a conduction path is present. Ground, or earth, is considered a neutral object that can accept or provide an undefined number of electrons. If the conduction path is broken, the neutral object has a net positive charge due to a deficiency of electrons.

Damaged or burned electrical components must be replaced immediately to prevent irreversible damage to electrical equipment.

Electrons flow from the negative terminal through a circuit toward the positive terminal of the voltage source. *Polarity* is the positive (+) or negative (–) state of an object. Many electrical devices have separate output terminals marked with a plus sign (+) to indicate the positive terminal, and a minus sign (–) to indicate the negative terminal. The correct polarity must be known when connecting equipment to a voltage source. Irreversible damage can occur if polarity is reversed.

Terms

Polarity is the positive (+) or negative (–) state of an object.

Length is the measurement of linear units.

BASIC UNITS OF MEASURE

Electrical properties are measured so that knowledge of electrical systems, components, devices, and circuitry can be obtained. Standard units of measure are required to properly communicate electrical properties to electrical system designers, installers, technicians, and end users.

The English and metric (SI) systems apply standard quantities to determine measurements. Standard quantities include length, mass, time, temperature, electric current, voltage, and resistance.

Conversion from one system to the other is accomplished by applying conversion factors. **See Figure 1-14.**

METRIC/ENGLISH CONVERSION FACTORS	
Metric to English Units	**English to Metric Units**
1 m = 39.37 in.	1 yd = 0.914 m
1 m = 3.28 ft	1 ft = 0.3048 m
1 km = 0.6214 mi	1 in. = 0.0254 m
1 cm = 0.3937 in.	1 in. = 2.54 cm
1 kg = 2.20 lb	1 lb = 0.454 kg
1 g = 0.03527 oz	1 oz = 28.3527 g

Figure 1-14. Conversion between English and metric units is accomplished by applying conversion factors.

Tech Tip

The meter is the metric system standard measurement unit of length. A meter is the length of the path travelled by light in a vacuum during a time interval of 1/299792458 of a second. One meter equals 10 dm (decimeters), 100 cm (centimeters), or 1000 mm (millimeters).

Length

Length is the measurement of linear units. Standard units of length measurement are the inch (in.), foot (ft), yard (yd), and mile (mi) in the English system and the millimeter (mm), centimeter (cm), meter (m), and kilometer (km) in the metric system. English units can be converted to metric units. For example, 1 m is equal to approximately 39.37 in.

Mass

Weight is a measure of gravity or the force of the Earth's attraction to a body. *Mass* is the measurement of matter contained in an object. The mass of an object never changes, whereas the weight of an object changes in relationship to its distance from the Earth's center. Various types of mechanical scales are used to measure weight. Spring balances are often used for light objects. Hanging scales, beam-balance scales, and platform scales are used for heavier objects. Beam-balance scales use beams with calibrated weights on one side of the beam. **See Figure 1-15.** Standard units of weight measurement in the English system are the ounce (oz), pound (lb), and ton (t). The standard unit of weight measurement in the metric system is the gram (g). For example, 1 cm^3 of water equals 1 g. However, because the gram is small and difficult to work with, the kilogram (kg) is sometimes used.

Tech Tip

The kilogram is the metric system standard measurement unit of weight and equals approximately 2.2 lb.

BEAM-BALANCE SCALES

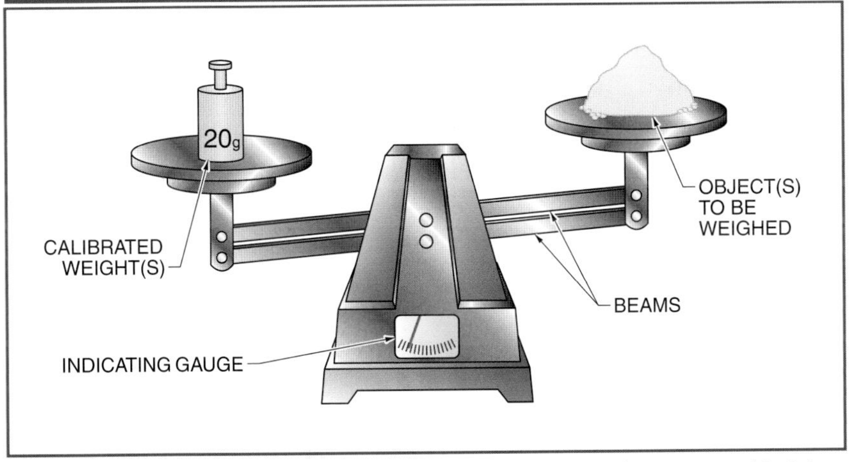

Figure 1-15. Mechanical scales, such as beam-balance scales, are used to take weight measurements.

Terms

Weight is a measure of gravity or the force of the Earth's attraction to a body.

Mass is the measurement of matter contained in an object.

A **second** is a measured duration of time based on cycles of radiation measured with a spectrometer.

Time

A *second* is a measured duration of time based on cycles of radiation measured with a spectrometer. The second is used in the English and the metric system as the base unit of time. The second is commonly used for time measurement and for scientific purposes. Units of time are proportionally related. **See Figure 1-16.**

UNITS OF TIME AND THEIR RELATIONSHIP	
Second (s)	60 s = 1 min
Minute (min)	60 min = 1 hr
Hour (hr)	24 hr = 1 d
Day (d)	365.26 d = 1 y
Year (y)	100 y = 1 century

Figure 1-16. The second is used in both the English and metric systems as the base unit of time.

Terms

Temperature is the measurement of the intensity of heat.

Thermal conductivity is the higher the amount of heat produced, the faster the temperature increases in the body being heated.

Temperature

Temperature is the measurement of the intensity of heat. All heating elements produce heat. Temperature increases when heat is produced. The amount of increase in temperature depends on the amount of heat produced, the mass of the body being heated, and the material of which the heated body is made.

Thermal conductivity is the higher the amount of heat produced, the faster the temperature increases in the body being heated. The larger the body to be heated, the slower the temperature rise of the body. Faster heat transfer occurs in materials that are better conductors of heat. The rate a material conducts heat depends on the thermal conductivity of the material. **See Figure 1-17.**

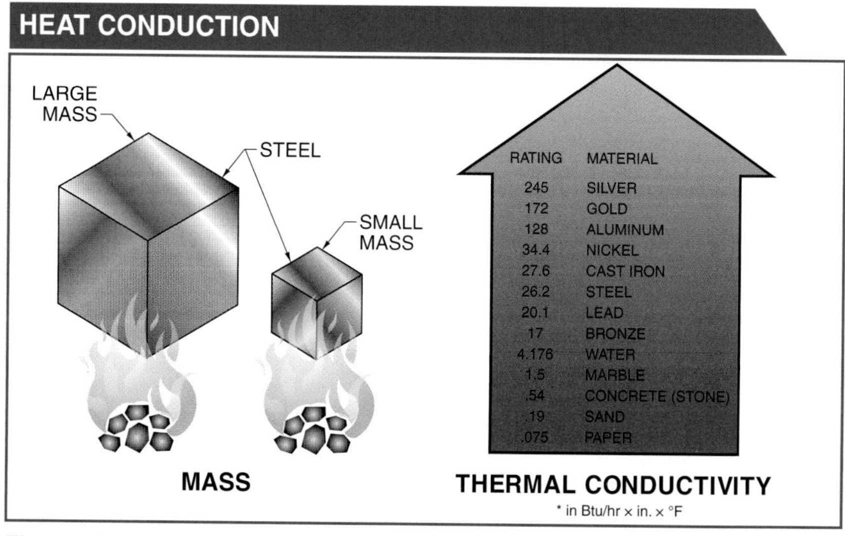

Figure 1-17. The rate a material conducts heat depends on the thermal conductivity of the material.

The metric system base unit of temperature is the kelvin. The kelvin (°K) is mainly used for scientific measurements. For practical applications, Celsius (°C) is used in the metric system and Fahrenheit (°F) is used in the English system. Converting one unit to the other is required because Celsius and Fahrenheit are both used when taking electrical measurements. **See Figure 1-18.** To convert a Fahrenheit temperature reading to Celsius, subtract 32 from the Fahrenheit reading and divide by 1.8:

$$°C = \frac{(°F - 32)}{1.8}$$

where

$°C$ = degrees Celsius

$°F$ = degrees Fahrenheit

32 = difference between bases

1.8 = ratio between bases

Example: What is the Celsius equivalent of 90°F?

$$°C = \frac{(°F - 32)}{1.8}$$

$$°C = \frac{(90 - 32)}{1.8}$$

$$°C = \frac{58}{1.8}$$

$$°C = \mathbf{32.22°C}$$

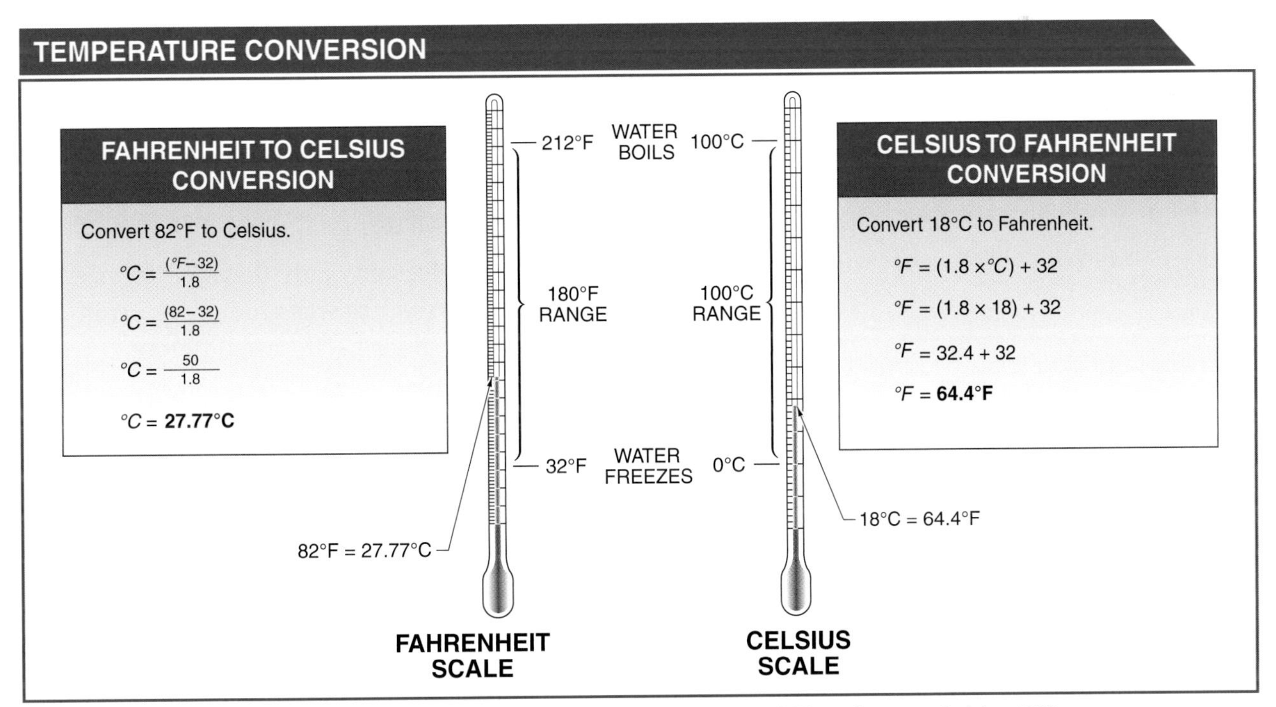

Figure 1-18. Temperature is commonly expressed in degrees Fahrenheit (°F) or degrees Celsius (°C).

To convert a Celsius temperature reading to Fahrenheit, multiply 1.8 by the Celsius reading and add 32:

$°F = (1.8 \times °C) + 32$

where

$°F$ = degrees Fahrenheit

1.8 = ratio between bases

$°C$ = degrees Celsius

32 = difference between bases

Example: What is the Fahrenheit equivalent of 100°C?

$°F = (1.8 \times °C) + 32$

$°F = (1.8 \times 100) + 32$

$°F = 180 + 32$

$°F =$ **212°F**

Electric Current

Current (I) is the amount of electrons flowing through an electrical circuit. Current flows through a circuit when a source of power is connected to a device that uses electricity. The metric system base unit for electrical current is the ampere (A). An *ampere* is 1 coulomb of electrons passing a given point in one second. One *coulomb* is 6.241×10^{18} electrons. The more power a load requires, the larger the amount of current.

Current may be direct or alternating. *Direct current (DC)* is current that flows in only one direction. Direct current flows in any circuit connected to a power supply producing a DC voltage. *Alternating current (AC)* is current that reverses its direction of flow at regular intervals. Alternating current flows in any circuit connected to a power supply producing an AC voltage. The flow of current in a conductor is from the negative terminal to the positive terminal.

Early scientists believed that electrons only flowed from positive to negative. Later, when atomic structure was studied, electron flow from negative to positive was introduced. The two different theories are referred to as conventional current flow and electron current flow. *Conventional current flow* is current flow from positive to negative. *Electron current flow* is current flow from negative to positive. Both conventional current flow and electron current flow are used when designing electrical circuits and systems. Conventional current flow is typically used by electrical engineers when designing circuits for small engines and automotive, marine, construction, and aviation systems. Electron current flow is typically used when designing solid-state circuits and systems. The flow of electrons (current) is measured in amperes (A). **See Figure 1-19.**

Terms

Current (I) is the amount of electrons flowing through an electrical circuit.

An **ampere** is 1 coulomb of electrons passing a given point in one second.

One **coulomb** is 6.241×10^{18} electrons.

Direct current (DC) is current that flows in only one direction.

Alternating current (AC) is current that reverses its direction of flow at regular intervals.

Conventional current flow is current flow from positive to negative.

Electron current flow is current flow from negative to positive.

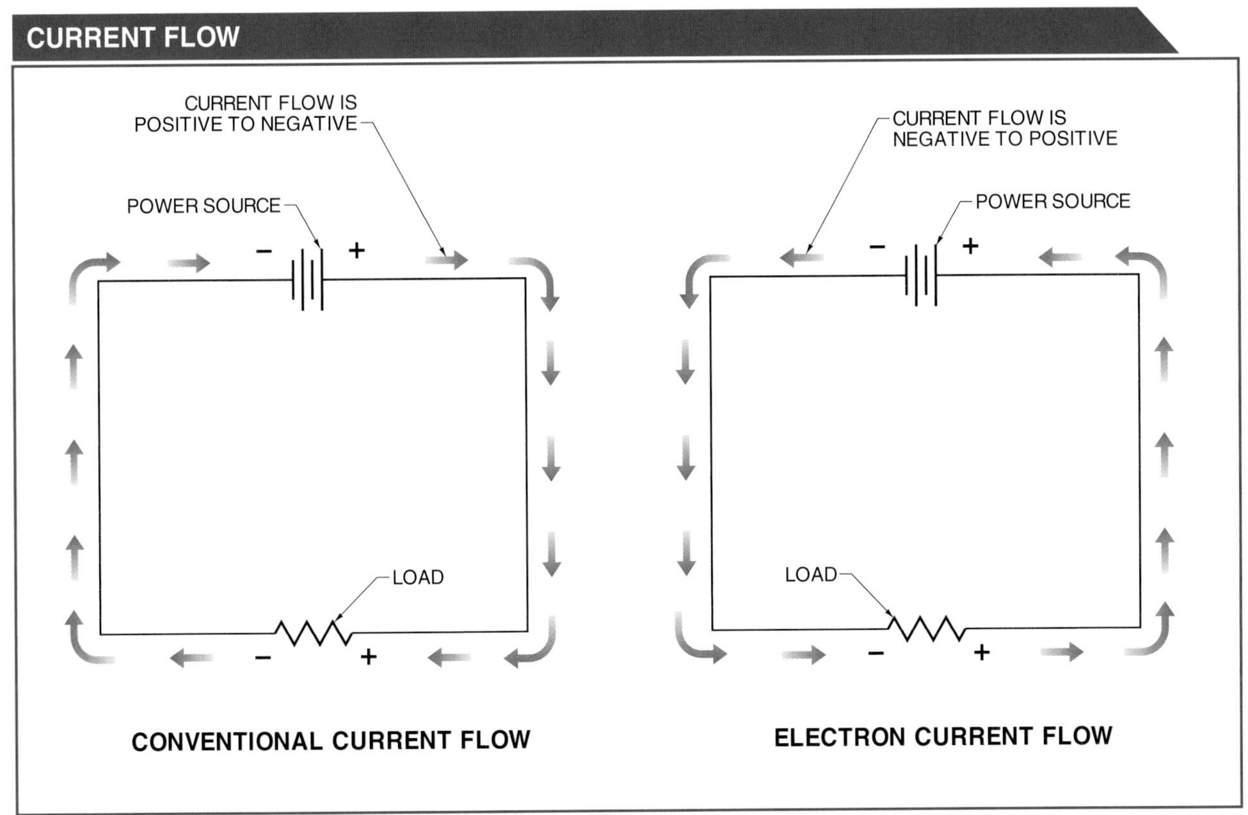

Figure 1-19. Conventional current flow is current flow from positive to negative. Electron current flow is current flow from negative to positive.

Voltage

Voltage (V) is the amount of electrical pressure in a circuit. Voltage is measured in volts (V). All electrical circuits must have a source of power to produce work. The source of power used depends on the application and the amount of power required. All sources of power produce a set voltage level or voltage range. Voltage is also known as electromotive force (EMF) or potential difference. DC voltage is produced by batteries, thermocouples, photocells, and rectified AC voltage supplies and can vary from almost pure DC voltage to full-wave DC voltage. **See Figure 1-20.** DC voltage is typically used in portable equipment (automobiles, golf carts, flashlights, cameras, etc.). AC voltage is used in residential, commercial, and industrial lighting and power distribution systems.

Terms

Voltage (V) is the amount of electrical pressure in a circuit.

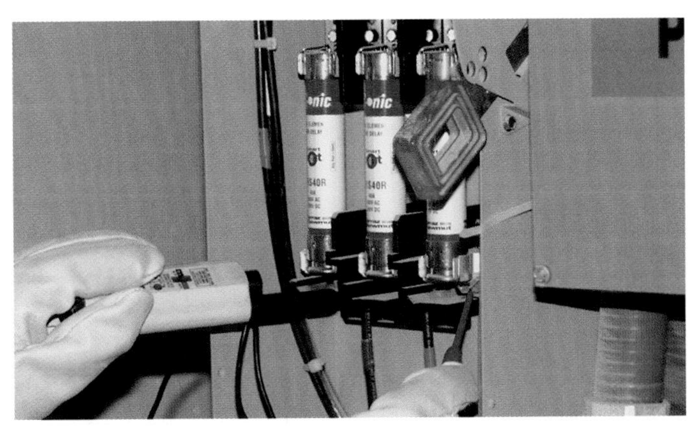

A voltage tester is an electrical test instrument that indicates the approximate voltage amount and the type of voltage (AC or DC) in a circuit.

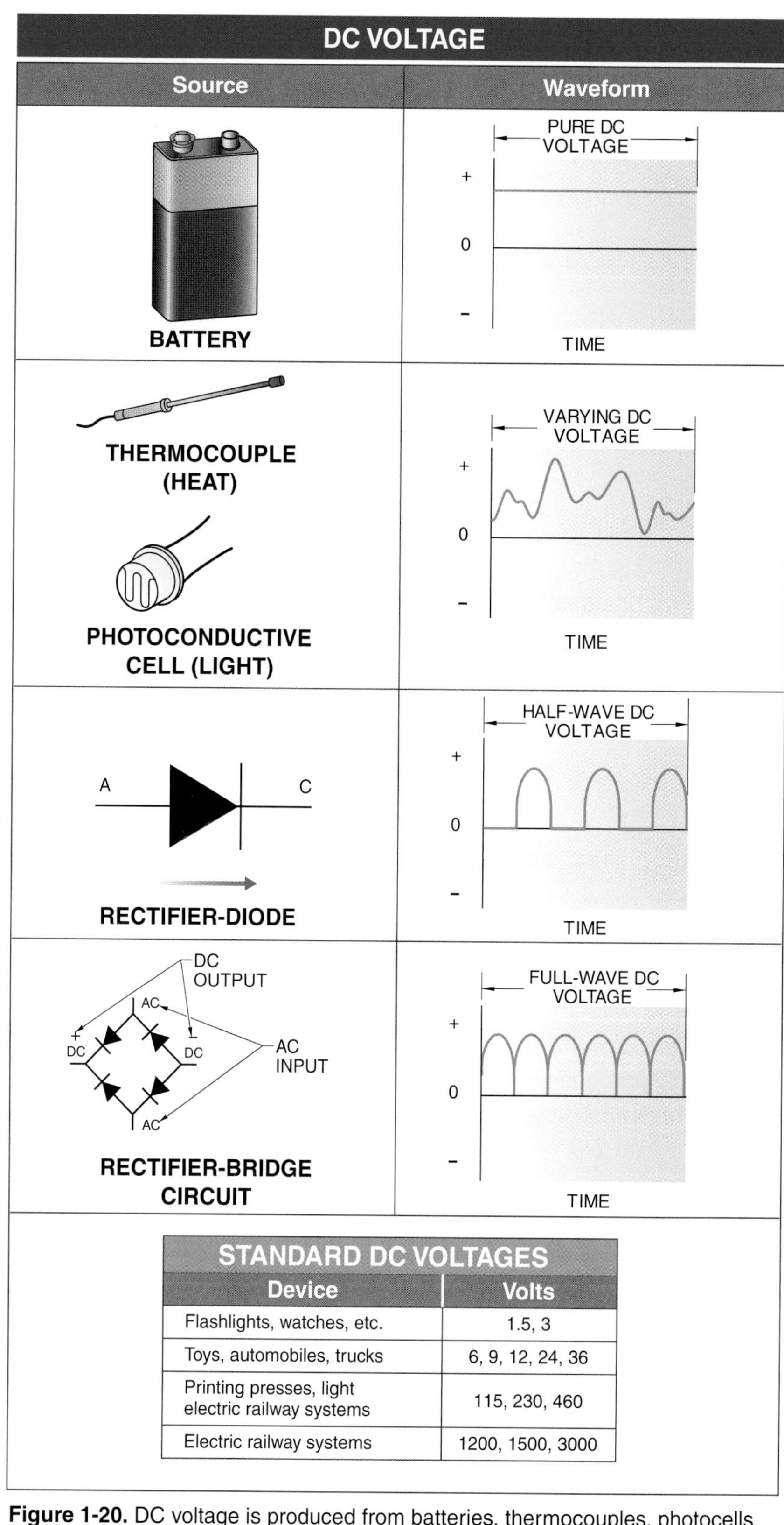

STANDARD DC VOLTAGES

Device	Volts
Flashlights, watches, etc.	1.5, 3
Toys, automobiles, trucks	6, 9, 12, 24, 36
Printing presses, light electric railway systems	115, 230, 460
Electric railway systems	1200, 1500, 3000

Figure 1-20. DC voltage is produced from batteries, thermocouples, photocells, and rectified AC voltage supplies and can vary from almost pure DC voltage to full-wave DC voltage.

Voltage is produced any time there is an excess of electrons at one terminal of a voltage source and a deficiency of electrons at the other terminal. The greater the difference in the number of electrons between the terminals, the higher the voltage produced.

The amount of voltage required in a device depends on the application. **See Figure 1-21.** For example, powering a flashlight requires a low voltage level. However, producing a picture on a TV screen requires an extremely high voltage level.

COMMON VOLTAGES	
Device	**Volts**
Flashlight battery (AAA, AA, C, D)	1.5
Automobile battery	12.0
Golf cart	36.0
Refrigerator, TV, VCR	115.0
Central air conditioner	230.0
Industrial motor	460.0
TV picture tube	25,000.0
High-tension power line	up to 500,000.0

Figure 1-21. The amount of voltage required in a device depends on the application.

Resistance

Resistance (R) is the opposition to the flow of electrons. Resistance is measured in ohms. The Greek symbol omega (Ω) is used to represent ohms. Resistance limits the flow of electrons in an electrical circuit. Resistances can be used in an electrical circuit for protection, operation, or current control, or in combination with other circuit components. The higher the resistance, the lower the flow of electrons. Likewise, the lower the resistance, the higher the flow of electrons. **See Figure 1-22.**

Terms

Resistance (R) is the opposition to the flow of electrons.

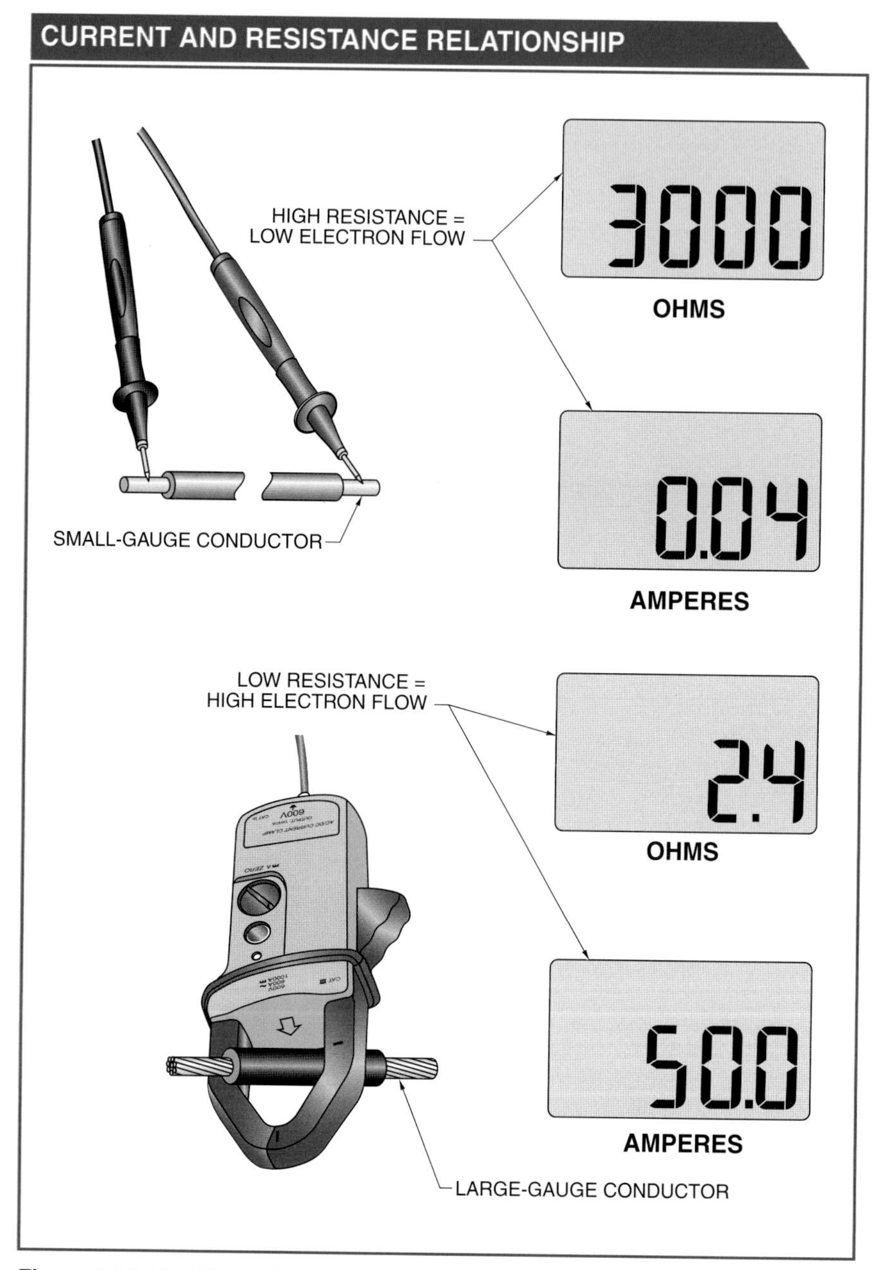

Figure 1-22. The higher the resistance, the lower the flow of electrons. Likewise, the lower the resistance, the higher the flow of electrons.

DERIVED UNITS OF MEASURE

Derived units of measure have algebraic relationships to base units. These derived units are necessary to relate electrical properties to physical values. The metric system is typically used in the majority of all countries. Derived units in the English system are primarily used in the United States. Other common units of physical measurements in the cgs (centimeter-gram-second) system are derived from the seven basic units. **See Figure 1-23.** For example, the dyne is a unit of measure of force. The gram (1/1000 kg), centimeter (1/100 m), and the second are used to define the dyne. The dyne is used to define the erg, a unit of work and the erg is used to define power.

METRIC DERIVED UNITS OF MEASUREMENT	
Quantity	**Unit**
Area	square centimeter (cm^2)
Volume	cubic centimeter (cm^3)
Speed or velocity	centimeter/second (cm/s)
Acceleration	centimeter/second/second (cm/s^2)
Force	dyne ($g \bullet cm/s^2$)
Pressure	dyne/squared centimeter ($dyne/cm^2$)
Energy and work	erg ($dyne \bullet cm = g \bullet cm^2/s^2$)
Power	erg/second (erg/s)
Luminous flux	lumen ($cd \bullet sr$)
Illuminance	lux (lm/m^2)
Density	gram/cubic centimeter (g/cm^3)

Figure 1-23. Common units of physical measurements in the cgs system are derived from the basic units of measurement.

The newton (N) is also a unit for force. The newton uses larger metric units, the meter and the kilogram. This is known as the mks (meter-kilogram-second) system. The unit for energy and work is the joule (J). There are also derived units in the English system. **See Figure 1-24.** With the exception of the ampere, which is one of the seven basic units of the metric system, all other units of measure of electrical units are derived from basic units. **See Appendix.**

Terms

Work (W) is the application of a force over a distance.

Force (F) is any cause that changes the position, motion, direction, or shape of an object.

Mechanical resistance is any force that tends to hinder the movement of an object.

Work

Work (W) is the application of a force over a distance. *Force (F)* is any cause that changes the position, motion, direction, or shape of an object. Work is accomplished when force overcomes mechanical resistance. *Mechanical resistance* is any force that tends to hinder the movement of an object. No work is produced if an applied force does not cause motion.

ENGLISH DERIVED UNITS OF MEASUREMENT	
Quantity	**Unit**
Area	square foot (ft^2)
Volume	cubic foot (ft^3)
Speed or velocity	feet/second (ft/s)
Acceleration	feet/second^2 (ft/s^2)
Force	pound-mass (lb_f)
Pressure	pound-force/inches squared (lb/in^2)
Energy and work	foot-pound-force (ft • lb_f)
Power	foot-pound-force/second (ft • lb/s)
Luminous flux	lumen (cd • sr)
Illuminance	footcandle (lm/ft^2)
Density	pounds/cubic feet (lb/ft^3)
Energy	electro volt ($1.603 \cdot 10^{-12}$)
Luminance	foot-lambert (cd/ft^2)

Figure 1-24. Derived units in the English system are primarily used in the United States.

The amount of work produced is determined by multiplying the force applied over the distance that an object is moved. **See Figure 1-25.** Thus, work is measured in pound-feet (lb-ft). Work produced is calculated by applying the following formula:

$W = F \times D$

where

W = work (in lb-ft)

F = force (in lb)

D = distance (in ft)

Example: How much work is accomplished when lifting a 72 lb pallet vertically from the floor to the top of a 3′ high workbench?

$W = F \times D$

$W = 72 \text{ lb} \times 3 \text{ ft}$

$W = \mathbf{216\ lb\text{-}ft}$

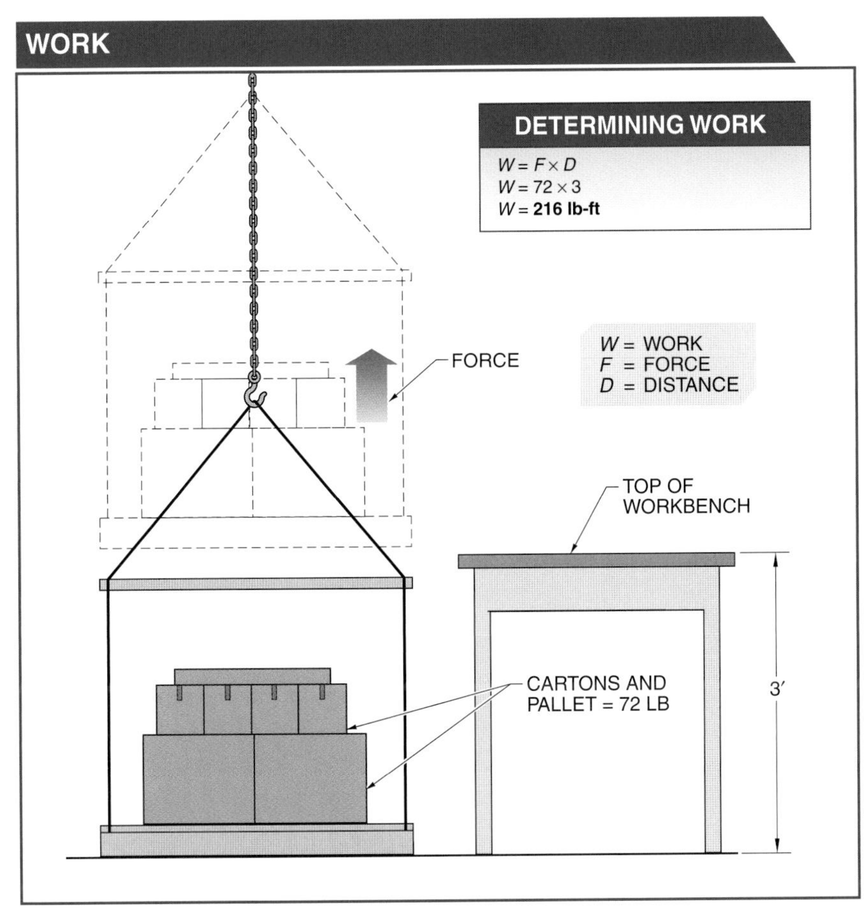

Figure 1-25. Work equals force times distance.

Resistance must be overcome to perform work. More work is required if the pallet is heavier, the distance longer, or a combination of the two. Less work is required if the pallet is lighter, the distance shorter, or a combination of the two.

Energy

Energy is the ability to do work and is measured in the same units as work. Energy units are erg, joule, volt, watt-hour, foot-pound, or foot-poundal, depending on the system of measurement being used.

Potential energy is stored energy a body has due to its position, chemical state, or condition. If a weight is raised two meters vertically, it has stored potential energy. If the same weight is moved two meters horizontally, it does not have any additional potential energy because its vertical height change is zero. The potential energy of a body equals the force times the distance it is raised. Potential energy is calculated by applying the following formula:

Terms

Energy is the ability to do work and is measured in the same units as work.

Potential energy is stored energy a body has due to its position, chemical state, or condition.

Tech Tip

The term energy comes from the Greek word "energeia," which means "activity."

$PE = F \times h$

where

PE = potential energy (in lb-ft)

F = force (in lb)

h = height (in ft)

Example: What is the potential energy of a 50 lb stone raised 20 ft?

$PE = F \times h$

$PE = 50 \text{ lb} \times 20 \text{ ft}$

$PE = \mathbf{1000\ lb\text{-}ft}$

If a weight is released and allowed to fall, the potential energy is changed to kinetic energy. *Kinetic energy* is the energy of motion. It is caused by a mass that is in motion. A mass in motion can perform work. Kinetic energy is calculated by applying the following formula:

$$KE = \frac{m \times v^2}{2}$$

where

KE = kinetic energy (in lb-ft²/s²)

m = mass (in lb)

v = velocity (in ft/s)

Terms

Kinetic energy is the energy of motion.

Power is the rate of doing work or using energy.

True power (P_T) is the actual power used in an electrical circuit.

Example: What is the kinetic energy of a 4 lb mass traveling at a speed of 2 ft/s?

$$KE = \frac{m \times v^2}{2}$$

$$KE = \frac{4 \text{ lb} \times (2 \text{ ft/s})^2}{2}$$

$$KE = \frac{4 \text{ lb} \times 4 \text{ ft}^2/\text{s}^2}{2}$$

$$KE = \frac{16 \text{ lb-ft}^2/\text{s}^2}{2}$$

$$KE = \mathbf{8\ lb\text{-}ft^2/s^2}$$

Power

Electrical energy is converted into another form of energy any time electrons flow in a circuit. Electrical energy may be converted into sound (speakers), rotary motion (motors), light (lamps), linear motion (solenoids), and heat (heating elements). **See Figure 1-26.** *Power* is the rate of doing work or using energy. Power in an electrical circuit may be expressed as either true power or apparent power. *True power* (P_T) is the actual power used in an electrical circuit.

True power is expressed in watts (W). True power is calculated by applying the following formula:

$P_T = I^2 \times R$

where

P_T = true power (in W)

I = total circuit current (in A)

R = total circuit resistance (in Ω)

ELECTRICAL ENERGY CONVERSION

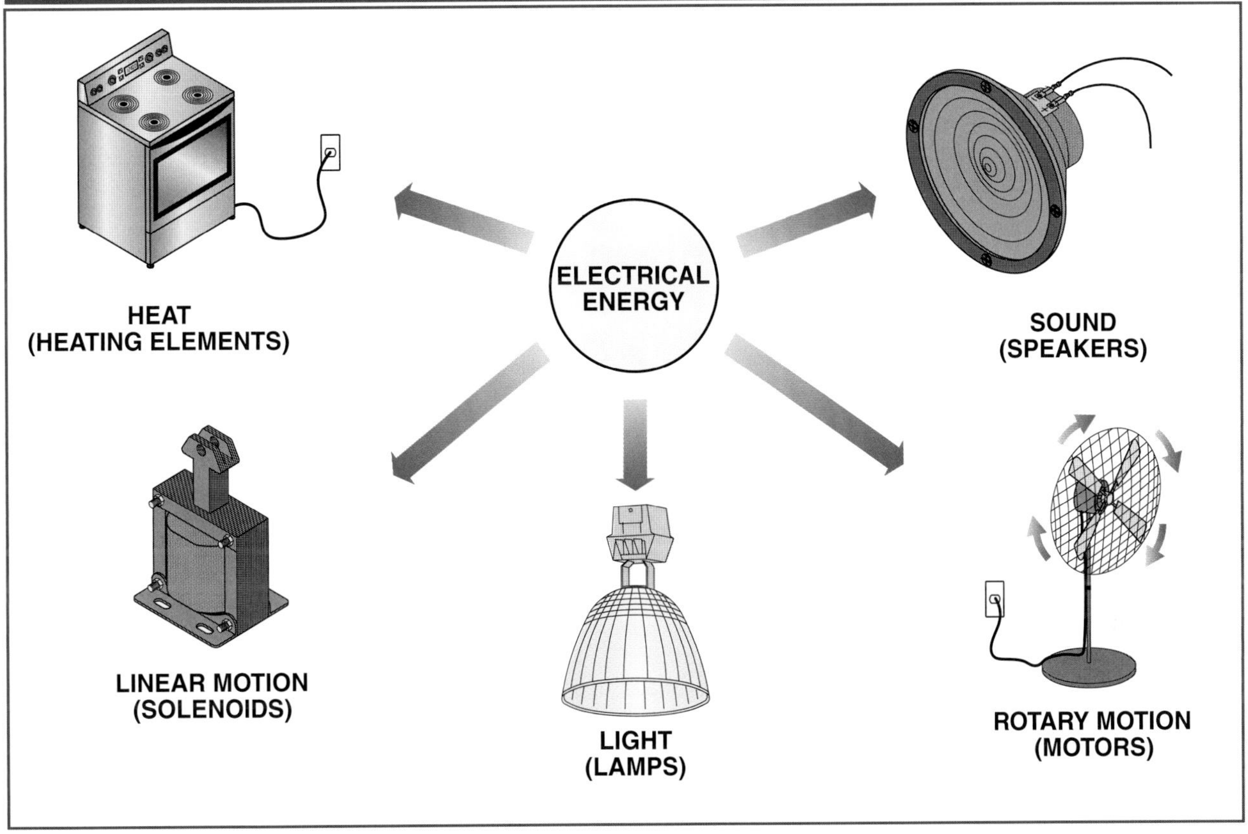

Figure 1-26. Electrical energy may be converted into sound, rotary motion, light, linear motion, and heat.

Example: What is the true power in a circuit that has a total current of 0.05 A and a total resistance of 1000 Ω?

$P_T = I^2 \times R$

$P_T = (0.05\ \text{A})^2 \times 1000\ \Omega$

$P_T = 0.0025\ \text{A} \times 1000\ \Omega$

$P_T = \mathbf{2.5\ W}$

Apparent power (P_A) is the product of the voltage and current in a circuit calculated without considering the phase shift that may be present between the voltage and current in the circuit.

Terms

Apparent power (P_A) is the product of the voltage and current in a circuit calculated without considering the phase shift that may be present between the voltage and current in the circuit.

Terms

Phase shift is the state in which voltage and current in an AC circuit reach their maximum amplitudes and zero levels at different times.

Mechanical power is the rate of doing work through mechanical means.

Phase shift is the state in which voltage and current in an AC circuit reach their maximum amplitudes and zero levels at different times. Apparent power is stated in volt amps (VA) or kilovolt amps (kVA). True power is always less than apparent power in any circuit in which there is a phase shift between voltage and current. **See Figure 1-27.** Apparent power is calculated by applying the following formula:

$P_A = V \times I$

where

P_A = apparent power (in VA)

V = measured voltage (in V)

I = measured current (in A)

POWER EXPRESSION METHODS

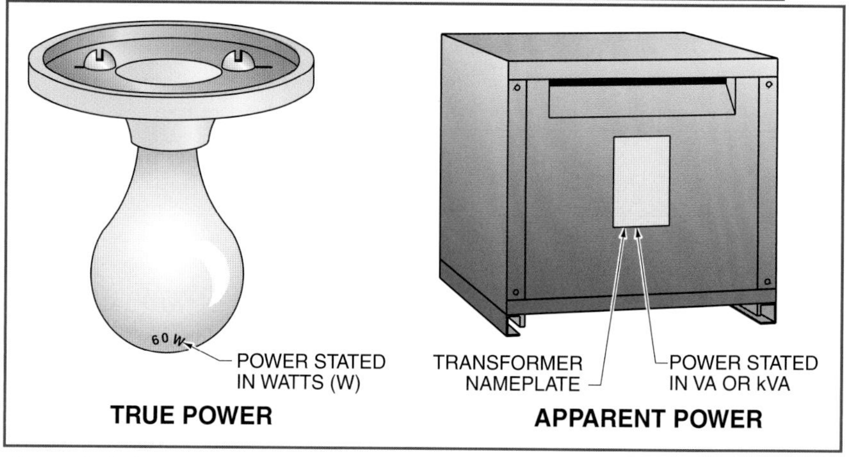

Figure 1-27. True power is the actual power used in an electrical circuit. Apparent power is the product of the voltage and current in a circuit calculated without considering the phase shift that may be present between the voltage and current.

Example: What is the apparent power in a circuit that has a measured voltage of 12 V and a measured current of 0.025 A?

$P_A = V \times I$

$P_A = 12\text{ V} \times 0.025\text{ A}$

$P_A = \mathbf{0.3\ VA}$

Mechanical power is the rate of doing work through mechanical means. Mechanical power is calculated by applying the following formula:

$$P_M = \frac{W}{t}$$

where

P_M = power (in lb-ft/sec)

W = work (in lb-ft)

t = time (in sec)

Example: What is the power required to lift a 5 lb load a distance of 1 ft in 5 sec?

$$P_M = \frac{W}{t}$$

$$P_M = \frac{5 \text{ lb-ft}}{5 \text{ sec}}$$

$$P_M = \mathbf{1\ lb\text{-}ft/sec}$$

Terms

A **watt (W)** is equal to one joule per second in the metric system.

One **horsepower (HP)** is equal to 746 W or 550 lb-ft per second (33,000 lb-ft per minute).

The watt is the basic unit of electrical power, and the horsepower (HP) is the basic unit of mechanical power. A *watt* (*W*) is equal to one joule per second in the metric system. In the English system, the foot-pound/second is used to express power. Because the horsepower is an English system unit used with motors, conversion between English and metric units is required. One *horsepower* (*HP*) is equal to 746 W or 550 lb-ft per second (33,000 lb-ft per minute). **See Figure 1-28.** Horsepower is calculated by applying the following formula:

$$HP = \frac{P_M}{550}$$

where

HP = horsepower

P_M = mechanical power (in lb-ft/sec)

550 = constant

HORSEPOWER

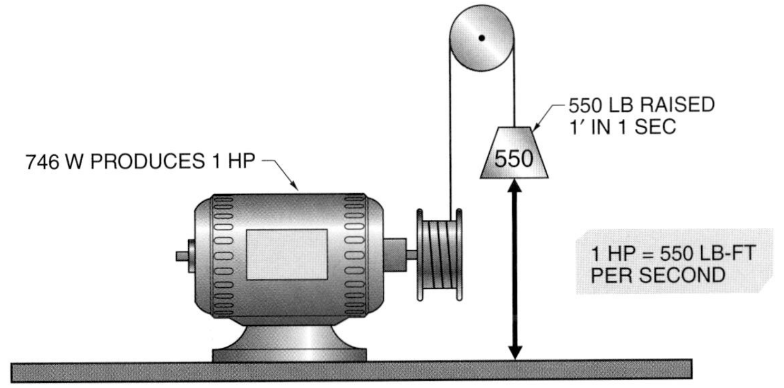

MECHANICAL ENERGY	ELECTRICAL ENERGY
½ HP	373 W
1 HP	746 W
2 HP	1492 W
5 HP	3730 W
100 HP	74,600 W

Figure 1-28. Motor power is rated in horsepower and watts.

Example: What is the horsepower required to lift 275 lb a distance of 1′ in 1 sec?

$$HP = \frac{P_M}{550}$$

$$HP = \frac{275}{550}$$

$$HP = \mathbf{0.5\ HP}$$

Tech Tip

A coulomb is a very large unit of measure. It requires nine billion newtons of force (approximately two billion pounds) to hold a positive coulomb (+C) 1 m from a negative coulomb (–C). A typically charged cloud about to produce a lightning bolt has a charge of about 30 coulombs.

Terms

Power loss is the difference between the power input and power output.

The **efficiency** of a system is the ratio of the power output to the power input.

Efficiency

Electrical and mechanical systems experience power loss when they require more input power than they can deliver to a load. *Power loss* is the difference between the power input and power output. In mechanical systems, power loss is due to friction. In electrical systems, power loss is due to electrical resistance (heat loss) in a circuit.

The *efficiency* of a system is the ratio of the power output to the power input. Efficiency is always less than one. Efficiency is most often used as a percentage. It is calculated by applying the following formula:

$$Eff = \frac{Power_{OUT}}{Power_{IN}} \times 100$$

where

Eff = efficiency (in %)

$Power_{OUT}$ = power output

$Power_{IN}$ = power input

Example: What is the efficiency of a system in which an audio amplifier can deliver 150 W of power to the load and the input power required is 200 W?

$$Eff = \frac{Power_{OUT}}{Power_{IN}} \times 100$$

$$Eff = \frac{150}{200} \times 100$$

$$Eff = 0.75 \times 100$$

$$Eff = \mathbf{75\%}$$

SCIENTIFIC NOTATION

The use of the powers of ten (scientific notation) is a simple method that can be used to express both extremely small and large numbers. **See Appendix.** Measurements of electricity often involve these numbers. For example, resistance is typically represented in megohms (1,000,000 = 10^6), and capacitance is often represented in picofarads ($1/100{,}000{,}000{,}000$ = 10^{-12}). Frequency is represented by gigahertz (1,000,000,000 = 10^9), and inductance is represented by millihenrys ($1/1000$ = 10^{-3}). Depending on the application, electrical potential can be represented by kilovolts (1000 = 10^3) and electrical current can be represented by microamperes ($1/1{,}000{,}000$ = 10^{-6}). Because these numbers involve a great number of zeros, it is much easier to express them using scientific notation. Modern test equipment will display a one- to four-digit value along with a unit symbol. For example, a resistance measurement on a digital multimeter (DMM) is displayed as 10 MΩ, rather than 10^7 Ω or 10,000,000 Ω. **See Figure 1-29.**

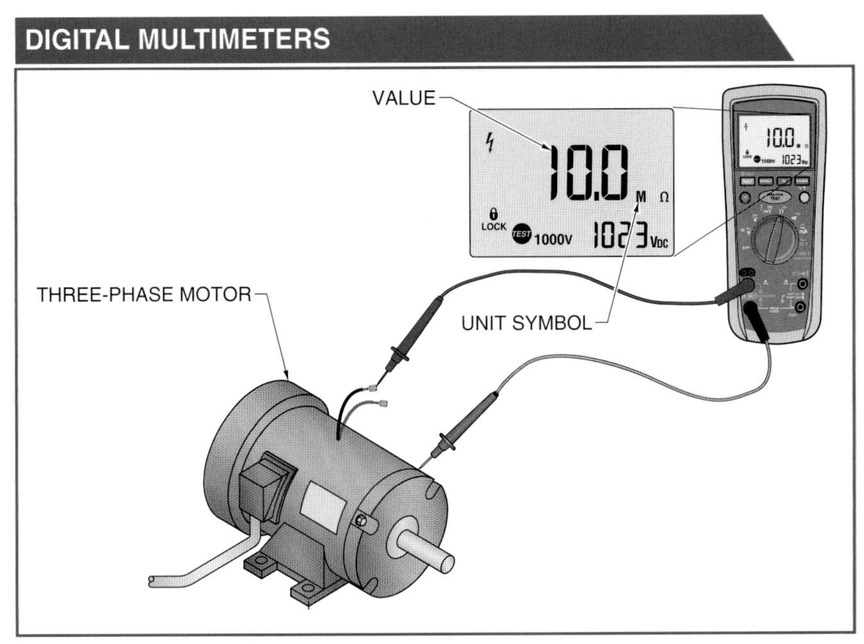

Figure 1-29. Modern test equipment displays a one- to four-digit value along with a unit symbol.

SUMMARY

An understanding of basic electricity principles, units of measure, physical breakdown of electricity, and safety principles is vital when beginning the study of electricity. Basic units of measure when working with electrical circuits and components include length, mass, time, temperature, current, light intensity, amount of substance, voltage, and resistance. Derived units of measure include work, energy, power, and efficiency. Conductors are materials that allow the flow of electrons while insulators are materials that resist, or limit, the flow of electrons.

ATPeResources.com/QuickLinks
Access Code: 232263

APPLICATION: INCANDESCENT LIGHT BULB CURRENT...

Background: Incandescent light bulbs have been commonly available for more than 100 years. Incandescent light bulbs are being phased out because about 95% of the consumed power is converted into heat rather than light. Incandescent light bulbs are being replaced by compact fluorescent lights (CFLs) and light-emitting diodes (LEDs) because of their efficiency.

Incandescent light bulbs are mostly resistive elements and are rated based on wattage. In the United States, these types of light bulbs are powered by 120 VAC. To find the amount of current used by an incandescent light bulb, the apparent power formula can be used. In this case, the apparent power in volt-amps is also equal to watts. The apparent power formula is the following:

$P_A = V \times I$
where
P_A = apparent power (in W)
V = volts (in V)
I = current (in A)

Example: Find the current required by a 40 W incandescent light bulb in a 120 VAC circuit.

$P_A = V \times I$
$40 \text{ W} = 120 \text{ VAC} \times I$

Therefore,

$$I = \frac{40 \text{ W}}{120 \text{ VAC}}$$

$I = 0.33 \text{ A}$

When multiple incandescent light bulbs are used on the same circuit, the total current required is the number of light bulbs times the current of one light bulb. The total current of multiple light bulbs on one circuit is calculated by applying the following formula:

$I_T = N_{LB} \times I_{LB}$
where
I_T = total current (in A)
N_{LB} = number of light bulbs
I_{LB} = current for one light bulb

Example: What is the total current of five 40 W light bulbs that use 0.33 A each?

$I_T = N_{LB} \times I_{LB}$
$I_T = 5 \times 0.33 \text{ A}$
$I_T = 1.65 \text{ A}$

Key Points: True power, apparent power, current

Problem: How many 100 W light bulbs can be connected to a 15 A circuit?

...APPLICATION: INCANDESCENT LIGHT BULB CURRENT

Solution: First, the amount of current needed by a single 100 W incandescent light bulb is calculated.

$$P_A = V \times I$$
$$100\text{ W} = 120\text{ VAC} \times I$$

Therefore,

$$I = \frac{100\text{ W}}{120\text{ VAC}}$$
$$I = 0.83\text{ A}$$

Once the current needed by a single 100 W light bulb is known, the maximum circuit current is divided by the individual light bulb current to find the maximum number of light bulbs that can be connected to a 15 A circuit as follows:

$$N_{LB} = \frac{I_{MAX}}{I_{LB}}$$

where
N_{LB} = number of light bulbs
I_{MAX} = maximum circuit current (in A)
I_{LB} = current for one light bulb

Therefore,

$$N_{LB} = \frac{I_{MAX}}{I_{LB}}$$
$$N_{LB} = \frac{15\text{ A}}{0.83\text{ A}}$$
$$N_{LB} = 18$$

Chapter Review

1. What is an electrostatic field?
2. What applications would require an electrician to have knowledge of element symbols?
3. How is the element carbon used with electrical components and devices?
4. What factor determines the rate in which a material conducts heat?
5. Why are gold-plated contacts not used in high-current applications?
6. What is an element?
7. Explain why aluminum is commonly used as a conductor in overhead power lines.
8. Per the NEC®, which material must conductors be composed of, unless otherwise specified?
9. When does conduction electrification occur?
10. Explain why most electrical contacts contain silver.
11. How many electron shells do most materials used in electronic and electrical applications contain?
12. What is the difference between conventional and electron flow?
13. What is a molecule?
14. How is work determined?
15. When is air gap used as an insulator?
16. What material is most commonly used as a conductor?
17. What are the three types of electrification?
18. List four types of electrical energy.
19. Considering bare, covered, or insulated conductors, which is the most commonly used in electrical systems?
20. Explain the difference between valance electrons in conductors and insulators.
21. What is used to determine an atom's weight and atomic number?
22. Why must the correct polarity be known when connecting equipment to a voltage source?
23. What are two additional terms used to identify voltage?
24. What are two applications in which static electricity has some limited practical use?
25. List three factors that affect the resistance of conductors.

2 Electrical Safety

OBJECTIVES

- Explain electrical safety rules and how they apply to working with electric circuits and equipment.
- Define the purpose of the *National Electrical Code®* (NEC®).
- Explain the effects of electrical shock.
- Describe proper grounding systems.
- List the different types of personal protective equipment (PPE).
- Explain the purpose of lockout/tagout systems and how they are used.
- List and explain the different classes of fires.
- Define hazardous locations.

An understanding of basic electrical safety rules and practices, proper equipment usage, equipment grounding, and the applicable codes and standards helps prevent accidents and injuries. In addition, an understanding of the correct use of personal protective equipment (PPE) for the working environment helps eliminate or reduce the severity of an injury. Working in the electrical field requires a thorough knowledge of PPE that prevents electrical shock and reduces the severity of an accident, the testing and maintenance of PPE, lockout-tagout procedures, electrical fire prevention, and hazardous locations.

ELECTRICAL SAFETY

Terms

The **National Fire Protection Association (NFPA)** is a national organization that provides guidance in assessing the hazards of the products of combustion.

Technicians must work safely at all times. Following electrical safety rules is required when working with electrical and electronic equipment to help prevent injuries caused by electrical energy sources. Electrical safety has been advanced by the efforts of the National Fire Protection Association (NFPA), the Occupational Safety and Health Administration (OSHA), and state safety laws. The *National Fire Protection Association (NFPA)* is a national organization that provides guidance in assessing the hazards of the products of combustion. The NFPA sponsors the development of the National Electrical Code® (NEC®).

National Electrical Code®

The National Electrical Code® (NEC®) is one of the most widely used and recognized electrical consensus standards in the world. The purpose of the NEC® is to protect people and property from hazards that arise from the use of electricity. Improper procedures when working with electricity can cause permanent injury or death. Many city, county, state, and federal agencies use the NEC® to set requirements for electrical installations. The NEC® is updated every three years. Electrical safety rules include the following:

- Always comply with the NEC®, state, and local codes.
- Use Underwriters Laboratories® (UL) approved equipment, components, and test equipment.
- Before removing any fuse from a circuit, be sure the switch for the circuit is open or disconnected. When removing fuses, use an approved fuse puller and break contact on the line side of the circuit first. When installing fuses, install the fuse first into the load side of the fuse clip, then into the line side.
- Inspect and test grounding systems for proper operation. Ground any conductive component or element that is not energized.
- Turn OFF, lock out, and tag out any circuit that does not need to be energized when maintenance is being performed.
- Always use personal protective equipment (PPE) and safety equipment.
- Perform the appropriate task required during an emergency situation.
- Use only a Class C fire extinguisher on electrical equipment. The color blue inside a circle identifies a Class C fire extinguisher.
- Always work with another individual when working in a dangerous area, on dangerous equipment, or with high voltages.
- Do not work when tired or taking medication that causes drowsiness unless specifically authorized by a physician.

- Do not work in poorly lighted areas.
- Ensure there are no atmospheric hazards such as flammable dust or vapor in the area.
- Use one hand when working on a live circuit to reduce the chance of an electrical shock passing through the heart and lungs.
- Never bypass fuses, circuit breakers, or any other safety device.

Electrical Shock

Electrical shock is a condition that results any time a body becomes part of an electrical circuit. Electrical shock effects vary from a mild sensation, to paralysis, to death. Also, severe burns may occur where current enters and exits the body. The severity of an electrical shock depends on the amount of electric current in milliamps (mA) that flows through the body, the length of time the body is exposed to the current, the path the current takes through the body, and the physical size and condition of the body through which the current passes. These differing current values indicate a ground fault in which the missing return current may be taking an alternate path to ground through an operator. This could result in serious injury or even death. Ground-fault circuit interrupters (GFCIs) are used to help prevent electrocutions and react quickly to a current imbalance, usually in less than 1/10 of a second. GFCIs should be installed where electrical circuits may accidentally come into contact with water, such as in kitchens, bathrooms, laundry rooms, and garages. GFCIs do not protect individuals from line contact hazards. **See Figure 2-1.**

Terms

Electrical shock is a condition that results any time a body becomes part of an electrical circuit.

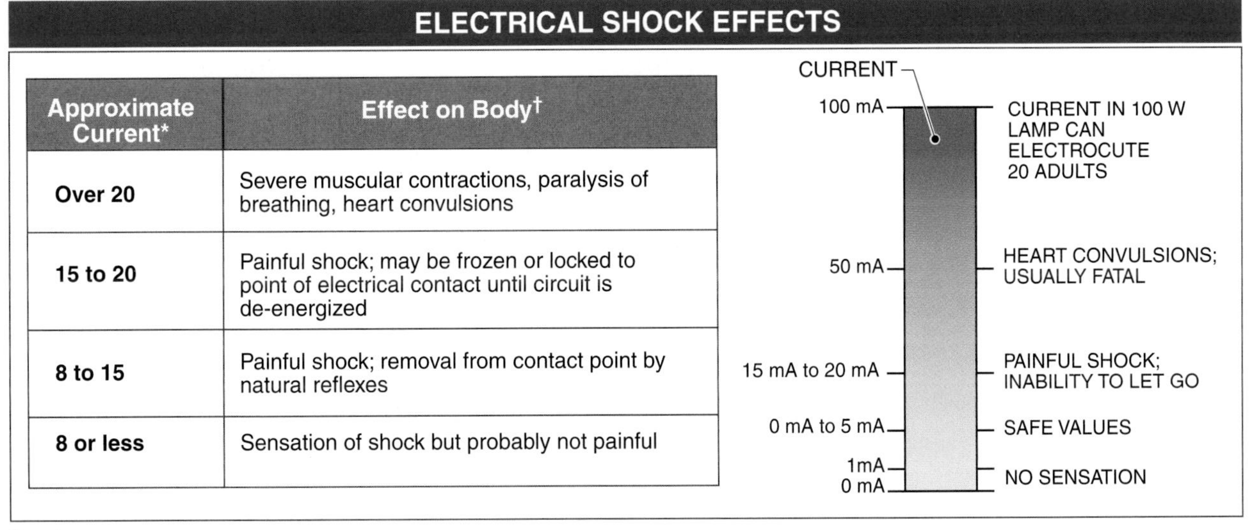

ELECTRICAL SHOCK EFFECTS

Approximate Current*	Effect on Body†
Over 20	Severe muscular contractions, paralysis of breathing, heart convulsions
15 to 20	Painful shock; may be frozen or locked to point of electrical contact until circuit is de-energized
8 to 15	Painful shock; removal from contact point by natural reflexes
8 or less	Sensation of shock but probably not painful

* in mA
† effects vary depending on time, path, amount of exposure, and condition of body

Figure 2-1. Electrical shock is a condition that results any time a body becomes part of an electrical circuit.

Individuals who work on or near electrical equipment must have respect for all voltages, have knowledge of the principles of electricity, and follow safe work procedures. All technicians should be encouraged to take a basic course in cardiopulmonary resuscitation (CPR) so they can aid a coworker in emergencies.

During an electrical shock, the body of a person becomes part of an electrical circuit. The resistance of a person's body to the flow of current varies. Sweaty hands have less resistance than dry hands. Standing on a wet floor provides less resistance than standing on a dry floor. Lower resistance increases current flow. As the current increases, the severity of the electrical shock increases.

If a person is receiving an electrical shock, power must be removed as quickly as possible. If the disconnect switch for the equipment circuit is nearby and can be operated safely, the power must be shut OFF. If power cannot be removed quickly, the victim must be removed from contact with the live parts. Insulated protective equipment, such as a hot stick, rubber gloves, wood poles, or plastic pipes, can be used to separate the victim from the energized circuit. When free from the electrical hazard, the victim must receive immediate medical attention.

Terms

Grounding is the connection of all exposed non-current-carrying metal parts to earth.

Grounding

Grounding is the connection of all exposed non-current-carrying metal parts to earth. Grounding provides a direct path to earth for unwanted fault current and safeguards equipment and personnel against the hazards of electrical shock. Unwanted current may exist because of insulation failure or because a current-carrying conductor makes contact with a non-current-carrying part of the system. Grounding is accomplished by connecting a circuit to a metal underground water pipe, a metal frame of a building, a grounding conductor, a ground ring, or a concrete-encased electrode. **See Figure 2-2.** To prevent problems, the grounding path must be as short as possible and of sufficient diameter (as recommended by the NEC®), must never be fused or switched, must be a permanent part of the electrical circuit, and must be continuous and uninterrupted from the electrical circuit to the grounding electrode.

Fluke Corporation

Earth-ground resistance testers are used to verify the integrity of grounding systems.

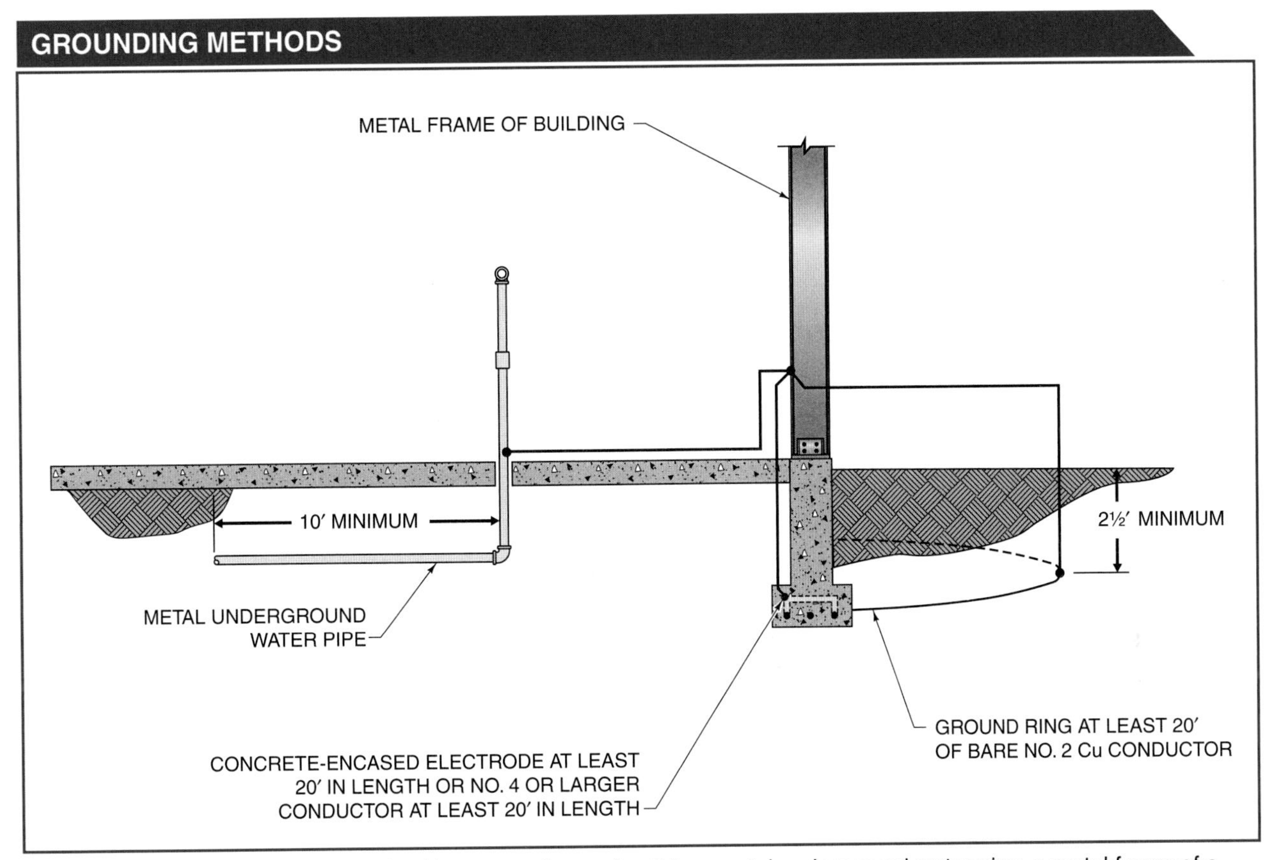

Figure 2-2. Grounding is accomplished by connecting a circuit to a metal underground water pipe, a metal frame of a building, a ground ring, or a concrete-encased electrode.

Personal Protective Equipment

Personal protective equipment (PPE) is clothing and/or equipment worn by a technician to reduce the possibility of injury in the work area. Minimum personal protective equipment includes protective clothing, head protection, eye protection, and hand protection. Additional equipment such as protective footwear, ear protection, or electrical insulating matting may be required for certain tasks. **See Figure 2-3.** The use of PPE is required whenever work is performed on or near energized exposed electrical circuits. The National Fire Protection Association standard, NFPA 70E, *Standard for Electrical Safety in the Workplace,* addresses "electrical safety requirements for employee workplaces that are necessary for the safeguarding of employees in their pursuit of gainful employment." For maximum safety, PPE must be used as recommended in NFPA 70E, OSHA Standard Part 1910, Subpart 1, *Personal Protective Equipment* (1910.132 through 1910.138) and other applicable safety mandates. Per NFPA 70E, the type of PPE required depends on the voltage and where the work is being performed.

Terms

Personal protective equipment (PPE) is clothing and/or equipment worn by a technician to reduce the possibility of injury in the work area.

PERSONAL PROTECTIVE EQUIPMENT

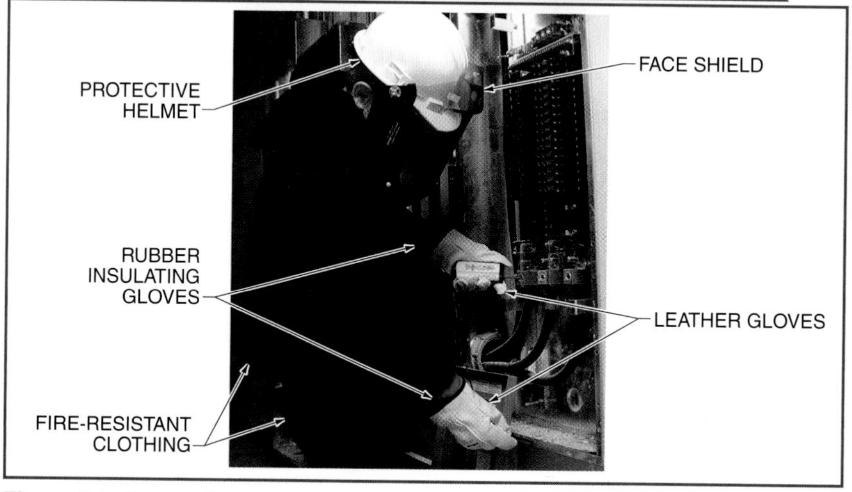

Figure 2-3. Personal protective equipment (PPE) is clothing and/or equipment worn by a technician to reduce the possibility of injury in the work area.

Protective Clothing. Sparks from an electrical circuit can cause a fire. Clothing made of synthetic materials such as nylon, polyester, or rayon must never be worn because they can burn and melt to the skin. The minimum PPE requirement for clothing is an untreated natural material long-sleeve shirt and long pants. Some applications require higher levels of protective clothing.

Approved flame-resistant (FR) clothing must be worn in conjunction with rubber insulating gloves for protection from electrical arcs when performing certain operations on or near energized equipment or circuits. FR clothing must be kept as clean and sanitary as practical and must be inspected prior to each use. Defective clothing must be removed from service immediately and replaced.

Head Protection. Head protection requires using a protective helmet. A *protective helmet* is a hard hat that is used in the workplace to prevent injury from the impact of falling and flying objects and from electrical shock. Protective helmets resist penetration and absorb impact force. Protective helmet shells are made of durable, lightweight materials. A lining keeps the shell away from the head to absorb shock and provide ventilation. Protective helmets are identified by class of protection against specific hazardous conditions.

Terms

A **protective helmet** is a hard hat that is used in the workplace to prevent injury from the impact of falling and flying objects and from electrical shock.

Class G, E, and C helmets are used in construction and in industrial applications. Class G protective helmets protect against low-voltage shock and burns and impact hazards, and are commonly used in construction and manufacturing facilities. Class E protective helmets protect against high-voltage shock and burns, impact hazards, and penetration by falling or flying objects. Class C protective helmets are manufactured with light materials yet provide adequate impact protection.

Eye Protection. Eye protection is worn to prevent eye or face injuries caused by flying particles, contact arcing, and radiant energy. Eye protection must comply with OSHA 29 CFR 1910.133, *Eye and Face Protection.* Eye protection standards are specified in ANSI Z87.1, *Occupational and Educational Eye and Face Protection.* Eye protection includes safety glasses, face shields, and arc-rated hoods. **See Figure 2-4.** *Safety glasses* are an eye protection device with special impact-resistant glass or plastic lenses, reinforced frames, and side shields. The plastic frames are designed to keep the lenses secured in the frame if an impact occurs. Plastic frames also minimize the shock hazard when working with electrical equipment. Side shields provide additional protection from flying objects. Tinted-lens safety glasses protect against low-voltage arc hazards.

Terms

Safety glasses are an eye protection device with special impact-resistant glass or plastic lenses, reinforced frames, and side shields.

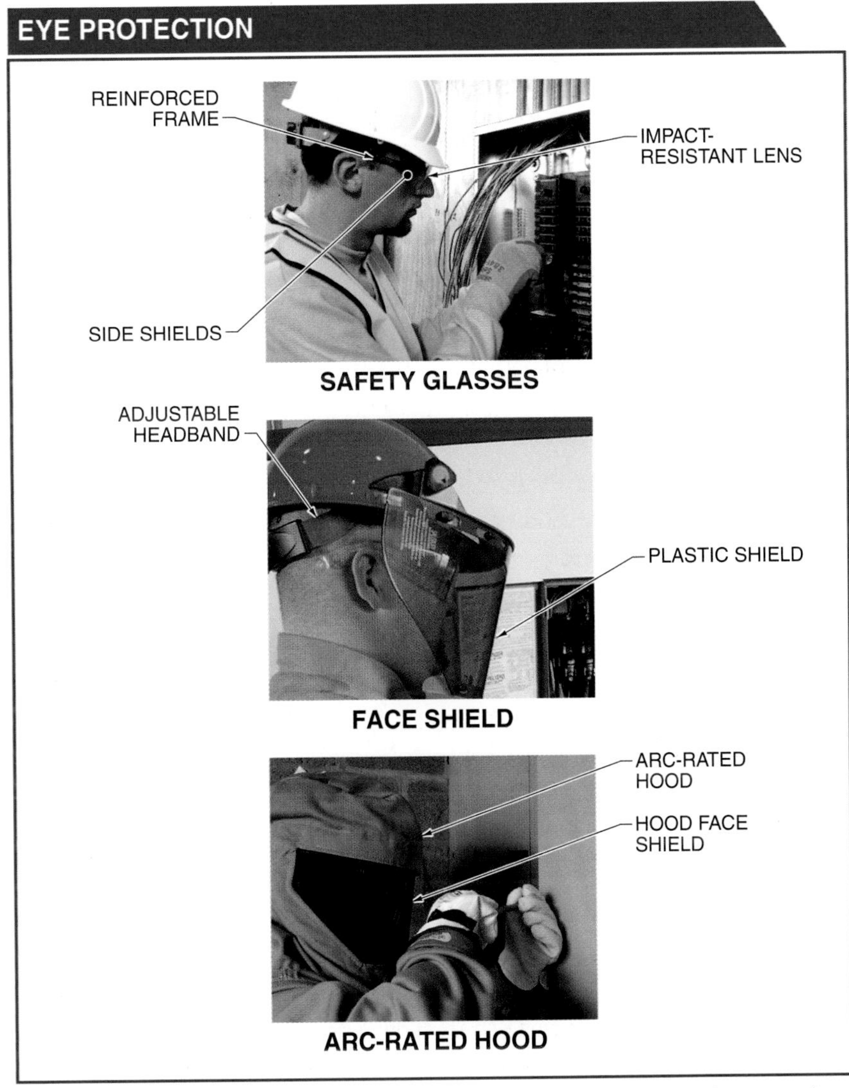

Figure 2-4. Eye protection includes safety glasses, face shields, and arc-rated hoods.

Terms

An **arc-rated face shield** is any eye and face protection device that covers the entire face with a plastic shield and is used for protection from flying objects.

An **arc-rated hood** is an eye and face protection device that consists of a flame-resistant hood and face shield.

Rubber insulating gloves are gloves made of latex rubber and are used to provide maximum insulation from electrical shock.

Leather protectors are gloves worn over rubber insulating gloves to prevent penetration of the rubber insulating gloves and to provide added protection against electrical shock.

An *arc-rated face shield* is any eye and face protection device that covers the entire face with a plastic shield and is used for protection from flying objects. Tinted face shields protect against low-voltage arc hazards.An *arc-rated hood* is an eye and face protection device that consists of a flame-resistant hood and face shield. Safety glasses, face shields, and arc-rated hoods must be properly maintained to provide protection and clear visibility.

Hand Protection. Hand protection includes gloves worn to prevent injuries to hands from cuts or electrical shock. The appropriate hand protection required is determined by the duration, frequency, and degree of the hazard to the hands. *Rubber insulating gloves* are gloves made of latex rubber and are used to provide maximum insulation from electrical shock. Rubber insulating gloves are rated and labeled for maximum voltage allowed, such as 26,500 V. *Leather protectors* are gloves worn over rubber insulating gloves to prevent penetration of the rubber insulating gloves and to provide added protection against electrical shock. Rubber insulating gloves and leather protectors must always be used together. **See Figure 2-5.**

The entire surface of a rubber insulating glove must be field tested (visual inspection and air test) before each use. In addition, rubber insulating gloves must also be laboratory tested by an approved laboratory every six months. Visual inspection of a rubber insulating glove is performed by stretching a small area of the glove (particularly the fingertips) and checking for defects such as punctures or pin holes, embedded or foreign material, deep scratches or cracks, cuts or snags, or deterioration caused by oil, heat, grease, insulating compounds, or any other substance that may harm the rubber.

Rubber insulating gloves must also be air tested when there is cause to suspect damage. The entire surface of the glove must be inspected by rolling the cuff tightly toward the palm so that air is trapped inside the glove, or by using a mechanical inflation device. The glove is then examined for punctures and other defects. Puncture detection is made easier by listening for escaping air while holding the glove to the face or ear. Some brands of rubber insulating gloves are available with two color layers. When one color layer becomes visible, the color serves as an indication to the user that the gloves must be replaced.

Proper care of leather protectors is essential for user safety. Leather protectors should be checked for cuts, tears, holes, abrasions, defective or worn stitching, oil contamination, and any other condition that might prevent them from adequately protecting rubber insulating gloves. Any substance that could physically damage rubber insulating gloves must be removed before using protector gloves.

RUBBER INSULATING GLOVE CLASSES		
Class	**Maximum Use Voltage***	**Color Of Label**
00	500	Beige
0	1000	Red
1	7500	White
2	17,000	Yellow
3	26,500	Green
4	36,000	Orange

* in V

LEATHER PROTECTORS

RUBBER INSULATING GLOVES

Figure 2-5. Hand protection includes gloves worn to prevent injuries to hands from cuts or electrical shock.

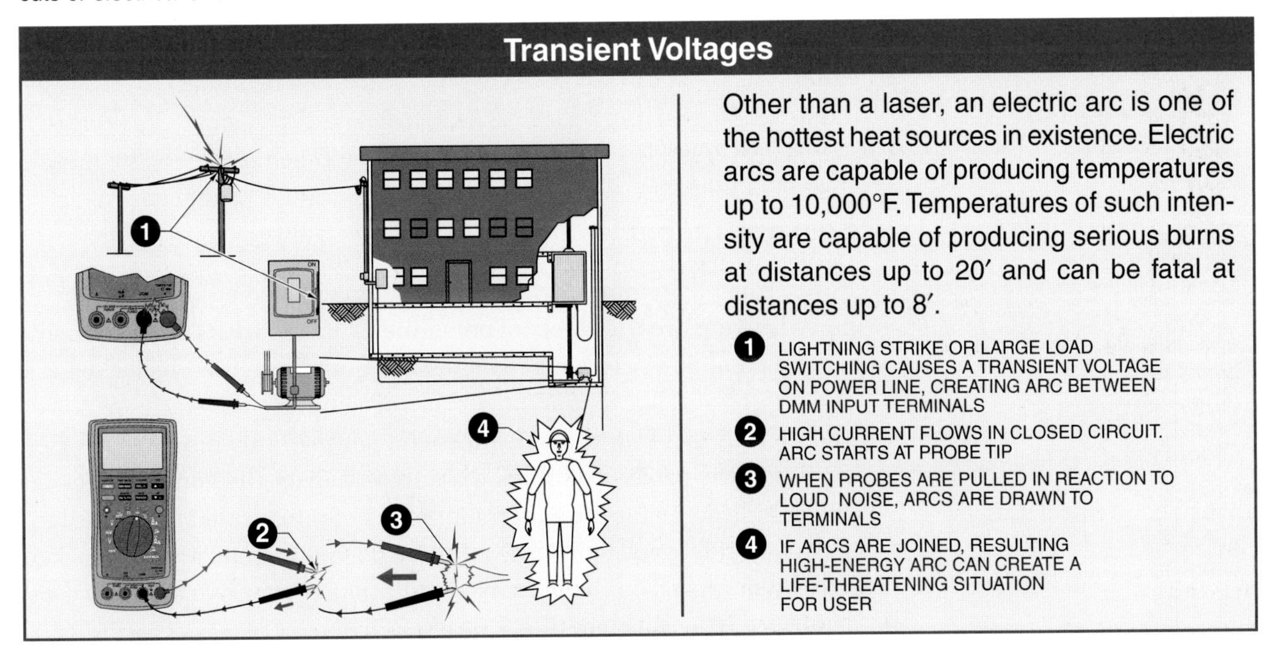

Electrical Insulating Matting. Electrical insulating matting is a personal protective device that provides electricians protection from electrical shock when working on energized electrical circuits. Dielectric black fluted rubberized insulating matting is specifically designed for use in front of open cabinets or high-voltage equipment and is used to protect electricians when working with voltages that are over 50 V. Electrical insulating matting is classified by material thickness, material width, and maximum working voltage. **See Figure 2-6.**

ELECTRICAL INSULATING MATTING RATINGS

Safety Standard	Material Thickness		Material Width (in.)	Test Voltage	Maximum Working Voltage
	Inches	Millimeters			
BS921*	0.236	6	36	11,000	450
BS921*	0.236	6	48	11,000	450
BS921*	0.354	9	36	15,000	650
BS921*	0.354	9	48	15,000	650
VDE0680†	0.118	3	39	10,000	1000
ASTM D178‡	0.236	6	24	25,000	17,000
ASTM D178‡	0.236	6	30	25,000	17,000
ASTM D178‡	0.236	6	36	25,000	17,000
ASTM D178‡	0.236	6	48	25,000	17,000

* BSI—British Standards Institute
† VDE—Verband Deutscher Elektrotechniker Testing and Certification Institute
‡ ASTM International

Figure 2-6. Electrical insulating matting is classified by material thickness, material width, and maximum working voltage.

Terms

Lockout is the process of removing the source of power and installing a lock that prevents the power from being turned ON.

Tagout is the process of placing a danger tag on the source of power, which indicates that the equipment may not be operated until the danger tag is removed.

Lockout/Tagout

Power must be removed when equipment is inspected, serviced, or repaired. To ensure the safety of personnel working with the equipment, all electrical power is removed and the equipment must be locked out and tagged out. *Lockout* is the process of removing the source of power and installing a lock that prevents the power from being turned ON. *Tagout* is the process of placing a danger tag on the source of power, which indicates that the equipment may not be operated until the danger tag is removed. Per OSHA standards, equipment must be locked out and/or tagged out before any installation, preventive maintenance, or service is performed. **See Figure 2-7.**

LOCKOUT/TAGOUT

Panduit Corp.

Figure 2-7. Per OSHA standards, equipment must be locked out and/or tagged out before any installation, preventive maintenance, or service is performed.

A danger tag has the same importance and purpose as a lock and is used alone only when a lock does not fit the disconnect device. A danger tag is attached at the disconnect device with a tag tie or equivalent and has space for the technician's name, craft, and other company-required information. A danger tag must withstand the elements and expected atmosphere for the maximum period of time that exposure is expected. Lockout/tagout is used in the following circumstances:

- power is not required to be ON to a piece of equipment to perform a task
- machine guards or other safety devices are removed or bypassed
- the possibility exists of being injured or caught in moving machinery
- jammed equipment is being cleared
- the danger exists of being injured if equipment power is turned ON

Terms

A **qualified person** is a person who has knowledge and skills related to the construction and operation of electrical equipment and has received appropriate safety training.

Lockout and tagouts do not by themselves remove power from a machine or its circuitry. OSHA provides a standard procedure for equipment lockout/tagout. Lockout is performed and tags are attached only after the equipment is turned OFF and tested.

Any person other than the authorized person who installed the lockout/tagout, except in an emergency, must not remove the lockout/tagout. In an emergency, only a qualified person may remove a lockout/tagout and only upon notification of the authorized person. A *qualified person* is a person who has knowledge and skills related to the construction and operation of electrical equipment and has received appropriate safety training. When more than one technician is required to perform a task on a piece of equipment, each technician must place a lockout and/or tagout on any energy-isolating device(s). A multiple lockout/tagout device (hasp) must be used because energy-isolating devices typically cannot accept more than one lockout/tagout. A list of company rules and procedures should be given to authorized personnel and any person who may be affected by a lockout/tagout.

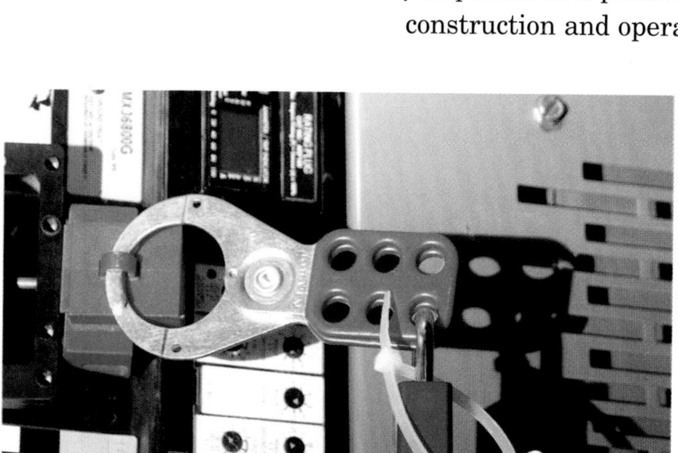

A multiple-hasp lockout/tagout device is required in situations where more than one technician performs work on a piece of electrical equipment.

Fire Safety

Fire safety requires following established procedures to reduce or eliminate conditions that could cause a fire. Guidelines in assessing hazards of the products of combustion are provided by the NFPA. Technicians must take responsibility for preventing the conditions that could result in a fire. This includes proper use and storage of combustible materials, immediate cleanup of combustible spills, and proper precautions when working with electrical hazards.

Prior to starting electrical work, all individuals should be advised of the location of the nearest telephone and fire alarm reporting station for summoning emergency medical assistance. When telephones are not available, two-way radios or some other means of rapid communication must be used. A procedure to evacuate the premises and account for all personnel in the event of a fire must be in place and practiced on a regular basis.

All facilities must have a fire safety plan. A fire safety plan establishes procedures that must be followed if a fire occurs. The fire safety plan lists the locations of the main electrical breaker, fire main, exits, fire alarms, and fire extinguishers for each area of a facility.

Incident Energy and Distance

Incident energy decreases by the inverse square of the distance as a person moves away from an arc source. This describes the relationship between incident energy and distance and also indicates the importance of working distance. The reverse is also true: incident energy increases by the square of the distance as a person gets closer to an arc source.

If an arc flash hazard warning label specifies a level of incident energy at a working distance of 18″, changes in distance will affect the incident energy a person receives. This is important to keep in mind if a worker moves closer to an exposed energized conductor in order to see through his or her bifocal glasses. Without realizing it, the worker will have reduced the effectiveness of his or her arc-rated clothing and PPE.

Below are a few examples of how the incident energy increases as distance decreases. The following calculations were performed using IEEE 1584-based software:

- At a working distance of 18″, the incident energy is listed on the label as 8.0 cal/cm^2.
- At a distance of 12″, the incident energy is 15.7 cal/cm^2.
- At a distance of 6″, the incident energy is 49.2 cal/cm^2.

This is an example of an instance when a worker would probably wear arc-rated PPE and equipment rated at 10 cal/cm^2 because it is commonly available and would properly protected the worker for the specified working distance of 18″. However, simply by decreasing the working distance, the worker might receive arc-related burns when he or she should not have received any.

Classes of Fires. The five classes of fires are Class A, Class B, Class C, Class D, and Class K. Class A fires include burning wood, paper, textiles, and other ordinary combustible materials containing carbon. Class B fires include burning oil, gas, grease, paint, and other liquids that convert to a gas when heated. Class C fires include burning electrical devices, motors, and transformers. Class D is a specialized class of fires including burning metals such as zirconium, titanium, magnesium, sodium, and potassium. Class K fires include burning grease in commercial cooking equipment. Fire extinguishers are selected for the class of fire based on the combustibility of the material. **See Figure 2-8.**

Emergency Exit Plan and Procedures

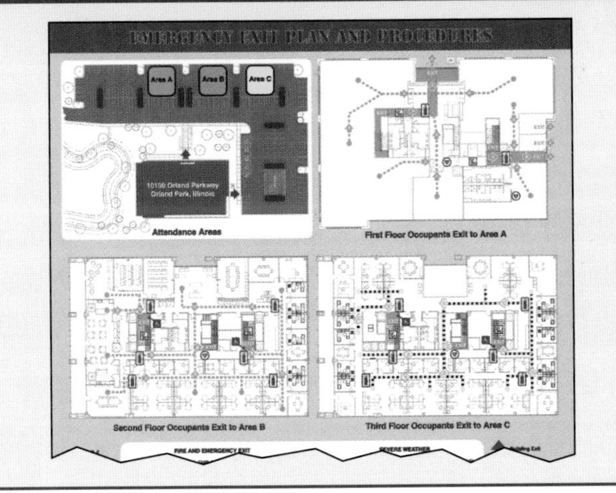

A fire safety plan is a part of an overall emergency exit plan and procedures. All commercial and industrial facilities should have a posted emergency procedure plan that includes emergency exits, fire extinguisher locations, severe weather procedures, first aid stations, and emergency contact information such as local police and fire department phone numbers. All individuals and visitors who are located within the facility should perform a regular review of this information.

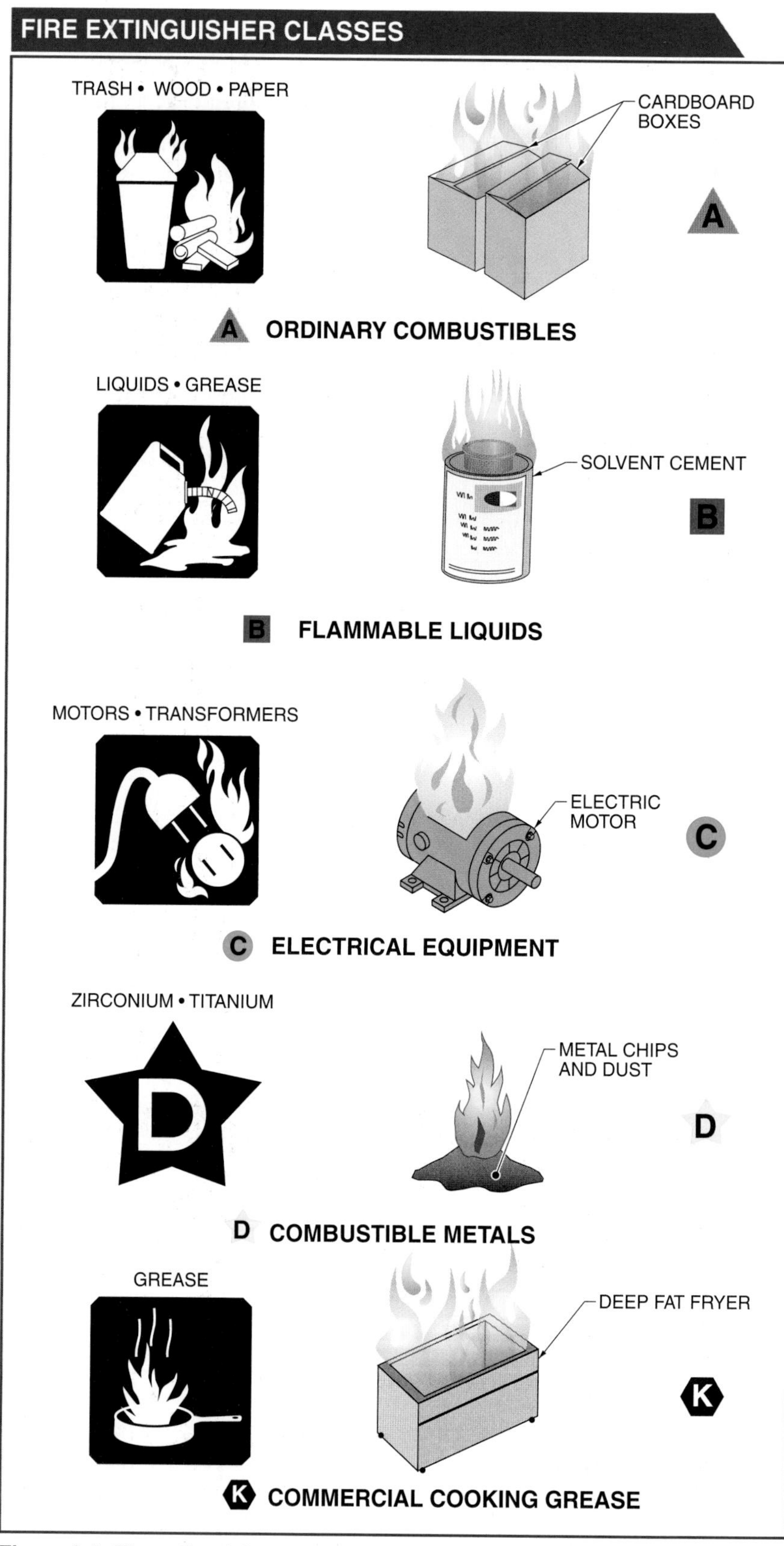

Figure 2-8. Fire extinguishers are selected for the class of fire based on the combustibility of the material.

Fuel, heat, and oxygen are the elements required to start and sustain a fire. A fire extinguishes itself when any one of the three elements is removed. Fire extinguishers operate by removing one or more of these elements, such as when cooling the burning materials or displacing the oxygen from the fire area. The instructions for use should be understood before using a fire extinguisher, and the correct fire extinguisher must be used for the class of fire.

Always refer to OSHA and NEC® guidelines when determining the required PPE for performing work with various types of electrical equipment.

Hazardous Locations

Article 500 of the NEC® and Article 440 of the NFPA cover hazardous locations. **See Figure 2-9.** Any hazardous location requires a maximum of safety and adherence to local, state, and federal guidelines and laws as well as in-plant safety rules. Locations can be hazardous due to proximity to potentially dangerous equipment, toxic or explosive atmospheres, or the possibility of entrapment.

Areas around exposed electrical equipment, moving mechanical equipment, or equipment at extreme temperatures or pressures may be considered hazardous. These areas may be indicated with danger or caution signs, and a hazard boundary may be marked by color-coded tape on floor or walls. Anyone working within the boundary must wear the correct PPE and follow proper procedures for that location's hazards.

Determining Classifications of Hazardous Locations

Designers and installers of electrical systems in hazardous locations have a responsibility to ensure that the installation of any electrical system does not jeopardize the safety of the personnel working in the area or the public at large. Often, the determination of the classification for a particular area can be a very complex issue. Designers and installers must determine the correct classification of the area. If sufficient information is not available to determine the correct classification, the owners of the building or structure where the classified area is located, their engineering support staff, and the authority having jurisdiction must be consulted before designing or installing electrical systems for these areas.

HAZARDOUS LOCATIONS—ARTICLE 500

Hazardous Location – A location where there is an increased risk of fire or explosion due to the presence of flammable gases, vapors, liquids, combustible dusts, or easily-ignitable fibers or flyings.

Location – A position or site.

Flammable – Capable of being easily ignited and of burning quickly.

Gas – A fluid (such as air) that has no independent shape or volume but tends to expand indefinitely.

Vapor – A substance in the gaseous state as distinguished from the solid or liquid state.

Liquid – A fluid (such as water) that has no independent shape but has a definite volume. A liquid does not expand indefinitely and is only slightly compressible.

Combustible – Capable of burning.

Ignitable – Capable of being set on fire.

Fiber – A thread or piece of material.

Flyings – Small particles of material.

Dust – Fine particles of matter.

Classes	Likelihood that a flammable or combustible concentration is present
I	Sufficient quantities of flammable gases and vapors present in air to cause an explosion or ignite hazardous materials
II	Sufficient quantities of combustible dust are present in air to cause an explosion or ignite hazardous materials
III	Easily ignitable fibers or flyings are present in air, but not in a sufficient quantity to cause an explosion or ignite hazardous materials

Divisions	Location containing hazardous substances
1	Hazardous location in which hazardous substance is normally present in air in sufficient quantities to cause an explosion or ignite hazardous materials
2	Hazardous location in which hazardous substance is not normally present in air in sufficient quantities to cause an explosion or ignite hazardous materials

Class I Division I:

- Spray booth interiors
- Areas adjacent to spraying or painting operations using volatile flammable solvents
- Open tanks or vats of volatile flammable liquids
- Drying or evaporation rooms for flammable vents
- Areas where fats and oils extraction equipment using flammable solvents is operated
- Cleaning and dyeing plant rooms that use flammable liquids that do not contain adequate ventilation
- Refrigeration or freezer interiors that store flammable materials
- All other locations where sufficient ignitable quantities of flammable gases or vapors are likely to occur during routine operations

Class II Division I:

- Grain and grain products
- Pulverized sugar and cocoa
- Dried egg and milk powders
- Pulverized spices
- Starch and pastes
- Potato and wood flour
- Oil meal from beans and seeds
- Dried hay
- Any other organic materials that may produce combustible dusts during their use or handling

Class III Division I:

- Portions of rayon, cotton, or other textile mills
- Manufacturing and processing plants for combustible fibers, cotton gins, and cotton seed mills
- Flax processing plants
- Clothing manufacturing plants
- Woodworking plants
- Other establishments involving similar hazardous processes or conditions

HAZARDOUS LOCATIONS		
Class	**Group**	**Material**
I	A	Acetylene
	B	Hydrogen, butadiene, ethylene oxide, propylene oxide
	C	Carbon monoxide, ether, ethylene, hydrogen sulfide, morpholine, cyclopropane
	D	Gasoline, benzene, butane, propane, alcohol, acetone, ammonia, vinyl chloride
II	E	Metal dusts
	F	Carbon black, coke dust, coal
	G	Grain dust, flour, starch, sugar, plastics
III	No groups	Wood chips, cotton, flax, and nylon

Figure 2-9. Article 500 of the NEC® covers hazardous locations.

Life-threatening atmospheres can develop because of oxygen deficiency or the presence of combustible and/or toxic gases. Oxygen deficiency is caused by the displacement of oxygen by leaking gases or vapors, the combustion or oxidation process, oxygen absorbed by the vessel or product stored, and/or oxygen consumed by bacterial action. Oxygen-deficient atmospheres in confined spaces can cause life-threatening conditions. **See Figure 2-10.** Leaking gases such as methane, carbon monoxide, carbon dioxide, and hydrogen sulfide commonly cause combustible or toxic atmospheres. An increase in the oxygen level above the normal 21% further increases the explosive potential of combustible gases. Finely ground materials including carbon, grain, fibers, metals, and plastics can also cause explosive atmospheres. The use of electrical equipment in areas where explosion hazards are present can lead to an explosion and fire.

POTENTIAL EFFECTS OF OXYGEN-DEFICIENT ATMOSPHERES*

Oxygen Content†	Effects and Symptoms‡
19.5	Minimum permissible oxygen level
15 to 19.5	Decreased ability to work strenuously; may impair condition and induce early symptoms in persons with coronary, pulmonary, or circulatory problems
12 to 14	Respiration exertion and pulse increase; impaired coordination, perception, and judgement
10 to 11	Respiration further increases in rate and depth; poor judgement; lips turn blue
8 to 9	Mental failure, fainting, unconsciousness, ashen face, blue lips, nausea and vomiting
6 to 7	Eight minutes, 100% fatal; 6 minutes 50% fatal; 4 to 5 minutes, recovery with treatment
4 to 5	Coma in 40 seconds, convulsions, respiration ceases, death

* Values are approximate and vary with state of health and physical activities
† % by volume
‡ at atmospheric pressure

Bacharach, Inc.

Figure 2-10. Oxygen-deficient atmospheres in confined spaces create life-threatening conditions.

Terms

A **confined space** is a space large enough that an employee can physically enter and perform assigned work, but has limited or restricted means for entry and exit, and is not designed for continuous employee occupancy.

Spaces that are small, crowded with obstructions, or have limited access are considered hazardous. A *confined space* is a space large enough that an employee can physically enter and perform assigned work, but has limited or restricted means for entry and exit, and is not designed for continuous employee occupancy. Confined spaces include storage tanks, process vessels, boilers, ventilation or exhaust ducts, sewers, underground utility vaults, pipelines, and open top spaces more than 4′ in depth such as pits and ditches. Confined spaces are also particularly susceptible to developing toxic or explosive atmospheres. Confined space permits are required for work in confined spaces and include procedures for monitoring the atmosphere and posting additional personnel outside the space to assist in an emergency.

SUMMARY

Electrical safety includes using personal protective equipment (PPE) when working on or near energized circuits, grounding equipment, taking precautions against electrical shock, using lockout/tagout procedures, and working safely in hazardous locations. Electrical safety practices, as described by the NEC®, must always be followed when performing work on or near energized electrical and electronic equipment.

APPLICATION: ELECRICAL SHOCK HAZARD

Background: Working with electricity can be hazardous. The effects of electricity on the human body vary based on the amount of current and amount of time exposed to the current. **See Figure 1.** The body begins to be affected by electricity at exposures above 5 mA.

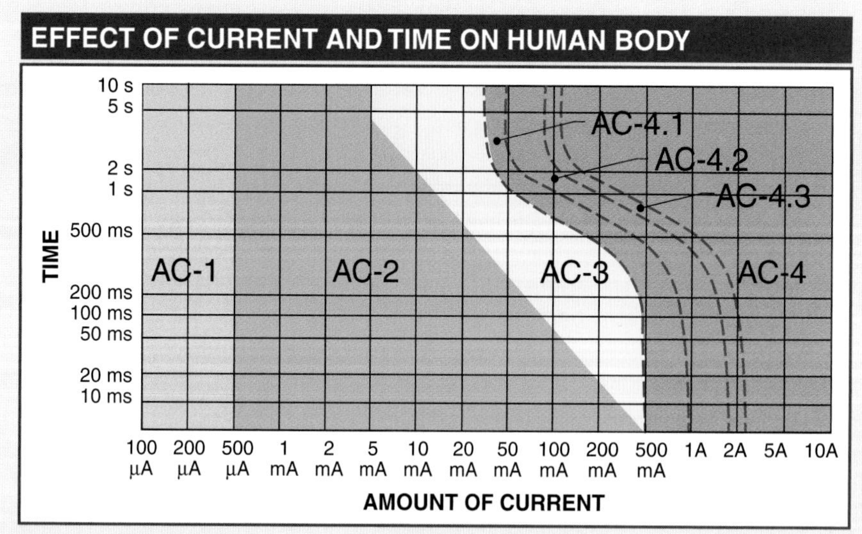

Figure 1. The effects of electricity on the human body vary based on the amount of current and amount of time exposed to the current.

APPLICATION: ELECTRICAL SHOCK HAZARD...

Based on current (I) and time exposure (T), alternating current passing through the human body has different detrimental effects as defined in International Electrical Commission (IEC) publication 60479-1. The following are the effects of electricity on the body at various levels of exposure:

- AC-1: imperceptible
- AC-2: perceptible but no muscle reaction
- AC-3: muscle contraction with reversible effects
- AC-4: possible irreversible effects
- AC-4.1: up to 5% probability of ventricular fibrillation
- AC-4.2: 5% to 50% probability of fibrillation
- AC-4.3: over 50% probability of fibrillation

The amount of current that flows through a body is dependent on the voltage and body resistance. Resistance of the human body to the flow of electricity can vary. In dry environments, the human body has a resistance of about 100,000 Ω. Resistance can be 1000 Ω or less if the skin is wet or broken.

Ohm's law is a simple formula that relates voltage, resistance, and current. Based on Ohm's law, when resistance decreases, current increases. To find the current, apply the following formula based on Ohm's law:

$$I = \frac{V}{R}$$

where
I = current (in A)
V = voltage (in V)
R = resistance (in Ω)

For example, how much current flows through the body if a person contacts 120 VAC in a dry environment?

$$I = \frac{V}{R}$$

$$I = \frac{120\ VAC}{100{,}000\ \Omega}$$

$$I = 1.2\ \text{mA}$$

Key Points: Ohm's law

Problem: If, after a heavy exercise workout, someone covered with sweat contacts 120 VAC, what level of exposure would they be susceptible to?

Solution: A human body that is wet has a resistance of 1000 Ω or less. The current, based on 1000 Ω, is found by applying the formula based on Ohm's law.

...APPLICATION: ELECTRICAL SHOCK HAZARD

$$I = \frac{V}{R}$$

$$I = \frac{120\ VAC}{1000\ \Omega}$$

$$I = 120\ \text{mA}$$

At 120 mA, the amount of time exposed to the current is a critical factor. In 500 ms, there is a probability of ventricular fibrillation.

Chapter Review

1. What is the purpose of the NEC®?
2. Explain where dielectric black fluted rubberized insulating matting is used and what voltages it is designed to provide protection from.
3. What three elements are required to start and sustain a fire?
4. What is considered the normal oxygen level as a percentage?
5. How is grounding accomplished?
6. List five areas that are considered to be confined spaces.
7. Per OSHA requirements, what action must be performed before any installation, preventive maintenance, or service is performed?
8. List the five classes of fires.
9. Explain how a visual inspection of rubber insulating gloves is performed.
10. Who is a qualified person?

3 Resistance

Resistance in an electrical circuit is the opposition to the flow of electrons. Heat is produced and electrical power consumed when current meets resistance. A resistor is an electrical device designed to introduce a specific amount of resistance into a circuit. Different applications call for different types of resistors. Resistors can be in series, parallel, or series/parallel combinations to meet the resistance required by an application. Standard analog and digital test equipment are available to measure resistance. Factors that affect resistance include type of material, conductor length, conductor cross-sectional area, and temperature.

OBJECTIVES

- Explain how resistance affects electrical circuits and what factors cause resistance to change.
- List and describe the different types of resistors.
- Explain the different types of resistor applications.
- Describe how resistance in a circuit is measured.

Tech Tip

The unit of resistance, the ohm, is named after German physicist Georg S. Ohm (1787–1854) who demonstrated the relationship between current, voltage, and resistance.

RESISTANCE

Resistance is the opposition to the flow of electrons. Heat is produced and electrical energy consumed when current meets resistance. The greater the resistance of a resistor the greater the decrease in circuit current for a constant voltage.

The electrical unit of resistance is the ohm. An *ohm* is the amount of resistance in a circuit when 1 ampere (A) flows with 1 volt (V) applied. The Greek letter omega (Ω) is used to represent the ohm, and the letter R is used to designate resistive components in a circuit. When there is more than one resistor in a circuit, the resistors are numbered (R_1, R_2, R_3, etc.). Symbols, numbers, and letters can be used to identify resistors within a circuit diagram. **See Figure 3-1.**

Terms

Resistance is the opposition to the flow of electrons.

An **ohm** is the amount of resistance in a circuit when 1 ampere (A) flows with 1 volt (V) applied.

Ohm's law is the relationship between voltage (V), current (I), and resistance (R) in an electrical circuit.

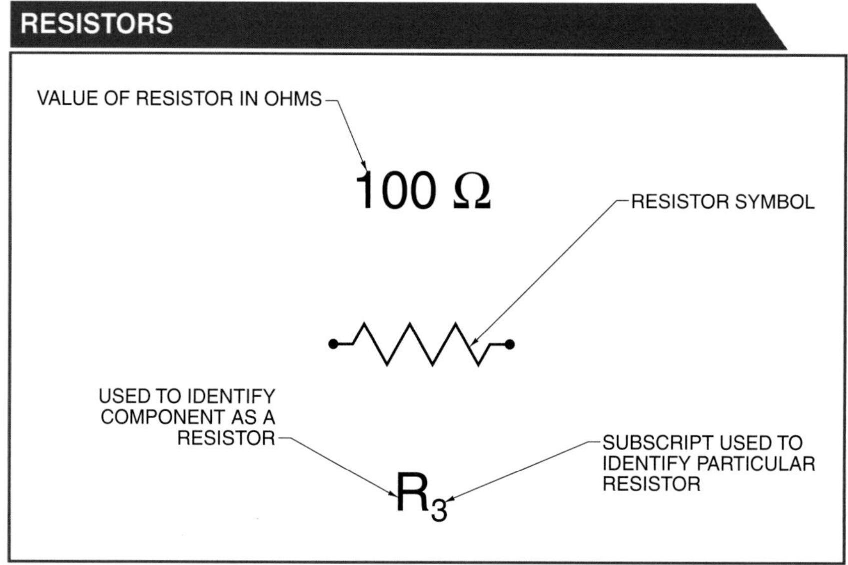

Figure 3-1. Symbols, numbers, and letters can be used to identify resistors in a circuit diagram.

Ohm's law is the relationship between voltage (*V*), current (*I*), and resistance (*R*) in an electrical circuit. Ohm's law is one of the most important concepts in the study of electricity. Ohm's law is expressed by the following mathematical equation:

$$R = \frac{V}{I}$$

where

R = resistance (in Ω)

V = voltage (in V)

I = current (in A)

Example: What is the resistance in a circuit with a current of 2 A and a voltage of 10 V?

$$R = \frac{V}{I}$$

$$R = \frac{10\text{ V}}{2\text{ A}}$$

$$R = \mathbf{5\ \Omega}$$

Tech Tip

At one time, the standard for the ohm was an artifact: a column of mercury having a weight of 14.4521 g, a uniform cross-sectional area of 1 mm^2, a length of 106.3 cm, and a temperature of 0°C. The artifact produced resistance to current in an electrical circuit. This artifact contained all the factors that affect the electrical resistance of a circuit. The artifact had a resistance of 1 Ω.

Factors Affecting Resistance

The four factors affecting the resistance of a substance are type of material, conductor length, conductor cross-sectional area, and temperature. A conductor can be a single wire or combination of wires suitable for carrying electric current.

Type of Material. The amount of resistance that a substance presents to the flow of electrons depends on the type of material. This is because certain materials have more free electrons than other materials. Materials with a large number of free electrons have low resistance, resulting in high current flow. Materials with few free electrons have high resistance, resulting in low current flow.

Terms

Resistivity (specific resistance) is the resistance of a material of a specific cubic size.

The resistance of a material is listed as its resistivity (specific resistance). *Resistivity (specific resistance)* is the resistance of a material of a specific cubic size. The differences in the resistivity of materials allows for a comparison of the ability of different materials to carry an electrical current. The Greek letter rho (ρ) is used to designate resistivity. The constant symbol K is also used to designate resistivity.

One cubic centimeter (cm^3) has been selected as the standard size of a resistivity sample. Using this standard size, the resistivity of annealed copper at 20°C is 1.72 × 10^{-6} (0.0000172) Ω/cm^3. **See Figure 3-2.** Substituting another material for a material that has the same physical dimensions will result in different levels of electron flow. The voltage, however, will remain constant.

Resistivity can be calculated using the metric or English system. However, the meter-kilogram-second (mks) system is officially used. Under this system, each side of the cube 1 cm in length is used to determine the SI unit of resistivity. The 1 cm cube basis is used to determine resistivity

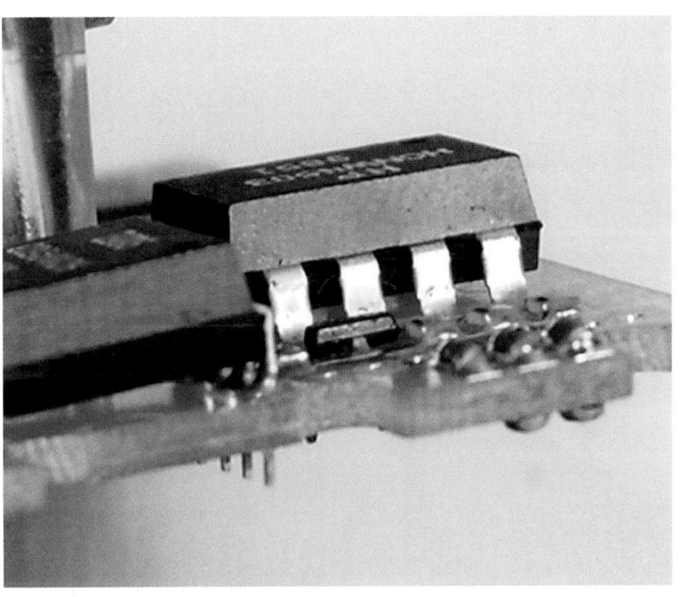

A network resistor package contains many resistors and is typically used where there is a space limitation.

because the official definition of the ohm is given in terms of voltage and current. The resistivity of a material is found by applying the following formula:

$$\rho = \frac{\text{volts per unit length}}{\text{amperes per unit of cross-sectional area}}$$

$$\rho = \frac{\text{volts/meter}}{\text{amperes/meter}^2}$$

ρ = volts/meter × meters^2/amperes

ρ = volts/amperes × meters

ρ = ohm-meter

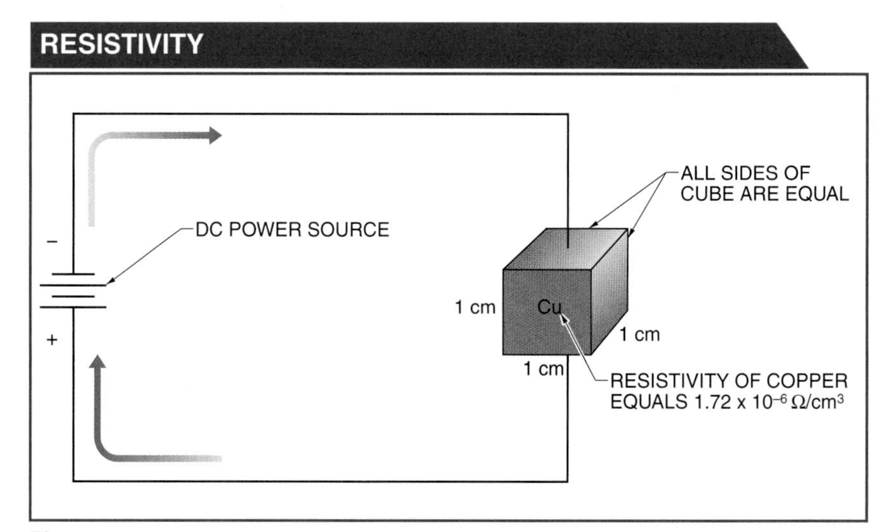

Figure 3-2. Resistivity (specific resistance) is the resistance of a material at a specific cube size.

The ohm-meter is the standard SI unit for resistivity and the symbol is Ω-m.

The resistivity of many common electrical conductors is well known. **See Figure 3-3.** The conductor material with the lowest resistivity is silver. For economic reasons, however, copper and aluminum are used for the majority of applications.

Two factors that affect resistivity, length and cross-sectional area, can be combined into a single equation. This equation is used to find the resistance of a given material with specific dimensions. The units of measure must be consistent and normally the mks system is used. Resistance is calculated by applying the following formula:

$$R = \rho \times \frac{L}{A}$$

where

R = resistance (in Ω)
ρ = resistivity (in Ω-m)
L = length (in m)
A = cross-sectional area (in mm^2)

MATERIAL RESISTIVITY VARIANCE	
Conductor Material	**Ohm-Meter (Ω-m)***
Silver	1.64×10^{-8}
Copper (annealed)	1.72×10^{-8}
Aluminum	2.83×10^{-8}
Tungsten	5.5×10^{-8}
Iron (pure)	12×10^{-8}
Nickel	7.8×10^{-8}
Constantan	49×10^{-8}
Nichrome II	110×10^{-8}

*at 20°C

Figure 3-3. Resistivity varies based on the type of material.

Example: A 100 m aluminum conductor has a cross-sectional area of 2.5 mm^2 (2.5 mm^2 = 2.5×10^{-6} m^2). The resistivity of aluminum is 2.83 $\times 10^{-8}$ W-m. One meter squared is equal to 106 mm^2 (1 m = 1000 mm; 1 m^2 = 106 mm^2). What is the resistance of the conductor?

$$R = \rho \times \frac{L}{A}$$

$$R = 2.83 \times 10^{-8}\ \Omega - \text{m} \times \frac{100\ \text{m}}{2.5 \times 10^{-6}\ \text{m}^2}$$

$$R = 2.83 \times 10^{-8}\ \Omega - \text{m} \times \frac{4 \times 10^{6}}{\text{m}}$$

$$R = \mathbf{0.1132\ \Omega}$$

Terms

Relative resistance is the comparison of the resistance of a given material to the resistance of copper.

Relative resistance is the comparison of the resistance of a given material to the resistance of copper. Although copper is not the best conductor, it is usually the material of choice for conductors in electrical circuits. For a relative comparison of resistance, copper is assigned a value of one. **See Figure 3-4.**

Silver is a better conductor than copper and has a relative resistance of 0.92. Gold has 1.38 times the resistance of copper, and Nichrome has 60 times the resistance. A good insulator such as glass or mica has a relative resistance at least 1,000,000 times that of copper.

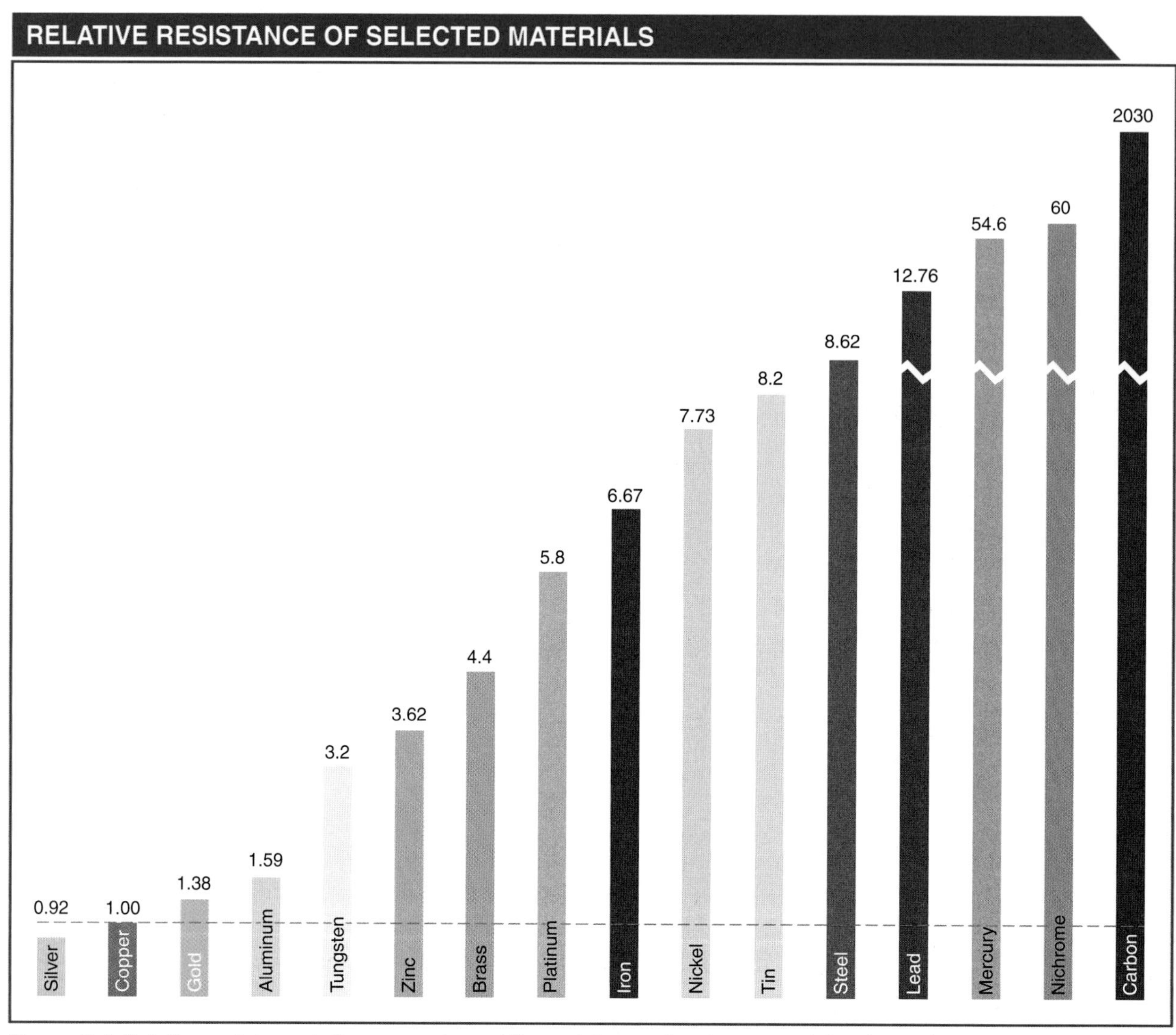

Figure 3-4. Relative resistance is the comparison of the resistance of a given material to the resistance of copper.

Terms

Conductance (G) is the ability of voltage to produce electron flow through a resistance.

Conductance (G) is the ability of voltage to produce electron flow through a resistance. Conductance is measured in siemens (S) in the SI system, where it is defined as ampere per volt (A/V). The conductance of a substance is the reciprocal of its resistance. Conductance is calculated by applying the following formula:

$$G = \frac{1}{R}$$

where

G = conductance (in S)

1 = constant

R = resistance (in Ω)

Example: What is the conductance of a 5 Ω resistor?

$$G = \frac{1}{R}$$

$$G = \frac{1}{5\ \Omega}$$

$$G = \mathbf{0.2\ S}$$

Terms

Relative conductance is the ability of a specific conductor to carry electrons as compared to the ability of a copper conductor to carry electrons.

Relative conductance is the ability of a specific conductor to carry electrons as compared to the ability of a copper conductor to carry electrons. Copper has a relative conductance of 1. The relative conductance of other metals is equal to the reciprocal of their relative resistance. Silver has a conductance of 1.08 and is the only metal with a relative conductance greater than 1.

Conductor Length. The resistance of a conductor is directly proportional to its length. The relationship between the resistance of a conductor and its length can be shown using a graph. **See Figure 3-5.** For example, if 1000′ of a given conductor has 1 Ω of resistance, then 2000′ has 2 Ω, and 3000′ has 3 Ω. This relationship can be expressed mathematically as the following direct proportion:

$$\frac{R_1}{R_2} = \frac{L_1}{L_2}$$

where

R_1 = initial resistance (in Ω)

R_2 = final resistance (in Ω)

L_1 = initial length (in ft)

L_2 = final length (in ft)

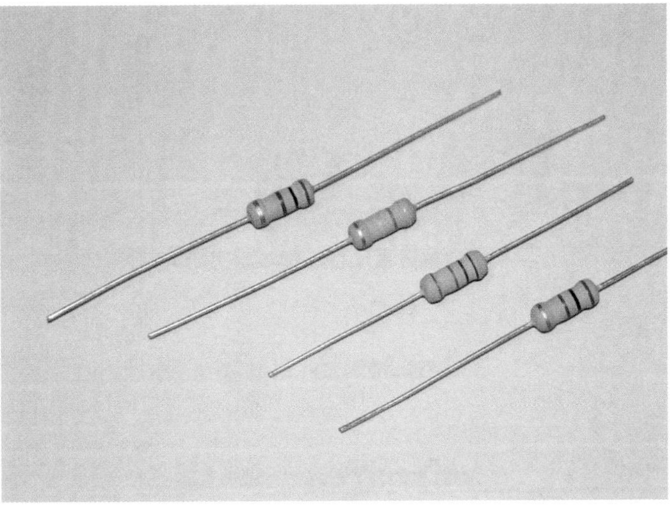

The resistance value of small resistors is easily identified by color bands on the resistors.

Example: A 500′ conductor has 2 Ω of resistance. What is the resistance of a conductor 125′ in length?

$$\frac{R_1}{R_2} = \frac{L_1}{L_2}$$

$$\frac{2}{R_2} = \frac{500'}{125'}$$

$$500' \times R_2 = 2 \times 125'$$

$$R_2 = \frac{2 \times 125'}{500'}$$

$$R_2 = \frac{250'}{500'}$$

$$R_2 = \mathbf{0.5\ \Omega}$$

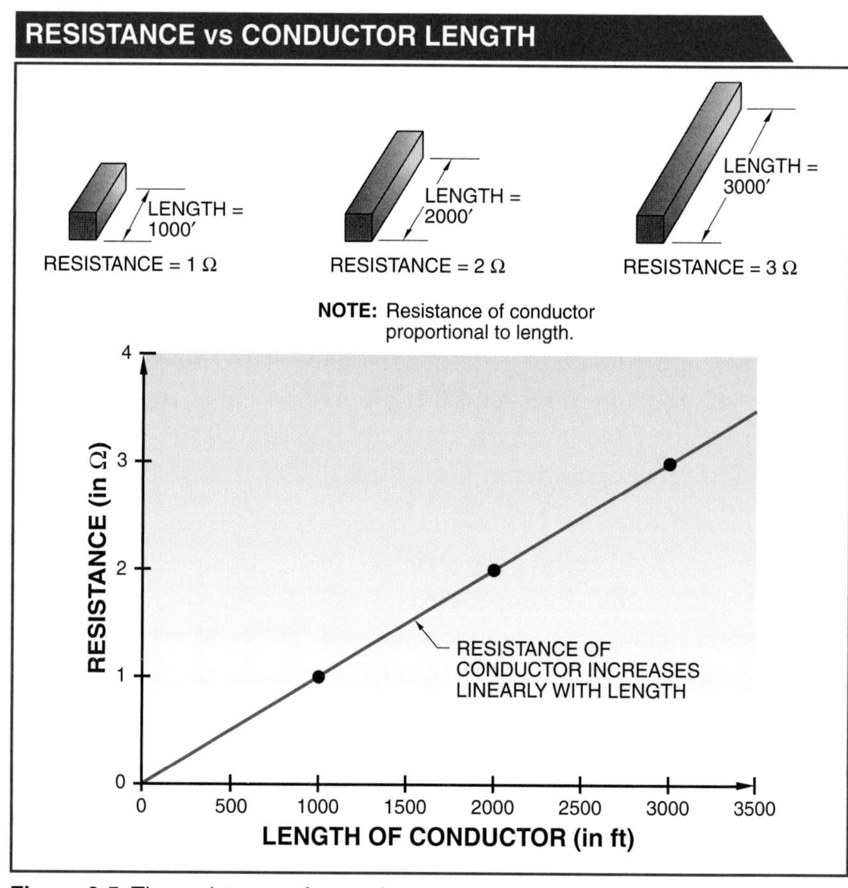

Figure 3-5. The resistance of a conductor increases linearly with its length.

Conductor Cross-Sectional Area. The resistance of a conductor is inversely proportional to its cross-sectional area. If a given length of a conductor with a cross-sectional area of 1 cm^2 has a resistance of 10 Ω, then the same conductor with a cross-sectional area of 2 cm^2 has a resistance of 5 Ω. If the cross-sectional area is increased to 4 cm^2, then the resistance decreases to 2.5 Ω. **See Figure 3-6.**

The relationship between the resistance of a conductor and its cross-sectional area can be shown using a graph. Unlike the graph showing the relationship between resistance and length, this graph does not depict a linear relationship. As the cross-sectional area approaches zero, the resistance of the conductor approaches infinity. Conversely, as the cross-sectional area becomes very large, the resistance of the conductor approaches 0 Ω. This relationship can be expressed mathematically as the following inverse proportion:

$$\frac{R_1}{R_2} = \frac{A_2}{A_1}$$

where

R_1 = initial resistance (in Ω)
R_2 = final resistance (in Ω)
A_2 = final area (in cm^2)
A_1 = initial area (in cm^2)

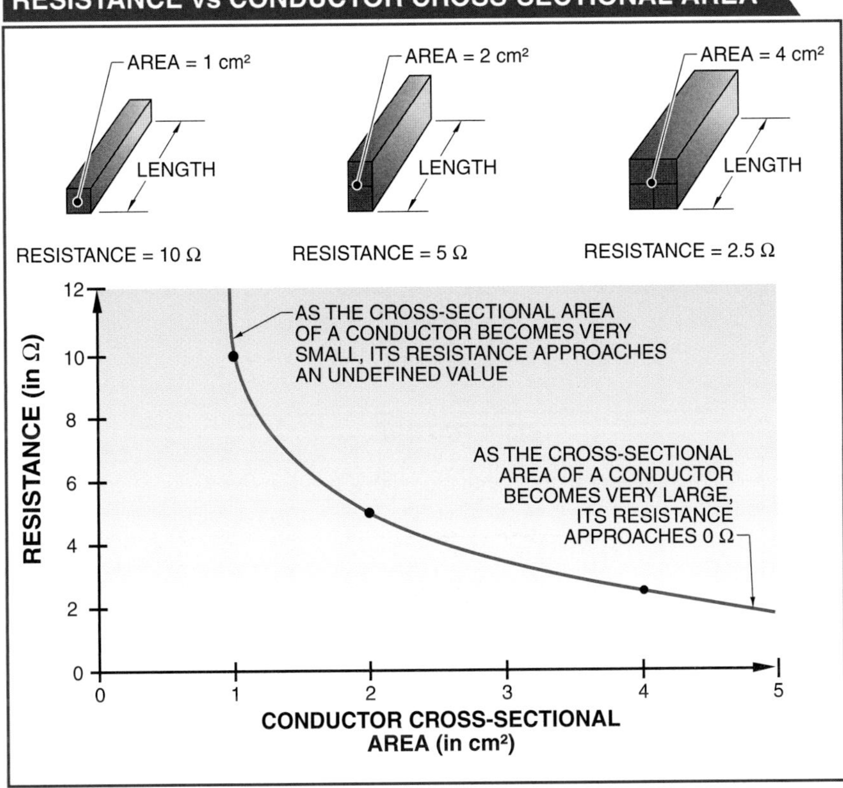

Figure 3-6. The resistance of a conductor is inversely proportional to its cross-sectional area.

Example: A square conductor of a given length has a resistance of 27 Ω. Each side of the square is equal to 1 mm. What is the resistance of the conductor if each side of the conductor is increased to 3 mm?

$$\frac{R_1}{R_2} = \frac{A_2}{A_1}$$

$$\frac{27\ \Omega}{R_2} = \frac{0.003 \times 0.003}{0.001 \times 0.001}$$

$$\frac{27\ \Omega}{R_2} = \frac{9 \times 10^{-6}}{1 \times 10^{-6}}$$

$$\frac{27\ \Omega}{R_2} = \frac{9}{1}$$

$$9(R_2) = 27\ \Omega$$

$$R_2 = \mathbf{3\ \Omega}$$

Terms

A **positive temperature coefficient** is the result of an increase in the resistance of a material with an increase in temperature.

A **negative temperature coefficient** is the result of a decrease in the resistance of a material with an increase in temperature.

Temperature. The electrical resistance of a material changes with temperature. Heat increases the resistance of most metals and their alloys. These materials have a positive temperature coefficient. A *positive temperature coefficient* is the result of an increase in the resistance of a material with an increase in temperature. **See Figure 3-7.**

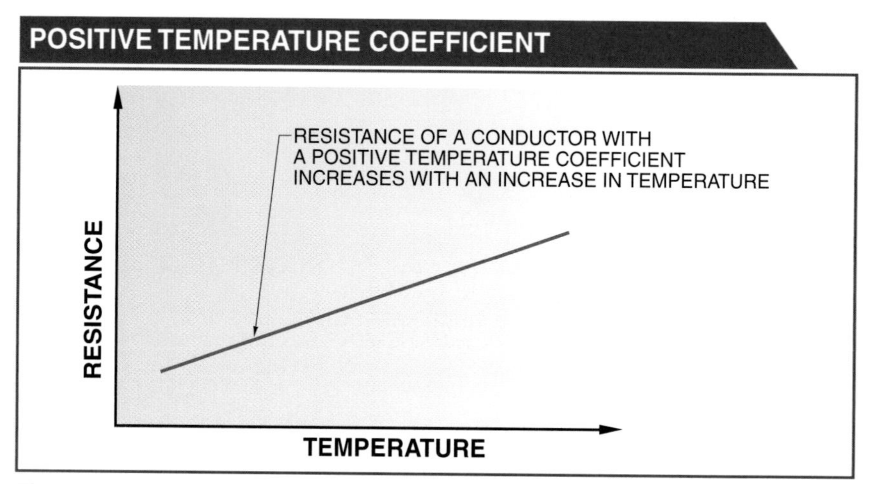

Figure 3-7. The resistance of a conductor with a positive temperature coefficient increases with an increase in temperature.

Although the values can range from 1% to approximately 6%, most metals have about a 4% increase in resistance for every 10°C increase in temperature. For example, the resistance of a piece of pure annealed copper wire is 1 Ω at 0°C and increases at a rate of about 0.4% for each degree Celsius increase. Aluminum bronze has about a 0.1% increase; iron has about a 0.6% increase. The temperature coefficient is expressed as a change in ohms per million ohms per degree Celsius.

The resistance of some materials decreases with an increase in temperature. These materials have a negative temperature coefficient. A *negative temperature coefficient* is the result of a decrease in the resistance of a material with an increase in temperature. **See Figure 3-8.**

Carbon and some semiconductor materials have a negative temperature coefficient. Carbon has a negative temperature coefficient of 500 ppm/°C. This means that carbon has a resistance of 1,000,000 Ω at 20°C, a resistance of 999,500 Ω at 21°C, and a resistance of 1,000,500 Ω at 19°C.

The temperature coefficient is represented in mathematical equations by the Greek letter alpha (α). Temperature coefficients of various materials can be obtained from tables. Nickel has one of the highest temperature coefficients. **See Figure 3-9.** The resistance of a conductor can be calculated for any temperature by applying the following formula:

$R_T = R_{20} \times [1 + \alpha (T_t - T_{20})]$

where

R_T = resistance of conductor at change in temperature

R_{20} = resistance at 20°C

1 = constant

α = temperature coefficient at 20°C (in $^1/_{°C}$)

T_t = final temperature (in °C)

T_{20} = coefficient temperature base (in °C)

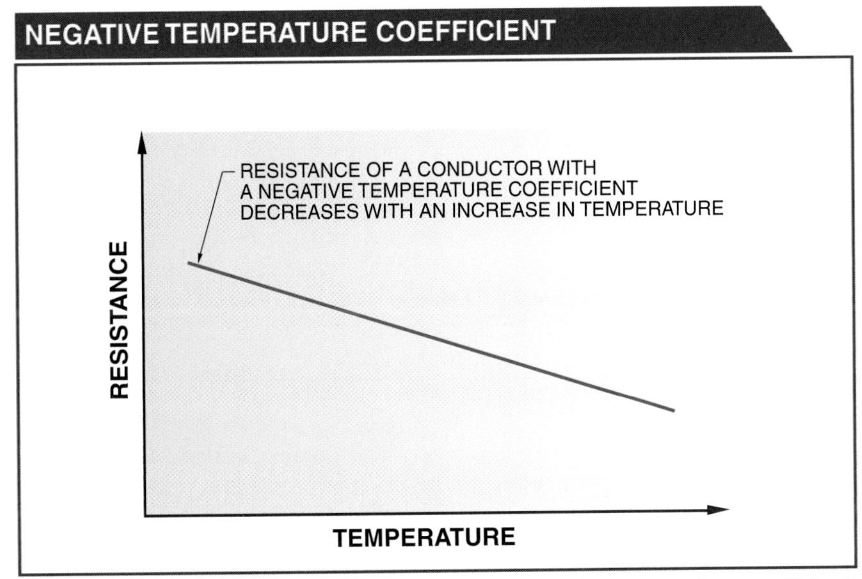

Figure 3-8. The resistance of a conductor with a negative temperature coefficient decreases with an increase in temperature.

Example: A copper conductor has 1.588 Ω of resistance at 20°C. What is the resistance of the conductor at 70°C?

$R_T = R_{20} \times [1 + \alpha (T_t - T_{20})]$

$$R_T = 1.588\ \Omega \times [1 + \frac{3930 \times 10^{-6}}{°C}(70°C - 20°C)]$$

$$R_T = 1.588\ \Omega \times (1 + \frac{0.00393}{°C} \times 50°C$$

$R_T = 1.588\ \Omega \times (1 + 0.1965)$

$R_T = 1.588\ \Omega \times 1.1965$

R_T = **1.900 Ω**

This relationship can also be expressed by modifying the physical characteristics equation for the resistance of a conductor. This formula becomes the following formula, where ΔT is the change in temperature:

$$R_T = \rho \times \frac{L}{A} \times (1 + \alpha \Delta T)$$

TEMPERATURE COEFFICIENTS	
Conductor Material	**Temperature Coefficient***
Aluminum	3900
Carbon	−500
Constantan (55% Cu, 45% Ni)	18
Copper (annealed)	3930
Iron (pure)	550
Manganin® (84% Cu, 12% Mn, 4% Ni)	20†
Nichrome I	170
Nickel	4700
Silver	3800
Tungsten	4500

*In parts per million @ 20°C

†Some tables show this material as having different temperature coefficients at temperatures between 0°C and 100°C

Figure 3-9. Most conductors have a positive temperature coefficient.

The resistance change of copper due to a temperature change is approximately 4% for each 10°C change in temperature. Generally, these small changes in resistance are not significant in a circuit as compared with the resistance of the load, and they can be ignored when making calculations. When temperature is a factor in wiring power circuits, the National Electrical Code® specifies the use of large conductors.

However, in some instances small changes in resistance can be of significance and must be addressed. For example, a small change in the circuit resistance of a measuring device such as an ammeter can cause a large error in the indicated value. In the design of these devices, alloys such as constantan or manganin, which have very low temperature coefficients, are used for the wiring of sensitive parts of the circuit. Constantan has a resistance change of 18 parts per million (ppm) for each degree Celsius, and manganin has a resistance change of 20 ppm for each degree Celsius, as compared to a change of 3930 ppm for copper wire.

The temperature coefficient equation is derived from experimental data. The equation, however, is not useful over very wide temperature ranges. For example, near absolute zero, the properties of some metals change radically. These metals are classified as superconductors. A *superconductor* is a material that exhibits zero resistance and imposes no opposition to electron flow nor consumes power at temperatures close to absolute zero (0°K, –273.15°C, –459.67°F).

Tech Tip

J. R. Brown devised the Brown & Sharpe (B&S) gauge for wire in 1857. The wire sizes are set based on successive steps in the wire-drawing process during manufacturing. The wire sizes are retrogressive, with smaller numbers having a larger wire diameter. This system was adopted and then replaced by the American Wire Gauge (AWG) for electrical and electronic wiring.

Wire Gauge

Wire gauge sizes are based on a simple mathematical law. The American Wire Gauge (AWG) system begins by assigning the number 36 to a small wire with a 0.0050″ diameter. The largest size of single strand wire, number 6/0 (pronounced six aught), has a diameter of 0.5800″. Intermediate diameters are determined by a geometric progression. **See Appendix.**

Terms

A **superconductor** is a material that exhibits zero resistance and imposes no opposition to electron flow nor consumes power at temperatures close to absolute zero (0°K, –273.15°C, –459.67°F).

Several useful mathematical relationships exist when working with wire gauge sizes. The first of these relationships is that with an increase of three gauge sizes, the cross-sectional area and weight of the wire doubles and the resistance is halved. The second is that an increase in six gauge sizes results in a diameter of twice the size. For example, 26 AWG wire has a diameter of 15.9 mils and 20 AWG wire has a diameter of 32 mils. A *mil* is equal to 0.001″. Finally, for an increase of ten gauge sizes, the cross-sectional area and weight increases ten times, and the resistance is reduced to one-tenth that of the smaller wire.

A knowledge of these relationships along with the fact that 10 AWG wire has a resistance of 1 Ω per 1000′, a diameter of 0.1″, and weighs 31.4 lb per 1000′, allows one to approximate data about copper wire of other AWG sizes. The smaller the AWG number, the greater the cross-sectional area and the heavier the wire. **See Figure 3-10.**

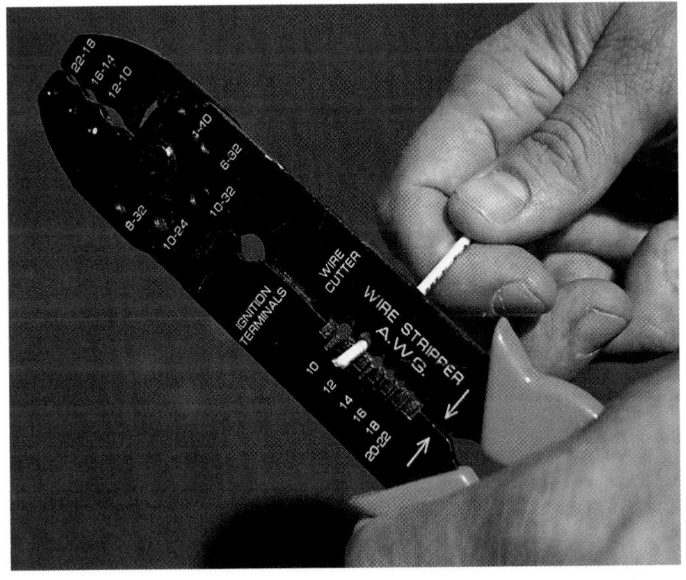

A wire stripper is used to strip insulation wires sized from 22 AWG to 10 AWG.

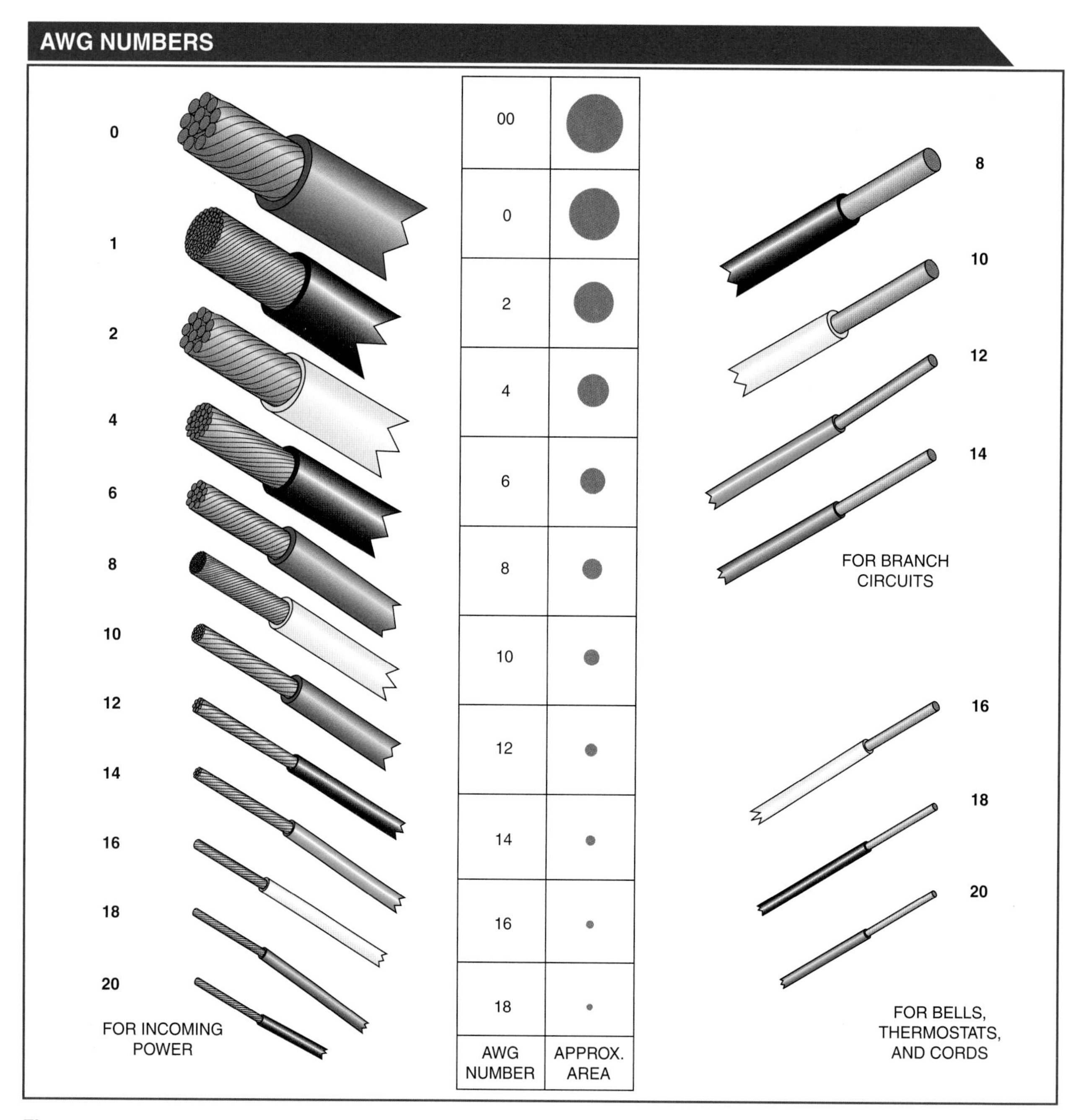

Figure 3-10. The smaller the AWG number, the greater the cross-sectional area and the heavier the wire.

The diameter of wire is given in mils (thousandths of an inch). Because the diameter of a round conductor may be only a fraction of an inch, it is expressed in mils to avoid the use of decimals. The inch unit is converted to mils by multiplying both the unit of width and the unit of length by a factor of 1000. For example, a wire with a diameter of 0.050″ (50 thousandths of an inch) is expressed as 50 mils.

The square mil is a convenient unit for expressing the cross-sectional area of conductors. A square mil is equal to the area of a square measuring one mil on each side. When calculating the square mils of a square conductor, the two sides of the conductor are multiplied together. **See Figure 3-11.**

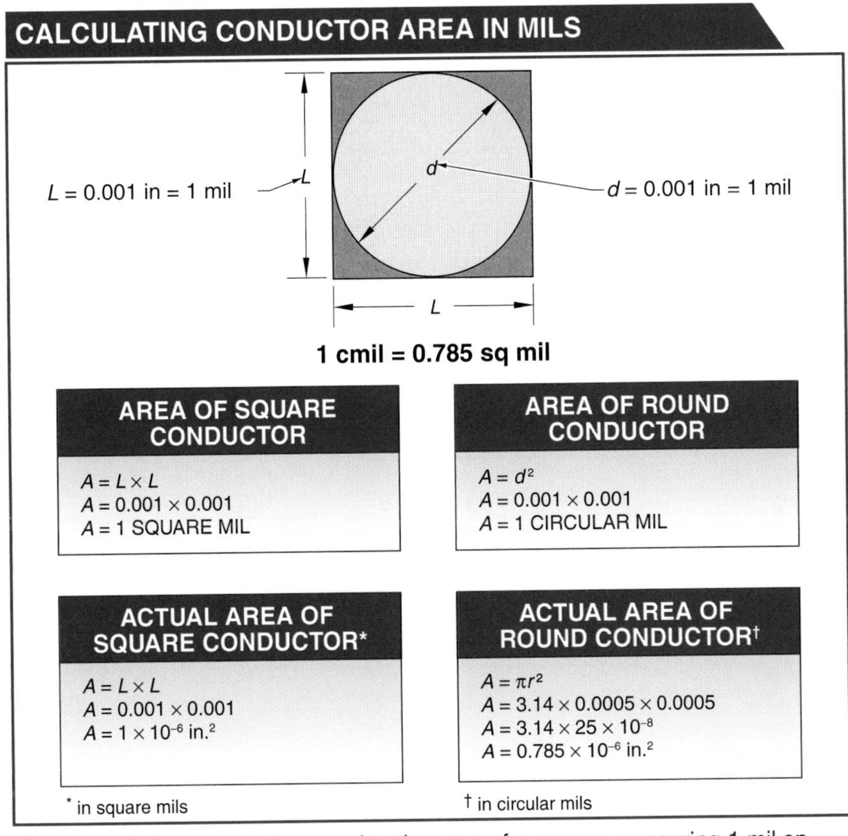

Figure 3-11. A square mil is equal to the area of a square measuring 1 mil on each side. A circular mil is an area of a circle with a diameter of 1 mil.

The cross-sectional area of a round conductor is given in circular mils. A *circular mil (cmil)* is an area of a circle with a diameter of 1 mil. It is the standard unit of wire cross-sectional area used by the AWG and the English wire tables. The circular mil area of a circular conductor is calculated by applying the following formula:

$CM = d^2$

where

CM = circular mil

d = diameter

Terms

A **circular mil (cmil)** is an area of a circle with a diameter of 1 mil.

Example: What is the circular mil area of a conductor with a diameter of 50 mils?

$CM = d^2$

$CM = 50^2$

$CM = \mathbf{2500}$

Wire Resistance

When computing the resistance of a section of wire, its resistivity, length, and cross-sectional area must be known. The AWG uses a resistivity given in ohms per circular mil per foot (Ω/cmil/ft). Understanding the resistivity of annealed copper wire and aluminum wire is important for the electrician as they are the common types of metals used in conductors. The resistivity of common conductor materials should be committed to memory. **See Figure 3-12.**

RESISTIVITY OF COMMON MATERIALS

Material	Resistivity*
Silver	9.6
Copper (annealed)	10.4
Aluminum	17
Tungsten	34
Brass	42
Nickel	60
Iron	61
Manganin®	264
Constantan	294
Cast Iron	435
Nichrome	675
Carbon	22,000

*in Ω per cmil per foot

Figure 3-12. The resistivity of common materials can be used along with the physical characteristics of a conductor to determine its resistance.

Terms

A **stranded conductor** is a conductor composed of several strands of solid wire wrapped together to make a single conductor.

As the circular mil area of a wire increases and the AWG numerical value decreases, the maximum allowable current-carrying capacity of the wire increases. Conductors normally have insulation and are referred to as insulated wires. Although the term wire refers just to the metal, it is generally understood to include insulation. A *stranded conductor* is a conductor composed of several strands of solid wire wrapped together to make a single conductor.

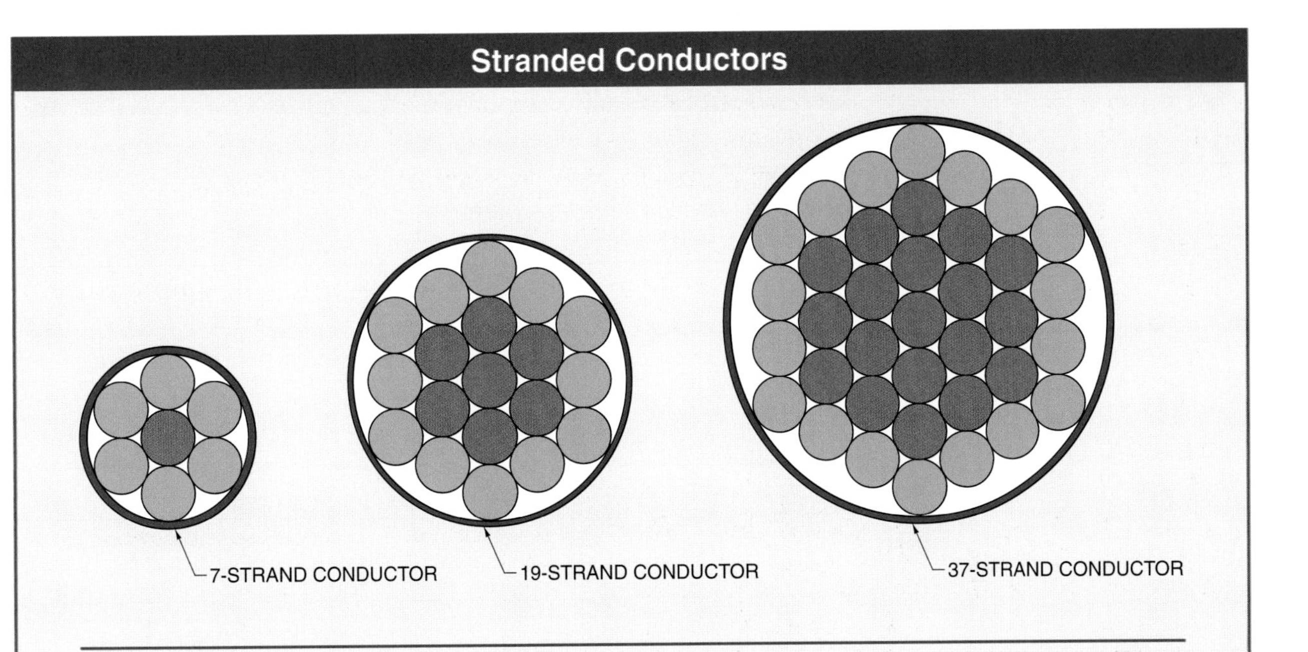

Conductors are stranded mainly to increase their flexibility. Twisting the conductor strands tightly together causes a cancellation of magnetic fields created by unwanted eddy currents. The wire strands in cables are arranged in the following order: The first layer of strands around the center conductor is made up of 6 conductors. The second layer is made up of 12 additional conductors. The third layer is made up of 18 additional conductors, and so on, in increments of six.

Thus, standard cables are composed of 7 strands, 19 strands, 37 strands, and so on, in continuing fixed increments. The overall flexibility of a cable can be increased by further stranding of the individual strands.

The maximum allowable current-carrying capacity of each wire gauge is different based on the type of insulation used. **See Figure 3-13.** This is due to the ability of the insulated wire to transfer the heat generated by electron flow through the resistance of the wire. Different types of insulating materials have different heat transfer ability. Excessive heat can melt conductor insulation. In extreme cases, such as a circuit fault, the conductor may vaporize. *Ampacity* is the maximum amount of current a conductor can carry continuously without exceeding its temperature rating. In Article 310 of the National Electrical Code® (NEC®), tables provide the maximum operating temperatures and ampacity for conductors. Tables are provided for single and multiple conductors for various types of insulation, operating conditions, and environments.

Terms

Ampacity is the maximum amount of current a conductor can carry continuously without exceeding its temperature rating.

SINGLE COPPER CONDUCTOR CURRENT-CARRYING CAPACITY*		
AWG	Insulation Type	
	Rubber-Thermoplastic	Heat-Moisture Resistant
0000	300	370
000	260	320
00	225	275
0	195	235
1	165	205
2	140	175
3	120	150
4	105	130
6	80	100
8	55	70
10	40	55
12	25	40
14	20	30

*in A at ambient temperatures below 30°C

Figure 3-13. The current-carrying capacity of each wire gauge is different based on the type of insulation used.

Terms

A **resistor** is an electrical device designed to introduce a specific amount of resistance into a circuit.

A **fixed resistor** is a resistor that has only one resistance value.

RESISTORS

A *resistor* is an electrical device designed to introduce a specific amount of resistance into a circuit. Electrical resistance is caused by the collision of electrons as they flow through a resistive material. The collisions convert electrical energy into another form of energy such as heat or light.

Resistors are classified as fixed or variable. Resistors are normally described by electrical resistance (in ohms), tolerance (percentage of stated value that is acceptable for an individual resistor), and power rating (ability to dissipate heat in watts). Resistors are also described by their method of construction (carbon composition, wire wound, or metal film) and circuit application.

Fixed Resistors

A *fixed resistor* is a resistor that has only one resistance value. The resistance value cannot be changed or adjusted. Fixed resistors include carbon composition, wire-wound, and metal film resistors.

Carbon Composition Resistors. A *carbon composition resistor* is a resistor that is constructed using carbon graphite mixed with clay. Many electronic devices require resistors with high resistance values that pass currents of only a few milliamps or microamps. The carbon composition resistor is the most frequently used type of resistor. A *joule* (J) is the unit of energy measurement. Carbon composition resistors are normally used when the conversion of electrical energy to heat is at a rate of one or two joules per second.

The resistance element of a carbon composition resistor consists of finely ground carbon mixed with an insulating resin composition (binder). The ratio of carbon to the insulating composition, along with the resistor dimensions, determines the resistance value. Tinned (covered with a thin layer of solder cover) copper pigtail leads are embedded in the composition and run through a plastic jacket that protects the element. **See Figure 3-14.** Typical applications for carbon composition resistors include power supplies and welding control units.

Terms

A **carbon composition resistor** is a resistor that is constructed using carbon graphite mixed with clay.

A **joule (J)** is the unit of energy measurement.

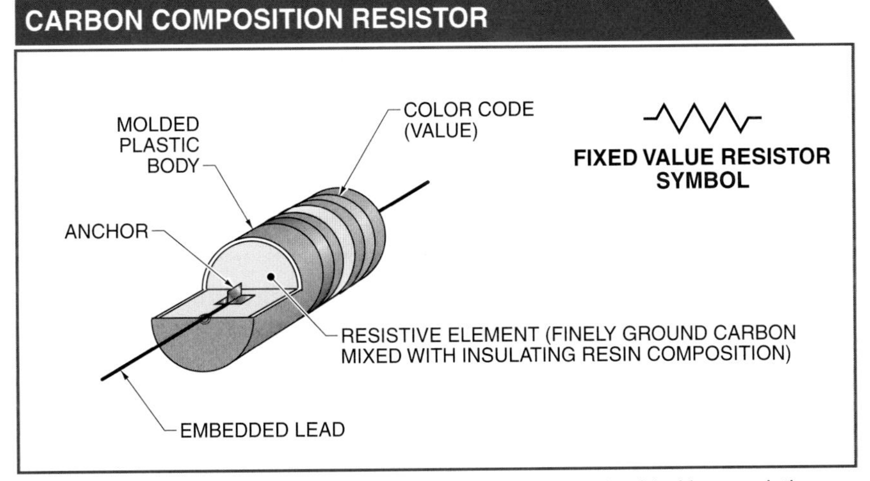

Figure 3-14. A carbon composition resistor has leads embedded in a resistive element within its molded body.

The resistance-temperature relationship of a carbon composition resistor is not linear as with other types of resistors. The mixture of the carbon, which has a negative temperature coefficient, and the binder, which has a positive temperature coefficient, produces a resistance curve that increases above and below 25°C. **See Figure 3-15.** However, this change is dependent on the physical size of the resistor and its power (wattage) rating. If the resistor is operated within its power rating, the variation is small enough that the carbon composition resistor behaves as if the resistance-temperature relationship is linear.

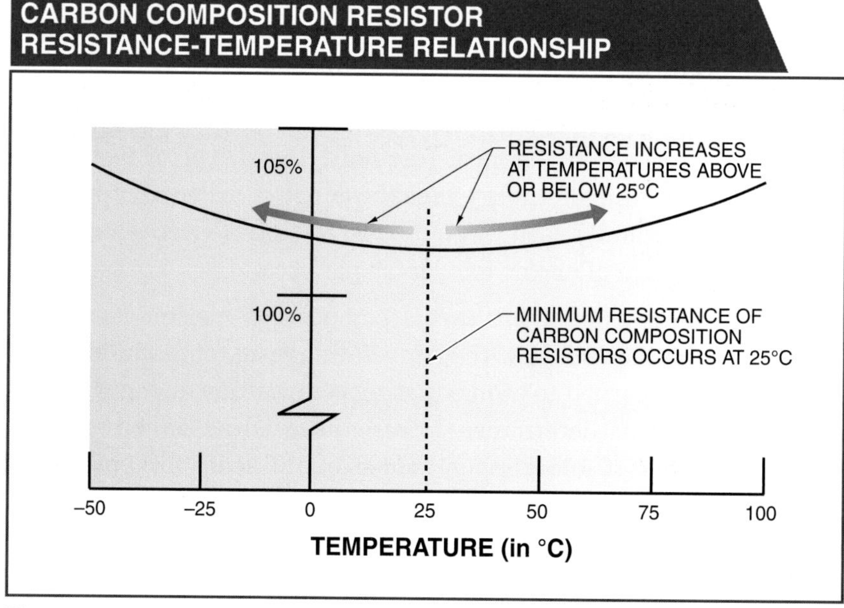

Figure 3-15. The resistance of a carbon composition resistor varies nonlinearly over a range of temperatures.

Tech Tip

With many electrical designs, a derating factor is used to extend component life. For example, a design may require a resistor to dissipate 1W; however, the designer may select a 2 W resistor based on a 50% derating requirement.

No relationship exists between a resistor's resistance value and its power rating. Carbon composition resistors are available in a wide range of resistance values, from less than 1 Ω to more than 100 MΩ, with power ratings of $\frac{1}{10}$ W to 2 W. The power rating of a carbon composition resistor is estimated based on the size of the resistor. In many electrical designs, a derating factor is used to extend the life of a component. For example, a design may require a resistor to dissipate 1 W but the designer may select a 2 W resistor based on a 50% derating requirement. **See Figure 3-16.**

Resistor Color Codes. Resistor value is indicated by a series of color bands placed on the resistor. Resistor color coding is the preferred method of marking the values on low-wattage resistors. By using the color coding method with the bands completely encircling the resistor, the value is always visible without having to move the resistor.

A resistor may have between two and four color bands with one color band that represents a power of 10 multiplier and one color band that represents resistor tolerance. **See Figure 3-17.** For example, a resistor has red, blue, yellow, and silver color bands. The red band represents a value of 2. The blue band represents a value of 6. The yellow band represents a multiplier of 10,000. The silver band represents a tolerance of ±10%. This resistor has a resistance of 260,000 Ω ±26,000 Ω.

CARBON COMPOSITION RESISTOR CHARACTERISTICS							
Rated Power*	Dimensions†			Maximum Rated Voltage‡	Maximum Overload Voltage‡	Resistance Range	Resistance Tolerance§
	L	D					
0.25	6.3	2.4		250	400	2.2 Ω – 22 MΩ	±5/10
0.5	9.5	3.6		350	700	2.2 Ω – 22 MΩ	±5/10
1	14.3	5.7		500	1000	2.2 Ω – 22 MΩ	±10

* in W
† in mm
‡ in V
§ in %

Figure 3-16. The power rating (in watts) of carbon composition resistors is based on the physical size of the resistor.

Because it is difficult to mass-produce resistors that have an exact value, manufacturers specify a tolerance by which resistors may vary from their stated value. Tolerance is specified as a plus/minus percentage. Common values of tolerance are ±1%, ±2%, ±5%, and ±10%. Resistors with no tolerance band have a tolerance of ±20%. Tolerances of less than ±0.01% are also possible. Precision resistors and high-power resistors normally have their values and tolerance printed on them in place of the color code.

Many characteristics of carbon composition resistors make them preferable for use in electrical and electronic circuits. However, an undesirable characteristic is that the resistance value of carbon composition resistors increases with age. In addition, the power ratings (in watts) of these resistors must match the power rating of the circuit in which they are used. If carbon composition resistors are overheated, their resistance is permanently changed, normally with a decrease in resistance value. Severe overheating results in the total destruction of the resistor with the possibility of either an open (no current flow) or a short (excessive current flow) circuit.

Although fixed-value resistors are available with almost any resistance value, they are typically manufactured in standard values. Low-tolerance resistors are produced with a wide range of resistance values, and high-tolerance resistors are produced with a narrow range of resistance values.

Wire-Wound Resistors. A *wire-wound resistor* is a resistor constructed using various types of wire wound on an insulating form. The wire is made of alloys such as nickel-chromium or copper-nickel. Tinned leads or terminal lugs are brought out at each end. The entire assembly, except for the terminals, is coated with an insulating material. The exact resistance of wire-wound resistors is determined

Terms

A **wire-wound resistor** is a resistor constructed using various types of wire wound on an insulating form.

by the type of material, the length, and the circular mil area of the wire used to form the resistor. Wire is often used in the manufacture of resistors with a power rating over 2 W.

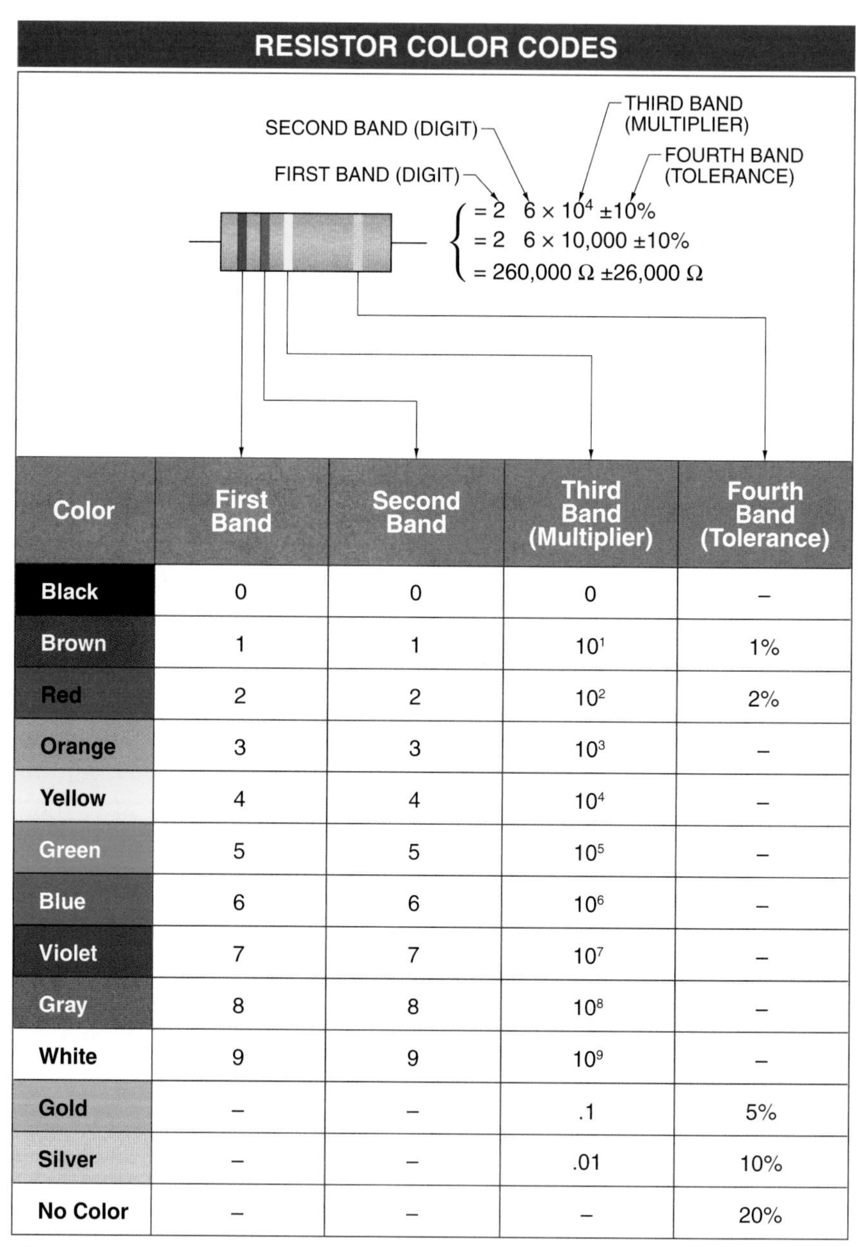

Color	First Band	Second Band	Third Band (Multiplier)	Fourth Band (Tolerance)
Black	0	0	0	–
Brown	1	1	10^1	1%
Red	2	2	10^2	2%
Orange	3	3	10^3	–
Yellow	4	4	10^4	–
Green	5	5	10^5	–
Blue	6	6	10^6	–
Violet	7	7	10^7	–
Gray	8	8	10^8	–
White	9	9	10^9	–
Gold	–	–	.1	5%
Silver	–	–	.01	10%
No Color	–	–	–	20%

Figure 3-17. A color code is used to specify resistor resistance and tolerance. Resistors can have up to six bands identifying resistance and tolerance.

Resistors used in electrical measurement instruments must be highly accurate. The tolerance rating of resistors used in these devices is usually ±1% or less. They can, however, have a wide range of power ratings.

Metal Film Resistors. A *metal film resistor* is a resistor that consists of a metal film of a selected metal alloy or carbon deposited in a vacuum on a small ceramic cylinder to which leads are attached. **See Figure 3-18.** The resistance of a metal film resistor is determined by the type of material used to form the film and the thickness and length of the film. The thickness of the deposit can be controlled, with normal thicknesses between 0.1″ and 10 millionths of an inch. The film may be etched and adjusted using a laser to obtain the proper resistance. A connecting lead is attached at each end, and the whole assembly is covered with an insulating coating. Although they are more expensive than carbon composition or wire-wound resistors, metal film resistors have generally replaced wire-wound resistors for applications that require a high degree of accuracy. Metal film resistors are particularly useful in high-voltage and high-frequency circuits.

Tech Tip

Early resistors consisted of a long, thin, insulated wire of an alloy with a very low temperature coefficient. Great care was taken in manufacturing this wire to ensure a uniform cross-sectional area so that the resistance was a linear function of the length and no hot spots developed when current passed through it. This wire was spooled on a small bobbin with terminals attached to the bobbin.

METAL FILM RESISTORS

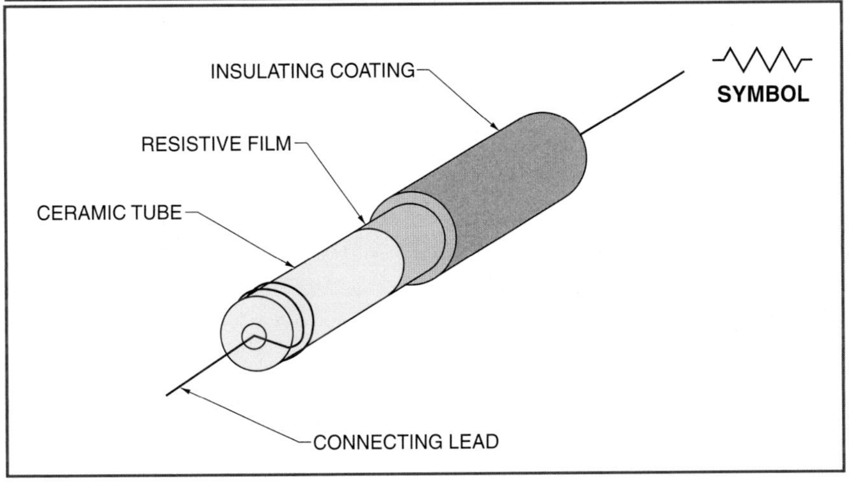

Figure 3-18. A metal film resistor consists of a metal film around a ceramic cylinder.

Terms

A **metal film resistor** is a resistor that consists of a metal film of a selected metal alloy or carbon deposited in a vacuum on a small ceramic cylinder to which leads are attached.

Metal film resistors can be manufactured with resistance values ranging from milliohms to gigaohms. Tolerances of 1⁄10 of a percent can be achieved. Metal film resistors exhibit very little change in resistance value with age. Care should be taken when soldering a metal film resistor into a circuit. Care should also be taken to avoid connecting it to a circuit that exceeds the resistor's wattage rating. Exposure to very low temperatures can also damage metal film resistors. The accumulative effect of these factors may result in a change from 1% to 5% of the resistance value. Power ratings range from a fraction of a watt to hundreds of watts. A resistor with 5 or 6 color bands is a metal film resistor. The fifth color band represents quality and the sixth color band represents temperature coefficient.

A metal film resistor is used to help control the flow of current in a transformer.

The temperature coefficient for metal film resistors manufactured using carbon deposits is usually negative. The range is between 0.01%/°C and 0.05%/°C for low-resistance values and slightly higher for high-resistance values. Mixing a small amount of boron with the carbon improves the temperature coefficient to a range of 0.005%/°C to 0.02%/°C.

A cermet film resistor is another form of film resistor. *Cermet* is a mixture of fine particles of glass or ceramic and powdered metals such as silver, platinum, or gold. Metal oxides and carbon are also used. The mixture is made into a paste and deposited as a spiral ribbon on a base (substrate) of glass, ceramic, or alumina. Leads are attached and the assembly is fired in a kiln and fused. This assembly is then covered with an insulating cover. Cermet is used in fixed and variable resistors. Cermet has a low temperature coefficient and is extremely stable and durable. **See Figure 3-19.**

Terms

Cermet is a mixture of fine particles of glass or ceramic and powdered metals such as silver, platinum, or gold.

A **variable resistor** is a resistor whose resistance value can be changed to any value within its range.

An **adjustable resistor** is a resistor whose resistance value can be changed by moving one or more contacting elements.

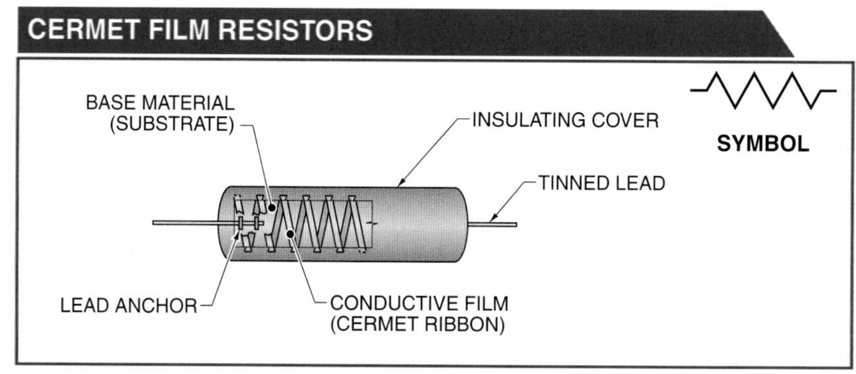

Figure 3-19. A cermet resistor has a ribbon of cermet wrapped around a base material.

Variable Resistors

A *variable resistor* is a resistor whose resistance value can be changed to any value within its range. Variable resistors are used when frequent adjustments are required, such as when changing the sound level on a television set. An *adjustable resistor* is a resistor whose resistance value can be changed by moving one or more contacting elements. Adjustable resistors are used in high-wattage circuits where the need to make an adjustment is infrequent. Limiting the output current of a power source to an absolute maximum value is an application of an adjustable resistor.

The resistance element of variable resistors may be carbon composition, wire wound, or deposited film. Variable resistors may be low-power or high-power and are manufactured with a wide range of resistance values. Variable resistors can be changed from a resistance of 0 Ω to the stated resistance value of the device. The resistance of a variable resistor can be changed by turning a shaft attached to a movable arm.

A potentiometer is a variable resistor used to provide a variable voltage in a circuit.

Although most variable resistors have shafts that can be turned less than 360° to obtain the full range of resistance, some precision variable resistors have geared mechanisms that require ten full turns to obtain the full range. Variable resistors include potentiometers and rheostats.

Terms

A **potentiometer** is a low-power three-terminal variable resistor.

A **rheostat** is a two-terminal variable resistor.

Potentiometers. A *potentiometer* is a low-power three-terminal variable resistor. Potentiometers are used for volume control in radios and amplifiers. Conductive plastics are often used in potentiometers as their resistive element. Plastics are economical to manufacture and have relatively long lives. The resistive elements are made by combining carbon powder with a plastic resin. This mixture is applied to a substrate such as ceramic. Plastics are used in potentiometers that have high usage.

Potentiometers are normally used as voltage dividers. In a voltage divider, the full voltage is applied across the total resistance with a portion of the full voltage available between the movable arm and one end of the potentiometer. A common mechanical arrangement of two potentiometers in one assembly uses two separate shafts to control each potentiometer. **See Figure 3-20.** The range of the back potentiometer is controlled by the inner shaft. The front potentiometer is controlled by the outer shaft. This assembly is smaller than if two separate potentiometers were used.

Rheostats. A *rheostat* is a two-terminal variable resistor. Rheostats are current-limiting devices and normally have a power rating over 2 W. Rheostats are used for controlling the light level of incandescent lamps. A potentiometer can be wired as a rheostat if only one end of the resistance element and the movable arm are connected to the circuit. When using a potentiometer as a rheostat, the power rating must be sufficient for the application. Rheostats are often wire wound and have higher power ratings than potentiometers. **See Figure 3-21.**

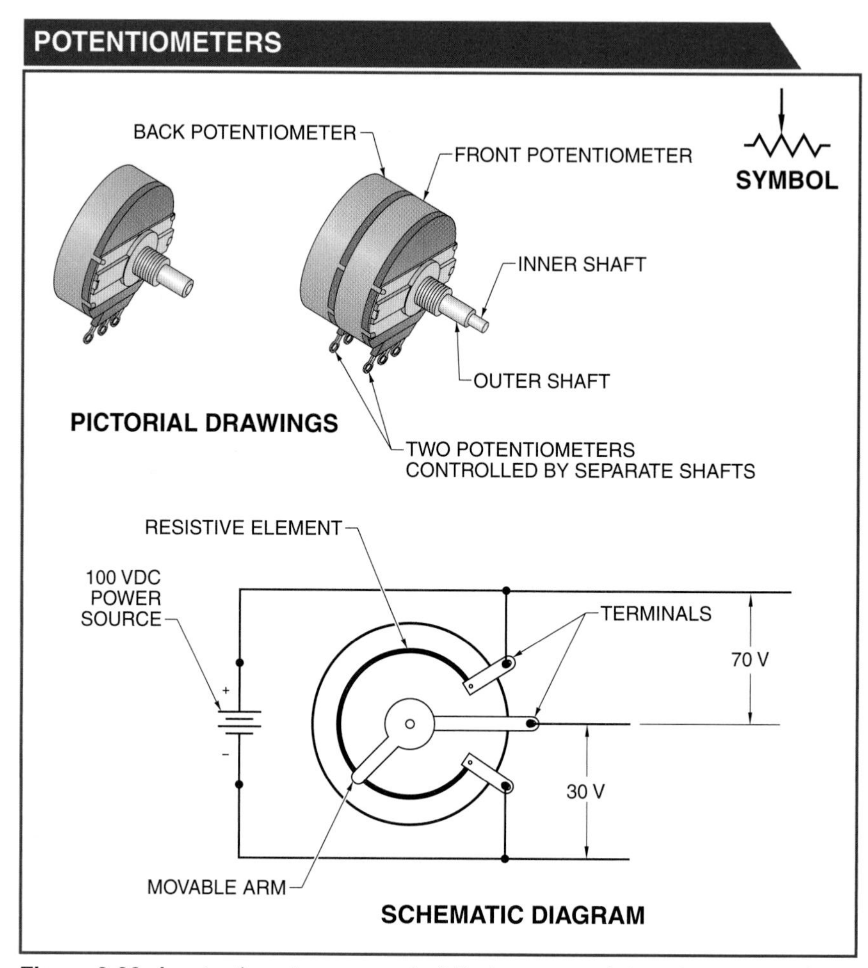

Figure 3-20. A potentiometer uses a shaft that moves an internal arm to set the resistance.

The relationship between resistance and shaft position in most rheostats is linear (resistance varies in direct proportion with the number of degrees the shaft is turned). However, in some rheostats, the relationship between resistance and shaft position is nonlinear (the rate of change in resistance varies as the shaft is turned). A different amount of resistance change occurs for equal degrees of change of the shaft.

Nonlinear rheostats and potentiometers are often used as volume and tone controls on high-fidelity audio amplifiers. These devices are selected to obtain an apparently uniform change in the level of sound. When a volume control is set at its midpoint, it would appear that one-half the power of the sound output is available as compared to full power when the control is in full volume position. This is not actually the case because a linear increase in voltage does not cause the same percentage increase in audio output. By tapering the control, the sound output increases or decreases by the same percentage as the change in the volume control.

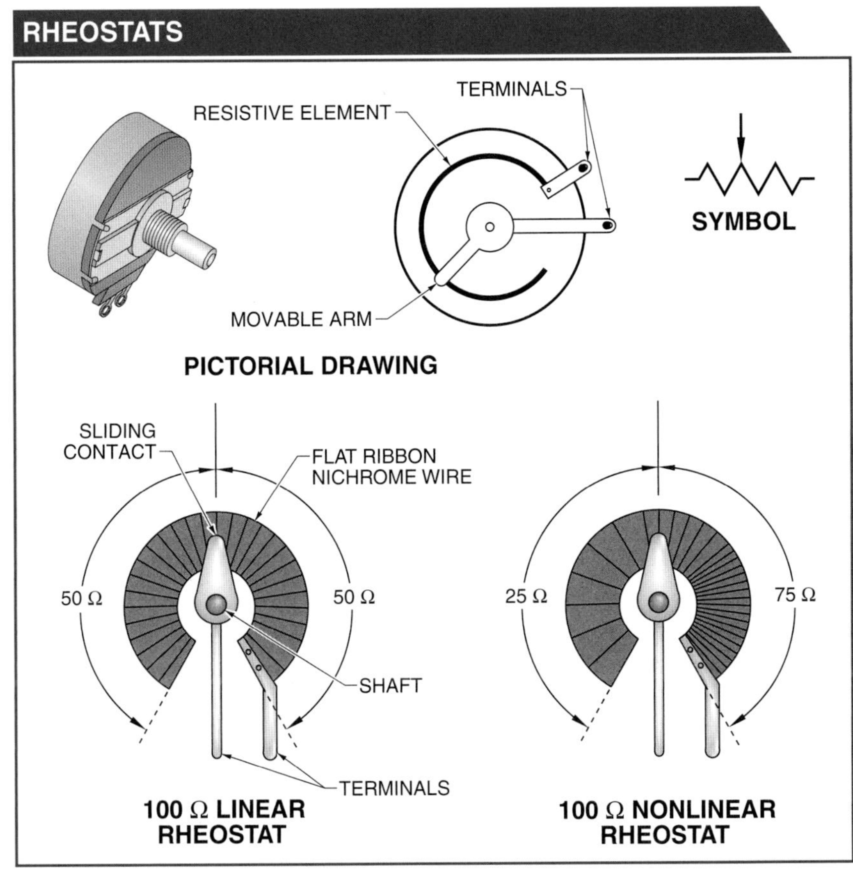

Figure 3-21. A rheostat is a two-terminal variable resistor. Rheostats are typically able to handle large amounts of power.

Ballast Resistors

A *ballast resistor* is a resistor whose resistance increases and decreases rapidly with changes in current. An increase in current causes a ballast resistor's resistance to increase, and a decrease in current causes its resistance to decrease. The current in the circuit remains constant due to this action. Ballast resistors make use of the fact that most materials have a positive temperature coefficient.

Ballast resistors are used to compensate for changes of voltages in power lines that are connected to electronic equipment sensitive to fluctuations. Ballast resistors are also used in electronic circuits that require a constant current to a given component. Most helium-neon lasers must use a ballast resistor in order to work properly.

Both the voltage and the power rating must be considered when selecting a ballast resistor for a particular application. The most reliable type is wire wound. Carbon composition ballast resistors can be used for low-current applications below 4.5 mA. Ballast resistors should be derated by at least 50%.

Terms

A **ballast resistor** is a resistor whose resistance increases and decreases rapidly with changes in current.

Power Resistors

A *power resistor* is a resistor with a power rating over 2 W. Power resistors are normally manufactured with a hollow center so that air flowing through them can help dissipate heat. This cooling effect extends the life of the resistor. Wire-wound power resistors using a ribbon made of a metal alloy such as nickel-chromium-iron wrapped around a ceramic support can dissipate hundreds of watts.

Power resistors may have multiple taps or may be adjustable. **See Figure 3-22.** Tapped resistors may have one or more taps. One tap provides three different values of resistance. Two taps provide six possible values of resistance. On an adjustable power resistor, the resistance can be changed by loosening the screw and nut assembly that keeps the indent in close contact with the exposed wire. Moving the assembly allows for selection of the resistance, from 0 Ω to the maximum value of the resistor. An arrow on any electrical symbol indicates the adjustability of the device.

Tech Tip

A good practice is to use a resistor with twice the wattage rating of that calculated for the component for normal circuit operation. This practice is not always followed in the manufacturing of consumer products due to cost.

Terms

A **power resistor** is a resistor with a power rating over 2 W.

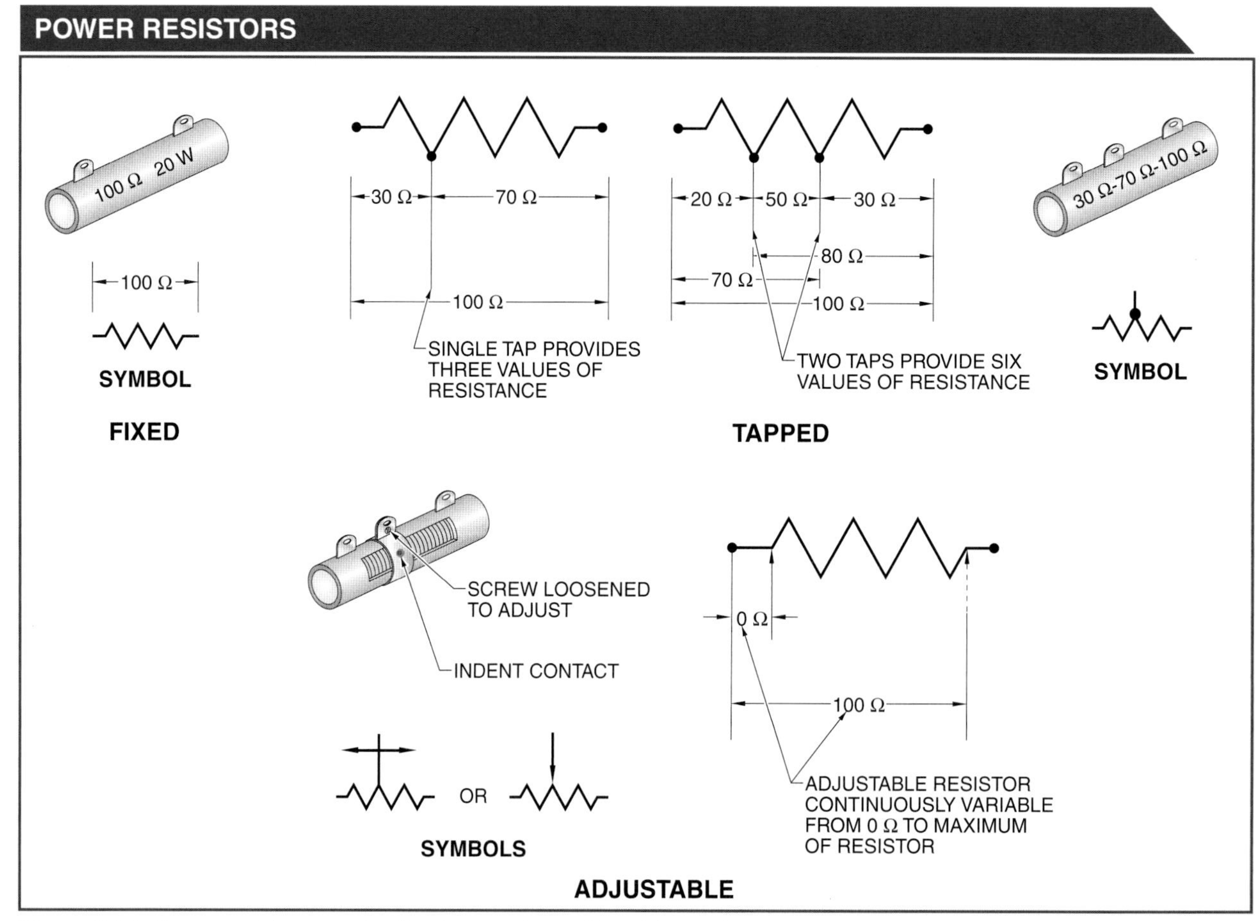

Figure 3-22. Power resistors may be fixed, tapped, or adjustable.

Resistor Networks

With the introduction of the printed circuit board, it became practice to place several components in a single package, and the number of discrete components included in any given device was reduced with each new development. Single integrated circuits have been developed that have more than a million transistors as well as the necessary resistors, capacitors, and inductors needed for proper operation of the device. Resistor networks have been developed that contain multiple resistors that have different physical arrangements within a single device. **See Figure 3-23.**

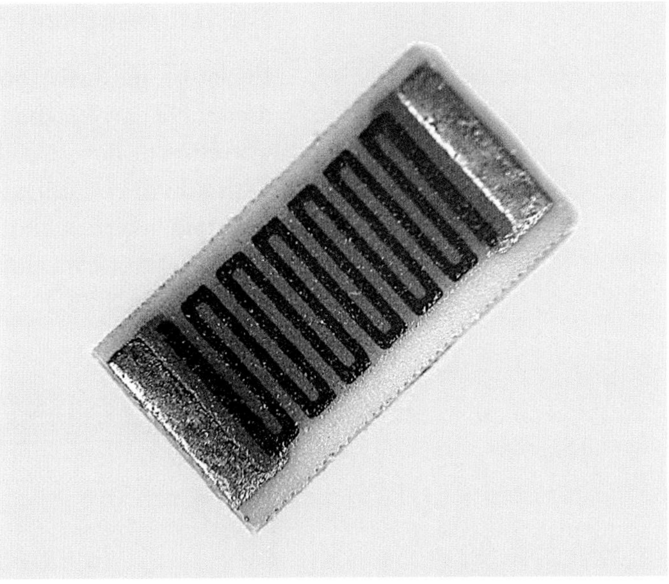

International Manufacturing Services, Inc.

Thick film high-power chip resistors are used for applications that require high power dissipation in a small package.

Resistor networks are available in a wide variety of shapes and sizes. Some resistor networks have heat sinks. A *heat sink* is a device that conducts and dissipates heat away from an electrical component. A heat sink consists of a block of metal that transfers the heat from the resistor to the chassis. Heat sinks are often used with solid-state devices, such as transistors and integrated circuits, to help remove excessive heat.

Terms

A **heat sink** is a device that conducts and dissipates heat away from an electrical component.

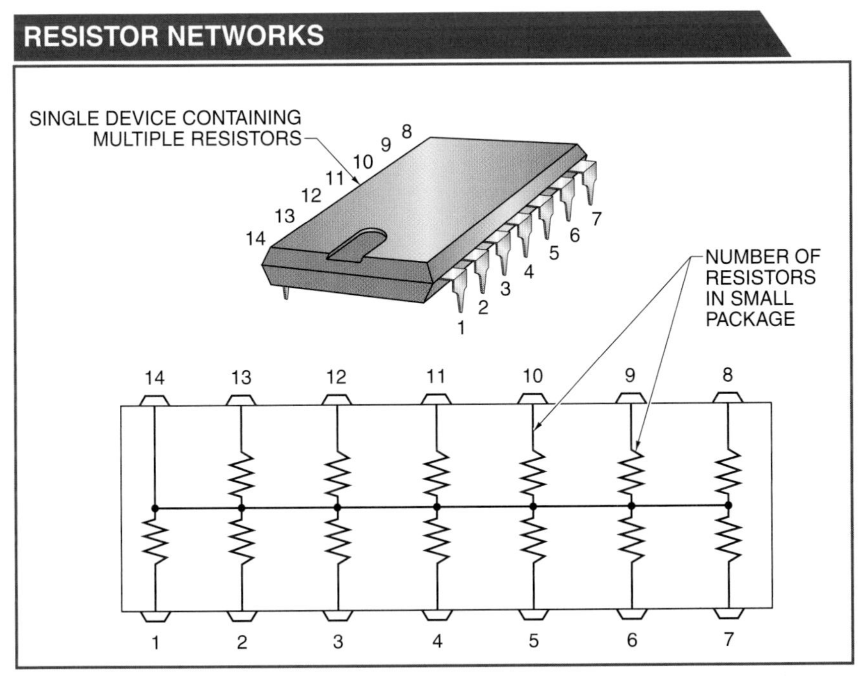

Figure 3-23. A resistor network consists of a number of resistors in a small package.

Resistor Applications

Resistors perform specific functions in a circuit. Resistors limit the amount of current that flows in a circuit and are selected by their values. The current flow in a circuit, over a range of voltages, is low in a circuit with a high resistance and high in a circuit with a low resistance. For example, a circuit that contains a 10 kΩ resistor has less current flow than a circuit containing a 5 kΩ resistor with the same voltage level. **See Figure 3-24.**

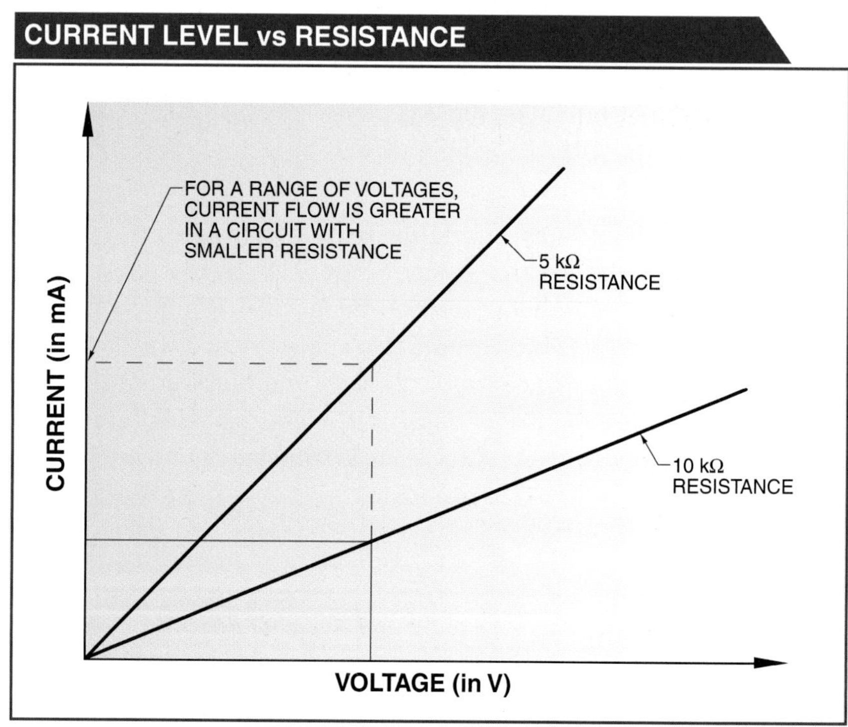

Figure 3-24. In a circuit, high resistance reduces the amount of current flow, regardless of the amount of voltage.

Resistors are also used to reduce the source voltage to the required voltage for a given load. For example, if a device requires a 6 V motor in an automobile that has a 12 V battery, a resistor can be used to reduce the voltage level to the level required. **See Figure 3-25.** If a range of voltages is required, a rheostat can be used in place of the resistor to provide an adjustable output voltage. For example, a rheostat will allow control of a fan motor speed.

Several different voltage levels are often required for proper electronic/electrical circuit operation. Using a single voltage source, resistors can be used as a voltage divider network to provide the required voltages. **See Figure 3-26.** Potentiometers are resistors that act as voltage dividers.

VOLTAGE LEVEL vs RESISTANCE

RESISTOR REDUCES 12 VDC POWER SOURCE TO REQUIRED 6 V FOR MOTOR
12 VDC POWER SOURCE
+
−
6 V MOTOR

Figure 3-25. Resistors can be used to reduce the source voltage to the voltage required by a load.

VOLTAGE DIVIDER NETWORK

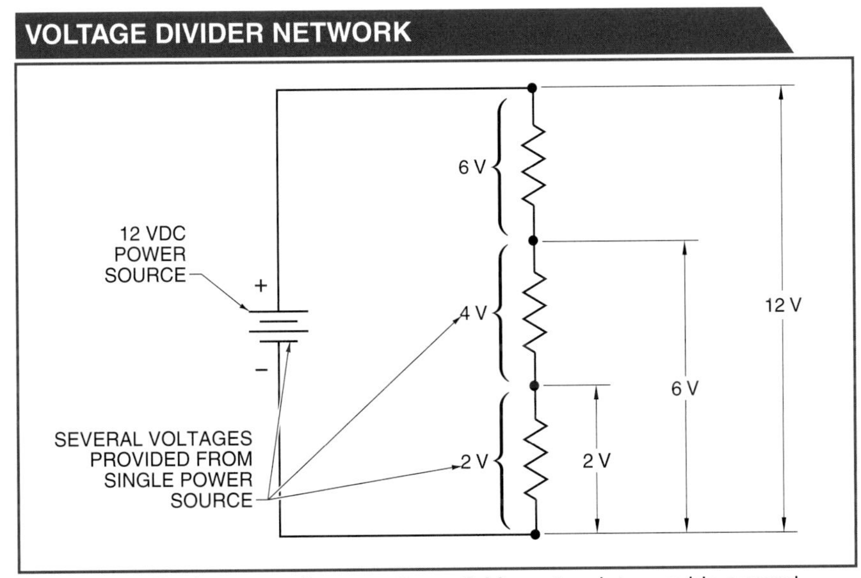

Figure 3-26. Resistors can form a voltage divider network to provide several voltages from a single source voltage.

A *capacitor* is an electronic device used to store an electric charge. Capacitors can store enough electrical energy to do bodily harm to anyone that contacts them. The NEC® requires the automatic discharge of capacitors to protect personnel working on some systems. A *bleeder resistor* is a resistor that is used to discharge a capacitor. **See Figure 3-27.** This safety feature is required for high-voltage applications where capacitors are used. For example, the capacitor charges to the value of the source voltage when the switch is in position 1. When the capacitor is disconnected from the source voltage by the switch, it is automatically connected to the bleeder resistor in position 2, causing a safe discharge.

Terms

A **capacitor** is an electronic device used to store an electric charge.

A **bleeder resistor** is a resistor that is used to discharge a capacitor.

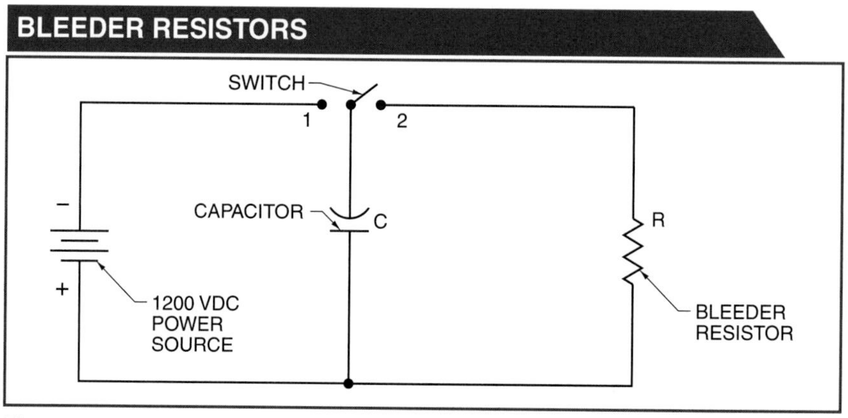

BLEEDER RESISTORS

Figure 3-27. To protect personnel, a bleeder resistor is used to automatically discharge capacitors in high-voltage applications.

Resistors are also used in devices to produce heat. Appliances such as electric stoves, toasters, waffle irons, and clothes dryers use resistive elements to produce heat for various applications. For example, a waffle iron includes a resistance heating element that heats the upper section and a resistance heating element that heats the lower section. An ON/OFF temperature control switch is used to turn the heating elements ON and control the temperature. **See Figure 3-28.**

Series and Parallel Resistors

Terms

A **series circuit** is a circuit that has only one current path.

A *series circuit* is a circuit that has only one current path. The total resistance of resistors connected in series equals the sum of the individual resistors. For example, if 100′ of wire with a resistance of 1 Ω is connected with another 100′ of wire with a resistance of 1 Ω, the length has doubled while the material and circular mil area remain the same. Since resistance is directly proportional to length, the resistance of the two lengths of wire is 2 Ω. The total resistance of the number of resistors (N) connected in series is calculated by applying the following formula:

$$R_T = R_1 + R_2 + R_3 + \ldots + R_N$$

where

R_T = total resistance

R_N = number of resistors connected in series

Example: What is the total resistance of three 1 kΩ resistors connected in series? **See Figure 3-29.**

$R_T = R_1 + R_2 + R_3$

$R_T = 1000\ \Omega + 1000\ \Omega + 1000\ \Omega$

$R_T = \mathbf{3000\ \Omega} = \mathbf{3\ k\Omega}$

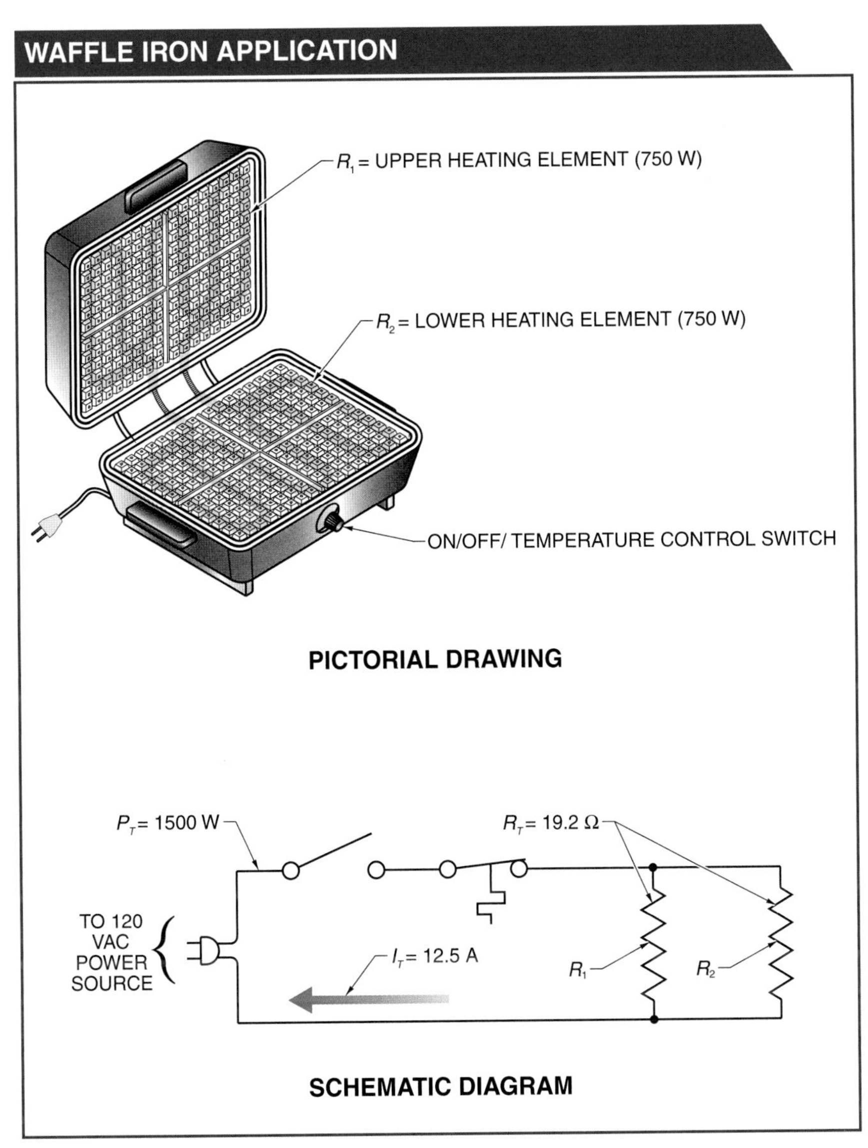

Figure 3-28. A waffle iron uses resistors to produce heat.

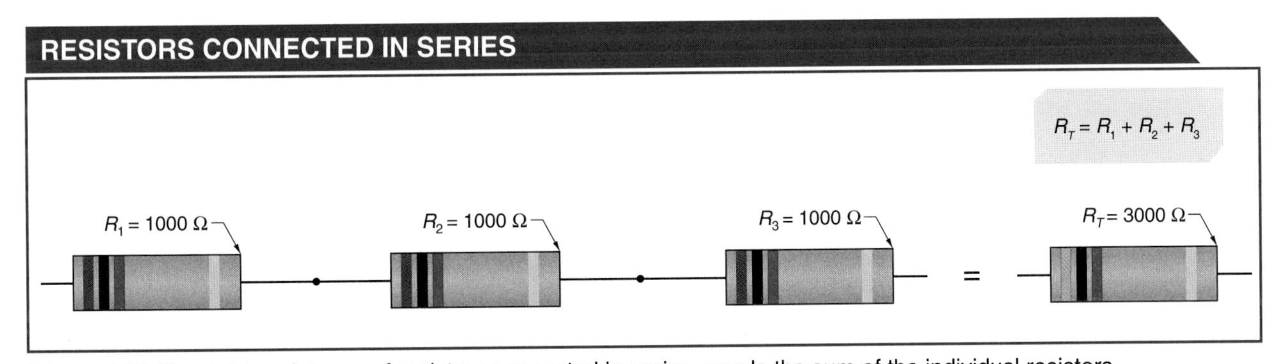

Figure 3-29. The total resistance of resistors connected in series equals the sum of the individual resistors.

Terms

A **parallel circuit** is a circuit that has more than one current path.

A *parallel circuit* is a circuit that has more than one current path. When resistors are connected in parallel with each other, their total resistance is less than the smallest resistance value. For example, if 100′ of wire with 1 Ω resistance is connected in series with an identical piece of wire, the length doubles and so does the resistance. When the same pieces of wire are connected in parallel, their total length remains 100′, but their circular mil areas increase. With double the circular mil area, the resistance is reduced to one-half its previous value (0.5 Ω). When multiple resistors of the same value are connected in parallel, the result is a resistance that is $1/n$ the resistance of one resistor. The total resistance of any number of resistors connected in parallel is calculated by applying the following formula:

$$R_T = \frac{R}{N}$$

where

R_T = total resistance

R = common resistor value

N = number of resistors of same value

Example: What is the total resistance of three 1.5 kΩ resistors connected in parallel? **See Figure 3-30.**

$$R_T = \frac{R}{N}$$

$$R_T = \frac{1500\ \Omega}{3}$$

$$R_T = \mathbf{500\ \Omega\ (0.5\ k\Omega)}$$

RESISTORS CONNECTED IN PARALLEL

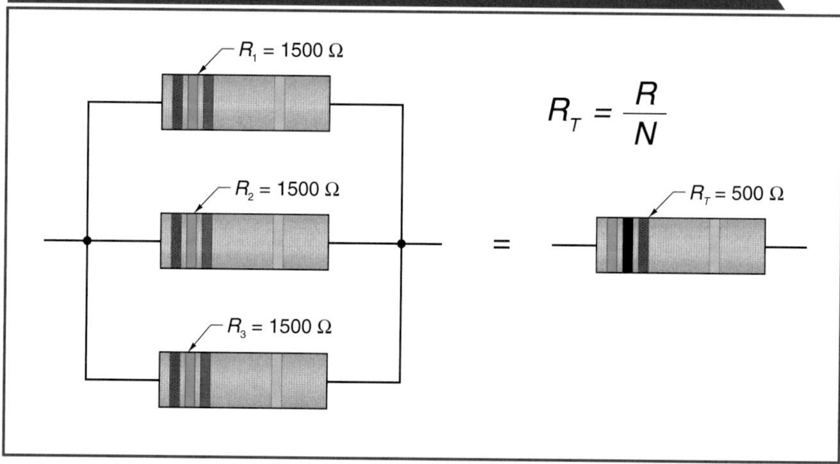

Figure 3-30. When three resistors of equal value are connected in parallel, their total value is equal to the value of one resistor divided by the number of resistors.

The total resistance of two resistors with unlike values connected in parallel is calculated by applying the following product over sum equation:

$$R_T = \frac{R_1 \times R_2}{R_1 + R_2}$$

Example: What is the total resistance in a circuit containing a 3 Ω resistor and a 6 Ω resistor connected in parallel?

$$R_T = \frac{R_1 \times R_2}{R_1 + R_2}$$

$$R_T = \frac{3\ \Omega \times 6\ \Omega}{3\ \Omega + 6\ \Omega}$$

$$R_T = \frac{18}{9}$$

$$R_T = \mathbf{2\ \Omega}$$

The total resistance of more than two resistors with unlike values connected in parallel is calculated by applying the following formula:

$$R_T = \frac{1}{\frac{1}{R_1} + \frac{1}{R_2} + \ldots + \frac{1}{R_N}}$$

Example: What is the total resistance in a circuit containing resistors of 6 Ω, 3 Ω, and 2 Ω connected in parallel?

$$R_T = \frac{1}{\frac{1}{R_1} + \frac{1}{R_2} + \frac{1}{R_3}}$$

$$R_T = \frac{1}{\frac{1}{6\ \Omega} + \frac{1}{3\ \Omega} + \frac{1}{2\ \Omega}}$$

$$R_T = \frac{1}{\frac{1}{6} + \frac{2}{6} + \frac{3}{6}}$$

$$R_T = \frac{1}{\frac{6}{6}}$$

$$R_T = \frac{1}{1}$$

$$R_T = \mathbf{1\ \Omega}$$

The total resistance of resistors of different values connected in parallel can also be found by using a calculator and determining decimal fractions. The decimal fractions can be added together and then divided into the numerator:

$$R_T = \frac{1}{\frac{1}{R_1}+\frac{1}{R_2}+\frac{1}{R_3}}$$

$$R_T = \frac{1}{\frac{1}{6\ \Omega}+\frac{1}{3\ \Omega}+\frac{1}{2\ \Omega}}$$

$$R_T = \frac{1}{0.167+0.333+0.5}$$

$$R_T = \frac{1}{1}$$

$$R_T = \mathbf{1\ \Omega}$$

Resistance Measurement

Terms

Continuity is the presence of a complete path for current flow.

A **continuity checker** is an instrument that indicates an open or closed circuit in a circuit in which all power is OFF.

The actual resistance value is not always required when taking measurements. *Continuity* is the presence of a complete path for current flow. A *continuity checker* is an instrument that indicates an open or closed circuit in a circuit in which all power is off. A continuity checker can be used to measure the integrity of a circuit or component. **See Figure 3-31.** A continuity checker consists of a battery, buzzer, and test leads and is often a function on a test instrument such as a DMM or ammeter. Continuity checkers are used to test switches, fuses, and grounds and to identify individual conductors in a multiwire cable.

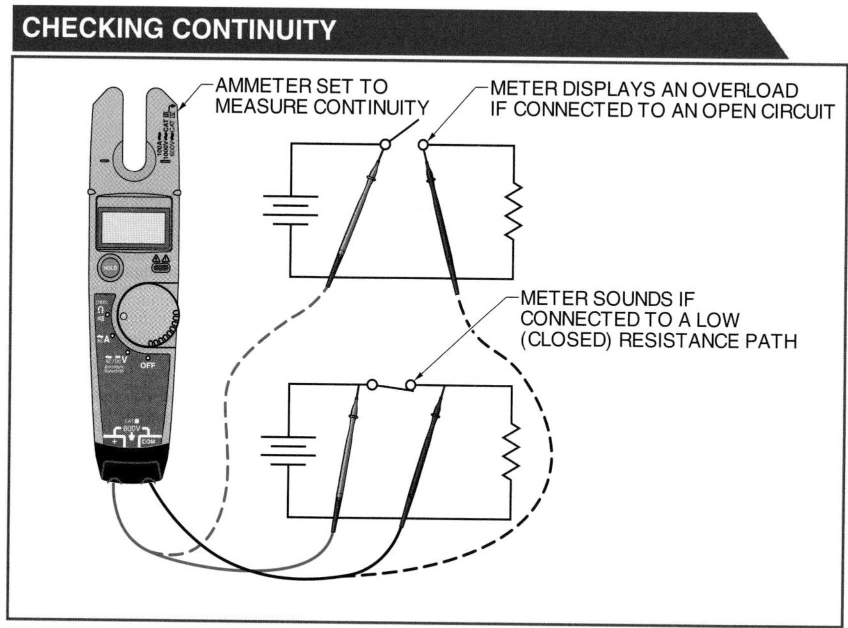

Figure 3-31. Continuity is the presence of a complete path for current flow and can be checked using an ammeter set to measure continuity.

When testing a switch, the meter will buzz when the switch is in the closed position and remain off when the switch in the open position. Most continuity checkers provide an audible indication when there is a complete circuit path.

Continuity checkers are only good for checking low-resistance circuits. Continuity checkers do not provide the actual resistance value of a circuit. An ohmmeter or multimeter is required if the circuit resistance value is needed.

An *ohmmeter* is an instrument used to measure the continuity and resistance of a circuit or component. With the exception of ohmmeters used to measure only continuity, they seldom exist as single-function devices. An ohmmeter that measures both continuity and resistance is usually built into a multimeter. A *multimeter* is a test tool used to measure two or more electrical values. Multimeters commonly measure voltage, current, and resistance. They can be either analog or digital devices.

Analog Multimeters. An *analog multimeter* is an electromechanical device that indicates readings by the mechanical motion of a pointer. **See Figure 3-32.** A multimeter that measures resistance is required to find exact resistance values. An analog multimeter, when set to measure resistance, allows testing of a wide range of resistance values. Analog multimeters have several scales for reading voltage, current, and resistance.

The basic circuit of an analog multimeter set to measure resistance is the same as that of a continuity checker except that the pointer movement replaces the lamp or audible indicator. The circuit includes a battery, pointer movement, a zero adjust rheostat, and an internal multiplier resistor (R_X) in series (one current path) with each other.

The resistance scale on an analog multimeter is nonlinear. The resistance scale covers a range from 0 Ω to infinity (∞). The numbers become crowded on the left-hand side of the scale, and it becomes difficult to obtain an accurate measurement. To increase accuracy, the range switch is changed so that the pointer reads the scale in the upper right two-thirds of the range of the pointer.

An analog multimeter must be read from perpendicular (90°) to scale face. A parallax error is introduced if the scale is viewed at an angle. A *parallax error* is the inaccuracy created by the difference in apparent position of an analog meter pointer when viewed from different angles not directly perpendicular to the pointer and scale. Some analog meters have a mirror on the scale that helps with the perpendicular alignment. When a mirror is present, the eye is aligned so that the pointer's reflection is not observed.

Terms

An **ohmmeter** is an instrument used to measure the continuity and resistance of a circuit or component.

A **multimeter** is a test tool used to measure two or more electrical values.

An **analog multimeter** is an electromechanical device that indicates readings by the mechanical motion of a pointer.

A **parallax error** is the inaccuracy created by the difference in apparent position of an analog meter pointer when viewed from different angles not directly perpendicular to the pointer and scale.

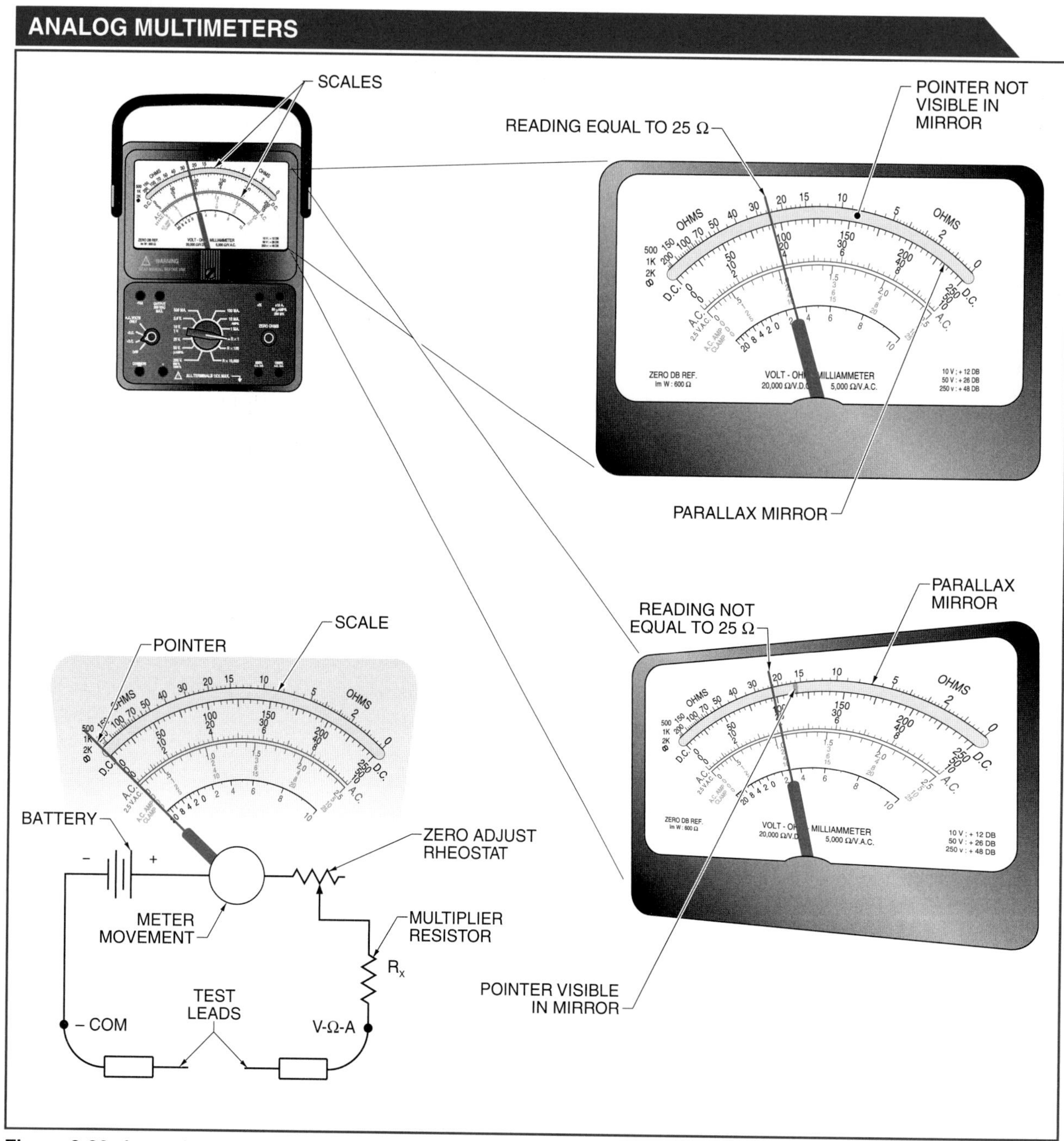

Figure 3-32. An analog multimeter indicates readings by the mechanical motion of a pointer.

The pointer will sometimes be between the major marks on the scale. When this occurs, the percentage of distance between the major marks is estimated, converted to a number, and added to the value of the next lowest mark. The inaccuracy introduced by this estimate is referred to as an interpolation error.

When the test probes are open, the pointer remains at the left-hand side of the scale indicating infinite value (∞). Under this condition, no current flows through the meter movement. When the test leads are shorted, the needle shows full scale deflection (0 Ω). **See Figure 3-33.** The zero adjust rheostat provides a means to compensate for changes in battery potential and slight changes in the values of circuit components. The zero adjust rheostat should be used to align the meter pointer to the 0 Ω mark. A center scale deflection occurs on the meter if the resistor being measured is equal to the multiplier resistor (R_x) plus the resistance of the rheostat.

The multiplier resistor is used to change the range of the meter. If a 12 Ω resistor is being measured, the multiplier is set to × 1. If a 120 Ω resistor is being measured, the multiplier is set to × 10 range and the value shown on the scale is multiplied by 10.

An analog multimeter has its own power source. It does not rely on circuit power to obtain a measurement. The first step in taking any resistance measurement is to remove all power from the circuit. If this is not done, the resistance measurement will be inaccurate and the meter may be damaged or destroyed.

An analog multimeter also has less accuracy when measuring voltages in sensitive circuits due to their low impedance inputs. The low impedance has greater impact on the circuit than the high impedance inputs found on digital multimeters.

Digital Multimeters. A *digital multimeter (DMM)* is an electronic device that indicates readings as numerical values. Digital multimeters have replaced analog multimeters in most electrical work. A DMM performs the same functions as an analog multimeter; however, DMMs are more durable and have higher accuracy than analog multimeters. They eliminate parallax and interpolation errors and often possess features that analog meters do not. For example, most DMMs measure alternating current (AC). They normally also have overload protection, which is automatic on some models. When the overload is removed, the meter resets itself so that the operator can continue taking measurements.

Digital multimeters are designed to measure a wide range of resistances. No single meter covers all ranges, but DMMs will normally measure resistance as low as 0.1 Ω and as high as 300 MΩ. For resistance measurements, many DMMs use a battery voltage of less than 0.3 VDC. This voltage is below the voltage that causes most solid-state devices to conduct current. A DMM has a function switch to set the measurement and range, a DC/AC switch, test lead connection jacks, and a numerical display. **See Figure 3-34.**

Terms

A **digital multimeter (DMM)** is an electronic device that indicates readings as numerical values.

Tech Tip

An analog multimeter is less accurate when measuring voltages in sensitive circuits due to their low impedance inputs. The low impedance has a greater impact on the circuit than high-impedance inputs for a digital multimeter.

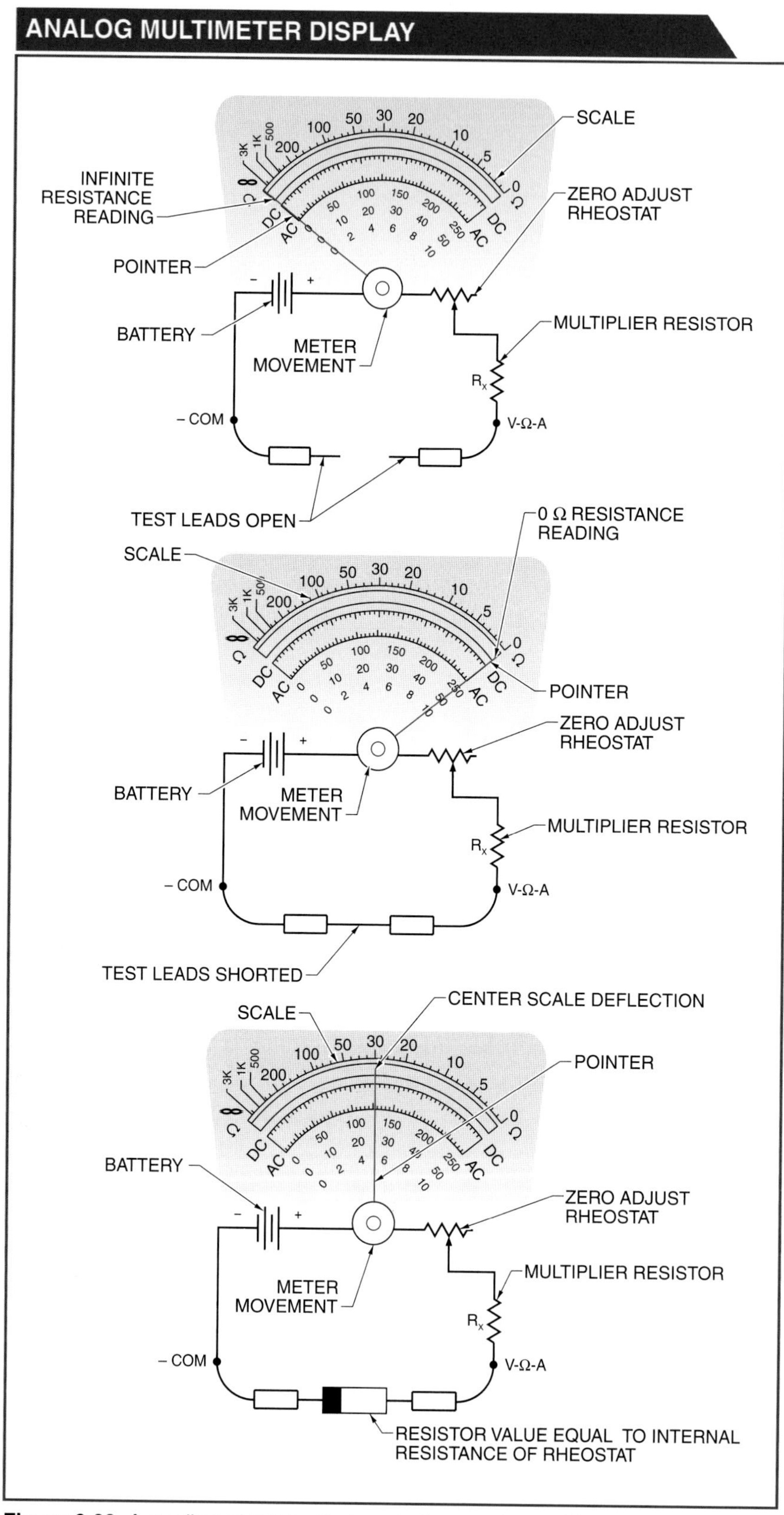

Figure 3-33. A reading of 0 Ω is displayed when analog multimeter test leads are shorted.

DIGITAL MULTIMETER FUNCTIONS

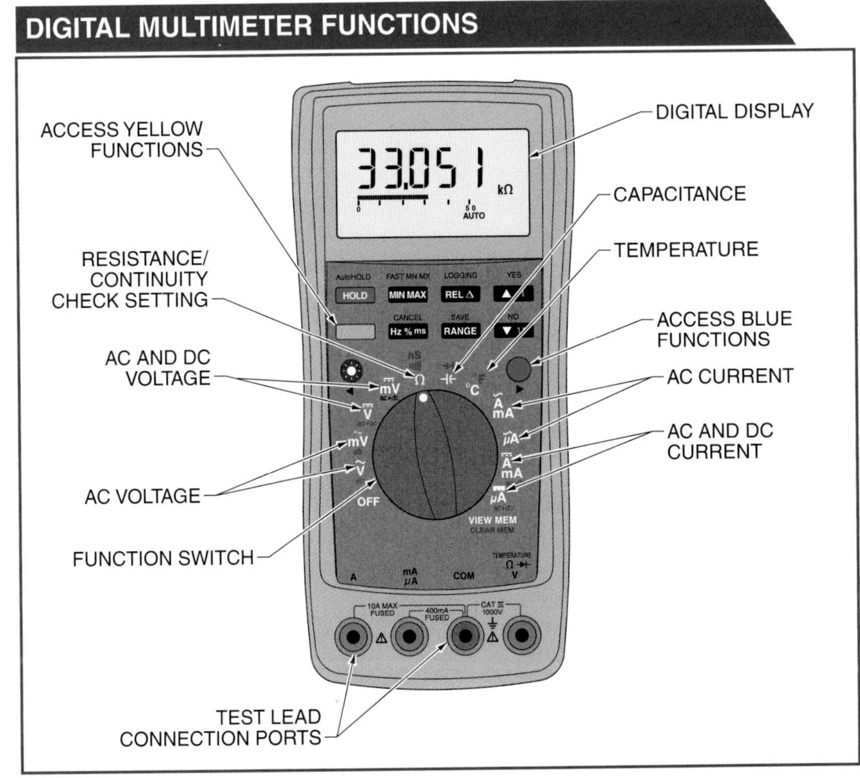

Figure 3-34. A digital multimeter has a function switch to set the measurement and range, a DC/AC switch, test lead connection jacks, and a numerical display.

Terms

An **autoranging DMM** is a DMM that automatically adjusts to a higher range setting if the range is not high enough.

A **megohmmeter** is a device that detects insulation deterioration by measuring high resistance values under high-voltage conditions.

Digital multimeters may have several resistance range settings or may be autoranging. An *autoranging DMM* is a DMM that automatically adjusts to a higher range setting if the range is not high enough. A buzzer may sound when continuity is established. The buzzer eliminates the need for the operator to read the value on the meter numerical display. The resistance value that prevents the buzzer from sounding varies between meters. To ensure safe operation, always consult the users manual before using any DMM.

Megohmmeters. Ordinary ohmmeters cannot take measurements of extremely high resistance. A *megohmmeter* is a device that detects insulation deterioration by measuring high resistance values under high-voltage conditions. **See Figure 3-35.** Megohmmeters are often referred to as Meggers® or insulation resistance testers (IR testers).

Fluke Corporation

An insulation multimeter is a test instrument specifically designed for taking resistance measurements ranging from 0.1 MΩ to 2 GΩ.

MEGOHMMETERS

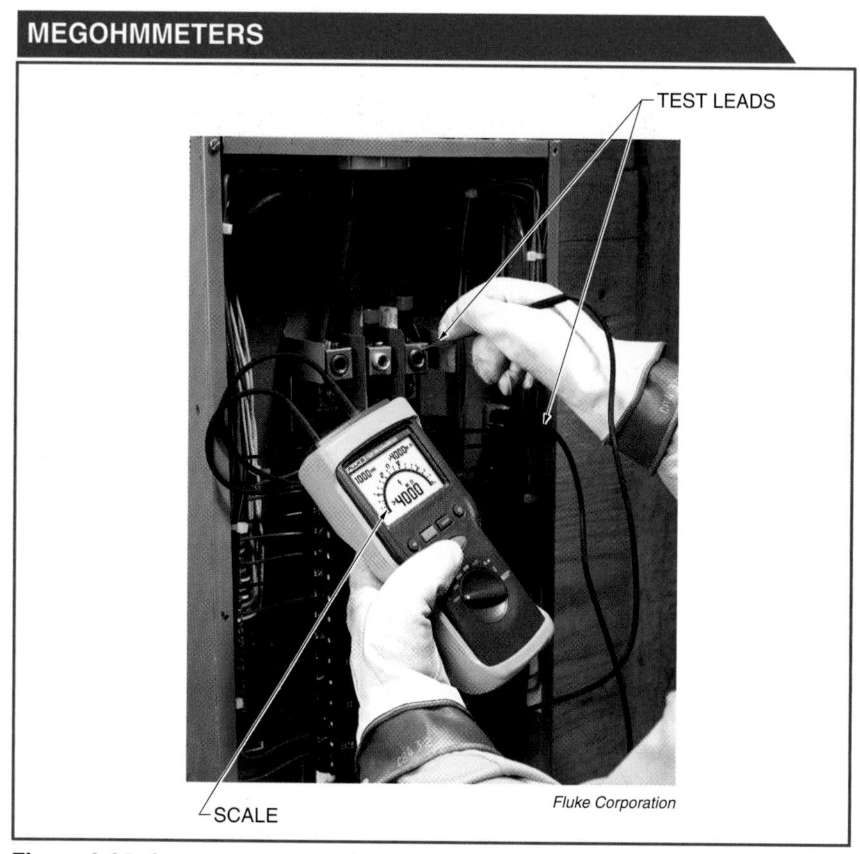

Figure 3-35. A megohmmeter is used to measure extremely high resistance.

In general, megohmmeters use a more powerful battery source than that used by ordinary multimeters. These tests are normally conducted using 500 V and 1000 V sources with the scale on the meter calibrated in megohms. Any material conducts electrons if enough voltage is applied across it. The minimum resistance in megohms (R_{min}) of the insulation of a power cable as specified by the Institute of Electrical and Electronics Engineers (IEEE) Standard 43 can be determined by the following equation:

$$R_{min} = kV + 1$$

where

R_{min} = minimum resistance (in MΩ)

kV = applied voltage (in kV)

1 = constant

Example: What is the minimum resistance of the insulation of a power cable that carries 2000 V?

$$R_{min} = kV + 1$$

$$R_{min} = 2 + 1$$

$$R_{min} = \mathbf{3\ M\Omega}$$

The absolute minimum resistance of the power cable insulation is 1 MΩ. For each kV the insulation must withstand, an additional 1 MΩ is added. These values vary depending on the authority setting the requirement. For example, an insulation resistance of 0.5 MΩ is allowed for lighting circuits by the U.S. Navy's NAVSEA command, whereas 1 MΩ is required for power circuits.

SUMMARY

Resistance is the opposition to the flow of electrons in an electrical circuit. The resistance of any mass is dependent upon its material makeup and its physical dimensions. Resistance increases as conductor length increases but decreases as conductor cross-sectional area increases. Temperature can also cause a change in resistance.

Increasing the resistance in a circuit results in a decreased flow of electrons. Heat is produced and electrical power consumed when current meets resistance, causing a decrease in circuit power.

Resistors are devices designed to present electrical resistance. These devices are rated in ohms and wattage. Resistors are produced in a wide range of values and shapes. Some resistors use color coding to indicate their values. Other resistors have their values printed on them. The different types of resistors include carbon composition, wire wound, and metal film. Each of these has advantages and disadvantages.

The total resistance of resistors connected in series equals the sum of the individual resistances. When resistors are connected in parallel, their total resistance is always less than the resistance of the resistor with the lowest value.

Some resistors are variable or adjustable. Variable resistors have a shaft connected to a variable arm that contacts the surface of the resistor. The resistor can be adjusted by turning the shaft. Variable resistors are normally classified as potentiometers or rheostats. Potentiometers have three terminals and are used to divide voltage. Rheostats have two terminals and are used to control current. Adjustable resistors are usually wire wound with a sliding contact that must be unfastened and then fastened at the selected resistance.

Test instruments used to measure resistance include analog multimeters and digital multimeters. Special instruments called megohmmeters are designed to measure very large resistances, such as that found with conductor insulation.

APPLICATION: RESISTANCE...

Background: Resistors are two-terminal circuit elements that are used to limit current and reduce voltage. Resistors come in many standard sizes. The standard resistor values are given in general by the following:

$$R = d \times 10^{\frac{i}{N}}$$

where
R = resistance (in Ω)
d = decade multiplier
$i = 0, 1, 2, \ldots, N - 1$
N = tolerance (*Note:* 1% = 96, 2% = 48, 5% = 24, 10% = 12)

The value of R is rounded to the proper number of significant digits based on the tolerance, which is 3 digits for 1% and 2% and 2 digits for 5% and 10%. Therefore, 10% resistor values in the 100s are the following:

$R = 100 \times 10^{i/12}$; $i = 0$ to 11
R = 100, 120, 150, 180, 220, 270, 330, 390, 470, 560, 680, 820

Note: Most of the values match the formula results with the correct number of significant digits. Some of the values are conventional rather than calculated (e.g., calculated 260 versus conventional 270).

Since resistors come in standard sizes, sometimes a circuit designer must use resistors connected in series and/or parallel combinations to achieve the desired resistance. For example, if a resistance value of 110 Ω is needed, a designer can use two 220 Ω resistors connected in parallel to give the proper resistance. Another design would be to use a 10 Ω resistor connected in series with a 100 Ω resistor.

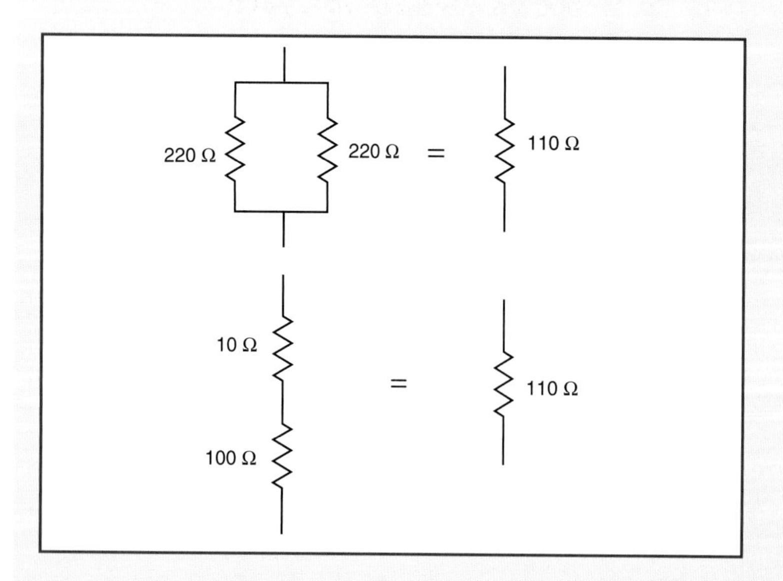

...APPLICATION: RESISTANCE...

When combining resistors of the same tolerance, the resultant resistance is still within the individual resistor tolerance. This is a worst-case scenario for tolerance since it is possible to get multiple resistors at one end of the tolerance scale. In reality, if one measured all resistors built in a given lot, the overall tolerance follows a normal distribution curve with a mean (μ) and standard deviation (σ). For a normal distribution curve, 68.2% of a resistor lot is within ±1 of the mean, and 95.4% of the resistors are within ±2σ of the mean.

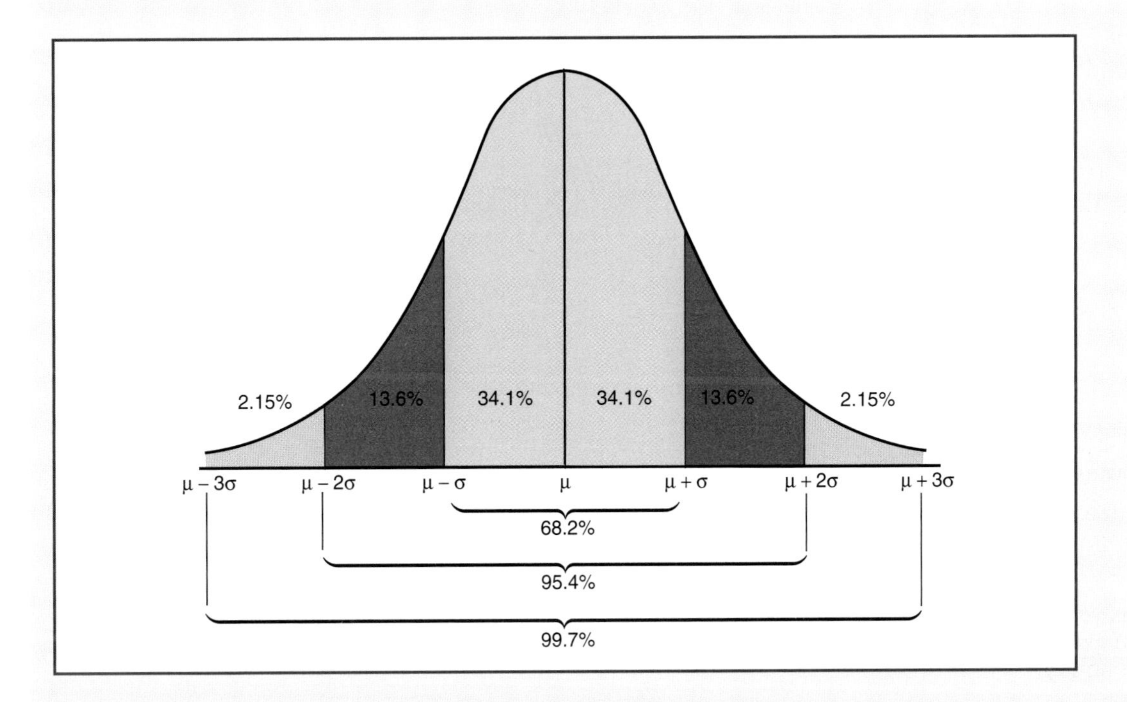

Statically, the combined resistance tolerance is actually less than the individual resistor tolerance. The combined resistance final tolerance is found by applying the following formula:

$$C_{RT} = \frac{I_{RT}}{\sqrt{N}}$$

where
C_{RT} = combined resistance tolerance (in %)
I_{RT} = individual resistance tolerance (in %)
N = number of resistors

Combining five 5% resistors, the expected resistance tolerance is the following:

$$C_{RT} = \frac{5\%}{\sqrt{5}}$$

$$C_{RT} = 2.2\%$$

Key Points: Resistance

...APPLICATION: RESISTANCE...

Problem: For a circuit design, a resistance of 300 Ω is required within a tolerance of 4% using only 10% resistors.

Solution: The minimum number of 10% resistors that is required is set by the required tolerance of 4%. The number of resistors needed is as follows:

$$C_{RT} = \frac{I_{RT}}{\sqrt{N}}$$

$$4\% = \frac{10\%}{\sqrt{N}}$$

$$N = \left(\frac{10\%}{4\%}\right)^2$$

$$N = 6.25$$

Seven is the minimum number of 10% tolerance resistors needed to reduce the circuit tolerance to 4%.

Standard 10% resistor values are the following:
10, 12, 15, 18, 22, 27, 33, 39, 47, 56, 68, 82
100, 120, 150, 180, 220, 270, 330, 390, 470, 560, 680, 820
1000, 1200, 1500, 1800, 2200, 2700, 3300, 3900, 4700, 5600, 6800, 8200
10,000, 12,000, 15,000, 18,000, 22,000, 27,000, 33,000, 39,000, 47,000, 56,000, 68,000, 82,000

For series resistors, the total resistance is the following:

$$R = R_1 + R_2 + \ldots R_N$$

For parallel resistors, the total resistance is the following:

$$R = \frac{1}{\left(\frac{1}{R_1} + \frac{1}{R_2} + \ldots + \frac{1}{R_N}\right)}$$

Using two 150 Ω resistors in series meets the 300 Ω requirement but does not reduce the tolerance to below 4%. The two resistors only reduce the tolerance to 7%.

For the same resistance value in parallel, the overall resistance is half. So, having two 600 Ω resistors in parallel is equal to having 300 Ω. Four 150 Ω resistors in series is equal to 600 Ω of resistance.

$$R = 150 + 150 + 150 + 150$$

$$R = 600$$

By adding another 600 Ω resistor (four 150 Ω resistors in series) the resistance becomes the following:

$$R = \frac{1}{\left(\frac{1}{600} + \frac{1}{600}\right)}$$

$$R = 300\ \Omega$$

...APPLICATION: RESISTANCE

The total number of 10% resistors used is eight, for an overall tolerance of ±3.5%.

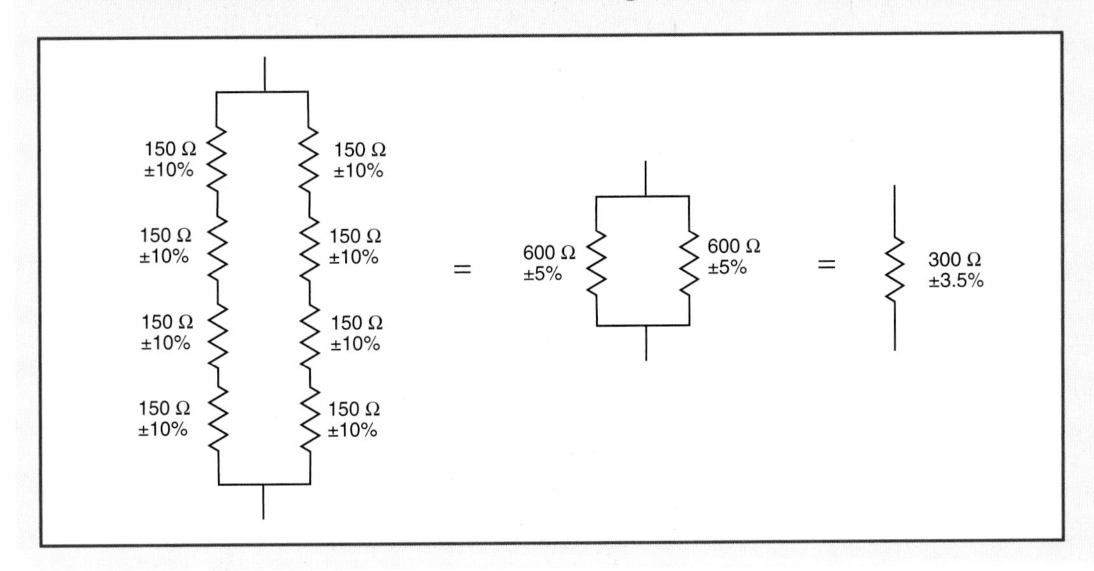

Chapter Review

1. What is the normal resistance measurement range for most DMMs?
2. What is a superconductor?
3. How does heat affect the resistance of most metals and their alloys?
4. Why is it important for electricians to understand the resistivity of annealed copper and aluminum?
5. What property allows voltage the ability to produce electron flow through a resistance?
6. What is the total resistance (in kΩ) of a group of 1000 Ω, 1200 Ω, and 1500 Ω resistors connected in series?
7. What are two common types of variable resistors?
8. What are the four factors that affect the resistance of a substance?
9. What do the fifth and sixth color bands represent on a metal film resistor?
10. What test instrument is used to detect insulation deterioration by measuring high resistance values under high-voltage conditions?
11. What are ballast resistors used for?
12. What is the only metal that has a relative conductance greater than 1?
13. What is the total resistance of four 2.5 kΩ resistors connected in parallel?
14. What type of resistor is used to discharge a capacitor?
15. How does an increase in the resistance of a circuit resistor affect circuit current for a constant voltage?
16. What are the two main classifications of resistors?
17. What is a negative temperature coefficient?
18. How does an increase of three wire gauge sizes affect the cross-sectional area, weight, and resistance of a wire?
19. What are two main advantages of using cermet in fixed resistors?
20. What is the standard SI unit for resistivity?

4

Voltage Sources

OBJECTIVES

- List and explain the different types of voltage sources.
- Explain how pressure affects voltage.
- Describe the different effects of light.
- Explain how heat affects voltage within a circuit.
- Explain chemical action and chemical device applications.
- Describe magnetism, magnetic processes, magnetic theory, and electromagnetism.
- List and describe magnetic device applications.

In order to have current flow in an electrical circuit, there must be a source of potential energy to move electrons in and out of their valence shells. The six sources of potential energy in electrical circuits are friction, pressure, light, heat, chemical action, and magnetism. Energy is used to produce electricity. Voltage is the amount of potential energy difference within an electrical circuit.

VOLTAGE SOURCES

Potential energy is produced by using a force to move electrons in and out of their valence shells. The six electrical sources of this energy are friction, pressure, light, heat, chemical action, and magnetism. **See Figure 4-1.** Magnetism is the most common usable source of electrical energy. Chemical action is the next most common usable source.

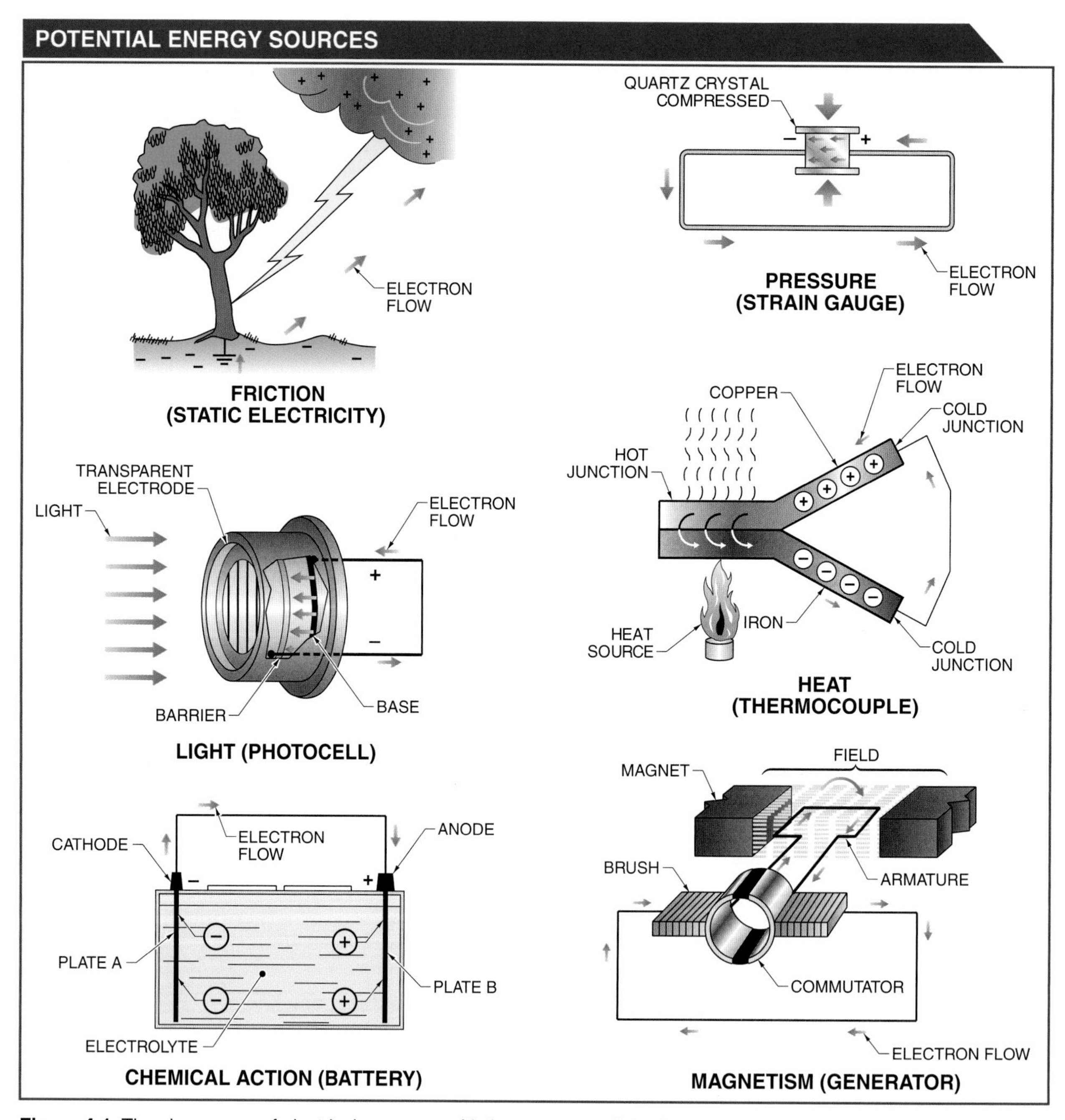

Figure 4-1. The six sources of electrical energy are friction, pressure, light, heat, chemical action, and magnetism.

Friction

The *triboelectric effect* is the generation of a static charge by friction. *Friction* is the resistance to motion that occurs when two surfaces move against each other. A *static charge* is an electrical charge at rest. Friction causes electrons from one source to flow to another material to create negative and positive static charges. A *negative static charge* is the accumulation of excessive electrons on a body. A *positive static charge* is a deficiency of electrons on a body. For example, a glass rod, when rubbed with a silk cloth, gives up some of its electrons to the silk. In this process, the silk becomes negatively charged, and the glass rod becomes positively charged. Friction is seldom used for producing electrical energy.

When a person with a negative static charge contacts an object with a positive static charge, all of the excess electrons flow (jump) to the object. All methods of electrification require electrons to move from one body to another. *Electrostatic discharge (ESD)* is the movement of electrons from a source to an object.

Electrostatic discharges as small as 100 V have been known to damage semiconductors. People normally do not feel an electrostatic discharge until the discharge reaches 3000 V. Precautions must be taken to discharge static charges before handling semiconductor devices. A wrist strap attached to a static line that is grounded will discharge static electricity from the body. **See Figure 4-2.**

Tech Tip

Static electricity is the oldest known form of electrical energy. The Greeks reported the effects of static electricity as long ago as 600 BC.

Terms

The **triboelectric effect** is the generation of a static charge by friction.

Friction is the resistance to motion that occurs when two surfaces move against each other.

A **static charge** is an electrical charge at rest.

A **negative static charge** is the accumulation of excessive electrons on a body.

A **positive static charge** is a deficiency of electrons on a body.

Electrostatic discharge (ESD) is the movement of electrons from a source to an object.

ELECTROSTATIC DISCHARGE

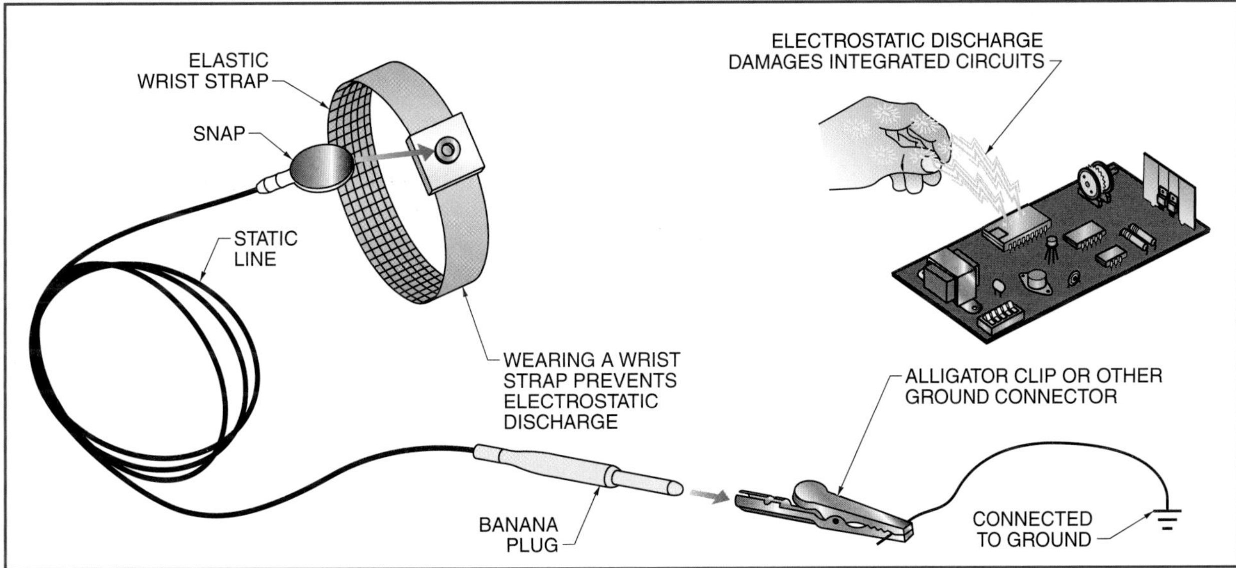

Figure 4-2. A wrist strap connected to ground is used to prevent electrostatic discharges from damaging electronic components.

Terms

Lightning is a transient, high-current electrostatic discharge that occurs in the atmosphere.

A **lightning stroke** is the initial electrostatic discharge and return discharge.

Small electrostatic discharges (static shock) are unpleasant and may be costly but are not life threatening. On the other hand, an aircraft in flight accumulates a static charge due to the friction between its surface and the passing air. These static charges may cause interference with radio communications and, under certain circumstances, can cause physical damage. When refueling aircraft, a static line is attached between the aircraft and ground to discharge any static charge buildup before beginning the fueling process. This eliminates the possibility of an explosion due to an electrostatic discharge.

Lightning. *Lightning* is a transient, high-current electrostatic discharge that occurs in the atmosphere. Lightning produces a very high flow of electrical current. A *lightning stroke* is the initial electrostatic discharge and return discharge. Lightning normally has a duration of less than one second and is composed of many distinct strokes that occur in such rapid succession that they appear as a single flash of light.

Lightning

Approximately 2000 thunderstorms may exist worldwide at any given time. In the United States, about 100 to 200 persons are killed and several hundred injured by lightning each year. In addition, lightning ground strikes in the United States cause property damage of several hundred million dollars yearly. Lightning also starts an estimated 10,000 forest fires every year.

Before lightning can occur, an electrical charge that can break down the insulation quality of the air must accumulate. Most experts believe that this electrical charge forms as different kinds of ice interact in a cloud. At the same time, updrafts of warm air and downdrafts of cold air flowing through ice crystals or droplets of water in the cloud cause a negative charge to form at the bottom of the cloud and a positive charge to form at the top. Thunderstorm clouds are generally negatively charged at the base and positively charged in higher regions. **See Figure 4-3.** Under certain conditions, this charge may be reversed or distributed differently in the cloud.

Although lightning may occur between clouds, between clouds and the ground, or between clouds and the clear air above them, over half of all lightning flashes occur within a single cloud.

ELECTRICAL CHARGES

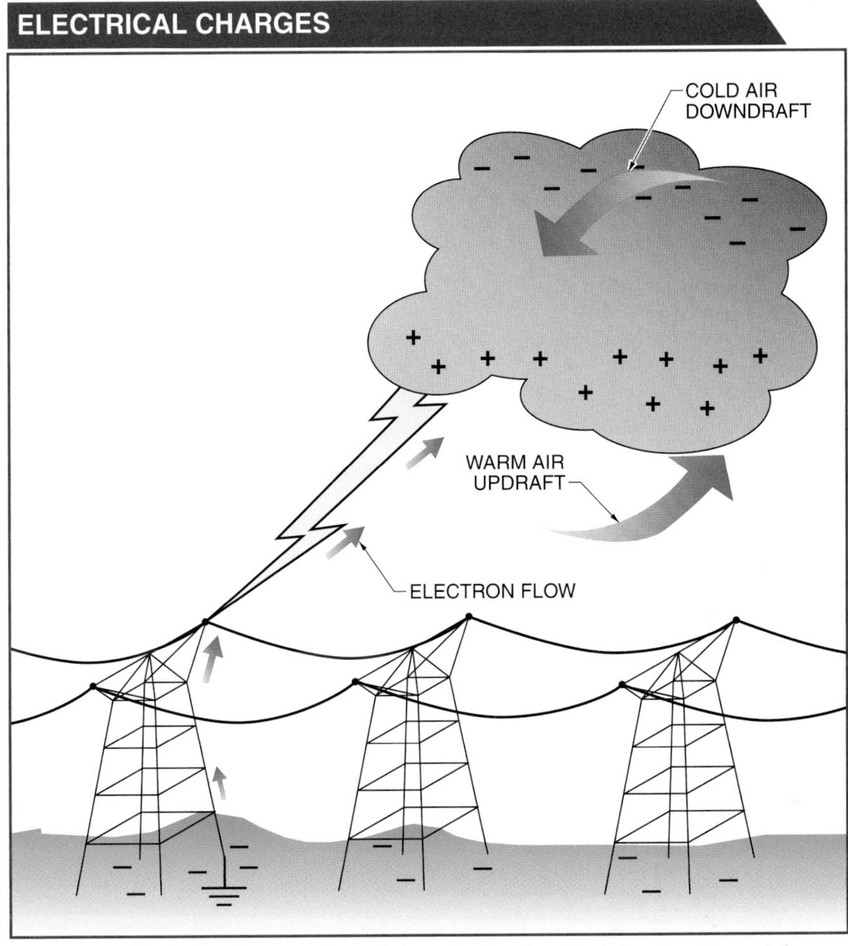

Figure 4-3. Updrafts of warm air and downdrafts of cold air flowing through ice crystals cause a negative charge to form at the bottom of a cloud and a positive charge to form at the top.

Lightning Protection. Benjamin Franklin, during his experiments with lightning, observed that lightning strikes normally occur at points of high elevation. Based on his observations, Franklin concluded that lightning was a form of electricity. He invented the first lightning rod system in 1752. The lightning rod system does not prevent lightning strokes. Rather, it provides a means of controlling the path of the electrostatic discharge with conductors. These conductors provide a low resistance path for the electron flow than that provided by the structure under protection.

High-voltage surge arrestors are used to protect electrical power generation equipment from power surges caused by lightning strikes.

To control lightning strikes, air terminals (copper rods) are fastened vertically at the apex of buildings to intercept electrostatic discharges. Electrostatic discharges are directed from the air terminals through heavy copper conductors to ground (earth). The copper rods must be of sufficient height for the lightning to strike them before it strikes the building. The area of protection of a lightning rod is a cone-shaped space that has a base radius equal to the height of the rod. It is possible, however, for lightning to strike within the protected area. **See Figure 4-4.**

LIGHTNING RODS

Figure 4-4. A lightning rod system protects a cone-shaped area that has a base radius equal to the height of the rod.

Lightning rods are not designed to take a direct hit from a lightning stroke. The current from a lightning stroke is so great that it can destroy the system. Instead, a lightning rod system is designed to slowly neutralize the charge by providing a conductive path for electrons. These currents are normally within the design capabilities of the system. This prevents the buildup of a large enough potential needed to break down the air and start the lightning stroke.

Power lines are often shielded by an overhead conductor that is grounded at fixed intervals. Grounding masts are often used with this conductor. The system works in the same manner as a lightning rod. In addition, electric power and communication lines are normally protected with surge and lightning arresters. These devices may be air gaps that are not affected by the normal potential of the conductors under protection. However, during a voltage surge or lightning strike, an air gap breaks down and acts as a conductor to ground. This keeps the potential of the protected conductors at a safe level.

Once the high potential has been neutralized, the air gap returns to its high resistance state. The distribution of power and communications are instantly restored.

Static Electricity Applications. Electrostatic discharge is not used in most applications of static electricity. Instead, the ability of charges to repel or attract each other is of greater use. This ability is used in electrostatic precipitators, cathode ray tubes, and manufacturing processes.

An *electrostatic precipitator* is a device that uses electricity to remove particles from flue gases. The operating principle of an electrostatic precipitator is that unlike charges attract. In an electrostatic precipitator, contaminated flue gas containing particles is passed through a grid containing positively charged electrodes and negatively charged plates. The particles become positively charged as they pass by the electrodes. Voltage potentials as high as 120 kV are often necessary to charge the particles. On each side of the electrodes are negatively charged collector plates. The particles are repelled by the electrodes because their charges are the same, and they are attracted to the collector plates because their charges are different. The collector plates are periodically vibrated to shake off the particles. The clean flue gas passes out of the precipitator into the smokestack or clean air return plenum with more than 99% of the contaminants removed. **See Figure 4-5.**

Terms

An **electrostatic precipitator** is a device that uses electricity to remove particles from flue gases.

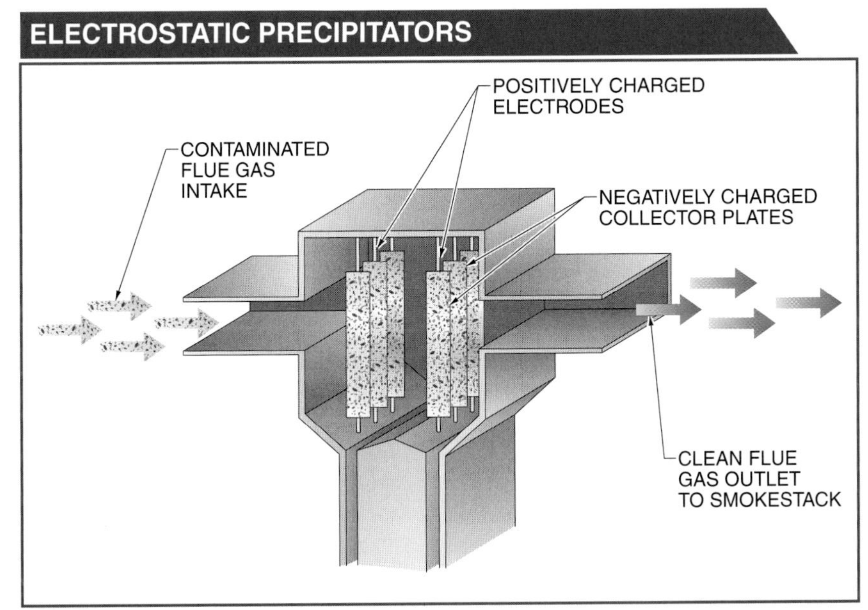

Figure 4-5. An electrostatic precipitator uses static electricity to remove contaminants from flue gases before releasing the gases into the atmosphere.

Tech Tip

Piezo (pronounced pee-ay'-zo) is derived from the Greek word *piezen,* meaning to squeeze or apply pressure.

Static electricity is also used in various manufacturing processes. Electrostatic spray painting is one application that has been particularly successful. In the electrostatic spray painting process, paint is charged as it leaves the spray gun with a charge opposite in polarity to the object being painted. The opposite charges of the paint and object attract each other. This process results in a uniform coat of paint on the object. It is particularly useful for painting irregular surfaces. In addition, little paint is lost to the area surrounding the charged object.

Electrostatic Spraying

SPRAYED MATERIAL WITH OPPOSITE CHARGE OF SPRAYED COMPONENTS

GROUNDED COMPONENTS TO BE SPRAYED

Electrostatic spraying of liquid material, usually paint, was first performed on automotive production lines. New technology uses lower voltages when spraying materials, allowing for the use of materials other than paint for a variety of applications. These applications include pest control, room sanitization, food processing, decontamination of equipment and personnel, mold removal, medical sterilization, and other applications that require a thin, uniform coating of material. A complete finishing or coating system typically consists of a pretreatment or washing unit, an industrial oven for drying, a coating application subsystem (paint spraying or powder coating), an oven or UV curing unit, a material handling subsystem, booth or enclosure, and a waste treatment or reclamation subsystem. Reclamation of excess material is a major advantage of using this type of system.

Terms

Piezoelectricity is the generation of voltage by the application of pressure.

The **piezoelectric effect** is the electrical polarization of some materials when mechanically strained.

The manufacturing of sandpaper also uses electrostatic principles. With sandpaper, an adhesive is applied to the underside of the paper. Negatively charged abrasive particles are attracted to the underside due to the positively charged plates located on the top side of the paper. **See Figure 4-6.** The sandpaper is dried by a heater and cut to size. The deposit on the paper is uniform, with very little waste material.

Pressure

Piezoelectricity is the generation of voltage by the application of pressure. The *piezoelectric effect* is the electrical polarization of some materials when mechanically strained. The voltage potential produced by the pressure on a crystal is directly proportional to the amount of strain applied. If a crystal is stretched, the voltage potential across the crystal changes polarity. For example, points on a crystal that are positive when pressure is applied change to negative when the crystal is stretched. **See Figure 4-7.**

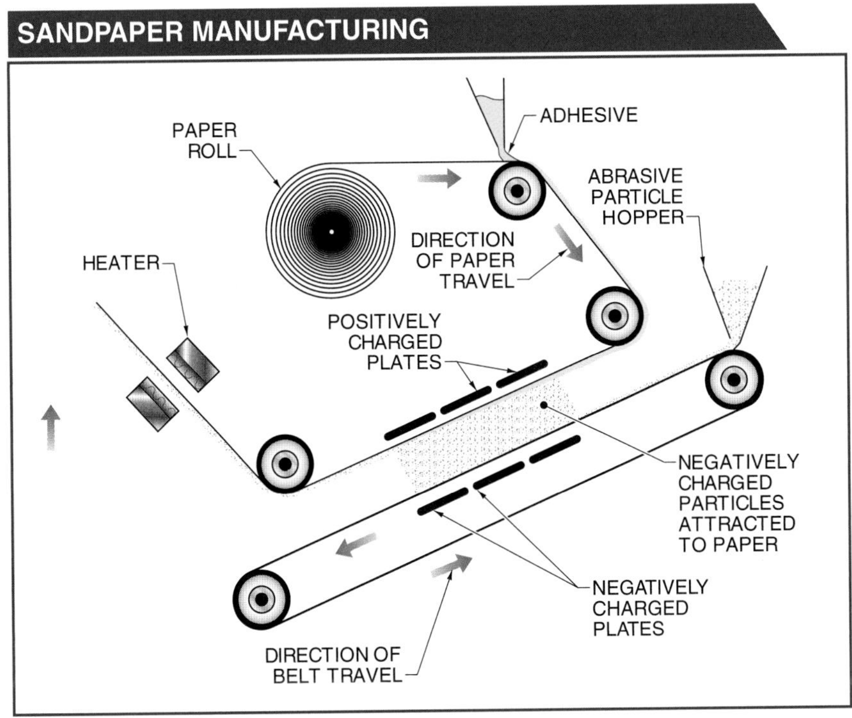

Figure 4-6. In the manufacturing of sandpaper, negatively charged abrasive particles are attracted to the underside of the paper due to the positively charged plates located on the top side of the paper.

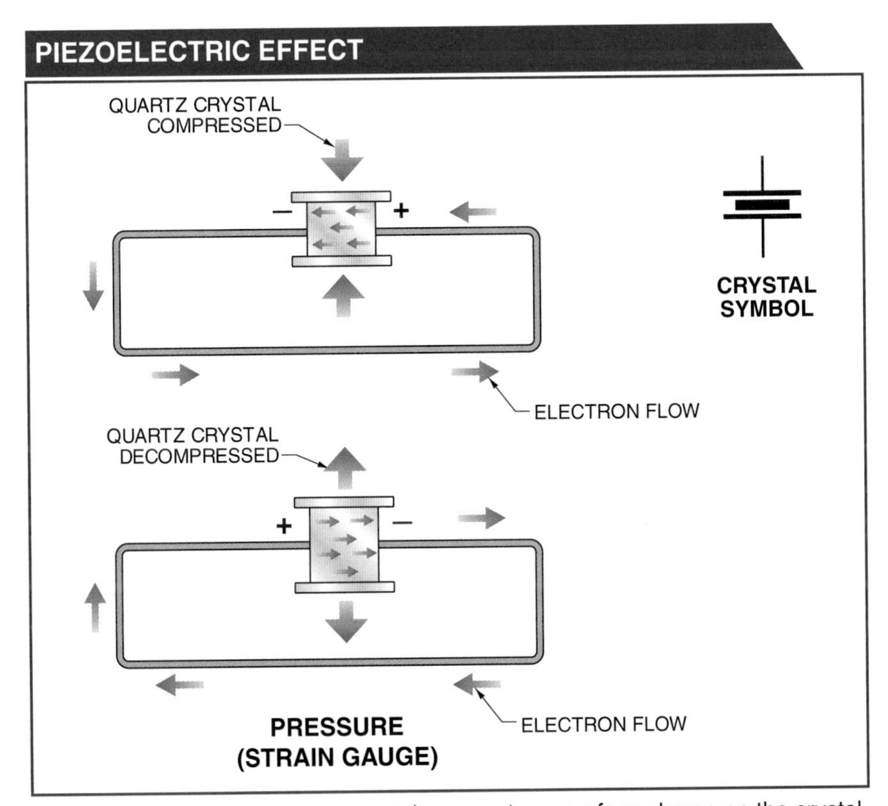

Figure 4-7. Pressure on some crystals generates a surface charge on the crystal. The amount of charge is proportional to the amount of pressure.

Electrons flow in one direction in a circuit connected to a crystal when the crystal is compressed and in the opposite direction when the crystal is decompressed. Also, the electron flow within the crystal in both cases is from the positive terminal to the negative terminal. The positive terminal has a deficiency of electrons; the negative terminal has an excess of electrons. If either the compression or decompression force is held constant, electron flow continues until these two charges are equalized, and then it ceases.

Pressure can be applied to a crystal by compression, stretching, twisting, or bending. The method of applying pressure depends on the crystal material. Some materials respond best to bending pressure, while others respond best to twisting pressure. The force of the pressure acts on the atoms of the crystal to force the electrons out of their orbits.

Piezoelectric Device Applications. A *transducer* is a device that is actuated by energy from one system and supplies energy in the same or different form to a second system. A *piezoelectric transducer* is a transducer that operates based on the interaction between the deformation of certain materials and an electric charge. For example, an electric conversion transducer may change an electric signal from one frequency to another frequency.

Terms

A **transducer** is a device that is actuated by energy from one system and supplies energy in the same or different form to a second system.

A **piezoelectric transducer** is a transducer that operates based on the interaction between the deformation of certain materials and an electric charge.

Piezoelectric transducers are used in microphones, watch timing circuits, radio and television station transmitters and receivers, ultrasound equipment, strain and pressure gauges, piezoelectric translators (PZTs), and piezoelectric motors. Ultrasonic transducers are used in medical equipment. In addition, crystals make excellent filters. Specific frequencies can either be passed to a load or rejected.

Tech Tip

In 1880, Pierre and Jacques Curie demonstrated the connection between the application of pressure on certain crystals (tourmaline, quartz, topaz, cane sugar, and Rochelle salt) and the generation of a surface charge across the crystals. The findings of Pierre and Jacques Curie regarding piezoelectricity were remarkable in that their discovery was made using only tinfoil, glue, wire, magnets, and a jeweler's saw used to cut the crystals.

A piezoelectric (crystal) microphone converts the pressure of sound waves into electric signals. The pressure from the sound waves causes the crystal material within the microphone to compress and decompress, producing the signals. **See Figure 4-8.** If an audio electric signal is applied to the crystal, the microphone becomes an earphone. In the earphone application, the applied electric signal causes the crystal to compress and decompress, creating sound.

Piezoelectricity is also used in strain and pressure gauges. Strain and pressure gauges consist of very fine electrical wires attached to the crystal. When the crystal is compressed or stretched, the potential generated is proportional to the force and can be measured with a voltmeter. These devices are used to measure the vibration of machinery and structures.

During the 1980s, piezoelectric translators (PZTs) and piezoelectric motors were developed. Both the translators and motors can make micropositioning adjustments due to the small movement of their crystals when voltage is applied. Piezoelectric motors are lightweight, are low-speed, and have a high torque (turning force). Because of the

Tech Tip

The invention of the piezoelectric motor is normally credited to H. V. Barth. Much of the development of this motor is also attributed to Akio Kumada, Toshiiku Sashida, and Sudnyika Ueka.

piezoelectric effect, these motors operate free of vibration and are used in applications such as precisely positioning a telescope. Their disadvantage is that they have a short life span.

PIEZOELECTRIC (CRYSTAL) MICROPHONE

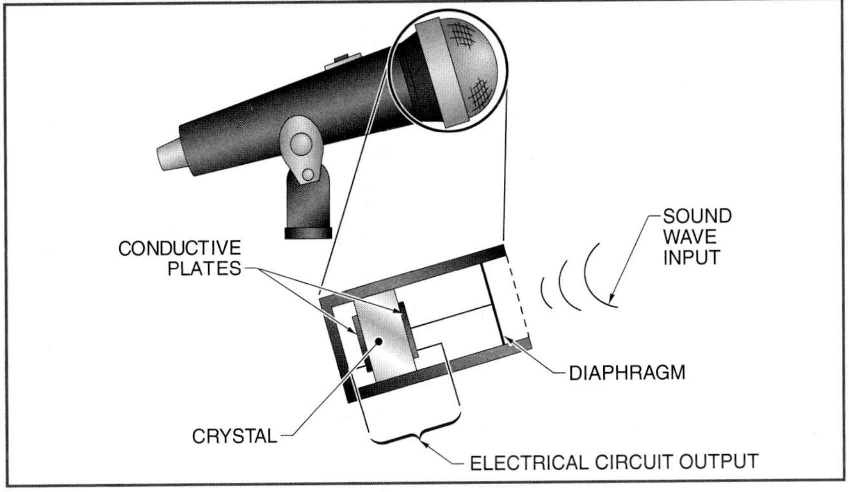

Figure 4-8. A piezoelectric (crystal) microphone converts the pressure of sound waves into electric signals.

Tech Tip

Although the piezoelectric effect was known for many years, it was the mid-1930s before any practical applications were developed. Research, development, and the perfection of large single quartz crystals were encouraged during World War II due to the need for communications devices such as transmitters and receivers.

Light

Certain materials such as cesium, selenium, cadmium, lead sulfide, silver oxide, copper oxide, potassium, and sodium emit electrons when exposed to light. The *photoelectric effect* is the conversion of light energy to electrical energy. The first clue to this phenomenon was discovered in 1887 by Heinrich Hertz, the father of modern radio. He observed that a spark could jump across a large gap if the gap was illuminated by ultraviolet light. Understanding of the photoelectric effect was furthered by Wilhelm Hallwachs. Hallwachs found that when ultraviolet light fell on a negatively charged metal plate, the plate lost its charge. When the plate was charged positively and exposed to ultraviolet light, no apparent change was observed. The plate emitted electrons when struck by light, thereby becoming positive. It was about 1897 when Joseph J. Thomson observed the emission of electrons when ultraviolet light fell on a metal surface. A *photoelectron* is an electron freed by light energy.

Photovoltaic Effect. The *photovoltaic effect* is the production of a voltage potential caused by the absorption of photons across a junction region of a semiconductor. A *photovoltaic cell* is a semiconductor device that converts light energy directly to electrical energy. A photovoltaic cell is known as a solar cell when the light source is sunlight. A photovoltaic cell consists of a layer of selenium sandwiched between a base plate conductor and a barrier layer, with a transparent cover placed on top of the barrier layer. **See Figure 4-9.**

Terms

The **photoelectric effect** is the conversion of light energy to electrical energy.

A **photoelectron** is an electron freed by light energy.

The **photovoltaic effect** is the production of a voltage potential caused by the absorption of photons across a junction region of a semiconductor.

A **photovoltaic cell** is a semiconductor device that converts light energy directly to electrical energy.

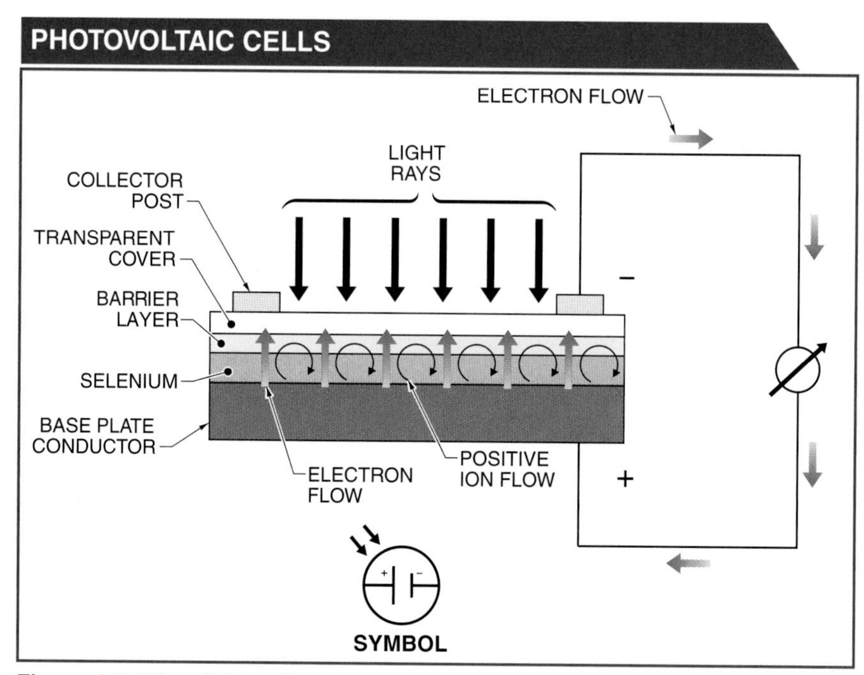

Figure 4-9. When light strikes the barrier layer in a photovoltaic cell, selenium electrons gain energy and move from the valence ring across the barrier layer to accumulate on the transparent cover.

When light passes through the transparent cover of a photovoltaic cell and strikes the barrier layer, electrons in the selenium atoms gain energy and move from the valence ring across the barrier layer to accumulate on the transparent cover. A voltage potential now exists between the collector post (–) in contact with the transparent cover and the base plate conductor (+) in contact with the selenium. The collector post has a negative polarity because it has an excess of electrons. The base plate conductor has a positive polarity because the selenium has given up electrons and positive ions remain. The barrier layer allows electron flow in only one direction so that the electrons cannot return directly to the selenium compound.

When an external circuit is connected to a photovoltaic cell, electrons flow from the negative terminal to the positive terminal (base plate conductor) of the cell. Inside the cell, the electrons move from the positive terminal to the negative terminal. Also, a flow consisting of the positive selenium ions moves toward the positive terminal. At this terminal, the ions regain their electrical balance by combining with the electrons flowing to that point from the external circuit.

Photovoltaic cells can be connected in series to provide a high voltage, in parallel to provide a high current, or in combination to provide a high current and high voltage. Photovoltaic cells can be used in remote locations where electrical power and phone service are not available.

Photoemissive Effect. The *photoemissive effect* is the release of electrons from a material (normally metal) by radiant energy (light) striking the surface of metal. Metals contain free electrons that can serve as mobile charges. A surface energy barrier binds these electrons to the metal. However, this binding force can be overcome, and electrons emitted into space if the energy level of the electrons are raised.

Photons striking the surface of a metal with energy levels less than the binding force of the electrons of the metal produce no emission. Photons striking the surface of a metal with energy levels greater than the binding force of the electrons of the metal produce an emission. Photons disappear once they have given up their energy. The level of emission from a photoemissive cell is directly proportional to the amount of light.

If this process takes place in a vacuum, the electrons that are emitted due to the photons of light can be collected by a positive plate (anode). The emitter of electrons (light-sensitive material) is called the cathode. **See Figure 4-10.**

Terms

The **photoemissive effect** is the release of electrons from a material (normally metal) by radiant energy (light) striking the surface of metal.

PHOTOEMISSIVE CELLS

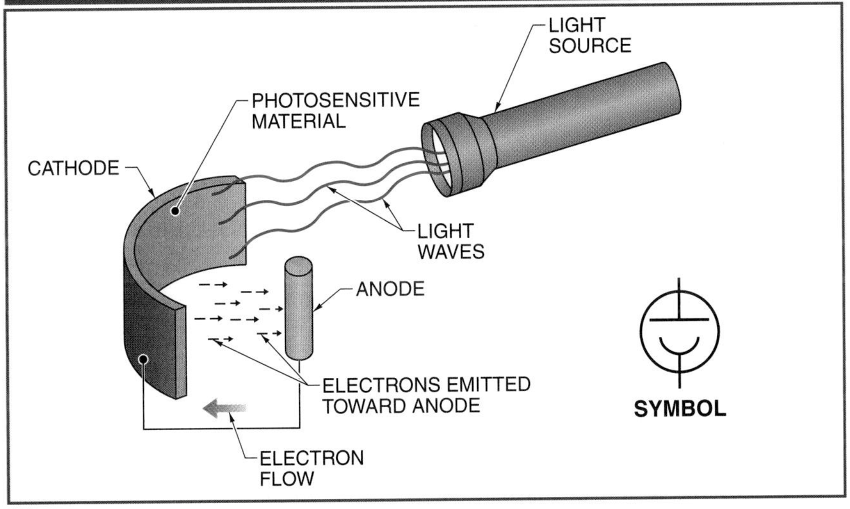

Figure 4-10. A photoemissive cell emits electrons when light energy is focused on its cathode.

Electrons are directed at the anode by the curved surface of the cathode. The more intense the light, the greater the number of photoelectrons emitted. A usable electron flow results if a wire is connected between the anode and the dark side of the cathode. The electrons pass through the dark side of the cathode and replace the electrons in the photosensitive material that were emitted due to the light. This electron flow is enhanced in a working circuit by the introduction of a battery with the negative battery terminal connected to the cathode and the positive battery terminal connected to the anode. A load connected in series is added to the circuit so the changes in current can be detected and used.

Terms

The **photoconductive effect** is a change in the electric conductivity of a solid or liquid due to light striking the material.

A **photoconductive cell (photocell)** is a transducer that conducts current when energized by light.

Photoconductive Effect. The *photoconductive effect* is a change in the electric conductivity of a solid or liquid due to light striking the material. The oldest photoelectric device (developed in the late 1800s) is the photoconductive cell. A *photoconductive cell (photocell)* is a transducer that conducts current when energized by light. Photocells are used to detect and measure radiant energy (light). Current increases with the intensity of light because resistance decreases. Photoconductive cells are also known as photoresistors. The materials used to construct a photoresistor have high resistance (hundreds of thousands of ohms) when in total darkness. The resistance of the cell decreases to a few ohms (100 Ω or less) when exposed to light. **See Figure 4-11.**

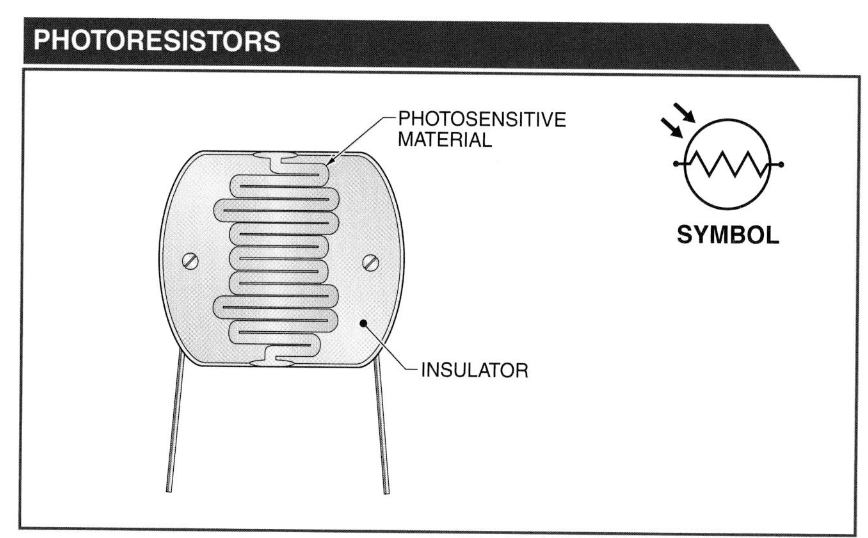

Figure 4-11. A photoresistor has a high resistance (hundreds of thousands of ohms) when in total darkness and low resistance (100 Ω or less) when exposed to light.

Photoresistors are constructed with a thin layer of photosensitive material deposited on an insulator such as ceramic. Typically, calcium sulfide, telluride, cadmium selenide, or cadmium sulfide are used as the photosensitive materials. Lead sulfide, lead selenide, and lead telluride are sensitive to infrared radiation, whereas cadmium sulfide has sensitivity to visible light.

Photoelectric Device Applications. Photoelectric devices have many applications in modern technology. One of the high profile applications is the use of photovoltaic cells in arrays to power satellites. Global positioning system (GPS) satellites use solar cells to generate power to recharge their batteries. Without solar power, space exploration would be greatly hampered. **See Figure 4-12.**

PHOTOVOLTAIC CELL SATELLITE APPLICATION

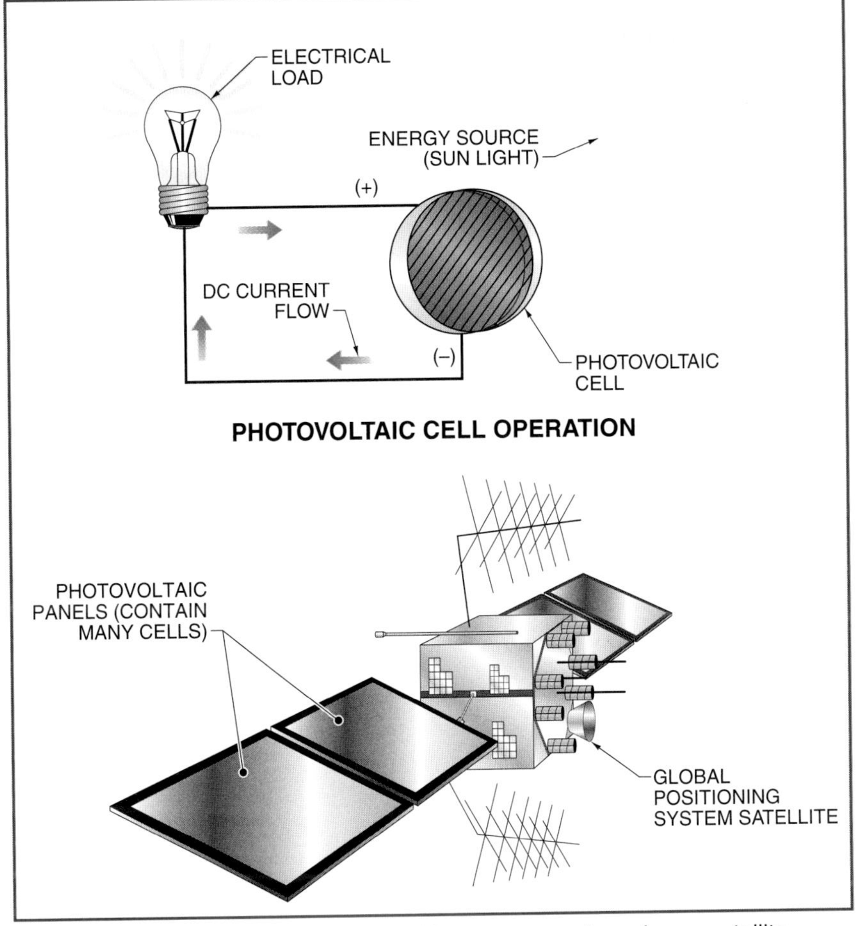

PHOTOVOLTAIC CELL OPERATION

Figure 4-12. Photovoltaic cells are used in outer space to recharge satellite batteries.

Economical applications of low-power photovoltaic cells have been incorporated in several devices for many years. Photovoltaic cells are used in light meters, watches, calculators, and cameras. Some calculators are entirely powered by solar cells. In addition, some cameras use photocells as a light meter that automatically adjusts for the light level focused on the film.

Photovoltaic cells can be made to operate in the infrared and ultraviolet regions of the electromagnetic spectrum. The voltage potential caused by the level of infrared radiation can be calibrated on a temperature scale. Instruments designed using this feature can provide a safe way of measuring the temperatures of the connections in electrical panels and at other electrical points where high voltages and/or high currents may present a danger of electrical shock. A high temperature reading indicates a dirty or loose connection.

Photoconductive cells have many applications. Photoconductive cells are used in the automatic control of streetlights that turn ON at dusk and OFF when a certain light level is achieved. Headlights of some automobiles automatically dim when their photoconductive cells sense light. Photoconductive cells are also used to control elevator doors and are used in some types of smoke detectors. They can also be used as sensors for counters on production lines. **See Figure 4-13.** Photoconductive cells are used in the price scanners at checkout stations in retail stores and with automatic doors.

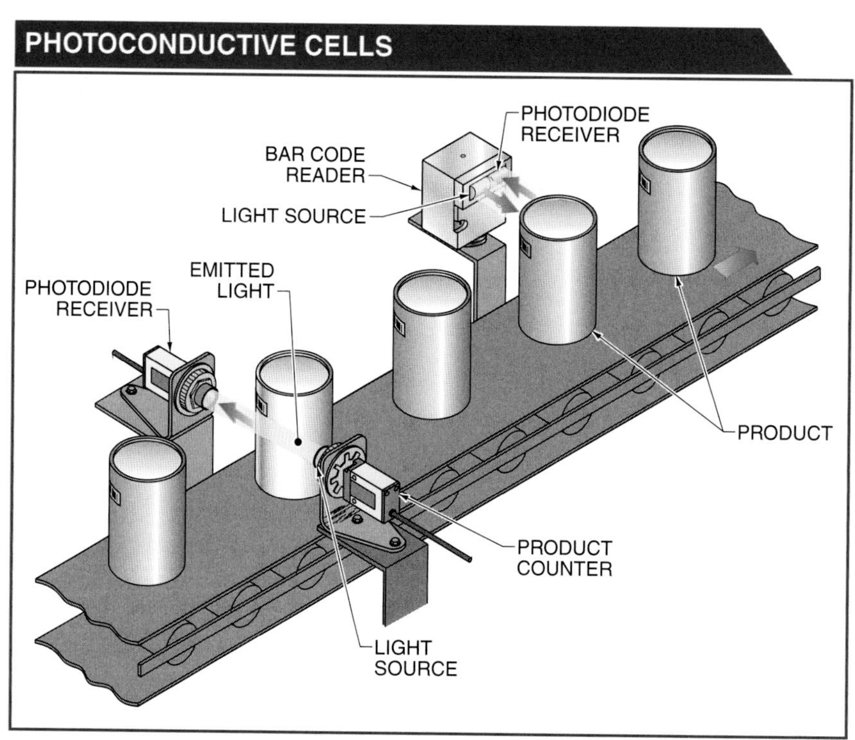

Figure 4-13. Photoconductive cells (photodiodes) are often used on production lines for counting products.

Terms

A **light-emitting diode (LED)** is a semiconductor device that emits light when an electrical current is present.

Light can be emitted from certain materials when an electrical current is present. A *light-emitting diode (LED)* is a semiconductor device that emits light when an electrical current is present. Combinations of the elements gallium and arsenic with silicon or germanium are used to construct LEDs. Light-emitting diodes have been designed to emit almost any desired color. Plastic lenses can be used to obtain additional colors. **See Figure 4-14.**

Light-emitting diodes are inexpensive, consume very little power, and have an unlimited life expectancy. These factors make them ideal alternatives to the small incandescent light bulbs used as indicators in digital clocks, radios, calculators, and a variety of appliances and other electronic devices. Light-emitting diodes are frequently used in the seven-segment alphanumerical displays on calculators and other electrical/electronic equipment.

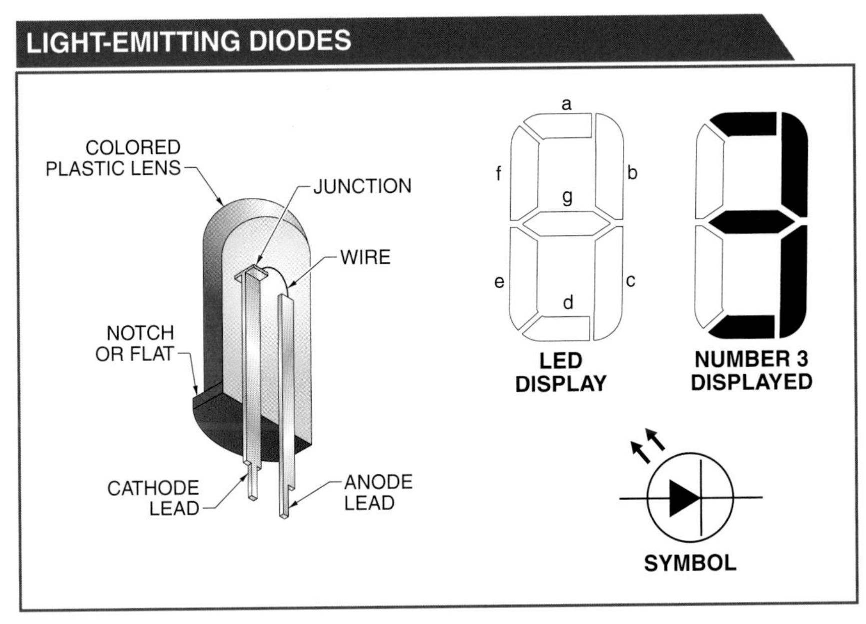

Figure 4-14. Light-emitting diodes are arranged in seven-segment sections to display alphanumeric symbols.

The telecommunications industry is increasing its use of fiber-optic cable and light. Light and fiber-optic cable are replacing the copper wire and electrical current of the traditional transmission method. To convert between the traditional electric and light transmission methods, optocouplers are used. An *optocoupler* is a device that normally consists of an LED as the input stage and a photodiode or phototransistor as the output stage of a fiber-optic system. A *photodiode* is a diode that is switched ON and OFF by a light. A *phototransistor* is a device that combines the effect of a photodiode and the switching capability of a transistor. **See Figure 4-15.**

Optocouplers may be self-contained units used to electrically isolate two systems using light. A self-contained optocoupler is indicated by dotted lines on the symbols that enclose the LED and phototransistor. Optocouplers may also consist of separate devices with an optical fiber cable connecting them. The separate enclosures are indicated by two sets of dotted lines. The two sets of dotted lines enclose the transmitter (LED) and detector (phototransistor).

Terms

An **optocoupler** is a device that normally consists of an LED as the input stage and a photodiode or phototransistor as the output stage of a fiber-optic system.

A **photodiode** is a diode that is switched ON and OFF by a light.

A **phototransistor** is a device that combines the effect of a photodiode and the switching capability of a transistor.

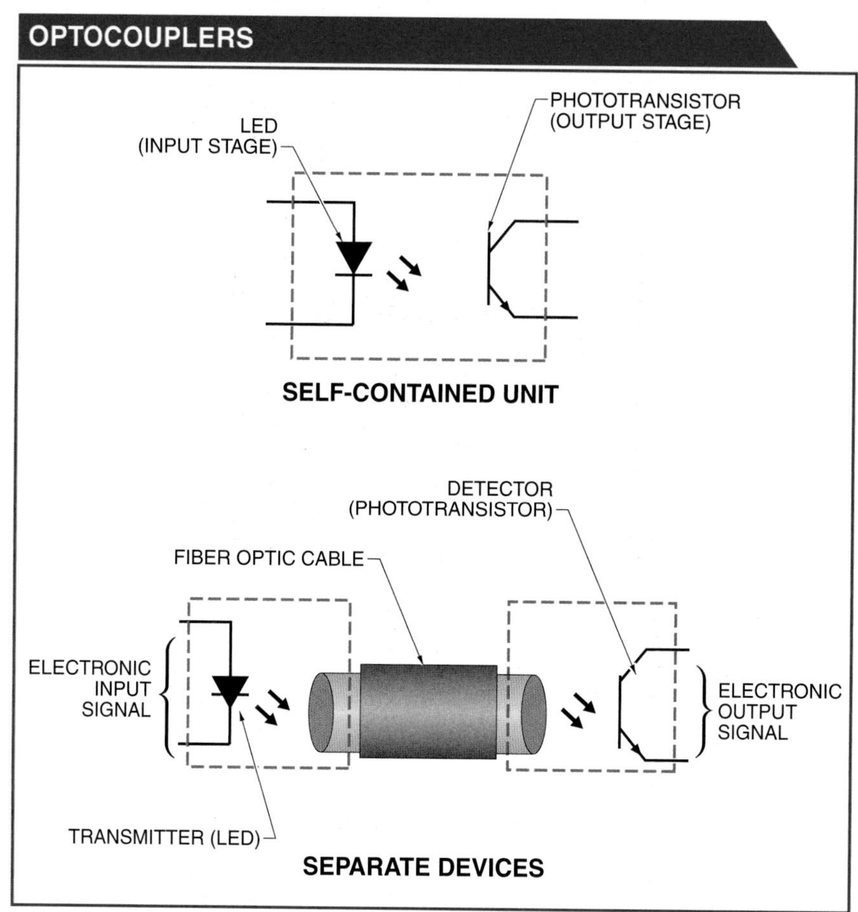

Figure 4-15. Optocouplers are used to interface between electrical and fiber-optic transmission systems.

Terms

Thermoelectricity is electrical energy produced by the action of heat.

The **Seebeck effect** is the production of a voltage potential from a temperature difference between two joined electrical conductors composed of different metals.

Heat

Heat can be used to generate potential energy. *Thermoelectricity* is electrical energy produced by the action of heat. The two thermoelectric effects are the Seebeck effect and the Peltier effect.

Seebeck Effect. The *Seebeck effect* is the production of a voltage potential from a temperature difference between two joined electrical conductors composed of different metals. The Seebeck effect occurs when the junction of two dissimilar metals with mobile charges is heated, and an electrical voltage occurs between the two open ends. This voltage is directly proportional to the temperature difference between the ambient temperature and the heat source. **See Figure 4-16.** The Seebeck effect is also referred to as the thermoelectric effect. The Seebeck effect occurs in liquid and solid materials. The materials include metals and semiconductors.

SEEBECK EFFECT

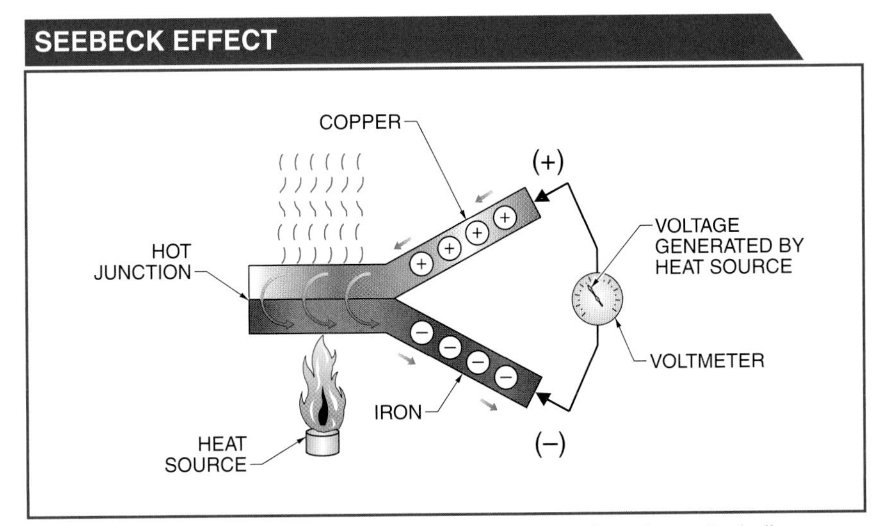

Figure 4-16. The Seebeck effect occurs when the junction of two dissimilar metals is heated and a voltage potential is generated between the two open ends.

Terms

The **Peltier effect** is the absorption or emission of heat at the junctions of two dissimilar metals when electrons flow through the junctions.

Peltier heat is the heat either emitted or absorbed at the junction of two dissimilar metals.

Peltier Effect. The *Peltier effect* is the absorption or emission of heat at the junctions of two dissimilar metals when electrons flow through the junctions. *Peltier heat* is the heat either emitted or absorbed at the junction of two dissimilar metals. Heat is either emitted or absorbed at the junctions depending on the direction of electron flow. **See Figure 4-17.** Peltier heat is different from Joule heat, which is the heat caused by current flow through a material.

Tech Tip

The Peltier effect is named after the French physicist Jean C. Peltier for his work on cooling and heating effects in unlike materials.

PELTIER EFFECT

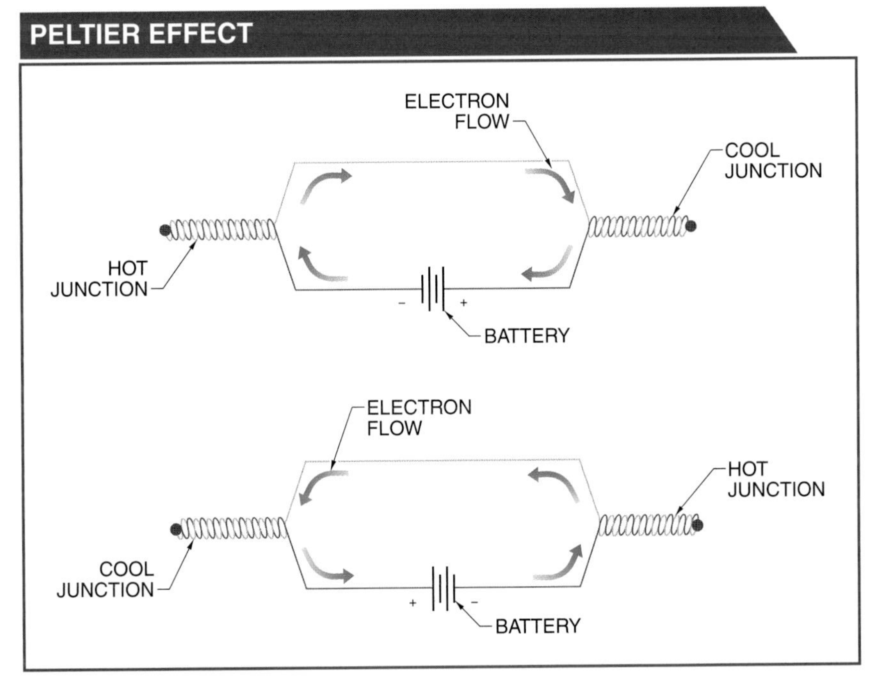

Figure 4-17. The Peltier effect occurs when heat is either emitted or absorbed at the junctions of two dissimilar metals depending on the direction of electron flow.

Terms

Thermionic emission is the release of electrons from a solid or liquid material based on the heat energy of the material.

With the Peltier effect, a net motion of charged particles exists. These charges transport energy. The energy levels differ in the two types of materials. As the charges move from one material to the other, they either give energy to or absorb energy from the junction. Whether a material gives energy or absorbs energy depends on the direction of electron flow.

Thermionic Emission. *Thermionic emission* is the release of electrons from a solid or liquid material based on the heat energy of the material. In 1883, Thomas Edison, while working on the development of the incandescent light bulb, inserted a metal plate close to the filament of the bulb. He then connected a meter to the plate and to the positive terminal of a battery. The negative terminal of the battery was connected to the bulb filament. Edison detected a current flow when the filament was heated. This surprised Edison because the circuit appeared open since the metal plate was not in contact with the filament. This phenomenon is referred to as the thermionic (Edison) effect. **See Figure 4-18.**

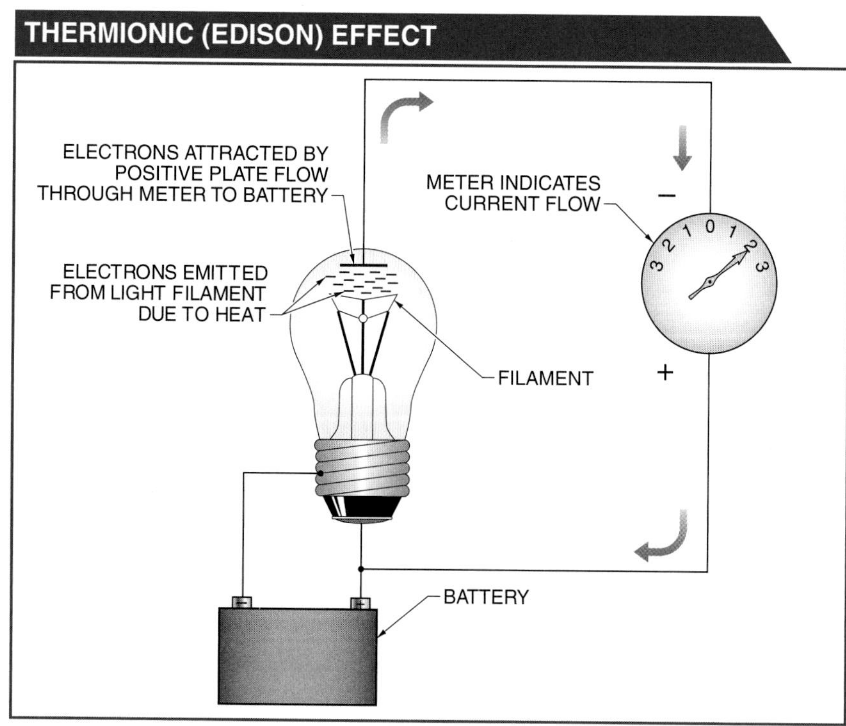

Figure 4-18. The thermionic (Edison) effect occurs when metal is heated and electrons are emitted into space.

Thermionic emission of electrons can be compared to boiling water. When water is placed in a container and heated, steam is given off. As the temperature rises to the boiling point, the amount of steam increases. If a metal is heated to the point of being red or white hot, electrons are emitted (boiled off) into space.

Most metals have many free electrons. Although these electrons are designated as free, they are only free to move from atom to atom within the lattice structure of the metal. When the surface of a solid is in contact with a liquid or gas, a thin film develops on the surface of the solid. Under normal conditions, this film prevents the electrons from escaping and is referred to as the surface barrier. Theoretically, in a perfect vacuum, electrons would not be bound to the metal. Because a perfect vacuum cannot be achieved, the surface barrier continues to exist.

Energy must be supplied to an electron for it to escape into a vacuum. This escape energy is called the work function of the particular metal. The unit of measure of the work function is the electron volt (eV). For example, the amount of kinetic energy required for an electron to escape from tungsten, the element used in an incandescent light bulb, is 4.53 eV. Tungsten produces approximately 7 mA per watt when heated to 2227°C. Other materials, such as the rare earth oxides, are more efficient at producing free electrons in a vacuum.

Thermoelectric Device Applications. The interaction of heat and electricity has many practical applications in modern technology. Thermal energy is used to generate electrical energy for many diverse purposes. Simultaneously, electron flow can generate heat or cause a cooling effect. Many thermoelectric applications use a thermocouple.

Thermocouple Applications

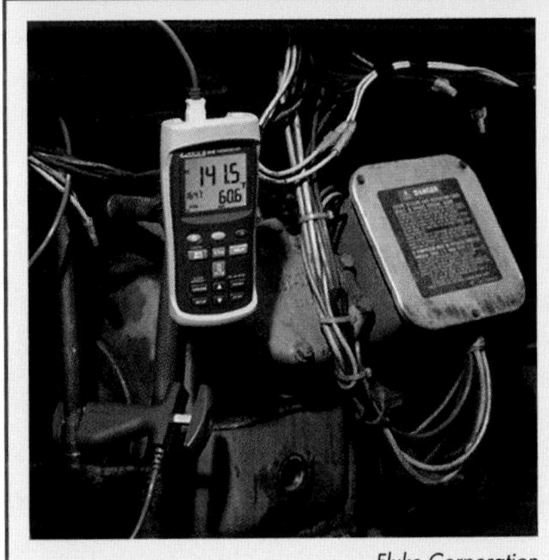

Fluke Corporation

Thermocouples are used in applications that require direct-contact temperature measurement. For example, a digital thermometer uses an external thermocouple as a temperature sensor. The thermocouple (sensor) is used to make contact with the component or component section that is under test. When the joined end is heated and the other end is at a lower temperature, an electric current is produced. The thermocouple reading is transferred to the digital display of the thermometer unit. For electric power measurements, the hot end heats by a resistor supplied with power from the power source to be measured. Thermocouple instruments typically have a wider measurement range and are more durable than other types of test instruments.

A *thermocouple* is a temperature sensor that consists of two dissimilar metals joined at the end where the heat is measured and produces a voltage output at the other end proportional to the measured temperature. Thermocouples are Seebeck devices. **See Figure 4-19.** If heat is applied to the junction, the voltage increases across the open ends. If the junction is cooled, the voltage decreases across the open ends. Thermocouples are useful in taking temperature measurements.

Terms

A **thermocouple** is a temperature sensor that consists of two dissimilar metals joined at the end where the heat is measured and produces a voltage output at the other end proportional to the measured temperature.

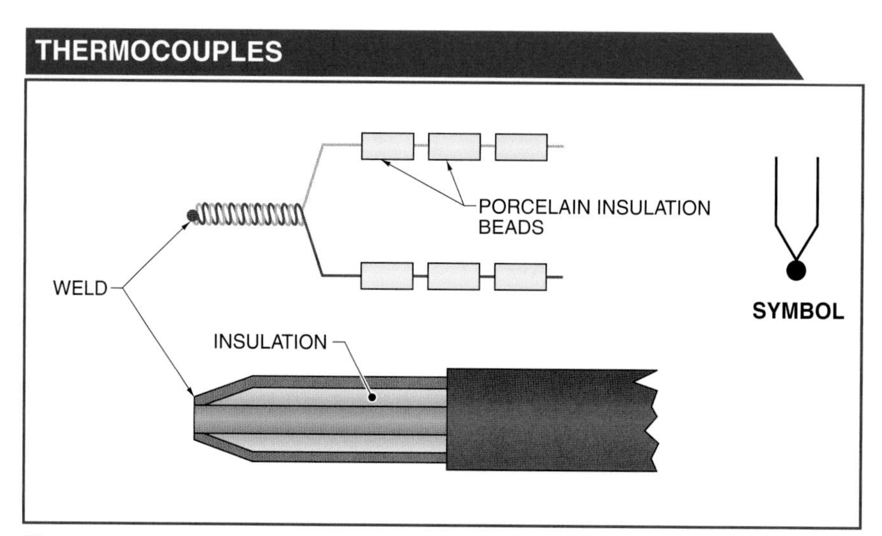

Figure 4-19. A thermocouple uses two dissimilar metals connected at a welded junction.

Terms

A **thermopile** is an array of thermocouples connected in series, parallel, or series/parallel combination to provide a higher voltage and/or current output than an individual thermocouple.

Porcelain insulation beads are used with some thermocouples so that they can be used for high-temperature applications. Thermocouples constructed for use as temperature probes consist of an inner conductor of metal welded at its end to an outer tube made of another kind of metal. Usually, a protective coating covers the probe to protect it from the environments in which it is used.

Thermocouples are designed for specific applications. Some thermocouples are designed to be immersed in liquids while others are designed for use in air. Others may be designed to operate in particular types of gases. The thermocouple head may be threaded for use in a tapped hole (well). A variety of other attachment methods are also available.

Different combinations of dissimilar metals are used in thermocouples. These metals include tungsten, platinum, rhodium, constantan, iron, copper, and alloys such as Alumel® (nickel-aluminum). Thermocouples are listed with an alphabetical code to designate the combination of metal used. For example, a J-type thermocouple is made of iron/constantan, and a T-type thermocouple is made of copper/constantan. A different potential per temperature degree is generated by each unique combination.

The most common application of thermoelectricity is the measurement of temperature. Thermocouples and thermopiles can be calibrated to measure a wide range of temperatures in very small increments, exceeding the performance capabilities of traditional mercury and alcohol thermometers. A *thermopile* is an array of thermocouples connected in series, parallel, or series/parallel combination to provide a higher voltage and/or current output than an individual thermocouple. The small voltage and current outputs of each thermocouple in a thermopile are added together, permitting a greater response to a given temperature change. For example, an antimony/bismuth thermopile with its unheated ends kept at a constant temperature can detect temperature changes of a hundred-millionth of a degree. Single temperature probes are commonly used to measure temperatures from –150°C to 1000°C. Specifications are available for thermocouples that measure temperatures as low as –200°C and as high as 1800°C.

A gas furnace may use a thermopile as an electrical source to activate its main gas valve. A thermocouple is an example of a Seebeck device. **Scc Figure 4-20.** The thermopile is located in the heat of the pilot light where it can generate the necessary potential energy to activate the main gas valve. If the pilot light goes out, the power is removed from the control circuit so that the main gas valve cannot be activated. This provides a safety feature in that the pilot light must be burning when the thermostat calls for heat, guaranteeing that gas supplied to the burner will be ignited.

THERMOPILE BURNER CONTROL APPLICATION

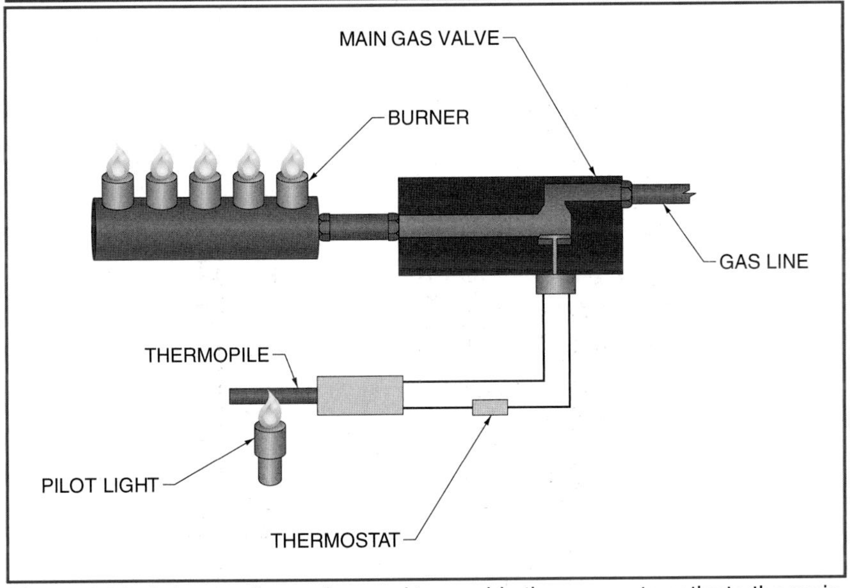

MAIN GAS VALVE
BURNER
GAS LINE
THERMOPILE
PILOT LIGHT
THERMOSTAT

Figure 4-20. A thermopile can be used to provide the energy to activate the main gas valve on a furnace.

Another device used in electrical and electronic circuits is the thermistor. A *thermistor* is a device that changes resistance in response to a change in temperature. Some thermistors are small enough to fit through the eye of a needle. Most thermistors are made of semiconductor materials. Manganese and nickel oxides are often used for thermistors. Thermistors are manufactured with positive and negative temperature coefficients and may be directly or indirectly heated. **See Figure 4-21.**

Terms

A **thermistor** is a device that changes resistance in response to a change in temperature.

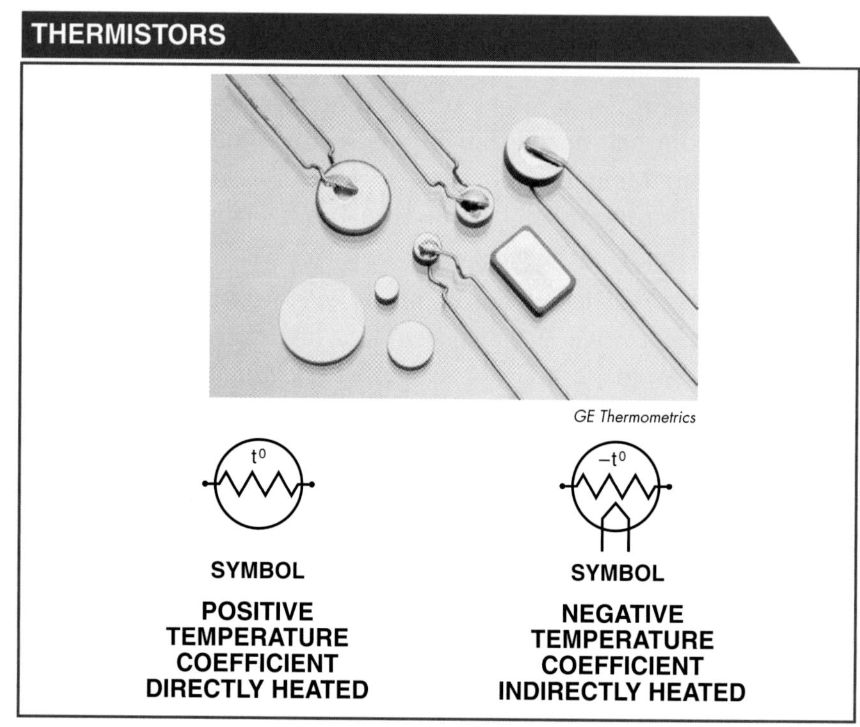

Figure 4-21. A thermistor is a device that changes resistance in response to a change in temperature.

Chemical Action

Chemical action was the first reliable and usable method devised for producing electrical energy. Luigi Galvani first discovered the process and published his results in 1792. In 1800, Allesandro Volta constructed the voltaic pile, the first battery, based on Galvani's findings. Volta documented his findings with the Royal Society of London. This ready and reliable source of electrical energy spurred research into the nature of electricity.

The voltaic pile consists of a series of silver discs separated from zinc discs by a porous material soaked in a solution of salt water. A conductor is connected at each end. The silver terminal is the positive electrode, and the zinc terminal is the negative electrode. Volta later developed a more efficient combination using a copper plate and a zinc plate in a lye solution. **See Figure 4-22.**

A *cell* is a device that produces electricity at a fixed voltage and current level. A cell consists of a combination of two metallic electrodes in an acid or alkaline solution. For example, copper and zinc in a lye solution is a chemical cell. A *battery* is a DC voltage source that converts chemical energy to electrical energy. A battery is formed when two or more cells are combined to provide a higher potential or current than a single cell provides. The voltaic pile is a battery. *Note:* The cells in the voltaic pile are combined in series.

Terms

A **cell** is a device that produces electricity at a fixed voltage and current level.

A **battery** is a DC voltage source that converts chemical energy to electrical energy.

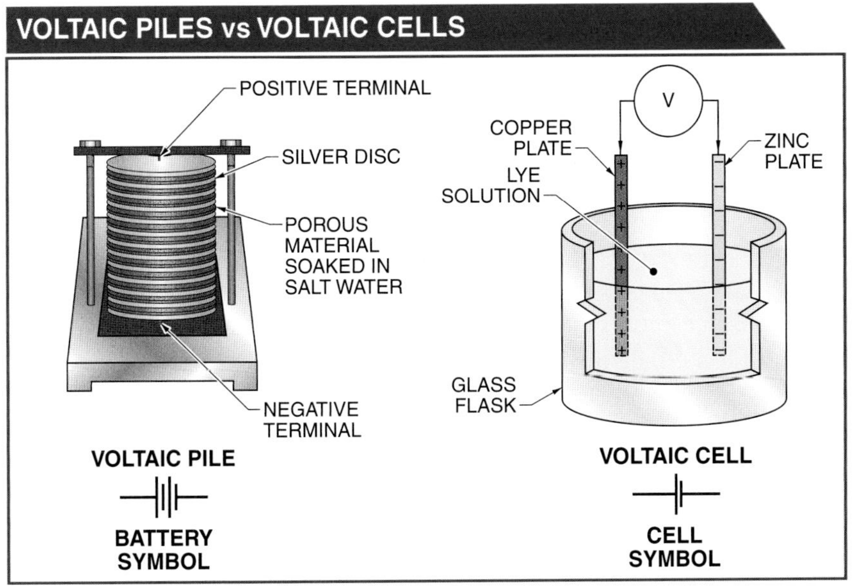

Figure 4-22. The voltaic pile was the first reliable and usable source of electrical energy. The voltaic cell uses a copper plate and a zinc plate in a lye solution.

When cells are connected in series, their voltages are added together. Cells connected in series will have the same current rating because the ability to deliver current to a load is limited to the current rating of the lowest cell. When cells are connected in parallel, their currents are added together. Cells connected in parallel will all have the same voltage rating because a battery has a discharge current that reduces the output voltage to the rated voltage of the lowest cell. For example, three 1.5 V/1.0 A cells produce 4.5 V/1.0 A when connected in series and 1.5 V/3.0 A when connected in parallel. Nine 1.5 V/1.0 A cells connected in a series/parallel connection produce 4.5 V/3.0 A. **See Figure 4-23.**

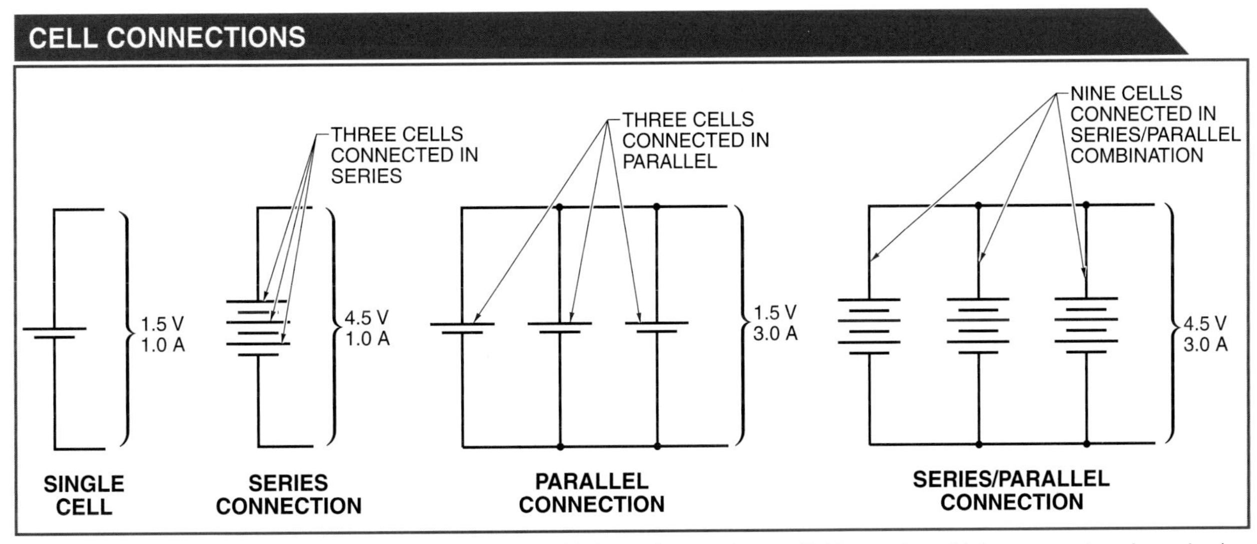

Figure 4-23. Cells may be wired in series to produce higher voltages, in parallel to produce higher current, or in series/parallel to provide higher voltage and current.

Terms

Ionization is the separation of atoms and molecules into particles that have electrical charges.

A chemical cell is formed when two electrodes of dissimilar metals are immersed in an electrolyte. The electrolyte may be an acid, alkaline, or salt solution. The purpose of the electrolyte is to refine or produce usable materials.

When an acid, such as sulfuric acid (H_2SO_4), is mixed in solution with water (H_2O), chemical ionization occurs. *Ionization* is the separation of atoms and molecules into particles that have electrical charges. These charged particles are called ions. The particles may be either negatively (excess of electrons) or positively (deficiency of electrons) charged. When water and sulfuric acid are mixed as an electrolyte, the sulfuric acid molecules split into three ions: two hydrogen ions (H+ H+), and one sulfate ion (SO–). An ionic balance is established in the solution when the number of negative and positive charges is equal. **See Figure 4-24.**

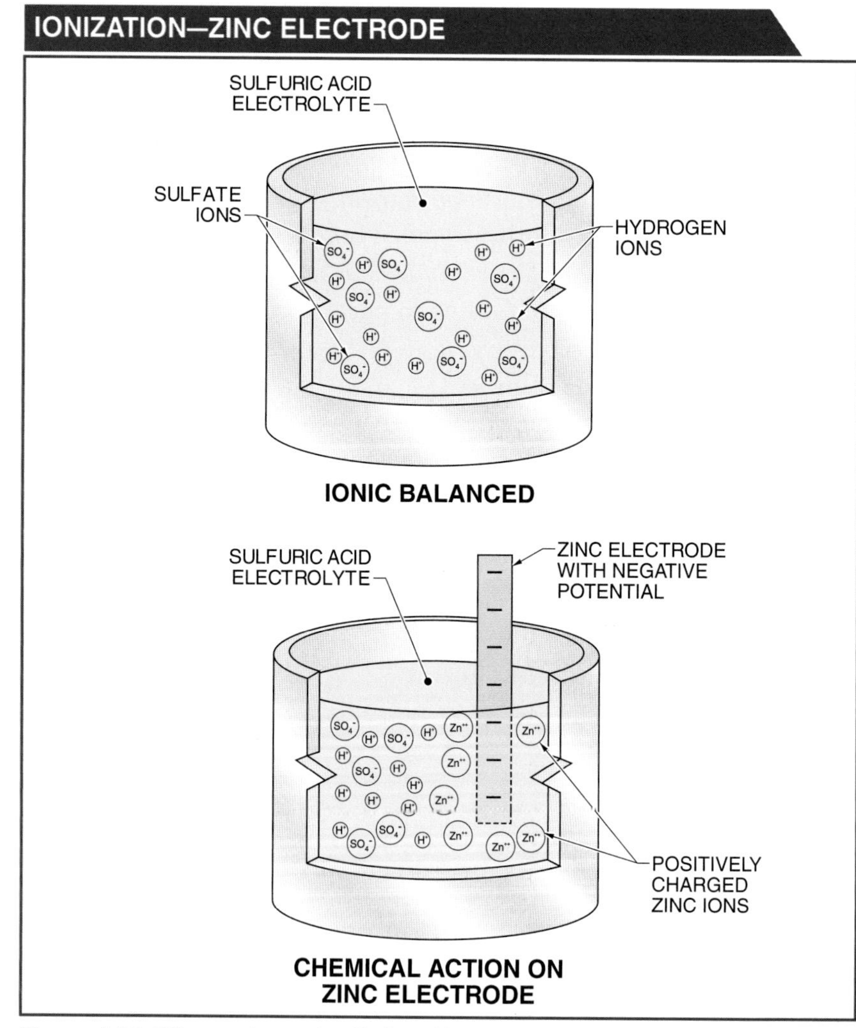

Figure 4-24. When water and sulfuric acid are mixed as an electrolyte, hydrogen and sulfate ions are formed. When a zinc electrode is added to the electrolyte, the zinc is dissolved as positively charged zinc ions, leaving the zinc electrode with a negative potential.

When a zinc (Zn) electrode is immersed in the electrolyte, the sulfuric acid dissolves some of the zinc into the solution, producing positively charged zinc ions (Zn++). For each zinc ion formed, two free electrons remain on the zinc electrode. In time, the zinc electrode becomes negatively charged while the electrolyte becomes positively charged. An equilibrium is established. The positive zinc ions collect near the negative zinc electrode.

When a copper electrode is immersed in the electrolyte, the positive hydrogen ions are attracted to the copper electrode where they receive an electron, neutralizing their charge. The neutral hydrogen atoms become atoms of hydrogen gas (H). The hydrogen gas collects around the copper electrode and can be seen as bubbles rising to the top of the solution. Due to the loss of electrons, the copper electrode becomes positively charged and an equilibrium is established. **See Figure 4-25.**

An electrical potential now exists between the copper and zinc electrodes. The amount of the potential depends on the metal used. Electrodes are chosen for chemical cells based on the ease with which one metal gives up electrons as compared with another metal. For a zinc-copper cell, the potential is approximately 1.08 V. Once this potential is achieved, the chemical action stops until a load is connected to the electrodes of the cell.

When a load is connected to the electrodes, electrons travel from the negative zinc electrode, through the load, and to the positive copper electrode. This upsets the chemical equilibrium of the cell. The zinc electrode now has a deficiency of electrons and the copper electrode has an excess of electrons. To compensate for this change, the positive hydrogen ions in the electrolyte flow to the copper electrode where they accept the excessive electrons and are neutralized. The neutral hydrogen ions travel up the copper electrode and leave the solution as a gas. Due to the loss of the negative particles on the zinc electrode, the negative sulfate molecules are attracted to the zinc electrode. Zinc atoms again enter the electrolyte. This replaces the negative balance of the zinc electrode.

Within the cell, the current carriers (ions) flow from positive to negative. In the external circuit, the current carriers flow from negative to positive. As long as the zinc is not consumed, the charges on the terminals are replaced as the load draws the current. Therefore, the potential remains almost constant.

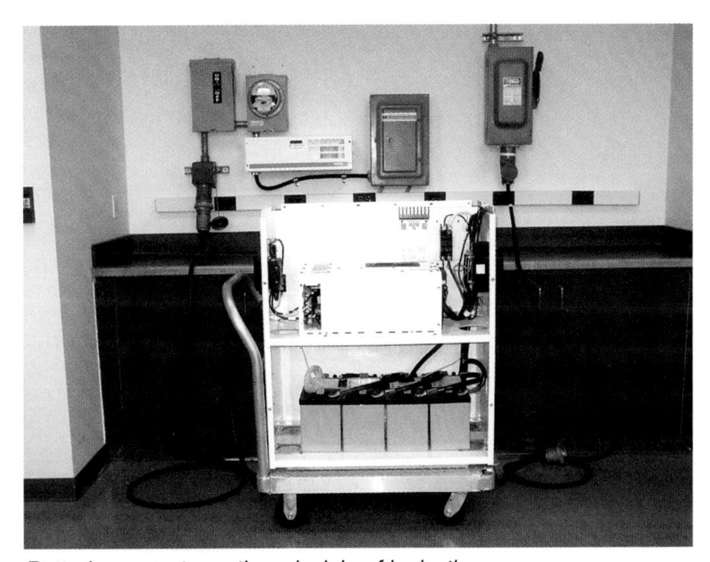

Batteries operate on the principle of ionization.

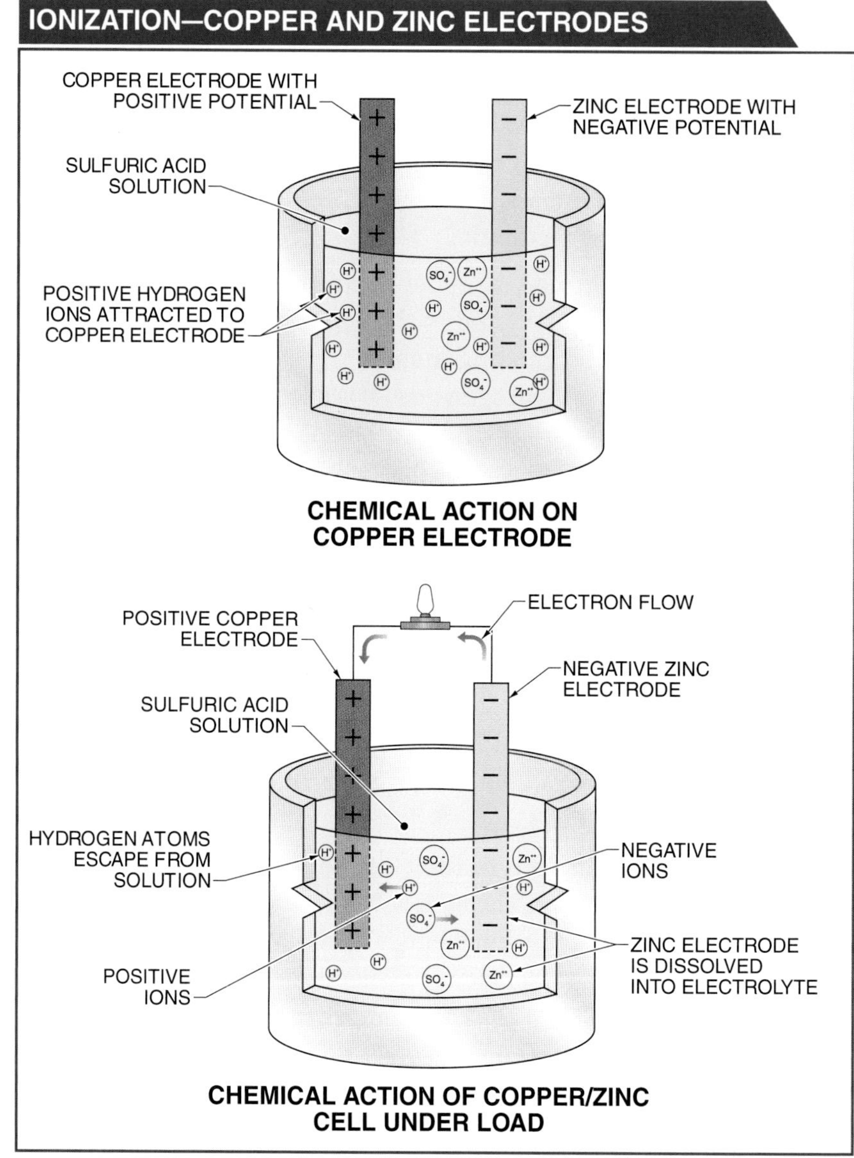

Figure 4-25. When a copper electrode is immersed in an electrolyte, electrons are transferred to the positive hydrogen ions, leaving the copper electrode with a positive potential. Connecting a load across the two electrodes allows electrons to flow from the zinc electrode to the copper electrode.

Chemical Device Applications. A cell or battery chemically stores and releases its electrical potential when a load is placed across its electrodes. In the 1830s, telegraph systems became a major consumer of batteries, making their manufacture a profitable commercial venture. By the 1870s, circuits were being produced for electric bells, an item with high consumer demand. By 1900, the flashlight had been invented and more than two million batteries were manufactured yearly. The growth of technology has continued to provide a high demand for batteries. Electrical energy produced by chemical action is used in a variety of applications in modern society. **See Figure 4-26.**

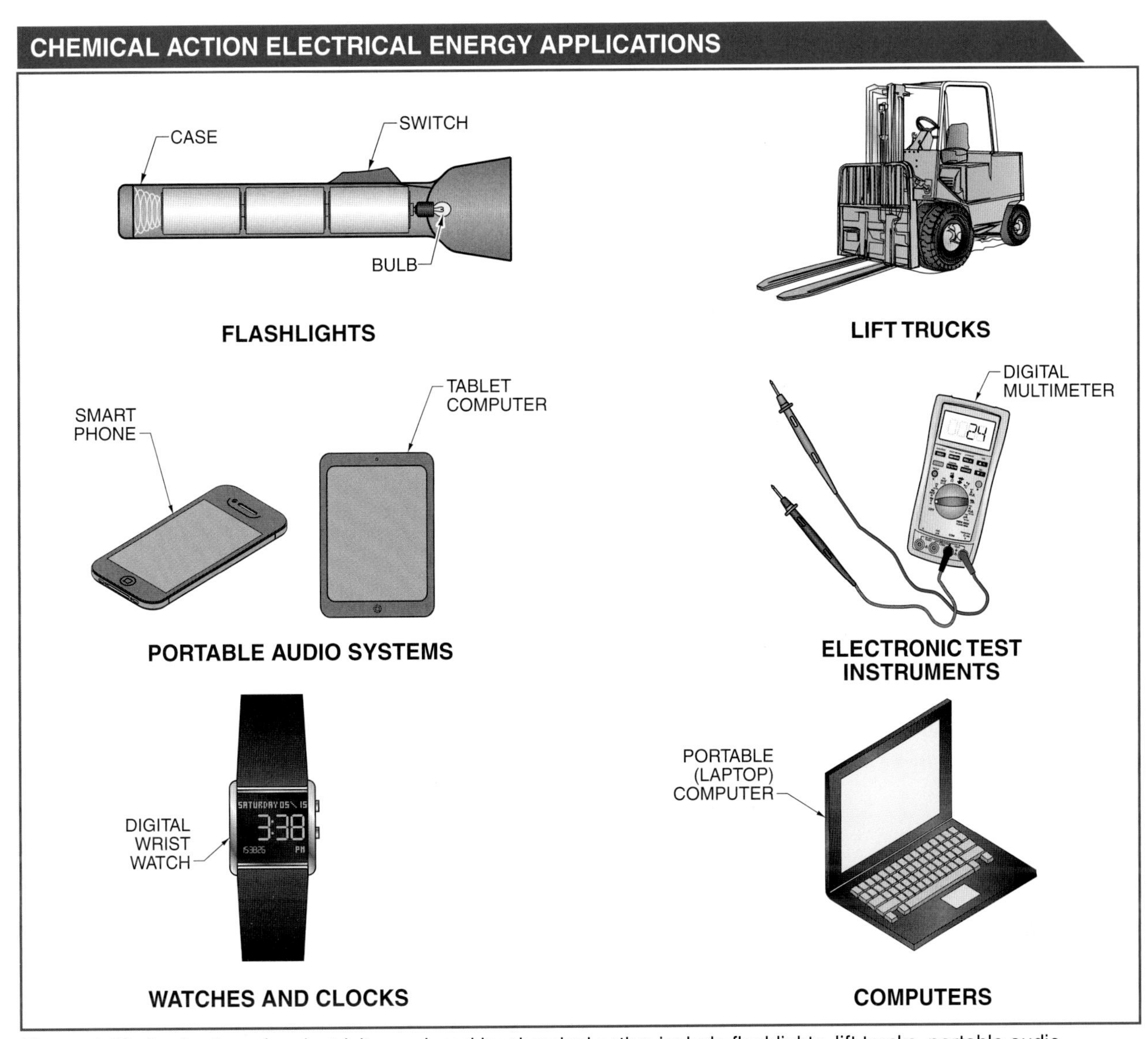

Figure 4-26. Applications for electricity produced by chemical action include flashlights, lift trucks, portable audio systems, electronic test instruments, watches, and computers.

DC Power Sources

Snorkel

Equipment such as golf carts, forklifts, aerial lifts, and electric cars require high voltage DC power sources. These devices normally use 6 VDC to 36 VDC power sources. Six volt or 12 V batteries are normally connected in series and/or parallel connections to produce the required power. Batteries are connected in series to increase the total voltage. Batteries are connected in parallel to increase the total current. A battery produces electricity through a chemical reaction between the battery plates and acid. Common battery ratings are 1.5 V, 6 V, 9 V, 12 V, 24 V, and 36 V. Batteries are a common source of electricity, especially in portable equipment.

Terms

A **magnet** is a substance that attracts iron and produces a magnetic field.

Magnetism is the invisible force exerted by a magnet.

A **paramagnetic material** is a partially magnetic material.

A **diamagnetic material** is a material only slightly repelled by a magnetic field.

The **north pole** of a magnet is the pole that points to the magnetic north pole of the Earth.

Most vehicles depend on batteries to start their gasoline and diesel engines. Forklifts used in areas where fumes from gasoline engines could be harmful are often battery powered. Research is continuing on an electric car that uses fuel cells which may one day replace fossil fuel vehicles. Portable audio systems such as CD players use batteries. Most portable electronic test instruments use batteries to power their electronic circuits. Common electric wristwatches and desktop computers rely on energy supplied by batteries to maintain date and time functions. Portable (laptop) computers can be operated entirely on battery power.

Magnetism

Magnetism is used as the primary source of electrical energy. Most commercially produced electrical power is generated using magnetism. Knowledge about magnetism dates to ancient times. Thales of Miletus (634–546 BC) recorded that iron was strangely attracted to a substance known as lodestone. Lodestone was discovered at Magnesia in Asia Minor and is now called magnetite. Magnetite is the naturally occurring magnet that we know as iron oxide (Fe_3O_4). Magnetite attracts articles that have iron in their molecular structure. In early times, the Chinese discovered that if a piece of magnetite were suspended on a string and left free to turn, it always aligned itself in a north-south direction.

A *magnet* is a substance that attracts iron and produces a magnetic field. *Magnetism* is the invisible force exerted by a magnet. Materials that are strongly attracted by a magnet are referred to as ferromagnetic materials. Besides iron, only nickel and cobalt occur naturally. However, alloys such as steel also exhibit magnetic characteristics.

Paramagnetic materials also exist. A *paramagnetic material* is a partially magnetic material. Aluminum, chromium, platinum, and air are paramagnetic materials. Paramagnetic materials are slightly attracted to a magnetic field. In addition, diamagnetic materials can be slightly magnetized under the influence of a strong magnetic field. A *diamagnetic material* is a material only slightly repelled by a magnetic field. Diamagnetic materials include copper, silver, gold, and mercury. All materials have some magnetic properties, but the magnetic characteristics of most materials are not easily observable. Paramagnetic and diamagnetic substances are usually placed in this nonmagnetic classification since these materials do not retain their magnetic properties once an external field is removed.

Every magnet has two poles, defined as a north pole and a south pole. The *north pole* of a magnet is the pole that points to the magnetic north pole of the Earth. **See Figure 4-27.**

NORTH POLE—MAGNETIC LOCATION

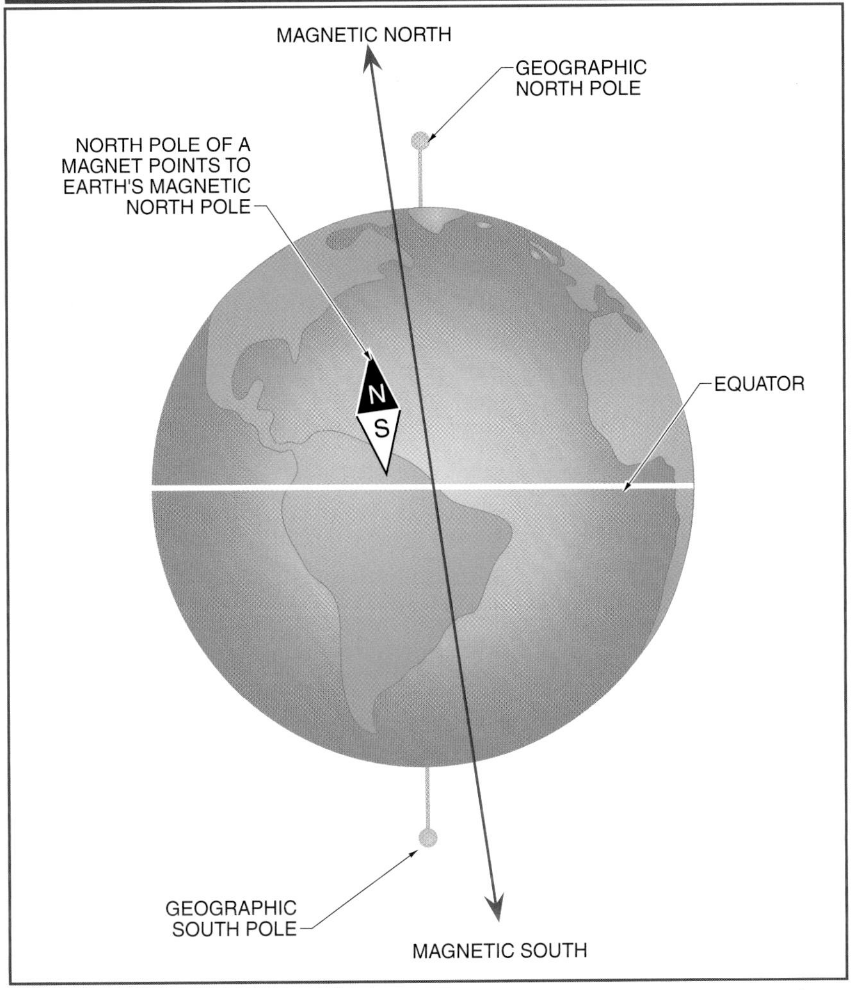

Figure 4-27. The north pole of a magnet is the pole that points to the magnetic north pole of the Earth.

Artificial Magnets. Naturally occurring magnets no longer have any practical value due to their small force. More powerful and conveniently shaped magnets can be produced. These artificial magnets are classified as either permanent (hard) magnets or temporary (soft) magnets. A *permanent magnet* is a magnet that retains its magnetism after the magnetizing force has been removed. Permanent magnets are made of special steels and alloys such as alnico, which consists principally of aluminum, nickel, and cobalt. Permanent magnets have a high degree of retentivity (ability to retain residual magnetism). A *temporary magnet* is a magnet that loses its magnetism when the magnetizing force is removed. Temporary magnets have low retentivity.

Terms

A **permanent magnet** is a magnet that retains its magnetism after the magnetizing force has been removed.

A **temporary magnet** is a magnet that loses its magnetism when the magnetizing force is removed.

Terms

Relative permeability is the ability of a material to conduct magnetic lines of force as compared to air, which has a permeability value of 1.

Alnico, hardened steels, and several other alloys are difficult to magnetize. These alloys are said to have low permeability. Once magnetized, however, these materials maintain their magnetic characteristics and force. Conversely, soft iron and annealed silicon steel have a high permeability; they are easy to magnetize. When the magnetizing force is removed, however, they retain only a small part of their magnetic force. The small amount of magnetism that remains in these materials is called residual magnetism. *Relative permeability* is the ability of a material to conduct magnetic lines of force as compared to air, which has a permeability value of 1. Paramagnetic substances have a permeability slightly greater than 1. Diamagnetic materials have a permeability of less than 1. Both permanent and temporary magnets have applications in modern technology.

Artificial magnets are manufactured in various shapes, sizes, and strengths depending on their intended use. Most magnets are bar, horseshoe, or ring magnets. **See Figure 4-28.** Bar magnets are often used to study the characteristics and effects of magnetism. Bar magnets may be square, rectangular, round, or other desired shapes. A horseshoe magnet takes its name from its shape. Horseshoe magnets have greater strength than bar magnets of the same size and material. Horseshoe magnets are often used in electrical meters. A closed ring magnet has its field confined to the ring. By cutting a slot in the ring, the magnetic field can be detected in the gap. Ring magnets were used as data storage devices in early computers. Temporary ring magnets are often used as magnetic shields.

Figure 4-28. Common manufactured magnet shapes include bar, horseshoe, and ring magnets.

Magnetic Poles. The attraction iron has to a bar magnet can be observed by placing the magnet into a pile of iron filings. When removed, most of the iron filings collect near both ends of the magnet. This shows that the strongest magnetic force exists at each end. The ends of the magnet are called its poles. All magnets have two poles. These are the north and south poles. Usually, the north pole is marked in some way by the original equipment manufacturer (OEM).

If a bar magnet is cut in half, each half has a north and a south pole. If a bar magnet is cut into thirds, each piece still maintains a north and a south pole. Regardless of the number of times a magnet is divided, each piece of the original magnet continues to exhibit a north and a south pole. A single pole magnet does not exist. If a north pole is present, then a magnet must also have at least one south pole. **See Figure 4-29.**

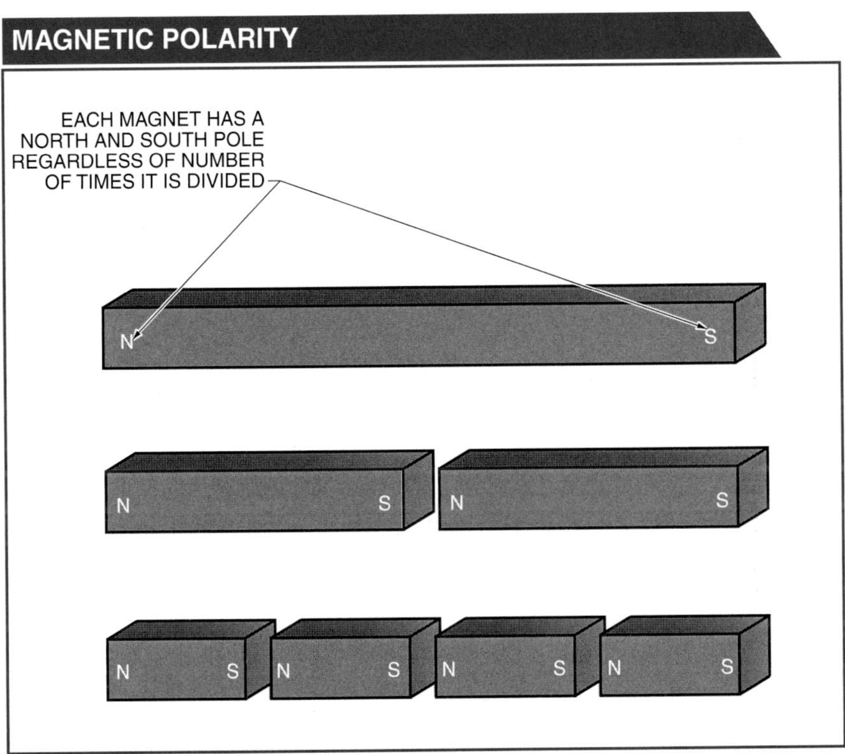

Figure 4-29. A magnet will always have a north and a south pole regardless of the number of times it is divided.

Magnetic Force. When two magnets are brought close to each other, an invisible force exists between them. This force may be one of attraction that pulls the magnets together, or it may be a force of repulsion that pushes the magnets apart. If one bar magnet is suspended by a string and another magnet is brought close, the force of attraction or repulsion can be determined. **See Figure 4-30.** Magnetic force is expressed in dynes. A *dyne* is a force that produces an acceleration of one centimeter per second per second on 1 gram of mass.

Terms

A **dyne** is a force that produces an acceleration of one centimeter per second per second on 1 gram of mass.

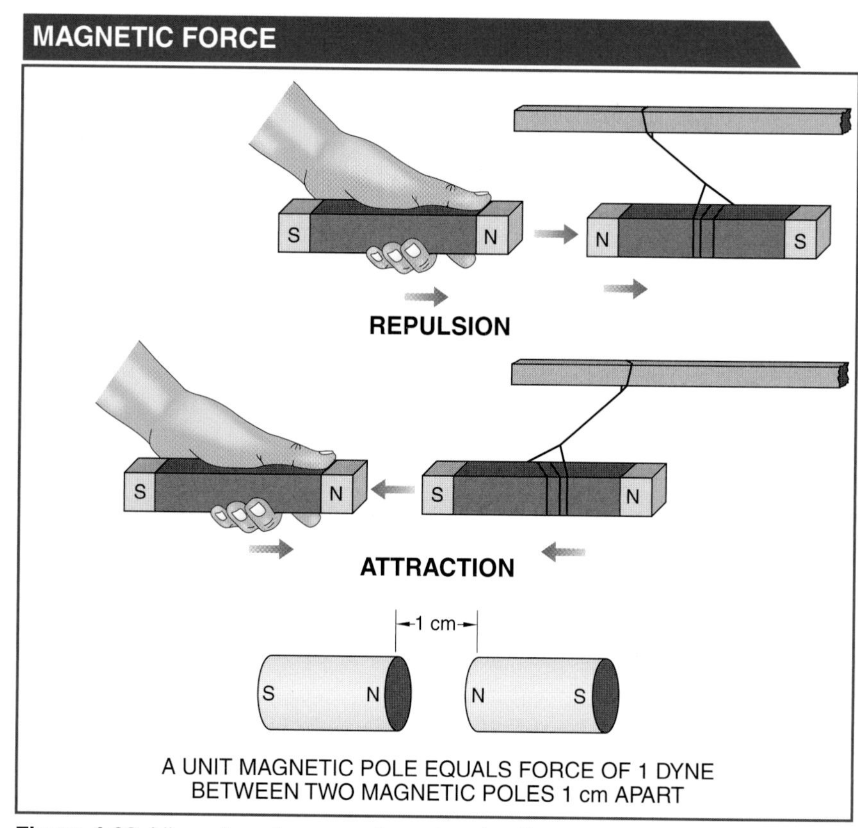

Figure 4-30. Like poles of a magnet repel and unlike poles of a magnet attract. A unit of magnetic force is equal to one dyne between the poles of two magnets separated by one centimeter.

The force between two magnetic poles is similar to the force that exists between two charges. In both cases, the force can be one of attraction or repulsion. In the 1780s, Charles-Augustin de Coulomb, a French physicist, developed the mathematical formula for calculating the force for either case. His law for magnetic poles states that the force between two magnetic poles is directly proportional to their magnetic strength and inversely proportional to the square of the distance between them. For example, if a force of 10 dynes exists between two magnetic poles separated by 1″ of air, a force of 160 dynes would exist if the poles were ¼″ apart. Based on the inverse square law, the force would increase 16 times. Conversely, if two magnetic poles were 2″ apart, a force of 2.5 dynes would exist. At twice the distance, the force would be one-fourth as strong. The equation for this relationship is as follows:

$$F = \frac{m_1 \times m_2}{d^2}$$

where

F = force (in dynes)

m_1 = strength of first magnetic pole (in unit magnetic poles)

m_2 = strength of second magnetic pole (in unit magnetic poles)

d^2 = distance between poles (in cm)

A *unit magnetic pole* is a force of one dyne between two magnetic poles separated by a distance of 1 cm. One pound-force is 444,822 dynes, making one dyne an extremely small force. The force between magnetic poles is affected by the medium between the poles. For a medium other than air, the permeability (μ) of the medium must be included in the calculation. Coulomb's law addresses a basic principle, but it is not commonly used to calculate magnetic force. The equation becomes:

$$F = \frac{m_1 \times m_2}{\mu d^2}$$

Example: What is the force between two magnetic poles separated by 10 cm in air if one pole has a strength of 30 unit magnetic poles and the other pole has a strength of 25 unit magnetic poles?

$$F = \frac{m_1 \times m_2}{\mu d^2}$$

$$F = \frac{30 \times 25}{(1)10^2}$$

$$F = \frac{750}{100}$$

$F =$ **7.5 dynes**

Magnetic Fields. A *magnetic field* is an invisible field produced by a current-carrying conductor, a permanent magnet, or the earth that develops a north and a south polarity. The English physicist Michael Faraday was the first scientist to visualize a magnetic field as a state of stress consisting of uniformly distributed lines of force (magnetic flux). *Magnetic lines of force* are the invisible lines of force that make up a magnetic field. **See Figure 4-31.** The magnetic field surrounding a magnet has a greater density at the poles and radiates out into the space surrounding the magnet in a symmetrical pattern.

Terms

A **unit magnetic pole** is a force of one dyne between two magnetic poles separated by a distance of 1 cm.

A **magnetic field** is an invisible field produced by a current-carrying conductor, a permanent magnet, or the earth that develops a north and a south polarity.

Magnetic lines of force are the invisible lines of force that make up a magnetic field.

MAGNETIC FIELDS

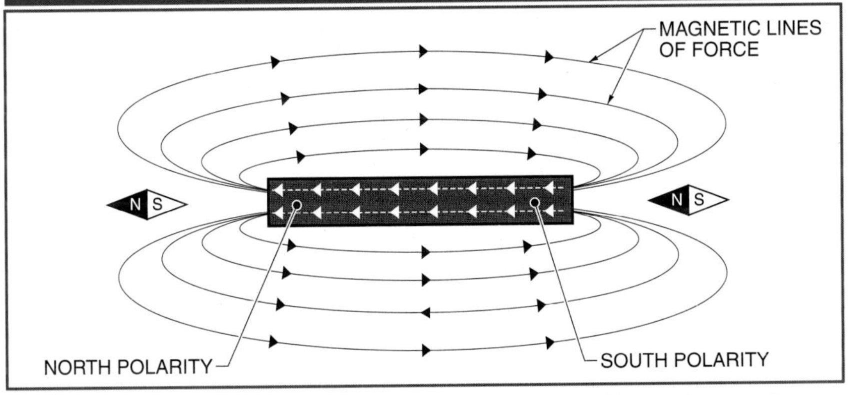

MAGNETIC LINES OF FORCE
NORTH POLARITY
SOUTH POLARITY

Figure 4-31. A magnetic field is the invisible field produced by a permanent magnet that develops a north and a south polarity.

Terms

Flux density is the measure of the magnetic lines of force per unit area taken at right angles to the direction of flux.

A **gauss** is a unit of flux density equal to one magnetic line of force per square centimeter.

Magnetic lines of force correspond to current in an electrical circuit, although unlike current, magnetic lines of force are not considered particles in motion, but as a field of force exerted in space, similar to the force of gravity. The more magnetic lines of force present in a given cross-sectional area, the greater the force and flux density. *Flux density* is the measure of the magnetic lines of force per unit area taken at right angles to the direction of flux. If the unit of length is a centimeter, then the flux density is measured in gauss. A *gauss* is a unit of flux density equal to one magnetic line of force per square centimeter. Flux density is greater at the poles of a magnet and spreads out over a larger area the farther the magnetic lines of force are from the poles. Flux density is expressed by the following equation:

$$B = \frac{\phi}{A}$$

where

B = flux density

ϕ = total number of magnetic lines of force

A = cross-sectional area (in cm^2)

Example: What is the flux density if 10,000 magnetic lines of force pass through an area of 1 sq in. (2.54 cm = 1 in.)?

$$B = \frac{\phi}{A}$$

$$B = \frac{10,000}{2.54 \times 2.54}$$

$$B = \mathbf{1550\ gauss}$$

The magnetic field surrounding a bar magnet can also be plotted using a compass. A compass aligns itself with the magnetic lines of force at each position. The compass needle rotates a full 360° as it is moved from one pole of the magnet to the other. The movement of the compass pointer also shows that magnetic lines of force have direction.

Just as the north pole of a magnet is defined in terms of the Earth's field, the direction of magnetic lines of force is also defined. By definition, the magnetic lines of force travel from the north pole to the south pole of the magnet. This is the polarity of the magnetic field. The magnetic lines of force travel from the south pole to the north pole inside the magnet to complete the loop.

Though magnetic lines of force are imaginary, they are useful in explaining magnetism and the interaction of electrons and magnetic fields. Magnetic lines of force represent the path along which a theoretically isolated magnetic pole would move from one pole to the other.

The magnetic lines of force can be compared to rubber bands in that they contract when the magnetic force is removed and expand when the magnetic force is increased. **See Figure 4-32.** Magnetic lines of force display the following characteristics:

- They are continuous.
- They push one another apart when traveling in the same direction.
- They cancel each other when traveling in opposite directions.
- They form the shortest loop possible.
- They never cross each other.
- They take the path of least reluctance.
- They pass through any type of material.
- They have elasticity.

Magnetic lines of force are continuous and form a complete loop. This means that a single magnetic line of force leaving the north pole of a magnet travels to the south pole and completes the loop back to the north pole through the magnetic material. This is similar to current flow in an electrical circuit where electrons leaving the negative terminal of an electrical source return to the positive terminal of the source. Within the electrical source, the electrons move from the positive pole to the negative pole. Neither the magnetic lines of force nor the electrons disappear.

Tech Tip

The fact that magnets always orient themselves in the same direction when free to turn shows that the Earth must also be a giant magnet. Otherwise, the same pole of a magnet would not always align in the same direction.

Magnetic lines of force traveling in the same direction repel one another. This is the reason like poles of magnets repel each other. The magnetic lines of force are more dense at the poles because the force on them is increased as they leave and approach their source.

Parallel magnetic lines of force traveling in opposite directions cancel each other. This is best explained using two current-carrying parallel conductors. The magnetic fields caused by electron flow run in opposite directions between the conductors. When the conductors are brought close enough for their fields to interact, the magnetic lines of force between the conductors cancel each other out. The resulting magnetic lines of force surround both conductors and pull them together.

Magnetic lines of force form the shortest loop possible. For example, when two unlike poles are opposite each other, the magnetic lines of force move them together. In addition, magnetic lines of force never cross each other.

Electromagnets, which use electric current and magnets, are used in metal processing facilities to move large amounts of magnetic materials, such as steel.

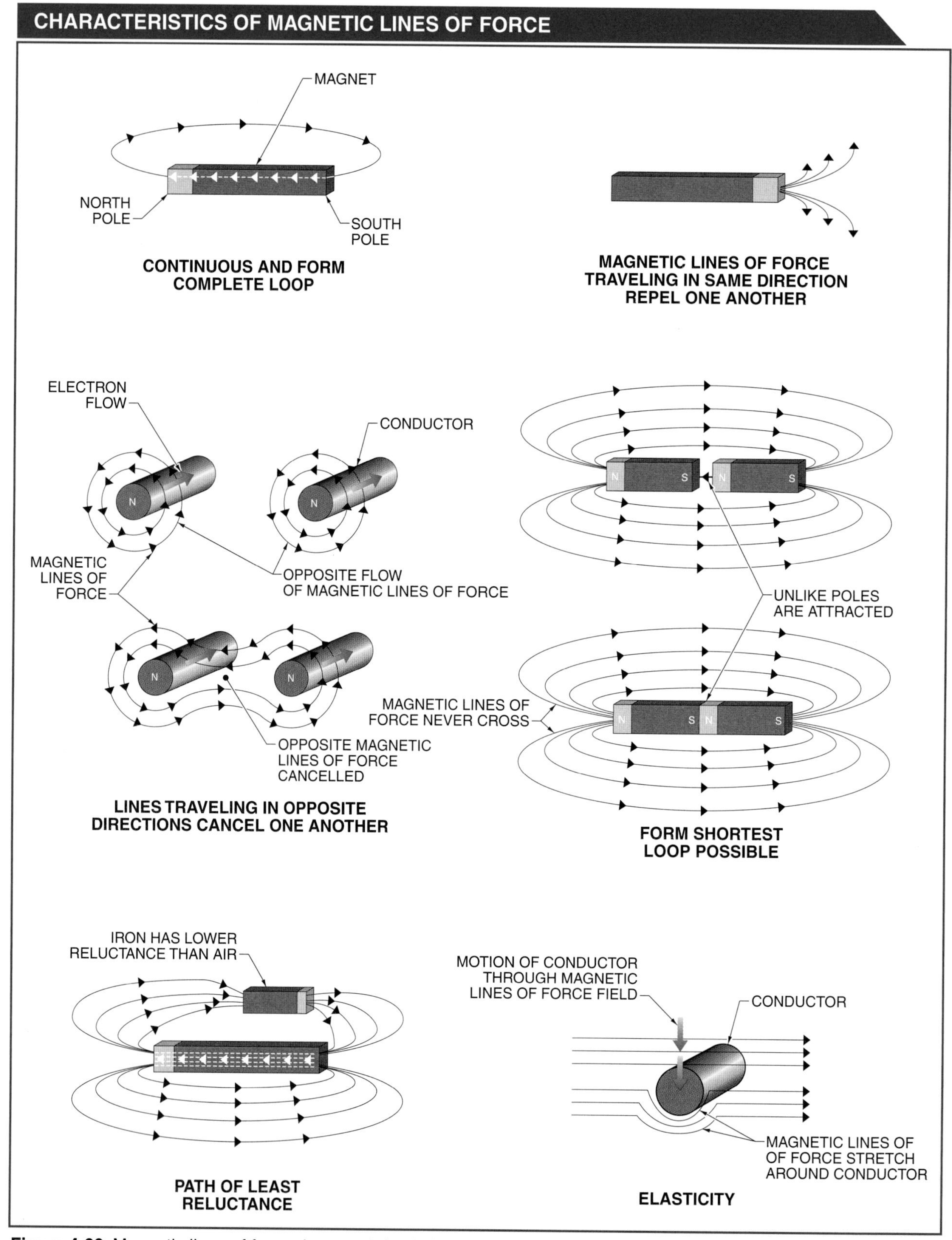

Figure 4-32. Magnetic lines of force characteristics help explain magnetism and the interaction of the electron with magnetic fields.

Magnetic lines of force take the path of least reluctance. *Reluctance* ($\Re$) is the property of an electric circuit that opposes magnetic lines of force. Reluctance is directly proportional to the length and inversely proportional to the permeability and cross-sectional area of the material through which the flux passes. Magnetic lines of force pass through any type of material, magnetic and nonmagnetic alike. The unit of reluctance, which has no given name, is expressed by the following equation:

$$\Re = \frac{l}{A \times \mu}$$

where

$\Re$ = reluctance

l = length (in cm)

A = cross-sectional area (in cm^2)

μ = permeability

Magnetic lines of force have elasticity. Magnetic lines of force stretch when a conductor passes through them and return to their original shape when the conductor has passed.

Magnetic lines of force can be harmful to some equipment. Because the magnetic lines of force pass through any type of material, shielding a conductor or object can be difficult. One method of shielding is to take advantage of the fact that magnetic lines of force take the path of least reluctance. This principle is often used to protect watch movements. The watch movement is protected from becoming magnetized by being enclosed in a low reluctance material that allows the magnetic field to bypass the watch movement.

Magnetic Theory. By the end of the 19th century, all known elements had been found to have some magnetic property. This includes substances such as wood, copper, glass, air, and water, which are classified as nonmagnetic. The magnetic properties of these materials are not easily observable under normal conditions.

The differences in the magnetic nature of matter are based on the atomic and molecular nature of matter. A clue was provided when it was discovered that if a magnet is cut into many parts, each part still maintains its magnetic properties with a north and south pole. If a magnet is divided to molecule size, each molecule has a north and south pole.

It is theorized that in a magnetic material, most of the molecules are aligned so that their poles point in the same direction. This forms regions of additive molecules in the material called magnetic domains. This is known as Weber's theory of magnetism. It can therefore be concluded that in a nonmagnetic material, the molecules are arranged in a random manner. **See Figure 4-33.**

Terms

Reluctance ($\Re$) is the property of an electric circuit that opposes magnetic lines of force.

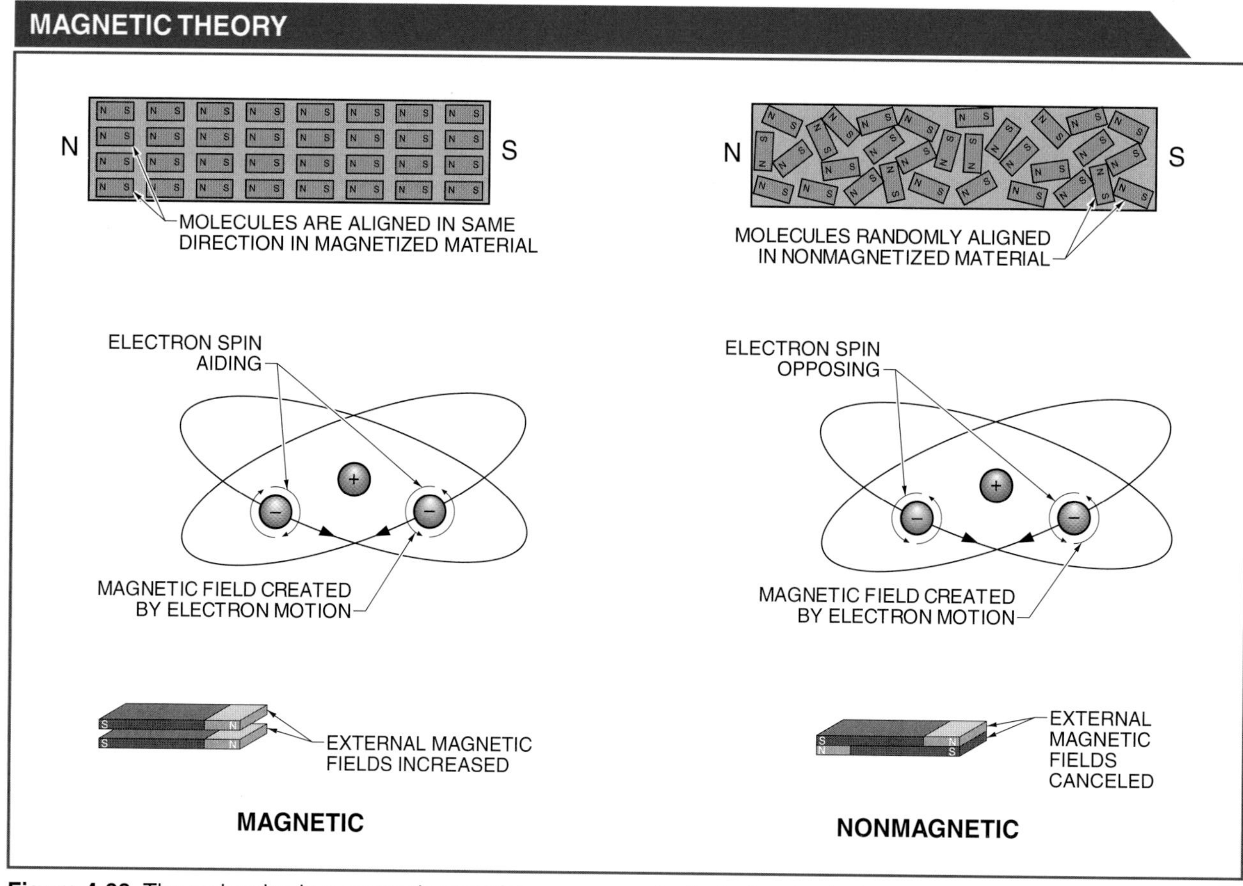

Figure 4-33. The molecules in a magnetic material are arranged with their poles in the same direction and the molecules in a nonmagnetic material are arranged randomly.

Tech Tip

Three points were discovered for the Earth's magnetic south pole using the dip pole method. Two of the points are located near the Amundsen-Scott South Pole Station. The other point is not even on the Antarctic continent. The real magnetic south pole was established as the point in the Antarctic Ocean at latitude 65.3° south and longitude 140° east.

The reason that some molecules align themselves in magnetic material and have a random arrangement in nonmagnetic material is explained by looking at an individual atom. The atomic theory of magnetism states that each atom of an element has a specific number of heavy positive protons in its nucleus. Around the nucleus, the same number of negative electrons revolve in discrete energy bands. Each electron in motion generates a magnetic field.

In a magnetized mass, paired electrons spin in such a direction that their magnetic fields are added together. This can be compared to two bar magnets placed with their north poles and south poles in the same direction. This condition creates a larger magnetic field than that of one magnet. In the atoms and molecules of most elements and compounds, the electrons are paired so that the effect of the magnetic field of one electron cancels that of the other. The effect is a nonmagnetic mass. This can be compared to two bar magnets, one with its north pole opposite the other's south pole.

According to this theory, it may appear that all atoms with odd numbers of electrons are magnetic. If atoms could be isolated, this might be seen to be true. However, atoms are found combined in molecular form and normally arrange themselves with eight valence electrons. This provides the balanced pairs that usually cause the formation of nonmagnetic materials. Materials composed of atoms and molecules having only paired electrons are diamagnetic. These atoms and molecules resist the influence of magnetic fields by setting up weak opposing fields. All materials, even paramagnetic substances and ferromagnetic metals, have some diamagnetic characteristics. Some electrons in these materials set up fields to cancel the intrusion of an outside field.

Materials that have an uneven number of electrons are normally paramagnetic. However, an atom can be paramagnetic in one compound and diamagnetic in another. For example, the oxygen-containing gases oxygen (O_2) and nitrous oxide (NO) are paramagnetic because some oxygen electrons are unpaired in these molecules. Nevertheless, in water (H_2O), all the oxygen electrons are paired within the oxygen atom. Water is therefore diamagnetic. This occurs because electrons have different motions in different kinds of molecules. Both diamagnetic and paramagnetic phenomena depend on the presence of an external magnetic field. When the external magnetic field is removed, diamagnetic and paramagnetic materials lose their magnetic characteristics because their electrons return to spins that cause their magnetic fields to cancel each other.

For some reason, this orderly process does not occur naturally in the ferromagnetic metals of iron, nickel, and cobalt. When atoms of these metals combine, they become ions and share valence electrons. Many valence electron spins do not cancel each other, resulting in a magnetic field. The molecules form regions of magnetic domains. Iron has four electrons that are unpaired. The magnetic fields of these electrons aid each other in forming a strong magnet.

Although iron is ferromagnetic, most pieces of iron are not magnets. This is because the rotations of the unpaired electrons are not aligned in the same direction. Domains are magnetized in different directions. Therefore, their magnetic fields cancel each other.

Electromagnetism. Magnetism and the electron are closely related to each other. They are considered to be expressions of a single force—the electromotive force. The three phenomena pertaining to electromagnetism are:

- Moving electric charges produce magnetic fields.
- Magnetic fields exert forces on moving electric charges.
- Changing magnetic fields in the presence of electric charges cause electrons to flow.

Tech Tip

Some soft magnets can be made into hard magnets. One process is to roll or hammer a hot piece of the soft magnet into a needle shape. While it is cooling, the soft magnet is placed in the field of an electromagnet or hard magnet and becomes a hard magnet.

In 1820, Danish physicist Hans Oersted noted that electron flow produces a magnetic field. He had a current-carrying conductor pointing in a north-south direction. A compass was near the wire. He noted that, in the absence of electron flow, the alignment of the wire and the alignment of the compass needle were the same. However, each time the circuit was closed to allow electron flow, the compass needle aligned itself at a right angle to the conductor. **See Figure 4-34.** When the direction of electron flow was reversed, the compass needle pointed in the opposite direction while maintaining a 90° displacement to the conductor.

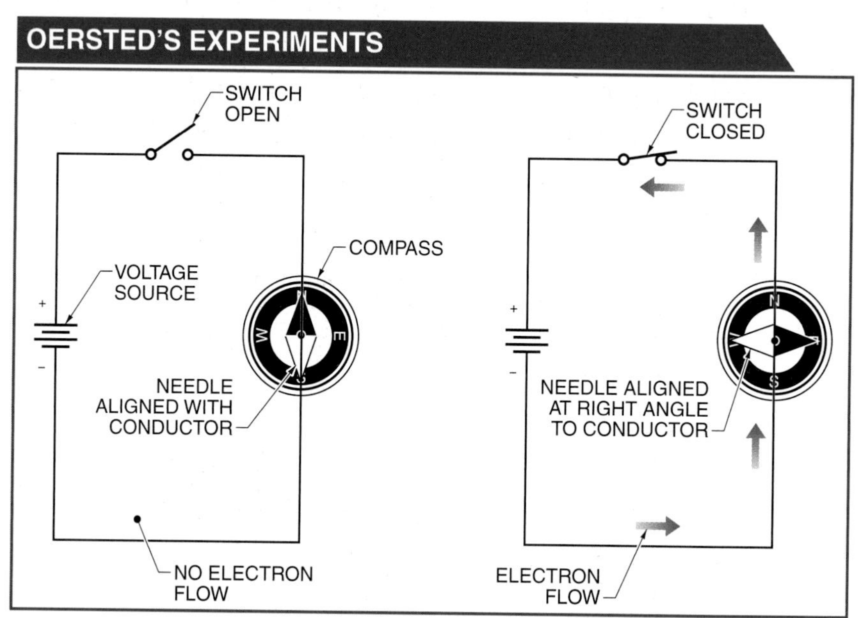

Figure 4-34. The needle of a compass aligns parallel to a conductor with no electron flow and perpendicular to a conductor with electron flow.

By definition, the direction of magnetic lines of force is from north to south. This means that the magnetic field of a current-carrying conductor enters the south pole of a compass needle and exits its north pole, causing the needle to align at 90° to the conductor. Based on this movement, it can be deduced that the magnetic lines of force must be concentric around the conductor. With the magnetic lines of force traveling in the same direction between two conductors, a force is created such that these fields oppose each other. **See Figure 4-35.** This results in a force that pushes the two conductors apart.

Oersted's findings led to speculation that the magnetic field of a magnet may interact with the magnetic field created by the flow of electrons through a conductor. Experiments by Michael Faraday demonstrated that the magnetic field of a magnet can cause a current-carrying conductor to move or rotate. **See Figure 4-36.** The magnet moves or rotates if the conductor is held in place and the magnet is free to move. These findings led to the development of the electric motor.

CONDUCTOR MAGNETIC FORCE

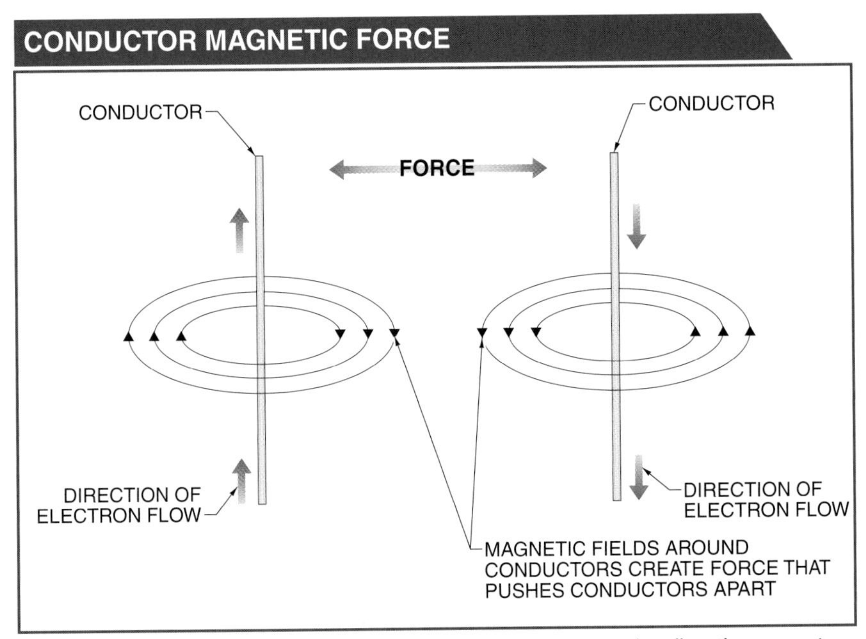

Figure 4-35. Two conductors with electrons flowing in opposite directions create a magnetic force that pushes the conductors apart.

MAGNETIC FIELD CREATED BY CONDUCTOR

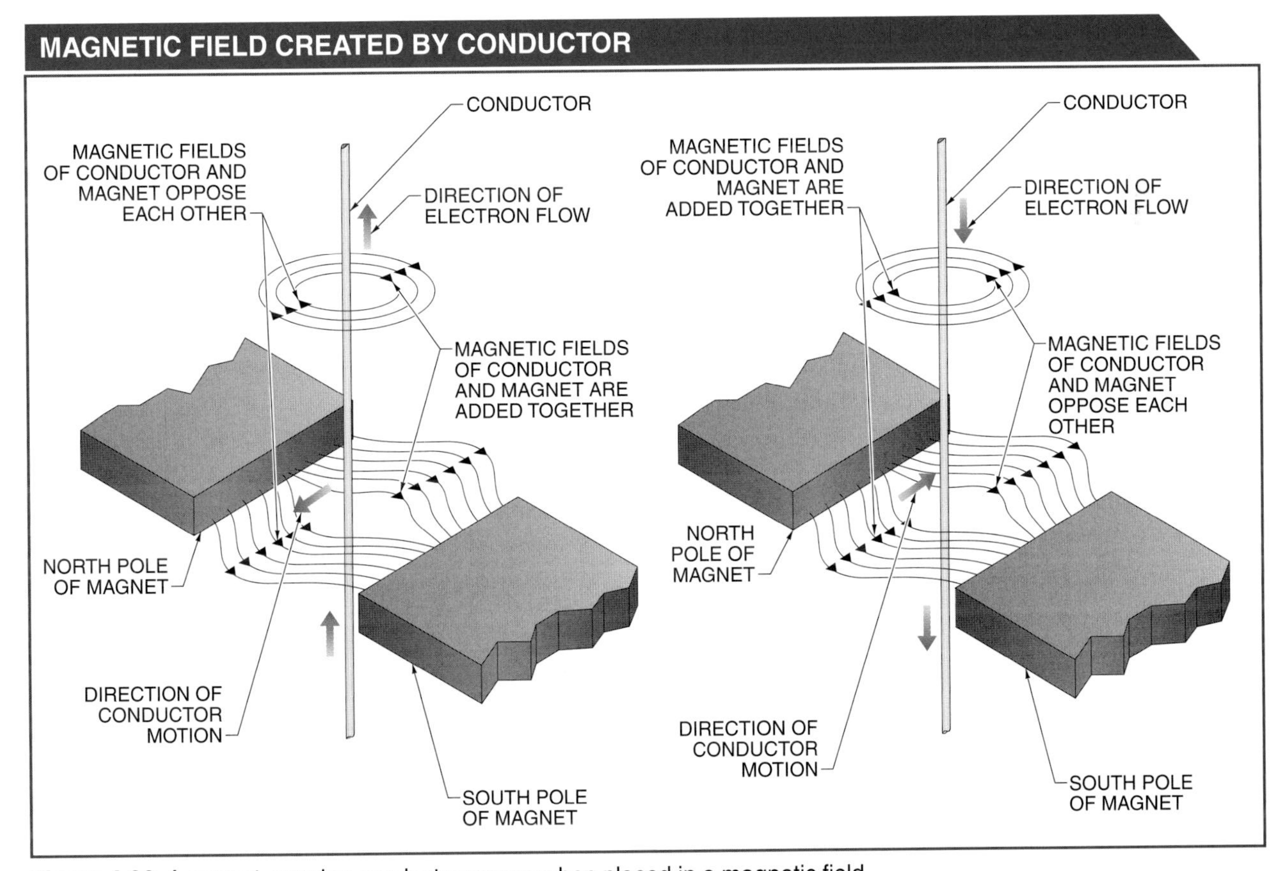

Figure 4-36. A current-carrying conductor moves when placed in a magnetic field.

In 1831, Faraday created electron flow in a conductor using the magnetic field of a permanent magnet. This proved that if there was relative motion between a magnet and a conductor, electron flow results in the conductor. **See Figure 4-37.** This was discovered concurrently by Joseph Henry and led to the development of the generator, which is the primary source of electrical energy production.

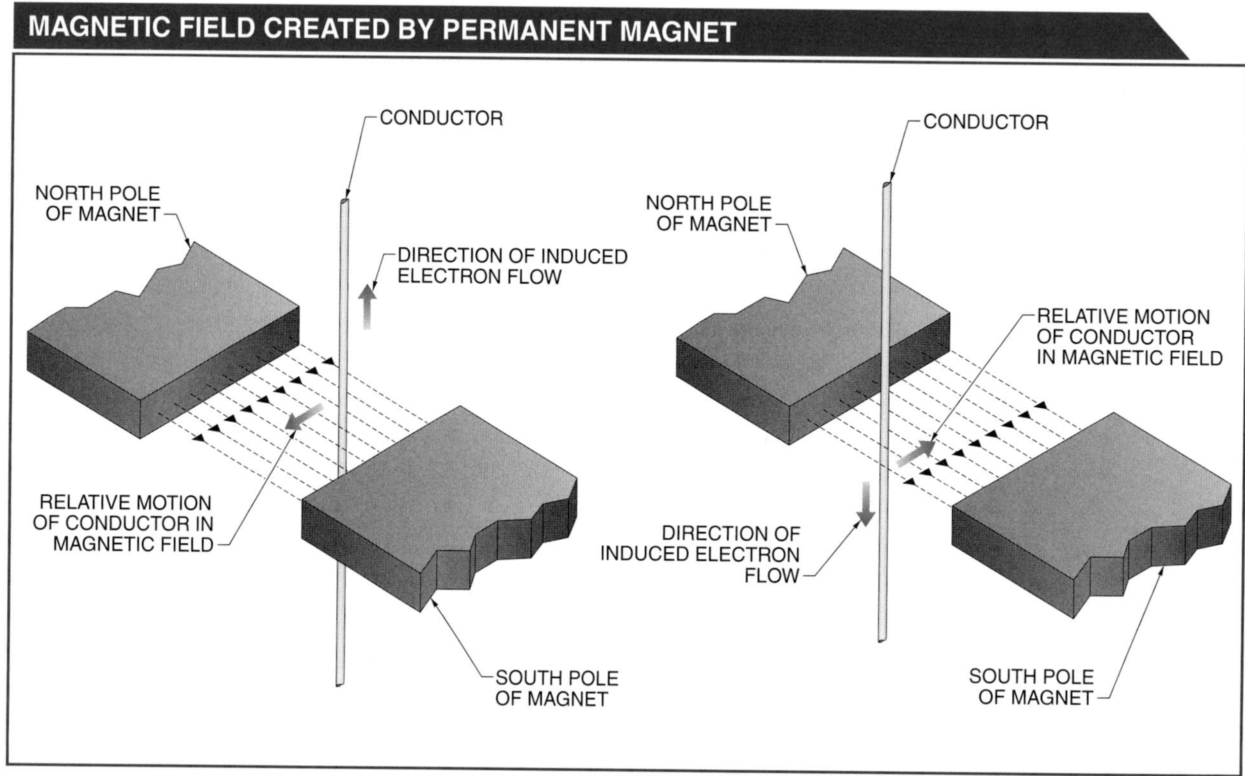

Figure 4-37. Electron flow is induced in a conductor when there is relative motion between the conductor and a magnetic field.

Faraday repeated the experiment using an electromagnet. In his experiments, he wound two coils of wire close together, connected a battery to the first coil, and connected a galvanometer to the second coil. The galvanometer was used to measure small amounts of electron flow. Faraday detected an electron flow with the galvanometer when the connection between the first coil and the battery was being made or broken. He noticed that the electron flow in the second coil flowed in one direction when the battery was being connected and flowed in the opposite direction when the battery was being disconnected. When electron flow was constant in the first coil, or when the battery was not connected, the galvanometer indicated zero current in the second coil. With this experiment, Faraday had invented the first electrical transformer. **See Figure 4-38.**

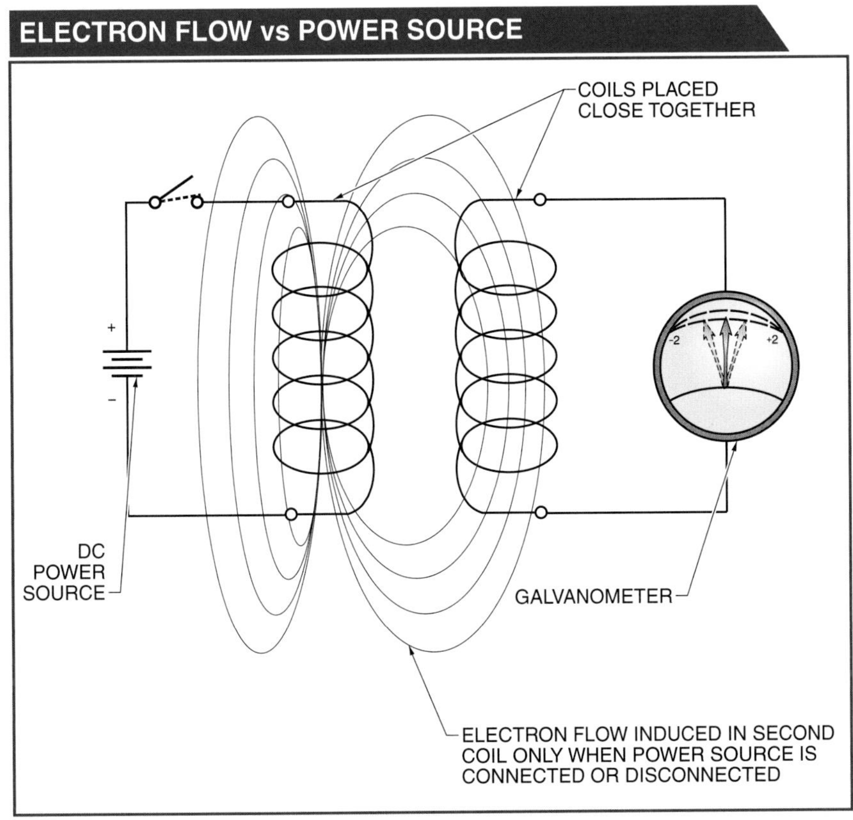

Figure 4-38. The needle of a galvanometer deflects when a battery is connected and disconnected. No deflection is observed when the switch is closed and the current is at a constant value.

Faraday developed a visual model of the magnetic field. He theorized that moving a magnet or an electrical charge disturbs the magnetic lines of force surrounding the magnet or charge. He concluded that this effect travels at a definite speed. In 1864, James Clerk Maxwell, a Scottish physicist, published his mathematical theory that the frequency of the changes in these fields could be treated as a wave motion and travel at the speed of light (300×10^6 m/s). Maxwell also came to the conclusion that light is an electromagnetic wave and all such waves consist of the growth and collapse of electric and magnetic fields at right angles to each other. The concentric magnetic lines of force can travel in the clockwise or counterclockwise direction depending on the direction of electron flow through a conductor.

Magnetism Applications. Magnetism is used in a wide variety of applications in modern industry. The ability of an electric current to create a magnetic field is used in generators, electric motors, power tools, and solenoids. **See Figure 4-39.**

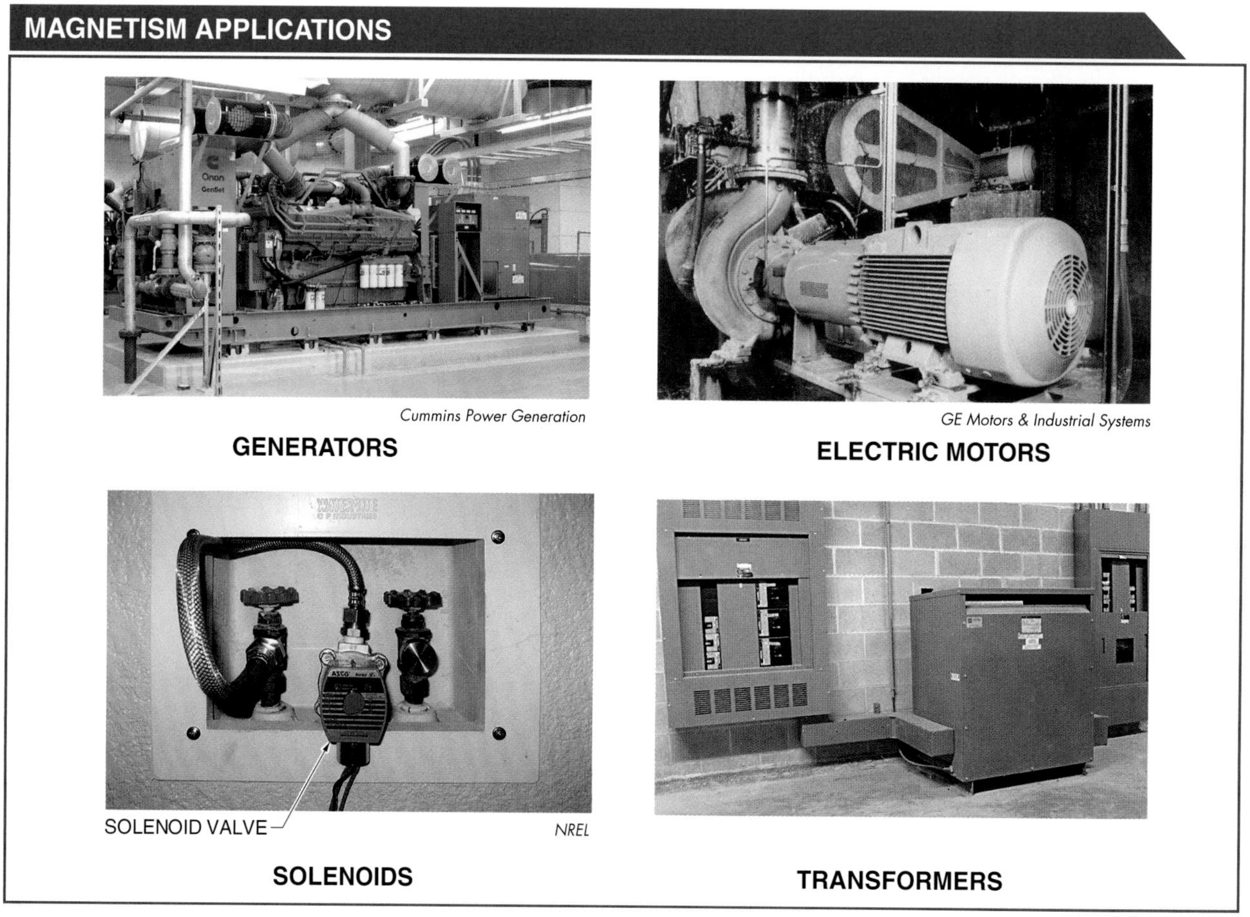

MAGNETISM APPLICATIONS

Cummins Power Generation

GENERATORS

GE Motors & Industrial Systems

ELECTRIC MOTORS

SOLENOID VALVE

NREL

SOLENOIDS

TRANSFORMERS

Figure 4-39. Magnetism is the most important source of electrical energy and is used in a wide variety of applications.

Terms

A **generator (alternator)** is a machine that converts mechanical energy into electrical energy by means of electromagnetic induction.

A **motor** is a rotating device that converts electrical energy into mechanical energy.

Regardless of the basic sources of potential energy that drives them, generators provide most of the electrical energy consumed. A *generator (alternator)* is a machine that converts mechanical energy into electrical energy by means of electromagnetic induction. Sources of potential energy used to drive generators include falling water, fossil fuels, nuclear energy, and electricity. A *motor* is a rotating device that converts electrical energy into mechanical energy. The torque (rotating mechanical force) on a motor shaft is used to produce work. An electric motor is used to drive a generator.

Electromagnetism made possible the invention of power tools. The variety of tools available include devices to drill, strike, saw, grind, sand, turn, cut, shape, and form materials. Power tools have taken much of the labor out of many tasks that previously could only be achieved through human power.

Magnetic radiation plays a large role in modern medicine. X-ray, CAT (computer-assisted tomography) scans, and MRI (magnetic resonance imaging) are a few of the applications. MRI began to be used in the early 1980s as a noninvasive way to see images of thin slices of the body. This is accomplished by measuring the characteristic magnetic behavior of specific nuclei in the water and fats of the body. These images help identify normal, damaged, and diseased tissue.

Another important use of electromagnetism is the generation of magnetic fields using electrical current. Electromagnets are used as substitutes for permanent magnets in many applications. Advantages of electromagnets over permanent magnets are that the strength of electromagnets can be varied by varying the amount of electron flow and the magnetism can be turned off by turning off the electron flow.

The advancement of the wave theory of electromagnetic radiation set the stage for modern communications. Electromagnetic fields of various frequencies make modern communications possible. Electron flow in a conductor results in a radiating magnetic field, and this field induces a current into any conductor it comes into contact with. Radio, television, and microwave transmissions of all types operate on this theory. Transmitters emit electromagnetic radiation from their antennas, and the signal induces the information into the antenna of the receiver.

SUMMARY

Electrical energy can be created from friction, pressure, light, heat, chemical action, and magnetism. Friction creates electrical energy when two different types of materials are rubbed together. Pressure applied to piezoelectric materials produces a voltage directly proportional to the amount of applied strain. Some materials emit electrons when exposed to light and are used as a source of electrical potential. Heat can also be used to generate electrical energy through the Seebeck or Peltier effect. Chemical action was the first reliable and usable method for producing electrical energy and is the basis for the common battery. Magnetism is the most common source of electrical energy. The relative motion between a magnet and a conductor is used to produce AC electricity.

APPLICATION: VOLTAGE SOURCES...

Background: Today, portable electronics are everywhere, and batteries are used to power these devices. Batteries are an example of how electrical energy is generated using a chemical action. Batteries have two or more cells. A cell is a device that produces electricity at a fixed voltage with limited current capacity. According to this definition, the small batteries sold to consumers are not batteries but cells. The different cell types have different voltages and current capacities and support different loads. **See Figure 1.**

BATTERY CELL TYPES			
Cell Type	Voltage*	Current Capacity†	Typical Load‡
D	1.5	13,000	200
C	1.5	6000	100
AA	1.5	2400	50
AAA	1.5	1000	10
9 V	9	500	15

* in V
† in mAh
‡ in mA

Figure 1. Different battery cell types have different voltages and current capacities and support different loads.

The different current capacities indicate how long a cell will last for a given load and is given in amp-hours (Ah). For example, if an AA cell was used with a 10 mA load, then the cell would last 240 hours (2400 mAh ÷ 10 mA = 240 hours). The typical load ensures maximum cell life. Higher loads reduce cell life below the listed capacity.

Key Points: Voltage source

Problem: A portable voltage source (battery) must be configured to support an application that requires 6.0 V with a load of 600 mA. The application also requires battery life greater than 100 hr. **See Figure 2.**

Solution: To start, select the cell configuration that provides the proper voltage. The cells being used provide 1.5 V each. To increase the voltage, several cells are connected in series. To meet the required voltage, the number of required series cells must be determined.

V_T = number of cells (N) × 1.5 V
$6\text{ V} = N \times 1.5\text{ V}$
$N = 6\text{ V} \div 1.5\text{ V}$
$N = 4$

...**APPLICATION:** VOLTAGE SOURCES...

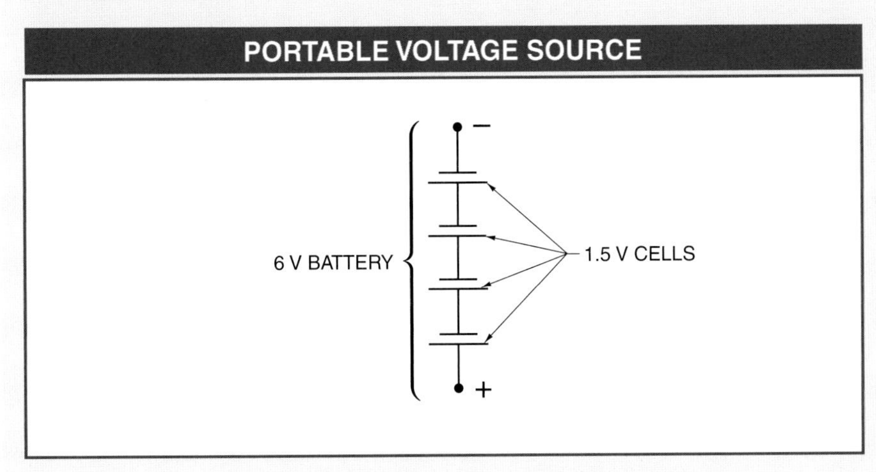

Figure 2. To increase the voltage, several cells are connected in series.

To create the required voltage source, a total of four 1.5 V cells are connected in series to form a 6 V battery. With the voltage sources connected in series, the current capability is limited to an individual cell. For example, if four AA cells are used, the 6 V battery has a maximum current capability of 2400 mAh, supporting a typical load up to 50 mA. The battery current capacity requirement is set by the current load and time. This is found by multiplying the current load and hours together for amp-hours.

Current capacity needed (in mAh) = current load (in mA) × hours (in h)
Current capacity needed (in mAh) = 600 mA × 100 h
Current capacity needed (in mAh) = 60,000 mAh

Since a single cell type cannot supply this current capacity, the capacity must be increased. To increase the battery current capacity, parallel batteries are added. Each parallel battery must match the original battery voltage. In this application, each parallel battery must be 6 V. To find the number of parallel batteries required (for a size D cell battery), the required capacity is divided by the battery capacity as follows:

$N = C_T \div C_B$
where
N = number of parallel batteries
C_T = total capacity required (in mAh)
C_B = battery capacity (in mAh)

For this application:
N = 60,000 mAh ÷ 13,000 mAh
N = 4.62
N = 5 parallel batteries

...APPLICATION: VOLTAGE SOURCES

Finally, maximum battery life must be verified by ensuring that the typical load is not exceeded. For a D cell battery, the typical load is 200 mA. Each parallel battery supports the typical load. To find the supported typical load, multiply the number of parallel batteries by the typical cell load.

$TLS = TCL \times N$
where
TLS = typical load support (in mA)
TCL = typical cell load (in mA)
N = number of parallel batteries

For this application:
$TLS = 200 \text{ mA} \times 5$
$TLS = 1000 \text{ mA}$

The D cell base battery in this application supports a typical load of 1000 mA, which is greater than the 600 mA requirement. So, the final D cell battery configuration required for this application is a total of 20 D cells (five parallel batteries with each battery containing four D cells connected in series). **See Figure 3.**

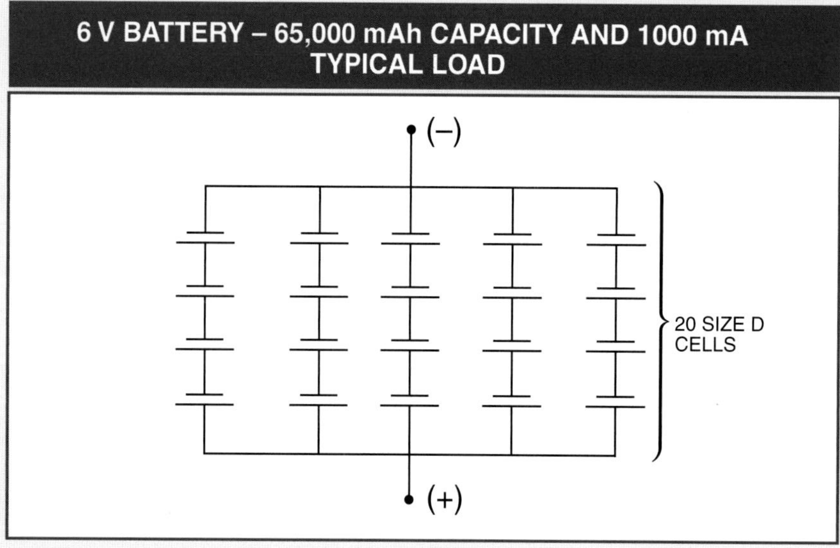

Figure 3. A total of 20 D cells is required for the battery configuration.

Chapter Review

1. Why do like poles of magnets repel one another?
2. How do global positioning systems (GPS) use solar cells?
3. List five applications for piezoelectric transducers.
4. Explain the differences between battery cells connected in series and battery cells connected in parallel.
5. At what point (in V) would a person feel an electrostatic discharge?
6. What is the purpose of using insulation beads with certain types of thermocouples?
7. List two advantages of light-emitting diodes (LEDs).
8. Explain the piezoelectric effect.
9. What is a gauss?
10. What does a high temperature reading on a solar cell indicate?
11. Explain the difference between a negative static charge and a positive static charge.
12. What are the two thermoelectric effects?
13. What are the two main advantages of electromagnets over permanent magnets?
14. How does a lightning rod system neutralize a charge?
15. Explain the difference between the Seebeck effect and the Peltier effect.
16. What is the most common application for the use of thermoelectricity?
17. List three phenomena that pertain to electromagnetism.
18. Explain the difference between paramagnetic and diamagnetic materials.
19. What is the photoelectric effect?
20. List five electrical sources of potential energy.
21. What are two common applications for photoconductive cells?
22. What is the operating principle of an electrostatic precipitator?
23. What is a transducer?
24. Explain why photovoltaic cells would be connected in series, parallel, or a series/parallel combination.
25. How does a lightning rod system work?

5

The Simple Circuit and Ohm's Law

A simple DC circuit consists of a voltage source, insulated conductors, and a load. Other circuit components such as switches (control devices) and fuses or circuit breakers (protective devices) may also be included. Ohm's law is used to determine the current, voltage, and resistance of a simple DC circuit. Ohm's law is one of the most valuable tools in the study of electricity.

OBJECTIVES

- Explain the concept of a simple circuit.
- Describe how conductors, switches, and loads are used in a simple circuit.
- Describe overcurrent and overcurrent protective devices.
- Explain how to take voltage and current measurements.
- Explain and perform calculations using Ohm's law.

Terms

An **electrical circuit** is an assemblage of conductors and electrical devices through which electrons flow.

THE SIMPLE CIRCUIT

An *electrical circuit* is an assemblage of conductors and electrical devices through which electrons flow. **See Figure 5-1.** Electrical circuits consist of a complete path for the flow of electrons between two or more points. The most fundamental electrical circuit is a simple circuit in which a single energy source supplies current to a single load through electrical conductors.

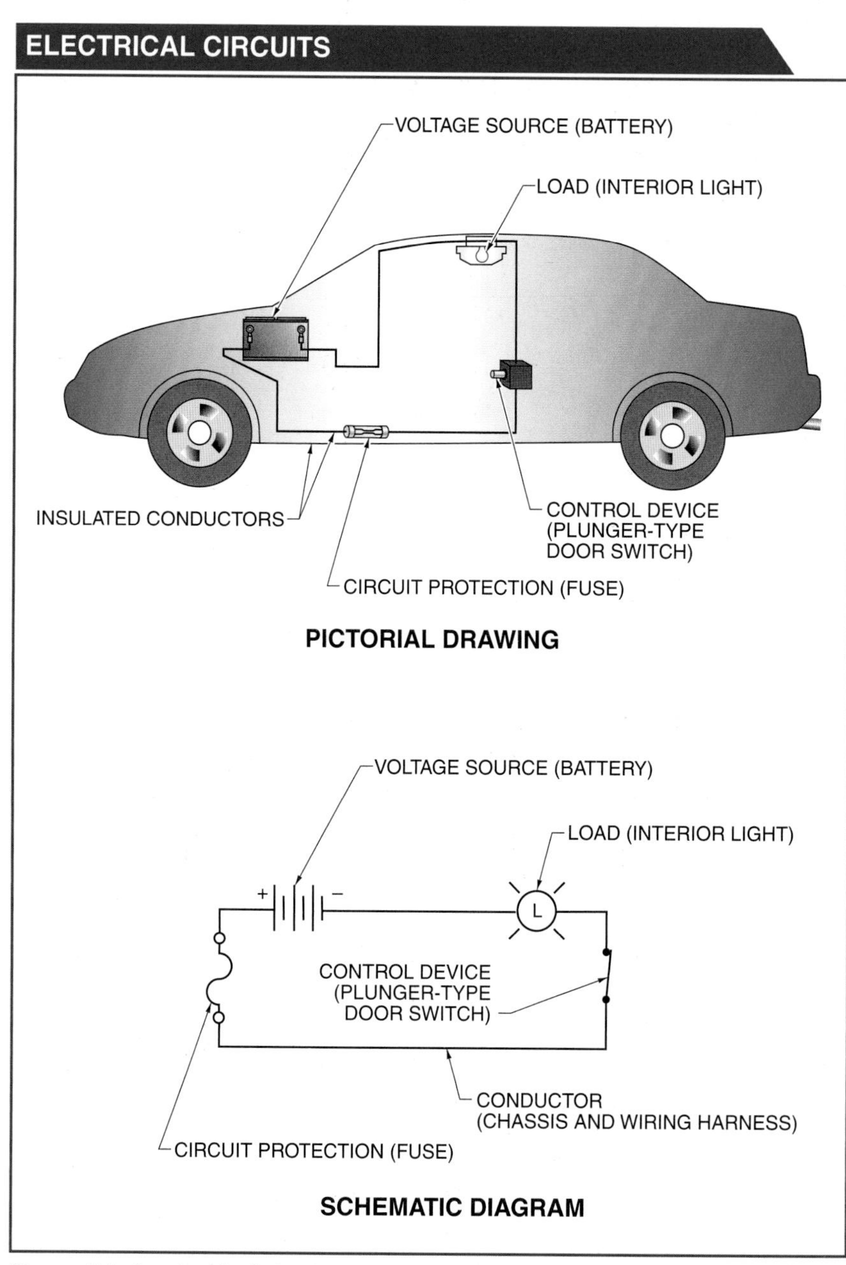

Figure 5-1. An electrical circuit consists of a voltage source, insulated conductors, a load, a switch, and a fuse.

All simple circuits must offer some opposition to the flow of electrons. Opposition may be high or low depending on the characteristics of the load. Simple circuits consist of the following three parts:

- a voltage source to produce electron flow
- insulated conductors to allow electrons to move from one point in the circuit to another and to confine the electron flow to the desired path through the conductors and intended components
- a load to limit the current level and to convert the electrical energy into another form of energy such as heat, light, or mechanical motion

A *closed circuit* is an electrical circuit in which electrons flow uninterrupted from the negative terminal of the voltage source, through the load, and back to the positive terminal of the voltage source. An *open circuit* is an electrical circuit in which the flow of electrons is interrupted. An open circuit has no electron flow.

Although control and protective devices are not required for a circuit to be complete, these devices may be required by electrical codes for the protection of persons and property. In most electrical circuits, a switch may be included for convenience, and a fuse or circuit breaker may be included for safety.

Terms

A **closed circuit** is an electrical circuit in which electrons flow uninterrupted from the negative terminal of the voltage source, through the load, and back to the positive terminal of the voltage source.

An **open circuit** is an electrical circuit in which the flow of electrons is interrupted.

A **conductor** is a material that has a low electrical resistance and permits electrons to move through it easily. Conductors are generally a single wire, a group of wires, or another material suitable for carrying electric current.

Conductors

A *conductor* is a material that has a low electrical resistance and permits electrons to move through it easily. Conductors are generally a single wire, a group of wires, or another material suitable for carrying electric current. Most single conductors are enclosed in an insulated cover to protect the conductor, increase safety, and meet electrical code requirements. Some single conductors, such as ground wires, may be bare. The schematic symbol for a conductor is a line connecting two devices in an electrical circuit. This symbol is the same for insulated or bare conductors. **See Figure 5-2.**

Unconnected conductors that cross each other are shown by the absence of a dot (node) or by using a loop over. The loop over method is the preferred method of showing that conductors are not connected. If conductors are connected, a dot is placed at their intersection. The dot method is the preferred method of showing that conductors are in physical contact with each other. A dot can also be used to show when one conductor terminates into another. Often, however, when one conductor terminates into another, the dot is eliminated. When a conductor is shielded, a dotted line is used in parallel with the conductor symbol.

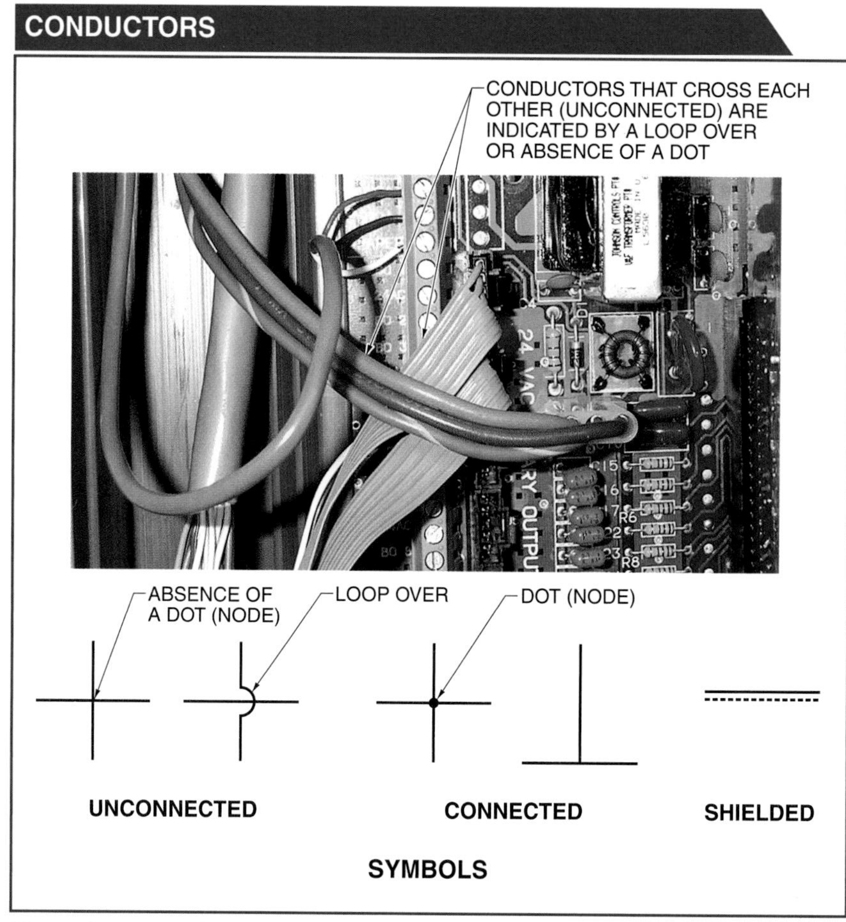

Figure 5-2. In a schematic or wiring diagram, conductors are shown as lines. Conductors that are connected often use a dot to indicate the connection.

Switches

A *switch* is a mechanical, electronic, or solid-state electrical device that is used to start, stop, or redirect the flow of electrons in an electrical circuit. Switches are added to a circuit as a control device. **See Figure 5-3.** Turning a lamp ON and OFF by using a switch is safer and more convenient than connecting and disconnecting a conductor. The switch is connected in series with the voltage source and load. There is only one current path in the circuit, so when the switch is open, current does not flow to the lamp.

Most switches are rated for the maximum current and voltage they can safely handle. Large switches may also be rated in horsepower. All switches use contacts to start or stop the flow of electrons in a circuit. A *contact* is the conducting part of a switch that operates with another conducting part to make or break a circuit.

Terms

A **switch** is a mechanical, electronic, or solid-state electrical device that is used to start, stop, or redirect the flow of electrons in an electrical circuit.

A **contact** is the conducting part of a switch that operates with another conducting part to make or break a circuit.

The position of the contacts (normally open or normally closed), number of poles (single-pole, double-pole), number of throws (single-throw, double-throw), and type of break (single-break, double-break) are used to describe switch contacts. **See Figure 5-4.**

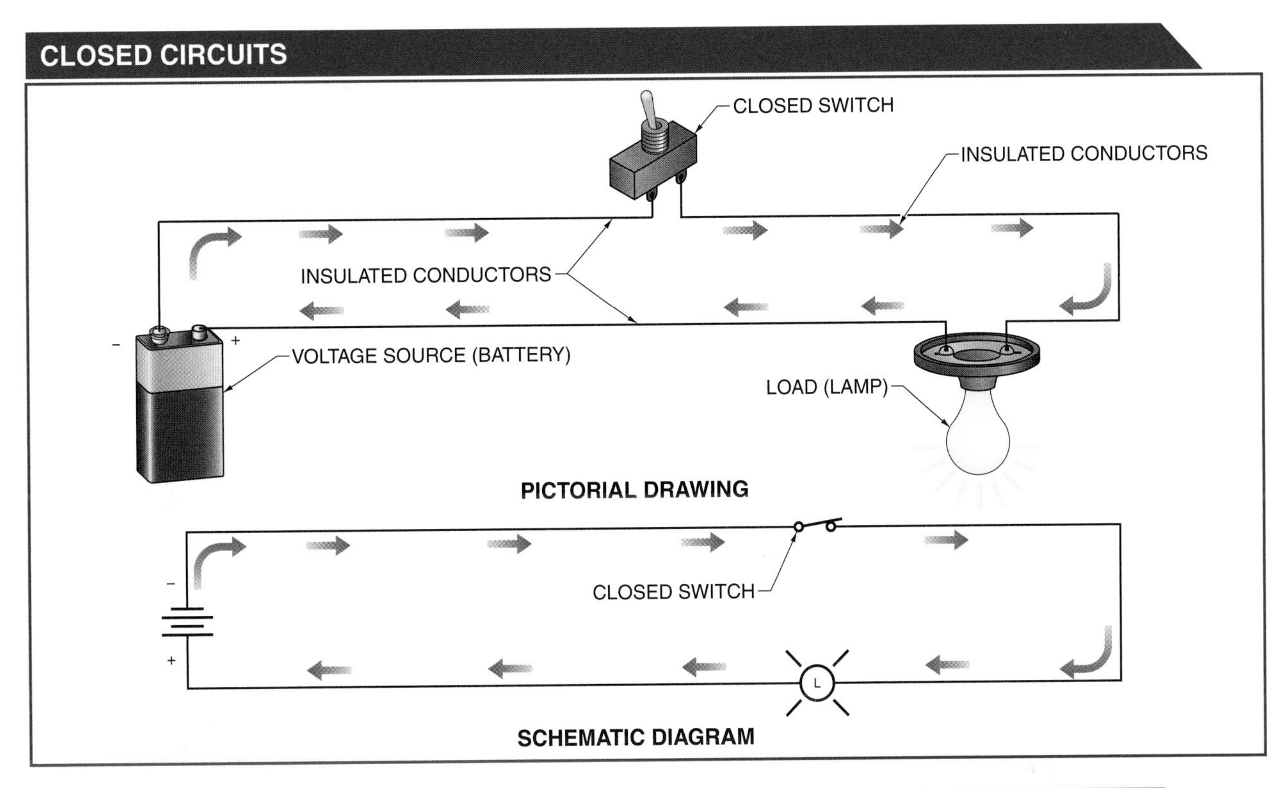

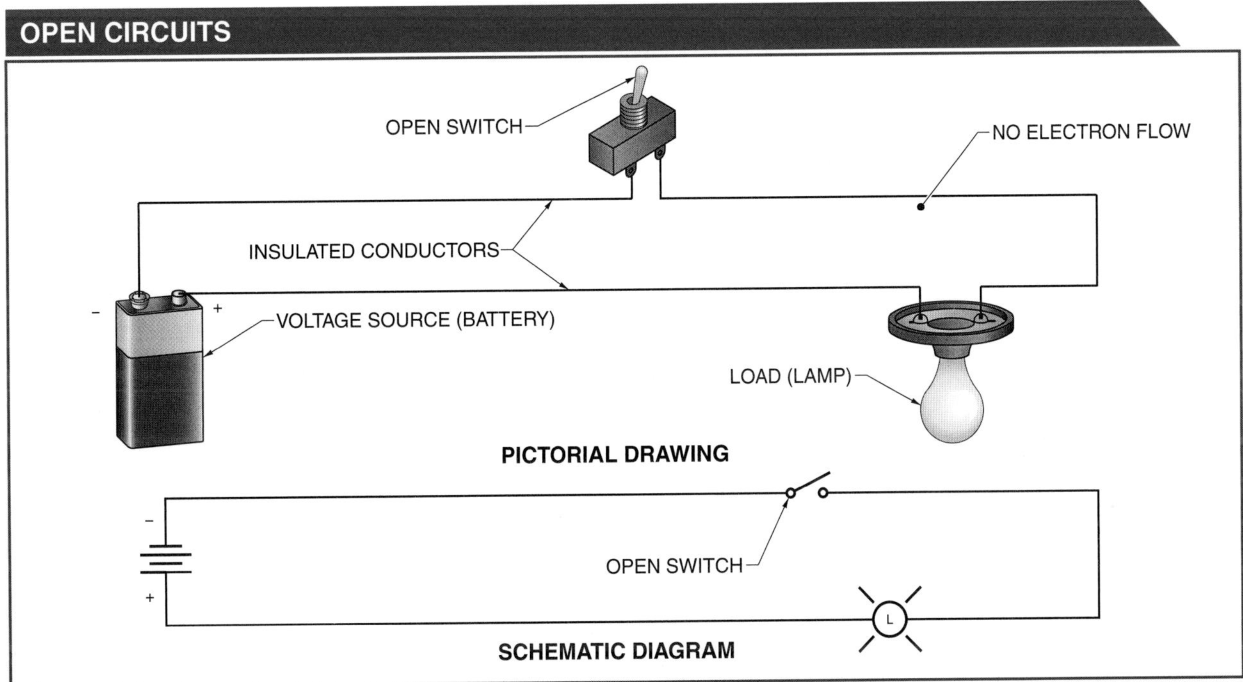

Figure 5-3. Switches are control devices and are used to close and open circuits safely.

SWITCH CONTACTS			
Abbreviation	**Symbol**	**Abbreviation**	**Symbol**
NO	MECHANICAL CONTACTS SOLID-STATE CONTACTS **NORMALLY OPEN**	**SPDT**	CLOSED CONTACT **SINGLE-POLE, DOUBLE-THROW, DOUBLE-BREAK**
NC	CONTACT CLOSED SOLID-STATE CONTACTS **NORMALLY CLOSED**	**DPST**	MECHANICALLY CONNECTED **DOUBLE-POLE, SINGLE-THROW, SINGLE-BREAK**
SPST or S	POLE TERMINALS **SINGLE-POLE, SINGLE-THROW, SINGLE-BREAK**	**DPDT**	COMMON TERMINALS MECHANICALLY CONNECTED **DOUBLE-POLE, DOUBLE-THROW, SINGLE-BREAK**
SPST	BREAK BOTH SIDES **SINGLE-POLE, SINGLE-THROW, DOUBLE-BREAK**	**DPDT**	MECHANICALLY CONNECTED **DOUBLE-POLE, DOUBLE-THROW, DOUBLE-BREAK**
SPDT OR S_3	**SINGLE-POLE, DOUBLE-THROW, SINGLE-BREAK**		COMMON TERMINAL **MULTIPLE-CONTACT SWITCH**

Figure 5-4. The position of the contacts, number of poles, number of throws, and type of break are used to describe switch contacts.

Terms

Normally open (NO) contacts are contacts that are open before being activated.

Normally closed (NC) contacts are contacts that are closed before being activated.

Position. The position of switch contacts refers to the contacts before they are activated. *Normally open (NO) contacts* are contacts that are open before being activated. When activated, normally open contacts are closed. *Normally closed (NC) contacts* are contacts that are closed before being activated. When activated, normally closed contacts are opened.

Any switch with a spring that holds its contacts in either the closed or open position has a normal position. The normal position is the position in which the spring places the switch. Contacts may or may not have a normal position. For example, most doorbell switches contain a spring that holds the contacts open; the switch contacts have a normal position. Switches used to turn lights on in a room do not contain a spring that holds the contacts open or closed; the switch contacts do not have a normal position.

Poles. A *pole* is the number of completely isolated circuits that a relay can switch. A single-pole (SP) contact can carry current through only one circuit at any one time. A double-pole (DP) contact can carry current through two circuits simultaneously. The two circuits are mechanically connected so that they open or close at the same time, while remaining electrically insulated from each other. This mechanical connection is represented in the symbol by a dashed line connecting the poles.

Throws. A *throw* is the number of closed contact positions per pole. Throws denote the number of different circuits that each individual pole is capable of controlling. If a pole controls only one circuit, it is a single throw (ST) pole. If a pole controls two circuits, it is a double throw (DT) pole.

An industrial pushbutton consists of one or more contact blocks (electrical contacts), a legend plate, and an operator.

Breaks. A *break* is a place on a switch contact that opens or closes a circuit. All contacts are either single break or double break. *Single-break contacts* are contacts that break the electrical circuit in one place. *Double-break contacts* are contacts that break the electrical circuit in two places.

Terms

A **pole** is the number of completely isolated circuits that a relay can switch.

A **throw** is the number of closed contact positions per pole.

A **break** is a place on a switch contact that opens or closes a circuit.

Single-break contacts are contacts that break the electrical circuit in one place.

Double-break contacts are contacts that break the electrical circuit in two places.

Switch Characteristics

Switches are available in many shapes. Switches can be activated manually, mechanically, or automatically. Once activated, the switch changes the position of the contacts. The contacts are used to start and stop the flow of electrons in a circuit. **See Figure 5-5.**

Manually operated switches include pushbuttons, selector switches, and foot switches. Manually operated switches are the most common type of switch used in electrical circuits. They are even used in automatically operated circuits to override automatic switches. The ability to override automatic switches is required when troubleshooting and in the event of an emergency. Common uses of manually operated switches include turning lights and appliances off and on. The symbol for a normally open manually operated switch shows the operator above the terminals. The symbol for a normally closed manually operated switch shows the operator below the terminals.

SWITCHES		
Device	**Abbreviation**	**Symbol**
Pushbuttons 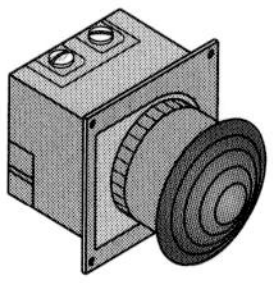	PB	TERMINALS MANUAL OPERATION **SINGLE CIRCUIT, NORMALLY OPEN**
	PB	**SINGLE CIRCUIT, NORMALLY CLOSED**
	PB	**DOUBLE-CIRCUIT, NORMALLY OPEN AND CLOSED**
	PB	MUSHROOM OPERATOR **MUSHROOM HEAD**
	PB	MECHANICAL LINK **MAINTAINED**
	PB/LT	R R = red G = green A = amber B = blue LAMP INSIDE OPERATOR **ILLUMINATED**
Selector Switches 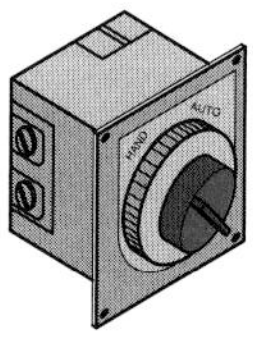	SELT SW or SS	RUN POSITION JOG POSITION JOG RUN A1 A2 J R A1 X A2 X CONTACTS OPEN "X" INDICATES CONTACTS CLOSED **TWO-POSITION**
	SELT SW or SS	J = jog R = run or reverse S = stop U = up D = down F = forward A = automatic M = manual STOP POSITION JOG STOP RUN A1 A2 J S R A1 X A2 X TRUTH TABLE FOR SWITCH **THREE-POSITION**
Foot Switches 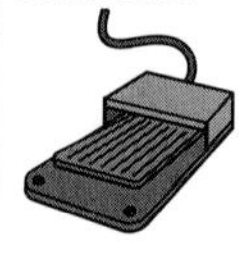	FTS	FOOT OPERATOR **NORMALLY OPEN**
	FTS	**NORMALLY CLOSED**

Device	**Abbreviation**	**Symbol**
Limit Switches 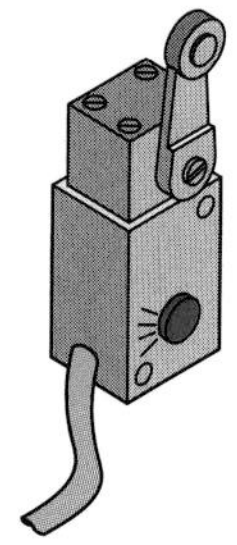	LS	MECHANICAL OPERATOR **NORMALLY OPEN**
	LS	OPERATOR IN CLOSED POSITION **NORMALLY OPEN, HELD CLOSED**
	LS	**NORMALLY CLOSED**
	LS	OPERATOR IN OPEN POSITION **NORMALLY CLOSED, HELD OPEN**
Proximity Switches 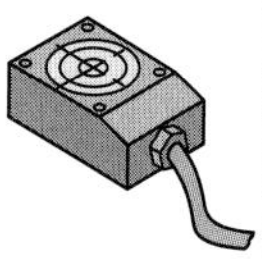	PR	SOLID-STATE OR **NORMALLY OPEN**
	PR	SOLID-STATE OR **NORMALLY CLOSED**
Temperature Switches 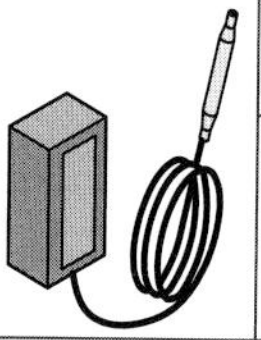	TAS or TH or THS	TEMPERATURE OPERATOR **NORMALLY OPEN**
	TEMP SW	**NORMALLY CLOSED**
Pressure Switches 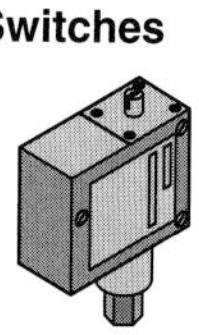	PS	PRESSURE OPERATOR **NORMALLY OPEN**
	PS	**NORMALLY CLOSED**
Flow Switches 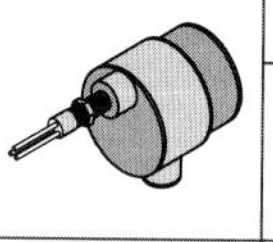	FLS	FLOW OPERATOR **NORMALLY OPEN**
	FLS	**NORMALLY CLOSED**
Level Switches 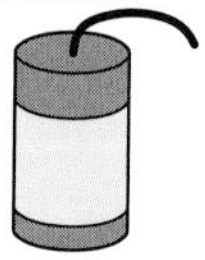	LS	LEVEL OPERATOR **NORMALLY OPEN**
	LS	**NORMALLY CLOSED**

Figure 5-5. Switches are available in many shapes and are often designated according to their use.

Mechanically operated switches such as limit, pressure, flow, and level switches detect the physical presence of an object and can be used as safety devices. For example, common uses of limit switches include determining if a door is closed (automobile, microwave, oven, etc.), if safety guards are in place, and if machine parts are properly placed. The symbol for a normally open mechanically operated switch shows the operator below the terminals. The symbol for a normally closed mechanically operated switch shows the operator above the terminals.

Electrically operated switches include proximity, temperature, flow, and level switches. Electrically operated switches operate electrical circuits with little or no manual input. Common uses of electrically operated switches include keeping rooms and ovens at a set temperature, basements dry, tanks full, pressure in a system, and products moving. The symbol for a normally open operated switch shows the operator below the terminals. The symbol for a normally closed operated switch shows the operator above the terminals.

Other common types of switches are used in homes and buildings to control lighting circuits. Common lighting circuit switches include two-way, three-way, and four-way switches. The switch used to control a lamp depends on the number of different locations the lamp must be controlled from. **See Figure 5-6.**

A *two-way switch* is a single-pole, single-throw (SPST) switch. A two-way switch may be placed in one of two different positions. A two-way switch allows electrons to flow through it in the ON position and does not allow electrons to flow through it in the OFF position. Two-way switches have ON and OFF position markings on them. Two-way switches are used to control a lamp from one location. Two-way switches are mounted so that ON is in the up position.

Terms

A **two-way switch** is a single-pole, single-throw (SPST) switch.

Four-Way Switches

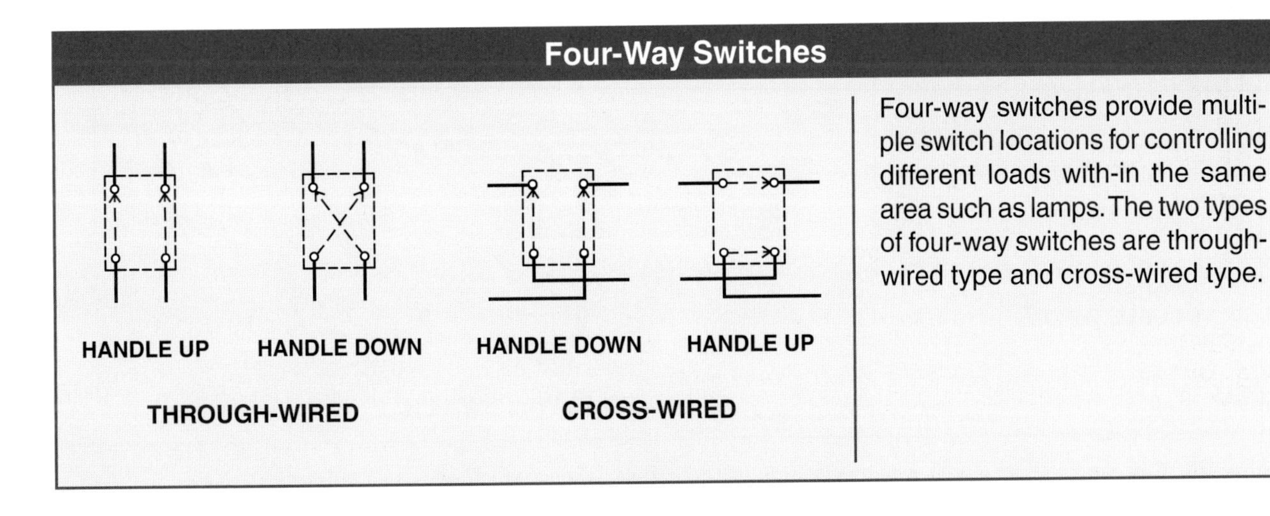

Four-way switches provide multiple switch locations for controlling different loads with-in the same area such as lamps. The two types of four-way switches are through-wired type and cross-wired type.

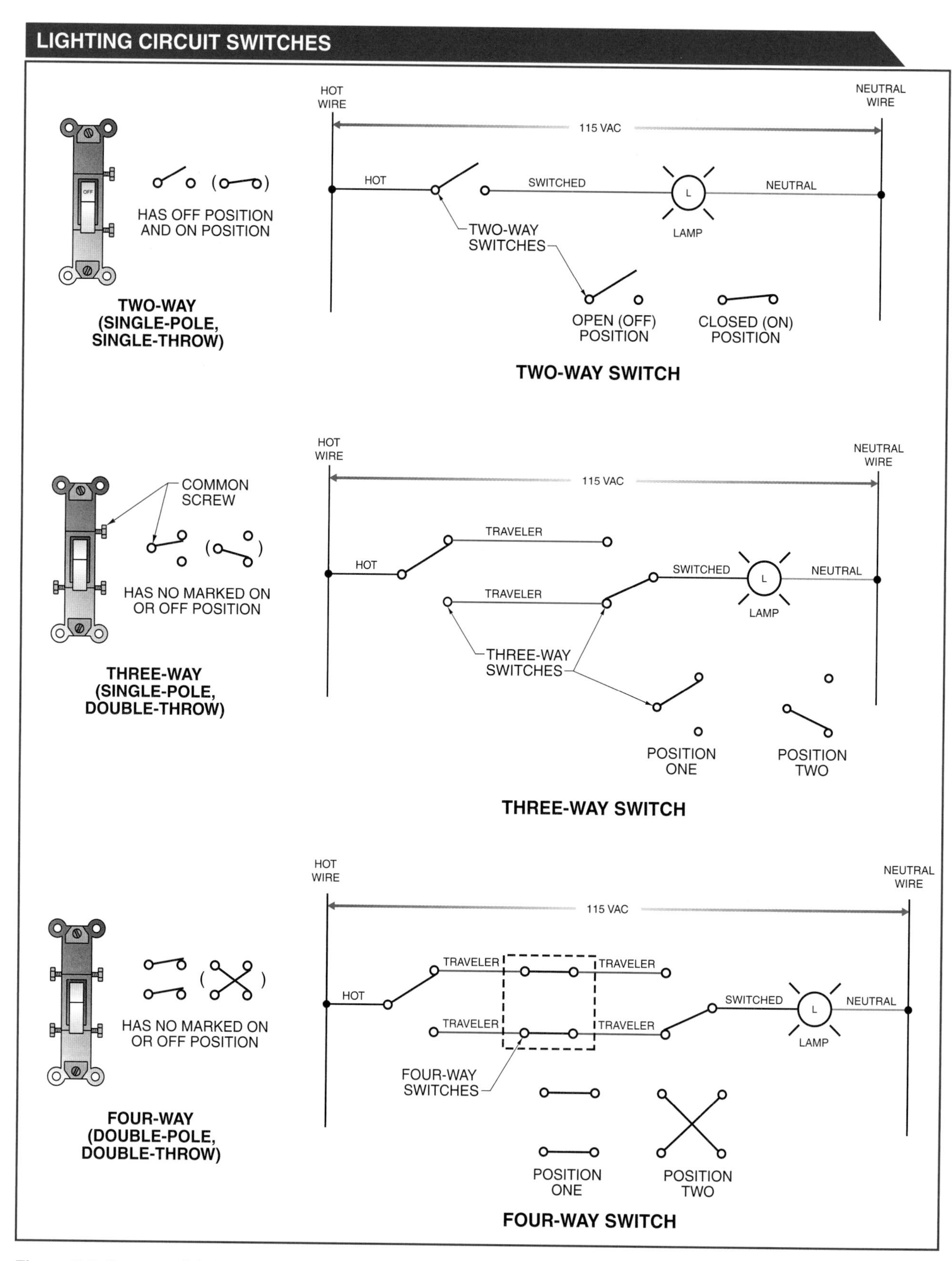

Figure 5-6. Common lighting circuit switches include two-way, three-way, and four-way switches.

A *three-way switch* is a single-pole, double-throw (SPDT) switch. A three-way switch has one common terminal and two traveler terminals. Three-way switches connect the common terminal to the top traveler terminal in one position and the common terminal to the bottom traveler terminal in the other position. Three-way switches do not have designated ON and OFF positions because the common terminal is always connected to one of the traveler terminals. Two three-way switches are used to control a lamp from two different locations.

A *four-way switch* is a double-pole, double-throw (DPDT) switch that changes the electrical connections inside the switch from straight to diagonal. Four-way switches do not have ON and OFF position markings on them. Four-way switches are used with three-way switches to control a lamp from three or more locations. When controlling a lamp from three locations, the four-way switch is placed between the two three-way switches.

Another type of switch is the rotary switch (wafer). A *rotary switch* is a switch that has one or more poles that can be connected to several positions. **See Figure 5-7.** A rotary switch is typically used to switch between several signal sources. For example, a listener can choose between listening to a radio station, an MP3 player, a compact disc, or a digital video disc (DVD) by rotating a selector switch on an audio receiver. More than one wafer controlled by a single shaft may be stacked to switch multiple circuits.

Terms

A **three-way switch** is a single-pole, double-throw (SPDT) switch.

A **four-way switch** is a double-pole, double-throw (DPDT) switch that changes the electrical connections inside the switch from straight to diagonal.

A **rotary switch** is a switch that has one or more poles that can be connected to several positions.

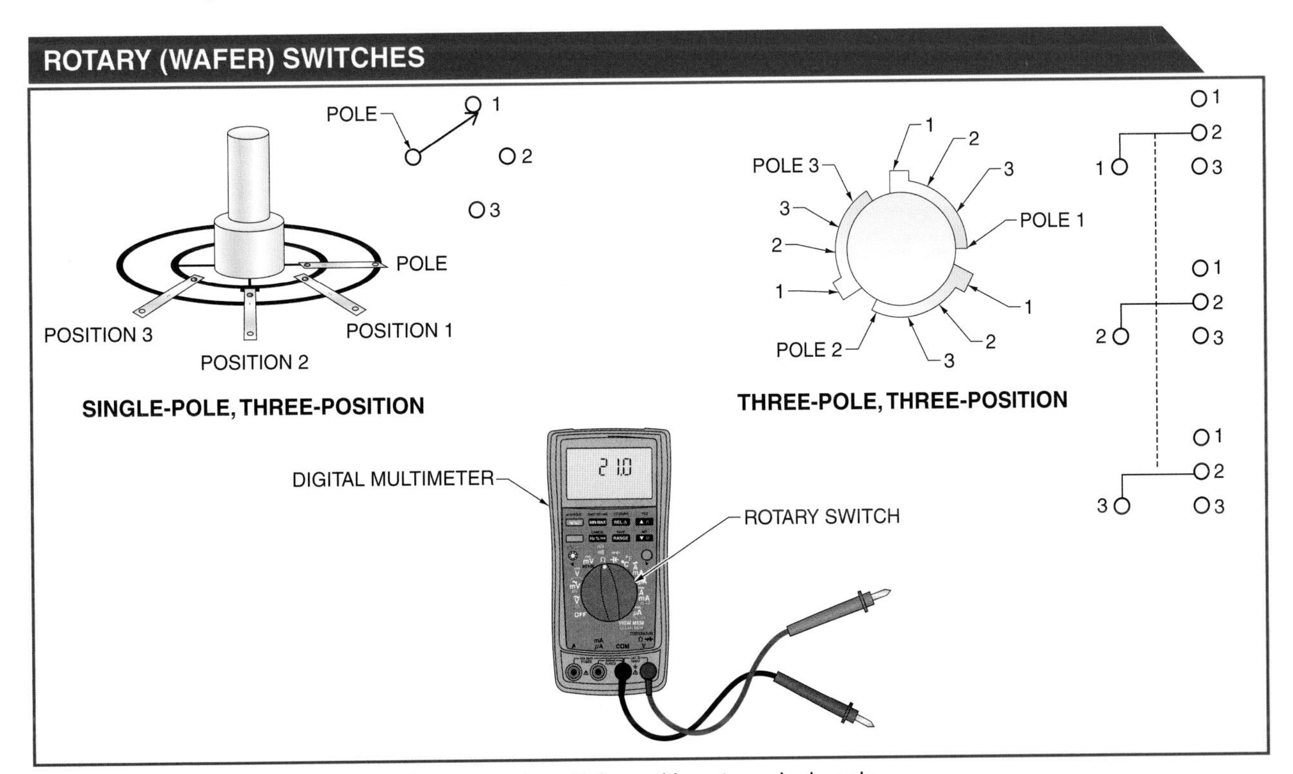

Figure 5-7. Rotary switches are used to connect multiple positions to a single pole.

Terms

A **load** is a device that converts electrical energy to motion, light, heat, or sound.

Loads

A *load* is a device that converts electrical energy to motion, light, heat, or sound. Common loads include motors (electrical energy to motion), lamps (electrical energy to light), heating elements (electrical energy to heat), and speakers (electrical energy to sound). **See Figure 5-8.**

LOADS

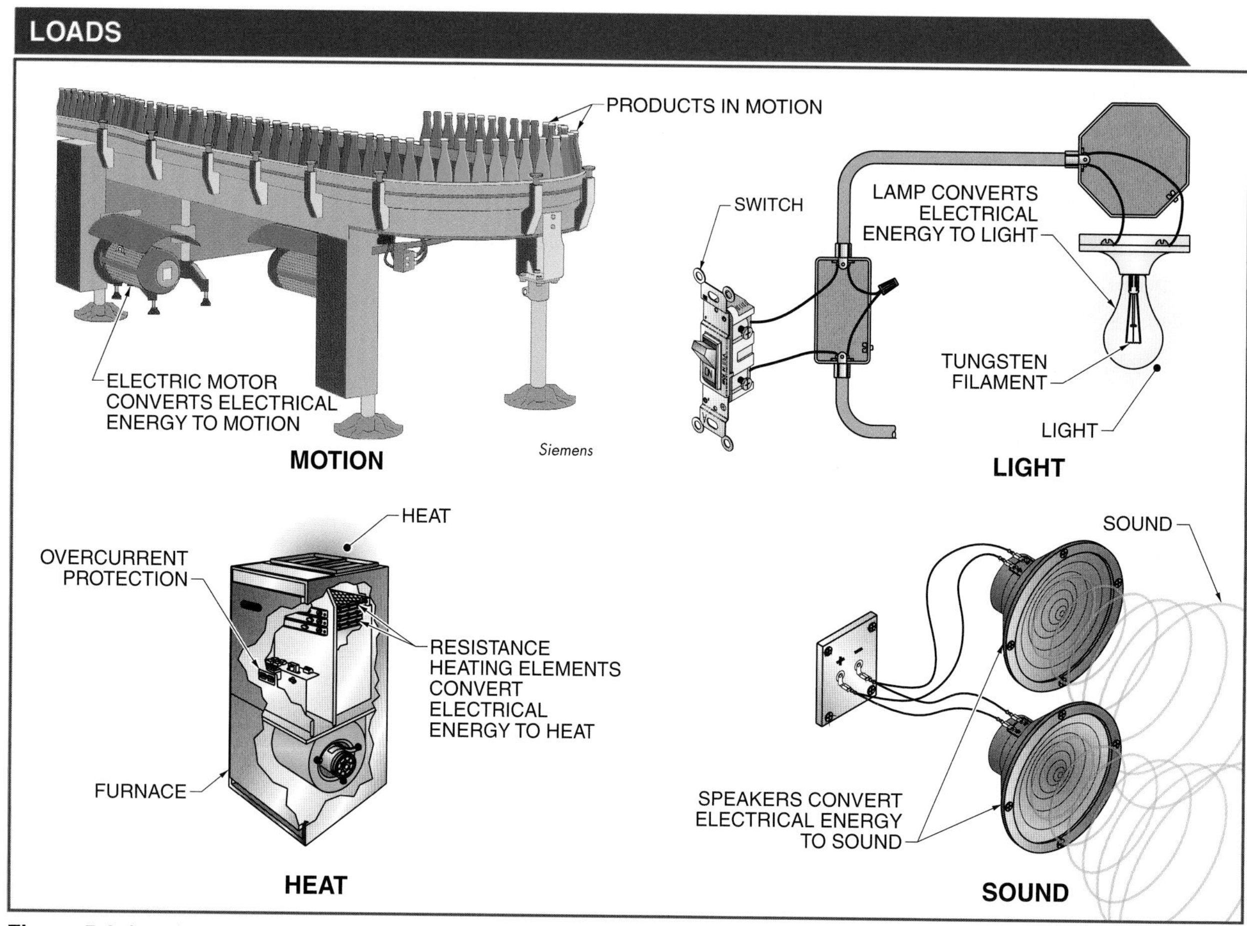

Figure 5-8. Loads convert electrical energy to another form of energy such as motion, light, heat, or sound.

An *electric motor* is a rotating device that converts electrical power into a rotating mechanical force. Motors are used in fans, compressors, cranes, conveyors, mixers, etc. In an electric motor, electrical energy is transformed into a rotating force that is used to perform work, such as moving products on a conveyor.

Terms

An **electric motor** is a rotating device that converts electrical power into a rotating mechanical force.

Incandescence is the emission of light energy from a substance when the substance is heated to a high temperature.

Incandescence is the emission of light energy from a substance when the substance is heated to a high temperature. An *incandescent lamp* is a device that produces light from the flow of current through a tungsten filament inside a sealed glass bulb. In an incandescent lamp, the tungsten filament is heated by electrical energy until it glows. The electrical energy is transformed to heat and then to light.

Light-emitting diodes and compact fluorescent lights are other devices that convert electrical energy into light. A *light-emitting diode (LED)* is a semiconductor device that emits light energy when current flows through it. A *compact fluorescent light (CFL)* is a fluorescent lamp that fits into an existing light bulb socket. LEDs and CFLs are lower-power and longer-life alternatives to incandescent light bulbs. CFLs are about 75% more energy efficient and last up to 10 times longer than incandescent light bulbs. LED light bulbs are about 85% more energy efficient and last up to 25 times longer than incandescent light bulbs.

An electric furnace produces heat from current flowing through resistance heating elements. A *resistance heating element* is a conductor that offers enough resistance to produce heat when connected to an electrical power supply. The lower the resistance is, the greater the heat produced. In a resistance heating element, electrical energy is transformed into heat that can be used for industrial processes or to heat building spaces.

Sound is energy that consists of pressure vibrations in the air. All sound is produced by a series of pressure vibrations originating from a vibrating object. A *speaker* is a device that converts electrical energy into vibrations (sound waves). A speaker consists of a coil of wire attached to a paper or polypropylene cone. A vibrating electric current passing through the coil creates a magnetic field that attracts or repels a large permanent magnet in the back of the speaker. This moves the cone back and forth producing sound waves.

Overcurrent

In a properly operating circuit, current flow is confined to the conductive paths provided by conductors and other components when the load is ON. Every load draws a normal amount of current when switched ON. The normal amount of current is the current level that the load, conductors, switches, and system is designed to safely handle. Under normal operating conditions, the current in a circuit should be equal to or less than the normal current level. However, at times an electrical circuit may have a higher-than-normal current flow (overcurrent). An *overcurrent* is a condition that exists in an electrical circuit when the normal load current is exceeded. An overcurrent condition exists during a short circuit or overload situation.

Short Circuits. A *short circuit* is any circuit in which current takes a shortcut around the normal path of current flow. A circuit may contain a partial short that causes an increased electron flow (overcurrent) or a dead short. A partial short may or may not cause damage depending on the ratings of the circuit components. A dead short may develop that completely removes the resistance of the load from the circuit. **See Figure 5-9.**

Terms

An **incandescent lamp** is a device that produces light from the flow of current through a tungsten filament inside a sealed glass bulb.

A **light-emitting diode** is a semiconductor device that emits light energy when current flows through it.

A **compact fluorescent light** is a fluorescent lamp that fits into an existing light bulb socket.

A **resistance heating element** is a conductor that offers enough resistance to produce heat when connected to an electrical power supply.

Sound is energy that consists of pressure vibrations in the air.

A **speaker** is a device that converts electrical energy into vibrations (sound waves).

An **overcurrent** is a condition that exists in an electrical circuit when the normal load current is exceeded.

A **short circuit** is any circuit in which current takes a shortcut around the normal path of current flow.

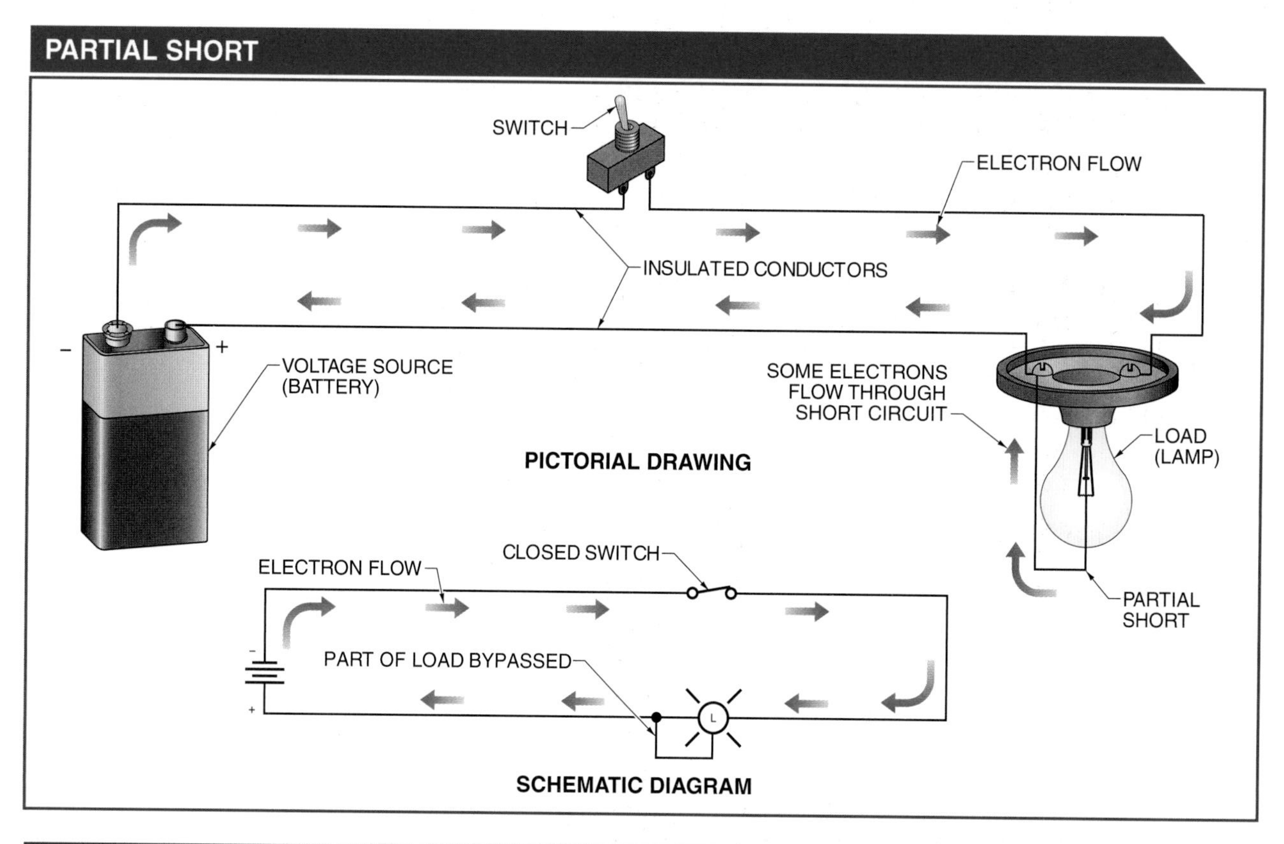

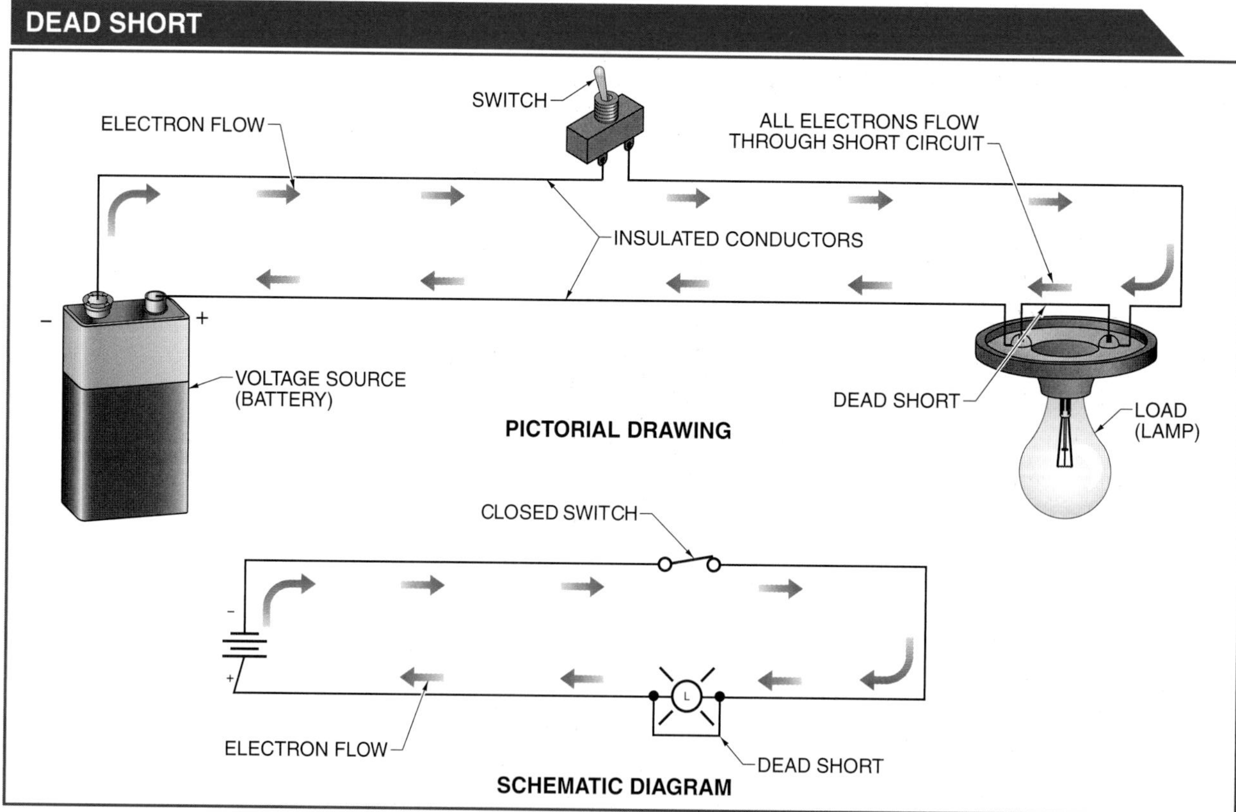

Figure 5-9. A short circuit has a resistance that is lower than the normal circuit resistance.

If the source has enough stored energy when a dead short occurs, circuit components will be damaged or explode. Switches can melt or vaporize. Conductors can over-heat and the insulation can burn off. The power source may also be damaged or destroyed when subjected to a dead short. Fires that result in loss of property and life can occur due to the temperatures generated by a partial short or a dead short. All circuits must be protected against a short circuit because short circuits are dangerous and damaging.

Overloads. An *overload* is a condition that occurs when circuit current rises above the current level at which the load and/or circuit is designed to operate. Overloads are caused by defective circuit components, overloaded equipment, or too many loads on one circuit. Overloaded equipment draws a higher than normal current based on the degree to which the equipment is overloaded. The more overloaded the equipment, the higher the current draw.

If an overload lasts for a brief time, the temperature rise is minimal and has little or no effect on the equipment or conductors. Sustained overloads, however, are destructive and must be prevented. A sustained overload results in the overheating of conductors, equipment, and any other components used in the circuit. Like short circuits, sustained overloads must be removed from the system. Unlike short circuits, overloads do not cause a sudden arc and do not have to be removed from the system immediately. If not removed within a short time period, however, overloads may cause a fire by overheating the equipment and conductors.

Overcurrent Protective Devices

An overcurrent protective device must be used to provide protection from short circuits and overloads to prevent the possible loss of property or life. An *overcurrent protective device (OCPD)* is a fuse or circuit breaker used to provide overcurrent protection in a circuit. Fuses and circuit breakers are OCPDs designed to automatically stop the flow of current in a circuit that has a short circuit or that is overloaded. **See Figure 5-10.**

An OCPD is placed in series with the voltage source and the load. This ensures that full circuit current flows though the OCPD. Overcurrent protective devices must be able to sense when a short circuit or excessive current exists, safely break (open) the circuit before any damage occurs, and have little effect when the circuit is operating normally. Overcurrent protective devices include fuses, circuit breakers, and overload relays.

Terms

An **overload** is a condition that occurs when circuit current rises above the current level at which the load and/or circuit is designed to operate.

An **overcurrent protective device (OCPD)** is a fuse or circuit breaker used to provide overcurrent protection in a circuit.

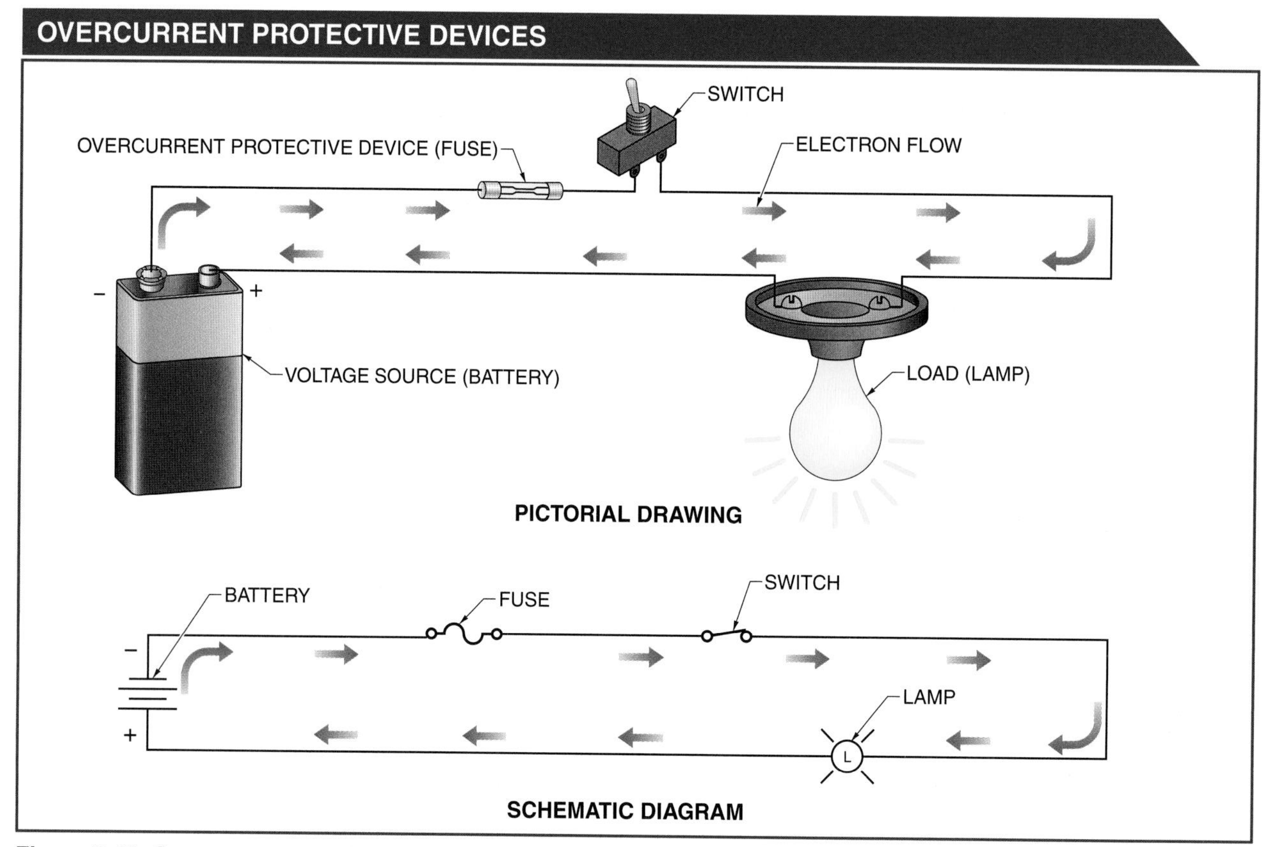

Figure 5-10. Overcurrent protective devices such as fuses are used to protect a circuit from a short circuit or overcurrent that can cause circuit damage.

Terms

A **fuse** is an overcurrent protective device with a fusible link that melts and opens a circuit when an overload condition or short occurs.

Fuses. A *fuse* is an overcurrent protective device with a fusible link that melts and opens a circuit when an overload condition or short occurs. The fusible link melts because the fuse is a metal that has a lower melting point than the copper of the conductor. For example, if copper and lead wire are connected in series, and a variable source of electrical energy is slowly adjusted to provide increments of higher current, both metals begin to heat due to the electron flow through them. Because lead has a higher resistance than copper, it gets hotter and melts before the copper melts. If the lead is alloyed with tin, it melts at an even lower temperature. By adjusting the physical dimensions of the alloy, it can be made to melt at specific current levels. This is the basis of fuse design. Therefore, a proper size fuse connected in series with a load opens before the current level can rise to a value high enough to harm circuit conductors or components.

Fuses have voltage, amperage, and interrupting ratings. The voltage rating determines the ability of a fuse to suppress any arc after the fuse opens. If a fuse with a voltage rating lower than circuit voltage is used, it is possible that the arc produced when the fuse opens will not be suppressed and the fuse will not quickly clear the fault. Most low-voltage systems contain fuses rated between 125 V and 600 V. When selecting a fuse, its voltage rating must be at least as great as the voltage source in the circuit it is protecting.

All fuses have an amperage rating. Normally, amperage ratings should not exceed the current-carrying capacity of the conductors in the circuit. For example, if the conductors are rated at 15 A, then the fuse size should not exceed 15 A. There are exceptions to this rule, however. The National Electrical Code® specifies fuse sizes for different types of loads. For example, 14 AWG wire is rated for 20 A, but its overcurrent protection shall not exceed 15 A. Fuse amperage ratings range from a fraction of an ampere to 6000 A.

Terms

The **interrupting rating** is the maximum amount of current that a fuse can safely interrupt without rupturing or arcing over.

Fuses must be able to clear faults under short-circuit conditions. Highly destructive energy can be present when a short circuit occurs. Short-circuit currents are only limited by the internal resistance of the voltage source and amount of resistance in the conductors up to the fault. This current, in the vast majority of cases, far exceeds the normal ampere rating of the fuse protecting the circuit. If the energy is high enough, the protection device may rupture in flames and vaporized metal. This vaporizing action can set up a conductive path around the fuse, which fails to clear the fault, and allows additional damage to occur. Because of this, it is important to consider the interrupting rating of a fuse. The *interrupting rating* is the maximum amount of current that a fuse can safely interrupt without rupturing or arcing over. Most modern current limiting fuses have interrupting ratings of 200,000 A or 300,000 A.

Fuse designs include cartridge and plug fuses. Fuses are available in various sizes and shapes. **See Figure 5-11.** Glass cartridge fuses may have a single wire or flat conductor as the fuse element. The flat conductor has less conducting area in the middle where the fuse element melts. A glass delayed action cartridge fuse opens only when the current is greater than its rating for a predetermined amount of time. Delayed action cartridge fuses are used for loads that have an initial current surge when power is applied. Some glass fuses have pigtails attached to them so that they can be soldered into a circuit. Glass fuses allow the observation of the condition of the fuse element without using an ohmmeter.

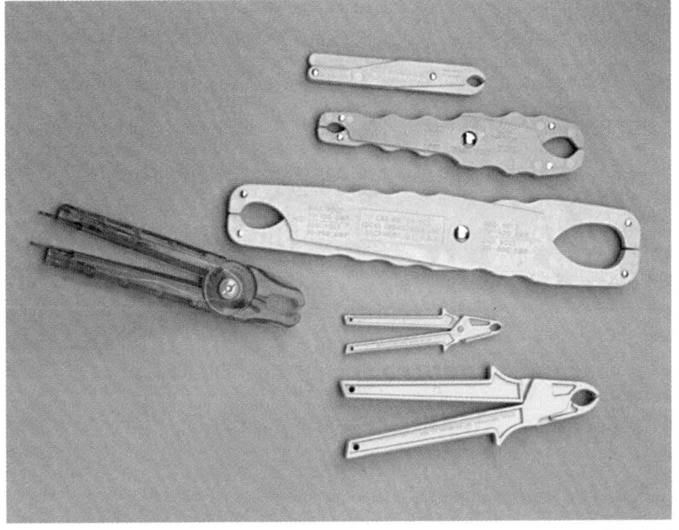

Gould Inc.

Fuse pullers are available in a variety of styles and sizes and are used to prevent slippage and electrical shock when removing cartridge-type fuses.

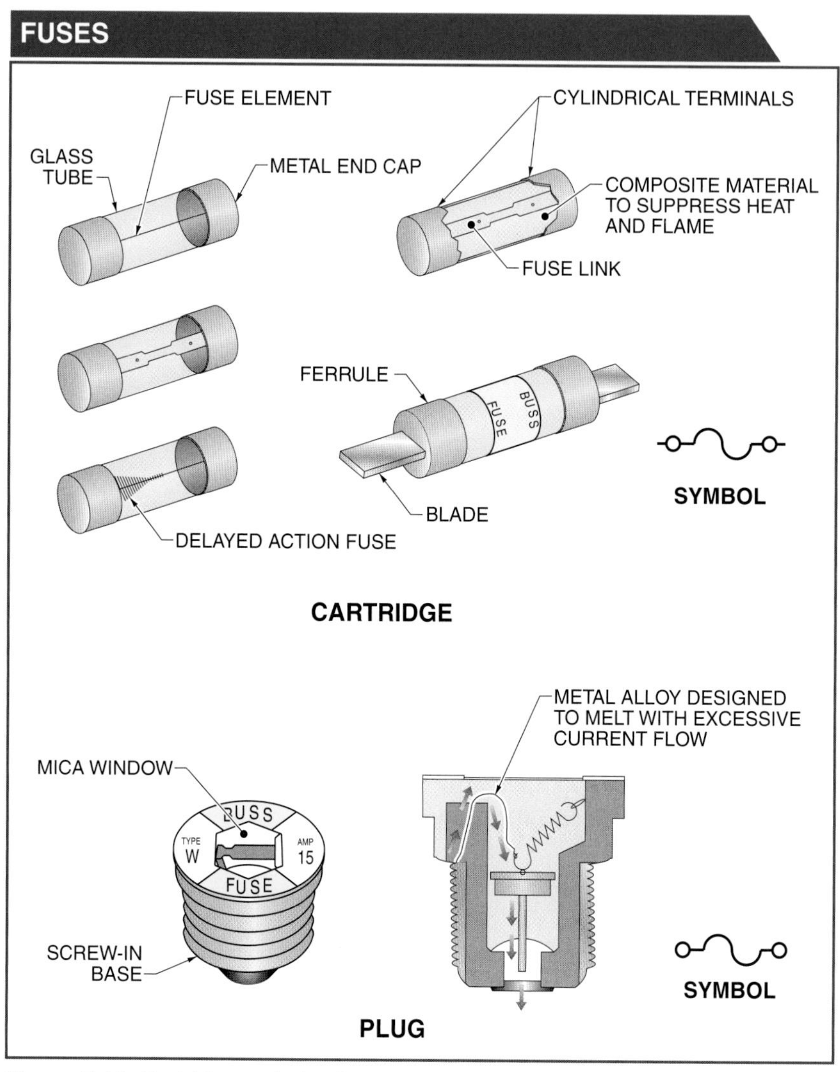

Figure 5-11. Cartridge and plug fuses may be surrounded with glass or encased in a composite material to suppress an arc or flame.

Terms

A **plug fuse** is a fuse that has a screw-in base like that of an incandescent lamp.

Some large cartridge fuse designs are completely enclosed and vary in size so that fuses with different values cannot be interchanged in a fuseholder. Large cartridge fuses may have cylindrical terminals or blade terminals. The cutaway of the fuse shows that a composite material surrounds the fuse link to suppress an arc or flame.

A *plug fuse* is a fuse that has a screw-in base like that of an incandescent lamp. Plug fuses and cartridge fuses were the primary protective devices used in most circuits before the development of the circuit breaker. Plug fuses normally have a mica window through which the condition of the fuse element can be observed. Some plug fuses have different size bases. The different size bases allow the fuse to be installed only in the type of circuit for which it is designed.

Circuit Breakers. A *circuit breaker* is an overcurrent protective device with a mechanical mechanism that manually or automatically opens a circuit when a short circuit or overload occurs. **See Figure 5-12.** Like fuses, circuit breakers are connected in series with circuit conductors. A circuit breaker opens and prevents current from flowing in a circuit when the current exceeds the rating of the circuit breaker. Circuit breakers contain a spring loaded electrical contact that opens the circuit. The spring opens and closes the contacts with a fast snap action. A circuit breaker does not have to be replaced each time the current rating is exceeded. Circuit breakers have replaced fuses in many applications. Circuit breakers have voltage, amperage, and interrupting ratings similar to fuses.

Terms

A **circuit breaker** is an overcurrent protective device with a mechanical mechanism that manually or automatically opens a circuit when a short circuit or overload occurs.

CIRCUIT BREAKERS

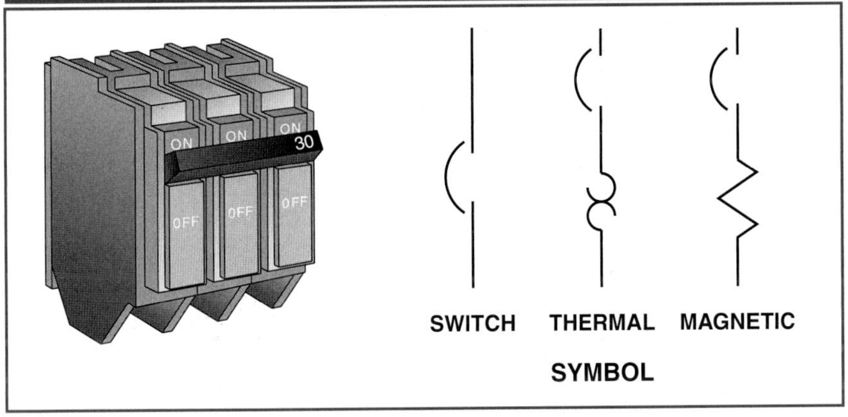

Figure 5-12. A circuit breaker is an overcurrent protective device that does not need to be replaced each time the circuit current rating is exceeded. Circuit breakers may be thermally or magnetically operated.

In addition to protecting a circuit from a short circuit or overload, a circuit breaker can be used to remove power from a specific circuit. An ON/OFF switch on the circuit breaker allows an individual to remove power from any circuit without removing the protective device. Excessive current causes the circuit breaker to trip to the OFF position and open the circuit. The circuit breaker is reset after the fault is cleared.

Circuit breakers may use thermal or magnetic methods to open a circuit when a short circuit or overload condition occurs. Thermal circuit breakers use a bimetallic strip attached to a latch mechanism. The bimetallic strip is made of two dissimilar metals that expand at different rates when heated. The bimetallic strip bends when heated and opens the contacts. **See Figure 5-13.** The bimetallic strip may be heated directly by circuit current or indirectly by the rise in temperature caused by an increase in circuit current.

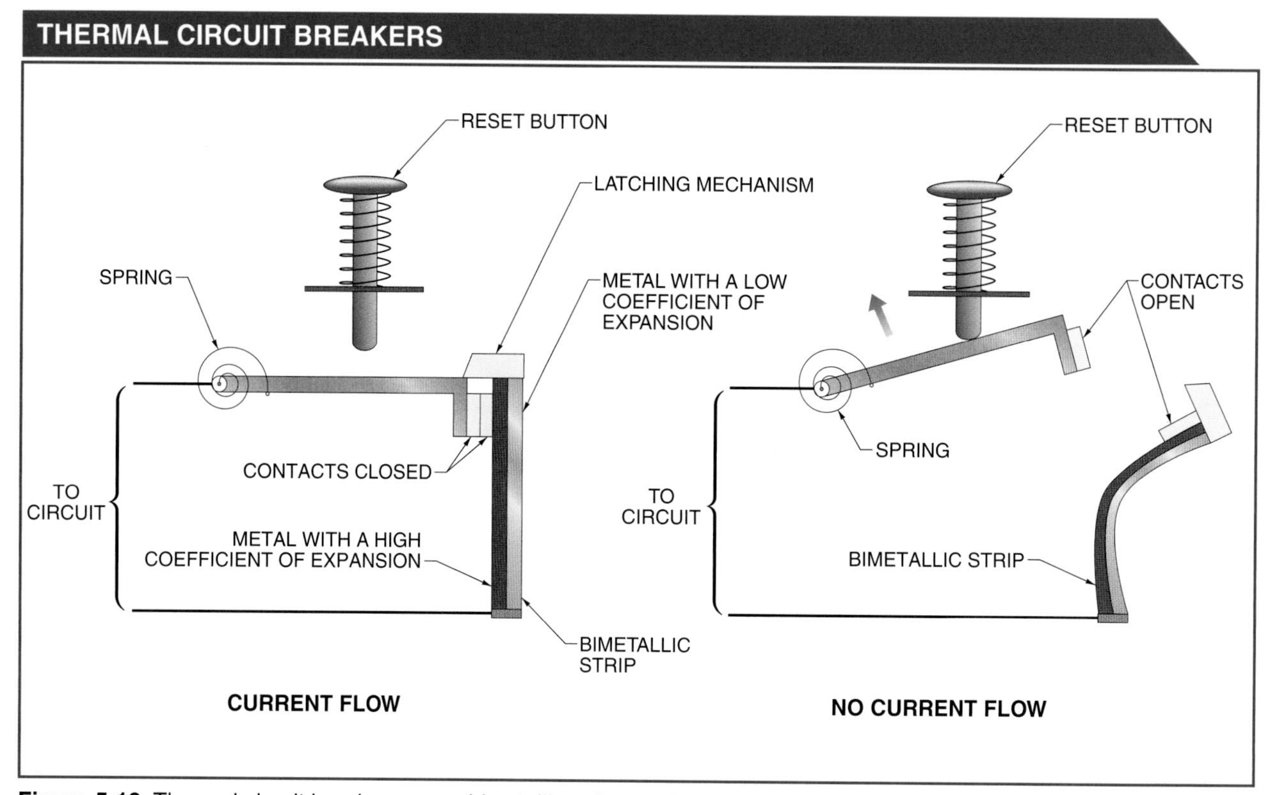

Figure 5-13. Thermal circuit breakers use a bimetallic strip attached to a latch mechanism to open the circuit when a short circuit or overload occurs.

Thermal circuit breakers are designed so that the bimetallic strip bends to release the contact under spring tension based on the amount of continuous current flow through it. The bimetallic strip must cool and return to its normal condition (size) at room temperature before the circuit breaker can be reset. Thermal protection of a circuit is not instantaneous. It requires time to heat the strip and for the strip to bend far enough to cause the contacts to snap open. A magnetic circuit breaker is used in applications where this delay can cause damage to a circuit. Thermal circuit breakers can be reset by pressing the pushbutton only after the bimetallic strip has cooled.

A magnetic circuit breaker uses an electromagnet coil and armature. In a magnetic circuit breaker, circuit current passes through the coil, producing a magnetic field. **See Figure 5-14.** Normal circuit current does not affect the armature. However, when circuit current creates a magnetic force that exceeds the spring force on the armature and the friction of the latching mechanism, the armature is pulled to the electromagnet coil. The spring-loaded contact arm is released, breaking the current flow to the load instantaneously. Magnetic circuit breakers can be manually reset immediately.

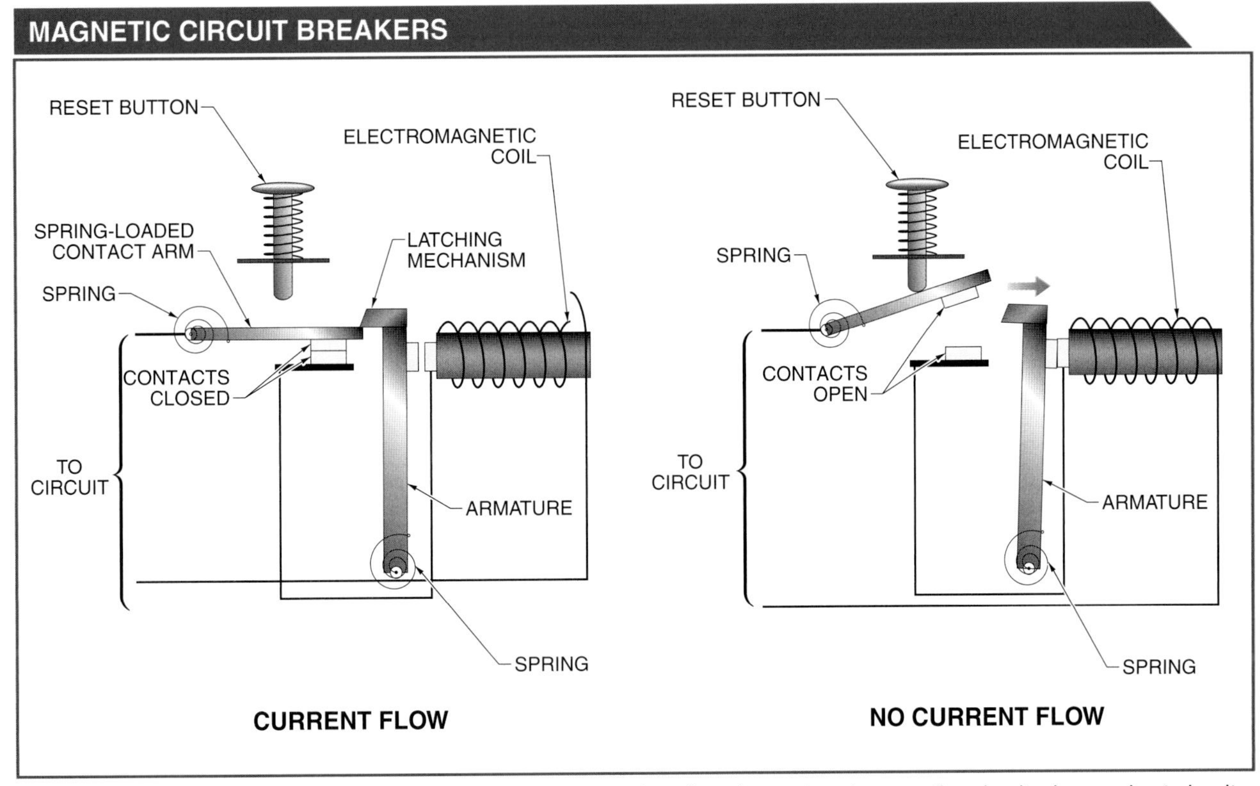

Figure 5-14. Magnetic circuit breakers use an electromagnetic coil and armature to open the circuit when a short circuit or overload occurs.

Most magnetic circuit breakers contain thermal and magnetic components. The thermal components protect the circuit against a constant overload current that is not of sufficient level to activate the magnetic components. This elevated current, if continuously present, can overheat components in the circuit over time. The magnetic components protect the circuit against high overload current or short-circuit current.

Overload Relays. An *overload relay* is an overload device that responds to electrical loads and operates at a preset value. Overload devices are required in motor circuits by the National Electrical Code®. Overload devices guard against continuous above-normal current levels that are not sufficient to blow a fuse or trip a circuit breaker but can cause damage if not corrected.

As with a circuit breaker, overload relays may use thermal or magnetic methods to open a circuit. These methods may be used separately or in combination. Overload relays differ from fuses and circuit breakers in that they are designed to operate indirectly in response to a relatively small amount of current. Circuit (line) current supplying the load does not flow through the sensing device of an overload relay.

Terms

An **overload relay** is an overload device that responds to electrical loads and operates at a preset value.

Thermal overload relays operate by opening a bimetallic strip upon a rise in temperature. Most thermal overload relays can be reset using a pushbutton. Automatic-reset thermal overload relays do not have a spring action that requires manual resetting of the contacts. The contacts open when the current level is exceeded for a given period of time. When the bimetallic strip cools, the contacts snap closed, and the circuit is automatically energized. Automatic-reset thermal overload relays are often used on motors but can be a safety hazard. To protect personnel and property, automatic-reset thermal overload relays should only be used on circuits that must be kept running. **See Figure 5-15.**

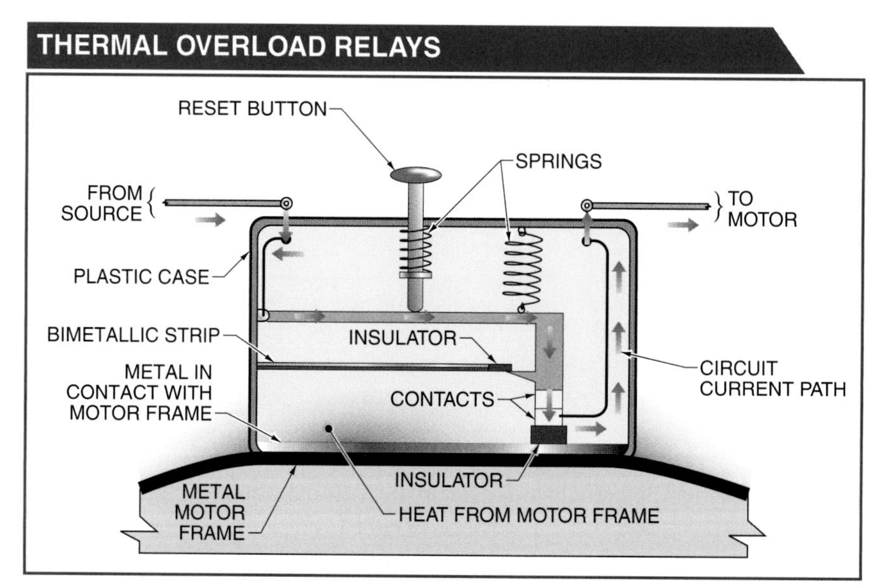

Figure 5-15. Thermal overload relay contacts open when the current level is exceeded for a given period of time. The temperature rise in the metal frame of the motor is used to heat the bimetallic strip.

Thermal overload relays are commonly fitted to the frame of a motor. The circuit (line) current flows through the closed contacts of the thermal overload relay. If the motor is overloaded, it draws a greater amount of current. This causes the temperature of the motor frame to increase. At some preselected temperature before damage is likely to occur, the bimetallic strip bends, releasing the latch on the upper contact mechanism. The insulated spring pulls the contacts apart, removing the current flow to the motor. Simultaneously, the spring pushes the reset button up. Once the bimetallic strip cools, the device can be returned to normal operation by pressing the reset button.

Overload relays are often designed as part of a contactor circuit. A *contactor* is a control device that uses a small input current to energize or de-energize the load connected to it. Contactors use a small input current to activate a relay that opens and closes the contacts in the load circuit. Switching a low input current that activates a relay is safer for an operator than switching a high load current. A single input current to the contactor can also be used to switch multiple output circuits. Many sets of contacts can be switched by a single armature.

An overload relay may be incorporated into a standard start/stop motor control circuit. In a standard motor control circuit, a motor starter or contactor coil controls a set of normally open contacts, and a set of normally closed overload relay contacts. **See Figure 5-16.**

The motor is started by pressing the start pushbutton. When the start pushbutton contacts close, electrons flow from the negative terminal of the source through the normally closed overload relay contacts, the relay coil, the start pushbutton contacts, the normally closed stop pushbutton contacts, and back to the positive terminal of the source. This activates the relay coil, which closes the normally open contacts in series with the source and motor and also closes the normally open holding contacts. The closing of the normally open contacts in series with the source and motor completes the circuit to the motor, allowing current to flow from the negative terminal of the source, through the motor windings and closed normally open contacts, and back to the positive terminal of the source. When the start pushbutton is released, the motor continues to run because the closed normally open holding contacts allow current to bypass the pushbutton and keep the circuit closed.

Terms

A **contactor** is a control device that uses a small input current to energize or de-energize the load connected to it.

MOTOR CONTROL CIRCUIT

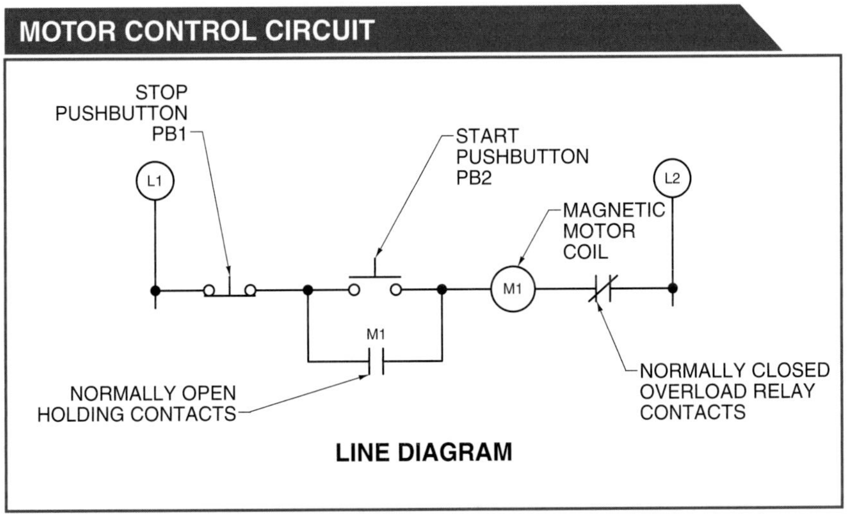

Figure 5-16. In a standard motor control circuit, a relay coil controls a set of normally open contacts and a set of normally closed overload relay contacts.

To stop the motor, the normally closed stop pushbutton is pressed. This opens the circuit to the relay coil, causing it to de-energize and return its contacts to their normally open condition. This opens the control circuit and simultaneously removes power from the motor.

The contacts of the overload relay are located in the control circuit and not in the motor power circuit. Because the control circuit requires only a low amount of current to operate, the contacts of the overload relay do not have to handle the high amount of current required to run the motor.

If the motor overheats while running, the normally closed overload relay contacts open, removing power from the control circuit. This opens contacts in the power circuit. When the power circuit opens, power is removed from the motor. The reset button on the overload relay must be pressed before the motor can be started again by pressing the start pushbutton.

Tech Tip

Always perform a voltage or current test on a known energy source prior to taking a test measurement on a piece of equipment.

VOLTAGE AND CURRENT MEASUREMENTS

Electrical measurements are commonly taken using a multimeter. Multimeters are test tools used to measure two or more electrical properties. Multimeters are used during electrical equipment installation and servicing. Multimeters commonly measure voltage, current, and resistance and can be either analog or digital devices. Digital multimeters (DMMs) are normally more accurate and commonly used than analog multimeters.

Precautions must be taken any time a multimeter is used to measure electrical properties. Multimeters should not be subjected to rough physical handling or electrical extremes. Instruction manuals should always be reviewed before any test instrument is used. The meter used for an application must meet or exceed accepted safety standards. Test leads must be in good condition and be able to withstand the voltage or current being measured.

DC Voltage Measurements

DC voltage is measured with a DMM using the following standard procedure. Always exercise caution when taking any circuit measurement. **See Figure 5-17.**

1. Set the function switch to DC voltage. If the DMM includes more than one DC setting, select the highest setting. This is particularly important when circuit voltage is unknown, although most modern DMMs are protected against an overvoltage condition.
2. Plug the black test lead into the common jack.
3. Plug the red test lead into the voltage jack.

4. Connect the test leads to the circuit. Voltage measurements are taken with the meter leads across the component (in parallel) under test. For correct polarity, connect the black test lead to the negative polarity test point and the red test lead to the positive polarity test point. If power is applied to a circuit and one side is grounded, the first test connection from the DMM should be made to the grounded side. The ground connection should be removed last when removing test leads from a circuit that has power applied to it.
5. Close the switch to apply power to the circuit.
6. Read the displayed voltage measurement. The displayed value must be less than the range selected. If the displayed value is within the range of a lower VDC setting, a more accurate measurement can be obtained by changing the setting. The meter must be disconnected from the circuit before changing the range setting. Most DMMs indicate an OL (overload) if the meter range setting is below the value being measured. An *autoranging DMM* is a DMM that automatically adjusts to a higher range setting if the range is not high enough. Accurate readings are best obtained by using the range setting that provides the best resolution without overloading the meter.
7. Turn circuit power OFF and remove the meter from the circuit. Turn the meter OFF to save the battery.

Terms

An **autoranging DMM** is a DMM that automatically adjusts to a higher range setting if the range is not high enough.

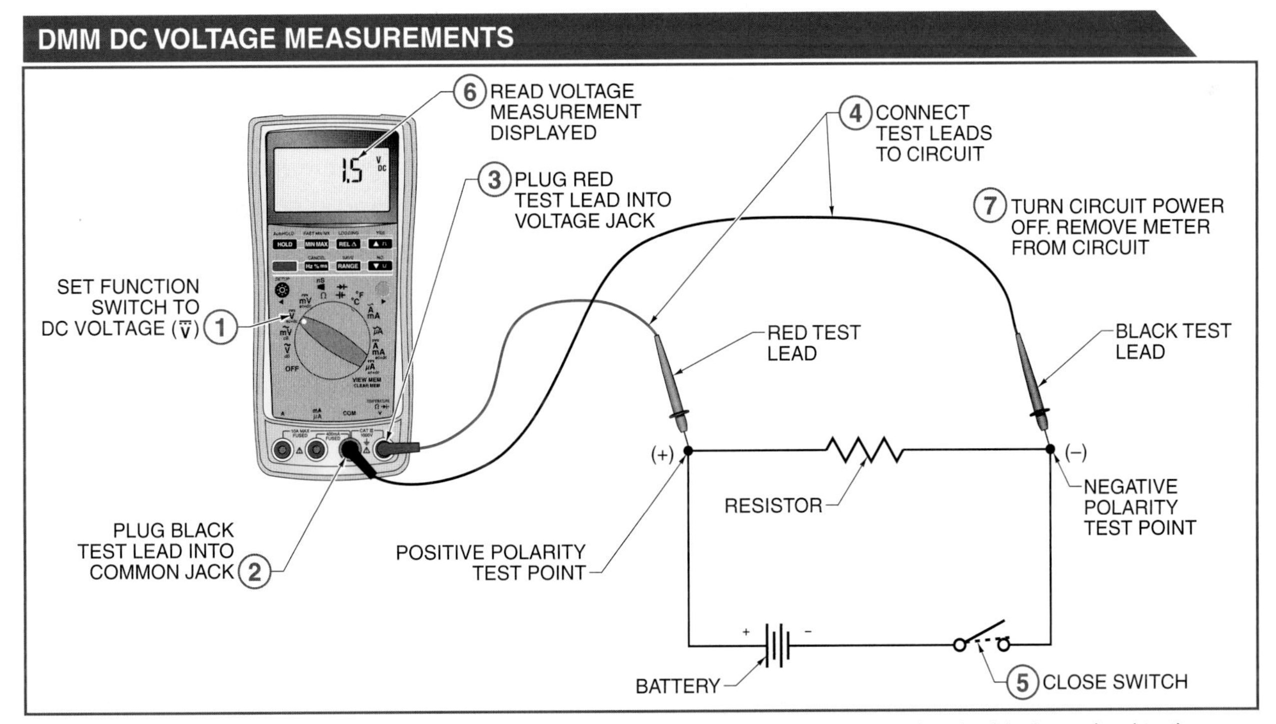

Figure 5-17. DC voltage measurements using a digital multimeter are taken by connecting the black test lead to the negative polarity test point and the red test lead to the positive polarity test point.

Voltage measurements can be made with an analog meter in a similar manner as with a digital multimeter. DC voltage is measured with an analog meter using the following standard procedure. Always refer to the instruction manual before using any meter. **See Figure 5-18.**

1. Set the DC polarity switch to the DC+ position.
2. Set the function/range switch to its highest voltage setting (DCV, 1000 V). If the circuit voltage is known, set the function/range switch to the known range.

CAUTION: The pointer of an analog meter can be driven off the scale and damaged if the circuit voltage is higher than the maximum value selected on the range switch.

3. Plug the black test lead into the common jack.
4. Plug the red test lead into the voltage jack.
5. Connect the test leads across the component. For correct polarity, connect the black test lead to the negative polarity test point and the red test lead to the positive polarity test point. The pointer of an analog meter can be driven off the scale with sufficient force to damage the instrument if the meter is connected in reverse polarity. Begin all measurements using the highest range setting to reduce the force and minimize any possible damage.
6. Close the switch to apply power to the circuit.
7. Read voltage measurement displayed on the AC DC scale. The 0–10 scale is used for the 1000 V, 100 V, 10 V, and 1 V ranges. The 0–30 scale is used for the 300 V, 30 V, and 3 V ranges.
8. Turn the power OFF and remove the meter from the circuit. Set the function/range switch to the OFF position.

Multimeters that are set to measure voltage present high opposition to current flow. For this reason, they draw a limited amount of current. Most digital multimeters have an input opposition to current flow of 10 MΩ for all voltage ranges except the 200 mV range, which has an opposition to current flow of 40 MΩ. Most analog multimeters have an opposition to current flow of 20 kΩ per volt on the DCV ranges and 5 kΩ per volt on the ACV ranges. This translates into 20 MΩ on the 1000 VDC range and 20 kΩ on the 1 VDC range.

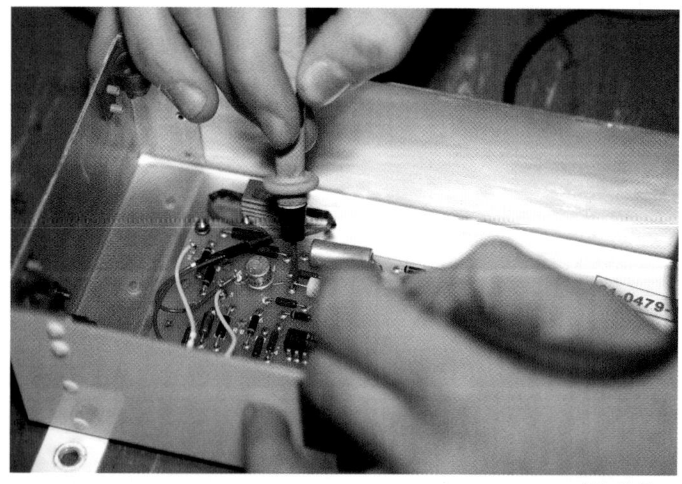

U.S. Air Force

In addition to testing electrical circuits, digital multimeters (DMMs) can also be used to test sensitive electronics equipment.

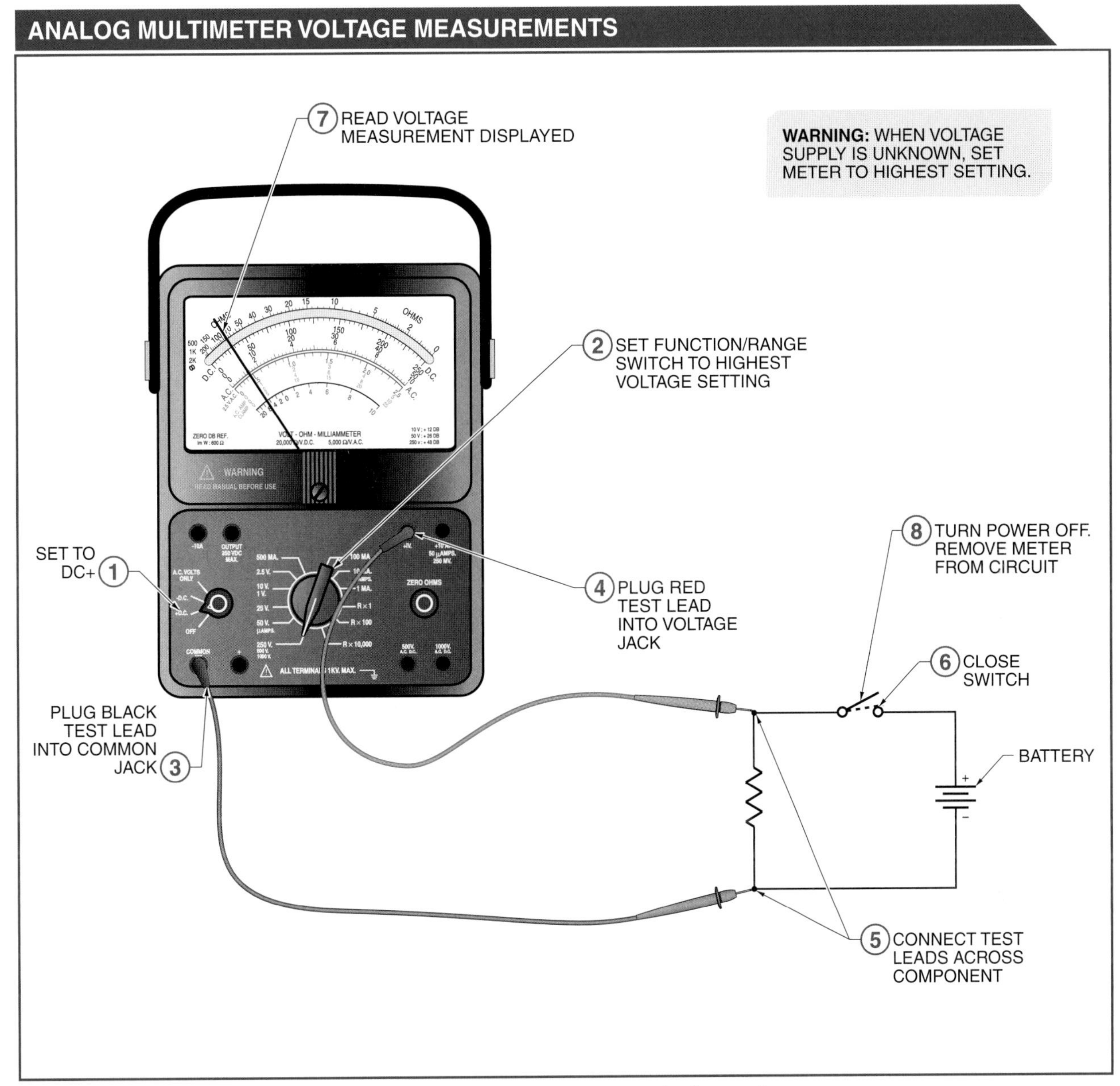

Figure 5-18. DC voltage is measured with an analog meter using standard procedures.

In some applications meter loading can occur. *Meter loading* is inaccurate readings due to the meter's resistance being in parallel with the component under test. For example, when reading a voltage of less than 1 V across a 1 MΩ resistor, a meter resistance of only 20 kΩ is placed across the resistor. The meter resistance can effectively cause the overall circuit resistance to change. This in turn causes the voltage drop to change to an inaccurate reading. To overcome this problem, a DMM can be used that has a high input resistance.

Terms

Meter loading is inaccurate readings due to the meter's resistance being in parallel with the component under test.

DC Current Measurements

To measure current flow through a component, a meter must be connected so that the total electron flow is through the meter circuit. This means that the meter must be connected in series with the component so that only one path for electron flow exists. Meters set to measure current must have a low resistance that does not substantially change the value of the current in the circuit. Direct current is measured with a DMM using the following standard procedure. Always turn the power to a circuit OFF before taking any measurements. **See Figure 5-19.**

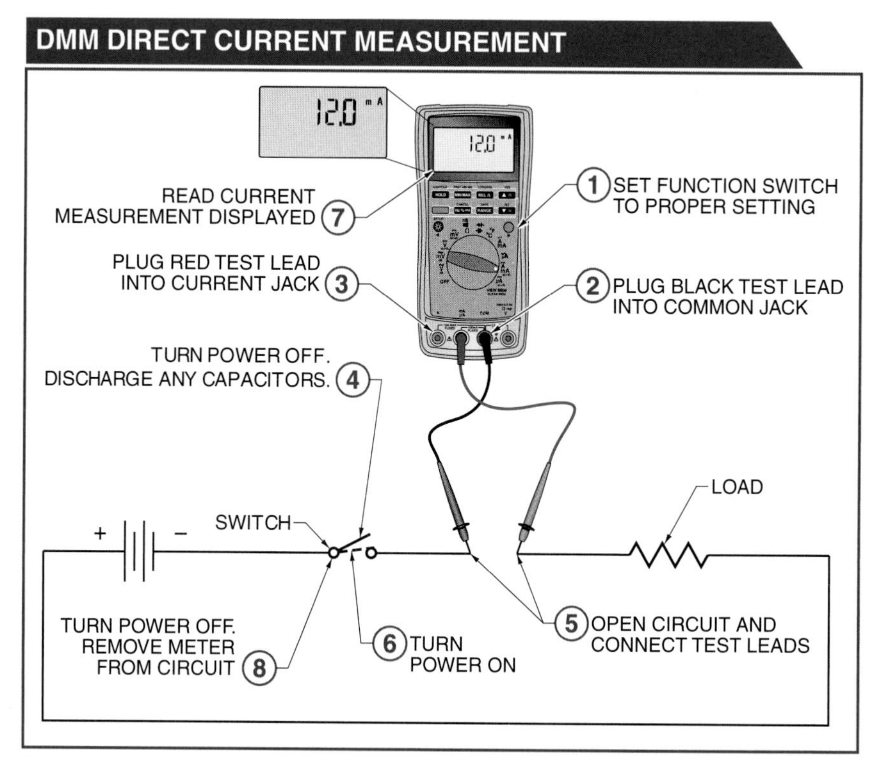

Figure 5-19. To measure current flow through a component, a meter must be connected so that the total electron flow is through the meter circuit.

1. Set the function switch to DC current position. The highest range setting must be used in circuits in which the current value is unknown. Most DMMs show an OL for an overload condition.
2. Plug the black test lead into the common jack.
3. Plug the red test lead into the current jack. The current jack may be marked A, or mA/mA.
4. Turn power to the circuit or device under test OFF and discharge all capacitors if possible.

5. Open the circuit so that the meter can be connected in series with the load. Attach the black (negative) test lead to the negative side of the opening. Connect the red (positive) test lead to the positive side of the opening. Reverse the black and red test leads if a negative sign appears to the left of the displayed value.
6. Turn the power ON to the circuit under test.
7. Read the current measurement displayed. If the value is below 1 A, a lower range can be used. To switch to a lower range, power must first be removed from the circuit. The red test lead is moved from the A jack to the mA jack and the function/range switch is set to the range indicated by the first reading. Power is returned to the circuit and the current measurement is taken.
8. Turn ciruit power OFF and remove the meter from the circuit. Turn meter OFF to save the battery.

Current measurements can also be taken using an analog multimeter. Direct current is measured with an analog multimeter using the following standard procedure. Always turn the power to the circuit OFF before taking any measurements. **See Figure 5-20.**

1. Set the polarity switch to the DC+ position.
2. Set the function/range switch to its highest current range. This reduces the chance that the pointer of the analog meter will be driven off the scale with sufficient force to damage the instrument.
3. Plug the black test lead into the common jack.
4. Plug the red test lead into the current jack.
5. Turn the power OFF to the circuit or device under test and discharge all capacitors if possible.
6. Open the circuit and connect the test leads to each side of the opening. The black test lead is connected to the negative polarity test point, and the red test lead is connected to the positive polarity test point. If the meter is connected in reverse polarity, the pointer of an analog meter can be driven off the scale with sufficient force to damage the instrument.
7. Turn the power ON to the circuit under test.
8. Read the current measurement displayed on the 1–10 AC-DC scale. The value on the scale is multiplied by 100 when the range is set to 1000 mA. The function/range switch may be changed to a lower range if the reading is less than 100 mA.
9. Turn circuit power OFF and remove the meter from the circuit. Set the function/range switch to the OFF position.

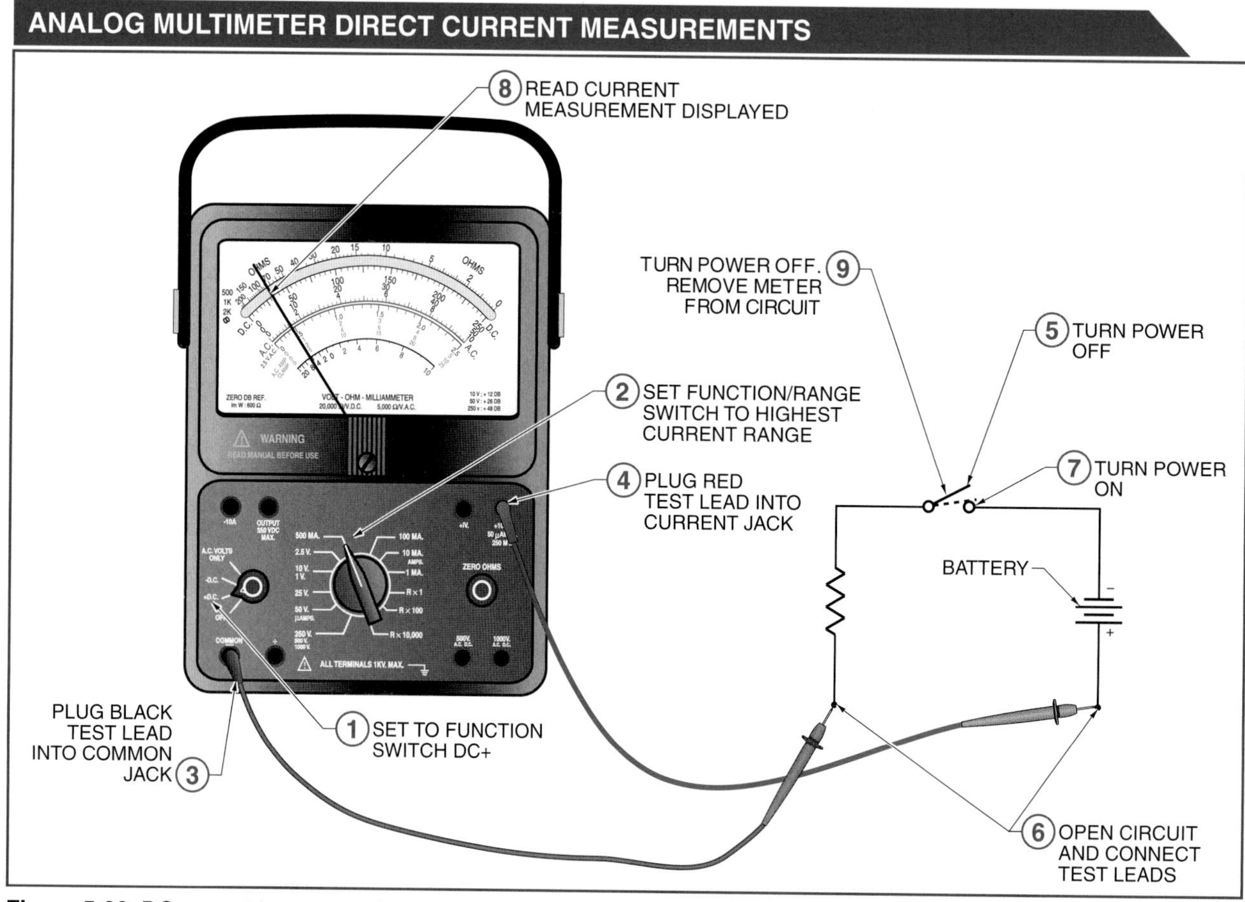

Figure 5-20. DC current is measured with an analog multimeter using standard procedures.

Clamp-On Ammeter Current Measurements. Clamp-on ammeters measure current in a circuit by measuring the strength of the magnetic field around a single conductor. Care should be taken to ensure that the meter does not pick up stray magnetic fields. Whenever possible, conductors under test should be separated from other surrounding conductors by a few inches. If this is not possible, several readings at different locations along the same conductor should be taken. Direct current is measured with a clamp-on ammeter or a DMM with a clamp on current probe accessory using the following standard procedure. **See Figure 5-21.**

1. Determine if the ammeter range is high enough to measure the maximum current that exists in the test circuit. If the ammeter range is not high enough, an accessory that has a high enough current rating or an ammeter with a higher range must be used.
2. Set the function switch to the proper current setting (600 A, 200 A, 10 A, 400 mA, etc.). If there is more than one current position or if the circuit current is unknown, the setting greater than the highest possible circuit current should be selected.

3. When required, plug the clamp-on current probe accessory into the multimeter. The black test lead of the clamp-on current probe accessory is plugged into the common jack. The red test lead is plugged into the mA jack for current measurement accessories that produce a current output. The red test lead is plugged into the voltage (V) jack for current measurement accessories that produce a voltage output. The current measurement accessories that produce current output are designed to measure AC only and generally deliver 1 mA to the meter for every 1 A of measured current (1 mA/A). Current accessories that produce a voltage output are designed to measure AC or DC current and deliver 1 mV to the meter for every 1 A of measured current (1 mV/A).
4. Open the jaws by pressing against the trigger.
5. Enclose one conductor in the jaws and ensure that the jaws are completely closed before taking readings.
6. Read the current measurement displayed.
7. Remove the clamp-on ammeter or clamp-on accessory from the circuit.

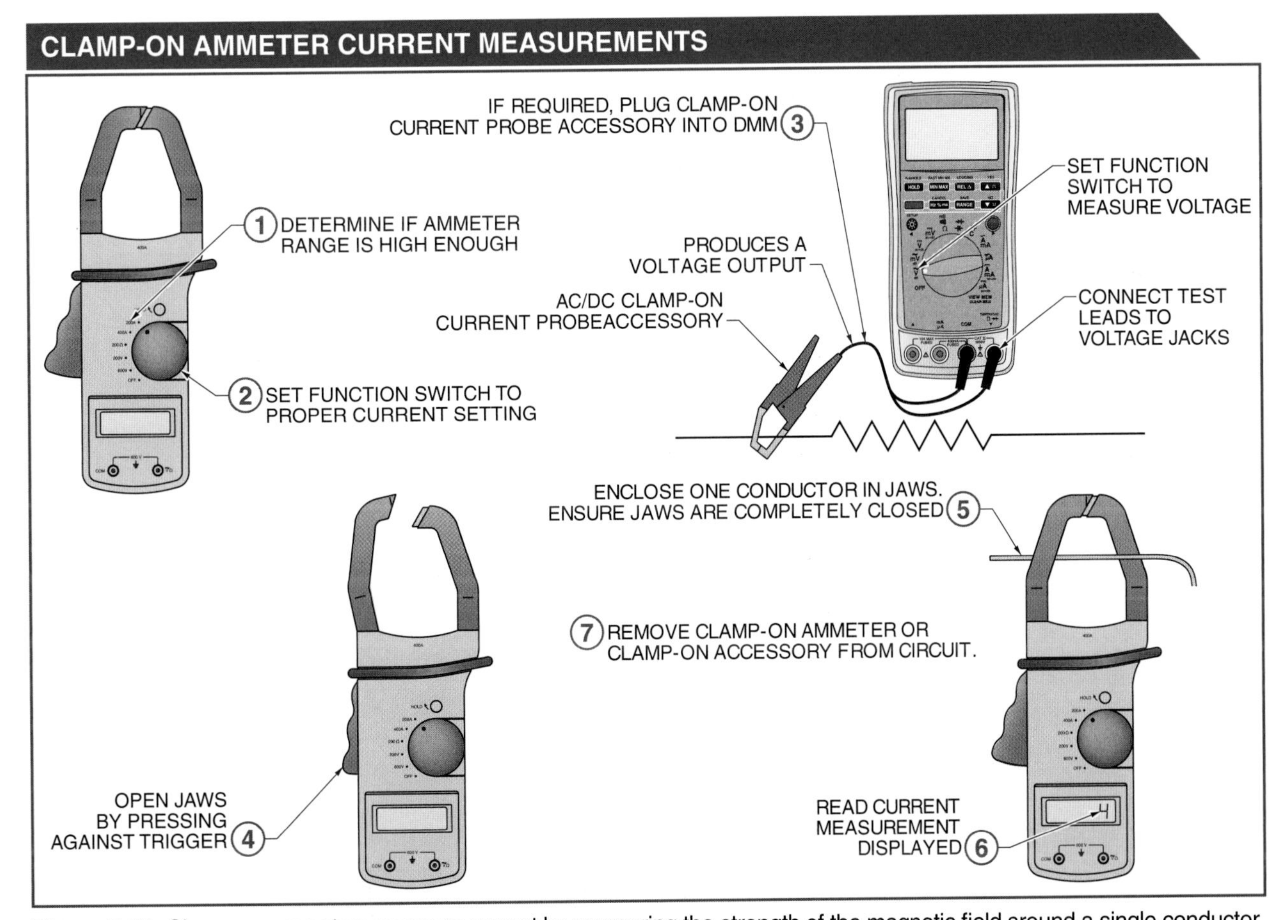

Figure 5-21. Clamp-on ammeters measure current by measuring the strength of the magnetic field around a single conductor.

Terms

Ohm's law is the relationship between voltage (V), current (I), and resistance (R) in an electrical circuit.

OHM'S LAW

In DC circuits, the higher the resistance is, the less the current flow. In addition, the higher the voltage is, the greater the current flow. The relationships between voltage, current, and resistance in a circuit were first discovered by Georg Simon Ohm during a series of experiments he conducted in the early 19th century.

Ohm's law is the relationship between voltage (V), current (I), and resistance (R) in an electrical circuit. Using Ohm's law, any value in this relationship can be found when the other two are known. The relationship between voltage, current, and resistance may be seen best in pie chart form. **See Figure 5-22.**

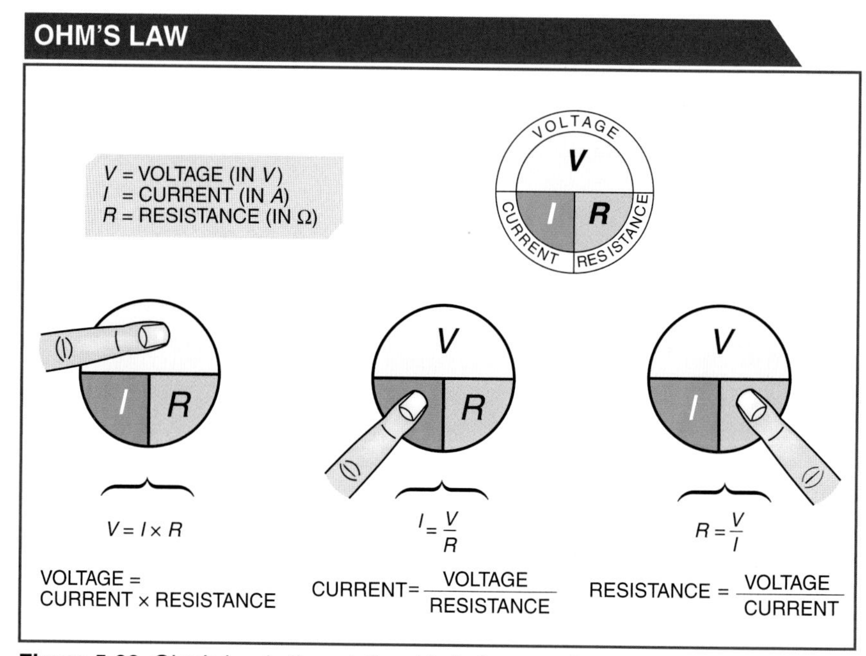

Figure 5-22. Ohm's law is the relationship between voltage, current, and resistance in an electrical circuit.

Determining Current

Ohm's law is used to calculate current in a DC circuit for various values of voltage and resistance. Ohm's law states that current in a circuit is equal to voltage divided by resistance. The current in a circuit is proportional to the amount of voltage and inversely proportional to the amount of resistance. If the voltage and resistance in a circuit are known, current can be calculated by applying the following formula:

$$I = \frac{V}{R}$$

where

I = current (in A)

V = voltage (in V)

R = resistance (in Ω)

Example: What is the current in a circuit that has a 2 Ω resistor connected to a 6 V supply?

$$I = \frac{V}{R}$$

$$I = \frac{6\ \text{V}}{2\ \text{A}}$$

$$\mathbf{I = 3\ A}$$

According to Ohm's law, if the resistance in a circuit is held constant and the voltage varied, the current can be determined for each value of voltage. The voltage/current curve is linear, which means that a specific change in voltage causes a specific change in current. A resistor is referred to as a linear load because of this straight line curve. **See Figure 5-23.** The voltage/current curve for a circuit having a fixed resistance of 2 Ω can be determined by applying the following formula:

$$I = \frac{V}{2\ \Omega}$$

If the voltage in a circuit is held constant and the resistance varied, the current can be determined for each value of resistance. The resistance/current curve is nonlinear, which means that current varies inversely with the amount of resistance. The circuit current decreases with an increase in resistance. The resistance/current curve for a circuit having a fixed voltage of 6 V can be determined by applying the following formula:

$$I = \frac{6\ \text{V}}{R}$$

If the voltage in a circuit is held constant and the resistance approaches 0 Ω, the current becomes very large. As the resistance approaches 0 Ω in a circuit that contains overcurrent protective devices, the fuse blows or circuit breaker trips, opening the circuit. In a circuit that does not contain overcurrent protective devices, the current is limited only by the internal resistance of the voltage source and the resistance of the conductors.

If the voltage in a circuit is held constant and the resistance continually increases, the current decreases, approaching 0 A. In an actual circuit, a switch is used to open the circuit, which increases the resistance to an undefined value, stopping the flow of current.

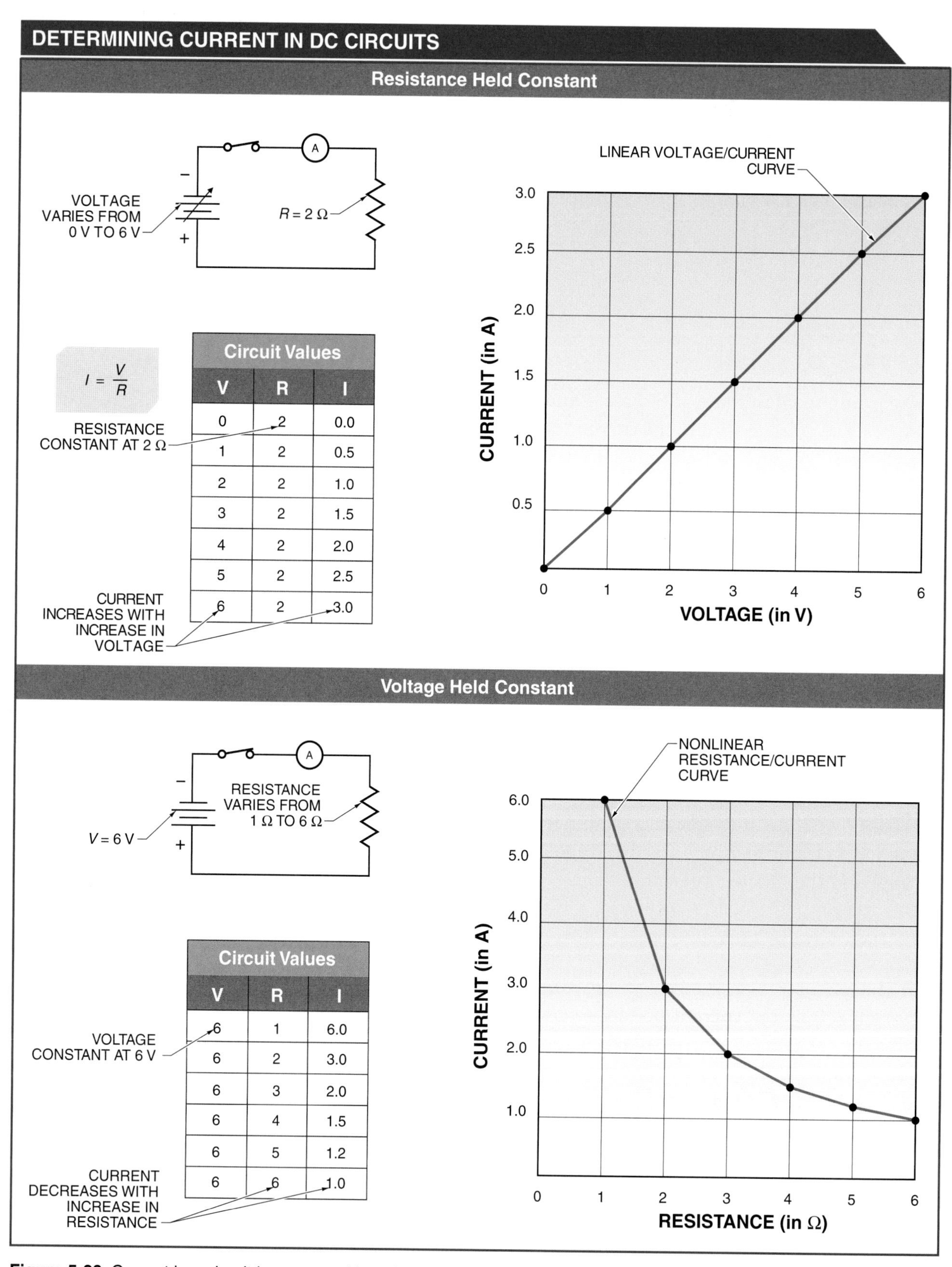

Circuit Values

V	R	I
0	2	0.0
1	2	0.5
2	2	1.0
3	2	1.5
4	2	2.0
5	2	2.5
6	2	3.0

Circuit Values

V	R	I
6	1	6.0
6	2	3.0
6	3	2.0
6	4	1.5
6	5	1.2
6	6	1.0

Figure 5-23. Current in a circuit increases with an increase in voltage and decreases with an increase in resistance.

Determining Voltage

Ohm's law is used to calculate voltage in a DC circuit for various values of current and resistance. Ohm's law states that voltage in a circuit is equal to current times resistance. If the current and resistance in a circuit are known, the voltage can be calculated by applying the following formula:

$$V = I \times R$$

where

V = voltage (in V)

I = current (in A)

R = resistance (in Ω)

Example: What is the voltage in a circuit with a 2 Ω resistor that draws 3 A?

$V = I \times R$

$V = 3\text{ A} \times 2\ \Omega$

$V = \mathbf{6\ V}$

According to Ohm's law, if the resistance in a circuit is held constant and the current varied, the voltage can be determined for each value of current. The current/voltage curve is linear, which means that the voltage drop across a resistor is directly proportional to the current flowing through it. **See Figure 5-24.**

If the current in a circuit is held constant and the resistance varied, the voltage can be determined for each value of resistance. The resistance/voltage curve is also linear which means that an increase in the resistance of a circuit with a constant current flow causes an increase in the voltage drop of the circuit.

Determining Resistance

Ohm's law is used to calculate resistance in a DC circuit for various values of voltage and current. Ohm's law states that resistance in a circuit is equal to voltage divided by current. If the voltage and current in a circuit are known, the resistance can be calculated by applying the following formula:

$$R = \frac{V}{I}$$

where

R = resistance (in Ω)

V = voltage (in V)

I = current (in A)

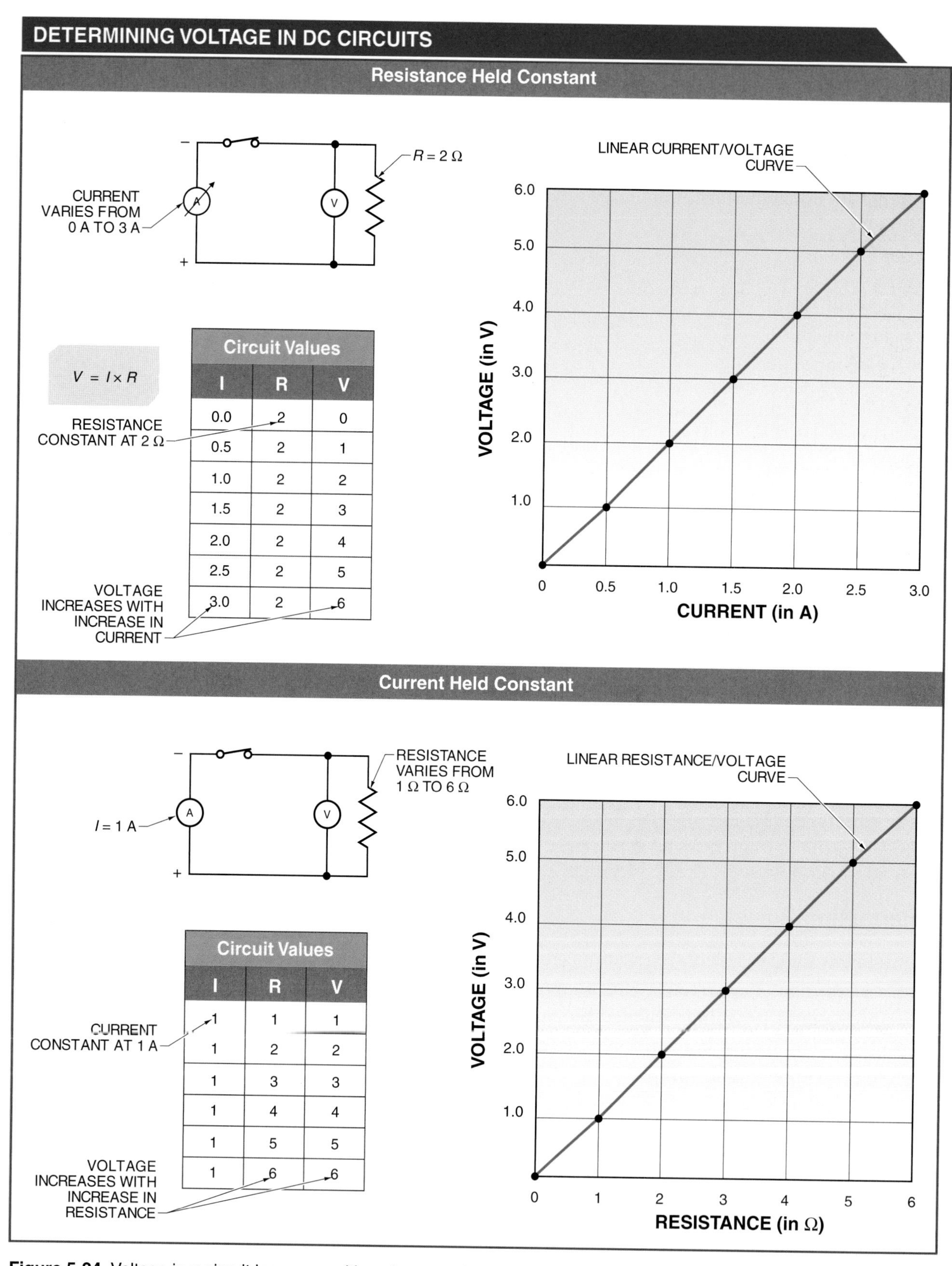

Figure 5-24. Voltage in a circuit increases with an increase in current and increases with an increase in resistance.

Example: What is the resistance of a circuit in which a load that draws 2 A is connected to a 6 V supply?

$$R = \frac{V}{I}$$

$$R = \frac{6\ \text{V}}{2\ \text{A}}$$

$$R = \mathbf{3\ \Omega}$$

According to Ohm's law, if the current in a circuit is held constant at 2 A and the voltage across it varied from 0 V to 6 V, the resistance value can be determined for each value of voltage. The voltage/resistance curve is linear, which means that with a constant current flow through a resistor, the value of the resistor must be increased to increase the voltage drop across it. **See Figure 5-25.**

If the voltage across a resistor is constant at 6 V and the current through the resistor is varied, the value of the resistor must also change to maintain the constant voltage. The resistance required to obtain this result is determined by dividing the voltage by the selected value of current. The current/resistance curve is nonlinear, which means that to maintain a constant 6 V across the resistor, the resistance must increase as the current decreases. In addition, with the voltage held constant and the current increasing toward an undefined value, the resistance approaches 0 Ω.

Terms

Power is the rate of doing work or using energy.

Determining Power

Power is the rate of doing work or using energy. In a mechanical system, work is done any time a force causes an object to move. Electrical force (voltage) of 1 V performs 1 joule of work for each coulomb. This relationship can be seen in the following equation:

$$volt = \frac{joule}{coulomb}$$

Electrical energy causes electron flow through a closed circuit. One ampere equals a flow of 1 coulomb in 1 second past a given point in a circuit. This relationship can be seen in the following equation:

$$ampere = \frac{coulomb}{second}$$

The same amount of work may be done in different amounts of time. For example, the same number of electrons may move from one point in a circuit to another in 1 second or in 1 minute depending on the rate at which they are moving. Although the same amount of work is done, the power is greater for the work done in one second than for the same work done in one minute.

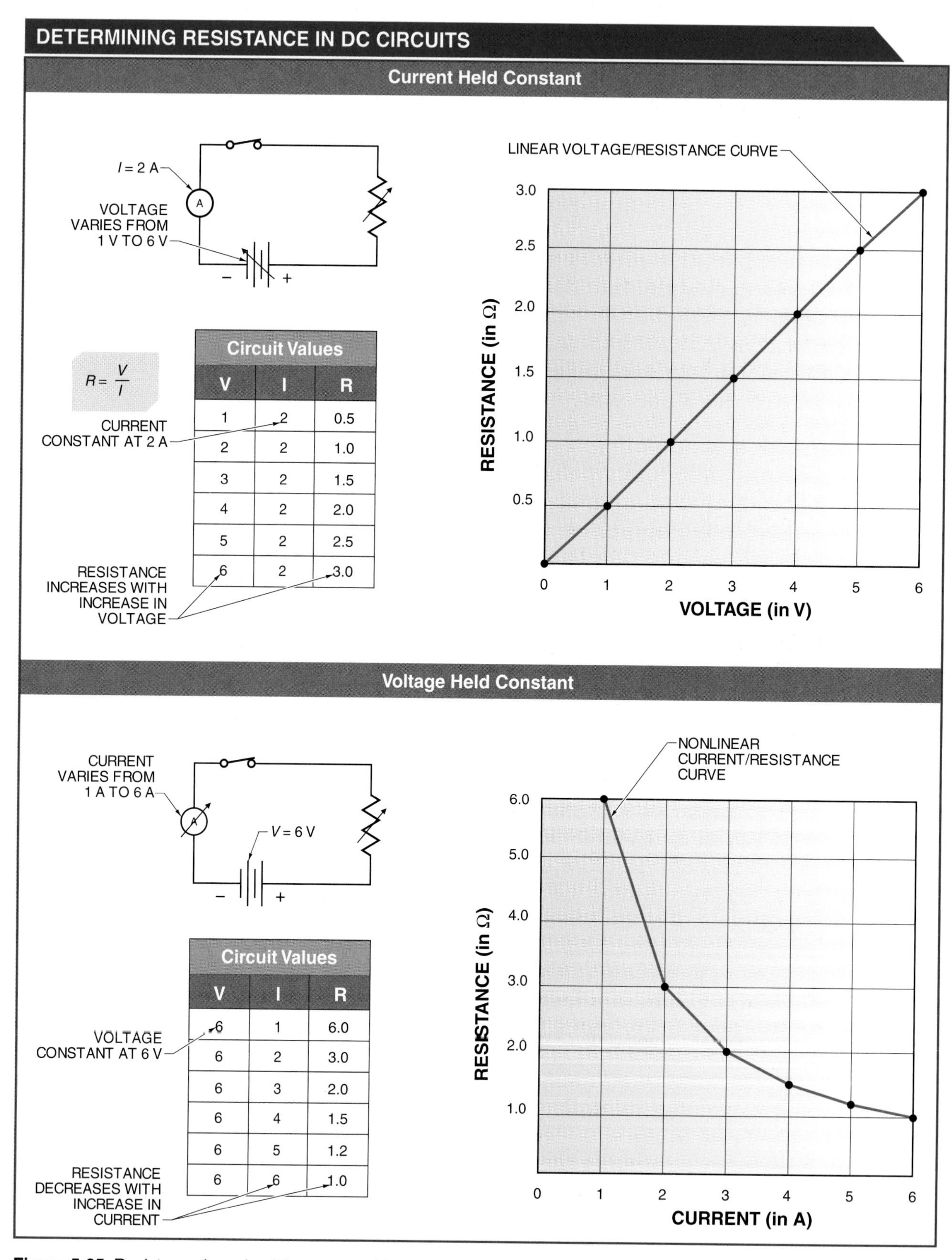

Figure 5-25. Resistance in a circuit increases with an increase in voltage and decreases with an increase in current.

No power is expended in a circuit with voltage present but without a flow of electrons. However, electrical power is used if the voltage causes electron flow. The unit of electrical power is the watt (W). One watt is equal to 1 V times 1 A. Electrical power (P) in a DC circuit can be calculated by applying the following equation:

$$P = I \times V = \frac{coulomb}{second} \times \frac{joule}{coulomb} = \frac{joule}{second}$$

The *power formula* is the relationship between power, voltage, and current in an electrical circuit. The power formula is often referred to as Watt's law. Any value in this relationship may be found when the other two values are known. The relationship between power, voltage, and current may be seen best in pie chart form. **See Figure 5-26.**

Terms

The **power formula** is the relationship between power, voltage, and current in an electrical circuit.

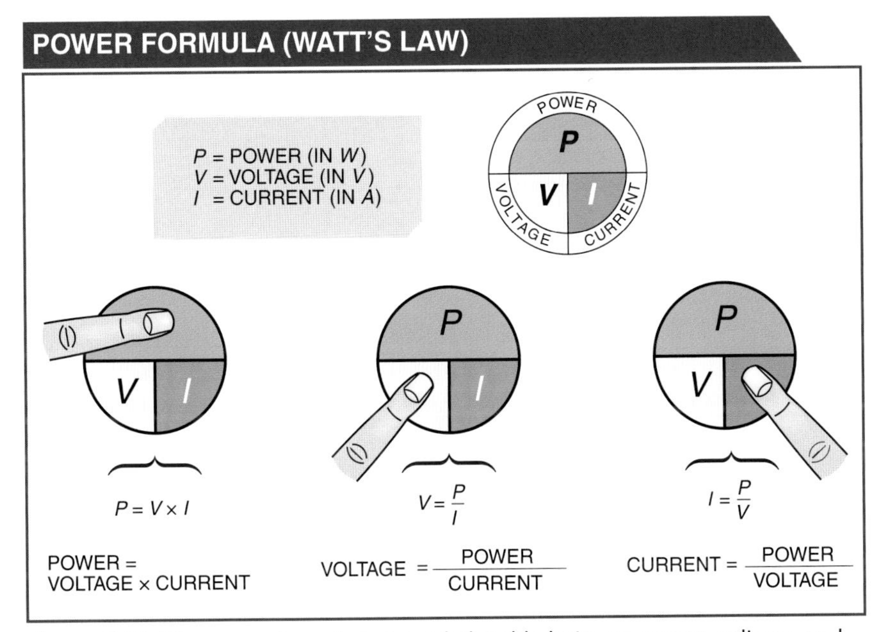

Figure 5-26. The power formula is the relationship between power, voltage, and current in an electrical circuit.

Power is directly related to voltage and current. Current flow through an electrical component either generates energy (a power source) or dissipates energy (a resistance, such as a lamp). Power is the rate at which energy is generated or consumed. The power supplied to a circuit must be consumed; therefore, power consumption must be equal to the power dissipated by a circuit. If the voltage in a circuit is varied from 0 V to 6 V and the resistance is held constant at 2 Ω, the current in the circuit is dependent on the value of the applied voltage. In this circuit, the current varies from 0 A to 3 A as calculated by Ohm's law. Power is calculated by multiplying current times voltage. The voltage/current/power curve is nonlinear. **See Figure 5-27.**

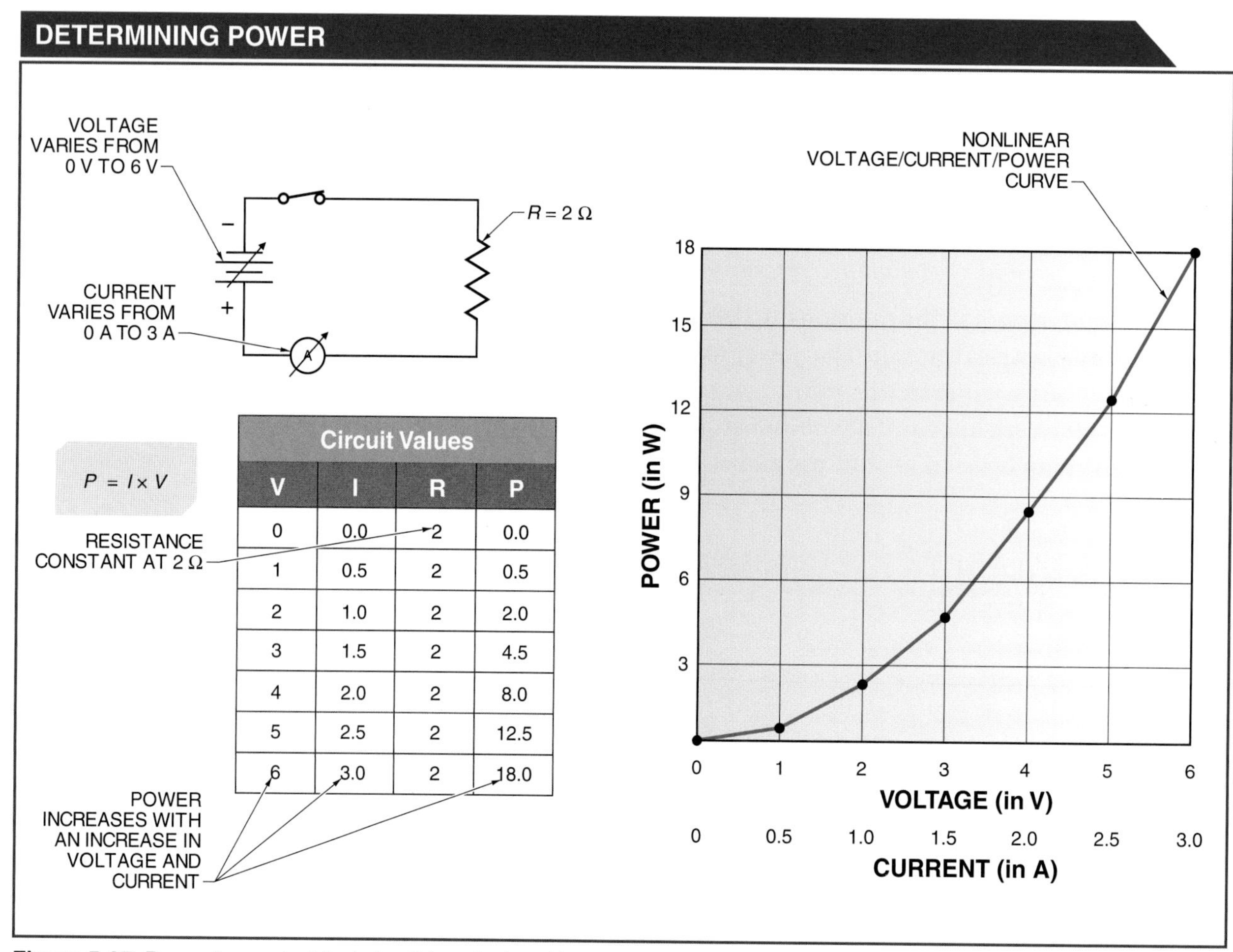

Circuit Values			
V	I	R	P
0	0.0	2	0.0
1	0.5	2	0.5
2	1.0	2	2.0
3	1.5	2	4.5
4	2.0	2	8.0
5	2.5	2	12.5
6	3.0	2	18.0

Figure 5-27. Power in an electrical circuit is calculated by multiplying current by voltage.

The power in a circuit increases based on a factor equal to the circuit voltage squared divided by the resistance, or by the circuit current squared times the resistance. This is true because when resistance is held constant, the current increases at the same rate as the voltage. For example, if the voltage is doubled, the current also doubles. Therefore, the power increases four times.

$(X)\text{P} = 2I \times 2V$

$(X)\text{P} = 4P$

$X = \mathbf{4}$

If the voltage is tripled, the current is also tripled, and the power increases nine times.

$(X)P = 3I \times 3V$

$(X)P = 9P$

$X = \mathbf{9}$

These relationships are derived by substituting voltage divided by resistance for current in the power formula:

$$P = I \times V$$

$$P = \frac{I}{R} \times V$$

$$P = \frac{V^2}{R}$$

These relationships can also be derived by substituting current times resistance for voltage in the power formula:

$$P = I \times V$$

$$P = I \times I \times R$$

$$P = I^2 \times R$$

When analyzing circuits in electronic equipment, most of the measurements that are taken involve voltage. As a result, the two values that are typically available for calculating power are voltage and resistance.

If the voltage in a circuit is fixed at 6 V, and resistance is varied from 0 Ω to 6 Ω, current also changes. In this circuit, the resistance/current/power curve is nonlinear. **See Figure 5-28.** Note that with voltage applied, and the resistance at 0 Ω, the current becomes undefined. The power consumed at this point is 0 W. Only resistance can consume power in an electrical circuit. Since no resistance is present, no power is consumed.

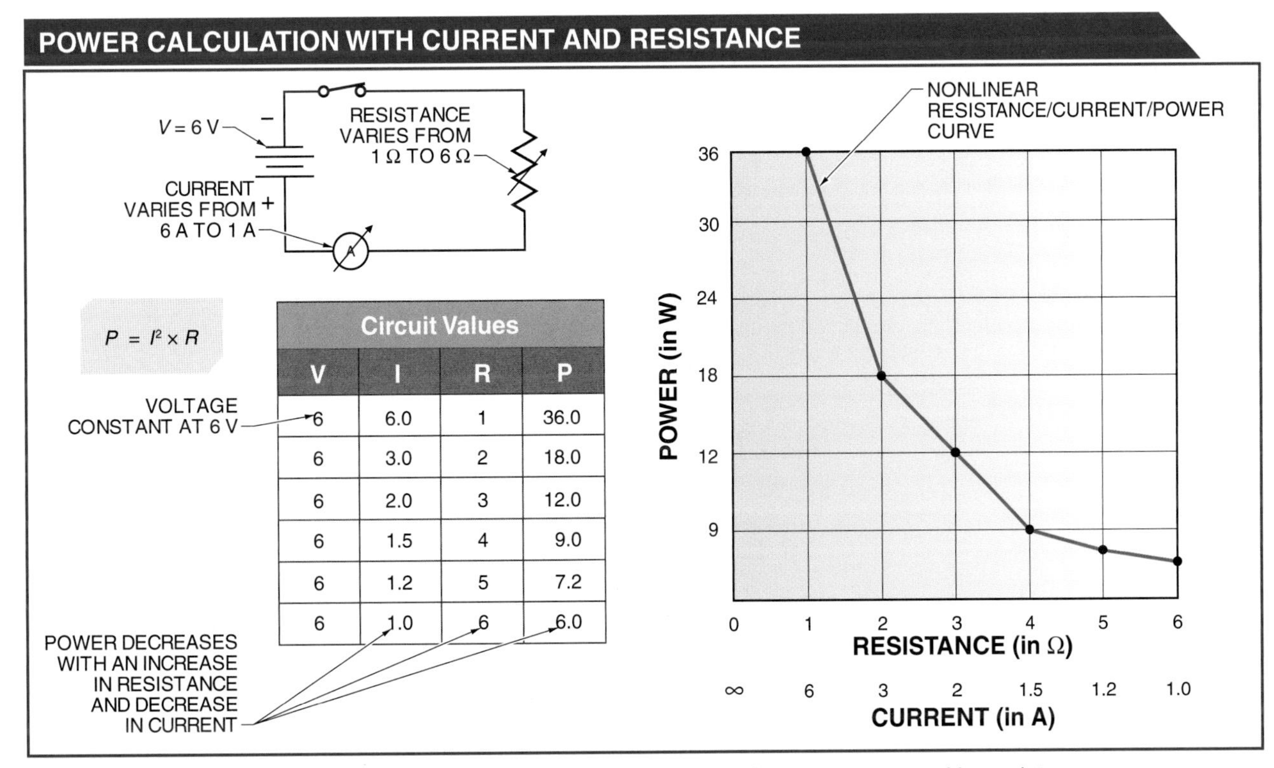

Circuit Values			
V	I	R	P
6	6.0	1	36.0
6	3.0	2	18.0
6	2.0	3	12.0
6	1.5	4	9.0
6	1.2	5	7.2
6	1.0	6	6.0

Figure 5-28. Power in an electrical circuit can be calculated by multiplying current squared by resistance.

If both the circuit voltage and resistance are varied to maintain a constant current, the resistance/voltage/power curve becomes linear. **See Figure 5-29.** At any point on the graph, if the voltage is divided by the resistance, the current is 1 A. It is only when the current is allowed to vary with the voltage that the power curve is nonlinear and varies at a square rate.

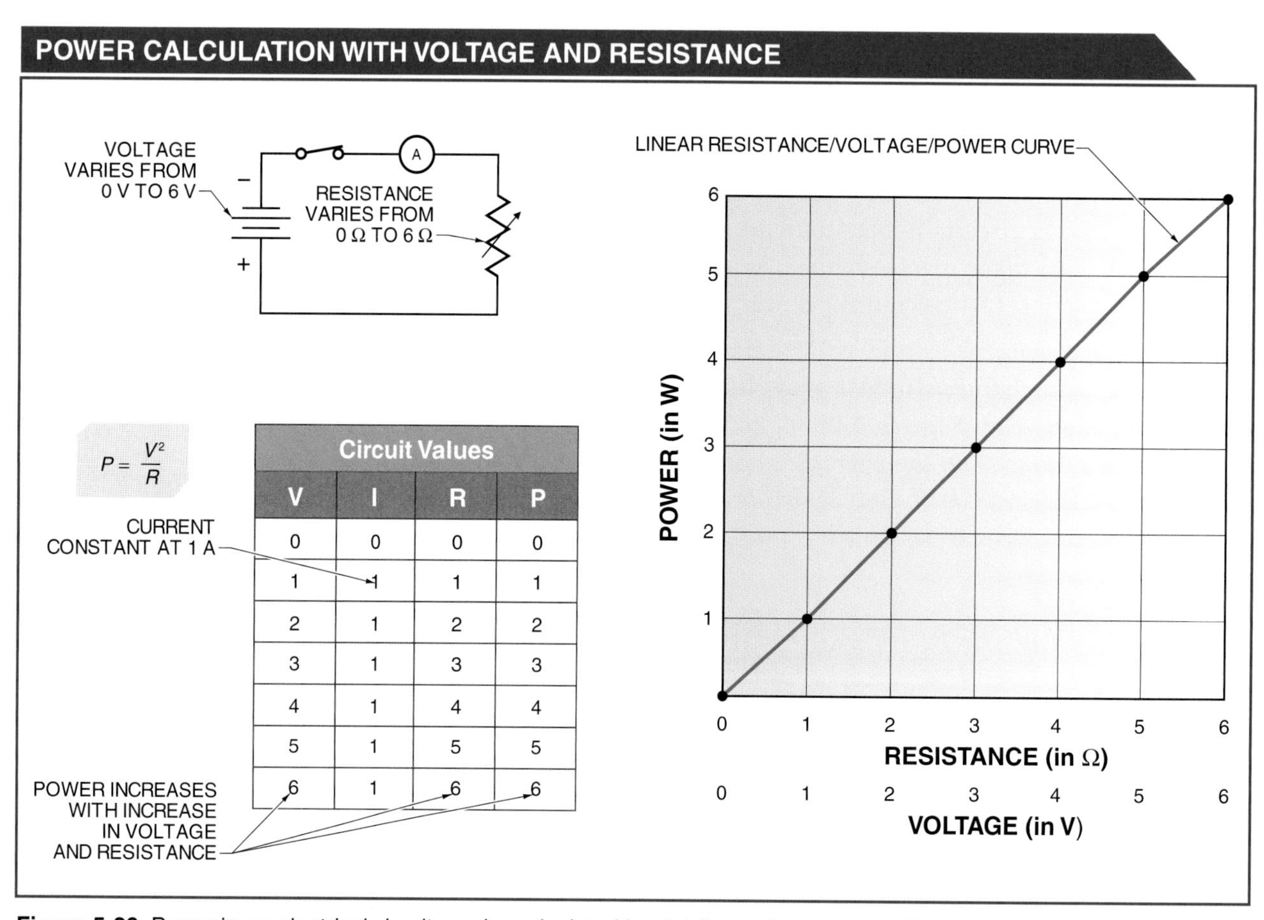

Circuit Values			
V	I	R	P
0	0	0	0
1	1	1	1
2	1	2	2
3	1	3	3
4	1	4	4
5	1	5	5
6	1	6	6

Figure 5-29. Power in an electrical circuit can be calculated by dividing voltage squared by resistance.

Ohm's law and the power formula can be combined to find any unknown value if two of the values are known. For example, if voltage and power are known, then the current and resistance can be determined. Through this combination, six basic formulas and six rearranged formulas are available. **See Appendix.** Three formulas are available to solve for each of the four basic values.

Note: No power can be consumed in a circuit unless there is current flow.

SUMMARY

A simple electrical circuit has a voltage source, conductors, and a load. In practice, most simple circuits also have a control device, such as a switch, and a protective device, such as a fuse or circuit breaker. Each component of a simple electrical circuit can differ in its appearance and how and when it can be used. For example, conductors may consist of thin wire that carries a small amount of current, or thick wire that carries a great amount of current. Conductors can be made of different metals such as copper or aluminum and may be insulated or bare. The different insulation materials allow a conductor to be used for certain applications or prohibit its use under certain environmental conditions. Conductors can be rectangular copper bars designed to carry high currents or a deposited film of copper on a printed circuit board that can carry only a small amount of current.

Ohm's law is the relationship between voltage, current, and resistance in an electrical circuit. The power formula is the relationship between power, voltage, and current in an electrical circuit. The unit of power is the watt. Ohm's law and the power formula can be combined mathematically to obtain twelve different formulas. Any time two of the values in a given formula are known, the unknown value can be determined.

APPLICATION: CURRENT LIMITING...

Background: In many applications, it is necessary to limit the amount of current in a circuit. For example, current limiting is required to properly operate a light-emitting diode (LED). An LED requires a minimum amount of current (~20 mA) to turn on, but allowing too much current (~30 mA) to flow through the LED will damage it.

For this application, a resistor needs to be selected to limit the LED current between 20 mA and 30 mA for a temperature range of 0°C to 40°C. The resistor temperature coefficient is –0.005°C, and the resistor has a tolerance of ±5%. The voltage source is a typical 9 V battery.

CURRENT LIMITING APPLICATION

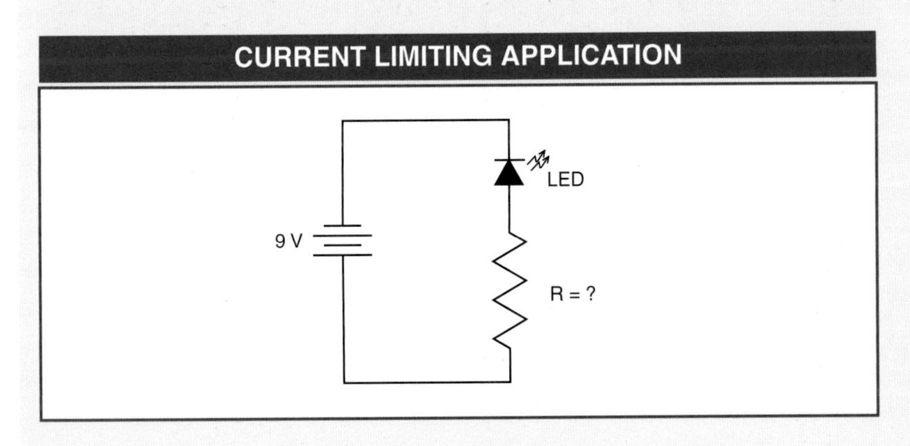

...APPLICATION: CURRENT LIMITING...

Key Points: Resistance; Ohm's law; Watt's law; simple series circuit

Solution: Since the current limiting resistor has a tolerance, the solution starts with a design to limit current at 25 mA, which is the center LED operating point. Using Ohm's law, the resistor value that limits current to 25 mA is the following:

$V = I \times R$
$V \div I = R$
$9\ V \div 0.025\ A = 360\ \Omega$

Applying the resistor tolerance of ±5% gives a resistance range of the following:

$360\ \Omega \times 95\%\ (-5\%) = 342\ \Omega$
$360\ \Omega \times 105\%\ (+5\%) = 378\ \Omega$

As part of the application, the impact of temperature on resistance must be analyzed. The selected resistor value is the resistance at 20°C. The effects of temperature can be found by using the temperature coefficient (α) of –0.005°C and calculating the changes at the temperature extremes. From this calculation, the minimum and maximum resistor values can be determined.

$R = R_{20}\ [1 + \alpha(T - T20)]$
$R = 342\ [1 + -0.005(40 - 20)] = 308\ \Omega$
$R = 342\ [1 + -0.005(0 - 20)] = 376\ \Omega$

$R = 378\ [1 + -0.005(40 - 20)] = 340\ \Omega$
$R = 378\ [1 + -0.005(0 - 20)] = 416\ \Omega$

So, for a 360 Ω, ±5% resistor operating between 0°C and 40°C, the minimum resistor value is 308 Ω and the maximum resistor value is 416 Ω. After determining the minimum and maximum resistances, it must be verified that the LED operating limits are met. Using Ohm's law again, the maximum and minimum LED currents can be calculated based on the resistor values. The minimum and maximum currents are as follows:

$V = I \times R$
$V \div R = I$
$9\ V \div 308\ \Omega = 0.029\ A\ (29\ mA)$
$9\ V \div 416\ \Omega = 0.022\ A\ (22\ mA)$

This resistor value works in this specific application since the currents are within the LED operating limits (between 20 mA and 30 mA). If the current fell outside the operating limits, a resistor with a tighter tolerance or one with a smaller temperature coefficient could be used. Finally, if a proper resistor could not be found, the operating temperature range could be restricted or a different LED with different operating currents could be selected.

In order to make the final resistor selection, the amount of power consumed by the resistor must be analyzed. Watt's law is used to calculate the power being consumed.

...APPLICATION: CURRENT LIMITING

$P = I^2 \times R$
$P = (0.029\ \text{A})^2 \times 308\ \Omega$
$P = 0.26\ \text{W}$

$P = (0.022\ \text{A})^2 \times 416\ \Omega$
$P = 0.20\ \text{W}$

Standard axial resistor power sizes are ⅛ W, ¼ W, ½ W, and 1 W. The maximum power is just over ¼ W, therefore, the next larger power size is required. So, for this LED application, a 360 Ω, ±5%, 0.5 W resistor properly limits the operating current, allowing this LED to emit light for temperatures from 0°C to 40°C.

Chapter Review

1. In what circuit are the contacts of the overload relay located?
2. List three types of electrically operated switches.
3. What is the basis of fuse design?
4. What is the voltage in a circuit with a 5 Ω resistor that draws 4 A?
5. What is the power formula?
6. What is the most common type of switch used in electrical circuits?
7. Explain why multimeters are designed to draw a limited amount of current.
8. With a thermal circuit breaker, what are the two main causes of heating of the bimetallic strip?
9. How are sustained overloads typically different from short circuits?
10. How are unconnected conductors that cross over one another represented in a schematic symbol?
11. What is the resistance of a circuit in which a load that draws 3 A is connected to a 24 V supply?
12. What is meter loading?
13. Are circuit breakers connected in series, in parallel, or in combination with circuit conductors?
14. What is the current in a circuit that has a 3 Ω resistor connected to a 12 V supply?
15. List the three most common types of overcurrent protective devices.
16. Explain the differences in protection between the thermal and magnetic components in a magnetic circuit breaker.
17. Explain the difference between a closed circuit and an open circuit.
18. What is the interrupting rating of most current limiting fuses?
19. List two common types of manually operated switches.
20. What is a resistance heating element?

6

DC Series Circuits

OBJECTIVES

- Describe the designation parameters for DC series circuits.
- Explain how current, resistance, voltage, and power affect DC series circuits.
- Describe ground reference.
- Explain maximum power transfer.

Ohm's law and the power formula are used to analyze direct current (DC) series circuits. For electrons to flow in a circuit, there must be a complete path for the electrons to move from the negative terminal of the power source, through the conductors, switches, protective device(s), loads, and back to the positive terminal of the power source.

In a DC series circuit, the current is the same at all points in the circuit. The voltage drop at any resistor is found by using Ohm's law. The total resistance is the sum of all individual resistors in a series circuit. The power at any point is determined by using the power formula. A ground reference (zero voltage point) can be placed anywhere in a series circuit. With DC circuits, the maximum power transfer occurs when the internal voltage source resistance is equal to the load resistance.

PARAMETER AND COMPONENT DESIGNATIONS

Each parameter and component of a circuit is assigned a designation that includes the alphabetic symbol for the parameter or component. If a circuit contains more than one component of the same type, each component is assigned a number. This can be seen with resistors numbered R_1, R_2, etc. **See Figure 6-1.**

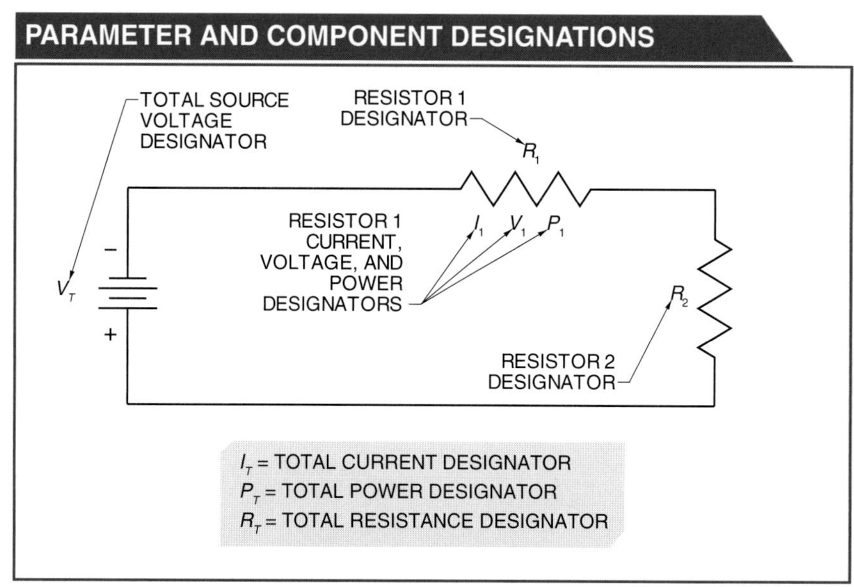

Figure 6-1. Parameters and components in complex circuits are described with designators consisting of letters and numbers.

The current flow through a resistor, the voltage drop across it, and the power consumed by it, receive the numerical subscript designation of the resistor. In complex circuits, the full designation of the component is sometimes used in the subscript. For example, the parameters of the components in a complex circuit may be given as I_{R1}, V_{R1}, and P_{R1} instead of I_1, V_1, and P_1.

Unlike a simple circuit, a complex circuit includes two or more resistive loads. In a complex circuit, the source voltage is designated V_T and the total circuit resistance designated R_T. The subscript T represents the total reading per the circuit. Total circuit current supplied by the power source and the total power consumed by the circuit are also assigned this subscript.

SERIES CIRCUITS

A *series circuit* is a circuit that has only one current path. This means that all the components in the circuit are connected end to end so that a complete, closed circuit is obtained. The flow of electrons stops if any component is open in a series circuit.

Terms

A **series circuit** is a circuit that has only one current path.

Current in Series Circuits

Current is the flow of electrons through an electrical circuit. Electrons are physical bodies that can be counted. The number of electrons that leave the negative terminal of a battery must return to the positive terminal. Since a series circuit has only one path, the flow of electrons must be the same at all points in the circuit. When a DMM set to measure current is inserted in series at every junction of the components in a series circuit, it reads the same current value at each junction. **See Figure 6-2.**

Terms

Current is the flow of electrons through an electrical circuit.

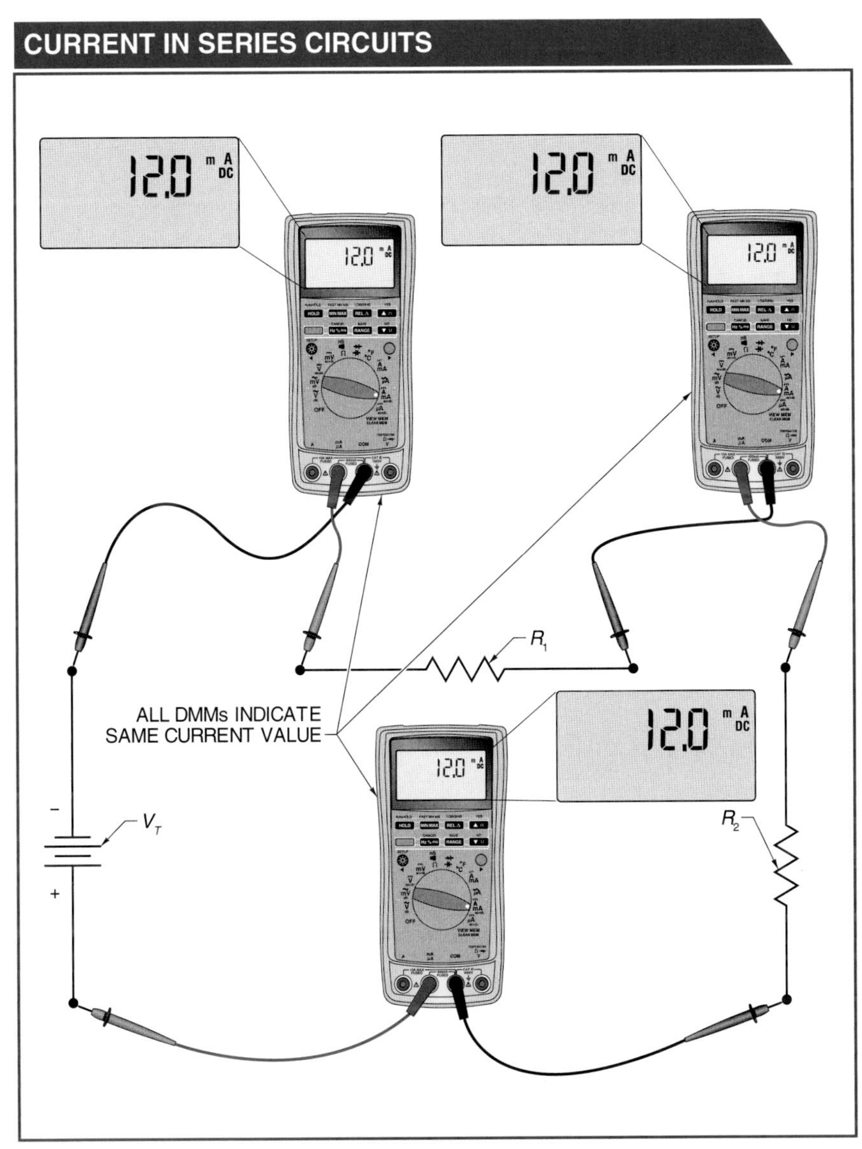

Figure 6-2. A DMM set to measure current can be used to show that the current in a series circuit is the same at any point in the circuit.

The total current in a series circuit is equal to the current through any component. Total current in a series circuit is calculated by applying the following formula:

$$I_T = I_1 = I_2 = \ldots = I_N$$

where

I_T = total current (in A)

I_1 = current through component 1 (in A)

I_2 = current through component 2 (in A)

I_N = current through component N (in A)

Example: What is the total current in a circuit that has 8 A flowing through resistor 1?

$$I_T = I_1$$

$$I_T = \mathbf{8\ A}$$

In a series circuit, if the current at any point is known, the current at any other point is also known. This fact is used to analyze DC series circuits.

Heating Circuit Current Calculation. Heat is produced any time electrons flow through a wire that has resistance. This method is used to produce heat in toasters, portable space heaters, hair dryers, and electric water heaters. When selecting a heating element for an application, the size of the heating element is typically stated in watts and is based on the application. The greater the required heat output, the higher the required wattage. Once a wattage size is selected, the total current is calculated by applying the power formula. The power formula is used to determine expected current values because most electrical equipment lists a voltage and power rating. **See Figure 6-3.** The listed power rating is given in watts (W) for most appliances or in horsepower (HP) for motors.

Example: What is the total current in a circuit when a 2000 W heating element is connected to a 115 V supply?

$$I_T = \frac{P_T}{V_T}$$

$$I_T = \frac{2000\ W}{115\ V}$$

$$I_T = \mathbf{17.39\ A}$$

Finding the current value is required because the water heater wire size, switch rating, and fuse (or circuit breaker) rating are all based on the amount of current in the circuit. Likewise, the conduit size is based on the wire size. Knowing the amount of current a load draws is essential to designing an electrical system.

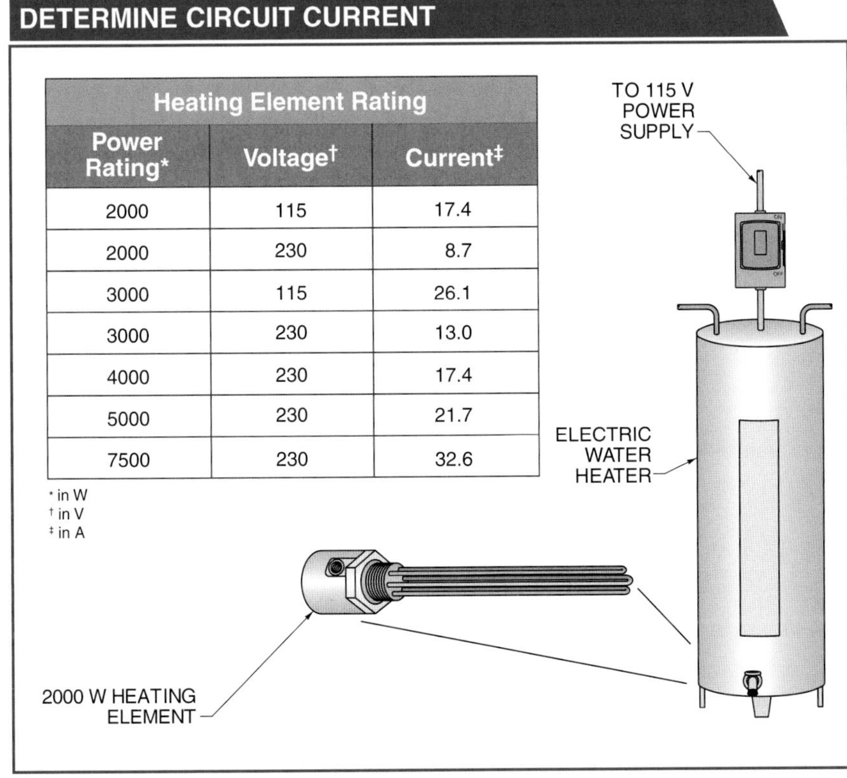

Heating Element Rating		
Power Rating*	Voltage†	Current‡
2000	115	17.4
2000	230	8.7
3000	115	26.1
3000	230	13.0
4000	230	17.4
5000	230	21.7
7500	230	32.6

* in W
† in V
‡ in A

Figure 6-3. The power formula is used to determine expected current values because most electrical equipment lists a voltage and power rating.

Resistance in Series Circuits

In a DC series circuit, total resistance is equal to the sum of the values of the resistors connected in series. **See Figure 6-4.** To calculate the total resistance in a DC series circuit, all resistance values are added together. Total resistance in a series circuit is calculated by applying the following formula:

$R_T = R_1 + R_2 + \ldots + R_N$

where

R_T = total resistance (in Ω)

R_1 = resistance of resistor 1 (in Ω)

R_2 = resistance of resistor 2 (in Ω)

R_N = resistance of resistor N (in Ω)

Example: What is the total resistance in a circuit containing a 2 Ω and a 4 Ω resistor connected in series?

$R_T = R_1 + R_2$

$R_T = 2\ \Omega + 4\ \Omega$

$R_T = \mathbf{6\ \Omega}$

RESISTANCE IN SERIES CIRCUITS

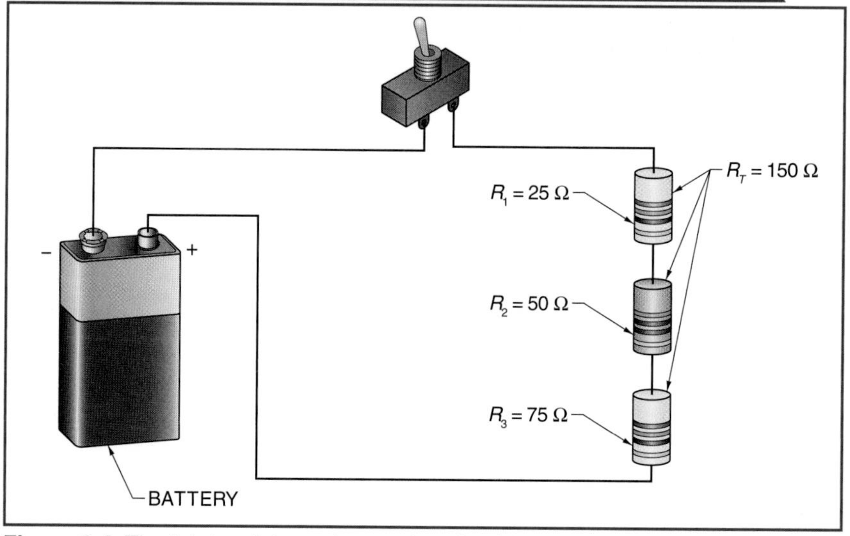

Figure 6-4. The total resistance in a series circuit is equal to the sum of the values of the resistors connected in series.

Total resistance in a series circuit can also be calculated by applying Ohm's law. By applying Ohm's law, if two values are known, the third can be determined.

Wire Resistance Calculation. In some applications, the resistance of the wire used in a circuit can have a significant effect on the voltage delivered to the load. To determine the voltage drop, the resistance of the wire is added to the resistance of the load.

Example: What is the voltage drop in a circuit that has a 100 Ω load connected to a 24 V voltage source and contains a total of 400′ of No. 20 AWG solid copper wire? **See Figure 6-5.**

The resistance of No. 20 AWG solid copper wire is approximately 11.9 Ω per 1000′ (0.0119 Ω/ft). The resistance for 400′ of No. 20 wire is 4.76 Ω (0.0119 Ω/ft × 400′ = 4.76 Ω). In a series circuit, total resistance is found by adding all the resistive elements. In this case, the total resistance is the resistance of the wire plus the load.

$$R_T = R_W + R_{LOAD}$$
$$R_T = 4.76\ \Omega + 100\ \Omega$$
$$R_T = \mathbf{104.76}\ \Omega$$

Once the total resistance is found, the total circuit current is calculated by applying Ohm's law

$$I_T = \frac{V_T}{R_T}$$

$$I_T = \frac{24\text{ V}}{104.76\ \Omega}$$

$$I_T = \mathbf{0.229\ A}$$

RESISTANCE APPLICATION

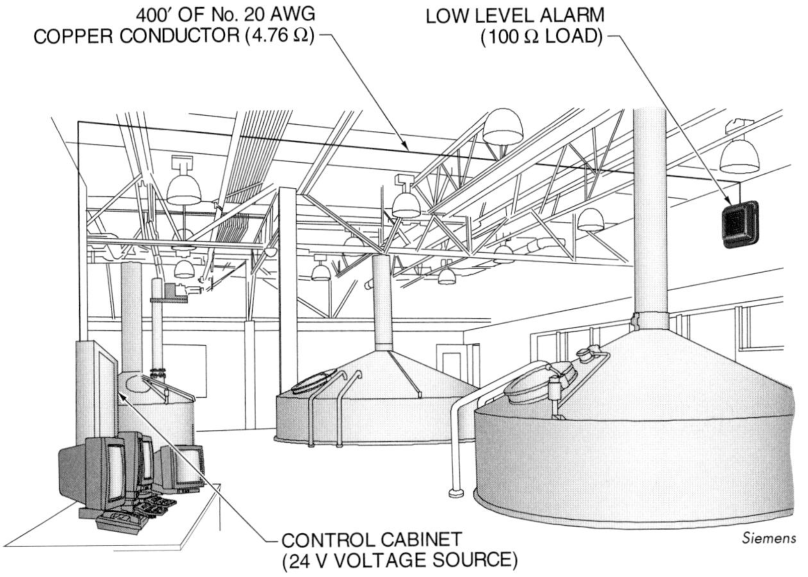

400′ OF No. 20 AWG COPPER CONDUCTOR (4.76 Ω)
LOW LEVEL ALARM (100 Ω LOAD)
CONTROL CABINET (24 V VOLTAGE SOURCE)
Siemens

Figure 6-5. The resistance of the conductors used in a circuit must be added to the load resistance in circuits that contain long wire or cable runs.

Once the total current is found, the voltage drop across the 100 Ω load can be calculated by applying Ohm's law.

$V_T = I_T \times R_{LOAD}$

$V_T = 0.229\ \text{A} \times 100\ \Omega$

$V_T = \mathbf{22.9\ V}$

The load consumes 22.9 V of the power supply's 24 V to the load because of the resistance of the copper wire. The copper wire consumes 1.1 V at the power supply's 24 V.

Voltage in Series Circuits

Ohm's law states that the voltage in a circuit is equal to the circuit current times the circuit resistance. The source voltage acts on the total resistance of the circuit. The total voltage in a series circuit is calculated by multiplying the circuit current by the circuit resistance.

Example: What is the supply voltage in a series circuit containing a total resistance of 6 Ω and total current of 8 A?

$V_T = I_T \times R_T$

$V_T = 8\ \text{A} \times 6\ \Omega$

$V_T = \mathbf{48\ V}$

Terms

Voltage drop is the amount of voltage consumed by a component as current passes through it.

Voltage drop is the amount of voltage consumed by a component as current passes through it. The voltage drop across a resistor is equal to the current flow through the resistor multiplied by the resistance of the resistor. Ohm's law can be used to calculate the voltage drop across any resistor. **See Figure 6-6.** The voltage drop across a resistor is calculated by applying the following formula:

$V = I \times R$

where

V = voltage drop of component (in V)

I = current through component (in A)

R = resistance of component (in Ω)

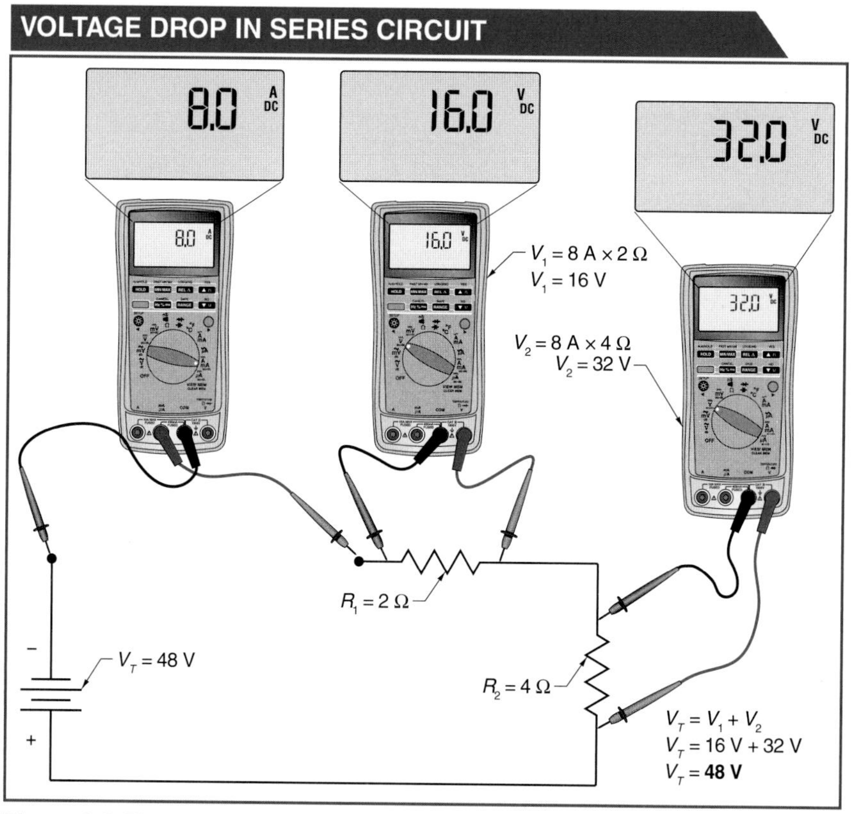

VOLTAGE DROP IN SERIES CIRCUIT

8.0 A DC

16.0 V DC

32.0 V DC

$V_1 = 8\ A \times 2\ \Omega$
$V_1 = 16\ V$

$V_2 = 8\ A \times 4\ \Omega$
$V_2 = 32\ V$

$R_1 = 2\ \Omega$

$V_T = 48\ V$

$R_2 = 4\ \Omega$

$V_T = V_1 + V_2$
$V_T = 16\ V + 32\ V$
$V_T = \mathbf{48\ V}$

Figure 6-6. The voltage drop across a resistor is equal to the current flow through the resistor times the resistance of the resistor.

Example: What is the voltage drop across a 2 Ω resistor in a circuit that has a total current of 8 A?

$V_1 = I_1 \times R_1$

$V_1 = 8\ A \times 2\ \Omega$

$V_1 = \mathbf{16\ V}$

The behavior of voltage in a series circuit gives rise to two fundamental concepts. First, since current is the same at any point in a series circuit, the largest voltage drop is across the largest resistance.

The voltage drop is proportional to the resistance value of the resistors in the circuit. Second, the sum of the voltage drops across series resistors is equal to the applied voltage. The sum of the voltage drops in a series circuit is calculated by applying the following formula:

$V_T = V_1 + V_2 + \ldots + V_N$

where

V_T = total circuit voltage drop (in V)

V_1 = voltage drop across load 1 (in V)

V_2 = voltage drop across load 2 (in V)

V_N = voltage drop across load N (in V)

Example: What is the total applied voltage of a series circuit containing 2 V, 2 V, and 8 V drops across three loads?

$V_T = V_1 + V_2 + V_3$

$V_T = 2\text{ V} + 2\text{ V} + 8\text{ V}$

$V_T = \mathbf{12\ V}$

These two concepts provide additional tools for circuit analysis. They are also useful for checking the accuracy of other calculations used in determining circuit parameters. For example, the sum of the voltage drops can be calculated to determine if it is equal to the applied voltage.

Minimum Allowable Voltage in Series Circuits. Conductors are connected in series with the source and the load and also exhibit a degree of resistance. The National Electrical Code® (NEC®) suggests that the conductors be sized so that they do not reduce the voltage at the load by more than 3% of the source voltage. The minimum allowable voltage at the load in a series circuit per the NEC® is calculated by applying the following formula:

$V_L = V_T - (0.03 \times V_T)$ or $V_L = V_T \times 0.97$

where

V_L = minimum allowable voltage at the load (in V)

V_T = total circuit voltage (in V)

0.03 = maximum allowable voltage drop per the NEC® (in %)

0.97 = minimum allowable voltage at load per the NEC® (in %)

Example: What is the minimum allowable voltage at the load of a 120 V circuit with a load resistance of 19 Ω and load current of 6 A?

$V_L = V_T - (0.03 \times V_T)$

$V_L = 120\text{ V} - (0.03 \times 120\text{ V})$

$V_L = 120\text{ V} - 3.6\text{ V}$

$V_L = \mathbf{116.4\ V}$

The minimum voltage at the load must be between 116.4 V and 120 V to comply with the NEC®. Actual voltage at the load can be calculated to determine if the circuit meets NEC® requirements by applying Ohm's law.

$V_L = I_L \times R_L$

$V_L = 6\text{ A} \times 19\ \Omega$

$V_L = \mathbf{114\ V}$

Because the actual voltage at the load is 114 V, it does not meet the minimum compliance of 116.4 V at the load. **See Figure 6-7.**

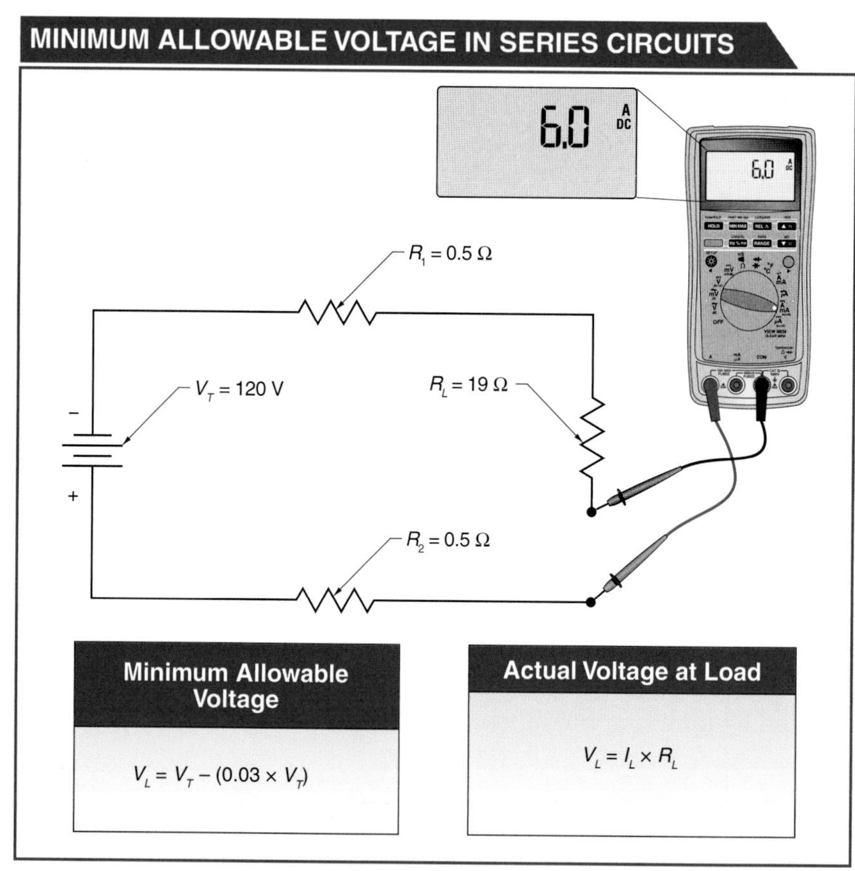

Figure 6-7. Per the NEC®, conductor must be sized so that they do not reduce the voltage at a load by more than 3% of the source voltage.

Power in Series Circuits

Power is produced when voltage is applied to a load and current flows through the load. The power produced is used to produce light (lamps), heat (heating elements), rotary motion (motors), or linear motion (solenoids).

More power is produced when a load's resistance is low or its applied voltage is high. Less power produced is produced when a load's resistance is high or its applied voltage is low. The amount of power produced is measured in watts (W).

The amount of power produced by a load is equal to the voltage drop across the load times the current through the load. The total power in a circuit is equal to the sum of the power produced by each load. **See Figure 6-8.** Total power in a series circuit when the power across each load is known or measured is calculated by applying the following formula:

$$P_T = P_1 + P_2 + P_3 + \ldots + P_N$$

where

P_T = total circuit power (in W)

P_1 = power of load 1 (in W)

P_2 = power of load 2 (in W)

P_3 = power of load 3 (in W)

P_N = power of load N (in W)

Example: What is the total power in a circuit if three loads are connected in series and each load produces 2 W?

$P_T = P_1 + P_2 + P_3$

$P_T = 2\text{ W} + 2\text{ W} + 2\text{ W}$

$P_T = \mathbf{6\text{ W}}$

POWER IN SERIES CIRCUITS

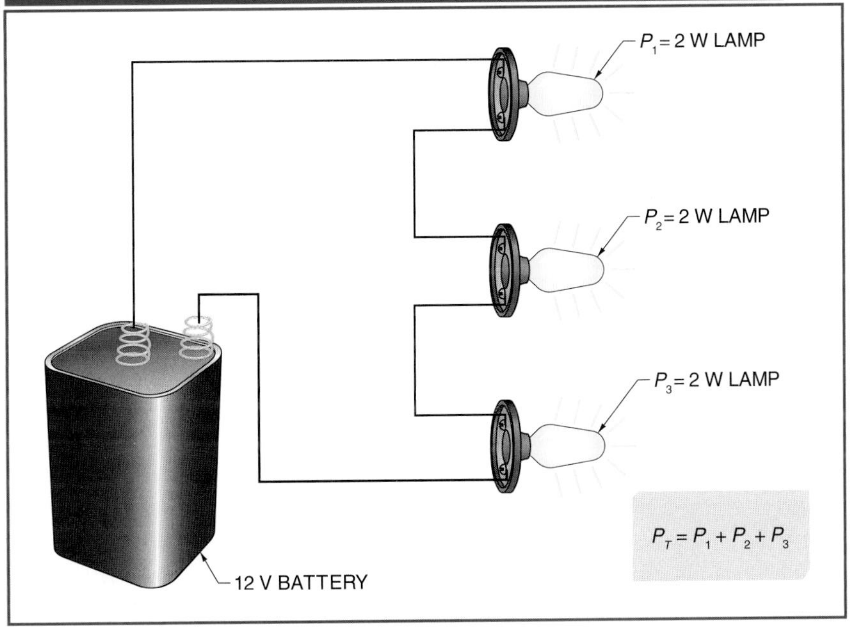

Figure 6-8. The total power in a series circuit is equal to the sum of the power produced by each load.

Total power in a series circuit can also be calculated by applying the power formula when the values of current, resistance, and voltage are known or can be measured. Total power in a series circuit when current and voltage are known is calculated by applying the following formula:

$P_T = I_T \times V_T$

where

P_T = total circuit power (in W)

I_T = total circuit current (in A)

V_T = total circuit voltage (in V)

Example: What is the total power in a 48 V circuit that has total current of 8 A?

$P_T = I_T \times V_T$

$P_T = 8\text{ A} \times 48\text{ V}$

$P_T = \mathbf{384\text{ W}}$

The power formula can also be used to calculate the power dissipated by a resistor when the current through the resistor and the voltage across it are known or can be measured. Power dissipated by a single resistor when the current through the resistor and the voltage across it are known is calculated by applying the following formula:

$P = I \times V$

where

P = power dissipated by resistor (in W)

I = current through resistor (in A)

V = voltage of resistor (in V)

DC Power Loss

TO VDC POWER SOURCE

TO VDC POWER SOURCE

N

The 3-Wire System

Thomas Edison addressed DC power loss problems with a three-wire system that is still in use today. The three-wire system allows a higher transmission voltage to service loads that require half of the source voltage between each side of the line. The potential between the neutral and the other two conductors in a three-wire system is equal. One line is negative in respect to the neutral, and the other line is positive. The neutral is neither positive nor negative. Usually, the neutral is white or grey in color and is connected to earth ground.

Maximum power loss on this system is no worse than with a two-wire system. Maximum power loss occurs when there is a load on one circuit but no load on the other. Minimum power loss occurs when the two circuits are balanced, or each circuit draws the same amount of current. Edison knew that higher transmission voltages could reduce losses and make the power distribution system more efficient. At that time, however, there were no economical means to step up and step down DC voltage.

Example: What is the power dissipated by a resistor with 4 A flowing through it and a voltage of 12 V?

$P = I \times V$

$P = 4\text{ A} \times 12\text{ V}$

$P = \mathbf{48\ W}$

Component Power Calculation. Calculating the amount of power a load consumes is necessary to properly size the load for the circuit parameters. Resistors are available with a variety of different power ratings. Resistor power ratings range from a fraction of a watt to many watts. Applying the power formula helps in selecting the correct resistor.

Example: What is the minimum power rating of a resistor for a DC series circuit that has a 12 V voltage source and 48 Ω of resistance? **See Figure 6-9.**

$$P_T = \frac{V_T^{\,2}}{R_T}$$

$$P_T = \frac{(12\text{ V})^2}{48\,\Omega}$$

$$P_T = \frac{144\text{ V}}{48\,\Omega}$$

$$P_T = \mathbf{3\ W}$$

COMPONENT POWER CALCULATION

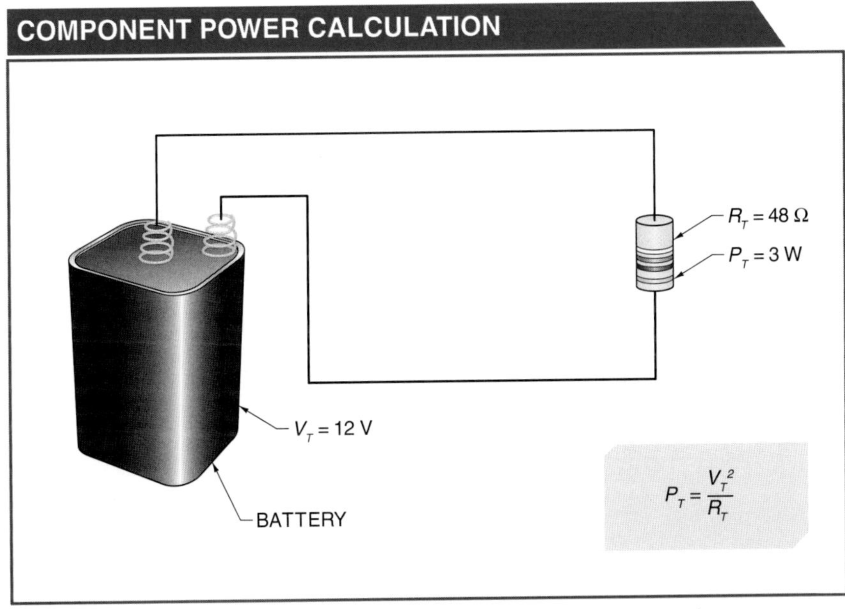

Figure 6-9. The power formula can be applied to find the amount of power consumed by a resistive load.

Terms

A **ground** is a low-resistance conducting connection between electrical and electronic circuits, equipment, and the earth.

A **voltage divider** is a circuit that is constructed by connecting resistors in series to produce a desired voltage drop across the resistors.

Ground Reference

Certain types of electronic systems require both a positive and a negative source of voltage for proper operation. Both polarities can be obtained from a single voltage source by using a ground reference. A *ground* is a low-resistance conducting connection between electrical and electronic circuits, equipment, and the earth. A ground is an acceptor of electrons and can be used as a voltage reference point in a circuit. A *voltage divider* is a circuit that is constructed by connecting resistors in series to produce a desired voltage drop across the resistors. The ground is used in the design of voltage dividers to provide both positive and negative voltage. For example, a circuit containing a 48 V supply is connected to two resistors connected in series. With respect to ground, the voltage across resistor 1 is −16 V, and the voltage across resistor 2 is +32 V. When measuring the voltage across resistor 1 with a DMM set to measure voltage, the positive lead is connected to ground. When measuring the voltage across resistor 2, the negative lead of the DMM is connected to ground. **See Figure 6-10.**

GROUND REFERENCE IN SERIES CIRCUITS

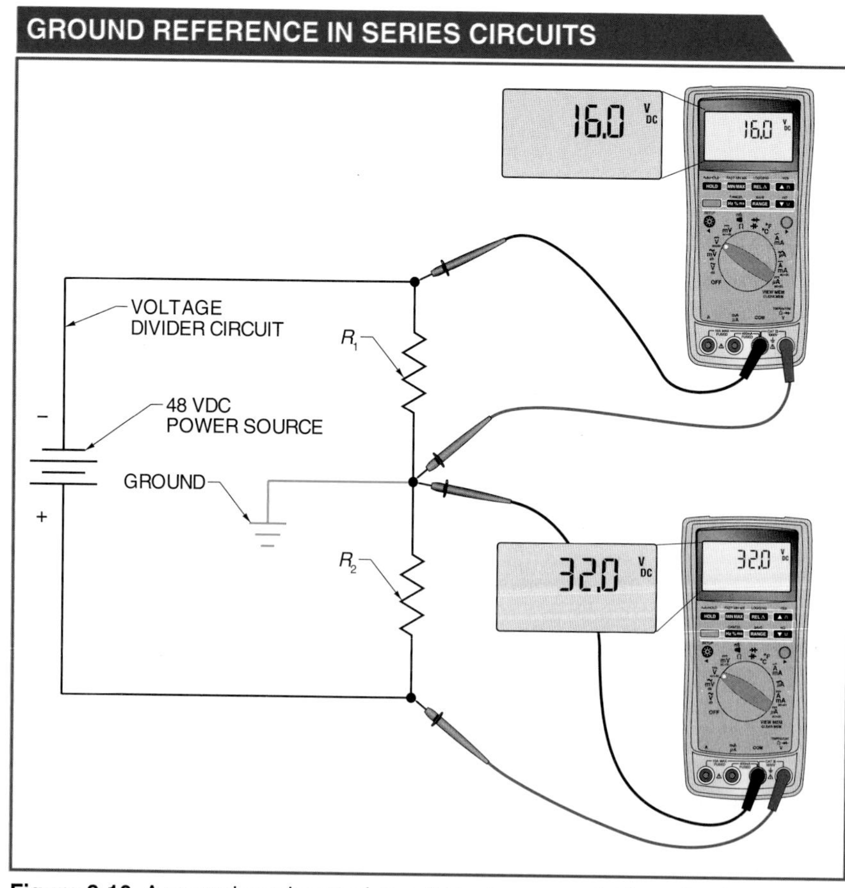

Figure 6-10. A ground can be used as a 0 V reference point in a circuit.

Maximum Power Transfer

All sources of electrical potential have internal resistance. Maximum power transfer occurs in DC circuits when the internal resistance of the power source (R_i) is equal to the resistance of the load (R_L). For example, 10 total power (P_T) calculations are taken from a circuit in which the resistance of the load (R_L) ranges from 0 Ω to 9 Ω. In this circuit, the battery has an internal resistance of 5 Ω. The internal resistance of the battery (R_i) is connected in series with the load, and these resistances are added to obtain the total circuit resistance (R_T). This value is divided into the source voltage to obtain the total circuit current for each load resistor value. The total circuit current is used to find the power dissipated through the internal resistance of the battery and load for each load resistance. **See Figure 6-11.**

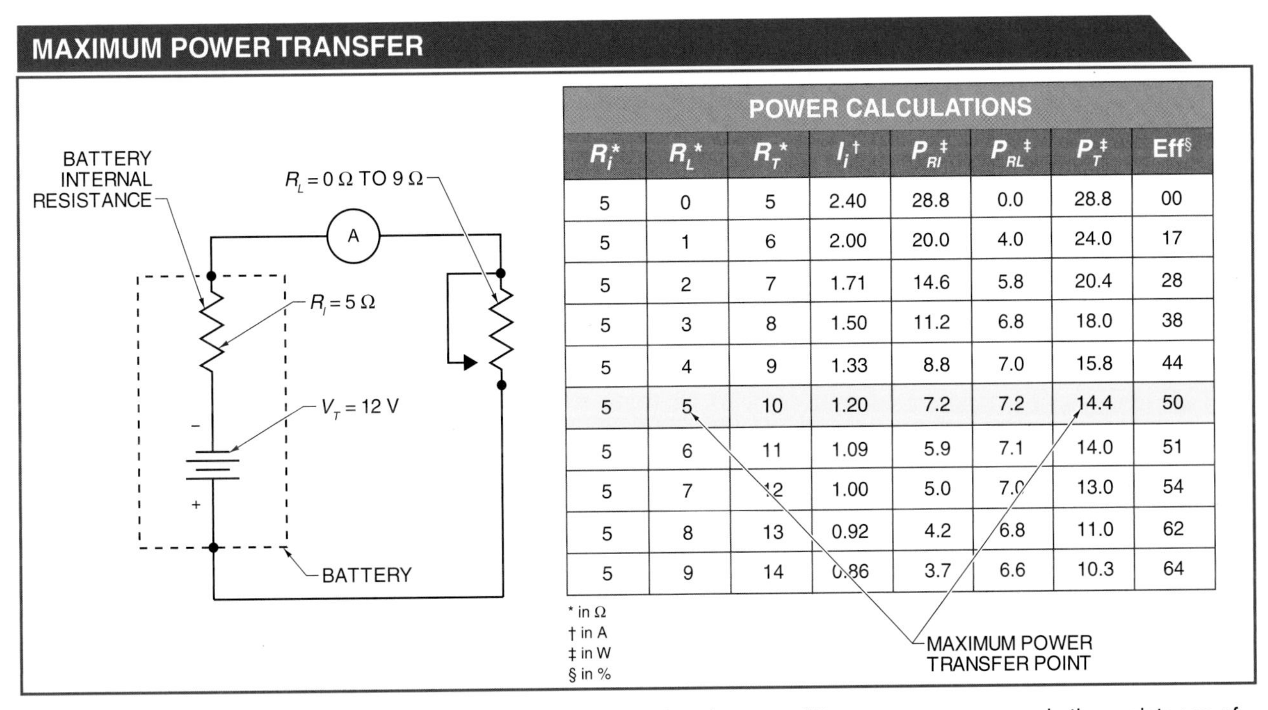

POWER CALCULATIONS

R_i*	R_L*	R_T*	I_i†	P_{Ri}‡	P_{RL}‡	P_T‡	Eff§
5	0	5	2.40	28.8	0.0	28.8	00
5	1	6	2.00	20.0	4.0	24.0	17
5	2	7	1.71	14.6	5.8	20.4	28
5	3	8	1.50	11.2	6.8	18.0	38
5	4	9	1.33	8.8	7.0	15.8	44
5	5	10	1.20	7.2	7.2	14.4	50
5	6	11	1.09	5.9	7.1	14.0	51
5	7	12	1.00	5.0	7.0	13.0	54
5	8	13	0.92	4.2	6.8	11.0	62
5	9	14	0.86	3.7	6.6	10.3	64

* in Ω
† in A
‡ in W
§ in %

Figure 6-11. Maximum power transfer occurs when the internal resistance of the power source equals the resistance of the load.

With a 0 Ω load resistance, the load does not dissipate any power (P_{RL}). The internal resistance of the source limits total current. The internal resistance of the battery dissipates the total power. As the resistance of the load is increased from 1 Ω to 5 Ω, the power delivered to the load by the source increases. Simultaneously, the power dissipated by the internal resistance of the battery decreases. As the resistance of the load is increased from 5 Ω to 9 Ω, the power dissipated by both the load and the internal resistance of the battery decreases. Power output from the source to the load peaks when the load resistance is 5 Ω. This is when the resistance of the load is equal to the internal resistance of the battery.

Note: The efficiency of this circuit is only 50% at the maximum power transfer point. A low efficiency is achieved with a load resistance below 5 Ω, and a high efficiency is achieved with a load resistance above 5 Ω. The percentage of efficiency for a given load is calculated by applying the following formula:

$$\%Eff = \frac{P_L}{P_T} \times 100$$

where

$\%Eff$ = circuit efficiency (in %)

P_L = power at load (in W)

P_T = total power output (in W)

Example: What is the efficiency of a circuit that has a total power of 200 W and a load power of 75 W?

$$\%Eff = \frac{75\text{ W}}{200\text{W}} \times 100$$

$$\%Eff = \mathbf{37.5}$$

A relationship between maximum power transfer and increased efficiency is evident. Where the amount of power involved is very large and high efficiency is required, the load resistance is increased in comparison with the source resistance. This relationship is typical with electrical power distribution systems that deliver power to commercial and residential customers.

If maximum power transfer between source and load is required, then the load resistance should match the source resistance. This relationship is typical with low power applications such as matching the output of an amplifier to a speaker. Whenever the load resistance is not matched to the source, less than maximum power is delivered to the load.

SUMMARY

A complex circuit includes two or more resistive loads. A series circuit has two or more components connected so that there is only one path for electron flow.

Current in a series circuit is the same at all points in the circuit. Resistor values add when connected in series. The sum of the voltage drops in a series circuit equal the source voltage. Total power dissipated by a circuit is equal to the sum of the power used by each resistor.

Opening a series circuit at any point stops the flow of electrons. If a short occurs, the current increases. Maximum transfer of power occurs when the internal resistance of the power source is equal to the resistance of the load. Also, relative potentials of both polarities can be obtained when a ground reference is used in a voltage divider network.

APPLICATION: VOLTAGE DIVIDERS...

Background: Resistors are two-terminal circuit elements that are used to limit current and reduce voltage. Because of this behavior, resistors can be used to create voltage dividers. Voltage dividers are used to reduce source voltage and are used in applications such as volume control, measurement circuits, bias circuits, and frequency filters.

The simplest form of a voltage divider uses a voltage source and two resistors.

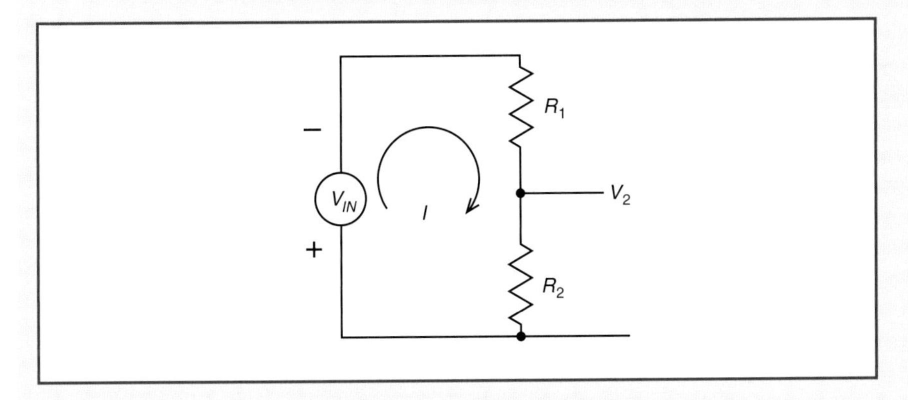

The voltage across R_2 is given by the following ratio:

$$V_2 = \frac{R_2}{(R_1 + R_2) \times V_{IN}}$$

In accordance with Ohm's law, the voltage V_2 is given by $I \times R_2$. The circuit current (I) is the source voltage (VIN) divided by the circuit total resistance, $R_1 + R_2$. The voltage divider formula is the following:

$$V_2 = R_2 \times I$$

where

$$I = \frac{V_{IN}}{R_1 + R_2}$$

Therefore,

$$V_2 = R_2 \times \frac{V_{IN}}{R_1 + R_2}$$

$$V_2 = \frac{R_2}{(R_1 + R_2) \times V_{IN}}$$

Key Points: Series DC circuit; Ohm's law

Problem: A system is being designed that turns on a light emitting diode (LED) when it is dark outside. A simple voltage divider sensing circuit is planned that uses a photo resistor or light-dependent resistor that controls voltage to a transistor that is used as the LED switch.

...APPLICATION: VOLTAGE DIVIDERS...

A photo resistor changes resistance based on the amount of ambient light. The dark resistance (R_D) may be 500 kΩ, while the illuminated resistance (R_I) is 25 kΩ. Other critical parameters are maximum applied voltage (150 V), continuous power dissipation (100 mW), and operating temperature (–30°C to +75°C).

Solution: The simple circuit to control the LED using a transistor is shown in Figure 2, which shows a simple circuit that uses a transistor to control current through the LED. Current flows between the C and E transistor terminals lighting the LED. When the transistor is off, the current path is broken turning off the LED.

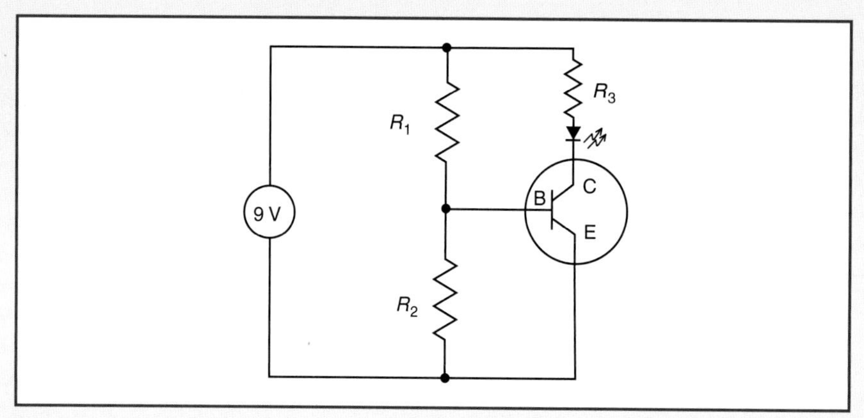

To turn on the transistor switch, the voltage at R_2 (V_2) must be greater than 0.7 V. To turn off the transistor switch, the V_2 voltage must be less than 0.7 V. A 9 V battery is being used for the voltage source. The value of the voltage at V_2 is given by using the voltage divider equation as follows:

$$V_2 = \frac{R_2}{(R_1 + R_2) \times V_{IN}}$$

A larger V_2 (> 0.7 V) is required when it's dark to turn on the transistor. When there is ambient light, V_2 must be less than 0.7 V to turn off the transistor. To start, set R_2 to be the photo-resistor. Therefore, the following equations for the circuit are as follows:

Transistor/LED on:

$$V_2 = \frac{R_2}{(R_1 + R_2) \times V_{IN}}$$

$$V_2 = \frac{R_D}{(R_1 + R_D) \times V_{IN}}$$

$$V_2 = \frac{0.7\text{ V} < 500\text{ k}\Omega}{(R_1 + 500\text{ k}\Omega) \times 9\text{ V}}$$

$$V_2 = R_1 < 5.928\text{ M}\Omega$$

...APPLICATION: VOLTAGE DIVIDERS...

Transistor/LED off:

$$V_2 = \frac{R_2}{(R_1 + R_2) \times V_{IN}}$$

$$V_2 = \frac{R_1}{(R_1 + R_I) \times V_{IN}}$$

$$V_2 = \frac{0.7\text{ V} > 25\text{ k}\Omega}{(R_1 + 25\text{ k}\Omega) \times 9\text{ V}}$$

$$V_2 = R_1 > 296\text{ k}\Omega$$

The solution using the voltage divider equation provides a large range for the R_1 resistor values that provide the proper voltage over the required conditions. One final equation to help select the final R_1 value is the maximum power dissipation of the photo resistor, which is 100 mW. Without R_1 changing, the maximum current flow is when the photo-resistor resistance is the smallest, which is at 25 kΩ. The power dissipation equation is as follows:

Maximum power dissipation > $I^2 \times R_2$

where

$$I = \frac{V_{IN}}{R_1 + R_2}$$

The current is reduced when R_1 increases since V_{IN} and R_2 are fixed. Selecting R_1 as 1 MΩ should satisfy all equations. Substituting R_I with 1 MΩ produces the following equations:

Light on:

$$I = 0.7\text{ V} < \frac{R_D}{R_1 + R_D} \times V_{IN}$$

$$I = 0.7\text{ V} < \frac{500\text{ k}\Omega}{R_1 + 500\text{ k}\Omega} \times 9\text{ V}$$

$$I = \frac{500\text{ k}\Omega}{1000\text{ k}\Omega + 500\text{ k}\Omega} \times 9\text{ V}$$

$$I = 3\text{ V} > 0.7\text{ V}$$

Light off:

$$I = 0.7\text{ V} > \frac{R_1}{R_1 + R_1} \times V_{IN}$$

$$I = 0.7\text{ V} > \frac{25\text{ k}\Omega}{R_1 + 25\text{ k}\Omega} \times 9\text{ V}$$

$$I = \frac{25\text{ k}\Omega}{1000\text{ k}\Omega + 25\text{ k}\Omega} \times 9\text{ V}$$

$$I = 0.2\text{ V} < 0.7\text{ V}$$

...APPLICATION: VOLTAGE DIVIDERS

Power dissipation:

$I = 100 \text{ mW} > I^2 \times R_2$

$$I = \frac{V_{IN}}{R_1 + R_2}$$

Light on:

$$I = \frac{9 \text{ V}}{1000 \text{ k}\Omega + 500 \text{ k}\Omega}$$

$I = 6.0\ \mu\text{A}$

$I = 6.0\ \mu\text{A}^2 \times 500 \text{ k}\Omega$

$I = 18\ \mu\text{W} < 100 \text{ mW}$

Light off:

$$I = \frac{9 \text{ V}}{1000 \text{ k}\Omega + 25 \text{ k}\Omega}$$

$I = 8.78\ \mu\text{A}$

$I = 8.78\ \mu\text{A}^2 \times 25 \text{ k}\Omega$

$I = 1.9\ \mu\text{W} < 100 \text{ mW}$

By selecting R_1 as the fixed resistor at 1 MΩ and R_2 as the photo resistor, the transistor can be controlled to turn on the LED at night and turn off the LED during the day. The amount of power being dissipated by the photo resistor has been verified to be within the limits of the device.

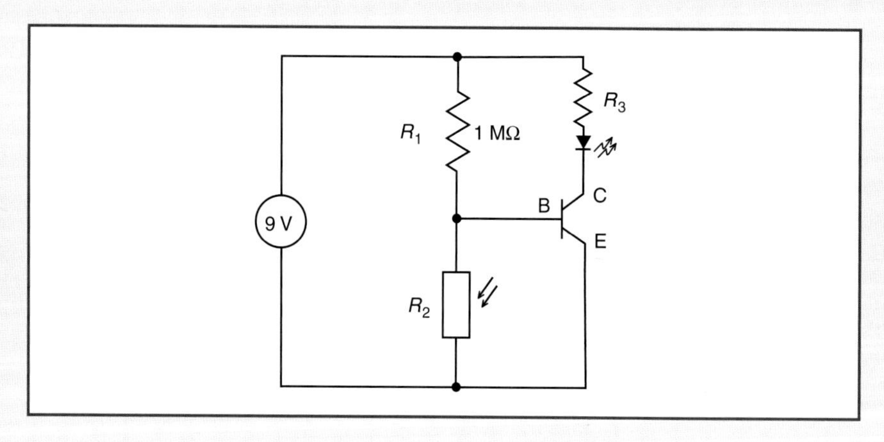

Chapter Review

1. What is a voltage divider?
2. What is the total resistance in a circuit containing a 2 Ω and 5 Ω resistor connected in series?
3. What is the minimum allowable voltage at the load of a 135 V circuit with a load resistance of 15 Ω and a load current of 8 A?
4. What is voltage drop?
5. What is a series circuit?
6. What is the total power in a 120 V circuit that has a total current of 15 A?
7. What is the suggested conductor sizing principle by the NEC® concerning load reduction?
8. What is the voltage drop in a circuit that has a 50 Ω load connected to a 120 V voltage source and contains a total of 20′ of No. 20 AWG solid copper wire?
9. What is the total power in a circuit that has four 3 W loads connected in series?
10. What is known about the flow of electrons in a series circuit as opposed to other types of circuits?
11. What is the supply voltage in a circuit containing a total resistance of 4 Ω and a total current of 3.5 A?
12. What is the minimum power rating of a resistor for a DC series circuit that has a 120 V voltage source and 64 Ω of resistance?
13. How is a complex circuit different from a simple circuit?
14. What is the total current in a circuit when a 200 W resistor is connected to a 120 V power supply?
15. What is the voltage drop across a 5 Ω resistor in a circuit that has a total current of 12 A?

NREL

7

DC Parallel Circuits

OBJECTIVES

- Explain the concept of a DC parallel circuit.
- Describe how voltage, current, resistance, and power affect DC parallel circuits.
- List the different DC parallel circuit applications.
- Describe the differences between open and short circuits.

Ohm's law and the power formula are used to analyze direct current (DC) parallel circuits. For current to flow in a circuit, there must be a complete path for the electrons to move from the negative terminal of the power source, through the conductors, switches, protective device(s), and loads, and back to the positive terminal of the power source.

The total voltage in a circuit containing parallel-connected loads is the same as the voltage across each load. The voltage drop across each load remains the same if parallel loads are added or removed. The total current in a circuit containing parallel-connected loads equals the sum of the current through all loads. The total current increases if loads are added in parallel and decreases if loads are removed. The total resistance in a parallel circuit is less than the smallest resistance value. The total resistance decreases if loads are added in parallel and increases if loads are removed.

Terms

A **parallel circuit** is a circuit that contains two or more loads and has more than one path for electron flow.

PARALLEL CIRCUITS

A *parallel circuit* is a circuit that contains two or more loads and has more than one path for electron flow. In a DC parallel circuit, the electrons flow in one direction. Each load in parallel is called a branch circuit. Parallel circuits are typically used for power distribution applications. **See Figure 7-1.** For example, the lighting systems in a temporary lighting application are connected in parallel so that each load has the same voltage, but each can draw a different amount of current.

PARALLEL CIRCUITS

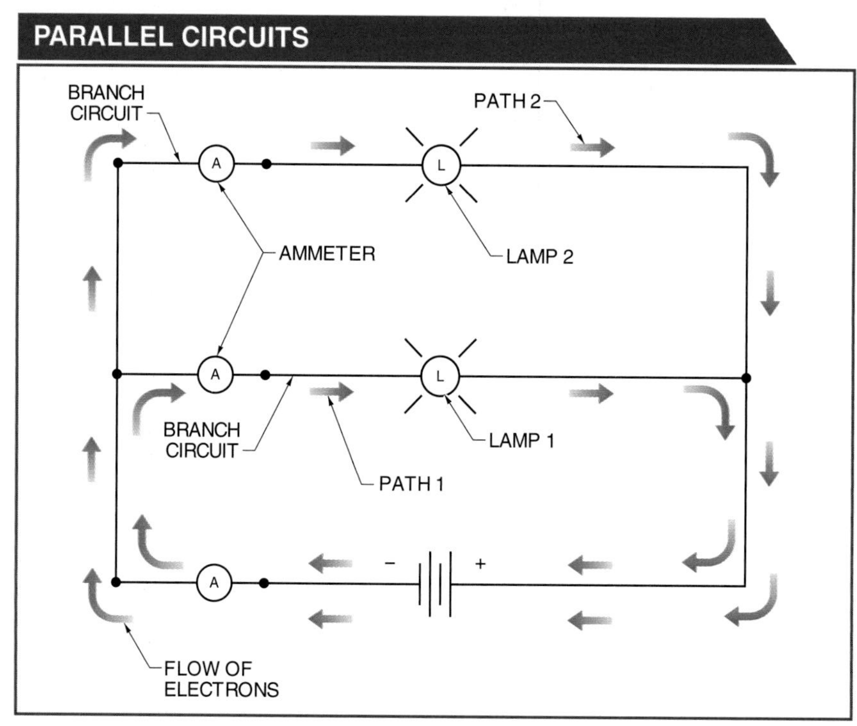

Figure 7-1. A parallel circuit has more than one current path for electron flow and is typically used for power distribution applications.

Voltage in Parallel Circuits

In a parallel circuit, the voltage drop across each load is the same. The voltage across each load remains the same if parallel loads are added or removed. Voltage measurements taken across the power source and each resistor show the same value. **See Figure 7-2.** Because voltage is the same, it becomes the reference point for analyzing all parallel circuits. Total voltage in a parallel circuit when the voltage drop across a load is known or measured is calculated by applying the following formula:

$$V_T = V_1 = V_2 = \ldots = V_N$$

where
V_T = total applied voltage (in V)
V_1 = voltage across load 1 (in V)
V_2 = voltage across load 2 (in V)
V_N = voltage across any branch circuit (in V)

Example: What is the total applied voltage in a circuit if the voltage drop across two loads connected in parallel is 12 V?
$V_T = V_1 = V_2$
V_T = 12 V = 12 V
V_T = **12 V**

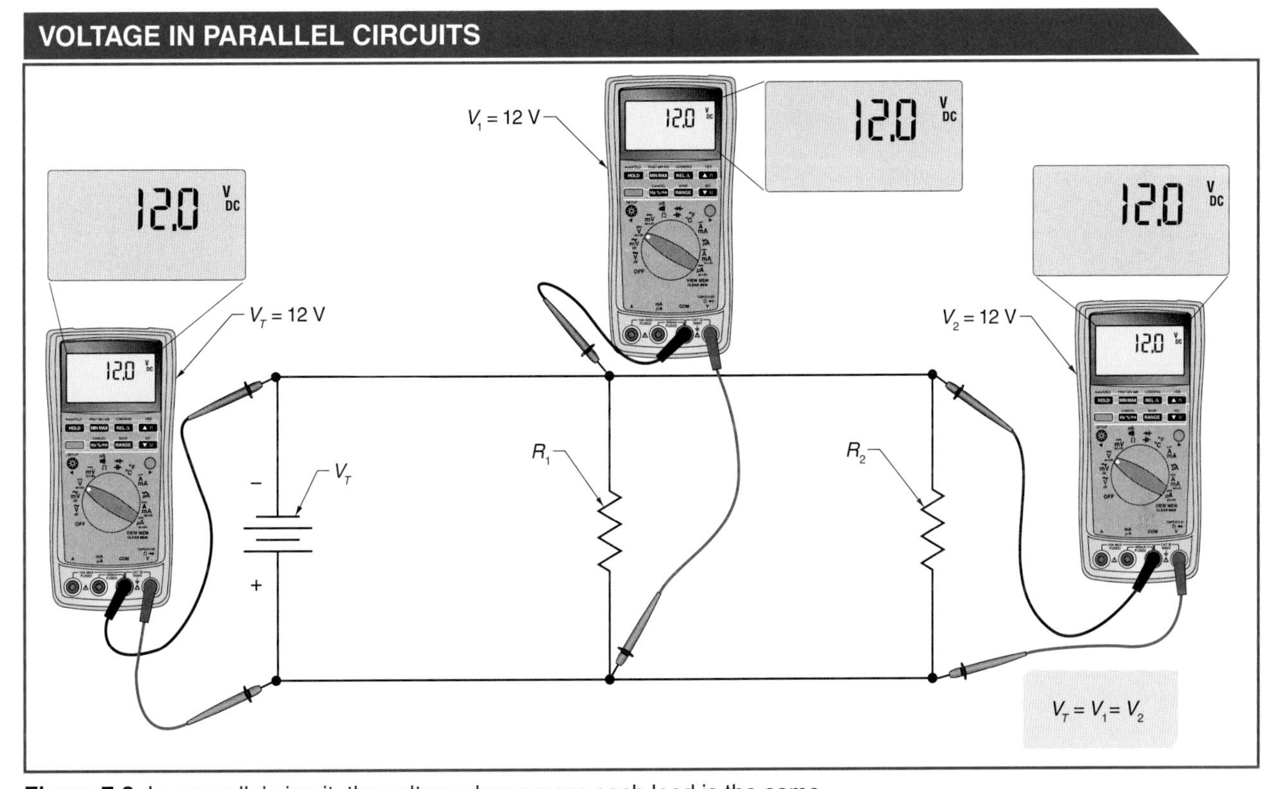

Figure 7-2. In a parallel circuit, the voltage drop across each load is the same.

Ohm's law can be also used to calculate total circuit voltage as well as the voltage across any resistor by multiplying the current in a branch circuit by the resistance of the branch. When the value of a resistor and current flowing through it are known or measured, the voltage across the resistor is calculated by applying the following formula:

$$V_N = I_N \times R_N$$

where

V_N = voltage across any branch circuit (in V)

I_N = current flow through any resistor (in A)

R_N = resistance of any resistor (in Ω)

Example: What is the voltage in a parallel circuit branch that has a resistor of 1.2 kΩ and a current of 5 mA flowing through it? **See Figure 7-3.**

$V_2 = I_2 \times R_2$

$V_2 = 0.005\ \text{A} \times 1200\ \Omega$

$V_2 = \mathbf{6\ V}$

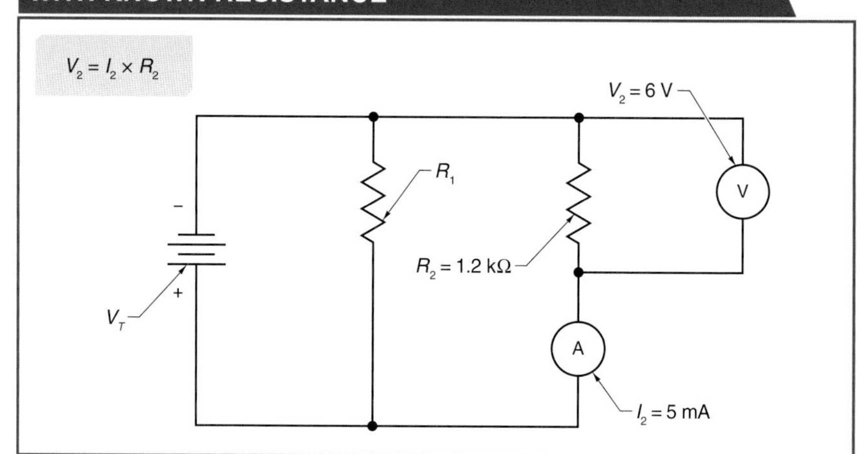

Figure 7-3. Ohm's law can be used to calculate the voltage drop across any branch of a parallel circuit when the resistance and current flow through the resistor are known or measured.

Current in Parallel Circuits

Total current in a circuit containing parallel-connected loads equals the sum of the current through all the loads. Total current increases if loads are added in parallel and decreases if loads are removed. Total current in a parallel circuit is calculated by applying the following formula:

$I_T = I_1 + I_2 + I_3 + I_4 + \ldots + I_N$

where

I_T = total circuit current (in A)

I_1 = current through load 1 (in A)

I_2 = current through load 2 (in A)

I_3 = current through load 3 (in A)

I_4 = current through load 4 (in A)

I_N = current through load N (in A)

Example: What is the total current in a circuit containing four loads connected in parallel if the current through the four loads is 0.833 A, 6.25 A, 8.33 A and 1.2 A? **See Figure 7-4.**

$I_T = I_1 + I_2 + I_3 + I_4$

$I_T = 0.833\text{ A} + 6.25\text{ A} + 8.33\text{ A} + 1.2\text{ A}$

$I_T = \mathbf{16.613\text{ A}}$

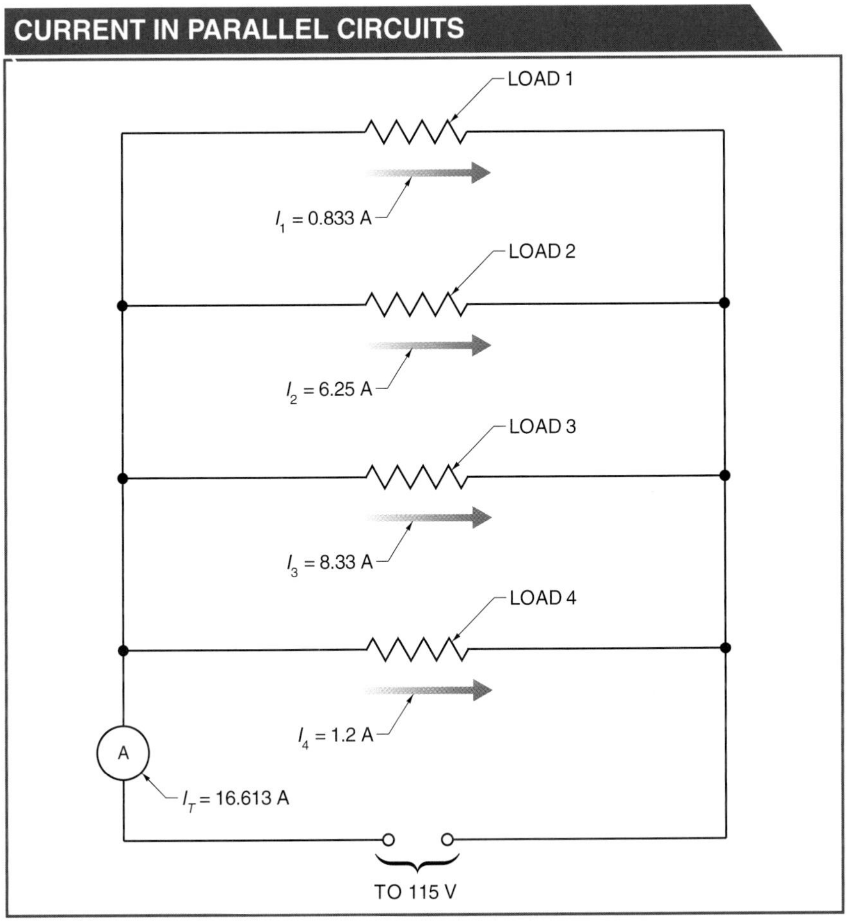

CURRENT IN PARALLEL CIRCUITS

Figure 7-4. Total current in a parallel circuit equals the sum of the current through all the loads.

If the voltage and resistance values are known, the current in any branch can be calculated by applying Ohm's law. Current flow through a branch circuit when the voltage and resistance values are known or measured is calculated by applying the following formula:

$$I_N = \frac{V_N}{R_N}$$

where

I_N = current through any branch (in A)

V_N = voltage through any branch (in V)

R_N = resistance through any branch (in Ω)

Example: What is the total current flow and the current flow through each branch of a circuit that has a total voltage of 6 V connected in parallel with a 1.2 kΩ resistor and a 2.4 kΩ resistor? **See Figure 7-5.**

Current flow through branch circuit 1:

$$I_1 = \frac{V_1}{R_1}$$

$$I_1 = \frac{6\text{ V}}{1200\,\Omega}$$

$$I_1 = 0.005\text{ A (5 mA)}$$

CALCULATING TOTAL PARALLEL CIRCUIT CURRENT

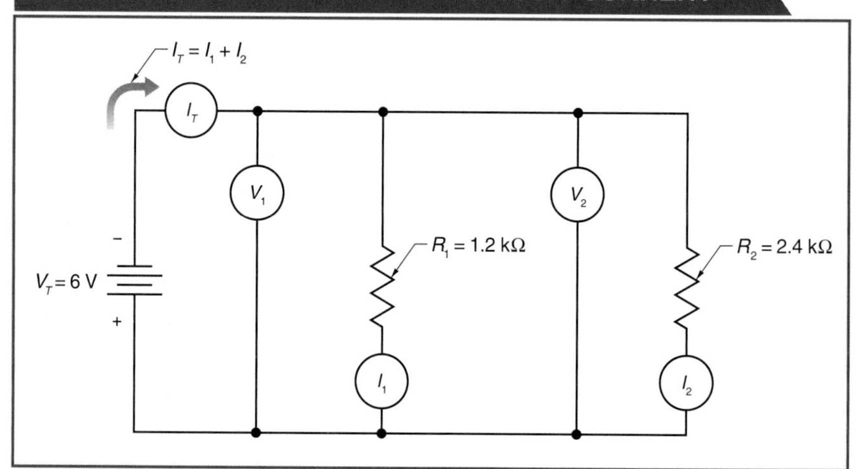

Figure 7-5. The total current in a parallel circuit can be found by applying Ohm's law (to calculate the current through each branch) and then summing all of the branch currents.

Current flow through branch circuit 2:

$$I_2 = \frac{V_2}{R_2}$$

$$I_2 = \frac{6\text{ V}}{2400\,\Omega}$$

$$I_2 = 0.0025\text{ A (2.5 mA)}$$

Total current flow:

$$I_T = I_1 + I_2$$
$$I_T = 0.005\text{ A} + 0.0025\text{ A}$$
$$I_T = \mathbf{0.0075\text{ A (7.5 mA)}}$$

Parallel Circuit Current Application

Current distribution in parallel-connected circuits equals the sum of the current through all branch circuits. Total current increases if parallel loads are added and decreases if parallel loads are removed. For example, a fan circuit may have an indicator lamp connected in parallel to provide a visual indication of fan operation. The current through the fan motor and indicator lamp are added to determine the total circuit current. Knowing the total circuit current allows the selection of properly sized electrical conductors and components. **See Figure 7-6.**

Example: What is the total current in a parallel circuit if the current in each branch is 0.2 A and 0.05 A?

$I_T = I_1 + I_2$

$I_T = 0.2\text{ A} + 0.05\text{ A}$

$I_T = \mathbf{0.25\text{ A}}$

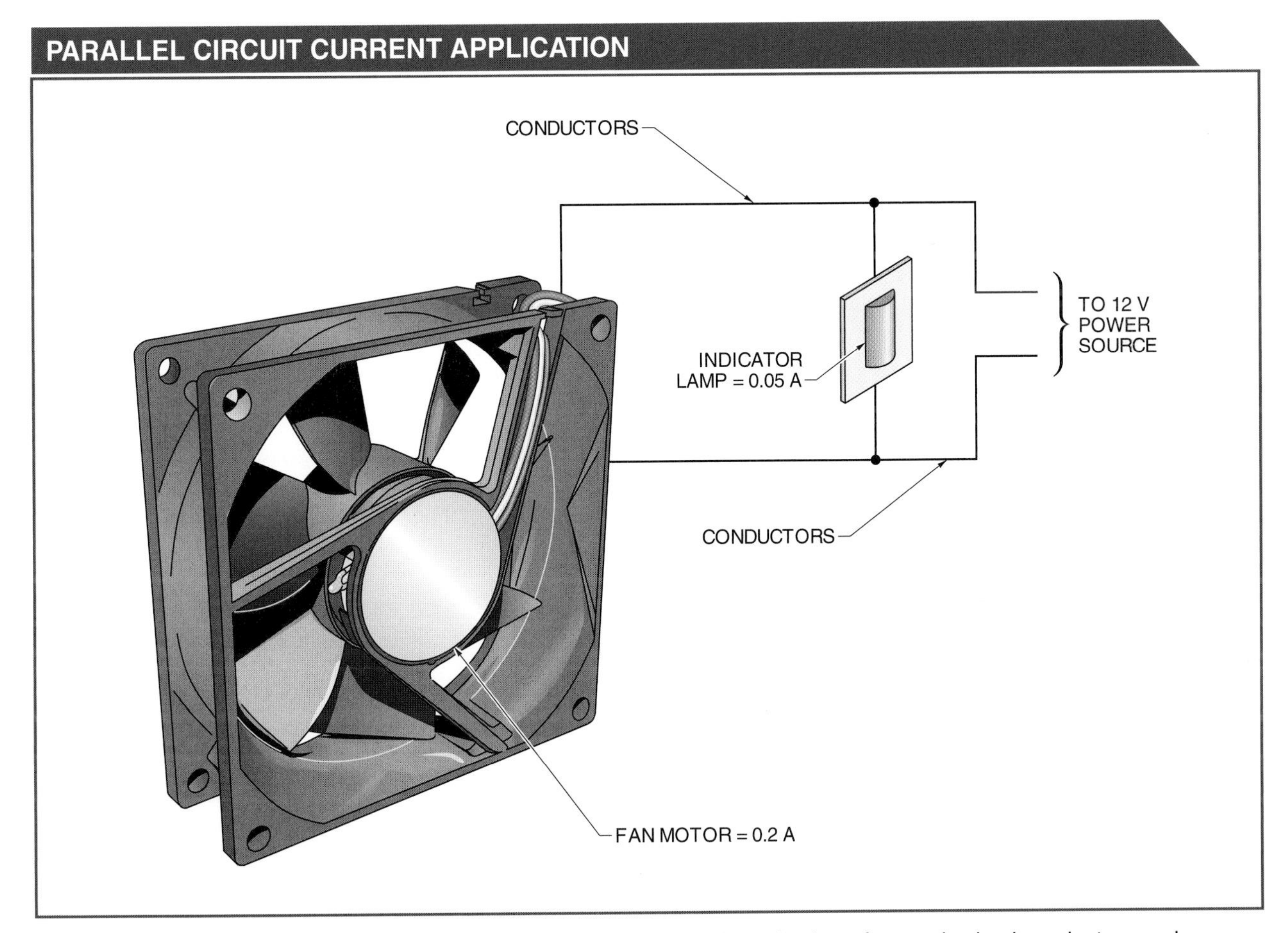

Figure 7-6. Knowing the total current in a parallel circuit allows for the selection of properly sized conductors and protection devices.

Terms

Resistance is the opposition to electron flow.

Resistance in Parallel Circuits

Resistance is the opposition to electron flow. The total resistance in a circuit containing parallel-connected loads is less than the smallest resistance value. The total resistance is lower because the resistors connected in parallel act as additional current pathways. Adding resistors in parallel results in a higher current for the same applied voltage: the resistance is reduced and the conductance improved. **See Figure 7-7.** Total resistance in a circuit containing two resistors connected in parallel is calculated by applying the "product over sum" method:

$$R_T = \frac{(R_1 \times R_2)}{(R_1 + R_2)}$$

where

R_T = total resistance (in Ω)

R_1 = resistance 1 (in Ω)

R_2 = resistance 2 (in Ω)

RESISTANCE IN PARALLEL CIRCUITS

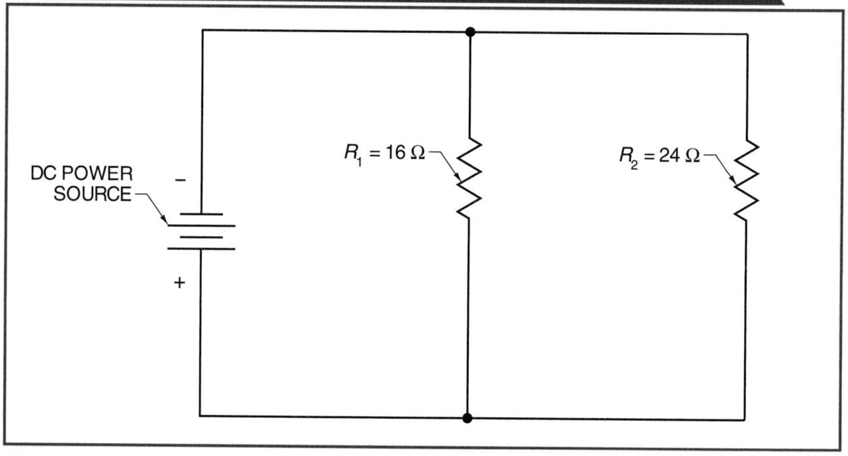

Figure 7-7. The total resistance in a circuit containing parallel-connected loads is less than the smallest resistance value.

Example: What is the total resistance in a circuit containing resistors of 16 Ω and 24 Ω connected in parallel?

$$R_T = \frac{(R_1 \times R_2)}{(R_1 + R_2)}$$

$$R_T = \frac{(16\,\Omega \times 24\,\Omega)}{(16\,\Omega + 24\,\Omega)}$$

$$R_T = \frac{384\,\Omega}{40\,\Omega}$$

$R_T = \mathbf{9.6}\ \Omega$

The product over sum method can also be used to calculate the equivalent resistance in a circuit containing more than two resistors connected in parallel. This equivalent resistance value is combined with the next selected resistance value to calculate the next equivalent resistance. This method is continued until a single equivalent resistance has been calculated.

The total resistance in a circuit containing more than two resistors connected in parallel may also be calculated by applying the following formula:

$$R_T = \frac{1}{\left\{\left(\frac{1}{R_1}\right)+\left(\frac{1}{R_2}\right)+\left(\frac{1}{R_3}\right)+\ldots+\left(\frac{1}{R_N}\right)\right\}}$$

where

R_T = total resistance (in Ω)
R_1 = resistance in circuit 1 (in Ω)
R_2 = resistance in circuit 2 (in Ω)
R_3 = resistance in circuit 3 (in Ω)
R_N = resistance in circuit N (in Ω)

Example: What is the total resistance in a circuit containing resistors of 16 Ω, 24 Ω, and 48 Ω connected in parallel?

$$R_T = \frac{1}{\left\{\left(\frac{1}{R_1}\right)+\left(\frac{1}{R_2}\right)+\left(\frac{1}{R_3}\right)\right\}}$$

$$R_T = \frac{1}{\left\{\left(\frac{1}{16\,\Omega}\right)+\left(\frac{1}{24\,\Omega}\right)+\left(\frac{1}{48\,\Omega}\right)\right\}}$$

$$R_T = \frac{1}{0.0625+0.04167+0.02083}$$

$$R_T = \frac{1}{0.125}$$

$R_T = \mathbf{8}\ \Omega$

Placing identical conductors in parallel results in double the electron flow and one-half the resistance compared to when only one conductor is connected to the source. For example, a single conductor has a diameter large enough to allow a single electron to flow past any point. If voltage is held constant and a second conductor of the same size is added in parallel, then two electrons flow past a given point simultaneously, doubling the electron flow. **See Figure 7-8.** Since the electron flow is doubled and voltage held constant, the total circuit resistance is reduced by 50%. Total circuit resistance in a parallel circuit when voltage and current are known or measured can be calculated by applying Ohm's law:

$$R_T = \frac{V_T}{I_T}$$

where

R_T = total resistance of circuit (in Ω)

V_T = total voltage of circuit (in V)

I_T = total current of circuit (in A)

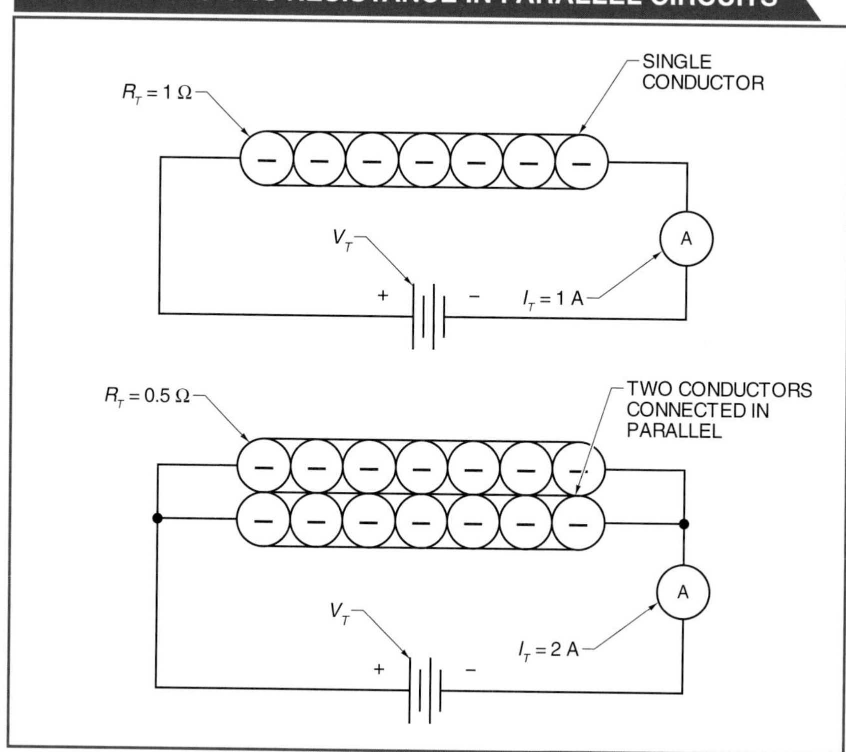

Figure 7-8. Placing identical conductors in parallel results in double the electron flow and one-half the resistance compared to when only one conductor is connected to the source.

Example: What is the total resistance in a parallel circuit that has a total current of 2.5 mA and a total voltage of 6 V?

$$R_T = \frac{V_T}{I_T}$$

$$R_T = \frac{6\text{ V}}{0.0025\text{ A}}$$

$$\mathbf{R_T = 2400\ \Omega\ (2.4\ k\Omega)}$$

If a second resistor of 1.2 kΩ is added in parallel with the 2.4 kΩ resistor, total circuit resistance decreases and total circuit current increases. **See Figure 7-9.** In this circuit, with the switch open, 2.5 mA flows through the 2.4 kΩ resistor. By closing the switch, the 1.2 kΩ resistor is added to the circuit in parallel. The new total resistance of the circuit can be calculated by applying the product over sum method.

Example: What is the total resistance of a circuit with resistors of 2.4 kΩ and 1.2 kΩ connected in parallel?

$$R_T = \frac{R_1 \times R_2}{R_1 + R_2}$$

$$R_T = \frac{2400\ \Omega \times 1200\ \Omega}{2400\ \Omega + 1200\ \Omega}$$

$$R_T = \frac{2,880,000\ \Omega}{3600\ \Omega}$$

$$\mathbf{R_T = 800\ \Omega}$$

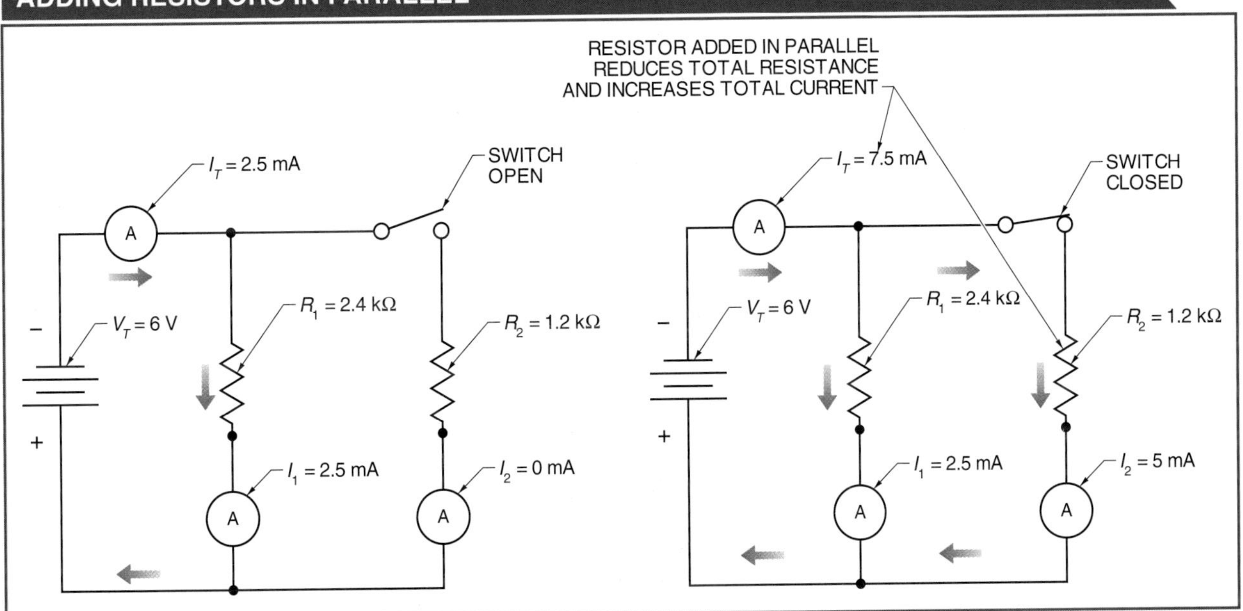

Figure 7-9. Adding a resistor across a load results in a higher current and a reduced circuit resistance.

The reduced resistance allows a greater current flow. Current flow through the branch circuit is calculated by applying Ohm's law. In this branch circuit with the switch closed, 5 mA flows through the 1.2 kΩ resistor.

$$I_2 = \frac{V_2}{R_2}$$

$$I_2 = \frac{6\text{ V}}{1200\,\Omega}$$

I_2 = **0.005 A (5 mA)**

The total circuit current is now 7.5 mA.

$I_T = I_1 + I_2$

I_T = 0.0025 A + 0.005 A

I_T = **0.0075 A (7.5 mA)**

The total resistance of any number of resistors connected in parallel can also be calculated by applying the following formula:

$$R_T = \frac{1}{S_1 + S_2 + S_3 + \ldots + S_N}$$

where

R_T = total resistance (in Ω)

S_1 = conductance of branch 1 (in S)

S_2 = conductance of branch 2 (in S)

S_2 = conductance of branch 3 (in S)

S_N = conductance of branch N (in S)

This formula is referred to as the reciprocal method because reciprocals of the branch resistance are added. This reflects the increase of the current flow (conductance) of the circuit. When only two resistors are in parallel, the reciprocal method reduces to the product over the sum of the resistor values.

Example: What is the total resistance in a parallel circuit that has resistor values of 1000 Ω, 2000 Ω, 4000 Ω, and 8000 Ω?

Conductance of resistor 1:

$$S_1 = \frac{1}{R_1}$$

$$S_1 = \frac{1}{1000}$$

S_1 = 0.001

Conductance of resistor 2:

$$S_2 = \frac{1}{R_2}$$

$$S_2 = \frac{1}{2000\ \Omega}$$

$S_2 = 0.0005$ S

Conductance of resistor 3:

$$S_3 = \frac{1}{R_3}$$

$$S_3 = \frac{1}{4000\ \Omega}$$

$S_3 = 0.00025$ S

Conductance of resistor 4:

$$S_4 = \frac{1}{R_4}$$

$$S_4 = \frac{1}{8000\ \Omega}$$

$S_4 = 0.000125$ S

Total resistance is equal to 533.33 Ω.

$$R_T = \frac{1}{S_1 + S_2 + S_3 + S_4}$$

$$R_T = \frac{1}{0.001\ \text{S} + 0.0005\ \text{S} + 0.00025\ \text{S} + 0.000125\ \text{S}}$$

$$R_T = \frac{1}{0.001875}$$

$R_T =$ **533.33** Ω

Use of these methods helps determine the resistance of parallel branch circuits and the entire system as well as verify test meter readings.

Ballast Resistor Calculation

In some applications, parallel circuits are used when a constant current flow from the voltage source is needed. In these circuits, ballast resistors are added in parallel to produce the constant flow of current. The ballast resistor size is determined by the parallel voltage divided by the amount of desired current. Ballast resistor size is calculated by applying the following formula:

$$R_B = \frac{V_T}{I_B}$$

where

R_B = ballast resistor value (in Ω)

V_T = voltage across ballast resistor (in V)

I_B = current through ballast resistor (in A)

Example: What size ballast resistor is required in a circuit with a 50 V source and a circuit load of 1000 Ω in which a constant current of 200 mA is needed? **See Figure 7-10.**

1. Calculate circuit load current without ballast resistor:

$$I_1 = \frac{V_T}{R_1}$$

$$I_1 = \frac{50\text{ V}}{1000\ \Omega}$$

$$I_1 = 0.05\text{ A}$$

2. Calculate current required for ballast resistor:

$$I_B = I_T - I_1$$
$$I_B = 0.2\text{ A} - 0.05\text{ A}$$
$$I_B = 0.15\text{ A}$$

3. Calculate resistance of ballast resistor:

$$R_B = \frac{V_T}{I_B}$$

$$R_B = \frac{50\text{ V}}{0.15\text{ A}}$$

$$R_B = \mathbf{333.33\ \Omega}$$

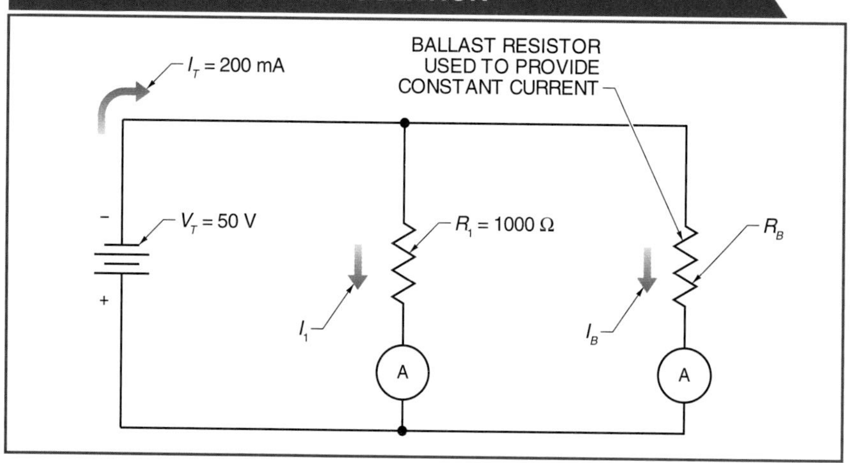

Figure 7-10. Ballast resistors are added in parallel to produce a constant flow of current.

Power in Parallel Circuits

Power is produced when voltage is applied to a load and current flows through the load. The power produced is used to produce light, heat, rotary motion, or linear motion. The lower the resistance of the load or the higher the applied voltage, the more power produced. The higher the resistance of the load or the lower the applied voltage, the less power produced. **See Figure 7-11.** Power is measured in watts. Total power in a parallel circuit is equal to the sum of the power produced by each load. Total power in a parallel circuit when the power across each load is known or measured is calculated by applying the following formula:

$$P_T = P_1 + P_2 + P_3 + \ldots + P_N$$

where

P_T = total circuit power (in W)
P_1 = power of load 1 (in W)
P_2 = power of load 2 (in W)
P_3 = power of load 3 (in W)
P_4 = power of load 4 (in W)
P_N = power of load N (in W)

Example: What is the total circuit power if four loads are connected in parallel and the loads produce 100 W, 750 W, 1000 W, and 150 W?

$$P_T = P_1 + P_2 + P_3 + P_4$$
$$P_T = 100\ \text{W} + 750\ \text{W} + 1000\ \text{W} + 150\ \text{W}$$
$$P_T = \mathbf{2000\ W}$$

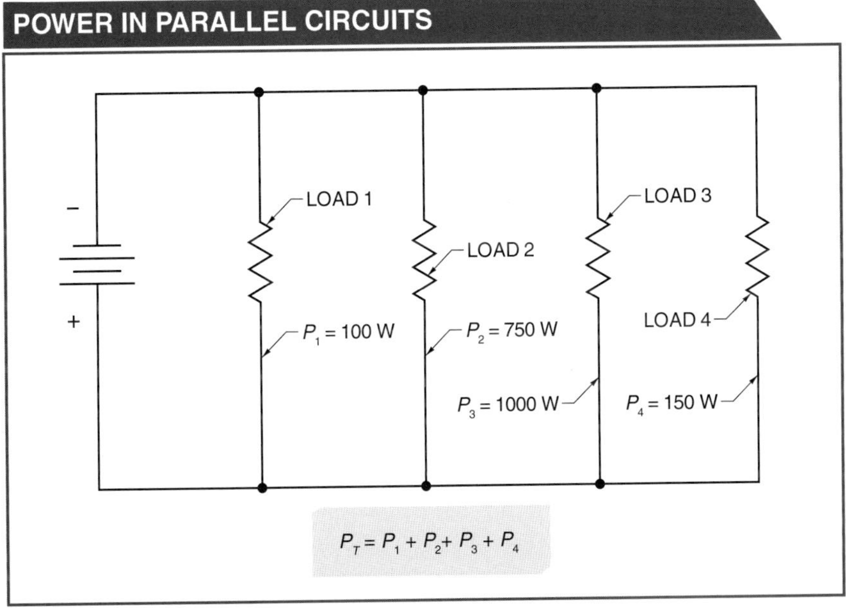

Figure 7-11. Total power in a circuit is equal to the sum of the power produced by each load.

Total power in a circuit is equal to the total voltage multiplied by total circuit current. In addition, the power produced by a load is equal to the voltage drop across the load times the current through the load. Total power in a circuit when voltage and current are known or measured, is calculated by applying the power formula:

$P_T = V_T \times I_T$

where

P_T = total power (in W)

V_T = total voltage (in V)

I_T = total current (in A)

Example: What is the power in a 115 V circuit with a current flow of 17.391 A?

$P_T = V_T \times I_T$

$P_T = 115 \text{ V} \times 17.391 \text{ A}$

$P_T = \mathbf{2000 \text{ W}}$

The total circuit current can be calculated by dividing total circuit power by total circuit voltage. Total current in a parallel circuit when total circuit power and total circuit voltage are known or measured is also calculated by applying the power formula:

$$I_T = \frac{P_T}{V_T}$$

where

I_T = total current (in A)

P_T = total power (in W)

V_T = total voltage (in V)

Example: What is the total current in a parallel circuit that has 115 V applied to it and produces 2000 W of power?

$$I_T = \frac{P_T}{V_T}$$

$$I_T = \frac{2000 \text{ W}}{115 \text{ V}}$$

$I_T = \mathbf{17.391 \text{ A}}$

Lighting Circuit Power Application

Power is calculated to determine the total power consumption in a circuit prior to the installation of additional loads on the same circuit. For example, four lamps with different wattages are connected in a parallel circuit. **See Figure 7-12.** Before the installation of additional loads, the power consumed is calculated to determine if the circuit can safely handle the additional loads. Once the total power is found, the total system current is calculated to ensure proper circuit operation.

LIGHTING CIRCUIT POWER APPLICATION

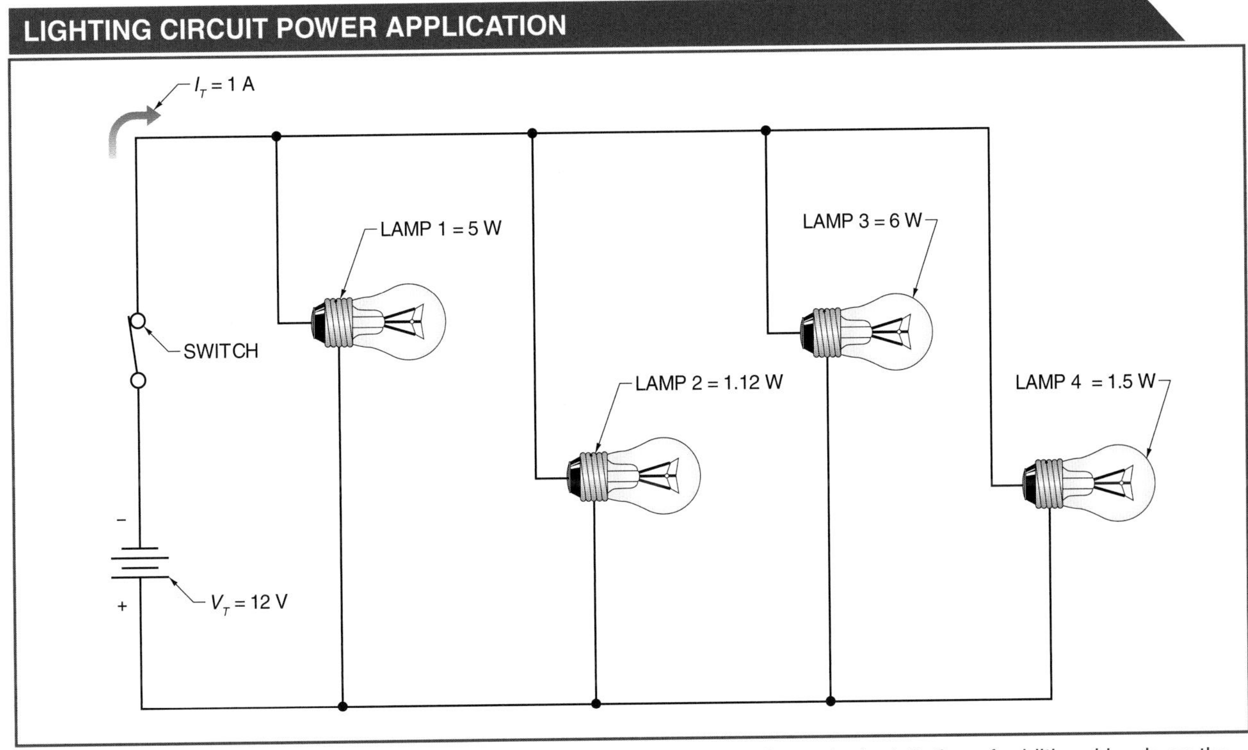

Figure 7-12. Total power consumption on a circuit must be determined prior to the installation of additional loads on the same circuit.

Example: Can a 1 A, 12 V power source support four separate lamps of 5 W, 1.12 W, 6 W, and 1.5 W that are connected in parallel?

1. Calculate total power:

$$P_T = P_1 + P_2 + P_3 + P_4$$

$$P_T = 5\text{ W} + 1.12\text{ W} + 6\text{ W} + 1.5\text{ W}$$

$$P_T = 13.62\text{ W}$$

2. Calculate total current:

$$I_T = \frac{P_T}{V_T}$$

$$I_T = \frac{13.62\text{ W}}{12\text{ V}}$$

$$I_T = \mathbf{1.135\ A}$$

This circuit cannot support the four separate lamps because the total current required by the four lamps is 1.135 A and maximum current from the power source is 1 A.

Terms

An **open circuit** is an electrical circuit in which the flow of electrons is interrupted.

A **short circuit** is any circuit in which current takes a shortcut around the normal path of current flow.

Open and Short Circuits

An *open circuit* is an electrical circuit in which the flow of electrons is interrupted. In an open circuit, two parts of the circuit are not connected by any device or conductor, opening the path of electron flow.

Unlike a series circuit, where an open at any point in the circuit causes the electron flow to stop in the entire circuit, the location of an open in a parallel circuit determines whether or not a branch or the entire circuit fails. Typical parallel circuit failures could be the result of a problem with the voltage source, such as a tripped circuit breaker, blown fuse, or disconnected power source. If the voltage source is found to be working properly, other possible causes must be checked.

Fluke Corporation

Power quality meters are used to maintain power systems, troubleshoot power problems, and diagnose equipment failures.

An open circuit causes zero current flow. Zero current flow in a circuit indicates that there is an open between the power source and a load or switch in the circuit. Another cause of zero current flow would be all loads being open. For example, if a main switch is open, then power is removed from all loads. If the main switch is closed, then power is supplied to all the loads and each load can be opened or closed without affecting the operation of other loads on the circuit. Using parallel power distribution permits various loads to be connected to the same power source. Parallel-connected loads can be controlled as a group or as individual devices.

A *short circuit* is any circuit in which current takes a shortcut around the normal path of current flow. The current through a short circuit can be any value because the resistance of the short circuit is 0 Ω.

If one component in a parallel circuit has a short circuit, the total resistance of the circuit is reduced to 0 Ω. For example, if there are three devices connected to a circuit and one of the devices short circuits, the total current in the circuit increases to an undefined value. Under these conditions, the conductors and their insulation are destroyed due to overheating. Overload protection devices are required in all circuits to prevent damage due to overheating.

SUMMARY

Parallel circuits have components connected in parallel across the same voltage source. In a DC parallel circuit, current flows in one direction. Voltage is the same across all the components that are in parallel, and source current is equal to the sum of the currents in the parallel branches. Current through a branch circuit is inversely proportional to the values of the resistors in the other branches, that is, the largest current flows through the smallest resistance and the smallest current flows through the largest resistance.

Total resistance of a parallel circuit is equal to the reciprocal of the sum of the reciprocals of the individual resistors. Adding a branch to a parallel circuit decreases total resistance and increases total current while removing a branch from a parallel circuit increases total resistance and decreases total current. A short-circuited branch circuit has a resistance of zero and an open branch circuit has infinite resistance. Total power consumed in a parallel circuit is equal to the sum of the power consumed by each component.

Parallel circuits are typically used in applications in which individually controlled loads require the same voltage but draw a different amount of current. Ohm's law and the power formula can be used to determine voltage, current, power, and resistance values in a parallel circuit.

APPLICATION: AMMETERS...

Background: An ammeter is a test instrument that measures current. To take measurements, the ammeter is placed in series with the current. **See Figure 1.** A basic ammeter design is based on a current divider. A current divider is a parallel circuit that allows current to travel through multiple paths. An ammeter has a maximum full scale current. A parallel shunt path is used to limit the amount of current that flows through the ammeter.

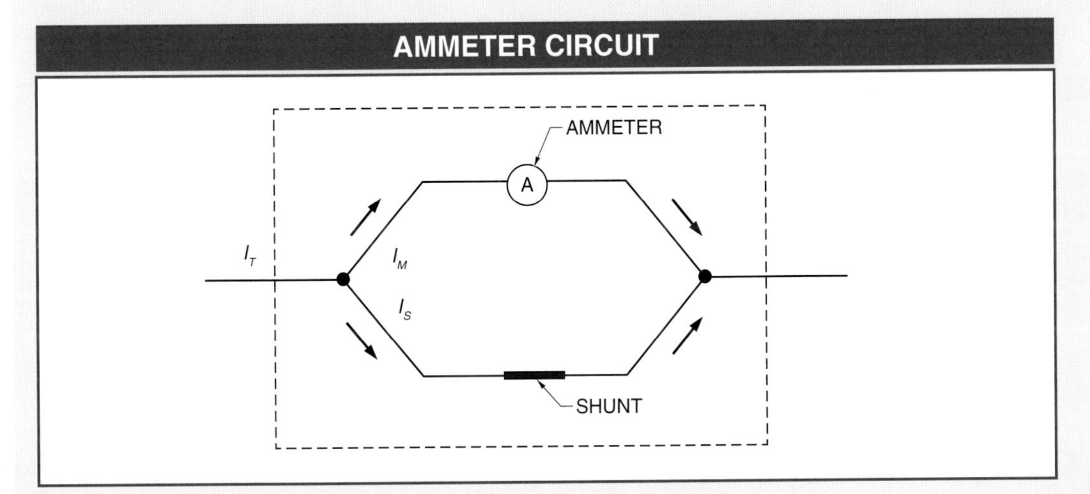

Figure 1. To take measurements, an ammeter is placed in series with current.

...APPLICATION: AMMETERS...

To allow an ammeter to measure current over a wide range, the shunt resistor's reading changes, based on the maximum amount of current that is being measured. An ammeter typically has a switch to select the maximum current. The ammeter characteristics are resistance and full scale current. For example, an ammeter may be 500 Ω with a full scale current of 1mA. In the circuit, the ammeter is a 500 Ω resistor and full needle deflection (maximum reading) occurs when 1 mA is flowing through the meter.

In a parallel circuit, voltage is used as the reference. The voltage across the ammeter and shunt resistor is the same. The design formulas start with the voltage across the ammeter and shunt resistor as follows:

$V_M = I_M \times R_M;$
$V_S = I_S \times R_S$
where
V_M = meter voltage (in V)
I_M = meter current (in A)
R_M = meter resistance (in Ω)
V_S = shunt voltage (in V)
I_S = shunt current (in A)
R_S = shunt resistance (in Ω)

In a parallel circuit, the voltage is the same across all parallel components. Also, the sum of the current through each path equals the total input current.

$V_M = V_S;$
$I_M \times R_M = I_S \times R_S;$
$I_T = I_M + I_S;$
$I_S = I_T - I_M$
where
I_T = total circuit current (in A)

When the maximum current is known along with the ammeter resistance and full scale current, proper shunt resistance can be calculated. The shunt resistance for the maximum current, allows the ammeter to achieve full needle deflection.

$I_M \times R_M = (I_T - I_M) \times R_S;$

$$R_S = \frac{I_M \times R_M}{I_T - I_M}$$

Key Points: Parallel DC circuits; current divider; Ohm's law

Problem: Design an ammeter that is able to measure five different currents: 1 mA, 10 mA, 50 mA, 100 mA, and 1 A. **See Figure 2.** The ammeter in use has a resistance of 100 Ω, and full scale is 50 μA.

...APPLICATION: AMMETERS...

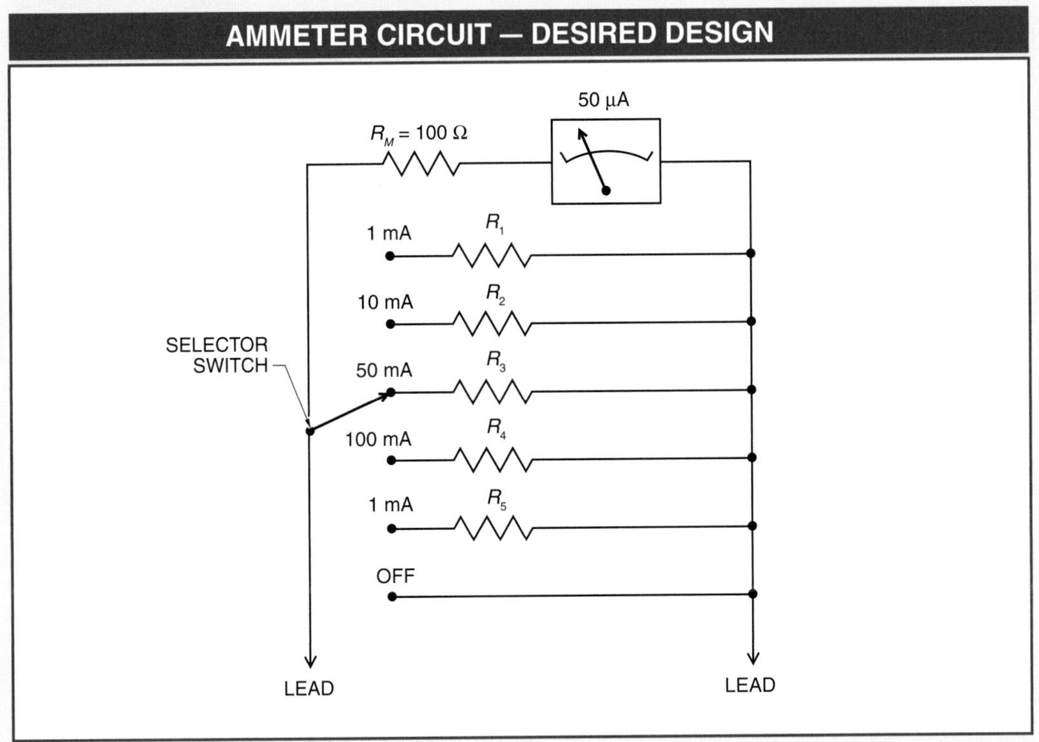

Figure 2. The ammeter must be able to measure five different currents: 1 mA, 10 mA, 50 mA, 100 mA, and 1 A.

Solution: The shunt resistor values to support the measurement of the five different maximum currents must be determined. Using the shunt resistor formula, each resistor value can be calculated. **See Figure 3.**

CURRENT VS. SHUNT RESISTANCE	
Maximum Current (I_T)	**Shunt Resistance (R_S)**
1 mA	R_1 = 5.263 Ω
10 mA	R_2 = 0.503 Ω
50 mA	R_3 = 0.100 Ω
100 mA	R_4 = 0.050 Ω
1A	R_5 = 0.005 Ω

Figure 3. Using the shunt resistor formula, each resistor value can be calculated.

$$R_S = \frac{I_M \times R_M}{I_T - I_M}$$

$R_M = 100\ \Omega$
$I_M = 50\ \mu A$

...APPLICATION: AMMETERS...

$$R_S = \frac{50\ \mu A \times 100\ \Omega}{I_T - 50\ \mu A}$$

$$R_S = \frac{0.005\ A \times 100\ \Omega}{I_T - 50\ \mu A}$$

With a 100 mA maximum current setting, a parallel circuit is created. **See Figure 4.**

AMMETER WITH DC PARALLEL CIRCUIT

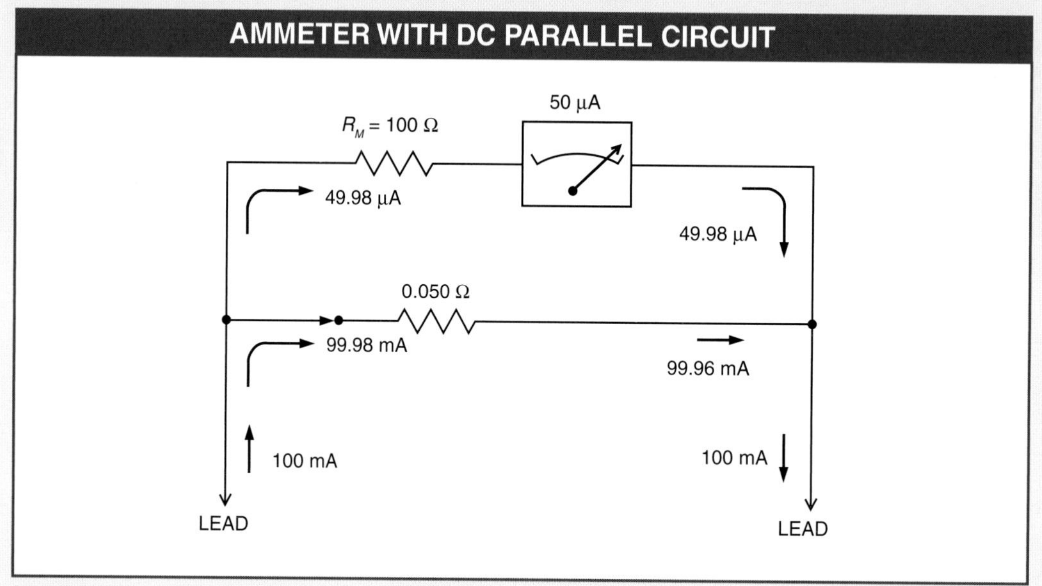

Figure 4. The voltage across the parallel circuit is the total current times the parallel resistance. The parallel resistance is the meter resistance in parallel with the shunt resistor.

The voltage across the parallel circuit is the total current times the parallel resistance. The parallel resistance is the meter resistance in parallel with the shunt resistor.

$$R = \frac{R_M}{R_S}$$

$$R = \frac{1}{\frac{1}{R_M} + \frac{1}{R_S}}$$

$$R = \frac{1}{\frac{1}{100\ \Omega} + \frac{1}{0.050\ \Omega}}$$

$$R = 0.049975\ \Omega$$

$$V = 100\ mA \times 0.049975\ \Omega$$

$$V = 0.004998\ V$$

...APPLICATION: AMMETERS

$$I_M = V \times R_M$$

$$I_M = \frac{0.004998 \text{ V}}{100 \ \Omega}$$

$$I_M = 49.98 \ \mu\text{A}$$

$$I_S = V \times R_S$$

$$I_S = \frac{0.004998 \text{ V}}{0.050 \ \Omega}$$

$$I_S = 99.96 \text{ mA}$$

Checking the circuit shows that most of the current flows through the shunt. When input current is 100 mA, the current causes the ammeter to be at full scale.

This particular ammeter design uses a simple DC parallel circuit to control the maximum current through the meter. It also measures currents greater than the meter can withstand alone.

Chapter Review

1. What is the total resistance of a circuit with resistors of 1.5 kΩ and 3.2 kΩ connected in parallel?
2. What is the total current in a parallel circuit if the current in each branch is 0.8 A and 0.02 A?
3. What is the total current in a circuit containing four loads connected in parallel if the current through the four loads is 0.565 A, 6.77 A, 3.75 A, and 5.14 A?
4. What is an open circuit and what causes the circuit to be open?
5. What is the total resistance in a parallel circuit that has a total current of 5 mA and a total voltage of 8 V?
6. Why can current in a short circuit be of any value but voltage is always 0 V?
7. What is the total resistance in a circuit containing resistors of 8 Ω and 5 Ω connected in parallel?
8. What is the total circuit power of four loads connected in parallel that produce 50 W, 250 W, 1225 W, and 800 W?
9. How is voltage drop affected in a parallel circuit?
10. What is the total current flow and current flow through each branch of a circuit that has a total voltage of 3 V connected in parallel with a 2.2 kΩ resistor and a 1 kΩ resistor?
11. What is the total applied voltage in a circuit if the voltage drop across three loads connected in parallel is 24 V?
12. What is the total resistance in a circuit containing resistors of 2.3 Ω, 6 Ω, and 8.5 Ω connected in parallel?
13. What is the total current in a parallel circuit that has 135 V applied to it and produces 1800 W of power?
14. What is a parallel circuit?
15. What is the voltage in a parallel circuit branch that has a resistor of 3.3 kΩ and current of 8 mA flowing through it?

DC Series/Parallel Circuits

OBJECTIVES

- Explain how resistance, current, voltage, and power affect DC series/parallel circuits.
- Perform calculations to determine resistance, current, voltage, and power in DC series/parallel circuits.
- Explain resistance, current, voltage drop, and power applications in DC series/parallel circuits.

In many direct current (DC) applications, a combination of both series and parallel circuits are used. Techniques for analyzing series/parallel circuits use a combination of series and parallel methods. To calculate unknown values in series/parallel circuits, separate calculations for series-connected and parallel-connected circuits are used. Most electronic circuits contain a combination of series/parallel circuits.

Terms

A **series/parallel circuit** is a circuit that contains a combination of both series-connected and parallel-connected components.

SERIES/PARALLEL CIRCUITS

A *series/parallel circuit* is a circuit that contains a combination of both series-connected and parallel-connected components. **See Figure 8-1.** Series/parallel circuits are often referred to as combination circuits. While some electronic circuits consist of either series-connected components only or parallel-connected components only, most electronic circuits are a combination of series-connected components and parallel-connected components. Switches, loads, meters, fuses, circuit breakers, and other electrical components or devices can be connected together in a series/parallel circuit.

DC SERIES/PARALLEL CIRCUIT

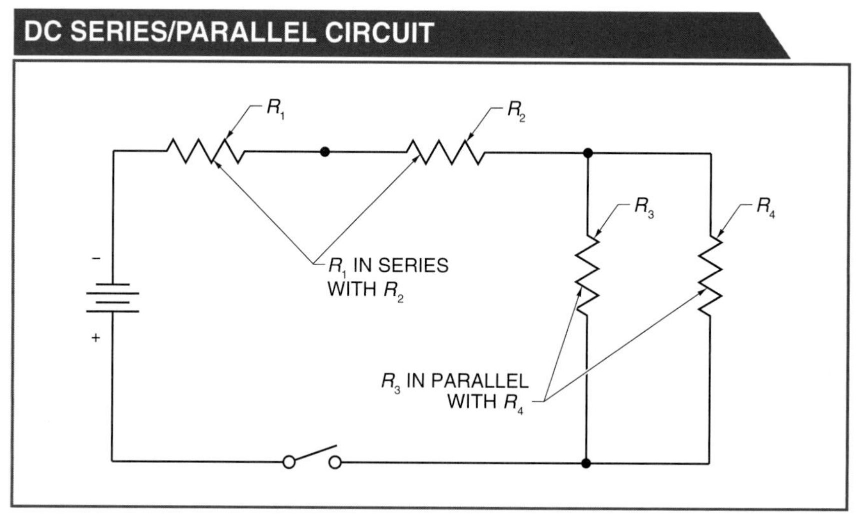

Figure 8-1. A series/parallel circuit contains both series-connected and parallel-connected components.

Safety precautions must be taken when working with series/parallel circuits because current can flow in part of a parallel circuit even though another part of the circuit is turned OFF. In most electrical circuits, loads such as lamps, motors, and solenoids are normally connected in parallel. Devices such as switches, fuses, and circuit breakers used to control and monitor current flow through the loads are connected in series. A digital multimeter (DMM) set to measure voltage is connected in parallel when taking voltage measurements. A DMM set to measure current is connected in series when taking current measurements. Understanding and recognizing series-connected and parallel-connected components and circuits enables a technician to take proper measurements, calculate circuit values, and make circuit modifications.

All DC voltage sources have a positive and a negative terminal (connection point). The positive and negative terminals establish polarity in the circuit. All points in a DC series/parallel circuit have polarity. Each component or device in a DC series/parallel circuit has a positive polarity side and a negative polarity side. The side of the component or device closest to the positive voltage terminal is the positive polarity side and the side of the component or device closest to the negative voltage terminal is the negative polarity side. **See Figure 8-2.**

DC SERIES/PARALLEL CIRCUIT POLARITY

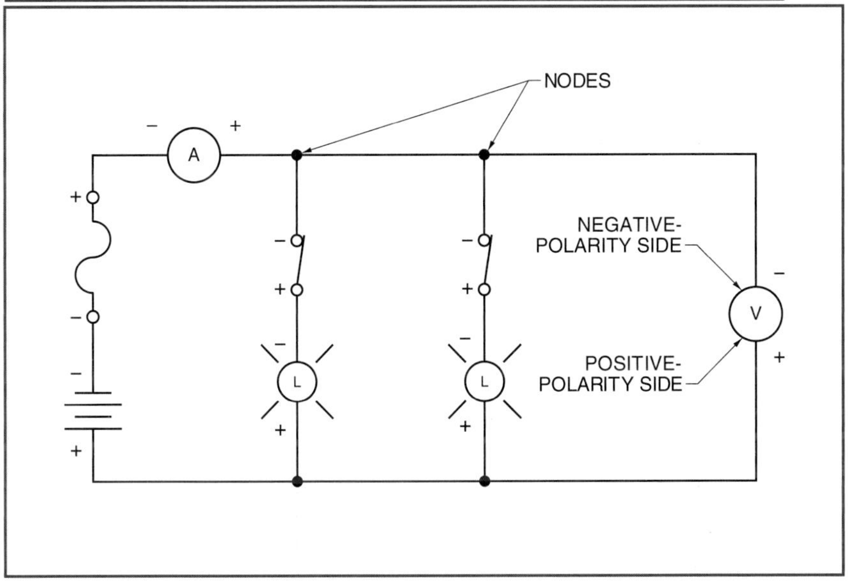

NODES
A
V
L
L
NEGATIVE-POLARITY SIDE
POSITIVE-POLARITY SIDE

Figure 8-2. Each component in a DC series/parallel circuit has a positive polarity side and a negative polarity side.

A *node* is a junction point in an electrical circuit where connections are made between different circuit paths. When working with series/parallel circuits, the current path is traced to the first node where the current is split off into either series or parallel-connected components. A series/parallel circuit has components that split off at the node from a series-connected path to a parallel-connected path.

Switches can be connected both in series and in parallel in a DC circuit. When switches in a circuit are connected in a series/parallel combination, two or more switches must be closed before current can flow. At least one series-connected switch or all parallel-connected switches must be opened to stop current flow. **See Figure 8-3.**

Terms

A **node** is a junction point in an electrical circuit where connections are made between different circuit paths.

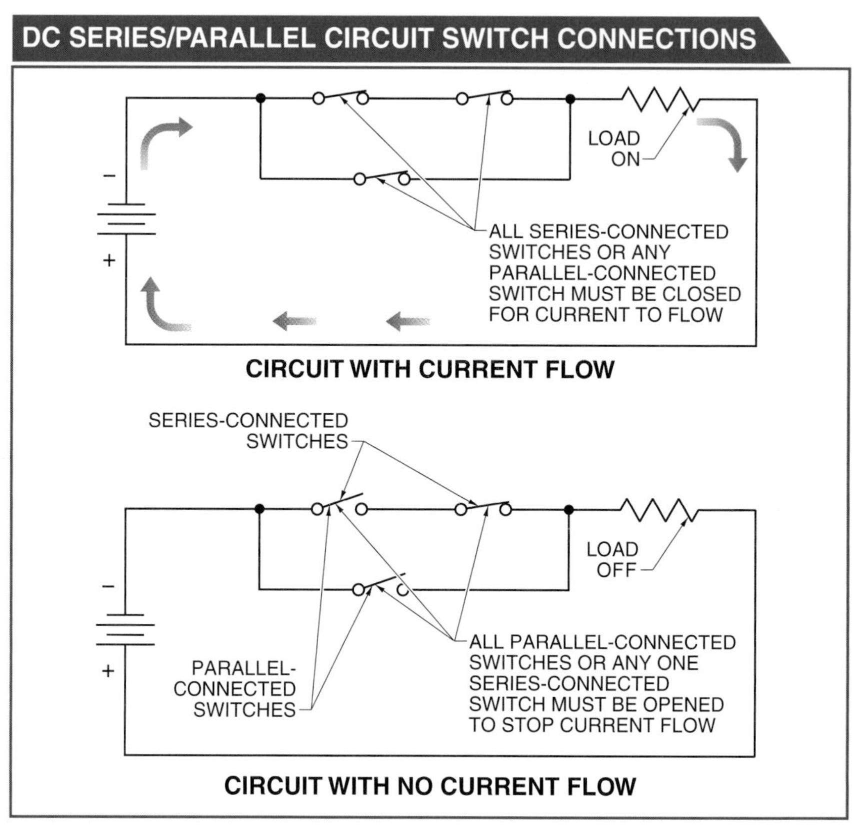

Figure 8-3. At least one series-connected switch or all parallel-connected switches must be opened to stop current flow.

ABB Inc., Drives & Power Electronics

A DC compound motor is an application where magnetic fields are connected in both series and parallel with the rotating part of the motor.

To solve for unknown values such as resistance, current, voltage, and power in a series/parallel circuit, the rules for both series-connected and parallel-connected circuits are applied. In series circuits, total current is equal at every point in the circuit, total resistance is equal to the sum of individual resistances, the sum of the voltage drops across all resistances is equal to the source voltage, and total power is equal to the sum of all loads in the circuit. In parallel circuits, total current is equal to the sum of all branch currents, total resistance is equal to the reciprocal of the sum of the reciprocal of individual resistances, voltage drop across any branch is equal to the source voltage, and power is equal to the sum of all loads in the circuit. **See Figure 8-4.**

DC SERIES/PARALLEL CIRCUIT RULES		
Value	**Series Circuit**	**Parallel Circuit**
Current	Only one current path and total current is same at all parts of circuit $I_T = I_1 = I_2 = I_3 = \ldots = I_N$	More than one current path and total current is equal to sum of current in the branch circuits $I_T = I_1 + I_2 + I_3 + \ldots + I_N$
Resistance	Total resistance is equal to sum of resistor values $R_T = R_1 + R_2 + R_3 + \ldots + R_N$	Total resistance is equal to reciprocal of sum of individual resistances $R = \frac{1}{\frac{1}{R_1} + \frac{1}{R_2} + \frac{1}{R_3} + \ldots + \frac{1}{R_N}}$
Voltage drop	Sum of voltage drops across resistors equals source voltage $V_T = V_1 + V_2 + V_3 + \ldots + V_N$	Voltage drop across each branch is equal to source voltage $V_T = V_1 = V_2 = V_3 = \ldots = V_N$
Power	Power consumed in series or parallel circuits is equal to sum of power consumed by resistors $P_T = P_1 + P_2 + P_3 + \ldots + P_N$	

Figure 8-4. When calculating values in DC series/parallel circuits, the rules of individual series circuits and individual parallel circuits are applied.

Resistance in Series/Parallel Circuits

A load such as a heating element or lamp has resistance in a circuit. Resistances and loads are often connected in a series/parallel combination. A circuit can contain any number of individual resistances connected in any number of different series/parallel circuit combinations. Total resistance in a series/parallel circuit is equal to one equivalent total resistance value. The equivalent total resistance value in a circuit containing series/parallel-connected resistances equals the sum of the series-connected resistances and the reciprocal of the sum of the reciprocal parallel-connected resistances. Total resistance in a circuit containing two resistances connected in parallel and two resistances connected in series can be calculated by applying the following formula:

$$R_T = \left(\frac{R_{P1} \times R_{P2}}{R_{P1} + R_{P2}}\right) + R_{S1} + R_{S2}$$

where

R_T = total resistance (in Ω)

R_{P1} = parallel resistance 1 (in Ω)

R_{P2} = parallel resistance 2 (in Ω)

R_{S1} = series resistance 1 (in Ω)

R_{S2} = series resistance 2 (in Ω)

Example: What is the total resistance in a series/parallel circuit that contains resistances of 10 Ω and 2500 Ω connected in series and resistances of 1000 Ω and 200 Ω connected in parallel?

$$R_T = \left(\frac{R_{P1} \times R_{P2}}{R_{P1} + R_{P2}}\right) + R_{S1} + R_{S2}$$

$$R_T = \left(\frac{1000\ \Omega \times 200\ \Omega}{1000\ \Omega + 200\ \Omega}\right) + 10\ \Omega + 2500\ \Omega$$

$$R_T = \left(\frac{200,000\ \Omega}{1200\ \Omega}\right) + 2510\ \Omega$$

$$R_T = 166.67\ \Omega + 2510\ \Omega$$

$$R_T = \mathbf{2676.67}\ \Omega$$

Total resistance in a series/parallel circuit containing multiple series/parallel combinations can be calculated by breaking down the circuit into its basic series and parallel parts. To calculate an equivalent resistance value, the formula for a series circuit is applied to the series part(s) and the formula for the parallel circuit is applied to the parallel part(s) of a series/parallel circuit. This process is continued until a single equivalent total resistance value is achieved. **See Figure 8-5.**

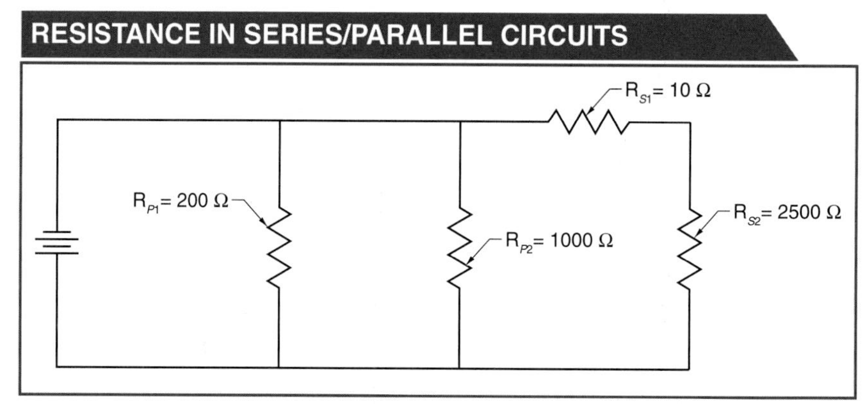

Figure 8-5. Total resistance in a series/parallel circuit containing multiple series/parallel combinations can be calculated by breaking down the circuit into its basic series and parallel parts.

For example, to calculate the total resistance in a series/parallel circuit containing multiple series/parallel combinations, the series-connected resistances and parallel-connected resistances are combined into a single resistance value. The equivalent resistance value of two resistances connected in series can be calculated by applying the following procedure:

1. Combine resistances connected in series into a single resistance value. Resistances connected in series are combined into a single resistance value by applying the following formula:

$R_{ST1} = R_{S1} + R_{S2}$

where

R_{ST1} = total series resistance 1 (in Ω)

R_{S1} = series resistance 1 (in Ω)

R_{S2} = series resistance 2 (in Ω)

2. Combine resistances connected in parallel into a single resistance value. Two resistances connected in parallel are combined into a single resistance value by applying the following formula:

$$R_{PT1} = \frac{R_{P1} \times R_{P2}}{R_{P1} + R_{P2}}$$

where

R_{PT1} = total parallel resistance 1 (in Ω)

R_{P1} = parallel resistance 1 (in Ω)

R_{P2} = parallel resistance 2 (in Ω)

Note: Three or more resistances connected in parallel are combined into a single resistance value by applying the following formula:

$$R_{PT1} = \frac{1}{\frac{1}{R_{P1}} + \frac{1}{R_{P2}} + \frac{1}{R_{P3}} + \ldots + \frac{1}{R_{PN}}}$$

where

R_{PN} = parallel resistance N (in Ω)

3. Combine the sum of series resistance values with the sum of the parallel resistance values. The sum of the series resistance values is combined with the sum of the parallel resistance values by applying the following formula:

$R_T = R_{ST1} + R_{PT2}$

Note: In larger circuits, this process must be applied until all sections of the circuit have been combined. **See Figure 8-6.**

DC Circuit Boards

Henkel Corporation

Electronic circuit boards are used in many DC applications including portable radios, smart phones, tablets, and other electronic devices. Because of the small size of these components, the circuit boards are designed to have densely arranged series, parallel, and series/parallel circuits, which makes visual inspection difficult. Using test instruments such as DMMs, ammeters, and megohmmeters to test for voltage, current, capacitance, and resistance is required when testing circuit board components.

CALCULATING RESISTANCE IN DC SERIES/PARALLEL CIRCUITS

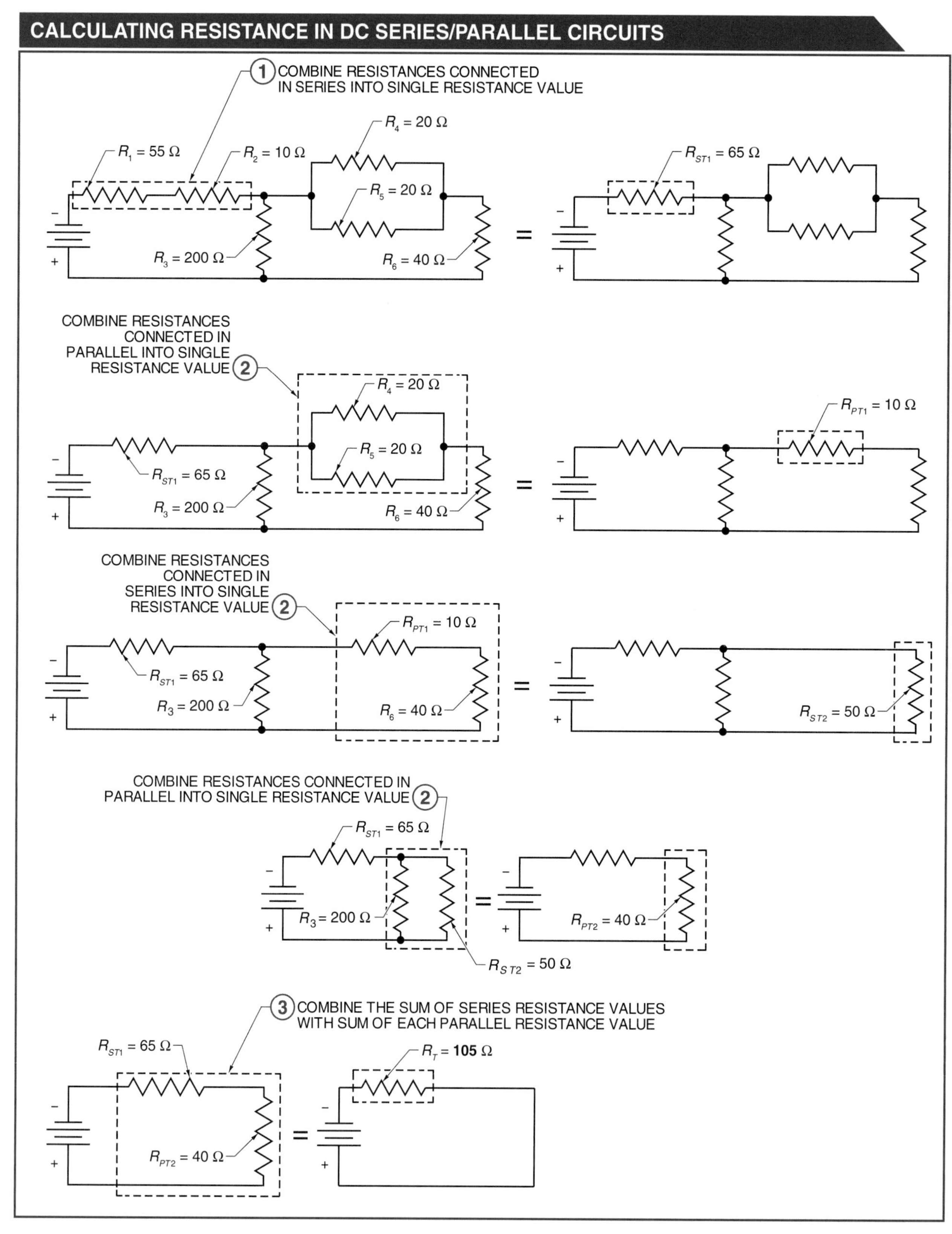

Figure 8-6. To find the total resistance in a series/parallel circuit containing multiple series/parallel combinations, the circuit is broken down to basic series and parallel parts.

Example: What is the total resistance in a circuit that has resistances of 55 Ω (R_{S1}), 10 Ω (R_{S2}), 40 Ω (R_{S3}), 20 Ω (R_{P1}), 20 Ω (R_{P2}), and 200 Ω (R_{P3}), connected in a series/parallel combination?

1. Combine resistances connected in series into a single resistance value.

$R_{ST1} = R_{S1} + R_{S2}$
$R_{ST1} = 55\ \Omega + 10\ \Omega$
$R_{ST1} = 65\ \Omega$

2. Combine resistances connected in parallel into a single resistance value.

$$R_{PT1} = \frac{R_{P1} \times R_{P2}}{R_{P1} + R_{P2}}$$

$$R_{PT1} = \frac{20\ \Omega \times 20\ \Omega}{20\ \Omega + 20\ \Omega}$$

$$R_{PT1} = \frac{400\ \Omega}{40\ \Omega}$$

$$R_{PT1} = 10\ \Omega$$

Combine resistances connected in series into a single resistance value.

$R_{ST2} = R_{PT1} + R_{S3}$
$R_{ST2} = 10\ \Omega + 40\ \Omega$
$R_{ST2} = 50\ \Omega$

Combine resistances connected in parallel into a single resistance value.

$$R_{PT2} = \frac{R_{ST2} \times R_{P3}}{R_{ST2} + R_{P3}}$$

$$R_{PT2} = \frac{50\ \Omega \times 200\ \Omega}{50\ \Omega + 200\ \Omega}$$

$$R_{PT2} = \frac{10{,}000\ \Omega}{250\ \Omega}$$

$$R_{PT2} = 40\ \Omega$$

3. Combine the sum of series resistance values with the sum of each parallel resistance value.

$R_T = R_{ST1} + R_{PT2}$
$R_T = 65\ \Omega + 40\ \Omega$
$R_T = \mathbf{105\ \Omega}$

Resistance Application. Frequently, the only known values in a series/parallel circuit are the values of the resistances and the source voltage. Typically, the parallel branches in the series/parallel circuit are reduced to their equivalent values of resistance.

The equivalent values of resistance become series resistance with other series resistances in the circuit. The resistance values are added to calculate the equivalent total resistance of the circuit. Other circuit parameters such as current, voltage drop, and total power can be calculated using equivalent total resistance of a circuit.

Example: What is the equivalent total resistance in a circuit containing 27 Ω (RP1) and 39 Ω (RP2) resistances connected in parallel and 12 Ω (RS1) and 18 Ω (RS2) resistances connected in series?

$$R_T = \left(\frac{R_{P1} \times R_{P2}}{R_{P1} + R_{P2}}\right) + R_{S1} + R_{S2}$$

$$R_T = \left(\frac{27\ \Omega \times 39\ \Omega}{27\ \Omega + 39\ \Omega}\right) + 12\ \Omega + 18\ \Omega$$

$$R_T = \left(\frac{1053\ \Omega}{66\ \Omega}\right) + 30\ \Omega$$

$$R_T = 15.95\ \Omega + 30\ \Omega$$

$$R_T = \mathbf{45.95\ \Omega}$$

Current in Series/Parallel Circuits

Current in series/parallel circuits follows the same laws as current in a series circuit and current in a parallel circuit. Current is equal in each series branch of a series/parallel circuit and equal to the sum of the current flow through each parallel branch of a series/parallel circuit. **See Figure 8-7.** To calculate total current in a series/parallel circuit, apply the following procedure:

1. Calculate resistance in series branch. Resistance in a series branch can be calculated by applying the following formula:

   $R_T = R_1 + R_2$

   where

   R_T = total resistance (in Ω)

   R_1 = resistance 1 (in Ω)

   R_2 = resistance 2 (in Ω)

2. Calculate current in series branch. Current in a series branch can be calculated by applying the following formula:

   $$I_S = \frac{V_S}{R_S}$$

   where

   I_S = series branch current (in A)

   V_S = series branch voltage (in V)

   R_S = series branch resistance (in Ω)

3. Calculate total current. Total current can be calculated by applying the following formula:

$I_T = I_{P1} + I_{P2}$

where

I_T = total circuit current (in A)

I_{P1} = parallel branch current 1 (in A)

I_{P2} = parallel branch current 2 (in A)

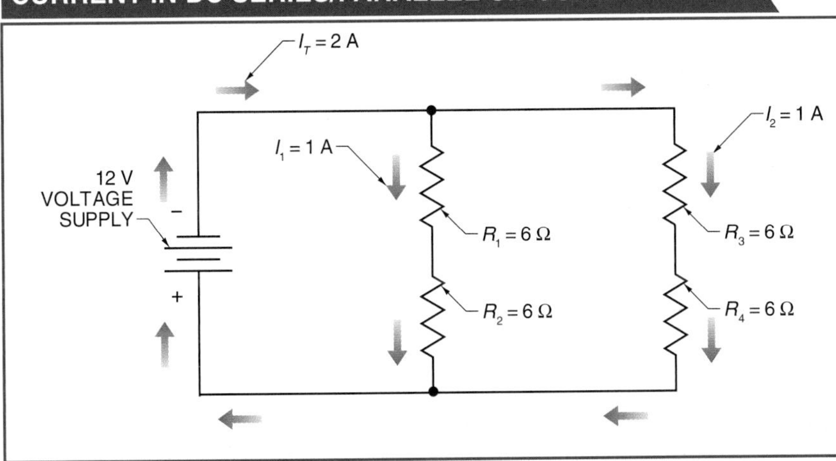

Figure 8-7. Current is equal to the sum of the current flow through each parallel branch of a series/parallel circuit.

Example: What is the total current in a 12 V circuit with two parallel branches containing two 6 Ω resistances connected in series?

1. Calculate resistance in series branch.

$R_T = R_1 + R_2$

$R_T = 6\ \Omega + 6\ \Omega$

$R_T = 12\ \Omega$

2. Calculate current in series branch.

$$I_S = \frac{V_S}{R_S}$$

$$I_S = \frac{12\ \text{V}}{12\ \Omega}$$

$I_S = 1\ \text{A}$

3. Calculate total current.

$I_T = I_{P1} + I_{P2}$

$I_T = 1\ \text{A} + 1\ \text{A}$

$I_T = \mathbf{2\ A}$

U.S. Navy

A digital multimeter can be used to determine current flow when working with electronic circuits.

Terms

A **shunt resistor** is a parallel-connected resistor used to increase the range of an ammeter.

Current flow through two 6 Ω resistances connected in series across a 12 V source voltage is 1 A at any point in a series circuit. Total current flow through two parallel-connected branches that each have 1 A of current flow through them is 2 A.

Current Application. An ammeter is used to measure the amount of current in a circuit. In an analog ammeter, the current is displayed along calibrated scales using a pointer (indicator). A *shunt resistor* is a parallel-connected resistor used to increase the range of an ammeter. The ammeter display has a resistance of 500 Ω and the amount of current required for full scale deflection of the pointer is 50 μA. Since the current to be measured may be greater than the current required for full scale deflection, one or more shunt resistors are used to limit the amount of current that flows through the ammeter display. Different shunt resistor combinations can be selected based on the amount of current being measured. **See Figure 8-8.**

For example, total current to be measured is 1 mA. Since full scale deflection limits the ammeter current to 50 μA, the remainder of the current must flow through the shunt resistors. The total current is equal to the sum of the branch currents, which is 950 μA (1000 μA − 50 μA = 950 μA).

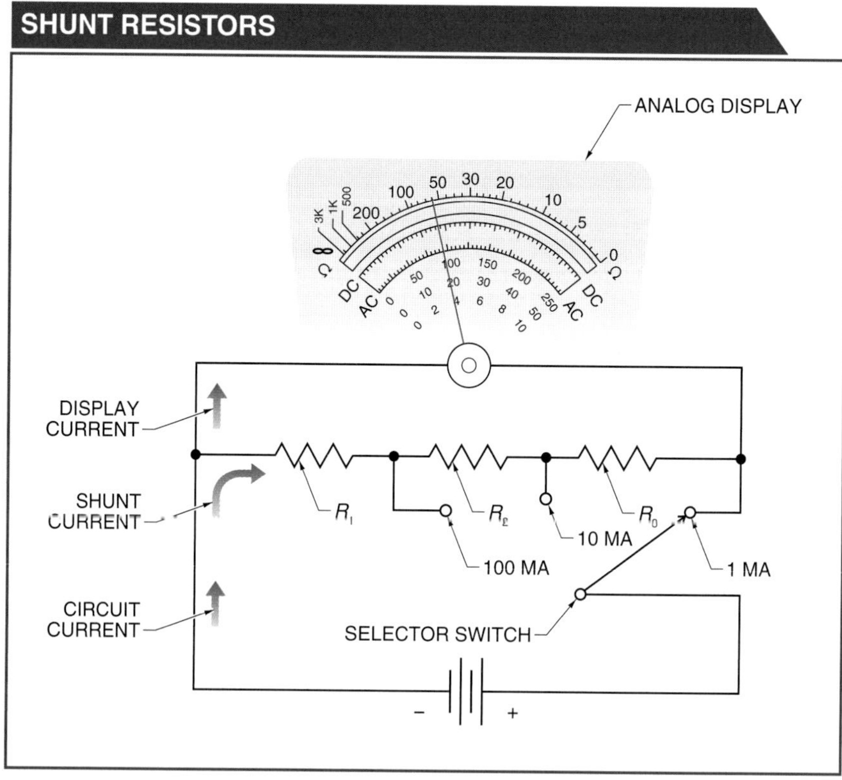

Figure 8-8. An analog ammeter contains different shunt resistor configurations to limit the current flow though the ammeter.

Since the shunt resistor is connected in parallel with the ammeter resistance, the voltage across the shunt resistors is equal to the voltage across the ammeter and can be calculated by applying the following formula:

$$V_{SH} = V_M$$

where

V_{SH} = voltage across shunt resistor (in V)

V_M = voltage across ammeter (in V)

Example: What is the voltage across an ammeter that has 12.2 V across a shunt resistor?

$V_{SH} = V_M$

$V_{SH} = 12.2\text{ V}$

$V_M = \mathbf{12.2\text{ V}}$

Shunt resistance can be calculated when the values of ammeter current and shunt current are known. Shunt resistance is equal to ammeter current multiplied by ammeter resistance, divided by shunt current, and is calculated by applying the following formula:

$$R_{SH} = \frac{I_M \times R_M}{I_{SH}}$$

where

R_{SH} = shunt resistance (in Ω)

I_M = ammeter current (in μA)

R_M = ammeter resistance (in Ω)

I_{SH} = current through shunt resistor (in μA)

Example: What is the shunt resistance in an ammeter that has ammeter current of 50 μA, ammeter resistance of 500 Ω, and current through the shunt resistor of 950 μA?

$$R_{SH} = \frac{50\ \mu\text{A} \times 500\ \Omega}{950\ \mu\text{A}}$$

$$R_{SH} = \frac{25{,}000\ \mu\text{V}}{950\ \mu\text{A}}$$

$$R_{SH} = \mathbf{26.32\ \Omega}$$

Terms

A **current divider** is a parallel circuit that divides the current proportionally between the branches.

The total current in a parallel circuit is inversely proportional to the size of the shunt resistors, as seen in this example, where 50 μA of current flows through a 500 Ω shunt resistor and 950 μA of current flows through a 26.32 Ω shunt resistor. This type of circuit is also referred to as a current divider. A *current divider* is a parallel circuit that divides the current proportionally between the branches. When using an analog ammeter, the resistance values for R_1 and $R_{1,2}$ can be calculated by applying the following formula (current divider formula):

$$R_N = \frac{I_M \times (R_M + R_{SH})}{I_T}$$

where

R_N = ammeter resistance at setting N (in Ω)
I_M = ammeter current (in μA)
R_M = ammeter resistance (in Ω)
R_{SH} = shunt resistance (in Ω)
I_T = total circuit current (in μA)

Example: What are the resistance values of an analog ammeter set at 100 mA and 10 mA respectively, with ammeter current of 50 μA, ammeter resistance of 500 Ω, shunt resistance of 26.32 Ω, and total current of 100,000 μA at the 100 mA setting and 10,000 μA at the 10 mA setting?

$$R_1 = \frac{50\ \mu A \times (500\ \Omega + 26.32\ \Omega)}{100{,}000\ \mu A}$$

$$R_1 = \frac{50\ \mu A \times 526.32\ \Omega}{100{,}000\ \mu A}$$

$$R_1 = \frac{26{,}316\ \mu V}{100{,}000\ \mu A}$$

$$R_1 = 0.263\ \Omega$$

$$R_{1,2} = \frac{50\ \mu A \times (500\ \Omega + 26.32\ \Omega)}{10{,}000\ \mu A}$$

$$R_{1,2} = \frac{50\ \mu A \times 526.32\ \Omega}{10{,}000\ \mu A}$$

$$R_{1,2} = \frac{26{,}316\ \mu V}{10{,}000\ \mu A}$$

$$R_{1,2} = 2.632\ \Omega$$

Because the shunt resistors R_1, R_2, and R_3 are connected in series, the values of resistances R_2 and R_3 are calculated by applying the following formulas:

$$R_2 = R_{1,2} - R_1$$
$$R_2 = 2.632\ \Omega - 0.263\ \Omega$$
$$R_2 = 2.369\ \Omega$$

$$R_3 = R_{SH} - R_{1,2}$$
$$R_3 = 26.32\ \Omega - 2.63\ \Omega$$
$$R_3 = 23.69\ \Omega$$

Resistance calculations can be verified by applying the following formula:

$R_{SH} = R_1 + R_2 + R_3$

$R_{SH} = 0.263\ \Omega + 2.369\ \Omega + 23.69\ \Omega$

$R_{SH} = \mathbf{26.32\ \Omega}$

Voltage in Series/Parallel Circuits

The total source voltage across resistances connected in a series/parallel combination is divided across the individual resistances. The higher the resistance of any one resistance or equivalent parallel resistance, the higher the voltage drop. To calculate the voltage drop across each resistance in a series/parallel circuit, the total circuit resistance and total circuit current are calculated first. To calculate voltage drop across a resistance in a series/parallel circuit, apply the following procedure:

DC electronic circuits are used in applications such as electric rail systems.

1. Calculate total resistance. Total resistance in a circuit containing two resistances connected in parallel and two resistances connected in series can be calculated by applying the following formula:

$$R_T = \left(\frac{R_{P1} \times R_{P2}}{R_{P1} + R_{P2}}\right) + R_{S1} + R_{S2}$$

where
R_T = total resistance (in Ω)
R_{P1} = parallel resistance 1 (in Ω)
R_{P2} = parallel resistance 2 (in Ω)
R_{S1} = series resistance 1 (in Ω)
R_{S2} = series resistance 2 (in Ω)

2. Calculate total current. Total circuit current can be calculated by applying the following formula:

$$I_T = \frac{V_T}{R_T}$$

where
I_T = total circuit current (in A)
V_T = total circuit voltage (in V)
R_T = total circuit resistance (in Ω)

3. Calculate voltage drop across each resistance. Voltage drop across each resistance can be calculated by applying the following formula:

   $V_N = I_T \times R_N$

   where

   V_N = voltage drop across resistance N (in V)

   I_T = total circuit current (in A)

   R_N = resistance N (in Ω)

Example: What is the voltage drop across each resistance in a series/parallel 12 V circuit that contains two 120 Ω (R_{P1}, R_{P2}) resistances connected in parallel and 20 Ω (R_{S1}) and 40 Ω (R_{S2}) resistances connected in series?

1. Calculate total resistance.

$$R_T = \left(\frac{R_{P1} \times R_{P2}}{R_{P1} + R_{P2}}\right) + R_{S1} + R_{S2}$$

$$R_T = \left(\frac{120\ \Omega \times 120\ \Omega}{120\ \Omega + 120\ \Omega}\right) + 20\ \Omega + 40\ \Omega$$

$$R_T = \left(\frac{14{,}400\ \Omega}{240\ \Omega}\right) + 60\ \Omega$$

$$R_T = 60\ \Omega + 60\ \Omega$$

$$R_T = 120\ \Omega$$

2. Calculate total current.

$$I_T = \frac{V_T}{R_T}$$

$$I_T = \frac{12\ \text{V}}{120\ \Omega}$$

$$I_T = 0.1\ \text{A}$$

3. Calculate voltage drop across each resistance.

   Calculate voltage drop across R_{S1}.

   $V_{S1} = I_1 \times R_{S1}$

   $V_{S1} = 0.1\ \text{A} \times 20\ \Omega$

   $V_{S1} = \mathbf{2\ V}$

   Calculate voltage drop across R_{S2}.

   $V_{S2} = I_2 \times R_{S2}$

   $V_{S2} = 0.1\ \text{A} \times 40\ \Omega$

   $V_{S2} = \mathbf{4\ V}$

   Calculate voltage drop across $R_{P1,2}$.

   $V_{P1,2} = I_{P1,2} \times R_{P1,2}$

   $V_{P1,2} = 0.1\ \text{A} \times 60\ \Omega$

   $V_{P1,2} = \mathbf{6\ V}$

Voltage drop across the 20 Ω resistance equals 2 V, voltage drop across the 40 Ω resistance equals 4 V and voltage drop across the 60 Ω resistance (equivalent resistance of two 120 Ω resistances connected in parallel) equals 6 V. Total voltage drop is equal to the source voltage (12 V).

Voltage Drop Application. Voltage drop in a circuit is calculated to ensure proper conductor size in a circuit. Per the NEC®, circuit conductors should be of sufficient size such that the voltage drop in the conductors is not more than 3% of the source voltage. When a voltage drop is present within circuit conductors, operating voltage at electrical equipment is less than the output voltage of the source. Inductive loads such as motors or ballasts that operate at voltage below their rating can overheat, resulting in shorter operating life as well as increased cost. **See Figure 8-9.** Maximum conductor resistance in a circuit is calculated to ensure that the NEC® 3% voltage drop requirement for a circuit is met. Maximum conductor resistance in a circuit is calculated by applying the following procedure:

1. Calculate total current. Total current can be calculated by applying the following formula:

   $I_T = I_1 + I_2 + I_3 + I_4$

   where

   I_T = total current (in A)

   I_1 = current 1 (in A)

   I_2 = current 2 (in A)

   I_3 = current 3 (in A)

   I_4 = current 4 (in A)

2. Calculate maximum voltage drop. Maximum voltage drop is equal to 3% of total voltage and can be calculated by applying the following formula:

   $V_D = V_T \times 0.03$

   where

   V_D = maximum voltage drop (in V)

   V_T = total voltage (in V)

   0.03 = constant (per NEC®)

3. Calculate maximum conductor resistance. Maximum conductor resistance can be calculated by applying the following formula (Ohm's law):

   $$R_{MAX} = \frac{V_D}{I_T}$$

   where

   R_{MAX} = maximum conductor resistance (in Ω)

   V_D = maximum voltage drop (in V)

   I_T = total current (in A)

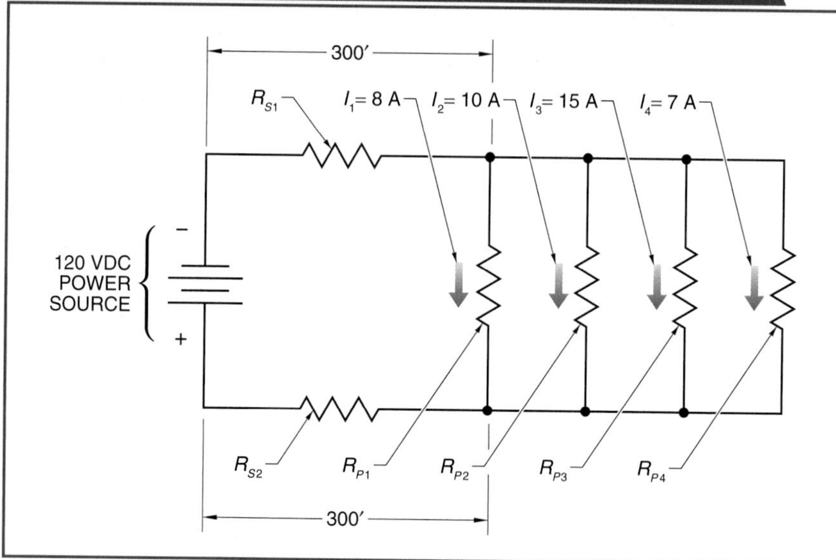

Figure 8-9. Voltage drop is calculated to ensure loads are operated at the voltage level for which they were designed.

Example: What is the maximum conductor resistance that meets the NEC® 3% voltage drop suggestion for a circuit that has a 120 V voltage source 300′ from four parallel-connected loads of 8 A, 10 A, 15 A, and 7 A?

1. Calculate total current.

$I_T = I_1 + I_2 + I_3 + I_4$

$I_T = 8\text{ A} + 10\text{ A} + 15\text{ A} + 7\text{ A}$

$I_T = 40\text{ A}$

2. Calculate maximum voltage drop.

$V_D = V_T \times 0.03$

$V_D = 120\text{ V} \times 0.03$

$V_D = 3.6\text{ V}$

3. Calculate maximum conductor resistance.

$$R_{MAX} = \frac{V_D}{I_T}$$

$$R_{MAX} = \frac{3.6\text{ V}}{40\text{ A}}$$

$$R_{MAX} = \mathbf{0.09\ \Omega}$$

Maximum conductor resistance of 0.09 Ω equals total resistance for 600′ (300′ × 2) of cable. The cable length is doubled in this application because the voltage source is 300′ from the loads, causing the total amount of cable used in the entire circuit to be 600′.

To find the proper conductor size to meet the NEC® 3% requirement, the resistance must be less than or equal to 0.09 Ω over 600′. Generally, conductor resistance is specified in ohms per 1000′. Converting the resistance of any conductor length to 1000′ is calculated by applying the following formula:

$$R_{1000} = \frac{R_L \times 1000'}{L}$$

where

R_{1000} = resistance per 1000′ of conductor (in Ω)

R_L = conductor resistance (in Ω)

L = conductor length (in ft)

Example: What is the minimum conductor size that meets 0.09 Ω over 600′ at 65°C?

$$R_{1000} = \frac{R_L \times 1000'}{L}$$

$$R_{1000} = \frac{0.09\ \Omega \times 1000'}{600'}$$

$$R_{1000} = 0.15\ \Omega$$

Size AWG 1 annealed solid copper wire has a resistance of 0.146 Ω per 1000′ at 65°C. Therefore, the minimum conductor size that meets 0.09 Ω over 600′ at 65°C is AWG 1.

Power in Series/Parallel Circuits

Power in a series/parallel circuit is produced when current flows through a resistance. The higher the resistance or amount of current, the more power produced. The lower the resistance or amount of current, the less power produced. Total power produced in a circuit is equal to the sum of the power produced by each resistance (load). **See Figure 8-10.** Total power in a series/parallel circuit can be calculated by applying the following procedure:

1. Calculate total resistance. Total resistance can be calculated by applying the following formula:

$$R_T = R_{S1} + R_{S2} + \left(\frac{1}{\frac{1}{R_{P1}} + \frac{1}{R_{P2}} + \frac{1}{R_{P3}}} \right)$$

where

R_T = total resistance (in Ω)

R_{S1} = series resistance 1 (in Ω)

R_{S2} = series resistance 2 (in Ω)

R_{P1} = parallel resistance 1 (in Ω)
R_{P2} = parallel resistance 2 (in Ω)
R_{P3} = parallel resistance 3 (in Ω)

2. Calculate total current. Total current can be calculated by applying the following formula (Ohm's law):

$$I_T = \frac{V_T}{R_T}$$

where
I_T = total current (in A)
V_T = total voltage (in V)
R_T = total resistance (in Ω)

3. Calculate voltage drop. Voltage drop can be calculated by applying the following formula:

$$V_D = (I_T \times R_{S1}) + (I_T \times R_{S2})$$

where
V_D = voltage drop (in V)
I_T = total current (in A)
R_{S1} = series resistance 1 (in Ω)
R_{S2} = series resistance 2 (in Ω)

4. Calculate parallel voltage. Parallel voltage can be calculated by applying the following formula:

$$V_P = V_T - V_D$$

where
V_P = parallel voltage (in V)
V_T = total voltage (in V)
V_D = voltage drop (in V)

5. Calculate power though each series-connected load. Power through each series-connected load can be calculated by applying the following formula:

$$P_{SN} = I_{SN}^{\ 2} \times R_{SN}$$

where
P_{SN} = power through series-connected load N (in W)
I_{SN} = current through series-connected load N (in A)
R_{SN} = resistance through series-connected load N (in Ω)

6. Calculate power through each parallel-connected load. Power through each parallel-connected load can be calculated by applying the following formula:

$$P_{PN} = \frac{V_{PN}^{\ 2}}{R_{PN}}$$

where

P_{PN} = power through parallel-connected load N (in W)
V_{PN} = voltage through parallel-connected load N (in V)
R_{PN} = resistance through parallel-connected load N (in Ω)

7. Calculate total power. Total power can be calculated by applying the following formula:

$$P_T = P_{S1} + \ldots + P_{SN} + P_{P1} + \ldots + P_{PN}$$

where

P_T = total power (in W)
P_{S1} = power through series-connected load 1
P_{SN} = power through series-connected load N
P_{P1} = power through parallel-connected load 1
P_{PN} = power through parallel-connected load N

8. Confirm value of total power by application of power formula (Ohm's law). Power formula can be calculated by applying the following formula:

$$P_T = V_T \times I_T$$

where

P_T = total power (in W)
V_T = total voltage (in V)
I_T = total current (in A)

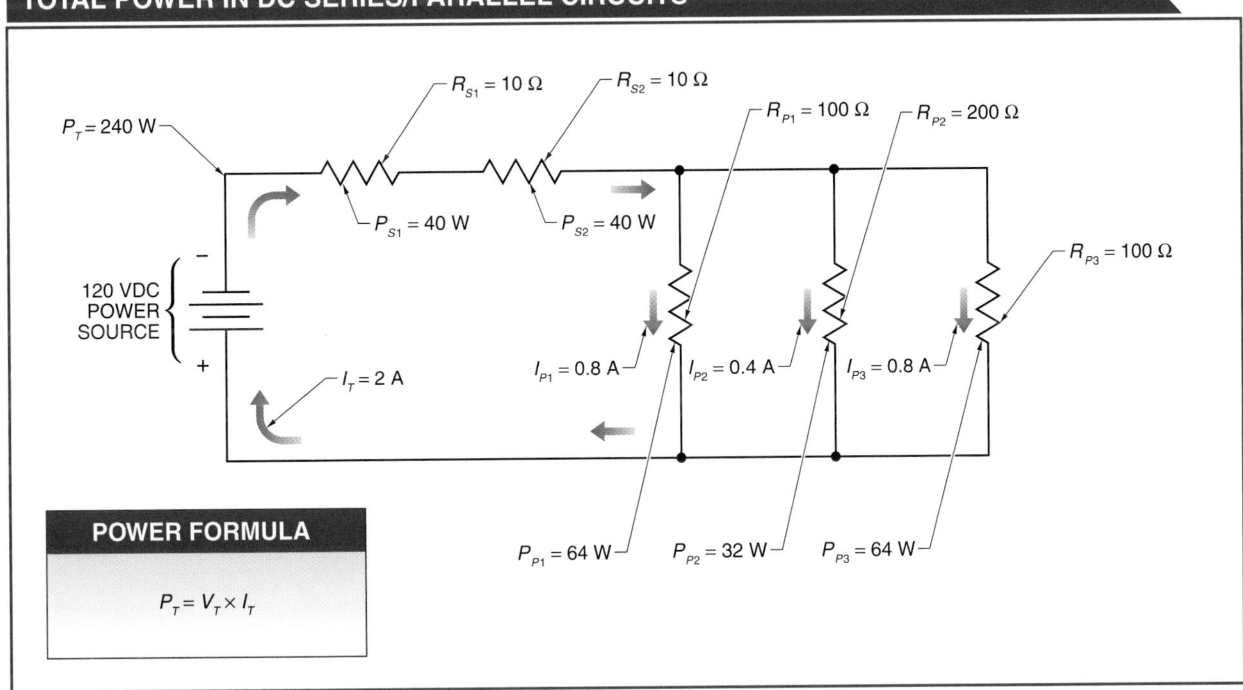

Figure 8-10. Total power in a series/parallel circuit is equal to the sum of the power produced by each resistance (load).

Example: What is the total power in a 120 V circuit that has two 10 Ω loads connected in series (R_{S1} and R_{S2}) with three loads of 100 Ω, 200 Ω, and 100 Ω connected in parallel (R_{P1}, R_{P2}, and R_{P3})?

1. Calculate total resistance.

$$R_T = R_{S1} + R_{S2} + \left(\frac{1}{\frac{1}{R_{P1}} + \frac{1}{R_{P2}} + \frac{1}{R_{P3}}} \right)$$

$$R_T = 10\ \Omega + 10\ \Omega + \left(\frac{1}{\frac{1}{100\ \Omega} + \frac{1}{200\ \Omega} + \frac{1}{100\ \Omega}} \right)$$

$$R_T = 10\ \Omega + 10\ \Omega + \left(\frac{1}{0.010 + 0.005 + 0.010} \right) \Omega$$

$$R_T = 10\ \Omega + 10\ \Omega + \left(\frac{1}{0.025} \right) \Omega$$

$$R_T = 10\ \Omega + 10\ \Omega + 40\ \Omega$$

$$R_T = 60\ \Omega$$

2. Calculate total current.

$$I_T = \frac{V_T}{R_T}$$

$$I_T = \frac{120\ \text{V}}{60\ \Omega}$$

$$I_T = 2\ \text{A}$$

3. Calculate voltage drop.

$$V_D = (I_T \times R_{S1}) + (I_T \times R_{S2})$$

$$V_D = (2\ \text{A} \times 10\ \Omega) + (2\ \text{A} \times 10\ \Omega)$$

$$V_D = 20\ \text{V} + 20\ \text{V}$$

$$V_D = 40\ \text{V}$$

4. Calculate parallel voltage.

$$V_P = V_T - V_D$$

$$V_P = 120\ \text{V} - 40\ \text{V}$$

$$V_P = 80\ \text{V}$$

5. Calculate power though each series-connected load.

$$P_{SN} = I_{SN}^{\ 2} \times R_{SN}$$

$$P_{S1} = I_{S1}^{\ 2} \times R_{S1}$$

$$P_{S1} = (2\ \text{A})^2 \times 10\ \Omega$$

$$P_{S1} = 4\ \text{A} \times 10\ \Omega$$

$$P_{S1} = 40\ \text{W}$$

$$P_{S2} = I_{S2}^{\ 2} \times R_{S2}$$

$P_{S2} = (2\text{ A})^2 \times 10\ \Omega$
$P_{S2} = 4\text{ A} \times 10\ \Omega$
$P_{S2} = 40\text{ W}$

6. Calculate power through each parallel-connected load.

$$P_{PN} = \frac{V_{PN}^{\ 2}}{R_{PN}}$$

$$P_{P1} = \frac{V_{P1}^{\ 2}}{R_{P1}}$$

$$P_{P1} = \frac{80\text{ V}^2}{100\ \Omega}$$

$$P_{P1} = \frac{6400\text{ V}}{100\ \Omega}$$

$$P_{P1} = 64\text{ W}$$

$$P_{P2} = \frac{V_{P2}^{\ 2}}{R_{P2}}$$

$$P_{P2} = \frac{80\text{ V}^2}{200\ \Omega}$$

$$P_{P2} = \frac{6400\text{ V}}{200\ \Omega}$$

$$P_{P2} = 32\text{ W}$$

$$P_{P3} = \frac{V_{P3}^{\ 2}}{R_{P3}}$$

$$P_{P3} = \frac{80\text{ V}^2}{100\ \Omega}$$

$$P_{P3} = \frac{6400\text{ V}}{100\ \Omega}$$

$$P_{P3} = 64\text{ W}$$

7. Calculate total power.

$P_T = P_{S1} + \ldots + P_{SN} + P_{P1} + \ldots + P_{PN}$
$P_T = P_{S1} + P_{S2} + P_{P1} + P_{P2} + P_{P3}$
$P_T = 40\text{ W} + 40\text{ W} + 64\text{ W} + 32\text{ W} + 64\text{ W}$
$P_T = \mathbf{240\ W}$

8. Confirm value of total power by application of Ohm's law.

$P_T = V_T \times I_T$
$P_T = 120\text{ V} \times 2\text{ A}$
$P_T = \mathbf{240\ W}$

Power Application. When total power and source voltage values are known, other variables, such as current, can be calculated. **See Figure 8-11.** For example, in a 120 V circuit lighting application, four lights with wattages of 60 W (P_1), 100 W (P_2), 40 W (P_3), and 100 W (P_4) are connected in both series and parallel. Total power and source voltage are used to calculate total system current.

CURRENT AND POWER IN DC SERIES/PARALLEL CIRCUITS

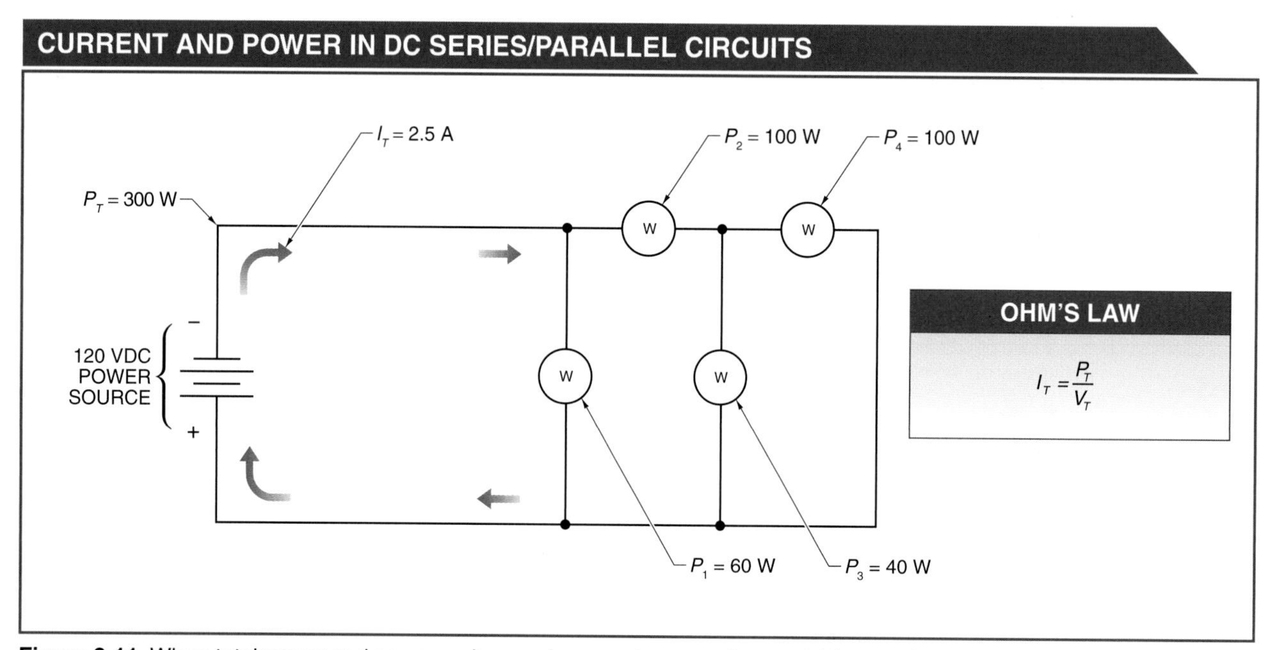

Figure 8-11. When total power and source voltage values are known, other variables, such as current, can be calculated.

Example: What is the total amount of current supplied by the source voltage?

$$P_T = V_T \times I_T$$

$$P_T = P_1 + P_2 + P_3 + P_4$$
$$P_T = 60\text{ W} + 100\text{ W} + 40\text{ W} + 100\text{ W}$$
$$P_T = 300\text{ W}$$

$$I_T = \frac{P_T}{V_T}$$
$$I_T = \frac{300\text{ W}}{120\text{ V}}$$
$$I_T = \mathbf{2.5\ A}$$

SUMMARY

Combination circuits contain elements of both series and parallel connections. The rules for solving for unknown parameters in series and parallel circuits are used along with Ohm's law and the power formula to analyze series/parallel circuits. To solve for unknown values in a series/parallel circuit, separate calculations are used for series-connected and parallel-connected circuits. In series circuits, total resistance is the sum of individual resistances, total current is the same at any point in the circuit, and the sum of the voltage drops is equal to the source voltage. In parallel circuits, total resistance is equal to the reciprocal of the sum of the reciprocal of the individual resistances, total current is equal to the sum of all branch currents, and voltage drop across any branch is equal to the source voltage. Total power is equal to the sum of the power consumed by individual resistances in both series-connected and parallel-connected branches. Reducing a series/parallel circuit to two or more simple circuits is required when performing calculations.

APPLICATION: LOADED UNBALANCED BRIDGE CIRCUITS...

Background: Bridge circuits are commonly found in resistive sensing applications. In this type of application, the sensor changes resistance based on the condition being measured such as temperature or strain. This change in resistance is measured using a bridge circuit.

A balanced bridge, such as a Wheatstone bridge, has balanced legs and is used to measure small changes in resistance. Small changes in the sensor resistance are measured as a voltage difference across the bridge points B and C. From this voltage difference, the sensor value is determined. **See Figure 1.**

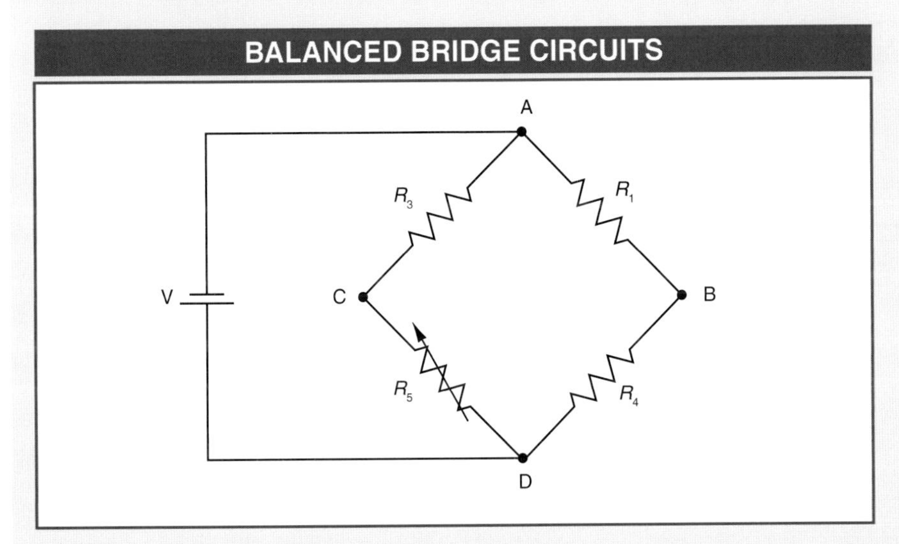

Figure 1. A balanced bridge has balanced legs and is used to measure small changes in resistance.

...APPLICATION: LOADED UNBALANCED BRIDGE CIRCUITS...

An unbalanced bridge does not have balanced legs. A loaded unbalanced bridge has a resistance across the bridge points B and C. This resistance can represent a measurement device such as a meter or another type of measurement element. **See Figure 2.**

UNBALANCED BRIDGE CIRCUITS

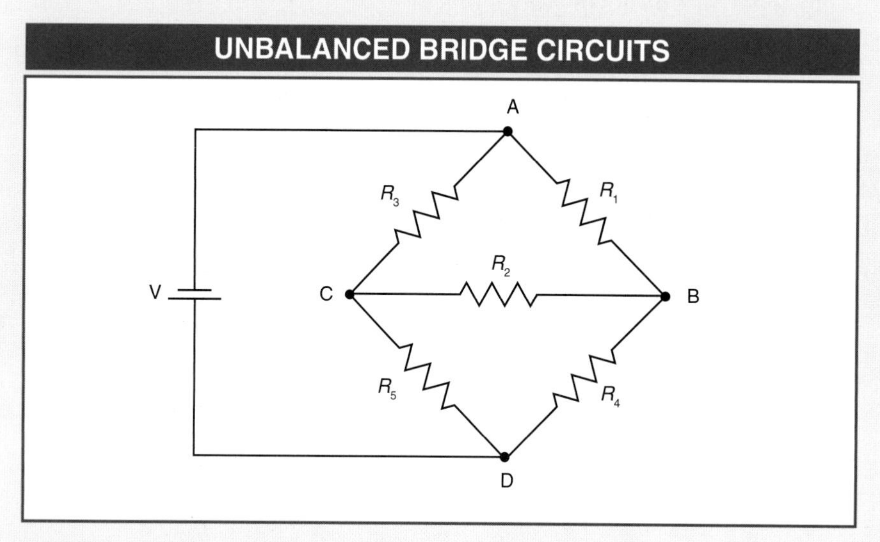

Figure 2. An unbalanced bridge does not have balanced legs.

Bridge circuits are difficult to analyze due to multiple current paths and the inability to clearly combine series and parallel resistors. To help analyze loaded unbalanced bridge circuits, a common transformation is used. Looking closely at the loaded unbalanced bridge circuit, a delta resistor combination is formed by the top three resistors. This delta configuration can be transformed into a wye configuration that simplifies the circuit. **See Figure 3.**

UNBALANCED BRIDGE WITH DELTA AND WYE CONFIGURATIONS

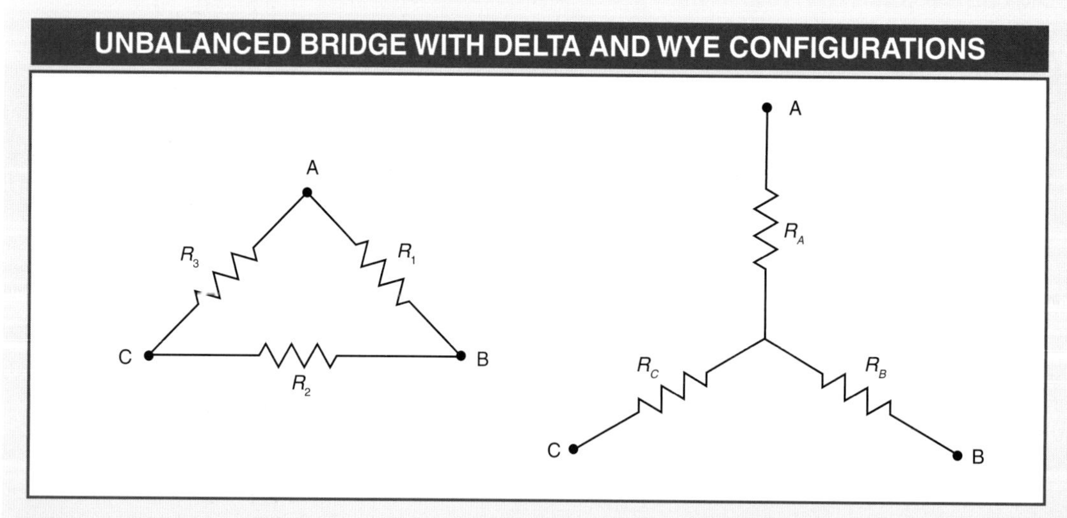

Figure 3. A delta configuration can be transformed into a wye configuration that simplifies the unbalanced bridge circuit.

...APPLICATION: LOADED UNBALANCED BRIDGE CIRCUITS...

To transform the delta resistor network into a wye network, the resistance values across each point, A, B, and C, are calculated for both the delta and wye resistor networks. Since the resistance is equivalent across each point, a set of transformation equations can be created.

The delta resistances are calculated by applying the following formulas:

$$R_{AB}(\Delta) = R_1 \parallel R_2 + R_3 = \frac{R_1 \times (R_2 + R_3)}{R_1 + R_2 + R_3}$$

$$R_{BC}(\Delta) = R_2 \parallel R_1 + R_3 = \frac{R_2 \times (R_1 + R_3)}{R_1 + R_2 + R_3}$$

$$R_{CA}(\Delta) = R_3 \parallel R_1 + R_2 = \frac{R_3 \times (R_1 + R_2)}{R_1 + R_2 + R_3}$$

Note: R_1 is parallel to R_2 and R_3.

The wye resistances are calculated by applying the following formulas:

$R_{AB}(\mathrm{Y}) = R_A + R_B$
$R_{BC}(\mathrm{Y}) = R_B + R_C$
$R_{CA}(\mathrm{Y}) = R_C + R_A$

The delta and wye resistor calculations can be equated by applying the following formulas:

$$R_{AB}(\Delta) = R_{AB}(\mathrm{Y})$$

$$R_{AB}(\Delta) = R_A + R_B$$

$$R_{AB}(\Delta) = \frac{R_1 \times (R_2 + R_3)}{R_1 + R_2 + R_3}$$

$$R_{BC}(\Delta) = R_{BC}(\mathrm{Y})$$

$$R_{BC}(\Delta) = R_B + R_C$$

$$R_{BC}(\Delta) = \frac{R_2 \times (R_1 + R_3)}{R_1 + R_2 + R_3}$$

$$R_{CA}(\Delta) = R_{CA}(\mathrm{Y})$$

$$R_{CA}(\Delta) = R_C + R_A$$

$$R_{CA}(\Delta) = \frac{R_3 \times (R_1 + R_2)}{R_1 + R_2 + R_3}$$

Adding all three equations together gives the following:

$$2 \times (R_A + R_B + R_C) = \frac{2 \times [(R_1 \times R_2) + (R_1 \times R_3) + (R_2 \times R_3)]}{R_1 + R_2 + R_3}$$

or

$$R_A + R_B + R_C = \frac{(R_1 \times R_2) + (R_1 \times R_3) + (R_2 \times R_3)}{R_1 + R_2 + R_3}$$

...APPLICATION: LOADED UNBALANCED BRIDGE CIRCUITS...

Subtracting individual equations from the above equation gives a separate equation for each wye resistor, R_A, R_B, and R_C as follows:

$$R_A = \frac{R_1 \times R_3}{R_1 + R_2 + R_3}$$

$$R_B = \frac{R_1 \times R_2}{R_1 + R_2 + R_3}$$

$$R_C = \frac{R_2 \times R_3}{R_1 + R_2 + R_3}$$

These are the final equations needed to convert a delta network into a wye configuration.

Key Points: DC series and parallel resistance; delta-to-wye transformers

Problem: Find the voltage across and the current through R_2. **See Figure 4.**

LOADED UNBALANCED BRIDGE — DELTA RESISTOR NETWORK

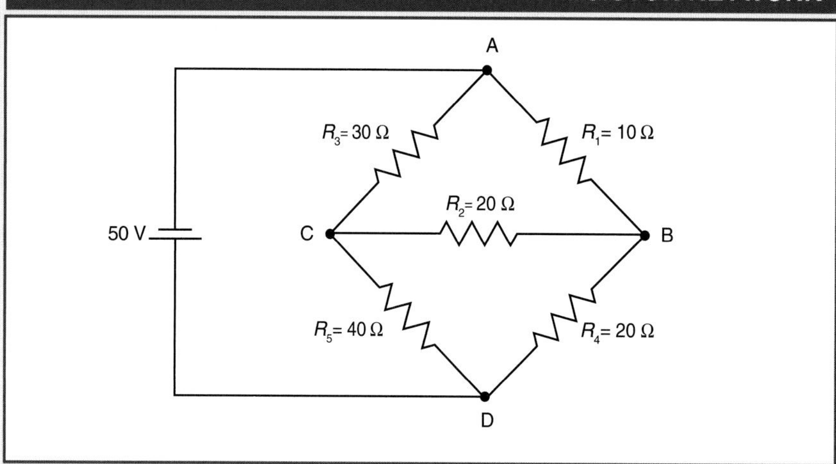

Figure 4. The voltage across and the current through R_2 must be found.

Solution: To find the voltage across and current through R_2, the voltage V_{BC} ($V_B - V_C$) needs to be calculated. The loaded unbalanced bridge circuit shown in Figure 4 can be simplified by changing the delta resistor configuration formed by R_1, R_2, and R_3 into a wye configuration. Calculate the wye resistor values by applying the following formula:

$$R_A = \frac{R_1 \times R_3}{R_1 + R_2 + R_3}$$

$$R_A = \frac{10\ \Omega \times 30\ \Omega}{60\ \Omega}$$

$$R_A = 5\ \Omega$$

...APPLICATION: LOADED UNBALANCED BRIDGE CIRCUITS...

$$R_B = \frac{R_1 \times R_2}{R_1 + R_2 + R_3}$$

$$R_B = \frac{10\ \Omega \times 20\ \Omega}{60\ \Omega}$$

$$R_B = 3.33\ \Omega$$

$$R_C = \frac{R_2 \times R_3}{R_1 + R_2 + R_3}$$

$$R_C = \frac{20\ \Omega \times 30\ \Omega}{60\ \Omega}$$

$$R_C = 10\ \Omega$$

The circuit is redrawn to replace the delta resistor network with a wye resistor network with the calculated values. **See Figure 5.**

LOADED UNBALANCED BRIDGE — WYE RESISTOR NETWORK

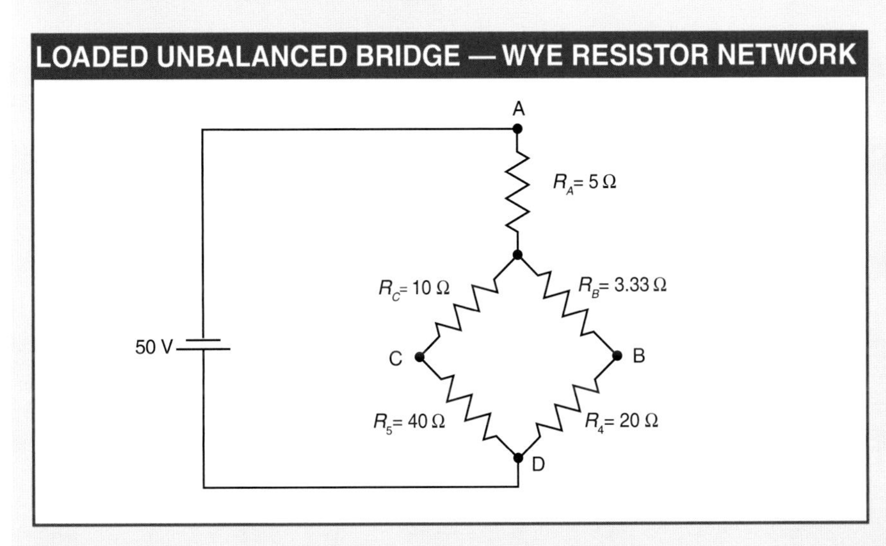

Figure 5. Loaded unbalanced bridge with a wye resistor network with the calculated values.

To find the voltage at points B and C, the voltage drop across R_4 and R_5 are needed since $V_B = V_{R4}$ and $V_C = V_{R5}$. To find the voltage drop across each resistor, the current through each resistor is used. The currents through R_4 and R_5 are different and set by the individual branch current. To find the branch currents, the voltage drop across R_A is needed.

The current through R_A is the total circuit current, which is found using Ohm's law and the total circuit voltage and resistance. **See Figure 6.** The total circuit resistance is equal to the following:

$$R_T = R_A + \frac{R_B + R_4}{R_C + R_5}$$

$$R_T = 5\ \Omega + \frac{3.33\ \Omega + 20\ \Omega}{10\ \Omega + 40\ \Omega}$$

...APPLICATION: LOADED UNBALANCED BRIDGE CIRCUITS...

$$R_T = 5\ \Omega + \frac{23.33\ \Omega}{50\ \Omega}$$

$$R_T = 20.91\ \Omega$$

TOTAL CIRCUIT VOLTAGE AND RESISTANCE

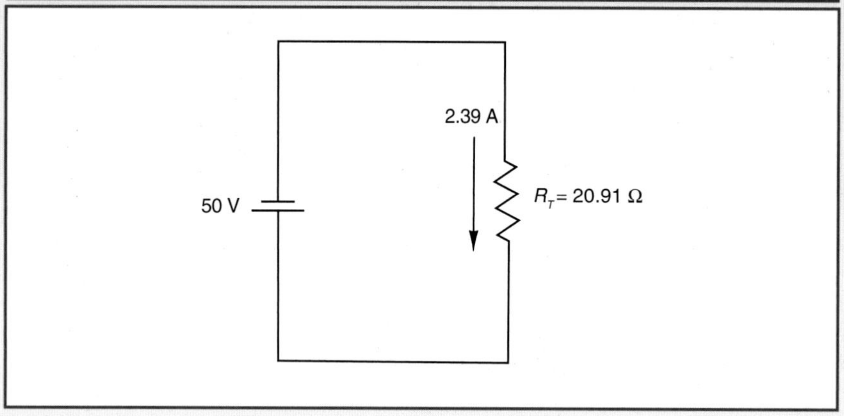

Figure 6. Total circuit current is found using Ohm's law and the total circuit voltage and resistance.

Total circuit current is found using Ohm's law as follows:

$$I_T = \frac{V_{IN}}{R_T}$$

where
V_{IN} = circuit voltage (in V)
R_T = total circuit resistance (in Ω)
I_T = total circuit current (in A)

Therefore,

$$I_T = \frac{50\ V}{20.91\ \Omega}$$

$$I_T = 2.39\ A$$

Total circuit current is equal to 2.39 A. **See Figure 7.**

The voltage drop through R_A is equal to the following:

$V_{RA} = R_T \times I_T$
$V_{RA} = 20.91\ \Omega \times 2.39\ A$
$V_{RA} = 11.95\ V$

...APPLICATION: LOADED UNBALANCED BRIDGE CIRCUITS...

TOTAL CIRCUIT CURRENT

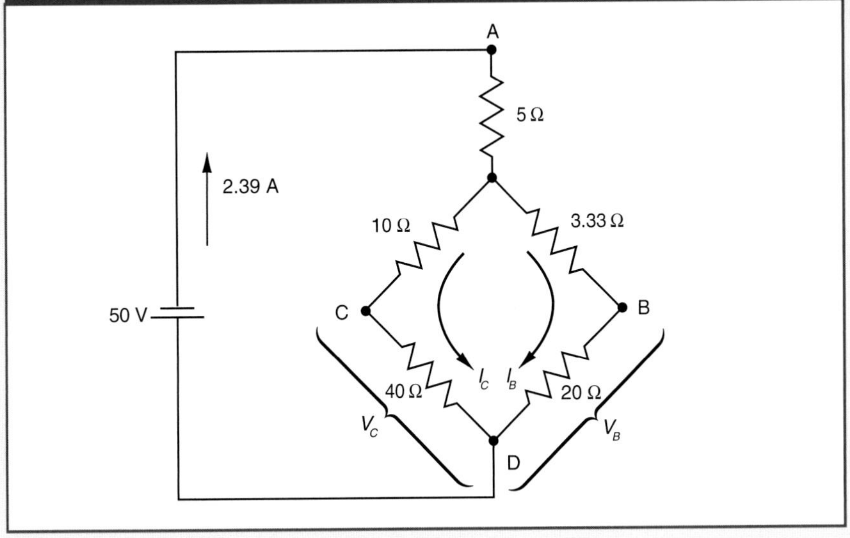

Figure 7. Total circuit current is equal to 2.39 A.

To find the voltage across resistors R_4 and R_5, the current in each branch is needed. Each branch is a parallel circuit, therefore the voltage across each branch and the branch resistance are used to find the branch current. The voltage across each branch is the input voltage minus the voltage drop through R_A. The branch voltage (V_B) equals 50 V – 11.95 V = 38.05 V. The current in each branch is calculated by applying the following formula:

$$I = \frac{V_B}{R}$$

where
I = branch current (in A)
V_B = branch voltage (in V)
R = branch resistance (in Ω)

Therefore,

$$I_B = \frac{38.05\text{ V}}{R_B + R_4}$$

$$I_B = \frac{38.05\text{ V}}{3.33\ \Omega + 20\ \Omega}$$

$$I_B = 1.63\text{ A}$$

$$I_C = \frac{38.05\text{ V}}{R_C + R_5}$$

$$I_C = \frac{38.05\text{ V}}{10\ \Omega + 40\ \Omega}$$

$$I_C = 0.76\text{ A}$$

...APPLICATION: LOADED UNBALANCED BRIDGE CIRCUITS

Ohm's law is used to calculate the voltage across R_4 and R_5 as follows:

$V_B = I_B \times R_4$
$V_B = 1.63 \text{ A} \times 20\ \Omega$
$V_B = 32.6 \text{ V}$

$V_C = I_C \times R_5$
$V_C = 0.76 \text{ A} \times 40\ \Omega$
$V_C = 30.4 \text{ V}$

Voltage V_{BC} is equal to the following:

$V_{BC} = V_B - V_C$
$V_{BC} = 32.6 \text{ V} - 30.4 \text{ V}$
$V_{BC} = 2.2 \text{ V}$

Finally, the current through R_2 is found by using Ohm's law with the voltage V_{BC} and R_2 as follows:

$$I_{R2} = \frac{V_{BC}}{R_2}$$

$$I_{R2} = \frac{2.2 \text{ V}}{20\ \Omega}$$

$$I_{R2} = 0.11 \text{ A}$$

The analysis of a loaded unbalanced bridge circuit was simplified by transforming a delta resistor network into a wye network. **See Figure 8.** This simple transformation allowed resistors to be combined into series and parallel combinations that aided in the circuit analysis.

LOADED UNBALANCED BRIDGE — SIMPLIFIED

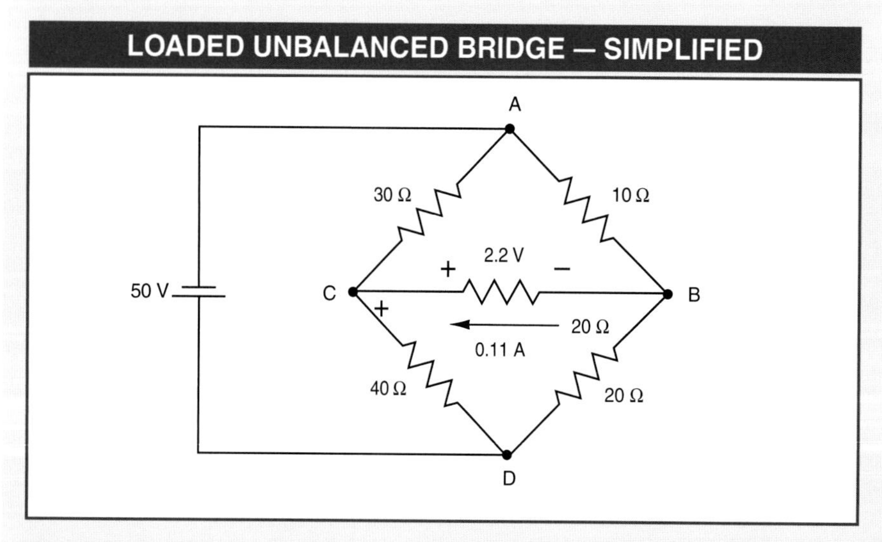

Figure 8. The analysis of a loaded unbalanced bridge circuit was simplified by transforming a delta resistor network into a wye network.

Chapter Review

1. What is the shunt resistance of an ammeter that has an ammeter current of 30 μA, ammeter resistance of 100 Ω, and current through the shunt resistor of 350 μA?
2. What must occur with switches before two or more switches can be connected in series/parallel circuits to allow current flow?
3. What is the total amount of current supplied by a 550 W circuit with a source voltage of 135 V?
4. What is true about series/parallel circuits regarding safety precautions and current?
5. What is the equivalent total resistance in a circuit containing resistances of 22 Ω (R_{P1}) and 7 Ω (R_{P2}) connected in parallel and resistances of 5 Ω (R_{S1}) and 24 Ω (R_{S2}) connected in series?
6. What is a current divider?
7. What is a node?
8. What is the voltage across an ammeter that has 14.6 V across a shunt resistor?
9. What is the total resistance in a series/parallel circuit that contains resistances of 8 Ω and 1200 Ω connected in series and resistances of 100 Ω and 20 Ω connected in parallel?
10. What is a commonly used term for series/parallel circuit?

NREL

9

Complex Network Analysis Techniques

OBJECTIVES

- Explain Kirchoff's voltage and current laws.
- Explain two-voltage-source T-circuits.
- Describe a bridge circuit.
- Explain the differences between the superposition, Thevenin's, and Norton's theorems and describe their applications.
- Describe Thevenizing circuits.

Complex networks are circuits in which the load is not connected in series or in parallel with the voltage source. Some complex networks have more than one voltage source. Ohm's law is used to analyze simple networks. To analyze complex networks, other techniques and methods are needed. These techniques and methods are based on concepts similar to Ohm's law and include Kirchhoff's laws, the superposition theorem, Thevenin's theorem, and Norton's theorem. These techniques and methods are additional tools used to understand circuit behavior.

Terms

Kirchhoff's current law (KCL) is a scientific law used in the evaluation of electrical circuits and states that the algebraic sum of all currents entering a branch point (junction) in a circuit must equal zero.

Potential is the voltage at a point in a circuit with respect to another point in the same circuit.

Potential difference is the algebraic difference in potential (voltage) between two different points in a circuit.

KIRCHHOFF'S LAWS

Kirchhoff's laws are referred to as Kirchhoff's current law (Kirchhoff's 1st law) and Kirchhoff's voltage law (Kirchhoff's 2nd law). Kirchhoff's laws are used by engineers and technicians to understand and evaluate complex network behavior.

Kirchhoff's Current Law

Kirchhoff's current law (KCL) is a scientific law used in the evaluation of electrical circuits and states that the algebraic sum of all currents entering a branch point (junction) in a circuit must equal zero. KCL is calculated by applying the following formula:

$0\text{ A} = I_1 + I_2 + I_3 + \ldots + I_N$

where

I_1 = current through branch 1 (in A)

I_2 = current through branch 2 (in A)

I_3 = current through branch 3 (in A)

I_N = current through branch N (in A)

The sign of each current indicates if the current is entering or leaving the junction. Currents entering the branch have positive values and currents leaving the branch have negative values.

Example: What is the current through branch 1, if the current through branch 2 equals 1.5 A and the current through branch 3 equals 1.0 A?

$0\text{ A} = I_1 + I_2 + I_3$

$0\text{ A} = I_1 + 1.5\text{ A} + 1.0\text{ A}$

$-I_1 = 1.5\text{ A} + 1.0\text{ A}$

$I_1 = \mathbf{-2.5\text{ A}}$

Because the current entering the junction in branches 2 and 3 equals 2.5 A, current leaving the junction through branch 1 must be equal to –2.5 A, according to KCL.

Current flow is the flow of electrons. The current flow exiting a source voltage will return to the same source voltage. The value of current entering a junction must equal the value of current exiting the same junction. **See Figure 9-1.**

Potential is the voltage at a point in a circuit with respect to another point in the same circuit. *Potential difference* is the algebraic difference in potential (voltage) between two different points in a circuit. Electrons cannot be stored or collected in a conductor because a potential could develop opposing the source voltage and current flow could cease, causing an open circuit. A potential exists across an open circuit that is equal to the voltage source causing current flow to be interrupted. Whether a single current enters a junction with three exit paths, or three currents are entering a junction with one exit path, current exiting the junction must always equal current entering the junction. A *nodal equation* is a network equation based on Kirchhoff's current law. Nodal equations are calculated by applying the following formula:

$I_{IN} = I_{OUT}$

where

I_{IN} = current entering a junction (in A)

I_{OUT} = current exiting a junction (in A)

Terms

A **nodal equation** is a network equation based on Kirchhoff's current law.

KIRCHHOFF'S CURRENT LAW

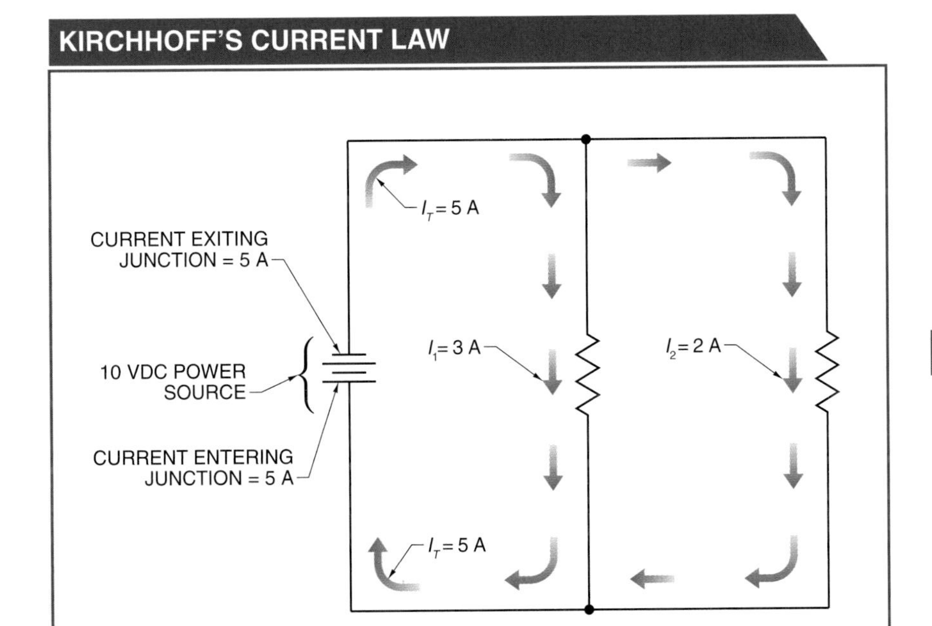

Figure 9-1. Kirchhoff's current law states that the amount of current entering a junction equals the amount of current exiting the same junction.

Tech Tip

In 1847, German physicist Gustav Kirchhoff discovered several properties of the behavior of voltage and current in electric circuits. He proposed two conclusions based on his experiments. These conclusions became known as Kirchhoff's laws and are referred to as Kirchhoff's current law and Kirchhoff's voltage law.

Example: What is the current exiting junction 1 if current entering junction 1 is equal to 8 A?

$I_{IN} = I_{OUT}$

$8\text{ A} = I_{OUT}$

$I_{OUT} = \mathbf{8\text{ A}}$

KCL is seldom used as a stand-alone nodal equation. Typically, KCL is used to provide values of equivalent currents that can be substituted in voltage equations to simplify solutions to complex circuit problems.

Terms

Kirchhoff's voltage law (KVL) is a scientific law used in the evaluation of electrical circuits and states that the algebraic sum of the voltages around any closed-loop circuit must equal zero.

Kirchhoff's Voltage Law

Kirchhoff's voltage law (KVL) is a scientific law used in the evaluation of electrical circuits and states that the algebraic sum of the voltages around any closed-loop circuit must equal zero. KVL is calculated by applying the following formula:

$$0\text{ V} = V_1 + V_2 + V_3 + \ldots + V_N$$

where

V_1 = voltage 1 (in V)

V_2 = voltage 2 (in V)

V_3 = voltage 3 (in V)

V_N = voltage N (in V)

The sign of each voltage indicates if the voltage increases the loop voltage or decreases the loop voltage. When using KVL to evaluate a circuit, if the loop enters the positive terminal, the voltage is positive. If the loop enters the negative terminal, the voltage is negative.

Example: What is voltage 1 in a closed loop, if voltage 2 equals 15 V and voltage 3 equals 7 V?

$$0\text{ V} = V_1 + V_2 + V_3$$

$$0\text{ V} = V_1 + 15\text{ V} + 7\text{ V}$$

$$-V_1 = 15\text{ V} + 7\text{ V}$$

$$V_1 = \mathbf{-22\text{ V}}$$

If loop voltages 2 and 3 equal 22 V, then voltage 1 is equal to −22 V, since KVL states that the closed-loop voltage must be zero.

KVL can be used in the evaluation of all voltage sources and voltage drops in a circuit. KVL can be applied any time there is voltage present in a closed-loop circuit. For example, if only resistance is present in a closed-loop circuit, then there must be a complete path for current flow, which causes voltage drops in the closed-loop circuit. Connecting a voltage source in the closed loop is not required. For example, in a circuit with four resistances, when the voltage drops across three of the resistors voltages are known, KVL can be used to calculate the voltage drop across the fourth resistance. **See Figure 9-2.** Starting at the positive side of resistance 2 (R_2) and proceeding in a clockwise direction, voltage drop is calculated by applying the following formula:

$$0\text{ V} = V_{R2} + V_{R1} - V_{R3} - V_{R4}$$

where

V_{R1} = voltage across resistance 1 (in V)

V_{R2} = voltage across resistance 2 (in V)

V_{R3} = voltage across resistance 3 (in V)

V_{R4} = voltage across resistance 4 (in V)

KIRCHHOFF'S VOLTAGE LAW

$V_{R1} = 3\text{ V}$

$V_{R2} = 6\text{ V}$

$V_{R3} = 4\text{ V}$

$V_{R4} = ?$

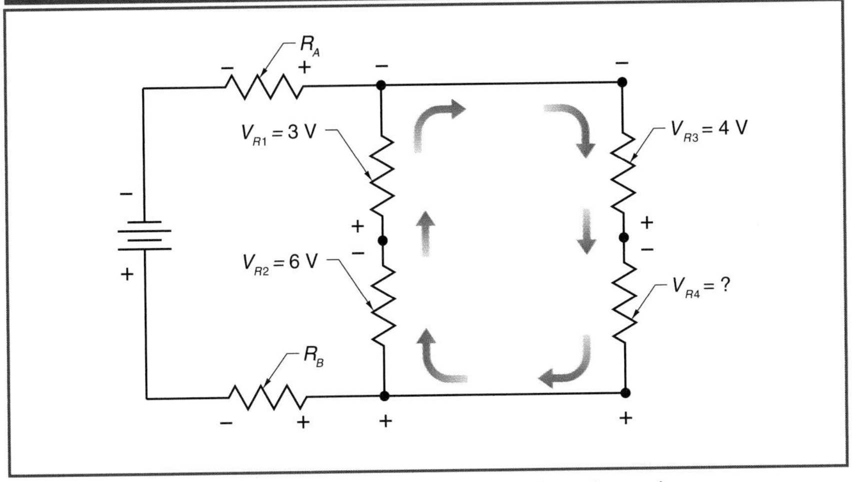

Figure 9-2. In a circuit with four resistances, when the voltage drops across three of the resistances are known, Kirchhoff's voltage law can be used to find the voltage drop across the fourth resistance.

Example: What is the R_4 voltage drop if the R_1 voltage drop equals 3 V, the R_2 voltage drop equals 6 V, and the R_3 voltage drop equals 4 V?

$0\text{ V} = V_{R2} + V_{R1} - V_{R3} - V_{R4}$

$0\text{ V} = 6\text{ V} + 3\text{ V} - 4\text{ V} - V_{R4}$

$0\text{ V} = 5\text{ V} - V_{R4}$

$V_{R4} = \mathbf{5\ V}$

Terms

A **two-voltage-source T-circuit** is a T-shaped circuit with two voltage sources.

TWO-VOLTAGE-SOURCE T-CIRCUITS

A *two-voltage-source T-circuit* is a T-shaped circuit with two voltage sources. Two-voltage-source T-circuits are a common design for electrical circuits. **See Figure 9-3.** Kirchhoff's laws are used to calculate unknown values in a T-circuit. To calculate for unknown values in a two-voltage-source T-circuit using Kirchhoff's laws, apply the following procedure:

1. Calculate voltage for all closed-loop circuits. Voltage for a closed-loop circuit is calculated by applying the following formula (KVL):

   $0\text{ V} = V_T$

   where

   V_T = the algebraic sum of voltages around a closed-loop circuit

2. Calculate all unknown current values. The current through the load resistor is a combination of the current in loop A and the current in loop B. The load resistor current is calculated by applying the following formula:

$I_L = I_1 + I_2$

where

I_L = current through load resistor (in A)

I_1 = current in loop A (in A)

I_2 = current in loop B (in A)

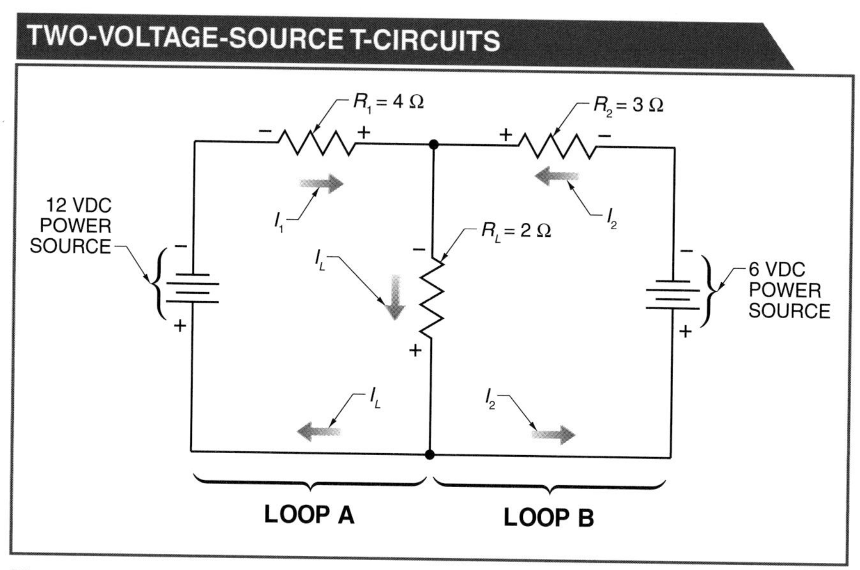

Figure 9-3. A two-voltage-source T-circuit is a T-shaped circuit that has two voltage sources.

Example: What is the voltage drop for a 2 Ω load resistor that is connected between a 12 V closed-loop circuit (loop A) with a resistance of 4 Ω and a 6 V closed-loop circuit (loop B) with a resistance of 3 Ω?

1. Calculate voltage for all closed-loop circuits.

Calculate voltage for loop A.

$0 = V_T$

$0 = 12 - R_1I_1 - R_LI_L$

$0 = 12 - 4I_1 - 2I_L$

$12 = 4I_1 + 2I_L$

Calculate voltage for loop B.

$0 = V_T$

$0 = 6 - R_2I_2 - R_LI_L$

$0 = 6 - 3I_2 - 2I_L$

$6 = 3I_2 + 2I_L$

Note: At this point there are two equations and three unknown variables (I_1, I_2, and I_L). To solve simultaneous equations, it is necessary to have at least as many equations as the number of unknown variables.

2. Calculate unknown current values through application of Kirchhoff's current law.

Solve for current I_2 for 12 V power source.

$12 = 4I_1 + 2(I_1 + I_2)$

$12 = 4I_1 + 2I_1 + 2I_2$

$12 = 6I_1 + 2I_2$

Solve for current I_2 for 6 V power source.

$6 = 3I_2 + 2(I_1 + I_2)$

$6 = 3I_2 + 2I_1 + 2I_2$

$6 = 2I_1 + 5I_2$

Multiply loop 2 by 3 to cancel I_1.

$18 = 6I_1 + 15I_2$

Multiply loop 1 by –1 to cancel I_1.

Add values for both 12 V power source and 6 V power source.

$$18 = 6I_1 + 15I_2$$

$$\frac{-12}{6} = \frac{-6I_1 - 2I_2}{13I_2}$$

$$I_2 = \frac{6}{13}$$

$$I_2 = 0.462 \text{ A}$$

Substitute current I_2 to solve for current I_1.

$12 = 6I_1 + 2I_2$

$12 = 6I_1 + 2(0.462)$

$12 = 6I_1 + 0.924$

$11.076 = 6I_1$

$$I_1 = \frac{11.076}{6}$$

$I_1 = 1.846 \text{ A}$

Use Kirchhoff's current law to solve for I_L.

$I_L = I_1 + I_2$

$I_L = 1.846 \text{ A} + 0.462 \text{ A}$

$I_L = \mathbf{2.308\ A}$

Terms

A **Wheatstone bridge** is a circuit used to take precise measurements of resistance.

A **balanced resistance bridge** is a Wheatstone bridge with the resistances adjusted so there is zero potential across the bridge.

BRIDGE CIRCUITS

Bridge circuits are typically used as measuring and control circuits. A *Wheatstone bridge* is a circuit used to take precise measurements of resistance. A Wheatstone bridge schematic diagram is typically drawn with four resistances connected in a diamond-shaped configuration. **See Figure 9-4.** A Wheatstone bridge operates on the principle that the sum of the voltage drops in a series circuit must equal the value of the voltage source. A *balanced resistance bridge* is a Wheatstone

Terms

An **unbalanced resistance bridge** is a Wheatstone bridge with fixed resistances and the voltage across the bridge proportional to the temperature of the variable resistor.

bridge with the resistances adjusted so there is zero potential across the bridge. The variable resistor R_3 is adjusted to match the resistance of resistor R_4 in order to balance the bridge. The value of R_4 can be determined when R_1, R_2, and R_3 are known. An *unbalanced resistance bridge* is a Wheatstone bridge with fixed resistances and the voltage across the bridge proportional to the temperature of the variable resistor. The variable resistor is adjusted to balance the circuit again. For example, when the temperature of the variable resistor changes, the resistance increases and the bridge is no longer balanced. The voltmeter now registers a potential across the bridge. The change in resistance is proportional to the change in temperature.

WHEATSTONE BRIDGE CIRCUITS

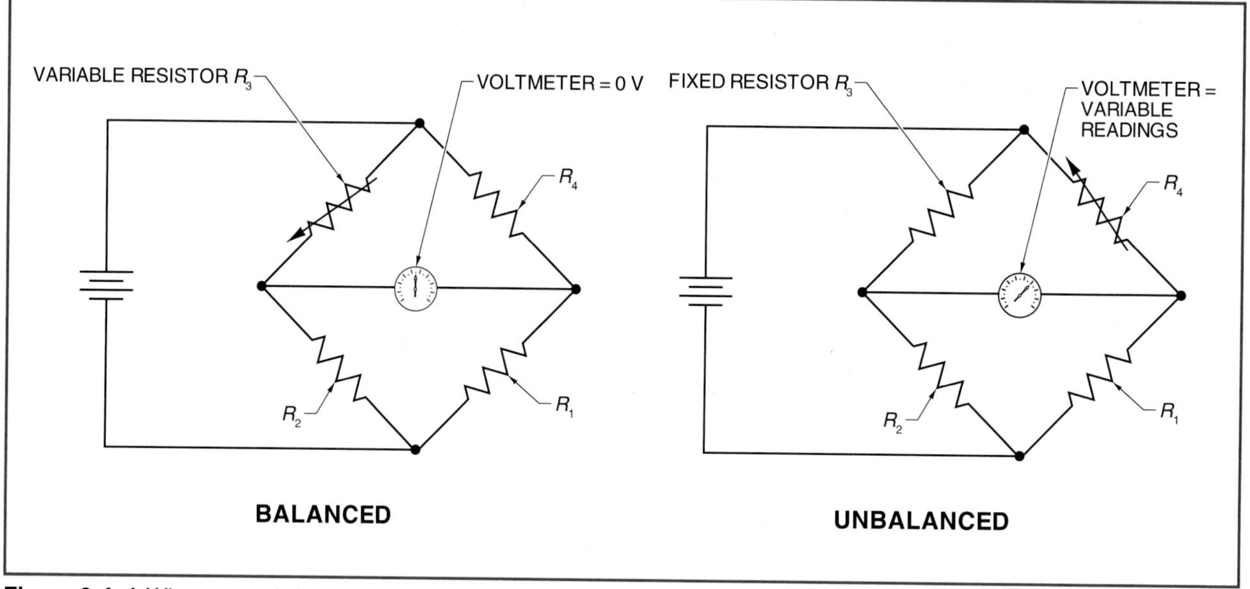

Figure 9-4. A Wheatstone bridge schematic diagram is typically drawn with four resistances connected in a diamond-shaped configuration.

Terms

The **superposition theorem** is a method of analyzing a circuit with multiple voltage sources by analyzing one voltage source at a time and combining the individual effects to determine circuit parameters.

SUPERPOSITION THEOREM

The *superposition theorem* is a method of analyzing a circuit with multiple voltage sources by analyzing one voltage source at a time and combining the individual effects to determine circuit parameters.

When using the superposition theorem, the effect of each voltage source is analyzed individually and then combined to understand circuit parameters. **See Figure 9-5.** Superposition theorem calculations can be performed by using the rules for series and parallel circuits, Kirchhoff's laws, and/or Ohm's law. The current values from each voltage source are then added together to obtain the current flow through each component.

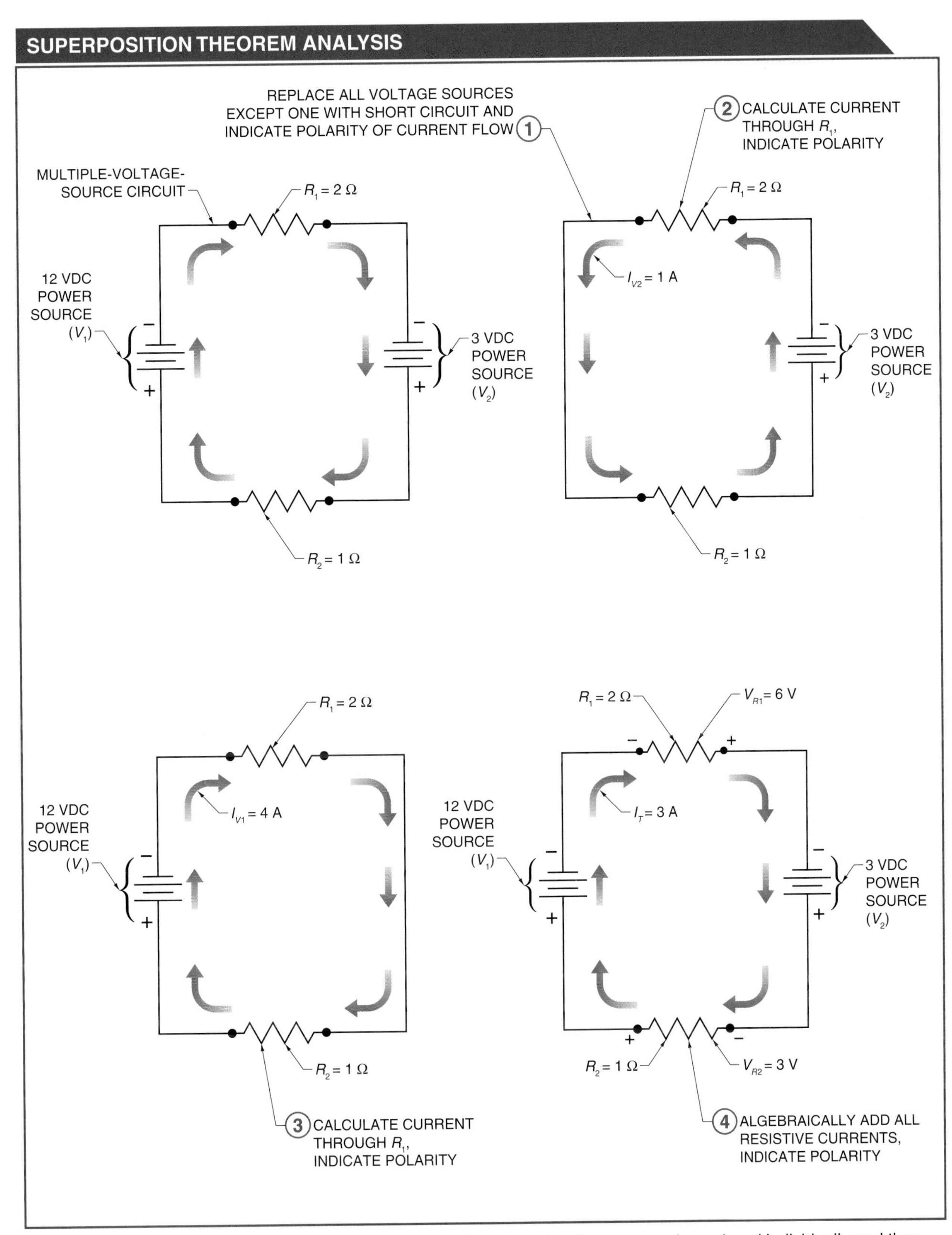

Figure 9-5. When using the superposition theorem, the effect of each voltage source is analyzed individually and then combined to determine circuit parameters.

The superposition theorem is used to analyze multiple-voltage-source circuits by applying the following procedure:

1. Replace all voltage sources but one with a short circuit and indicate polarity of current flow.
2. Calculate current through each resistance using Ohm's law and indicate polarity of current flow.

   Current through each resistance is calculated by applying the following formula:

   $$I = \frac{V}{R}$$

   where

   I = resistive current (in A)

   V = voltage source (in V)

   R = equivalent resistance (in Ω)
3. Repeat steps 1 and 2 for all voltage sources.
4. Algebraically add all resistive currents and indicate polarity. Resistive currents are algebraically added by applying the following formula:

   $$I_T = I_{V1} + I_{V2} + \ldots + I_{VN}$$

   where

   I_T = total circuit current (in A)

   I_{V1} = current from voltage source 1 (in A)

   I_{V2} = current from voltage source 2 (in A)

   I_{VN} = current from voltage source N (in A)

Example: What is the total current in a multiple-voltage-source circuit with voltage source values of 12 V and 3 V and resistance values of 2 Ω and 1 Ω?

1. Replace all voltage sources but one with a short circuit and indicate polarity of current flow.
2. Calculate current through each resistance and indicate polarity of current flow.

   Calculate voltage 1 current.

   $$I_{V1} = \frac{V_1}{R_1 + R_2}$$

   $$I_{V1} = \frac{12\text{ V}}{2\ \Omega + 1\ \Omega}$$

   $$I_{V1} = \frac{12\text{ V}}{3\ \Omega}$$

   $$I_{V1} = 4\text{ A}$$

3. Repeat steps 1 and 2 for all voltage sources.

$$I_{V2} = -\frac{V_2}{R_1 + R_2}$$

$$I_{V2} = -\frac{3\text{ V}}{2\ \Omega + 1\ \Omega}$$

$$I_{V2} = -\frac{3\text{ V}}{3\ \Omega}$$

$$I_{V2} = -1\text{ A}$$

Note: The current flow of voltage source 2 is in the opposite direction as the current flow from voltage source 1, so the current value is negative.

4. Algebraically add all resistive currents and indicate polarity.

$$I_T = I_{V1} + I_{V2} + \ldots + I_{VN}$$

$$I_T = 4\text{ A} + (-1\text{ A})$$

$$I_T = \mathbf{3\ A}$$

KVL can be used to verify this solution. KVL states that the sum of the voltages in a closed loop must equal zero. For this circuit, Ohm's law is used to determine the voltage drop in each resistor. The following formula is used:

$$0\text{ V} = V_1 - V_{R1} - V_2 - V_{R2}$$

$$0\text{ V} = 12\text{ V} - (2\ \Omega \times 3\text{ A}) - 3\text{ V} - (1\ \Omega \times 3\text{ A})$$

$$0\text{ V} = 12\text{ V} - 6\text{ V} - 3\text{ V} - 3\text{ V}$$

$$0\text{ V} = 0\text{ V}$$

The superposition theorem can be used to solve for circuit parameters when there are more than two voltage sources. Steps 1 and 2 are repeated for each voltage source.

Tech Tip

French electrical engineer Leon Charles Thevenin developed Thevenin's theorem in 1882 by performing experiments and conducting research involving Ohm's law and Kirchhoff's laws. Thevenin's work is applied as Thevenin's theorem, which is used to calculate current values in complex electrical circuits.

THEVENIN'S THEOREM

Thevenin's theorem is a method of circuit analysis that reduces a complex circuit to one voltage source with a series resistance. A voltage source can be analyzed with one of two different considerations. A voltage source can be considered as an ideal voltage source with no internal resistance, or it can be considered a practical voltage source. An ideal voltage source is theoretical, while a practical voltage source has a resistance connected in series with it, causing its output voltage to change with changes in the load. **See Figure 9-6.** Thevenin's theorem states that any linear circuit can be viewed as a two-terminal network consisting of an open-circuit voltage with a series resistor.

Terms

Thevenin's theorem is a method of circuit analysis that reduces a complex circuit to one voltage source with a series resistance.

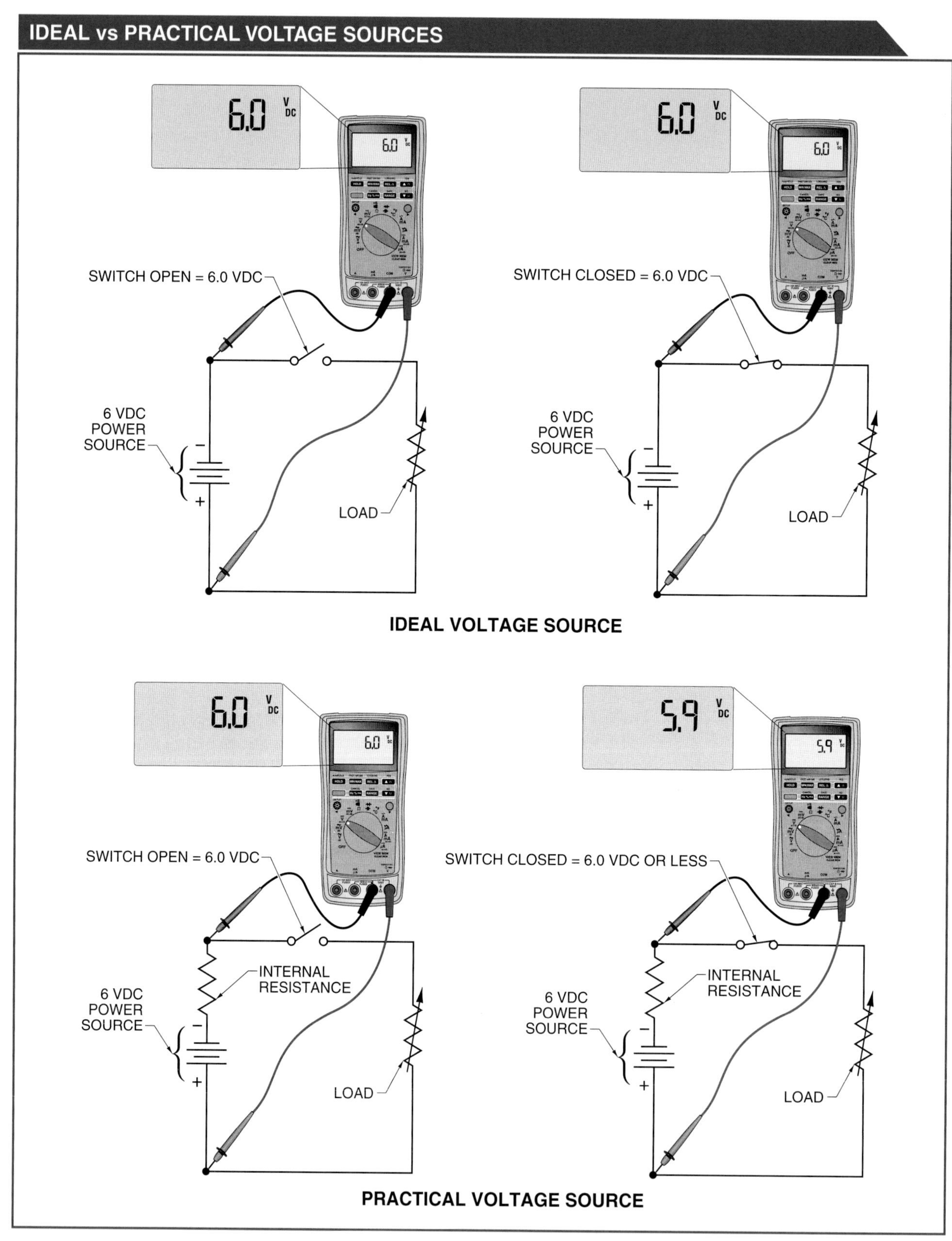

Figure 9-6. An ideal voltage source is theoretical, while a practical voltage source has internal resistance connected in series with it, causing its output voltage to change with changes in the load.

For example, with an ideal 6 V voltage source, with or without a load applied, the amount of voltage remains constant at 6 V. Regardless of how small the resistance is made to increase current flow, the amount of voltage does not change. The ideal voltage source is used when making current calculations where voltage and resistance are known, because in most cases, as long as the current drain on the voltage source does not exceed its current rating, the values obtained using an ideal voltage source are acceptable. The difference between an ideal voltage source and a practical voltage source is typically set at a 1% change in voltage. This change in voltage is typically less than that caused by rounding of numbers in calculations.

All voltage sources have some internal resistance in series with the constant voltage. When the internal resistance becomes a significant part of the resistive load, its value must be considered in voltage and current calculations for the load. With Thevenin's theorem, the entire circuit (with the exception of the load) is considered part of the voltage source. A Thevenin's equivalent circuit has a voltage source with a series resistor (R_{TH}). **See Figure 9-7.**

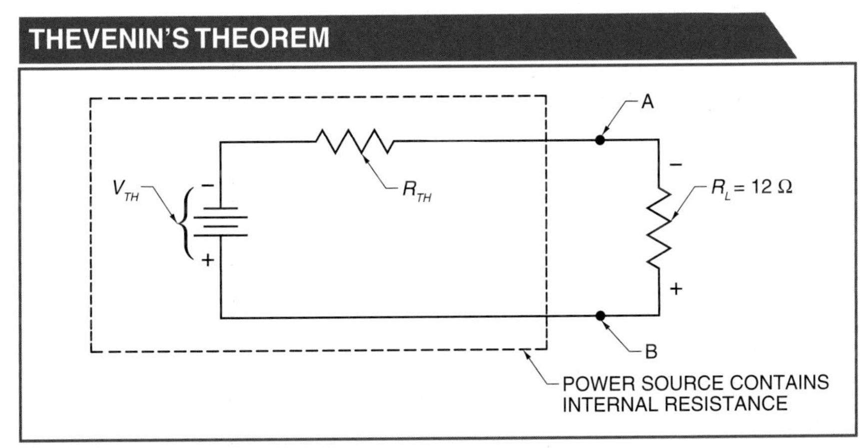

Figure 9-7. When performing circuit analysis with Thevenin's theorem, the entire circuit, with the exception of the load, is considered part of the power (voltage) source.

Note: Open-circuit voltage is always greater than when the voltage source is under load. The more current drawn from the voltage source, the lower the voltage between points A and B. When a short circuit is connected across these points, there is maximum current flow, but the total voltage is dropped across the internal resistance.

When Thevenin's theorem is applied, any circuit can be viewed as a two-voltage-source network consisting of an open-circuit voltage in series with a resistance. These values are known as Thevenin's open-circuit voltage (V_{TH}) and Thevenin's equivalent resistance (R_{TH}).

Joseph Henry

American scientist and inventor Joseph Henry was responsible for numerous inventions and discovered several major principles of electromagnetism, including self-inductance. He also developed a highly efficient electromagnet. Henry invented the first electrical relay in which a small current controlled a much larger current. Henry's work led to the development of the practical telegraph, transformers, and modern radios.

Henry discovered the magnetic principle of induction at about the same time as Faraday, although Faraday is often credited for the discovery. Henry never patented his devices because he believed that scientific discoveries should be used for the common benefit of all humanity. The unit of electrical inductance, the Henry, is named in his honor.

Thevenizing a Circuit

Terms

Thevenizing is the process of simplifying a complex circuit into an equivalent circuit containing a single voltage source and resistance connected in series.

Thevenizing a circuit is the process of simplifying a complex circuit into an equivalent circuit containing a single voltage source and resistance connected in series. **See Figure 9-8.** Thevenizing a circuit can be performed on a Wheatstone bridge circuit by applying the following procedure:

1. Remove load resistance from circuit.
2. Calculate voltage between points A and B. Voltage between points A and B is calculated by applying the following formulas:

Calculate voltage for R_2. Voltage for R_2 is calculated by applying the following formula:

$$V_{R2} = \frac{R_2}{R_1 + R_2} V_P$$

where

V_{R2} = voltage through resistance 2 (in V)
R_2 = resistance 2 (in Ω)
R_1 = resistance 1 (in Ω)
V_P = power source voltage (in V)

Calculate voltage for R_4. Voltage for R_4 is calculated by applying the following formula:

$$V_{R4} = \frac{R_4}{R_3 + R_4} V_P$$

where

V_{R4} = voltage through resistance 4 (in V)
R_4 = resistance 4 (in Ω)
R_3 = resistance 3 (in Ω)
V_P = power source voltage (in V)

Calculate voltage between points A and B (Thevenin's equivalent voltage). Voltage between points A and B is calculated by applying the following formula:

$$V_{AB} = V_{R2} - V_{R4}$$

where

V_{AB} = voltage between points A and B (in V)
V_{R2} = voltage through resistance 2 (in V)
V_{R4} = voltage through resistance 4 (in V)

3. Redraw circuit to show voltage source removed from circuit and a short circuit between points C and D.
4. Calculate Thevenin's resistance. Thevenin's resistance is calculated by applying the following formula:

$$R_{TH} = \frac{1}{\left(\frac{1}{R_1}\right)+\left(\frac{1}{R_2}\right)} + \frac{1}{\left(\frac{1}{R_3}\right)+\left(\frac{1}{R_4}\right)}$$

where

R_{TH} = Thevenin's resistance (in Ω)
R_1 = resistance 1 (in Ω)
R_2 = resistance 2 (in Ω)
R_3 = resistance 3 (in Ω)
R_4 = resistance 4 (in Ω)

5. Redraw simplified circuit to show Thevenin's equivalent voltage and resistance.

Note: Voltage between points A and B (V_{AB}) is also equal to Thevenin's equivalent voltage (V_{TH}).

6. Add load resistance removed from step 1.
7. Calculate load voltage. Load voltage is calculated by applying the following formula:

$$V_L = \frac{R_L}{R_L + R_{TH}} \times V_{TH}$$

where

V_L = voltage through load (in V)
R_L = resistance through load (in Ω)
R_{TH} = Thevenin's resistance (in Ω)
V_{TH} = Thevenin's voltage (in V)

8. Calculate load current. Load current is calculated by applying the following formula:

$$I_L = \frac{V_L}{R_L}$$

where

I_L = current through load (in mA)
R_L = resistance through load (in Ω)

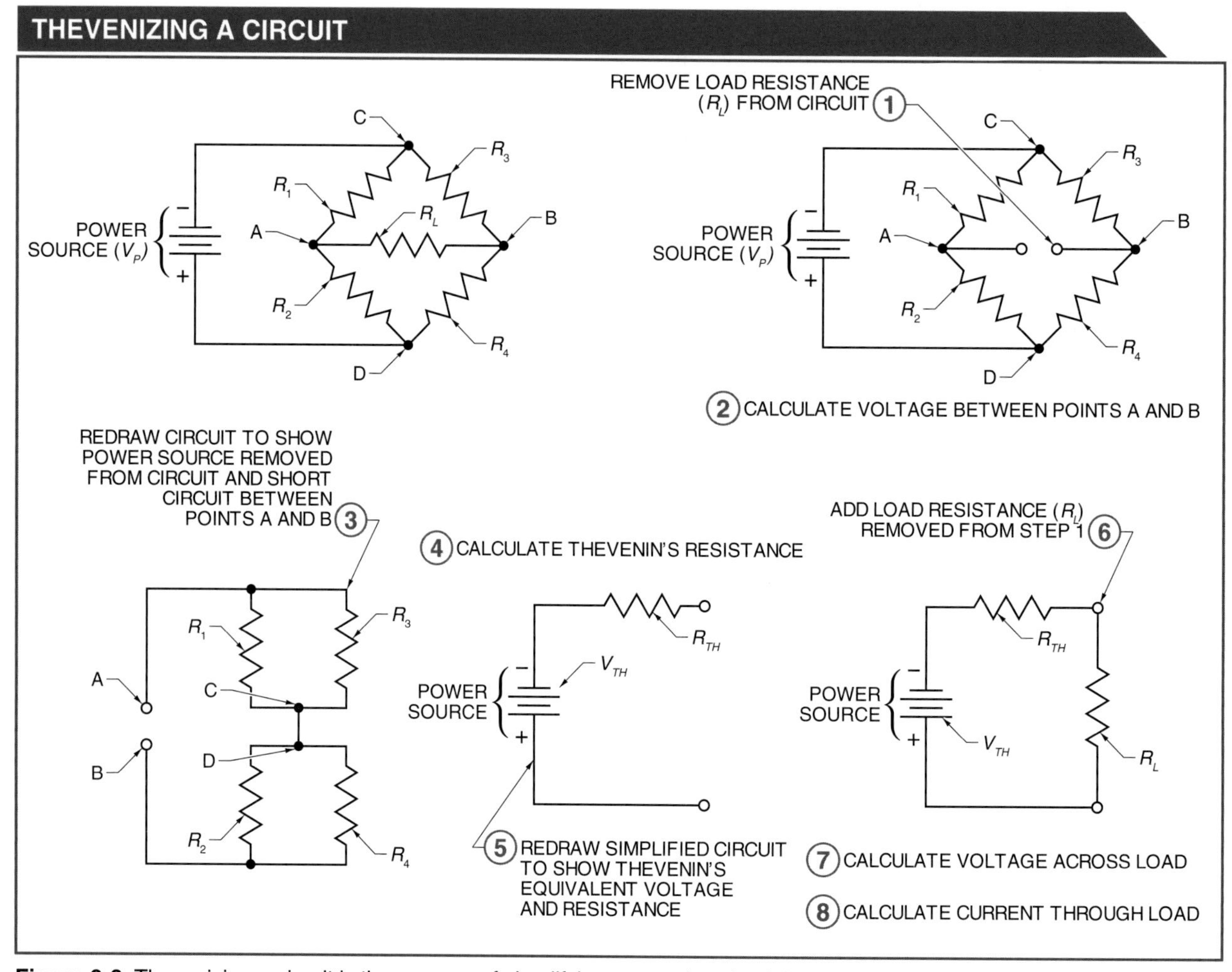

Figure 9-8. Thevenizing a circuit is the process of simplifying a complex circuit into an equivalent circuit that contains a single voltage source and resistor connected in series.

Example: Use Thevenin's theorem to analyze a Wheatstone bridge circuit with a load resistance placed across the output. **See Figure 9-9.**

1. Remove load resistance from circuit.
2. Calculate voltage between points A and B.

Calculate voltage for R_2.

$$V_{R2} = \left(\frac{R_2}{R_1 + R_2}\right)V_P$$

$$V_{R2} = \left(\frac{600\ \Omega}{1500\ \Omega + 600\ \Omega}\right)12\ \text{V}$$

$$V_{R2} = \left(\frac{600\ \Omega}{2100\ \Omega}\right)12\ \text{V}$$

$$V_{R2} = (0.28)\ 12\ \text{V}$$

$$V_{R2} = 3.43\ \text{V}$$

Calculate voltage for R_4.

$$V_{R4} = \left(\frac{R_4}{R_3 + R_4}\right)V_P$$

$$V_{R4} = \left(\frac{240\ \Omega}{2400\ \Omega + 240\ \Omega}\right)12\ \text{V}$$

$$V_{R4} = \left(\frac{240\ \Omega}{2640\ \Omega}\right)12\ \text{V}$$

$$V_{R4} = (0.091)\ 12\ \text{V}$$

$$V_{R4} = 1.09\ \text{V}$$

Calculate voltage between points A and B (Thevenin's equivalent voltage).

$$V_{AB} = V_{R2} - V_{R4}$$

$$V_{AB} = 3.43\ \text{V} - 1.09\ \text{V}$$

$$V_{AB} = 2.34\ \text{V}$$

3. Redraw circuit to show voltage source removed from circuit and a short circuit between points C and D.
4. Calculate Thevenin's resistance.

$$R_{TH} = \frac{1}{\left(\frac{1}{R_1}\right)+\left(\frac{1}{R_2}\right)} + \frac{1}{\left(\frac{1}{R_3}\right)+\left(\frac{1}{R_4}\right)}$$

$$R_{TH} = \frac{1}{\left(\frac{1}{1500\ \Omega}\right)+\left(\frac{1}{600\ \Omega}\right)} + \frac{1}{\left(\frac{1}{2400\ \Omega}\right)+\left(\frac{1}{240\ \Omega}\right)}$$

$$R_{TH} = \frac{1}{0.00067 + 0.00167} + \frac{1}{0.00042 + 0.0042}$$

$$R_{TH} = \frac{1}{0.0023} + \frac{1}{0.0046}$$

$$R_{TH} = 434.78 + 217.39$$

$$R_{TH} = 652.17\ \Omega$$

5. Redraw simplified circuit to show Thevenin's equivalent voltage and resistance.
6. Add load resistance removed from step 1 to simplified circuit drawn in step 5.
7. Calculate load voltage.

$$V_L = \frac{R_L}{R_L + R_{TH}} \times V_{TH}$$

$$V_L = \frac{1500\ \Omega}{1500\ \Omega + 652.17\ \Omega} \times 2.34\ \text{V}$$

$$V_L = \frac{1500\ \Omega}{2152.17\ \Omega} \times 2.34\ \text{V}$$

$$V_L = 1.63\ \text{V}$$

8. Calculate load current.

$$I_L = \frac{V_L}{R_L}$$

$$I_L = \frac{1.63\ \text{V}}{1500\ \Omega}$$

$$I_L = \mathbf{1.1\ mA}$$

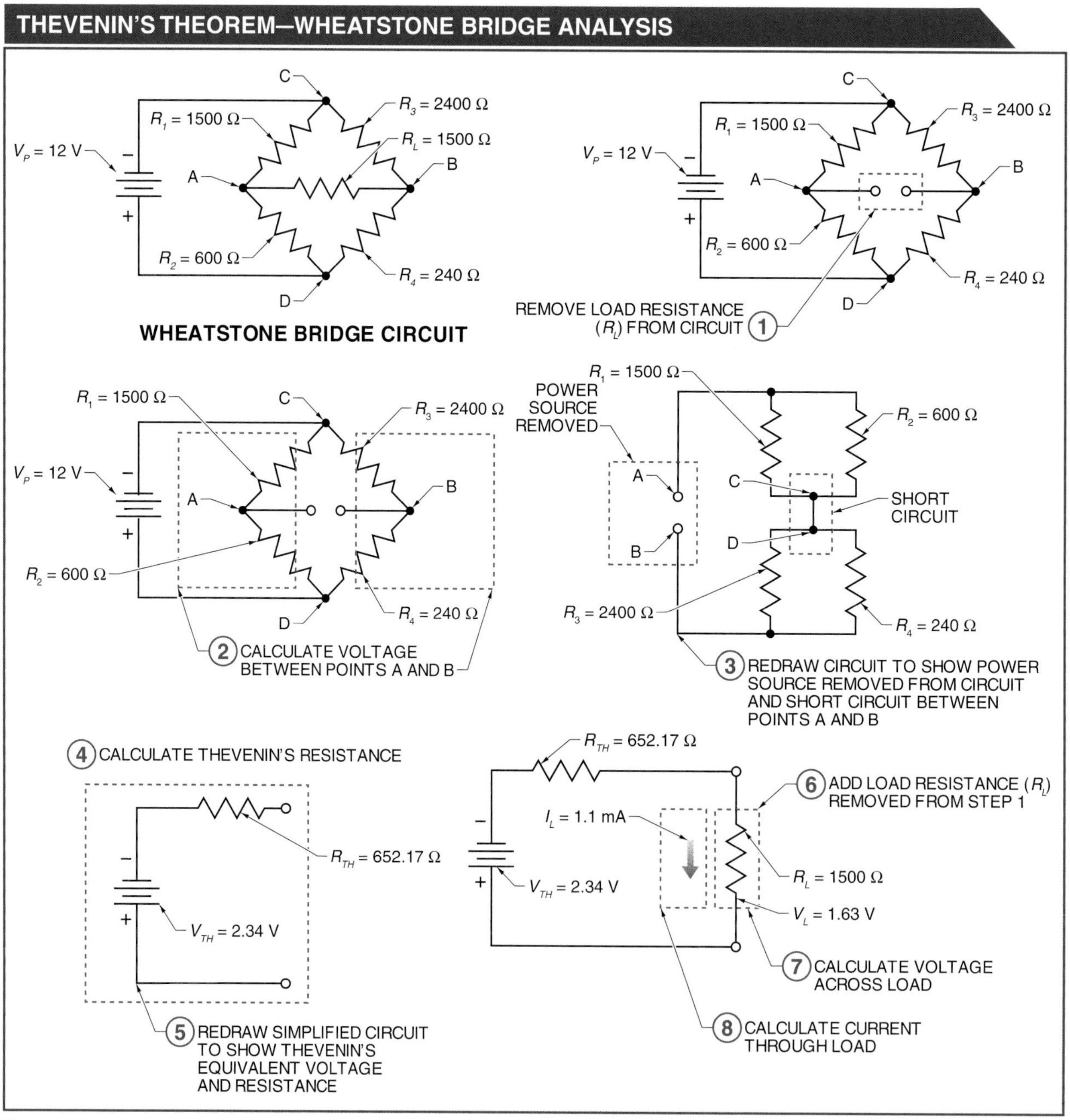

THEVENIN'S THEOREM—WHEATSTONE BRIDGE ANALYSIS

Figure 9-9. Thevenin's theorem can be used to analyze a complex series-parallel circuit such as a Wheatstone bridge with a load resistance across the output.

NORTON'S THEOREM

Norton's theorem is a method of circuit analysis that reduces complex and basic circuits into a circuit with one current source and a parallel resistance. Norton's theorem can be applied to any circuit and is used to reduce a circuit into a single current source and a single parallel-connected resistance. **See Figure 9-10.**

Norton's equivalent current and resistance in a basic are calculated by applying the following procedure:

1. Place short circuit across load points A and B.
2. Calculate Norton's equivalent circuit current. Norton's equivalent circuit current is calculated by applying the following formula:

   $$I_{NT} = \frac{V_P}{R_1}$$

   where
   I_{NT} = Norton's equivalent current (in A)
   V_P = Power source voltage (in V)
   R_1 = resistance 1 (in Ω)
3. Remove short circuit across points A and B and replace voltage source with short circuit.
4. Calculate Norton's equivalent circuit resistance. Norton's equivalent circuit resistance is calculated by applying the following formula:

   $$R_{NT} = \frac{R_1 \times R_2}{R_1 + R_2}$$

   where
   R_{NT} = Norton's equivalent resistance (in Ω)
   R_1 = resistance 1 (in Ω)
   R_2 = resistance 2 (in Ω)
5. Redraw circuit and replace short circuit across power source with Norton's equivalent current and connect in parallel with Norton's equivalent resistance.

Network Theorem Applications

Network theorems such as Thevenin's theorem and Norton's theorem are typically referenced when analyzing circuits with multiple loads. Network theorems are used to reduce complex, multiple-load circuits to simple circuits and can be used to analyze any electronic circuit.

NORTON'S THEOREM

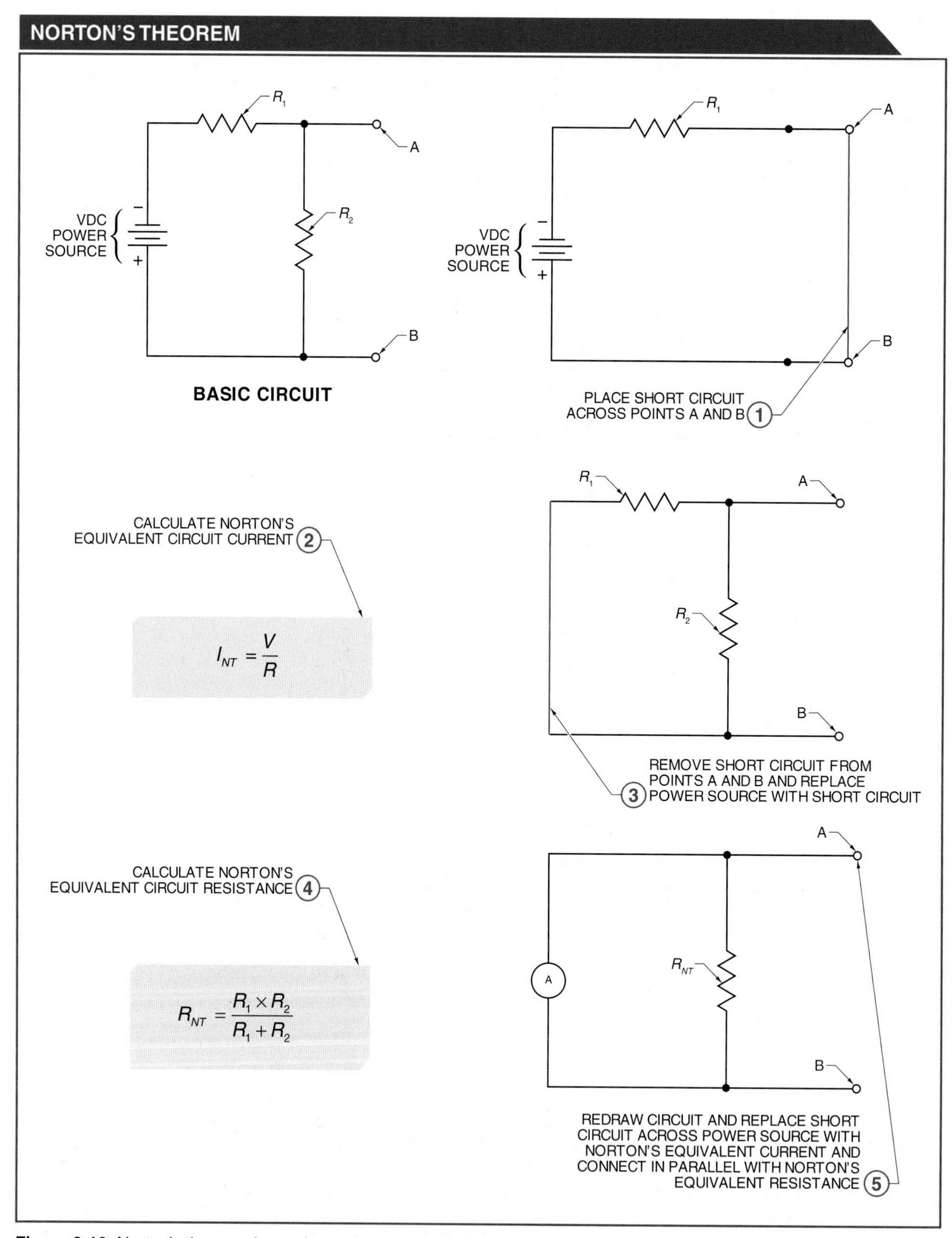

Figure 9-10. Norton's theorem is used to reduce a complex circuit into a single current source and a single parallel-connected resistance.

Example: What is Norton's equivalent current and resistance in a basic circuit with a 36 VDC power source and resistances of 4 Ω and 6 Ω connected in series? **See Figure 9-11.**

1. Place short circuit across points A and B.
2. Calculate Norton's equivalent circuit current.

$$I_{NT} = \frac{V_P}{R_1}$$

$$I_{NT} = \frac{36\text{ V}}{4\ \Omega}$$

$$I_{NT} = 9\text{ A}$$

3. Remove short circuit across points A and B and replace power source with short circuit.
4. Calculate Norton's equivalent circuit resistance.

$$R_{NT} = \frac{R_1 \times R_2}{R_1 + R_2}$$

$$R_{NT} = \frac{4\ \Omega \times 6\ \Omega}{4\ \Omega + 6\ \Omega}$$

$$R_{NT} = \frac{24\ \Omega}{10\ \Omega}$$

$$R_{NT} = \mathbf{2.4\ \Omega}$$

5. Redraw circuit and replace short circuit across power source with Norton's equivalent current and connect in parallel with Norton's equivalent resistance.

Any resistance value can now be connected across points A and B and other electrical parameters can be calculated accordingly. While most circuits can be analyzed with Ohm's law, Kirchhoff's voltage law (KVL), and Kirchhoff's current law (KCL), circuits that contain multiple voltage sources cannot be analyzed with Ohm's law and are difficult to analyze with KVL and KCL. Thevenin's and Norton's theorems are used with these types of circuits. While Thevenin's theorem is used to simplify a circuit into an equivalent circuit that contains a single power source and a series resistance, Norton's theorem is used to simplify a circuit into an equivalent circuit that contains a single current source and a parallel resistance.

Milwaukee Electric Tool Corporation

Battery packs used with cordless electric hand tools have output changes with changes in load and are examples of practical voltage sources.

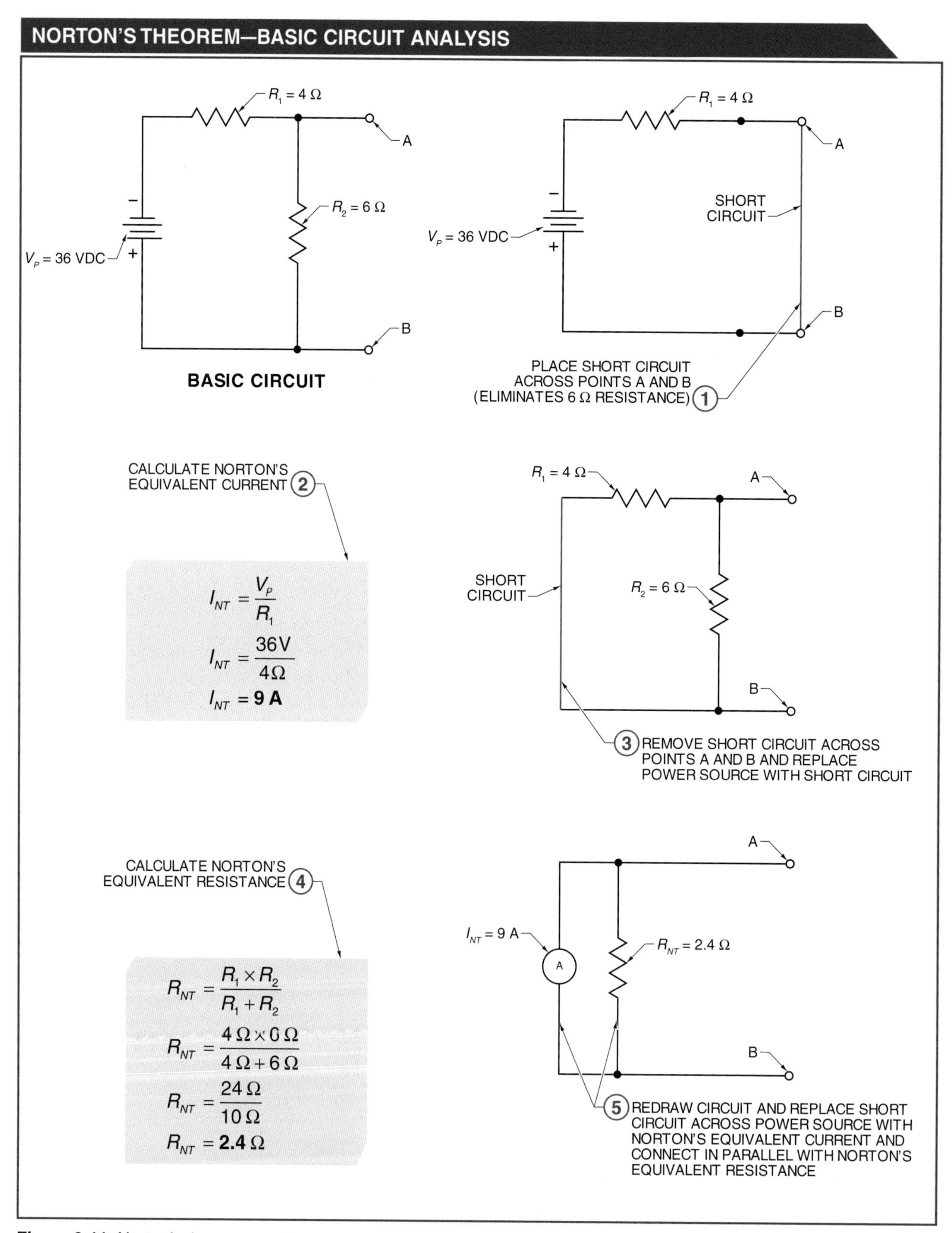

Figure 9-11. Norton's theorem can be used to simplify a circuit and to calculate electrical parameters such as current and resistance.

SUMMARY

Most circuits can be analyzed by using Kirchhoff's laws, the superposition theorem, Thevenin's theorem, or Norton's theorem. Kirchhoff's laws state that closed-loop voltage equals zero and that the current into a junction equals the current out of that junction. Kirchhoff's laws can be used to determine all circuit parameters.

The superposition theorem is only applicable in multiple-voltage-source circuits and is used to calculate circuit parameters by analyzing one voltage source at a time. Performing calculations with the superposition theorem requires the application of Ohm's law and the rules for series and parallel circuits. Final circuit parameters are found by summing the individual voltage source contributions. As with Kirchhoff's laws, all circuit parameters can be determined using the superposition theorem.

Thevenin's theorem is used to simplify complex circuits into a single source and a series resistor. Once a Thevenin's equivalent circuit is developed, only a single calculation using a two-resistor series circuit with a voltage drop across each resistor is required to calculate the voltage across the load resistance.

Norton's theorem is used to simplify complex circuits into a single current source with a parallel resistor. With Norton's theorem, only one calculation is necessary, by using two parallel resistors that divide the circuit current into two paths to calculate the current flow through any load.

Thevenin's and Norton's theorems can be used with a single voltage source or multiple voltage sources. These theorems do not lend themselves to solving for all circuit parameters, and other techniques may be required to develop the equivalent circuits. The benefit of these theorems is that they greatly simplify the process of determining the voltage or current through a load.

APPLICATION: ZENER DIODE VOLTAGE REGULATORS...

Background: Voltage regulators are used to provide a stable and constant voltage to a load. Voltage regulators are used in applications that require a voltage source other than the available voltage. For example, an LED lightbulb does not directly work with the same 120 VAC as a regular incandescent lightbulb. Internal to the LED lightbulb is a voltage regulator that converts 120 VAC to a DC voltage usable by the LED elements.

Voltage regulators are used for both AC and DC voltages. **See Figure 1.** A simple DC voltage regulator uses a semiconductor device, such as a zener diode, and creates a voltage reference based on current flow through the device. The remaining current, the shunt path, flows to the load that is attached in parallel with the voltage regulator.

...APPLICATION: ZENER DIODE VOLTAGE REGULATORS...

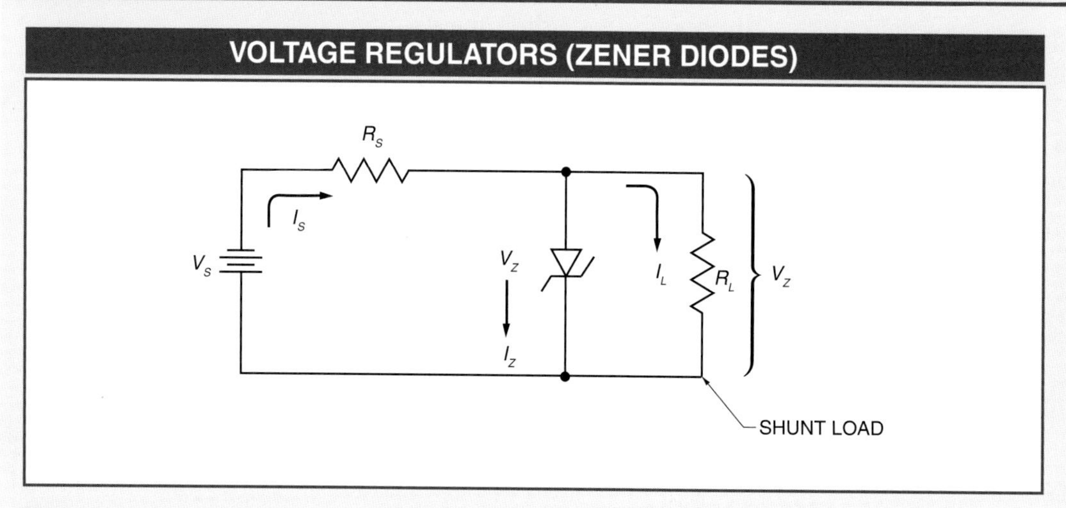

Figure 1. A simple DC voltage regulator uses a semiconductor device, such as a zener diode, and creates a voltage reference based on current flow through the device.

A zener diode voltage regulator provides a constant voltage when the diode is reverse-biased with a minimum amount of current and regulates the voltage up to a maximum current. **See Figure 2.** Exceeding the maximum current causes the zener diode to fail. When the zener diode is forward biased, it operates like a normal diode.

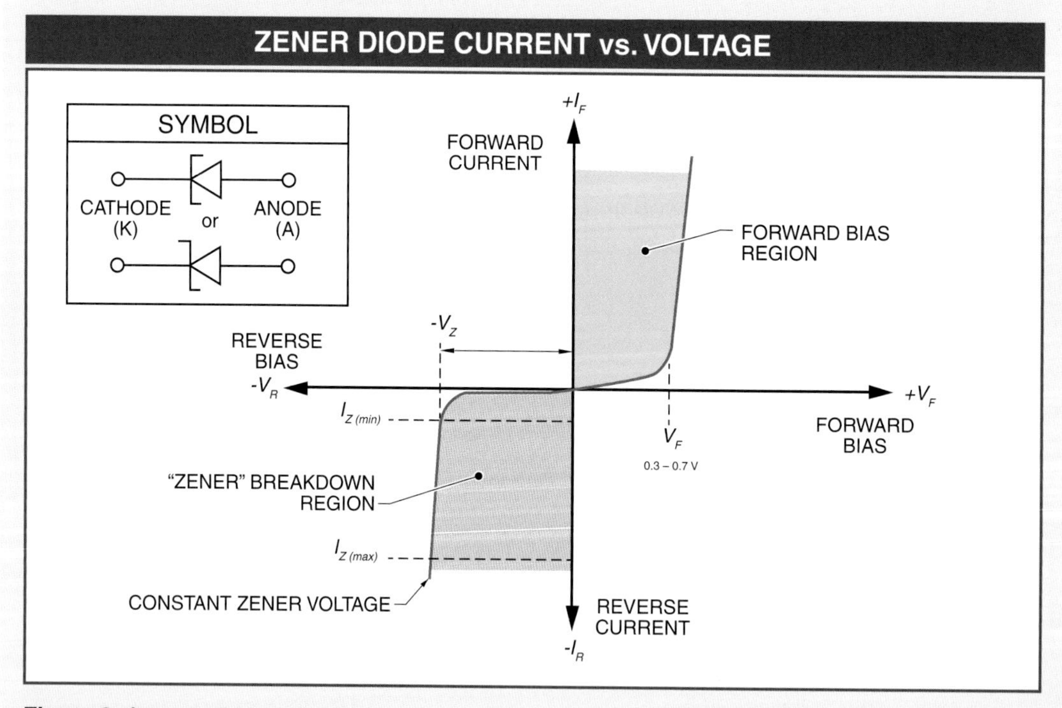

Figure 2. A zener diode voltage regulator provides a constant voltage when the diode is reverse-biased with a minimum amount of current and regulates the voltage up to a maximum current.

...APPLICATION: ZENER DIODE VOLTAGE REGULATORS...

Zener diodes are available in a variety of wattages and voltages. Zener diodes rated for 1 W have voltages ranging from 3.3 VDC to 75 VDC. The zener diode is used in a series/parallel circuit. The circuit has a voltage source greater than the zener voltage, a series current limiting resistor, and the zener diode. A current limiting resistor is used to protect the zener diode from excessive current. In parallel with the zener diode is the load. Zener diodes added to voltage regulator applications are also used as waveform clippers and voltage shifters. **See Figure 3.**

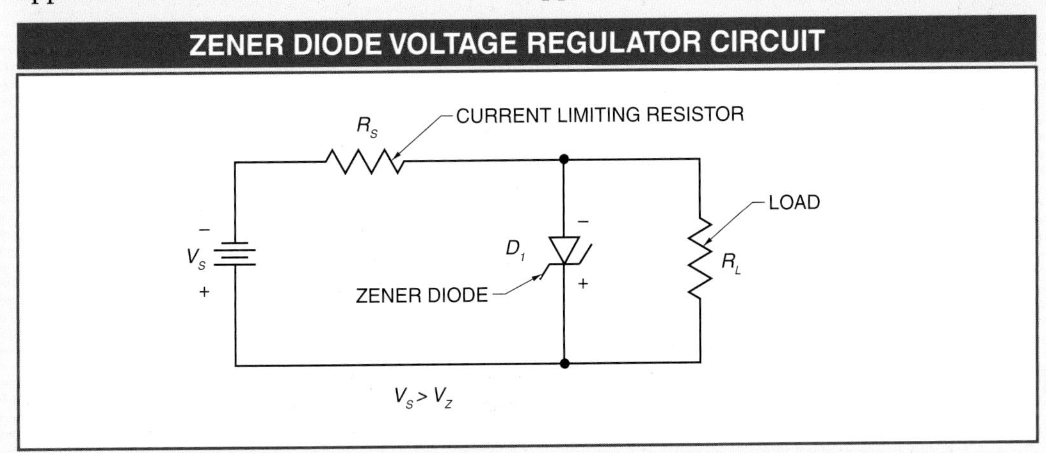

Figure 3. A current limiting resistor is used to protect the zener diode from excessive current.

Key Points: Kirchhoff's current and voltage laws; Watt's law

Problem: Given a source voltage of 12 VDC, design a 5.1 VDC voltage regulator using a 1 W zener diode. Design the voltage regulator to operate without a load and determine the maximum load (minimum resistive load) the voltage regulator can support.

Solution: A standard voltage of the 1 W zener diode is 5.1 V. **See Figure 4.** The operating electrical characteristics are as follows:

Type	V_Z (V)	I_{ZT} (mA)	Z_{ZT} (Ω)	Z_{ZK}(Ω)	I_{ZK} (mA)	I_R(μA)	V_R(v)
1N4733A	5.1	49	7.0	550	1.0	10	1.0

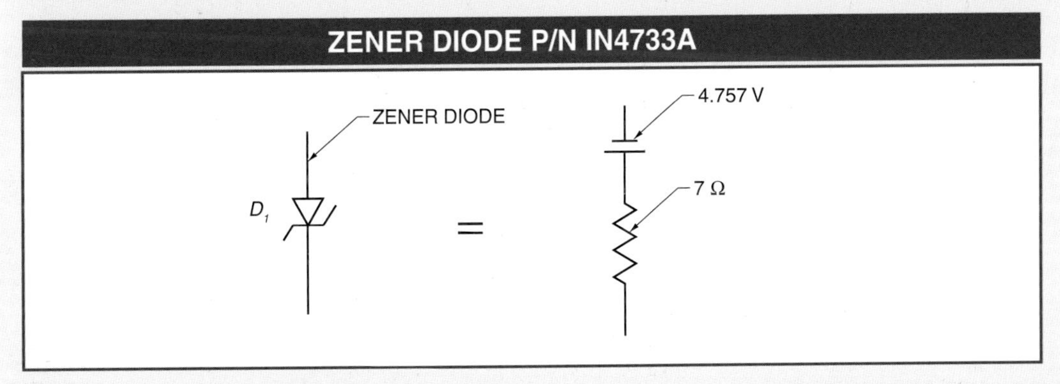

Figure 4. The operating electrical characteristics of the voltage regulator part number IN4733A can be used to determine the maximum load it can support.

...APPLICATION: ZENER DIODE VOLTAGE REGULATORS...

The linear model (approximation) of a zener diode is a voltage source and resistor. The resistor value, or impedance, at the test current is 7.0 Ω (Z_{ZT}) and is specified in the operating electrical characteristics. **See Figure 5.** To find the zener diode voltage source (V_{Z0}), apply the following formula:

$V_Z = V_{Z0} + I_{ZT} \times Z_{ZT}$
where
V_Z = voltage at test current (in V)
V_{Z0} = voltage source (in V)
I_{ZT} = test current (in A)
Z_{ZT} = impedance at test current (in Ω)

The formula is rearranged to solve for V_{Z0} as follows:
$5.1\text{ V} = V_{Z0} + (0.049\text{ A} \times 7.0\ \Omega)$
$V_{Z0} = 5.1\text{ V} - (0.049\text{ A} \times 7.0\ \Omega)$
$V_{Z0} = 5.1\text{ V} - 0.343\text{ V}$
$V_{Z0} = 4.757\text{ V}$

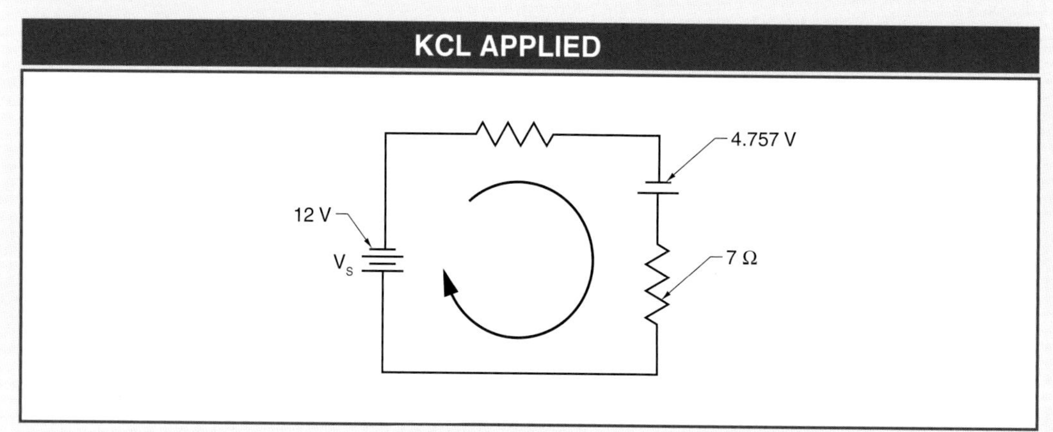

Figure 5. The linear model (approximation) of a zener diode is a voltage source and resistor.

The first step in the circuit analysis, after finding the parameters of the zener diode, is to determine the current limiting resistor value, R_S. This resistor limits the circuit current, and its value is calculated when the circuit load is infinity (no load). In this condition, the entire circuit current is flowing through the zener diode. The maximum zener diode current is limited to P_d, which is 1 W. To find R_S, the maximum current (I_{MAX}) in the circuit is first determined using the power formula. The power formula is applied as follows:

$P_d = I_{MAX} \times V_Z$
$1\text{ W} = I_{MAX} \times V_Z$
where
$V_Z = V_{Z0} + I_{MAX} \times Z_{ZT}$

Therefore,
$1\text{ W} = I_{MAX} \times (V_{Z0} + I_{MAX} \times Z_{ZT})$
$1\text{ W} = I_{MAX} \times (Z_{ZT} \times I_{MAX} + V_{Z0})$
$1\text{ W} = Z_{ZT} \times I_{MAX}^2 + V_{Z0} \times I_{MAX}$

...APPLICATION: ZENER DIODE VOLTAGE REGULATORS...

The equation is set to zero and becomes the following:

$0 = Z_{ZT} \times I_{MAX}{}^2 + V_{Z0} \times I_{MAX} - 1$

$0 = 7 \times I_{MAX}{}^2 + 4.757 \times I_{MAX} - 1$

The quadratic formula is used to solve for I_{MAX}:

$$x = \frac{-b \pm \sqrt{b^2 - 4ac}}{2a}$$

where

$a = 7$

$b = 4.757$

$c = -1$

$x = I_{MAX}$

$$I_{max} = \frac{-4.757 \pm \sqrt{4.757^2 - 4 \times 7 \times -1}}{2 \times 7}$$

$$I_{max} = \frac{-4.757 \pm \sqrt{22.629 + 28}}{14}$$

$$I_{max} = \frac{-4.757 \pm \sqrt{50.629}}{14}$$

$$I_{max} = \frac{-4.757 \pm 7.115}{14}$$

$$I_{max} = 0.168 \text{ A}$$

The next step in finding R_s is to determine the amount of circuit current by using a circuit source voltage (V_S) of 12 V and KVL. This law states that the sum of the voltages in a loop is equal to zero as follows:

$$0 = V_S - I_{MAX} \times R_S - V_{Z0} - I_{MAX} \times Z_{ZT}$$

The formula is rearranged to solve for R_S as follows:

$$V_S - V_{Z0} = I_{MAX} \times R_s + I_{MAX} \times Z_{ZT}$$

$$V_S - V_{Z0} = I_{MAX} \times (R_s + Z_{ZT})$$

$$12 \text{ V} - 4.757 \text{ V} = 0.168 \text{ A} \times (R_s + 7.0 \ \Omega)$$

$$7.243 \text{ V} = 0.168 \text{ A} \times (R_s + 7.0 \ \Omega)$$

$$R_s = \frac{7.243 \text{ V}}{0.168 \text{ A} - 7.0 \ \Omega}$$

$$R_s = 36 \ \Omega$$

...**APPLICATION:** ZENER DIODE VOLTAGE REGULATORS...

Now that the circuit values are calculated, the next step in the analysis is to find the minimum resistive load (R_L) that allows the circuit to operate properly. R_L can be found using Kirchhoff's laws: KCL and KVL.

For the zener diode to properly regulate voltage, the minimum current that must flow through the zener diode is 1.0 mA (I_{ZK}). The remaining circuit current flows through the resistive load. Under this condition, the minimum resistive load can be calculated using a combination of KCL and KVL. Using KCL, the relationship between the different loop currents and load current is the following:

$$0 = I_1 - I_L - I_2$$
$$I_L = I_1 - I_2$$

The current in loop I_2 is the zener diode current. The minimum zener diode current is the following:

$$I_2 = I_{ZK}$$
$$I_2 = 0.001 \text{ A}$$

Using KVL, the voltages in each current loop are set equal to zero.

Loop I_1:

$$0 = V_S - R_S \times I_1 - R_L \times I_L$$
$$0 = 12 \text{ V} - 36 \ \Omega \times I_1 - R_L \times I_L$$
$$12 \text{ V} = 36 \ \Omega \times I_1 + R_L \times (I_1 - I_2)$$
$$12 \text{ V} = 36 \ \Omega \times I_1 + R_L \times (I_1 - 0.001 \text{ A})$$
$$12 \text{ V} = 36 \ \Omega \times I_1 + R_L \times I_1 - R_L \times 0.001 \text{ A}$$

Loop I_2:

$$0 = -V_{Z0} - R_Z \times I_2 + R_L \times I_L$$
$$0 = -V_{Z0} - R_Z \times I_2 + R_L \times (I_1 - I_2)$$
$$0 = -4.757 \text{ V} - 7 \ \Omega \times 0.001 \text{ A} + R_L \times (I_1 - 0.001 \text{ A})$$
$$4.764 \text{ V} = R_L \times I_1 - R_L \times 0.001 \text{ A}$$

The value of I_1 is found by subtracting the two equations as follows:

$$12.000 \text{ V} = 36 \ \Omega \times I_1 + R_L \times I_L - R_L \times 0.001 \text{ A}$$
$$-4.764 \text{ V} = -(R_L \times I_L) + R_L \times 0.001 \text{ A}$$
$$7.233 \text{ V} = 36 \ \Omega \times I_1$$
$$I_1 = 0.201 \text{ A}$$

...APPLICATION: ZENER DIODE VOLTAGE REGULATORS

The value of R_L is found by using the equation for loop I_2 and substituting for I_1:

$$4.764\ \text{V} = R_L \times I_1 - R_L \times 0.001\ \text{A}$$
$$4.764\ \text{V} = R_L \times (I_1 - 0.001\ \text{A})$$
$$R_L = \frac{4.767\ \text{V}}{(0.201\ \text{A} - 0.001\ \text{A})}$$
$$R_L = \mathbf{23.835\ \Omega \approx 24\ \Omega}$$

Therefore, the minimum resistive load the circuit supports is 24 Ω. For the zener diode to operate properly in the circuit, the zener voltage must be greater than 4.764 V ($V_{ZO} + Z_{ZT} \times I_{ZK}$). Thus, the minimum zener voltage is 4.764 V.

In this application, a voltage regulator is designed using a zener diode. The circuit is designed such that the zener diode operates within its parameter with or without a load. The minimum resistive load supported by the zener diode is 24 Ω found using Kirchhoff's laws.

Chapter Review

1. What is the difference between an ideal voltage source and practical voltage source?
2. Explain the difference between a balanced and unbalanced resistance bridge.
3. Why is it impossible for electrons to be stored or collected in a conductor?
4. Explain Norton's theorem.
5. How are values of voltage and current entering and exiting a circuit identified in KVL and KCL?
6. What is Thevenin's theorem?
7. How is KCL typically used in an equation?
8. What is "Thevenizing a circuit"?
9. What is the most common use of KVL and KCL?
10. What is a Wheatstone bridge?
11. Which is greater: open circuit voltage or voltage under load?
12. What is potential?
13. What is voltage 1 in a closed loop if voltage 2 is equal to 24 V and voltage 3 is equal to 12 V?
14. How is the superposition theorem used to analyze voltage?
15. What is the current entering a junction if current exiting a junction is equal to 1.325 A?

10

Electromagnetism

OBJECTIVES

- Explain how conductors and coils form electromagnets.
- Describe magnetic circuit properties such as magnetizing force intensity, permeability, hysteresis, reluctivity, and permeance.
- Explain the concept of electromagnetic induction.
- Explain Lenz's law and how it applies to electromagnetism.
- Describe the operating principles of DC generators and their key components.
- Describe how DC voltage is generated.

One of the greatest discoveries in the field of electricity is the discovery of electromagnetism. The primary uses of electromagnetism are in motors and the generation of electrical power, which have proven to be great enablers for other innovations.

Basic electricity and magnetism are closely related to each other. Both are considered expressions of a single force, the electromagnetic force. Describing one without the other is difficult. Through understanding the relationship between electricity and magnetism, the application of electromagnetic principles can be understood. When studying electromagnetism, three phenomena are noted. The three phenomena are moving electric charges produce magnetic fields, changing magnetic fields in the presence of electric charges cause a current to flow, and magnetic fields exert forces on moving electric charges.

Terms

Electromagnetism is the magnetism produced when electricity passes through a conductor.

ELECTROMAGNETISM

Electromagnetism is the magnetism produced when electricity passes through a conductor. When there is no current flow in a conductor, electrons move in a random manner. When voltage is applied to a conductor causing current flow, all of the electrons move in one direction. This causes their magnetic fields to align with each other. The magnetic fields combine and extend outside the conductor. **See Figure 10-1.** Electromagnetism affects magnetic field strength and distance, conductors, coils, and electromagnets.

Magnetic Field Strength and Distance

Theoretically, the magnetic field around the circumference of a conductor extends an infinite amount into space. If a compass is placed near the magnetic field surrounding a conductor, the compass will indicate the polarity of the field. However, the further the distance the compass is moved away from the conductor, the less strength the field has on the compass. At some point, there is no effect, and the compass will align with the magnetic field of the earth.

If the current flow through the conductor is increased, a magnetic field can again be detected at the greater distance. This demonstrates the relationship between the level of current flow in a conductor and the perpendicular distance of a magnetic field from the conductor.

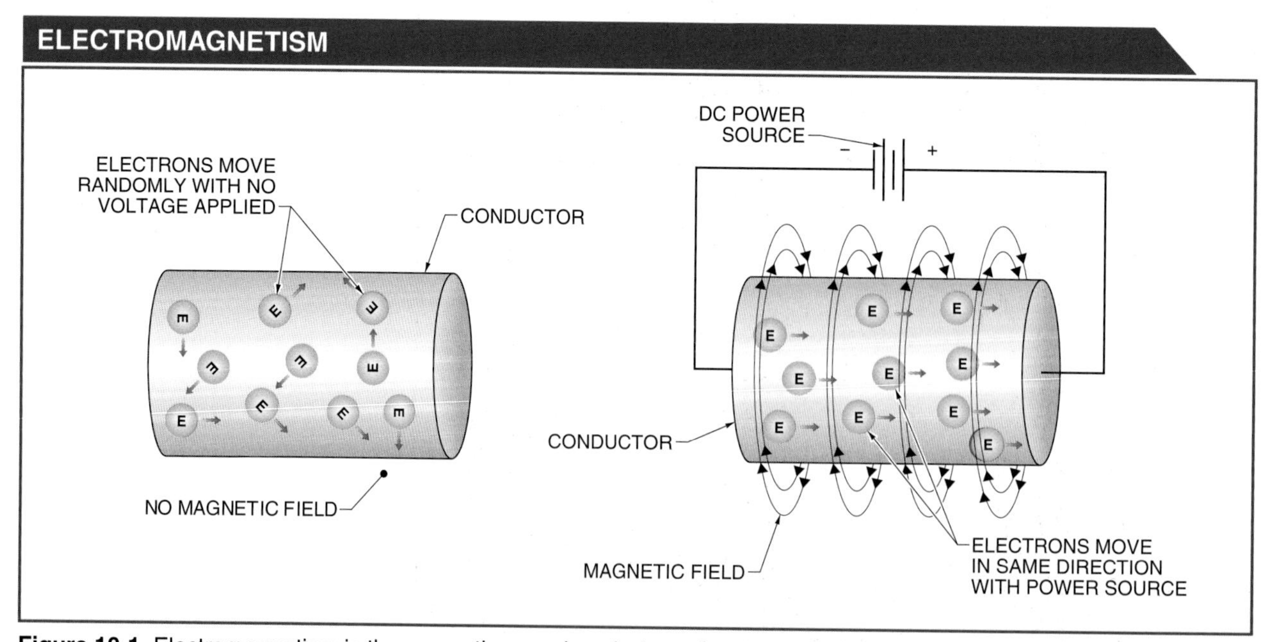

Figure 10-1. Electromagnetism is the magnetism produced when electricity passes through a conductor.

The magnetic field stretches the entire length of the conductor with concentric rings of magnetic force. Because the magnetic field from a single conductor consists of concentric circles radiating out from the center of the conductor, it is difficult to magnetically shield the field using any type of known material, including materials such as those used in electrical conduit and wire insulation.

Plasma in Electronics

Many conductors are nonmetallic, including plasma. Plasma is an electrically conductive collection of charged particles that responds collectively to electromagnetic forces. Plasma is typically in the form of a gas cloud or a charged ion beam, but it may also take the form of microscopic dust and grains. A common electronics application of plasma is a plasma television (plasma-screen television). A plasma television uses a flat-panel screen where light is created by phosphors excited by a plasma discharge between two flat panels of glass. Although lighter and thinner than standard cathode ray tube (CRT) television screens, plasma television screens have a usable life of about 100,000 hours, which corresponds to the life of the gas between the panels of the screen.

Conductors

A *conductor* is a material that has low electrical resistance and permits electrons to move through it easily. When two parallel current-carrying conductors have current flow in opposite directions, magnetic lines of force traveling in the same direction between the conductors are created. Magnetic lines of force traveling in the same direction repel each other. This results in a force that tends to push the two conductors apart. If the current flow is reversed in one conductor, the direction of the magnetic lines of force surrounding the same conductor also reverses. Magnetic lines of force traveling in the same direction repel each other while magnetic lines of force traveling in the opposite direction attract each other. **See Figure 10-2.**

Current-carrying conductors that are perpendicular to each other produce magnetic fields that attempt to align parallel with each other so that the current flow is in the same direction. The magnetic lines of force always move from a strong magnetic field toward a weak magnetic field. The direction of movement is the result of all magnetic forces acting on the conductor. The direction of the magnetic lines of force is from north to south, which means that the magnetic field of the current-carrying conductor enters the south pole of the conductor and exits at its north pole. This movement causes the needle of a compass to align 90° to the conductor and the magnetic lines of force to be concentric around the circumference of the conductor.

Terms

A **conductor** is a material that has low electrical resistance and permits electrons to move through it easily.

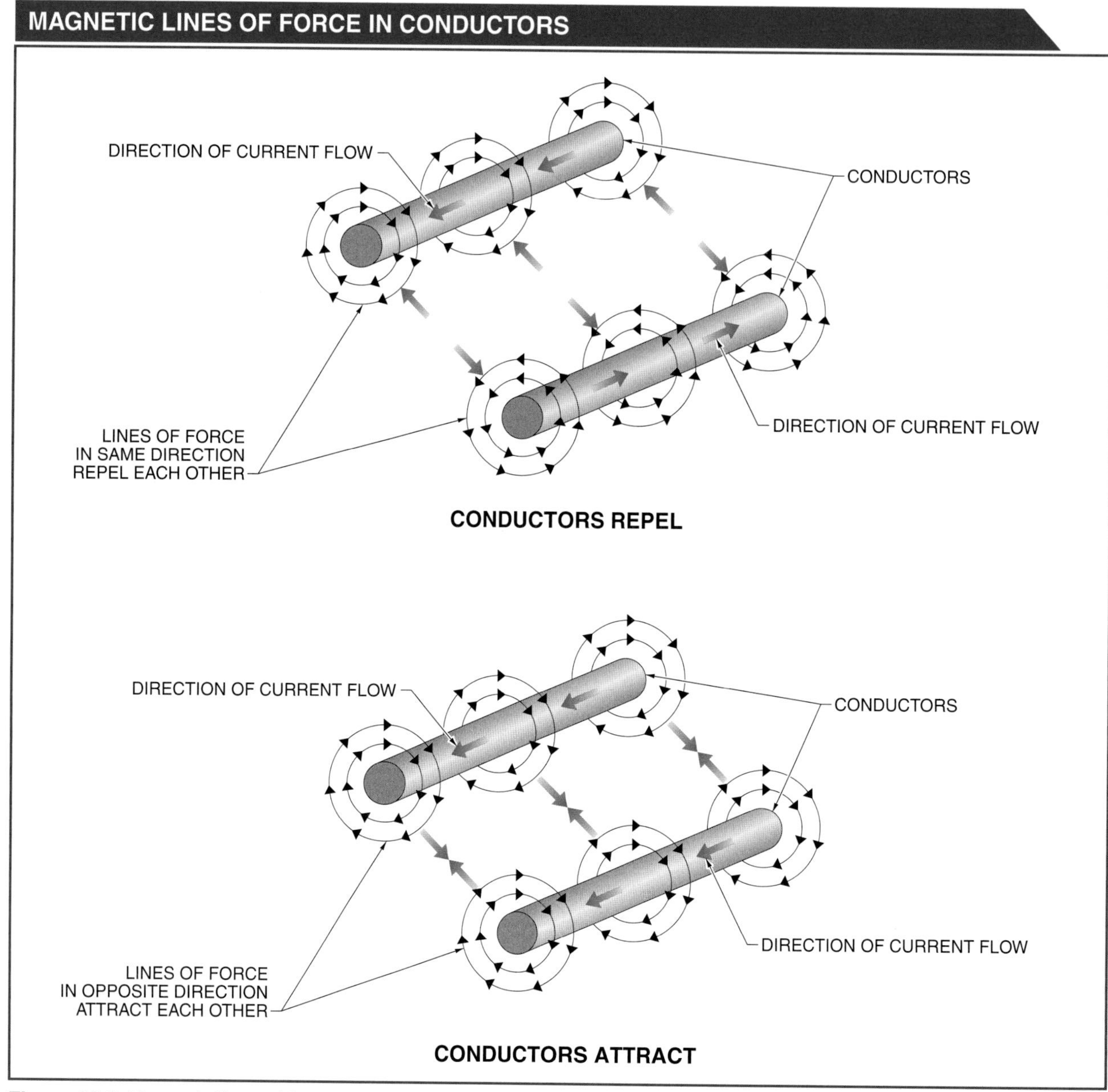

Figure 10-2. Magnetic lines of force traveling in the same direction repel each other while magnetic lines of force traveling in the opposite direction attract each other.

The left-hand rule for conductors can be used to determine the direction of magnetic lines of force around a conductor. The left-hand rule for conductors states that when a conductor is wrapped with the left hand with the thumb in the direction of the current flow, the fingers point in the direction of the magnetic lines of force. The direction of current is from negative to positive and can be determined by the polarity of the source. **See Figure 10-3.**

LEFT-HAND RULE FOR CONDUCTORS

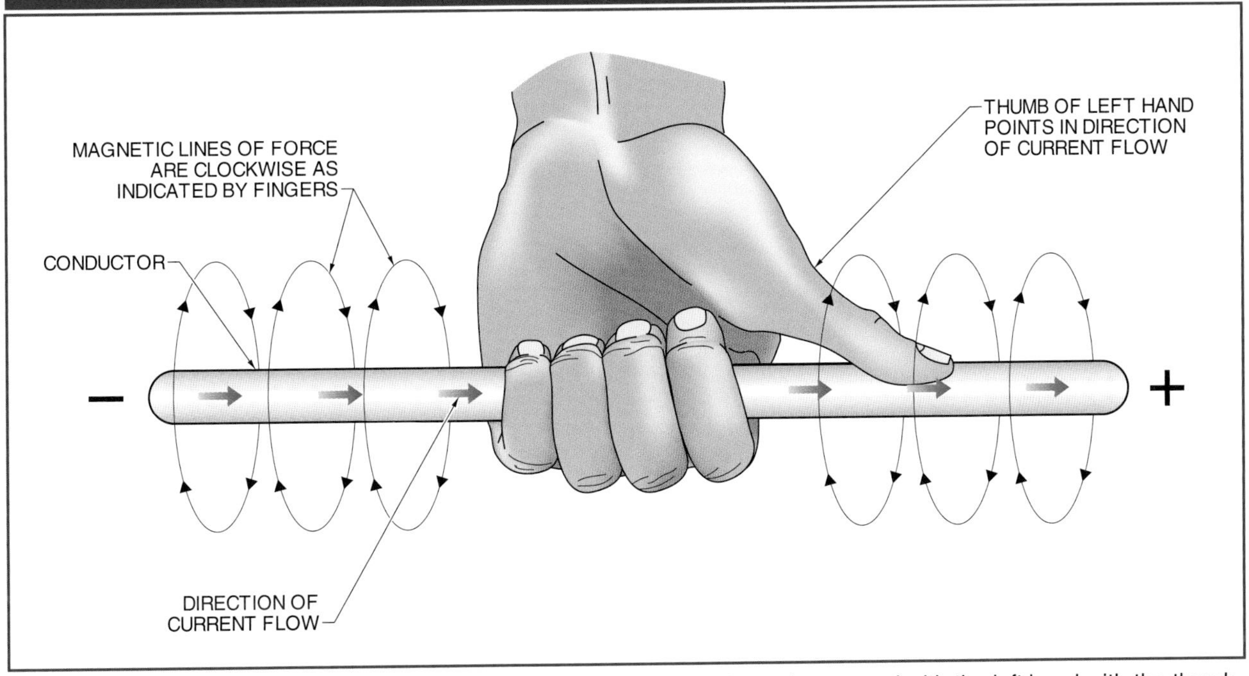

Figure 10-3. The left-hand rule for conductors states that when a conductor is wrapped with the left hand with the thumb in the direction of the current flow, the fingers point in the direction of the magnetic lines of force.

When a straight, current-carrying conductor produces a magnetic field, the magnetic lines of force are concentric around the conductor, without detectable magnetic poles. This is indicated by moving a compass from below the conductor to above the conductor and observing the rotation of the compass needle. As the compass is moved in a circular motion around the circumference of the conductor, the needle swings to align with the magnetic lines of force perpendicular to the conductor.

By forming a loop in a current-carrying conductor, an electromagnet with a north and south pole is formed. The south pole of the electromagnet attracts the north pole of the compass. **See Figure 10-4.**

Terms

A **coil** is a circular wound wire (winding) consisting of insulated conductors arranged to produce lines of magnetic flux.

Coils

A *coil* is a circular wound wire (winding) consisting of insulated conductors arranged to produce lines of magnetic flux. A coil offers considerable opposition to the passage of AC but very little opposition to DC. Coils may be wound loosely or tightly. The tighter the turns of a coil, the stronger the magnetic field surrounding the coil. **See Figure 10-5.**

Tech Tip

The French physicist André M. Ampère discovered that if a straight current-carrying conductor is formed into a loop (turn), the magnetic lines of force are given a polarity.

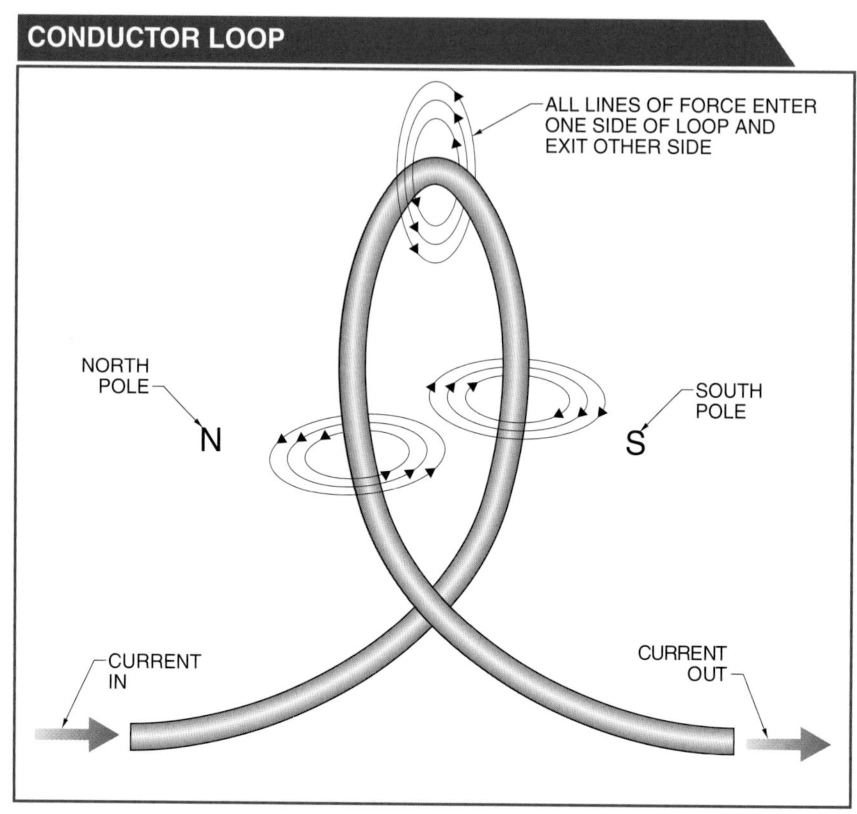

Figure 10-4. By forming a loop in a current-carrying conductor, an electromagnet with a north and south pole is formed.

Because there is current flow in the same direction in all of the turns of a coil, the magnetic field produced is similar to the magnetic field produced by parallel conductors with current flow in the same direction. Magnetic lines of force are created between adjacent segments of the coil that oppose each other and weaken the magnetic field between the turns. By winding the coil turns tightly at the bottom of the coil, the magnetic field is made stronger, with more magnetic lines of force encircling the entire coil. Under this condition, the individual concentric magnetic force lines around the conductor join to encircle the entire coil.

When current flow is present through the coil, the coil has a magnetic field comparable to a permanent magnet. The polarity of a coil is determined by applying the left-hand rule for coils. To apply the left-hand rule for coils, grasp the coil with the left hand so that the fingers encircle the coil in the direction of the current flow through the coil and extend the thumb at a right angle to the fingers. The thumb points toward the north pole of the coil. If the direction of current is reversed, the polarity of the magnetic field also reverses. Several factors affect the strength of a coil's magnetic field including the number of conductor turns, amount of current flow through the coil, ratio of coil's length to its width, and type of coil core material.

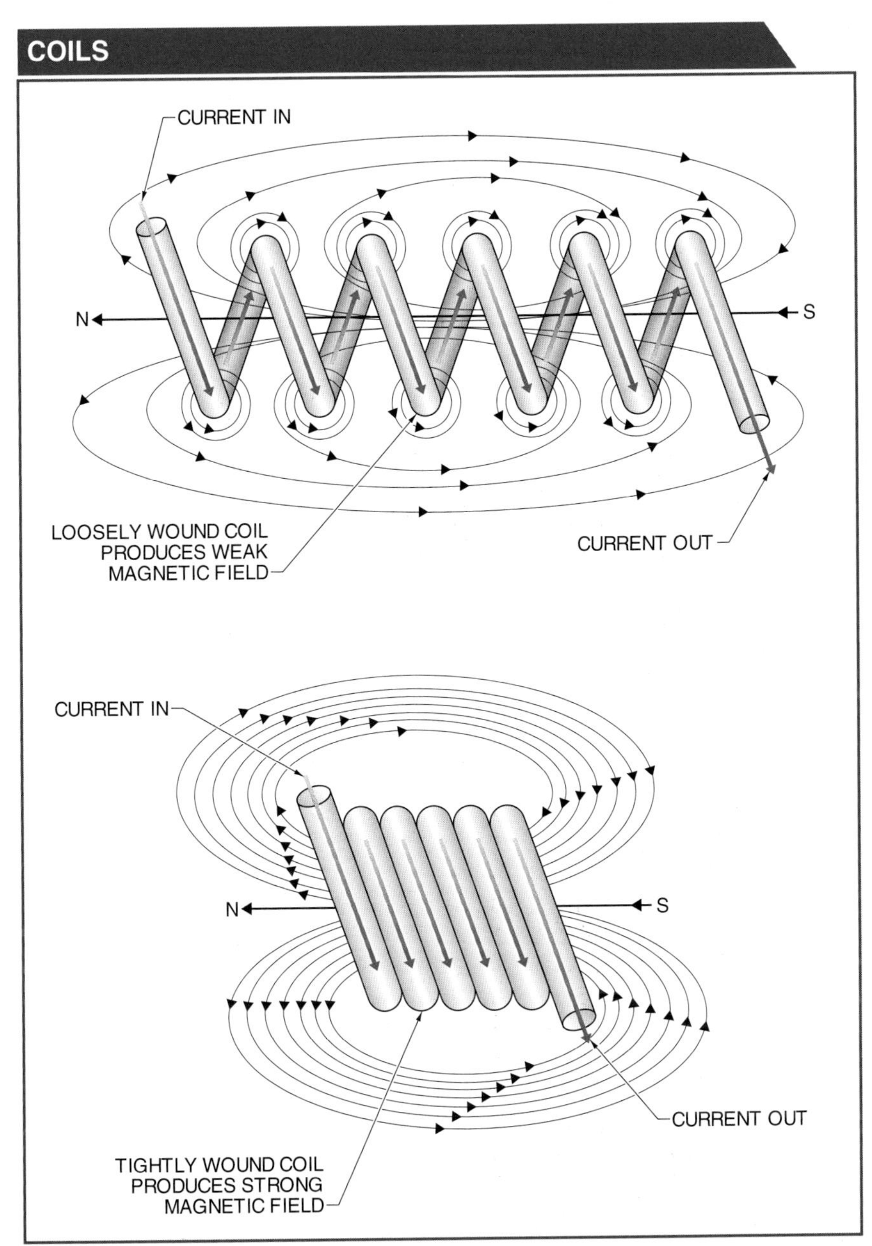

Figure 10-5. Coils may be wound loose or tight. When the turns are tighter, the magnetic field surrounding the coils is stronger.

Electromagnets. *Electromagnets* are coils that have a core material made of soft iron. The core material enhances the strength of the magnetic field because the soft iron provides a better path for the magnetic lines of force than air and concentrates them into a smaller area. Once current is removed from the coil, the electromagnet loses its magnetic characteristics because soft iron has very little residual magnetism. The left-hand rule for coils is used to determine the polarity of an electromagnet's magnetic field.

Terms

Electromagnets are coils that have a core material made of soft iron.

Terms

A **magnetic circuit** is the path or paths taken by the magnetic lines of force leaving the north pole of a magnet and returning to the south pole of the magnet.

MAGNETIC CIRCUIT PROPERTIES

A *magnetic circuit* is the path or paths taken by the magnetic lines of force leaving the north pole of a magnet and returning to the south pole of the magnet. Many electrical devices, such as transformers and motors, depend on magnetism for proper operation. In order for these devices to perform their functions satisfactorily, exact magnetic circuit properties must be determined. Magnetic circuits must have the correct strength, have the correct paths, and be made in suitable shapes and of suitable materials. Magnetic circuit paths may be series, parallel, or series/parallel. **See Figure 10-6.**

MAGNETIC CIRCUIT PATHS

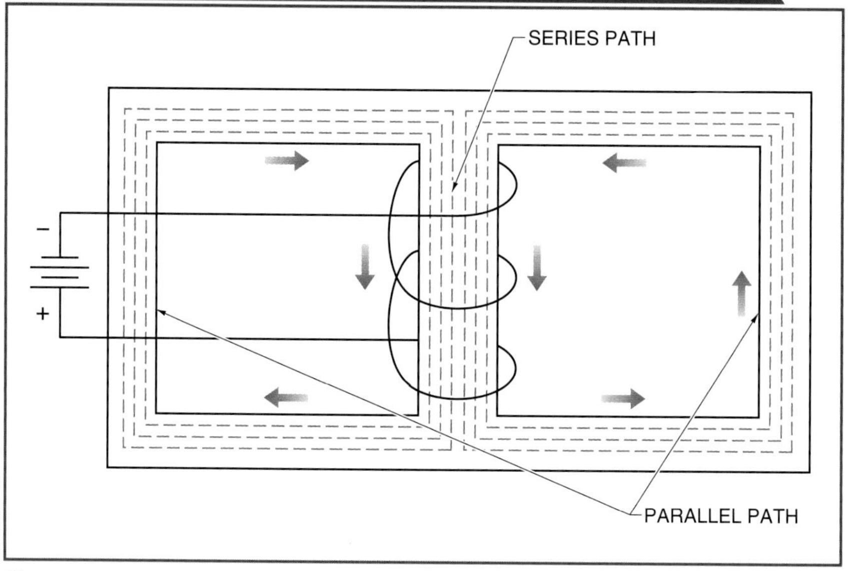

Figure 10-6. Magnetic circuits can have either series, parallel, or series/parallel paths.

Rowland's Law

Tech Tip

The American physicist Henry A. Rowland proved that electrostatic charges in motion produce a magnetic field similar to the flow of electrons (current) in a conductor. He also defined the value of the ohm (Ω).

Any flux line that leaves the north pole of a magnet must return to the south pole of the same magnet. Each flux line is continuous. Magnetic circuit operating conditions are determined by calculating circuit conditions and parameters such as magnetic lines of force (magnetic flux), magnetomotive force, reluctance, intensity of magnetizing force, permeability, hysteresis, reluctivity, and permeance. These conditions and properties are calculated by applying Rowland's law.

Ohm's law is used to find the level of current flow in an electrical circuit. It states that the current flow (I) in a circuit is directly proportional to applied voltage (V) and inversely proportional to resistance (R). Rowland's law for magnetic circuits is similar to Ohm's law.

Rowland's law states that the number of magnetic lines of force (Φ) is directly proportional to the magnetomotive force (F_m) and inversely proportional to the reluctance (R_m) of the circuit. Although Ohm's law and Rowland's law are similar, a comparison shows that each law has separate parameters for flow, force, and opposition. **See Figure 10-7.**

COMPARISON OF OHM'S LAW AND ROWLAND'S LAW		
	Electric Circuit	**Magnetic Circuit**
Ohm's law	$I = \frac{V}{R}$	—
Rowland's law	—	$\Phi = \frac{F_m}{R_m}$
Flow units	ampere, I, A	flux, Φ, maxwells, weber
Force units	volt, V, emf, E	F_m, mmf, gilbert, ampere-turn
Opposition units	ohm, R, Ω	R_m, rel, gilberts/maxwell, ampere-turn/weber

Figure 10-7. Ohm's law is used with electric circuits while Rowland's law is used with magnetic circuits. Each law has separate parameters for flow, force, and opposition.

Magnetic Flux. *Magnetic flux (Φ)* is a measure of magnetic induction represented by magnetic lines of force. The *maxwell (Mx)* is the centimeter-gram-second (cgs) unit of flux and is defined as one line of flux. A maxwell is often referred to as a line of force. The *weber (Wb)* is the unit of magnetic lines of force in the meter-kilogram-second (mks) measurement system, or Système International (SI). The mks system is based on the meter and kilogram instead of the centimeter and gram in the cgs system. Magnetic lines of force are calculated by applying the following formula:

$$Wb = 10^{-8} \times Mx$$

where

Wb = total magnetic lines of force

Mx = 1 magnetic line of force

10^{-8} = constant

Terms

Magnetic flux (Φ) is a measure of magnetic induction represented by magnetic lines of force.

The **maxwell (Mx)** is the centimeter-gram-second (cgs) unit of flux and is defined as one line of flux.

The **weber (Wb)** is the unit of magnetic lines of force in the meter-kilogram-second (mks) measurement system, or Système International (SI).

Example: What is the number of magnetic lines of force for 2,500,000 Mx?

$Wb = 10^{-8} \times Mx$

$Wb = 10^{-8} \times 2{,}500{,}000$

$Wb = \mathbf{0.025}$

Terms

Magnetomotive force (mmf) is the current through a coil multiplied by the number of turns on the coil.

A **gilbert (Gb)** is the unit of measure in the cgs system for mmf.

Reluctance is the property of an electric circuit that opposes magnetic lines of force.

Magnetomotive Force. *Magnetomotive force (mmf)* is the current through a coil multiplied by the number of turns on the coil. Magnetomotive force in a magnetic circuit operates on a similar principle as voltage (emf) in an electric circuit and is the force that produces the magnetic lines of force in the magnetic circuit. A *gilbert (Gb)* is the unit of measure in the cgs system for mmf. For coils with a length that is ten times or greater than its diameter, the mmf is calculated by applying the following formula:

$F_m = 1.257 \times I \times N$

where

F_m = total circuit mmf (in Gb)

1.257 = constant of length to diameter ratio

I = circuit current (in A)

N = number of conductor turns

Example: What is the total mmf for a circuit that has 4 A of current flow and 200 conductor turns?

$F_m = 1.257 \times I \times N$

$F_m = 1.257 \times 4.0 \times 200$

$F_m = \mathbf{1005.6\ Gb}$

When the length of a coil is less than ten times its diameter, the constant 1.257 is dropped due to the change in the ratio of length to diameter. The mmf is then equal to the current multiplied by the number of turns. The unit of force then becomes the ampere-turn as used in the SI system.

Reluctance. *Reluctance* is the property of an electric circuit that opposes magnetic lines of force. Reluctance for a magnetic material varies with flux density. The unit of reluctance is the opposition to flow presented by 1 cm^3 of air to the passage of magnetic lines of force. There is no widely accepted unit for reluctance although the "rel" is commonly used. Reluctance in magnetic circuits is similar to resistance in electrical circuits. Reluctance is calculated by applying the following formula:

$$R_m = \frac{F_m}{\Phi}$$

where

R_m = total reluctance (in rels)

F_m = total mmf (in Gb)

Φ = number of magnetic lines of force (in Mx)

Example: What is the reluctance in a circuit with 4,500,000 Mx and a total circuit mmf of 2 Gb?

$$R_m = \frac{F_m}{\Phi}$$

$$R_m = \frac{2 \text{ Gb}}{4{,}500{,}000 \text{ Mx}}$$

R_m = **0.0000004 rels**

Reluctance, like resistance, varies in direct proportion to the length of the core material and inversely with its area and permeability. *Permeability* (μ) is the ability of a material to carry magnetic lines of force.

Although the reluctance measured of a coil with an air gap is linear, reluctance measured using a magnetic substance is nonlinear. A permeability chart is used to obtain reluctance values. **See Appendix.** Air exhibits reluctance 50 times to 5000 times greater than that of magnetic materials. Nonmagnetic materials, such as paper and wood, have about the same value of reluctance (within ±1%) as air. Magnetic circuits that combine a magnetic material with an air gap are common. Similar to resistances in an electric circuit connected in series, reluctances add together in a magnetic circuit. **See Figure 10-8.** Reluctance in a magnetic circuit with air gap is calculated by applying the following formula:

$$R_{mT} = R_{m1} + R_{m2} + \ldots + R_{mN}$$

where

R_{mT} = total circuit reluctance (in rels)

R_{m1} = circuit 1 reluctance (in rels)

R_{m2} = circuit 2 reluctance (in rels)

R_{mN} = circuit N reluctance (in rels)

Terms

Permeability (μ) is the ability of a material to carry magnetic lines of force.

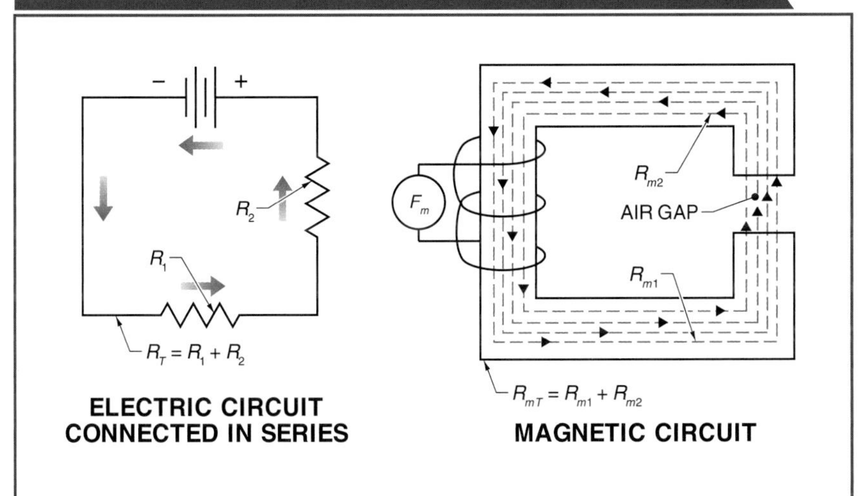

Figure 10-8. Similar to resistances in an electric circuit connected in series, reluctances add together in a magnetic circuit.

Terms

The **intensity of magnetizing force** measures the magnetic strength per unit length.

Example: What is the total circuit reluctance in a magnetic circuit that has 50 × 10^{-6} rels, 30 × 10^{-6} rels, and 20 × 10^{-6} rels in series?

$R_{mT} = R_{m1} + R_{m2} + R_{m3}$

$R_{mT} = 50 \times 10^{-6}$ rels + 30×10^{-6} rels + 20×10^{-6} rels

$R_{mT} = 100 \times 10^{-6}$ rels

$R_{mT} =$ **0.0001 rels**

Magnetizing Force Intensity

The *intensity of magnetizing force* measures the magnetic strength per unit length. High current produces high magnetic fields. The unit of measure of the intensity of magnetizing force is calculated by applying the following procedure:

1. Calculate magnetomotive force (mmf). The mmf is calculated by applying the following formula:

   $F_m = 1.257 \times I \times N$

2. Calculate the intensity of magnetizing force.

   $$H = \frac{F_m}{l}$$

   where

   F_m = total mmf (in Gb)

   1.257 = constant

   I = current (in A)

   N = number of turns

   H = intensity of magnetizing force (in Oe)

   l = unit length (in cm)

Example: What is the intensity of magnetizing force for a conductor carrying 100 mA with 1000 turns over a 10 cm length?

1. Calculate the mmf.

   $F_m = 1.257 \times I \times N$

   $F_m = 1.257 \times 0.100$ A × 1000 turns

   $F_m = 125.7$ Gb

2. Calculate the intensity of the magnetizing force.

   $$H = \frac{F_m}{l}$$

   $$H = \frac{125.7 \text{ Gb}}{10 \text{ cm}}$$

   $H =$ **12.57 Oe**

In the mks system, mmf is measured in ampere-turns, and length is measured in meters. The mks intensity of magnetizing force is measured in ampere-turn per meter units.

Example: What is the intensity of the magnetizing force in the msk system for a conductor carrying 100 mA with 1000 turns over a 10 cm length?

1. Calculate the mmf.

$F_m = I \times N$

$F_m = 0.100 \text{ A} \times 1000 \text{ turns}$

$F_m = 100 \text{ amp-turns}$

2. Calculate the intensity of the magnetizing force.

$$H = \frac{F_m}{l}$$

$$H = \frac{100 \text{ amp-turns}}{0.1 \text{ m}}$$

$H =$ **1000 amp-turns/m**

The magnetic intensity from a conductor can also be measured in oersteds. With a conductor, 1 Oe is the magnetic intensity at 1 cm from the center of a conductor carrying 5 A. The magnetic field intensity of a conductor is calculated by applying the following formula:

$$H = \frac{I}{5 \times r}$$

where

H = magnetic field intensity (in Oe)

I = current (in A)

5 = constant

r = distance from conductor center (in cm)

Example: What is the magnetic field intensity (in oersteds) from the center of a straight conductor carrying 2.25 at 15 cm?

$$H = \frac{I}{5 \times r}$$

$$H = \frac{2.25 \text{ A}}{5 \times 15 \text{ cm}}$$

$H =$ **0.03 Oe**

Magnetic field intensity at the center of a flat circular coil with current flowing through it is calculated by applying the following formula:

$$H = \frac{\pi \times N \times I}{5 \times r}$$

where

H = magnetic field intensity (in Oe)

π = ratio of coil circumference to diameter (3.14)

N = number of turns

I = current (in A)

r = radius of turns (in cm)

Example: What is the magnetic field intensity at the center of a flat circular coil of 500 turns that is 20 cm in diameter and carries 3 A?

$$H = \frac{\pi \times N \times I}{5 \times r}$$

$$H = \frac{3.14 \times 500 \text{ turns} \times 3 \text{ A}}{5 \times (0.5 \times 20) \text{ cm}}$$

$$H = \frac{4710}{50}$$

$H = \mathbf{94.2\ Oe}$

Permeability

Permeability is the ability of a material to carry magnetic lines of force. Permeability (μ) is defined as the ratio of flux density (B) in lines per square centimeter to the field intensity (H) in oersteds. A vacuum has a permeability of 1. The permeability of air is also designated as 1 but is actually slightly higher at 1.0000004.

A gauss is equal to one line of force per square centimeter. Gauss (G) measures the amount of density for magnetic lines of force (flux density). The Earth's flux density is about 1 G at the surface. Flux density is calculated by applying the following formula:

$$B = \frac{\Phi}{A}$$

where

B = flux density

Φ = total magnetic lines of force (in Mx)

A = total area of magnetic lines of force (in cm^2)

Example: What is the flux density of a cube that has lengths of 10 cm and magnetic lines of force of 1 Mx?

$$B = \frac{\Phi}{A}$$

$$B = \frac{1 \text{ Mx}}{10 \text{ cm} \times 10 \text{ cm}}$$

$$B = \frac{1 \text{ Mx}}{100 \text{ cm}}$$

$B = \mathbf{0.01\ G}$

As the magnetic intensity of a nonmagnetic substance is increased to obtain a greater flux density, the permeability of the nonmagnetic substance decreases. Permeability of a nonmagnetic substance is not a fixed value. Variation in the intensity of the magnetizing force (H)

applied to nonmagnetic substances results in a proportional increase or decrease in the flux density (B) and shows a linear relationship between B and H.

This relationship is not true for magnetic substances. The density of the magnetic lines of force of a magnetic substance depends on the permeability of the substance and the amount of ampere-turns per unit length. All magnetic substances have a different permeability, and the permeability curve is nonlinear. For example, the magnetization curves for soft iron, soft cast steel, and cast iron show a greater flux density for soft iron than for soft cast steel or cast iron. **See Figure 10-9.** The flux density rises quickly at the beginning of the magnetizing process and at a greatly reduced rate afterwards. At some point, a magnetic saturation occurs. Magnetic saturation is the point at which a further increase in current does not cause a further increase in the number of magnetic lines of force.

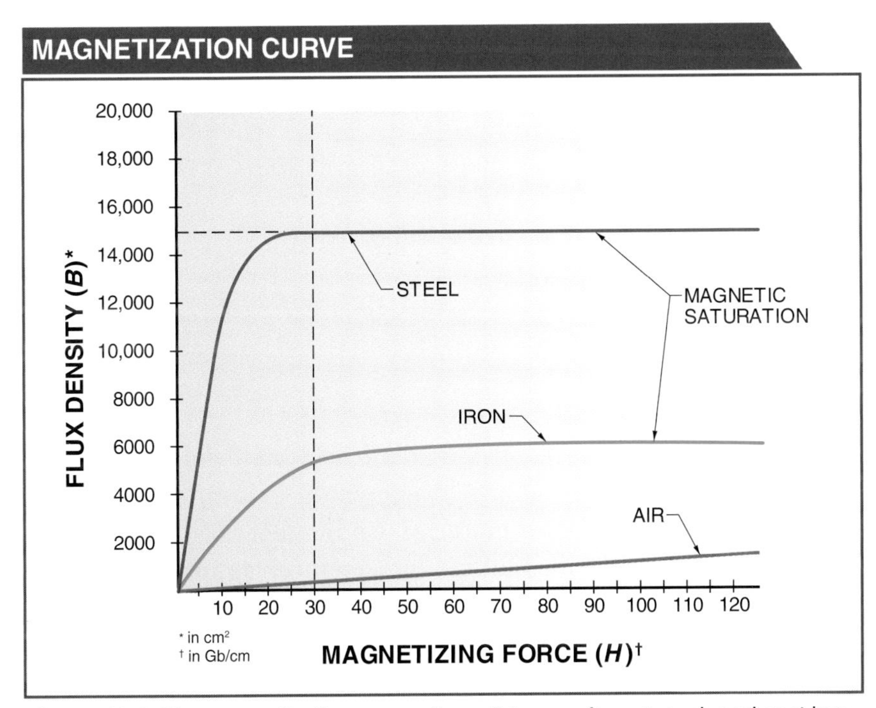

Figure 10-9. The magnetization curves for soft iron, soft cast steel, and cast iron show a greater flux density for soft iron than for soft cast steel or cast iron.

Terms

A **paramagnetic substance** is a substance with a permeability greater than 1.

A **diamagnetic substance** is a substance with permeability less than 1 and is repelled by either pole of a magnet.

A **ferromagnetic substance** is a substance that has permeability greater than 1 as well as greater than a paramagnetic substance.

A *paramagnetic substance* is a substance with a permeability greater than 1. A paramagnetic substance has a greater ability to be magnetized than air, but less ability than a ferromagnetic material. A *diamagnetic substance* is a substance with permeability less than 1 and is repelled by either pole of a magnet. A *ferromagnetic substance* is a substance that has permeability greater than 1 as well as greater than a paramagnetic substance.

A simple process determines whether a substance is paramagnetic, diamagnetic, or ferromagnetic. If the substance is suspended between the poles of a strong magnet and it aligns itself so that its longer dimension is in line with the field, then it is either ferromagnetic or paramagnetic. If the substance aligns itself so that it assumes a position perpendicular to the magnetic field, then the substance is diamagnetic. Permeability is calculated by applying the following formula:

$$\mu = \frac{B}{H}$$

where

μ = permeability (in G/Oe)

B = flux density (in G)

H = magnetizing force (in Oe)

Example: What is the permeability of a substance that has field intensity of 25 Oe and a flux density of 125,000 G?

$$\mu = \frac{B}{H}$$

$$\mu = \frac{125{,}000\ \text{G}}{25\ \text{Oe}}$$

$$\mu = \mathbf{5000\ G/Oe}$$

Hysteresis

Terms

Hysteresis is the property of ferromagnetic materials where the magnetic induction of a coil lags the magnetic field that is charging the coil.

Hysteresis is the property of ferromagnetic materials where the magnetic induction of a coil lags the magnetic field that is charging the coil. Each molecule of a material is a magnet with a north and a south pole. All molecules must align themselves in the same direction to obtain the maximum flux density. Once the molecules are aligned, flux density remains at the same value, causing the core to become saturated.

The time lag in magnetization of the core behind the value of H (magnetizing force) is caused by molecular agitation. That is, each molecule must align itself in one direction, and then, when the direction of current flow through the coil is reversed, it must align itself in the opposite direction. If the polarity of H is reversed slowly, very little energy is used. If the polarity of H is reversed rapidly, considerable heat is generated by the molecules bumping into each other in the core, resulting in a high-energy loss known as hysteresis loss. A hysteresis curve for a constant changing intensity of a magnetizing force is used to track hysteresis. **See Figure 10-10.**

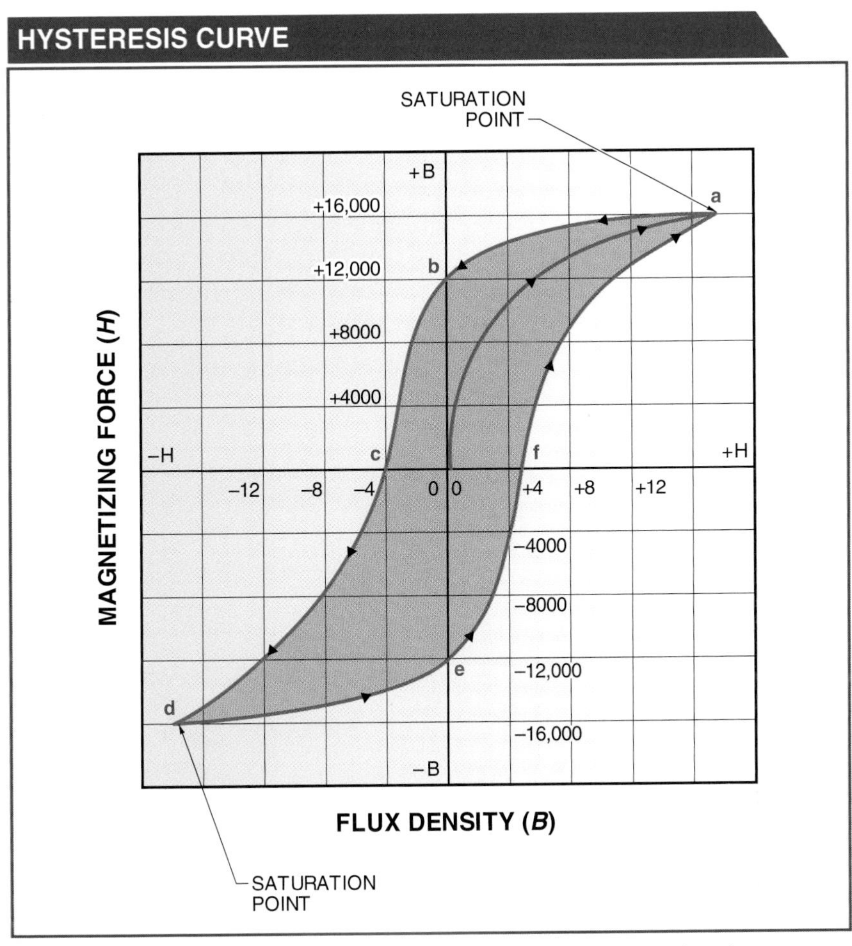

Figure 10-10. A hysteresis curve for a constantly changing intensity of a magnetizing force is used to track hysteresis.

With the core completely demagnetized, power is applied. The magnetizing force (H) is increased from the origin (0) to point a where the core material is saturated at +16,000 G. As magnetizing force is decreased to 0, the core material does not demagnetize at the same rate. At point b with 0 Oe magnetizing force applied, the flux density is still about +12,000 G. At this point, the current through the coil is reversed so that the core magnetizes in the opposite direction. The core is not fully demagnetized until magnetizing force of 4 Oe are reached at point c. As the magnetization process continues in the negative direction from point c to point d, the polarity of the core reverses and is fully saturated at magnetizing force 16,000 G. As the magnetizing force is reduced to 0, the core material still has −12,000 G at point e. Increasing magnetizing at this point results in demagnetizing the core at point f and saturating the core in the positive direction at point a.

Terms

Retentivity is the ability of the core material to retain its magnetism after the magnetizing force is removed.

Retentivity is the ability of the core material to retain its magnetism after the magnetizing force is removed. Retentivity is most noticeable in hard steel and least noticeable in soft iron. The value of the residual magnetism when magnetizing force is reduced to 0 depends on the type of core material and the value of the magnetizing force obtained. Core materials with the greatest retentivity, and therefore the greatest area enclosed by the hysteresis curve, have the greatest energy loss. Materials that have greater hysteresis curve areas typically have greater magnetic retentivity. For example, annealed steel has about one-half the magnetic retentivity as that of hard steel. **See Figure 10-11.**

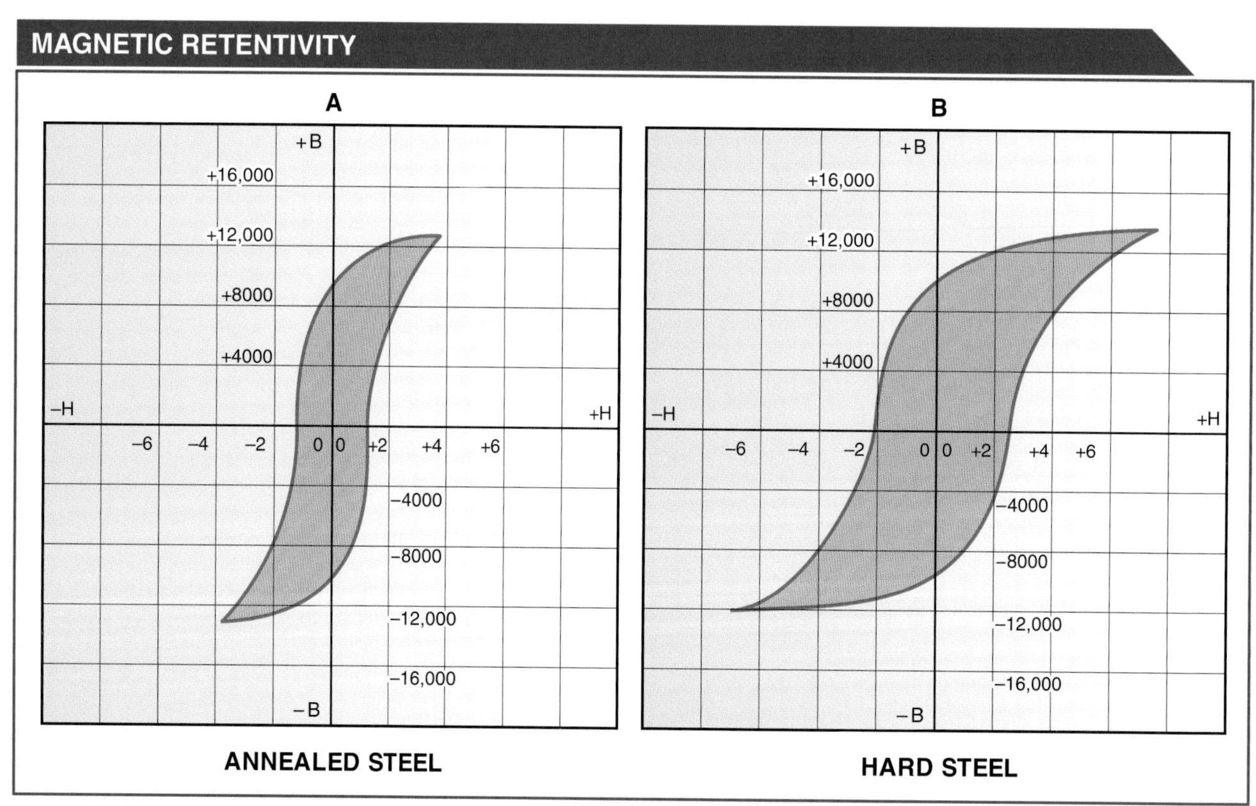

Figure 10-11. Annealed steel has about one-half the magnetic retentivity as that of hard steel.

Tech Tip

Because the materials are always saturated in a positive or a negative direction, they made excellent devices to store digital information in the 1950s and 1960s, although the memory density was small compared to that of modern technology.

Magnetic substances composed of magnesium, manganese, or iron have a rectangular hysteresis curve. **See Figure 10-12.** As magnetizing force is increased gradually in a positive direction, almost no change in the B occurs. At the point where magnetizing force reaches a critical point (a), the core becomes saturated. As magnetizing force is decreased, the core remains saturated in a positive direction until a critical point (b) is reached, causing the core to saturate in the negative direction. This cycle then repeats.

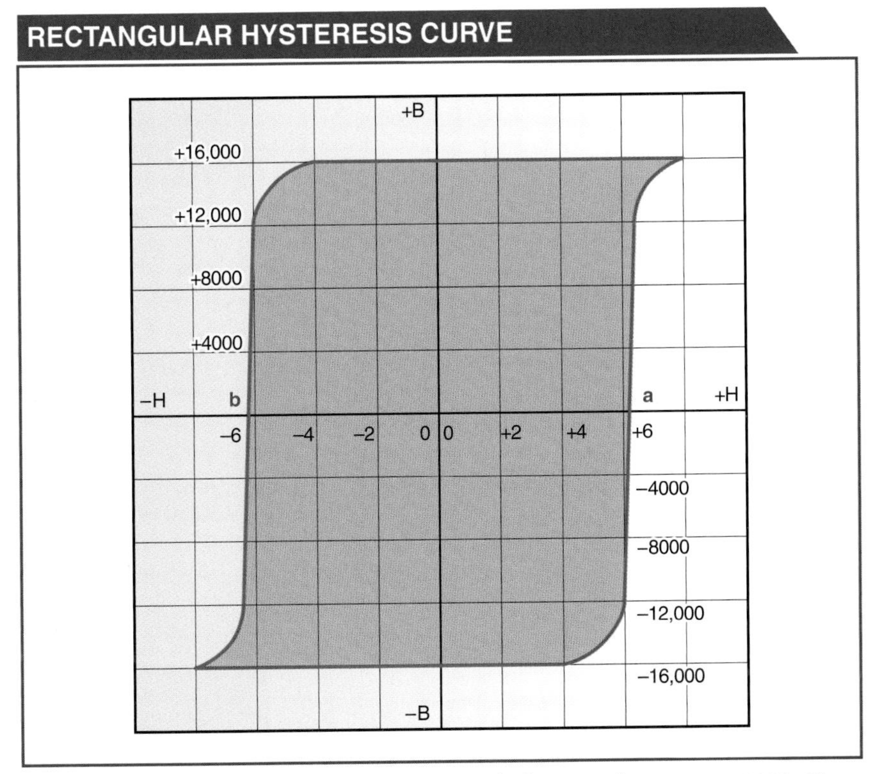

Figure 10-12. Magnetic substances composed of magnesium, manganese, or iron have a rectangular hysteresis curve.

Reluctivity

Reluctivity is the reluctance per cubic centimeter of a material. Reluctivity is sometimes used to compare the reluctance of various materials to air. For example, air has a reluctivity that is 1800 times that of iron. Reluctivity is the reciprocal of permeability, which is the ability to conduct magnetic lines of force. The larger the reluctivity, the greater the resistance to the magnetic lines of force.

Terms

Reluctivity is the reluctance per cubic centimeter of a material.

Electromagnets

In addition to applications in solenoids, relays, and clutches, electromagnets are used to lift heavy masses of magnetic materials. While permanent magnets have a magnetically retentive (hard) coil material, electromagnets have a nonmagnetically retentive (soft) coil material that exhibits magnetic properties only when current flows through the coil. By controlling current flow with a DC power source and a switch, electromagnets can be turned ON and OFF and used to move heavy magnetic materials, such as steel, from one location to another.

Terms

Permeance is the property of a core material that allows the passage of magnetic lines of force.

Electromagnetic induction is the production of a potential difference across a conductor when it is exposed to a varying magnetic field.

Inductance (L) is the property of a device or circuit that causes it to oppose a change in current due to energy stored in a magnetic field.

Permeance

Permeance is the property of a core material that allows the passage of magnetic lines of force. Permeance is compared with conductance in an electrical circuit and is the reciprocal of reluctance. Permeance is calculated by applying the following formula:

$$P_m = \frac{1}{R_m}$$

where

P_m = material permanence (in μ)

R_m = material reluctance (in rels)

Example: What is the permeance of a material with a reluctance of 44.44 x 10^{-6} rels?

$$P_m = \frac{1}{R_m}$$

$$P_m = \frac{1}{44.44 \times 10^{-6} \text{ rels}}$$

$$P_m = \frac{1}{0.00004444 \text{ rels}}$$

$$P_m = \mathbf{22502.25}\ \mu$$

ELECTROMAGNETIC INDUCTION

Electromagnetic induction is the production of a potential difference across a conductor when it is exposed to a varying magnetic field. *Inductance (L)* is the property of a device or circuit that causes it to oppose a change in current due to energy stored in a magnetic field. The current flowing in a coil produces a field that expands out of and surrounds the conductor. Energy is stored in that field. When the source voltage goes from peak to zero, the energy in the field is returned to the coil and converted back to electrical energy.

The three requirements for induction are a conductor, a magnetic field, and relative motion between the conductor and the field. The AC power flowing through the conductor generates an expanding and collapsing magnetic field.

The direction of the magnetic field around a conductor can be determined using the left-hand rule for conductors. When an electric line of force crosses a magnetic line of force, they are perpendicular to each other. When the conductor moves up through the stationary magnetic field, electron flow is toward the bottom side of the conductor. If the conductor moves down through the magnetic field, electron flow is toward the top side of the conductor. This process in each condition causes the electrons in the conductor to move to one end of the conductor, leaving a deficit of electrons at the opposite end of the conductor. The voltage is induced across the length of the conductor with its polarity set by the direction of motion through the magnetic field.

The polarity of current flow in the conductor is determined by application of the left-hand rule for generators. **See Figure 10-13.** In the left-hand rule for generators, the index finger should be pointed in the direction of the magnetic field. The thumb points in the direction of motion of the conductor through the field, and the middle finger, which should be extended out at 90° from the index finger, will indicate the direction of current flow in the conductor.

LEFT-HAND RULE FOR GENERATORS

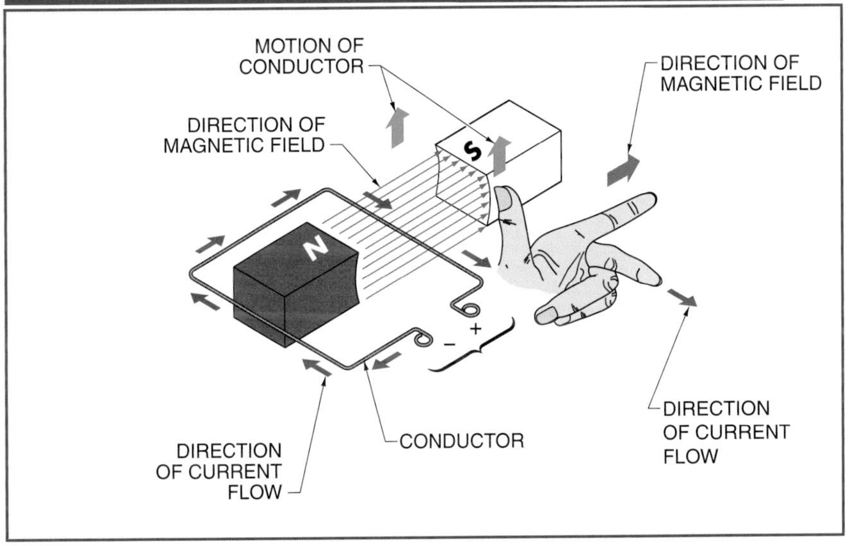

Figure 10-13. The polarity of current flow in the conductor is determined by application of the left-hand rule for generators.

The amount of electromagnetic induction in a conductor is determined by the number of coil turns (loops), strength of magnetic field, relative motion, and angle of cutting motion and is given by the following formula:

$$V = \frac{\Phi \times S}{10^8}$$

where

V = induced voltage (in V)

Φ = number of magnetic lines of force (in Mx)

S = speed of relative motion (cycles per second)

Example: What is the induced voltage when a conductor cuts 100,000,000 magnetic lines of force two times per second?

$$V = \frac{\Phi \times S}{10^8}$$

$$V = \frac{100{,}000{,}000 \text{ Mx} \times 2}{10^8}$$

$$V = \mathbf{2\ V}$$

Number of Coil Turns

When the number of coil turns is increased, the level of induced voltage is also increased. For example, if 0.5 V were induced into one coil turn, then 2 V would be induced into four coil turns. The level of induced voltage is directly proportional to the number of coil turns. **See Figure 10-14.** The number of coil turns is the multiplier when all other factors are held constant.

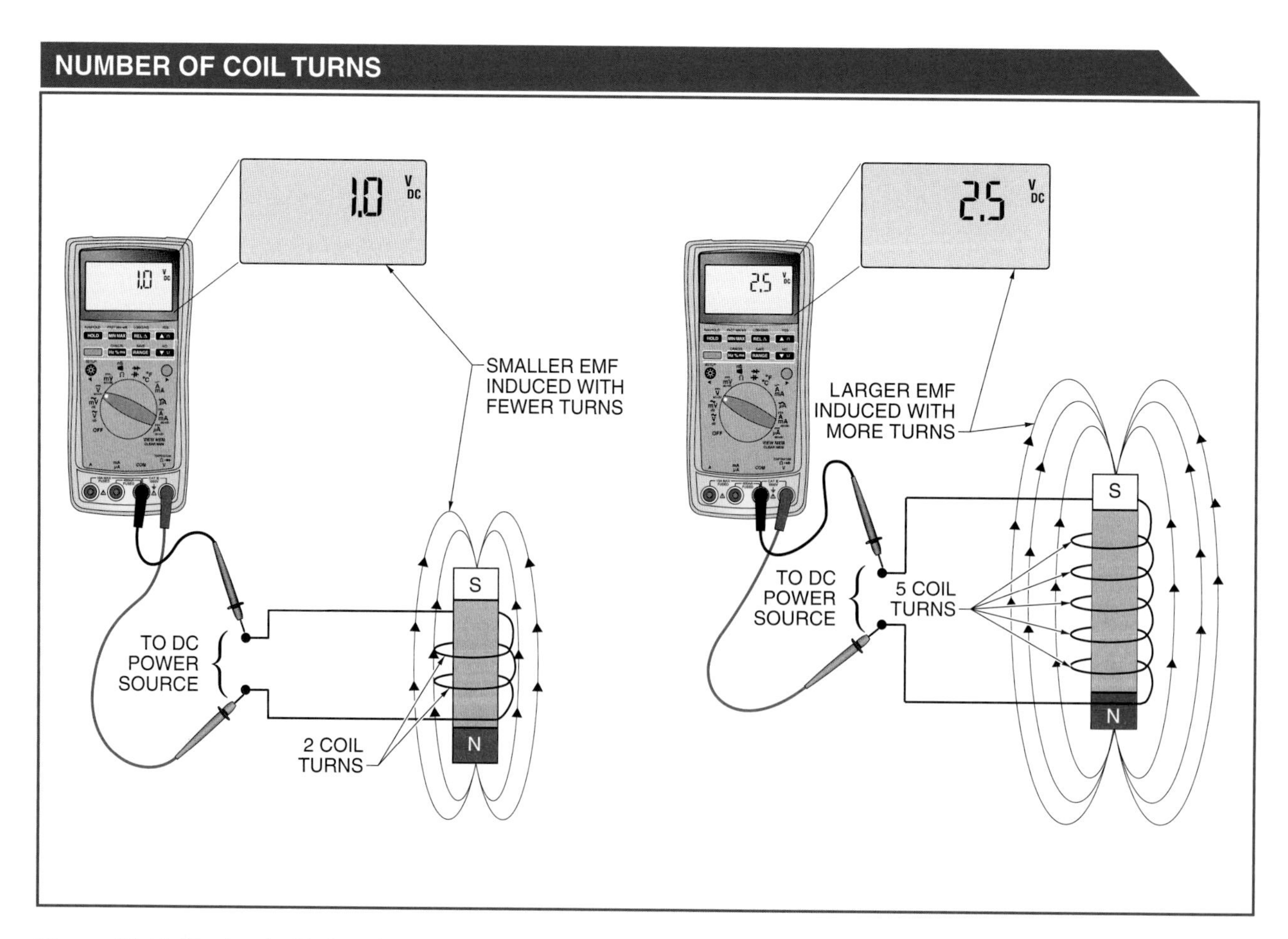

Figure 10-14. The level of induced voltage is directly proportional to the number of coil turns.

Strength of Magnetic Field

When permanent magnets are used as the source for magnetic lines of force to induce a voltage, the number of magnetic lines of force is fixed. Electromagnets are typically used to adjust the number of magnetic lines of force. **See Figure 10-15.** Increasing resistance in a conductor results in lower current flow and fewer magnetic lines of force. Decreasing resistance in a conductor results in higher current flow and more magnetic lines of force. With all other factors held constant, increasing the magnetic lines of force results in a higher voltage being induced into the conductor. There must be relative motion between the magnetic lines of force and the conductor in order for a voltage to be induced. If a conductor is resting in the flux field, or if the conductor and the flux field are both moving together in the same direction with the same velocity, no voltage is induced.

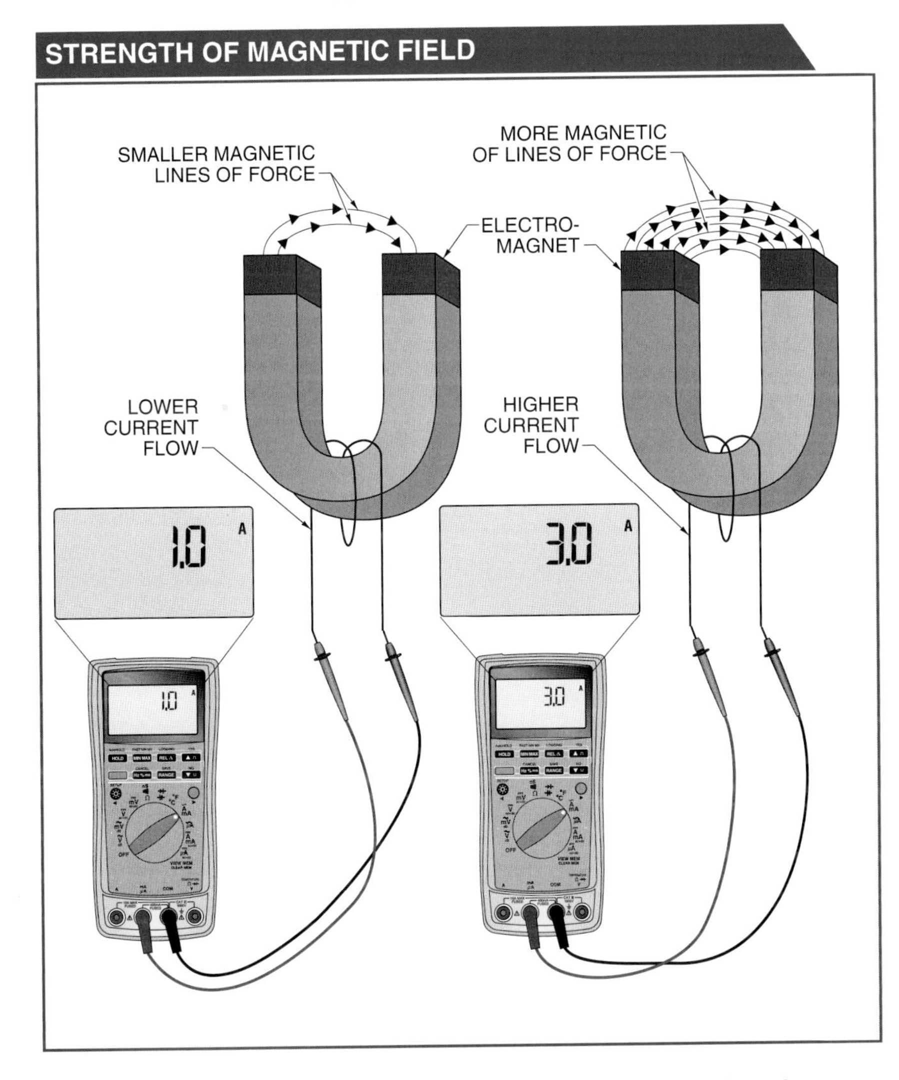

Figure 10-15. Electromagnets are typically used to adjust the number of magnetic lines of force.

Terms

Relative motion is the speed at which a conductor cuts a magnetic field.

Relative Motion. *Relative motion* is the speed at which a conductor cuts a magnetic field. When relative motion occurs between a field created by magnetic lines of force and a conductor, voltage is induced into the conductor. Increasing the speed at which the conductor cuts the magnetic field increases the induced voltage. The amount of induced voltage is directly proportional to the velocity of the cutting motion.

Angle of Cutting Motion. The angle at which the conductor cuts the magnetic field affects the amount of induced voltage. **See Figure 10-16.** When the conductor is moved parallel to the magnetic field, no magnetic lines of force are cut, and no voltage is induced. Maximum induction occurs when the conductor cuts the magnetic lines of force at 90°. At any angle between 0° and 90°, the induced voltage equals the sine of the angle times the maximum induced voltage.

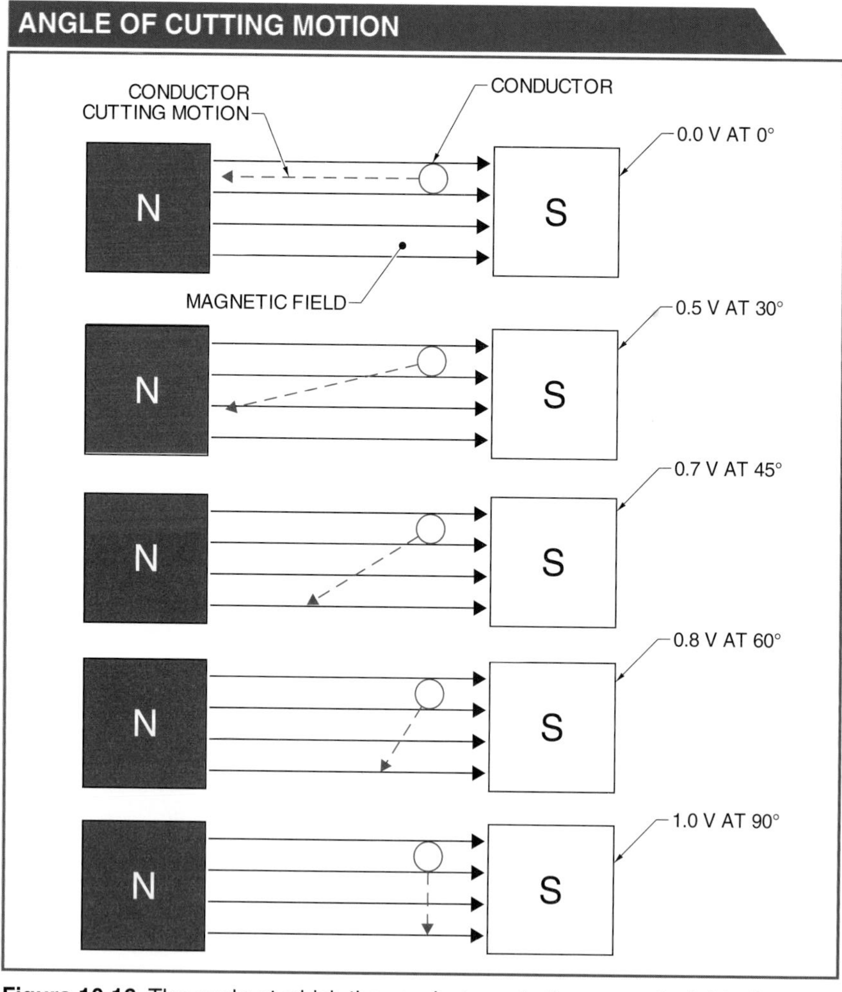

Figure 10-16. The angle at which the conductor cuts the magnetic field affects the amount of induced voltage.

LENZ'S LAW

Lenz's law is the basic principle used to determine the direction of an induced voltage or current. Lenz's law states that the voltage induced due to electromagnetic induction must be in such a direction that any current flow resulting from the induced voltage creates a magnetic field that opposes the force that caused the electrical energy. When a load is connected across a conductor, current results and power is consumed.

According to the law of conservation of energy, energy cannot be created or destroyed. Consequently, the electrical power produced must come from another form of energy. In this case, mechanical energy is converted to electrical energy. For a conductor to move through the magnetic field, work must be performed. The speed at which the conductor moves through the field determines the input power (work/time). When more electrical power is needed, the mechanical input power must increase to produce it. For example, Lenz's law can be demonstrated through the principle of operating a hand-cranked generator. Turning the crank without a load across the output of the generator is performed easily. If a load is placed across the output of the generator, turning the crank becomes more difficult. If the output of the generator is short-circuited, turning the crank smoothly is impossible.

Tech Tip

Experiments by the Russian physicist H. F. Emil Lenz led to the principle of Lenz's law, which is used to determine the direction of induced voltage.

Terms

Lenz's law is the basic principle used to determine the direction of an induced voltage or current.

A **generator (alternator)** is a machine that converts mechanical energy into electrical energy by means of electromagnetic induction.

A **direct current (DC) generator** is a generator that operates on the principle that when a conductor coil is rotated in a magnetic field, a voltage is induced in the coil.

DIRECT CURRENT GENERATORS

A *generator (alternator)* is a machine that converts mechanical energy into electrical energy by means of electromagnetic induction. A *direct current (DC) generator* is a generator that operates on the principle that when a conductor coil is rotated in a magnetic field, a voltage is induced in the coil. The amount of voltage induced in the coil is determined by the rate at which the coil is rotated in the magnetic field. When a conductor coil is rotated in a magnetic field at a constant rate, the voltage induced in the coil depends on the number of magnetic lines of force in the magnetic field at each given instant of time. DC generators consist of field windings, an armature, a commutator, brushes, and a frame. **See Figure 10-17.**

NREL

DC generators are typically installed in wind turbine systems.

DC GENERATORS

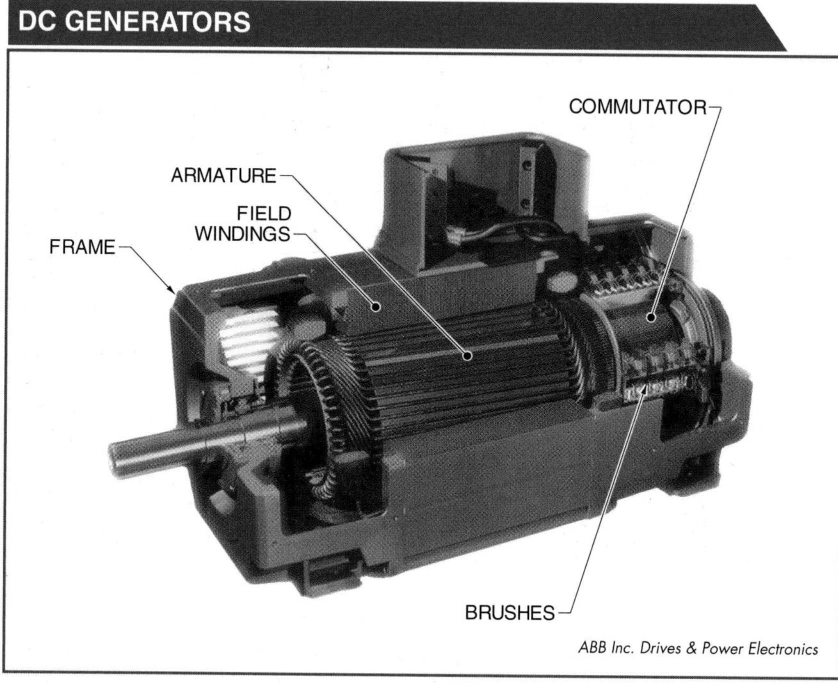

Figure 10-17. DC generators consist of field windings, an armature, a commutator, brushes, and a frame.

Terms

Field windings are magnets used to produce the magnetic field in a generator.

An **armature** is the current-carrying part of a DC generator.

Field Windings

Field windings are magnets used to produce the magnetic field in a generator. The magnetic field in a generator is produced by permanent magnets or electromagnets. Permanent magnets are used in very small machines referred to as magnetos. The disadvantages of a permanent magnet are that its magnetic lines of force decrease as it ages and its strength cannot be varied for control purposes. Most generators use electromagnets, which must be supplied with current. If the current for the field windings is supplied by an outside source (a battery or another generator), the generator is separately excited. If the generator supplies its own current for the field windings, the generator is referred to as self-excited. DC generators are typically self-excited.

Armature

An *armature* is the current-carrying part of a DC generator. A DC generator always has a rotating armature and a stationary field (field windings). The rotating armature may consist of many coils. Although increasing the number of coils reduces the ripples (pulsations) in the output voltage, it is impossible to remove the ripples completely. The armature core is made of sheet-steel laminations that are electrically isolated from each other by coatings of varnish. **See Figure 10-18.**

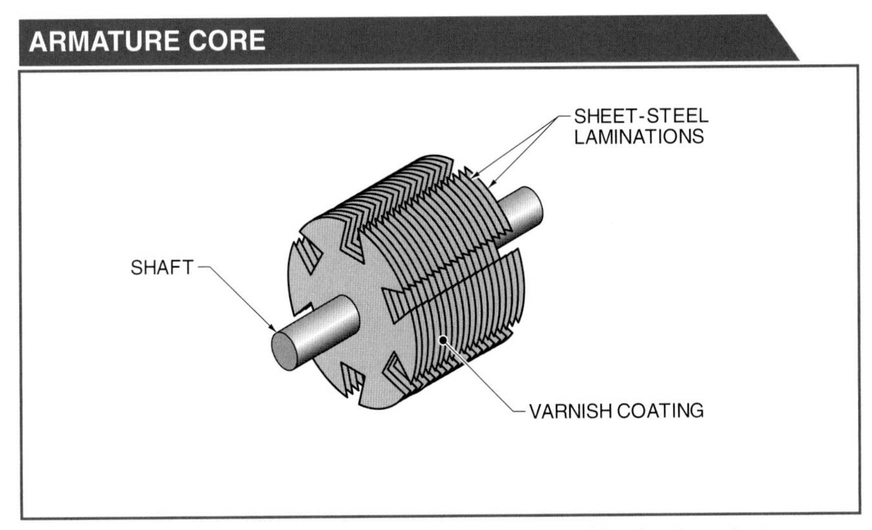

Figure 10-18. The armature core is made of sheet-steel laminations that are electrically isolated from each other by coatings of varnish.

Eddy current is unwanted current induced in the metal structure of a device due to the rate of change in the induced magnetic field. Sheet-steel laminations are used to prevent eddy current. The sheet-steel laminations increase the overall resistance to eddy current flow and limit heat buildup. Each lamination reduces eddy current loss to one-half of the power loss if no laminations are used. There is only ¼ of the loss with two laminations, and 1⁄16 the loss with four laminations.

The armature of high-power generators is often the stationary member (stator). The field then becomes the rotor of the generator. The reason for this design is that removing a very large current from the armature using brushes is not practical due to arcing. With the armature stationary, fixed wiring is used to remove high current. Brushes are used to apply the current to the field windings since the current needed for the magnetic field is very small as compared to the current supplied by the generator.

Terms

Eddy current is unwanted current induced in the metal structure of a device due to the rate of change in the induced magnetic field.

A **commutator** is the part of the armature that connects each armature coil to the brushes by using copper bars (commutator segments) that are insulated from each other with pieces of mica.

Commutator

A *commutator* is the part of the armature that connects each armature coil to the brushes by using copper bars (commutator segments) that are insulated from each other with pieces of mica. **See Figure 10-19.** Each end of a conductor is connected to a commutator segment. A voltage is induced in the conductor whenever the conductor cuts the magnetic lines of force of a magnetic field. The commutator segments reverse the connections to the brushes every half cycle. This maintains a constant polarity of output voltage produced by the generator.

COMMUTATORS

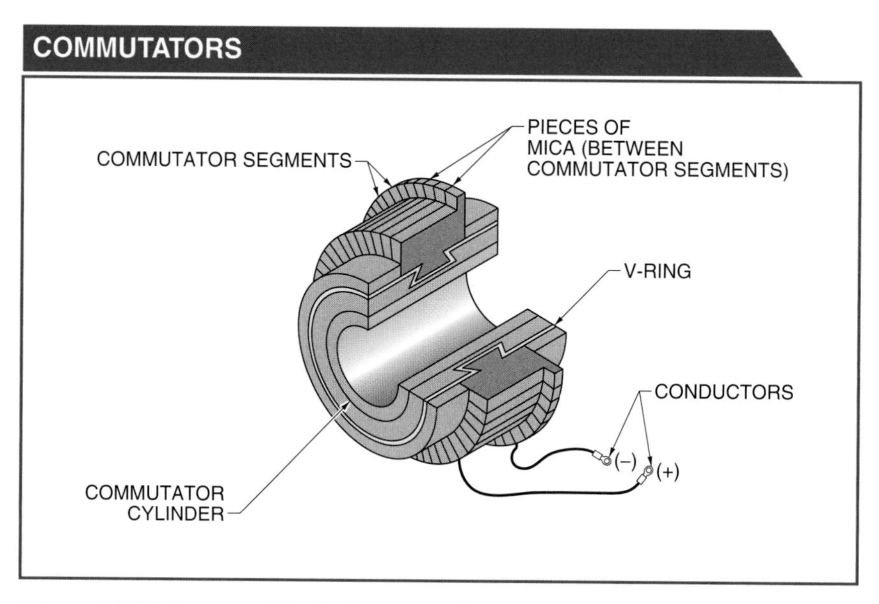

Figure 10-19. A commutator is the part of the armature that connects each armature coil to the brushes by using copper bars (commutator segments) that are insulated from each other with pieces of mica.

DC generators use wedge-shaped commutator segments. Voltage and number of poles in the generator determine the number of commutator segments. Commutator segments are held together by V-rings. Voltage between commutator segments is limited to 15 V to prevent arcing. Commutator segments are composed of hard-drawn copper assembled on a cylinder. Each segment is insulated from the cylinder and other segments by pieces of mica. The mica between the segments is undercut by 1⁄64″ to 1⁄8″ to prevent contact with the brushes.

Brushes

A brush is the sliding contact that rides against the commutator and is used to connect the armature to the external output circuit. Brushes are used to remove the electrical energy from the generator's commutator. Brushes are made of carbon or graphite materials and are usually easy to replace. A brush is held in a brush holder and is free to move up and down in the holder. The freedom allows the brush to follow irregularities in the surface of the commutator. Brush holders are mounted on the generator's frame, but they are electrically insulated from it. A spring placed behind a brush forces the brush to make contact with the commutator. The spring pressure is usually adjustable. A pigtail connects a brush to the power supply. A *pigtail* is an extended, flexible connection or a braided copper conductor.

Terms

A **brush** is the sliding contact that rides against the commutator and is used to connect the armature to the external output circuit.

A **pigtail** is an extended, flexible connection or a braided copper conductor.

Brush material must have good conductivity and be soft enough not to damage the copper segments of the commutator. Brush material is composed of a mixture of carbon and graphite for high-voltage machines and of graphite and metallic powder for low-voltage machines. The graphite in the brushes provides lubrication. No other lubrication should be used since it may cause electrical problems and equipment damage.

Electromagnets are used in wind turbine braking systems.

Brushes can usually be worn down to a fraction of their length before they need to be replaced. Because brushes wear down, constant pressure is applied with an adjustable tension spring so that the brushes can maintain contact with the commutator. **See Figure 10-20.** The pressure applied to the brushes is typically between 1.5 psi and 2 psi. Improper tension on the brush or misalignment with the axis of the commutator results in excessive arcing. Excessive arcing at the brushes greatly limits their life. Arcing at the brushes may be caused by improper spring tension; installation where contact with the full area of the commutator is not achieved; vibration or chatter during operation; dirty, worn, out-of-round, or loose commutator segments; defective solder connection between the coil and commutator segment; short-circuited armature windings; or an overloaded generator.

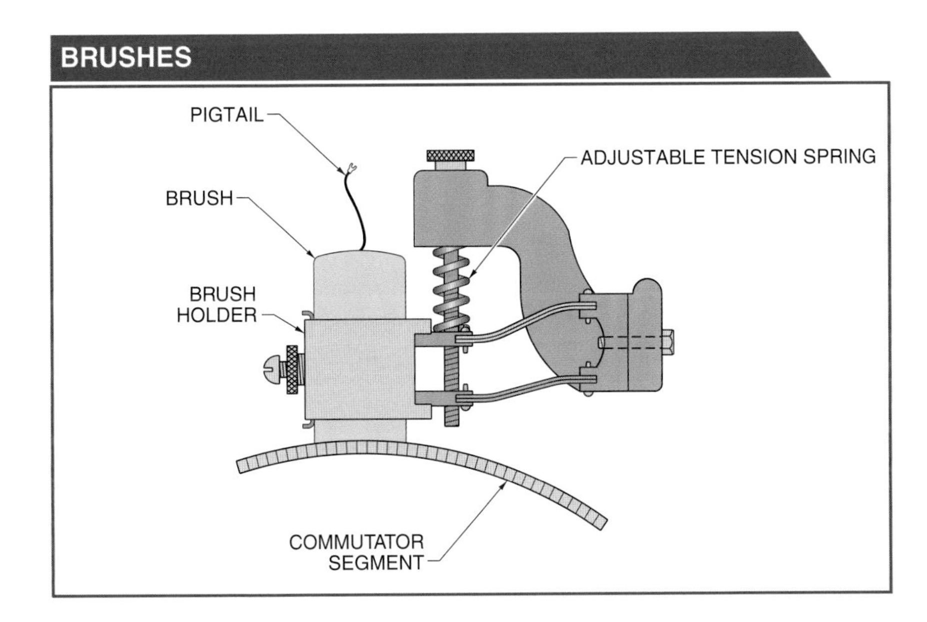

Figure 10-20. Because brushes wear down with use, constant pressure is applied with an adjustable tension spring so that the brushes can maintain contact with the commutator.

Terms

A **yoke** is the center portion of the frame in which the field coils are mounted.

The **end frames (end bells)** are the two end parts of the frame.

Frame

The frame of a generator provides mechanical support for the generator parts. Most frames consist of a yoke and two end frames. A *yoke* is the center portion of the frame in which the field coils are mounted. The *end frames (end bells)* are the two end parts of the frame. The end bells are attached to the yoke and support the armature shaft with bearings mounted in their frames. Some generators have provisions to lubricate the bearings through an opening in the end frame. The frame is typically composed of annealed steel that has good magnetic qualities that limit energy losses. A typical large generator may weigh several hundred pounds. A cooling fan is often connected inside or outside the bell housing at the drive shaft to pull air through the generator for cooling purposes.

GENERATION OF DC VOLTAGE

A DC generator is used to generate voltage and change alternating current to direct current. In a DC generator, as the red (top) end of the armature loop rotates in a clockwise direction 180° and cuts the magnetic lines of force, it generates the first 180° of the sine wave. **See Figure 10-21.** Brush 2 rides on the segment of the commutator attached to the top end of the loop, which cuts down through the magnetic lines of force.

As the loop passes through 180°, the red end of the loop begins cutting up through the magnetic lines of force and the other side of the loop moves down through the magnetic lines of force. This means that current has alternated (changed direction) in the loop.

Note: Brush 2 is still attached to the segment of the commutator that cuts down though the field. Brush 1 is still attached to the segment of the commutator that cuts up through the field. The effect of the commutator is that it has rectified (changed to DC) the alternating current. Current always flows in the same direction through the load.

Two cycles are generated for each 360° of rotation of the commutator. Therefore, the frequency of the current is twice that of an AC generator given the same number of poles and speed of rotation. The split in the commutator is designed so that it is aligned with the brushes at 0° and 180°. This is the amount of time that the loop is moving parallel to the flux field and no voltage is being induced. This reduces the chance of sparking when the brushes move from one segment of the commutator to the next segment. As the brushes move from one segment to the other, they short the coil.

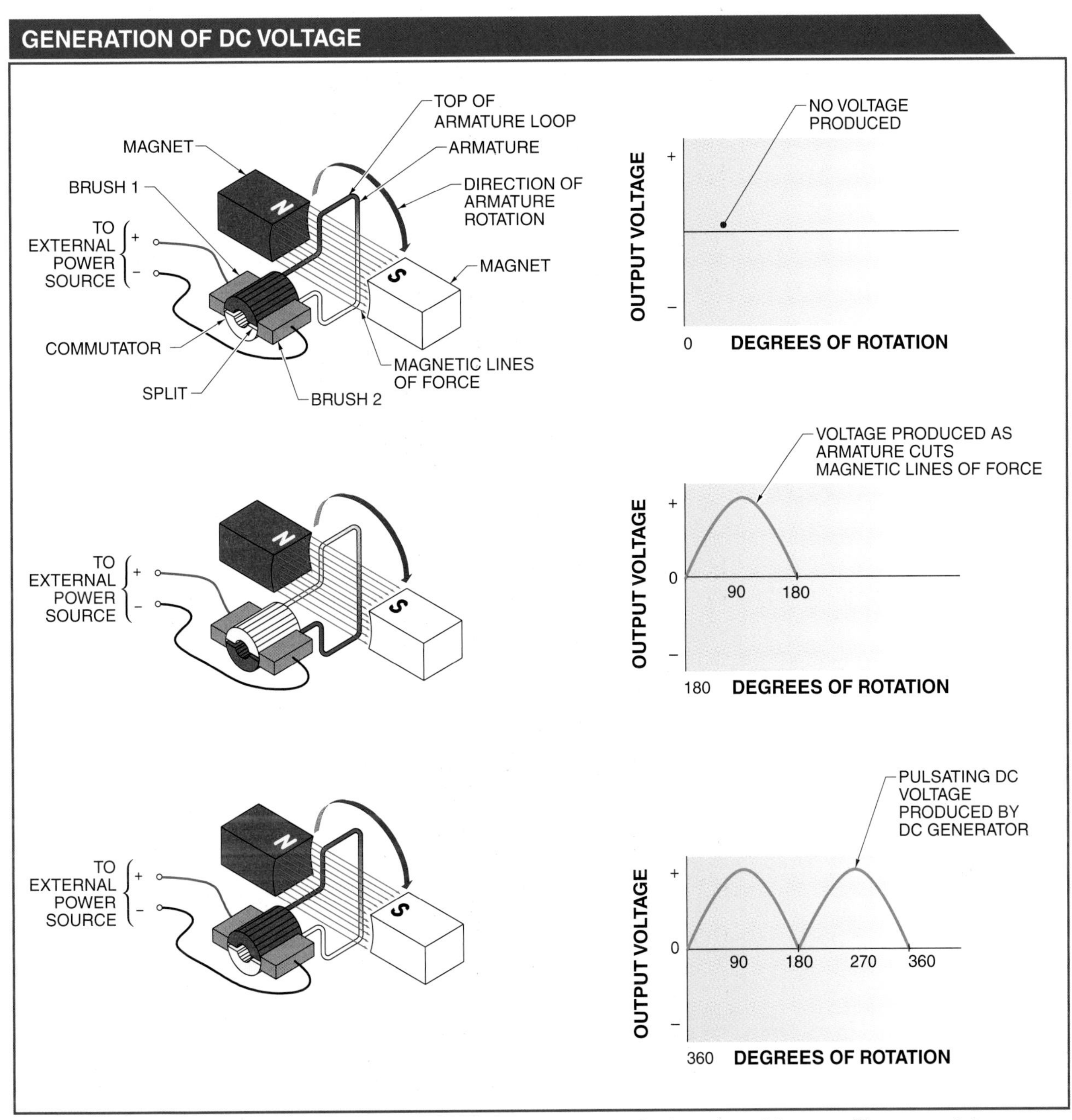

Figure 10-21. Two cycles of DC are generated for each 360° of mechanical rotation of the rotor. The commutator ensures that the current flow through the load is always in the same direction.

Effect of Additional Loops

With all conditions held constant, and an additional loop added to the coil, the emf is doubled. If more loops are added, the output voltage equals the number of loops multiplied by the emf induced into a single loop.

Terms

A **ripple** is a change in DC output voltage level.

A DC generator produces a ripple in the direct voltage. A *ripple* is a change in DC output voltage level. For most applications, the percent of ripple for a single loop is unacceptable. A second loop placed at 90° from the first loop can be added to the armature to reduce the severity of the ripple voltage. **See Figure 10-22.**

The commutator is now divided into four segments with only one set of brushes. With this arrangement, the voltage cannot fall any lower than that shown at point A on the graph. The value of this low point is 0.707 times V_{MAX}, as opposed to 0 V for the single loop. Ripple voltage is reduced, and the average voltage is increased using two loops. Further improvement on reducing ripple voltage can be made by adding more loops and segments.

EFFECT OF ADDITIONAL LOOPS

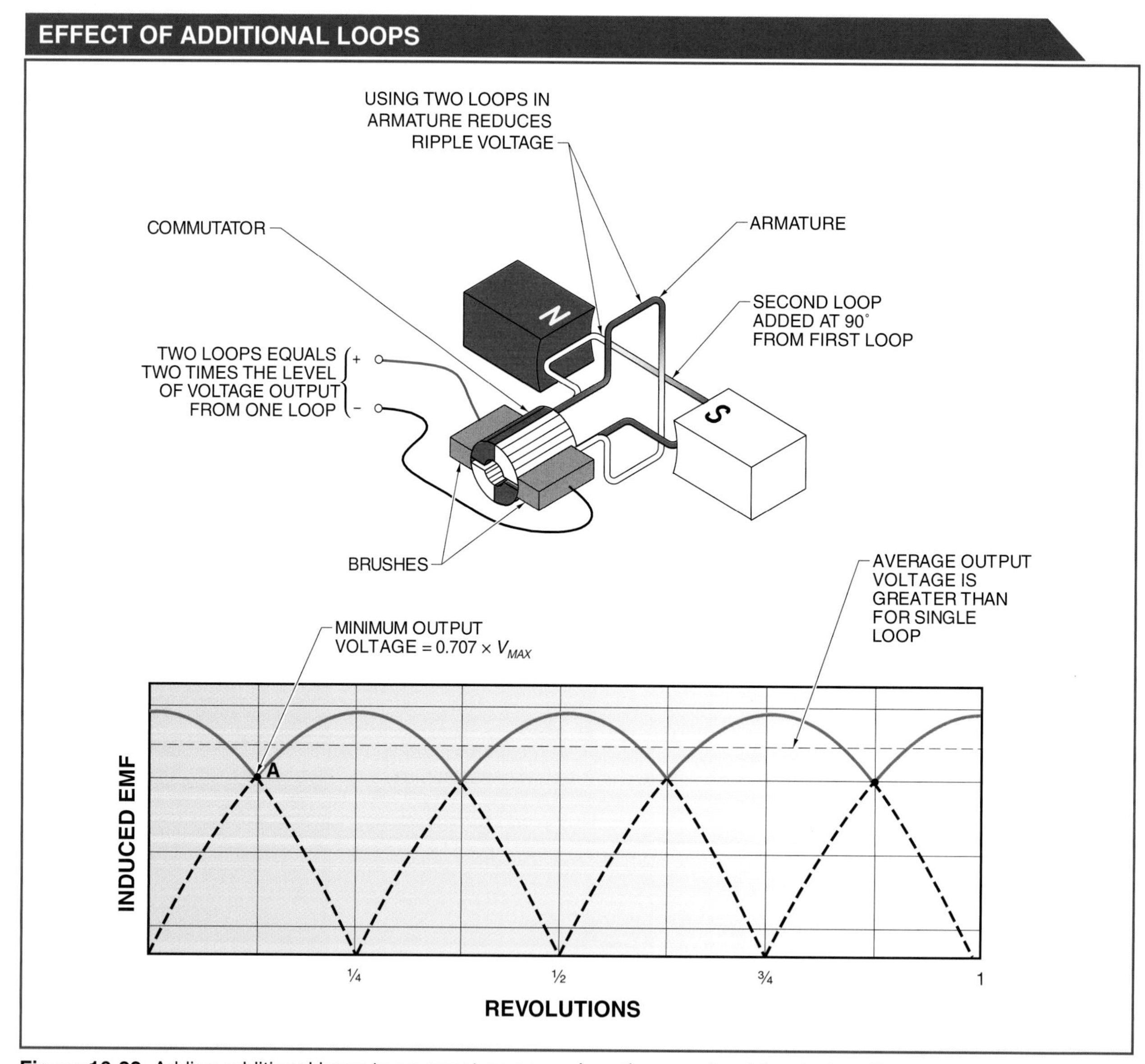

Figure 10-22. Adding additional loops to an armature can reduce the severity of the ripple voltage.

SUMMARY

Electromagnetism is the magnetism produced when electricity passes through a conductor. Electromagnetism is applied to a variety of electrical applications. The three principles of electromagnetism are as follows:

- moving electric charges produce magnetic fields
- changing magnetic fields in the presence of electric charges cause current to flow
- magnetic fields exert forces on moving electric charges

Rowland's law is used to determine magnetic lines of force, magnetomotive force, and reluctance. Magnetic circuit properties include intensity of magnetizing force, permeability, hysteresis, reluctivity, and permeance. Electromagnetic induction involves the number of conductor turns in a coil, strength of the magnetic field, speed of relative motion, and angle of cutting motion. Lenz's law is used to determine the speed of the induced voltage.

A DC generator is a machine that converts mechanical energy into electrical energy by means of electromagnetic induction. Direct current generator components are windings, an armature, a commutator, brushes, and a frame.

APPLICATION: RADIO FREQUENCY IDENTIFICATION ANTENNAS...

Background: Passive radio frequency identification (RFID) tags are commonly used to identify and track objects. These devices are similar in operation and function to barcodes, which are being replaced. Typically, an electronic RFID reader is used to get the unique information from an RFID tag. The reader uses an optical system to read the tag. The reader transmits electromagnetic energy that is used to power the RFID tag so that the identifying information can be transmitted back to the reader.

An RFID reader generates a magnetic field by running current through a conductor, which is the antenna. Ampère's law describes the relationship between the magnetic field around an electric current and the electric current. According to Ampère's law, the magnetic field strength is a function of the amount of current and the distance from the transmitting source. The formula for a conductor that is infinite in length is the following:

$$B = \frac{\mu_0 I}{2\pi r}$$

where

B = magnetic field (in T)
$\mu_0 = 4\pi \times 10^{-7}$ (in Tm/A) (constant for permeability of free space)
I = current (in A)
π = 3.14 (constant)
r = radial distance (in m)

...APPLICATION: RADIO FREQUENCY IDENTIFICATION ANTENNAS...

For a conductor of limited length, the physical relationship between the conductor and magnetic field is best illustrated through the use of a graph. The formula for the magnetic field strength for an RFID antenna includes the angles between the reader antenna and location of the RFID tag. The magnetic field strength is found by applying the following formula:

$$B = \frac{\mu_0 I}{4\pi r}(\cos\infty_2 - \cos\infty_1)$$

For the special case of an infinitely long conductor, ∞_1 is 180° and ∞_2 is 0°. Using these angles, the original formula (cos 0° – cos 180° = 2) is derived. **See Figure 1.**

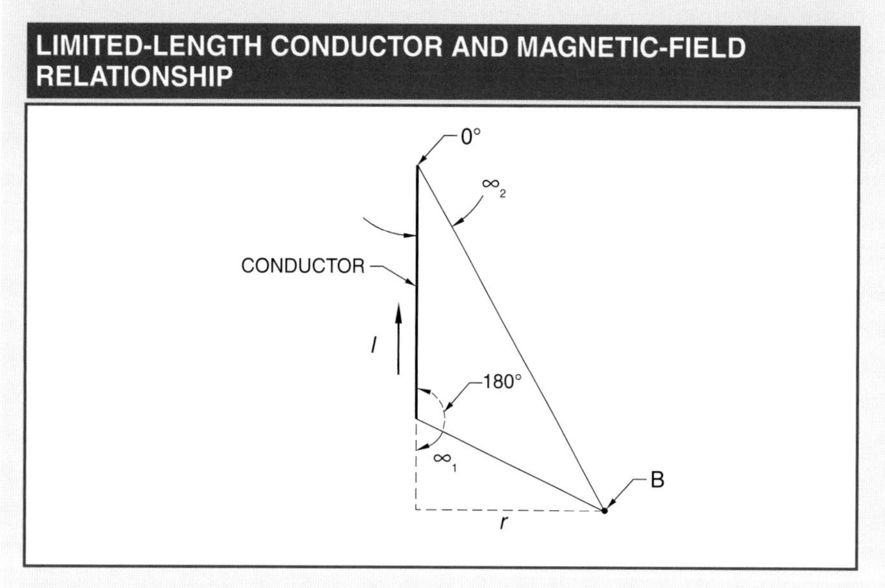

Figure 1. For a conductor of limited length, the physical relationship between the conductor and magnetic field is best illustrated through the use of a graph.

Key Points: Ampère's law; tesla (T); gauss (G)

Problem: An RFID tag (r) is located 0.87 m away from a reader antenna. The reader antenna is a straight conductor that is 1 m in length. The RFID tag is at the center of the conductor. Determine the magnetic field strength at the RFID tag.

Solution: With the RFID tag at the center of the conductor, an equilateral triangle is formed from the invisible planes created by the reader antenna and RFID tag. **See Figure 2.** With this type of triangle, all sides are of equal length and all interior angles are 60°. Angle ∞_1 is the difference between the angle of a straight line (180°) and the angle of the triangle (60°), which equals 120°. The magnetic field strength is as follows:

$$B = \frac{\mu_0 I}{4\pi r}(\cos\infty_2 - \cos\infty_1)$$

$$B = \frac{(4\pi \times 10^{-7}\,\text{Tm/A}) \times 1\,\text{A}}{4\pi \times 0.87\text{m}}(\cos 60° - \cos 120°)$$

...APPLICATION: RADIO FREQUENCY IDENTIFICATION ANTENNAS

$$B = \frac{1\times10^{-7}\,\text{T}}{0.87}\left[0.5-(-0.5)\right]$$

$$B = 1.15\times10^{-7}\,\text{T or } 0.00115\,\text{G}$$

RFID TAG AND READER ANTENNA

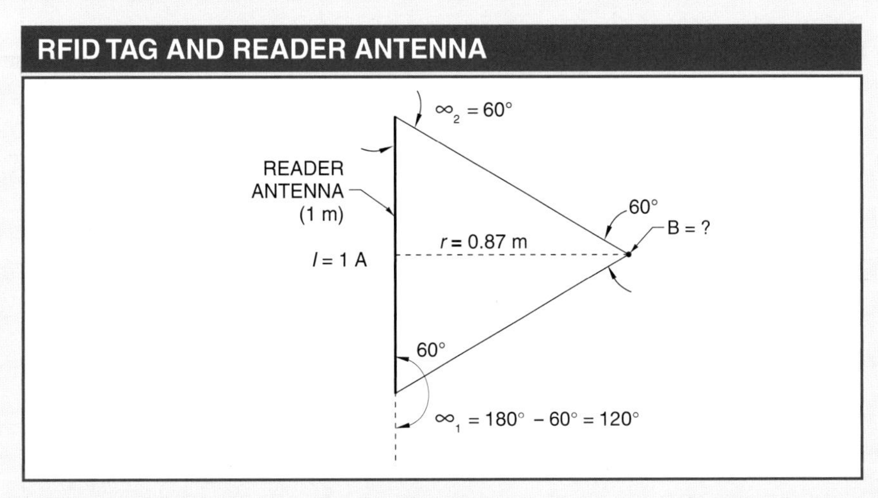

Figure 2. With the RFID tag at the center of the conductor, an equilateral triangle is formed from the invisible planes created by the reader antenna and the RFID tag.

RFID tags, rather than barcodes, are used to identify objects. RFID tags are passive and use electromagnetic induction to power and transmit unique tag information. The magnetic field strength can be small relative to the Earth's magnetic field, which is about 0.5 G.

Chapter Review

1. What design in metal components is used to prevent eddy currents?
2. What is another name for a maxwell?
3. In which materials is retentivity the most and least noticeable?
4. What is the total mmf for a circuit that has 5 A of current flow and 225 conductor turns?
5. Explain the left-hand rule for conductors.
6. What are the three requirements for induction?
7. What variable is reluctance in a magnetic circuit comparable to in an electrical circuit?
8. What is the purpose of only using graphite as lubrication between DC generator brushes?
9. What is inductance?
10. How do magnetic lines of force react when traveling in the same direction and traveling in the opposite direction?
11. What is the core material in an electromagnet?
12. What is the standard voltage limitation between commutator segments?
13. Explain permeability.
14. How are permanent magnets and electromagnets used with magnetic lines of force to induce voltage?
15. What is magnetomotive force?
16. How does reluctivity compare to the resistance of magnetic lines of force?
17. Explain the difference between paramagnetic, diamagnetic, and ferromagnetic substances.
18. List three factors that affect the strength of a coil's magnetic field.
19. What is magnetic flux?
20. What is electromagnetism?

11

DC Circuit Inductance

OBJECTIVES

- Explain inductance and describe factors that affect inductance.
- Describe inductor construction.
- List and describe the different inductor types.
- Describe resistive, ideal, and practical inductive circuits.
- Explain the inductive-resistive time constant and high-induced voltages.
- Explain self-inductance and mutual inductance.
- Define series- and parallel-connected inductors.

Inductance is one of the electrical properties that are present in all electrical circuits. The other properties that are present are resistance, which is the opposition to current flow, and capacitance, which is the ability to store a charge. Inductance is used in applications such as transformers, solenoids, automobile electronic systems, televisions, electric motors, and generators.

Terms

Inductance (L) is the property of a device or circuit that causes it to oppose a change in current due to energy stored in a magnetic field.

An **inductor** is a passive device that stores energy in the form of a magnetic field.

The **henry (H)** is the unit of inductance equal to the amount of voltage induced per the rate of current change.

A **solenoid** is an electric output device that converts electrical energy into linear mechanical force.

Tech Tip

The unit of measure for inductance, the henry, is named after the American scientist Joseph Henry, who discovered the property of self-induction.

INDUCTANCE

Inductance (L) is the property of a device or circuit that causes it to oppose a change in current due to energy stored in a magnetic field. An *inductor* is a passive device that stores energy in the form of a magnetic field. Inductors consist of a coil or conductor loop around a core. The *henry (H)* is the unit of inductance equal to the amount of voltage induced per the rate of current change. For example, if a change of 1 A in 1 sec causes an induced voltage of 1 V, then the inductor has 1 H of inductance. Induced voltage in a circuit is calculated by applying the following formula:

$$V = \frac{-L\Delta i}{\Delta t}$$

where

V = induced voltage (in V)

L = inductance (in H)

Δi = change in current (in A)

Δt = change in time (in sec)

Example: What is the induced voltage for an inductor of 120 µH with a current change of 100 mA in 10 µs (microseconds)?

$$V = \frac{-L\Delta i}{\Delta t}$$

$$V = \frac{-120 \times 10^{-6} \times 100 \times 10^{-3}}{10 \times 10^{-6}}$$

$$V = \frac{-0.00012 \times 0.100}{0.000010}$$

$$V = \frac{-0.000012}{0.000010}$$

$V = \mathbf{-1.2\ V}$

Inductors are manufactured with inductance values ranging from microhenrys to henrys. Higher-value inductors between 15 H and 30 H are normally limited to use in power supplies and audio amplifier applications. Inductance of 1 H or greater is unsuitable for most circuit applications involving frequencies higher than the audio range of 15 Hz to 20 kHz. Lower value inductors are used for higher-frequency applications such as in radio and television circuits. Inductive values are usually printed directly on the inductor or a label is attached to it. A maximum current rating is typically given for inductors of 15 H and greater.

Inductors are typically used in solenoids. A *solenoid* is an electric output device that converts electrical energy into linear mechanical force. For example, in DC applications, an electromagnet uses a solenoid to control lifting and moving heavy materials such as steel coils.

FACTORS AFFECTING INDUCTANCE

Factors that affect inductance include the number of conductor (coil) turns around the circumference of a core, the spacing between coil turns on a core, the core diameter (cross-sectional area), the shape of the core, the number of coil layers, the type of core material, and the core permeability. **See Figure 11-1.**

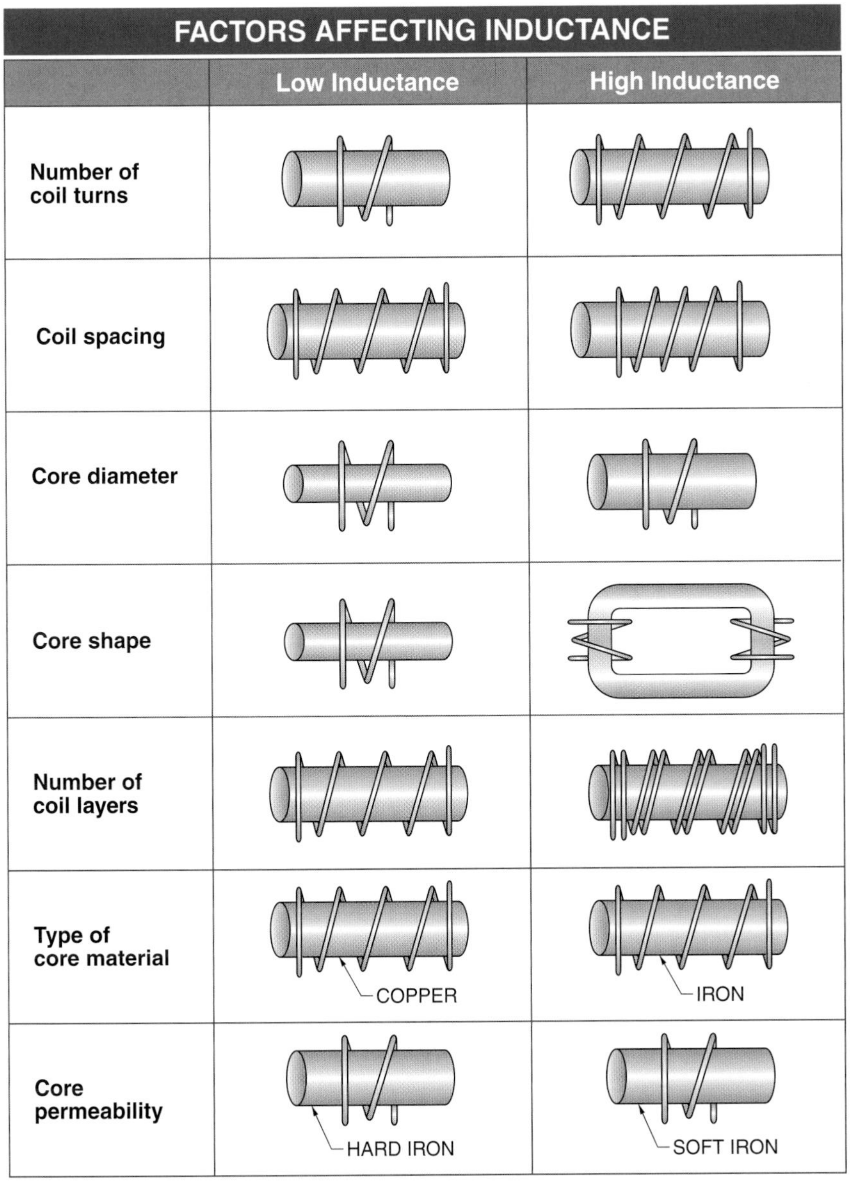

FACTORS AFFECTING INDUCTANCE		
	Low Inductance	**High Inductance**
Number of coil turns		
Coil spacing		
Core diameter		
Core shape		
Number of coil layers		
Type of core material	COPPER	IRON
Core permeability	HARD IRON	SOFT IRON

Figure 11-1. Factors that affect inductance include the number of conductor (coil) turns around the circumference of a core, spacing between coil turns, core diameter, core shape, number of coil layers, type of core material, and core permeability.

Tech Tip

Solenoids are used in DC applications such as automatic door locks in an automobile.

The higher the number of conductor turns around the circumference of a core, the higher the inductance. A smaller amount of space between conductor turns on a core results in higher flux density and higher inductance. Because a large-diameter core allows for a larger area of the coil to have closer spacing, a large-diameter core has more inductance than a small-diameter core. A core that is shaped to have the north and south poles as close as possible to one another also allows for best circulation of magnetic lines of force within the coil. For example, if coils are wound on a closed core, inductance is increased by about 2000 times over the inductance of a straight core. The greater the number of coil layers on a core, the smaller the amount of space between conductors, resulting in higher inductance. The lesser the number of coil layers on a core, the greater the amount of space between conductors, resulting in lower inductance.

Terms

Permeability is the ability of a material to carry magnetic lines of force.

The type of core material used affects inductance. A diamagnetic material, such as copper, has a weaker magnetic field than a ferrous material, such as iron, causing the magnetic core to have lower inductance. *Permeability* is the ability of a material to carry magnetic lines of force. Permeability may also be expressed as the ratio of the number of magnetic lines of force that pass through a given type of material as compared to the number of magnetic lines of force that pass through air of the same area.

DC electromagnets use solenoids to control the lifting of ferrous materials.

The magnetic field strength of a given inductor can be expressed as the ratio of the magnetic lines of force produced per the rate of change of the current flow through the coil. Inductance is a measure of the number of magnetic lines of force that cut or link the turns of a coil. If one hundred million (10^8) magnetic lines of force pass through one turn of a coil for a change of current flow of 1 A in one second, then 1 H is present. The inductance of a coil is calculated by applying the following formula:

$$L = \frac{N\Delta\Phi}{\Delta I \times 10^8}$$

where

L = inductance (in H)

N = number of coil turns

$\Delta\Phi$ = change in magnetic lines of force (in Mx)

ΔI = change in current (in A)

10^8 = constant to calculate L

Example: What is the inductance of a coil if the number of turns is increased from 1 to 4?

$$L = \frac{N\Delta\Phi}{\Delta I \times 10^8}$$

$$L = \frac{4 \times 10^8}{1 \times 10^8}$$

$$L = \frac{400{,}000{,}000}{100{,}000{,}000}$$

$L =$ **4 H**

Inductance increases in proportion to the squared value of the number of turns, the core permeability, and the cross-sectional area. A larger diameter, or cross-sectional area of the core, means that a greater length of the conductor is adjacent to the next turn. Inductance is inversely proportional to core length. The same number of turns on a longer-length core has a lower inductance than those on a shorter-length core. On the shorter core, the turns are closer together, so more magnetic lines of force cut adjacent turns.

For coils with iron cores, the permeability of the core is not constant. Permeability of the core depends on the size of the magnetizing current. At some point, core saturation occurs, and the flux density does not change with additional increases in the magnetizing force. Without a flux density change, no voltage is induced. The distance between turns affects the inductance of a coil. In a straight length of a conductor, very little distance exists between the magnetic lines of force between one part of the conductor and another. The inductance, therefore, is very small. Inductance is increased as the conductor is wound into coils. The method in which coils are wound affects the value of inductance. A single coil layer is loosely wound, limiting the magnetic lines of force linking the turns between the coils. Also, the lateral movement of the magnetic lines of force does not link flux effectively due to the single layer of turns. A coil design with multiple turns increases the flux linkage and, thus, the inductance of the coil.

Tech Tip

An inductor placed in the power line of an automobile radio allows the DC to pass and blocks any high-frequency AC (ignition noise and static) from entering the radio.

Science of Inductance

Most light electric rail systems have a DC motor that receives power through the rails, overhead power lines, or a third electrified rail. When the train accelerates, there is high current in the electrified rails, power lines, or third rail. At uniform speed, the current is small and steady. If the voltage is reduced for the purpose of stopping or slowing the train, the current remains for a short period of time, even without voltage present. This current is created by the collapsing magnetic fields in the motor coils, which are electrical inductors. As with mass in motion not being able to change instantly (Newton's third law of motion), the current in an electrical inductor also cannot change instantly.

INDUCTOR CONSTRUCTION

Inductors are constructed in various shapes and sizes with different coil styles. For applications in electrical and electronic equipment, the length of a coil is generally not greater than 10 times its diameter or width. Inductors can be identified by the various types of coil styles. Each inductor coil style has a related formula for determining its inductance, which is known as a Wheeler's formula. Wheeler's formulas are accurate to within a range of a few percent. To simplify the difference in coil styles, inductors are classified into four groups, with each group having its own formula for determining inductance. The four groups of coil styles are single-layer coil, multilayer coil, pancake coil, and toroid coil. **See Figure 11-2.**

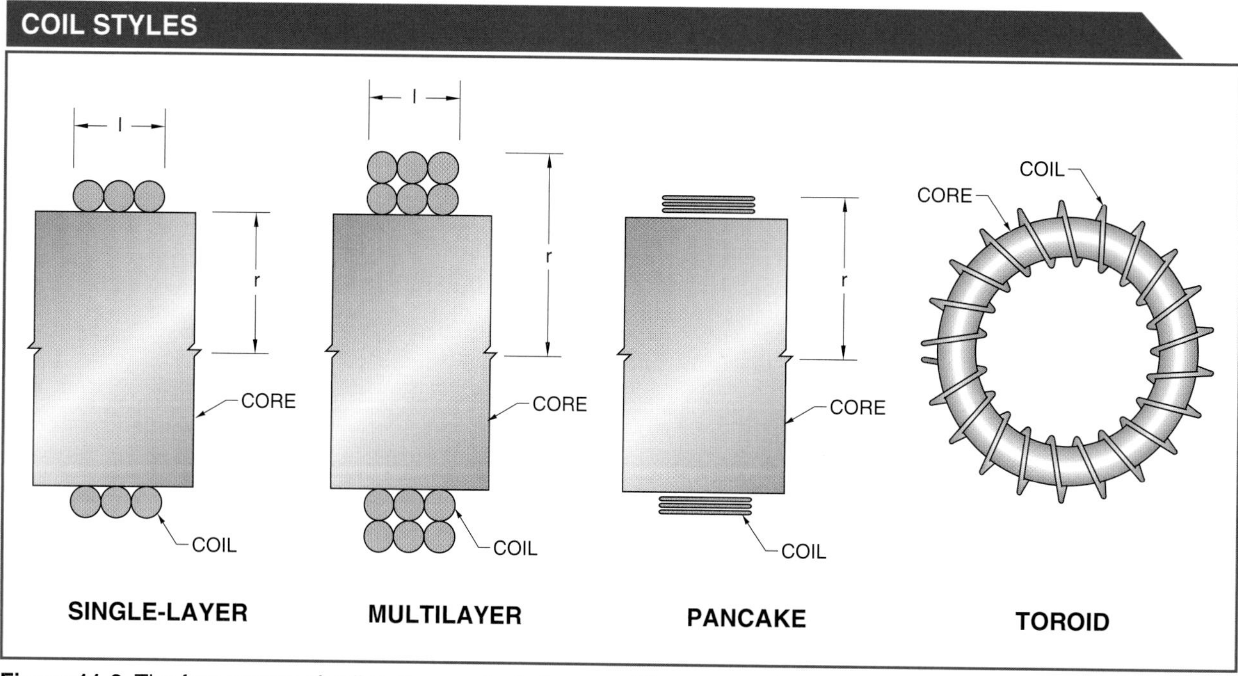

Figure 11-2. The four groups of coil styles are single-layer, multilayer, pancake, and toroid.

Terms

A **single-layer coil** is a coil that has only one layer of turns and is wound on an air-core form.

A general approximation of the inductance for any coil is calculated using the following formula:

$$L = \frac{1.26 \times N^2 \times \mu \times A}{l \times 10^8}$$

where

L = inductance (in H)

1.26 = constant

N = number of coil turns
μ = relative permeability
A = cross-sectional area (in cm^2)
l = core length (in cm)

Example: What is the inductance of a square coil that measures 50 cm in length with sides of 10 cm each, 100 turns, and a relative permeability of 3000?

$$L = \frac{1.26 \times N^2 \times \mu \times A}{l \times 10^8}$$

$$L = \frac{1.26 \times 100^2 \times 3000 \times 100}{50 \times 10^8}$$

$$L = \frac{3.78 \times 10^9}{50 \times 10^8}$$

L= **0.756 H**

Single-Layer Coils

A *single-layer coil* is a coil that has only one layer of turns and is wound on an air-core form. Single-layer coils are also referred to as solenoids. Inductance of a single-layer coil, where the length of the coil does not exceed its diameter, is calculated by applying the following formula:

$$L = \frac{(rN)^2}{9r + 10l}$$

where
L = inductance (in µH)
r = coil radius (in in.)
N = number of coil turns
l = coil length (in in.)

Example: What is the inductance of a single-layer coil that has a radius of 0.5″, a length of 4″, and 10 turns?

$$L = \frac{(rN)^2}{9r + 10l}$$

$$L = \frac{(0.5 \times 10)^2}{9 \times 0.5 + 10 \times 4}$$

$$L = \frac{25}{44.5}$$

L = **0.562 µH**

Terms

A **single-layer coil** is a coil that has only one layer of turns and is wound on an air-core form.

Tech Tip

In 1753, English scientist John Canton discovered the principle of electrostatic induction. Canton discovered that if a charged object was brought close to an uncharged object, the uncharged object would take on the charge of the charged object.

When the dimensions of the core are known, the number of turns required for a given amount of inductance is calculated by applying the following formula:

$$N = \frac{\sqrt{L(9r + 10l)}}{r}$$

where

N = number of coil turns

L = inductance (in μH)

r = coil radius (in in.)

l = coil length (in in.)

Example: What number of turns is required for a solenoid where the desired inductance is 10 μH, with a 0.25″ radius and a length of 3″?

$$N = \frac{\sqrt{L(9r + 10l)}}{r}$$

$$N = \frac{\sqrt{10\ \mu H\ (9 \times 0.25'' + 10 \times 3'')}}{0.25''}$$

$$N = \frac{\sqrt{10\ \mu H\ (2.25'' + 30'')}}{0.25''}$$

$$N = \frac{\sqrt{10\ \mu H\ (32.25'')}}{0.25''}$$

$$N = \frac{18}{0.25}$$

N = **72 turns**

Factors that must be taken into consideration when calculating inductance are the type and size of the coil and coil insulation. The diameter of the coil with insulation must be small enough that the required number of turns can be achieved in a single layer. The number of turns per inch depends on the type of coil insulation used. Inductor coil insulation types include polymeric coatings of polyvinyl acetal, polyester, or polyurethane. **See Appendix.** Once the calculation is made as to the number of turns required for a given inductance on a given core size, the total length of the coil needed for the required number of turns can be determined. A greater number of coil turns in an inductor produces a stronger magnetic field. The stronger the magnetic field generated by the coil, the greater the inductance value. The total coil length required for the number of turns is calculated by applying the following formula:

$$l = N \times (\pi \times d)$$

where

l = coil length (in in.)

N = number of coil turns

π = 3.14

d = core diameter (in in.)

Example: What length of coil wire is required for a solenoid of 200 turns with a radius of 0.25 ?

$l = N \times (\pi \times d)$

$l = 200 \times (3.14 \times 0.5'')$

$l = 200 \times 1.57$

$l = \mathbf{314''}$

Multilayer Coils

A *multilayer coil* is a coil that has more than one layer of turns and is wound on an air-core form. The overall depth of a multilayer coil will affect inductance. The greater the depth of a multilayer coil, the lower the inductance. Multilayer coils are used in applications where space limitation is required, such as mobile telephones. Inductance for multilayer coils is calculated by applying the following formula:

$$L = \frac{0.8(rN)^2}{6r + 9l + 10b}$$

where

L = inductance (in μH)

r = mean (average) radius (in in.)

N = number of coil turns

l = coil length (in in.)

b = coil depth (in in.)

Terms

A **multilayer coil** is a coil that has more than one layer of turns and is wound on an air-core form.

Example: What is the inductance of a multilayer coil of 50 turns on a 0.1″ radius air cylinder with a 1″ length and a coil depth of 0.5″?

$$L = \frac{0.8(rN)^2}{6r + 9l + 10b}$$

$$L = \frac{0.8 \times (0.1'' \times 50)^2}{6 \times 0.1 + 9 \times 1 + 10 \times 0.5}$$

$$L = \frac{0.8 \times 25}{0.6 + 9 + 5}$$

$$L = \frac{20}{14.6}$$

$L = \mathbf{1.37\ \mu H}$

Terms

A **pancake coil** is a short coil with only a few conductor turns along its length.

Pancake Coils

A *pancake coil* is a short coil with only a few conductor turns along its length. A pancake coil is also referred to as a flat spiral coil. A pancake coil is often wound with a thin, flat, rectangular metal ribbon conductor rather than with a circular conductor. Inductance for a pancake coil is calculated by applying the following formula:

$$L = \frac{(rN)^2}{8r + 11b}$$

L = inductance (in μH)

r = mean (average) radius (in in.)

N = number of coil turns

b = coil depth (in in.)

Example: What is the inductance of a pancake coil of 25 turns with a radius of 0.125″ and a coil depth of 0.2″?

$$L = \frac{(rN)^2}{8r + 11b}$$

$$L = \frac{(0.125 \times 25)^2}{8 \times 0.125 + 11 \times 0.2}$$

$$L = \frac{(3.125)^2}{1.000 + 2.200}$$

$$L = \frac{9.766}{3.200}$$

L = **3.05 μH**

Terms

A **toroid coil** is a doughnut-shaped coil with a closed circular or rectangular core.

Toroid Coils

A *toroid coil* is a doughnut-shaped coil with a closed circular or rectangular core. Toroid coils are used for such applications as deflection coils on cathode ray tubes used in sonar equipment, radar equipment, as transformers in power supplies, and in low-frequency radio applications. Unlike other types of coils, where the magnetic field passes through the air and the core material, the magnetic field of a toroid coil is confined to the core. These coils are wound as single and multilayer types depending on the application and have high inductance in relation to their size. **See Figure 11-3.**

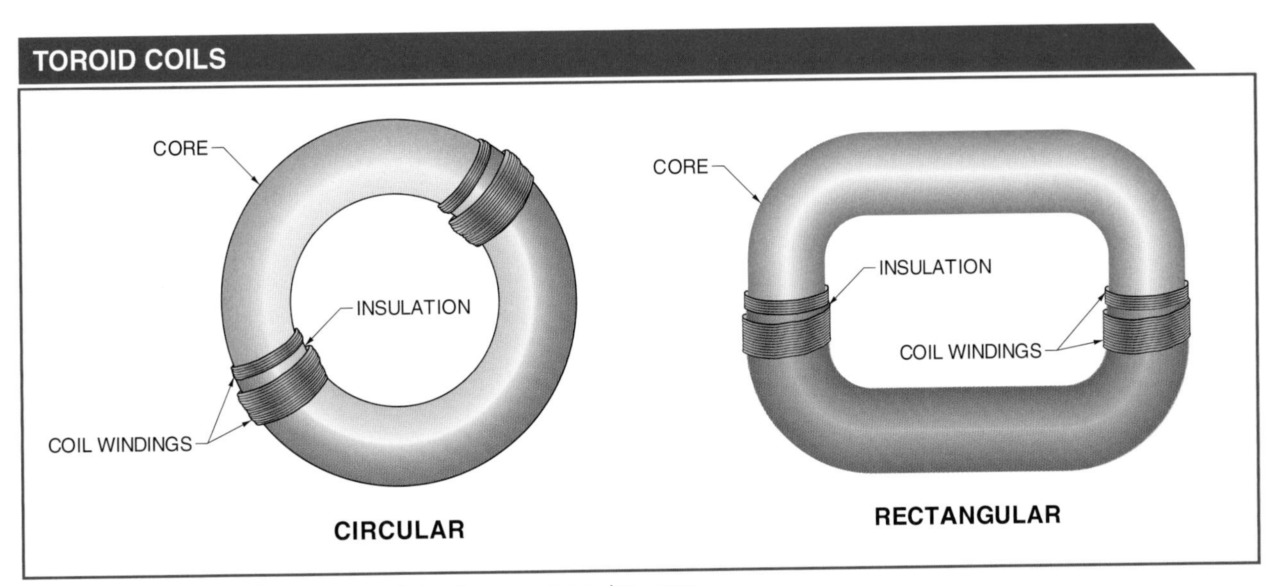

Figure 11-3. Toroid coils have closed circular or rectangular cores.

INDUCTOR TYPES

Inductor types include variable inductors, molded inductors, radio frequency chokes, shielded inductors, and noninductive coil inductors. In addition to inductance, inductors are also rated according to current, DC resistance, voltage, tolerance, and quality.

Variable Inductors

A *variable inductor* is an inductor in which the inductance is varied by a core that can be moved into and out of the center of the coil. **See Figure 11-4.** Variable inductors may be tapped in the same way as a tapped resistor. Typically, the inductance is varied by having a core that can be moved into and out of the center of the coil windings. This is often accomplished using threads and a screw adjustment. Variable inductors are used in applications such as radio tuners and amplifiers.

Molded Inductors

A *molded inductor* is an inductor that is encapsulated in plastic or other insulating materials with a conductor at each end for connecting or soldering into a circuit. Molded inductors are sometimes referred to as radial inductors. These inductors resemble resistors and have color codes to indicate their value. Molded inductors may have air, powdered iron, or ferrite cores. Molded inductors are available in various sizes with an inductance value identified by four or five colored bands on the inductor's case. **See Figure 11-5.**

Terms

A **variable inductor** is an inductor in which the inductance is varied by a core that can be moved into and out of the center of the coil.

A **molded inductor** is an inductor that is encapsulated in plastic or other insulating materials with a conductor at each end for connecting or soldering into a circuit.

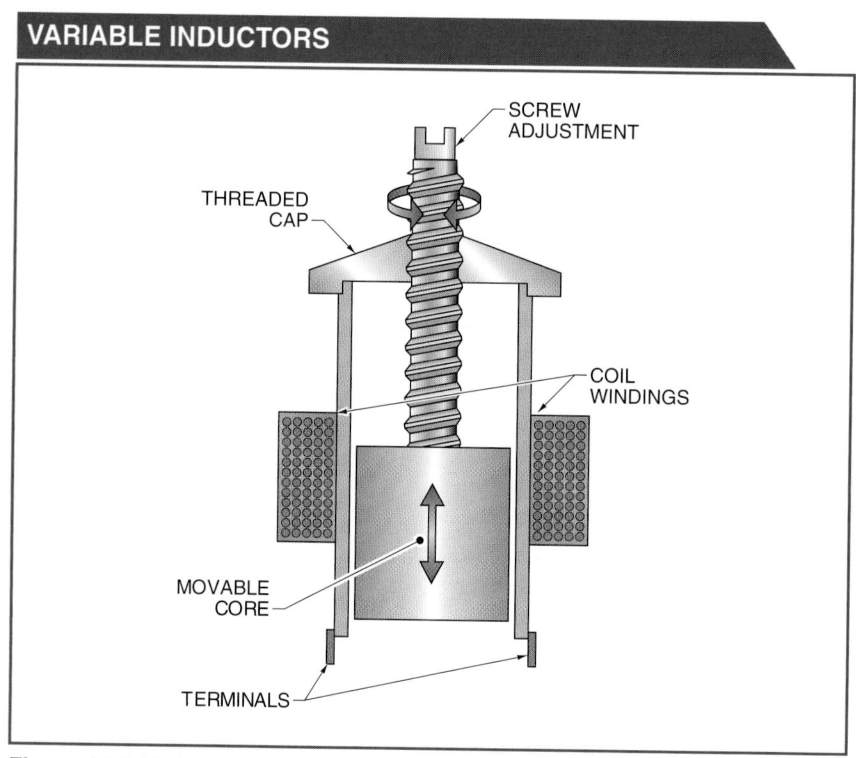

Figure 11-4. Variable inductors are adjusted with a core that can be moved into and out of the center of the coil windings.

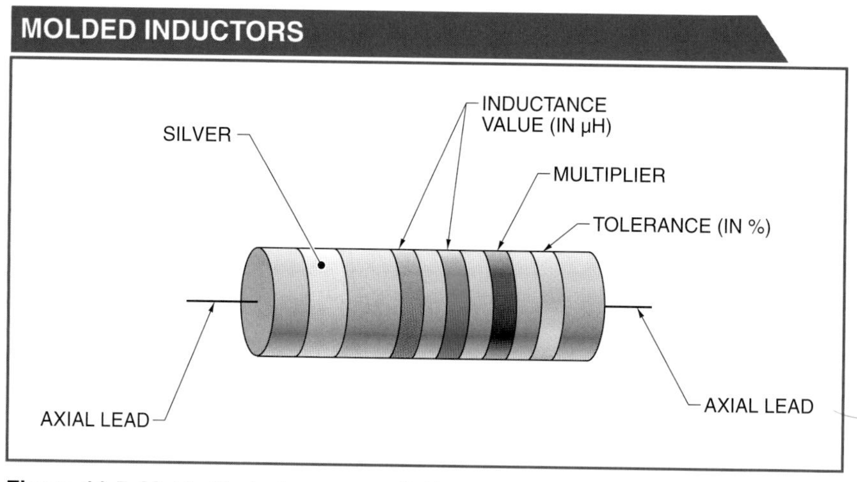

Figure 11-5. Molded inductors are available in various sizes with inductance value identified by four or five colored bands on the inductor case.

The first two bands on the inductor are the color-coded inductance value (in microhenrys). The third band is the color-coded multiplier and the color of the fourth band indicates inductance tolerance (as a percentage). Typically, molded inductors are used in electronic circuit applications.

Radio Frequency Chokes

A *radio frequency (RF) choke* is an inductor designed for high impedance over large frequency ranges. Radio frequency chokes were first used in early radio transmitters and are referred to as RF chokes. Radio frequency chokes are available in various sizes and forms. Radio frequency chokes have air, powdered iron, or ferrite cores, depending on the type of application and application frequency. Radio frequency chokes are unshielded for use in high-frequency applications. Typically, fine-gauge conductors of many turns are wrapped around a molded frame with connecting leads. Conductors at each end of the RF choke are soldered to the connecting leads to simplify soldering the RF choke into a circuit.

Radio frequency chokes also use a system of color-coded bands to identify inductance values. When either the first or second band of the three bands is gold, it represents the decimal point for inductance values less than 10. For inductance values of 10 or more, the first two bands represent inductance values, and the third band represents the multiplier. For small units, a series of four dots may be used instead of bands to indicate the inductance and tolerance. **See Appendix.**

Terms

A **radio frequency (RF) choke** is an inductor designed for high impedance over large frequency ranges.

Shielded Inductors

A *shielded inductor* is an inductor contained in a shield composed of low-reluctance magnetic material brought to ground potential. **See Figure 11-6.** Inductors are often shielded in electrical and electronic circuits to prevent the interaction of magnetic fields.

Terms

A **shielded inductor** is an inductor contained in a shield composed of low-reluctance magnetic material brought to ground potential.

SHIELDED INDUCTORS

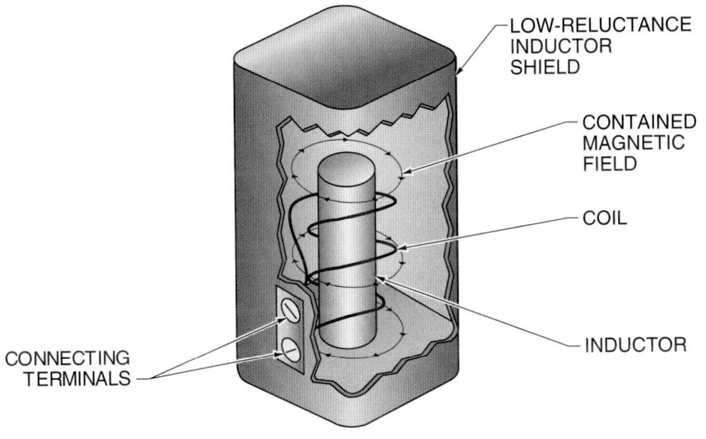

Figure 11-6. A shielded inductor is an inductor contained in a shield of low-reluctance magnetic material brought to ground potential. The shield prevents stray magnetic fields from influencing the inductor.

An inductor shield prevents the magnetic field of the coil from escaping to affect other circuits. At the same time, the inductor shield prevents stray magnetic fields from other circuits from entering and affecting the protected inductor. Shielded inductors are available in adjustable and fixed designs. Adjustable shielded inductors normally have an access hole for the inductor in the top or bottom of the case. Adjustable shielded inductors often have two coils, and two adjustable cores for tuning circuits. Fixed shielded inductors have screw-type connectors on one side for connection to a flat surface. Other fixed shielded inductors, such as those used on printed circuit boards, have rigid terminals at the bottom of the case so that they can be soldered into place.

Noninductive Coil Inductors

Terms

A **noninductive coil inductor** is an inductor that has coils wound in such a manner that the inductance created by one coil turn is canceled by an adjacent coil turn.

A *noninductive coil inductor* is an inductor that has coils wound in such a manner that the inductance created by one coil turn is canceled by an adjacent coil turn. Typically, inductance is introduced into a circuit to store electrical energy. However, certain applications require the circuit to be purely resistive in nature. For example, if a coil is wound around a mold to manufacture a resistor, inductance is also introduced. To overcome inductance, the resistor may be noninductive wound. Resistance without inductance is achieved by double-winding the coil onto the mold. **See Figure 11-7.** This results in current flow in the opposite direction in conductors next to each other, thereby canceling the magnetic effect of the current flow. Unwanted signals are sometimes caused by current flow though conductors connecting various components and devices in a circuit.

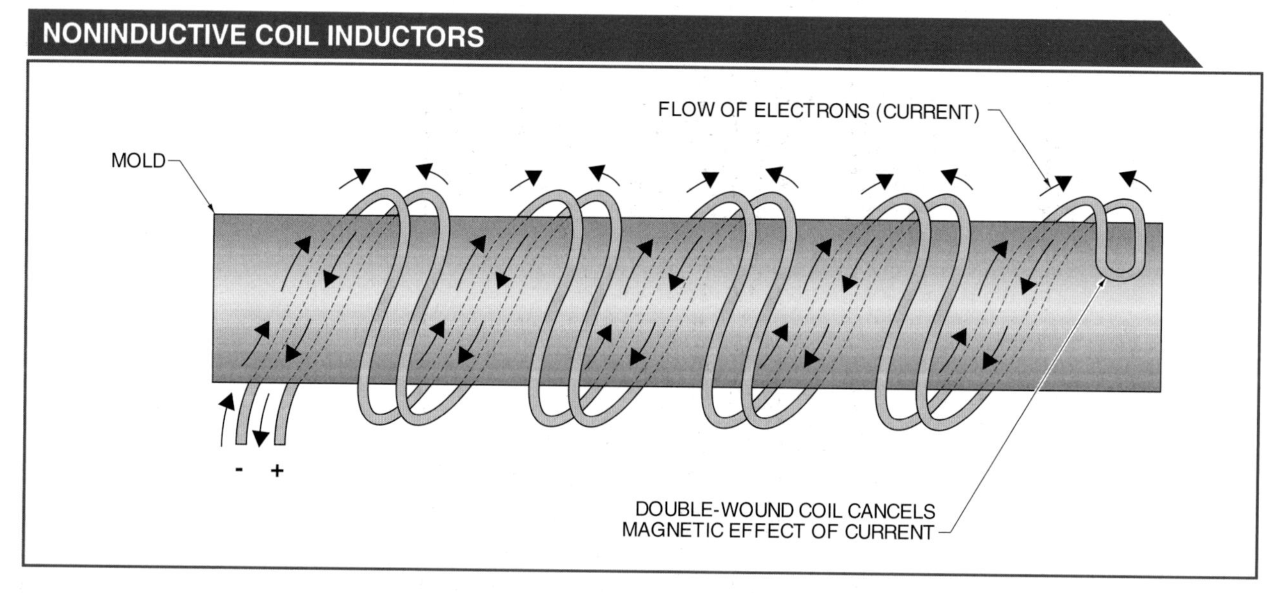

Figure 11-7. Resistance without inductance is achieved by double-winding the coil onto the mold.

Inductor Ratings

Typically, inductors are rated in terms of inductance measured in henrys. Inductors are also rated according to current, DC resistance, voltage, tolerance, and quality. The current rating indicates how much current the inductor coil can continuously carry without overheating. Inductance is higher at lower current levels and lower at higher current levels. The resistance of the coil is the DC resistance of the coil as measured with a DMM set to measure resistance.

Resistance of the coil affects the time constant and the quality of the coil. *Inductive reactance* is the opposition of an inductor to alternating current. Inductive reactance of the inductor coil is used in the design of tuned electronic circuits. Lowering DC resistance raises the inductive reactance of an inductor coil.

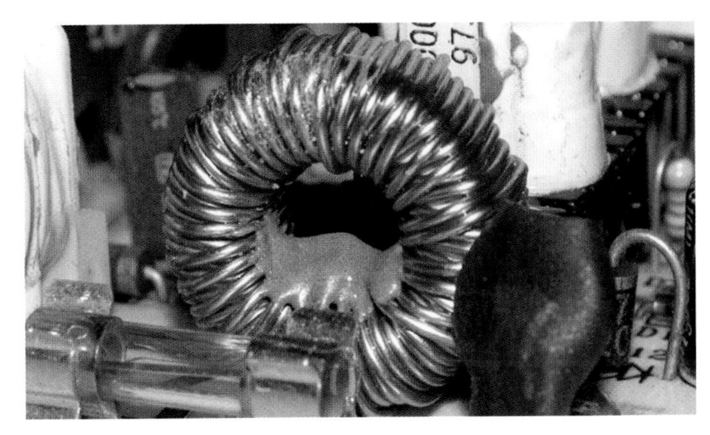

Toroidal inductors are used for smoothing current and storing energy in inductive filter circuits.

The voltage rating of an inductor signifies the maximum voltage that the insulation of the coil can continuously withstand without deteriorating. Exceeding the voltage rating of a coil may or may not cause instantaneous deterioration, but the life of the insulation decreases if this rating is continuously exceeded. Voltage ratings are used mostly with laminated iron-core inductors because they are used in power supplies. Laminated iron-core inductors often have their cores physically and electrically connected to ground. If the coil insulation deteriorates, there is the possibility that the conductor can touch the core and short to ground.

This type of inductor is typically used in power supplies with several hundred volts and high currents present in the coils. Inductors have tolerances of ±1% for precision types and up to 20% in types where tolerance is not as important. Inductors used in electrical and electronic equipment typically have tolerances of ±20%.

INDUCTOR-STORED ENERGY

An inductor stores energy and opposes any change in current through it. Energy is stored in a magnetic field created by the current flow through the coil. A *joule (J)* is the unit for energy measurement. Stored energy relates to the value of inductance and the instantaneous value of current and is calculated by applying the following formula:

$W = \frac{1}{2}LI^2$

where

Terms

Inductive reactance is the opposition of an inductor to alternating current.

A **joule (J)** is the unit for energy measurement.

W = energy (in J)

L = inductance (in H)

I = instantaneous value of current (in A)

Example: What is the stored energy for a 2 H inductor with an instantaneous current of 3 A?

$W = \frac{1}{2} LI^2$

$W = \frac{1}{2} \times 2 \text{ H} \times 3^2 \text{ A}$

$W = \mathbf{9\ J}$

Tech Tip

Inductance does not affect steady DC because without a change in current there is no change in the magnetic field.

Energy supplied by the current flow through the coil aligns the electron orbits of the atoms, which form magnetic fields. When current decreases, energy returns to the circuit as the magnetic field collapses into the coil. The returned energy allows the previously aligned orbital electrons to resume their random orbits. The amount of energy returned to the circuit depends on the amount of energy stored in the magnetic field and the time involved. For example, if 10 J are stored by the magnetic field, and the field collapses in one second, then 10 J/s (10 W) are returned to the circuit. If the field collapses in 1 millisecond, then 10,000 J/s (10 kW) are returned to the circuit. The faster the field collapses, the greater the power returned to the circuit.

Terms

A **resistive circuit** is an electrical circuit that contains only resistance.

An **ideal inductive circuit** is a circuit that has an ideal inductor connected to a battery through a switch.

RESISTIVE AND IDEAL INDUCTIVE CIRCUITS

A *resistive circuit* is an electrical circuit that contains only resistance. For example, a heating element is a resistive circuit. An *ideal inductive circuit* is a circuit that has an ideal inductor connected to a battery through a switch. An ideal inductive circuit is a theoretical circuit that is purely inductive. In order to fully understand a practical inductive circuit, it is important to understand and contrast a resistive circuit and an ideal inductive circuit.

Resistive Circuits

A resistive circuit has current flow from the negative terminal of the battery, through the resistor, and back to the positive terminal of the battery when the switch is closed. Current rises to its maximum value instantly when the switch is closed and falls instantly when the switch is opened. **See Figure 11-8.** The current achieves its maximum value of 5 A instantaneously. Very little is required for the current level to change from 0 A to 5 A. Current continues to flow at this level (5 A) until the switch is opened after 6 sec. Current then falls instantly back to 0 A. The resistive voltage has the same shape as the current since the voltage value is equal to the current value multiplied by the resistance value. In a resistive circuit, resistance merely opposes, or limits, current flow.

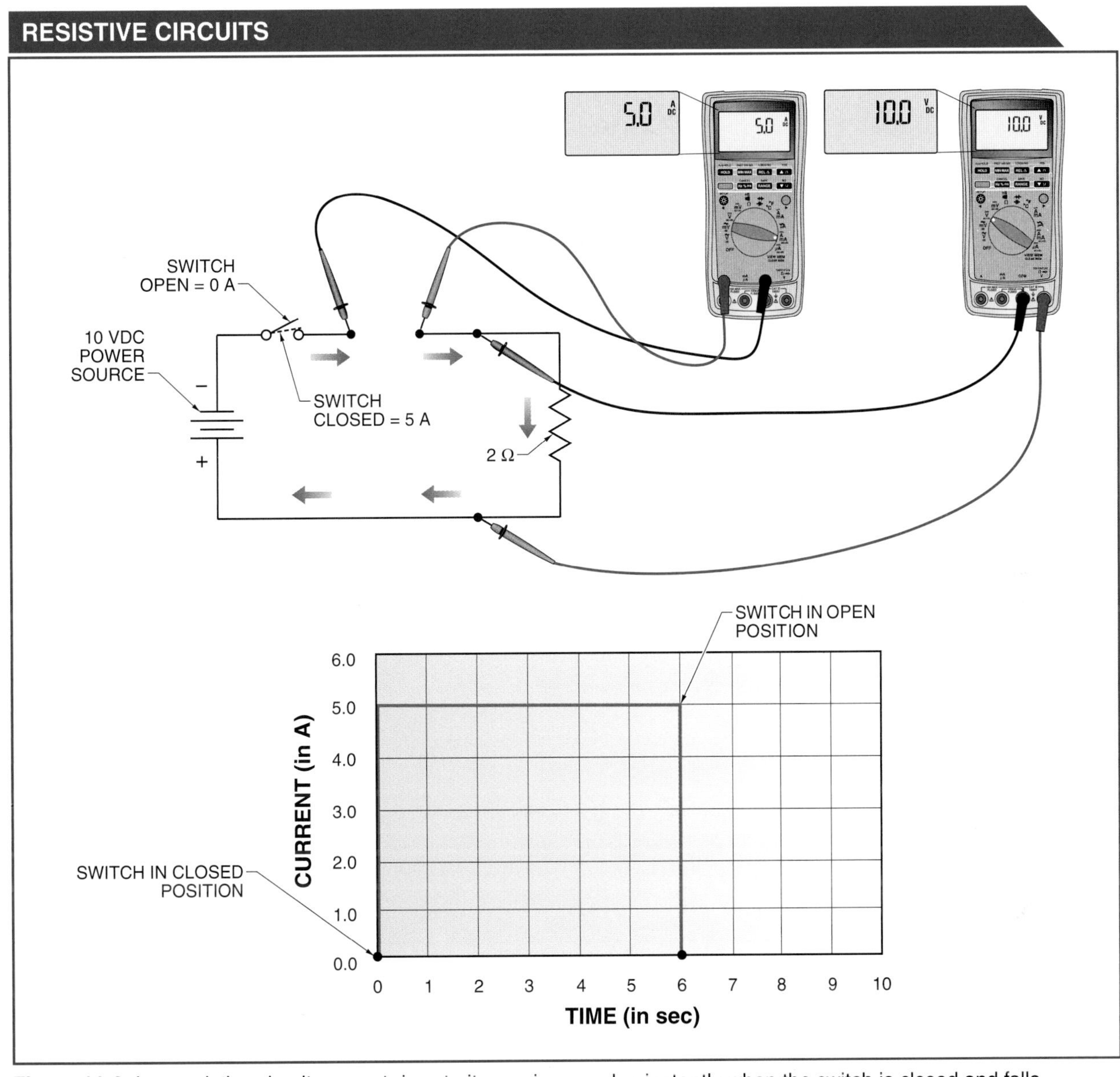

Figure 11-8. In a resistive circuit, current rises to its maximum value instantly when the switch is closed and falls instantly when the switch is opened.

Ideal Inductive Circuits

Ideal inductive circuits are circuits that have no resistance or capacitance and are purely inductive. In an ideal inductive circuit, current rises indefinitely at a linear rate to an undefined value. **See Figure 11-9.** The circuit is theoretical since the conductor used in an ideal inductor has no resistance. When the switch is placed in position 1, current does not increase instantly as in the case of the resistive circuit. Instead, the current increases linearly with time as long as the battery is connected to the inductor.

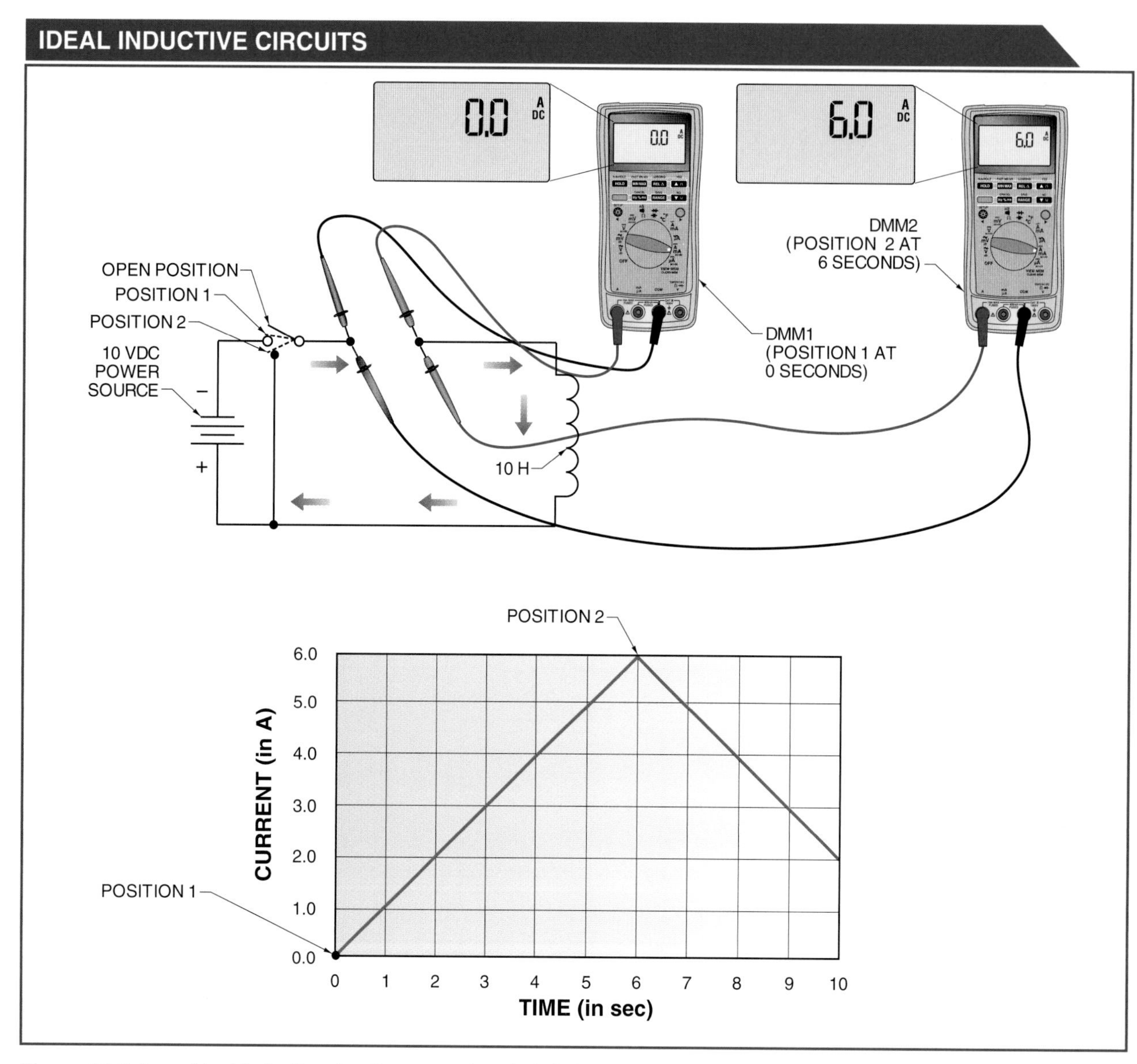

Figure 11-9. In an ideal inductive circuit, current rises indefinitely at a linear rate to an undefined value.

Terms

Electromotive force (EMF) is electrical pressure applied to a circuit.

Counter-electromotive force (CEMF) is the property of a conductor to oppose any change in current.

Electromotive force (EMF) is electrical pressure applied to a circuit. *Counter-electromotive force (CEMF)* is the property of a conductor to oppose any change in current. Counter-electromotive force is also referred to as back-electromotive force because the induced voltage is always in a direction that opposes the change in current. The value of inductance from CEMF is very low in DC and low-frequency AC circuits.

The inductive voltage opposes the change in the current in the circuit for the first 6 sec. In a purely inductive circuit, CEMF must always equal the source voltage. A change of 1 A/sec multiplied by inductance of 10 H is equal to the voltage of the battery. The opposing voltage remains inductive as long as the battery is connected to the circuit. Current continues to increase at this rate until the battery is removed from the circuit. As the current increases, the magnetic field around the inductor is strengthened. If the switch is changed instantly to position 2 at 6 sec, the battery is removed from the circuit.

With the battery removed, the magnetic field around the inductor begins to collapse, inducing a voltage that acts to keep the current flowing at the same level and in the same direction through the inductor. Because of the opposing voltage, the current cannot fall to zero instantly. Instead, it decreases at the same rate as it increased, 1 A/sec. Current flow continues until the magnetic field has completely collapsed. If the switch is in position 1 and then placed in the open position, a large inductive voltage is created. This is due to the rapid collapse of the current. Theoretically, an ideal inductive circuit is one in which the current rises and falls at a constant rate. Realistically, some resistance and capacitance are always present.

PRACTICAL INDUCTIVE CIRCUITS

A *practical inductive circuit* is a circuit that has both resistance and inductance. In a practical inductive circuit, the resistance and source voltage of a circuit limit maximum current. When circuit current is constant with a given source voltage, resistance in a practical inductive circuit can be determined using Ohm's law. The primary source of resistance in a practical inductive circuit is the resistance of the wire used in the coil. This resistance is equal to the resistivity of the wire multiplied by its length divided by the circular mil area.

Terms

A **practical inductive circuit** is a circuit that has both resistance and inductance.

For example, in a circuit that has resistance of 20 Ω connected in series with inductance of 10 H, the maximum current is determined by application of Ohm's law ($I = V/R$). In a circuit with pure resistance, current achieves maximum value instantly. In a circuit with pure inductance, current increases linearly for as long as the source voltage is applied.

In a series inductive-resistive circuit, the CEMF of the inductor opposes any change in current, thereby preventing instant change. At the same time, the current does not increase at a linear rate due to the resistances in series with the inductor. While current increases, three voltages are present in the circuit. These are the source voltage (V_T), inductive voltage (V_L), and resistive voltage drop (V_R). **See Figure 11-10.** The voltage drop across the resistor is equal to $I \times R$ and follows the rate of the current rise. When the switch is placed in position 1, the inductive voltage is equal to and opposite the source voltage. At this point the current is zero, but its rate of change is maximized and can be calculated by applying the following formula:

$$\frac{\Delta i}{\Delta t} = \frac{V_T}{L}$$

where

Δi = change in current (in A)

Δt = change in time (in sec)

V_T = source voltage (in V)

L = inductance (in H)

Example: What is the rate of change in current for a circuit with source voltage of 10 V and an inductance of 10 H?

$$\frac{\Delta i}{\Delta t} = \frac{V_T}{L}$$

$$\frac{\Delta i}{\Delta t} = \frac{10\text{ V}}{10\text{ H}}$$

$$\frac{\Delta i}{\Delta t} = 1\text{ A/sec}$$

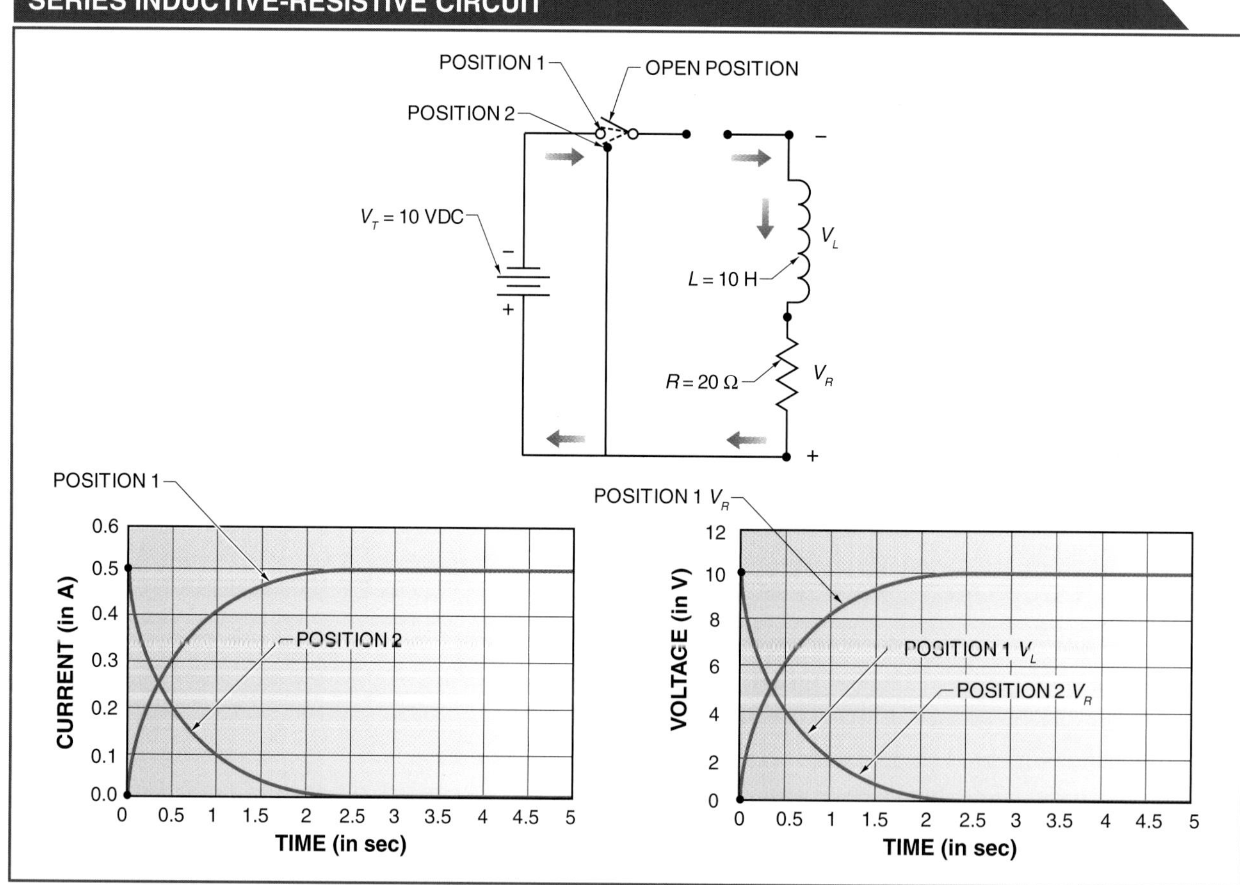

Figure 11-10. Voltages present in a series inductive-resistive circuit with increasing current include source voltage (V_T), inductive voltage (V_L), and resistive voltage drop (V_R).

There is no resistive voltage drop since there is no current flow. After the initial action, the source voltage causes current to increase in the circuit causing a resistive voltage drop and a decrease in CEMF. At any time, the sum of CEMF and $I \times R$ is equal to the source voltage. As current begins to flow, a resistive voltage drop develops. Inductive voltage decreases at the same rate. Because CEMF decreases, the current increases, as does resistive voltage drop. After the switch is closed, the current reaches its maximum value, and resistive voltage is equal to the source voltage. Current is constant at this point, limited only by the value of the resistance in the circuit. Since the current is no longer changing, no voltage is induced in the inductor.

When the switch is changed to position 2 the battery is removed from the circuit, discharging the energy stored in the magnetic field of the coil. The magnetic field surrounding the coil collapses through the coil and initially induces an inductive voltage of 10 V that opposes the decreasing current. Therefore, the current cannot fall instantly from 0.5 A to 0 A, as was the case of the resistive circuit. Nor does current decrease at the linear rate of the purely inductive circuit. As the magnetic field continues to collapse, the rate decreases and a smaller voltage is induced. Because the same values of circuit components and devices are used both in the charge circuit and the discharge circuit, the same amount of time is required for the current to reach its maximum and minimum values.

INDUCTIVE-RESISTIVE TIME CONSTANT

The *time constant* is the time required for the current in an inductive-resistive circuit to reach 63.2% of its maximum value after power is applied to the circuit, or to decrease by 63.2% (to 36.8% of maximum power) when the power is removed from the circuit. **See Figure 11-11.**

For each time constant, the current changes another 63.2% of the remaining value. The rate of change in current generated by this process is known as an exponential curve. In practical inductive-resistive circuits, the ratio of inductance to resistance is seldom greater than unity (1). Therefore, time constant is typically expressed in milliseconds (ms) or microseconds (μs). When the inductance and resistance of a circuit are known, the time constant for the circuit can be calculated by applying the following formula:

$$\tau = \frac{L}{R}$$

where

τ = time constant (in s)

L = inductance (in H)

R = resistance (in Ω)

Terms

The **time constant** is the time required for the current in an inductive-resistive circuit to reach 63.2% of its maximum value after power is applied to the circuit, or to decrease by 63.2% (to 36.8% of maximum power) when the power is removed from the circuit.

Example: What is the time constant for a circuit with an inductance of 10 H and a resistance of 20 Ω?

$$\tau = \frac{L}{R}$$

$$\tau = \frac{10\ \text{H}}{20\ \Omega}$$

$$\tau = \mathbf{0.5\ s}$$

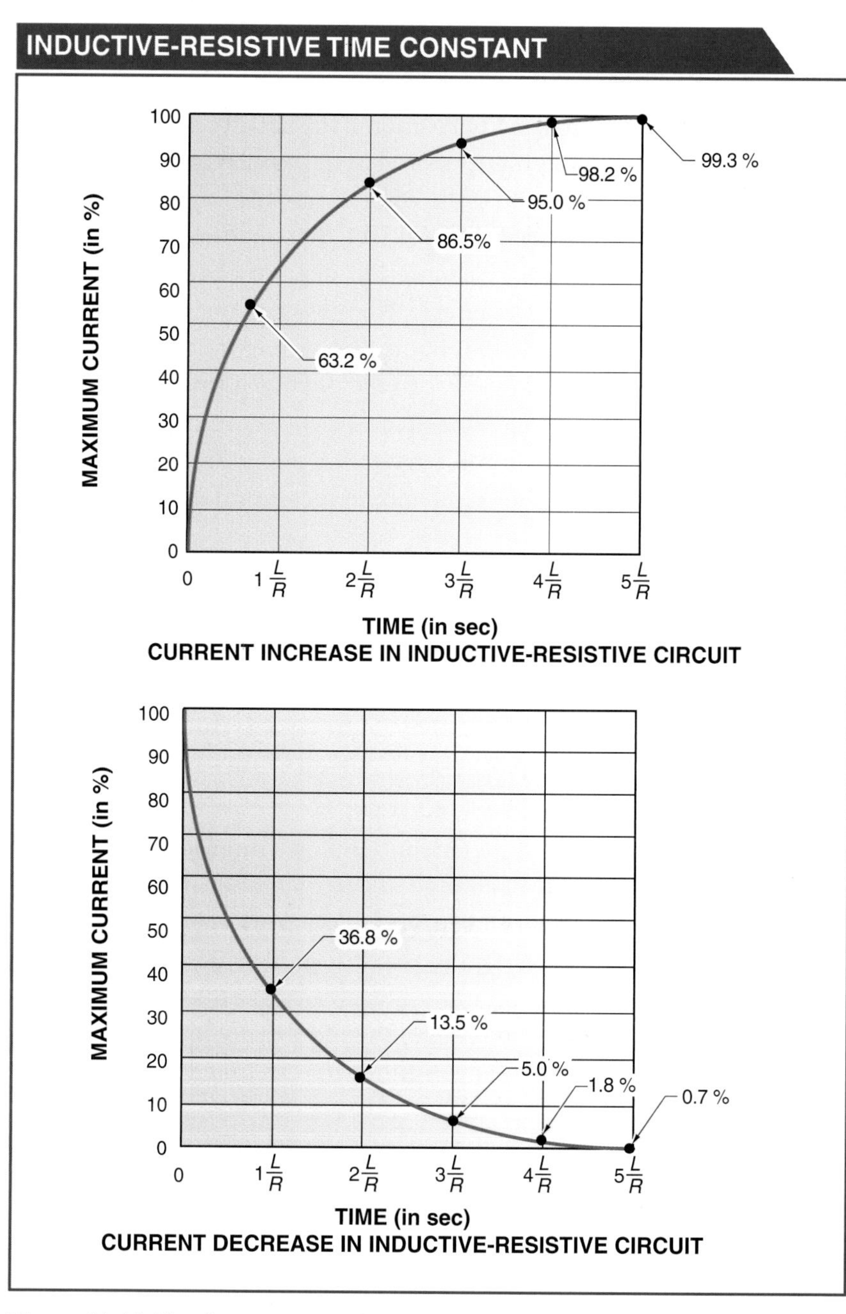

Figure 11-11. The time constant of an inductive-resistive circuit is the time required for the current to increase or decrease by 63.2% of its value.

Once the time constant is known, the time required for the current to reach its maximum value when the circuit is energized, or to decrease to 0 A when the circuit is de-energized, can be determined. Theoretically, the current never reaches its maximum value when power is applied to the inductive-resistive circuit, nor does it decrease to 0 A when power is removed. The reason for this is that only 63.2% of the current is used during each time constant. Total change in current occurs within five time constant cycles.

In the inductive-resistive circuit, the series current and inductor voltage can be described mathematically. To find the equation that describes the circuit current, Kirchhoff's voltage law is used. This law states that the voltage around a closed loop is equal to zero.

$$V_T - V_L - V_R = 0$$
$$V_T = V_L + V_R$$

where
V_T = source voltage (in V)
V_L = inductor voltage (in V)
V_R = resistor voltage (in V)

The voltage at the inductor is given by the following:

$$V_L = L\frac{\Delta i}{\Delta t}$$

where
V_L = inductor voltage (in V)
L = inductance (in H)
Δi = change in current (in A)
Δt = change in time (in sec)

The resistor voltage is simply current times the resistance as follows:

$$V_R = i \times R$$

Substituting for V_L and V_R, the equation becomes the following:

$$V_T = L\frac{\Delta i}{\Delta i} + i \times R$$

This equation is a first-order differential equation, which has a common solution. The solution for i is the following:

$$i(t) = \frac{V_T}{R}(1 - e^{-Rt/L})$$

where
$i(t)$ = current at time t (in A)
V_T = source voltage (in V)

NREL

DC inductive-resistive circuits are found in the electronic systems in electric vehicles.

R = resistance (in Ω)
1 = constant
e = natural log (2.718)
t = time (in sec)
L = inductance (in H)

Example: What is the current for every 0.5 sec for a circuit with a 10 V voltage source, 10 H inductor, and 20 Ω resistor?

$$i(t) = \frac{V_T}{R}(1 - e^{-Rt/L})$$

$$i(t) = \frac{10\text{ V}}{20\ \Omega}(1 - e^{-20\Omega(t/10\text{ H})})$$

$$i(t) = 0.5(1 - e^{-2t})$$

INDUCTOR CURRENT vs TIME

Time*	$i(t)$†	Number of Time Constants	Percentage of Current
0.0	0.000	0	0.0%
0.5	0.316	1	63.2%
1.0	0.432	2	86.5%
1.5	0.475	3	95.0%
2.0	0.491	4	98.2%
2.5	0.497	5	99.3%
3.0	0.499	6	99.8%
3.5	0.500	7	99.9%
4.0	0.500	8	100.0%
4.5	0.500	9	100.0%
5.0	0.500	10	100.0%

* in sec
† in A

The equation for the inductor voltage that changes over time is the following:

$$V_L(t) = V_T(e^{-Rt/L})$$

where

$V_L(t)$ = inductor voltage at time t (in V)

V_T = source voltage (in V)

R = resistance (in Ω)

e = natural log (2.718)

t = time (in seconds)

L = inductance (in H)

Example: What is the inductor voltage for every 0.5 sec for a circuit with a 10 V voltage source, 10 H inductor, and 20 Ω resistor?

$$V_L(t) = V_T(e^{-Rt/L})$$

$$V_L(t) = 10(e^{-20\Omega t/10\text{ H}})$$

$$V_L(t) = 10\,(e^{-2t})$$

INDUCTOR VOLTAGE vs TIME

Time*	$V_L(t)$	Number of Time Constants	Percentage of Current
0.0	10.000	0	100.0%
0.5	3.679	1	36.8%
1.0	1.353	2	13.5%
1.5	0.498	3	5.0%
2.0	0.183	4	1.8%
2.5	0.067	5	0.7%
3.0	0.025	6	0.2%
3.5	0.009	7	0.1%
4.0	0.003	8	0.0%
4.5	0.001	9	0.0%
5.0	0.000	10	0.0%

* in sec

HIGH INDUCED VOLTAGES

High induced voltage is an energy release caused by the coil in an inductive circuit. *Inductive kick* is the energy release in a coil anytime the inductive current is abruptly stopped or reversed. The energy contained in a collapsing magnetic field surrounding a coil must be dissipated slowly and evenly within the circuit. The high induced voltage created in the dissipation process is sufficient to create the arc (current flow) across switch contacts, creating heat from the stored energy. The energy from the inductive kick of a coil can cause electrical shock hazards, damage switch contacts, or damage the insulation of a coil. Abrupt removal of power from a circuit often causes coils to fail because of inductive kick. Safety precautions should always be taken not to open the field windings rapidly in motors and generators.

Terms

High induced voltage is an energy release caused by the coil in an inductive circuit.

Inductive kick is the energy release in a coil anytime the inductive current is abruptly stopped or reversed.

SELF-INDUCTANCE

Self-inductance is the property of a conductor to induce voltage within itself due to changes in current. A coiled conductor has more inductance than a straight conductor. Current flow through a conductor causes a concentric magnetic field to radiate from the center of the conductor. As current increases or decreases, the strength of the magnetic field changes in direct proportion, and voltage is induced into the conductor that opposes the change in current. **See Figure 11-12.**

Counter-electromotive force always opposes current change in a conductor. As source voltage increases, current increases, and the magnetic field expands outward from the center of the conductor. An expanding magnetic field induces voltage that opposes the source voltage, delaying the rise in current. As source voltage decreases, current decreases, and the magnetic field collapses around the conductor. A changing magnetic field induces CEMF, which aids the source voltage in keeping the current at its previous level of flow. Polarity of CEMF can be determined by applying the left-hand rule for generators.

Terms

Self-inductance is the property of a conductor to induce voltage within itself due to changes in current.

MUTUAL INDUCTANCE

Mutual inductance is the effect of one coil inducing a voltage into another coil. *Mutual induction* is the ability of an inductor in one circuit to induce a voltage in another circuit. When a second coil, not physically connected to a voltage source, is placed within the changing flux field of a self-inducting coil, a voltage is induced in the second coil. When coils are arranged in this manner, they have

Terms

Mutual inductance is the effect of one coil inducing a voltage into another coil.

Mutual induction is the ability of an inductor in one circuit to induce a voltage in another circuit.

mutual induction. The *coefficient of coupling* is the percentage of magnetic lines of force created by one coil that cut the windings of the second coil. **See Figure 11-13.** Mutual inductance is measured in henrys and is calculated by applying the following formula:

$$M = k\sqrt{L_1 \times L_2}$$

where

M = mutual inductance (in H)

k = coefficient of coupling (in %)

L_1 = inductance 1 (in H)

L_2 = inductance 2 (in H)

Terms

The **coefficient of coupling** is the percentage of magnetic lines of force created by one coil that cut the windings of the second coil.

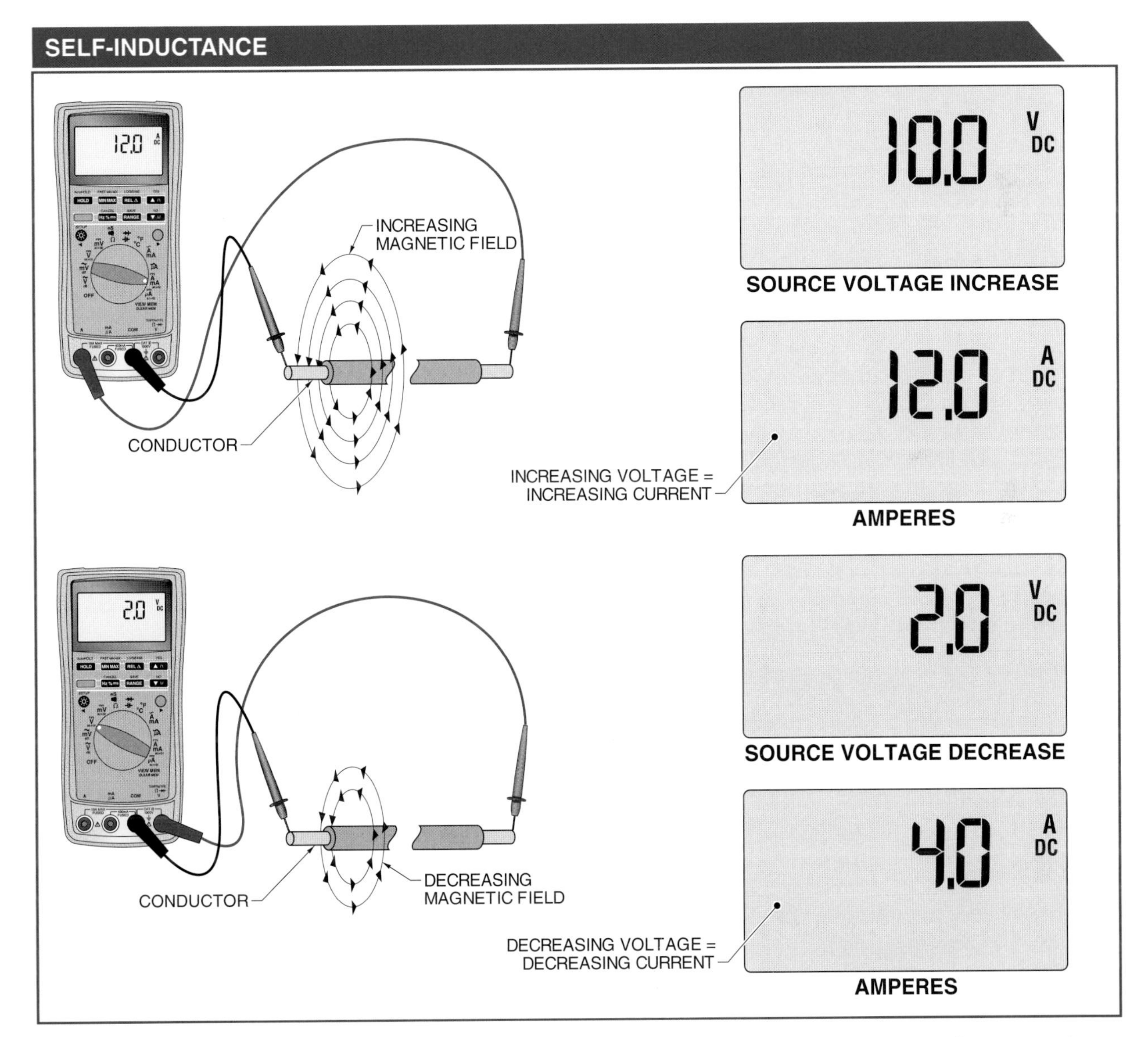

Figure 11-12. With self-inductance, as current changes, the strength of the magnetic field changes in direct proportion, and voltage is induced in the conductor that opposes the change in current.

COEFFICIENT OF COUPLING

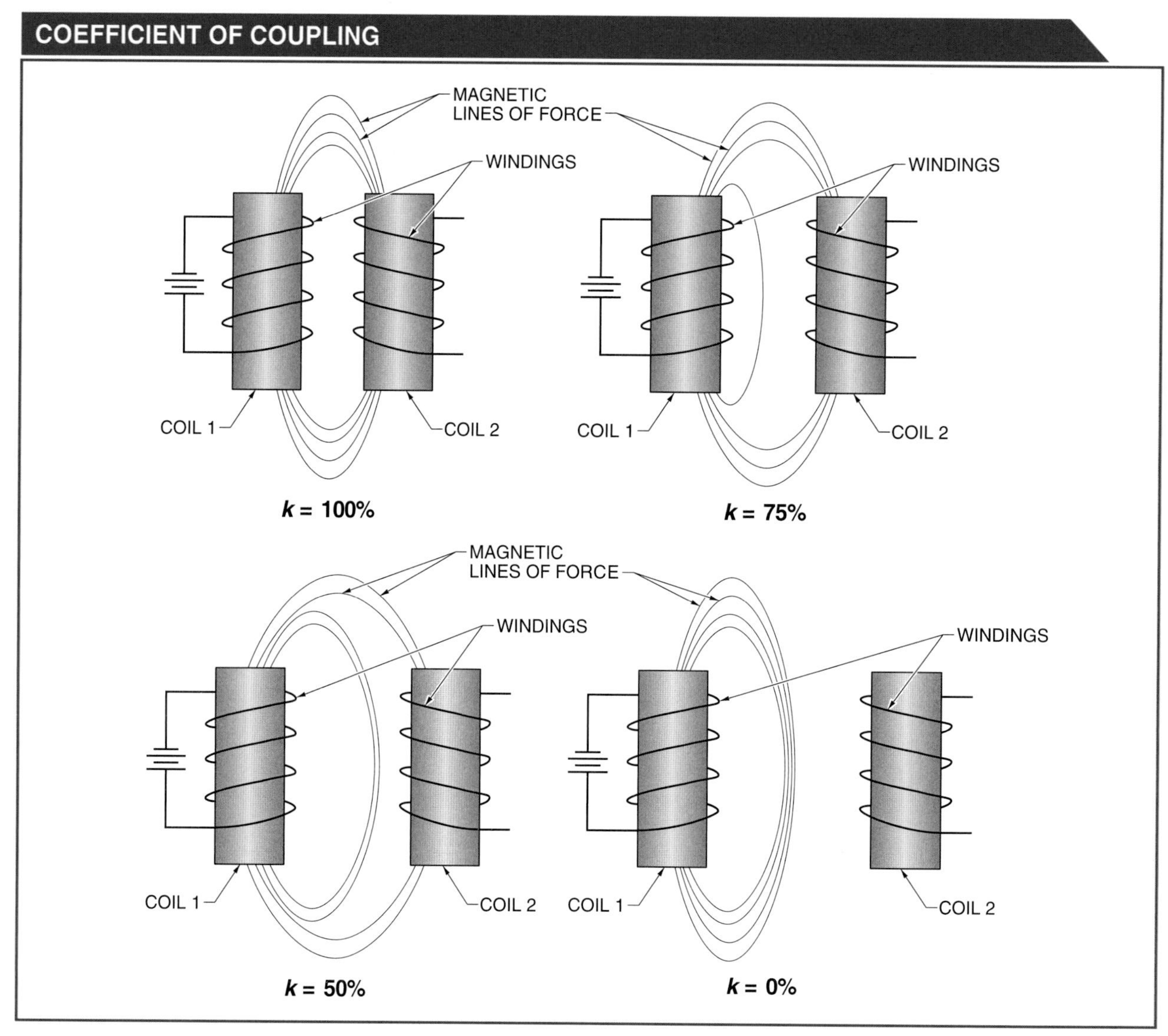

Figure 11-13. The coefficient of coupling (k) is the percentage of magnetic lines of force created by one coil that cut the windings of a second coil.

Example: What is the mutual inductance of two coils, each with an inductance of 2 H, with a coefficient of coupling of 75%?

$$M = k\sqrt{L_1 \times L_2}$$

$$M = 0.75\sqrt{2\text{ H} \times 2\text{ H}}$$

$$M = 0.75 \times 2$$

$$M = \mathbf{1.5\ H}$$

The coefficient of coupling is dependent upon the physical characteristics of the coils, the distance between the coils, and the permeability of the core. If all of the magnetic lines of force produced on one coil cut the windings of the second coil, then the coefficient of coupling is 1, or 100%. In actual electrical applications it is nearly impossible to attain a 100% coefficient of coupling.

The inductive voltage in a coil is dependent upon the number of coil turns. Inductive voltage in each coil is due to the change in the magnetic field. Inductive voltages add to provide the total voltage. The more coil turns, the higher the inductive voltage. Inductive voltage is also dependent upon the magnetic lines of force linking the coils and the rate of change in the magnetic field created by the lines of force.

INDUCTORS CONNECTED IN SERIES

Inductors are often connected in series. Each inductor opposes the change in current in the circuit. Since the inductors are connected in series with each other, the current flow is the same through all of them. To satisfy Kirchhoff's voltage law, the sum of the voltage drops across inductors connected in series must equal the source voltage. When the magnetic fields of two inductors aid each other, their mutual inductance (M) must be added together to their individual values to obtain the total inductance for the circuit. When no mutual induction exists between the inductors, their values add together in the same manner as resistors connected in series add together. Total inductance for inductors connected in series is calculated by applying the following formula:

$L_T = L_1 + L_2 + L_3 + \ldots + L_N$

where

L_T = total inductance (in H)

L_1 = inductance 1 (in H)

L_2 = inductance 2 (in H)

L_3 = inductance 3 (in H)

L_N = inductance N (in H)

Tech Tip

Inductors are connected in series in a DC circuit in order to increase total inductance.

Example: What is the total inductance of three inductors connected in series with inductance values of 1 H, 2 H, and 3 H?

$L_T = L_1 + L_2 + L_3 + \ldots + L_N$

$L_T = 1\text{ H} + 2\text{ H} + 3\text{ H}$

$L_T = \mathbf{6\ H}$

This formula is modified if mutual induction exists between inductors. The total induction under this condition is either greater or less than the sum of inductor values. If the value is greater than the sum of inductor values, then the net magnetic fields are aiding each other. If the value is less than the sum of inductor values, then the net magnetic fields are opposing each other. **See Figure 11-14.** Total induction for two inductors connected in series with aiding magnetic fields is calculated by applying the following formula:

$$L_T = \left(L_1 + k\sqrt{L_1 \times L_2}\right) + \left(L_2 + k\sqrt{L_1 \times L_2}\right)$$

where

L_T = total inductance (in H)

k = coefficient of coupling (in %)

L_1 = inductance 1 (in H)

L_2 = inductance 2 (in H)

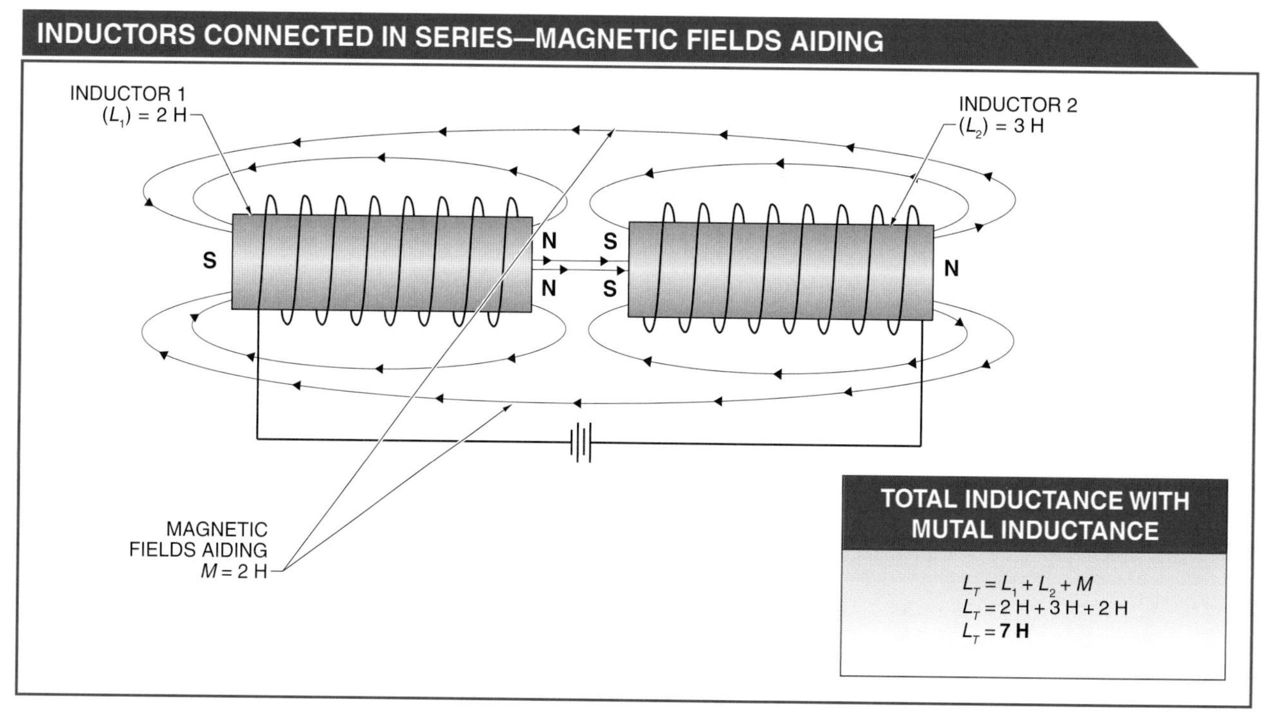

Figure 11-14. When the magnetic fields of two inductors are aiding, their mutual inductance (M) must be added to their individual values to obtain the total inductance for the circuit.

Example: What is the total inductance of two inductors connected in series with inductance values of 1 H and aiding magnetic fields with a coefficient coupling of 75%?

$$L_T = \left(L_1 + k\sqrt{L_1 \times L_2}\right) + \left(L_2 + k\sqrt{L_1 \times L_2}\right)$$

$$L_T = \left(1\text{ H} + 0.75\sqrt{1\text{ H} \times 1\text{ H}}\right) + \left(1\text{ H} + 0.75\sqrt{1\text{ H} \times 1\text{ H}}\right)$$

L_T = (1 H + 0.75) + (1 H + 0.75)

L_T = 1.75 + 1.75

L_T = **3.5 H**

Total induction for two inductors connected in series with opposing magnetic fields is calculated by applying the following formula:

$$L_T = \left(L_1 - k\sqrt{L_1 \times L_2}\right) + \left(L_2 - k\sqrt{L_1 \times L_2}\right)$$

where

L_T = total inductance (in H)
L_1 = inductance 1 (in H)
k = coefficient of coupling (in %)
L_2 = inductance 2 (in H)

Example: What is the total inductance of two 1 H inductors connected in series that have opposing magnetic fields with a coefficient coupling of 75%?

$$L_T = \left(L_1 - k\sqrt{L_1 \times L_2}\right) + \left(L_2 - k\sqrt{L_1 \times L_2}\right)$$

$$L_T = \left(1\text{ H} - 0.75\sqrt{1\text{ H} \times 1\text{ H}}\right) + \left(1\text{ H} - 0.75\sqrt{1\text{ H} \times 1\text{ H}}\right)$$

$$L_T = (1\text{ H} - 0.75) + (1\text{ H} - 0.75)$$

$$L_T = 0.25 + 0.25$$

$$L_T = \mathbf{0.5\ H}$$

A *variometer* is an inductor that consists of two coils, one fixed and one movable. A variometer operates on the principle that the magnetic fields of two inductors can be adjusted to aid or to oppose each other. A variometer has an outer fixed coil connected in series with an inner movable coil. **See Figure 11-15.** The total inductance is at a maximum when the inner coil is in a position that puts its magnetic field in phase with the field of the outer coil. As the inner coil is rotated through 180°, the fields are in opposition and the overall inductance is at a minimum. The inside coil can be adjusted using a movable shaft so that the relative position of the two coils can be varied between the points of maximum aid and maximum opposition to the magnetic fields. When the coils are perpendicular to each other, the total impedance is equal to the sum of the inductance of the two coils. Variometers are often used to tune circuits for radio transmissions.

Terms

A **variometer** is an inductor that consists of two coils, one fixed and one movable.

Tech Tip

Portable variometers are used by operators of lightweight aircraft, such as hangliders and paragliders, to measure the angle at which an aircraft is ascending or descending.

VARIOMETERS

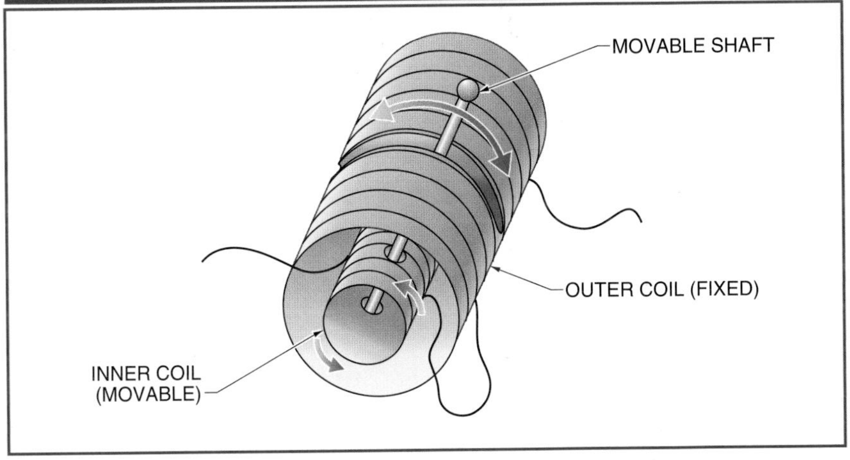

Figure 11-15. A variometer has an outer fixed coil connected in series with an inner movable coil.

Tech Tip

Inductors are connected in parallel in a DC circuit in order to reduce total inductance.

INDUCTORS CONNECTED IN PARALLEL

When inductors are connected in parallel in a circuit, the total circuit current increases. Each inductor individually opposes any change in current flow through its branch. The greatest amount of current flow is through the branch with the smallest value of inductance. When inductors are connected in parallel, total inductance of a circuit can be calculated using a method similar to that for calculating total resistance of a parallel circuit. If there is no mutual inductance between inductors, total inductance for inductors connected in parallel is calculated by applying the following formula:

$$L_T = \frac{1}{\frac{1}{L_1} + \frac{1}{L_2} + \frac{1}{L_3} + \ldots + \frac{1}{L_N}}$$

where

L_T = total inductance (in H)

L_1 = inductance 1 (in H)

L_2 = inductance 2 (in H)

L_3 = inductance 3 (in H)

L_N = inductance N (in H)

Example: What is the total inductance of three inductors connected in parallel with inductance values of 2 H, 4 H, and 1 H and no mutual inductance?

$$L_T = \frac{1}{\frac{1}{L_1} + \frac{1}{L_2} + \frac{1}{L_3} + \ldots + \frac{1}{L_N}}$$

$$L_T = \frac{1}{\frac{1}{2\text{ H}} + \frac{1}{4\text{ H}} + \frac{1}{1\text{ H}}}$$

$$L_T = \frac{1}{0.50\text{ H} + 0.25\text{ H} + 1\text{ H}}$$

L_T = **0.57 H**

When there are only two inductors in parallel, total inductance for inductors connected in parallel is calculated by applying the following formula:

$$L_T = \frac{L_1 \times L_2}{L_1 + L_2}$$

where

L_T = total inductance (in H)

L_1 = inductance 1 (in H)

L_2 = inductance 2 (in H)

Measuring Inductance

While electrical parameters such as voltage, current, resistance, and capacitance are easily measured with a variety of test instruments, inductance values have historically been calculated using mathematical formulas that rely on the dimensions of the coil and core and the number of coil turns. Most electrical test instruments do not have a function that measures inductance. To measure inductance accurately, a specialty meter such as an LCR (inductance-capacitance-resistance) meter or an LC (inductance-capacitance) meter is required. Inductance-capacitance meters are less costly than LCR meters and typically measure inductance with a resolution of 1 mH and accuracy of ±1%. LCR meters have more functions than LC meters. LCR meter types include hand-held, portable, and fixed. Typical measurement functions that are considered when using LCR meters include resistance, impedance measurement range and accuracy, capacitance measurement range and accuracy, and inductance measurement range and accuracy.

Example: What is the total inductance of two inductors connected in parallel with inductance values of 2 H and 4 H and no mutual inductance?

$$L_T = \frac{L_1 \times L_2}{L_1 + L_2}$$

$$L_T = \frac{2\text{ H} \times 4\text{ H}}{2\text{ H} + 4\text{ H}}$$

$$L_T = \frac{8\text{ H}}{6\text{ H}}$$

L_T = **1.33 H**

This calculation is used to reduce parallel inductive circuits, two inductors at a time, to obtain an equivalent inductance. Equivalent inductances are used until the final total inductance is calculated.

As with inductors connected in series, inductors connected in parallel may have aiding or opposing magnetic fields, which increases or decreases the total inductance. **See Figure 11-16.** When inductors have aiding mutual induction, total induction is calculated by applying the following formula:

$$L_T = \frac{1}{\left(\frac{1}{L_1 + M}\right) + \left(\frac{1}{L_2 + M}\right)}$$

where

L_T = total inductance (in H)

M = mutual inductance (in H)

L_1 = inductance 1 (in H)

L_2 = inductance 2 (in H)

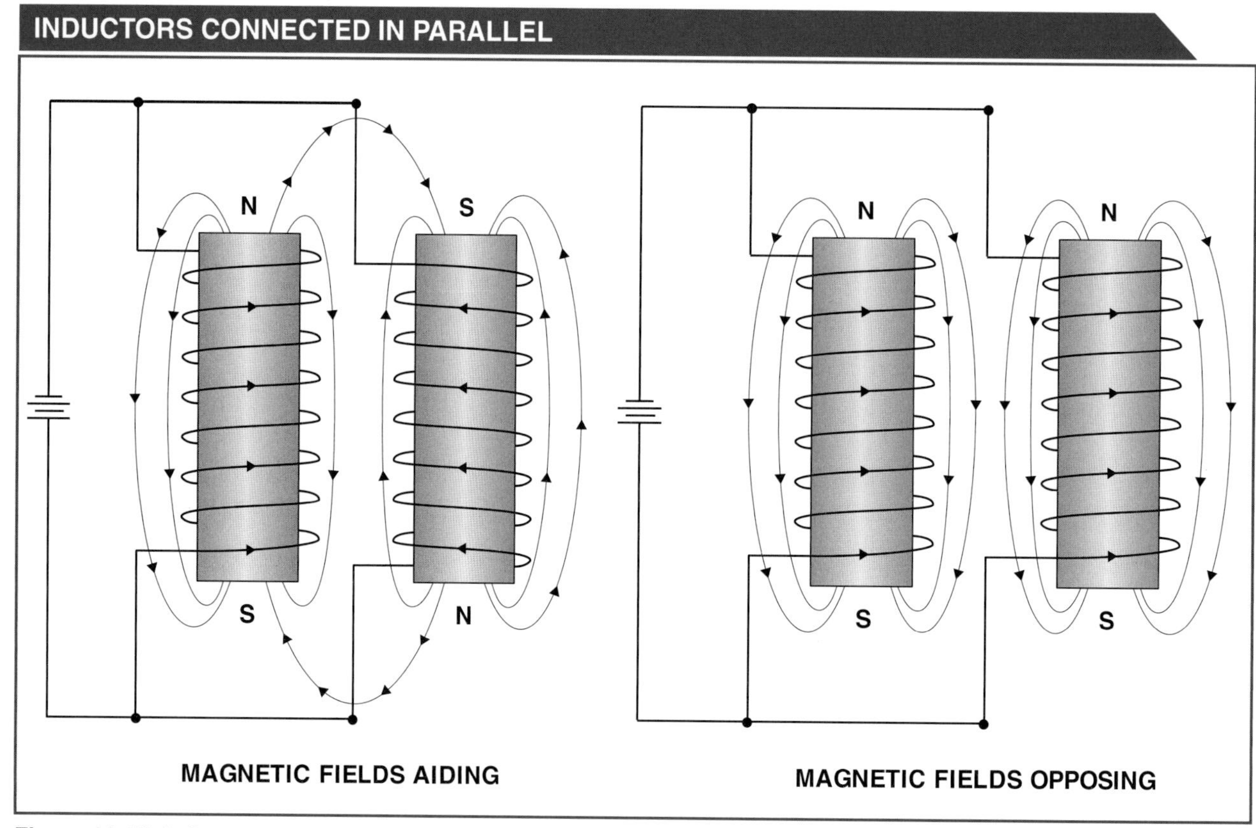

Figure 11-16. Inductors connected in parallel have magnetic fields that may aid or oppose one another.

Example: What is the total inductance of two inductors connected in parallel with inductance values of 0.1 H and 0.5 H, with mutual inductance 0.112 H?

$$L_T = \frac{1}{\left(\frac{1}{L_1 + M}\right) + \left(\frac{1}{L_2 + M}\right)}$$

$$L_T = \frac{1}{\left(\frac{1}{0.1\ \text{H} + 0.112\ \text{H}}\right) + \left(\frac{1}{0.5\ \text{H} + 0.112\ \text{H}}\right)}$$

$$L_T = \frac{1}{\frac{1}{0.212\ \text{H}} + \frac{1}{0.612\ \text{H}}}$$

$$L_T = \frac{1}{4.717 + 1.634}$$

$$L_T = \frac{1}{6.351}$$

L_T = **0.157 H**

When inductors have opposing mutual induction, total induction is calculated by applying the following formula:

$$L_T = \frac{1}{\left(\frac{1}{L_1 - M}\right) + \left(\frac{1}{L_2 - M}\right)}$$

where

L_T = total inductance (in H)

M = mutual inductance (in H)

L_1 = inductance 1 (in H)

L_2 = inductance 2 (in H)

Example: What is the total inductance of two inductors connected in parallel with inductance values of 2 H and 6 H, with opposing mutual inductance 0.35 H?

$$L_T = \frac{1}{\left(\frac{1}{L_1 - M}\right) + \left(\frac{1}{L_2 - M}\right)}$$

$$L_T = \frac{1}{\left(\frac{1}{2\text{ H} - 0.35\text{ H}}\right) + \left(\frac{1}{6\text{ H} - 0.35\text{ H}}\right)}$$

$$L_T = \frac{1}{\frac{1}{1.65\text{ H}} + \frac{1}{5.65\text{ H}}}$$

$$L_T = \frac{1}{0.606 + 0.177}$$

$$L_T = \frac{1}{0.783}$$

L_T = **1.277 H**

SUMMARY

Inductance is the property of a circuit that opposes any change in current. Inductors are devices intentionally connected in circuits to perform this function and are available in many shapes, forms, and sizes. The physical construction of an inductor often depends on the circuit application and requirements.

The henry is the unit of measurement for inductance and is defined as the voltage induced for a rate of change of current. Inductance time constants show how inductance and resistance in a circuit affect circuit current. Current rises at an exponential rate despite the time involved for the current to reach maximum or minimum values. Self-induced voltage (self-induction) exists only when the current is changing.

Energy is measured in joules and is stored in the magnetic field of the coil. Abruptly opening the circuit can cause this energy to be expended over a very short time, causing a high voltage to occur across the open contacts. This high voltage, referred to as inductive kick, can have considerable power that can be hazardous to personnel and equipment.

Mutual induction occurs when the magnetic field of one coil cuts through a second coil. Energy is transferred between coils through mutual induction. When mutual induction occurs, the magnetic fields of the coils may be additive or subtractive. When the magnetic fields aid each other, the total induction in the circuit increases. When the magnetic fields oppose each other, the total induction in the circuit decreases. When inductors are connected in series, their inductances add similarly to resistances in series. The total inductance of a circuit with inductors connected in parallel can be determined using the reciprocal formula of the formula used to find resistance for resistors connected in parallel.

APPLICATION: BOOST-CONVERTER SWITCHING-MODE POWER SUPPLY...

Background: Power supplies are commonly found in most electronic equipment such as televisions, radios, and computers. Power supplies are used to convert electrical power into a form that is usable for the application. They are used to convert AC power into DC power or convert DC power into a different DC voltage. DC-to-DC power supplies may be used to decrease voltage, invert voltage, or increase voltage.

Two common power supply designs are linear power supplies and switching-mode power supplies (SMPSs). SMPSs are used when higher efficiency is needed or a smaller size is required. Linear regulators are mainly used for circuit simplicity and because they generate little electrical noise that could affect other circuits.

In applications in which higher voltage is needed, a boost-converter SMPS can be used. A boost converter is also known as a step-up converter. A boost-converter SMPS changes a DC voltage into a higher DC voltage. Power is not created by an SMPS. So, if there is perfect efficiency (100%), the input power equals the output power. For example, if the output voltage for a specific application is increased four times, then the output current of that same application is a quarter of the input current.

...APPLICATION: BOOST-CONVERTER SWITCHING-MODE POWER SUPPLY...

The primary component used to increase the voltage in a boost-converter SMPS is an inductor. The inductor is charged by an input voltage and discharged to the load at a periodic rate. This switching process, once it reaches steady state, increases the input voltage to the output.

The basic boost-converter SMPS has an input voltage (V_{IN}), series inductor (L), ideal switch (S), and output voltage (V_{OUT}). The output voltage has a capacitor (C) and a load (R). **See Figure 1.** The ideal switch changes between position 1 and 2 at a period rate with a duty cycle. The duty cycle is a percentage of time (D) the switch is in position 1 and the remainder of time the switch is in position 2 ($1 - D$).

BOOST CONVERTER SWITCHING-MODE POWER SUPPLIES

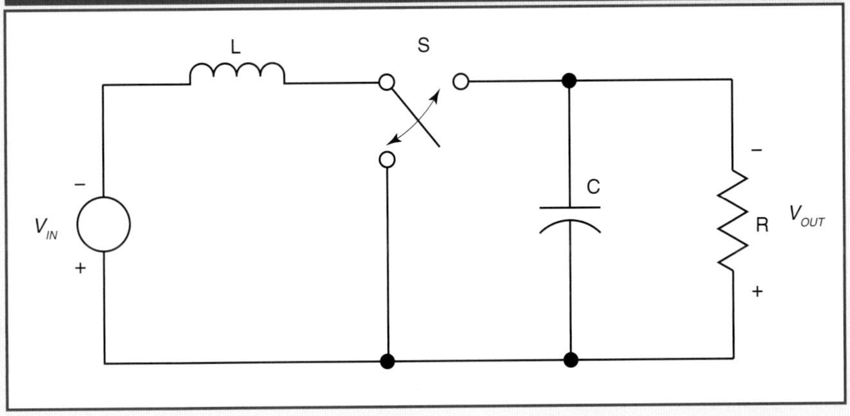

Figure 1. A basic boost-converter SMPS has an input voltage, series inductor, ideal switch, and output voltage. The output voltage has a capacitor and a load.

When the switch is in position 1, the inductor is charged by the input voltage. The inductor voltage (V_L) is equal to V_{IN}. The inductor voltage (V_L) is found by applying the following formula:

$$V_L = \frac{L_L di}{dt}$$

where
V_L = inductor voltage (in V)
L_L = inductor inductance (in H)
di = change in current (in A)
dt = change in time (in sec)

The inductor current increases at a rate of V_L divided by L_L for the time the switch is in position 1. **See Figure 2.** This continues until the switch is moved to position 2.

...APPLICATION: BOOST-CONVERTER SWITCHING-MODE POWER SUPPLY...

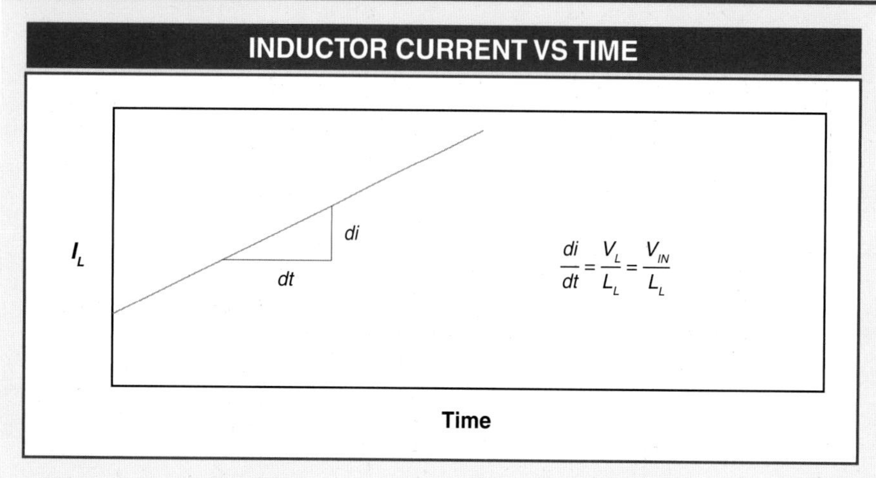

Figure 2. The inductor current increases at a rate of V_L divided by L_L for the time the switch is in position 1.

When the switch is in position 1, the output capacitor provides current to the load. A capacitor can be viewed as a battery that can be charged quickly but holds a limited charge. **See Figure 3.** The current provided by the capacitor (I_C) is equal to V_{OUT} divided by R.

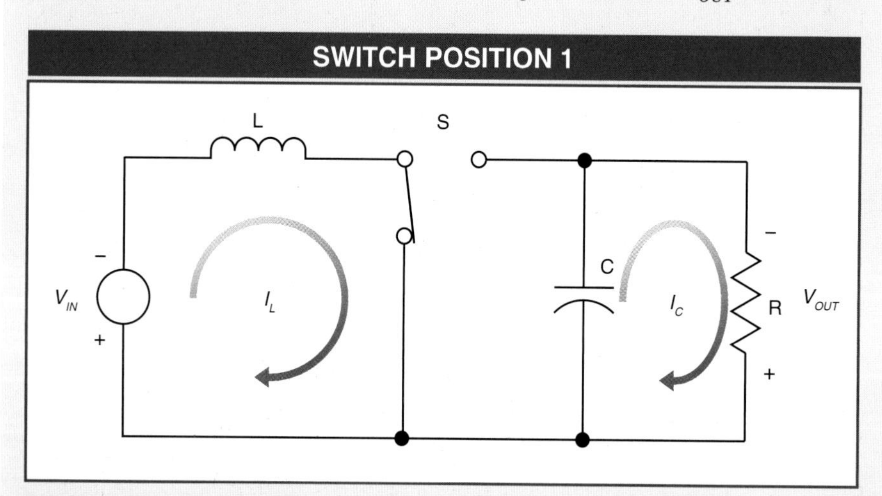

Figure 3. When the switch is in position 1, the output capacitor provides current to the load.

When the switch is in position 2, the inductor current is discharged to both the capacitor and load. **See Figure 4.** The inductor voltage is the difference between the input voltage and output voltage ($V_{IN} - V_{OUT}$). The inductor current (I_L) is found by applying the following formula:

$$V_{IN} - V_{OUT} = \frac{L_L di}{dt}$$

where
V_{IN} = input voltage (in V)
V_{OUT} = output voltage (in V)
L_L = inductor inductance (in H)
di = change in current (in A)
dt = change in time (in sec)

...APPLICATION: BOOST-CONVERTER SWITCHING-MODE POWER SUPPLY...

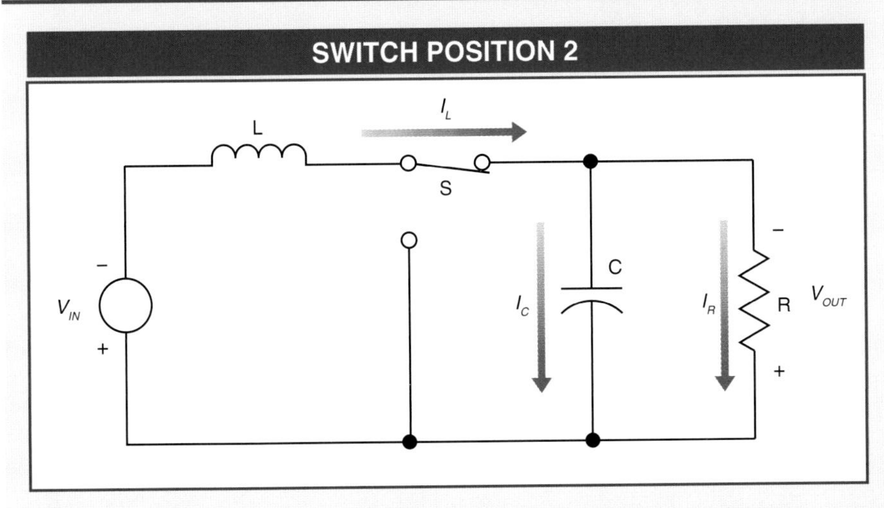

Figure 4. When the switch is in position 2, the inductor current is discharged to both the capacitor and load.

Therefore, the inductor current decreases at a rate of V_{IN} minus V_{OUT} divided by L_L [$(V_{IN} - V_{OUT}) \div L_L$] when the switch is in position 2. **See Figure 5.** While the switch is in position 2, the inductor provides current to both the load and capacitor. The current provided to the capacitor is found by using Kirchoff's current law (KCL) as follows:

$$0 = I_L - I_C - I_R$$

or

$$0 = \frac{I_L - I_C - V_{OUT}}{R}$$

or

$$I_C = \frac{I_L V_{OUT}}{R}$$

where
I_C = capacitor current (in A)
I_L = inductor current (in A)
I_R = resistor (load) current (in A)
V_{OUT} = output voltage (in V)
R = load resistance (in Ω)

...APPLICATION: BOOST-CONVERTER SWITCHING-MODE POWER SUPPLY...

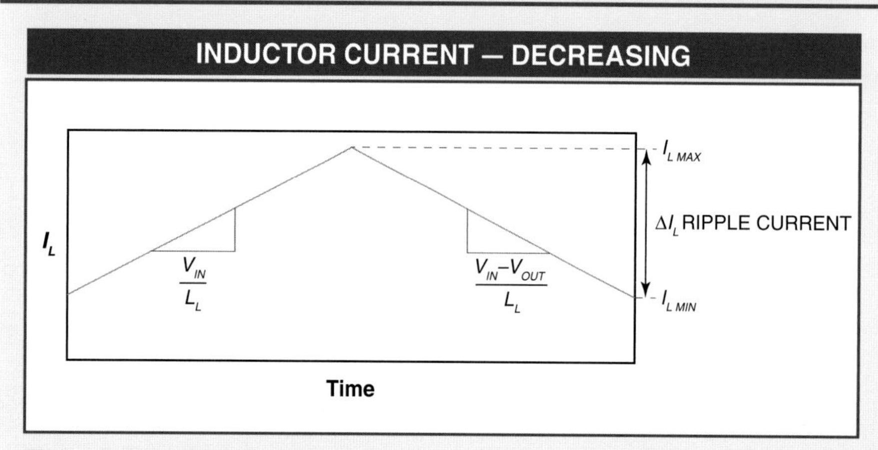

Figure 5. The inductor current decreases at a rate of V_{IN} minus V_{OUT} divided by L_L when the switch is in position 2.

A boost converter SMPS can operate in two different modes. One mode is continuous mode in which the inductor current never returns to zero. The switching rate is such that before the inductor current completely discharges (switch position 2), the inductor is recharged (switch position 1). The other mode is the discontinuous mode in which the inductor is fully discharged, and current returns to zero. In continuous mode, the boost-converter SMPS output voltage is only dependent on the duty cycle. In discontinuous mode, the output voltage is dependent on the inductor value as well as several other factors. This application only examines simplified boost-converter SMPS continuous-mode operation.

Inductors in a steady-state condition follow the volt-second balance condition in which the inductor average current does not change over time. If the volt-second balance condition is not met, then the current continues to either increase or decrease. Using the formulas and the volt-second principle, the relationship between voltage input and output can be derived.

$$V_{L\text{-}AVG} = V_{L1} + V_{L2}$$

Therefore,

$$V_{L\text{-}AVG} = V_{IN} \times D + (V_{IN} - V_{OUT}) \times (1 - D)$$

Therefore,

$$0 = V_{IN} \times D + (V_{IN} - V_{OUT}) \times (1 - D)$$

Therefore,

$$V_{OUT} = \frac{V_{IN}}{(1-D)}$$

where

$V_{L\text{-}AVG}$ = inductor average voltage (in V)
V_{L1} = inductor voltage when switch is in position 1 (in V)
V_{L2} = inductor voltage when switch is in position 2 (in V)
V_{IN} = input voltage (in V)
V_{OUT} = output voltage (in V)
D = percentage of time switch is in position 1 (in %)

...APPLICATION: BOOST-CONVERTER SWITCHING-MODE POWER SUPPLY...

D can be varied to increase the output voltage. However, the realistic practical limit is a maximum output voltage increase of about five times or an 80% duty cycle (*D*). **See Figure 6.**

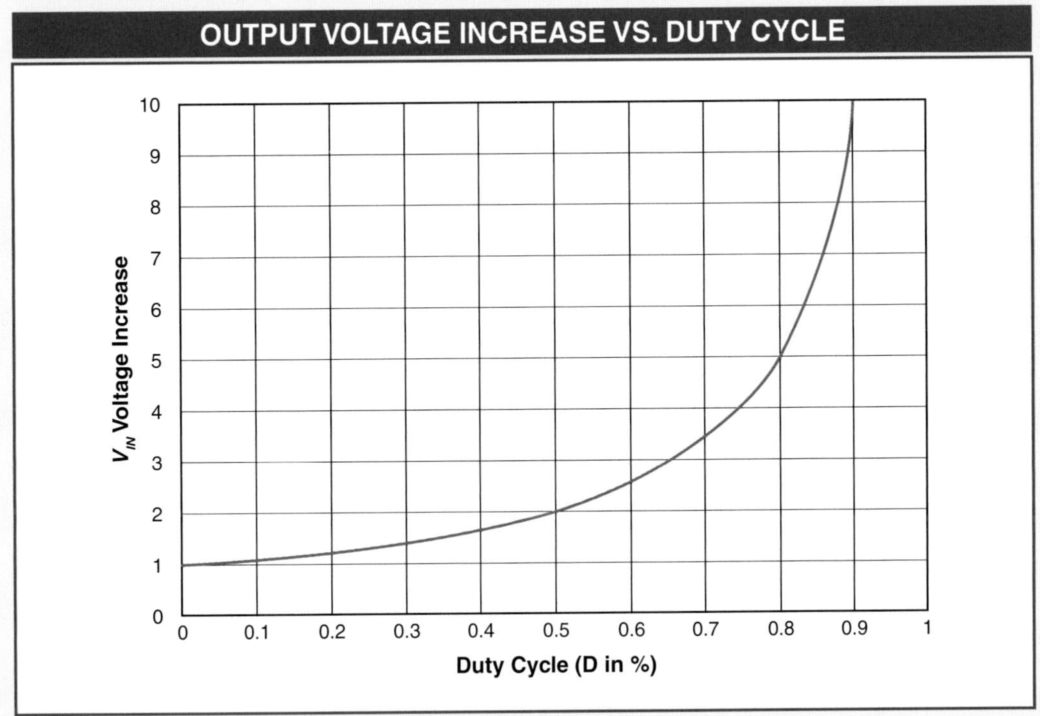

Figure 6. The realistic practical limit of maximum output voltage is an increase of about five times or an 80% duty cycle (D).

Based on the formulas, the inductor value does not affect the output voltage since the analysis is for continuous mode operation. To find the inductor value, an estimate of the inductor ripple current is needed. The inductor ripple current is the difference between minimum inductor current and maximum inductor current. The maximum inductor current is reached when the boost-converter SMPS switches from switch position 1 to 2. The minimum inductor current is reached when the boost-converter SMPS switches from switch position 2 to 1.

Inductor ripple current should be about 20% to 40% of the output current. If the output current is 2 amps, then the desired inductor ripple current is 0.6 amps.

Once the inductor ripple current is determined, the inductor value can be calculated using the basic inductor formula that relates inductor value, voltage, and change in current. The inductor ripple current is the change in current (*di*) and the charging or discharging change in time (*dt*). If the charging cycle is used, then the value *D* is used for the duty cycle. If the discharging cycle is used, then $1 - D$ is used for the duty cycle.

...APPLICATION: BOOST-CONVERTER SWITCHING-MODE POWER SUPPLY...

For charging cycle:

$$L_L = \frac{V_L \times \left(\frac{D}{f_s}\right)}{\Delta I_L}$$

For discharging cycle:

$$L_L = \frac{V_L \times \left(\frac{1-D}{f_s}\right)}{\Delta I_L}$$

where
L_L = inductor inductance (in H)
V_L = inductor voltage (in V)
D = switch duty cycle charge/discharge (in %)
f_S = switch frequency (in Hz)
ΔI_L = inductor ripple current (in A)

Key Points: DC inductor steady-state operation

Problem: A new Bluetooth® hands-free circuit is being designed for a car manufacturer and requires 24 VDC at 1 A. The available voltage from the car battery is only 12 VDC. An SMPS needs to be designed that operates at 100,000 Hz, uses the car battery input voltage, and provides the needed voltage and current for the Bluetooth® hands-free circuit.

Solution: Since the design requires a voltage increase, a boost-converter SMPS will be designed. The parameters for the boost-converter SMPS are the following:

V_{IN} = 12 VDC
I_{IN} = ___ A
V_{OUT} = 24 VDC
I_{OUT} = 1 A
f_S = 100,000 Hz
D = ___%
L_L = ___ H
ΔI_L = ___ A

Since the application is a boost converter SMPS with no losses, power is conserved between the input and output. Since the output and input power are equal, the input current (I_{IN}) can be calculated as follows:

$$P_{IN} = P_{OUT}$$

...APPLICATION: BOOST-CONVERTER SWITCHING-MODE POWER SUPPLY...

Therefore,

$$V_{IN} \times I_{IN} = V_{OUT} \times I_{OUT}$$
$$12\text{ VDC} \times I_{IN} = 24\text{ VDC} \times 1\text{ A}$$
$$I_{IN} = \frac{(24\text{ VDC} \times 1\text{ A})}{12\text{ VDC}}$$
$$I_{IN} = 2\text{ A}$$

The input and output voltages have been provided along with the switching frequency. From these values, the switch position 1 duty cycle can be determined as follows:

$$V_{OUT} = \frac{V_{IN}}{(1-D)}$$
$$24\text{ VDC} = \frac{12\text{ VDC}}{(1-D)}$$
$$1-D = \frac{12\text{ VDC}}{24\text{ VDC}}$$
$$D = 1-\frac{12\text{ VDC}}{24\text{ VDC}}$$
$$D = 1-0.5$$
$$D = 0.5\text{ or }50\%$$

To find the inductor value, the inductor ripple current is calculated. The inductor ripple current (ΔI_L) should be 20% to 40% of the output current of 1 amp. So, the inductor ripple current is 0.3 A. The inductor value is calculated during the charge cycle when V_L is equal to V_{IN} of 12 VDC. So, the inductor value is the following:

$$L_L = V_L \times \frac{\left(\frac{D}{f_s}\right)}{\Delta I_L}$$
$$L_L = 12\text{ VDC} \times \frac{\left(\frac{0.5}{100{,}000\text{ Hz}}\right)}{0.3\text{ A}}$$
$$L_L = 200\ \mu\text{H}$$

...APPLICATION: BOOST-CONVERTER SWITCHING-MODE POWER SUPPLY

The calculation for the inductor charge cycle can also be used with the discharge cycle. The inductor voltage during the discharge cycle is V_{IN} minus V_{OUT}, which is equal to 12 VDC. Therefore, the inductor value is the following:

$$L_L = V_L \times \frac{\left[\frac{(1-D)}{f_s}\right]}{\Delta I_L}$$

$$L_L = 12\text{ VDC} \times \frac{\left(\frac{0.5}{100{,}000\text{ Hz}}\right)}{0.3\text{ A}}$$

$$L_L = 200\ \mu\text{H}$$

In this application, a boost-converter SMPS was designed to change 12 VDC at 2 A into 24 VDC at 1 A. The boost-converter SMPS uses a 200 μH inductor that is switched using a 50% duty cycle at 100,000 Hz switch frequency. SMPSs have a higher level of complexity compared to linear regulators, but SMPSs consume less power and are generally smaller in size, making them idea for lightweight applications such as portable electronics. **See Figure 7.**

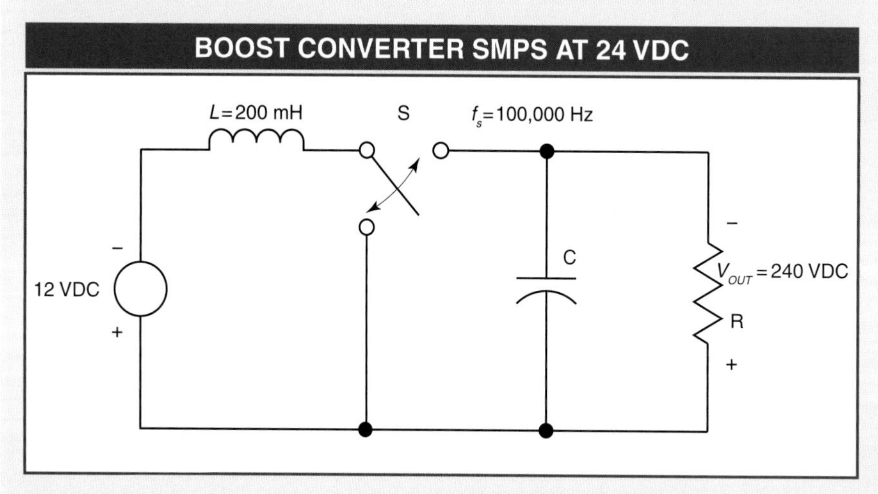

Figure 7. In this application, a boost-converter SMPS was designed to change 12 VDC at 2 A into 24 VDC at 1 A.

Chapter Review

1. What is mutual inductance?
2. Why is counter-electromotive force sometimes referred to as back-electromotive force?
3. How can permeability be expressed as a ratio?
4. What length of coil wire is required for a solenoid of 150 turns with a radius of 0.375″?
5. How can inductor values be quickly located?
6. List five DC inductor ratings.
7. What is the total inductance value of three inductors connected in parallel with inductance values of 3 H, 3 H, and 2 H and no mutual inductance?
8. What is the induced voltage for an inductor of 75 μH with a current change of 60 mA in 5 μs (microseconds)?
9. What tolerances do inductors used in electrical and electronic circuits typically have?
10. How does connecting inductors in parallel affect total circuit current?
11. What is the mutual inductance of two coils, each with an inductance of 4 H, with a coefficient of coupling of 65%?
12. List four types of inductors.
13. Explain why a large-diameter core has more inductance than a small-diameter core.
14. What principle does a variometer operate on?
15. What is a noninductive coil inductor?
16. What is another term commonly associated with single-layer coils?
17. What is inductance?
18. What is the total inductance value of two inductors connected in parallel with inductance values of 3 H and 5 H and no mutual inductance?
19. What circuit knowledge is required to understand a practical inductive circuit?
20. How is a pancake coil different from a standard coil?
21. What is the time constant for a circuit with an inductance of 20 H and a resistance of 15 Ω?
22. List two common applications for inductors with values that range from 15 H to 30 H.
23. What is the standard length of the coil used in construction of inductors used in electrical and electronic equipment?
24. What type of application are molded inductors typically used in?
25. List five factors that affect inductance.

12

DC Circuit Capacitance

OBJECTIVES

- Explain capacitor construction, symbols, and principles of operation.
- Explain how capacitors are rated.
- List factors affecting DC capacitance.
- List capacitor types and classifications.
- Describe capacitor specifications and identification codes for the different types of capacitors.
- Explain the resistive-capacitive time constant and how stored energy can be present in a capacitor.
- Explain DC capacitors connected in series, parallel, and series-parallel.
- Describe capacitor losses.
- Explain how DC capacitors are tested.

All electrical circuits, no matter how complex, rely on three basic properties: resistance, inductance, and capacitance. Resistance is the property that opposes the flow of current, and inductance is the characteristic that opposes changes in current. Capacitance is the ability of a component or circuit to store energy in the form of an electrical charge. Capacitors are as common as resistors and inductors in electrical and electronic circuits. Capacitors are used in ignition systems, fluorescent lamp starters, motors, radio and television circuits, sensors, filters, power factor correction circuits, and control circuits of all types.

Terms

A **capacitor** is an electrical device designed to store electrical energy by means of an electrostatic field.

Capacitance (C) is the ability of a component or circuit to store energy in the form of an electrical charge.

Plates are the conductors in a capacitor.

Dielectric is a nonconductor of DC current. Dielectric material is used as the electric insulation between the plates of a capacitor.

CAPACITOR CONSTRUCTION AND SYMBOLS

A *capacitor* is an electrical device designed to store electrical energy by means of an electrostatic field. *Capacitance (C)* is the ability of a component or circuit to store energy in the form of an electrical charge.

A capacitor is constructed of two parallel conductors separated by an electrical insulator. **See Figure 12-1.** *Plates* are the conductors in a capacitor. *Dielectric* is a nonconductor of DC current. Dielectric material is used as the electric insulation between the plates of a capacitor. Several different symbols are used in schematic diagrams to represent the various types of capacitors. **See Appendix.**

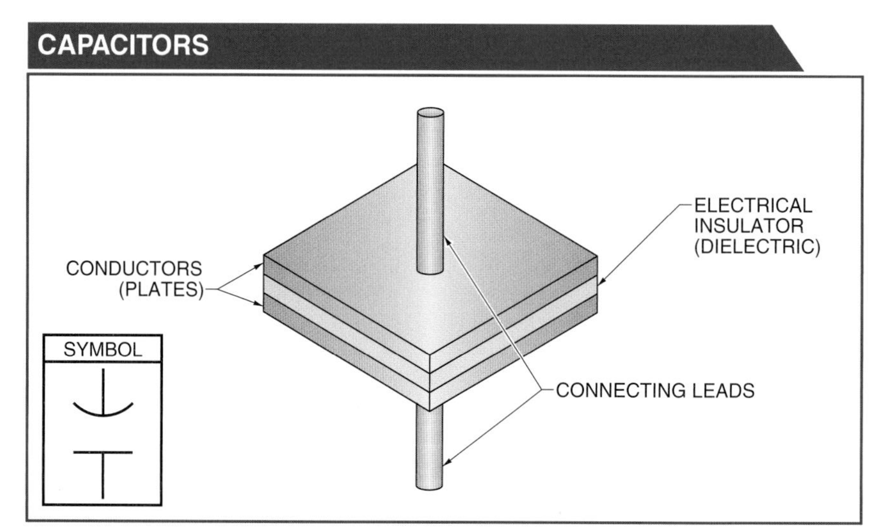

Figure 12-1. A capacitor is constructed using two conductors separated by an electrical insulator (dielectric).

A polarized capacitor is a capacitor that has separate negative and positive connection leads. A nonpolarized capacitor has connection leads that can be either positive or negative. Observing the polarity of the voltage is not necessary when connecting a nonpolarized capacitor in a circuit. However, polarity must be observed with a polarized capacitor, because mismatched polarities on the connection leads of a polarized capacitor will damage the capacitor and cause an explosion, potentially harming anyone within the vicinity of the capacitor.

The end of the capacitor, represented by the curved portion of the symbol, is typically connected to the lower-potential connection lead.

This connection lead is often at ground potential or connected to the grounded chassis of the device. The lead indicated by the curved portion is connected to the conductor inside the capacitor. On a capacitor, this conductor is indicated by a ring or mark on that end. This practice helps to prevent short circuits due to insulation damage. By having the higher voltage on the inside conductor, the dielectric and the outer insulating material of the capacitor must both be breached for a short circuit to occur. If the higher voltage is on the outside plate, only the insulation of the capacitor case is damaged.

A *variable capacitor* is a capacitor that varies in capacitance value. Variable capacitors can also be designed to change values by mechanical means or by varying the voltage across certain types of capacitors.

Terms

A **variable capacitor** is a capacitor that varies in capacitance value.

Leakage current is the small amount of current that flows through insulation.

Displacement current is current that exists in addition to normal current.

PRINCIPLES OF CAPACITOR OPERATION

Capacitors operate on a few basic principles of operation that determine current flow, voltage, and resistance. For example, a DC circuit with three separate switch positions can be used to demonstrate capacitor operation principles. **See Figure 12-2.** With the switch in position 2, there is no current flow in the circuit. The capacitive voltage is 0 V. If the switch is changed to position 1, the power source is connected through R_1 to the capacitor. Electrons flow from the negative terminal of the power source and collect on the plate of the capacitor, giving that plate a net negative charge. The electrons are prevented from flowing through the capacitor due to the dielectric between the plates.

Because of the force of the electrostatic field, an equal number of electrons leave the other capacitor plate and flow back to the positive terminal of the power source. This leaves the plate with a deficiency of electrons and a positive potential. The charges on the plates develop voltage. Current continues to flow through R_1 until capacitive voltage equals the source voltage. At this point, current cannot flow in either direction. The dielectric between the plates prevents current flow.

To discharge the capacitor, the switch is moved to position 3. This puts R_2 across the charged capacitor. Current now flows from the negative terminal of the capacitor, through R_2, and back to the positive terminal of the capacitor. A DMM set to measure current shows maximum current when the switch is first closed, decreasing exponentially as capacitive voltage decreases.

Leakage current is the small amount of current that flows through insulation. Leakage current is typically considered undesirable because it flows through the dielectric. *Displacement current* is current that exists in addition to normal current. Displacement current is proportional to the rate of change of the electric field in the capacitor.

Tech Tip

Capacitance and the capacitor were discovered independently and concurrently by Prussian scientist Ewald Georg von Kleist and Dutch physicist Peiter van Musschenbroek in 1745 while both were studying the effects of electrostatic fields.

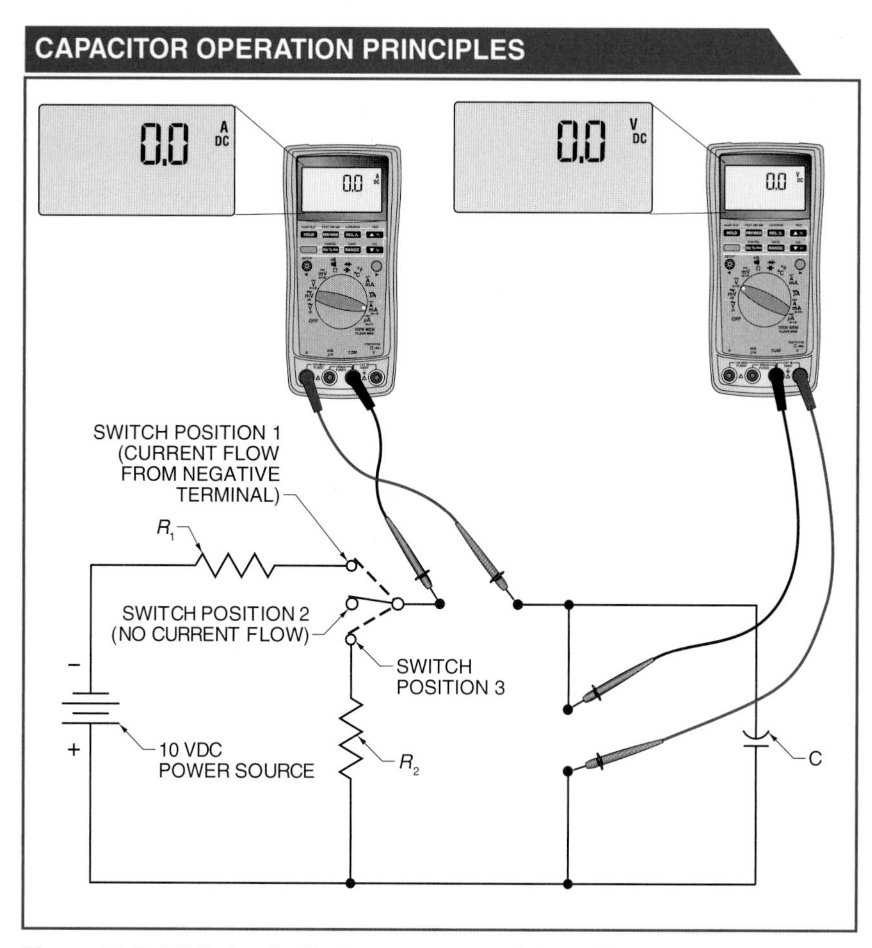

Figure 12-2. A DC circuit with three separate switch positions can be used to demonstrate capacitor operation principles.

In a capacitive circuit with DC voltage applied, current flow ceases when the capacitive voltage equals the value of the source voltage. In a resistive circuit with DC voltage applied, current continues to flow even though the resistive voltage is equal to the source voltage. In both cases, Kirchhoff's law (the algebraic sum of all voltages in a closed loop circuit is equal to 0) is satisfied. The dielectric prevents electrons from flowing between the plates of the capacitor. The capacitor plates are much larger than the cross-sectional area of the connecting conductors. This allows room for the electrons to disperse over the surface of the plates. Because of the larger area of the plates, an abundance of free electrons are on both plates when no electrical force is applied to it.

When each of the plates is the same size and no electrical force is present, the number of free electrons on each is nearly equal. When electrical force is applied, electrons begin to collect on the plate connected to the negative terminal of the source. An electrostatic field passes through the dielectric. The accumulation of electrons on the negatively charged plate repels an equal number of electrons from the positively charged plate. Electrons from the positively charged plate of the capacitor are simultaneously attracted back to the positive terminal of the power source. With a charged capacitor, the electron orbits become stretched toward the positively charged plate. **See Figure 12-3.**

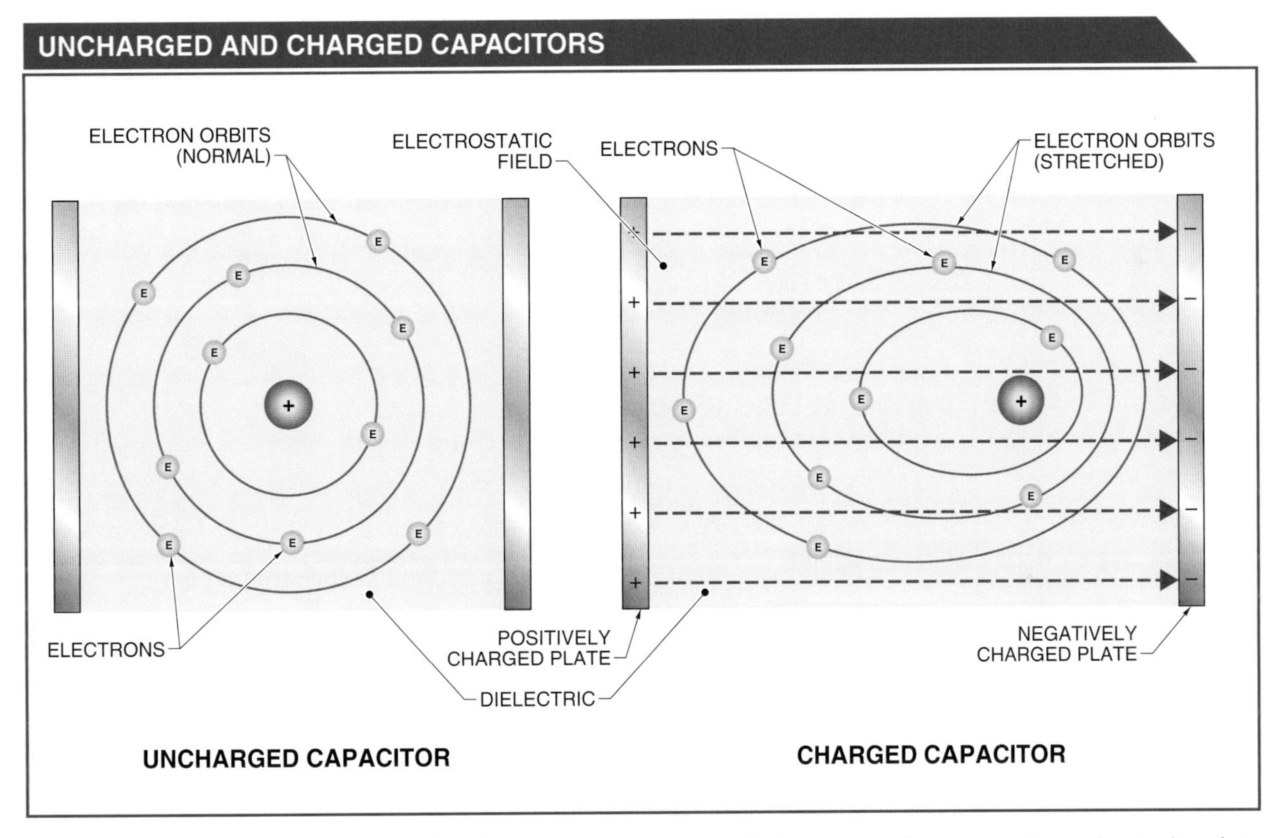

Figure 12-3. With an unchanged capacitor, electron orbits are in a perfect circular motion. With a charged capacitor, the electron orbits become stretched toward the positively charged plate.

Polarization of the molecules in the dielectric also has the effect of moving the charges on the plates closer together. Therefore, the capacitance is increased by the action of the molecules in the dielectric. Energy is required to distort the electron orbits. This energy is transferred from the electrostatic field to the electrons. Since energy cannot be destroyed, the energy required to elongate the electron orbits can be recovered when the electrons return to their normal orbits.

Terms

A **coulomb (C)** is the quantity of electric charge that passes any cross-section of a conductor in one second when the current is maintained at 1 A.

A **farad (F)** is equal to the capacitance of the accumulated charge of one coulomb for each volt applied.

Dielectric breakdown is the breakdown of insulation between the plates of a capacitor.

CAPACITOR RATINGS

Capacitors are rated according to the amount of capacitance and their voltage-handling capability. The amount of capacitance in a circuit is defined as the ratio of charge to volts. A *coulomb (C)* is the quantity of electric charge that passes any cross-section of a conductor in one second when the current is maintained at 1 A. A *farad (F)* is equal to the capacitance of the accumulated charge of one coulomb for each volt applied. A farad is a large unit that is too large to be used for an average capacitor application. However, modern techniques allow for the production of high-capacitance capacitors with low-voltage ratings that have ratings in farads. Most capacitors have capacitance units in either microfarads (μF) or picofarads (pF). The picofarad is typically used in place of the micro-microfarad (μμF), although the micro-microfarad may be found in older schematic diagrams or on older components still in use. Capacitance is calculated by applying the following formula:

$$C = \frac{Q}{V}$$

where

C = circuit capacitance (in F)

Q = circuit charge (in C)

V = circuit voltage (in V)

Example: What is the capacitance for 1 C of charge for 1 V?

$$C = \frac{Q}{V}$$

$$C = \frac{1\text{ C}}{1\text{ V}}$$

$$C = \mathbf{1\ F}$$

If a capacitor charges at a rate of 1 V each second with a charging current of 1 A, then the capacitor has a capacitance of 1 F.

Dielectric breakdown is the breakdown of insulation between the plates of a capacitor. Dielectric breakdown is the result of the orbital electrons in the dielectric becoming free current-carrying electrons. If too much capacitive voltage is present, the insulating material between the plates may fail, causing dielectric breakdown.

When dielectric breakdown occurs, the capacitor either arcs over (electric arc is present between the contacts) or short-circuits, causing high leakage current to develop. Damage caused by dielectric breakdown is permanent for most capacitors, and damaged capacitors must be replaced. Capacitors with gaseous or liquid dielectrics are an exception. Dielectric breakdown that occurs through air or oil insulation is temporary. These materials recover their dielectric strength once voltage is removed from the circuit.

Capacitor short-circuits can result in damage to other parts of the circuit. To prevent additional problems, care should be taken to ensure that the proper voltage rating is used when replacing capacitors in a circuit. For example, a nonpolarized capacitor rated at 120 VDC is not safe to use in a 120 VAC circuit. Alternating current voltage has peak values of 170 V, exceeding the 120 V rating of the capacitor.

Working voltage is the maximum voltage that can be safely applied to a capacitor. Typical working voltage values range from 1 V to 2000 V. *Dielectric strength* is the ability of a dielectric material to withstand voltage applied across it during a breakdown. Different materials have different dielectric strengths. **See Figure 12-4.** Dielectric strength is expressed in volts per unit of dielectric thickness. Dielectric strength depends on the thickness of the dielectric. The thicker the dielectric, the greater the voltage it can withstand.

Terms

Working voltage is the maximum voltage that can be safely applied to a capacitor.

Dielectric strength is the ability of a dielectric material to withstand voltage applied across it during a breakdown.

DIELECTRIC STRENGTH	
Material	**Dielectric Strength***
Air	20 to 80
HDPE	500
Castor oil	370 to 630
Formica®	450
Glass	200 to 250
Hard rubber	450
Lucite	480 to 500
Mica	2000
Paper (beeswax)	1800
Paper (paraffin)	1200 to 1250
Porcelain	40 to 750

* in V/mil

Figure 12-4. Different materials have different dielectric strengths.

FACTORS AFFECTING CAPACITANCE

The three factors that affect the capacitance of a circuit are the surface area of the conductive plates, thickness of the dielectric material, and the dielectric constant of the dielectric material. **See Figure 12-5.** Although the farad is defined in terms of the charge and the voltage, these factors do not change capacitance unless the voltage rating of the capacitor is exceeded. An increase in voltage results in a proportionally larger charge.

FACTORS AFFECTING CAPACITANCE

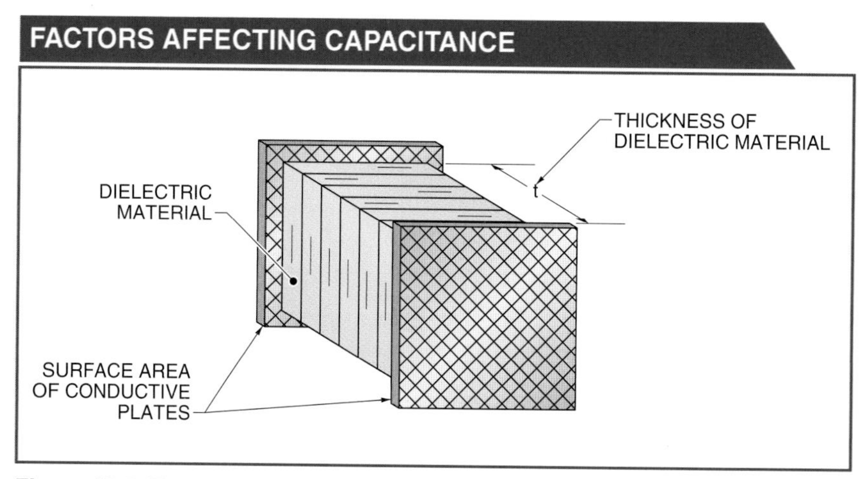

Figure 12-5. The three factors that affect the capacitance of a circuit are the surface area of the conductive plates, the thickness of the dielectric material, and the dielectric constant of the dielectric material.

Surface Area of Capacitor Plates

The surface area of capacitor plates affects total capacitance of a capacitor. When two capacitors have equally thick dielectric made of the same type of material, capacitance can be changed by changing the size of the capacitor plates. Capacitance is directly proportional to the size of the capacitor plates. The larger the capacitor plates, the more charge they can hold.

As capacitor surface area increases, capacitance value increases proportionally. **See Figure 12-6.** Doubling the plate dimensions results in four times the capacitance, and quadrupling the dimensions results in sixteen times the capacitance. For example, if three capacitors have the same dielectric and amount of space between the capacitor plates and a 1″ square capacitor has 1 μF of capacitance, then a 2″ square capacitor has 4 μF, and a 4″ square capacitor has 16 μF.

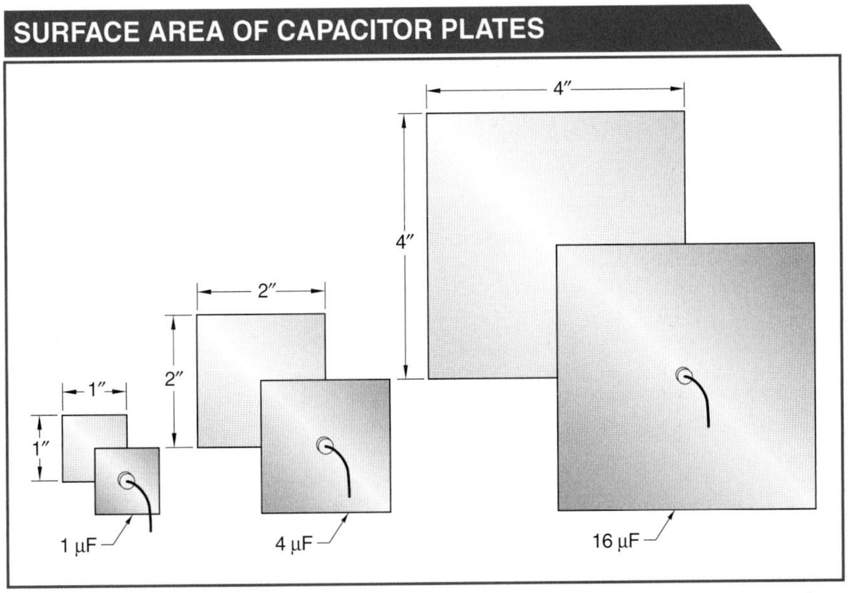

SURFACE AREA OF CAPACITOR PLATES

Figure 12-6. As capacitor surface area is increased, the value of capacitance is increased proportionally.

Thickness of Dielectric

The thickness of dielectric of a capacitor affects total capacitance. As a capacitor is electrified, electrons are repelled from the positive plate and accumulate on the negative plate. When more electrons are forced onto the negative plate, an equal number are repelled from the positive plate. As the positive plate becomes more positive, it attracts electrons to the negative plate. As the charge between the plates increases, so does the strength of the electrostatic field. Charges on each plate also affect the electrons in the dielectric. Electrons orbits are elongated away from the negative plate towards the positive plate. Reducing the thickness of the dielectric brings the capacitor plates closer together and increases capacitance. There are limitations on how close the capacitor plates can be brought together. The limitations on the distance between the capacitor plates determine the voltage rating of the capacitor. If the capacitor plates are too close together, electrons may arc across the gap, causing degradation of the dielectric and damage to the capacitor.

Degradation of the dielectric depends on the level of the voltage, distance between the plates, and the type of dielectric material. Any material will eventually degrade if a high enough voltage is applied. Low voltage ratings allow for less separation between the plates and a higher capacitance rating. Capacitors are available with ratings of several hundred volts with separation as small as 0.0005″ between the plates.

Doubling the capacitor plate area while moving the capacitor plates twice the distance apart counteracts the effect of the doubled capacitor plate area. **See Figure 12-7.** For example, capacitor A has a plate area of 8 cm^2 and capacitor B has a plate area of 16 cm^2. If both capacitors had equal distance between the plates, then capacitor B would have twice the capacitance of capacitor A. However, the doubled distance between plates negates the doubled plate area of capacitor B, so the two capacitors have equal capacitance.

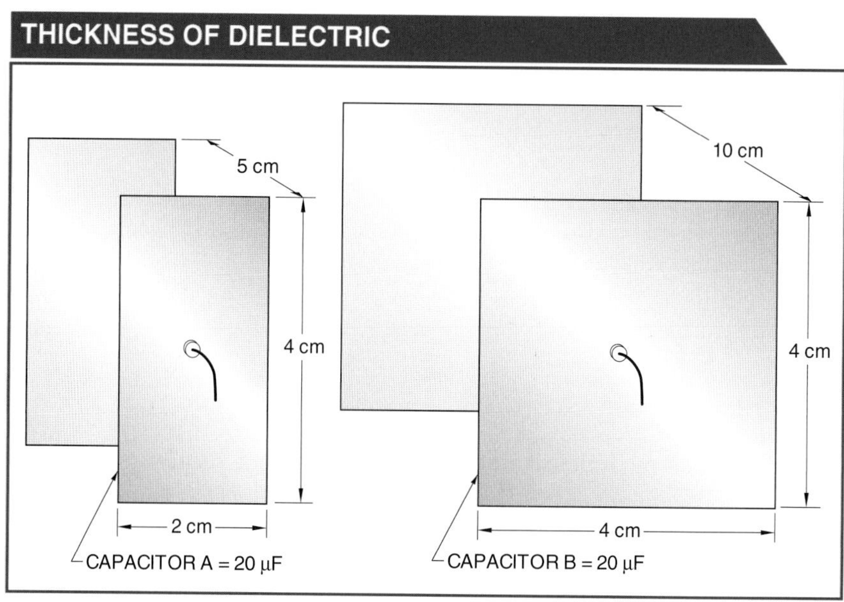

Figure 12-7. Doubling the capacitor plate size while moving the capacitor plates twice the distance apart results in unchanged capacitance.

Dielectric Constant

Terms

Dielectric constant (k) is the ratio of the capacitance of a capacitor with a given dielectric to the capacitance of a capacitor with air (vacuum) as the dielectric.

A *dielectric constant* (k) is the ratio of the capacitance of a capacitor with a given dielectric to the capacitance of a capacitor with air (vacuum) as the dielectric. The dielectric constant is the ability of a substance to convey the influence of an electric field. Dielectric constant is also referred to as specific inductive capacity or capacitivity. A dielectric constant is a rating of materials used as a dielectric in a capacitor.

The dielectric constant of a material is the ratio of the capacitance of that material to the capacitance of air (vacuum), which has a dielectric constant of 1. Cellulose, for example, has a dielectric constant of 7. Therefore, cellulose increases capacitance by seven times that of air. The average dielectric constants for dielectric materials used for capacitors differ based on the selected material. **See Figure 12-8.**

DIELECTRIC CONSTANT RATING	
Material	**Dielectric Constant***
Vacuum	1.0000
Air	1.0006
Paraffin	1.9 to 2.2
Teflon® (plastic)	2.0
Lucite®	2.4 to 3.0
Resin	2.5
Rubber	2.5 to 7.0
Paper	2.5 to 3.5
Oil	4.3 to 4.7
Glass	4.2 to 6.0
Polycarbonate	4.4
Mica	5.0 to 9.0
Ceramic	5.5 to 6.0
Cellulose	7.0
Mycalex®	8.0
Titanium	90 to 170
Barium-strontium titanate ceramic	7500

* at 1 kHz

Figure 12-8. The dielectric constants of capacitors differ based on the selected dielectric material.

The capacitance of parallel plates is significantly dependent on the type of insulating material used for the dielectric. Insulation is classified as either solid or fluid. Solid insulation can be further classified as either rigid or flexible. Different types of material have various levels of opposition to the passage of electrostatic fields. When the dielectric constant is known, capacitance is calculated by applying the following formula:

$$C = \frac{0.2249 \times k \times A}{t}$$

where

C = capacitance (in pF)

A = surface area of plates (in in.^{2})

t = thickness of dielectric (in in.)

k = dielectric constant

0.2249 = MKS to English conversion constant (in pF/in.)

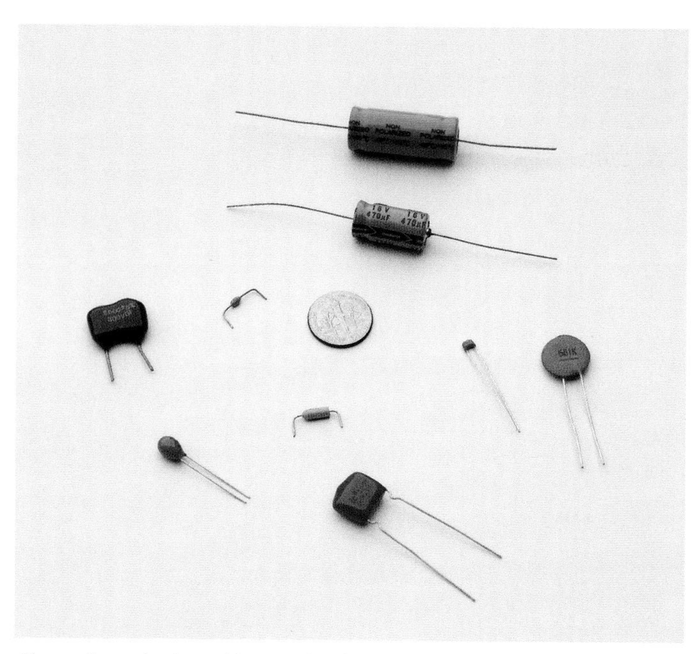

Capacitors designed for use in electronic applications are available in different shapes and sizes.

Example: What is the capacitance of a paper capacitor that has a dielectric constant of 3.0 with plates that are 2″ by 2″ and separated by 0.010″?

$$C = \frac{0.2249 \times k \times A}{t}$$

$$C = \frac{0.2249 \times 3.0 \times (2'' \times 2'')}{0.010''}$$

C = **270 pF**

Dielectric constant can vary considerably because of the capacitor manufacturing process. Dielectric constant should not be confused with dielectric strength. A requirement of high working voltage often precludes the use of an insulator with a high dielectric constant to increase the capacitance. Capacitor size typically increases as capacitance and voltage requirements increase.

CAPACITOR TYPES AND CLASSIFICATIONS

Capacitors are available in two types and with several different classifications. The two types of capacitors are fixed and variable. One method of classifying capacitors is by the type of dielectric used in their structure. Other classification methods may involve the structure of the element or capacitor polarization. Typical fixed capacitor types are paper, mica, ceramic, oil, or aluminum electrolytic capacitors. To keep the size of fixed capacitors as small as possible, special techniques are used in their design and manufacture, including rolling a series of plates separated by a dielectric. Usually, variable capacitors use either air or mica as the dielectric. Supercapacitors use thin metal film technology (TMF) and have more capacitance per unit area than conventional capacitors.

Paper Capacitors

A *paper capacitor* is a capacitor that has thin sheets of metallic foil, which act as plates, separated by waxed paper, which acts as the dielectric. The leads are connected to the foil. **See Figure 12-9.** The paper and foil are cut in long strips and alternated so that the dielectric is between the conductive materials. Typically, two or three strips of waxed paper are layered so that any small holes in the individual strips of waxed paper do not result in dielectric breakdown and cause the foil plates to short-circuit. The capacitor assembly is tightly wrapped around an insulated core and enclosed with a cardboard cover. It is then waxed or dipped in a paraffin coating to electrically insulate the capacitor. This process also prevents moisture, which causes increased leakage current and corrosion, from entering the capacitor assembly.

Terms

A **paper capacitor** is a capacitor that has thin sheets of metallic foil, which act as plates, separated by waxed paper, which acts as the dielectric.

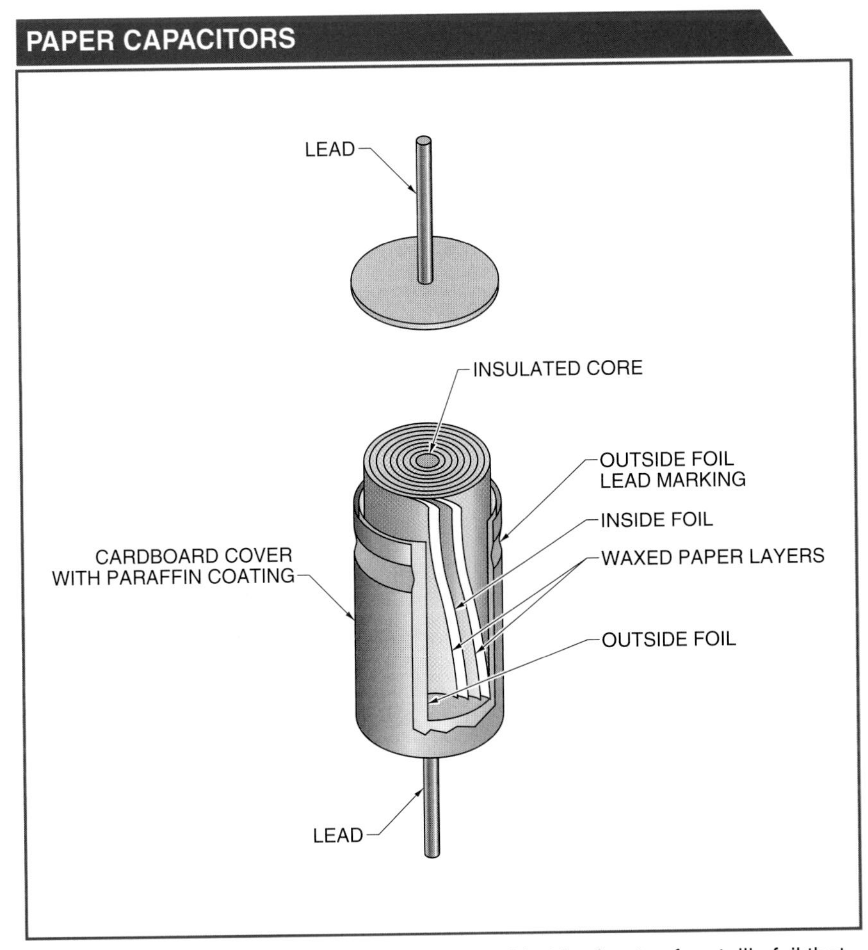

Figure 12-9. A paper capacitor is a capacitor with thin sheets of metallic foil that act as plates separated by waxed paper that acts as the dielectric.

Leads are fastened to the inside foil and the outside foil that make up the plates of the capacitor. The outside foil forms the exterior layer of the capacitor (inside the capacitor case) while the inside foil forms an interior layer of the capacitor. A single band is used to mark the lead connected to the outside foil at one end of the capacitor. This mark should not be confused with a polarity mark. The potential connected to the lead can be either negative or positive. When a capacitor is connected in a circuit, the marked end or lower potential point is normally connected to ground. If connected to ground, the outside foil acts as a shield against stray magnetic and electric fields.

Capacitance and voltage ratings are normally printed on paraffin-insulated, cardboard-covered paper capacitors. Although most paper capacitors are manufactured in a cylindrical design, they may be manufactured in other designs such as a flat square or rectangle. Paper capacitors are most commonly used in circuits where maximum voltages range from 600 V to 800 V. Their capacitance normally ranges from 300 pF to 8 μF. Paper capacitors are inexpensive when compared with many other types of capacitors and provide a good ratio of capacitance for size, but they have largely been replaced by capacitors that use plastic dielectrics. An application where paper capacitors are still in use is as antitheft protection devices on retail merchandise. Items have an attached coil and paper capacitor circuit that resonates when passed through alarm portals located near store entrance and exit locations. When an item is purchased, the cashier applies enough voltage to the capacitor to destroy it, and the circuit no longer resonates.

Mica Capacitors

Terms

A **mica capacitor** is a capacitor that is composed of metal foil plates separated by layers of dielectric material.

A *mica capacitor* is a capacitor that is composed of metal foil plates separated by layers of dielectric material. Mica capacitors are sometimes referred to as glass capacitors. The dielectric in a mica capacitor is tightly fitted between the plates. The entire assembly is enclosed in a molded rigid plastic case to protect the assembly. The plates on a mica capacitor are interconnected to increase their surface area. Leads are attached to each plate and brought through the case. Certain capacitors have their values printed on the case. A color code is typically used for capacitance and voltage ratings.

Mica has excellent dielectric strength and can withstand voltages up to 7.5 kV. Capacitance values of mica capacitors range from 50 pF to 0.02 μF. Mica capacitors tend to be highly stable with good reliability and are used in tuned circuit applications such as in oscillators and filters.

Ceramic Capacitors

A *ceramic capacitor* is a capacitor that uses ceramic-coated cylindrical tubes or flat metal disks as the capacitor and the dielectric. The tubes or disks consist of thin films of metal deposited on the dielectric. A break is made in the deposited metal on each tube or disk. Without this break, the film would cause a short circuit across the capacitor. The tube or disk is inserted into a metallic case and coated with ceramic. Leads are soldered onto each tube or disk and brought out through the ceramic-coated case. Ceramic capacitors have capacitance values that range from 1 pF to 0.01 μF. Ceramic capacitors have excellent breakdown voltage properties, with ratings as high as 30 kV. Ceramic capacitors are the most commonly used capacitors.

Ceramic capacitors are used in high-frequency applications such as computers and peripherals because of their construction, in communication products for their size, and in switched-mode power supply products for their low equivalent series resistance.

Terms

A **ceramic capacitor** is a capacitor that uses ceramic-coated cylindrical tubes or flat metal disks as the capacitor and the dielectric.

An **oil capacitor** is a capacitor that uses oil or oil-impregnated paper as a dielectric.

Tech Tip

Capacitor manufacturers can design ceramic capacitors with custom or proprietary ceramic compositions if stock ceramic capacitors are insufficient for a given application.

Oil Capacitors

An *oil capacitor* is a capacitor that uses oil or oil-impregnated paper as a dielectric. Oil capacitors are divided into two classes. Both types of oil capacitors are usually marked on the capacitor case for proper identification. The outside plate of an oil capacitor is usually marked with a dot or arrow.

The first class of oil capacitor uses oil-impregnated paper, which meets voltage requirements from 600 V to 2 kV and is used in applications with large currents. Oil-impregnated paper has a high dielectric constant that is ideal for high-capacitance applications. Oil-impregnated paper is used with other types of dielectrics to prevent arcing between the plates. If arcing does occur, the oil usually repairs any damage. It is for this reason that oil-impregnated capacitors are sometimes referred to as self-healing capacitors. This type of capacitor can be mistaken for the aluminum-enclosed paper type because their appearances are similar.

The second class of oil capacitor is the oil-filled type, where oil is used as the dielectric. Oil-filled capacitors are primarily used as filters in the output of high-voltage DC power supplies. They are also used in broadcast and radar power supplies, and other high-voltage DC supplies. Oil-filled capacitors are larger in size than oil-impregnated paper capacitors but have the same capacitance rating due to the difference in the dielectric constant. Oil-filled capacitors are sometimes referred to as high-energy storage capacitors.

Sometimes glass or porcelain is added to an oil-filled capacitor as extra insulation along with the oil. The glass or porcelain is used at the connection terminals. Capacitors with oil and glass or porcelain insulators have high current- and voltage-handling capabilities. Capacitance ratings are average, but voltage ratings are high and range from 15 kV to 20 kV. The glass or porcelain insulators used at the terminals distinguish these types of capacitors from other capacitor types.

Oil-filled capacitors are designed for work with high voltages and high currents because they normally discharge quickly and have a maximum peak current-discharge rating. High currents and high voltages put extreme stress on the plates and dielectric in oil-filled capacitors. Although a single instance of exceeding these ratings does not usually destroy the capacitor, exceeding the ratings repeatedly will cause premature failure. Most oil-filled capacitors have an energy rating along with capacitance, voltage, and current ratings. The energy rating specifies how many joules the capacitor stores when charged to its maximum voltage.

Chip Capacitors

Chip capacitors (surface-mount capacitors) are capacitors without leads. Chip capacitors are different than leaded capacitors in that chip capacitors do not have leads for connection purposes. Chip capacitors are generally designed as one of three types. The three main types of chip capacitors are film capacitors, multilayer capacitors, and single-layer chip capacitors. Film capacitors are constructed of single or multiple layers of plastic film dielectric wound alternately with metal foil electrodes. Multilayer chip capacitors are designed with several layers of dielectric, while single-layer chip capacitors have a single layer of dielectric material. Types of dielectric material used in chip capacitors include ceramic, ceramic composites, niobium, paraffin paper, polycarbonate, polyester, polypropylene, polystyrene, and tantalum.

Terms

An **aluminum electrolytic capacitor** is a capacitor composed of aluminum that contains an electrolyte.

An **ion** is an electrically charged atom.

An **electrolyte** is a nonmetallic electric conductor in which current is carried by the movement of ions.

Aluminum Electrolytic Capacitors

An *aluminum electrolytic capacitor* is a capacitor composed of aluminum that contains an electrolyte. An *ion* is an electrically charged atom. An *electrolyte* is a nonmetallic electric conductor in which current is carried by the movement of ions. Aluminum electrolytic capacitors are sometimes referred to simply as electrolytic capacitors. An electrolyte may be in the form of a liquid (wet electrolytic capacitors) or a paste (dry electrolytic capacitors). Most aluminum electrolytic capacitors are the dry type because of spillage problems with the wet type. A dry electrolytic capacitor does contain some moisture; if it were to become totally dry, the capacitor would not function. Internally, an aluminum electrolytic capacitor is constructed much like a paper capacitor, in that an aluminum capacitor has foil sheets rolled into a cylindrical cartridge.

Aluminum electrolytic capacitors are capacitors that consist of cathode aluminum foil, electrolytic capacitor paper, and a layer of aluminum oxide. The aluminum oxide layer acts as the dielectric and has rectifying properties. In addition, when in contact with the electrolyte, the dielectric possesses an isolative property, which prevents forward passage. The greater the purity of the aluminum oxide, the greater the dielectric of the aluminum electrolytic capacitor.

In aluminum electrolytic capacitors, an electrolytic process forms the dielectric. Typically, two aluminum electrodes are placed in an electrolyte of borax or carbonate. Gauze between the aluminum electrodes absorbs the electrolyte to provide the electrolysis that produces an oxide film. The electrolytes employed in wet aluminum electrolytic capacitors may be divided into two general classifications, simple aqueous solutions and polyhydric alcohol solutions (such as glycerol). Aqueous solutions generally consist of a solution of boric acid and ammonium borate or sodium borate. In cases where capacitors are used in applications requiring higher-than-normal temperatures, an electrolyte consisting of distilled water, boric acid, ammonium borate, and either glycerol or glycol is employed. An electrolyte containing either glycerol or glycol will not freeze at temperatures at which the simple aqueous electrolytes solidify, maintaining capacitor operation over a much larger temperature range.

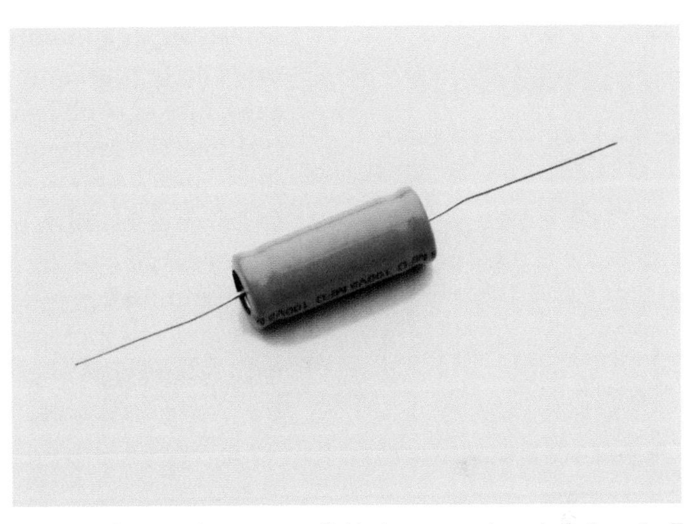

Electrolytic capacitors are available in a range from less than 1 μF up to 1,000,000 μF.

All styles of aluminum electrolytic capacitor are available in one of two off-the-shelf form factors. Custom or proprietary varieties are also available. The two widely available form factors are leaded capacitors and chip aluminum electrolytic capacitors. Leaded capacitors connect to circuits via leads. Chip aluminum electrolytic capacitors do not have leads and are also referred to as surface mount capacitors. Types of leads available for leaded aluminum electrolytic capacitors include axial leads, radial leads, flying leads, tab leads, screw leads, gull wing leads, and J-leads.

Aluminum electrolytic capacitors are manufactured as single units and multiple units. Geometrically shaped symbols printed on the body of the capacitor are used to identify the capacitance and voltage ratings of multiple-unit capacitors. **See Figure 12-10.** For example, a circle, a triangle, a square, and a semicircle may be used to designate the terminals on a multiple-unit capacitor with four terminals. Typically, uF is used to designate microfarads when the value is printed on the capacitor. The aluminum case acts as the negative terminal for all four capacitors.

Aluminum electrolytic capacitors have high leakage current compared to other types of capacitors. To prevent high leakage current, the polarity of an electrolytic capacitor must be determined before connecting to a voltage. With the aluminum case construction, the marked terminals are positive. On single units, the terminals are marked as positive or negative. If the polarity of the voltage is reversed on an aluminum electrolytic capacitor, high leakage current could occur and could cause a short circuit. A short-circuit condition can cause enough heat buildup in the capacitor to bring the electrolyte to a boil and can create enough pressure to cause an explosion. Since electrolytic capacitors are usually polarized, they are limited to DC circuits. Electrolytic capacitors have a capacitance range between 1.0 μF and 2.6 F, with a maximum voltage rating of 500 VDC. To obtain higher voltages, capacitors can be stacked in series. This arrangement increases the size of the capacitor so that it compares with other types of capacitors with higher capacitance ratings than the electrolytic type.

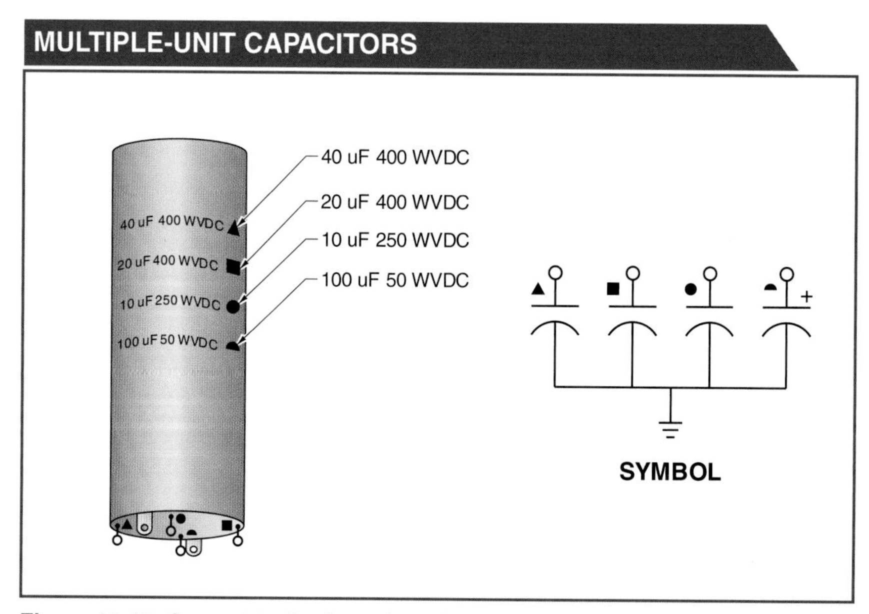

Figure 12-10. Geometrically shaped symbols are used to identify the capacitance and voltage ratings of a multiple-unit capacitor.

Variable Capacitors

A *variable capacitor* is a capacitor that varies in capacitance value. A variable capacitor is constructed so that the capacitance values within the capacitor can be changed. **See Figure 12-11.** A knob can be attached to the shaft and turned manually, or the capacitor may have an adjustment screw. A *rotor plate* is the movable plate of a variable capacitor. A *stator plate* is the stationary plate of a variable capacitor. As the two plates are intermeshed, capacitance is increased. When the plates are moved apart, less plate area is opposite each part, and capacitance is decreased.

Terms

A **variable capacitor** is a capacitor that varies in capacitance value.

A **rotor plate** is the movable plate of a variable capacitor.

A **stator plate** is the stationary plate of a variable capacitor.

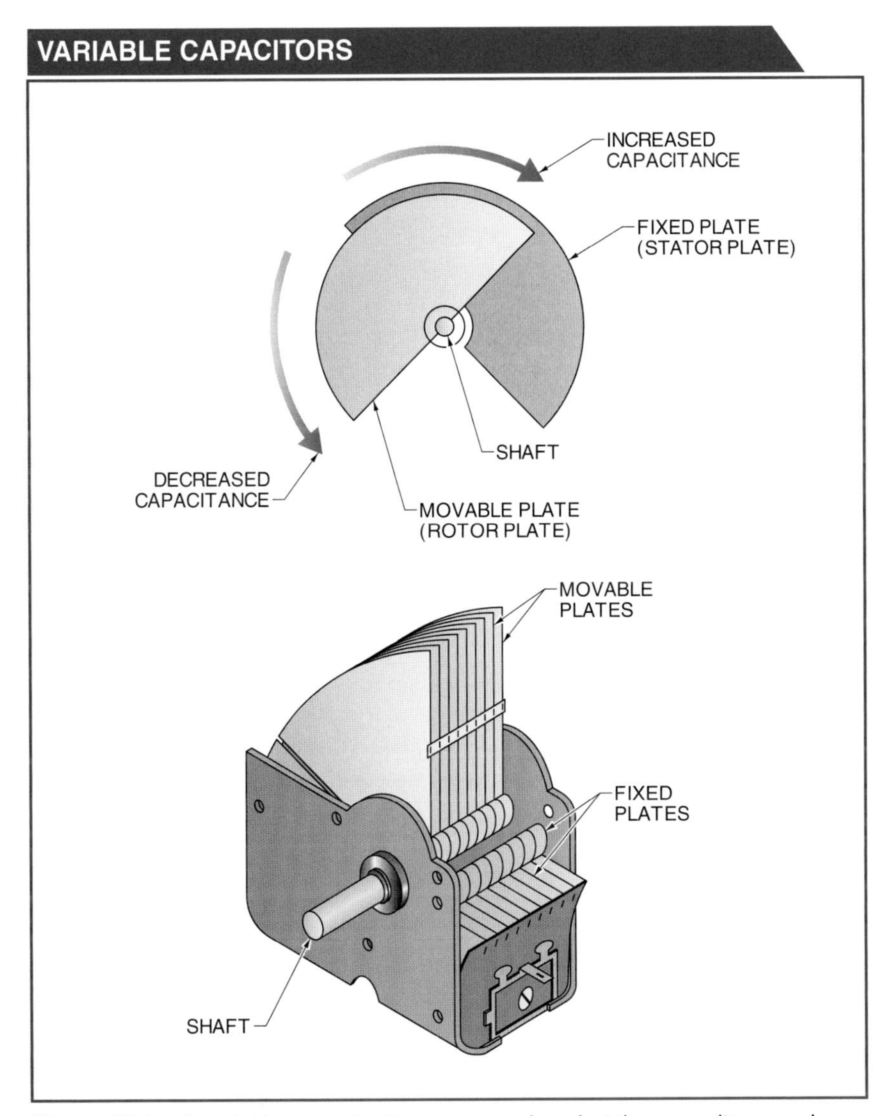

Figure 12-11. A variable capacitor is constructed so that the capacitance value of the capacitor can be changed.

A multiplate variable capacitor is typical of those used to select a frequency on a radio or to change the frequency of an audio or radio frequency generator. Capacitance of a variable capacitor is calculated by applying the following formula:

$$C = \frac{0.2249 \times A \times (N-1)}{t}$$

where

C = capacitance (in pF)

A = area of single plate (in in.^{2})

N = number of plates

t = distance between plates (in in.)

0.2249 = MKS to English conversion constant (in pF/in.)

Example: What is the maximum capacitance of a 10-plate variable capacitor, with each plate having an area of 0.5 in^2 and a distance of 0.1″ between the plates?

$$C = \frac{0.2249 \times A \times (N-1)}{t}$$

$$C = \frac{0.2249 \times 0.5 \text{ in}^2 \times (10-1)}{0.1''}$$

$$C = \frac{0.2249 \times 0.5 \times 9}{0.1}$$

$$C = \frac{1.012}{0.1}$$

C = **10.12 pF**

It is sometimes necessary to subtract the area of one plate when calculating the total area. For example, it requires two plates to obtain one area between plates. If there are three plates, then there is capacitance between two sets of plates. The two outside plates do not have capacitance between them when there are more than two plates. A *trimmer capacitor* is a capacitor used with a larger variable capacitor to fine-tune the overall capacitance of the circuit. Small mica trimmer capacitors are often attached to variable air capacitors. Trimmer capacitors typically have an adjustment screw. As the adjustment screw is tightened, the plates are forced closer together, increasing capacitance. A trimmer capacitor is used to calibrate the high-frequency end of a radio tuning range. A trimmer capacitor is connected in parallel with a larger tuning capacitor. The capacitance range of trimmer capacitors is only a few picofarads.

A *padder capacitor* is a small adjustable capacitor that is connected in series in the tuning circuit of a radio. A padder capacitor

Terms

A **trimmer capacitor** is a capacitor used with a larger variable capacitor to fine-tune the overall capacitance of the circuit.

A **padder capacitor** is a small adjustable capacitor that is connected in series in the tuning circuit of a radio.

is a variation of a trimmer capacitor and is used to calibrate the low-frequency end of a radio tuning range. In a padder capacitor, the rotor plate is moved with an adjustment screw. The adjustment screw does not move the plates closer together but varies the plate overlap with rotor and stator plates. The adjustment is continuous from minimum to maximum capacitance with a range in the lower picofarad values.

A *varicap capacitor* is a variable solid-state diode that operates under reverse bias. As the voltage is changed across the capacitor, its capacitance changes at a predictable, nonlinear rate. In effect, changes in the voltage cause the thickness of the insulation area between the cathode and the anode to vary, thereby changing the capacitance of the diode. Although varicap capacitors are designed for a particular range of capacitance for a given reverse voltage change, any solid-state diode can act as a varicap capacitor. Both high-voltage and low-voltage varicap capacitors are available. Varicap capacitors are used in radio and electronic circuits as replacements for metal-plate variable capacitors such as tuning, trimmer, and padder capacitors. Disadvantages of varicap capacitors include their nonlinear characteristics and noise generation during normal operation.

Terms

A **varicap capacitor** is a variable solid-state diode that operates under reverse bias.

A **supercapacitor** is a capacitor that offers high capacitance in a small volume.

Supercapacitors

A *supercapacitor* is a capacitor that offers high capacitance in a small volume. A supercapacitor has energy storage similar to a battery and has the discharge characteristics of a capacitor. In some cases, supercapacitors compete with batteries.

Variable capacitors have an adjustment mechanism to obtain specific capacitance values.

A supercapacitor utilizes the ion storage capacity of the electrode-electrolyte interface. A supercapacitor is used as a battery replacement because a supercapacitor has characteristics that are similar to a battery. **See Figure 12-12.** When an external DC power source is applied, an excess of electrons collects on the negative plate, and electrons leave the positive plate. This action is offset by the ions in the electrolyte. Because the separation of negative and positive charges in this type of system is in the order of molecular size, the capacitance per unit area is much larger than that of conventional electrolytic capacitors. Because of the proximity of the charges to each other, the limiting factor is the amount of working voltage.

SUPERCAPACITOR vs BATTERY COMPARISON				
Type	Supercapacitor	Primary Battery	Secondary Battery	
	Dynacap	Lithium	Nickel-Cadmium	Lithium Ion
Chargeable	Yes	No	Yes	Yes
Charge Time*	30 to 60	–	Hours	Hours
Voltage†	2.5 to 6.3	3	1.2	3
Cycles	>1,000,000	1	500 to 1000	500 to 1000
Weight	Light	Heavy	Heavy	Heavy
Landfill Pollution	No	Yes	Yes	Yes

* in sec
† in V

Figure 12-12. Supercapacitors are similar to batteries, which they are designed to replace.

Supercapacitors have several advantages over batteries. Unlike nickel-cadmium and lithium ion batteries, which require current limiting devices when charging, there is no current limit when charging a supercapacitor, greatly reducing recharging time. Where ordinary batteries can operate for 500 to 1000 operating cycles, for all practical purposes, a supercapacitor can operate for an unlimited number of operating cycles. Unlike a battery, a fully charged supercapacitor can have its terminals short-circuited without any harmful effects on the capacitor.

CAPACITOR SPECIFICATIONS

Capacitor specifications are required to ensure proper equipment operation. The minimum specifications required for general-use capacitors include capacitance rating, voltage rating, capacitance tolerance, capacitor type and classification, capacitor dimensions, capacitor shape, and capacitor lead configurations. Environmental specifications for capacitors include temperature coefficient and operating temperature range. Capacitor specifications are available from the capacitor manufacturer and should also be verified by performing capacitance tests with a DMM set to measure capacitance prior to installation in equipment.

CAPACITOR IDENTIFICATION CODES

Capacitor identification codes from three different standards systems are used when marking capacitors for identification purposes. These standards systems are the Joint Army-Navy (JAN), the Electronics Industries Association (EIA), and the International Electrotechnical Commission (IEC).

Markings with colors, codes, dots, and numbers in combination with letters are used to identify capacitor types and capacitance values. Some manufacturers produce capacitors with no markings or with only proprietary markings. These types of capacitors require a chart from the manufacturer to identify capacitor values. Electrolytic capacitors are clearly marked with their capacitance, polarity, and usually their working voltage. Some capacitor markings indicate capacitance values in microfarads or picofarads.

Capacitance and voltage ratings are normally printed on waxed or paraffin-insulated paper capacitors. Although most paper capacitors are manufactured in a tubular form, they may have other shapes. Certain paper capacitors are formed without a core and may have flat, square, or oblong plastic enclosures or different shapes of aluminum cases. Other types of capacitor identification codes include ceramic capacitor identification codes and dipped tantalum capacitor identification codes. Capacitor code systems include dot system identification codes and alphanumeric system identification codes.

Small capacitors are used with inductors and resistors in DC applications such as portable radio tuning circuits.

Tubular Paper Capacitor Identification Codes

Tubular paper capacitors are identified with a code consisting of six colored bands which is read from left to right starting with the first colored band closest to one end. **See Appendix.** The first three bands designate the value of the capacitor (in picofarads). The fourth band indicates the tolerance as a percentage of the capacitance. A capacitor with a ±10% tolerance can have capacitance up to 10% less or 10% more than the manufacturer's rated capacitance value. The last two color bands indicate the first two digits of the voltage rating and are read from right to left. The digits from the last two color bands are multiplied by 100 to obtain the working voltage of the capacitor. When the capacitor has a voltage rating less than 1 kV, the second voltage band is omitted, and the first digit is multiplied by 100.

Ceramic Capacitor Identification Codes

Ceramic capacitors are identified using a code consisting of five colored bands. The first band on the left is normally wider than the other four bands. The wide band indicates the temperature coefficient of the capacitor. Ceramic capacitors may have either negative or positive coefficients, given in parts per million per degree centigrade (ppm/°C). A minus sign is used to show the negative temperature coefficient, and a positive sign is used for the positive temperature coefficient. Some charts use N or P for this purpose. For example, N30 indicates –30°C and P30 indicates 30°C.

Following the temperature coefficient band are two digits and the multiplier for the capacitance. The fifth band indicates the tolerance. Tolerance is divided into two columns. One column uses a percentage value of the capacitance for values greater than 10 pF. The other column uses the difference in pF that the capacitance can vary from its color-coded value for values less than 10 pF. Ceramic capacitor color codes follow the same system as tubular paper color codes.

Terms

Tantalum is a hard metallic material that is resistant to acid. Tantalum is commonly used as a dielectric for capacitors.

A **dipped tantalum capacitor** is a small capacitor that has tantalum dielectric and resembles the head of a match in shape.

Dipped Tantalum Capacitor Identification Codes

Tantalum is a hard metallic material that is resistant to acid. Tantalum is commonly used as a dielectric for capacitors. A *dipped tantalum capacitor* is a small capacitor that has tantalum dielectric and resembles the head of a match in shape. Dipped tantalum capacitor identification codes are printed on the head of the capacitor. The capacitor body color indicates the first digit, followed by a narrow color band for the second digit. A wider band below the second digit color band is the multiplier. Capacitance is given in picofarads. A dot on top of the capacitor indicates the tolerance of the capacitor and a dot on its side indicates its working voltage. Dipped tantalum capacitors are used in small electronic components including portable telephones, pagers, personal computers, and automotive electronics. Dipped tantalum capacitors have high capacitance and low voltage ratings. **See Appendix.**

Dot-System Identification Codes

Dot-system identification codes are typically found on paper and mica capacitors. **See Appendix.** Sometimes dots are enclosed in an arrow to show the direction the code is to be read. The first dot in the upper left-hand corner indicates the type of capacitor. A black dot indicates a JAN mica capacitor, a white dot indicates an EIA mica capacitor, and a silver dot indicates that the capacitor is a paper type. Dot location is the same regardless of capacitor type.

Colored dots are read from left to right. Arrows are sometimes used to indicate the proper code sequence. The arrow typically points to the right.

A three-dot system indicates only the capacitance (in picofarads), while a five-dot system indicates the capacitance (in picofarads) using the first three dots. Voltage rating and the tolerance are provided by the remaining two dots. The six-dot system provides three digits for the capacitance in picofarads followed with the working voltage and the tolerance. An arrowhead indicates the direction in which the capacitance is read. The arrow encloses the first two digits and the multiplier for determining the capacitance in picofarads. The dot opposite the arrowhead is the tolerance. The dot opposite the tail of the arrow is the voltage rating. The dot between the tolerance and the voltage rating has no significance and is not used as part of the code.

Alphanumeric-System Identification Codes

Certain capacitors have values marked on them with alphanumeric identification codes. With many capacitors, units are not specified. Most capacitors have numbers that may be in units of either microfarads or picofarads. However, because the units of measurement are not specified for the values marked on many capacitors, capacitance values cannot be determined without manufacturer's specification sheets or direct measurement with a DMM set to measure capacitance.

Aluminum electrolytic capacitors and some tantalum capacitors are marked for capacitance, polarity, and voltage rating. Typically, a polarized tantalum capacitor has two sets of numbers and no other markings except the polarity of the plates. Sometimes the length of the leads indicates polarity, and the longer lead is the positive plate.

While the polarity is clear, the coding can be complex. For example, a silver mica capacitor is marked the military specifications CM 05 F D 271 J 03. The letters CM indicate that the package is molded plastic and numbers 05 designate the package size. The letter F stands for the capacitor's maximum capacitance change over a temperature range. The letter D is the voltage rating of 500 V. The number 271 indicates a capacitance of 270 pF. The letter J indicates a tolerance of ± 5%, and 03 is the style number. This system incorporates military specifications with IEC markings.

Some manufacturers use an alphanumeric system of digits and letters marked on the capacitor and have adopted the IEC standard for marking capacitors. **See Appendix.** This system is similar to the

Tech Tip

Two types of aluminum electrolytic capacitors are leaded capacitors and chip aluminum electrolytic capacitors. Leaded capacitors are used to connect to circuits via leads. Chip aluminum electrolytic capacitors (surface-mount capacitors) do not have leads.

color-code (band) method of identification, but uses numbers and letters rather than color bands. The IEC code uses three digits followed by a single letter. In some cases, two manufacturer-specific letters may precede the digits. These letters may indicate a particular type or series of capacitor. For example, a capacitor is marked xx273M. The first digit is two, the second digit is seven, and the third digit indicates a multiplier of 10^3. This results in a value of 27,000 pF. The letter following the numbers indicates the tolerance. Typically, the letter M represents a tolerance of 20%. Tolerance for capacitors over 10 pF is given as a percentage of the value of capacitance, and tolerance for capacitors less than 10 pF is given as a variance of picofarads.

RESISTIVE-CAPACITIVE TIME CONSTANT

Terms

A **resistive-capacitive time constant** is the amount of time required for the capacitor to charge to 63.2% of the maximum voltage across a resistive-capacitive circuit.

A *resistive-capacitive time constant* is the amount of time required for the capacitor to charge to 63.2% of the maximum voltage across a resistive-capacitive circuit. Recall that Ohm's law states that resistive voltage is equal to resistive current times the value of the resistance. There is resistive voltage only as long as there is resistive current.

A capacitor is capable of storing a charge of electrons. When a capacitor is uncharged, the number of electrons on each plate is equal. When a voltage source is placed across the capacitor, additional electrons collect on one plate, and an equal number of electrons leave the opposite plate. This process continues until the capacitive voltage is equal to the source voltage across it. When the source voltage is removed, the charge remains on the plates of the capacitor until it is discharged. The time constant for a series resistive-capacitive circuit is calculated by applying the following formula:

$$\tau = RC$$

where

τ = time constant (in sec)

R = circuit resistance (in Ω)

C = circuit capacitance (in F)

Circuit boards used in electronics often contain groups of various types of capacitors.

Example: What is the time constant of a 1 F capacitor charged through a 1 Ω resistor?

$\tau = RC$

$\tau = 1\ \Omega \times 1\ \text{F}$

$\tau = \mathbf{1\ sec}$

In each of the following time constants, the capacitor charges to 63.2% of the difference between the maximum voltage and the capacitive voltage. Theoretically, a capacitor never charges to its full value. However, for all practical purposes, a capacitor charges and current decreases at an exponential rate, reaching full charge after five time constants. **See Figure 12-13.** At t_0, the current in the circuit is at its maximum level. It then decreases at an exponential rate as the capacitive voltage increases. After five time constants, the capacitive voltage is equal to the source voltage, and the current flow ceases.

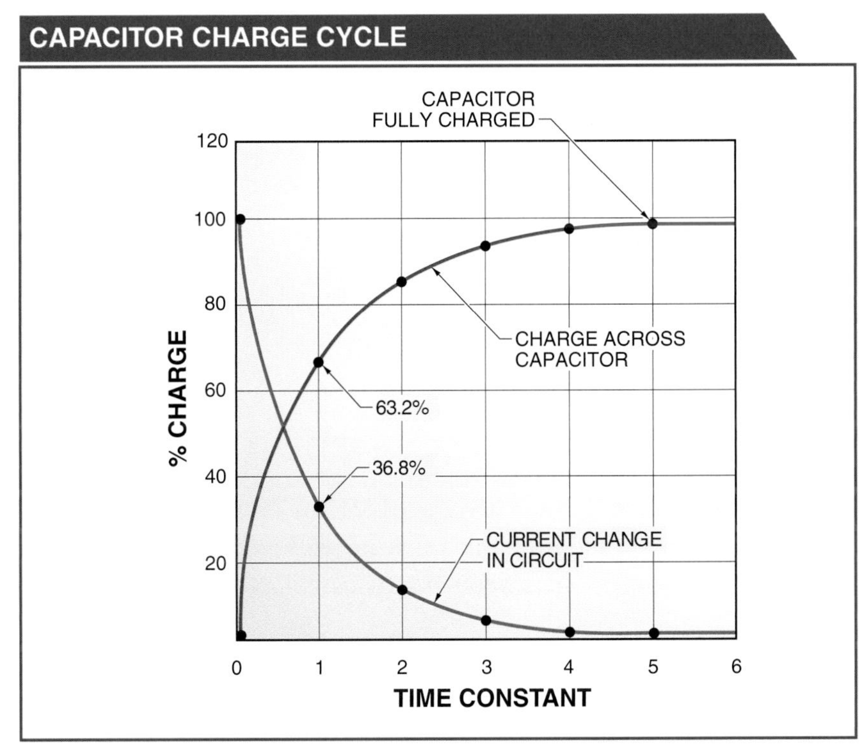

Figure 12-13. A capacitor charges and current decreases at an exponential rate, reaching full charge after five time constants.

A capacitor discharges at the same exponential rate as it charges. For example, in an application where a capacitor is used as a power source filter for steady DC output voltage, a small time constant is used for capacitor charging and a large time constant is used for capacitor discharging. **See Figure 12-14.** The power source resistance needs to be low to minimize the resistive-capacitive time constant, while the load resistance needs to be high to maximize the resistive-capacitive time constant.

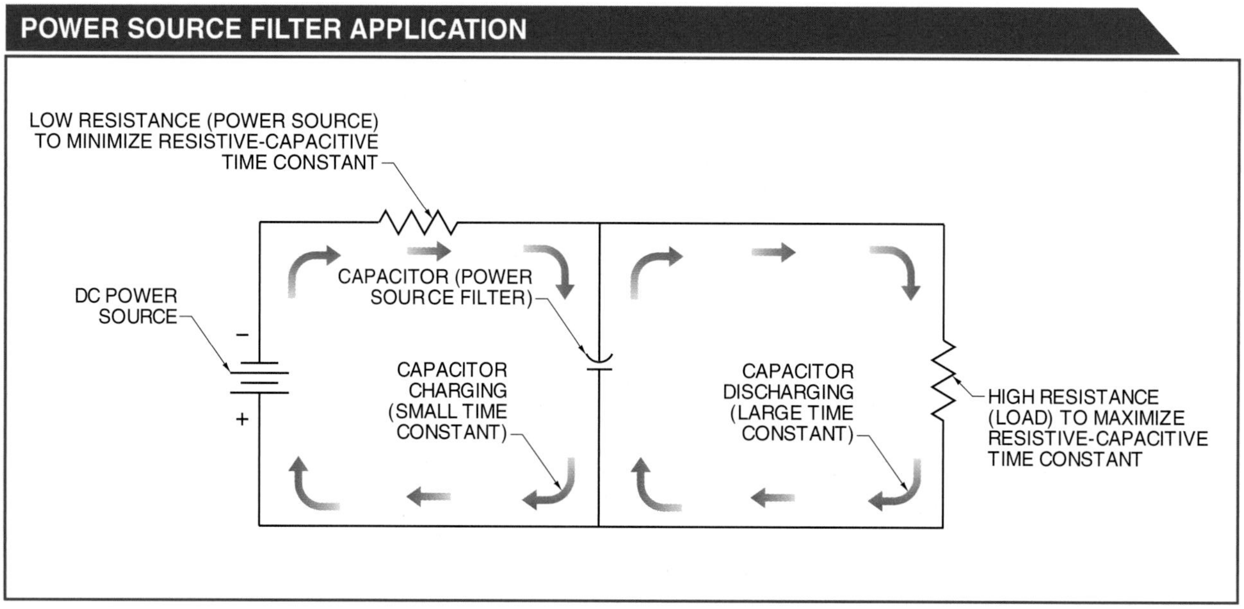

Figure 12-14. Power source resistance needs to be low to minimize the resistive-capacitive time constant, while load resistance needs to be high to maximize the resistive-capacitive time constant.

STORED ENERGY IN A CAPACITOR

The amount of stored energy in a capacitor is dependent on the amount of capacitance and the working voltage rating of the capacitor. Stored energy in a capacitor is calculated by applying the following formula:

$$W = \frac{1}{2} \times C \times V^2$$

where

W = stored energy (in J)

C = capacitance (in F)

V = volts (in V)

Example: What is the stored energy of a 250 μF capacitor rated at 120 V?

$$W = \frac{1}{2} \times C \times V^2$$

$$W = 0.50 \times (250 \text{ F} \times 10^{-6}) \times (120 \text{ V})^2$$

$$W = 0.50 \times 0.00025 \times 14{,}400$$

$$W = \mathbf{1.8 \text{ J}}$$

Stored energy of 1.8 J is not much energy. A 25 W lamp, for instance, uses 25 J every second. Some capacitors with high capacitance and voltage ratings can store several thousand joules of energy. Capacitors with high capacitance and voltage ratings are used for applications such as capacitor-discharge welding and with surge suppressors to prevent current surges on overhead power lines. For example, during the welding cycle, this type of capacitor can discharge a high current over a short time period supplying much of the current to perform the weld. It can then be recharged over a much longer time period for the next cycle.

CAPACITOR CONNECTIONS

As with resistors and inductors, capacitors can also be connected in various configurations. Capacitors can be connected together in parallel so that their capacitances add together. Capacitors can also be connected together in series so that the working voltage can be increased. For certain circuit applications, capacitors are connected in series/parallel combinations.

Capacitors Connected in Series

The capacitance of capacitors connected in series reduces capacitance, such as resistance for resistors connected in parallel. When capacitors are connected in series, their total value decreases in the same manner as values of resistance and inductance decrease when connected in parallel. The total capacitance is always less than the smallest capacitor connected in series. **See Figure 12-15.**

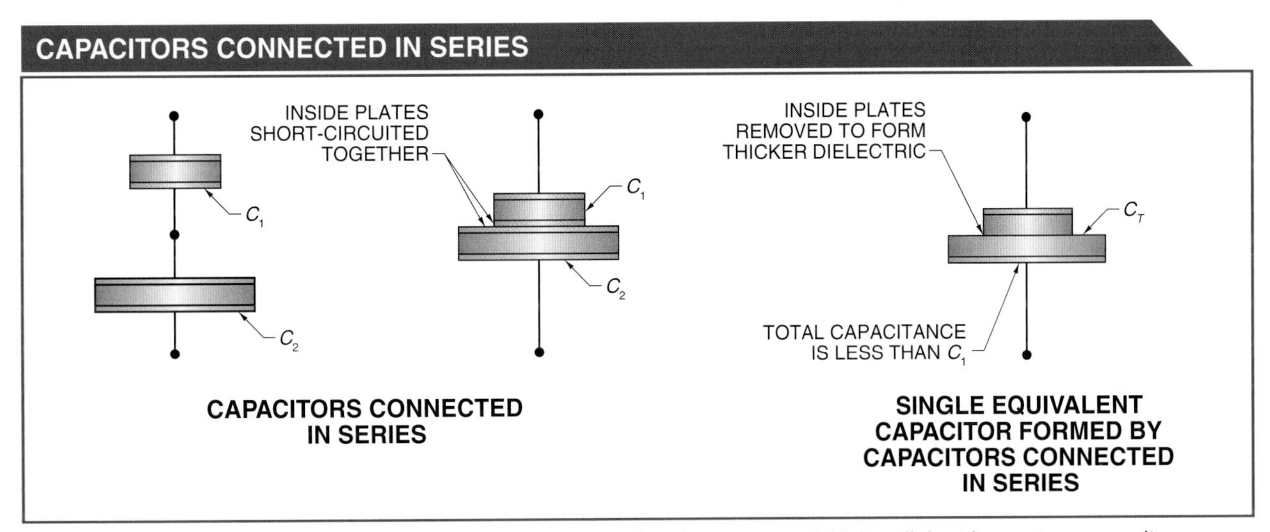

Figure 12-15. When connecting capacitors in series, a smaller plate area and thicker dielectric create a capacitance that is less than that of the smaller capacitor connected in the series.

For example, with two capacitors connected in series, the two inside plates are short-circuited together and have no capacitance between them. Therefore, the leads are shortened to bring the two plates together to make a single plate. The two plates, isolated between two dielectrics, serve no purpose and are removed to form a single equivalent capacitor. The properties of the single equivalent capacitor are such that the total plate area has effectively been reduced to that of the smaller capacitor. Simultaneously, the total thickness of the dielectric has been increased to the sum of the thickness of the dielectric of the two capacitors.

Capacitors connected in series are used to block DC voltage and pass AC signal voltage through the use of charge and discharge current, which helps in achieving coupling, bypassing, and filtering of AC signals. When combining capacitors connected in series, a smaller plate area and thicker dielectric create a capacitance that is less than that of the smaller capacitor connected in the series. Total capacitance for more than two capacitors connected in series is calculated by applying the following formula:

$$C_T = \frac{1}{\frac{1}{C_1} + \frac{1}{C_2} + \frac{1}{C_3} + \ldots + \frac{1}{C_N}}$$

where

C_T = total circuit capacitance (in µF)

C_1 = capacitance 1 (in µF)

C_2 = capacitance 2 (in µF)

C_3 = capacitance 3 (in µF)

C_N = capacitance N (in µF)

Example: What is total circuit capacitance of four capacitors connected in series with capacitances of 10 µF, 20 µF, 20 µF, and 25 µF?

$$C_T = \frac{1}{\frac{1}{C_1} + \frac{1}{C_2} + \frac{1}{C_3} + \ldots + \frac{1}{C_N}}$$

$$C_T = \frac{1}{\frac{1}{10\ \mu F} + \frac{1}{20\ \mu F} + \frac{1}{20\ \mu F} + \frac{1}{25\ \mu F}}$$

$$C_T = \frac{1}{0.1\ \mu F + 0.05\ \mu F + 0.05\ \mu F + 0.04\ \mu F}$$

C_T = **4.2 µF**

When two capacitors are connected in series, total capacitance is calculated by applying the following formula:

$$C_T = \frac{C_1 \times C_2}{C_1 + C_2}$$

Example: What is the total capacitance of two capacitors connected in series with capacitances of 10 μF at 100 V and 20 μF at 150 V?

$$C_T = \frac{C_1 \times C_2}{C_1 + C_2}$$

$$C_T = \frac{10\ \mu\text{F} \times 20\ \mu\text{F}}{10\ \mu\text{F} + 20\ \mu\text{F}}$$

$$C_T = \frac{200\ \mu\text{F}}{30\ \mu\text{F}}$$

C_T = **6.7 μF**

In a DC circuit, the working voltages of the two capacitors add together for a total value of 250 V because connecting the capacitors in series increases the dielectric thickness.

Ultracapacitors

Ultracapacitors are passive electronic components that store energy by separating positive and negative charges. Specifications for ultracapacitors include capacitance range, capacitance tolerance, total capacitance, working DC voltage, current, leakage current, power, and equivalent series resistance. Ultracapacitors are used in fuel cells for hybrid vehicles and robotic motion control systems.

Capacitors Connected in Parallel

The capacitance of capacitors connected in parallel adds in the same way as values of resistors connected in series. The capacitance of capacitors connected in parallel is calculated by applying the following formula:

$$C_T = C_1 + C_2 + C_3 + \ldots + C_N$$

When combining capacitors connected in parallel, dielectric thickness remains the same. Connecting capacitors in parallel results in an increased plate area with the same dielectric thickness. Increasing the plate area without increasing the dielectric thickness causes the total capacitance to be the sum of the capacitance of the individual capacitors. **See Figure 12-16.**

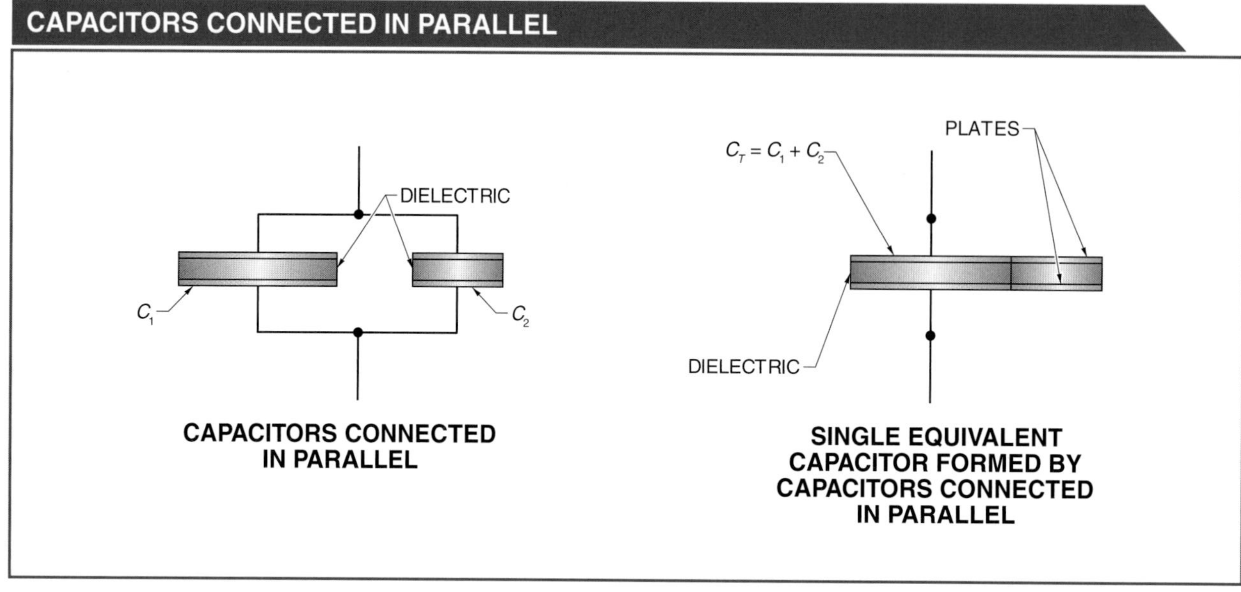

Figure 12-16. When connecting capacitors in parallel, increased plate area and the same dielectric thickness produce a total capacitance that is the sum of the capacitance of the individual capacitors.

Example: What is total circuit capacitance of four capacitors connected in parallel with capacitances of 10 μF, 20 μF, 20 μF, and 25 μF?

$C_T = C_1 + C_2 + C_3 + \ldots + C_N$

$C_T = 10\ \mu F + 20\ \mu F + 20\ \mu F + 25\ \mu F$

$C_T = \mathbf{75\ \mu F}$

The dielectric thickness remains the same for the four capacitors. Therefore, the working voltage remains the same. When capacitors with different working voltages are connected in parallel, the lowest working voltage needs to be observed. For example, if two capacitors have working voltages of 100 V and two capacitors have working voltages of 150 V, when connected in parallel, a 100 V working voltage needs to be observed for all four capacitors.

Capacitors Connected in Series/Parallel

Certain circuits can have capacitors connected in series/parallel combinations. When analyzing a circuit with capacitors connected both in series and parallel, the values of the parallel-connected capacitors are determined first and then added to the equivalent values of the series-connected capacitors. **See Figure 12-17.** Capacitance for a series/parallel capacitor circuit is calculated by applying the following procedure:

1. Combine parallel-connected capacitors into equivalent capacitance value(s). The capacitance value of two capacitors connected in parallel is calculated by applying the following formula:

   $C_{2,3} = C_2 + C_3$

   where

   $C_{2,3}$ = total parallel capacitance 2 and 3 (in μF)
   C_2 = parallel capacitance 2 (in μF)
   C_3 = parallel capacitance 3 (in μF)

2. Combine single equivalent capacitance value(s) from parallel-connected capacitors with values from series-connected capacitors into a single capacitance value. The capacitance value of two capacitors connected in series is calculated by applying the following formula:

   $$C_T = \frac{C_1 \times C_{2,3}}{C_1 + C_{2,3}}$$

   where

   C_T = total series capacitance 1 (in μF)
   C_1 = series capacitance 1 (in μF)
   $C_{2,3}$ = total parallel capacitance 2 and 3 (in μF)

CAPACITORS CONNECTED IN SERIES/PARALLEL

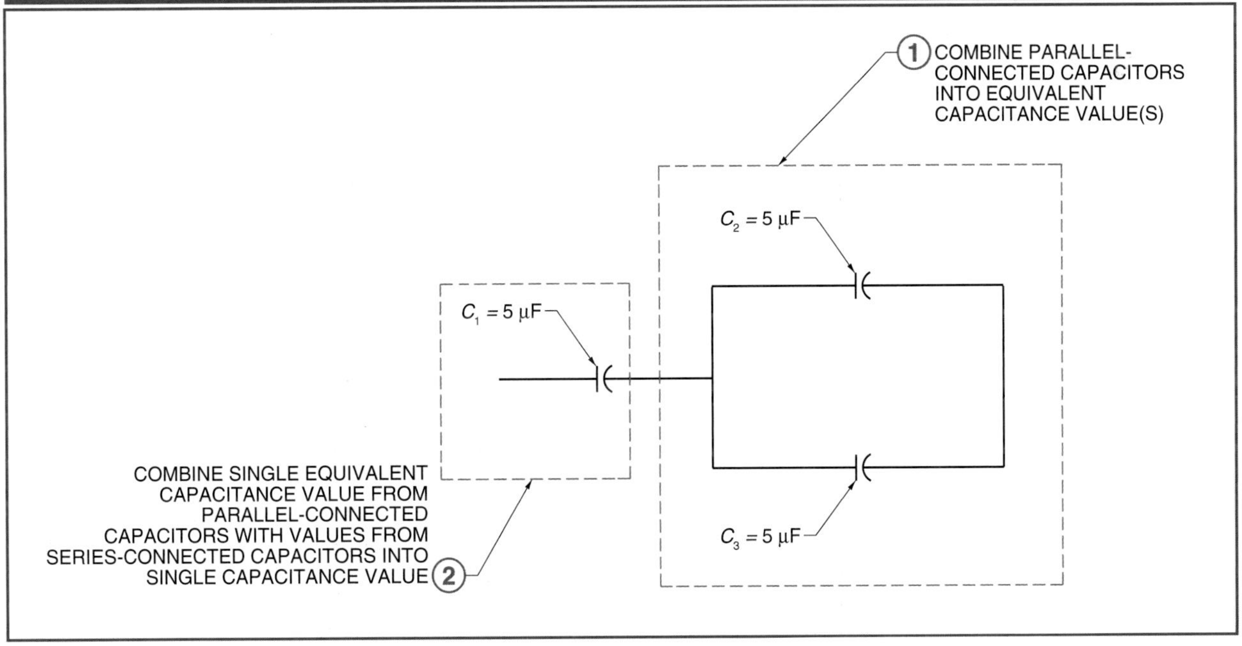

Figure 12-17. When analyzing a series/parallel capacitor circuit, the values for parallel-connected components should be determined first and then added to the values for series-connected components.

Example: What is the total capacitance of a circuit that has a 5 μF capacitor connected in series with two 5 μF capacitors connected in parallel?

1. Combine parallel-connected capacitors into equivalent capacitance value(s).

$C_{2,3} = C_2 + C_3$

$C_{2,3} = 5\ \mu F + 5\ \mu F$

$C_{2,3} = 10\ \mu F$

2. Combine single equivalent capacitance value(s) from parallel-connected capacitors with values from series-connected capacitors into a single capacitance value.

$$C_T = \frac{C_1 \times C_{2,3}}{C_1 + C_{2,3}}$$

$$C_T = \frac{5\ \mu F \times 10\ \mu F}{5\ \mu F + 10\ \mu F}$$

$$C_T = \frac{50\ \mu F}{15\ \mu F}$$

$C_T = \mathbf{3.3\ \mu F}$

Note: The final capacitance value must be less than the smallest series-connected capacitance value.

CAPACITOR LOSSES

Capacitor losses typically occur through the dielectric. Two basic types of losses occur. The first type of loss is dielectric leakage current. Dielectric leakage current results from the dielectric having high resistance. In time, a capacitor with voltage across the dielectric will discharge through the dielectric. Ordinarily, discharge current does not factor into circuit calculations. Electrolytic capacitors have higher leakage current than other types of capacitors. Anytime the amount of leakage current is high, a rapid loss of the charge occurs, causing an increase in temperature. A capacitor may overheat under this condition. Capacitors that overheat are considered damaged and should be replaced.

The second type of loss is dielectric hysteresis loss. Dielectric hysteresis loss can be compared with hysteresis loss in magnetic circuits. Dielectric hysteresis loss is the result of dielectric impurities in the dielectric. Dielectric impurities are caused by positive and negative charges that accumulate in certain regions of the dielectric. The positive and negative charges are caused by changes in the orientation of electron orbits due to rapid reversals of the polarity. Different dielectrics have different hysteresis losses. For example, a vacuum has the lowest hysteresis loss while solid materials, such as glass or wax, have the highest hysteresis losses. Losses in capacitors are much lower than losses in inductors.

TESTING CAPACITORS

An ohmmeter can be used to test the integrity of a capacitor. Since an ohmmeter has a source voltage, when it is connected across a discharged capacitor the capacitor will charge. The amount of time it takes for the capacitor to charge is dependent upon the resistance in the ohmmeter circuit and the value of the capacitor. A low reading on the ohmmeter display shows the charging condition. The reading then increases to show maximum resistance on the ohmmeter's scale. This test should be performed at the ohmmeter's highest resistance range to provide a longer time constant.

After the capacitor has charged, the ohmmeter measures the maximum resistance of the dielectric. For all types of capacitors except electrolytic capacitors, the ohmmeter displays its maximum resistance. If the maximum resistance value is not obtained, then the capacitor under test is defective. Short circuits cause the ohmmeter to indicate near 0 Ω. Aluminum electrolytic capacitors, with more leakage current than normal, may have a final reading as low as 100 kΩ. This is because the dielectric is a very thin layer of oxide that is not as good an insulator as the other materials used for dielectrics.

Polarity should be observed when testing aluminum electrolytic capacitors. If polarity is reversed, leakage is increased, and a good capacitor can be mistakenly classified as bad. If the reading does not change when the ohmmeter is connected to a discharged capacitor, the capacitor may be open. However, when performing resistance tests on capacitors that have low values of capacitance, the resistive-capacitive time constant may not be long enough to cause an ohmmeter deflection during the charge cycle. The ohmmeter should still show its maximum resistance at the end of the test. The amount of capacitance required to cause a deflection for an ohmmeter can be determined by experimentation. Typically, when there is deflection on the ohmmeter, and the reading falls back down to show maximum resistance, the capacitor is good. This includes a capacitance value that changes little over time for most types of capacitors. When there is no deflection and only the maximum resistance reading is displayed, it is possible that the capacitor is open. If a reading on absolute capacitance is required, a DMM set to measure capacitance is used to take capacitance readings.

SUMMARY

Capacitance is defined as the ability of a component or circuit to store energy in the form of an electrical charge. Capacitance in a circuit opposes changes in voltage. The unit of capacitance is the farad, although most capacitance values are expressed in microfarads or picofarads. The physical factors that effect capacitance are the surface area of the plates, the thickness of the dielectric, and the dielectric constant. The dielectric is used as the electric insulation between the plates of a capacitor. The type of dielectric and its thickness determine the working voltage of a capacitor. For the same size and classification of capacitors, higher voltage ratings result in lower capacitance, and lower voltage ratings result in higher capacitance.

Capacitors are either fixed-type or variable-type. A variable capacitor is a capacitor that varies in capacitance value. Capacitors may also be classified by the type of dielectric used. Series resistive-capacitive circuits are used as timing circuits.

Capacitors charge and discharge at an exponential rate. A resistive-capacitive time constant is the amount of time required for the capacitor to charge to 63.2% of the maximum voltage across a resistive-capacitive circuit. In one time constant, 63.2% of the capacitor's voltage is affected. It requires five time constants for a capacitor to charge completely or to discharge fully.

Capacitors can be connected in series, parallel, or in series/parallel. Total capacitance for series-connected capacitors can be determined using the same calculations as for total resistance of resistors connected in parallel. Total capacitance of series-connected capacitors will always be less than that of the smallest capacitor. Total capacitance of parallel-connected capacitors is equal to the sum of the capacitance and adds in the same manner as values of resistors connected in series. Total capacitance value of parallel-connected capacitors will always be greater than that of the largest capacitor. Total capacitance for capacitors connected in series/parallel can be calculated by applying the same rules as for calculating the values of capacitors connected in series and parallel. The parallel combinations should be calculated first, with the equivalent value(s) added to the series capacitors. The condition of a capacitor can often be determined using only an ohmmeter. If absolute capacitance readings are required, a DMM set to measure capacitance is used instead of an ohmmeter.

APPLICATION: SWITCHED-CAPACITOR VOLTAGE INVERTERS...

Background: Circuits that interact with sensors, circuits that use operational amplifiers, and data communication interfaces sometimes require an inverted voltage. When positive and negative voltages are needed, two voltage sources are cascaded with a center ground reference. **See Figure 1.**

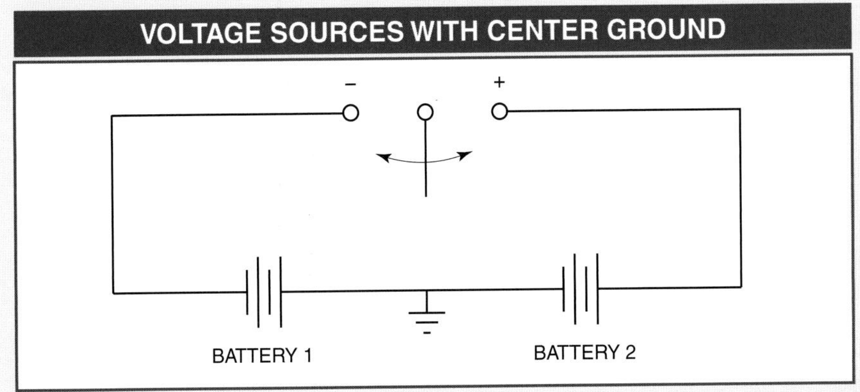

Figure 1. Dual voltage sources are often cascaded with a center ground reference.

In applications in which only a single voltage supply is available, a switched-capacitor voltage inverter can be used to generate the required inverted voltage such that the output voltage (V_{OUT}) is equal to the negative input voltage ($-V_{IN}$). A switched-capacitor voltage inverter has a charge-pump input capacitor (C_1) that is changed by the input voltage during the first half of the switching period. During the second half of the switching period, the input capacitor is disconnected from the input voltage, inverted, and connected to the output capacitor (C_2) and the load. **See Figure 2.** The average input current is approximately equal to the output current.

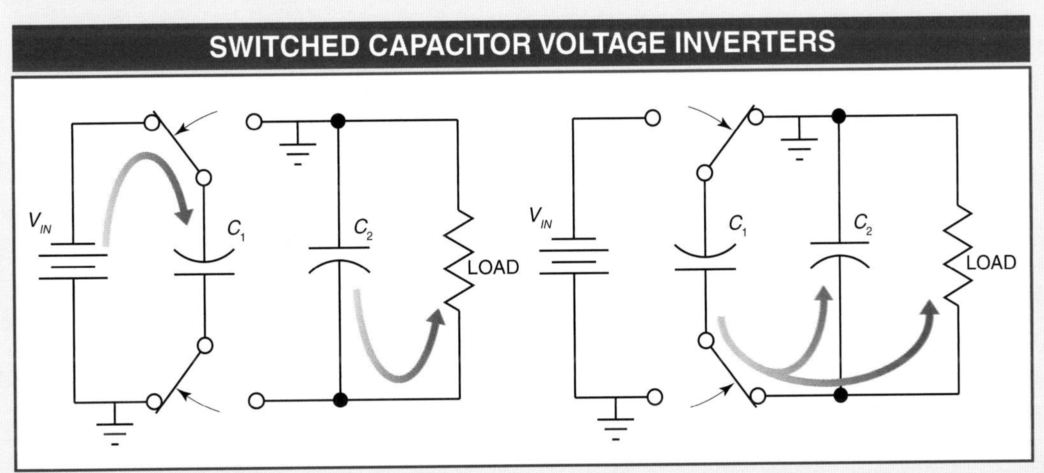

Figure 2. A switched-capacitor voltage inverter has a charge-pump input capacitor (*C*1) that is changed by the input voltage during the first half of the switching period.

Higher switching frequencies, or shorter switching periods, allow the use of smaller capacitors. The switching frequency usually has a 50% duty cycle for maximum charge transfer. This means during the switching period, the charge-pump capacitor is connected to the input voltage half of the time and to the output capacitor the other half of the time. The switching period is related to the switching frequency and is calculated by applying the following formula:

... APPLICATION: SWITCHED-CAPACITOR VOLTAGE INVERTERS...

$$T = \frac{1}{f}$$

where

T = switching period (in sec)
f = switching frequency (in Hz)

The amount of charge transferred between the charge-pump capacitor and the load capacitor is dependent on the load current and switching frequency. While the charge-pump capacitor is being charged by the input voltage, the load capacitor is providing the load current and the output voltage droops according to the following:

$$i = \frac{C \times dv}{dt}$$

or

$$dv = \frac{i \times dt}{C}$$

where

i = load current (in A)
C = load capacitor capacitance (in F)
dv = change in voltage (in V)
dt = change in time (in sec)

Note: The switching period is two times dt.

For example, if the load current is 100 mA with a capacitance of 100 μF, switching frequency of 100 kHz, and change in time of 5 μsec, then the voltage droop is the following:

$$dv = \frac{i \times dt}{C}$$

$$dv = \frac{0.100\text{ A} \times 5\text{ μsec}}{100\text{ μF}}$$

$$dv = 5\text{ mV}$$

When the charge-pump capacitor is connected to the output capacitor, a charge is transferred between the two capacitors. This charge transfer adheres to the conservation of charge law. Each capacitor (C_1 and C_2) has a charge (q) that is based on its capacitance and voltage. The formula to find the charge is the following:

$$q = CV$$

where

q = charge (in coulombs)
C = capacitance (in F)
V = voltage (in V)

When C_1 and C_2 are connected via the switch during the second half of the switching period, the voltage across C_1 and C_2 is the same and set by one of the following formulas:

... APPLICATION: SWITCHED-CAPACITOR VOLTAGE INVERTERS...

$$q_T = q_1 + q_2$$
$$q_T = C_1 \times V_1 + C_2 \times V_2$$
$$V_T = \frac{q_T}{C_1 + C_2}$$
$$V_T = \frac{C_1 \times V_1 + C_2 \times V_2}{C_1 + C_2}$$
$$V_T = \frac{\frac{C_1 \times V_1}{(C_1 + C_2) + C_2 \times V_2}}{C_1 + C_2}$$

where

q_T = total charge (in coulombs)
V_T = total output voltage (in V)
C_1 = capacitor 1 capacitance (in F)
V_1 = capacitor 1 voltage (in V)
C_2 = capacitor 2 capacitance (in F)
V_2 = capacitor 2 voltage (in V)

On startup, V_2 is zero. If C_1 and C_2 are equal, then V_T on the first switching cycle is equal to the following:

$$V_T = \frac{C_1 \times V_1 + C_2 \times V_2}{C_1 + C_2}$$
$$V_T = \frac{C_1 \times V_1}{C_1 + C_2}$$
$$V_T = \frac{C_1 \times V_1}{2 \times C_1; \text{when } C_1 = C_2}$$
$$V_T = \frac{1}{2} \times V_1$$

V_T is the new output voltage. On the second switching cycle, V_T is equal to V_2, assuming there is no change in voltage during the charge pump charging cycle, as follows:

$$V_T = \frac{C_1 \times V_1 + C_2 \times V_2}{C_1 + C_2}$$
$$V_T = \frac{1}{2} \times V_1 + \frac{1}{2} \times \frac{1}{2} \times V_1$$
$$V_T = \frac{1}{2} \times V_1 + \frac{1}{4} \times V_1$$
$$V_T = \frac{3}{4} \times V_1$$

The output voltage continues to increase over time until the maximum voltage (V_{IN}) is reached.

Key Points: Capacitance, conservation of charge law

Problem: An old switched-capacitor voltage inverter design is to be reused for a new application. **See Figure 3.** A table is used to calculate the output voltage (V_T) after each cycle. **See Figure 4.** *Note*: The inverted voltage value is ignored in the calculations to avoid confusion. The output voltage droop occurs when the output capacitor supplies the entire load current because it impacts the value of V_2 in the calculations. The key formulas are the following:

...APPLICATION: SWITCHED-CAPACITOR VOLTAGE INVERTERS...

SWITCHED-CAPACITOR VOLTAGE INVERTER ORIGINAL DESIGN

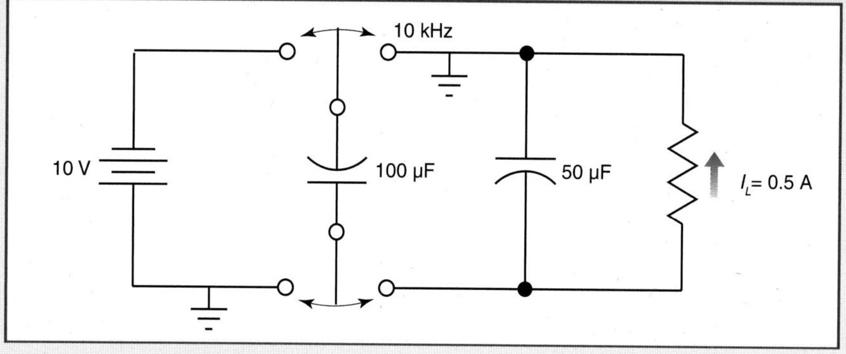

Figure 3. An original switched-capacitor voltage inverter design is to be reused for a new application.

Output voltage:

$$V_T = \frac{C_1 \times V_1 + C_2 \times V_2}{C_1 + C_2}$$

Output voltage droop:

$$dv = \frac{i \times dt}{C}$$

$$dt = \frac{\frac{1}{2} \times 1}{f}$$

$$V_2 = V_T - \text{droop (on next cycle)}$$

OUTPUT VOLTAGE

Cycle	V_1	C_1	V_2	C_2	V_T	Droop
1 (100 µsec)	10 VDC	100 µF	0.00 VDC	50 µf	6.67 VDC	0.50 VDC
2 (200 µsec)	10 VDC	100 µF	6.17 VDC	50 µf	8.72 VDC	0.50 VDC
3 (300 µsec)	10 VDC	100 µF	8.22 VDC	50 µf	9.41 VDC	0.50 VDC
4 (400 µsec)	10 VDC	100 µF	8.91 VDC	50 µf	9.64 VDC	0.50 VDC
5 (500 µsec)	10 VDC	100 µF	9.14 VDC	50 µf	9.71 VDC	0.50 VDC
6 (600 µsec)	10 VDC	100 µF	9.21 VDC	50 µf	9.74 VDC	0.50 VDC
7 (700 µsec)	**10 VDC**	**100 µF**	**9.24 VDC**	**50 µf**	**9.75 VDC**	**0.50 VDC**
8 (800 µsec)	10 VDC	100 µF	9.25 VDC	50 µf	9.75 VDC	0.50 VDC
9 (900 µsec)	10 VDC	100 µF	9.25 VDC	50 µf	9.75 VDC	0.50 VDC
10 (1 msec)	10 VDC	100 µF	9.25 VDC	50 µf	9.75 VDC	0.50 VDC

Figure 4. A table is used to calculate the output voltage (V_T) after each cycle.

...APPLICATION: SWITCHED-CAPACITOR VOLTAGE INVERTERS

How many switching cycles are needed before steady-state operation is achieved?

The number of switching cycles before steady-state voltage output is achieved is 7 (700 µsec).

What is the maximum steady-state output voltage?

The maximum steady-state output voltage is 9.75 VDC.

What is the minimum steady-state output voltage?

The minimum steady-state output voltage is 9.75 VDC. The overall output ripple is 0.5 VDC or about 5%.

What improvements can be made to increase the output voltage?

To increase the output voltage and reduce output ripple, the switching frequency can be increased from 10 kHz to 100 kHz. The maximum steady-state output voltage will be increased to 9.97 VDC with an overall output ripple of 0.05 VDC or less than 1%.

A switched-capacitor voltage inverter is a simple circuit that uses two capacitors and a switching network to generate an inverted voltage output. The basic circuit uses a capacitor that is switched periodically between an input voltage and output capacitor. This charge-pump capacitor is charged by the input voltage and transfers that charge to the inverted output capacitor that supplies current to the load.

Chapter Review

1. What are the three different standards systems used for identifying capacitors?
2. What units are most capacitors rated in?
3. Why is the dielectric breakdown that occurs through air or oil insulation temporary?
4. How are capacitors that overheat repaired?
5. What occurs when capacitor plates are too close together?
6. Where are oil-filled capacitors primarily used?
7. What is the capacitance for 5 C of charge for 6 V?
8. What is the time constant of a 5 F capacitor charged through a 3 Ω resistor?
9. What causes dielectric hysteresis loss?
10. Why is leakage current considered to be undesirable in a capacitor?
11. List three factors that affect the capacitance of a circuit.
12. What is the capacitance for 3 C of charge for 12 V?
13. What is a resistive-capacitive time constant?
14. What is a supercapacitor?
15. What is capacitance?
16. What is the main advantage of a supercapacitor over a battery?
17. List five types of leads used with aluminum electrolytic capacitors.
18. What is a dielectric constant?
19. What is dielectric breakdown in a capacitor the result of?
20. What is the time constant of a 0.5 F capacitor charged through a 5 Ω resistor?
21. What is the capacitance for 2 C of charge for 1 V?
22. What is used as the dielectric for variable capacitors?
23. Why must polarity be observed with a polarized capacitor?
24. How are capacitance and voltage ratings identified on multiple-unit capacitors?
25. When does current flow cease in a capacitive circuit with DC voltage applied?

13

AC Fundamentals

OBJECTIVES

- Distinguish between alternating current (AC) and direct current (DC).
- Explain AC terminology.
- Describe the operation of an AC generator.
- Explain the principles related to AC power generation, transmission, distribution, and conversion.

Alternating current (AC) is the voltage found in residential, commercial, and industrial locations. The direction of current flow in an AC system reverses periodically. Alternating current is generated at power plants and distributed to homes and businesses through a vast distribution network of high-voltage towers, step-up and step-down transformers, substations, and distribution transformers.

Knowledge of AC properties and related terminology is critical for working with AC power. It is important to understand the similarities and differences between AC and direct current (DC) power. Specialized terminology is used to describe the properties of AC power. Also, certain electrical phenomena occur only with AC. Finally, to fully understand AC power, it is important to know how AC is created and distributed for use.

ALTERNATING CURRENT

Terms

Alternating current (AC) is current that reverses its direction of flow at regular intervals.

Direct current (DC) is current that flows in only one direction.

Alternating current (AC) is current that reverses its direction of flow at regular intervals. Current first flows in one direction in the circuit, then changes direction and flows in the opposite direction. When polarity of the source voltage reverses direction, the current flow also reverses direction. As with DC waveforms, AC waveforms can have several different shapes, similar in appearance to DC waveforms. However, AC waveforms are different from DC waveforms in that they cross the 0 A baseline level when there is a reversal of current flow. **See Figure 13-1.** In telecommunications systems, AC has an advantage over DC because it radiates electromagnetic waves into space. AC creates changes in the electromagnetic field due to the reversing of the current. These changes in the electromagnetic field are used to change AC voltage levels, allowing information to be transmitted through space.

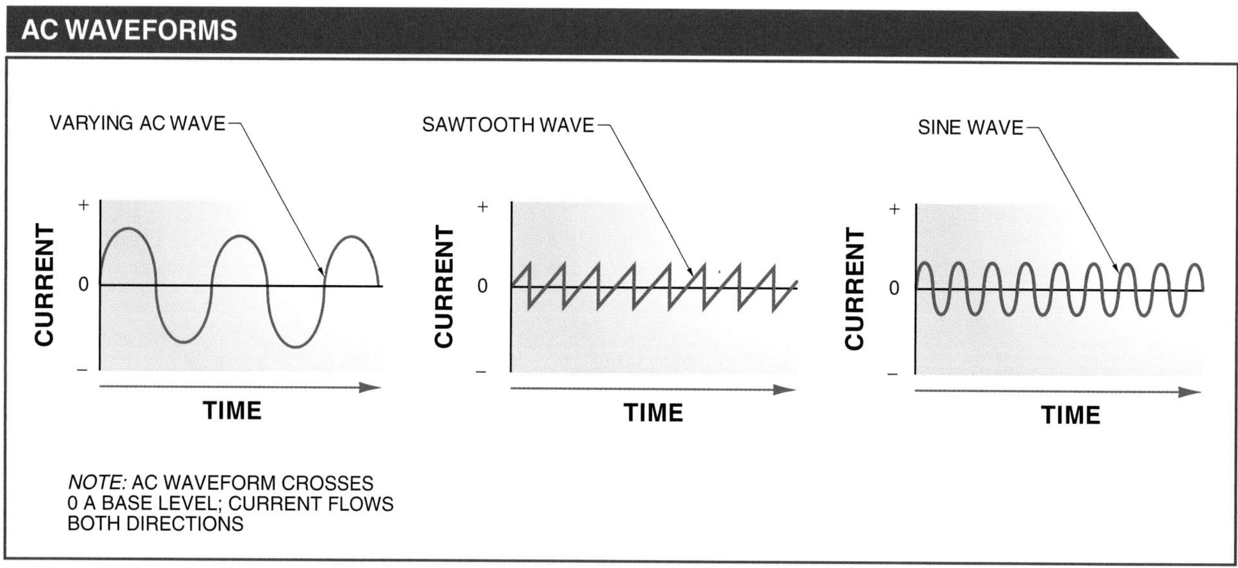

Figure 13-1. AC waveforms are different from DC waveforms in that AC waveforms cross the 0 A baseline level when there is a reversal of current flow.

Direct current (DC) is current that flows in only one direction. The polarity of current flow is from the negative terminal of the power source toward the positive terminal of the power source. DC waveforms can have several shapes including varying DC current waveforms, sawtooth waveforms, and sine waveforms. **See Figure 13-2.** Base levels of waveforms are at different levels. When none of these waveforms cross the 0 A baseline level, then current in the circuit flows in one direction and is DC.

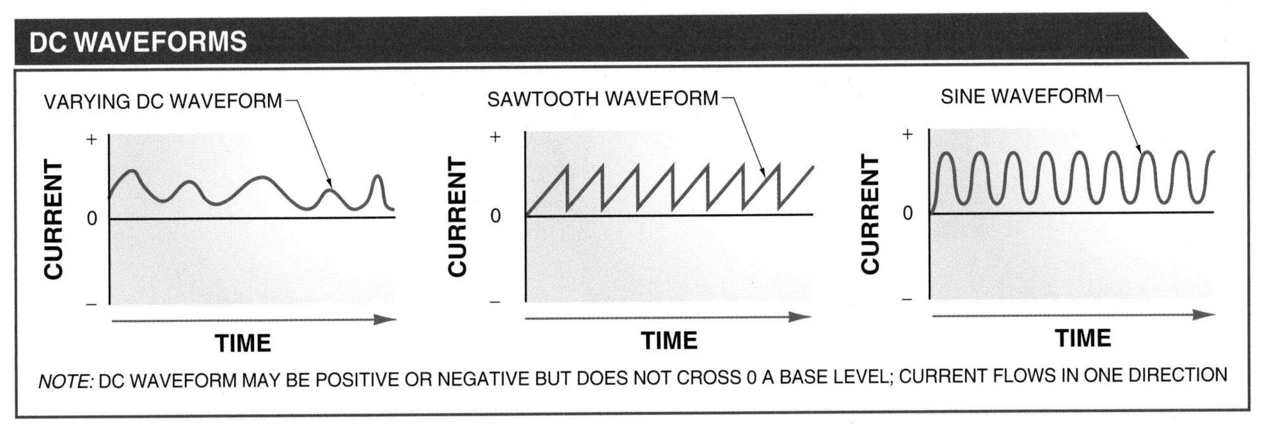

Figure 13-2. DC waveforms can have several shapes, including varying DC current waveforms, sawtooth waveforms, and sine waveforms.

AC differs from DC in that AC flow periodically reverses in direction. This reversal of current flow is an advantage of AC distribution systems over DC distribution systems. AC is more easily distributed at higher voltages and lower current, with greater efficiency in the delivery of power. Resistance and distribution currents result in lower efficiency in DC power distribution than AC power distribution. Although there are clear advantages of AC power distribution over DC power distribution, the use of high-voltage DC (HVDC) power systems has increased due to the development of semiconductor and control technology.

Terminology commonly used when working with AC circuits and equipment includes cycle, period, alternation, frequency, wavelength, skin effect, and phase. These terms are specific to AC systems and are used when designing, installing, and troubleshooting AC electrical and electronic systems.

Electromagnetic Interference

NORMAL WAVEFORM

ALTERED WAVEFORM DUE TO EMI

Cables used in electronic environments are often subject to electromagnetic interference (EMI). A possible solution to EMI problems is to use optical fiber cable. However, optical fiber cable is generally not as robust as copper wire cable. In addition, optical fiber is more expensive than copper cable. A properly balanced unshielded twisted pair provides significant immunity to EMI. A balanced line is a signal transmission cable in which two conductors carry the signal and create a balanced circuit alternating from positive to negative. The paired wires balance out each other's signals and help to nullify the effect of electromagnetic interference (EMI) from nearby wires or electromagnetic fields.

Another possible solution to EMI problems is to use shielded CAT 5e copper cable. This cable includes an aluminum foil shield that can provide significant immunity to EMI when properly grounded.

Terms

A **cycle** is one complete positive and negative alternation of a waveform.

A **period** is the time required to produce one complete cycle of a waveform.

An **alternation** is half of a cycle.

Cycle

A *cycle* is one complete positive and negative alternation of a waveform. For example, a sine wave cycle for voltage corresponds to 360° of mechanical rotation of an armature. The process of generating sine wave cycles for voltage continues as long as the armature of an AC generator is rotated in the magnetic field. For a two-pole generator, one cycle is generated for each 360° of mechanical rotation of the rotor. The induced voltage corresponds to the mechanical degree of rotation.

Period

A *period* is the time required to produce one complete cycle of a waveform. The period of a waveform is used in different ways to analyze a circuit. For example, the period is used to calculate the relationship between an AC waveform and the length of transmission line used to carry the signal. The relationship between period and transmission line length determines if the source current impedance or load impedance dominates circuit behavior. The shorter the period to transmission line length, the more likely the circuit is dominated by load impedance. A period is calculated by applying the following formula:

$$T = \frac{N}{f}$$

where
T = period (in sec)
N = number of cycles
f = frequency (in Hz)

Example: What is the period required to complete one cycle for a 60 Hz generator?

$$T = \frac{N}{f}$$

$$T = \frac{1}{60\ \text{Hz}}$$

$$T = \mathbf{0.016\ sec}$$

Alternation

An *alternation* is half of a cycle. An alternation begins each time a sine wave crosses the baseline and the direction of current flow reverses through the load. Positive alternations occur during the period the sine wave crosses the 0 A baseline in the positive direction and continue until it crosses the 0 A baseline in the negative direction.

Negative alternations occur during the period the sine wave crosses below the 0 A baseline (negative direction) and continue until it crosses above the 0 A baseline (positive direction) to start the next positive alternation. Each cycle in an AC sine wave has two alternations, one positive and one negative. **See Figure 13-3.**

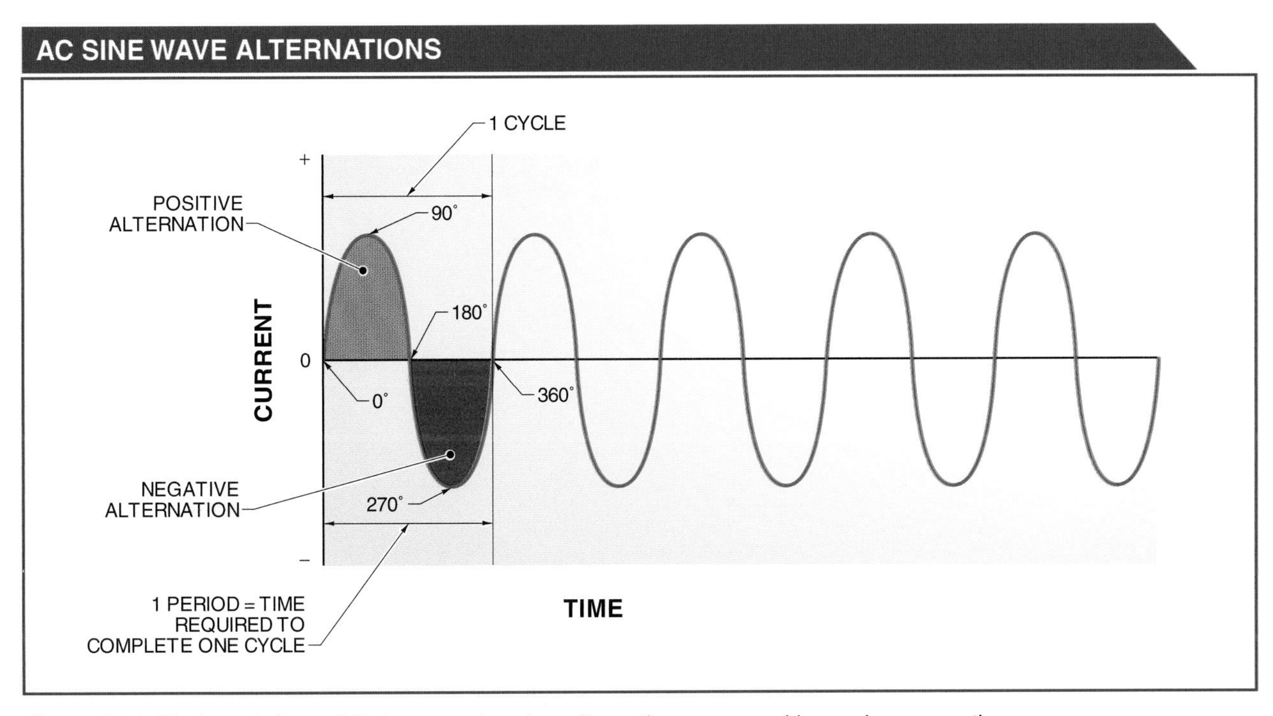

Figure 13-3. Each cycle in an AC sine wave has two alternations, one positive and one negative.

Frequency

Frequency is the number of cycles per second (cps) in an AC sine wave. The *hertz* is the international unit of frequency equal to 1 cycle per second. Frequency is also used to describe the power used in residential, commercial, and industrial facilities. For example, most power in North America operates at 60 Hz. Frequencies used for electromagnetic transmission are designated in bands. Frequency bands range from extremely low frequency (ELF) at 3 Hz to 30 Hz, to extremely high frequency (EHF) at 30 GHz to 300 GHz. **See Figure 13-4.**

In the United States, amplitude-modulated (AM) broadcasting operates at 540 kHz to 1600 kHz (1.6 MHz). Frequency-modulated (FM) broadcasting operates at 54 MHz to 216 MHz. Ultrahigh freq uency (UHF) broadcasting operates at 470 MHz to 890 MHz. Broadcast television frequencies range from channel 2 at 54 MHz (FM) to channel 83 at 890 MHz (UHF). Cable television (CATV) transmission uses a wide range of frequencies from 10 MHz to 1 GHz.

Terms

Frequency is the number of cycles per second (cps) in an AC sine wave.

The **hertz** is the international unit of frequency equal to 1 cycle per second.

FREQUENCY BANDS		
Frequency Band	**Frequency Range**	**Applications**
Extremely low (ELF)	3 Hz to 30 Hz	Biological usages, military communications
Superlow (SLF)	30 Hz to 300 Hz	AC power grids
Ultralow (ULF)	300 Hz to 3 kHz	Mine communications
Very low (VLF)	3 kHz to 30 kHz	Beacons, wireless monitors
Low (LF)	30 kHz to 300 kHz	Navigation, time signals
Medium (MF)	300 kHz to 3 MHz	AM broadcast
High (HF)	3 MHz to 30 MHz	Shortwave amateur radio
Very high (VHF)	30 MHz to 300 MHz	FM and TV broadcast
Ultrahigh (UHF)	300 MHz to 3 GHz	TV broadcast, mobile phones, wireless LAN, communications systems
Superhigh (SHF)	3 GHz to 30 GHz	Microwave, mobile phones, wireless LAN, radar
Extremely high (EHF)	30 GHz to 300 GHz	Radio astronomy, high-speed microwave radio

Figure 13-4. Frequency bands range from extremely low frequency (ELF) at 3 Hz to 30 Hz to extremely high frequency (EHF) at 30 GHz to 300 GHz.

Terms

A **wavelength** (λ) is the distance covered by one complete cycle of a given frequency of a signal as it passes through the air.

Wavelength

A *wavelength* (λ) is the distance covered by one complete cycle of a given frequency of a signal as it passes through the air. There is an inverse proportion between the wavelength of a signal and its frequency. Signals of different wavelengths fall into different frequency categories including radio waves, infrared light waves, visible light waves, X-ray light waves, and cosmic waves. **See Figure 13-5.** Electrical communication systems operate over a wide range of frequencies. These frequencies include systems in the audio range of 20 Hz to 20 kHz such as intercoms, telephones, and public address systems.

Wavelength is equal to the velocity of the wave divided by its frequency. When working with electrical signals, velocity is equal to the speed of light at 300×10^6 (300,000,000) meters per second. Wavelength becomes shorter as frequency is increased, and is measured in meters.

Wavelength can be calculated by applying the following formula:

$$\lambda = \frac{\upsilon}{f}$$

where

λ = wavelength (in m)

υ = velocity (in m/sec)

f = frequency (in Hz)

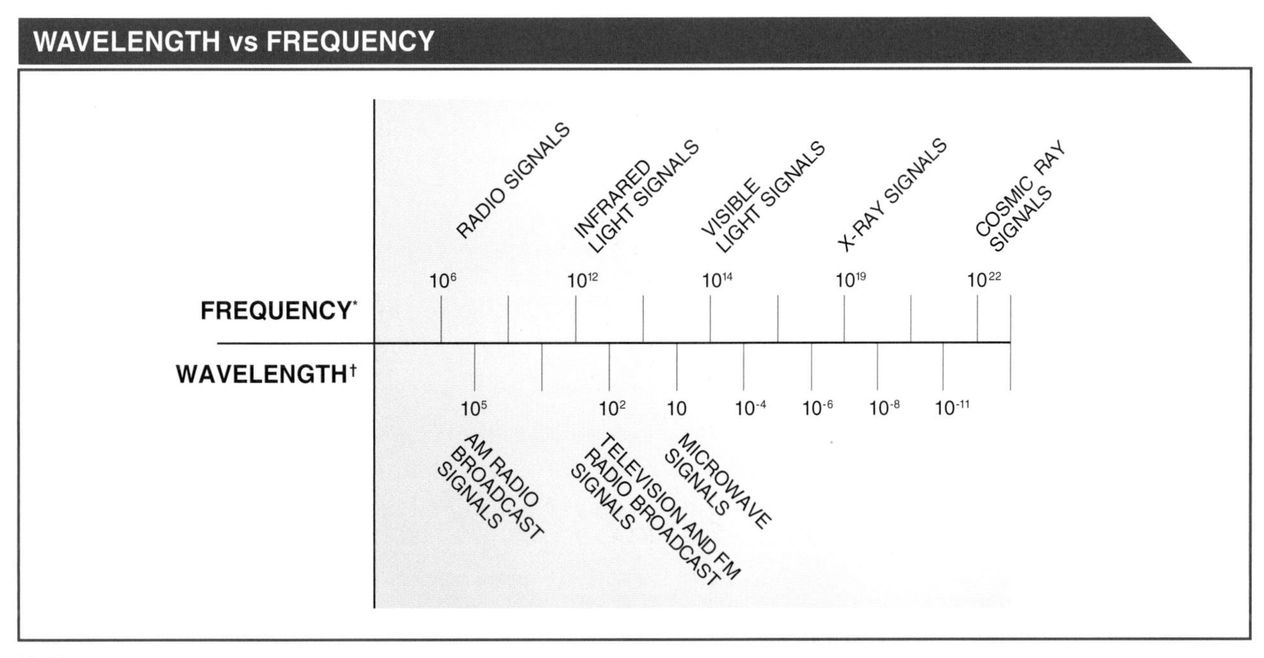

Figure 13-5. Signals of different wavelengths fall into different frequency categories, ranging from radio signals to cosmic signals.

Example: What is the wavelength of a transmitter that has a frequency of 1 MHz?

$$\lambda = \frac{\upsilon}{f}$$

$$\lambda = \frac{300{,}000{,}000 \text{ m/sec}}{1{,}000{,}000 \text{ Hz}}$$

λ = **300 m**

When the period of a sine wave is known, wavelength can be calculated by applying the following formula:

$$\lambda = \upsilon \times T$$

where

λ = wavelength (in m)

υ = velocity (in 300×10^6 m/sec)

T = period (in sec)

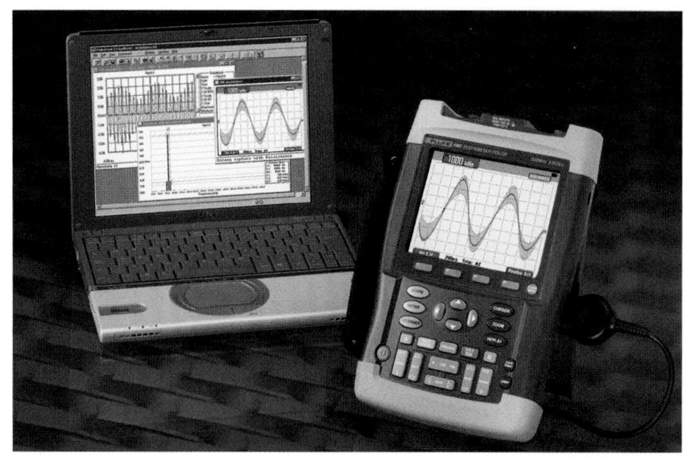

Fluke Corporation

Graphic-display test instruments provide a visual display of the frequency distribution in a circuit.

Example: What is the wavelength for a transmitter that has a period of 0.1667 sec?

$\lambda = \upsilon \times T$

$\lambda = 300{,}000{,}000 \text{ m/sec} \times 0.1667 \text{ sec}$

$\lambda = \mathbf{5{,}000{,}000 \text{ m}}$

Skin Effect

Skin effect is the effect that occurs in AC when more current flows near the outer surface (skin) of a conductor and less flows near the center of a conductor at higher frequencies. In both AC and DC circuits, the resistance value of a conductor opposes the current flow in the conductor. Skin effect increases as the frequency through the conductor increases. In DC, electron flow tends to be throughout the conductor, while in AC, skin effect tends to move electron flow to the surface of the conductor. **See Figure 13-6.**

Higher frequencies have greater skin effects. Since the electrons are not using the full cross-sectional area of the conductor, the resistance of the conductor has effectively been increased. Skin effect has little influence at low frequencies and is normally undetectable on power lines operating at frequencies of 50 Hz to 60 Hz.

Terms

Skin effect is the effect that occurs in AC when more current flows near the outer surface (skin) of a conductor and less flows near the center of a conductor at higher frequencies.

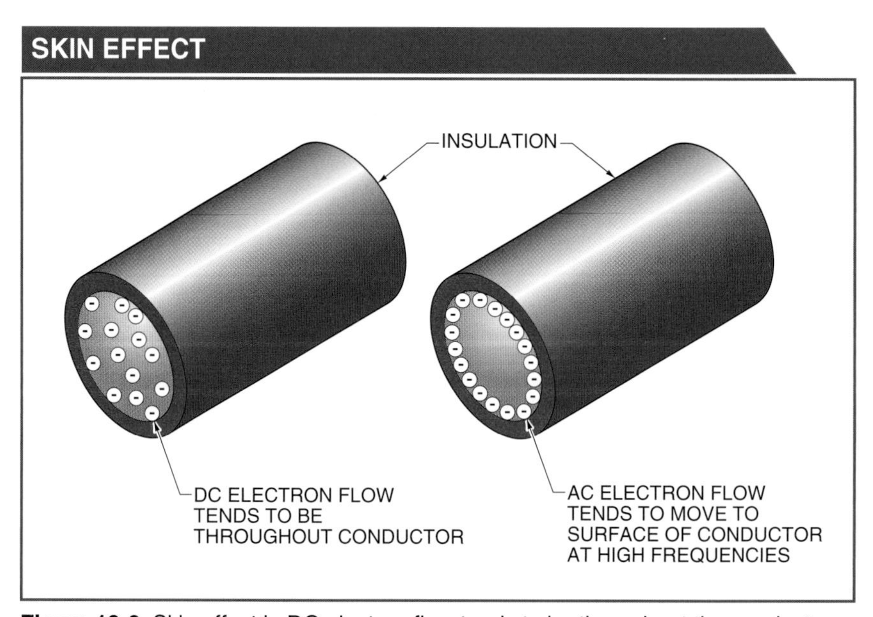

Figure 13-6. Skin effect in DC electron flow tends to be throughout the conductor, while skin effect in AC electron flow tends to move to the surface of the conductor.

A *stranded conductor* is a conductor composed of several strands of solid wire wrapped together to make a single conductor. To reduce skin effect in power cables and cords, stranded conductors are normally used rather than solid conductors. **See Figure 13-7.** A stranded conductor with the same circular-mil rating of a solid conductor has a greater surface area and flexibility than the solid conductor. Stranded conductors can also reduce undesirable eddy currents if the individual conductor strands are twisted tightly together. Twisting the conductor strands tightly together causes a cancellation of magnetic fields caused by eddy currents. Telecommunications conductors are twisted in this manner to reduce interference and crosstalk.

Terms

A **stranded conductor** is a conductor composed of several strands of solid wire wrapped together to make a single conductor.

Phase (ϕ) is the time relationship of a sine wave to a known time period.

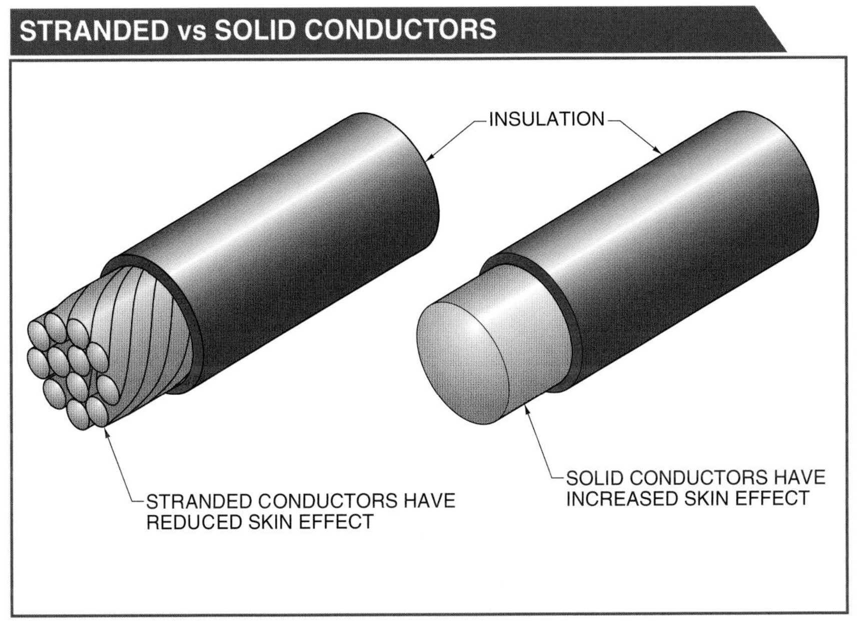

Figure 13-7. To reduce skin effect in power cables and cords, stranded conductors are normally used rather than solid conductors.

Phase

Phase (ϕ) is the time relationship of a sine wave to a known time period. Sine waves are often used to depict the relationships between two or more signals. In-phase sine waves have peak values and baseline crossings that occur at the same time, while out-of-phase sine waves have peak values and baseline crossings that occur at different times. **See Figure 13-8.** Sine waves with no time displacement are in phase. The phase angle between them is 0°. Both sine waves cross the baseline at the same time and move in the same direction. Maximum out-of-phase condition occurs when signals are 180° out of phase.

With a maximum out-of-phase condition, both signals cross the baseline at the same time, but move in opposite directions from one another. As one sine wave moves in a positive direction, the opposite sine wave moves in a negative direction. Sine waves that are 90° out of phase lag the opposite sine wave by 90°. For example, a three-phase (3ϕ) AC generator produces three AC outputs. Each of the three AC outputs are out of phase by 120° because there are three equally spaced windings in a 3ϕ AC generator.

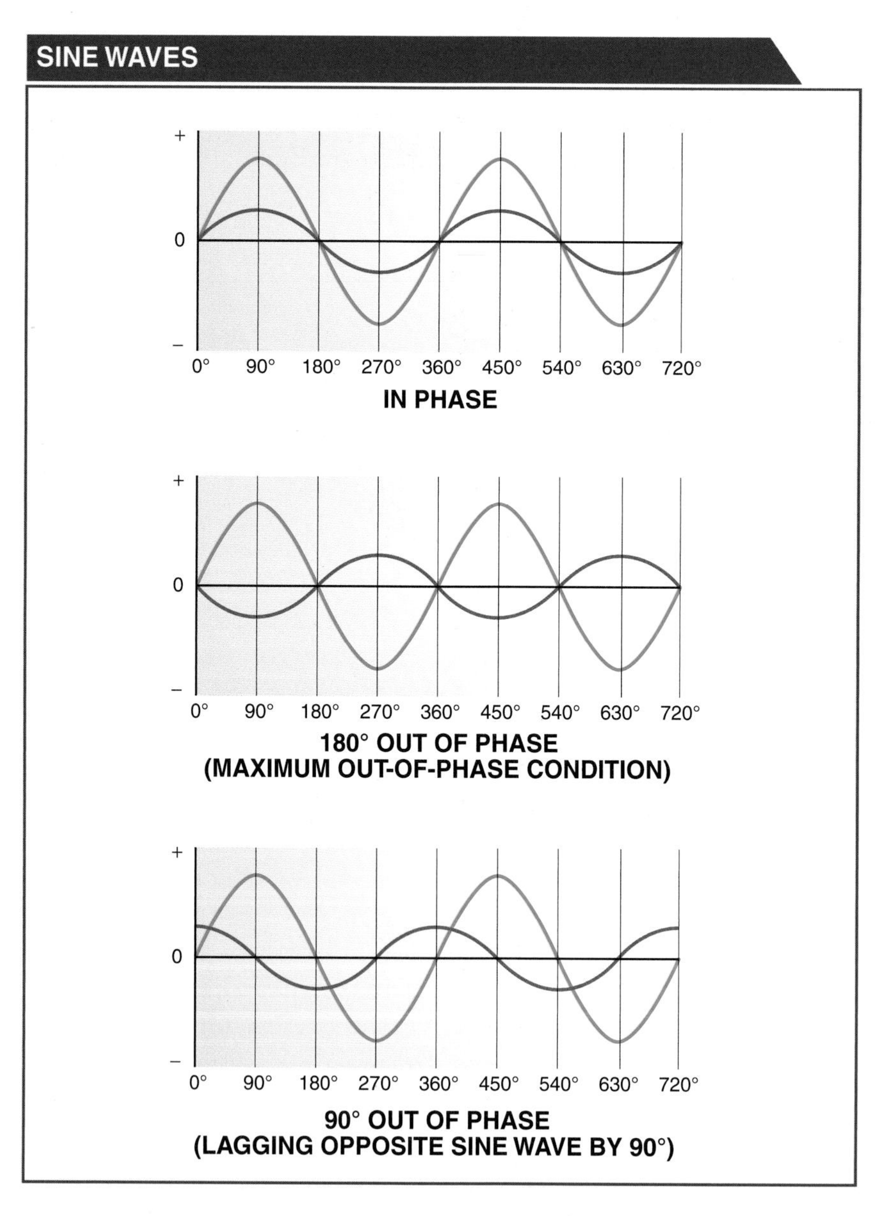

Figure 13-8. In-phase sine waves have peak values and baseline crossings that occur at the same time, while out-of-phase sine waves have peak values and baseline crossings that occur at different times.

SINGLE-PHASE AC GENERATORS

A single-phase (1ϕ) AC generator consists of a stationary magnet and a single winding that is rotated in the field. The rotating winding is referred to as the armature, and the stationary magnet is referred to as the stator. The armature is the part of the generator from which electric power is generated. When the armature is rotated in a magnetic field, a voltage is induced into the loop of wire. When a load is connected across the armature, current flows through the armature.

A *generator (alternator)* is a machine that converts mechanical energy into electrical energy by means of electromagnetic induction. The difference between a DC generator (dynamo) and an AC generator is that the DC generator has split rings (commutator), whereas the AC generator has slip rings. Each slip ring is insulated from the other and attached to one lead of the armature. Brushes ride on the slip rings to remove power from the generator. As the armature is rotated in the magnetic field, the voltage at each position is equal to the sine of the angle multiplied by the maximum voltage.

Each complete rotation of the armature in a 1ϕ AC generator produces one complete AC cycle. **See Figure 13-9.** The rotation speed of the armature determines the output frequency of an AC generator. In position A, before the armature begins to rotate in a clockwise direction, there is no voltage and no current in the external (load) circuit because the armature is not cutting across any magnetic lines of force (0° of rotation).

Terms

A **generator (alternator)** is a machine that converts mechanical energy into electrical energy by means of electromagnetic induction.

Early Applications of Electricity

The electrical industry essentially began in 1860, when English physicist Joseph Swan patented a low-voltage lamp. Swan's lamp was a low-voltage, battery-operated, high-current component, with low light output and a short life. Transmission losses due to the resistance of the wire and the high current required the power source to be close to the lamp. This condition ruled out the possibility of a central power station to supply the lamps. Frequent changes of batteries were necessary to keep the lamps operational. Swan's low-voltage lamp is used in modern applications such as flashlights and vehicle lighting systems, where the electrical source is a DC battery located close to the low-voltage, high-current lamps.

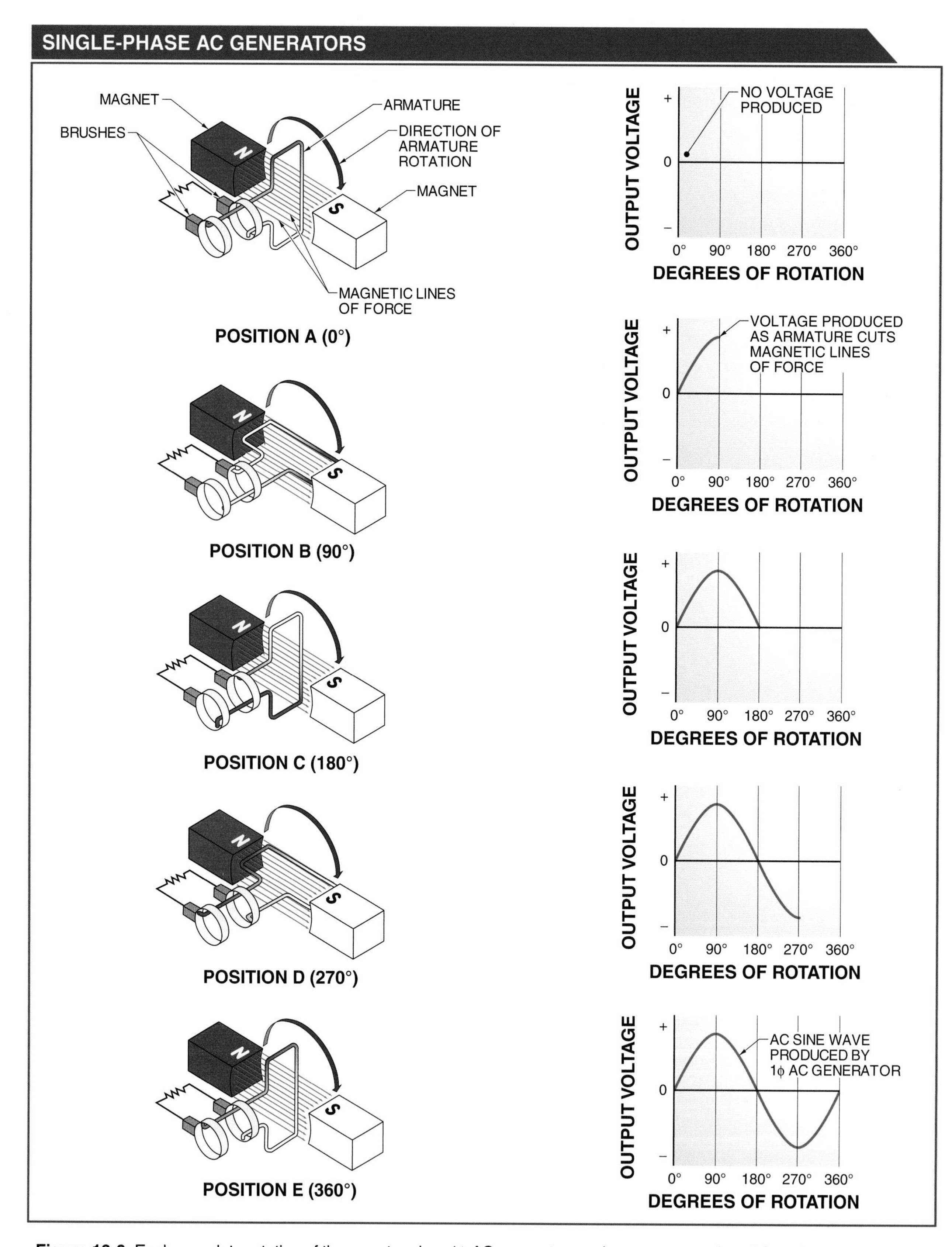

Figure 13-9. Each complete rotation of the armature in a 1ϕ AC generator produces one complete AC cycle.

As the armature rotates from position A to position B, each half of the armature cuts across the magnetic lines of force, producing current in the external circuit. The current increases from zero to its maximum value in one direction. This changing value of current is represented by the first quarter (90° of rotation) of the sine wave.

As the armature rotates from position B to position C, current continues in the same direction. The current decreases from its maximum value to zero. This changing value of current is represented by the second quarter (90° to 180° of rotation) of the sine wave.

As the armature continues to rotate to position D, each half of the coil cuts across the magnetic lines of force in the opposite direction. This changes the direction of the current. During this time, the current increases from zero to its maximum negative value. This changing value of current is shown by the third quarter (180° to 270° of rotation) of the sine wave.

As the armature continues to rotate to position E (position A), the current decreases to zero. This completes one 360° cycle of the sine wave.

AC Generator Output Frequency

AC generator output frequency is based on the number of poles and the rotational speed of the armature. AC generator output frequency is calculated by applying the following formula:

$$f = \frac{P \times S}{120}$$

where

f = frequency (in Hz)

P = number of poles

S = speed of armature rotation (in rpm)

120 = constant

Example: What is the output frequency of a two-pole AC generator rotating at 3600 rpm?

$$f = \frac{P \times S}{120}$$

$$f = \frac{2 \times 3600 \text{ rpm}}{120}$$

$$f = \frac{7200}{120}$$

f = **60 Hz**

Cummins Power Generation

Large-scale AC generators are used in locations such as hospitals, airports, and school or office campuses.

Early Light Transmission

American scientist and inventor Thomas Edison began research on the electric lamp in 1878 in an effort to improve upon a low-voltage lamp. Edison set out to develop a higher-resistance filament of 100 Ω or more, using a higher voltage and a lower current to produce incandescence. Comparing the two lamps, a 40 W/5 V Swan lamp required 8 A to emit light, while a 40 W lamp operated at 100 V requires only 0.40 A to emit light. Edison's high-voltage, low-current lamp had the advantage of lowering transmission losses on the conductors while providing the same level of illumination.

A conversion factor of 120 is used to convert to cycles per sec. By increasing the speed of rotation of the armature, EMF also increases. As the armature is rotated in a magnetic field, voltage is induced that is proportional to the sine wave of the angle at which the armature cuts the magnetic field. For example, with a two-pole AC generator, each mechanical degree of rotation equals one electrical degree. One complete revolution generates one AC cycle.

Beginning at position A, the armature rotates in a counterclockwise direction and moves parallel with the magnetic field. Because the armature does not cut the magnetic lines of force while moving parallel to the field, no voltage is induced. At position B, the armature has rotated mechanically 90° and is rotating perpendicularly to the magnetic lines of force. This condition results in the maximum number of magnetic lines of force being cut. A maximum voltage is induced that corresponds electrically to 90°.

As the armature rotates mechanically another 90° to position C, the armature rotates parallel to the magnetic lines of force, and no voltage is induced. This corresponds electrically to 180°. At position D, the armature cuts the maximum number of magnetic lines of force. A maximum negative voltage corresponds electrically to 270°. At position E, the armature has rotated mechanically 360°. At this point, the armature rotates parallel with the flux field and no voltage is induced, which corresponds electrically to 360°. Another 360° of mechanical rotation (second cycle) corresponds to electrical degrees 360° to 720°.

For a four-pole generator, two electrical degrees are produced for each mechanical degree of rotation. For a six-pole generator, three electrical degrees are produced for each mechanical degree of rotation. For an eight-pole generator, four electrical degrees are produced for each mechanical degree of rotation. The number of electrical degrees produced in a generator can be calculated by applying the following formula:

$$V = \frac{M \times P}{2}$$

where

V = electrical rotation (in °)

M = mechanical rotation (in °)

P = number of poles

Example: What number of electrical degrees is produced in a generator with 8 poles when the armature is rotated 90°?

$$V = \frac{M \times P}{2}$$

$$V = \frac{90° \times 8}{2}$$

V = **360° (1 cycle)**

Increasing the number of poles allows for a reduction in the speed of armature rotation. Lower armature rotation speed increases the life of the generator. If the operating frequency is reduced, the rotating speed is also reduced. The speed of armature rotation can be calculated by applying the following formula:

$$S = \frac{f \times 120}{P}$$

where

S = speed of armature rotation (in rpm)

f = frequency (in Hz)

120 = constant

P = number of poles

Example: What is the speed of armature rotation for a 60 Hz four-pole generator?

$$S = \frac{f \times 120}{P}$$

$$S = \frac{60\text{ Hz} \times 120}{4}$$

S = **1800 rpm**

In the United States and Canada, 60 Hz power is standard, and 50 Hz power is standard in Europe. Different power standards affect generator output. For example, an eight-pole generator that delivers an output frequency of 50 Hz requires a speed of 750 rpm, and an eight-pole generator that delivers an output frequency of 60 Hz requires a speed of 900 rpm.

The frequency output of an AC generator must be kept constant. Many types of loads are designed to operate at the specified frequency of the power system. A frequency that is too low can cause electric motors to run slow and draw excessive current. A frequency that is too high can cause electric motors to run too fast and overheat. Some devices, such as transformers, draw high current when the frequency is reduced, thereby lowering their efficiencies and reducing their ability to deliver rated capacities.

Generators with different frequencies cannot be interconnected without an elaborate industrial electronic system, such as a solid-state converter that converts the incoming AC power to DC power and synthesizes the output AC waveform. Mechanical systems are also used to perform AC to DC conversion.

AC POWER GENERATION

AC power is generated from electric power generators. Steam-powered turbines typically drive electric power generators. Steam-powered turbines receive their energy from either fossil fuel or nuclear energy. The generator in a steam-powered turbine system converts the mechanical energy of the turbine into electrical energy through magnetic induction. Typically, AC generators are high-voltage 3ϕ machines.

To obtain three phases, 3ϕ generators have three sets of armature windings that are located mechanically 120° apart. Generation of 3ϕ power is essentially the same as for 1ϕ power. A 3ϕ generator has three sets of equally spaced windings. Adjacent windings of each phase are spaced mechanically 60° apart. **See Figure 13-10.** Each phase is brought out to its own set of slip rings. The phases are generated so that phase 1 leads phase 2 by 120° and phase 3 by 240°.

With high-power AC generators that have high current output, the armature is often the stator (stationary part), and the field is rotated. With high-power AC generators, the power output is captured by fixed wiring attached to the armature. The stationary armature windings surround the rotating field. Fixed wiring overcomes the disadvantage of using brushes and slip rings, which limit power output. Excessive arcing occurs at the brushes and slip rings when high levels of current are present.

Fluke Corporation

Insulation testers are used to test the integrity of the insulation of conductors and windings in electric motors.

By combining the individual coils of each phase into a single winding in the stationary armature, it is easier to recognize the 120° separation between phases.

THREE-PHASE GENERATORS

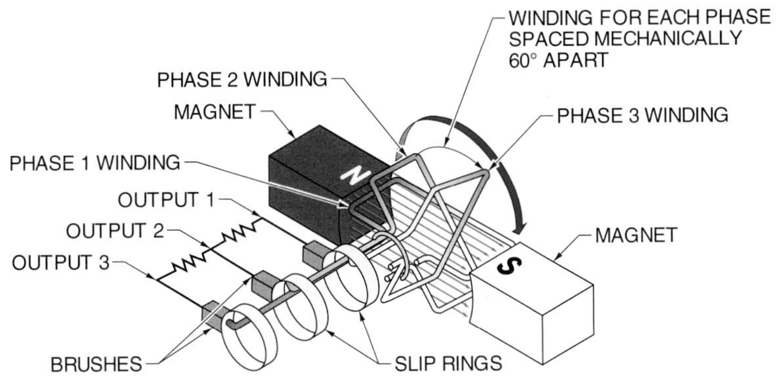

Figure 13-10. To obtain three phases, a 3ϕ generator has three sets of armature windings that are located mechanically 60° apart.

Typically, an electromagnet is used as the rotor in place of a four-pole permanent magnet. DC power is applied to the field coil through brushes and slip rings to create the field. The current required for the field is much lower than the current output of the generator. The strength of the magnetic field can be adjusted, thereby changing the level of the output voltage. On large generators such as those used for power generation, a DC generator armature is mounted on the shaft of the AC generator armature to supply the current to the field windings. This eliminates the need for brushes and slip rings. The strength of the electromagnet for small DC generators can be adjusted externally with fixed wiring to its field. Only magnetic coupling is required in this case.

A *volt ampere (VA)* is the unit of measure for apparent power. A volt ampere is the amount of current that a generator can deliver to the load at its rated voltage. Although electrical power generation is usually rated in megawatts (MW), AC generators are rated in volt amperes. In AC circuits, voltage and current are not always in phase, which results in less real or true power (W) being dissipated in a circuit containing inductance or capacitance than the product of the source voltage times the total amperes.

Terms

A **volt ampere (VA)** is the unit of measure for apparent power. A volt ampere is the amount of current that a generator can deliver to the load at its rated voltage.

Regardless of the type of load that is applied to it, an AC generator has internal resistance that generates heat when current is removed. This heat can become excessive and destroy the AC generator if its current rating is exceeded. Output voltages of commercial AC generators are typically in the range of 2 kV through 4 kV, but can have output voltages as high as 22 kV.

AC POWER TRANSMISSION AND DISTRIBUTION

Terms

A **transformer** is an electrical device that uses electromagnetism to change voltage from one level to another or to isolate one voltage from another.

AC power is transmitted and distributed to all parts of the electrical utility system from the power-generating plant to customer service-entrance equipment. A *transformer* is an electrical device that uses electromagnetism to change voltage from one level to another or to isolate one voltage from another. Transformers are used to change high voltage levels from the power supplier (utility company) to lower voltage levels required by the end user. A power transmission and distribution system commonly includes the following parts:

Tech Tip

There are currently about 60 systems worldwide that convert AC power to DC power for transmission over long distances.

- Step-up transformers. The generated voltage is stepped up to transmission voltage level, usually between 12.47 kV and 245 kV.
- Power plant transmission lines. The 12.47 kV to 245 kV power plant transmission lines deliver power to the transmission substations.
- Power transmission substations. The voltage is transformed to a lower primary (feeder) voltage, usually between 4.16 kV and 34.5 kV.
- Primary transmission lines. The 4.16 kV to 34.5 kV primary transmission lines deliver power to the distribution substations and heavy industry.
- Distribution substations. The voltage is transformed down to utilization voltages, ranging from 480 V to 4.16 kV.
- Distribution lines. Power is carried from the distribution substation along the street or rear lot lines to the final step-down transformers.
- Final step-down transformers. Voltage is transformed to the required voltage, such as 480 V or 120/240 V. The final step-down transformers may be installed on poles, grade-level pads, or in underground vaults. The secondary of the final step-down transformer is connected to service drops that deliver power to service-entrance equipment. **See Figure 13-11.**

A utility company power transmission and distribution system delivers power to industrial, commercial, and residential customers. The number and size of transformers used to step down (reduce) the voltage before the customer power distribution system depends on customer power requirements. For example, in a typical heavy industrial facility, the electricity may be delivered directly from a transmission substation to an outside transformer vault. Service-entrance conductors are routed from the outside transformer vault through an outdoor busway to a metered switchboard. Power is then fed through circuit breakers in the panelboard and routed through busways to power distribution panels and busways with plug-in sections to the points of use.

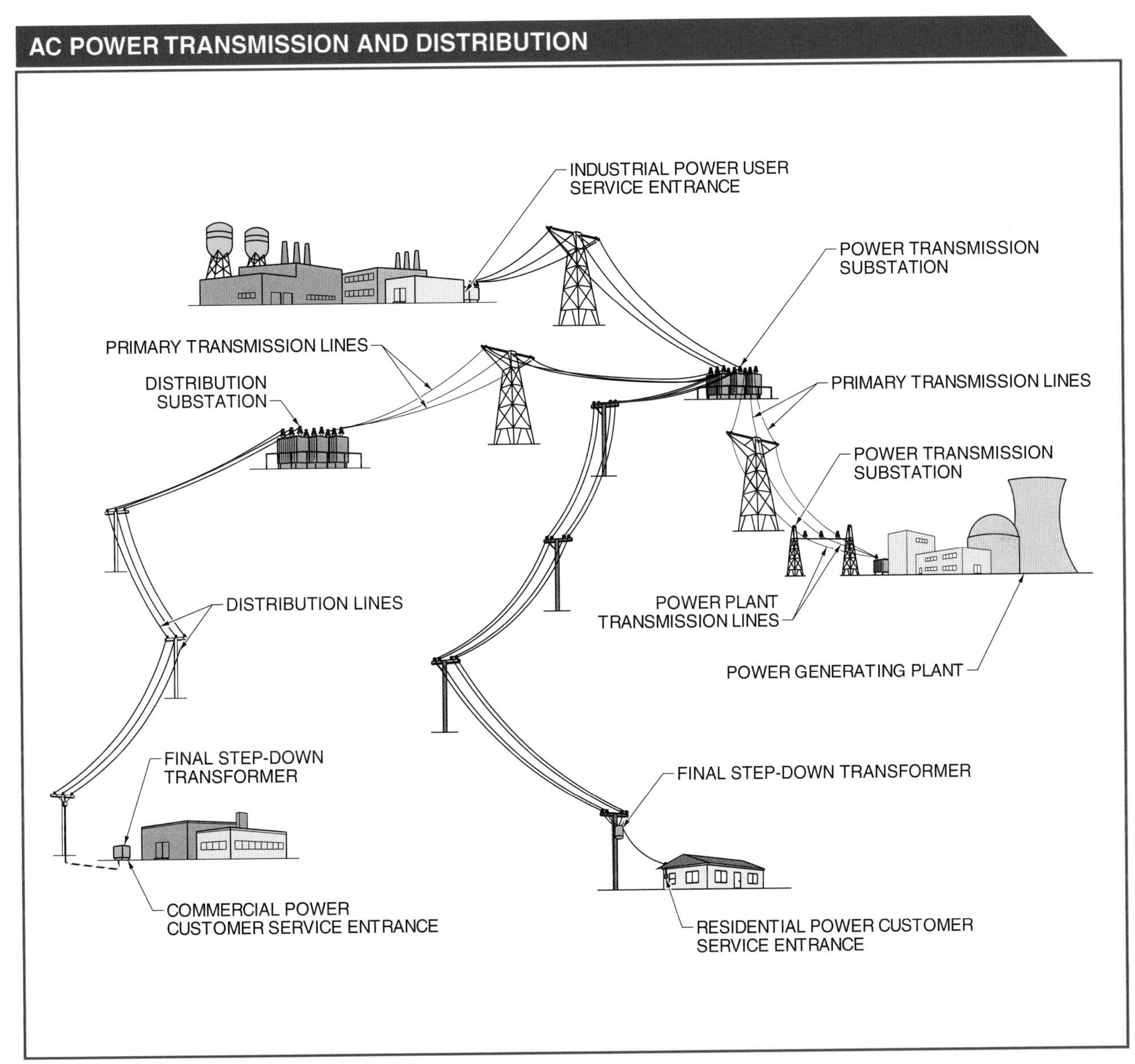

Figure 13-11. AC power is transmitted and distributed to all parts of the electrical utility system, from the power-generating plant to customer service-entrance equipment.

Tech Tip

Of the available global hydro energy of 15,000 TWhr/year, only about 2000 TWhr/year are actually in production. This means that Central and Western Europe, where hydro energy resources are nearly fully utilized, can begin receiving hydroelectric power from regions such as central Africa.

Depending on customer needs, the power distribution system delivers power at standard voltage levels and fixed current ratings to set points such as receptacles. Common voltage levels include 120 V, 208 V, 240 V, 277 V, and 480 V.

High transmission voltages are reduced to lower voltages at primary distribution transformer substations. Primary distribution voltage is usually 23 kV, but may be other voltage values. This level of voltage is supplied to industrial and commercial end users that have a high demand. End users have their own transformer substations to step the voltage down for local distribution. In all cases, these are 3ϕ four-wire systems, with the fourth wire as the grounded conductor. Primary distribution voltages are stepped down to 2.3 kV to 4.1 kV for local distribution in residential areas. These voltages are then stepped down to 120/240 V, 1ϕ, three-wire drop to residential end users.

Terms

A **utilization voltage** is a secondary voltage of transformers applied to rated loads.

A *utilization voltage* is a secondary voltage of transformers applied to rated loads. A utilization voltage is used to provide power to light commercial or residential components. Most household lamps, appliances, and heaters specify 120 V or 240 V.

High-Voltage Direct Current (HVDC)

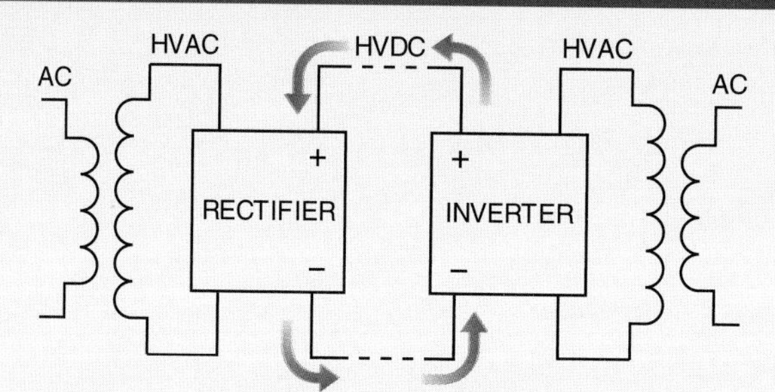

In the late eighteen hundreds, DC power distribution was used primarily for lighting. Due to the difficulties of varying DC voltage, the distribution was generally low voltage, around 100 VDC. To minimize losses, large conductors were used to distribute power, although they were costly and limited the distance power could be transmitted. Eventually, AC power distribution overtook DC power distribution due to the invention of the iron core transformer. Using an inexpensive transformer, voltages could be stepped up or down to reduce conductor size and improve transmission efficiencies over long distances.

During the twentieth century, electronic components were developed that permitted high-voltage rectification and inversion, making high-voltage direct-current (HVDC) power distribution possible. In 1954, the world's first HVDC transmission system was commissioned for a 20 MW system using an undersea cable that operated at 100 kV. A typical HVDC power distribution system converts AC to DC using a rectifier. Once the power has been transmitted, the DC is converted back to AC using an inverter.

Recently, there has been renewed interest in HVDC power distribution systems, especially in the renewable energy industry. HVDC power distribution sustains fewer losses than AC over long distances. The primary applications of HVDC include connections between unsynchronized power grids, systems that use underground and/or underwater connections greater than 50 km, and above ground connections exceeding 800 km.

AC POWER CONVERSION

AC power can be easily stepped up or stepped down for distribution by using transformers. AC is also converted into DC for use in many electronic applications. Modern technology also permits the conversion of AC power into high-voltage DC (HVDC) for power distribution.

A *converter* is a device that changes AC power to DC power. An *inverter* is a device that changes DC power into AC power. At the location where power is to be used, an inverter changes the HVDC back to AC power, to be stepped down using a transformer. Modern converters also have the ability to interconnect systems of different frequencies.

For example, a 50 Hz power supply system can be interconnected with a 60 Hz power supply system to share the same power grid. HVDC transmission has other advantages over HVAC transmission. For example, HVAC transmission requires four conductors. A conductor is required for each phase along with a grounded conductor. For the same amount of power transmission, HVDC transmission requires only two conductors. By using two conductors rather than four conductors, conductor resistance is reduced by 50%. In addition to cost savings in material and line loss reduction, transmission towers are also less costly to maintain due to the lower weight load generated by fewer conductors. Losses due to eddy currents and skin effect are also eliminated. With lower conversion costs, HVDC transmission is less expensive than HVAC over long distances.

Rectification is the process of changing AC to DC. Rectification is typically used to create DC voltage from AC voltage. An *ignitron tube* is a single-anode pool tube in which an igniter is used to initiate the cathode spot before each conducting period. **See Figure 13-12.** An ignitron converts AC to DC. Ignitrons are capable of withstanding high voltages without breakdown and have the ability to conduct very high currents. Ignitron tubes with ratings of 50 kV at 700,000 A were developed for high-current AC rectification for large industrial facilities. Devices within an ignitron include an igniter connector, igniter, water-cooling tube, anode connector, anode, mercury-pool, and a cathode connector. Ignitrons are sometimes water-cooled to remove heat generated during operation.

Anode voltage is normally applied during operation. This voltage may be either DC or AC. For an ignitron to conduct electricity, the anode must be positive. During the AC cycle, the positive alternation must be present. The mercury is vaporized each cycle of operation through an igniter, which is in contact with the mercury-pool rectifier tube. Precise timing of conduction is achieved with the control circuit vaporizing the mercury pool into a gas. The ignitron continues to conduct until the positive potential is removed from the anode.

Terms

A **converter** is a device that changes AC power to DC power.

An **inverter** is a device that changes DC power into AC power.

Rectification is the process of changing AC to DC. Rectification is typically used to create DC voltage from AC voltage.

An **ignitron tube** is a single-anode pool tube in which an igniter is used to initiate the cathode spot before each conducting period.

In a converter circuit, ignitrons are wired in series so that voltages higher than the tube rating can be used. If greater current capacity is required, the ignitrons can also be wired in parallel. Ignitrons are characterized by their ability to withstand currents several times their rated value for several cycles.

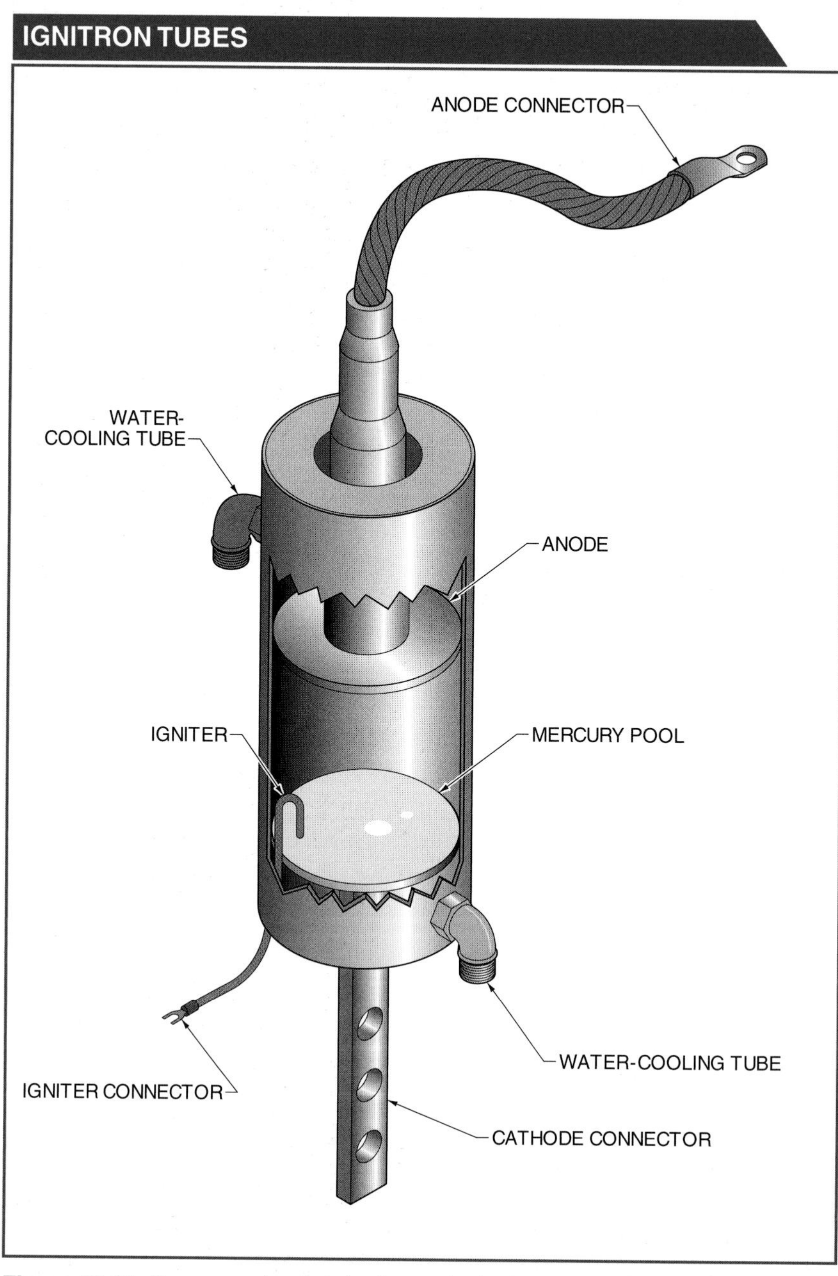

Figure 13-12. Components of an ignitron tube include an igniter connector, igniter, anode connector, anode, mercury-pool, water-cooling tube, and a cathode connector.

Ignitrons are manufactured and used in place of semiconductors in certain applications. Because ignitrons can switch several hundred kA of current and withstand high voltage levels, typical ignitron applications include the switching of high-capacitance capacitor banks during electromagnetic forming, electrohydraulic forming, or emergency short-circuiting of high-voltage power sources.

A *diode* is an electronic device that allows current to flow in only one direction. An *anode* is a positive lead of a diode. A *cathode* is a negative lead of a diode. A diode allows current to flow only when a positive voltage is applied to the anode (forward bias). The diode does not allow current to flow when the anode and cathode have the same polarity or if a positive voltage is applied to the cathode (reverse bias).

A *thyristor* is a solid-state switching device that switches current ON using a quick pulse of control current. Thyristor types include the silicon-controlled rectifier (SCR) and the triac. Thyristors are typically used in power control applications such as rectification. Thyristors substantially reduce the size and complexity of HVDC converter stations.

A *silicon-controlled rectifier (SCR)* is a solid-state rectifier with the ability to rapidly switch currents. The gate pulse on an SCR serves the same purpose as the igniter on the ignitron. Operation of the SCR is identical to that of the ignitron. A gate pulse is necessary for the SCR to conduct and the anode must be positive with respect to the cathode. To stop conduction, the positive potential on the anode must be removed. Silicon-controlled rectifiers use three electrodes for normal operation. The three SCR electrodes are the anode, cathode, and gate. The three terminals of the triac are the main terminal 1 (MT_1), main terminal 2 (MT_2), and gate. **See Figure 13-13.** Silicon-controlled rectifiers are typically used in applications such as power control systems, battery chargers, and protection circuits.

Silicon-controlled rectifiers are also used with high-voltage and current applications, often to control AC where the change of current flow causes the SCR to switch OFF and ON automatically.

Terms

A **diode** is an electronic device that allows current to flow in only one direction.

An **anode** is a positive lead of a diode.

A **cathode** is a negative lead of a diode.

A **thyristor** is a solid-state switching device that switches current ON using a quick pulse of control current.

A **silicon-controlled rectifier (SCR)** is a solid-state rectifier with the ability to rapidly switch currents.

Early AC Power Systems

American scientist Nikola Tesla, who at one time worked for Thomas Edison, was instrumental in the development of AC power systems. Tesla invented many AC components and devices along with the polyphase generator. American inventor George Westinghouse purchased the rights to Tesla's patents. Westinghouse set out to establish AC as the primary source of electrical power. Westinghouse liked the idea of supplying AC power over long distances using large generators, which was not possible with DC power generation. In 1893, both Edison and Westinghouse bid against each other to provide electrical lighting for the Columbian Exhibition and World's Fair in Chicago, Illinois. Westinghouse won the bid, bringing AC power to the attention of the public.

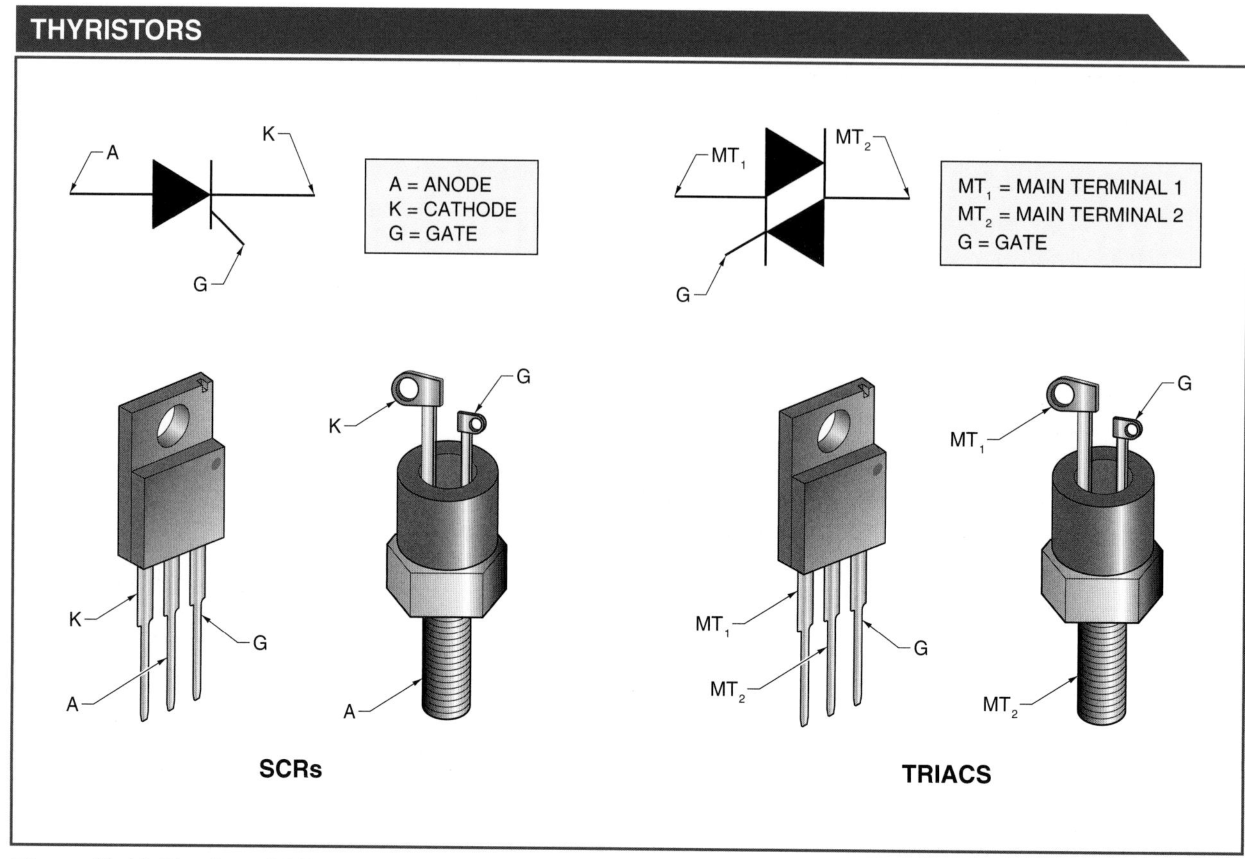

Figure 13-13. The three SCR electrodes are the anode, cathode, and gate, and the three terminals of the triac are the main terminal 1 (MT_1), main terminal 2 (MT_2), and gate.

Terms

Forward breakover voltage is the voltage required to switch an SCR into a conductive state from a nonconductive state.

A **harmonic** is a frequency that is an integer (whole number) multiple (second, third, fourth, fifth, etc.) of the fundamental frequency.

The gate serves as the control point for an SCR. The SCR differs from a semiconductor diode in that it does not pass significant current, even when forward biased, unless the anode voltage equals or exceeds the forward breakover voltage. *Forward breakover voltage* is the voltage required to switch an SCR into a conductive state from a nonconductive state. An SCR switches ON and becomes highly conductive when the forward breakover voltage is reached. An SCR is unique because the gate current is used to reduce the level of breakover voltage necessary for the SCR to conduct.

Low-current SCRs can operate with an anode current of less than 1 mA. High-current SCRs can handle load currents in the hundreds of amperes. The size of an SCR increases with an increase in its current rating. Silicon-controlled rectifiers are used in numerous applications such as power switching, speed controllers for motors, power supplies, protection circuits, and electric automobiles.

A *harmonic* is a frequency that is an integer (whole number) multiple (second, third, fourth, fifth, etc.) of the fundamental frequency. Harmonics can be either voltage harmonics or current harmonics.

Voltage and current harmonics can affect each other. For example, current harmonics can produce voltage harmonics. Each harmonic has a name (number) or order (such as fifth order harmonic), frequency, and sequence. Harmonics are generated any time a nonlinear device, such as a thyristor, is switched. For example, the second harmonic of 60 Hz is 120 Hz, the third harmonic of 60 Hz is 180 Hz, and this pattern continues for as many harmonics as are generated. Some waveforms (ideal sine waves) contain only the fundamental frequency.

An *even harmonic* is an even multiple of the fundamental frequency (second, fourth, sixth, etc.). An *odd harmonic* is an odd multiple of the fundamental frequency (third, fifth, seventh, etc.). Harmonics are also classified by their sequence when related to three-phase electrical systems. A *sequence* is the relationship between a harmonic and the fundamental frequency. Positive sequence harmonics are undesirable because they can overheat conductors and transformers due to the addition of the waveforms. Negative sequence harmonics weaken rotating magnetic fields, creating potential torque problems with motors. Zero sequence harmonics, known as triplens, which are every third harmonic, are in-phase and add to voltages and/or currents in neutral wire that cause heating effects. Waveforms may include odd harmonics, even harmonics, or both odd and even harmonics, in addition to the fundamental frequency. **See Figure 13-14.**

When DC is converted back to AC, the thyristor control unit triggers each unit in the proper sequence to generate the desired harmonic frequency. These pulses normally occur 50 or 60 times per second, but may occur as few as 15 times per second to provide 15 Hz for some light rail systems. Any frequency can be generated. Optical pulses are used to gate the thyristors, which allows pulses to be generated at ground potential and connected to a thyristor that may be 500 kV above ground potential.

Terms

An **even harmonic** is an even multiple of the fundamental frequency (second, fourth, sixth, etc.).

An **odd harmonic** is an odd multiple of the fundamental frequency (third, fifth, seventh, etc.)

A **sequence** is the relationship between a harmonic and the fundamental frequency.

Early Power Distribution Systems

After Thomas Edison developed a marketable high-voltage incandescent lamp, he turned his energies to creating a system to produce and distribute electrical power to supply the lamps. A central power supply was possible because of the higher efficiency and lower current needed to bring a lamp to incandescence. By 1880, Edison had a central power station in London, England, using DC generators to supply electricity for his lamps. In 1882, the Pearl Street Station opened in New York City and initially served 85 customers, with a total of 400 lamps, using a two-wire feeder-main system. This plant eventually supplied DC power to one square mile of Manhattan. By the late 1880s, electric motors changed power distribution from a service primarily used to power nighttime lighting to a 24-hour-a-day service with increasing demand by industry and transportation. By the end of the 1880s, small power stations were strategically located in many U.S. cities. Each of these power stations serviced only a few city blocks.

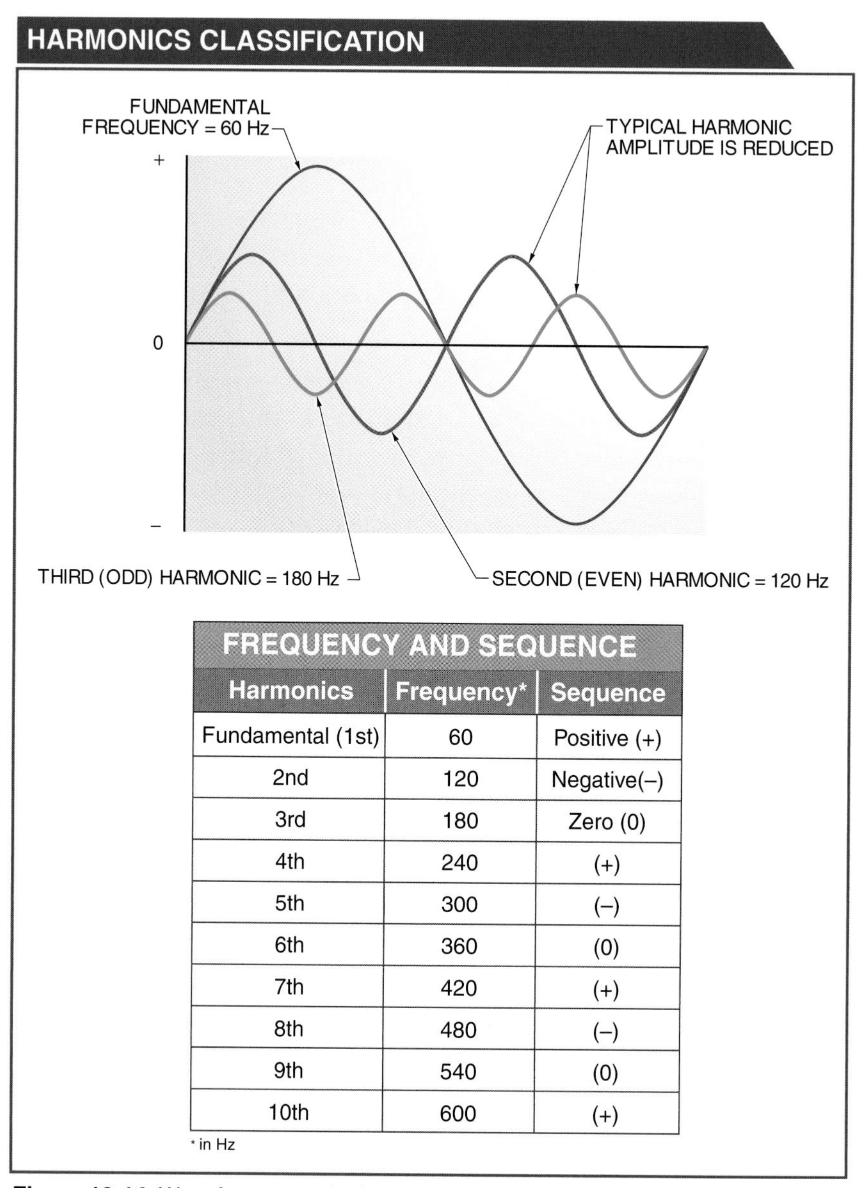

FREQUENCY AND SEQUENCE		
Harmonics	Frequency*	Sequence
Fundamental (1st)	60	Positive (+)
2nd	120	Negative(–)
3rd	180	Zero (0)
4th	240	(+)
5th	300	(–)
6th	360	(0)
7th	420	(+)
8th	480	(–)
9th	540	(0)
10th	600	(+)

* in Hz

Figure 13-14. Waveforms may include odd harmonics, even harmonics, or both odd and even harmonics, in addition to the fundamental frequency.

DC motors are used for applications ranging from large motors used in elevator systems to small motors used in computers and electronic equipment. AC motors are used for applications ranging from residential appliances to industrial production and process equipment. High performance levels are obtained from AC polyphase motors by applying a variable frequency to control their speed. To obtain a variable frequency, the 60 Hz frequency must first be converted to DC and then back to AC. The DC power source is typically located in the motor control circuit. Also, a fluorescent lamp has to rectify the distributed AC power to each lamp into DC power. The DC is then converted back into a higher frequency to power the lamps.

Changing AC to DC is easily accomplished with modern semiconductor devices. Changing 3ϕ AC power is more economical than using 1ϕ power for this purpose. In large facilities, AC power can be changed at a central location and then distributed to those loads that require DC power.

Tech Tip

The largest HVDC transmission project using ignitrons began operation in 1970 between the cities of Dallas, Oregon and Los Angeles, California, a distance of 1476 km (917 miles). This system first operated at 400 kVDC and delivered 1400 MW. This voltage was soon increased to 800 kVDC with a capacity of 1600 MW. The current system now operates at 1 MVDC and has a capacity of 3100 MW.

SUMMARY

Alternating current (AC) fundamentals are used to explain the differences between direct current (DC) and alternating current electricity. DC always flows in one direction in a circuit, and AC periodically reverses its direction. Terminology used with AC systems includes cycle, alternation, frequency, period, wavelength, skin effect, and phase.

A 1ϕ AC generator consists of a stationary magnet and a single winding that is rotated in the magnetic field created by the stationary magnet. The rotating winding is referred to as the armature, and the stationary magnet is referred to as the stator. The armature is the part of the generator from which electric power is removed. Increasing the number of poles allows for a reduction in the speed of armature rotation. Lower armature speed increases the life of the generator. If the operating frequency is reduced, the rotating speed is also reduced. In the United States and Canada 60 Hz power is standard and 50 Hz power is standard in Europe.

AC power is derived from electric power generators. A converter is a device that changes AC power to DC power. An inverter is a device that changes DC power to AC power. At the location of usage, high-voltage DC power (HVDC) is changed by an inverter back to AC power, to be stepped down using a transformer. Modern converters also have the ability to interconnect systems of different frequencies.

Changing AC to DC is accomplished with modern semiconductor devices such as thyristors and SCRs. Rectifying 3ϕ AC power is more economical than using 1ϕ power for this purpose. In large facilities, AC power can be rectified at a central location and then distributed to those loads that require DC power.

ATPeResources.com/QuickLinks
Access Code: 232263

APPLICATION: HARMONICS...

Background: The number of nonlinear power-line loads has grown significantly over the last decade. This growth is due to the increased use of variable-speed drives, electronic ballasts, computers, and other electronic equipment that use solid-state switch-mode power supplies to convert AC to DC. The increase in nonlinear loads creates harmonics on the power lines due to a variable need for current.

Harmonics are voltages and currents that are related to the fundamental frequency. For US power distribution, the fundamental frequency is 60 Hz. Harmonics of 60 Hz (first/fundamental) are 120 Hz (second), 180 Hz (third), 240 Hz (fourth), and 300 Hz (fifth).

Excessive power-grid harmonics can lead to power quality issues such as conductor overheating, reduced capacitor life, false circuit-breaker tripping, inefficient generator operation, and telephone-line interference.

To determine the characteristic harmonics, the number of rectifiers (pulses) used to convert AC to DC must be known. Once the number of rectifiers is known, the characteristic harmonics can be determined by applying the following formula:

$$H = N \times P \pm 1$$

where
H = harmonic number
N = integer (1, 2, 3, ...)
P = number of rectifiers

For each integer (N), two harmonics are calculated: first by adding 1 and second by subtracting 1 from the product of the given integer and the number of rectifiers. The harmonic frequencies are calculated by multiplying the harmonic number by the fundamental power-line frequency, which is 60 Hz in the United States.

Key Points: Power-line harmonics
Problem: In the United States, three-phase (3ϕ) power is converted from AC to DC by using six rectifiers. **See Figure 1.** What are the characteristic harmonics and their frequencies?

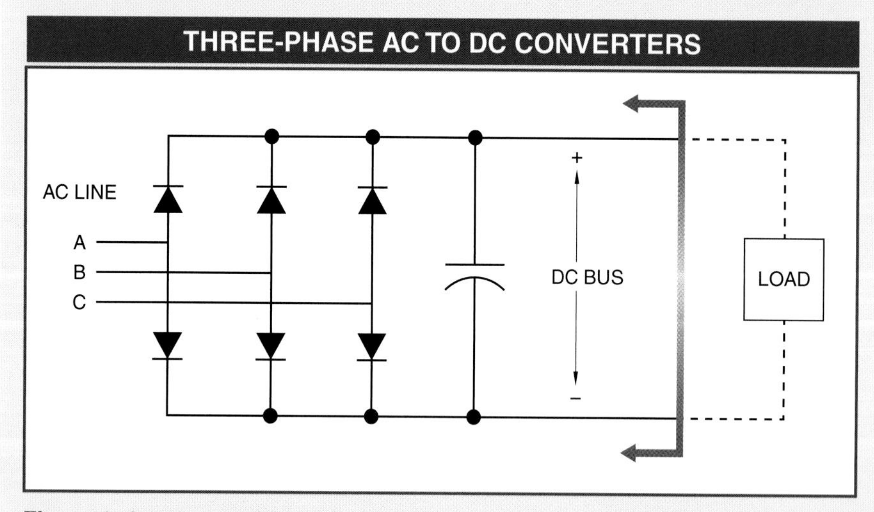

Figure 1. A typical 3ϕ AC to DC converter uses six rectifiers to perform the conversion.

...APPLICATION: HARMONICS

Solution: Using the characteristic harmonics formula, the harmonics are calculated and then multiplied by 60 Hz to find the frequencies.

$$H = N \times P \pm 1$$

For the six-rectifier configuration, the 2nd, 3rd, and 4th harmonics are unknown as well as the 6th, 8th, 9th, and 10th. The relative amplitude between harmonics reduces by half when the frequency is doubled. **See Figure 2.** For example, if harmonic 1 is 100 V, then harmonic 2 is 50 V, and harmonic 4 is 25 V.

HARMONIC FREQUENCIES

N	±1	Harmonic Number (*H*)	Frequency*
1	−1	5th	300
1	+1	7th	420
2	−1	11th	660
2	+1	13th	780
3	−1	17th	1020
3	+1	19th	1140
4	−1	23rd	1380
4	+1	25th	1500
5	−1	29th	1740
5	+1	31st	1860

* in Hz

Figure 2. The relative amplitude between harmonics reduces by half when the frequency is doubled.

Many methods can be used to reduce harmonics, including the following:

- reducing the number of nonlinear loads to less than 30%
- using a 12-pulse converter to increase the harmonic number, which reduces the overall impact but does not eliminate the harmonics
- using harmonic tap filters to eliminate harmonic currents
- using delta-delta and delta-wye transformers with an equal number of nonlinear loads (the configuration acts as a 12-pulse converter)

Chapter Review

1. How is the amplitude of a waveform plotted?
2. How does DC compare to AC?
3. What type of current periodically reverses its direction of flow?
4. Explain why a source voltage with a sine wave does not always produce an alternating current.
5. What is the advantage of a high-voltage lamp over a low-voltage lamp of the same wattage?
6. What is an electromagnetic device that changes mechanical energy into electrical energy?
7. What is an electromagnetic device that changes electrical energy into mechanical energy?
8. With generators, which component has electrical energy removed from it?
9. What is another term for a DC generator?
10. What is a complete set of voltages that repeats called?
11. What occurs when the direction of current reverses in a circuit?
12. What is the unit of measure for frequency?
13. What is the time required for one cycle called?
14. What is the frequency range of the UHF band?
15. What is the tendency of electrons to move to the surface of a conductor as the frequency is increased known as?
16. What are eddy currents?
17. What is the time displacement of two signals of the same frequency known as?
18. How can the frequency of an AC generator be increased?
19. If a generator has eight poles, how many electrical degrees are generated for each mechanical degree the rotor rotates?
20. How is most AC power generated?
21. Is there a tendency to raise or lower distribution voltage?
22. Does HVDC transmission require more or less copper than HVAC systems?
23. Why are positive sequence harmonics considered undesirable?
24. How much power does a 10 Ω resistor dissipate with 25 A flowing through it?
25. What is the current flow through a lamp rated 100 W and 120 V?
26. What is the period for 1 MHz?
27. How many meters are in the wavelength of a 1000 Hz signal?
28. What is the frequency of an eight-pole generator turning at 1200 rpm?
29. What is the speed of a 50 Hz, six-pole generator?
30. Find the secondary current of a transformer that has V_P = 2100 V, I_P = 25 A, and V_S = 240 V.

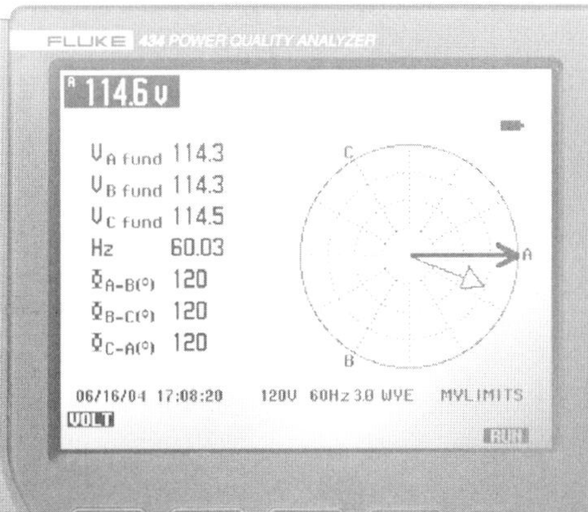

14

Vectors and Phase Relationships

OBJECTIVES

- Define vector.
- Define phasor.
- Describe vector mathematical and graphical calculation methods.
- Discuss vector addition.
- Explain vector subtraction.

Alternating current (AC) circuit parameters change constantly. These changes are a result of the AC generator and circuit components. Comprehending signals in constant change that are out of phase with each other is difficult. To aid with this process, graphic representations are made of various AC circuit parameters. These parameters are often represented graphically as vectors.

Vectors are used to represent AC magnitude and phase information. The different methods used to solve vector operations include graphical and mathematical methods. Solving and understanding vector operations is helpful when using electrical test instruments such as digital multimeters and power quality meters.

Terms

A **vector** is a quantity involving direction and magnitude (scalar value) represented by a straight line with an arrowhead on one end.

VECTORS

A *vector* is a quantity involving direction and magnitude (scalar value) represented by a straight line with an arrowhead on one end. A vector is used to represent the changing value of a sine wave. The length of the line represents the magnitude, or quantity, of the vector. This is the same as a scalar value such as the length of a ruler. However, a vector value differs from a scalar value in that the vector value also has direction. An arrowhead indicates the direction of the vector. Vectors are used to represent such factors as force, direction, and velocity.

Scalar diagrams represent only magnitude, while vectors represent direction and magnitude. **See Figure 14-1.** For example, a scalar diagram can be used to represent an automobile that is moving west (direction) at 50 mph (magnitude). An automobile that is moving west at 50 mph is considered a vector value because both direction and magnitude are represented.

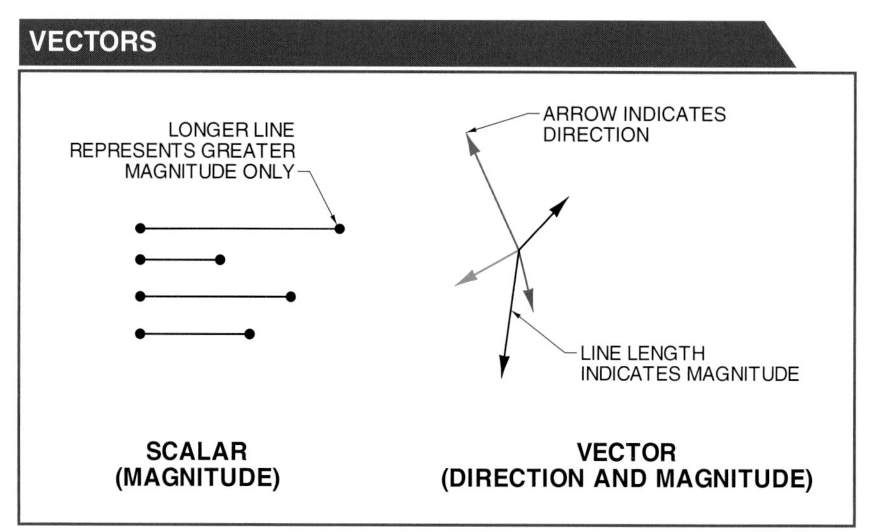

Figure 14-1. Scalar diagrams represent only magnitude, while vectors represent direction and magnitude.

Terms

A **phasor** is a vector in which length represents the magnitude of an electrical parameter and direction represents angle theta in electrical degrees.

PHASORS

A *phasor* is a vector in which length represents the magnitude of an electrical parameter and direction represents angle theta in electrical degrees. The term phasor is used in AC analysis to replace the term vector because it shows both amplitude and phase. Phasors are used to illustrate that the position of the line changes with time as the sine wave moves from 0° to 360°.

Phasors can be represented in graphical format. For example, an x-y graph can be used to represent the output voltage of a single-phase

(1ϕ) generator. The phasor is rotated conventionally in a counterclockwise direction. An x-y graph uses the x-axis as the horizontal axis and the y-axis as the vertical axis. **See Figure 14-2.** The origin (x, y = 0, 0) of the graph is listed by value and separated by a comma.

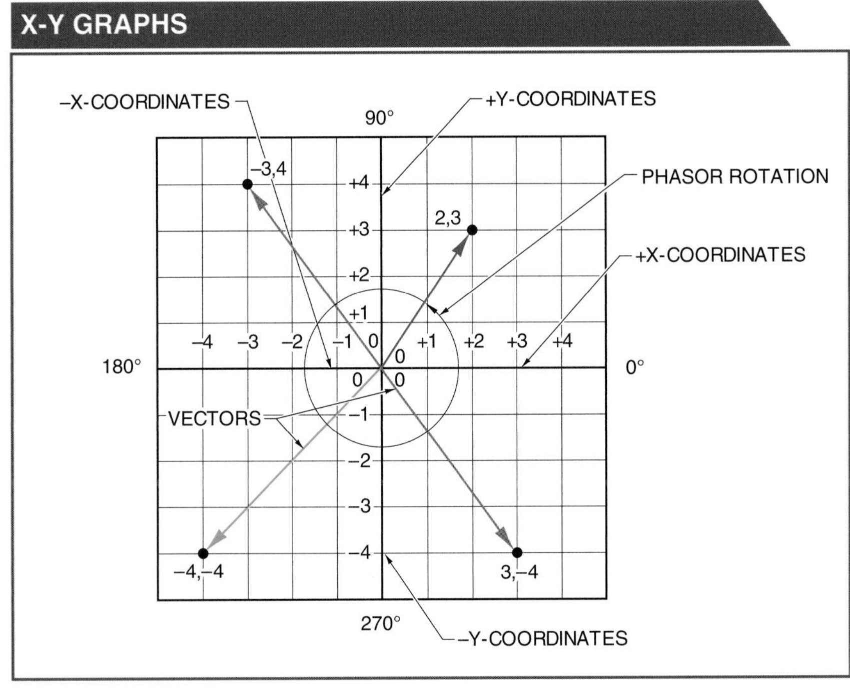

Figure 14-2. An x-y graph uses the x-axis as the horizontal axis and the y-axis as the vertical axis.

Coordinates on the x-axis that are right of the y-axis are positive, and coordinates of the x-axis that are left of the y-axis are negative. Coordinates on the y-axis that are above the x-axis are positive, and coordinates on the y-axis that are below the x-axis are negative.

Using these coordinates, any vector or phasor originating at the origin can be drawn when the x and y coordinates are known. The x-coordinate is listed first. A comma and the value of the y-coordinate follow the x-coordinate number. X-Y graphs are typically continued beyond the values shown on the x-axis and y-axis indefinitely in each direction (negative and positive). For example, a plot of (–4, 3) is shown on a scale with axes that extend beyond these variables in both negative and positive directions.

Phasors can also be constructed in rectangular coordinates. Rectangular coordinates are typically written in vector algebra format. For example, using vector algebra format, rectangular coordinates are written (2, j3), (–3, j4), (–4, –j2), and (3, –j4). The first digit in the algebraic expression is the x coordinate, and the second digit, or y coordinate, is identified with the letter j in front of the coordinate value.

Phasors move with time. For example, a phasor moves with time when the phase of the signal is changing. At any point, the phasor movement and instantaneous phase relationship between signals can be seen. A vector can be used to represent the instantaneous value of a phasor. For example, phasor graphs can be used to represent the voltage output of a three-phase (3ϕ) generator. **See Figure 14-3.** Three-phase generators have their output voltages separated by 120°.

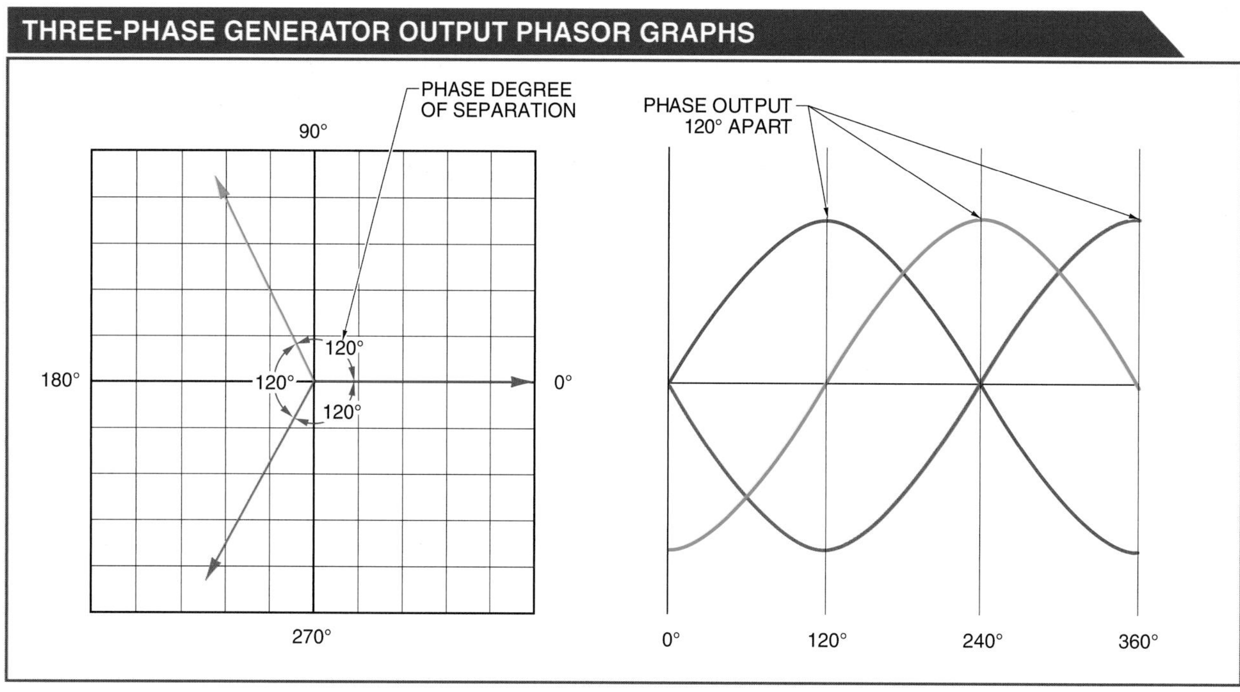

Figure 14-3. Phasor graphs can be used to represent the output voltage of 3ϕ generators.

The Greek letter theta (θ) is used to designate the angle between the reference axis and the vector. Subscripts are used when more than one vector is present. Vectors have the same direction when they are parallel to each other and the line that joins their origins does not intersect the line that joins their termination point (arrowhead).

Vectors are equal when they have the same magnitude and the same direction. When using vectors, any given vector can be replaced by a vector of equal value (having same direction and magnitude). The angle of the vector is positive (+) when it is rotated from the reference line in a counterclockwise direction. Conversely, the vector angle is negative (–) if it is rotated from the reference line in a clockwise direction.

VECTOR CALCULATION METHODS

Vector calculation methods are used to solve for AC values that require the need to combine multiple phasors into a single vector. Vector calculation methods include mathematical methods, (trigonometry), and graphical methods, (vector diagram calculation method and the parallelogram calculation method), which are performed with a scale and protractor.

Mathematical Calculation Methods

Mathematical vector calculation methods are used to perform vector calculations when a high level of accuracy is required. Mathematical vector calculation methods utilize trigonometry, as well as basic addition, subtraction, multiplication, and division functions, to perform the majority of calculations. *Trigonometry* is a branch of mathematics in which the relationships between the lengths of the sides of a triangle and the angles of a triangle are used to perform calculations.

A *right triangle* is a triangle with a right angle. A *right angle* is an angle that is 90°. A right angle is formed when two lines are perpendicular to each other. An *acute angle* is an angle that is less than 90°. Trigonometric functions are the ratios of the lengths of the sides of a triangle with respect to one of the acute angles of a triangle. Trigonometric functions are used to calculate both the length and the angle of the vector.

Trigonometric functions most commonly used in the solution of parameters for AC circuits include sine, cosine, and tangent. The *hypotenuse* is the side of a right triangle that is opposite the right angle. The *sine of a right triangle* (*sin*) represents the ratio of the length of the side opposite an acute angle to the length of the hypotenuse. The *cosine of a right triangle* (*cos*) represents the ratio of the length of the side adjacent to an acute angle to the length of the hypotenuse. The *tangent of a right triangle* (*tan*) represents the ratio of lengths of the sides opposite and adjacent to an acute angle. Trigonometric functions are normally abbreviated sin, cos, and tan, respectively. **See Appendix.** Regardless of the magnitude of the lengths of the sides of a triangle, the ratio of triangle sides is always the same for any given angle.

Trigonometric functions can be used to solve for phasor values. **See Figure 14-4.** To simplify performing calculations with trigonometric functions, x (horizontal coordinate) is substituted for the opposite side; y (vertical coordinate) is substituted for the adjacent side; and z (resultant vector) is substituted for the hypotenuse.

For example, a rotating vector (z) always has a value of 1.000. Trigonometric functions can be developed using a rotating vector with a value of 1.000. **See Figure 14-5.** At 0°, a rotating vector is equal to its x-coordinate, and the y-coordinate is 0.000. Therefore, both sine θ and tangent θ have a value of 0.000, and cosine θ has a value of 1.000.

Terms

Trigonometry is a branch of mathematics in which the relationships between the lengths of the sides of a triangle and the angles of a triangle are used to perform calculations.

A **right triangle** is a triangle with a right angle.

A **right angle** is an angle that is 90°.

An **acute angle** is an angle that is less than 90°.

The **hypotenuse** is the side of a right triangle that is opposite the right angle.

The **sine of a right triangle** (*sin*) represents the ratio of the length of the side opposite an acute angle to the length of the hypotenuse.

The **cosine of a right triangle** (*cos*) represents the ratio of the length of the side adjacent to an acute angle to the length of the hypotenuse.

The **tangent of a right triangle** (*tan*) represents the ratio of lengths of the sides opposite and adjacent to an acute angle.

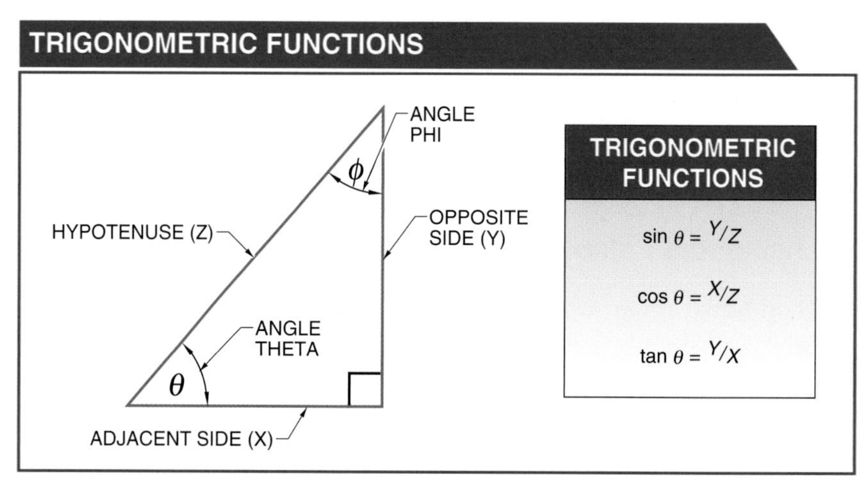

Figure 14-4. Trigonometric functions can be used to solve for phasor values.

TRIGONOMETRIC FUNCTIONS WITH ROTATING VECTOR

Vector Type	0°	30°	45°	60°	90°
Z	1.000	1.000	1.000	1.000	1.000
cos	1.000	0.866	0.707	0.500	0.000
sin	0.000	0.500	0.707	0.866	1.000
tan	0.000	0.577	1.000	1.732	∞
	0°	30°	45°	60°	90°
VECTOR	VECTOR VALUES	VECTOR VALUES	VECTOR VALUES	VECTOR VALUES	VECTOR VALUES
(Z) ⟶	1.000	1.000	1.000	1.000	1.000
(X) ⟶	1.000	0.866	0.707	0.500	0.000
(Y) ⟶	0.000	0.500	0.707	0.866	1.000

Figure 14-5. Trigonometric functions can be developed using a rotating vector with a value of 1.000.

As a vector rotates in a counterclockwise direction, the x-coordinate decreases from 1.000, and the y-coordinate increases toward its maximum value of 1.000. At 30°, cosine θ is equal to 0.866 and sine θ is equal to 0.500. Tangent θ is equal to 0.500 divided by 0.866, or 0.577.

At 45°, the x-coordinate is equal to the y-coordinate. Tangent θ is 1.000, and cosine θ is equal to sine θ with a value of 0.707. When the vector is at 60°, cosine θ has a value of 0.500, and sine θ has a value of 0.866. Tangent θ is equal to 0.866 divided by 0.500, or 1.732. When the vector is 90°, the value of sine θ is its maximum value of 1.000, cosine θ is its minimum value of 0.000, and tangent θ is an undefined number (∞).

Angle theta (θ) can be calculated by using the arcsine ($\sin^{-1}$), the arccos ($\cos^{-1}$), or the arctan ($\tan^{-1}$) by applying the following formulas:

$$\theta = \sin^{-1}\left(\frac{y}{z}\right)$$

or

$$\theta = \cos^{-1}\left(\frac{x}{z}\right)$$

or

$$\theta = \tan^{-1}\left(\frac{y}{x}\right)$$

where

θ = angle theta (in °)

x = vector 1 (in V)

y = vector 2 (in V)

z = resultant vector (in V)

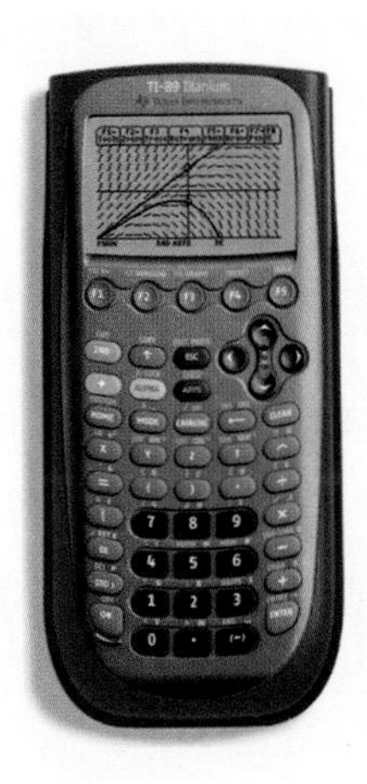

Texas Instruments

Scientific and graphing calculators can be used when performing vector calculations.

Example: What is the angle θ for a 2ϕ generator with two 100 V vectors and a resultant vector of 141.4 V?

$$\theta = \sin^{-1}\left(\frac{y}{z}\right)$$

$$\theta = \sin^{-1}\left(\frac{100\text{ V}}{141.4\text{ V}}\right)$$

$$\theta = \sin^{-1} 0.707$$

$$\theta = \mathbf{45°}$$

or

$$\theta = \cos^{-1}\left(\frac{x}{z}\right)$$

$$\theta = \cos^{-1}\left(\frac{100\text{ V}}{141.4\text{ V}}\right)$$

$$\theta = \cos^{-1} 0.707$$

$$\theta = \mathbf{45°}$$

or

$$\theta = \tan^{-1}\left(\frac{y}{x}\right)$$

$$\theta = \tan^{-1}\left(\frac{100\text{ V}}{100\text{ V}}\right)$$

$$\theta = \tan^{-1} 1.000$$

$$\boldsymbol{\theta = 45°}$$

With a 2ϕ generator, the angle is equal to 45°, with sine and cosine equal to 0.707 and tangent equal to 1.000. Once the value of each of the trigonometric functions is calculated, the values are located in the trigonometric function table with the corresponding angle. **See Appendix.**

Graphical Calculation Methods

Graphical calculation methods are used to perform vector calculations when a high level of accuracy is not required. Graphical methods include the vector diagram calculation method and the parallelogram calculation method. With the vector diagram calculation method, the vectors are graphically connected head to tail to calculate the resultant vector. With the parallelogram calculation method, a line drawn parallel to each vector results in a perpendicular intersection point used to calculate the resultant vector. The parallelogram calculation method operates on two vectors at a time. The vector diagram calculation method can operate on multiple vectors at a time.

Terms

The **vector diagram calculation method** is a graphical method of calculation that uses vector values to perform vector calculations.

A **resultant vector** is the single vector that is calculated by drawing a vector from the tail of the first vector to the head of the last vector.

Vector Diagram Calculation Method. The *vector diagram calculation method* is a graphical method of calculation that uses vector values to perform vector calculations. **See Figure 14-6.** The vector diagram calculation method connects the heads and tails of the individual vectors. The vector diagram calculation method is sometimes referred to as the vector triangle method. A *resultant vector* is the single vector that is calculated by drawing a vector from the tail of the first vector to the head of the last vector. To perform vector calculations by using the vector diagram calculation method, apply the following procedure:

1. Connect the head of one vector to the tail of the other vector.
2. Draw the resultant vector from tail of the first vector to head of the last vector.
3. Measure the resultant vector by using the scale (provided).
4. Measure the resultant vector angle (θ) by using a protractor (not provided).

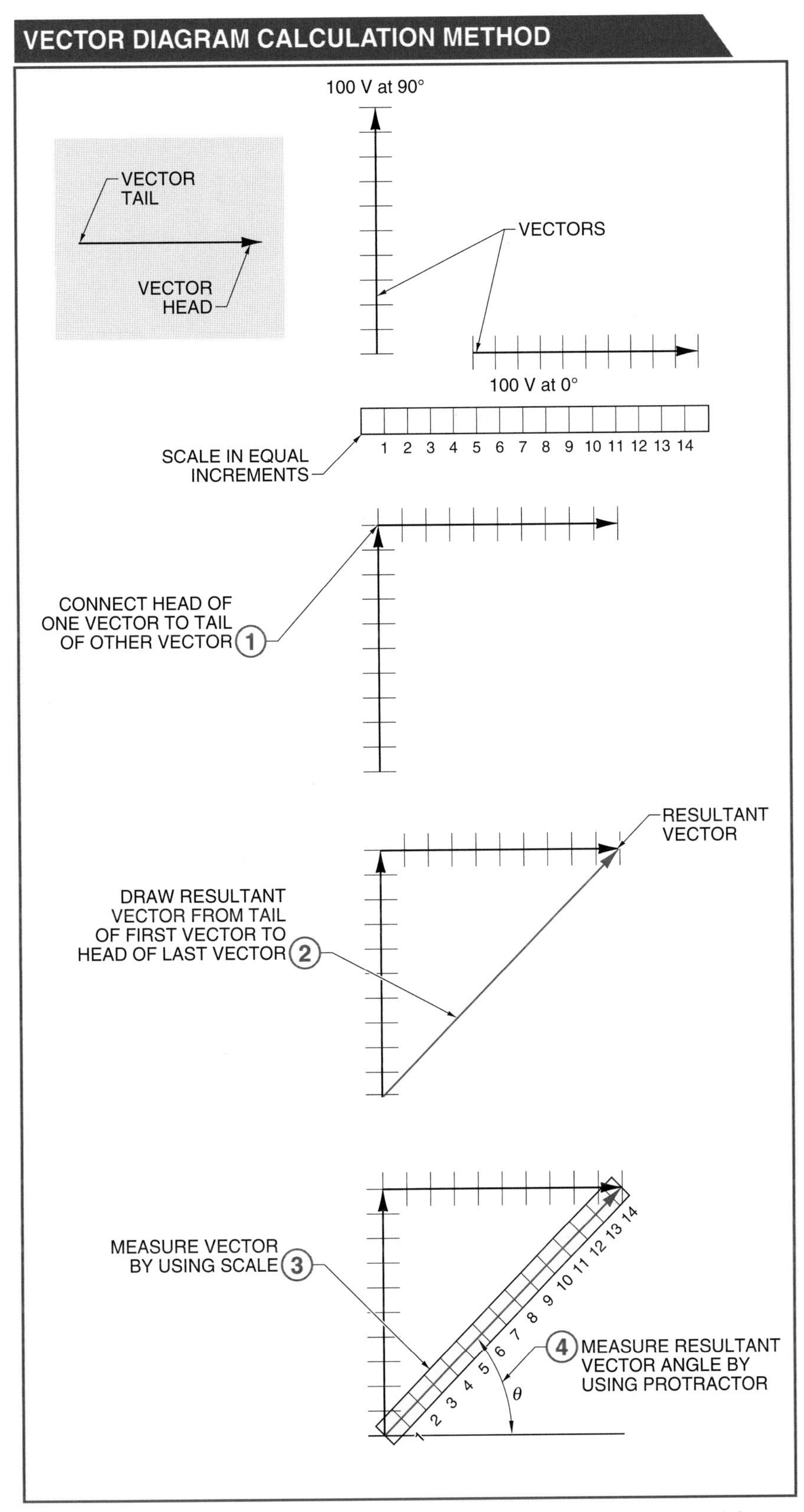

Figure 14-6. The vector diagram calculation method is a graphical method that connects vector values head-to-tail to perform vector calculations.

Terms

The **parallelogram calculation method** is a graphical method of calculation that uses parallelogram values to perform vector calculations.

A **parallelogram** is a four-sided plane figure with opposite sides parallel and equal.

Parallelogram Calculation Method. The *parallelogram calculation method* is a graphical method of calculation that uses parallelogram values to perform vector calculations. **See Figure 14-7.** A *parallelogram* is a four-sided plane figure with opposite sides parallel and equal. This method calculates the resultant vector by calculating the difference between parallel vectors. The difference between the parallelogram calculation method and the vector diagram calculation method is that the parallelogram calculation method operates on two vectors at a time. To perform vector calculations by using the parallelogram calculation method, apply the following procedure:

1. Draw a vector that is parallel to vector B from head of vector A.
2. Draw a vector that is parallel to vector A from head of vector B.
3. Draw resultant vector from origin to point of intersection of lines from first two steps.
4. Measure resultant vector by using the scale (provided).
5. Measure resultant vector angle (θ) by using a protractor (not provided).

VECTOR ADDITION

When solving for AC values, vectors and phasor values are often added to produce a single vector. Both graphical and mathematical methods can be used to calculate the resultant vector. Vectors can be added when vectors are in the same direction, in the opposite direction, and when there are specific vectors that are greater or less than 90°.

Techniques used to add scalar quantities are used when adding vectors. Scalar quantities, such as mass, length, and time, are added algebraically. When adding vectors, mathematical signs preceding vector quantities are considered.

For example, in DC circuits current flow into a junction is equal to current flow out of the same junction. Currents that are in phase with each other are added as scalar quantities and calculated by applying the following formula:

$$I_1 = I_2 + I_3$$

where

I_1 = current 1 flow out of the junction (in A)

I_2 = current 2 flow into the junction (in A)

I_3 = current 3 flow into the junction (in A)

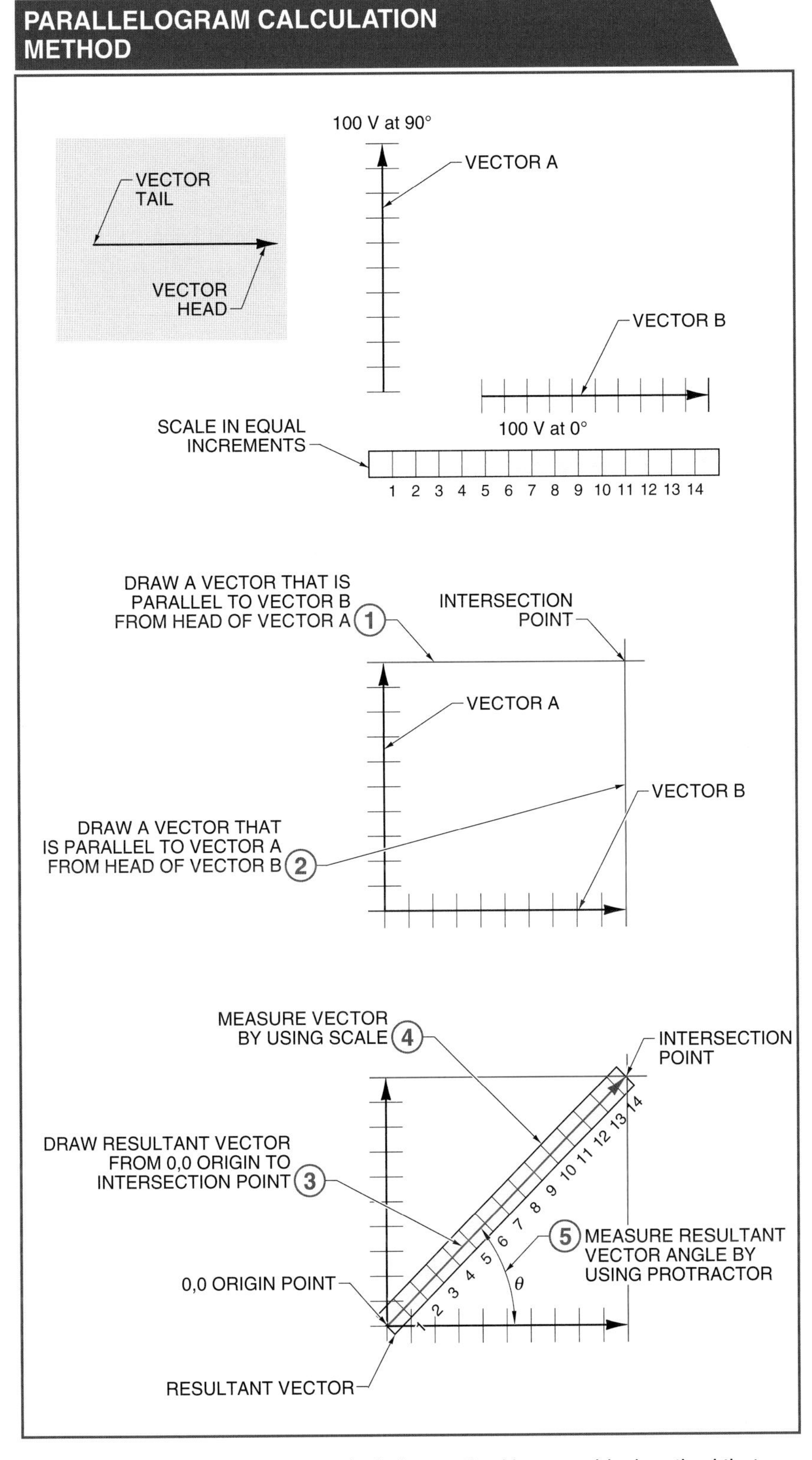

Figure 14-7. The parallelogram calculation method is a graphical method that uses a parallelogram to perform vector calculations.

Example: What is the current flow out of the junction for current I1 if the two currents entering the junction are equal to 3 A (I2) and 2 A (I3)?

$I_1 = I_2 + I_3$

$I_1 = 3\text{ A} + 2\text{ A}$

$I_1 = \mathbf{5\text{ A}}$

Using the same equation, when any two currents are known, the third current is calculated by applying the following formula:

$I_2 = I_1 - I_3$

Example: What is the current through I2 if the current through I1 is equal to 5 A and the current through I3 is equal to 2 A?

$I_2 = I_1 - I_3$

$I_2 = 5\text{ A} - 2\text{ A}$

$I_2 = \mathbf{3\text{ A}}$

Adding Vectors with Same Direction

Vectors with the same direction can be added using graphical methods. For example, vector A has three units at 90° and vector B has two units at 90°. The vectors can be added graphically by placing the tail (origin) of vector B at the head of vector A. The two vectors are combined to form a resultant vector by measuring the two vectors, to obtain five units at 90°. Parallel vectors in the same direction, regardless of angle, add together as scalar values. **See Figure 14-8.**

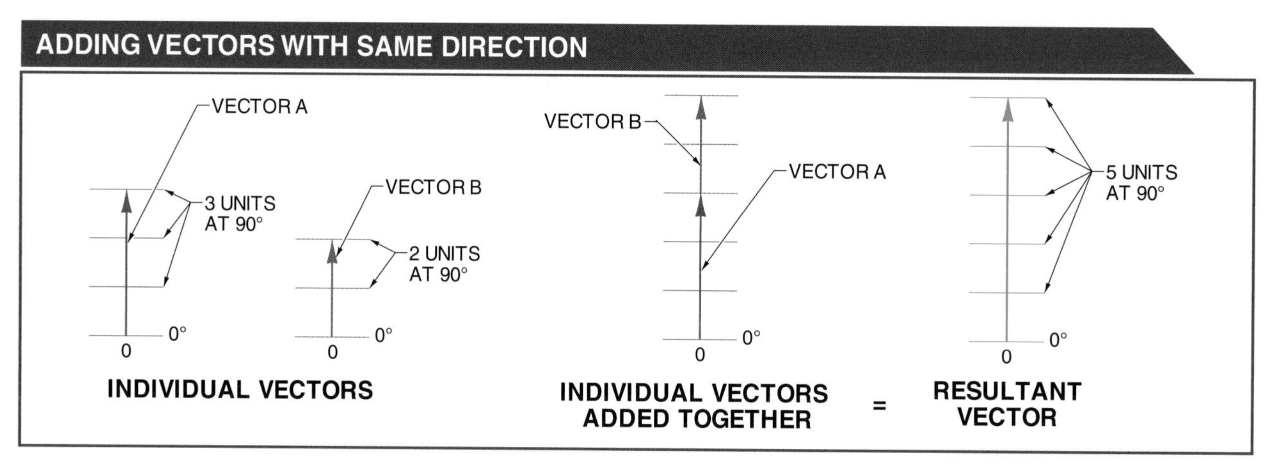

Figure 14-8. Parallel vectors with the same direction, regardless of angle, add together as scalar values.

Vectors can be used to represent parameters such as voltage in a circuit. For example, when two batteries are connected in series, the vectors have the same direction and the voltages add. For instance, a 3.6 V battery connected so that its positive terminal is connected to the negative terminal of a 1.5 V battery has a resultant vector equal to 5.1 V.

Adding Vectors with Opposite Directions

Vectors with opposite directions can be added using graphical methods such as the vector diagram calculation method. **See Figure 14-9.** For example, if vector A has three units at 270°, and vector B has two units at 90°, the vectors can be added graphically by placing the tail of vector B at the head of vector A. Measuring the resultant vector shows one unit at –90°.

For example, batteries that oppose each other are similar to vectors that add in opposite directions. If a 3.6 V battery at 0° opposes a 1.5 V battery at 180°, the resulting voltage equals 2.1 V at 0°.

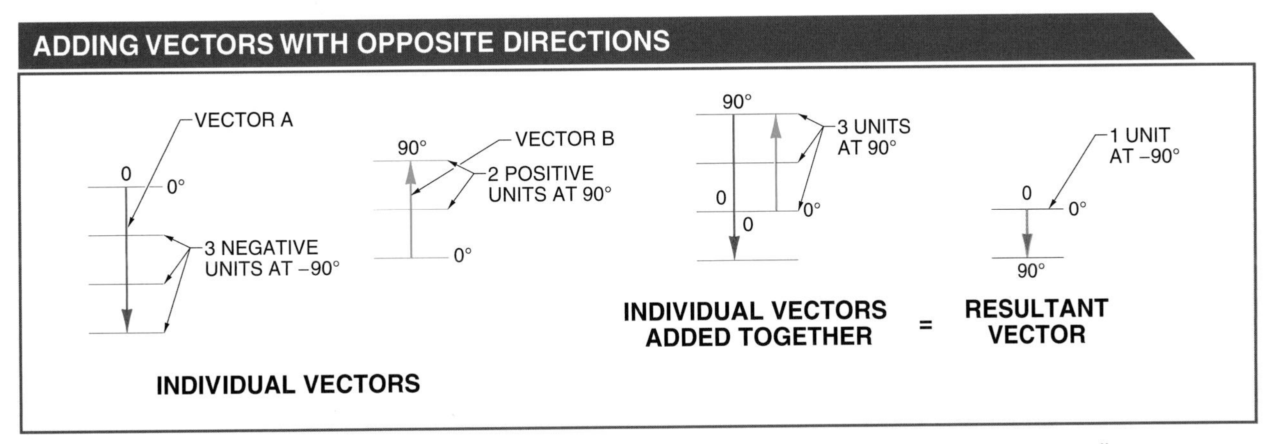

Figure 14-9. Vectors with opposite directions can be added using graphical methods such as the vector diagram calculation method.

Adding Vectors 90° Apart

In AC circuits, circuit parameters are sometimes 90° apart. For example, a 2ϕ generator has voltages that are 90° apart. Two vectors of equal length are used to represent these voltages, one at 90° and the other at 0°. Graphical or mathematical methods can be used to solve for the resultant vector.

For example, to solve for the resultant vector using the vector diagram calculation method, a scale is used to represent two vectors of 100 V each. Each unit of the scale represents 10 V. To calculate the voltage between phases of a generator using the vector diagram calculation method, apply the following procedure:

1. Connect the head of one vector to the tail of the other vector. Vector order is unimportant.
2. Draw the resultant vector from the opposite tail to the opposite head of each vector.
3. Measure the resultant vector by using the scale (provided).
4. Measure the resultant vector angle by using a protractor (not provided).

To calculate the resultant vector graphically, the resultant vector must be measured with a scale, while the resultant vector angle is measured by using a protractor. **See Figure 14-10.** For a scale with 14 units at 10 V per unit, the resultant vector is equal to 140 V. A protractor can be used to verify that the resultant vector angle is approximately 45°.

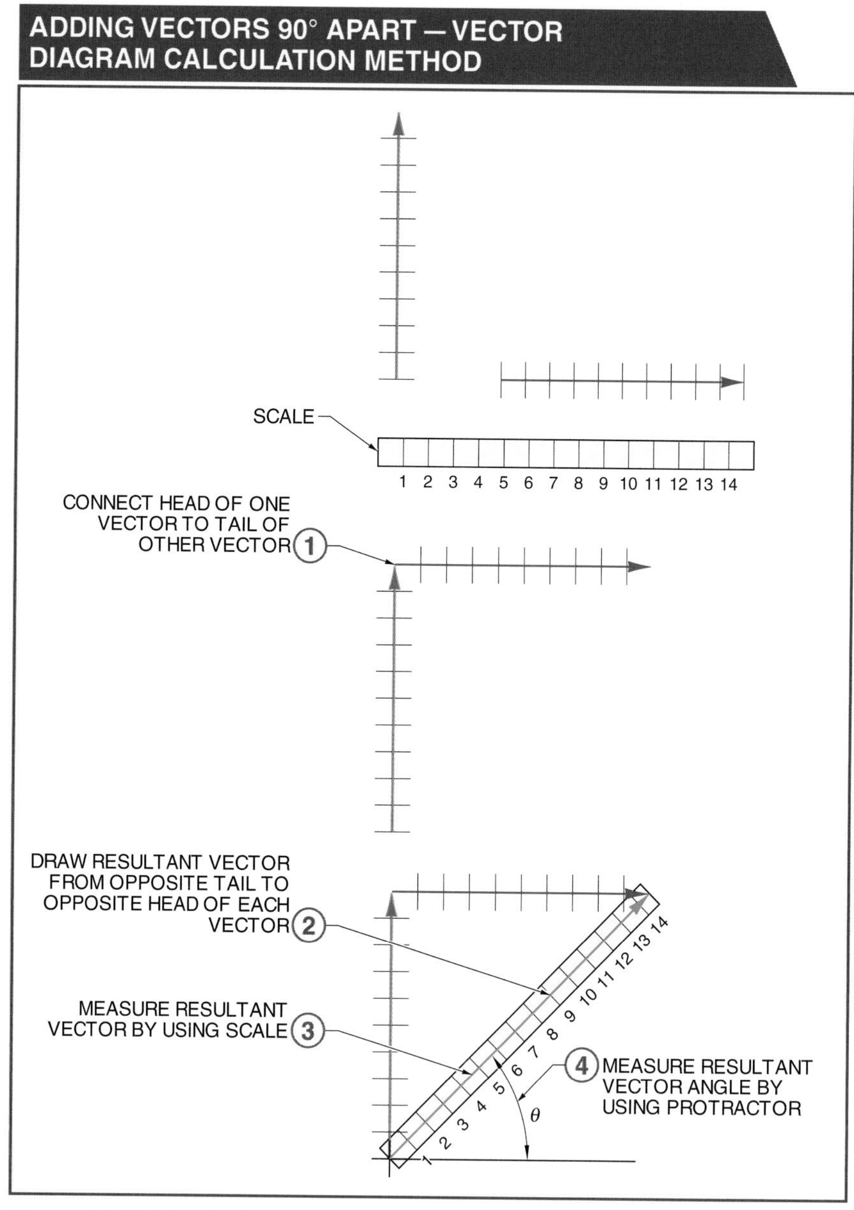

Figure 14-10. To calculate a resultant vector graphically, the resultant vector must be measured with a scale, while the angle is measured by using a protractor.

Graphical vector solutions often provide good visual records of the results. However, the values of length and angle are only estimates and may not be acceptable answers for some purposes. Mathematical methods are used to provide answers that are more accurate.

Mathematical methods are based on the Pythagorean theorem. The *Pythagorean theorem* states that the square of the hypotenuse of a right triangle is equal to the sum of the squares of the other two sides. The Pythagorean theorem is a mathematical method used to solve for vector values. The Pythagorean theorem can be used to calculate the voltage (resultant vector) between the phases (vectors) of a 2ϕ generator. **See Figure 14-11.** The resultant vector and angle theta are calculated by applying the following procedure:

1. Calculate resultant vector. The resultant vector is calculated by applying the following formula (Pythagorean theorem):

   $$V_R = \sqrt{V_A^2 + V_B^2}$$

   where

   V_R = hypotenuse of triangle (resultant vector)
   V_A = horizontal side of triangle (x-vector)
   V_B = vertical side of triangle (y-vector)

2. Calculate angle theta. Angle theta is calculated by applying the following formula:

   $$\theta = \sin^{-1}\left(\frac{V_B}{V_R}\right)$$

   where

   θ = angle theta (in °)
   V_B = vertical side of triangle (y-vector)
   V_R = hypotenuse of triangle (resultant vector)

Terms

The Pythagorean theorem states that the square of the hypotenuse of a right triangle is equal to the sum of the squares of the other two sides.

PYTHAGOREAN THEOREM

$$V_R = \sqrt{V_A^2 + V_B^2}$$

RESULTANT VECTOR (V_R = 141.4 V)
Y-VECTOR (V_B = 100 V)
θ
X-VECTOR (V_A = 100 V)

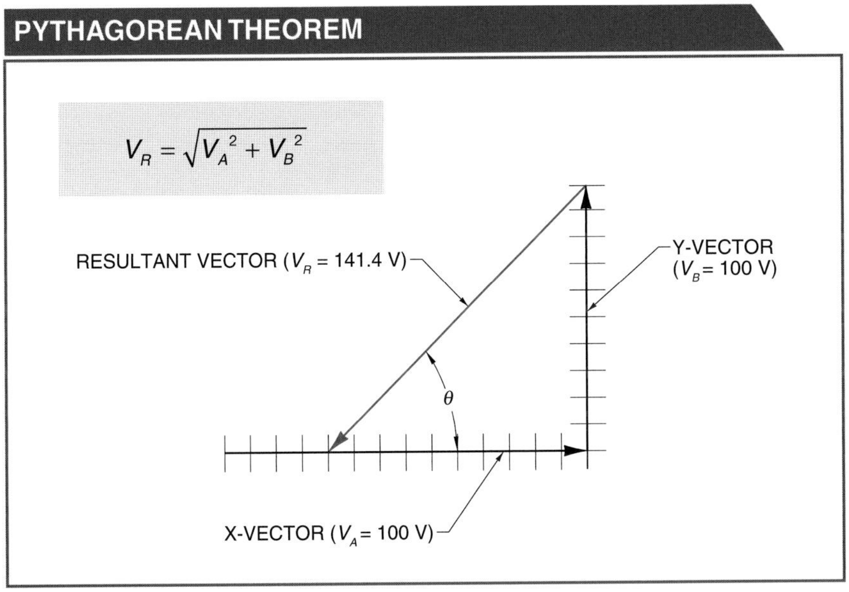

Figure 14-11. The Pythagorean theorem can be used to calculate the voltage (resultant vector) between the phases (vectors) of a 2ϕ generator.

Example: What is the maximum voltage and angle theta when both phases of a 100 V 2ϕ generator are summed?

1. Calculate resultant vector.

$$V_R = \sqrt{V_A^2 + V_B^2}$$

$$V_R = \sqrt{100\ V^2 + 100\ V^2}$$

$$V_R = \sqrt{20{,}000}$$

$$V_R = \mathbf{141.4\ V}$$

2. Calculate angle theta.

$$\theta = \sin^{-1}\left(\frac{V_B}{V_R}\right)$$

$$\theta = \sin^{-1}\left(\frac{100\text{ V}}{141.4\text{ V}}\right)$$

$$\theta = \sin^{-1} 0.707$$

$$\theta = \mathbf{45°}$$

When the y-vector (VA) is placed tail to head to the x-vector (VB), the voltage between phases, the resultant vector (VR), can be calculated. The resultant vector demonstrates the maximum voltage developed in a 2ϕ generator circuit when both phases are added together.

Adding Vectors Less than 90° Apart

Vectors less than 90° apart can be added graphically or mathematically. To solve graphically, the vector diagram calculation method can be used to calculate the final resultant vector. **See Figure 14-12.** To solve using the vector diagram calculation method, apply the following procedure:

1. Start at the origin (0, 0) and connect vector A head to vector B tail.
2. Connect vector B head to vector C tail.
3. Draw resultant vector (R_{ABC}) from vector A tail to vector C head.
4. Measure the resultant vector angle by using a protractor (not provided).

Note: If three or more vectors are to be added, the vector diagram calculation method of addition is continued until all vectors have been included. For the parallelogram graphical method, two vectors are added together to calculate an intermediate resultant vector. That resultant vector is then added with the next vector to create another resultant vector, until all vectors have been added.

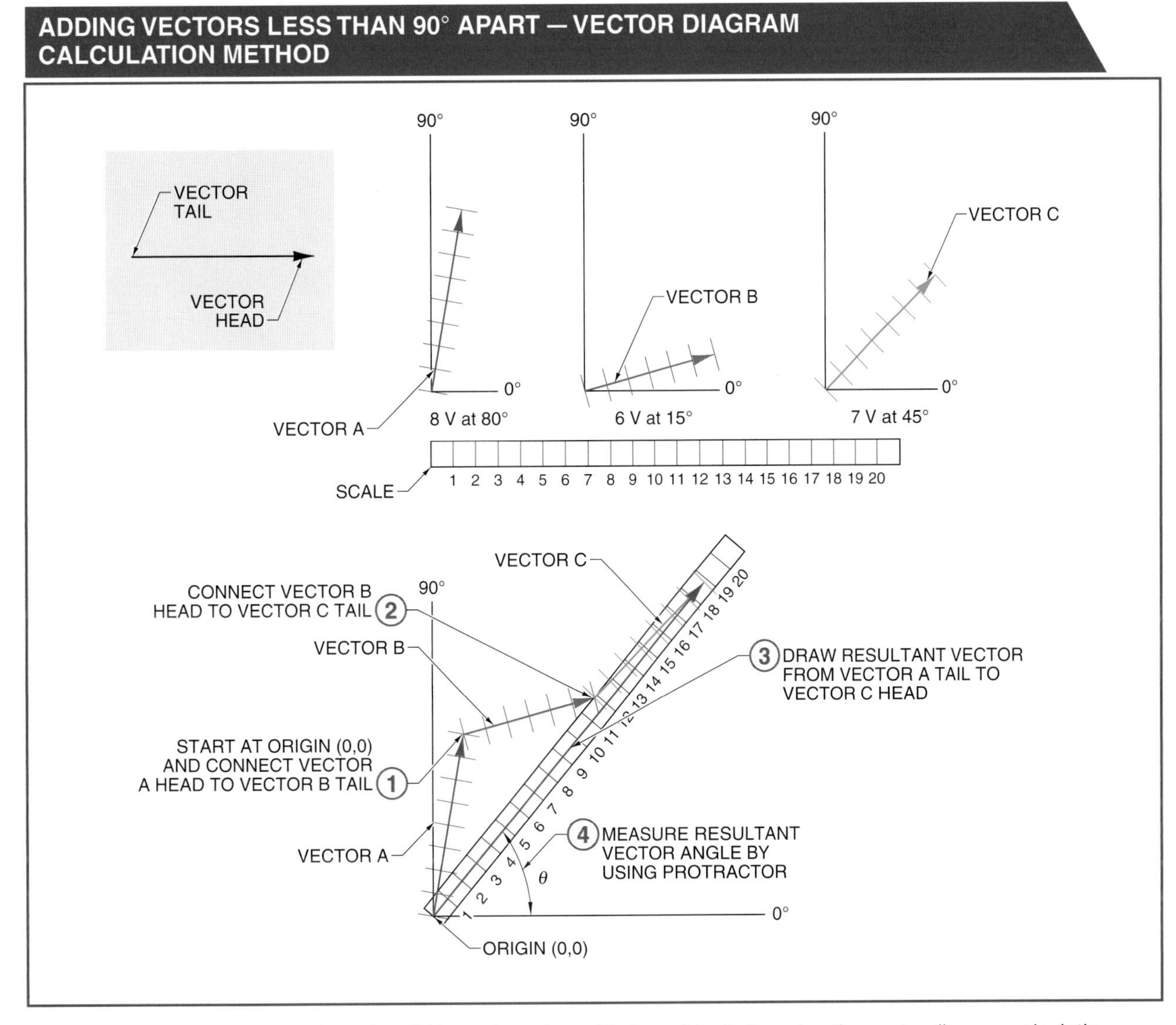

Figure 14-12. Vectors that are less than 90° apart can be added graphically by using the vector diagram calculation method.

When two or more vectors are not in-phase with each other, they cannot be added directly. Also, the Pythagorean theorem does not apply when vectors are not at right angles to each other. When vectors are not at right angles to each other, trigonometric functions are required to solve mathematically for the resultant vector.

For example, any vector can be broken into vertical and horizontal components by using the length of the vector and the sine and cosine values of the resultant vector angle. **See Figure 14-13.** When using the sine and cosine functions, each vector is reduced to its equivalent vertical and horizontal components. The equivalent vertical and horizontal components are calculated by applying the following procedure:

1. Calculate vertical component values. Vertical component values are calculated by applying the following formula:

   $V_N = \sin\theta \times R$

   where

   V_N = vertical vector value (in V)

   R = resultant vector value (in V)

   $\sin\theta$ = sine of resultant vector phase (in °)

2. Calculate horizontal component values. Horizontal component values are calculated by applying the following formula:

   $H_N = \cos\theta \times R$

   where

   H_N = horizontal vector value (in V)

   $\cos\theta$ = cosine of resultant vector phase (in °)

   R = resultant vector value (in V)

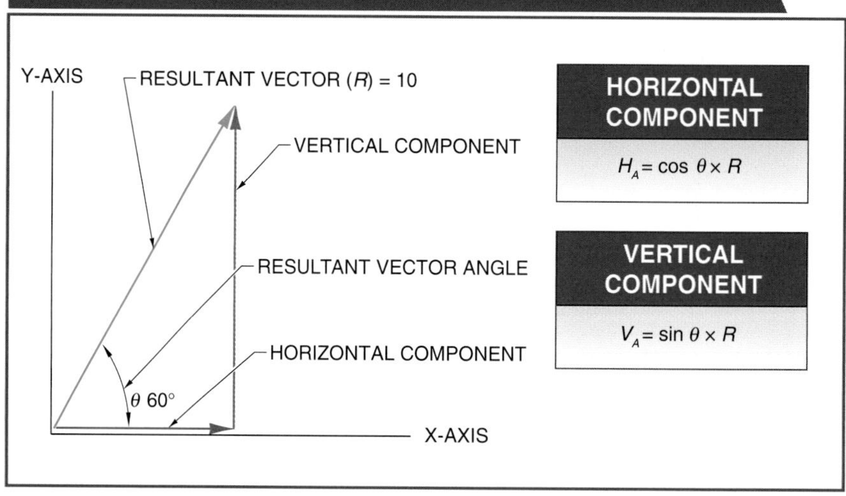

Figure 14-13. Any vector can be broken into vertical and horizontal components by using the length of the vector and sine and cosine values of the resultant vector angle.

Example: What are the horizontal and vertical values of vector A at 8 V and 80°, vector B at 6 V and 15°, and vector C at 7 V and 45°?

1. Calculate vertical component values.

   $V_A = \sin\theta \times R$

   $V_A = \sin 80° \times 8\text{ V}$

   $V_A = 0.9848 \times 8\text{ V}$

   $V_A = 7.88\text{ V}$

$V_B = \sin\theta \times R$
$V_B = \sin 15° \times 6\text{ V}$
$V_B = 0.2588 \times 6\text{ V}$
$V_B = 1.55\text{ V}$

$V_C = \sin\theta \times R$
$V_C = \sin 45° \times 7\text{ V}$
$V_C = 0.7071 \times 7\text{ V}$
$V_C = 4.95\text{ V}$

2. Calculate horizontal component values.

$H_A = \cos\theta \times R$
$H_A = \cos 80° \times 8\text{ V}$
$H_A = 0.1736 \times 8\text{ V}$
$H_A = 1.39\text{ V}$

$H_B = \cos\theta \times R$
$H_B = \cos 15° \times 6\text{ V}$
$H_B = 0.9659 \times 6\text{ V}$
$H_B = 5.79\text{ V}$

$H_C = \cos\theta \times R$
$H_C = \cos 45° \times 7\text{ V}$
$H_C = 0.7071 \times 7\text{ V}$
$H_C = 4.95\text{ V}$

Once the horizontal and vertical values are calculated, all horizontal and vertical components can be added together in a scalar manner since they are in-phase with each other.

$V_{ABC} = V_A + V_B + V_C$
$V_{ABC} = 7.88\text{ V} + 1.55\text{ V} + 4.95\text{ V}$
$V_{ABC} = \mathbf{14.38\text{ V}}$

$H_{ABC} = H_A + H_B + H_C$
$H_{ABC} = 1.39\text{ V} + 5.79\text{ V} + 4.95\text{ V}$
$H_{ABC} = \mathbf{12.13\text{ V}}$

This results in two sides of a right triangle. The resultant vector (hypotenuse) is calculated by applying the following formula:

$$R = \sqrt{H_T^{\,2} + V_T^{\,2}}$$

where
R = resultant vector (in V)
H_T = total value of horizontal vector components (in V)
V_T = total value of vertical vector components (in V)

Example: What is the resultant vector of two sides of a right triangle with a horizontal component of 12.13 V and a vertical component of 14.38 V?

$$R = \sqrt{H_T^{\,2} + V_T^{\,2}}$$

$$R = \sqrt{12.13\text{ V}^2 + 14.38\text{ V}^2}$$

$$R = \sqrt{147.14 + 206.78}$$

$$R = \sqrt{353.92}$$

$$R = \mathbf{18.81\ V}$$

The angle of the resultant vector is calculated by applying the following formula:

$$\theta = \tan^{-1}\left(\frac{V_R}{H_R}\right)$$

where

θ = vector angle (in °)

V_R = vertical resultant vector (in V)

H_R = horizontal resultant vector (in V)

Example: What is the vector angle with a horizontal component of 12.13 V and a vertical component of 14.38 V?

$$\theta = \tan^{-1}\left(\frac{V_R}{H_R}\right)$$

$$\theta = \tan^{-1}\left(\frac{14.38\text{ V}}{12.13\text{ V}}\right)$$

$$\theta = \mathbf{49.85°}$$

VECTOR SUBTRACTION

As with vector addition, vector subtraction can be performed with both graphical and mathematical methods. When performing vector subtraction with the vector diagram calculation method, the subtracting vector is rotated 180° in a counterclockwise direction and then added to the opposite vector. The addition of these vectors may be done algebraically if they are parallel to each other. If the vectors are 90° apart, the Pythagorean theorem is used to calculate the resultant vector. Graphical solutions are possible using either vector diagram or parallelogram calculation methods. For the best accuracy, mathematical methods should be used.

Vector subtraction when using mathematical formulas is calculated by applying the following procedure:

1. Rotate vector B 180°.
2. Add vector A and vector B.
3. Calculate the resultant vector length. Resultant vector length is calculated by applying the following formula:

   $$R_{A-B} = \sqrt{x^2 + y^2}$$

   where

   R_{A-B} = resultant vector length

   x = x-axis vector

   y = y-axis vector
4. Calculate angle theta. Angle theta is calculated by applying the following formula:

   $$\theta = \tan^{-1}\left(\frac{y}{x}\right)$$

Example: What is the resultant vector magnitude and angle θ when subtracting vector B at 4 V and 0° (4, 0) from vector A at 3 V and 90° (0, 3)?

1. Rotate vector B 180°. This creates vector B at 4 V and 180°, (–4, 0).
2. Add vector A and vector B. The mathematical result is (–4 + 0) + (0 + 3) = (–4, 3).
3. Calculate the resultant vector length.

   $$R_{A-B} = \sqrt{x^2 + y^2}$$

   $$R_{A-B} = \sqrt{-4^2 + 3^2}$$

   $$R_{A-B} = \sqrt{16 + 9}$$

   $$R_{A-B} = \sqrt{25}$$

   $$R_{A-B} = \mathbf{5}$$
4. Calculate angle theta.

   $$\theta = \tan^{-1}\left(\frac{y}{-x}\right)$$

   $$\theta = \tan^{-1}\left(\frac{3}{-4}\right)$$

   $$\theta = \tan^{-1}(-0.7500)$$

   $$\theta = \mathbf{-36.87°}$$

Tech Tip

Angle theta is measured clockwise from the 0° line. This is shown mathematically with a negative value of the tangent. The angle from the 0° line in a counterclockwise direction is calculated as follows:
360° – 36.87° = 323.13°

Terms

An **abscissa** is the x-value of a trigonometric function.

An **ordinate** is the y-value of a trigonometric function.

A **radius vector** is the radius value of a trigonometric function.

The resultant vector magnitude is always a positive number. When x or y is a negative value, the value is squared. The value of RA–B is equal to 5 units at 323.13°.

Additional terms used when working with trigonometric functions include abscissa, ordinate, and radius vector. An *abscissa* is the x-value of a trigonometric function. An *ordinate* is the y-value of a trigonometric function. A *radius vector* is the radius value of a trigonometric function.

Determining if a trigonometric function is a positive or negative value depends on the quadrant location of the rotating vector diagram. **See Figure 14-14.** The x and y coordinates are typically in the center of a rotating vector. The rotating vector is always a constant positive number equal to unity. The cosine and sine waves are to the left of the coordinates, and the tangent wave is to the right of the coordinates.

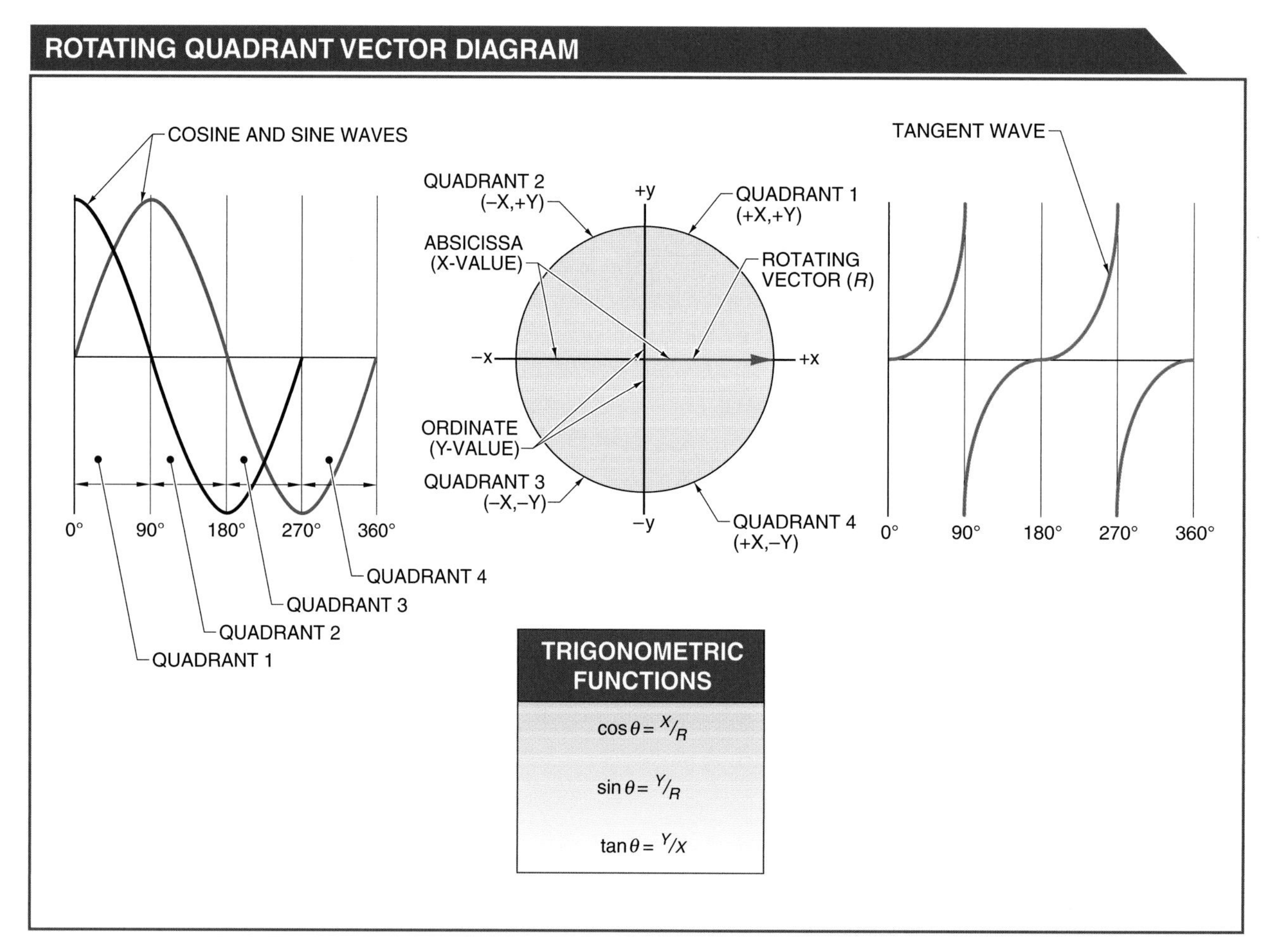

Figure 14-14. The quadrant location of a rotating vector diagram can be used to calculate whether a trigonometric function is positive or negative.

Subtracting Vectors Less than 90° Apart

Vectors less than 90° apart can be subtracted using graphical and mathematical methods. Each type of method produces the same result, although the mathematical method of vector subtraction is more accurate. Vector subtraction is identical to vector addition except that the number to be subtracted is rotated 180° before the addition. **See Figure 14-15.**

The mathematical method of vector subtraction involves several steps. When subtracting vectors that are fewer than 90° apart from one another with the mathematical method of vector subtraction, apply the following procedure:

1. Rotate vector A 180°.
2. Solve mathematically for vector A vertical and horizontal components by applying the following formula:

   $V_A = \sin \theta_A \times V_{AM}$

   and

   $H_A = \cos \theta_A \times V_{AM}$

   where

   θ_A = angle of vector A from horizontal
   V_A = vector A vertical component
   H_A = vector A horizontal component
   V_{AM} = vector A magnitude (in V)
3. Solve mathematically for vector B vertical and horizontal components by applying the following formula:

   $V_B = \sin \theta_B \times V_{BM}$

   and

   $H_B = \cos \theta_B \times V_{BM}$

   where

   θ_B = angle of vector B from horizontal
   H_B = vector B horizontal component
   V_B = vector B vertical component
   V_{BM} = vector B magnitude (in V)
4. Add vertical components together to form V_{B-A}.

   $V_{B-A} = V_B + V_A$
5. Add horizontal components together to form H_{B-A}.

   $H_{B-A} = H_B + H_A$
6. Calculate angle θ_{B-M}. Angle θ_{B-M} is calculated by applying the following formula:

   $$\theta_{B-A} = \tan^{-1}\left(\frac{V_{B-A}}{H_{B-A}}\right)$$

7. Calculate resultant vector magnitude by using sine function, cosine function, or Pythagorean theorem. The resultant vector magnitude can be calculated by applying one of the following formulas:

$$Z_{B-A} = \frac{V_{B-A}}{\sin \theta_{B-A}}$$

or

$$Z_{B-A} = \frac{H_{B-A}}{\cos \theta_{B-A}}$$

or

$$Z_{B-A} = \sqrt{V_{B-A}{}^2 + H_{B-A}{}^2}$$

where

Z_{B-A} = resultant vector magnitude
V_{B-A} = vertical vector components
H_{B-A} = horizontal vector components

Example: What is the resultant vector when vector A at 6 V and 70° is subtracted from vector B at 8 V and 25°? **See Figure 14-15.**

1. Rotate vector A 180°.
   Vector A magnitude = 6 V
   Vector A phase = 180° + 70°
   Vector A phase = 250°

2. Solve mathematically for vector A vertical and horizontal components.
   $V_A = \sin \theta_A \times V_{AM}$
   $V_A = \sin 250° \times 6\text{ V}$
   $V_A = -0.9397 \times 6\text{ V}$
   $V_A = -5.64\text{ V}$

   $H_A = \cos \theta_A \times V_{AM}$
   $H_A = \cos 250° \times 6\text{ V}$
   $H_A = -0.3420 \times 6\text{ V}$
   $H_A = -2.05\text{ V}$

3. Solve mathematically for vector B vertical and horizontal components.
   $V_B = \sin \theta_B \times V_{BM}$
   $V_B = \sin 25° \times 8\text{ V}$
   $V_B = 0.4226 \times 8\text{ V}$
   $V_B = 3.38\text{ V}$

   $H_B = \cos \theta \times V_{BM}$
   $H_B = \cos 25° \times 8\text{ V}$
   $H_B = 0.9063 \times 8\text{ V}$
   $H_B = 7.25\text{ V}$

4. Add vertical components together to form V_{B-A}.

$V_{B-A} = V_A + V_B$

$V_{B-A} = -5.64 \text{ V} + 3.38 \text{ V}$

$V_{B-A} = -2.26 \text{ V}$

5. Add horizontal components together to form H_{B-A}.

$H_{B-A} = H_A + H_B$

$H_{B-A} = -2.05 \text{ V} + 7.25 \text{ V}$

$H_{B-A} = 5.2 \text{ V}$

6. Calculate angle θ_{B-A}.

$$\theta_{B-A} = \tan^{-1}\left(\frac{V_{B-A}}{H_{B-A}}\right)$$

$$\theta_{B-A} = \tan^{-1}\left(\frac{-2.26 \text{ V}}{5.2 \text{ V}}\right)$$

$$\theta_{B-A} = -23.5°$$

Measuring from the 0° line the counterclockwise angle is:

$\theta = 360° - 23.5°$

$\theta = 336.5°$

7. Calculate resultant vector magnitude. Apply the sine or cosine function to calculate the value of the resultant vector or use the Pythagorean theorem.

$$Z_{B-A} = \frac{V_{B-A}}{\sin \theta_{B-A}}$$

$$Z_{B-A} = \frac{V_{B-A}}{\sin(-23.5°)}$$

$$Z_{B-A} = \frac{-2.26 \text{ V}}{-0.3986}$$

$Z_{B-A} = \mathbf{5.67 \text{ V}}$

or

$$Z_{B-A} = \frac{H_{B-A}}{\cos \theta_{B-A}}$$

$$Z_{B-A} = \frac{5.2 \text{ V}}{\cos(-23.5°)}$$

$$Z_{B-A} = \frac{5.2 \text{ V}}{0.9170}$$

$Z_{B-A} = \mathbf{5.67}$

or

$$Z_{B-A} = \sqrt{{V_{B-A}}^2 + {H_{B-A}}^2}$$

$$Z_{B-A} = \sqrt{5.2^2 + (-2.26)^2}$$

$$Z_{B-A} = \sqrt{27.04 + 5.11}$$

$$Z_{B-A} = \sqrt{32.15}$$

$$Z_{B-A} = \mathbf{5.67\ V}$$

The resultant vector is equal to 5.67 V at 336.5° (360° – 23.5° = 336.5°).

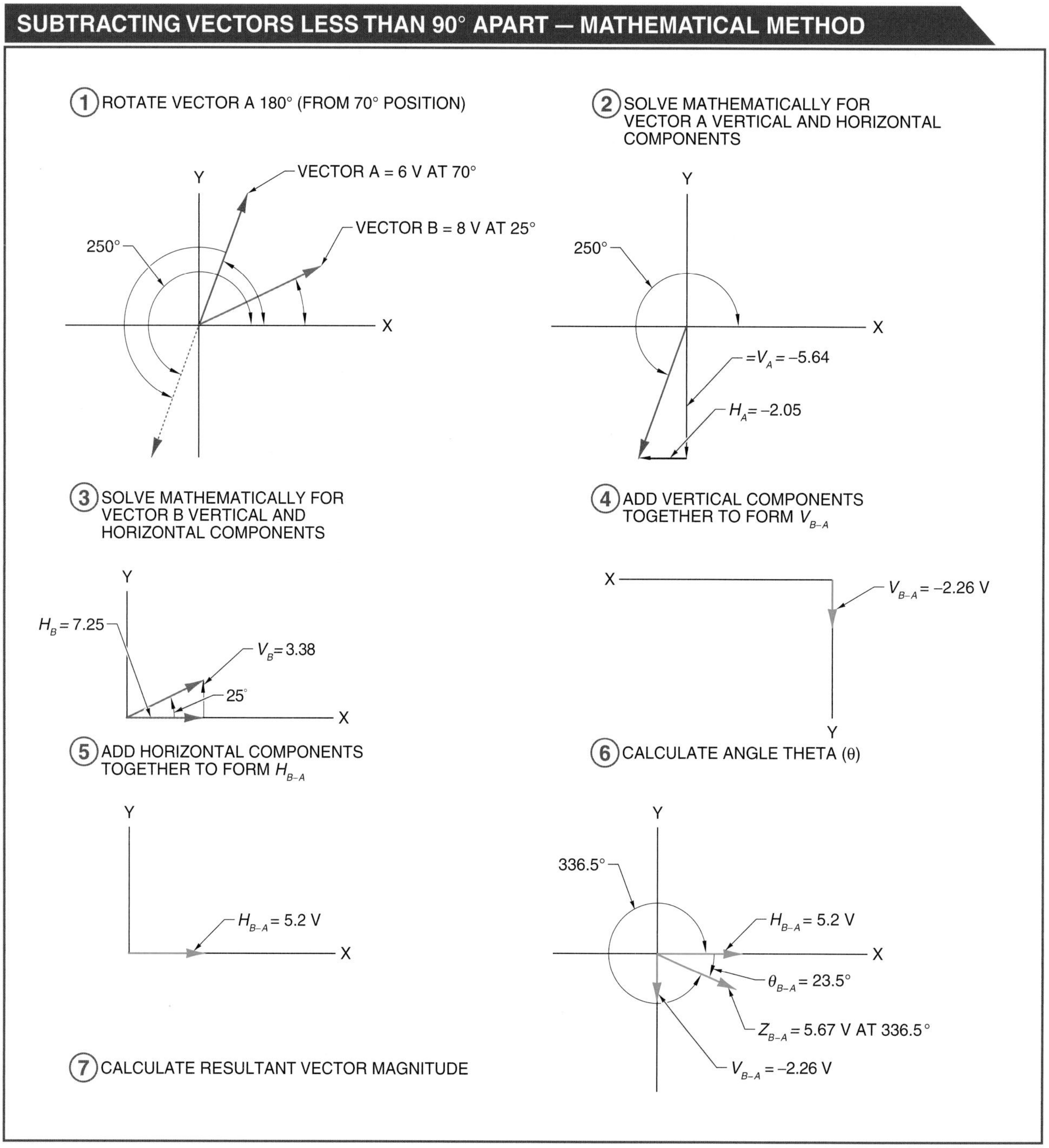

SUBTRACTING VECTORS LESS THAN 90° APART — MATHEMATICAL METHOD

1 ROTATE VECTOR A 180° (FROM 70° POSITION)

VECTOR A = 6 V AT 70°
VECTOR B = 8 V AT 25°
250°

2 SOLVE MATHEMATICALLY FOR VECTOR A VERTICAL AND HORIZONTAL COMPONENTS

250°
$=V_A = -5.64$
$H_A = -2.05$

3 SOLVE MATHEMATICALLY FOR VECTOR B VERTICAL AND HORIZONTAL COMPONENTS

$H_B = 7.25$
$V_B = 3.38$
25°

4 ADD VERTICAL COMPONENTS TOGETHER TO FORM V_{B-A}

$V_{B-A} = -2.26$ V

5 ADD HORIZONTAL COMPONENTS TOGETHER TO FORM H_{B-A}

$H_{B-A} = 5.2$ V

6 CALCULATE ANGLE THETA (θ)

336.5°
$H_{B-A} = 5.2$ V
$\theta_{B-A} = 23.5°$
$Z_{B-A} = 5.67$ V AT 336.5°
$V_{B-A} = -2.26$ V

7 CALCULATE RESULTANT VECTOR MAGNITUDE

Figure 14-15. Subtracting vectors less than 90° apart using the mathematical method involves several steps.

SUMMARY

Vectors are graphic figures that can be used to represent various AC circuit parameters. A vector is a scalar quantity that has direction. Vectors can be added or subtracted. Vector analysis is the primary tool in understanding how parameters of an AC circuit change. Vector analysis can take two forms, graphical and mathematical. Two graphical methods available for the solution of vector values are the vector diagram calculation method and the parallelogram calculation method. With the vector diagram calculation method, vectors are arranged head to tail with the tail portion of the first vector connected to the head portion of the last vector to obtain the resultant vector. With the parallelogram calculation method, lines are drawn parallel to each of the two vectors. The resultant vector is drawn from the origin of the two vectors to the point of intersection of the constructed lines. In both cases, the angle must be estimated or measured with a protractor. Graphical methods can be used when a high degree of accuracy is not required.

Mathematical methods are used when a high degree of accuracy is required. Mathematical methods require use of the Pythagorean theorem and trigonometric functions to solve for the magnitude and direction of the resultant vector. Typically, the horizontal method decomposes each vector into horizontal and vertical components. Horizontal and Vertical components are used to find the resultant vector magnitude by using Pythagorean theorem and angle theta by using trigonometric functions.

APPLICATION: THE j OPERATOR...

Background: When analyzing AC circuits, phasors are used to analyze the phase relationships between voltage and current. In AC circuits, components such as inductors and capacitors introduce changes in the phase relationship between voltage and current. For complex circuits, phasors can be difficult to use. Complex numbers can be used instead to simplify the analysis.

The lowercase letter "j" represents $\sqrt{(-1)}$. For circuit analysis, the value of j is used to indicate a phase advance of 90°. A value of –j is used to indicate a phase lag of 90°. The relationship between sin and cos is 90°. **See Figure 1.**

...APPLICATION: THE j OPERATOR...

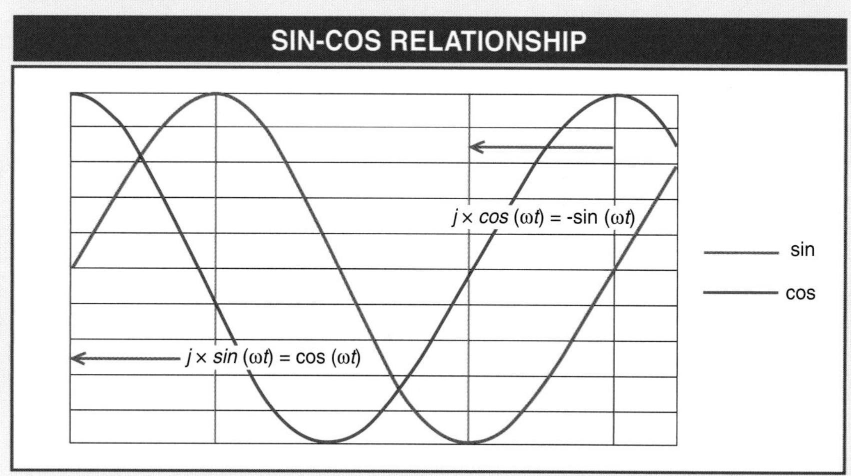

Figure 1. Cosine leads sine by 90°.

Therefore, the following statements are true:

$j \times sin(\omega t) = \cos(\omega t)$

$j \times cos(\omega t) = -\sin(\omega t)$

$-j \times sin(\omega t) = -\cos(\omega t)$

$-j \times cos(\omega t) = \sin(\omega t)$

where

$j = \sqrt{(-1)}$ (constant)

ω = angular frequency (in radians per sec)

t = time (in sec)

Angular frequency is defined as the following:

$\omega = 2\pi \times \mathrm{f}$

where

ω = angular frequency (in radians per sec)

2π = 6.28 (constant)

f = frequency (in Hz)

When analyzing circuit voltage, Ohm's law is used for AC circuits. For AC circuits, the resistance is replaced with impedance (Z).

$V = I \times Z$

where

V = voltage (in V)

I = current (in A)

Z = impedance (in Ω)

The AC impedance in terms of the j operator of the different circuit elements are the following:

resistor $Z = R$

inductor $Z = j\omega L$

capacitor $Z = \dfrac{1}{(j\omega C)}$

...APPLICATION: THE j OPERATOR

where
Z = impedance (in Ω)
R = resistance (in Ω)
L = inductance (in H)
C = capacitance (in F)
$j = \sqrt{(-1)}$ (constant)
ω = angular frequency (in radians per sec)

Example: What is the AC voltage across a resistor (R) and inductor (L) when the current through each device is $sin\ (\omega t)$?

$$V = I \times Z$$
$$V = sin\ (\omega t) \times R$$
$$V = R \times sin\ (\omega t)$$

$$V = sin\ (\omega t) \times j\omega L$$
$$V = \omega \times L \times j \times sin\ (\omega t)$$
$$V = \omega \times L \times \cos\ (\omega t)$$

As the frequency increases, the angular frequency (ω) increases, and the voltage across the inductor increases. The voltage across the resistor remains the same. Since the current remains the same, the inductor impedance must increase over frequency.

Key Points: Complex numbers

Problem: What is the AC voltage across a capacitor when the current is $sin\ (\omega t)$?

Solution: Using Ohm's law and the impedance of the capacitor, the voltage is calculated as follows:

$$V = I \times Z$$
$$V = \frac{sin(\omega t) \times 1}{(j\omega C)}$$
$$V = \frac{sin(\omega t) \times -j}{(\omega C)}$$
$$V = \frac{-1 \times j \times sin(\omega t)}{(\omega C)}$$
$$V = \frac{-1 \times cos(\omega t)}{(\omega C)}$$
$$V = \frac{-cos(\omega t)}{(\omega C)}$$

As the frequency increases, the angular frequency (ω) increases, and the voltage across the capacitor decreases. Since the current remains the same, the capacitor impedance must decrease over frequency.

Chapter Review

1. What is a radius vector?
2. How is vector subtraction different from vector addition?
3. List the three most common trigonometric functions used in the solution of parameters for AC circuits.
4. What is a resultant vector?
5. How does a vector value differ from a scalar value?
6. What is an ordinate?
7. How is the term "phasor" used in AC analysis?
8. What is an abscissa?
9. What is the Pythagorean theorem?
10. What is the angle θ for a 2ϕ generator with two 120 V vectors and a resultant vector of 153.8 V?
11. What is a phasor?
12. How is the Pythagorean theorem used when vectors are not at right angles to one another?
13. When would vector calculations be used to solve for AC values?
14. What is the difference between the parallelogram and vector diagram calculation methods?
15. What is a vector used to represent?

15

Resistive AC Circuits

OBJECTIVES

- List the instantaneous values of AC.
- Define the different values of AC voltage and current and describe the relationships between AC values.
- Explain resistance in AC circuits.
- Describe the differences between series, parallel, and series/parallel resistive AC circuits.

Unlike direct current (DC), alternating current (AC) periodically changes its direction of current flow. In a resistive AC circuit with constant voltage applied, the instantaneous, peak, average, and effective values are all different. Therefore, when making calculations in a resistive AC circuit, it is first necessary to define each of these values.

As with DC circuits, AC circuit power is based on voltage and current. To calculate the equivalent DC power dissipation for an AC circuit, the circuit effective value parameters are used rather than peak or average values. Series, parallel, and series-parallel (combination) AC resistive circuits use Ohm' s law to calculate voltage and current values.

INSTANTANEOUS VALUES OF AC PARAMETERS

Instantaneous AC parameters in a resistive circuit are the values of voltage and/or current at a given point of time. In an AC circuit, the voltage and current continually change value at a periodic rate. For example, a conductor rotating 360° through a magnetic field generates a sine wave that demonstrates changing values in an AC circuit. **See Figure 15-1.** In the generation of a sine wave, 0 V is induced into a rotating conductor at 0° and at 180°. This is because the conductor is moving parallel to the magnetic field and not cutting the field. Maximum positive voltage is induced at 90° rotation, and maximum negative voltage is induced at 270° rotation. At these points, the conductor is perpendicular to the magnetic flux field and cuts the maximum number of magnetic lines of force.

INSTANTANEOUS VALUES OF AC VOLTAGE

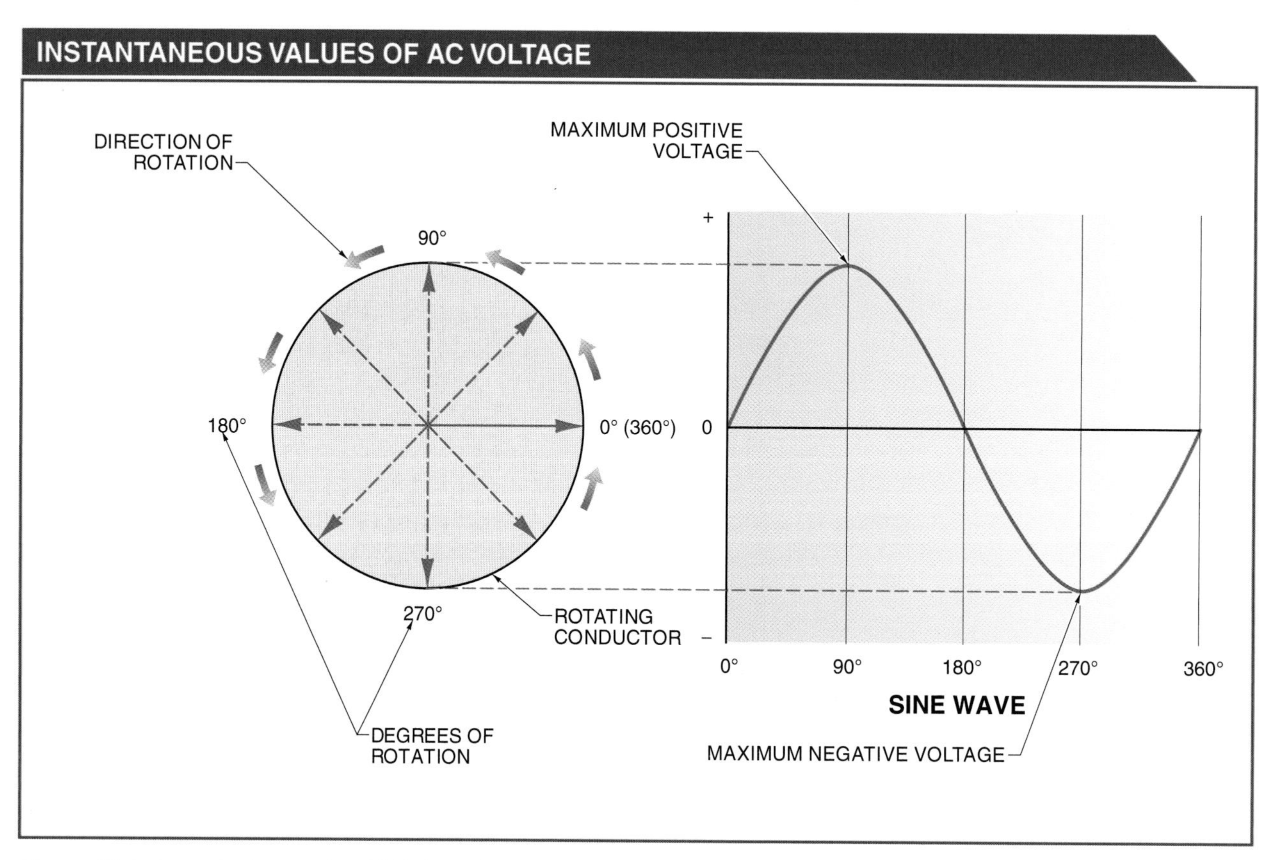

Figure 15-1. A conductor rotating 360° through a magnetic field generates a sine wave that demonstrates changing values in an AC circuit.

Besides maximum and minimum voltage points, voltage values at certain angles or phases are often used. These angles include 30° (50% of maximum voltage), 45° (70.7% of maximum voltage), and 60° (86.6% of maximum voltage). The voltage percentage value is based on the sine of the angle of rotation. Sometimes voltage is needed for an intermediate angle not listed. Because a sine wave follows a trigonometric sine function as the vector rotates, a simple equation yields any instantaneous value for any rotation angle. The instantaneous value of AC voltage can be calculated by applying the following formula:

$V = VP \times \sin\theta$

where

V = instantaneous voltage (in V)

VP = maximum peak voltage at 90° (in V)

θ = angle theta (in °)

Terms

The **peak value** of an AC voltage or current in a resistive circuit is the maximum instantaneous value of either the positive or negative alternation of the sine wave.

Example: What is the instantaneous value of voltage in a resistive AC circuit with an angle theta (θ) of 35° and a maximum generated voltage of 170 V?

$V = VP \times \sin\theta$

$V = 170\ \text{V} \times \sin 35°$

$V = 170\ \text{V} \times 0.5736$

$V = \mathbf{97.5\ V}$

The value of sine at 35° is 0.5736. **See Appendix.** Voltage at any given point during the rotation of the vector can be calculated by multiplying the sine of the vector rotation angle by the maximum peak voltage.

VALUES OF ALTERNATING VOLTAGE AND CURRENT

Values of AC voltage and current in a resistive circuit may be given as peak value, average value, or effective value. An AC voltage is described completely when one of these values is given along with the frequency of the sine wave. For example, the common value of voltage used in any home in the United States is 120 VAC effective with a frequency of 60 Hz. Since an AC sine wave for voltage or current has many different and instantaneous values throughout the cycle, it is common practice to define the relationships between the different AC values.

Peak Value

The *peak value* of an AC voltage or current in a resistive circuit is the maximum instantaneous value of either the positive or negative alternation of the sine wave. Positive peak value of an AC voltage is above the 0 V baseline and occurs at 90°, and negative peak value of an AC voltage is below the 0 V baseline and occurs at 270°. **See Figure 15-2.**

PEAK VALUE

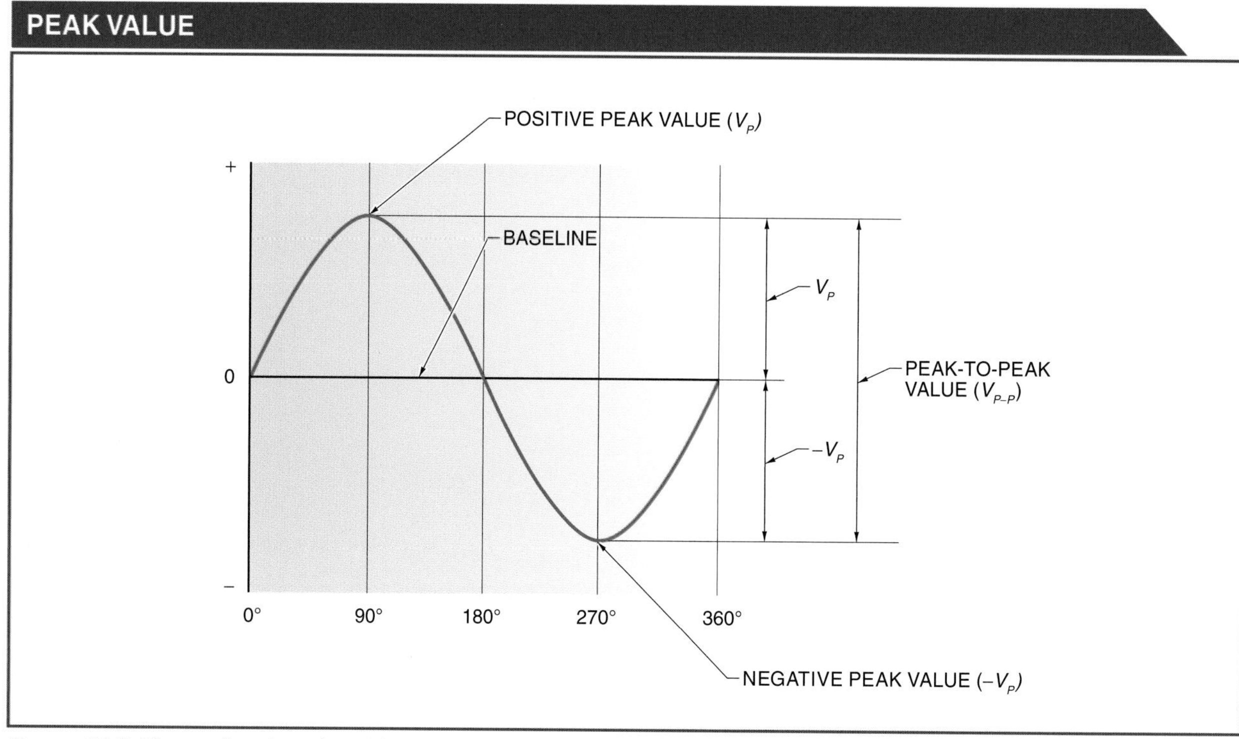

Figure 15-2. The peak value of an AC voltage or current is an instantaneous maximum value that occurs at 90° and 270°. Positive peak value occurs at 90° and negative peak value occurs at 270°.

Terms

A **peak-to-peak value** is the value measured from the maximum positive alternation to the maximum negative alternation of a sine wave.

A *peak-to-peak value* is the value measured from the maximum positive alternation to the maximum negative alternation of a sine wave. Voltages and currents in an AC resistive circuit are sometimes given in their peak-to-peak values. Because a pure sine wave has equal values above and below the 0 V baseline, the positive and negative peaks are equal. When either the positive or negative peak value of voltage is known, then the peak-to-peak value of the voltage can be calculated by applying the following formula:

$$V_{P-P} = 2 \times V_P$$

where

V_{P-P} = AC peak-to-peak value (in V)

V_P = AC positive peak value (in V)

Example: What is the peak-to-peak voltage value of a resistive AC circuit if positive peak voltage value is equal to 170 V?

$$V_{P-P} = 2 \times V_P$$

$$V_{P-P} = 2 \times 170 \text{ V}$$

$$V_{P-P} = \mathbf{340\ V}$$

In this example, only voltage is used. Formulas for voltage apply equally to alternating current, with peak current (I_P) and peak-to-peak current (I_{P-P}) substituted for V_P and V_{P-P}.

Peak values of voltage are often used with electrostatic devices such as a cathode ray tube (CRT). For example, a capacitor charges to the peak voltage. Deflection of the electron beam in a CRT computer monitor is responsive to the peak-to-peak voltages. Typically, the vertical scale on an oscilloscope is calibrated in peak-to-peak or peak values.

Average Value

The *average value* of an AC voltage or current in a resistive AC circuit is the mathematical mean of all instantaneous values in a sine wave. **See Figure 15-3.** For certain AC applications, the average value of a sine wave is used. Either voltage or current can be represented. Because the wave is symmetrical, when the average value above the baseline is averaged with the average value below the baseline, the value is zero. Therefore, when the average value is used, it relates to the value of one alternation, or half-cycle, of a sine wave. Either a positive or negative alternation can be used.

Terms

The **average value** of an AC voltage or current in a resistive AC circuit is the mathematical mean of all instantaneous values in a sine wave.

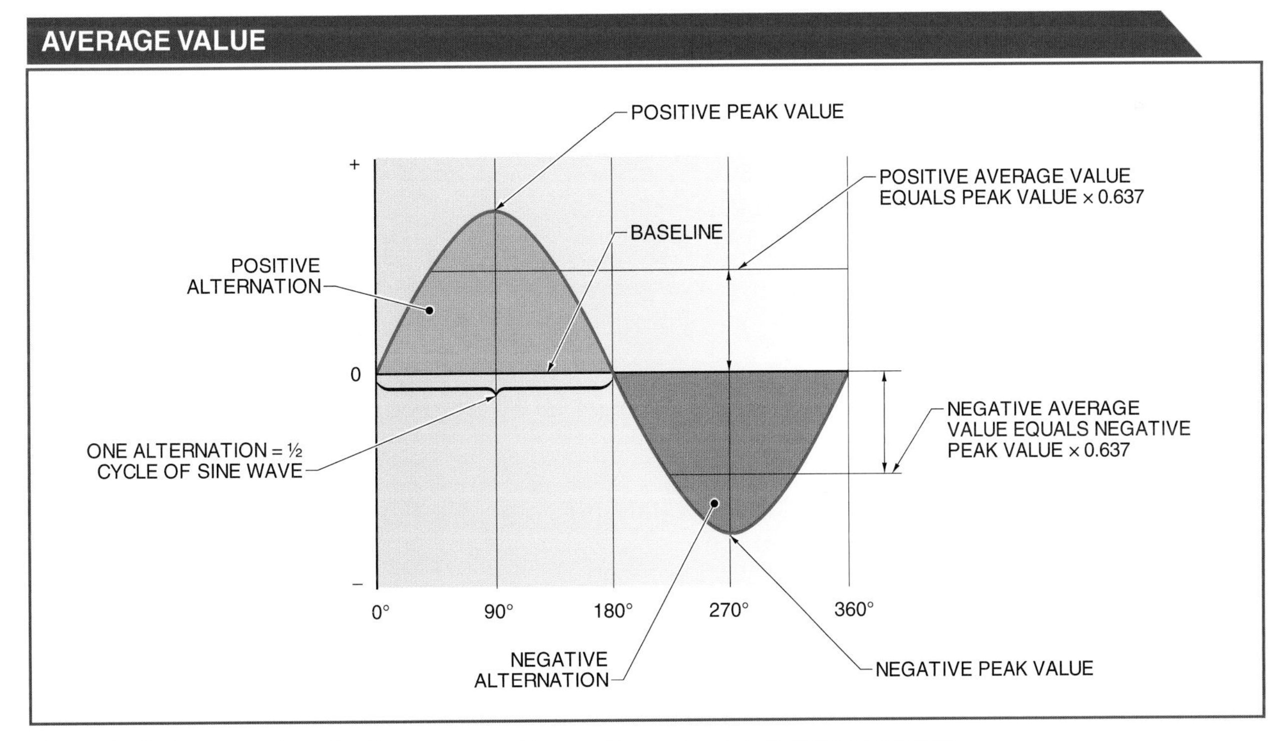

Figure 15-3. Average value of voltage or current is equal to peak value multiplied by 0.637.

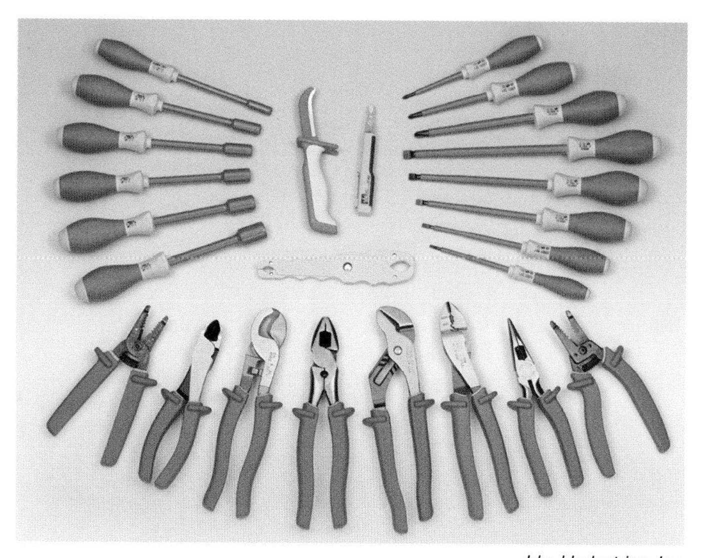

Ideal Industries, Inc.

The insulation rating of an electrician's insulated hand tool is based on the peak voltage its insulation material can withstand.

The average voltage of a sine wave is the average of all instantaneous values for one alternation. That is, if the values of the sine function were added for each degree from 0° to 90°, with the sum divided by 90, an approximation of the average value would be calculated. The value of average voltage is calculated to be equal to the instantaneous peak value multiplied by a factor of 0.637. Average voltage can be calculated by applying the following formula:

$V_{AVG} = V_P \times 0.637$

where

V_{AVG} = average voltage value (in V)

V_P = peak value (in V)

0.637 = conversion factor from V_P to V_{AVG}

Example: What is the average voltage value if the peak voltage value is equal to 170 V?

$V_{AVG} = V_P \times 0.637$

$V_{AVG} = 170 \text{ V} \times 0.637$

$V_{AVG} = \mathbf{108.3 \text{ V}}$

If the average voltage value is known, the peak value can be calculated by applying the following formula:

$$V_P = \frac{V_{AVG}}{0.637}$$

where

V_P = peak value (in V)

V_{AVG} = average voltage value (in V)

0.637 = conversion factor from V_P to V_{AVG}

Example: What is the peak voltage value of a resistive AC circuit if the average value of the same circuit is equal to 108.3 V?

$$V_P = \frac{V_{AVG}}{0.637}$$

$$V_P = \frac{108.3 \text{ V}}{0.637}$$

$V_P = \mathbf{170 \text{ V}}$

Effective Value

The *effective value* of an AC voltage or current is the value of a sine wave that produces the same amount of heat in a pure resistive circuit as DC of the same value. Units of AC voltage and current share a common basis with units of DC voltage and current. The common basis is that the effective value of 1 A of AC causes the same heating effect as 1 A of DC.

Terms

The **effective value** of an AC voltage or current is the value of a sine wave that produces the same amount of heat in a pure resistive circuit as DC of the same value.

When AC flows through a resistance, the power dissipated is constantly changing with the AC sine wave. The power dissipated at any instant can be calculated by applying the following formula:

$P_D = I^2 \times R$

where

P_D = power dissipated (in W)

I = circuit current (in A)

R = circuit resistance (in Ω)

Example: What is the power dissipated in a 4 A circuit with a resistance of 10 Ω?

$P_D = I^2 \times R$

$P_D = (4\text{ A})^2 \times 10\ \Omega$

$P_D = 16\text{ A} \times 10\ \Omega$

$P_D =$ **160 W**

Power dissipated is always a positive value, even when the current value is negative. The value of I^2 is always positive because a negative value multiplied by a negative value yields a positive value.

The *root-mean-square voltage value (rms value)* is the voltage value of a sine wave that produces the same amount of heat in a pure resistive circuit as DC of the same value. The rms value is also known as the mathematical factor for the effective value.

Terms

The **root-mean-square voltage value (rms value)** is the voltage value of a sine wave that produces the same amount of heat in a pure resistive circuit as DC of the same value.

When calculating the rms value, the effective value of the current or voltage sine wave is shown to be equal to peak value divided by the square root of 2. A current sine wave can be calculated by applying the following formula:

$$I_{EFF} = \frac{I_P}{\sqrt{2}}$$

where

I_{EFF} = current effective value (in A)

I_P = current peak value (in A)

$\sqrt{2} = 1.414$

Since the reciprocal value of 1.414 is 0.707, a current sine wave can also be calculated by applying the following formula:

$I_{EFF} = 0.707 \times I_P$

When calculating voltage sine waves, V_{EFF} is substituted for I_{EFF}.

Example: What is the effective voltage if peak voltage is equal to 230 V?

$V_{EFF} = 0.707 \times V_P$

$V_{EFF} = 0.707 \times 230 \text{ V}$

$V_{EFF} = \mathbf{162.6\ V}$

The average value of AC has no relationship to the average value of DC, because an AC sine wave is constantly changing—from zero to maximum positive value and back to zero, to maximum negative value and back to zero. This allows for two heating and cooling cycles for each cycle of the AC sine wave. With DC, the heating is constant, and no cooling occurs. Therefore, an average DC current causes greater heat than an average AC current.

In an AC circuit, effective value must be greater than the average value to provide the same heating effect of DC of equal value. Mathematically, this is because effective value is calculated by taking the square root of the average of all the instantaneous values squared while the average value is calculated by simply taking the mean of all the instantaneous values. Positive effective value is equal to peak value multiplied by 0.707. Negative effective value is equal to the peak value multiplied by –0.707. **See Figure 15-4.**

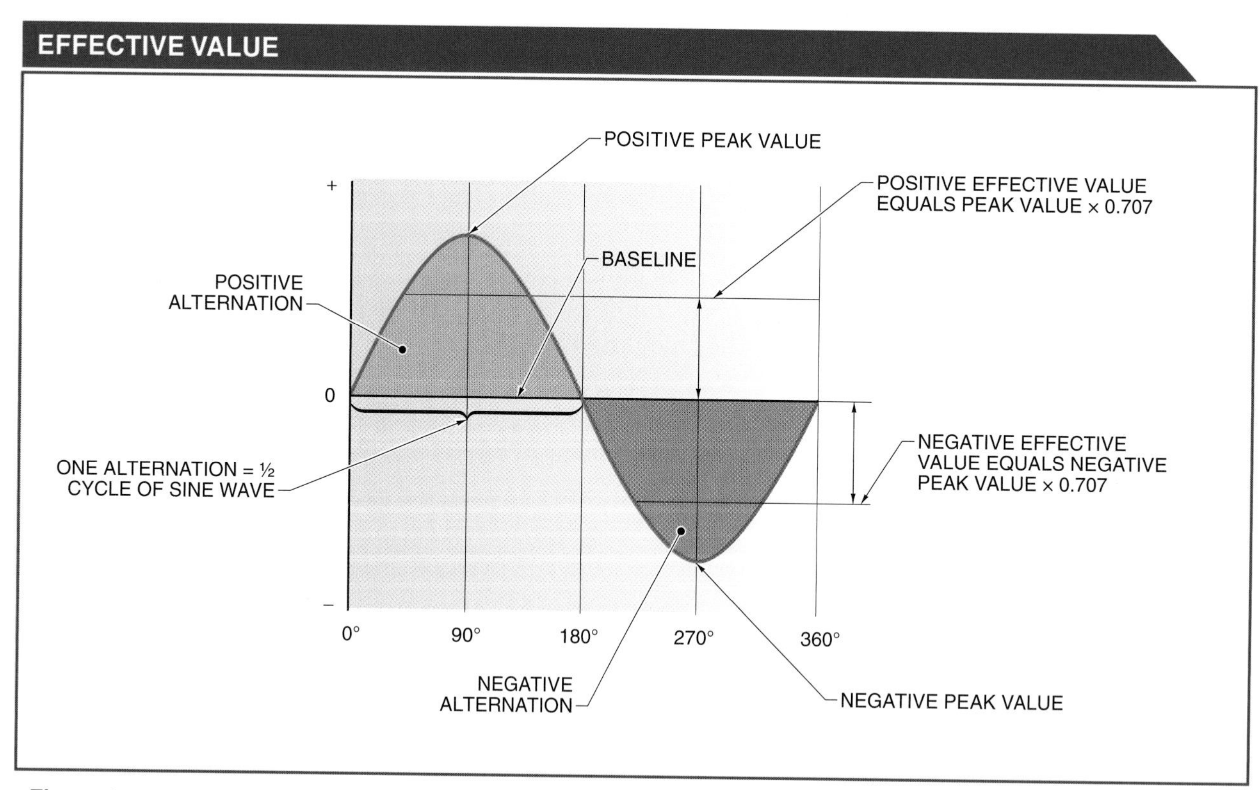

Figure 15-4. Effective value of voltage or current is equal to peak value multiplied by 0.707.

Note: When an AC voltage or current rating is given without any qualifiers such as peak or average, it is implied that the voltage or current rating is the effective value. For example, an appliance is rated at 120 VAC/10 A/60 Hz. The effective values are 120 VAC and 10 A.

Relationships Between AC Values

Relationships between AC values in a resistive AC circuit can vary greatly. For example, peak values are considerably greater than average and effective values. In a resistive AC circuit, the components and devices must be designed to withstand peak values, even though they only occur twice each cycle. The dielectric strength of insulators must have voltage ratings that exceed the peak voltage. In certain circumstances, the voltage rating must be greater than the peak-to-peak value.

Certain devices are used to help control varying values in resistive AC circuits. For example, a rectifier changes AC to DC by allowing the current to flow in only one direction. In a rectifier circuit, the diode is rated for the peak-to-peak voltage, while the capacitor is rated for peak voltage. **See Figure 15-5.** Controlling AC values is accomplished with a diode that allows current to flow through it only when its anode portion is positive in respect to its cathode portion. With a capacitor in the circuit, only the positive alternation is developed across the load resistance. Without a capacitor in the circuit, only the positive alternation is visible across the load resistance with an oscilloscope.

A *filter* is a selective network of resistances, capacitances, and inductances that provides little opposition to certain frequencies or DC while blocking or attenuating other frequencies. For example, when a filter capacitor is replaced, it charges during the first half of the sine wave. The filter discharges through the load during the second half of the positive cycle and during the time the cycle is in its negative alternation. The capacitor smooths out the large ripple of the rectified wave.

A *filter capacitor* is a capacitor utilized with inductors and/or resistors for controlling harmonics problems in the power source, such as reducing voltage distortion due to large rectifier loads. *Ripple voltage* is the portion of the output voltage of a power source that is related in frequency to the power source.

A *bleeder resistor* is a resistor that is used to discharge a capacitor. This reduces changes in the ripple voltage when current is delivered to the load. A bleeder resistor is used to protect equipment from excessive voltages if the load is removed or substantially reduced.

Terms

A **filter** is a selective network of resistances, capacitances, and inductances that provides little opposition to certain frequencies or DC while blocking or attenuating other frequencies.

A **filter capacitor** is a capacitor utilized with inductors and/or resistors for controlling harmonics problems in the power source, such as reducing voltage distortion due to large rectifier loads.

Ripple voltage is the portion of the output voltage of a power source that is related in frequency to the power source.

A **bleeder resistor** is a resistor that is used to discharge a capacitor.

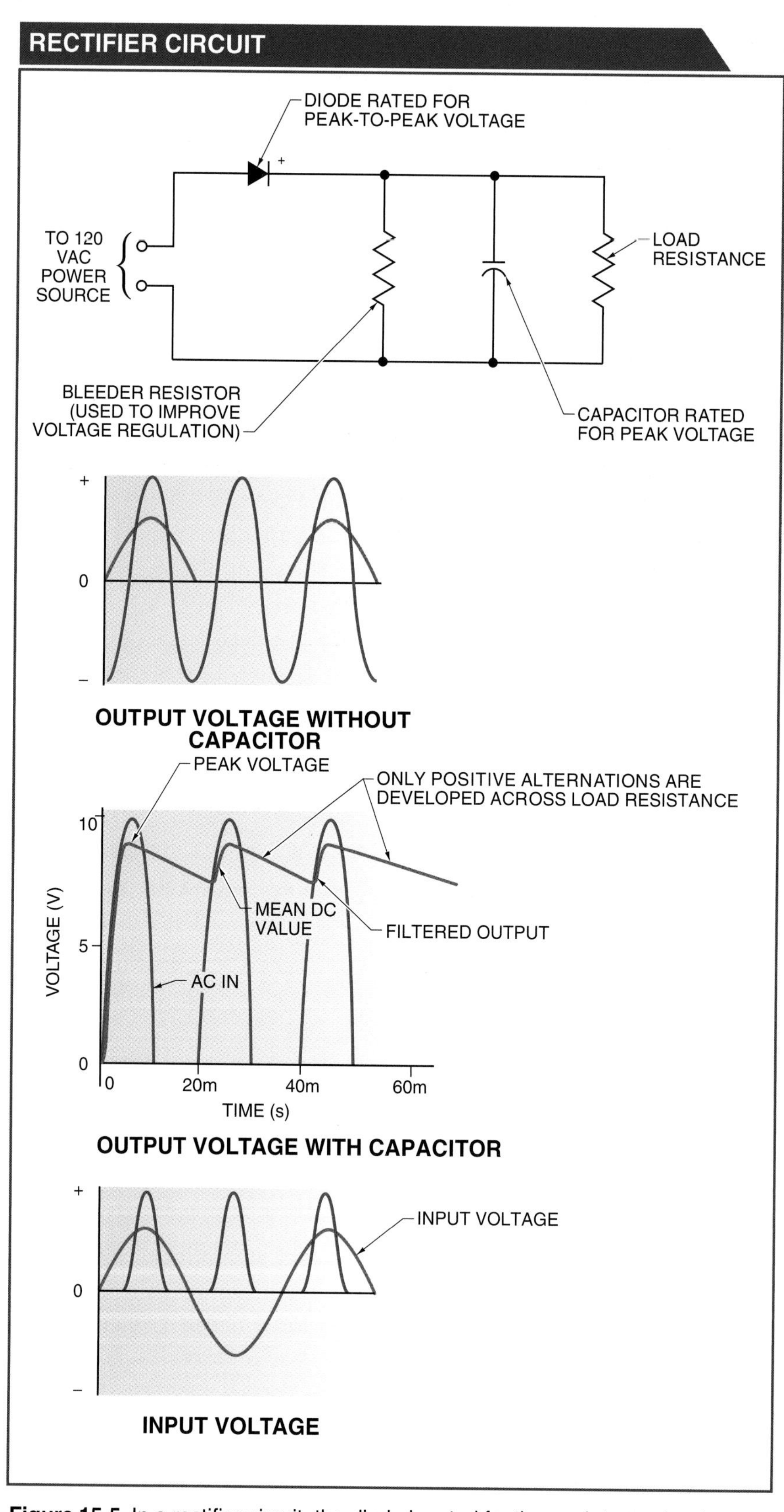

Figure 15-5. In a rectifier circuit, the diode is rated for the peak-to-peak voltage, while the capacitor is rated for peak voltage.

A bleeder resistor gets its name from another of its purposes, which is to "bleed" the residual charge off of a capacitor when the power source is turned OFF. Bleeding off the residual charge prevents accidental shock and equipment damage while work is performed on equipment that is not energized.

When current flows through a diode, electrons are supplied by the positive plate of the capacitor and flow through the bleeder resistor. The capacitor charges the anode portion of the diode to the peak value of the sine wave.

The peak-to-peak voltage is applied across the diode. During the negative half-cycle, the DC voltage on the cathode is maintained by the charged capacitor. The DC voltage plus the negative peak voltage across the anode make the voltage across the diode nearly equal to the peak-to-peak value. The rectifier must therefore be rated at or above the peak-to-peak voltage, and the capacitor must be rated above the peak value. Average value, effective value, peak value, and peak-to-peak value can be converted to other values by applying multiplication conversion factors. **See Figure 15-6.**

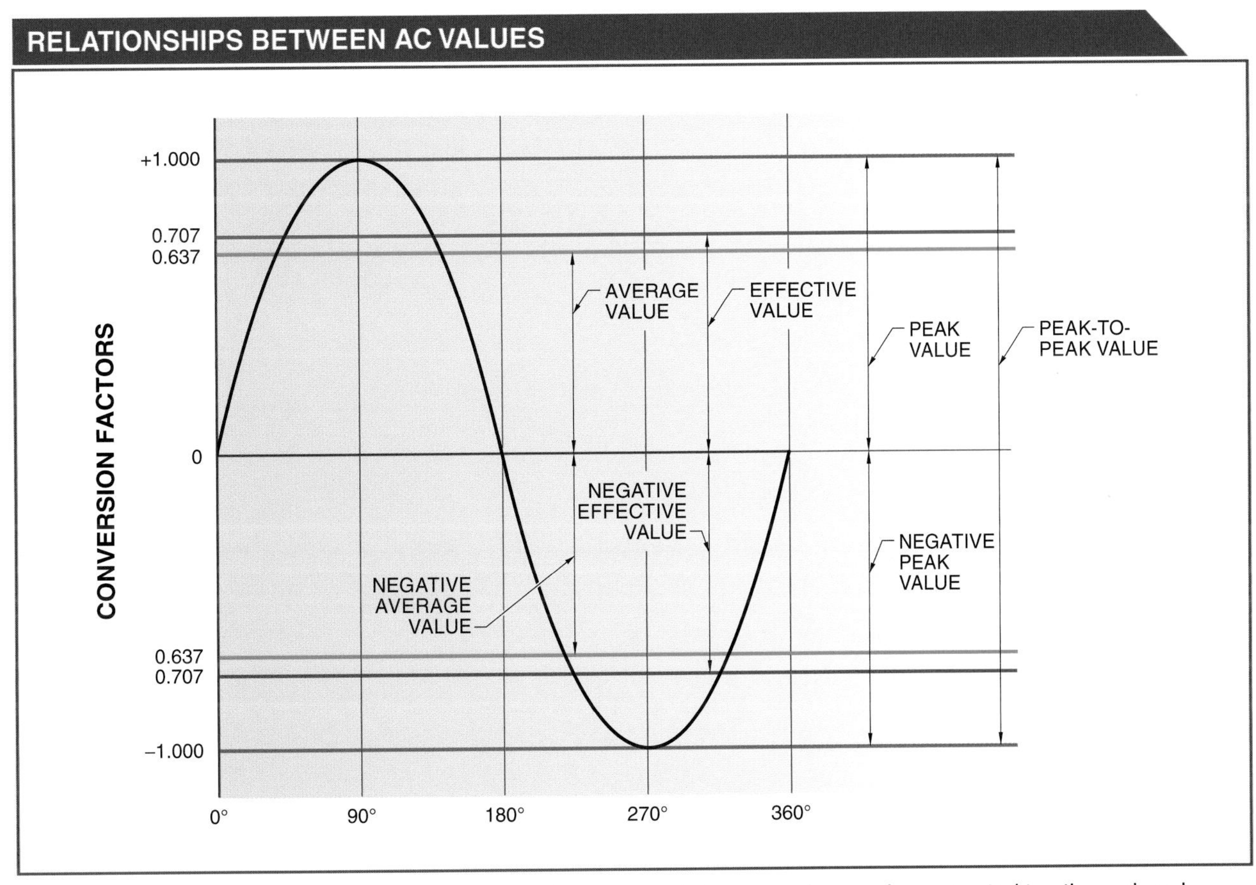

Figure 15-6. Average value, effective value, peak value, and peak-to-peak value can be converted to other values by applying multiplication conversion factors.

RESISTANCE IN AC CIRCUITS

Resistance in AC circuits merely opposes the flow of current in a circuit. A resistive AC circuit with effective voltage and current present can be treated in the same way as a DC resistive circuit. Any time an AC value is given without specifying peak, peak-to-peak, average, or effective values, it is assumed to be an effective value.

When viewing a schematic diagram that has representations of voltage and current sine waves and shows the phase relationship between them, the voltage sine wave represents instantaneous values supplied by an AC generator. The current sine wave is derived from the voltage sine wave and represents the instantaneous voltage points divided by a resistance. These sine waves are in phase with each other and are always in the same direction, crossing the baseline together. No time difference exists between them. In a pure resistive AC circuit, voltage and current are always in phase and cross the baseline together. **See Figure 15-7.**

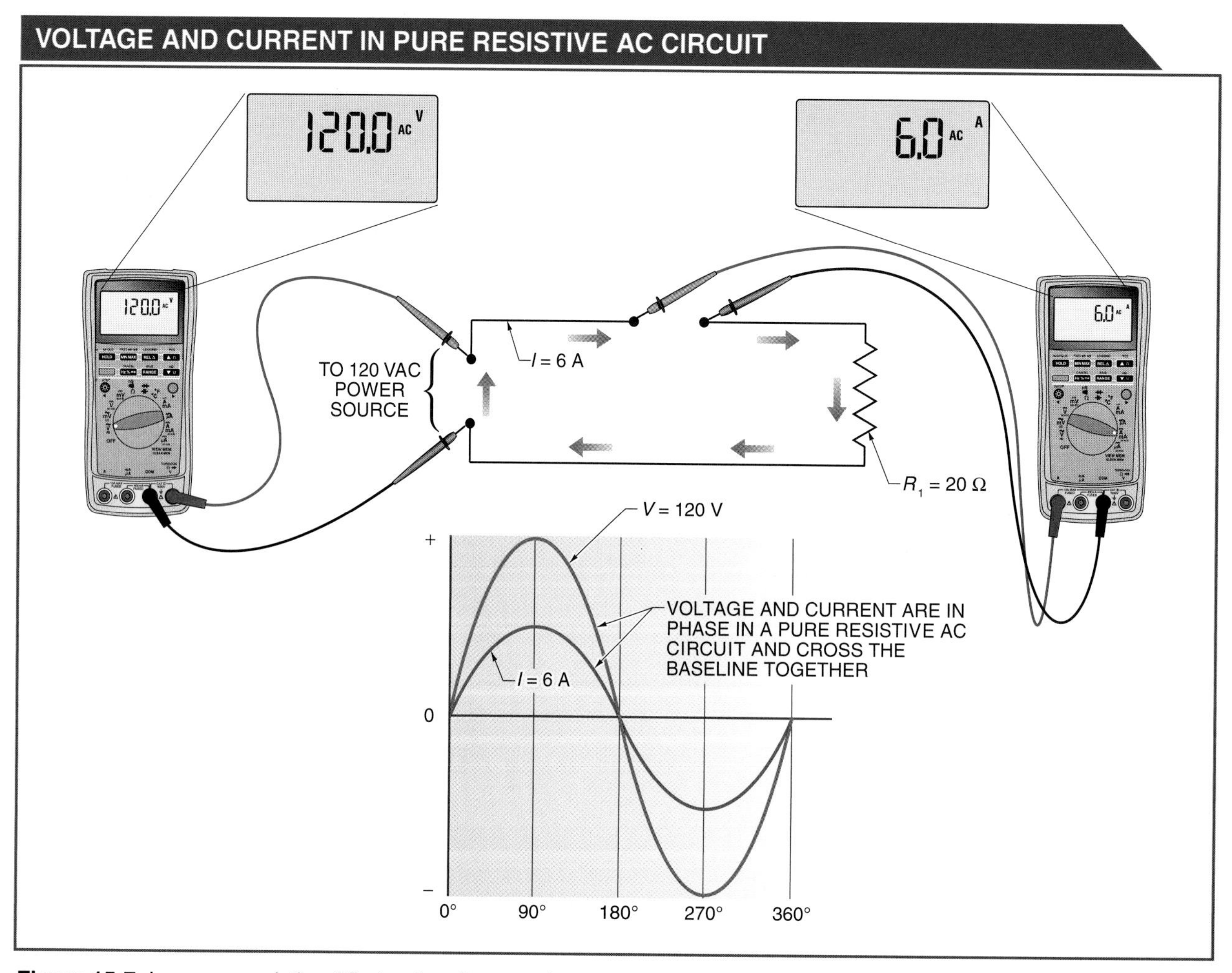

Figure 15-7. In a pure resistive AC circuit, voltage and current are always in phase and cross the baseline together.

A *power curve* is a graph that shows the changes in circuit power as AC parameters change over time. A power curve can be constructed by multiplying the instantaneous voltage value by the instantaneous current value. In an AC resistive circuit, instantaneous current and voltage are in phase while power is always a positive value. **See Figure 15-8.** Power in an AC resistive circuit can be calculated by applying the following formula:

$P = V \times I$

where

P = power (in W)

V = instantaneous voltage (in V)

I = instantaneous current (in A)

Terms

A **power curve** is a graph that shows the changes in circuit power as AC parameters change over time.

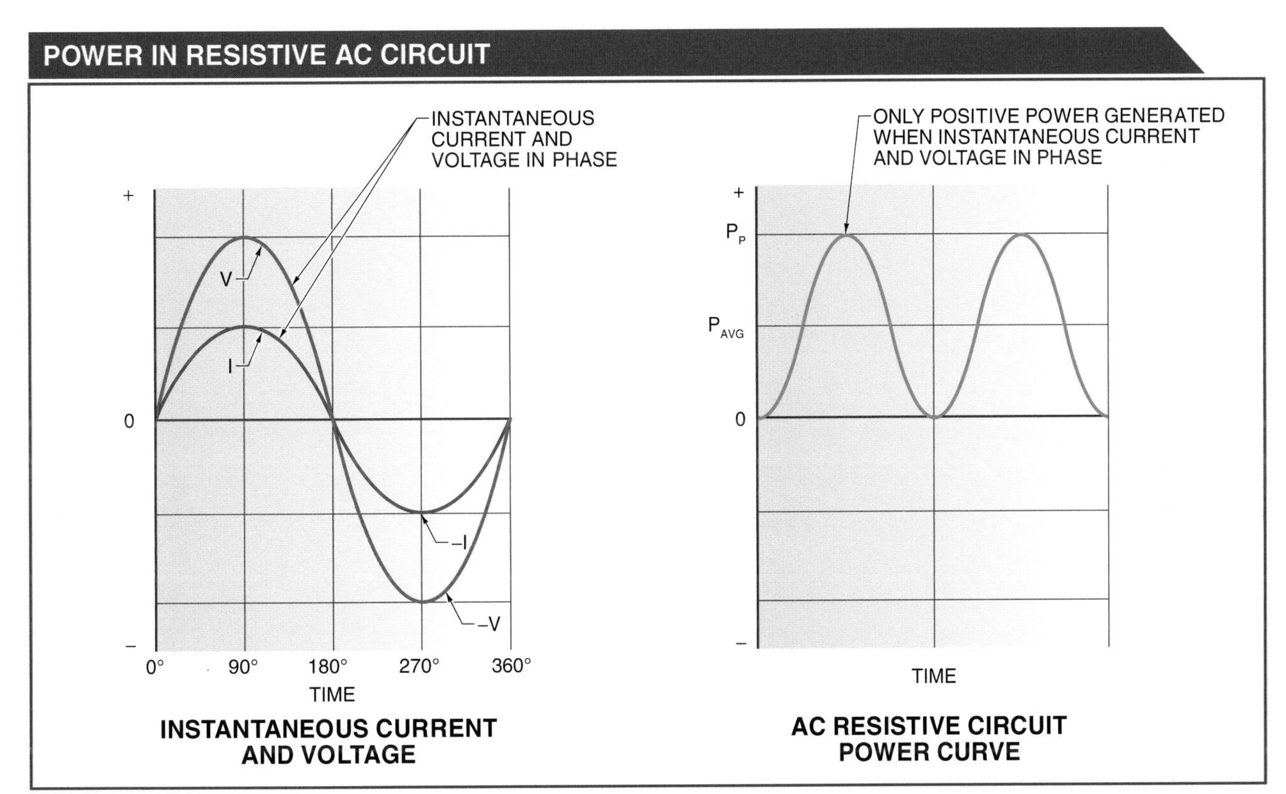

Figure 15-8. In a resistive AC circuit, instantaneous current and voltage are in phase while power is always a positive value.

Example: What is the power in an AC resistive circuit when voltage is –60 V and current is –3 A?

$P = V \times I$

$P = -60\text{ V} \times -3\text{ A}$

$P = \mathbf{180\text{ W}}$

When voltage and current are at their peak values, power in an AC resistive circuit can be calculated by applying the following procedure:

1. Calculate peak voltage. Peak voltage can be calculated by applying the following formula:

   $$V_P = \frac{V_{EFF}}{0.707}$$

   where
   V_P = peak voltage (in V)
   V_{EFF} = effective voltage (in V)
2. Calculate peak current. Peak current can be calculated by applying the following formula:

   $$I_P = \frac{I_{EFF}}{0.707}$$

   where
   I_P = peak current (in A)
   I_{EFF} = effective current (in A)
3. Calculate peak power. Peak power can be calculated by applying the following formula:
   $P_P = V_P \times I_P$
   where
   P_P = peak power (in W)
   V_P = peak voltage (in V)
   I_P = peak current (in A)

Example: What is the peak power in an AC resistive circuit when effective voltage is 120 V and effective current is 6 A?

1. Calculate peak voltage.

   $$V_P = \frac{V_{EFF}}{0.707}$$

   $$V_P = \frac{120\text{ V}}{0.707}$$

   $V_P = 170\text{ V}$
2. Calculate peak current.

   $$I_P = \frac{I_{EFF}}{0.707}$$

   $$I_P = \frac{6\text{ A}}{0.707}$$

   $I_P = 8.5\text{ A}$
3. Calculate peak power.
   $P_P = V_P \times I_P$
   $P_P = 170\text{ V} \times 8.5\text{ A}$
   $P_P = \mathbf{1445\ W}$

Power is positive and reaches its maximum value at 90° and 270°. At 0°, 180°, and 360°, both the voltage and current are zero. Likewise, power falls to 0 W at these points. Power in an AC resistive circuit varies between 0 W and its maximum value at twice the frequency of the source voltage.

Average power in an AC resistive circuit is indicated by a dotted line through the power curve. Average power can be calculated by applying the following formula:

$$P_{AVG} = \frac{P_P}{2}$$

where

P_{AVG} = average power (in W)

P_P = peak power (in W)

Example: What is the average power in an AC resistive circuit when peak power is 1445 W?

$$P_{AVG} = \frac{P_P}{2}$$

$$P_{AVG} = \frac{1445\ \text{W}}{2}$$

$$P_{AVG} = \mathbf{722.5\ W}$$

If voltage and current are separated, average power can be calculated by applying the following formula:

$$P_{AVG} = \frac{V_P}{\sqrt{2}} \times \frac{I_P}{\sqrt{2}}$$

where

P_{AVG} = average power (in W)

V_P = peak voltage (in V)

I_P = peak current (in A)

$\sqrt{2} = 1.414$

Example: What is the average power in a circuit with a peak voltage of 170 V and peak current of 8.5 A?

$$P_{AVG} = \frac{V_P}{\sqrt{2}} \times \frac{I_P}{\sqrt{2}}$$

$$P_{AVG} = \frac{V_P}{1.414} \times \frac{I_P}{1.414}$$

$$P_{AVG} = \frac{170\ \text{V}}{1.414} \times \frac{8.5\ \text{A}}{1.414}$$

$$P_{AVG} = 120.201 \times 6.010$$

$$P_{AVG} = \mathbf{722.5\ W}$$

Understanding the relationship between power, voltage, and current is required when analyzing resistance in AC circuits. At times only voltage and wattage of each load are known. To calculate the power dissipation in an AC circuit, voltage and current are used. Power dissipated in an AC resistive circuit can be calculated by applying the following formula:

$$P_D = V_{EFF} \times I_{EFF}$$

where

P_D = power dissipation (in W)

V_{EFF} = effective voltage (in V)

I_{EFF} = effective current (in A)

Example: What is the power dissipation in an AC circuit that has effective voltage of 120 V and effective current of 0.5 A?

$$P_D = V_{EFF} \times I_{EFF}$$

$$P_D = 120 \text{ V} \times 0.5 \text{ A}$$

$$P_D = \mathbf{60\ W}$$

This formula is essentially the same as the power formula for a DC circuit. Thus, the effective values of current and voltage in an AC resistive circuit dissipate the same amount of power (heat) as those of a DC resistive circuit having equal values of voltage and current.

SERIES RESISTIVE AC CIRCUITS

Series resistive AC circuits are identical to series resistive DC circuits except for the difference in power source. In an AC circuit, current flow reverses direction periodically over time. A series circuit is a circuit in which there is only one current path. Therefore, total current is equal to the current flow through any component or device. The value of the current flow into a point in a series circuit is equal to the current flow away from that point. Current is used as the reference for all circuit parameters. In a series resistive AC circuit, individual resistances and voltages add together, while current is the same at any point in the circuit. **See Figure 15-9.**

As with a DC circuit, resistances connected in series add together in an AC circuit. Resistance in a series resistive AC circuit can be calculated by applying the following formula:

$$R_T = R_1 + R_2 + R_3 + \ldots + R_N$$

where

R_T = total resistance (in Ω)

R_1 = resistance 1 (in Ω)

R_2 = resistance 2 (in Ω)

R_3 = resistance 3 (in Ω)

R_N = resistance N (in Ω)

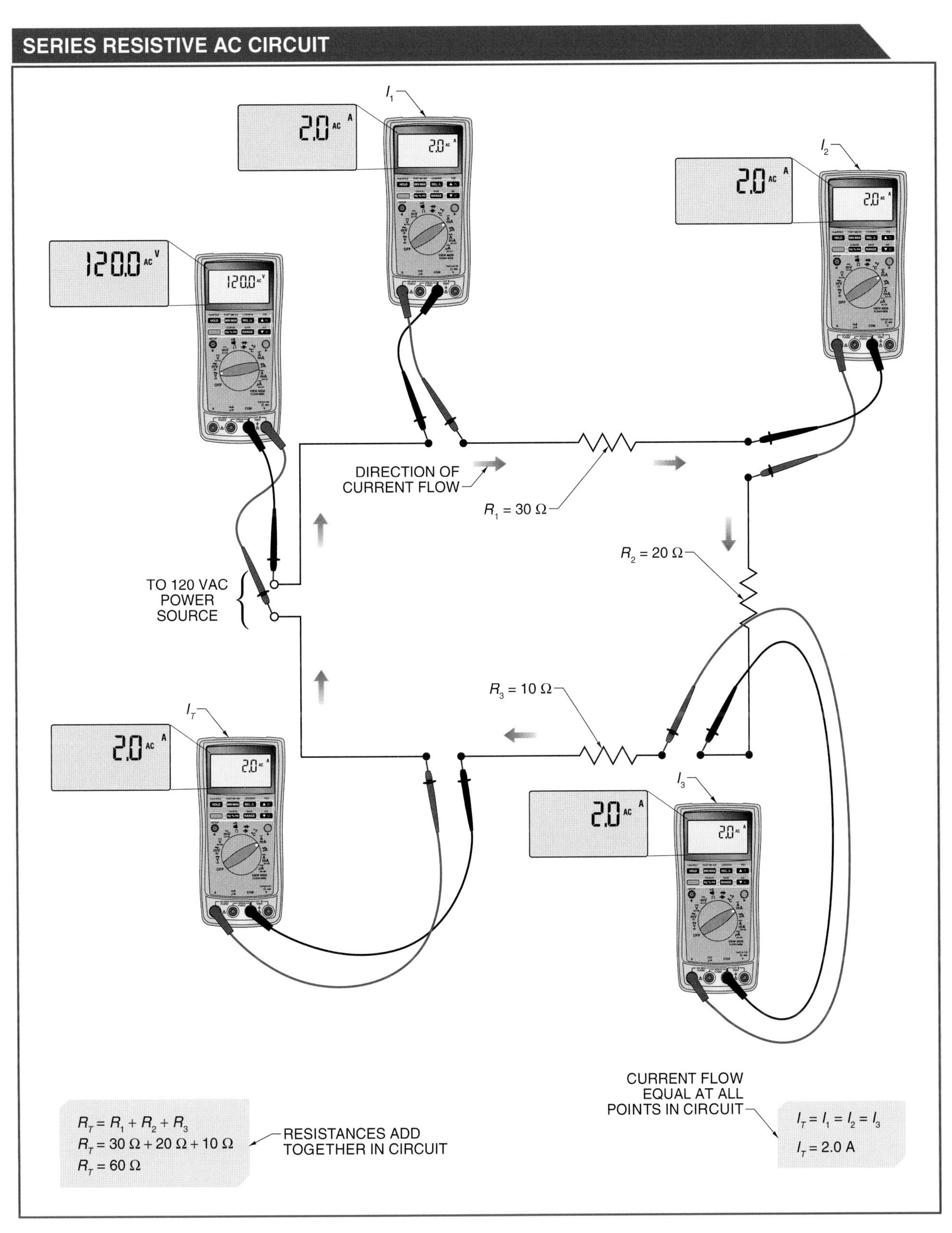

Figure 15-9. In a series resistive AC circuit, individual resistances and voltages add together, while current is equal at any point in the circuit.

Example: What is the total resistance in a series resistive AC circuit containing 30 Ω, 20 Ω, and 10 Ω resistances connected in series?

$R_T = R_1 + R_2 + R_3 + \ldots + R_N$

$R_T = 30\ \Omega + 20\ \Omega + 10\ \Omega$

$R_T = \mathbf{60\ \Omega}$

Ohm's law can be used to calculatc total current. Total current can be calculated by applying the following formula:

$$I_T = \frac{V_T}{R_T}$$

where

I_T = total current (in A)

V_T = total voltage (in V)

R_T = total resistance (in Ω)

Example: What is the total current in a 120 VAC circuit that has a total resistance of 60 Ω?

$$I_T = \frac{V_T}{R_T}$$

$$I_T = \frac{120\ \text{VAC}}{60\ \Omega}$$

$I_T = \mathbf{2\ A}$

Total current in a series resistive AC circuit is equal to the current through any of its components or devices. Total current in a series circuit can be calculated by applying the following formula:

$I_T = I_1 = I_2 = \ldots = I_N$

where

I_T = total current (in A)

I_1 = current 1 (in A)

I_2 = current 2 (in A)

I_N = current N (in A)

Example: What is the total current in a series resistive AC circuit that has 2 A of current flow through component 2?

$I_T = I_1 = I_2 = \ldots = I_N$

$I_T = I_2$

$I_T = \mathbf{2\ A}$

Voltage drop across each resistance in a series resistive AC circuit can be calculated by applying the following formula (Ohm's law):

$V_{DN} = I_N \times R_N$

where

V_{DN} = voltage drop N (in V)

I_N = branch current N (in A)
R_N = resistance N (in Ω)

Example: What is the voltage drop across a 20 Ω resistance in a series resistive AC branch circuit that has current of 2 A?

$V_{DN} = I_N \times R_N$
$V_{D1} = I_1 \times R_1$
$V_{D1} = 2\text{ A} \times 20\ \Omega$
$V_{D1} = \mathbf{40\ V}$

In a series resistive AC circuit, the sum of the voltage drops is equal to the source voltage. The sum of the voltage drops in a series resistive AC circuit can be calculated by applying the following formula:

$V_{DT} = V_{D1} + V_{D2} + \ldots + V_{DN}$

where

V_{DT} = total circuit voltage drop (in V)
V_{D1} = voltage drop 1 (in V)
V_{D2} = voltage drop 2 (in V)
V_{DN} = voltage drop N (in V)

Example: What is the total source voltage of a series resistive AC circuit containing 60 V, 40 V, and 20 V drops across three loads?

$V_{DT} = V_{D1} + V_{D2} + \ldots + V_{DN}$
$V_{DT} = V_{D1} + V_{D2} + V_{D3}$
$V_{DT} = 60\text{ V} + 40\text{ V} + 20\text{ V}$
$V_{DT} = \mathbf{120\ V}$

As with a DC series circuit, total power dissipated by a series resistive AC circuit can be calculated by applying the following formula:

$P_D = I_T \times V_T$

where

P_D = power dissipation (in W)
I_T = total current (in A)
V_T = total voltage (in V)

Example: What is the total power dissipated in a 120 V circuit that has 2 A of total current flow through it?

$P_D = I_T \times V_T$
$P_D = 2\text{ A} \times 120\text{ V}$
$P_D = \mathbf{240\ W}$

Analysis of a series resistive AC circuit is identical to the analysis of a series resistive DC circuit. Resistance merely opposes current flow. No shift in the phase occurs between current and voltage in a series resistive AC circuit. When the type of AC source voltage is not specified, it is assumed that the voltage is the effective value.

PARALLEL RESISTIVE AC CIRCUITS

Parallel resistive AC circuits are identical to parallel DC resistive circuits except for the difference in power source. Current flow in an AC circuit reverses direction periodically over time. A parallel resistive AC circuit has more than one current path. In a parallel resistive AC circuit, the voltage is the same across all components and devices. Consequently, voltage is used as a reference when analyzing parallel resistive AC circuits.

When analyzing a parallel resistive AC circuit, individual currents add together. Total resistance in a parallel resistive circuit is equal to total voltage divided by total current. **See Figure 15-10.**

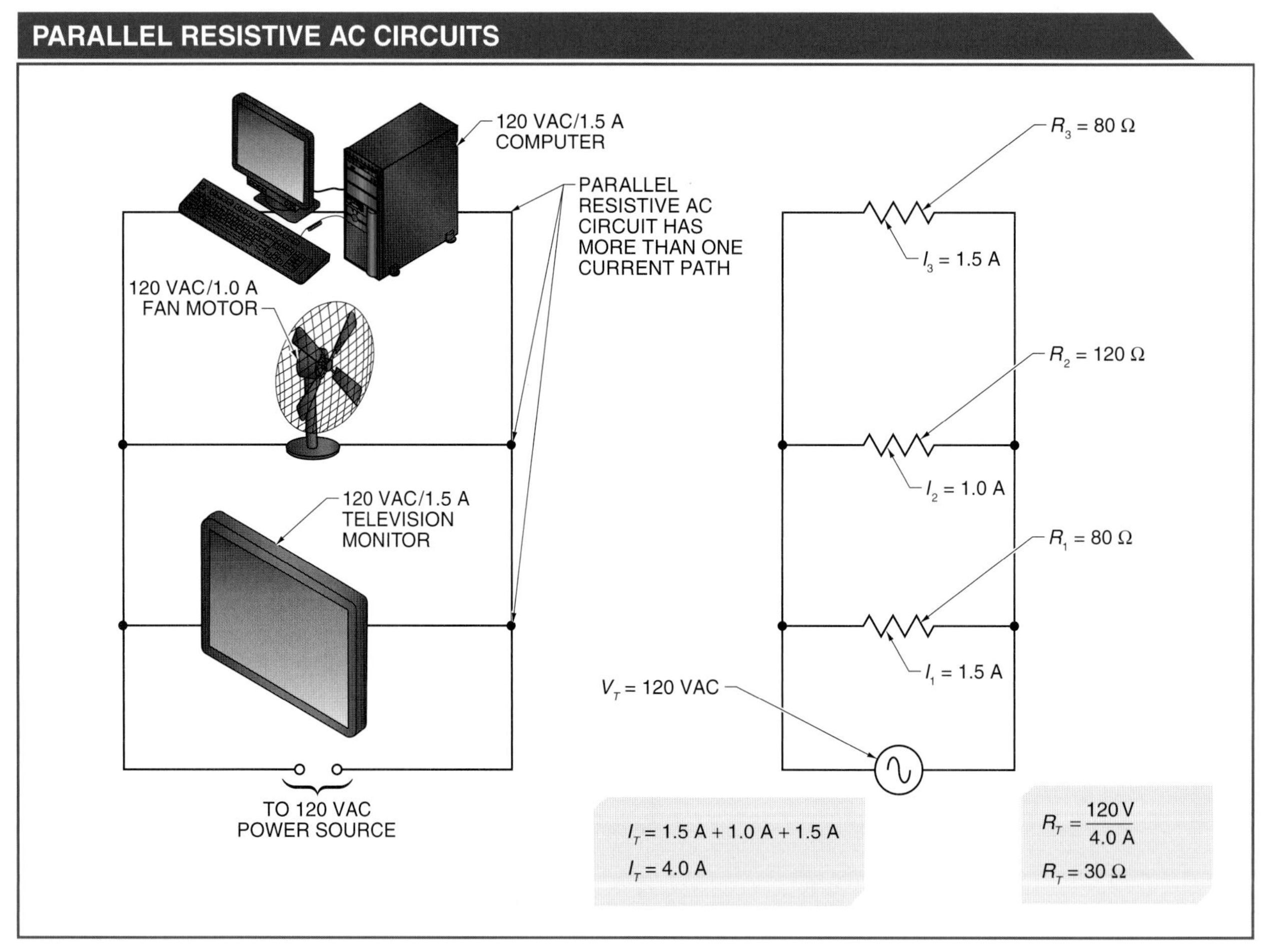

Figure 15-10. In a parallel resistive AC circuit, individual currents add together and voltage is the same across each load in the circuit. Total resistance is equal to total voltage divided by total current.

When resistances are added together in a series resistive circuit, the total circuit resistance increases. The opposite circuit condition occurs in a parallel resistive circuit. When resistances are added together in a parallel resistive circuit, the total circuit resistance decreases. When the load voltage is known or measured, total voltage in a parallel resistive AC circuit can be calculated by applying the following formula:

$$V_T = V_1 = V_2 = \ldots = V_N$$

where

V_T = total source voltage (in V)

V_1 = voltage 1 (in V)

V_2 = voltage 2 (in V)

V_N = voltage N (in V)

Example: What is the total voltage in a parallel resistive AC circuit that has 1000 V across load 3?

$$V_T = V_1 = V_2 = \ldots = V_N$$

$$V_T = V_1 = V_2 = V_3$$

$$V_T = V_3$$

$$V_T = \mathbf{1000\ V}$$

When the source voltage and each resistance in a circuit are known, each resistive current can be calculated by applying the following formula:

$$I_N = \frac{V_T}{R_N}$$

where

I_N = resistive current N (in A)

V_T = source voltage (in V)

R_N = resistance N (in Ω)

Example: What is the current through each branch of a 120 VAC circuit that has resistances of 80 Ω, 120 Ω, and 80 Ω connected in parallel?

$$I_N = \frac{V_T}{R_N}$$

$$I_1 = \frac{120\ V}{80\ \Omega}$$

$$I_1 = \mathbf{1.5\ A}$$

$$I_2 = \frac{120\ V}{120\ \Omega}$$

$$I_2 = \mathbf{1.0\ A}$$

$$I_3 = \frac{120\ V}{80\ \Omega}$$

$$I_3 = \mathbf{1.5\ A}$$

Terms

A **series/parallel circuit** is a circuit that contains a combination of both series-connected and parallel-connected components.

Total current in a parallel resistive AC circuit can be calculated by applying the following formula:

$$I_T = I_1 + I_2 + \ldots + I_N$$

where

I_T = total circuit current (in A)

I_1 – current 1 (in A)

I_2 = current 2 (in A)

I_N = current N (in A)

Example: What is the total current in a parallel resistive AC circuit containing three loads connected in parallel if the current through each of the loads is 1.5 A, 1.0 A, and 1.5 A?

$$I_T = I_1 + I_2 + \ldots + I_N$$

$$I_T = I_1 + I_2 + I_3$$

$$I_T = 1.5\text{ A} + 1\text{ A} + 1.5\text{ A}$$

$$I_T = \mathbf{4\text{ A}}$$

Total resistance of a parallel resistive AC circuit can be calculated by applying the following formula:

$$R_T = \frac{V_T}{I_T}$$

where

R_T = total resistance (in Ω)

V_T = total voltage (in V)

I_T = total current (in A)

Example: What is the total resistance of a 120 VAC parallel resistive circuit with a total current flow of 4 A?

$$R_T = \frac{V_T}{I_T}$$

$$R_T = \frac{120\text{ V}}{4\text{ A}}$$

$$R_T = \mathbf{30\ \Omega}$$

Total power dissipated by a parallel resistive AC circuit can be calculated by applying the following formula:

$$P_D = I_T \times V_T$$

where

P_D = power dissipation (in W)

I_T = total current (in A)

V_T = total voltage (in V)

Example: What is the total power dissipated in a parallel resistive 120 VAC circuit with total current of 4 A?

$P_D = I_T \times V_T$

$P_D = 4\ \text{A} \times 120\ \text{V}$

$P_D = \mathbf{480\ W}$

Note: In a resistive circuit, all currents are in phase with each other and with the source voltage.

Megohmmeters

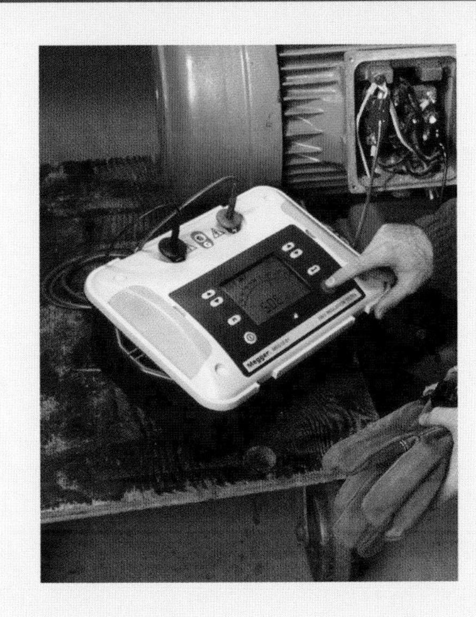

A megohmmeter is a meter that is used to measure the integrity of the insulation of individual conductors, motor windings, or transformer windings. A megohmmeter produces a high DC voltage into the conductor or motor winding being tested and measures the leakage current through the insulation to determine the resistance of the insulation. A megohmmeter measurement is displayed in megohms, which are calculated using Ohm's law.

Megohmmeters are manufactured in a variety of styles. Most megohmmeters have a function switch or selector switch for choosing the appropriate test voltage. The megohmmeter display can be an analog display or a digital display. The power source of a megohmmeter can be generated by a hand crank, battery power, or 120 VAC. Some models have dual power sources, such as a hand crank and 120 VAC.

SERIES/PARALLEL RESISTIVE AC CIRCUITS

A series/parallel circuit is a circuit that contains a combination of both series-connected and parallel-connected components. When analyzing these circuits where values of the resistances are known, the first step is to reduce all parallel resistances to a single value. The result is a circuit containing only series resistances. Adding these values results in the total resistance of the circuit. Dividing total resistance into the source voltage yields the total circuit current. The steps used to perform an analysis of a resistive AC circuit are similar to the steps used to perform an analysis of a resistive DC circuit. **See Figure 15-11.** An analysis of a series/parallel AC resistive circuit can be performed by applying the following procedure:

1. Combine resistances connected in series into a single resistance value. A single resistance value can be calculated by applying the following formula:

   $R_{ST} = R_1 + \ldots + R_N$

   where

   R_{ST} = total series resistance (in Ω)

   R_1 = resistance 1 (in Ω)

   R_N = resistance N (in Ω)

2. Combine two resistances connected in parallel into a single resistance value. Parallel resistance can be calculated by applying the following formula:

   $$R_{PT} = \frac{R_2 \times R_3}{R_2 + R_3}$$

   where

   R_{PT} = total parallel resistance (in Ω)

   R_2 = resistance 2 (in Ω)

   R_3 = resistance 3 (in Ω)

3. Combine series and parallel resistances into a single (total) resistance value. Total resistance can be calculated by applying the following formula:

   $R_T = R_{ST} + R_{PT}$

   where

   R_T = total resistance (in Ω)

   R_{ST} = total series resistance (in Ω)

   R_{PT} = total parallel resistance (in Ω)

4. Calculate total current. Total current can be calculated by applying the following formula:

   $$I_T = \frac{V_T}{R_T}$$

   where

   I_T = total current (in A)

   V_T = total voltage (in V)

   R_T = total resistance (in Ω)

5. Calculate voltage drop across the circuit. Voltage drop across the circuit can be calculated by applying the following formula:

   $V_{DN} = I_T \times R_N$

   where

   V_{DN} = voltage drop across resistance N (in V)

   I_T = total current (in A)

   R_N = resistance N (in Ω)

6. Verify accuracy of calculations by application of Kirchhoff's voltage and current laws. Kirchhoff's voltage and current laws state that algebraic sum of the voltages and currents around the closed loop circuit must equal zero. Kirchhoff's laws are calculated by applying the following formulas:

   $V_N = 0$

   and

   $I_N = 0$

   where

   V_N = algebraic sum of all voltages (in V)

   and

   I_N = algebraic sum of all currents (in A)

7. Calculate total power consumed in the circuit. Total power consumed in a circuit can be calculated by applying the following formula:

   $P_T = I_T \times V_T$

   where

   P_T = total circuit power (in W)

   I_T = total circuit current (in A)

   V_T = total circuit voltage (in V)

8. Calculate the power dissipated in each load. Power dissipated in each load can be calculated by applying the following formula:

   $P_{DN} = I_N \times V_N$

   where

   P_{DN} = power dissipation in circuit N (in W)

   I_N = current in circuit N (in A)

   V_N = voltage in circuit N (in V)

Outdoor stadium lighting is an example of applications where parallel-connected AC components are used.

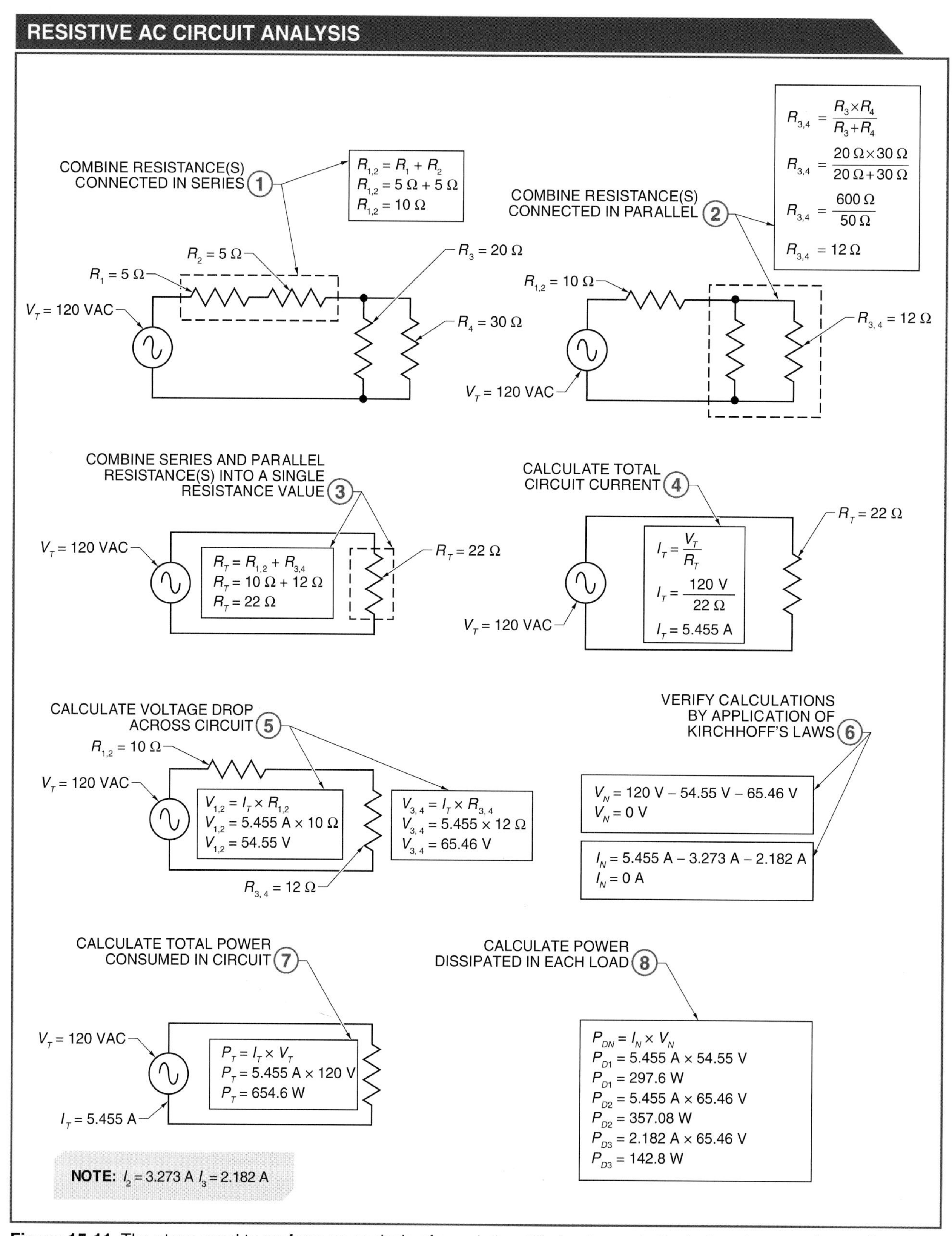

Figure 15-11. The steps used to perform an analysis of a resistive AC circuit are similar to the steps used to perform an analysis of a resistive DC circuit.

Example: What is the power dissipated in each load in a 120 VAC series/parallel circuit that has a series resistance of 10 Ω and parallel resistances of 20 Ω and 30 Ω?

1. Combine resistances connected in series into a single resistance value.

   $R_{ST} = R_1 + R_2$

   $R_{ST} = \mathbf{10\ \Omega}$

2. Combine resistances connected in parallel into a single resistance value.

   $$R_{PT} = \frac{R_3 \times R_4}{R_3 + R_4}$$

   $$R_{PT} = \frac{20\ \Omega \times 30\ \Omega}{20\ \Omega + 30\ \Omega}$$

   $$R_{PT} = \frac{600\ \Omega}{50\ \Omega}$$

   $R_{PT} = \mathbf{12\ \Omega}$

3. Combine series and parallel resistances into a single (total) resistance value.

   $R_T = R_{ST} + R_{PT}$

   $R_T = 10\ \Omega + 12\ \Omega$

   $R_T = \mathbf{22\ \Omega}$

4. Calculate total current.

   $$I_T = \frac{V_T}{R_T}$$

   $$I_T = \frac{120\ \text{V}}{22\ \Omega}$$

   $I_T = \mathbf{5.455\ A}$

5. Calculate voltage drop across the circuit.

   $V_{D1} = I_T \times R_1$

   $V_{D1} = 5.455\ \text{A} \times 10\ \Omega$

   $V_{D1} = \mathbf{54.55\ V}$

   $V_{D2,3} = I_T \times R_{2,3}$

   $V_{D2,3} = 5.455\ \text{A} \times 12\ \Omega$

   $V_{D2,3} = \mathbf{65.46\ V}$

Total current flows through R_1 so that I_T and I_1 are equal. The total current divides between the parallel resistances R_2 and R_3, with the larger current passing through R_2, which is the smaller of the two resistances. Since the voltage and resistance are known, the simplest way of calculating these two currents is to apply Ohm's law.

$$I_2 = \frac{V_2}{R_2}$$

$$I_2 = \frac{65.46 \text{ V}}{20\ \Omega}$$

$I_2 = \mathbf{3.273\ A}$

$$I_3 = \frac{V_3}{R_3}$$

$$I_3 = \frac{65.46 \text{ V}}{30\ \Omega}$$

$I_3 = \mathbf{2.182\ A}$

$I_T = I_2 + I_3$
$I_T = 3.273 \text{ A} + 2.182 \text{ A}$
$I_T = \mathbf{5.455\ A}$

6. Verify accuracy of calculations by application of Kirchhoff's voltage and current laws. Kirchhoff's laws can be used to check the accuracy of the calculations. The algebraic sum of the voltages around the closed loops must equal 0.
$V_N = V_T - V_1 - V_2$
$V_N = 120 \text{ V} - 54.55 \text{ V} - 65.46 \text{ V}$
$V_N = \mathbf{0\ V}$

The algebraic sum of the currents around the closed loops must equal 0.
$I_N = I_1 - I_2 - I_3$
$I_N = 5.455 \text{ A} - 3.273 \text{ A} - 2.182 \text{ A}$
$I_N = \mathbf{0\ A}$

7. Calculate total power consumed in the circuit.
$P_T = I_T \times V_T$
$P_T = 5.455 \text{ A} \times 120 \text{ V}$
$P_T = \mathbf{654.6\ W}$

8. Calculate the power dissipated in each load.
$P_{DN} = I_N \times V_N$

$P_{D1} = I_1 \times V_1$
$P_{D1} = 5.455 \text{ A} \times 54.55 \text{ V}$
$P_{D1} = \mathbf{297.6\ W}$

$P_{D2} = I_2 \times V_2$
$P_{D2} = 3.273 \text{ A} \times 65.46 \text{ V}$
$P_{D2} = \mathbf{214.3\ W}$

$P_{D3} = I_3 \times V_3$
$P_{D3} = 2.182 \text{ A} \times 65.46 \text{ V}$
$P_{D3} = \mathbf{142.8\ W}$

This example demonstrates the basic technique for solving series/parallel AC resistive circuits. In more complex circuits that have additional series and/or parallel loads, the same approach is applied by adding series elements and combining parallel elements.

SUMMARY

A resistive AC circuit can be treated identically to a resistive DC circuit provided the effective value of AC is used when performing calculations. In a resistive AC circuit, however, values other than effective value are sometimes required when performing calculations. These values include instantaneous value, average value, peak value, and peak-to-peak value. Resistive AC circuit values can be converted from one value to another by performing mathematical calculations. When a current or voltage is given without the value being specified, it is assumed to be the effective value. Resistive AC circuits can be connected in series, parallel, or series/parallel combinations. In a resistive AC circuit, voltage and current are always in-phase with each other.

APPLICATION: AC RESISTIVE POWER...

Background: In a resistive DC circuit, the power dissipated in a resistor can be calculated using several different methods. Power is typically calculated using the following formula:

$P = I^2 \times R$
where
P = power (in W)
I = current (in A)
R = resistance (in Ω)

By using Ohm's law, several other power formulas can be derived. Additional formulas used to calculate the power dissipated include the following:

$P = V \times I$
and
$$P = \frac{V^2}{R}$$

In a resistive AC circuit, power is also dissipated in the form of heat. The AC power dissipation formulas are the same as in a DC circuit. To find the equivalent power dissipation between an AC and DC circuit, the AC root-mean-square (rms) voltage or current value is used. The relationship between rms and peak values is as follows:

...APPLICATION: AC RESISTIVE POWER...

$V_{rms} = V_{peak} \times 0.707$
where
V_{rms} = volts rms (in V)
V_{peak} = volts peak (in V)
$0.707 = \text{constant}\left(\frac{1}{\sqrt{2}}\right)$

Key Points: AC power dissipation

Problem: A DC circuit with a voltage source of 100 V and a resistive load of 20 Ω is used as a heater. If the resistor value is changed to 40 Ω, what is the peak AC voltage needed to generate the same amount of heat as in the DC circuit?

Solution: First, DC power is calculated as follows:

$$P = \frac{V^2}{R}$$

$$P = \frac{(100\text{ VDC})^2}{20\ \Omega}$$

$$P = 500\text{ W}$$

Next, the rms AC voltage for the equivalent heat dissipation is found using the same power dissipation formula and solving for voltage as follows:

$$P = \frac{V^2}{R}$$

$$500\text{ W} = \frac{V^2}{40\ \Omega}$$

$$V^2 = 500\text{ W} \times 40\ \Omega$$

$$V = \sqrt{500\text{ W} \times 40\ \Omega}$$

$$V = \sqrt{20{,}000}$$

$$V = 141.4$$

The found AC voltage is in rms. To find the peak voltage value, the following formula is used:

$$V_{rms} = V_{peak} \times 0.707$$

$$141.4\text{ V} = V_{peak} \times 0.707$$

$$V_{peak} = \frac{141.4\text{ V}}{0.707}$$

$$V_{peak} = 200\text{ V}$$

...APPLICATION: AC RESISTIVE POWER

In resistive AC and DC circuits, heat is dissipated by resistors in the circuits. When comparing AC and DC power dissipation, the AC rms voltage or current is used in the comparison. The AC rms value is also known as the effective value.

Chapter Review...

1. What is the total current in a parallel resistive AC circuit containing three loads connected in parallel if the current through each of the loads is 0.5 A, 0.7 A, and 1.1 A?
2. What is the total resistance of a 12 VAC parallel resistive circuit with a total current flow of 0.8 A?
3. What is the total power dissipated in a parallel resistive 208 VAC circuit with total current of 2.2 A?
4. What is the peak power in an AC resistive circuit when effective voltage is 208 V and effective current is 10 A?
5. What is the average power in an AC resistive circuit when peak power is 4145 W?
6. What is the average power in a circuit with a peak voltage of 145 V and peak current of 6 A?
7. What is a peak-to-peak value?
8. What is a filter?
9. What is the power dissipation in an AC circuit that has effective voltage of 208 V and effective current of 1.5 A?
10. What are instantaneous AC parameters in a resistive circuit?
11. What is the current through each branch of a 135 VAC circuit that has resistances of 10 Ω, 12 Ω, and 8 Ω connected in parallel?
12. What is the peak-to-peak voltage value of a resistive AC circuit if the positive peak voltage value is equal to 145 V?
13. What are the three voltage and current value designations in a resistive circuit?
14. What is the total resistance in a series resistive AC circuit containing 5 Ω, 5 Ω, and 8 Ω resistances connected in series?
15. In the generation of a sine wave, why is 0 V induced into a rotating conductor at 0° and at 180°?

...Chapter Review

16. What is the total current in a 135 VAC circuit that has a total resistance of 40 Ω?

17. What is the power dissipated in a 6 A circuit with a resistance of 3.3 Ω?

18. What is the total current in a series resistive AC circuit that has 1.2 A of current flow through component 2?

19. What is a bleeder resistor used for?

20. What is the voltage drop across a 10 Ω resistance in a series resistive AC branch circuit that has current of 3 A?

21. What is the most common voltage and frequency used in residential installations in the United States?

22. What is the average voltage value if the peak voltage value is equal to 145 V?

23. What is the peak voltage value of a resistive AC circuit if the average value of the same circuit in Question 22 is equal to 92.4 V?

24. What is the effective voltage if peak voltage is equal to 135 V?

25. What is the instantaneous value of voltage in a resistive AC circuit with an angle theta (θ) of 47° and a maximum generated voltage of 170 V?

26. What is the total source voltage of a series resistive AC circuit containing 3 V, 6 V, and 10 V drops across three loads?

27. What is the total power dissipated in a 208 V circuit that has 5 A of total current flow through it?

28. How is a power curve constructed?

29. What is the common basis between the units of AC voltage and current and the units of DC voltage and current?

30. What is the total voltage in a parallel resistive AC circuit that has 3.3 V across load 3?

16

Inductive AC Circuits

Square D | Schneider Electric

OBJECTIVES

- Describe inductive AC circuits and the factors that affect inductive reactance.
- Explain current and voltage phase relationships in inductive AC circuits.
- Explain the differences between series and parallel inductive-resistive AC circuits.
- Explain how impedance, reactive power, apparent power, power factor, frequency, and inductance affect series inductive-resistive circuits.
- Describe how current, impedance, frequency, and inductance affect parallel inductive-resistive circuits.

Inductance is the ability of a changing magnetic field to induce voltage in a conductor. Inductance is also the characteristic of an electrical circuit that opposes any change in circuit current. These two characteristics are tied together. The induced voltage of an inductor is always in such a direction that it opposes any change in current flow.

When inductance is present in an AC circuit, the alternating current produces effects not found in a DC circuit. These differences are caused by the characteristics of an AC signal frequency and the characteristics of an inductor. Different applications call for different levels of inductance. Inductors can be connected in series or in parallel and can be connected with resistances to form inductive-resistive circuits. Techniques for analyzing inductive-resistive series and parallel circuits for voltage, current, impedance, and power are used when working with these types of circuits.

INDUCTIVE CIRCUITS

Terms

A **pure inductive circuit** is an AC circuit with an inductor and no resistance.

Inductive circuits have an AC power source with current and voltage across an inductive load. **See Figure 16-1.** A *pure inductive circuit* is an AC circuit with an inductor and no resistance. When no resistance is present, the total source voltage is across the inductive load. The source voltage and the inductive voltage are equal and in phase with each other. Although a pure inductive circuit is strictly theoretical, the concept is used to gain understanding of circuit parameters such as inductive reactance, current, phase relationships between current and voltage, and power in inductive AC circuits.

INDUCTIVE CIRCUITS

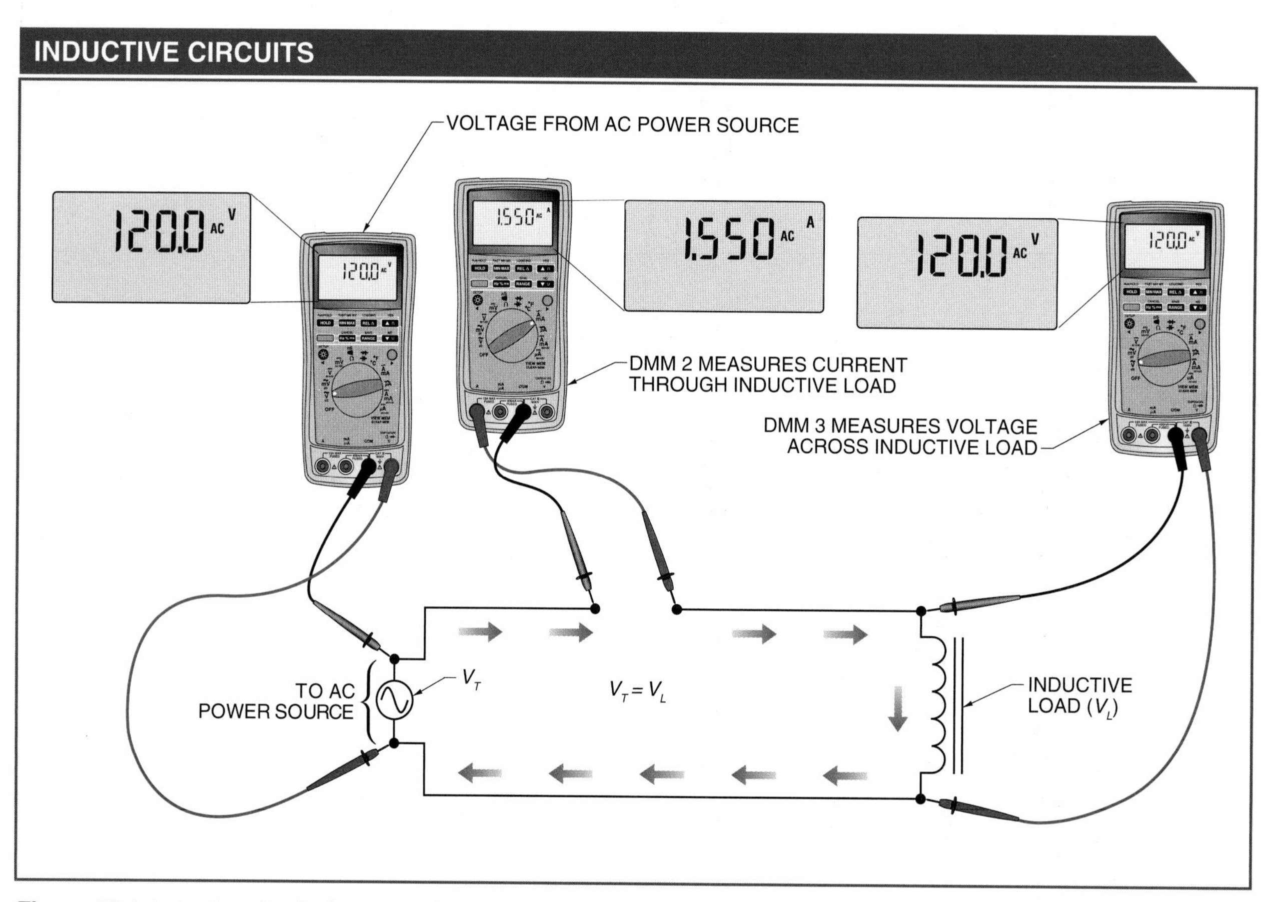

Figure 16-1. Inductive circuits have an AC power source with current and voltage across an inductive load.

Although inductance has little effect in a DC circuit, it has a significant effect in an AC circuit. **See Figure 16-2.** When there is inductance in an AC circuit, the AC produces effects not calculated in DC circuits, such as reduced lamp brilliance and increased resistance, due to the AC frequency. For example, the differences caused by inductance can be shown in a circuit containing a 120 V/100 W rated lamp and an inductor connected in series. The lamp is supplied by two power sources, one at 120 VDC and another at 120 VAC. Placing the double-pole, double-throw (DPDT) switch to position 1 connects the DC power source to the circuit, which gives the lamp normal brilliance. When the switch is changed to position 2, the AC power source is connected to the circuit. In this configuration, the lamp has reduced brilliance.

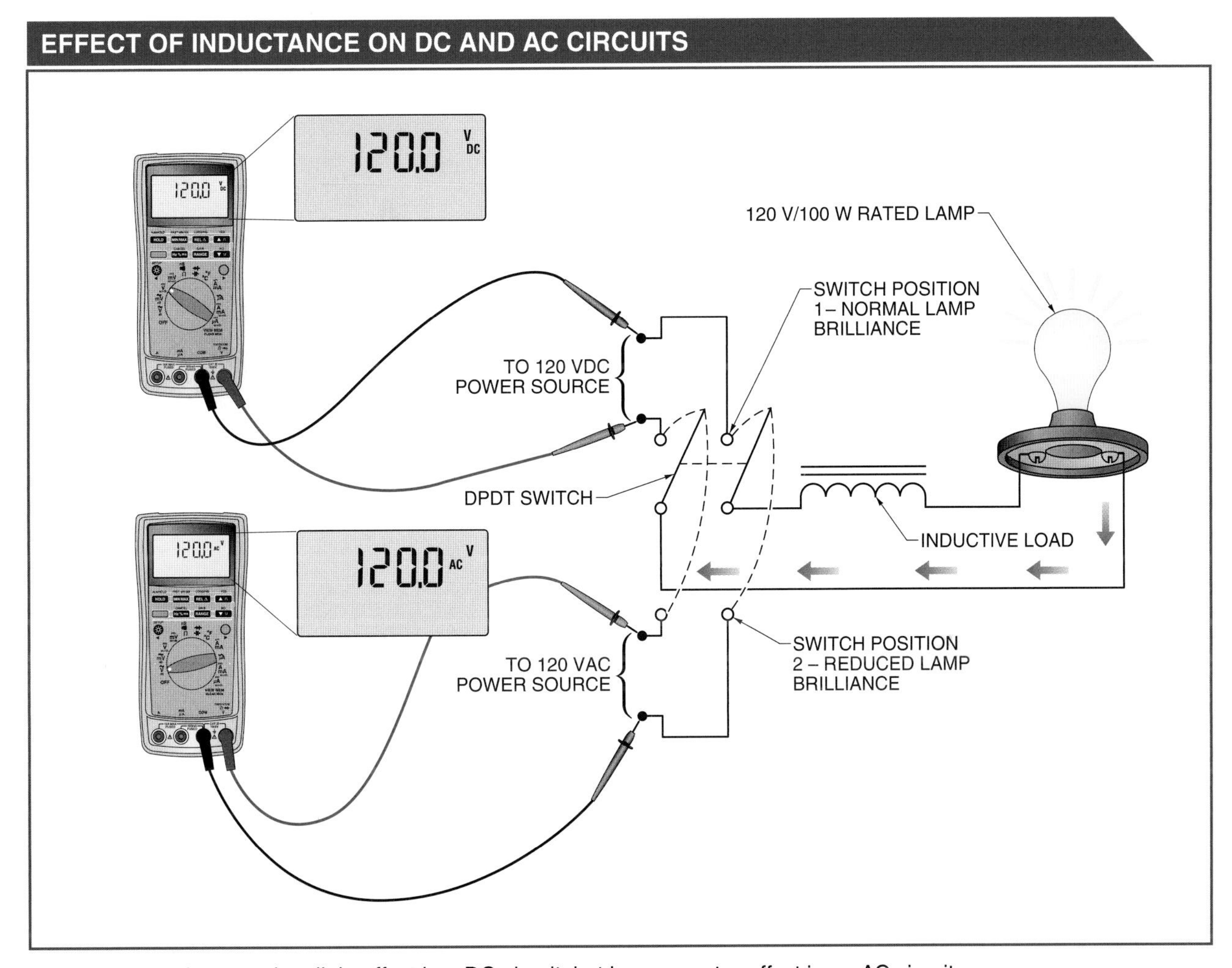

Figure 16-2. Inductance has little effect in a DC circuit, but has a greater effect in an AC circuit.

The failure of the lamp to obtain full brilliance when the switch is changed to the AC power source (position 2) indicates an increase in the opposition to current flow. Interaction between the inductance and the AC power source adds to the resistance of the lamp to cause a lower level of current flow than that which occurs with the DC power source. The opposition of the inductor to current flow is affected by the amount of inductance and the AC frequency in the circuit.

Inductive Reactance

Terms

Inductive reactance is the opposition of an inductor to alternating current.

Susceptance is the ability of a reactive circuit to conduct current and is measured in siemens (S).

Inductive reactance is the opposition of an inductor to alternating current. Opposition to alternating current is caused by the counter electromotive force (CEMF) developed by the inductor. CEMF is created by the change in current, which is caused by a constantly changing magnetic field that cuts the inductor. **See Figure 16-3.** *Susceptance* is the ability of a reactive circuit to conduct current and is measured in siemens (S). Susceptance is the reciprocal of the reactive component of impedance.

The level of the inductive voltage across an inductor due to change in current is equal to and opposite the source voltage and is calculated by applying the following formula:

$$V_L = I_L \times X_L$$

where

V_L = inductive voltage (in V)

I_L = inductive current (in A)

X_L = inductive reactance (in Ω)

INDUCTIVE REACTANCE IN AC CIRCUITS

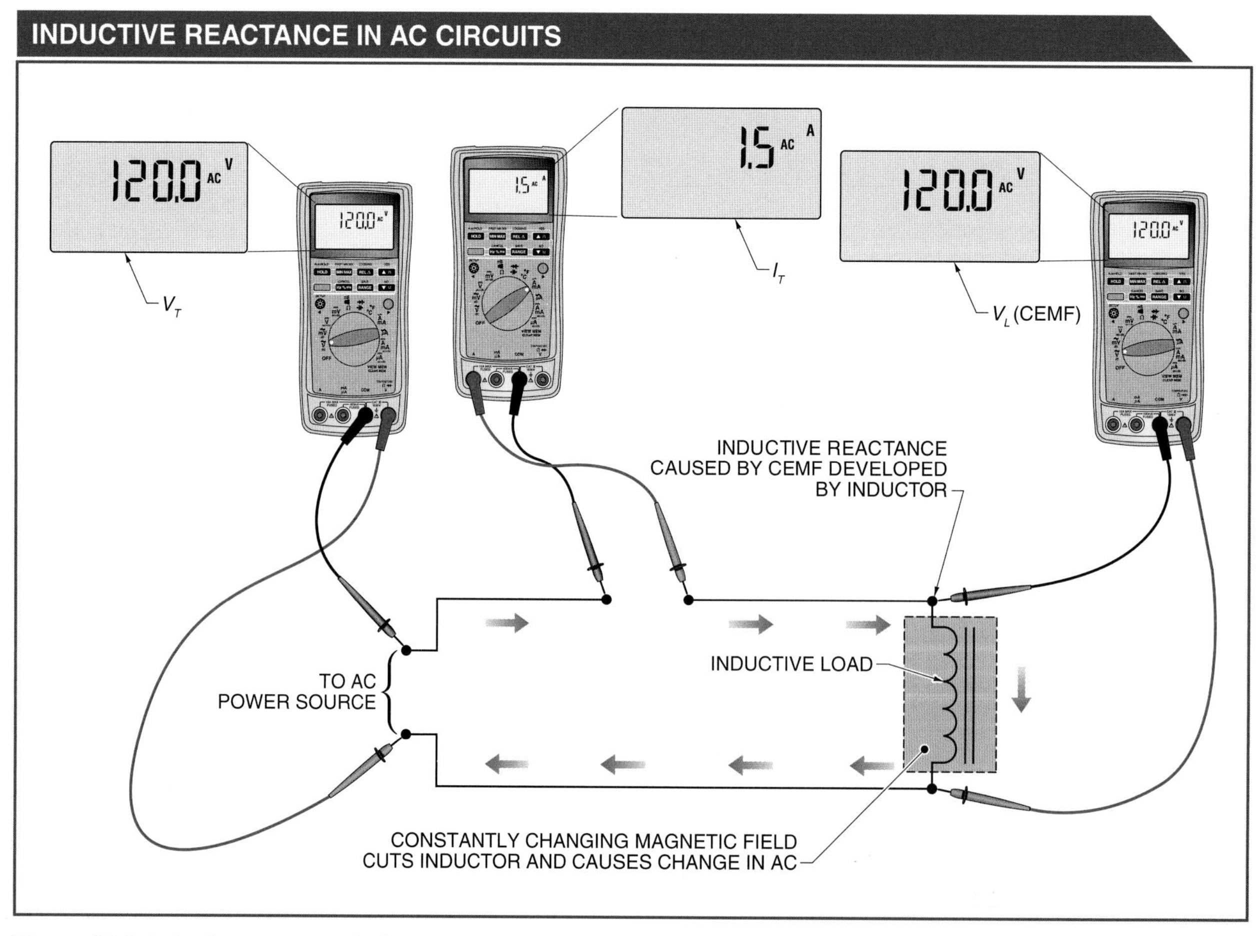

Figure 16-3. Inductive reactance is the opposition to AC flow.

Example: What is the inductive voltage of an inductive AC circuit that has an inductive current of 4 A and inductive reactance of 30 Ω?

$V_L = I_L \times X_L$

$V_L = 4\text{ A} \times 30\ \Omega$

$V_L = \mathbf{120\ V}$

A closed-loop circuit is a circuit in which current flows in a continuous loop. According to Kirchhoff's voltage law, the algebraic sum of the voltage in a closed-loop circuit is equal to zero. The algebraic sum of the voltage in a closed-loop circuit is calculated by applying the following formula:

$0 = V_T - V_L$

where

V_T = source voltage (in V)

V_L = inductive voltage (in V)

Example: What is the inductive voltage of a closed-loop circuit with a source voltage of 240 V?

$0\text{ V} = V_T - V_L$

$0\text{ V} = 240\text{ V} - V_L$

$V_L = \mathbf{240\ V}$

Tech Tip

A simple way to remember the relationship between source voltage and inductive current is by remembering the phrase "ELI the ICE man." The acronym "ELI" means that voltage (*E*) in an inductor (*L*) leads current (*I*). The acronym "ICE" means that current (*I*) in a capacitor (*C*) leads voltage (*E*).

Inductor Testing

Inductors are tested for three types of problems: open circuits, complete short circuits, and partial short circuits. Open circuits are the most common inductor problem and can be detected using an ohmmeter or a DMM set to measure resistance. An acceptable reading is typically between 0 Ω and 500 Ω. For complete or partial short circuits, it is best to use a capacitance and inductance analyzer to check the functionality of capacitors and inductors in a circuit.

In an inductive AC circuit, CEMF is 180° out of phase with the source voltage. Source voltage is greatest when inductive current makes its most rapid change as it crosses the baseline. Inductive current leads CEMF by 90° and lags source voltage by 90°. **See Figure 16-4.**

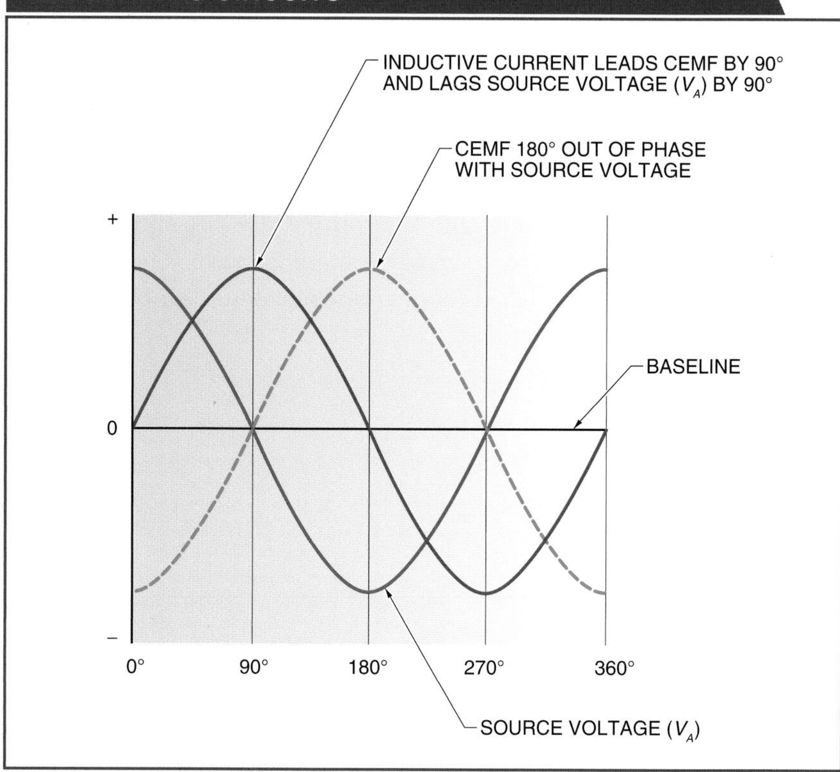

Figure 16-4. In an inductive AC circuit, counter-electromotive force (CEMF) is the greatest when inductive current crosses the baseline.

Using a DMM to measure inductive voltage and current in an inductive AC circuit, the inductive reactance can be calculated by applying the following formula:

$$X_L = \frac{V_L}{I_L}$$

where

X_L = inductive reactance (in Ω)

V_L = inductive voltage (in V)

I_L = inductive current (in A)

Example: What is the inductive reactance in an inductive AC circuit if a DMM indicates a voltage of 120 V and a current of 2 A?

$$X_L = \frac{V_L}{I_L}$$

$$X_L = \frac{120\text{ V}}{2\text{ A}}$$

$$X_L = \mathbf{60\ \Omega}$$

Factors Affecting Inductive Reactance. Factors affecting inductive reactance are the size of the inductor and the AC frequency. In a DC circuit, the inductive voltage is proportional to the size of the inductor. An inductor with a high value of inductance is more easily magnetized than an inductor with a low value of inductance. In an AC circuit, the magnetic field rate of change is constant and is based on the signal frequency.

The greater the frequency, the greater the inductor-induced voltage, and therefore the greater the inductive reactance. Inductive reactance in an inductive AC circuit is affected by the frequency and inductance. Inductive reactance changes directly in proportion with changes in either frequency or inductance. **See Figure 16-5.** Inductive reactance is calculated by applying the following formula:

$X_L = 2\pi \times f \times L$

where

X_L = inductive reactance (in Ω)

2π = 6.28

f = frequency (in Hz)

L = inductance (in H)

EFFECT OF INDUCTIVE REACTANCE ON FREQUENCY AND INDUCTANCE

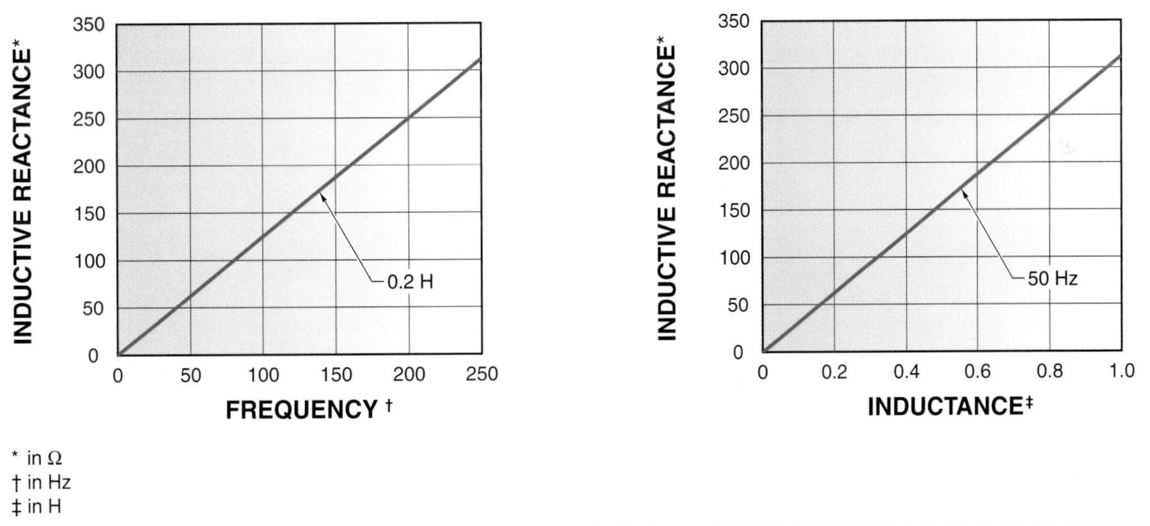

INDUCTIVE REACTANCE*
FREQUENCY†
0.2 H
INDUCTIVE REACTANCE*
INDUCTANCE‡
50 Hz

* in Ω
† in Hz
‡ in H

Figure 16-5. Inductive reactance changes directly in proportion with changes in either frequency or inductance.

Example: What is the inductive reactance in an inductive AC circuit that contains a 60 Hz AC lamp with an inductance value of 0.4 H?

$X_L = 2\pi \times f \times L$

$X_L = 6.28 \times 60 \text{ Hz} \times 0.4 \text{ H}$

$X_L = \mathbf{150.7\ \Omega}$

Inductance can be held constant with the frequency varied, or frequency can be held constant with the inductance varied. Inductive reactance increases linearly with an increase in frequency and with an increase in inductance.

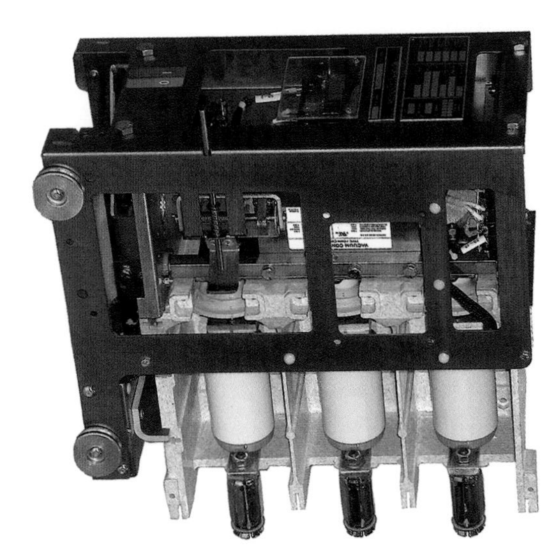

Square D/Schneider Electric

Inductive components are present in devices used for motor control, such as contactors.

In an AC circuit, if frequency or inductance is varied, the inductive reactance changes. With power circuits, inductance is typically the only factor that varies since the power service provider supplies a constant power line frequency that seldom changes. However, power line frequency may be converted to a range of frequencies for tasks such as controlling motor speeds. In electronic circuits, inductance is typically constant and frequency varies. Under normal conditions, inductance does not change unless it is from a variable inductor. This information can be used to confirm that a lamp has normal brilliance when a DC power source is applied to the same circuit. The frequency of the circuit with the DC power source is equal to 0 Hz because the current does not alternate as in an AC circuit.

Example: What is the inductive reactance of a lamp with a frequency of 0 Hz and inductance of 0.4 H?

$X_L = 2\pi \times f \times L$

$X_L = 6.28 \times 0 \text{ Hz} \times 0.4 \text{ H}$

$X_L = \mathbf{0}\ \Omega$

With DC power applied, the voltage drop across the inductor is 0 V. The total line voltage is across the lamp, producing normal brilliance.

Example: What is inductive reactance of a lamp with frequency of 60 Hz and inductance of 0.4 H?

$X_L = 2\pi \times f \times L$

$X_L = 6.28 \times 60 \text{ Hz} \times 0.4 \text{ H}$

$X_L = \mathbf{150.72}\ \Omega$

With AC power applied, the inductor has a resistive value that causes a voltage drop. The voltage drop causes less current flow through the lamp. Therefore, the lamp consumes less power and produces less brilliance. **See Figure 16-6.**

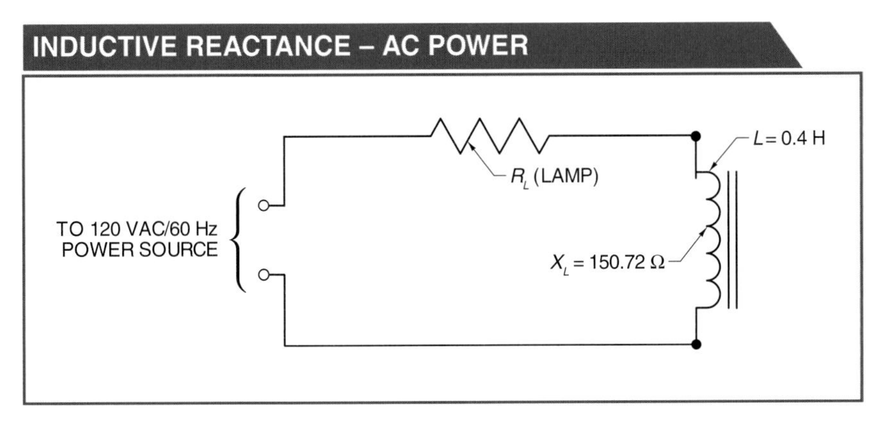

Figure 16-6. In a 120 VAC/60 Hz AC circuit, an inductance (*L*) of 0.4 H causes an inductive reactance (X_L) of 150.72 Ω, which reduces total loop current.

Impedance is the total opposition of any combination of resistance, inductive reactance, and capacitive reactance offered to the flow of alternating current. The value of impedance is always greater than that of reactance or resistance. If either resistance or reactance has a larger value in comparison to the other, the impedance value will be closer to the larger value. For example, in a simple AC circuit that includes a lamp and an inductor, impedance is the combination of inductive reactance and lamp resistance. Resistance and impedance are calculated by applying the following procedure:

1. Calculate component lamp resistance. Component resistance is calculated by applying the following formula:

$$R_T = \frac{V_A^{\ 2}}{P_L}$$

where
R_T = component resistance (in Ω)
V_A = source voltage (in V)
P_L = component power (in W)

2. Calculate impedance. Total impedance is calculated by applying the following formula:

$$Z_T = \sqrt{R_T^{\ 2} + X_L^{\ 2}}$$

where
Z_T = total impedance (in Ω)
R_T = total resistance (in Ω)
X_L = inductive reactance (in Ω)

Terms

Impedance is the total opposition of any combination of resistance, inductive reactance, and capacitive reactance offered to the flow of alternating current.

Note: The resistance and series inductive reactance cannot be directly added together because the lamp and inductor voltage are 90° out of phase. $Z_T = R_T$ at $0° + X_L$ at 90°. **See Figure 16-7.**

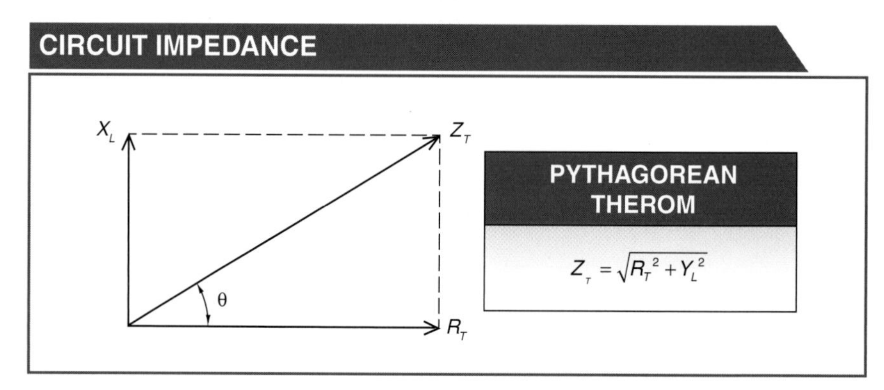

Figure 16-7. To find circuit impedance, the Pythagorean theorem is used.

Example: What is the total resistance and impedance of a 120 VAC inductive circuit with a 100 W lamp and inductive reactance of 150.7 Ω?

1. Calculate component resistance.

$$R_T = \frac{V_A^{\ 2}}{P_L}$$

$$R_T = \frac{(120\text{ V})^2}{100\text{ W}}$$

$$R_T = \frac{14{,}400\text{ V}}{100\text{ W}}$$

$$R_T = 144\ \Omega$$

2. Calculate impedance.

$$Z_T = \sqrt{R_T^{\ 2} + X_L^{\ 2}}$$

$$Z_T = \sqrt{(144\ \Omega)^2 + (150.7\ \Omega)^2}$$

$$Z_T = \sqrt{20{,}736\ \Omega^2 + 22{,}710\ \Omega^2}$$

$$Z_T = \sqrt{43{,}446\ \Omega^2}$$

$$Z_T = \mathbf{208\ \Omega}$$

The impedance value is closer to the inductive reactance value. Total circuit current can be calculated by using Ohm's law, where the impedance replaces resistance. Total circuit current is calculated by applying the following formula:

$$I_T = \frac{V_A}{Z_T}$$

where

I_T = total circuit current (in A)
V_A = source voltage (in V)
Z_T = total impedance (in Ω)

Example: What is the total circuit current in a 120 V circuit that has impedance of 208 Ω?

$$I_T = \frac{V_A}{Z_T}$$

$$I_T = \frac{120\text{ V}}{208\ \Omega}$$

I_T = **0.57 A**

Increased impedance due to inductance reduces the current and lowers the power dissipated in the lamp. Power dissipation in a resistive circuit is calculated by applying the following formula:

$P_D = I_T^2 \times R_N$
where
P_D = power dissipation (in W)
I_T = total current (in A)
R_N = component N resistance (in Ω)

Example: What is the power dissipation in a lamp with total current of 0.57 A and component resistance of 144 Ω?

$P_D = I_T^2 \times R_N$
$P_D = (0.57\text{ A})^2 \times 144\ \Omega$
$P_D = 0.325\text{ A} \times 144\ \Omega$
P_D = **47 W**

In an inductive AC circuit, an inductor limits the power dissipated by the lamp to less than its rated value because of inductive reactance, which is not present in a DC circuit. These values can change because of the changing resistance of the lamp. The lamp filament has a positive temperature coefficient because its filament has a much higher resistance at incandescence than it does when at ambient temperature.

Current in Inductive AC Circuits

Current in an inductive AC circuit can be calculated once the inductive reactance and the source voltage are known. Current can be calculated in this type of circuit by modifying Ohm's law to substitute inductive reactance (X_L) for resistance (R). Current in a pure inductive circuit where the inductive reactance and voltage are known is calculated by applying the following formula:

$$I_L = \frac{V_L}{X_L}$$

where
I_L = inductive current (in A)
V_L = inductive voltage (in V)
X_L = inductive reactance (in Ω)

Example: What is the current in a 120 VAC inductive circuit with an inductive reactance of 150.7 Ω?

$$I_L = \frac{V_L}{X_L}$$

$$I_L = \frac{120 \text{ VAC}}{150.7 \ \Omega}$$

$$I_L = \mathbf{0.796 \ A}$$

Inductive voltage is equal to 120 VAC. When the source voltage and inductive reactance are known, inductive current can be calculated by application of Ohm's law. **See Figure 16-8.**

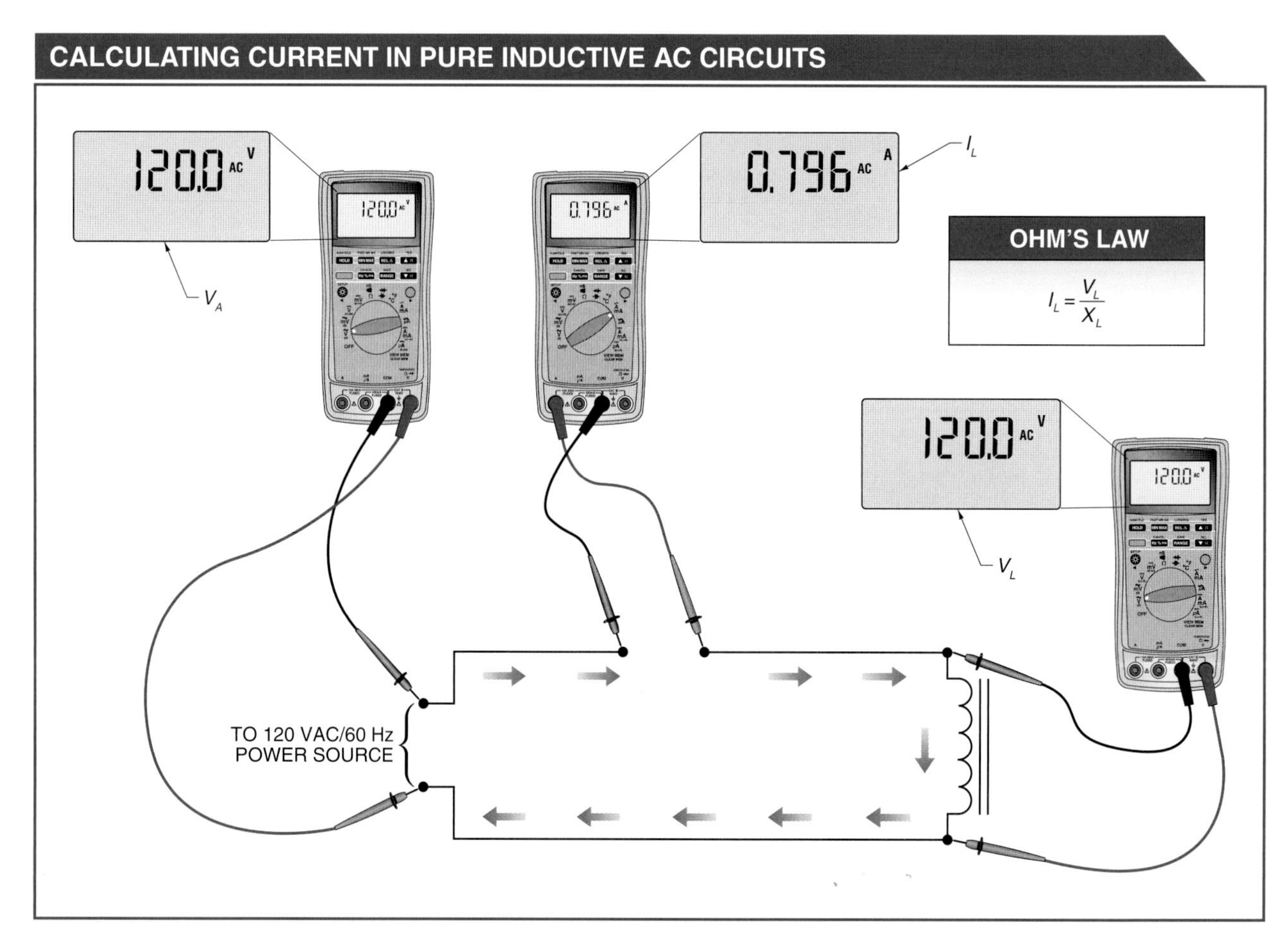

Figure 16-8. When source voltage and inductive reactance are known, current in a pure inductive AC circuit can be calculated by using Ohm's law.

Current and Voltage Phase Relationships in Inductive AC Circuits

Current and voltage phase relationships in an inductive AC circuit are similar to current and voltage phase relationships in an inductive DC circuit. At the beginning of quadrant 1 of the sine wave (0°), inductive voltage is at its maximum negative point and opposes the source voltage, which is at its maximum positive point. Inductive current is 0 A. At the end of quadrant 1, source voltage is equal to 0 V and inductive current is at its maximum value. Inductive current achieves its maximum positive value (90°) after the source voltage achieves its maximum positive value. As source voltage falls to its maximum negative value in the quadrant 2, inductive current falls from its maximum positive value toward 0 A. This process continues through quadrant 4, with inductive current always lagging the source voltage by 90°. **See Figure 16-9.**

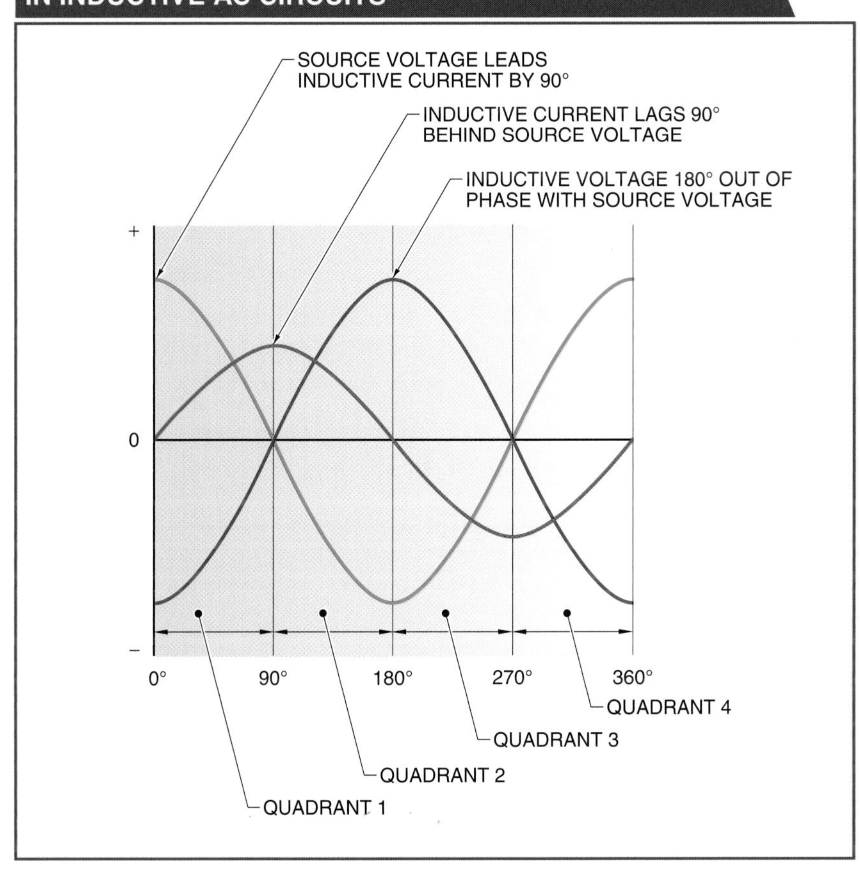

Figure 16-9. Inductive voltage is 180° out of phase with source voltage, and inductive current leads inductive voltage by 90°.

Power in Inductive AC Circuits

Terms

Apparent power is the product of the voltage and current in a circuit, calculated without considering the phase shift that may be present between the voltages and current in a circuit.

True power is the actual power used in an electrical circuit.

Reactive power is power supplied to a reactive load.

Power in an inductive AC circuit has several components, including apparent power, true power, and reactive power. *Apparent power* is the product of the voltage and current in a circuit, calculated without considering the phase shift that may be present between the voltages and current in a circuit. Apparent power is measured in volt amperes (VA). *True power* is the actual power used in an electrical circuit. True power is measured in watts (W). *Reactive power* is power supplied to a reactive load. Reactive power is measured in voltamperes reactive (VAR). Apparent power represents a load or circuit that includes both true power and reactive power. Apparent power in an inductive circuit is calculated by applying the following formula:

$P_A = I_T \times V_A$

where

P_A = apparent power (in VA)

I_T = total current (in A)

V_A = source voltage (in V)

Example: What is the apparent power in an inductive circuit that has total current of 0.796 A and source voltage of 120 VAC?

$P_A = I_T \times V_A$

$P_A = 0.796 \text{ A} \times 120 \text{ VAC}$

$P_A = \mathbf{95.5\ VA}$

Only resistance can dissipate power in an electrical circuit. When a circuit is inductive, an inductor requires energy to create the expanding magnetic field around it. Energy is stored in the magnetic field. When the magnetic field collapses around the inductor, the energy from the magnetic field returns to the circuit.

Apparent power in an inductive AC circuit can be seen in a changing waveform. **See Figure 16-10.** In quadrants 1 and 3, the inductor consumes power from the circuit. In quadrants 2 and 4, energy returns to the circuit. Consumed power is always positive, and power returned to the circuit is always negative.

In quadrants 1 and 3, current and source voltage are both positive values causing the inductor to consume energy from the circuit. A positive value multiplied by a positive value, or a negative value multiplied by a negative value, yields a positive value. In quadrants 2 and 4, voltage and current have opposite values causing the inductor to return energy to the circuit. A positive value multiplied by a negative value, or a negative value multiplied by a positive value, yields a negative value.

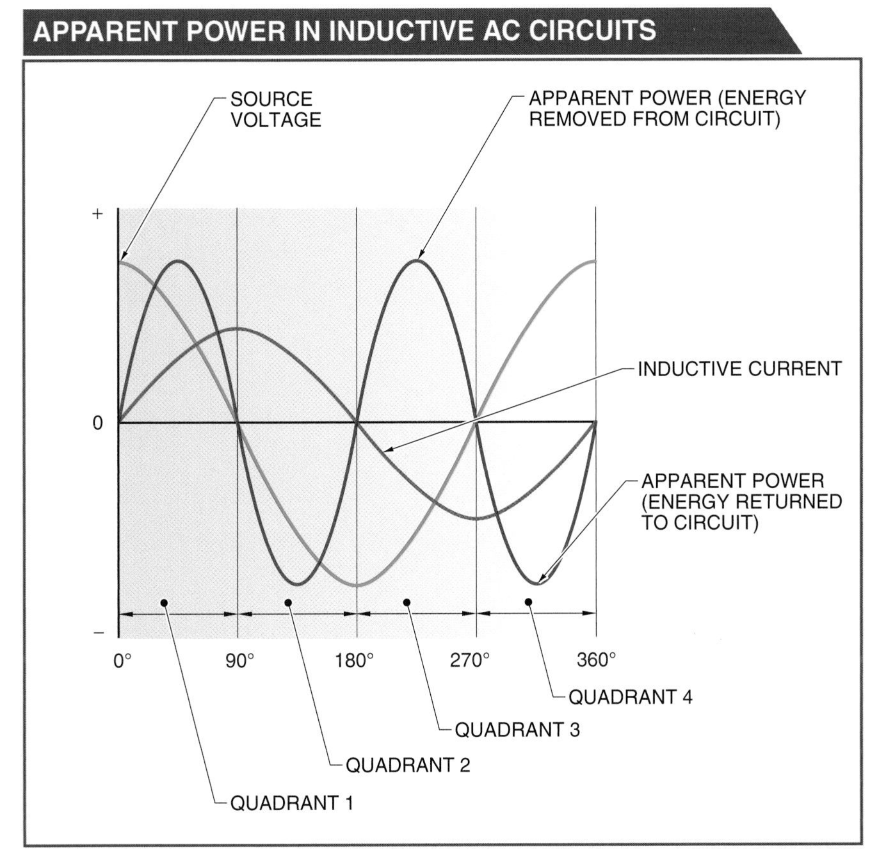

Figure 16-10. Apparent power in an inductive circuit can be seen in a changing sine wave.

SERIES INDUCTIVE-RESISTIVE AC CIRCUITS

Theoretically, in a series inductive-resistive AC circuit, no power is consumed when only inductance is present in a circuit. Practically, though, since inductors are composed of conductor coils, and conductors have resistance, all inductors have some resistance. This resistance is in series with the inductance. In a live circuit, it is not possible to separately measure inductive voltage and inductive resistance because the inductor is the resistive element.

When resistance is low in relation to inductive reactance, the inductor has pure inductance because the resistance has little impact on the overall impedance of the circuit. In applications where the inductive reactance is 10 or more times greater than its resistance, the resistance is not a factor and is normally ignored.

Problems in series inductive-resistive circuits may be detected because of other noticeable conditions. A problem in a series inductive-resistive circuit causes the current to be either too high or too low or to have a value of 0 A.

Terms

A **quality factor** is the ratio of the inductive reactance to the resistance of an inductor.

A **voltage phasor** is a vector diagram used to show the magnitude and phase of voltage in a series inductive-resistive AC circuit.

Inductors are sometimes compared as to their ability to approach perfect inductance. A *quality factor* is the ratio of the inductive reactance to the resistance of an inductor. Quality factor is commonly abbreviated as "Q" and does not have a unit of measure. The quality factor of an inductor is calculated by applying the following formula:

$$Q = \frac{X_L}{R_L}$$

where

Q = quality factor

X_L = inductive reactance (in Ω)

R_L = inductive resistance (in Ω)

Example: What is the quality factor of an inductor that has an inductive reactance of 50 Ω and an inductive resistance of 2 Ω?

$$Q = \frac{X_L}{R_L}$$

$$Q = \frac{50\ \Omega}{2\ \Omega}$$

$$Q = \mathbf{25}$$

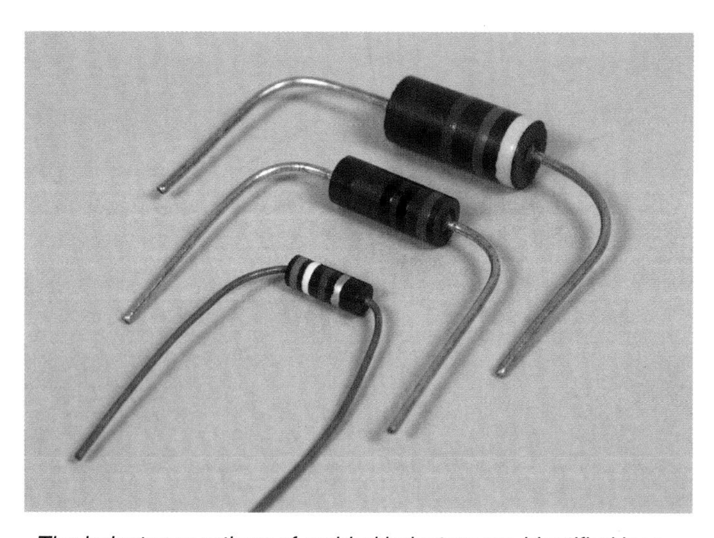

The inductance ratings of molded inductors are identified by a series of color bands.

Proper analysis of series inductive-resistive AC circuits involves separate analysis of voltage phasors, impedance, reactive power, power factor, and frequency.

Voltage Phasors in Series Inductive-Resistive AC Circuits

A *voltage phasor* is a vector diagram used to show the magnitude and phase of voltage in a series inductive-resistive AC circuit. When analyzing an inductive-resistive circuit, voltage phasors are used to calculate the effective voltage and phase difference between the inductive and resistive voltages.

Inductive devices such as AC motors are the equivalent of series inductive-resistive circuits. **See Figure 16-11.** In any series circuit, the current is the same at all points of the circuit. For this reason, current is used as a reference point for analyzing a series inductive-resistive AC circuit. A reference point is the starting point for the circuit analysis and is a known quantity for the circuit. Waveforms, as seen on a power quality meter or an oscilloscope, are used to demonstrate the relationship between the resistive voltage drop and inductor in this type of circuit. The resistive voltage drop is in phase with the current. Resistance opposes the flow of current. The inductive voltage drop leads the current by 90°. Inductance opposes any change in current. Problems or changes in inductive-resistive circuits also cause angle theta to change in value.

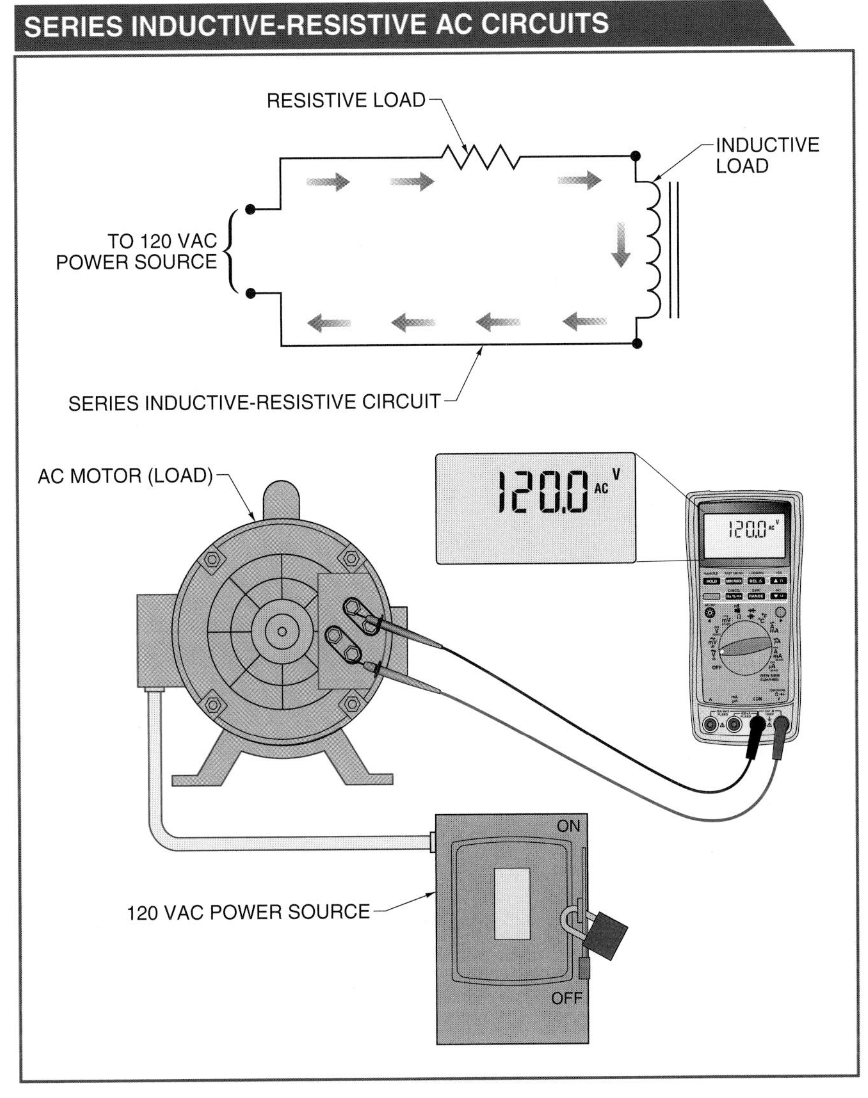

Figure 16-11. Inductive devices such as AC motors are the equivalent of series inductive-resistive AC circuits.

Kirchhoff's voltage law can be applied to series inductive-resistive AC circuits. When voltages across components are added together, their sum equals the source voltage. Source voltage is calculated by applying the following formula:

$$V_A = \sqrt{V_R^2 + V_L^2}$$

where

V_A = source voltage (in V)

V_R = resistive voltage drop (in V)

V_L = inductive voltage drop (in V)

Example: What is the source voltage in a series inductive-resistive AC circuit that includes a 120 VAC lamp and an inductor with a voltage drop of 208 V?

$$V_A = \sqrt{V_R^2 + V_L^2}$$

$$V_A = \sqrt{120\ V^2 + 208\ V^2}$$

$$V_A = \sqrt{14{,}400 + 43{,}264}$$

$$V_A = \sqrt{57{,}664}$$

$$V_A = \mathbf{240\ V}$$

Since the resistive and inductive voltages are 90° out of phase, they cannot be added together directly. These voltage vectors form a right triangle. When inductive voltage and resistive voltage are known, a graphical method, such as the parallelogram method, can also be used to calculate the source voltage. **See Figure 16-12.**

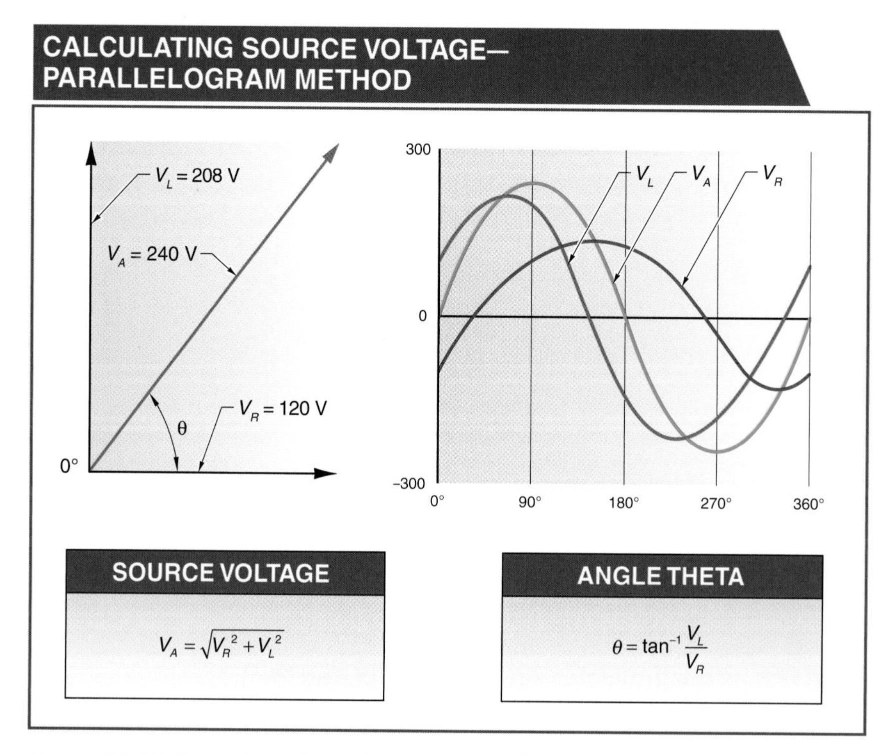

Figure 16-12. When inductive voltage and resistive voltage are known, the parallelogram method can be used to calculate the source voltage.

Angle theta is the phase difference between the source voltage and resistive voltage. Angle theta is sometimes referred to as phase angle. The greater the ratio of inductive voltage drop to resistive voltage drop, the greater the angle theta. Angle theta is calculated by applying the following formula:

$$\theta = \tan^{-1}\left(\frac{V_L}{V_R}\right)$$

where

θ = angle theta (in °)

$\tan^{-1}$ = arctangent

V_L = inductive voltage (in V)

V_R = resistive voltage (in V)

Example: What is the angle theta in an inductive-resistive AC circuit that includes both a 120 VAC lamp and an inductive load with a voltage drop of 208 V?

$$\theta = \tan^{-1}\left(\frac{V_L}{V_R}\right)$$

$$\theta = \tan^{-1}\left(\frac{208\text{ V}}{120\text{ V}}\right)$$

$$\theta = \tan^{-1} 1.73$$

$$\theta = \mathbf{60°}$$

Terms

Angle theta is the phase difference between the source voltage and resistive voltage.

Impedance in Series Inductive-Resistive AC Circuits

Impedance is the total opposition of any combination of resistance, inductive reactance, and capacitive reactance offered to the flow of alternating current. Resistance and inductive reactance combine in a series inductive-resistive AC circuit to oppose the flow of current. The unit of impedance is the ohm (Ω). *Admittance* is the ability of a circuit containing both resistance and reactance to conduct current. The unit of admittance is the siemen (S), which is the same unit as is used for conductance and susceptance.

When inductive current and the voltages across the lamp (resistive voltage) and the inductor (inductive voltage) are known, the impedance of the circuit can be calculated. **See Figure 16-13.** Impedance of a series inductive-resistive AC circuit containing a lamp and an inductor is calculated by applying the following procedure:

1. Determine component resistance. Component resistance is calculated by applying the following formula:

$$R_N = \frac{V_R}{I_R}$$

Terms

Admittance is the ability of a circuit containing both resistance and reactance to conduct current.

where

R_N = component N resistance (in Ω)

V_R = resistive voltage (in V)

I_R = resistive current (in A)

2. Determine circuit inductive reactance. Circuit inductive reactance is calculated by applying the following formula:

$$X_L = \frac{V_L}{I_L}$$

where

X_L = inductive reactance (in Ω)

V_L = inductive voltage (in V)

I_L = inductive current (in A)

3. Determine circuit impedance. Circuit impedance is calculated by applying the following formula:

$$Z = \sqrt{R_N^{\ 2} + X_L^{\ 2}}$$

where

Z = circuit impedance (in Ω)

R_N = component N resistance (in Ω)

X_L = inductive reactance (in Ω)

Note: Since the voltages are divided by a common factor (current), the impedance angle is the same as the voltage angle.

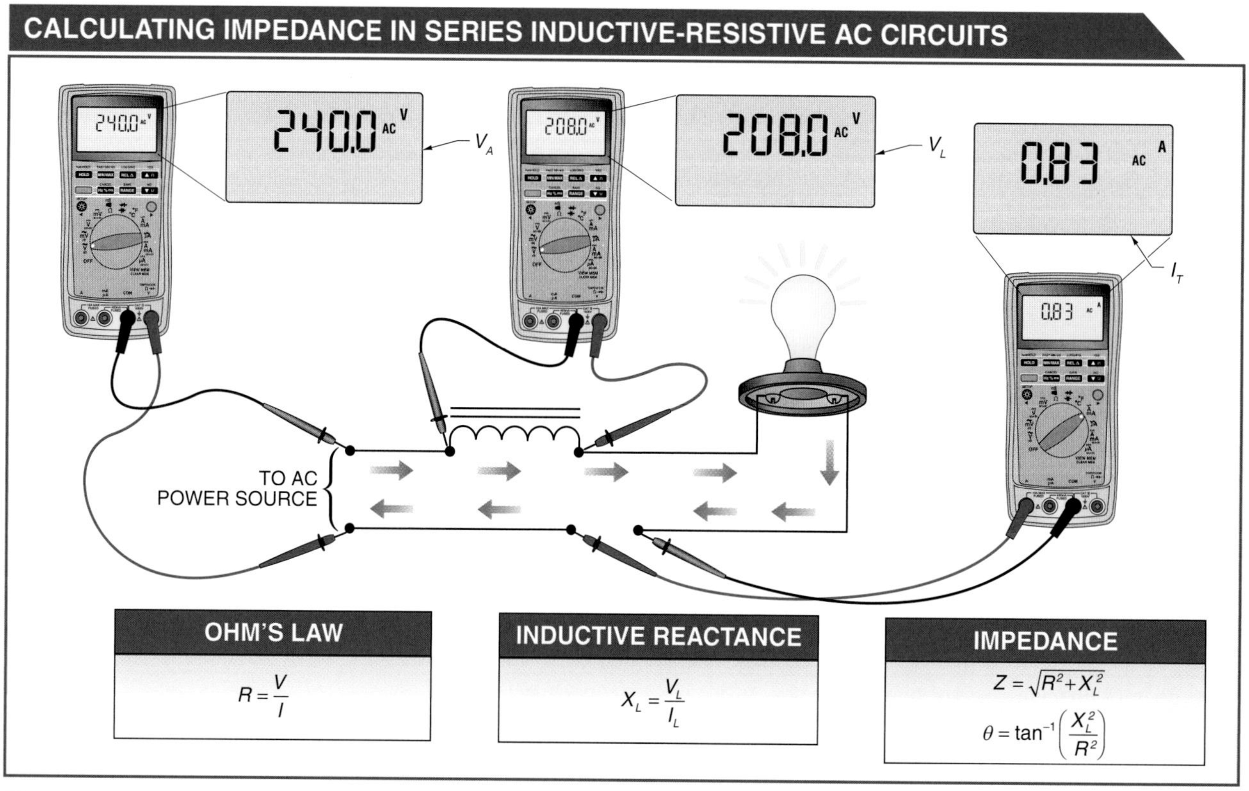

CALCULATING IMPEDANCE IN SERIES INDUCTIVE-RESISTIVE AC CIRCUITS

Figure 16-13. When the inductive voltages and the current across the lamp and the inductor are known, the impedance of the circuit can be calculated.

Example: What is the impedance of a 0.83 A series inductive-resistive AC circuit that has 120 V across a lamp and 208 V across an inductor?

1. Determine component resistance.

$$R_N = \frac{V_R}{I_R}$$

$$R_N = \frac{120\text{ V}}{0.83\text{ A}}$$

$$R_N = 144\ \Omega$$

2. Determine circuit inductive reactance.

$$X_L = \frac{V_L}{I_L}$$

$$X_L = \frac{208\text{ V}}{0.83\text{ A}}$$

$$X_L = 250\ \Omega$$

3. Determine circuit impedance.

$$Z = \sqrt{R_N^{\,2} + X_L^{\,2}}$$

$$Z = \sqrt{(144\ \Omega)^2 + (250\ \Omega)^2}$$

$$Z = \sqrt{20{,}736\ \Omega + 62{,}500\ \Omega}$$

$$Z = \sqrt{83{,}236\ \Omega}$$

$$Z = \mathbf{288.5}\ \Omega$$

Reactive Power in Series Inductive-Resistive AC Circuits

Reactive power is power supplied to a reactive load. The unit of reactive power is voltamperes reactive (VAR) rather than watts (W), which are used with true power. True power represents a pure resistive component or load, and reactive power represents a pure reactive component or load. The difference between true power and reactive power is that true power supplied to a resistive load is used to perform work or is dissipated as heat. Reactive power supplied to a reactive component such as an inductor averages out to zero and is not converted to work. In a series inductive-resistive AC circuit, reactive power is calculated by applying the following formula:

$$P_{VAR} = X_L \times I_T^{\,2}$$

where

P_{VAR} = reactive power (in VAR)

X_L = inductive reactance (in Ω)

I_T = series circuit current (in A)

Terms

Reactive power is power supplied to a reactive load.

Example: What is the reactive power of a series inductive-resistive AC circuit with an inductive reactance of 250 Ω and current of 2 A?

$P_{VAR} = X_L \times I_T^2$

$P_{VAR} = 250\ \Omega \times 2\ A^2$

$P_{VAR} = \mathbf{1000\ VAR}$

Almost all AC circuits include impedance in the form of inductive reactance and/or capacitive reactance. Inductive reactance is by far the more common, since all motors, transformers, solenoids, and inductors have inductive reactance. Inductive reactance or capacitive reactance causes voltage and current to be out of phase because of their opposition to current or voltage changes.

In any AC circuit, measuring circuit voltage and current with a DMM is not useful in determining true power because the voltage measurement is multiplied by the current measurement. True power cannot be calculated because there is no consideration for the phase shift between voltage and current. A power quality meter must be used to measure reactive power and true power in an AC circuit. **See Figure 16-14.**

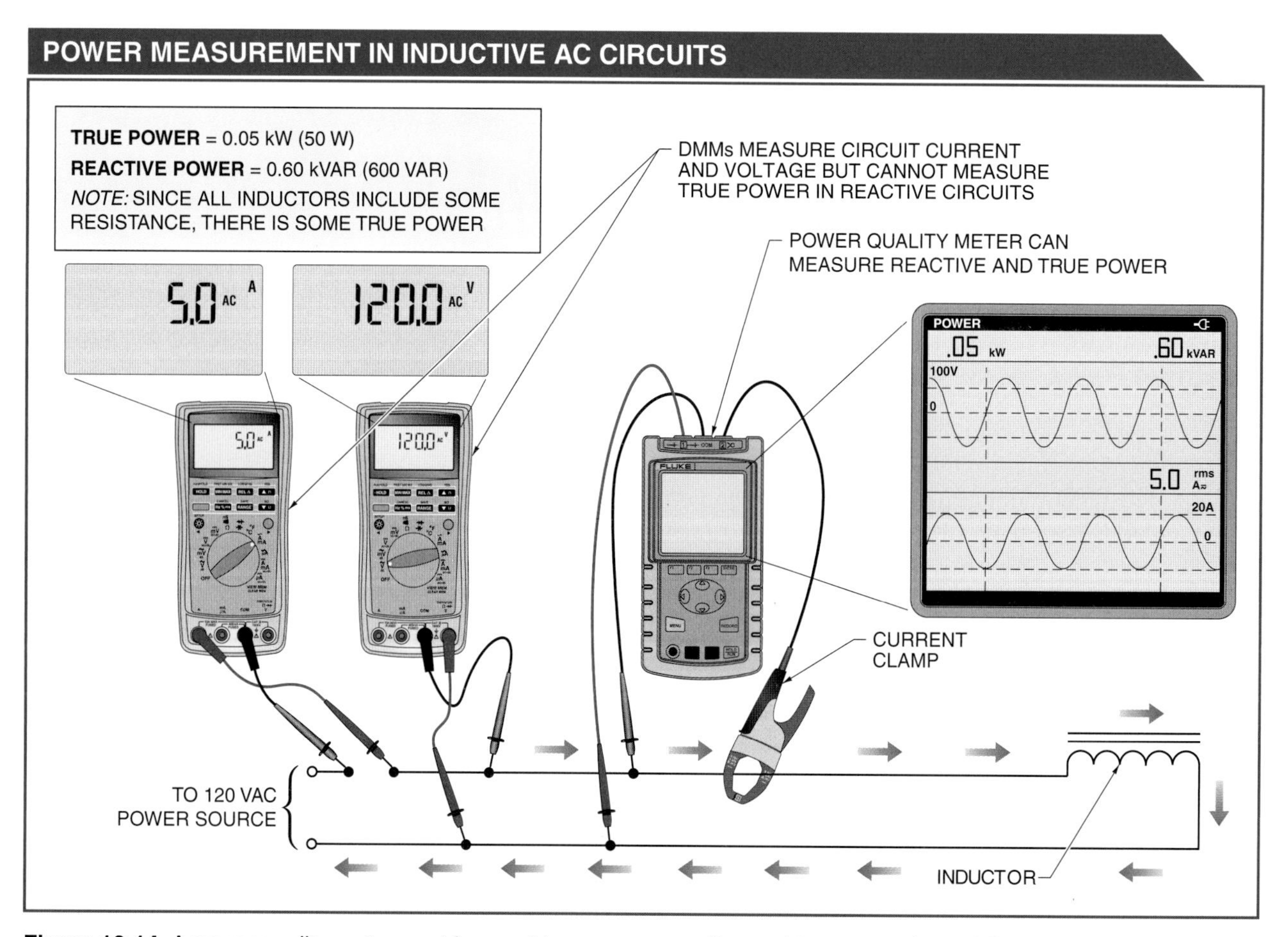

Figure 16-14. A power quality meter must be used to measure reactive and true power in an AC circuit.

In an inductive-resistive AC circuit, the amount of reactive power is high compared to circuit true power. This is because almost all of the power is reactive and continuously alternates back and forth, and there is no net power loss or gain. The power alternates back and forth because magnetic energy is alternately stored and released by an inductor connected to an AC power supply. The vector for an inductive circuit shows that current lags voltage by 90°. There is some true power in any inductive-resistive AC circuit because all inductors have some resistance. True power in an inductive-resistive AC circuit is converted to heat in the inductor due to the resistance of the inductor.

Although there is little true power in a pure reactive circuit, there is still current in the circuit. This current must be considered when sizing conductors, transformers, and all power distribution equipment used to deliver power to the circuit. Reactive power is a necessary part of any AC circuit that makes use of magnetic fields, such as found in solenoids, motors, or transformers.

Apparent Power in Series Inductive-Resistive AC Circuits

Apparent power is the vector combination of true power and reactive power. Apparent power can be calculated from true power and reactive power. As with voltage and impedance, true and reactive power vectors form a right triangle. When true power and reactive power are known, apparent power is calculated by applying the following formula:

$$P_A = \sqrt{P_T^2 + P_{VAR}^2}$$

where:

P_A = apparent power (in VA)
P_T = true power (in W)
P_{VAR} = reactive power (in VAR)

Example: What is the apparent power of a series inductive/resistive AC circuit with true power of 5 W and reactive power of 40 VAR?

$$P_A = \sqrt{P_T^2 + P_{VAR}^2}$$

$$P_A = \sqrt{5\ W^2 + 40\ W^2}$$

$$P_A = \sqrt{25\ W + 1600\ W}$$

$$P_A = \sqrt{1625}$$

$$P_A = \mathbf{40.3\ VA}$$

Terms

Power factor is the ratio of the true power to apparent power.

Power Factor in Series Inductive-Resistive AC Circuits

Power factor is the ratio of the true power to apparent power. In a series resistive AC circuit, true power is equal to apparent power. In AC circuits containing reactance, true power dissipated by resistance is always less than the apparent power in the circuit. Power factor is typically provided as a percentage by multiplying by 100%. For example, a 0.50 power factor is written as a power factor of 50%. Power factor in a series inductive-resistive AC circuit is calculated by applying the following formula:

$$PF = \frac{P_T}{P_A}$$

where

PF = power factor (in %)

P_T = true power (in W)

P_A = apparent power (in VA)

Example: What is the power factor in a series inductive-resistive AC circuit that has true power of 5 W and apparent power of 40.3 VA?

$$PF = \frac{P_T}{P_A}$$

$$PF = \frac{5.0\ \text{W}}{40.3\ \text{VA}}$$

PF = **12% (0.12)**

The range of the power factor is from 1 to 0. Current and source voltage are in phase when the load is purely resistive and the power factor is 1. When current lags the source voltage by 90°, power factor is 0.

Most power line circuits are inductive due to motors and other inductive loads sharing the line with the resistive loads. These circuits have a lagging power factor since the current lags the source voltage in inductive circuits.

Terms

Unity power factor is a power factor equal to 1.

Poor power factor is the condition of a circuit in which power factor is less than 90%.

Unity power factor is a power factor equal to 1. *Poor power factor* is the condition of a circuit in which power factor is less than 90%. Poor power factor increases line loss because a greater level of current flows through the conductors. Poor power factor can result in heat damage to insulation and other circuit components and typically requires installation of larger conductors to meet the requirements of the National Electrical Code® (NEC®).

Under normal conditions in power circuits, unity power factor is the most desirable condition. That is, all power delivered to the load is used by the load. When poor power factor occurs, inefficiencies are increased. Generators and transformers must have a high rating to supply the volt amperes, not just the true power. Power factor is typically measured in series/parallel circuits having resistance, inductance, and capacitance.

Frequency in Series Inductive-Resistive AC Circuits

Although electricians in the United States and Canada work mostly with a frequency of 60 Hz in series inductive-resistive AC circuits, there are situations where another frequency may be encountered. Utility power frequencies in various countries range from 25 Hz to 60 Hz. Most European countries use systems that require 50 Hz power. Some applications may also require a frequency other than 60 Hz. For example, some aircraft alternators have a 400 Hz output because of weight considerations since 400 Hz systems require smaller inductors, transformers, and motors.

Many types of equipment are manufactured in other countries, purchased by end users in the United States, and installed and operated at 60 Hz. Equipment manufactured in countries other than the United States may have a standard operating frequency of 50 Hz. Frequency changes affect parameters of a series inductive-resistive AC circuit such as impedance, voltage, and power. **See Figure 16-15.** Frequency changes do not affect the resistance or the source voltage. These values remain constant.

The value of inductive reactance in a series inductive-resistive AC circuit is doubled when the frequency value is doubled, and its value is reduced by half when the frequency value is reduced by half. Increasing the frequency causes impedance to increase and current to decrease. The maximum impedance for a series inductive-resistive circuit is undefined because as frequency continually increases toward infinity, impedance also approaches infinity. Lowering the frequency causes impedance to decrease and current to increase.

Inductive voltage increases with increased frequency and decreases with decreased frequency. In all cases, the vector sum of the inductive voltage and the resistive voltage must equal the source voltage. At infinite frequency, the total source voltage is dropped across the inductor. At 0 Hz, total source voltage is dropped across the resistor. When a ratio of 10 to 1 exists between the inductive reactance and the resistance, the circuit is usually considered a pure inductive or pure resistive circuit, depending on which of the two is larger. This is because the component vector impact is very small by comparison.

FREQUENCY CHANGES IN INDUCTIVE-RESISTIVE AC CIRCUITS

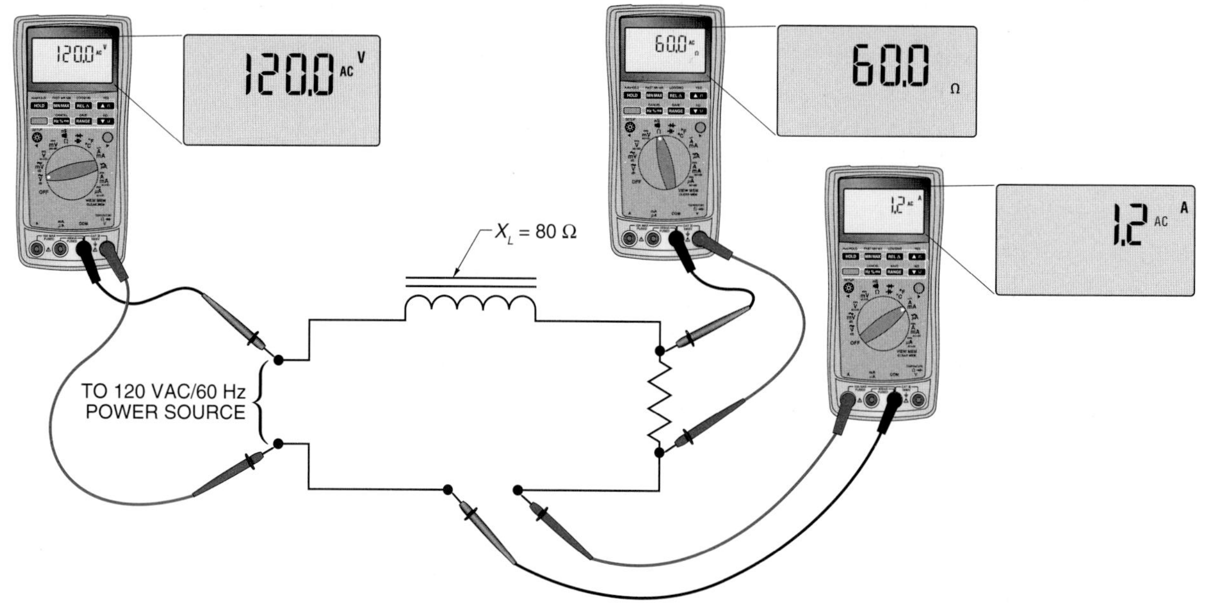

120.0 AC V

60.0 Ω

1.2 AC A

X_L = 80 Ω

TO 120 VAC/60 Hz POWER SOURCE

EFFECTS OF FREQUENCY IN AC CIRCUITS

Frequency*	Impedance	Voltage	Power
120	Z = 171 Ω; X_L = 160 Ω; θ; θ = 69.4°; R = 60 Ω	V_A = 120 V; V_L = 112.3 V; θ; θ = 69.4°; V_R = 42 V	V_A = 84 VA; VAR = 79 VAR; θ; θ = 69.4°; P_T = 30 W
60	Z = 100 Ω; X_L = 80 Ω; θ; θ = 53.13°; R = 60 Ω	V_A = 120 V; V_L = 96 V; θ; θ = 53.13°; V_R = 72 V	V_A = 144 VA; VAR = 115 VAR; θ; θ = 53.13°; P_T = 86 W
30	Z = 72 Ω; X_L = 40 Ω; θ; θ = 33.7°; R = 60 Ω	V_A = 120 V; V_L = 66 V; θ; θ = 33.7°; V_R = 99 V	V_A = 199 VA; VAR = 110 VAR; θ; θ = 33.7°; P_T = 165 W

* in Hz

Figure 16-15. Frequency changes affect series inductive-resistive AC circuit parameters such as impedance, voltage, and power.

Power vectors can be used to show that the ratio of true power to apparent power decreases with an increase in frequency and increases with a decrease in frequency. This is reflected in the change in angle theta, which is larger at the high frequency than at the low frequency. The cosine of angle theta is equal to the power factor, ranging from 1 at 0° to 0 at 90°. Power factor is at 100% at a frequency of 0 Hz and at 0% when the frequency is very large. Frequency is considered undefined when it is infinite.

Because inductive reactance is also directly proportional to circuit inductance, the effects caused by a change in frequency, such as changes to impedance, voltage, and power, will also occur when inductance is changed in a series inductive-resistive circuit.

Inductance in Series Inductive-Resistive AC Circuits

Inductive reactance is directly proportional to the frequency and the value of inductance in a series inductive-resistive AC circuit. An increase in frequency results in an increase in inductive reactance, impedance, angle theta, and the voltage drop across the inductor. Current decreases along with voltage drop across the resistor. At the same time the power factor of the circuit is decreased. A decrease in frequency results in the opposite effect of each of the circuit parameters. Because inductive reactance is also proportional to circuit inductance, the same effects will occur when the inductance of a series inductive-resistive circuit is changed.

PARALLEL INDUCTIVE-RESISTIVE AC CIRCUITS

Voltage is the same across all components connected in parallel in inductive-resistive AC circuits. For this reason, voltage is used as a reference point for analyzing parallel inductive-resistive AC circuits. Although voltage is the same across all components in a parallel inductive-resistive AC circuit, resistive voltage is in phase with resistive current, and inductive voltage is 90° out of phase with inductive current, because the CEMF generated by the inductor opposes changes in current. Sine waves are used to show relationships between source voltage and current in a parallel inductive-resistive circuit. **See Figure 16-16.**

Waveform diagrams are used to show the source voltage beginning at 0° and ending at 360°. Source voltage is used as the reference point at 0° in the vector diagram. Resistance opposes current flow, and resistive current is in phase with the source voltage. Inductance opposes any change in current, and inductive current lags voltage by 90°. This places inductive current at 270° or –90°, relative to the inductive voltage.

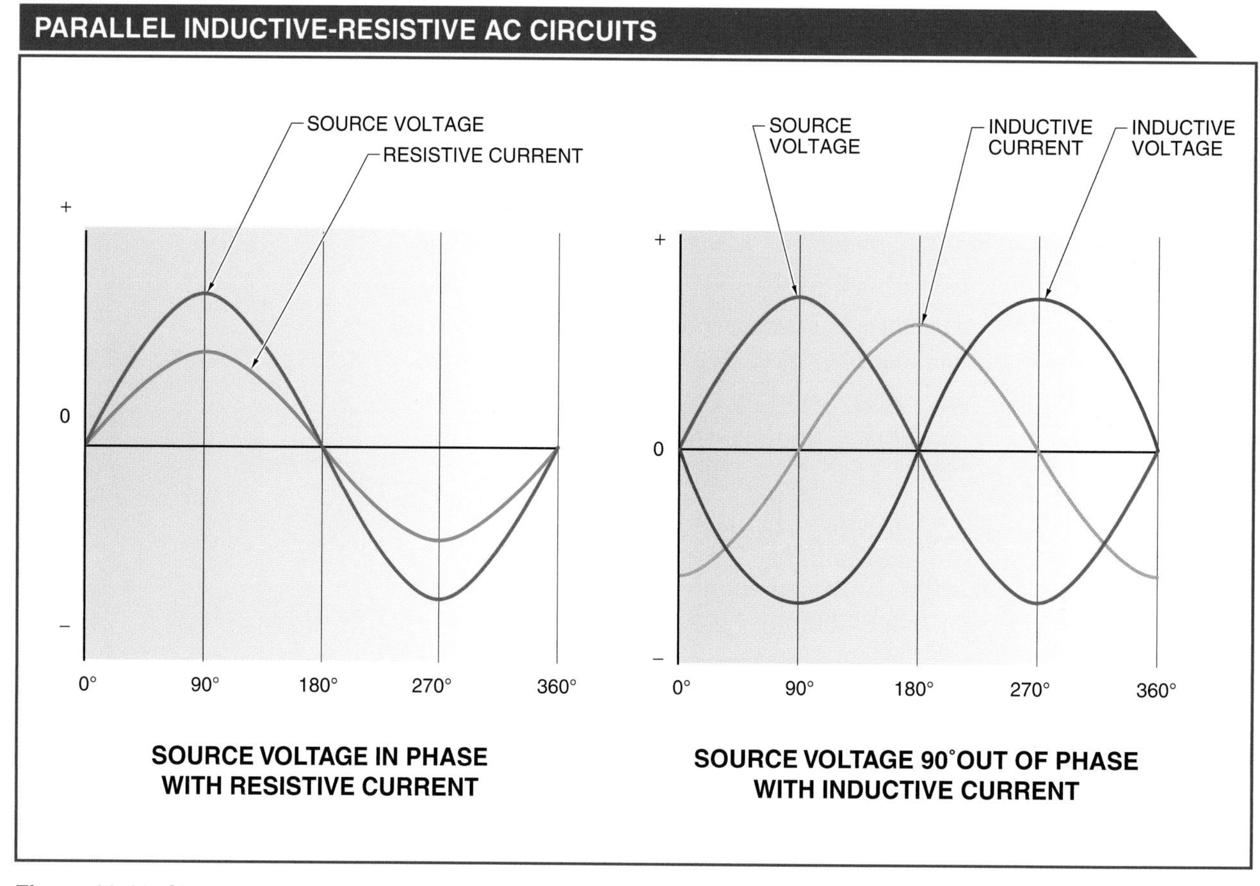

Figure 16-16. Sine waves are used to show relationships between source voltage and current in a parallel inductive-resistive circuit.

Circuit parameters typically analyzed in parallel inductive-resistive circuits include current, impedance, frequency, and inductance. Since the reference point for the parallel circuit is voltage, current is used to calculate many of these parameters.

Current in Parallel Inductive-Resistive AC Circuits

In a parallel inductive-resistive AC circuit, the algebraic sum of the two currents is not equal to the source current. The Pythagorean theorem is used to calculate source current. Source current is calculated by applying the following formula:

$$I_T = \sqrt{I_R^2 + I_L^2}$$

where

I_T = source current (in A)

I_R = resistive current (in A)

I_L = inductive current (in A)

Example: What is the source current in a parallel inductive-resistive AC circuit with a resistive current of 4 A and an inductive current of –3 A?

$$I_T = \sqrt{I_R^{\,2} + I_L^{\,2}}$$

$$I_T = \sqrt{(4\ \text{A})^2 + (-3\ \text{A})^2}$$

$$I_T = \sqrt{16\ \text{A} + 9\ \text{A}}$$

$$I_T = \sqrt{25\ \text{A}}$$

$$I_T = \mathbf{5\ A}$$

Angle theta is calculated by applying the following formula:

$$\theta = \tan^{-1}\left(\frac{I_L}{I_R}\right)$$

where

θ = angle theta (in °)

I_L = inductive current (in A)

I_R = resistive current (in A)

Example: What is the angle theta in a parallel inductive-resistive AC circuit with inductive current of –3 A and resistive current of 4 A?

$$\theta = \tan^{-1}\left(\frac{I_L}{I_R}\right)$$

$$\theta = \tan^{-1}\left(\frac{-3\ \text{A}}{4\ \text{A}}\right)$$

$$\theta = \tan^{-1}(-0.75)$$

$$\theta = \mathbf{-36.87}°$$

Note: Angle theta is a negative value because the source current lags the source voltage at 0°.

Because angle theta must be calculated to satisfy the conditions for a vector diagram, trigonometric functions can be used in place of the Pythagorean theorem to determine the magnitude of the current. The cosine of the angle is equal to the ratio of the adjacent side to the hypotenuse. **See Appendix.** When angle theta is known, current is calculated by applying the following formula:

$$I_T = \frac{I_R}{\cos\theta}$$

where

I_T = source current (in A)

I_R = resistive current (in A)

θ = angle theta

Example: What is the source current in a parallel inductive-resistive AC circuit with a resistive current of 4 A and an angle theta of −36.87°?

$$I_T = \frac{I_R}{\cos\theta}$$

$$I_T = \frac{4\text{ A}}{\cos(-36.87°)}$$

$$I_T = \frac{4\text{ A}}{0.800}$$

$$I_T = \mathbf{5\text{ A}}$$

Impedance in Parallel Inductive-Resistive AC Circuits

When values of total current and source voltage are known, the simplest method of determining impedance in a parallel inductive-resistive AC circuit is by application of Ohm's law. **See Figure 16-17.** Impedance in a parallel inductive-resistive AC circuit is calculated by applying the following formula:

$$Z = \frac{V_A}{I_T}$$

where

Z = impedance (in Ω)

V_A = source voltage (in V)

I_T = total current (in A)

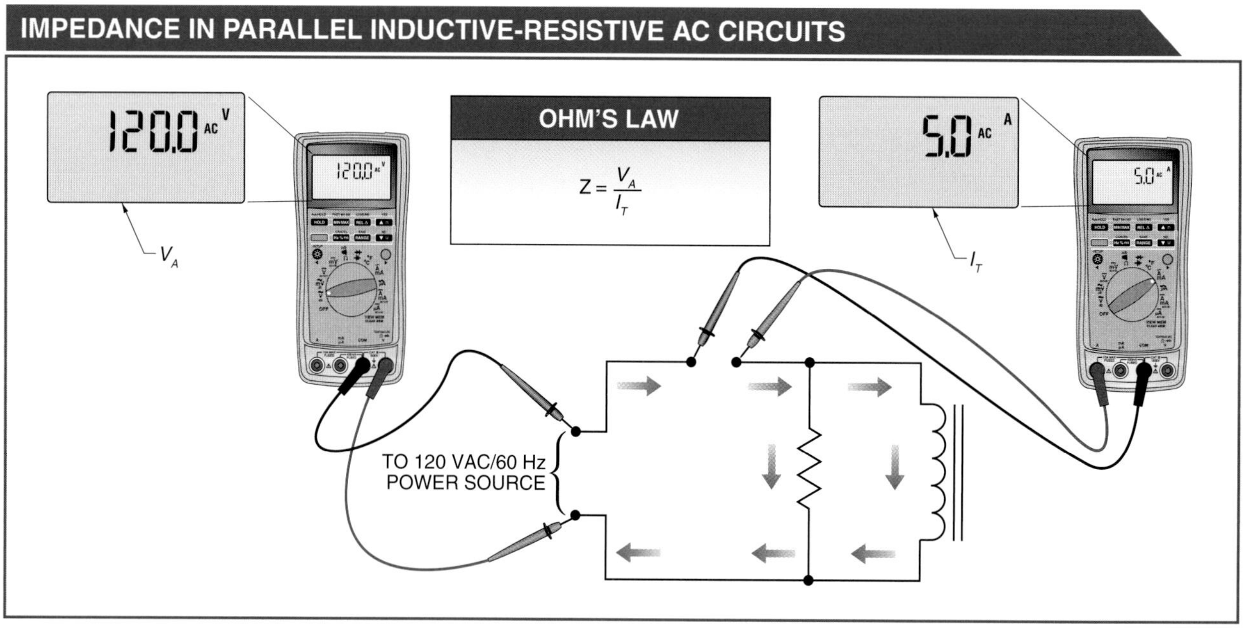

IMPEDANCE IN PARALLEL INDUCTIVE-RESISTIVE AC CIRCUITS

Figure 16-17. When source voltage and total current are known, impedance in a parallel inductive-resistive AC circuit can be calculated using Ohm's law.

Example: What is the impedance of a parallel inductive-resistive AC circuit with a source voltage of 120 VAC and total current of 5 A?

$$Z = \frac{V_A}{I_T}$$

$$Z = \frac{120\text{ V}}{5\text{ A}}$$

$$Z = \mathbf{24\ \Omega}$$

In a series inductive-resistive circuit, total impedance is calculated by adding the resistance and inductive reactance vectors together as is done for resistances connected in series. For a parallel inductive-resistive AC circuit, total impedance is calculated by using the product-over-sum method as used with resistances connected in parallel. For a parallel inductive-resistive AC circuit, total impedance is calculated by applying the following formula:

$$Z_T = \frac{R_T \times X_L}{\sqrt{{R_T}^2 + {X_T}^2}}$$

where

Z_T = total impedance (in Ω)

R_T = total resistance (in Ω)

X_L = inductive reactance (in Ω)

Example: What is the impedance of a parallel inductive-resistive AC circuit with a resistance of 60 Ω, and an inductive reactance of 80 Ω?

$$Z_T = \frac{R_T \times X_L}{\sqrt{{R_T}^2 + {X_T}^2}}$$

$$Z_T = \frac{60\ \Omega \text{ at } 0° \times 80\ \Omega \text{ at } 90°}{\sqrt{(60\ \Omega \text{ at } 0°)^2 + (80\ \Omega \text{ at } 90°)^2}}$$

$$Z_T = \frac{4800\ \Omega \text{ at } 90°}{\sqrt{10{,}000\ \Omega \text{ at } 53°}}$$

$$Z_T = \mathbf{48}\ \Omega \text{ at } 37°$$

Frequency in Parallel Inductive-Resistive AC Circuits

As with a series inductive-resistive AC circuit, a change in frequency in a parallel inductive-resistive AC circuit also causes changes in circuit parameters such as current and power. Because voltage is constant across the components of a parallel inductive-resistive circuit, vectors representing values of current and power can be used to show changes for 120 Hz, 60 Hz, and 30 Hz frequencies. **See Figure 16-18.**

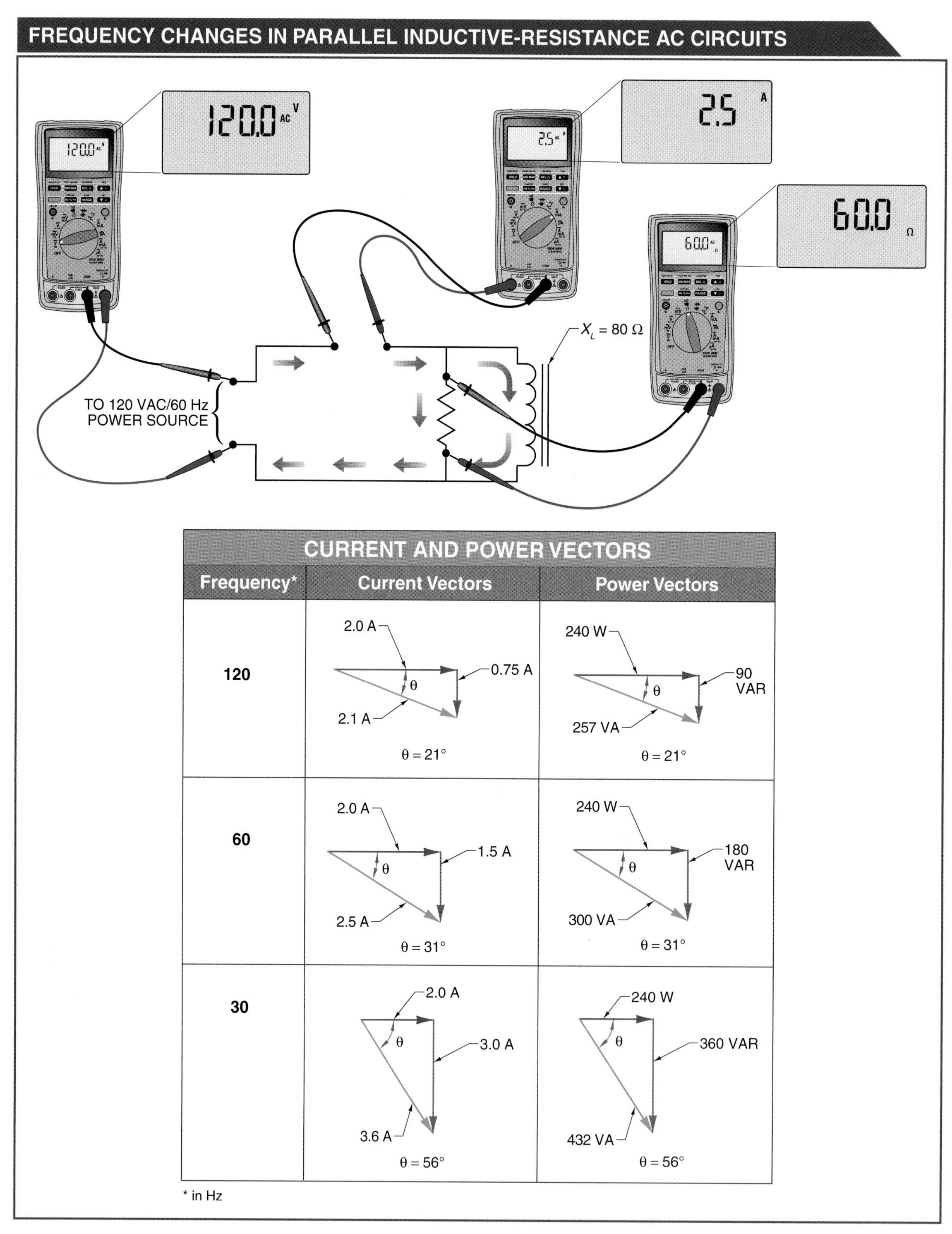

Figure 16-18. Because voltage is constant across the components of a parallel inductive-resistive circuit, vector diagrams of current and power can be used to show changes for 120 Hz, 60 Hz, and 30 Hz frequencies.

In a parallel inductive-resistive AC circuit, frequency does not affect the resistance; therefore, the resistive current remains constant. However, a frequency change causes a change in the inductive reactance. An increase in frequency causes the inductive current to decrease. With less inductive current, total current is reduced, and the impedance of the circuit is increased. Maximum impedance of a parallel inductive-resistive circuit is the value of the resistance when no measurable current passes through the inductor. This occurs when the frequency is very high. Minimum impedance occurs when the frequency is at 0 Hz and there is a very large amount of inductive current.

A parallel inductive-resistive AC circuit is usually considered a pure inductive circuit when inductive current is ten times the resistive current. If resistive current is ten times the inductive current, the circuit is usually considered a pure resistive circuit.

Because resistive current does not change, the true power dissipated in the circuit remains constant. Since inductive current varies with frequency, the reactive power of the circuit changes, as well as the apparent power. Frequency, therefore, causes changes in the power factor. The greater the frequency in a parallel inductive-resistive circuit, the greater the power factor.

Inductance in Parallel Inductive-Resistive AC Circuits

Inductance in a parallel inductive-resistive AC circuit acts similarly to frequency in the same type of circuit. An increase in frequency for a parallel inductive-resistive AC circuit results in an increase in inductive reactance and circuit impedance. Inductive current decreases while resistive current remains constant. Because inductive current decreases relative to resistive current, the total current is lower and less inductive. Lower current results in a greater power factor. Decreasing the frequency results in increasing values in each of the circuit parameters. Because inductive reactance is directly proportional to circuit inductance, the effects that occur with a change in frequency also occur when inductance is changed. Changes in circuit parameters in both series and parallel inductive-resistive AC circuits vary with an increase or decrease in frequency or inductance. **See Figure 16-19.** Changes in frequency or inductance do not affect resistive current, resistive voltage, or inductive voltage in parallel circuits.

Impedance values are calculated in AC applications such as personal computer monitors.

FREQUENCY AND INDUCTANCE CHANGES IN INDUCTIVE-RESISTIVE AC CIRCUITS				
Parameter	Increase in Frequency or Inductance		Decrease in Frequency or Inductance	
	Series	Parallel	Series	Parallel
X_L	↑	↑	↓	↓
I_T	↓	↓	↑	↑
θ	↑	↓	↓	↑
Z	↑	↑	↓	↓
V_R	↓	–	↑	–
V_L	↑	–	↓	–
I_R	↓	–	↑	–
I_L	↓	↓	↑	↑
cos θ	↓	↑	↑	↓

Figure 16-19. Changes in circuit parameters in series and parallel inductive-resistive AC circuits vary with an increase or decrease in frequency or inductance.

SUMMARY

Inductive reactance is the opposition of an inductor to AC. Inductive reactance is affected by the inductance and frequency in an AC circuit. Inductive voltage is based on the amount of change in current over a period. Inductive voltage is 180° out of phase with the source voltage. Inductive current is calculated by application of Ohm's law once the inductive voltage and inductive reactance are known. Current lags the source voltage by 90°. Theoretically, no power is consumed in a pure inductive AC circuit because the energy used to create the magnetic field is stored and returned to the circuit when the magnetic field collapses.

Apparent power in an inductive-resistive AC circuit is calculated in the same manner as with a DC circuit and is measured in voltamperes (VA). Changes to circuit parameters in inductive-resistive AC circuits can be compared for changes in frequency or inductance. These changes in circuit parameters include:

- In a series inductive-resistive AC circuit, an increase in inductance or frequency results in an increase in inductive reactance, impedance, and cosine of θ. Total current varies inversely with these changes.

- In a parallel inductive-resistive AC circuit, an increase in inductance or frequency results in an increase in inductive reactance and impedance. The line current and cosine of θ vary inversely with these changes.
- In all types of series circuits, the voltage leads the current and has a leading angle theta. In all types of parallel circuits, the current lags the voltage and has a lagging angle theta.
- Power factor, as reflected by the cosine of θ, decreases with an increase in frequency or inductance in the series circuit. Power factor increases in a parallel circuit. A decrease in frequency or inductance causes an inverse effect in these circuits.

As with other circuits, there are series and parallel inductive-resistive AC circuits. In a series inductive-resistive AC circuit, the reference point is current, and in a parallel inductive-resistive AC circuit, the reference point is voltage.

APPLICATION: INDUCTOR LOW-PASS FILTERS...

Background: Inductors are circuit elements that block AC voltage. In an inductor, the reactance increases when the frequency increases. When an inductor is placed in series with a resistor, a special circuit called a low-pass filter is created. **See Figure 1.**

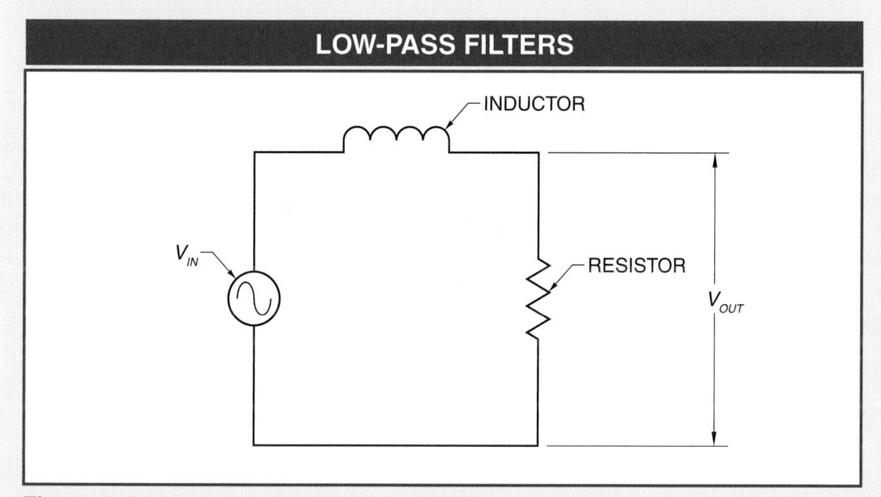

Figure 1. A simple low-pass filter is created with an inductor and series resistor.

At low frequencies, the inductor has little reactance. Therefore, most of the input voltage is dropped across the resistor. An increase in inductor reactance at higher frequencies increases the voltage drop across the inductor, which reduces the voltage drop across the resistor.

...APPLICATION: INDUCTOR LOW-PASS FILTERS...

Key Points: inductor reactance; low-pass filter

Problem: AC voltage is driving a 75 Ω resistive load through an inductor, as shown in **Figure 1**. What is the inductance if the desired voltage across the resistive load is 50% of the source AC voltage at 1 MHz? What is the voltage ratio (V_{out}/V_{in}) across the 75 Ω resistor for frequencies from 1 Hz to 100 MHz, at a power of 10?

Solution: The circuit is a simple voltage divider, with Z_1 impedance as the inductive reactance and Z_2 impedance as the resistor. **See Figure 2.**

VOLTAGE DIVIDERS

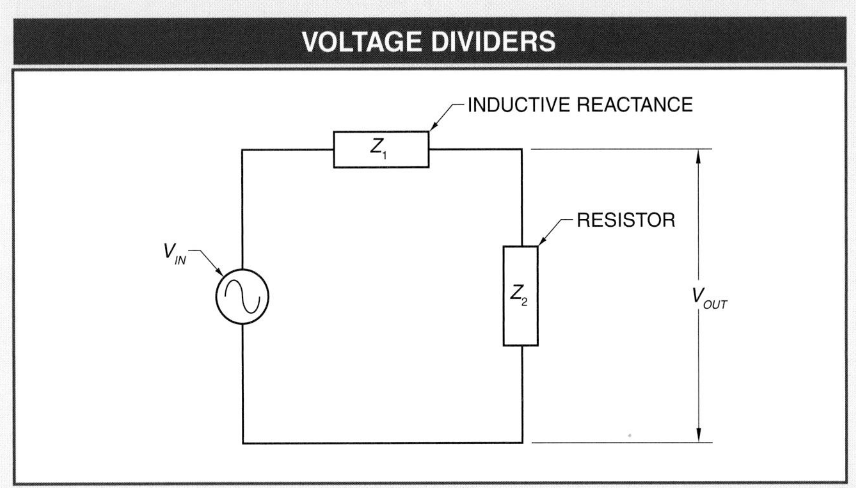

Figure 2. A simple voltage divider is formed by two impedances in series with a voltage source.

The voltage divider is calculated by applying the following formula:

$$V_{out} = \frac{Z_2}{Z_1 + Z_2} \times V_{in}$$

where
V_{out} = voltage output across Z_2 (in V)
V_{in} = voltage input (in V)
Z_1 = impedance of element 1 (in Ω)
Z_2 = impedance of element 2 (in Ω)

The impedances of the inductor and resistor cannot be added directly because the voltages are 90° out of phase. To add the inductor and resistor impedances, apply the following formula:

$$Z_I + Z_R = \sqrt{X_L^2 + R^2}$$

where
Z_I = inductor impedance (in Ω)
Z_R = resistor impedance (in Ω)
X_L = inductive reactance (in Ω)
R = resistance (in Ω)

...APPLICATION: INDUCTOR LOW-PASS FILTERS...

The inductor impedance is found by using the following inductor reactance formula:

$$X_L = 2 \times \pi \times f \times L$$

where
X_L = inductive reactance (in Ω)
π = 3.14 (constant)
f = frequency (in Hz)
L = inductance (in H)

To find the inductance (L) that reduces the output voltage to 50% of the input voltage, the voltage divider and inductor reactance formulas are combined.

$$V_{out} = \frac{R}{\sqrt{X_L^2 + R^2}} \times V_{in}$$

$$0.5 = \frac{R}{\sqrt{X_L^2 + R^2}}$$

$$\sqrt{X_L^2 + R^2} = \frac{R}{0.5}$$

$$X_L^2 + R^2 = \left(\frac{R}{0.5}\right)^2$$

$$X_L = \sqrt{\left(\frac{R}{0.5}\right)^2 - R^2}$$

$$2 \times \pi \times f \times L = \sqrt{\left(\frac{R}{0.5}\right)^2 - R^2}$$

$$L = \frac{\sqrt{\left(\frac{R}{0.5}\right)^2 - R^2}}{2 \times \pi \times f}$$

$$L = \frac{\sqrt{\left(\frac{75}{0.5}\right)^2 - 75^2}}{2 \times \pi \times 1{,}000{,}000}$$

The inductance is 20.7 μH. To find the voltages at different frequencies, the voltage divider and inductive reactance formulas are used with the calculated value for inductance (L).

$$\frac{V_{out}}{V_{in}} = \frac{R}{\sqrt{X_L^2 + R^2}}$$

$$\frac{V_{out}}{V_{in}} = \frac{75\ \Omega}{(2 \times \pi \times f \times 20.7\ \text{mH})\Omega^2 + 75\ \Omega^2}$$

...APPLICATION: INDUCTOR LOW-PASS FILTERS

Various voltage ratios are determined by substituting different values for the frequency (*f*). **See Figure 3.**

LOW-PASS FILTER FREQUENCIES	
Frequency (*f*)*	Voltage Ratios $\left(\frac{V_{OUT}}{V_{IN}}\right)^{\dagger}$
1	100
10	100
100	100
1000	100
10,000	100
100,000	99
1,000,000	50
10,000,000	6
100,000,000	1

* in Hz
† in %

Figure 3. A series inductor-resistor circuit forms a low-pass filter.

In a low-pass filter, at low frequencies, the input voltage is transferred unchanged to the output. At high frequencies, the output voltage is reduced. In this circuit, at 1000 Hz, all of the input voltage is present across the load. At 1 MHz, 50% of the input voltage is across the load. Only 1% of the input voltage is present across the load at 100 MHz.

Chapter Review

1. Why is inductance the only factor that varies with power circuits?
2. What is the unit of measure for reactive power?
3. What is the inductive voltage of an inductive AC circuit that has an inductive current of 8 A and inductive reactance of 40 Ω?
4. What is the unit of measure for admittance?
5. Explain poor power factor.
6. What is the most common frequency of inductive-resistive AC circuits used in North American countries?
7. What is the inductive voltage of a closed-loop circuit with a source voltage of 120 V?
8. How must reactive power and true power be measured in an AC circuit?
9. What is quality factor?
10. What is susceptance and its unit of measure?
11. What is the inductive voltage of an inductive AC circuit that has an inductive current of 2 A and inductive reactance of 15.5 Ω?
12. What is the unit of measure for apparent power?
13. What is the inductive voltage of a closed-loop circuit with a source voltage of 208 V?
14. Which common electrical inductive device is the equivalent of a series inductive-resistive circuit?
15. What is impedance?
16. What is the unit of measure for true power?
17. What is the inductive voltage of a closed-loop circuit with a source voltage of 135.5 V?
18. How are pure inductive circuits used by electrical and electronics technicians?
19. What two factors affect inductive reactance?
20. What is the inductive voltage of an inductive AC circuit that has an inductive current of 0.5 A and inductive reactance of 25 Ω?

17

Capacitive AC Circuits

OBJECTIVES

- Describe pure capacitive circuits and the relationships between voltage, current, capacitive reactance, and power.
- Develop an understanding of series resistive-capacitive circuits and how such circuits are affected by current, impedance, power, frequency, and capacitance.
- Analyze series resistive-capacitive circuits using vector diagrams.
- Develop an understanding of parallel resistive-capacitive circuits and how such circuits are affected by current, impedance, frequency, and capacitance.
- Explain series/parallel resistive-capacitive circuits.

A capacitor is an electrical device that stores electrical energy by means of an electrostatic field. The charge on a capacitor consists of an excess of electrons on one plate and an equally large deficiency of electrons on another plate. Capacitance opposes any change in the voltage in a circuit. In a DC circuit when a capacitor is connected in series with a source, current flows in the circuit until the capacitor is charged to a voltage that is equal and opposite to the voltage source.

The source voltage in an AC circuit is always changing. At the same time, the capacitor is continually charging in one direction, discharging, and then charging in the opposite direction. Because of this process, an alternating current flows continuously in the circuit.

Terms

A **pure capacitive circuit** is a theoretical circuit without any resistance.

PURE CAPACITIVE CIRCUITS

A *pure capacitive circuit* is a theoretical circuit without any resistance. A pure capacitive circuit is an ideal circuit that has only an AC power source and a single capacitive load. In a pure capacitive circuit, current leads the source voltage by 90°. The stored charge across the capacitance lags the source voltage by 180°. **See Figure 17-1.** Although a pure capacitive circuit is strictly theoretical, the concept is used to gain understanding of capacitive reactance, of relationships between current, voltage, and capacitive reactance, and of power in a capacitive circuit.

PURE CAPACITIVE CIRCUITS...

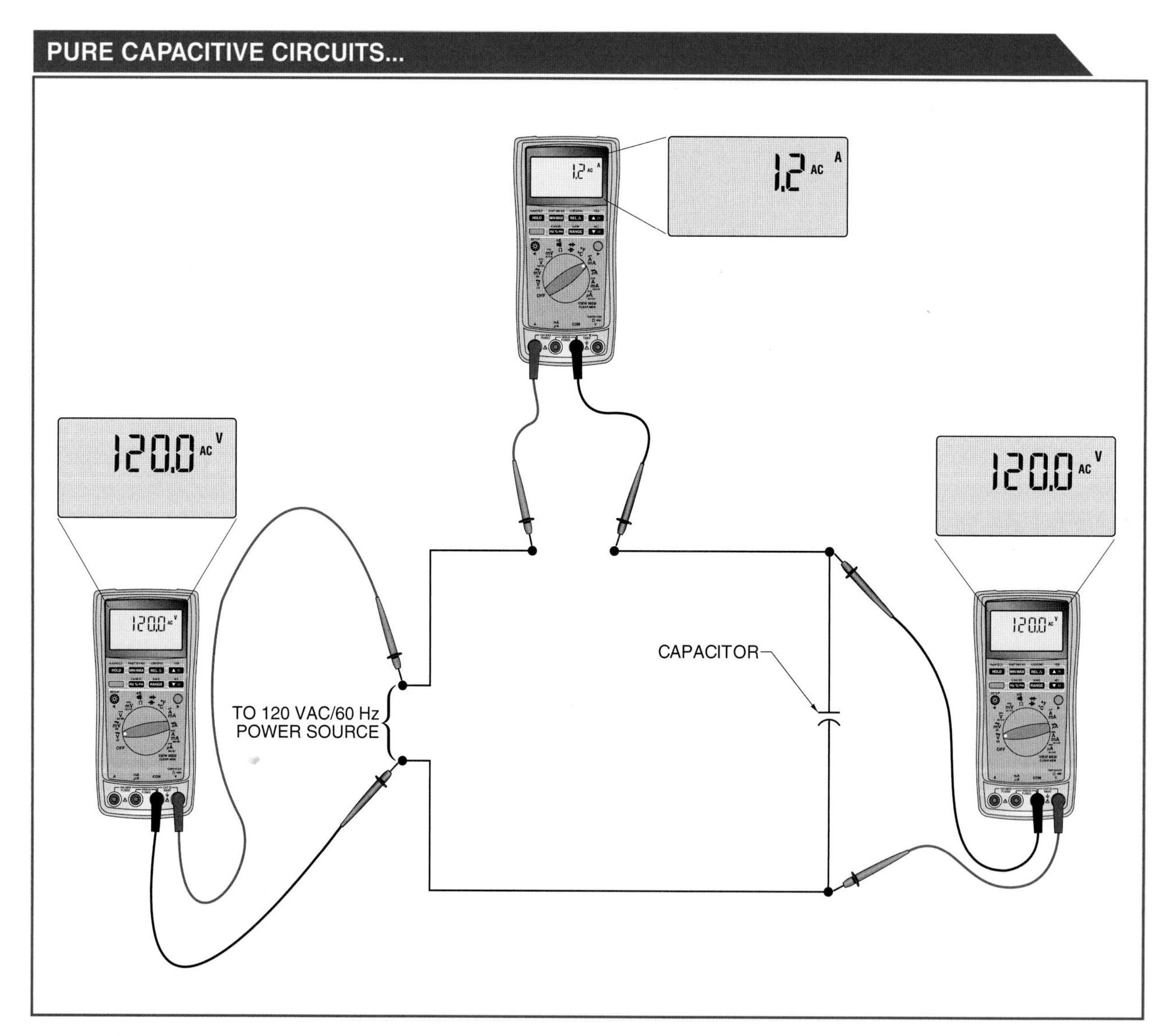

Figure 17-1 . . .

... PURE CAPACITIVE CIRCUITS

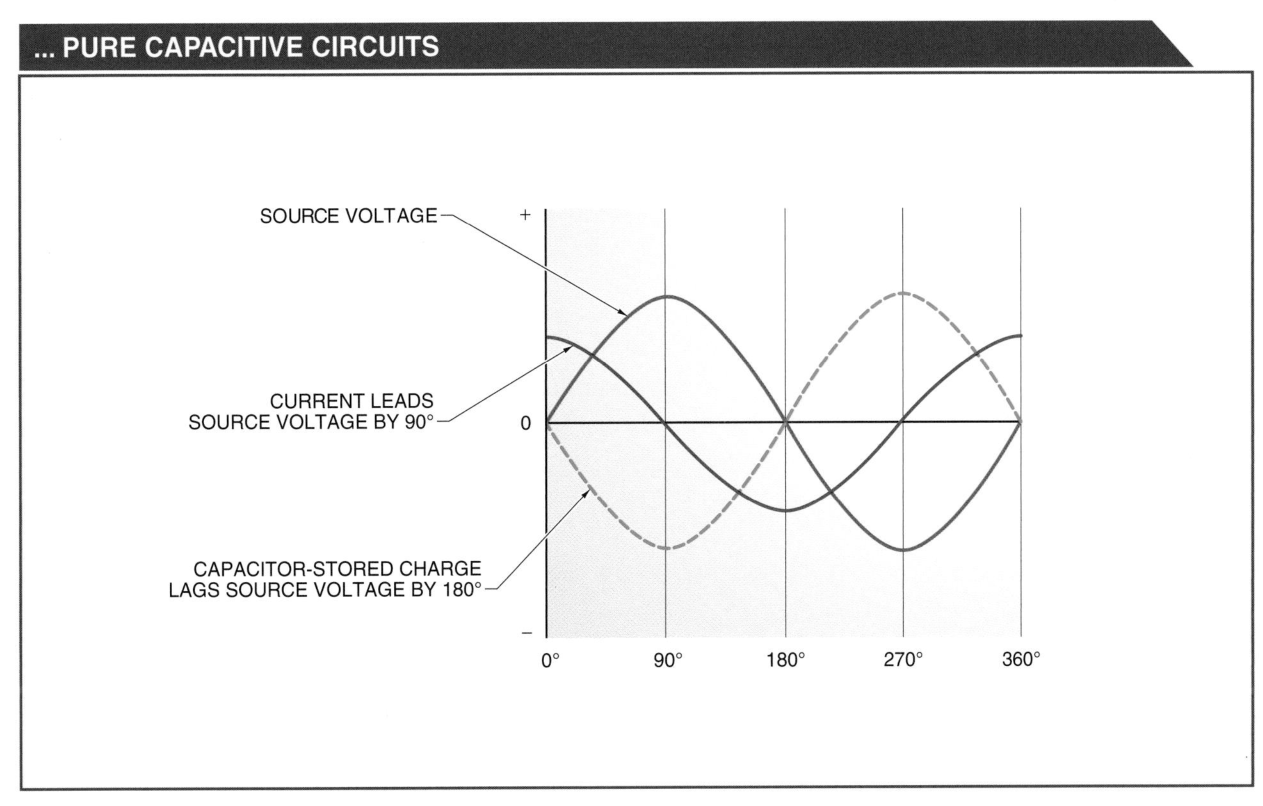

. . . Figure 17-1. In a pure capacitive circuit, the current leads the source voltage by 90°, and the capacitor-stored charge lags the source voltage by 180°.

Capacitive Reactance in Pure Capacitive Circuits

Capacitive reactance is the opposition to current flow by a capacitor (capacitance). When capacitance is added in series to an AC circuit, the circuit current is reduced due to the capacitive reactance. Capacitive reactance, as with resistance and inductive reactance, is measured in ohms (Ω). However, capacitive reactance acts differently than does resistance and is the opposite of inductive reactance because capacitance opposes any changes in voltage. A capacitor causes opposition to current flow due to the charge collecting on its plates that opposes the change of voltage in the circuit. When the voltage and current are known, capacitive reactance is calculated by applying the following formula:

$$X_C = \frac{V_C}{I_C}$$

where

X_C = capacitive reactance (in Ω)

V_C = capacitive voltage (in V)

I_C = capacitive current (in A)

Terms

Capacitive reactance is the opposition to current flow by a capacitor (capacitance).

Example: What is the capacitive reactance in an AC circuit with a capacitive voltage of 120 V and capacitive current of 2 A?

$$X_C = \frac{V_C}{I_C}$$

$$X_C = \frac{120\ \text{V}}{2\ \text{A}}$$

$$X_C = \mathbf{60\ \Omega}$$

As source voltage rises from 0 V to a finite value, current immediately rises to its maximum value since, theoretically, there is no opposition to current flow in an ideal circuit. Electrons flow to one plate of the capacitor and leave the opposite plate to flow back to the source. This causes the capacitor to charge in an opposite direction as the source voltage, which opposes the changing current. At all times in a closed-loop circuit, capacitive voltage is equal to the source voltage. Capacitive voltage complies with Kirchhoff's voltage law and is calculated by applying the following formula:

$$0\ \text{V} = V_A - V_C$$

where

V_A = source voltage (in V)

V_C = capacitive voltage (in V)

Example: What is the capacitive voltage of a closed-loop circuit with a source voltage of 240 V?

$$0\ \text{V} = V_A - V_C$$

$$0\ \text{V} = 240\ \text{V} - V_C$$

$$V_C = \mathbf{240\ V}$$

When capacitive voltage is equal to the source voltage at 90°, current flow in the circuit is 0 A. At 90°, source voltage begins to decrease. At the same time, the capacitor, which opposes any change in voltage, begins to discharge (when there is no resistance). Line current reversal occurs at the point where the current wave crosses the 0 A baseline. As source voltage decreases, discharge current from the capacitor increases, until it reaches its maximum negative value (180°). At 180°, source voltage and capacitive voltage are equal to 0 V.

Source voltage continues in the negative direction, and the capacitor charges in the reverse direction. At 270°, the capacitive voltage is equal to and opposite the source voltage, and current is at 0 A. This process continues to 360°, where the current returns to its maximum level and both voltages are at their minimum levels.

The ability of a capacitance to oppose changes in voltage by charging and discharging causes the opposition to current flow. The faster the current changes, the larger the capacitance and the lower the capacitive reactance. Therefore, capacitive reactance is inversely proportional to frequency (in Hz) and capacitance (in F). When frequency and capacitance are known, capacitive reactance is calculated by applying the following formula:

$$X_C = \frac{1}{2\pi \times f \times C}$$

where

X_C = capacitive reactance (in Ω)

$2\pi = 6.28$

f = frequency (in Hz)

C = capacitance (in F)

Example: What is the capacitive reactance of a 1.0 μF (1×10^{-6} F) capacitance at a frequency of 100 Hz?

$$X_C = \frac{1}{2\pi \times f \times C}$$

$$X_C = \frac{1}{6.28 \times 100 \text{ Hz} \times (1 \times 10^{-6}) F}$$

$$X_C = \frac{1}{628 \times 0.000001}$$

$$X_C = \frac{1}{0.000628}$$

X_C = **1592.4** Ω

If either frequency or capacitance is doubled, then capacitive reactance is reduced by 50%. Conversely, if either the frequency or the capacitance is reduced by 50%, then capacitive reactance is doubled.

Voltage, Current, and Capacitive Reactance Relationships

A pure capacitive AC circuit is used to analyze and understand voltage, current, and capacitive reactance relationships. In a pure capacitive AC circuit, when voltage, frequency, and capacitance are known, other circuit parameters can be calculated. **See Figure 17-2.** Current in a pure capacitive AC circuit is calculated by applying the following procedure:

1. Calculate the capacitive reactance of the circuit. Capacitive reactance is calculated by applying the following formula:

   $$X_C = \frac{1}{2\pi \times f \times C}$$

   where

X_C = capacitive reactance (in Ω)
2π = 6.28
f = frequency (in Hz)
C = capacitance (in F)

2. Calculate total current. Total current is calculated by applying the following formula:

$$I_T = \frac{V_A}{X_C}$$

where
I_T = total current (in A)
V_A = source voltage (in V)
X_C = capacitive reactance (in Ω)

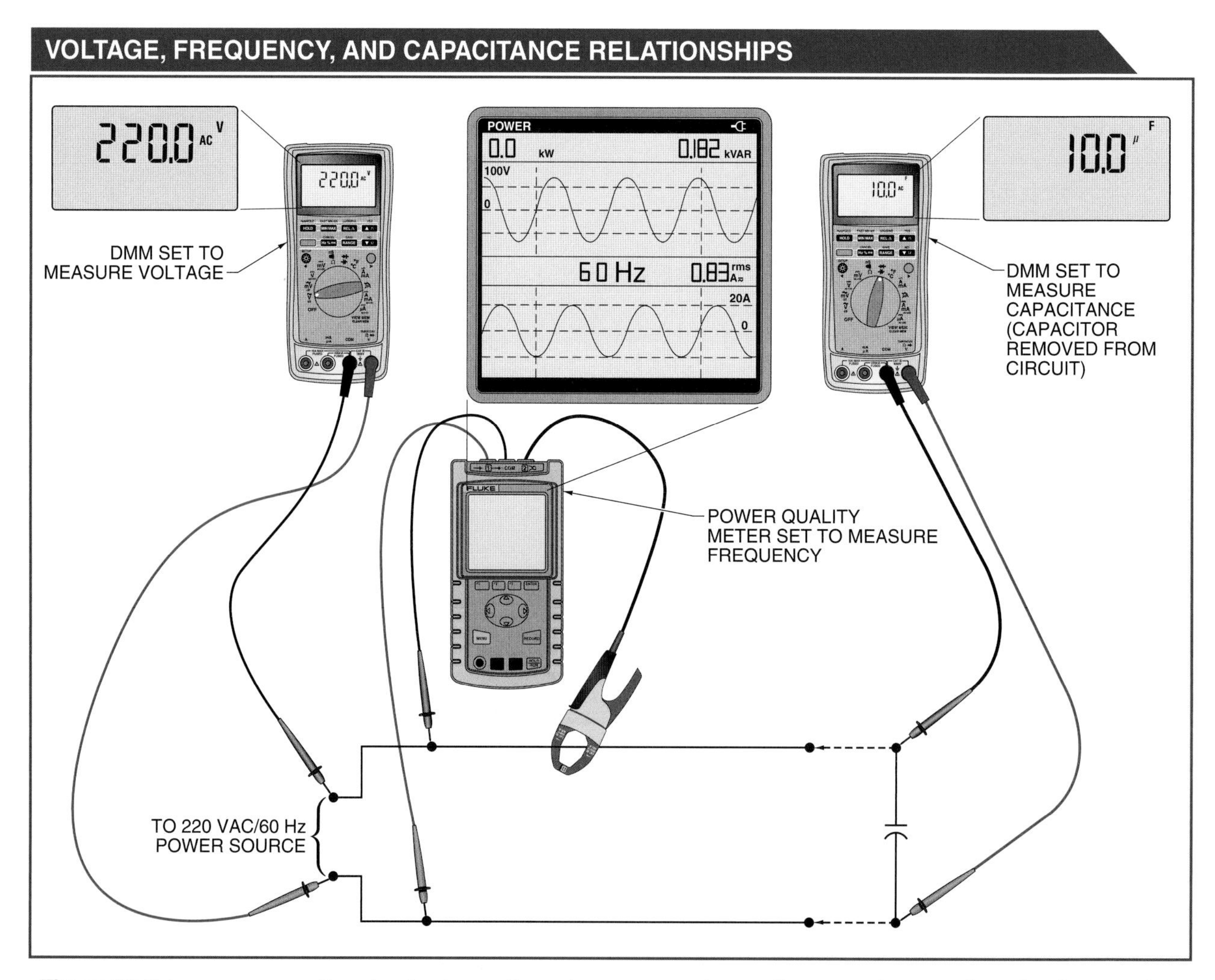

Figure 17-2. In a pure capacitive circuit, when voltage, frequency, and capacitance are known, other circuit parameters can be calculated.

Testing Capacitors

Testing capacitors in low-voltage applications is typically performed using a DMM set to measure resistance or an ohmmeter. Testing capacitors with capacitance values of 1pF to 20 F is performed with a capacitance tester. Capacitance testers are used to check for defective capacitors and can perform tests for dielectric absorption, capacitor value change, capacitor leakage, and equivalent series resistance.

Dielectric absorption is checked to ensure that the capacitor's dielectric is discharging enough to not effect the capacitor's properties. Capacitor value change is tested to ensure that a capacitor's value does not change by more than 15% over a time period of 12 months or more. Capacitor leakage is tested to ensure that there are not defects with the dielectric that may cause excess current to "leak" from the capacitor. Equivalent series resistance is tested to ensure that the amount of series resistance between axial leads or lead-to-plate connections of the capacitor is such that the capacitance value does not change.

Example: What is the total current in a pure capacitive circuit that has a 10 μF capacitance connected to a 220 VAC/60 Hz power source?

1. Calculate capacitive reactance of the circuit.

$$X_C = \frac{1}{2\pi \times f \times C}$$

$$X_C = \frac{1}{6.28 \times 60 \text{ Hz} \times \left(10 \times 10^{-6} \text{F}\right)}$$

$$X_C = \frac{1}{376.8 \times 0.00001}$$

$$X_C = \frac{1}{0.003768}$$

$$X_C = 265.4 \ \Omega$$

2. Calculate total current.

$$I_T = \frac{V_A}{X_C}$$

$$I_T = \frac{220 \text{ V}}{265.4 \ \Omega}$$

$$I_T = \mathbf{0.83\ A}$$

A reference point is used to determine different parameters of the circuit. In a simple series circuit, current is the reference point. Source voltage is directly across the capacitive load. Circuit current leads the source voltage by 90°, and current must flow before the capacitor can charge.

In a resistive circuit, circuit current is in phase with the source voltage. In a capacitive circuit, circuit current leads the source voltage by 90°. **See Figure 17-3.**

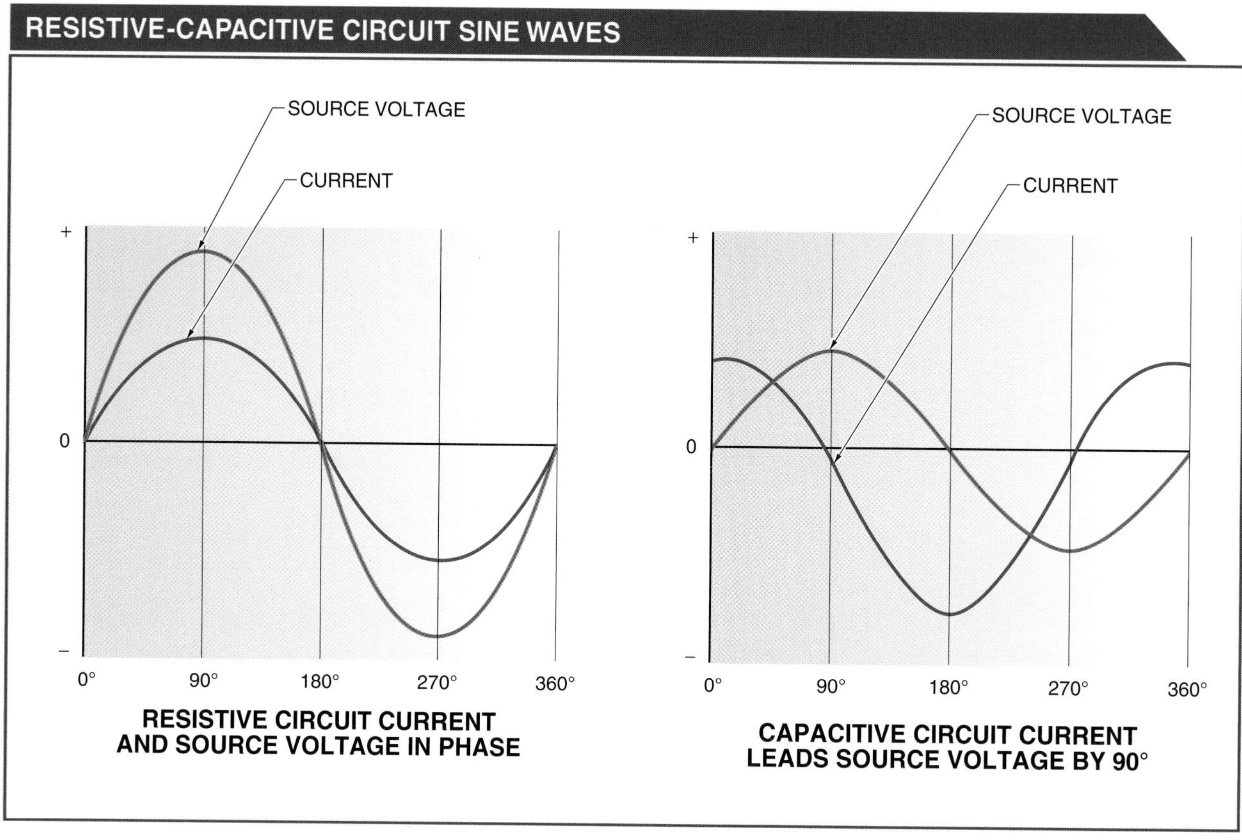

Figure 17-3. In a resistive circuit, circuit current is in phase with the source voltage. In a capacitive circuit, circuit current leads the source voltage by 90°.

Power in Pure Capacitive Circuits

Power in a pure capacitive circuit is equivalent to reactive power in a capacitive circuit. **See Figure 17-4.** Reactive power in a capacitive circuit is calculated by applying the following formula (power formula):

$P_{VAR} = I_C \times V_C$

where

P_{VAR} = reactive power (in VAR)

I_C = capacitive current (in A)

V_C = capacitive voltage (in V)

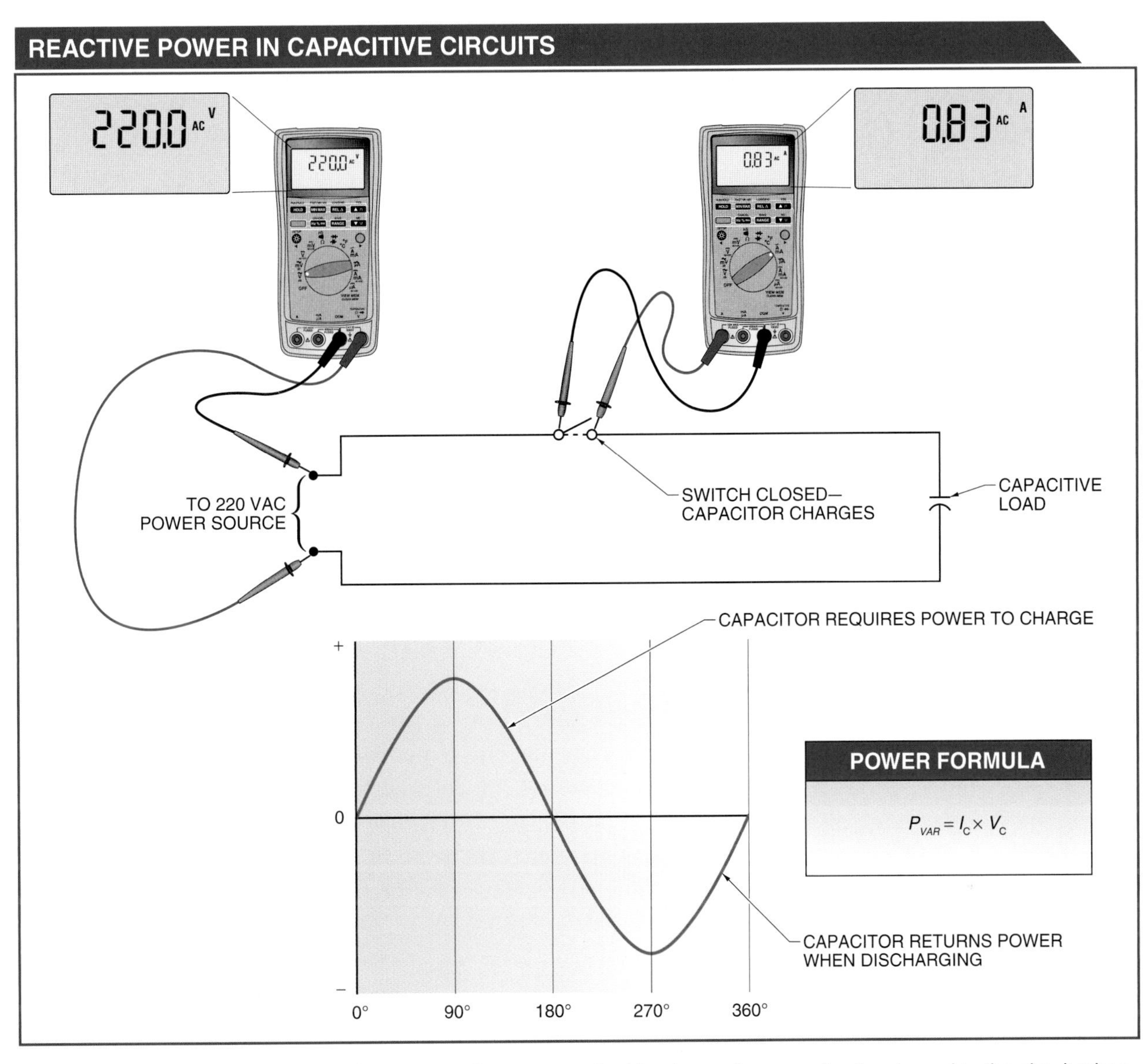

Figure 17-4. In a capacitive circuit, the amount of energy required to charge the capacitor is returned to the circuit when the capacitor discharges. Reactive power is calculated by using the power formula.

Example: What is the value of reactive power in a capacitive series circuit that has total current of 0.83 A and source voltage of 220 V?

$P_{VAR} = I_C \times V_C$

$P_{VAR} = 0.83\text{ A} \times 220\text{ V}$

$P_{VAR} = \mathbf{182.6\ VAR}$

Only resistance can dissipate true power. In a pure capacitive circuit, the capacitor uses energy to charge. When the capacitor discharges, the same amount of energy is returned to the circuit. In the case of an ideal capacitive series circuit, the reactive power is the same as apparent power.

In the first and third quadrants, the capacitor draws energy from the circuit. In the second and fourth quadrants, the capacitor discharges and returns power to the circuit. Consumed power has a positive value, and power returned to the circuit has a negative value.

In the first and third quadrants, the signs for the current and source voltage are the same. A positive multiplied by a positive in the first quadrant, or a negative multiplied by a negative in the third quadrant, yields a positive value. A negative multiplied by a positive in the second quadrant, or a positive multiplied by a negative in the fourth quadrant, yields a negative value.

SERIES RESISTIVE-CAPACITIVE CIRCUITS

Although there is always some resistance present in a circuit, a pure capacitive circuit can be approximated. For example, a series resistive-capacitive circuit always has some resistance present due to the resistance of the capacitor leads, conductors carrying current to the capacitor, and the internal resistance of the voltage source. All of these resistances are in series with the capacitor. If capacitive reactance is ten times that of the resistance, the circuit acts as a pure capacitive circuit. This is not the case in most practical circuits because capacitive reactance can change with frequency. In turn, the ratio of resistance to capacitance also changes. Analysis of series resistive-capacitive circuits involves separate analysis using values of current, impedance, power factor, frequency, and capacitance with vector diagrams.

Current in Series Resistive-Capacitive Circuits

Current in a series resistive-capacitive circuit is the same at all points for all components and devices. Voltage vector diagrams in series resistive-capacitive circuits are used to analyze the circuit. In any series resistive-capacitive circuit, the resistive current and the voltage drop across the resistor are in phase. At the same time, the voltage drop across the capacitor lags the total circuit current by 90°. **See Figure 17-5.** Because current is the same at any point in a series circuit, it is used as the reference point for other circuit parameters.

According to Kirchhoff's voltage law, the source voltage is equal to the sum of the voltage drops across the resistance and capacitance. Since these voltages are out of phase with one another, the source voltage is calculated by adding resistive and capacitive voltage vectors. Source voltage can be calculated by using the vector diagram method or mathematical method. Source voltage in a series resistive-capacitive circuit is calculated with the mathematical method by applying the following formula:

$$V_A = \sqrt{V_R^{\,2} + (-V_C)^2}$$

where

V_A = source voltage (in V)

V_R = resistive voltage (in V)

V_C = capacitive voltage (in V)

Note: V_C is negative because capacitive voltage lags current by 90°.

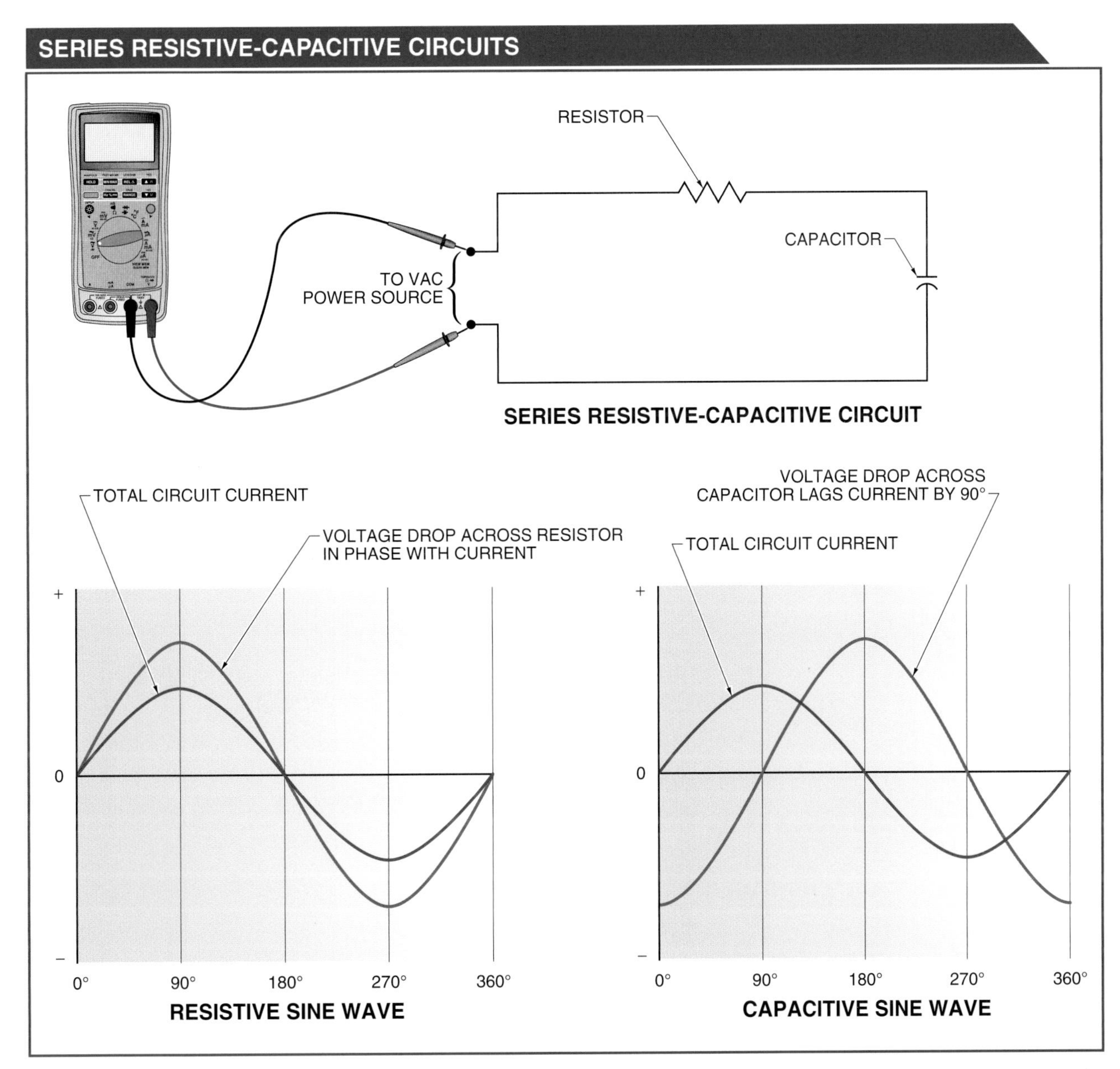

Figure 17-5. In a series resistive-capacitive circuit, the voltage drop across the resistor is in phase with the current, and the voltage drop across the capacitor lags the current by 90°.

Example: What is the source voltage in a series resistive-capacitive circuit that has resistive voltage of 208 V across a lamp and capacitive voltage of 120 V?

$$V_A = \sqrt{{V_R}^2 + (-V_C)^2}$$

$$V_A = \sqrt{(208\text{ V})^2 + (-120\text{ V})^2}$$

$$V_A = \sqrt{43{,}264 + 14{,}400}$$

$$V_A = \sqrt{57{,}664}$$

$$V_A = \mathbf{240.13\ V}$$

The vectors representing voltage drops across the lamp and the capacitor must be added together to obtain the value of the source voltage. Adding the voltages together yields a value of 328 V, whereas the actual value is only 240 V. The difference in values is caused by the voltage drop across the capacitor lagging the total circuit current by 90°.

The greater the ratio of voltage drop across the capacitance to voltage drop across the resistance, the greater the angle theta that the circuit current leads the source voltage. Angle theta (θ) is calculated by calculating the voltage ratio value and then determining the angle for which the tangent ratio is valid. **See Appendix.** Angle theta is calculated by applying the following formula:

$$\theta = \tan^{-1}\left(\frac{-V_C}{V_R}\right)$$

where

θ = angle theta (in °)

V_C = capacitive voltage (in V)

V_R = resistive voltage (in V)

Example: What is the angle theta in a series resistive-capacitive circuit that has resistive voltage of 208 V and capacitive voltage of 120 V?

$$\theta = \tan^{-1}\left(\frac{-V_C}{V_R}\right)$$

$$\theta = \tan^{-1}\left(\frac{-120\text{ V}}{208\text{ V}}\right)$$

$$\theta = \tan^{-1}(-0.5769)$$

$$\theta = \mathbf{-30°}$$

If angle theta is known, the cosine function can be used to calculate the value of the source voltage. **See Appendix.** When angle theta is known, source voltage is calculated by applying the following formula:

$$V_A = \frac{V_R}{\cos\theta}$$

where

V_A = source voltage (in V)

V_R = resistive voltage (in V)

θ = angle theta (in °)

Example: What is the source voltage in a series resistive-capacitive circuit that has resistive voltage of 208 V and an angle theta of −30°?

$$V_A = \frac{V_R}{\cos\theta}$$

$$V_A = \frac{208\text{ V}}{\cos(-30°)}$$

$$V_A = \frac{208\text{ V}}{0.866}$$

V_A = **240.19 V**

Impedance in Series Resistive-Capacitive Circuits

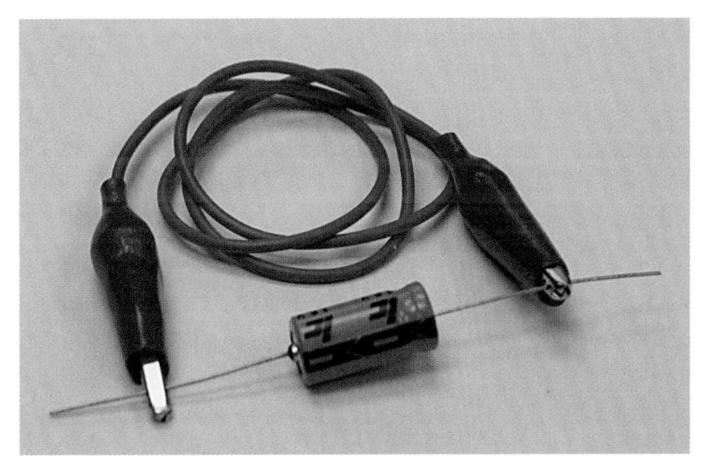

Capacitors are discharged of residual current by shorting the axial leads together.

Impedance in a series resistive-capacitive circuit is the combination of resistance and capacitive reactance to oppose the flow of current. When total current and source voltage in a series resistive-capacitive circuit are known, total impedance can be calculated by using a variation of the voltage, current, and impedance relationship formula. **See Appendix. See Figure 17-6.** Total impedance in a series resistive-capacitive circuit is calculated by applying the following procedure:

1. Calculate total current. Total current is calculated by applying the following formula:

   $$I_T = \frac{P_R}{V_R}$$

   where

   I_T = total current (in A)

   P_R = power consumed by resistive load (in W)

   V_R = resistive voltage (in V)

2. Calculate total impedance. Total impedance is calculated by applying the following formula:

$$Z_T = \frac{V_A}{I_T}$$

where

Z_T = circuit impedance (in Ω)

V_A = source voltage (in V)

I_T = total current (in A)

VOLTAGE, CURRENT, AND IMPEDANCE RELATIONSHIP FORMULA

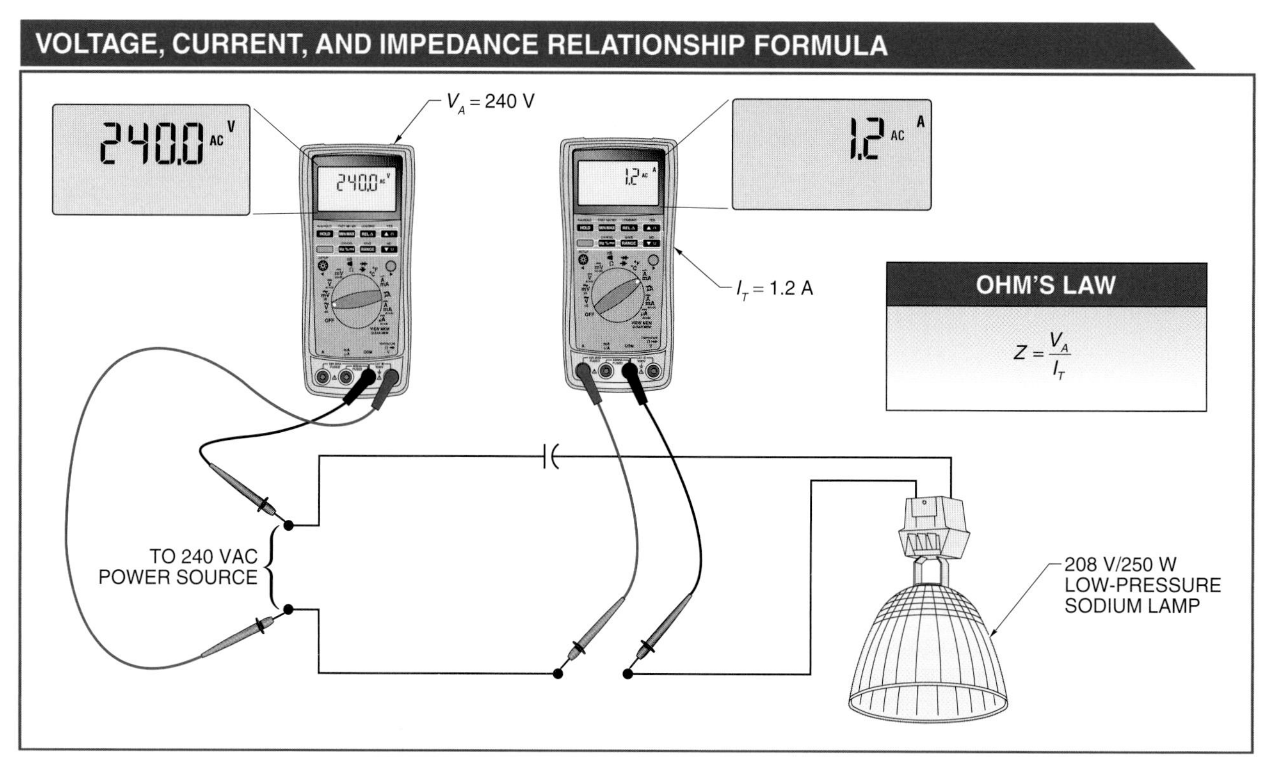

Figure 17-6. When the values of total current and source voltage are known, impedance in a series resistive-capacitive circuit can be calculated by using a variation of the voltage, current, and impedance relationship formula.

Example: What is the total impedance of a series resistive-capacitive circuit with a source voltage of 240 V and a lamp rated at 208 V/250 W?

1. Calculate total current.

$$I_T = \frac{P_R}{V_R}$$

$$I_T = \frac{250\text{ W}}{208\text{ V}}$$

$$I_T = 1.2\text{ A}$$

2. Calculate total impedance.

$$Z_T = \frac{V_A}{I_T}$$

$$Z_T = \frac{240 \text{ V}}{1.2 \text{ A}}$$

Z_T = **200 Ω**

When the values of circuit resistance and capacitive reactance are known, total impedance in a series resistive-capacitive circuit can be calculated by using the vector diagram method of calculation. **See Figure 17-7.** Using a vector diagram, total impedance is calculated by applying the following formula:

$$Z_T = \sqrt{R_T^2 + (-X_C)^2}$$

where

Z_T = total impedance (in Ω)

R_T = total resistance (in Ω)

X_C = capacitive reactance (in Ω)

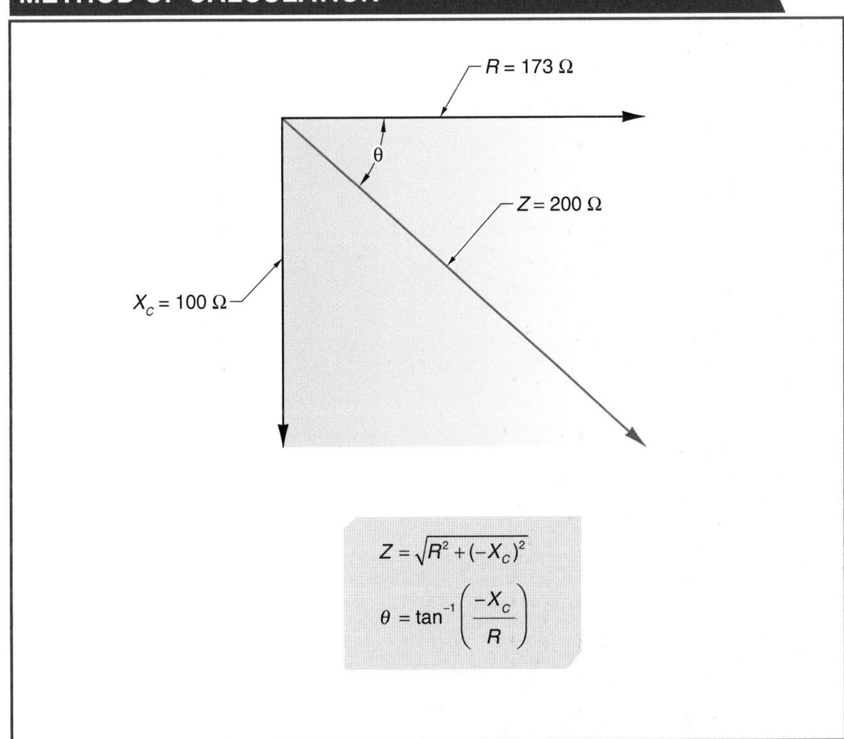

Figure 17-7. When the values of circuit resistance and capacitive reactance are known, impedance in a series resistive-capacitive circuit can be calculated by using the vector diagram method of calculation.

Example: What is the total impedance of a circuit that has a vector diagram indicating total resistance of 173 Ω and capacitive reactance of 100 Ω?

$$Z_T = \sqrt{R_T^2 + (-X_C)^2}$$

$$Z_T = \sqrt{173\ \Omega^2 + (-100\ \Omega)^2}$$

$$Z_T = \sqrt{29{,}929 + 10{,}000}$$

$$Z_T = \sqrt{39{,}929}$$

$$Z_T = \mathbf{199.82\ \Omega}$$

Tech Tip

Large capacitors that are used in televisions and other electronic equipment retain their charge indefinitely even after power is removed from the component. Capacitors that are removed from electronic equipment must be discharged to prevent damage to equipment and electrical shock.

Angle theta is the angle by which current leads the source voltage, as shown in a vector diagram. Angle theta can be calculated when the values of the impedance vector are known. Since the voltages in the voltage vector are divided by the common value of current, the resistive values of the impedance vector have the same ratio as that of the voltage vector. The value of the arctangent of the impedance vectors is angle theta. **See Appendix.** Angle theta is calculated by applying the following formula:

$$\theta = \tan^{-1}\left(\frac{-X_C}{R_T}\right)$$

where

θ = angle theta (in °)

X_C = capacitive reactance (in Ω)

R_T = total resistance (in Ω)

Example: What is the angle theta in a series resistive-capacitive circuit that has a vector diagram indicating total resistance of 173 Ω and capacitive reactance of 100 Ω?

$$\theta = \tan^{-1}\left(\frac{-X_C}{R_T}\right)$$

$$\theta = \tan^{-1}\left(\frac{-100}{173\ \Omega}\right)$$

$$\theta = \tan^{-1}(-0.5780)$$

$$\theta = \mathbf{-30°}$$

If angle theta is known, the cosine function can be used to confirm the value of impedance. When values of total resistance and angle theta are known, total impedance is calculated by applying the following formula:

$$Z_T = \frac{R_T}{\cos\theta}$$

where

Z_T = total impedance (in Ω)

R_T = total resistance (in Ω)

θ = angle theta (in °)

Example: What is the total impedance in a series resistive-capacitive circuit that has total resistance of 173 Ω and an angle theta of –30°?

$$Z_T = \frac{R_T}{\cos\theta}$$

$$Z_T = \frac{173\ \Omega}{\cos(-30°)}$$

$$Z_T = \frac{173\ \Omega}{0.866}$$

Z_T = **201.16** Ω

Ganged Capacitors

In certain applications such as radio tuning and broadcasting equipment, two or more variable capacitors are sometimes connected in series and must have their capacitance values vary simultaneously. In order to vary capacitance values of two or more variable capacitors from one control mechanism, capacitors are mechanically connected, or ganged, by running a common rotor through each capacitor. Ganged capacitors are always variable-type capacitors.

Power Vector Diagrams of Series Resistive-Capacitive Circuits

A power vector diagram of a series resistive-capacitive circuit can be used to calculate the apparent power of the circuit. When the values of source voltage and total current are known, apparent power in a series resistive-capacitive circuit can be calculated by multiplying the voltage and current. **See Figure 17-8.** Each voltage vector is multiplied by the circuit current value to calculate the true and reactive power. Apparent power in a series resistive-capacitive circuit is calculated by applying the following formula:

$$P_A = \sqrt{{P_{VAC}}^2 + {P_T}^2}$$

where

P_A = apparent power (in VA)

P_{VAC} = reactive power (in VAC)

P_T = true power (in W)

Example: What is the apparent power in a series resistive-capacitive circuit that has 144 VAC from a capacitor and dissipates 249.6 W of true power?

$$P_A = \sqrt{P_{VAC}^{\ 2} + P_T^{\ 2}}$$

$$P_A = \sqrt{144\text{ VAC}^2 + 249.6\text{ W}^2}$$

$$P_A = \sqrt{20{,}736 + 62{,}300}$$

$$P_A = \sqrt{83{,}086}$$

P_A = **288 VA**

True power in a series resistive-capacitive circuit is the actual power used in an electrical circuit. True power is used when calculating power factor. Only resistance can dissipate true power. True power in a series resistive-capacitive circuit is calculated by applying the following formula:

$$P_T = I_R \times V_R$$

where

P_T = true power (in W)

I_R = resistive current (in A)

V_R = resistive voltage (in V)

Example: What is the true power in a series resistive-capacitive circuit with resistive current of 1.2 A and resistive voltage of 208 V?

$P_T = I_R \times V_R$

P_T = 1.2 A × 208 V

P_T = **249.6 W**

Reactive power in a series resistive-capacitive circuit can be used with the true power vector to calculate the apparent power and is also used to calculate power factor corrections. Reactive power in a series resistive-capacitive circuit is calculated by applying the following formula:

$$P_{VAR} = I_C \times V_C$$

where

P_{VAR} = reactive power (in VAR)

I_C = capacitive current (in A)

V_C = capacitive voltage (in V)

Example: What is the reactive power in a series resistive-capacitive circuit with capacitive current of 1.2 A and capacitive voltage of 120 V?

$P_{VAR} = I_C \times V_C$

P_{VAR} = 1.2 A × 120 V

P_{VAR} = **144 VAR**

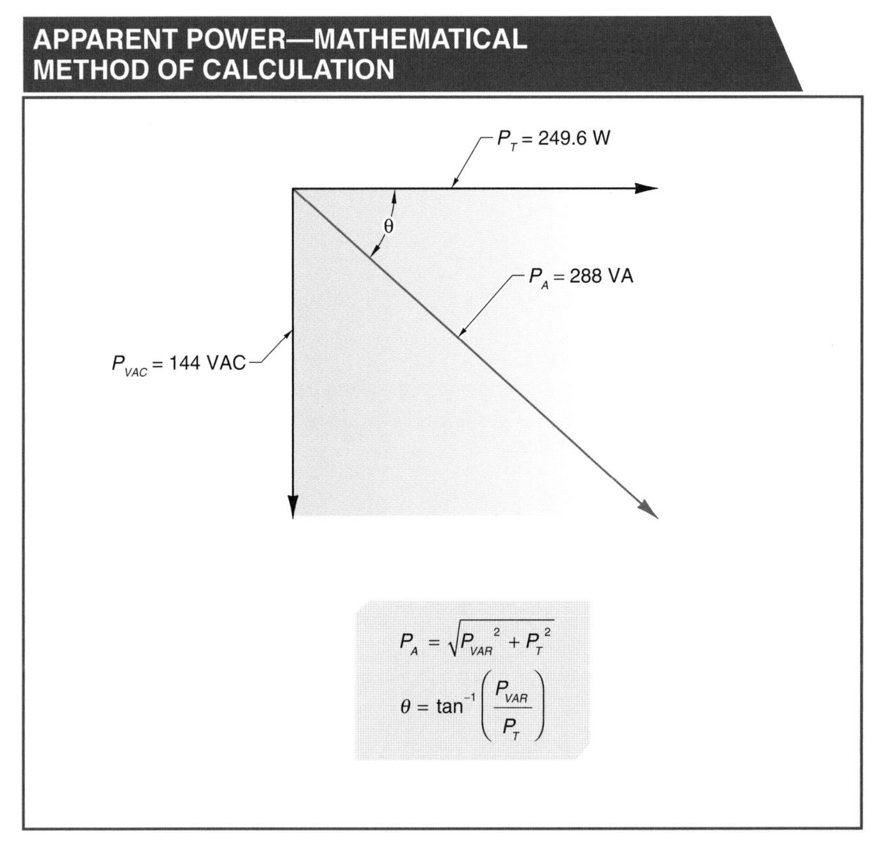

Figure 17-8. When the values of source voltage and total current are known, apparent power in a series resistive-capacitive circuit can be calculated by using the mathematical method of calculation.

Power Factor in Series Resistive-Capacitive Circuits

Power factor in a series resistive-capacitive circuit is similar to power in an inductive-resistive circuit because power factor is equal to the cosine of angle theta (cos θ) in both types of circuits. The difference between the two types of circuits is that in an inductive-resistive circuit, apparent power is lagging, indicating inductance (current lags source voltage). In a series resistive-capacitive circuit, apparent power leads current and source voltage, indicating capacitance (current leads source voltage). **See Figure 17-9.** Apparent power leads current and voltage in a series resistive-capacitive circuit because of the effect of capacitive voltage present in the circuit. The values of apparent power and true power affect the value of power factor. Power factor in a series resistive-capacitive circuit is calculated by applying the following formula:

$$PF = \frac{P_T}{P_A} \times 100$$

where

PF = power factor (in %)

P_T = true power (in W)

P_A = apparent power (in VA)

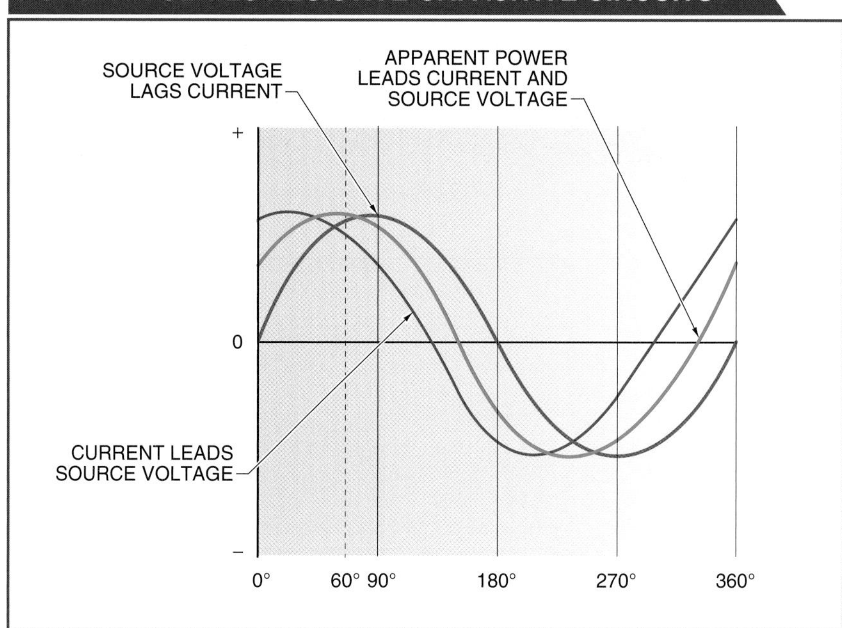

Figure 17-9. In a series resistive-capacitive circuit, current leads source voltage, indicating that the circuit is capacitive.

Example: What is the power factor in a series resistive-capacitive circuit that has true power of 249.6 W and apparent power of 288 VA?

$$PF = \frac{P_T}{P_A} \times 100$$

$$PF = \frac{249.6 \text{ W}}{288 \text{ VA}} \times 100$$

$$PF = 0.867 \times 100$$

PF = **86.7%**

Power factor is not an angular measure but a numerical ratio, with a value between 0 and 1, or 0% and 100%, that is also equal to the cosine of angle theta.

Frequency in Series Resistive-Capacitive Circuits

In a series resistive-capacitive circuit, changes in frequency can cause changes in circuit parameters such as impedance, voltage, power, and angle theta. **See Figure 17-10.** A change in frequency causes these parameters to change because of capacitive reactance. Resistance and source voltage are not affected by changes in frequencies. When frequency is changed, the only parameter directly affected is the capacitive reactance, while the other parameters are indirectly affected by the change.

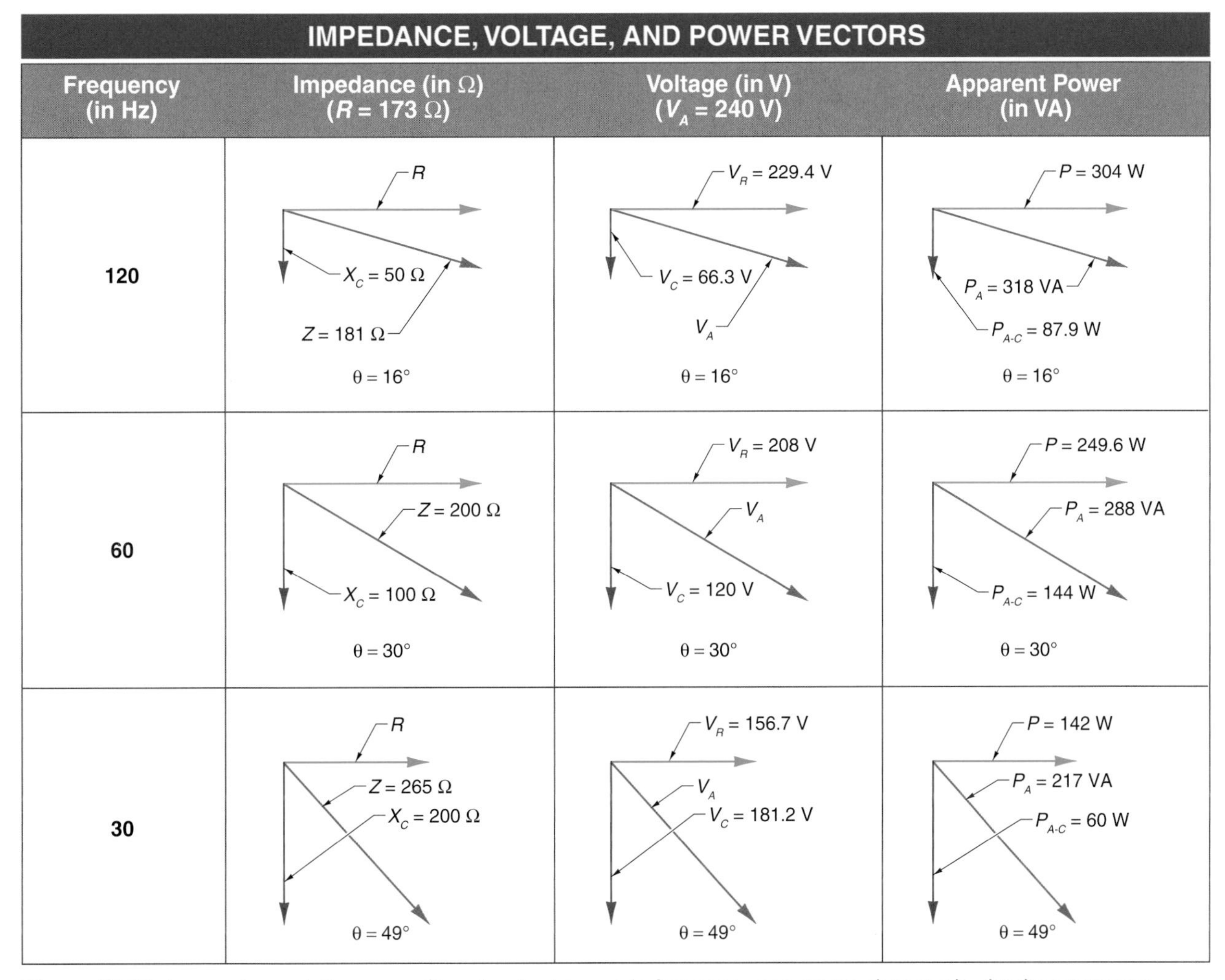

Figure 17-10. In a series resistive-capacitive circuit, changes in frequency can cause changes in circuit parameters such as impedance, voltage, apparent power, and angle theta.

For example, when frequency is doubled in a series resistive-capacitive circuit, capacitive reactance is reduced by 50%. The change in reactance reduces the impedance in the circuit. Since total impedance is reduced, source voltage is held constant, and total current increases. An increase in total current results in a greater voltage drop across the resistance, and more true power is dissipated. Angle theta becomes smaller, and the power factor of the circuit is increased.

When frequency is reduced, capacitive reactance is increased. An increase in capacitive reactance results in a higher circuit impedance and reduced circuit current. A reduction in total current causes less voltage to be dropped across the resistance, resulting in less true power dissipation. Angle theta becomes larger, and the power factor of the circuit is reduced.

Capacitance in Series Resistive-Capacitive Circuits

Capacitance in series resistive-capacitive circuits is inversely proportional to changes in frequency. Increasing the amount of capacitance reduces the amount of capacitive reactance. Reducing the amount of capacitance increases the amount of capacitive reactance. Capacitive reactance in a circuit is indirectly proportional to the frequency and the capacitance.

The value of capacitance can be calculated in a series resistive-capacitive circuit if the values of frequency and capacitive reactance are known. Capacitance in a series resistive-capacitive circuit is calculated by applying the following formula:

$$C = \frac{1}{2\pi \times f \times X_C}$$

where

C = capacitance (in F)

2π = 6.28 (in radians per cycle)

f = frequency (in Hz)

X_C = capacitive reactance (in Ω)

Example: What is the capacitance in a series resistive-capacitive circuit with a frequency of 60 Hz and a capacitive reactance of 100 Ω?

$$C = \frac{1}{2\pi \times f \times X_C}$$

$$C = \frac{1}{6.28 \times 60 \text{ Hz} \times 100 \ \Omega}$$

$$C = \frac{1}{37{,}680}$$

C = **2.7 × 10^{-5} F (27 μF)**

PARALLEL RESISTIVE-CAPACITIVE CIRCUITS

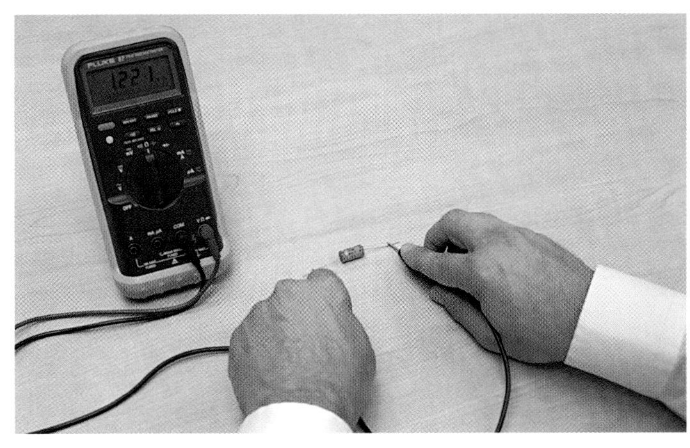

A DMM set to measure resistance is used to test capacitors with values of 0.5 µF and higher.

Voltage is the same across all components connected in parallel in resistive-capacitive circuits. For this reason, voltage is used as a reference point for analyzing parallel resistive-capacitive circuits. In a parallel resistive-capacitive circuit, resistive branch circuit current is in phase with voltage, and capacitive branch circuit current leads voltage by 90°. **See Figure 17-11.** Waveform diagrams are used to show the source voltage beginning at 0° and ending at 360°. Resistance opposes current flow, and resistive current is in phase with the source voltage. Capacitance opposes the change in voltage, and the current for the capacitor is shown leading the voltage by 90°. Parallel resistive-capacitive circuits are used in electronic AC applications such as computers, CD-ROM players, DVD players, DVR equipment, amplifiers, stereo receivers, oscilloscopes, and video production equipment. Factors that affect resistive-capacitive circuits include current, impedance, frequency, and capacitance.

Oscilloscopes

Oscilloscopes provide examples of both AC and DC parallel resistive-capacitive circuits. Oscilloscopes translate an electronic signal into a pattern or waveform on a screen. As the signal is traced across the screen, the waveform creates a signature of the signal's characteristics. Specifications for oscilloscopes include bandwidth, number of input channels, number of inputs, and screen resolution.

Oscilloscopes are available with many different features. Some types of oscilloscopes have a relay or switch output for limit detection or other types of signaling. Other types of oscilloscopes include portable oscilloscopes, which are powered by a replaceable or rechargeable battery and are designed to be used while held in one hand. Oscilloscopes that are rated for high-power applications can monitor and/or display currents and voltages associated with electrical power or high-power switching. Typically, these currents and voltages are much higher than standard sensor signal levels. Oscilloscopes can include a hard drive, nonvolatile memory, or on-board random-access memory (RAM) for data storage. Removable storage media devices such as tapes, diskettes, and PCMCIA cards are also available.

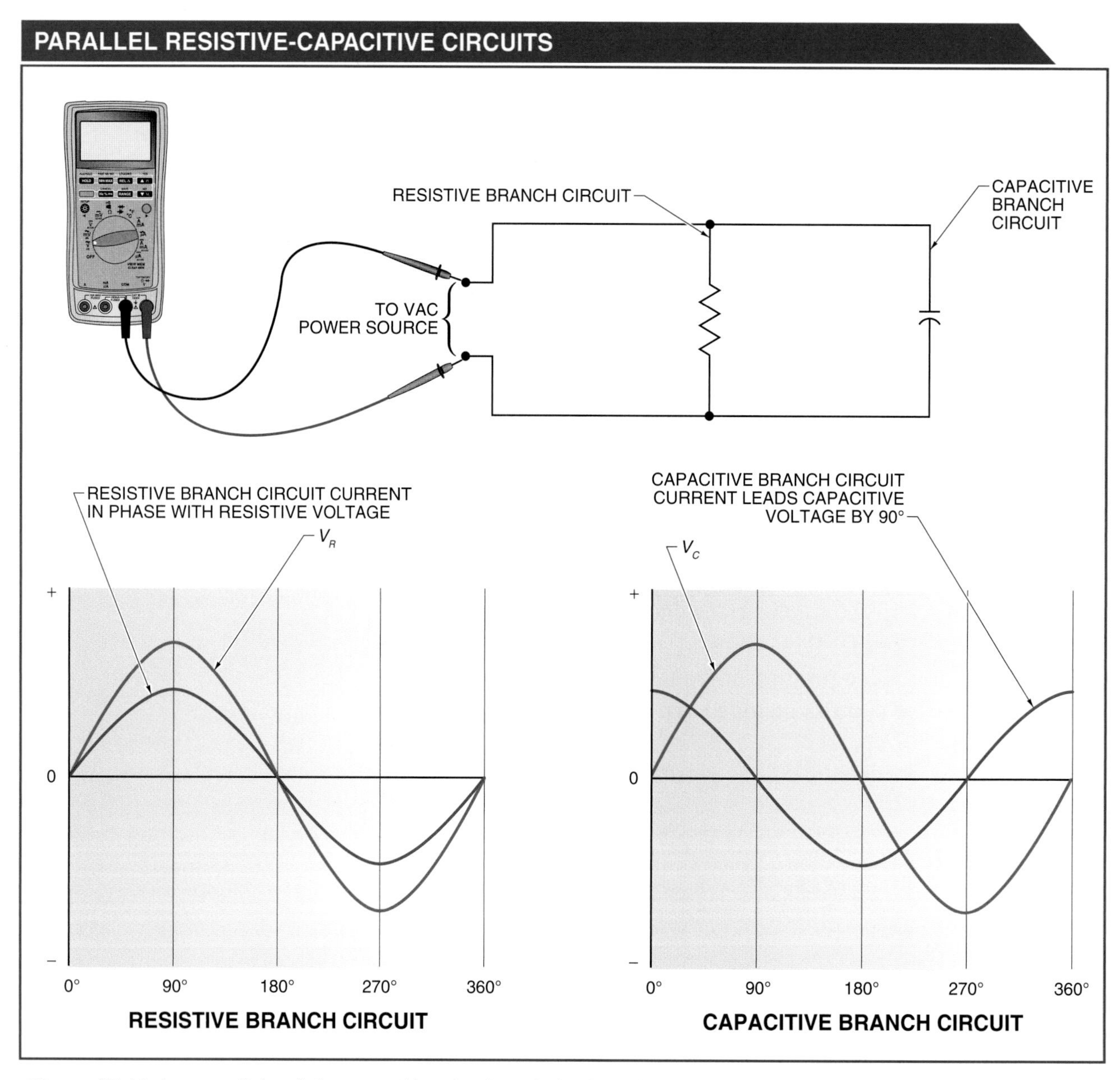

Figure 17-11. In a parallel resistive-capacitive circuit, resistive branch circuit current is in phase with voltage, and capacitive branch circuit current leads voltage by 90°.

Current in Parallel Resistive-Capacitive Circuits

In a parallel resistive-capacitive circuit, the algebraic sum of multiple currents is equal to the source current (total current). When the amount of current in each branch circuit is known, a current vector diagram can be constructed to solve for the value of the source current using the vector diagram method. **See Figure 17-12.** The Pythagorean theorem can also be used to calculate the source current. Source current is calculated by applying the following formula:

$$I_T = \sqrt{I_R^2 + I_C^2}$$

where

I_T = source current (in A)
I_R = resistive current (in A)
I_C = capacitive current (in A)

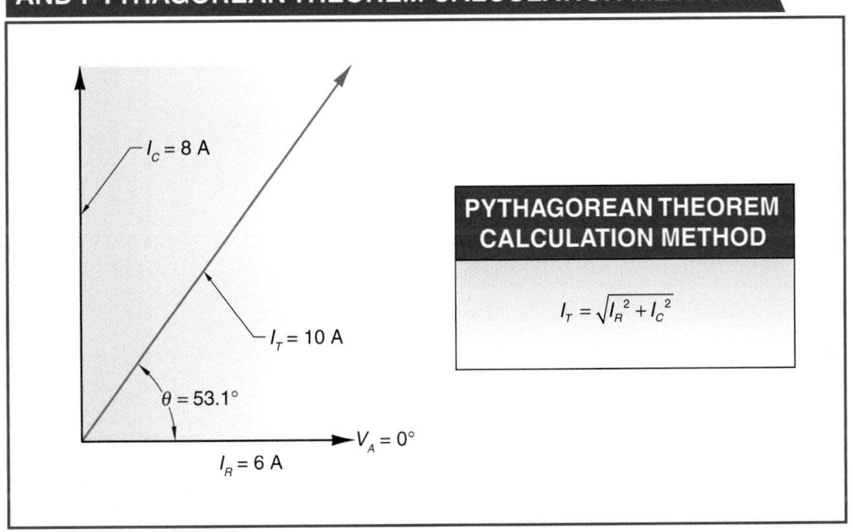

Figure 17-12. When the value of each branch circuit current is known, a current vector diagram can be constructed to solve for the value of the source current by using the vector diagram or pythagorean theorem calculation methods.

Example: What is the source current in a parallel resistive-capacitive circuit with resistive current of 6 A and capacitive current of 8 A?

$$I_T = \sqrt{I_R^2 + I_C^2}$$

$$I_T = \sqrt{(6\text{ A})^2 + (8\text{ A})^2}$$

$$I_T = \sqrt{36 + 64}$$

$$I_T = \sqrt{100}$$

$$I_T = \mathbf{10\ A}$$

Angle theta can be calculated by using the arctangent ($\tan^{-1}$). Angle theta is calculated by applying the following formula:

Tech Tip

When testing a capacitor with an ohmmeter or DMM, the voltage of the test instrument's source voltage should not exceed the voltage rating of the capacitor under test.

$$\theta = \tan^{-1}\left(\frac{I_C}{I_R}\right)$$

where

θ = angle theta (in °)

I_C = capacitive current (in A)

I_R = resistive current (in A)

Example: What is the angle theta in a parallel resistive-capacitive circuit with resistive current of 6 A and capacitive current of 8 A?

$$\theta = \tan^{-1}\left(\frac{I_C}{I_R}\right)$$

$$\theta = \tan^{-1}\left(\frac{8\text{ A}}{6\text{ A}}\right)$$

$$\theta = \tan^{-1}(1.333)$$

$$\theta = \mathbf{53.1°}$$

Note: Angle theta is a positive value since the circuit current of 10 A leads the source voltage by 53.1°.

Because the angle theta must be calculated to determine a vector, trigonometric functions can be used in place of the Pythagorean theorem to determine total current. Once angle theta is calculated as 53.1°, current can be calculated by applying the following formula:

$$I_T = \frac{I_R}{\cos\theta}$$

where

I_T = source current (in A)

I_R = resistive current (in A)

θ = angle theta (in °)

Example: What is the source current in a parallel resistive-capacitive circuit with a resistive current of 6 A and angle theta of 53.1°?

$$I_T = \frac{I_R}{\cos\theta}$$

$$I_T = \frac{6\text{ A}}{\cos(53.1°)}$$

$$I_T = \frac{6\text{ A}}{0.600}$$

$$I_T = \mathbf{10\ A}$$

Impedance in Parallel Resistive-Capacitive Circuits

When values of total current and source voltage are known, the simplest method of determining impedance in a parallel resistive-capacitive circuit is by application of Ohm's law. Ohm's law is used to calculate impedance by dividing the source voltage by the total circuit current. **See Figure 17-13.** Impedance in a parallel resistive-capacitive circuit is calculated by applying the following formula:

$$Z = \frac{V_A}{I_T}$$

where

Z = impedance (in Ω)

V_A = source voltage (in V)

I_T = total current (in A)

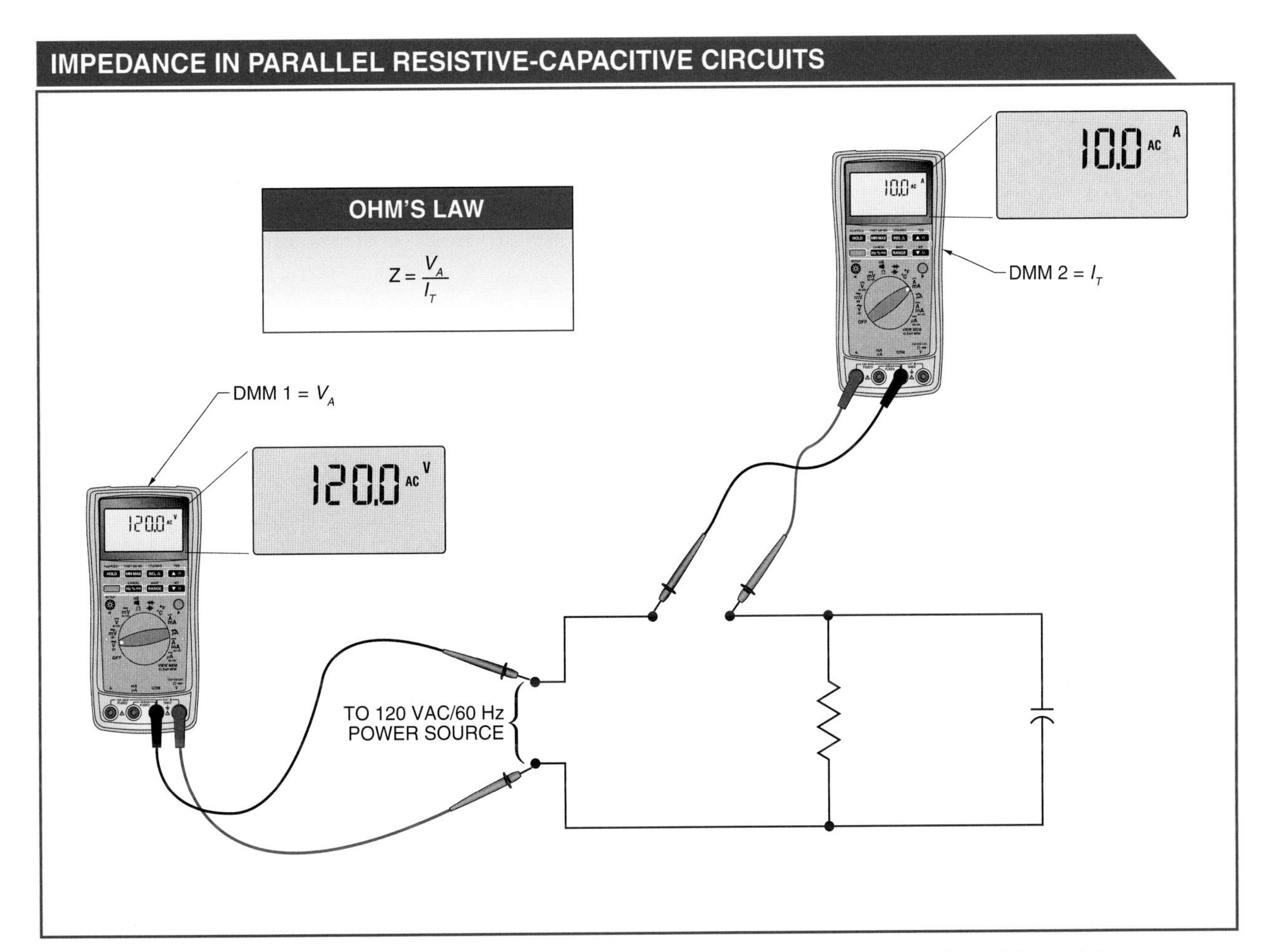

Figure 17-13. When values of total current and source voltage are known, the simplest method of determining impedance in a parallel resistive-capacitive circuit is by application of Ohm's law.

Example: What is the impedance of a parallel inductive-resistive circuit with a source voltage of 120 V and total current of 10 A that leads the source voltage by 53.1°?

$$Z = \frac{V_A}{I_T}$$

$$Z = \frac{120\text{ V}}{10\text{ A at }53.1°}$$

$Z =$ **12 Ω at –53.1°**

When the reactive and resistive values of the components are given, impedance in a parallel resistive-capacitive circuit can be calculated by using the product-over-sum method. Impedance in a parallel resistive-capacitive circuit is calculated with the product-over-sum method by applying the following formula:

$$Z = \frac{R \times X_C}{\sqrt{R^2 + X_C{}^2}}$$

where

Z = circuit impedance (in Ω)

R = circuit resistance (in Ω)

X_C = capacitive reactance (in Ω)

Example: What is the impedance of an AC circuit with a 20 Ω resistance connected in parallel with a capacitance that has reactance of 15 Ω?

$$Z = \frac{R \times X_C}{\sqrt{R^2 + X_C{}^2}}$$

$$Z = \frac{20\ \Omega\text{ at }0° \times 15\ \Omega\text{ at } -90°}{\sqrt{(20\ \Omega\text{ at }0°)^2 + (15\ \Omega\text{ at } -90°)^2}}$$

$$Z = \frac{20\ \Omega\text{ at }0° \times 15\ \Omega\text{ at } -90°}{\sqrt{400 + 225\text{ at } -36.9°}}$$

$$Z = \frac{20\ \Omega\text{ at }0° \times 15\ \Omega\text{ at } -90°}{\sqrt{625\text{ at } -36.9°}}$$

$$Z = \frac{300\ \Omega\text{ at } -90°}{25\ \Omega\text{ at } -36.9°}$$

$Z =$ **12 Ω at –53.1°**

Note: The different phase angles must be taken into consideration when reactance is present. This is done by using vector algebra. When multiplying vectors the angles are added, and when dividing vectors the angles are subtracted. When adding two vectors the angle is determined by using the arc tangent function.

Frequency in Parallel Resistive-Capacitive Circuits

In a parallel resistive-capacitive circuit, a change in frequency causes changes in circuit parameters such as capacitive reactance and reactive power. Frequency has no effect on resistance or source voltage. Therefore, resistive current remains constant. **See Figure 17-14.**

Since resistive current does not change, the true power in the circuit remains the same when the frequency is changed. Capacitive reactance changes each time the frequency changes. A change in capacitive reactance results in different capacitive current in each branch circuit and causes current to increase as frequency increases. A change in current in the capacitive branch circuit causes the line current to change, as well as the apparent power. Reactive power in the capacitive branch circuit changes in proportion to the line current change.

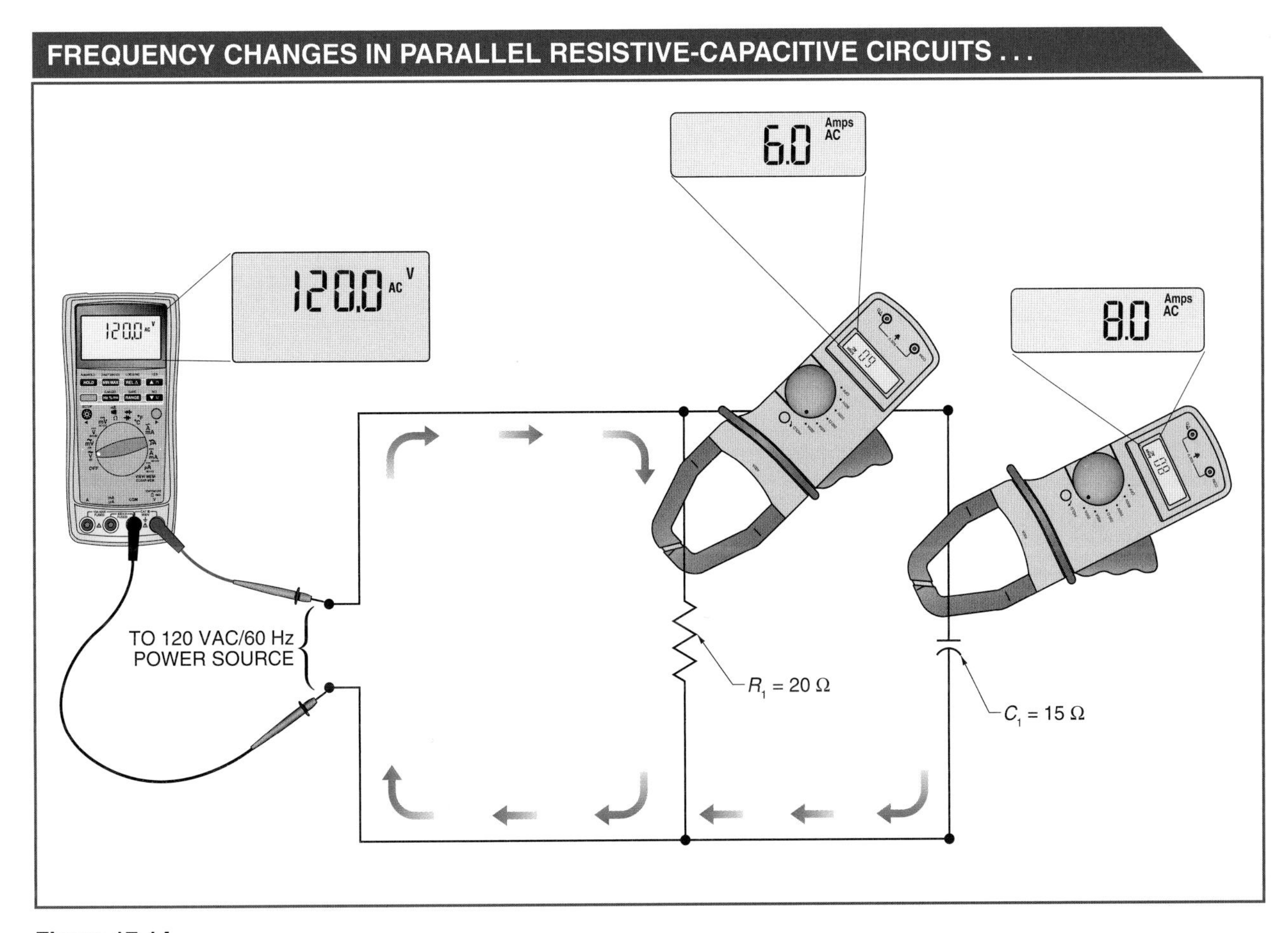

Figure 17-14. . .

. . . FREQUENCY CHANGES IN PARALLEL RESISTIVE-CAPACITIVE CIRCUITS

CURRENT AND POWER VECTORS		
Frequency (in Hz)	Current (at 6 A)	Power (at 120 W)
120	V_A = 120 V X_C = 7.5 Ω 17 A 16 A θ = 70° 6 A V_A	V_A = 120 V X_C = 7.5 Ω 2050 VA 1920 VAR θ = 70° 120 W V_A
60	V_A = 120 V X_C = 15 Ω 10 A 8 A θ = 55° 6 A V_A	V_A = 120 V X_C = 15 Ω 1200 VA 960 VAR θ = 55° 120 W V_A
30	V_A = 120 V X_C = 30 Ω 7 A 4 A θ = 30° 6 A V_A	V_A = 120 V X_C = 30 Ω 840 VA 480 VAR θ = 30° 120 W V_A

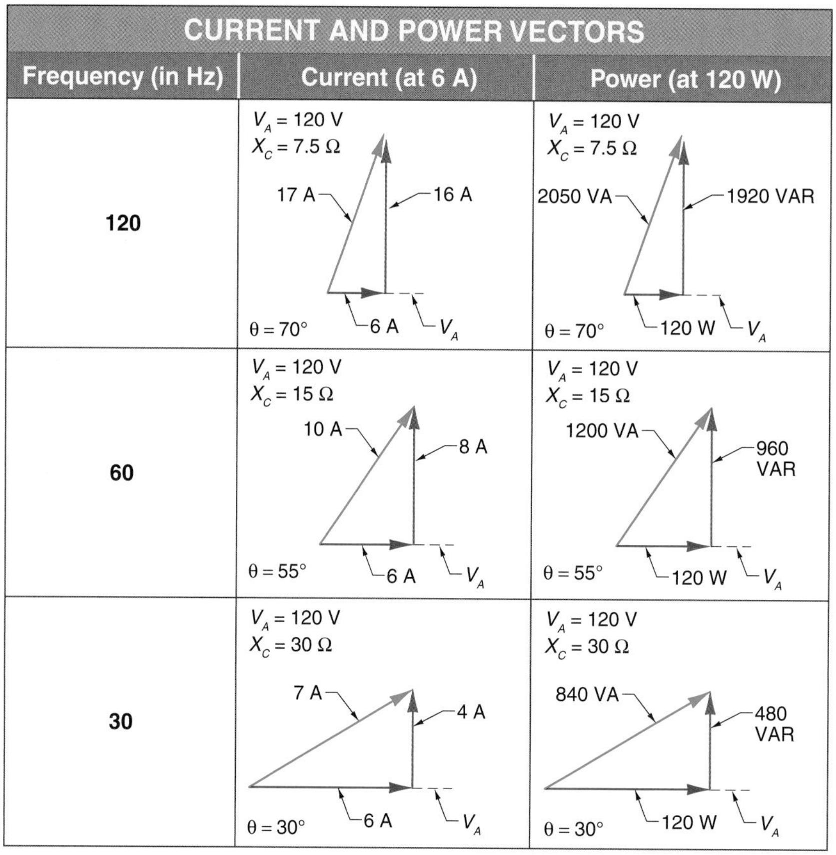

. . . Figure 17-14. In a parallel resistive-capacitive circuit, a change in frequency causes changes in circuit parameters such as capacitive reactance and VAC but does not affect resistance or source voltage.

Tech Tip

The most common type of electronic circuit is a resistor and capacitor connected in series (RC series circuit).

Capacitance in Parallel Resistive-Capacitive Circuits

Capacitance in parallel resistive-capacitive circuits is affected by changes in frequency and capacitive reactance. Capacitive reactance is inversely proportional to the frequency. Changes in capacitance will also change the capacitive reactance in an inversely proportional manner. For example, if the capacitive reactance of a circuit is 100 Ω at 60 Hz, it will change to 200 Ω at 30 Hz and 50 Ω at 120 Hz. In order to determine capacitance in a circuit, the values of frequency and capacitive reactance must be known. Capacitance in a parallel resistive-capacitive circuit is calculated by applying the following formula:

$$C = \frac{1}{2\pi \times f \times X_C}$$

where

C = capacitance (in F)

2π = 6.28 (in radians per cycle)

f = frequency (in Hz)

X_C = capacitive reactance (in Ω)

Example: What is the capacitance in a parallel resistive-capacitive circuit with a frequency of 60 Hz and a capacitive reactance of 15 Ω?

$$C = \frac{1}{2\pi \times f \times X_C}$$

$$C = \frac{1}{6.28 \times 60 \text{ Hz} \times 15 \text{ }\Omega}$$

$$C = \frac{1}{5652}$$

C = **177 μF**

Changes in circuit parameters in both series and parallel resistive-capacitive circuits vary with an increase or decrease in frequency or capacitance. **See Figure 17-15.**

FREQUENCY AND CAPACITANCE CHANGES IN RESISTIVE-CAPACITIVE CIRCUITS				
Parameter	Increase in Frequency or Capacitance		Decrease in Frequency or Capacitance	
	Series	Parallel	Series	Parallel
X_C	↓	↓	↑	↑
I_T	↑	↑	↓	↓
θ	↓	↑	↑	↓
Z	↓	↓	↑	↑
V_R	↑	–	↓	–
V_C	↓	–	↑	–
I_R	↑	–	↓	–
I_C	↑	↑	↓	↓
$\cos\theta$	↑	↓	↓	↑

Figure 17-15. Values of circuit parameters in both series and parallel resistive-capacitive circuits vary with an increase or decrease in frequency or capacitance.

SERIES/PARALLEL RESISTIVE-CAPACITIVE CIRCUITS

Series/parallel resistive-capacitive circuits are circuits that contain both resistances and capacitances connected in series and parallel. This type of circuit is used in many types of electronics applications. Series/parallel resistive-capacitive circuits are analyzed in a manner similar to that used for any series/parallel circuit. With a series/parallel resistive-capacitive circuit, the simplest manner in which to analyze the circuit is to reduce the circuit to a single load with a single power source. Parallel combinations are reduced to their series-equivalent circuits so that the components can be added together directly. In a series/parallel resistive-capacitive circuit, current is used as the reference point for series components, and voltage is used as the reference point across parallel branches. The reference point is used to analyze the circuit. In a series/parallel resistive-capacitive circuit, resistive voltage drop is in phase with current (I_R) while capacitive voltage drop lags current (I_C) by 90°. **See Figure 17-16.**

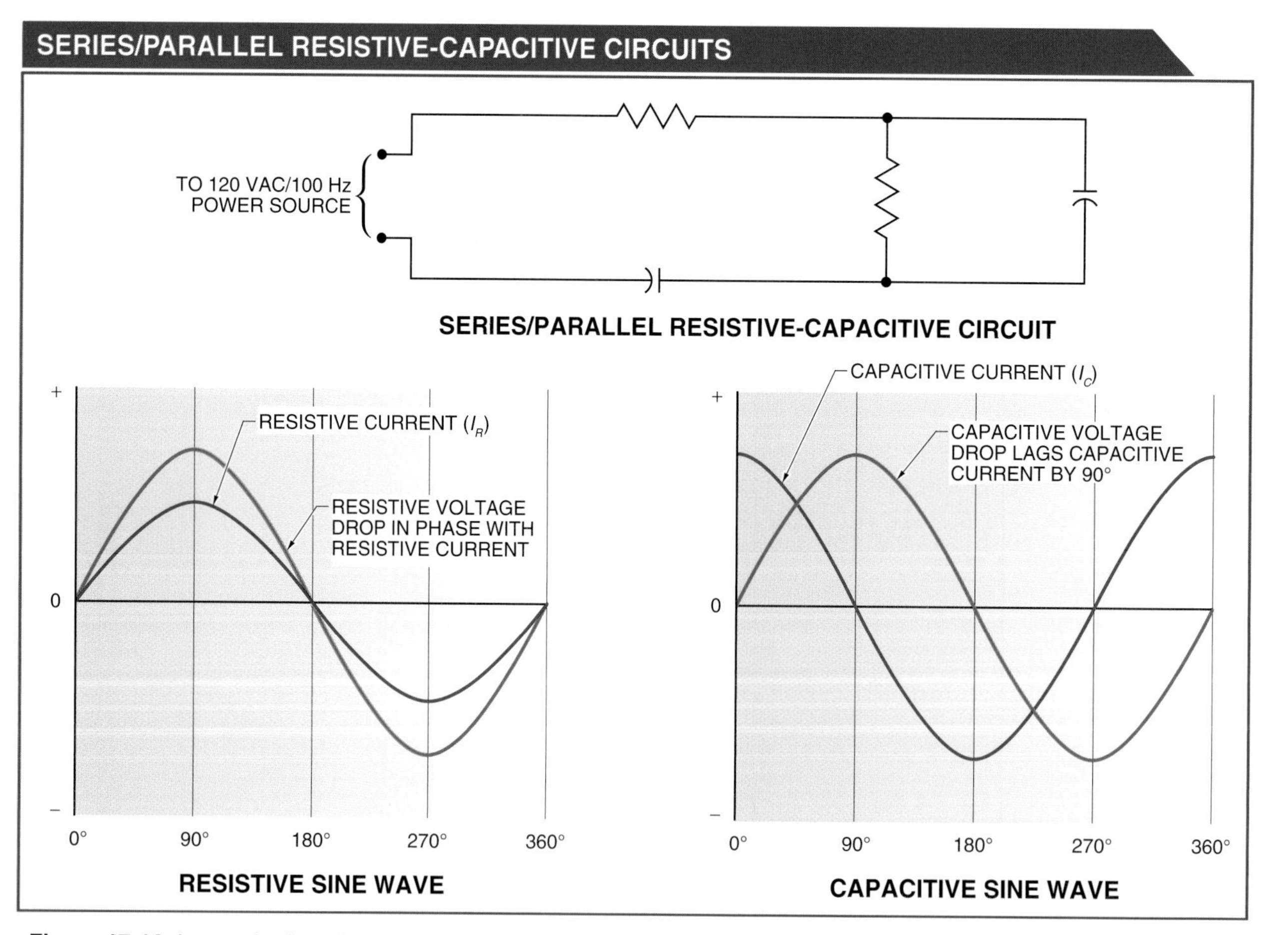

Figure 17-16. In a series/parallel resistive-capacitive circuit, resistive voltage drop is in phase with current (I_R), while capacitive voltage drop lags current (I_C) by 90°.

SUMMARY

Capacitance is the ability to store a charge and to oppose any change in voltage. In a capacitive circuit, circuit current leads the source voltage. Capacitive reactance is the opposition to current flow in an AC circuit due to the presence of capacitance. Capacitive reactance is inversely proportional to frequency and capacitance.

A pure capacitive circuit does not have any resistance in the circuit. A pure capacitive circuit has only an AC source and a single capacitive load. Capacitive reactance in a pure capacitive circuit is the total opposition to the current flow. In a pure capacitive circuit, current leads voltage by 90°. The amount of energy that is required to charge a capacitor is returned to the circuit. There is only apparent power in a pure capacitive circuit. No true power is consumed, and the power factor is equal to 0%. Only resistance consumes electrical power.

Resistive-capacitive circuits can be series, parallel, or series-parallel (combination) circuits. In a series circuit, current is used as the reference point. In a parallel circuit, voltage is used as the reference point. In a series-parallel circuit, current is used as the reference point for series components, and voltage is used as the reference point across parallel branches. The voltage drop across all resistances is in phase with the current through them. The voltage drop across all capacitances lags the current by 90°.

Terms

Bias tee is a three-port device that combines and separates radio frequency (RF) signals and direct current (DC) voltage.

APPLICATION: BIAS TEES...

Background: A *bias tee* is a three-port device that combines and separates radio frequency (RF) signals and direct current (DC) voltage. A bias tee is used to distribute a DC voltage with an alternating current (AC) signal. Some bias tee applications include remotely powering antenna amplifiers, cable equipment, Ethernet-connected equipment, and photodiodes. A bias tee consists of a capacitor and inductor. **See Figure 1.**

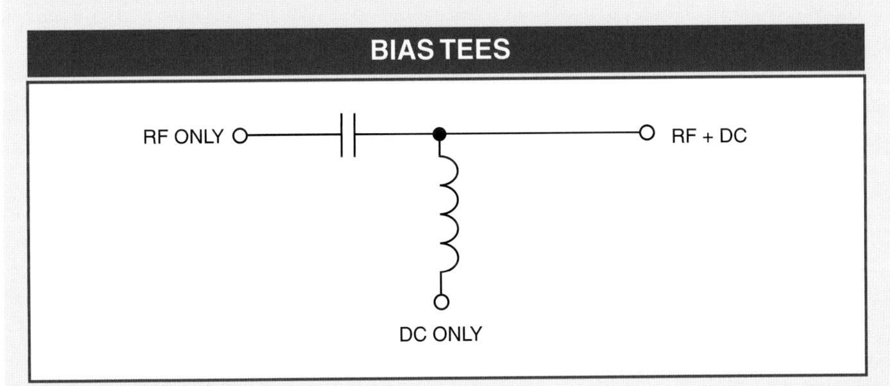

Figure 1. A bias tee is a three-port device that combines and separates radio frequency (RF) signals and direct current (DC) voltage.

...APPLICATION: BIAS TEES...

The bias tee capacitor is used to block DC on the RF-only port, while the inductor blocks AC on the DC-only port. The RF + DC port has the combined signals. Two bias tees are used to insert and remove DC from the AC signal path, which is typically a 50 Ω or 75 Ω coaxial cable. **See Figure 2.** Bias tees can also work over a range of AC frequencies.

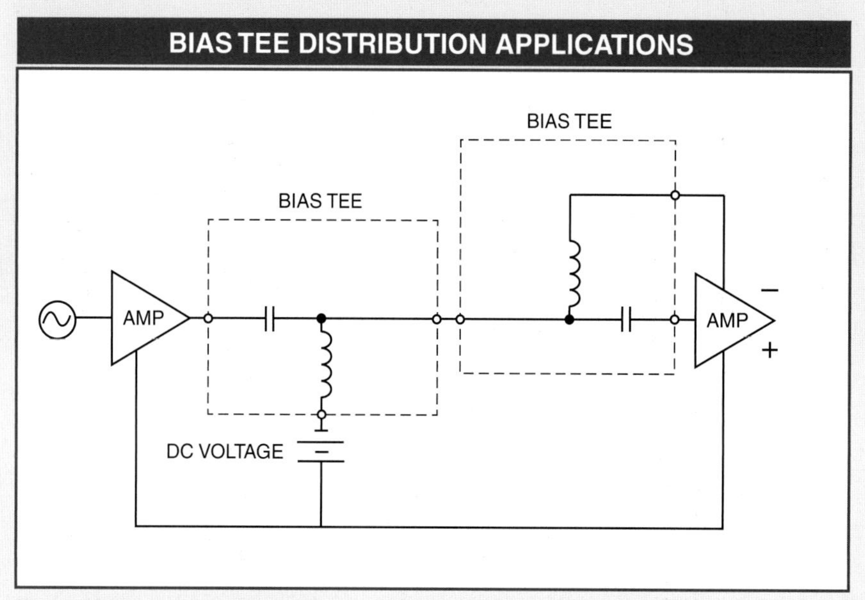

Figure 2. Bias tees are used to distribute AC signals and DC voltages to remote devices.

A bias tee is designed for the system characteristic impedance, Z_0. The capacitor impedance (X_C) is selected to be approximately 100 times smaller than the system characteristic impedance. The inductor impedance (X_L) is to be approximately 100 times larger than the system characteristic impedance. Since bias tees are sometimes designed to work over a range of frequencies, when calculating the cap active and inductive impedances, the lowest operating frequency is used to minimize AC signal impact.

Key Points: Capacitance reactance, inductor reactance

Problem: A 10 MHz AC signal with DC power must be transmitted over a cable television 75 Ω coaxial cable. Calculate the bias-tee capacitor and inductor values.

Solution: The coaxial cable characteristic impedance is 75 Ω. The desired capacitor reactance is 0.75 Ω (100 times smaller than Z_0), and the desired inductor reactance is 7500 Ω (100 times larger than Z_0). The lowest operating frequency is 10 MHz.

The capacitor value is found by using the following capacitive reactance formula:

$$X_C = \frac{1}{2 \times \pi \times f \times C}$$

where

X_C = capacitive reactance (in Ω)
π = 3.14 (constant)
f = frequency (in Hz)
C = capacitance (in F)

...APPLICATION: BIAS TEES

$$X_C = \frac{1}{2 \times \pi \times f \times C}$$

$$C = \frac{1}{2 \times \pi \times f \times X_C}$$

$$C = \frac{1}{6.28 \times 10{,}000{,}000 \text{ Hz} \times 0.75\ \Omega}$$

$$C = 21.2 \text{ pF}$$

The inductor value is found by using the following inductive reactance formula:

$$X_L = 2 \times \pi \times f \times L$$

where
X_L = inductive reactance (in Ω)
π = 3.14 (constant)
f = frequency (in Hz)
L = inductance (in H)

$$X_L = 2 \times \pi \times f \times L$$

$$L = \frac{X_L}{2 \times \pi \times f}$$

$$L = \frac{7500\ \Omega}{6.28 \times 10{,}000{,}000 \text{ Hz}}$$

$$L = 119.4\ \mu\text{H}$$

For a cable television coaxial cable that has a characteristic impedance of 75 Ω, the bias-tee network needed to transmit a 10 MHz AC signal with DC voltage uses a 21.2 pF capacitor and 119.4 μH inductor.

Chapter Review...

1. What is the impedance of a parallel inductive-resistive circuit with a source voltage of 208 V and total current of 12 A that leads the source voltage by 67.38°?
2. What is the impedance of an AC circuit with a 17 Ω resistance connected in parallel with a capacitance that has reactance of 10 Ω? (*Note:* Use source voltage vector from question 1.)
3. What is the source current in a parallel resistive-capacitive circuit with resistive current of 2 A and capacitive current of 6 A?
4. What is the source voltage in a series resistive-capacitive circuit that has resistive voltage of 135 V across a lamp and capacitive voltage of 111 V?
5. What is the capacitance in a parallel resistive-capacitive circuit with a frequency of 50 Hz and a capacitive reactance of 12 Ω?
6. Describe capacitive reactance.
7. What is the angle theta in a parallel resistive-capacitive circuit with resistive current of 5 A and capacitive current of 12 A?
8. What is the difference between a series resistive-capacitive circuit and an inductive-resistive circuit?
9. What is the value of reactive power in a capacitive series circuit that has total current of 0.67 A and source voltage of 220 V?
10. What is the source voltage in a series resistive-capacitive circuit that has resistive voltage of 120 V and an angle theta of –48.37°?
11. What is the capacitive reactance in an AC circuit with a capacitive voltage of 208 V and capacitive current of 6 A?
12. What is the apparent power in a series resistive-capacitive circuit that has 115 VAC from a capacitor and dissipates 227.9 W of true power?
13. What is the true power in a series resistive-capacitive circuit with resistive current of 5.7 A and resistive voltage of 240 V?
14. What is the reactive power in a series resistive-capacitive circuit with capacitive current of 5.7 A and capacitive voltage of 120 V?
15. What is the capacitive reactance of a 2.0 μF (1×10^{-6} F) capacitance at a frequency of 60 Hz?
16. What is the capacitance in a series resistive-capacitive circuit with a frequency of 100 Hz and a capacitive reactance of 150 Ω?
17. What is the total impedance of a series resistive-capacitive circuit with a source voltage of 135 V and a lamp rated at 120 V/75 W?
18. What is the total impedance of a circuit that has a vector diagram indicating total resistance of 224 Ω and capacitive reactance of 120 Ω?
19. Describe a pure capacitive circuit.
20. What is the capacitive voltage of a closed-loop circuit with a source voltage of 120 V?
21. What is the angle theta in a series resistive-capacitive circuit that has a vector diagram indicating total resistance of 224 Ω and capacitive reactance of 120 Ω?

...Chapter Review

22. What is the total impedance in a series resistive-capacitive circuit that has total resistance of 224 Ω and an angle theta of –28.18°?
23. What parameters are directly affected when frequency is changed in a series resistive-capacitive circuit?
24. What is the total current in a pure capacitive circuit that has a 20 μF capacitance connected to a 120 VAC/60 Hz power source?
25. What is the source current in a parallel resistive-capacitive circuit with a resistive current of 5 A and angle theta of 67.38°?
26. What is the angle theta in a series resistive-capacitive circuit that has resistive voltage of 120 V and capacitive voltage of 135 V?
27. What is the unit of measure for capacitive reactance and inductive reactance?
28. List four factors that affect resistive-capacitive circuits.
29. What is the power factor in a series resistive-capacitive circuit that has true power of 1368 W and apparent power of 256.88 VA?
30. What is the capacitance in a parallel resistive-capacitive circuit with a frequency of 50 Hz and a capacitive reactance of 12 Ω?

18

Inductive-Resistive-Capacitive Circuits

OBJECTIVES

- Perform series inductive-resistive-capacitive circuit analysis.
- Perform parallel inductive-resistive-capacitive circuit analysis.
- Perform series/parallel inductive-resistive-capacitive circuit analysis.

In most practical circuits, the properties of resistance, capacitance, and inductance are found. In every conductor there is resistance due to the substance, length, cross-sectional area, and temperature of the conductor. All commonly used conductors have resistance. Capacitance is present because the conductor is separated by insulation from other conductors or earth. Inductance is present along the length of the conductor due to the expanding and collapsing magnetic field caused by the current flow through the conductor.

Frequency, inductance, and capacitance are combined in AC circuits to achieve certain effects such as signal filtering or power factor correction. Both series and parallel inductive-resistive-capacitive circuits are analyzed using graphical and mathematical methods. These methods are combined to calculate inductive-resistive-capacitive series and parallel circuit parameters such as total line current, voltage drop across components, circuit impedance, component power, and power factor.

Terms

A **series inductive-resistive-capacitive (LRC) circuit** is a circuit where all inductive, resistive, and capacitive circuit elements are connected such that there is only one current path.

SERIES INDUCTIVE-RESISTIVE-CAPACITIVE CIRCUIT ANALYSIS

A *series inductive-resistive-capacitive (LRC) circuit* is a circuit where all inductive, resistive, and capacitive circuit elements are connected such that there is only one current path. When a load is a series inductive-resistive combination, current flow is initially opposed by the counter-electromotive force (CEMF) across an inductor. Current then decreases over five time constants to the minimum value, as limited by the resistance. When a load is a series resistive-capacitive combination, current flow is initially at its maximum value as limited by the resistance, and in five time constants decreases to 0 A, as the charge on the capacitor is equal to the source voltage.

In a series inductive AC circuit, line current lags the source voltage. When the circuit is capacitive, line current leads the source voltage. With these types of circuits, capacitance and inductance present an opposition to current flow known as reactance. Reactance is combined with resistance to calculate the impedance (total opposition to current flow) in an AC circuit.

In a series resistive AC circuit, the source voltage and line current are in phase with each other. A series resistive AC circuit is analyzed in the same manner as a series resistive DC circuit. If the effective value of AC is used, then the DC values and AC values of the circuit parameters are equal.

When analyzing a series circuit, current is used as a reference for all other circuit parameters. Current is the same at all points in the circuit. Series inductive-resistive-capacitive circuits are typically analyzed to calculate circuit parameters such as impedance, current flow, power dissipation, voltage drop, and power factor.

A series inductive-resistive-capacitive circuit is analyzed by applying the following procedure:

1. Calculate circuit inductive reactance. Circuit inductive reactance is calculated by applying the following formula:

   $X_L = 2\pi \times f \times L$

   where

   X_L = inductive reactance (in Ω)

   2π = 6.28 (in radians per cycle)

   f = frequency (in Hz)

   L = inductance (in H)

2. Calculate circuit capacitive reactance. Circuit capacitive reactance is calculated by applying the following formula:

   $$X_C = \frac{1}{2\pi \times f \times C}$$

where

X_C = capacitive reactance (in Ω)

2π = 6.28 (in radians per cycle)

f = frequency (in Hz)

C = capacitance (in F)

3. Calculate circuit impedance by combining the reactance and resistance vectors. The magnitude of the circuit impedance is calculated by applying the following formula:

$$Z = \sqrt{R^2 + (X_L - X_C)^2}$$

where

Z = circuit impedance (in Ω)

R = circuit resistance (in Ω)

X_L = inductive reactance (in Ω)

X_C = capacitive reactance (in Ω)

The angle of the circuit impedance is calculated by applying the following formula:

$$\theta = \tan^{-1}\left(\frac{X}{R}\right)$$

where

θ = angle theta (in °)

X = total reactance (in Ω)

R = resistance (in Ω)

4. Calculate circuit current. Circuit current is calculated by applying the following formula (Ohm's law):

$$I = \frac{V_A}{Z}$$

where

I = circuit current (in A)

V_A = source voltage (in V)

Z = circuit impedance (in Ω)

Air-core inductors are typically used in electronics applications that have series inductive-resistive-capacitive (LRC) circuits.

5. Calculate voltage drop across inductive, resistive, and capacitive components. Inductive, capacitive, and resistive voltage drops are calculated by applying the following formulas (derived from Ohm's law):

$V_L = I \times X_L$
$V_C = I \times X_C$
$V_R = I \times R$
where
V_L = inductive voltage drop (in V)
I = circuit current (in A)
X_L = inductive reactance (in Ω)
V_C = capacitive voltage drop (in V)
X_C = capacitive reactance (in Ω)
V_R = resistive voltage drop (in V)
R = circuit resistance (in Ω)

6. Calculate apparent power. Apparent power is calculated by applying the following formula:
   $P_{APP} = I \times V_A$
   where
   P_{APP} = apparent power (in VA)
   I = circuit current (in A)
   V_A = source voltage (in V)
7. Calculate reactive power. Reactive power is calculated by applying the following formula:
   $P_{VAR} = I \times V_L$
   $P_{VAR} = I \times V_C$
   where
   P_{VAR} = reactive power (in VAR)
   I = circuit current (in A)
   V_L = inductive voltage drop (in V)
   V_C = capacitive voltage drop (in V)
8. Calculate true power. True power is calculated by applying the following formula:
   $P_{TRUE} = I \times V_R$
   where
   P_{TRUE} = true power (in W)
   I = circuit current (in A)
   V_R = resistive voltage drop (in V)
9. Calculate power factor. Power factor is calculated by applying the following formula:
   $$PF = \frac{P_{TRUE}}{P_{APP}} \times 100\%$$
   where
   PF = power factor (in %)
   P_{TRUE} = true power (in W)
   P_{APP} = apparent power (in VA)

Without taking any measurements, values that are typically known in a series inductive-resistive-capacitive circuit are the frequency, source voltage, resistance, and the capacitance and inductance values of the electrical components. **See Figure 18-1.**

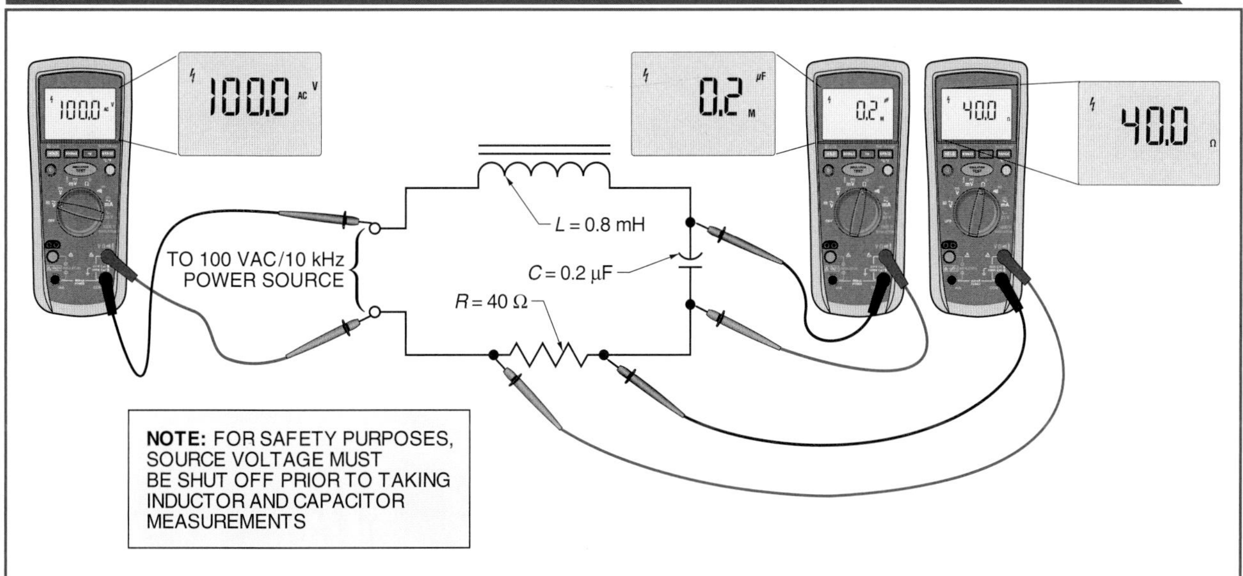

Figure 18-1. Values that are typically known in a series inductive-resistive-capacitive circuit are frequency, source voltage, resistance, and the capacitance and inductance values of the electrical components.

Example: What is the analysis of a series inductive-resistive-capacitive circuit with a frequency of 10 kHz, source voltage of 100 VAC, and a 0.8 mH inductor, 0.2 μF capacitor, and 40 Ω resistor?

1. Calculate circuit inductive reactance.

$X_L = 2\pi \times f \times L$

$X_L = 6.28 \times 10^4 \times 0.8 \times 10^{-3}$

$X_L = 50\ \Omega$

2. Calculate circuit capacitive reactance.

$$X_C = \frac{1}{2\pi \times f \times C}$$

$$X_C = \frac{1}{6.28 \times 10^4 \times 0.2 \times 10^{-6}}$$

$X_C = 80\ \Omega$

Since an inductive reactance and a capacitive reactance are 180° out of phase with each other, they can be added together. The inductive reactance is a positive value, and the capacitive reactance is a negative value, because of this phase relationship.

$X_L - X_C = 50\ \Omega - 80\ \Omega$

$X_L - X_C = -30\ \Omega$

Tech Tip

A series LRC circuit is inductive at higher frequencies and capacitive at lower frequencies. A parallel LRC circuit is capacitive at higher frequencies and inductive at lower frequencies.

Since X_C is greater than X_L, the circuit is capacitive, with the line current leading the source voltage.

3. Calculate circuit impedance.

$$Z = \sqrt{R^2 + (X_L - X_C)^2}$$

$$Z = \sqrt{40\ \Omega^2 + (50\ \Omega - 80\ \Omega)^2}$$

$$Z = \sqrt{40\ \Omega^2 + (-30\ \Omega^2)}$$

$$Z = \sqrt{1600 + 900}$$

$$Z = \sqrt{2500}$$

$$Z = 50\ \Omega$$

$$\theta = \tan^{-1}\left(\frac{X}{R}\right)$$

$$\theta = \tan^{-1}\left(\frac{-30\ \Omega}{40\ \Omega}\right)$$

$$\theta = \tan^{-1}(-0.75)$$

$$\theta = -36.9°$$

4. Calculate the circuit current using Ohm's law.

$$I = \frac{V_A}{Z}$$

$$I = \frac{100\text{ V}}{50\ \Omega \text{ at} - 36.9°}$$

$$I = 2\text{ A at } 36.9°$$

5. Calculate voltage drops across inductive, resistive, and capacitive components.

$$V_L = I \times X_L$$
$$V_L = 2\text{ A} \times 50\ \Omega$$
$$V_L = 100\text{ V}$$

$$V_R = I \times R$$
$$V_R = 2\text{ A} \times 40\ \Omega$$
$$V_R = 80\text{ V}$$

$$V_C = I \times X_C$$
$$V_C = 2\text{ A} \times 80\ \Omega$$
$$V_C = 160\text{ V}$$

In this example, inductive voltage drop equals the source voltage, and capacitive voltage drop exceeds the source voltage. Remember that each voltage is a vector. Kirchhoff's voltage law, which states that the sum of the voltage drops around a closed loop is equal to zero, is applicable.

$$V_A = \sqrt{V_R^2 + (V_L - V_C)^2}$$

$$V_A = \sqrt{(80\text{ V})^2 + (100\text{ V} - 160\text{ V})^2}$$

$$V_A = \sqrt{6400 + 3600}$$

$$V_A = \sqrt{10{,}000}$$

$$V_A = 100\text{ V}$$

When the individual voltages across each component are measured in the circuit, the calculated values are obtained. When inductive and capacitive voltage are measured, their combined value is equal to 60 V because the voltage vectors are 180° out of phase with each other. **See Figure 18-2.**

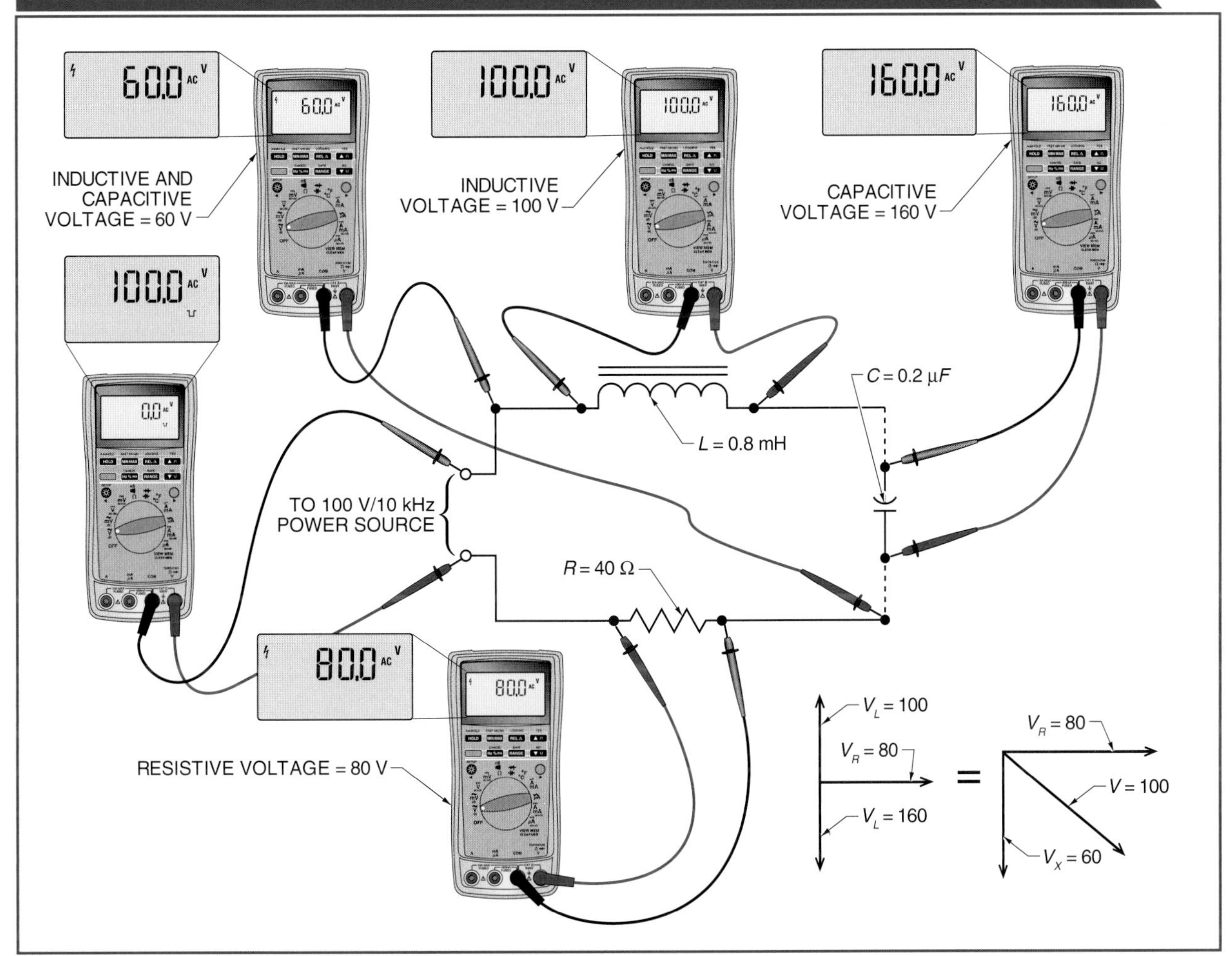

Figure 18-2. When inductive and capacitive voltage are measured, their combined value is equal to 60 V because the voltage vectors are 180° out of phase with each other.

6. Calculate apparent power.

$P_{APP} = I \times V_A$

$P_{APP} = 2\text{ A} \times 100\text{ V}$

$P_{APP} = 200\text{ VA}$

7. Calculate reactive power.

$P_{VAR} = I \times V_L$

$P_{VAR} = 2\text{ A} \times 100\text{ V}$

$P_{VAR} = 200\text{ VAR}$

$P_{VAR} = I \times V_C$

$P_{VAR} = 2\text{ A} \times 160\text{ V}$

$P_{VAR} = 320\text{ VAR}$

8. Calculate true power.

$P_{TRUE} = I \times V_R$

$P_{TRUE} = 2\text{ A} \times 80\text{ V}$

$P_{TRUE} = 160\text{ W}$

9. Calculate power factor.

$$PF = \frac{P_{TRUE}}{P_{APP}} \times 100\%$$

$$PF = \frac{160\text{ W}}{200\text{ VA}} \times 100\%$$

$PF = \mathbf{80\%}$

When analyzing a series inductive-resistive-capacitive circuit, a power vector diagram can be constructed using the power calculations. **See Figure 18-3.**

SERIES INDUCTIVE-RESISTIVE-CAPACTIVE CIRCUITS—POWER VECTOR DIAGRAMS

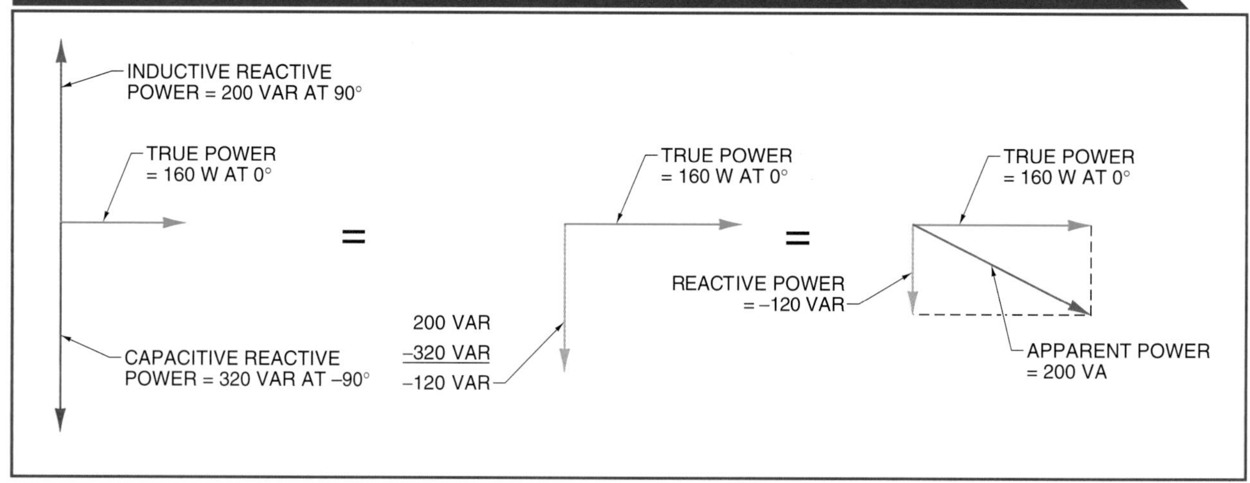

Figure 18-3. When analyzing a series inductive-resistive-capacitive circuit, a power vector diagram can be constructed by using power calculations.

At power line frequencies, the inductive and capacitive components of a conductor are not typically a factor. However, as frequency is increased, the effects of inductance and capacitance can become significant and must be considered. In addition, all three of these components are combined in AC series circuits to achieve desired effects such as filtering signals or power factor correction.

In a series circuit, if inductive reactance is larger than capacitive reactance, then the circuit is inductive. If, in a series circuit, capacitive reactance is larger than inductive reactance, then the circuit is capacitive. The circuit in the example is capacitive, so line voltage lags the source current. If the circuit were inductive, the line voltage would lead the source current.

PARALLEL INDUCTIVE-RESISTIVE-CAPACITIVE CIRCUIT ANALYSIS

Voltage in a parallel circuit is the same across all parallel components. For this reason, voltage is used as the reference in a parallel inductive-resistive-capacitive circuit. Parallel inductive-resistive-capacitive circuits behave differently than series inductive-resistive-capacitive circuits. For the same ohmic values as in a series circuit, the source current, impedance power, and power factor in a parallel circuit are different. Parallel inductive-resistive-capacitive circuits are typically analyzed to calculate the same operating parameters as in a series circuit. These parameters include source current, power, and power factor.

A parallel inductive-resistive-capacitive circuit is analyzed by applying the following procedure:

1. Calculate component reactance. Component reactance is calculated by applying the following formulas:

   $X_L = 2\pi \times f \times L$

   $X_C = \frac{1}{2\pi \times f \times C}$

   where

   X_L = inductive reactance (in Ω)

   X_C = capacitive reactance (in Ω)

   2π = 6.28 (in radians per cycle)

   f = frequency (in Hz)

   L = inductance (in H)

   C = capacitance (in F)

Fluke Corporation

Inductive-resistive-capacitive circuits are used to produce sine waves in electronic test instruments such as power quality meters.

2. Calculate each branch circuit current. Each branch circuit current is calculated by applying the following formula (Ohm's law):

$$I = \frac{V_A}{X_N}$$

where

I = inductive or capacitive branch circuit current (in A)

V_A = source voltage (in V)

X_N = inductive or capacitive reactance (in Ω)

Resistive branch circuit current is calculated by applying the following formula:

$$I_R = \frac{V_A}{R}$$

where

I_R = resistive branch circuit current (in A)

V_A = source voltage (in V)

R = resistance (in Ω)

3. Calculate total circuit current. Total circuit current is calculated by applying the following formula:

$$I_T = \sqrt{I_R{}^2 + (I_C - I_L)^2}$$

where

I_T = total circuit current (in A)

I_R = resistive branch circuit current (in A)

I_C = capacitive branch circuit current (in A)

I_L = inductive branch circuit current (in A)

4. Calculate angle theta. Angle theta is calculated by applying the following formula:

$$\theta = \tan^{-1}\left(\frac{I_{LC}}{I_R}\right)$$

where

θ = angle theta (in °)

I_{LC} = reactive branch circuit current (in A)

I_R = resistive branch circuit current (in A)

5. Calculate circuit impedance. Circuit impedance is calculated by applying the following formula (Ohm's law):

$$Z = \frac{V_A}{I_T}$$

where

Z = circuit impedance (in Ω)

V_A = source voltage (in V)

I_T = total circuit current (in A)

6. Calculate apparent power. Apparent power is calculated by applying the following formula:

$$P_{APP} = I \times V_A$$

where

P_{APP} = apparent power (in VA)

I = circuit current (in A)

V_A = source voltage (in V)

As seen in a series inductive-resistive-capacitive circuit, the apparent power in a parallel inductive-resistive-capacitive circuit is equal to the line current multiplied by the source voltage.

Tech Tip

In 1745, German scientist E. Georg von Kleist and University of Leiden (Leyden) professor Pieter van Musschenbroek each discovered that a glass vessel filled with water and charged by a friction source could store an electric charge. This device became known as the Leyden jar and was the first reported capacitor.

7. Calculate true power. True power is calculated by applying the following formula:

$$P_{TRUE} = I_R \times V_A$$

where

P_{TRUE} = true power (in W)

I_R = resistive current (in A)

V_A = source voltage (in V)

8. Calculate reactive power. Reactive power is calculated by applying the following formula:

$$P_{VAR} = I_L \times V_A$$

$$P_{VAR} = I_C \times V_A$$

where

P_{VAR} = reactive apparent power (in VAR)

I_L = inductive current (in A)

I_C = capacitive current (in A)

V_A = source voltage (in V)

9. Calculate power factor. Power factor is calculated by applying the following formula:

$$PF = \frac{P_{TRUE}}{P_{APP}} \times 100\%$$

where

PF = power factor (in %)

P_{TRUE} = true power (in W)

P_{APP} = apparent power (in VA)

As with a series inductive-resistive-capacitive circuit, the power factor in a parallel inductive-resistive-capacitive circuit is equal to the true power divided by the apparent power.

Without taking any measurements, values that are typically known in a parallel inductive-resistive-capacitive circuit are the source voltage and component values.

Example: What is the analysis of a parallel inductive-resistive-capacitive circuit that has a source voltage of 100 VAC at 10 kHz, an inductor of 0.8 mH, a capacitor of 0.2 μF, and a resistor of 40 Ω connected together in parallel?

Note: Using voltage as the reference, inductive branch circuit current lags voltage by 90°, capacitive branch circuit current leads voltage by 90°, and resistive branch circuit current is in phase with voltage. **See Figure 18-4.**

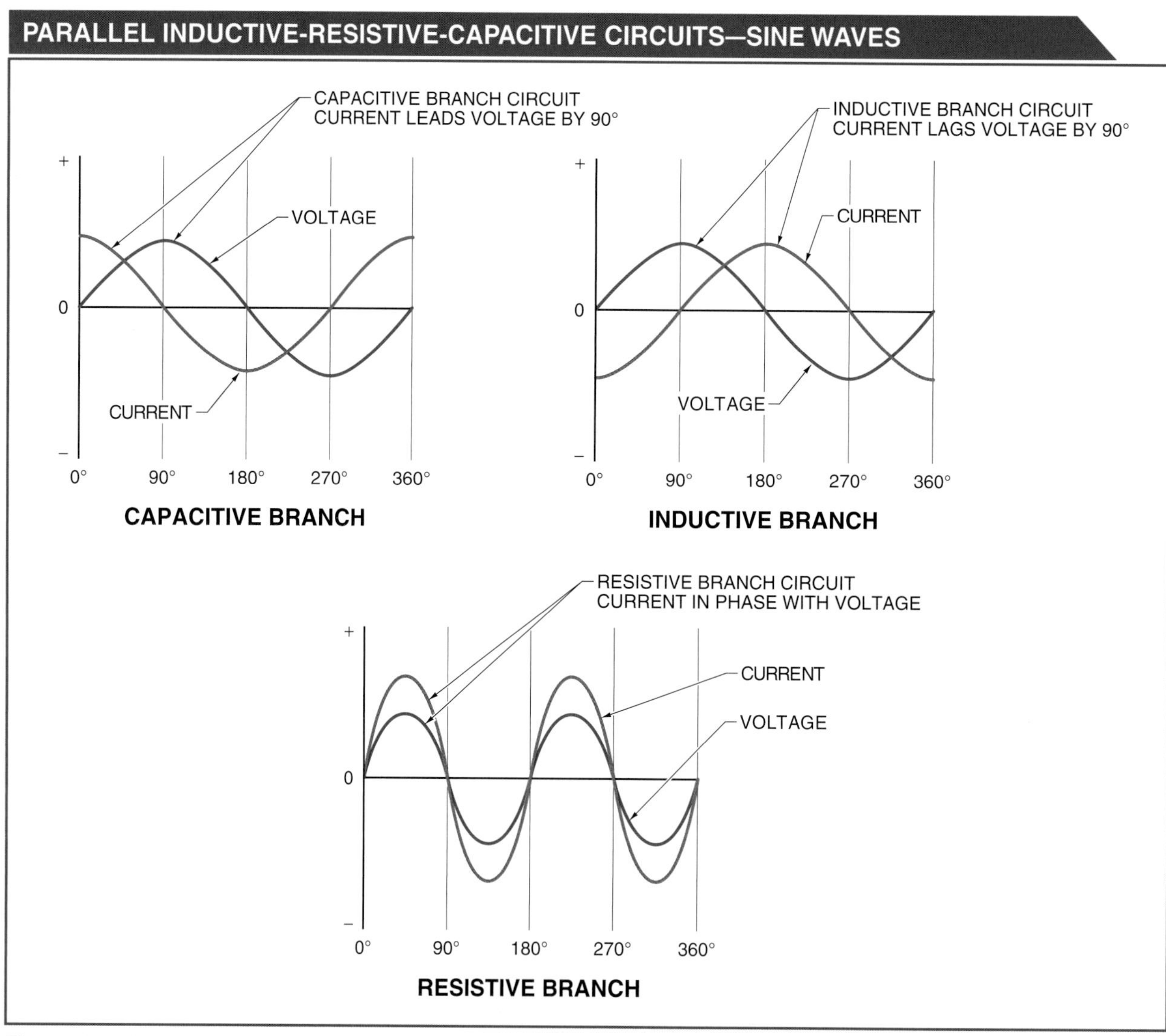

Figure 18-4. In a parallel inductive-resistive-capacitive circuit, inductive branch circuit current lags voltage by 90°, capacitive branch circuit current leads voltage by 90°, and resistive branch circuit current is in phase with voltage.

1. Calculate component reactance.

$X_L = 2\pi \times f \times L$

$X_L = 6.28 \times 10^4 \times 0.8\ \text{H} \times 10^{-3}$

$X_L = 50\ \Omega$

$$X_C = \frac{1}{2\pi \times f \times C}$$

$$X_C = \frac{1}{6.28 \times 10^4 \times 0.2\ \text{F} \times 10^{-6}}$$

$X_C = 80\ \Omega$

2. Calculate each branch circuit current.

Calculate inductive branch circuit current.

$$I_L = \frac{V_A}{X_L}$$

$$I_L = \frac{100\ \text{V}}{50\ \Omega \text{ at } 90^\circ}$$

$I_L = 2\ \text{A at } -90^\circ$

Calculate capacitive branch circuit current.

$$I_C = \frac{V_A}{X_C}$$

$$I_C = \frac{100\ \text{V}}{80\ \Omega \text{ at } -90^\circ}$$

$I_C = 1.25\ \text{A at } 90^\circ$

Calculate resistive branch circuit current.

$$I_R = \frac{V_A}{R}$$

$$I_R = \frac{100\ \text{V}}{40\ \Omega \text{ at } 0^\circ}$$

$I_R = 2.5\ \text{A at } 0^\circ$

3. Calculate total circuit current.

$$I_T = \sqrt{I_R^2 + (I_C - I_L)^2}$$

$$I_T = \sqrt{(2.5\ \text{A})^2 + (1.25\ \text{A} - 2\ \text{A})^2}$$

$$I_T = \sqrt{6.8125\ \text{A}}$$

$I_T = 2.61\ \text{A}$

Because the inductive and capacitive branch circuit currents are 180° out of phase, the smaller current reduces the larger current.

Tech Tip

An ideal inductor ($R_1 = 0\ \Omega$) is equivalent to a short circuit in a DC circuit once steady-circuit conditions have been established.

4. Calculate angle theta.

$$\theta = \tan^{-1}\left(\frac{I_{LC}}{I_R}\right)$$

$$\theta = \tan^{-1}\left(\frac{1.25 - 2\text{ A}}{2.5\text{ A}}\right)$$

$$\theta = \tan^{-1}\left(\frac{-0.75\text{ A}}{2.5\text{ A}}\right)$$

$$\theta = \tan^{-1} -0.30$$

$$\theta = -16.7^\circ$$

As with a series inductive-resistive-capacitive circuit, where the voltage across a component can exceed the source voltage, the current in either the capacitive-reactive or inductive-reactive branch can exceed the line current under certain conditions. For example, increasing or decreasing the frequency causes reactive branch circuit currents to change. Increasing the frequency causes current flow to increase in the capacitive branch and decrease in the inductive branch. Decreasing the frequency causes current flow to increase in the inductive branch and decrease in the capacitive branch. These changes are due to changes in reactive impedance.

5. Calculate circuit impedance.

$$Z = \frac{V_A}{I_T}$$

$$Z = \frac{100\text{ V}}{2.61\text{ A at } -16.7^\circ}$$

$$Z = 38.3\ \Omega \text{ at } 16.7^\circ$$

6. Calculate apparent power.

$$P_{APP} = I \times V_A$$

$$P_{APP} = 2.61\text{ A} \times 100\text{ V}$$

$$P_{APP} = 261\text{ VA}$$

7. Calculate true power.

$$P_{TRUE} = I_R \times V_A$$

$$P_{TRUE} = 2.5\text{ A} \times 100\text{ V}$$

$$P_{TRUE} = 250\text{ W}$$

8. Calculate reactive power.

$$P_{VAR} = I_L \times V_A$$

$$P_{VAR} = 2\text{ A} \times 100\text{ V}$$

$$P_{VAR} = 200\text{ VAR}$$

$$P_{VAR} = I_C \times V_A$$

$P_{VAR} = 1.25 \text{ V} \times 100 \text{ V}$

$P_{VAR} = 125 \text{ VAR}$

Reactive power values are 180° out of phase. Total reactive power is calculated by adding the inductive and capacitive power vectors together. When the reactive power values are added together, the result is equal to:

$VAR_C - VAR_L = 125 \text{ VAR} - 200 \text{ VAR} = 75 \text{ VAR at } -90°$

When analyzing a parallel inductive-resistive-capacitive circuit, a power vector diagram can be constructed to calculate apparent power. **See Figure 18-5.**

9. Calculate power factor.

$$PF = \frac{P_T}{P_A} \times 100\%$$

$$PF = \frac{250 \text{ W}}{261 \text{ VA}} \times 100\%$$

$$PF = \mathbf{96\%}$$

Honeywell

LRC circuits are commonly used in industrial components.

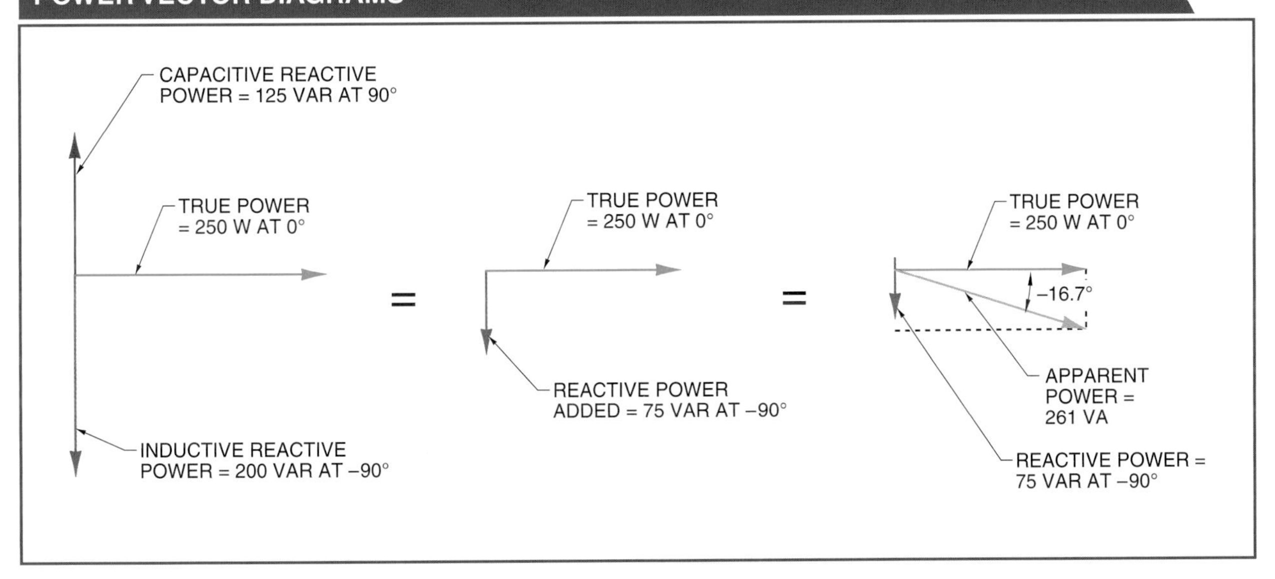

Figure 18-5. When analyzing a parallel inductive-resistive-capacitive circuit, a power vector diagram can be constructed to calculate apparent power.

SERIES/PARALLEL INDUCTIVE-RESISTIVE-CAPACITIVE CIRCUIT ANALYSIS

When analyzing parallel inductive-resistive-capacitive circuits, it is assumed that each branch is purely inductive, capacitive, or resistive. However, the two reactive branches may also have some resistance. The additional resistance in the reactive branches creates a series/parallel inductive-resistive-capacitive circuit. **See Figure 18-6.** Series/parallel inductive-resistive-capacitive circuits are typically analyzed to calculate the total circuit current, circuit power, and power factor. A series/parallel inductive-resistive-capacitive circuit is analyzed by applying the following procedure:

1. Calculate the reactance of each branch circuit, if not known. Branch circuit inductive reactance is calculated by applying the following formula:

   $X_L = 2\pi \times f \times L$

   where

   X_L = inductive reactance (in Ω)

   2π = 6.28 (in radians per cycle)

   f = frequency (in Hz)

   L = inductance (in H)

Branch circuit capacitive reactance is calculated by applying the following formula:

$$X_C = \frac{1}{2\pi \times f \times C}$$

where

X_C = capacitive reactance (in Ω)

2π = 6.28 (in radians per cycle)

f = frequency (in Hz)

C = capacitance (in F)

2. Calculate branch circuit impedances and angle theta. Inductive or capacitive impedance is calculated by applying the following formula:

   $$Z_N = \sqrt{X_N^2 + R_N^2}$$

   where

   Z_N = branch circuit impedance (in Ω)

   X_N = branch circuit inductive or capacitive reactance (in Ω)

   R_N = branch circuit resistance (in Ω)

Angle theta is calculated by applying the following formula:

$$\theta = \tan^{-1}\left(\frac{X_N}{R_N}\right)$$

SERIES/PARALLEL INDUCTIVE-RESISTIVE-CAPACITIVE CIRCUITS

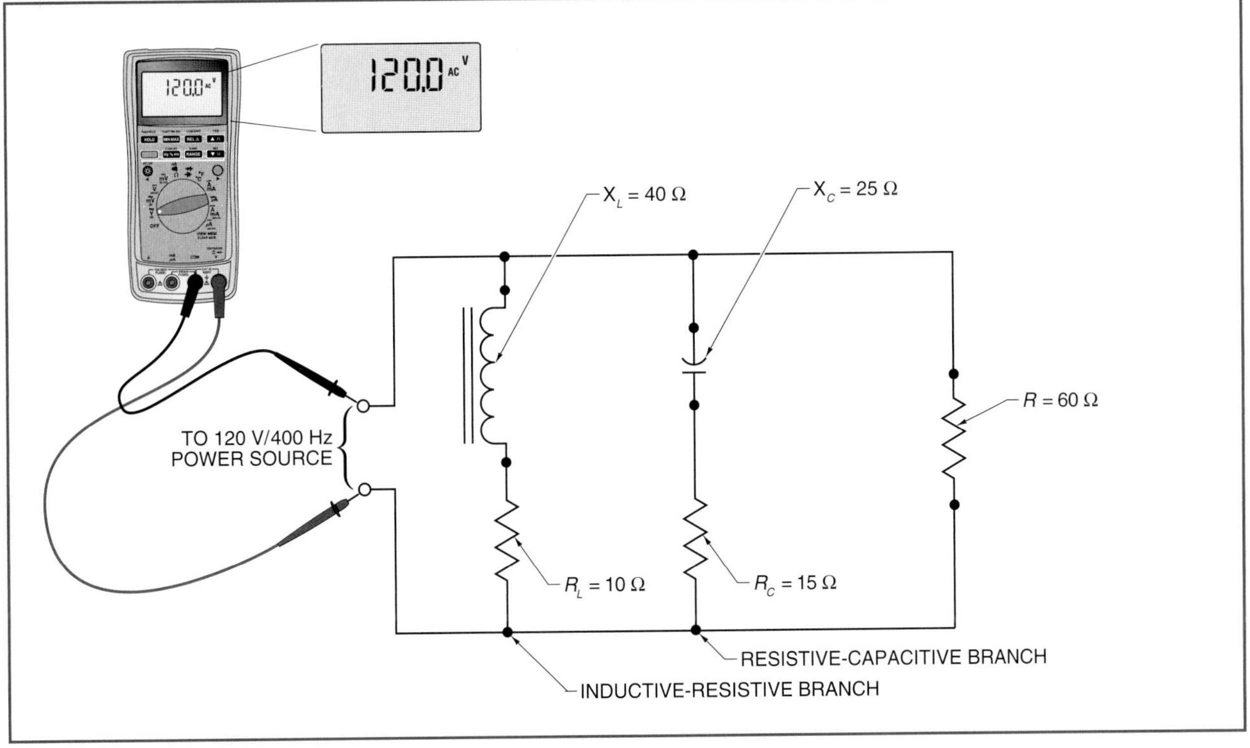

Figure 18-6. Because the parallel inductive and capacitive branches have some resistance, the circuit is considered a series/parallel inductive-resistive-capacitive circuit.

3. Calculate each branch circuit current. Inductive or capacitive branch circuit current is calculated by applying the following formula (Ohm's law):

$$I_N = \frac{V_A}{Z_N}$$

where

I_N = inductive or capacitive branch circuit current (in A)

V_A = source voltage (in V)

Z_N = inductive or capacitive branch circuit impedance (in Ω)

Resistive branch circuit current is calculated by applying the following formula:

$$I_R = \frac{V_A}{R}$$

where

I_R = branch circuit current (in A)

V_A = source voltage (in V)

R = resistance (in Ω)

Tech Tip

The unit of flux density, the gauss, is named in honor of German scientist and mathematician Carl Friedrich Gauss. In 1833, Gauss worked with Wilhelm Weber researching electricity and magnetism, leading to the development of the first electromagnetic telegraph. Samuel Morse later developed a similar telegraph system independently in the United States.

4. Develop vector diagrams of each branch circuit and reduce each branch circuit current to its vertical and horizontal components. Branch circuit current is reduced to its vertical component by applying the following formula:

   $V_N = \sin\theta \times I_N$

   where

   V_N = inductive or capacitive vertical component (in A)

   $\sin\theta$ = sine of angle theta (in °)

   I_N = inductive or capacitive current (in A)

Branch circuit current is reduced to its horizontal component by applying the following formula:

$H_N = \cos\theta \times I_N$

where

H_N = inductive or capacitive horizontal component (in A)

$\cos\theta$ = cosine of angle theta (in °)

I_N = inductive or capacitive current (in A)

5. Add vertical and horizontal vector values together.

Vertical vector values are added together by applying the following formula:

$V_T = V_C + V_L$

where

V_T = total vertical vector value (in A)

V_C = capacitive vertical vector value (in A)

V_L = inductive vertical vector value (in A)

Horizontal vector values are added together by applying the following formula:

$H_T = I_R + H_C + H_L$

where

H_T = horizontal vector total (in A)

I_R = resistive current (in A)

H_C = capacitive horizontal vector (in A)

H_L = inductive horizontal vector (in A)

6. Calculate total line current and angle theta. Total line current is calculated by applying the following formula (Pythagorean theorem):

   $$I_T = \sqrt{H_T^{\,2} + V_T^{\,2}}$$

   where

   I_T = total line current (in A)

   H_T = sum of horizontal components (in A)

   V_T = sum of vertical components (in A)

Angle theta is calculated by applying the following formula:

$$\theta = \tan^{-1}\left(\frac{V_T}{H_T}\right)$$

where

θ = angle theta (in °)

V_T = sum of vertical components (in A)

H_T = sum of horizontal components (in A)

7. Calculate total circuit impedance. Total circuit impedance is calculated by applying the following formula (Ohm's law):

$$Z_T = \frac{V_A}{I_T}$$

where

Z_T = total circuit impedance (in Ω)

V_A = source voltage (in V)

I_T = total circuit current (in A)

8. Calculate apparent power. Apparent power is calculated by applying the following formula:

$$P_{APP} = I_T \times V_A$$

where

P_{APP} = apparent power (in VA)

I_T = total source current (in A)

V_A = source voltage (in V)

9. Calculate true power. True power is calculated by applying the following formula:

$$P_{TRUE} = I_H \times V_A$$

where

P_{TRUE} = true power (in W)

I_H = horizontal component current (in A)

V_A = source voltage (in V)

10. Calculate reactive power. Reactive power is calculated by applying the following formula:

$$P_X = I_V \times V_A$$

where

P_X = reactive power (in VAR)

I_V = vertical component current (in A)

V_A = source voltage (in V)

11. Calculate power factor. Power factor is calculated by applying the following formula:

$$PF = \frac{P_{True}}{P_{App}} \times 100\%$$

Fluke Corporation

Digital multimeter measurements can be used to calculate LRC circuit parameters in electronic equipment.

where

PF = power factor (in %)

P_{True} = true power (in A)

P_{App} = apparent power (in VA)

Example: What is the analysis of a parallel inductive-resistive-capacitive circuit with a source voltage of 120 VAC at 400 Hz, with parallel inductance of 16 mH, capacitance of 16 µF, and resistance of 60 Ω?

Note: The inductive branch has 10 Ω of resistance and the capacitive branch has 15 Ω of resistance.

1. Calculate the reactance of each branch circuit.

$X_L = 2\pi \times f \times L$

$X_L = 6.28 \times 400 \times (16 \times 10^{-3})$

$X_L = 40\ \Omega$

$$X_C = \frac{1}{2\pi \times f \times C}$$

$$X_C = \frac{1}{6.28 \times 400\ \text{Hz} \times \left(16 \times 10^{-6}\right)}$$

$X_C = 25\ \Omega$

2. Calculate branch circuit impedances and angle theta.

Calculate inductive branch circuit impedance.

$$Z_N = \sqrt{{X_N}^2 + {R_N}^2}$$

$$Z_L = \sqrt{{X_L}^2 + {R_L}^2}$$

$$Z_L = \sqrt{(40\ \Omega)^2 + (10\ \Omega)^2}$$

$$Z_L = \sqrt{1600 + 100}$$

$$Z_L = \sqrt{1700}$$

$Z_L = 41.23\ \Omega$

Calculate capacitive branch circuit impedance.

$$Z_N = \sqrt{X_N^{\ 2} + R_N^{\ 2}}$$

$$Z_C = \sqrt{X_C^{\ 2} + R_C^{\ 2}}$$

$$Z_C = \sqrt{25\ \Omega^2 + 15\ \Omega^2}$$

$$Z_C = \sqrt{625 + 225}$$

$$Z_C = \sqrt{850}$$

$$Z_C = 29.15\ \Omega$$

Calculate angle theta for inductive branch circuit.

$$\theta_L = \tan^{-1}\left(\frac{X_L}{R_L}\right)$$

$$\theta_L = \tan^{-1}\left(\frac{40\ \Omega}{10\ \Omega}\right)$$

$$\theta_L = \tan^{-1} 4.000$$
$$\theta_L = 75.96°$$

Calculate angle theta for capacitive branch circuit.

$$\theta_C = \tan^{-1}\left(\frac{-X_C}{R_C}\right)$$

$$\theta_C = \tan^{-1}\left(\frac{-25\ \Omega}{15\ \Omega}\right)$$

$$\theta_C = \tan^{-1} -1.667$$
$$\theta_C = -59.04°$$

$$Z_L = 41.23\ \Omega \text{ at } 75.96°$$
$$Z_C = 29.15\ \Omega \text{ at } -59.04°$$

The rules for a series circuit are used to analyze each branch circuit. The resistive branch circuit is the simplest branch to analyze since it merely opposes current flow. Resistive branch circuit current is in phase with resistive branch circuit voltage.

> **Tech Tip**
>
> LRC circuits are typically used in radio and communication applications as part of tuning equipment.

3. Calculate each branch circuit current.

Calculate inductive branch circuit current.

$$I_L = \frac{V_A}{Z_L}$$

$$I_L = \frac{120\ \text{V}}{41.23\ \Omega \text{ at } 75.96°}$$

$$I_L = 2.91\ \text{A at } -75.96°$$

Calculate the capacitive branch circuit current.

$$I_C = \frac{V_A}{Z_C}$$

$$I_C = \frac{120 \text{ V}}{29.15 \ \Omega \text{ at } -59.04°}$$

$I_C = 4.11 \text{ A at } 59.04°$

Calculate resistive branch circuit current.

$$I_R = \frac{V_A}{R}$$

$$I_R = \frac{120 \text{ V}}{60 \ \Omega \text{ at } 0°}$$

$I_R = 2 \text{ A at } 0°$

Because the branch circuits are connected in parallel to each other, the currents are referenced to the voltage.

4. Develop vector diagrams for each branch circuit and reduce each current to its vertical and horizontal components.

Reduce current to its vertical inductive component.

$V_L = \sin \theta \times I_L$

$V_L = \sin -75.96° \times 2.91 \text{ A}$

$V_L = -0.9703 \times 2.91 \text{ A}$

$V_L = -2.82 \text{ A}$

Reduce current to its horizontal inductive component.

$H_L = \cos \theta \times I_L$

$H_L = \cos -75.96° \times 2.91 \text{ A}$

$H_L = 0.2425 \times 2.91 \text{ A}$

$H_L = 0.71 \text{ A}$

Reduce current to its vertical capacitive component.

$V_C = \sin \theta \times I_C$

$V_C = \sin 59.04° \times 4.11 \text{ A}$

$V_C = 0.8575 \times 4.11 \text{ A}$

$V_C = 3.52 \text{ A}$

Reduce current to its horizontal inductive component.

$H_C = \cos \theta \times I_C$

$H_C = \cos 59.04° \times 4.11 \text{ A}$

$H_C = 0.5144 \times 4.11 \text{ A}$

$H_C = 2.11 \text{ A}$

Each current is reduced to its vertical and horizontal components to be used to calculate the total line current. Horizontal and vertical vector components are added together directly. When analyzing a series/parallel inductive-resistive-capacitive circuit, current vector diagrams can be constructed using the current calculations. **See Figure 18-7.**

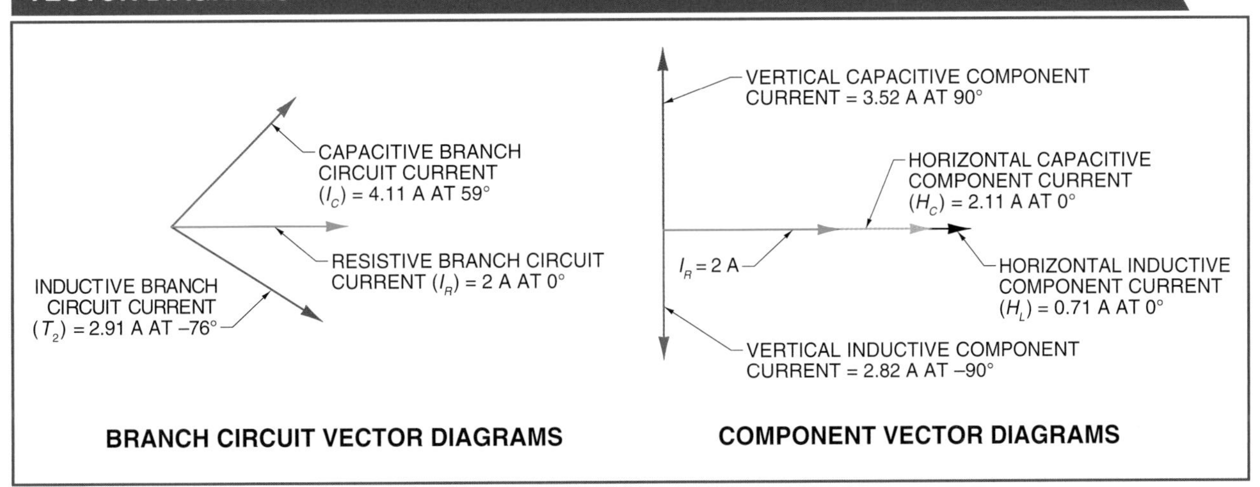

Figure 18-7. When analyzing a series/parallel inductive-resistive-capacitive circuit, current vector diagrams can be constructed using the current calculations.

5. Add vertical and horizontal vectors together.

Add vertical vectors together.

$V_T = V_C + V_L$

$V_T = 3.52 \text{ A} + (-2.82 \text{ A})$

$V_T = 0.7 \text{ A}$

Add horizontal vectors together.

$H_T = I_R + H_C + H_L$

$H_T = 2 \text{ A} + 2.12 \text{ A} + 0.7 \text{ A}$

$H_T = 4.82 \text{ A}$

The vertical components are added together to calculate the reactive current. A lagging current indicates an inductive circuit, while a leading current indicates a capacitive circuit.

6. Calculate total line current and angle theta.

$$I_T = \sqrt{H_T^{\,2} + V_T^{\,2}}$$

$$I_T = \sqrt{(4.82)\text{ A}^2 + (0.70)\text{ A}^2}$$

$$I_T = \sqrt{23.23 + 0.49}$$

$$I_T = \sqrt{23.72}$$

$$I_T = 4.87 \text{ A}$$

The vector diagram calculation method or the Pythagorean theorem can be used to calculate total current in a series/parallel inductive-resistive-capacitive circuit. **See Figure 18-8.**

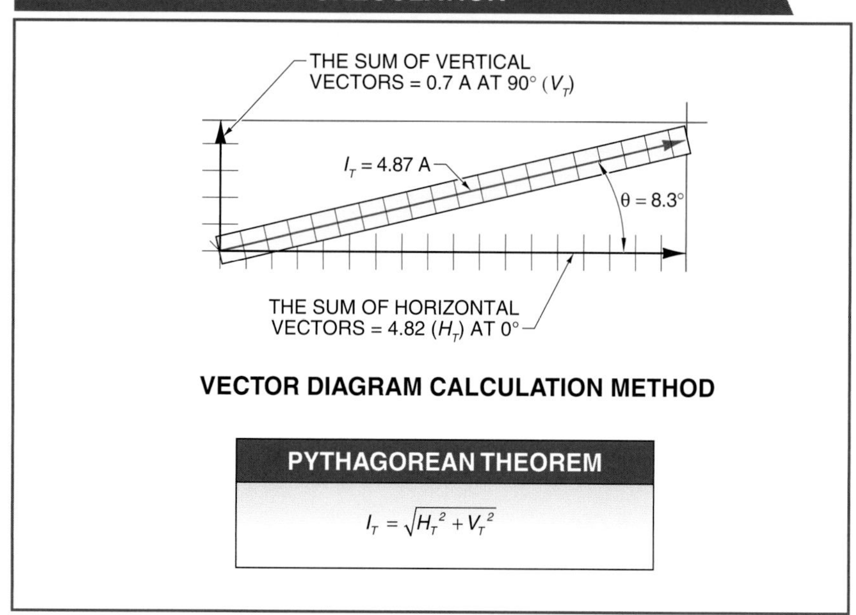

Figure 18-8. The vector diagram calculation method or the Pythagorean theorem can be used to calculate total current in a series/parallel inductive-resistive-capacitive circuit.

> **Tech Tip**
>
> An inductive probe is a fluid-level measuring instrument consisting of a sealed probe containing a coil, an electrical source that generates an alternating magnetic field, and circuitry to detect changes in inductance. The sealed probe is inserted into a vessel containing conductive material. As the fluid level changes, the magnetic field of the probe interacts with the conductive material. The interaction is detected by measuring the inductive reactance.

Calculate angle theta.

$$\theta = \tan^{-1}\left(\frac{V_T}{H_T}\right)$$

$$\theta = \tan^{-1}\left(\frac{0.7\text{ A}}{4.82\text{ A}}\right)$$

$$\theta = 8.3°$$

$$I_T = 4.87\text{ A at }8.3°$$

7. Calculate total circuit impedance.

$$Z_T = \frac{V_A}{I_T}$$

$$Z_T = \frac{120\text{ V}}{4.87\text{ A at }8.3°}$$

$$Z_T = 24.64\ \Omega\text{ at }-8.3°$$

Angle theta is a negative value because the line voltage lags the line current by 8.3°.

8. Calculate apparent power.

$P_A = I_T \times V_A$

$P_A = 4.87 \text{ A} \times 120 \text{ V}$

$P_A = 584.4 \text{ VA}$

9. Calculate true power.

$P_T = I_H \times V_A$

$P_T = 4.82 \text{ A} \times 120 \text{ V}$

$P_T = 578.4 \text{ W}$

10. Calculate reactive power.

$P_X = I_V \times V_A$

$P_X = 0.7 \text{ A} \times 120 \text{ V}$

$P_X = 84 \text{ VAR}$

11.

$$PF = \frac{P_T}{P_A} \times 100\%$$

$$PF = \frac{578.4 \text{ W}}{584.4 \text{ VA}} \times 100\%$$

$$PF = 99\%$$

SUMMARY

All circuits have inductance, resistance, and capacitance, to some extent. In power circuits with low frequencies, only the parameters that are intentionally introduced into the circuit are considered in the calculations. In circuits with high frequencies, this may not be the case because the effects of capacitance and inductance are more significant. Circuit current is the reference for analyzing a series circuit. The voltage across parallel branch circuits is the reference for analyzing a parallel circuit. Current is the same through all components of a series inductive-resistive-capacitive circuit, and the voltage is the same across all branch circuits of a parallel inductive-resistive-capacitive circuit. In a series/parallel inductive-resistive-capacitive circuit, the effects of nonideal inductors and capacitors have additional impacts that require a more complex analysis.

APPLICATION: MAXWELL-WIEN BRIDGES...

Background: Balanced bridges are commonly used to measure an unknown component value, such as resistance, capacitance, or inductance. In a balanced bridge, one or more components are adjusted until the voltage difference between the two parallel legs is zero. **See Figure 1.**

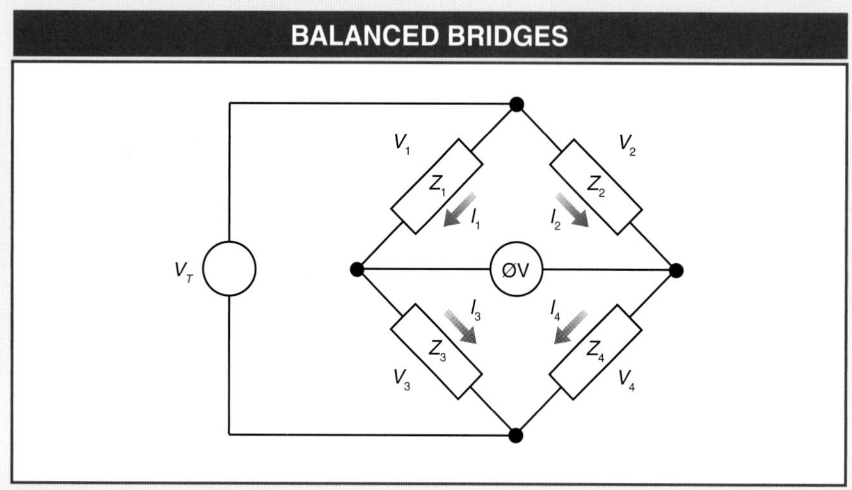

Figure 1. In a balanced bridge, one or more components are adjusted until the voltage difference between the two parallel legs is zero.

In analyzing a balanced bridge, several key equations are created. First, in the two parallel legs of the bridge, the voltage drops across Z_1 and Z_2 are equal as follows:

$$I_1 \times Z_1 = I_2 \times Z_2$$

where
I_X = current through impedance Z_X (in A)
Z_X = component X impedance (in Ω)

Also, the current through the impedances in each parallel leg is common as follows:

$$I_1 = I_3 = \frac{V_T}{Z_1 + Z_3}$$

$$I_2 = I_4 = \frac{V_T}{Z_2 + Z_4}$$

where
I_X = current through impedance Z_X (in A)
V_T = voltage across the bridge (in V)
Z_X = component X impedance (in Ω)

Using these equations, the impedance relationships can be determined as follows:

$$I_1 \times Z_1 = I_2 \times Z_2$$

...APPLICATION: MAXWELL-WIEN BRIDGES...

Substituting for I_1 and I_2, the equation yields the following:

$$\frac{V_T \times Z_1}{Z_1 + Z_3} = \frac{V_T \times Z_2}{Z_2 + Z_4}$$

$$\frac{Z_1}{Z_1 + Z_3} = \frac{Z_2}{Z_2 + Z_4}$$

$$Z_1 \times (Z_2 + Z_4) = Z_2 \times (Z_1 + Z_3)$$

$$Z_1 \times Z_2 + Z_1 \times Z_4 = Z_2 \times Z_1 + Z_2 \times Z_3$$

$$Z_1 \times Z_4 = Z_2 \times Z_3$$

Therefore, if Z_4 is unknown, it can be found by rearranging the equation as follows:

$$Z_4 = \frac{Z_2 \times Z_3}{Z_1}$$

The solution to finding the unknown impedance is independent of voltage and current and is completely based on known impedances.

A balanced bridge that uses only resistors is known as a Wheatstone bridge. **See Figure 2.** In a Wheatstone bridge, the values for R_2 and R_3 are known. R_1 is adjusted until the bridge is balanced. R_4 is found using the balanced bridge formula as follows:

$$Z_4 = \frac{Z_2 \times Z_3}{Z_1}$$

Therefore,

$$R_4 = \frac{R_2 \times R_3}{R_1}$$

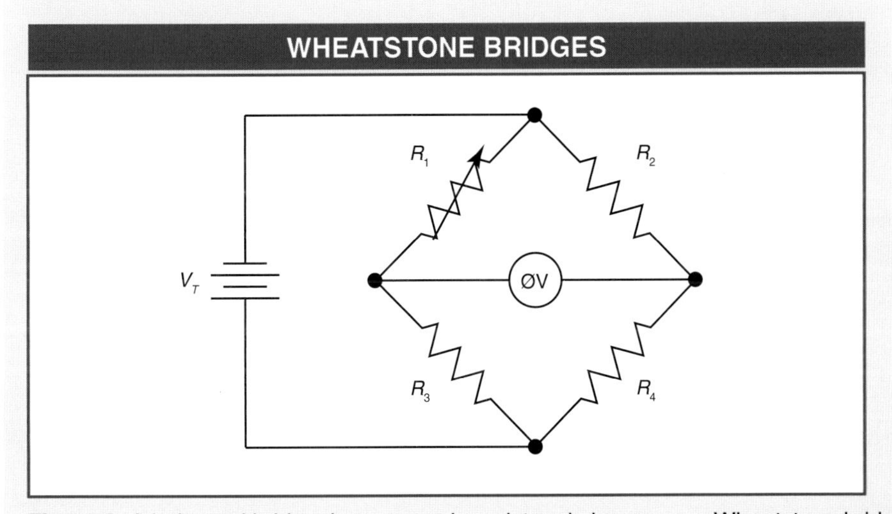

Figure 2. A balanced bridge that uses only resistors is known as a Wheatstone bridge.

...APPLICATION: MAXWELL-WIEN BRIDGES...

A Maxwell-Wien balanced bridge is used with capacitors and inductors in addition to resistors. When the bridge voltage is AC voltage, the capacitors and inductors can be used as bridge circuit elements and can be measured. The Maxwell-Wien bridge is used to find unknown inductance and inductor resistance values. **See Figure 3.** *Note:* If DC voltage is used, then the capacitor is open, the inductor is shorted, and the bridge becomes a Wheatstone bridge.

MAXWELL-WIEN BRIDGES

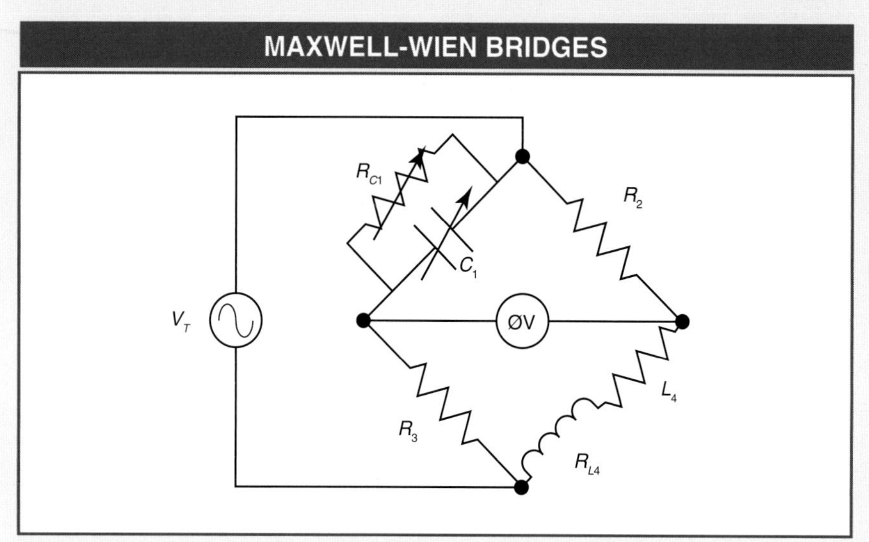

Figure 3. The Maxwell-Wien bridge is used to find unknown inductance and inductor resistance values.

The Maxwell-Wien bridge adjusts Z_1, which is an adjustable resistor connected in parallel with an adjustable capacitor, until the bridge is balanced. To find the inductance and inductor resistance, the balanced bridge equation is again used as follows:

$$Z_4 = \frac{Z_2 \times Z_3}{Z_1}$$

$$Z_4 = (Z_2 \times Z_3) \times \frac{1}{Z_1}$$

Z_1 is the impedance of the resistor in parallel with a capacitor. To simplify the impedance equation, $1 \div Z_1$ is used since it is needed to solve for Z_4.

$$Z_1 = \frac{1}{\left(\frac{1}{R_{C1\,at\,0^\circ}} + \frac{1}{X_{C1}}\right)}$$

$$\frac{1}{Z_1} = \frac{1}{R_{C1\,at\,0^\circ}} + \frac{1}{\left(\frac{1}{2 \times \pi \times f \times C_{1\,at\,-90^\circ}}\right)}$$

$$\frac{1}{Z_1} = \frac{1}{R_{C1\,at\,0^\circ}} + \frac{1}{2 \times \pi \times f \times C_{1\,at\,90^\circ}}$$

...APPLICATION: MAXWELL-WIEN BRIDGES...

where
Z_1 = impedance of R_{C1} in parallel with C_1 (in Ω)
R_{C1} = resistance (in Ω)
π = 3.14
f = frequency (in Hz)
C_1 = capacitance (in F)

Using the balanced bridge equation with the impedance values, the solution for Z_4 is the following:

$$Z_4 = \frac{Z_2 \times Z_3}{Z_1}$$

$$Z_4 = (Z_2 \times Z_3) \times \frac{1}{Z_1}$$

$$Z_4 = (R_2 \times R_3) \times \left(\frac{1}{R_{C1\,at\,0^\circ} + 2 \times \pi \times f \times C_{1\,at\,90^\circ}} \right)$$

$$Z_4 = \frac{R_2 \times R_3}{R_{C1\,at\,0^\circ} + (R_2 \times R_3) \times 2 \times \pi \times f \times C_{1\,at\,90^\circ}}$$

The value of Z_4 is the inductor resistance in series with the inductor reactance X_L. The Z_4 value is the following:

$$Z_4 = R_{L4\,at\,0^\circ} + 2 \times \pi \times f \times L_{4\,at\,90^\circ}$$

Setting the two equations equal to each other yields the following equation for Z_4:

$$R_{L4\,at\,0^\circ} + 2 \times \pi \times f \times L_{4\,at\,90^\circ} = \frac{R_2 \times R_3}{R_{C1at\,0^\circ} + (R_2 \times R_3) \times 2 \times \pi \times f \times C_{1\,at\,90^\circ}}$$

To find the inductance and inductor resistance, the final equations are formed for different phase angles at 0° and 90° as follows:

$$R_{L4\,at\,0^\circ} = \frac{R_2 \times R_3}{R_{C1\,at\,0^\circ}}$$

and

$$2 \times \pi \times f \times L_{4\,at\,90^\circ} = (R_2 \times R_3) + 2 \times \pi \times f \times C_{1\,at\,90^\circ}$$

Therefore,

$$R_{L4} = \frac{R_2 \times R_3}{R_{C1}}$$

and

$$L_4 = (R_2 \times R_3) \times C_1$$

where
R_X = resistance (in Ω)
L_4 = inductance (in H)
C_1 = capacitance (in C)

...APPLICATION: MAXWELL-WIEN BRIDGES

One advantage of a Maxwell-Wien balanced bridge is that, in theory, an adjustable inductor can be used in the bridge rather than a capacitor. However, calibration-grade capacitors are easier to make than similar inductors. Also, if the bridge only uses inductors, then mutual inductance between the inductors could introduce small measurement errors.

Key Points: balanced bridge; inductor reactance; capacitor reactance; phasor

Problem: A Maxwell-Wien balanced bridge is being used to find the inductance. The bridge is balanced with the adjustable resistor set to 100,000 Ω, and the adjustable capacitor is 0.15 µF. What is the inductance and inductor resistance? *Note:* The fixed balanced-bridge resistor values (R_2 and R_3) are each 5000 Ω.

Solution: Apply the Maxwell-Wien balanced bridge equations.

$$R_{L4} = \frac{R_2 \times R_3}{R_{C1}}$$

$$R_{L4} = \frac{5000\ \Omega \times 5000\ \Omega}{100{,}000\ \Omega}$$

$$R_{L4} = 250\ \Omega$$

$$L_4 = (R_2 \times R_3) \times C_1$$

$$L_4 = (5000\ \Omega \times 5000\ \Omega) \times 0.15 \times 10^{-6}\,\mathrm{F}$$

$$L_4 = 3.75\ \mathrm{H}$$

A Maxwell-Wien bridge is used to find the inductance. Once the adjustable resistor and capacitor create a balance across the bridge, two simple equations are used to determine the inductance and inductor resistance.

Chapter Review

1. What values of a parallel LRC circuit are typically known without taking measurements?
2. Why is voltage used as the reference in a parallel LRC circuit?
3. What does lagging current in a combination LRC circuit indicate?
4. What is the inductive and capacitive reactance of a series inductive-resistive-capacitive circuit with a frequency of 10 kHz, source voltage of 120 VAC, 0.6 mH inductor, 0.15 μF capacitor, and 25 Ω resistor?
5. What is a series LRC circuit?
6. What does leading current in a combination LRC circuit indicate?
7. What effect does increased frequency have on current in the inductive-reactive and capacitive-reactive branches of an LRC circuit?
8. What is the main purpose of analyzing a combination LRC circuit?
9. Why is current used as a reference in a series circuit?
10. List five values of an LRC circuit that are typically known without taking any measurements.
11. What is an ideal inductor?
12. What is the main purpose of analyzing a series LRC circuit?
13. What is the main difference in operation between series and parallel LRC circuits?
14. What is the main purpose of analyzing a parallel LRC circuit?
15. What is true power in a 5 A, 208 V parallel LRC circuit?

19

Resonance

OBJECTIVES

- Describe how frequency, impedance, current, voltage, power, quality factor, and bandwidth affect series resonant circuits.
- Describe how frequency, impedance, current, voltage, power, quality factor, and bandwidth affect parallel resonant circuits.
- Explain how low-pass, high-pass, band-pass, and band-reject filters are used in resonant circuits.
- Describe frequency filter networks.

Resonance is one of the most useful electrical concepts. Resonant circuits are used to create frequency filters that make modern communications possible. As with other circuits, series and parallel resonant circuits have properties that are used to create application-specific frequency filters. Frequency filters can also be made to pass either low-or high frequency signals. Frequency filters can be constructed to pass either low-or high-frequency filters. Frequency filters can also be made to pass or reject specific frequency ranges. Frequency filters are used in audio applications, power line conditioning, and communications. Resonant circuits have characteristics and properties that are used to create basic frequency filter designs.

Terms

Resonance is the condition where the inductive reactance (X_L) equals capacitive reactance (X_C) at a given frequency.

SERIES RESONANT CIRCUITS

A series resonant circuit has all circuit elements in series and the inductive and capacitance reactance are equal. *Resonance* is the condition where the inductive reactance (X_L) equals capacitive reactance (X_C) at a given frequency. **See Figure 19-1.** Major characteristics of a series resonant circuit include:

- minimum impedance ($Z = R$)
- maximum allowable current in circuit
- high inductive and capacitive voltages
- angle theta (θ) = 0°

Current is used as the reference point in a series resonant circuit. Typical applications for series resonant circuits include frequency filters for communications systems. Parameters that are typically measured in a series resonant circuit include impedance, current, voltage, power, quality factor, and bandwidth. These parameters are used to understand the operation of the circuit and to ensure that all components are properly sized to withstand potentially high inductive and capacitive voltages.

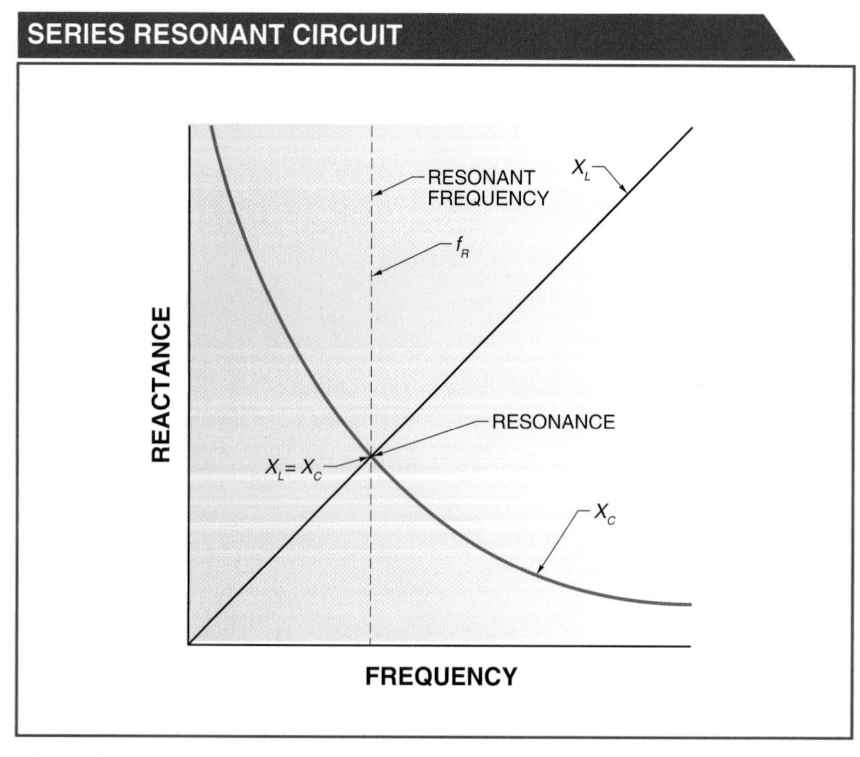

Figure 19-1. Resonance is the condition at a given frequency (f_R) where the inductive reactance (X_L) equals capacitive reactance (X_C).

Frequency in Series Resonant Circuits

Resonance occurs in series circuits when inductive reactance is equal to capacitive reactance. The point of resonance occurs at a certain frequency. Resonant frequency is calculated by applying the following formula:

$$f_R = \frac{1}{2\pi\sqrt{L \times C}}$$

where

f_R = resonant frequency (in Hz)

2π = 6.28 (in radians per cycle)

L = inductance (in H)

C = capacitance (in F)

Example: What is the resonant frequency of a series resonant circuit with inductance of 0.25 H and capacitance of 1 μF?

$$f_R = \frac{1}{2\pi\sqrt{L \times C}}$$

$$f_R = \frac{1}{6.28\sqrt{0.25\text{ H} \times 1\ \mu\text{F}}}$$

$$f_R = \frac{1}{6.28\sqrt{0.25 \times 0.000001}}$$

$$f_R = \frac{1}{6.28\sqrt{0.000000250}}$$

$$f_R = \frac{1}{6.28 \times 0.00050}$$

$$f_R = \frac{1}{0.00314}$$

f_R = **318.47 Hz**

Impedance in Series Resonant Circuits

Impedance in a series resonant circuit is equal to the vector sums of reactance and resistance. Because inductive and capacitive reactance are equal in a series resonant circuit, calculations of their values are not required to calculate impedance at resonance. Resonant series circuit impedance is equal to resonant series circuit resistance. Calculations of reactance are required to calculate the voltage drop across each component. Determining the voltage drop across each component ensures they have the proper voltage rating. The voltage drop across these components may exceed the source voltage. When the frequency and the values of inductance and capacitance are known, the inductive and capacitive reactance can be calculated.

Inductive reactance is calculated by applying the following formula:

$X_L = 2\pi \times f \times L$

where

X_L = inductive reactance (in Ω)

2π = 6.28 (in radians per cycle)

f = frequency (in Hz)

L = inductance (in H)

Example: What is the inductive reactance of a series resonant circuit with a frequency of 4.55 MHz and an inductance of 25 μH?

$X_L = 2\pi \times f \times L$

$X_L = 6.28 \times (4.55 \times 10^6 \text{ Hz}) \times (25 \times 10^{-6} \text{ H})$

$X_L = 6.28 \times (4{,}550{,}000 \text{ Hz}) \times (0.000025 \text{ H})$

$X_L = 6.28 \times 113.75$

$X_L = \mathbf{714\ \Omega}$

Capacitive reactance is calculated by applying the following formula:

$$X_C = \frac{1}{2\pi \times f \times C}$$

where

X_C = capacitive reactance (in Ω)

2π = 6.28 (in radians per cycle)

f = frequency (in Hz)

C = capacitance (in F)

Example: What is the capacitive reactance of a series resonant circuit with a frequency of 4.55 MHz and a capacitance of 49 pF?

$$X_C = \frac{1}{2\pi \times f \times C}$$

$$X_C = \frac{1}{6.28 \times (4.55 \times 10^6 \text{ Hz}) \times (49 \times 10^{-12} \text{ F})}$$

$$X_C = \frac{1}{6.28 \times (4{,}550{,}000 \text{ Hz}) \times (4.900000^{-11} \text{ F})}$$

$$X_C = \frac{1}{0.0014}$$

$X_C = \mathbf{714\ \Omega}$

Circuit impedance can be calculated when the inductive and capacitive reactance of a series resonant circuit are known. Circuit impedance is calculated by applying the following formula:

$$Z = \sqrt{R^2 + (X_L - X_C)^2}$$

where

Z = circuit impedance (in Ω)

R = circuit resistance (in Ω)

X_L = inductive reactance (in Ω)

X_C = capacitive reactance (in Ω)

Example: What is the impedance in a series resonant circuit with resistance of 200 Ω, inductive reactance of 714.35 Ω, and capacitive reactance of 714.28 Ω?

$$Z = \sqrt{R^2 + (X_L - X_C)^2}$$

$$Z = \sqrt{200\ \Omega^2 + (714\ \Omega - 714\ \Omega)^2}$$

$$Z = \sqrt{200\ \Omega^2 + 0\ \Omega^2}$$

$$Z = \sqrt{40{,}000}$$

Z = **200 Ω**

Using the equation for impedance for a series resonant circuit, the reactance cancels, leaving only resistance. An impedance vector diagram indicates that, at resonance, inductive reactance and capacitive reactance cancel, and total impedance is equal to total resistance. **See Figure 19-2.**

SERIES RESONANT CIRCUIT—VECTOR DIAGRAMS

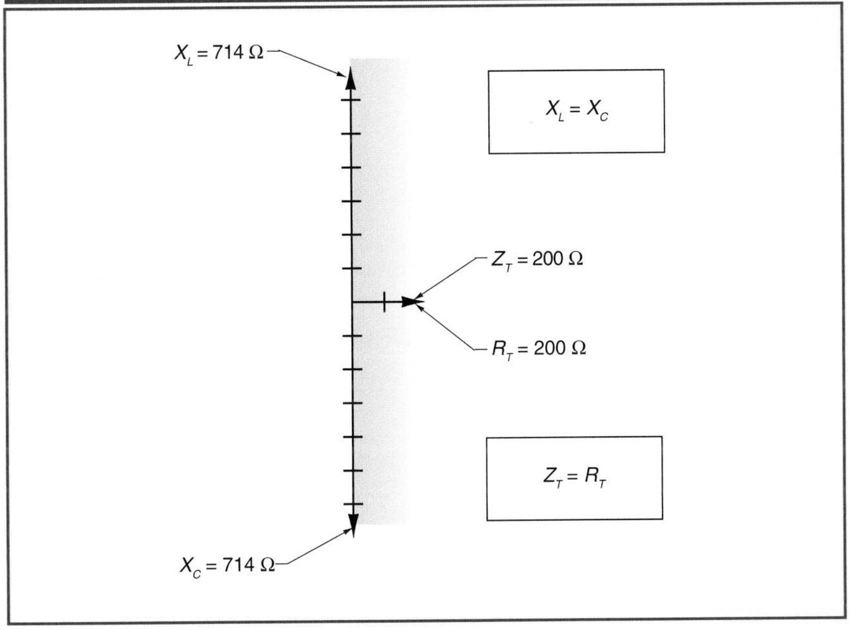

Figure 19-2. At resonance, an impedance vector diagram indicates that inductive reactance and capacitive reactance cancel, and total impedance is equal to total resistance.

Current in Series Resonant Circuits

Current in series resonant circuits is used to calculate the voltage drop across different circuit components. When the source voltage and impedance are known, Ohm's law is applied to calculate the circuit current in a series resonant circuit. Circuit current in a series resonant circuit is calculated by applying the following formula:

$$I_T = \frac{V_A}{Z}$$

where

I_T = circuit current (in A)

V_A = source voltage (in V)

Z = impedance (in Ω)

Example: What is the total current in a series resonant circuit that has source voltage of 100 V and impedance of 200 Ω?

$$I_T = \frac{V_A}{Z}$$

$$I_T = \frac{100 \text{ V}}{200 \text{ }\Omega}$$

$$I_T = \mathbf{0.5 \text{ A}}$$

Voltage in Series Resonant Circuits

Voltage drop across each component in a series resonant circuit is used to understand the proper component voltage rating. Excessive voltage may damage components. When values of total circuit current, total circuit resistance, and inductive and capacitive reactance for each component are known, the voltage drop across each component can be calculated. In a series resonant circuit, voltage drop across the inductor and capacitor can exceed source voltage. **See Figure 19-3.** This is because the current in the circuit is set only by the resistive element, which can create a low-impedance circuit relative to the inductive and capacitive reactance. Voltage drop across an inductor in a series resonant circuit is calculated by applying the following formula:

$$V_L = I_T \times X_L$$

where

V_L = voltage drop across inductor (in V)

I_T = total circuit current (in A)

X_L = inductive reactance (in Ω)

Voltage drop across a capacitor in a series resonant circuit is calculated by applying the following formula:

$V_C = I_T \times X_C$

where

V_C = voltage drop across capacitor (in V)

I_T = total circuit current (in A)

X_C = capacitive reactance (in Ω)

Voltage drop across a resistor in a series resonant circuit is calculated by applying the following formula:

$V_R = I_T \times R_T$

where

V_R = voltage drop across resistor (in V)

I_T = total circuit current (in A)

R_T = total circuit resistance (in Ω)

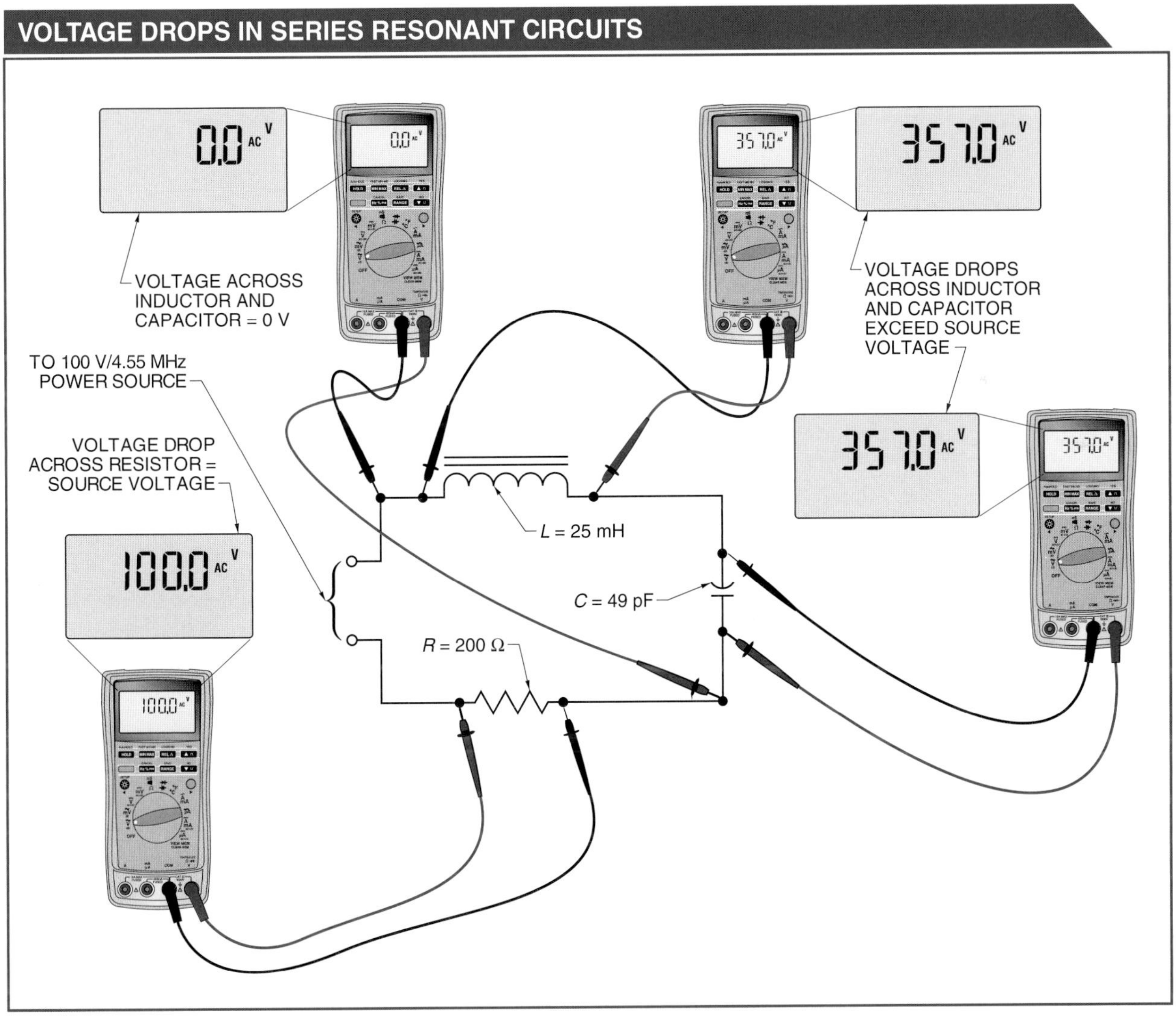

VOLTAGE DROPS IN SERIES RESONANT CIRCUITS

Figure 19-3. In a series resonant circuit, voltage drops across the inductor and capacitor can exceed the source voltage.

Example: What is the voltage drop across each component in a series resonant circuit with total current of 0.5 A, inductive and capacitive reactance of 714 Ω, and resistance of 200 Ω?

Calculate voltage drop across inductor.

$V_L = I_T \times X_L$

$V_L = 0.5\ A \times 714\ \Omega$

$V_L = \mathbf{357\ V}$

Calculate voltage drop across capacitor.

$V_C = I_T \times X_C$

$V_C = 0.5\ A \times 714\ \Omega$

$V_C = \mathbf{357\ V}$

Calculate voltage drop across resistor.

$V_R = I_T \times R_T$

$V_R = 0.5\ A \times 200\ \Omega$

$V_R = \mathbf{100\ V}$

Because the reactive component voltages can exceed the source voltage, the reactive components must have voltage ratings greater than the source voltage. If resistance is lower, then current flow is higher, and the voltages across the reactive components are higher. For example, if the resistance is changed to 100 Ω, then the inductive and capacitive voltage changes to 714 V.

Power in Series Resonant Circuits

Power in series resonant circuits can be used to indicate whether the circuit is at resonance. At resonant frequency, a series resonant circuit is neither capacitive nor inductive. When total circuit current, total circuit resistance, and the source voltage are known, apparent power, true power, and power factor can be calculated. Apparent power in a series resonant circuit is calculated by applying the following formula:

$P_A = I_T \times V_A$

where

P_A = apparent power (in VA)

I_T = total circuit current (in A)

V_A = source voltage (in V)

True power in a series resonant circuit is calculated by applying the following formula:

$P_T = I_T^2 \times R_T$

where

P_T = true power (in W)

I_T = total circuit current (in A)

R_T = total circuit resistance (in Ω)

Power factor in a series resonant circuit is calculated by applying the following formula:

$$PF = \frac{P_T}{P_A} \times 100\%$$

where

PF = power factor (in %)

P_T = true power (in W)

P_A = apparent power (in VA)

Example: What are the apparent power, true power, and power factor in a series resonant circuit with total circuit current of 0.5 A, source voltage of 100 V, and total circuit resistance of 200 Ω?

Calculate apparent power.

$P_A = I_T \times V_A$

$P_A = 0.5\ A \times 100\ V$

$P_A = \mathbf{50\ VA}$

Calculate true power.

$P_T = I_T^2 \times R_T$

$P_T = (0.5\ A)^2 \times 200\ \Omega$

$P_T = \mathbf{50\ W}$

Calculate power factor.

$$PF = \frac{P_T}{P_A} \times 100\%$$

$$PF = \frac{50\ W}{50\ VA} \times 100\%$$

$PF = \mathbf{100\%}$

Because the source voltage in the resonant circuit is in phase with the voltage drop across the resistor, current causes the voltage drop across the resistor. Line current is in phase with the source voltage. True power is equal to the apparent power, and power factor is equal to 100%.

In a series resonant circuit, maximum power occurs at resonance (f_R). **See Figure 19-4.** Tuning the frequency below resonance causes the circuit to become capacitive with a leading current. Tuning the frequency above resonance causes the circuit to become inductive with a lagging current.

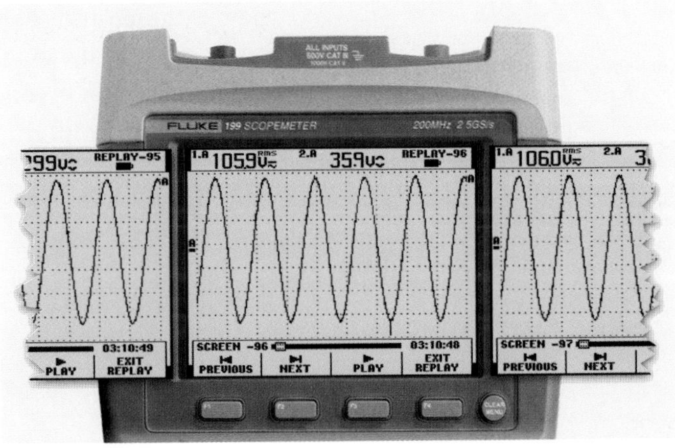

Fluke Corporation

Resonant circuits display oscillations when certain test instruments are used.

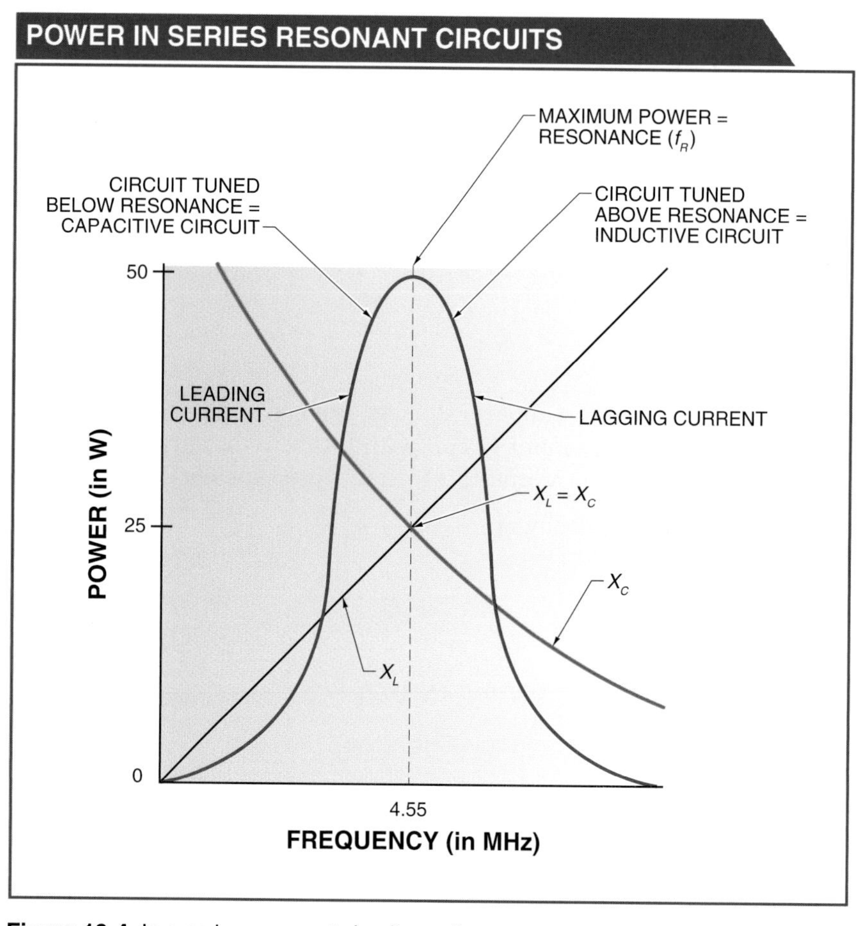

Figure 19-4. In a series resonant circuit, maximum power occurs at resonance (f_R).

Quality Factor in Series Resonant Circuits

An inductor in an AC circuit has a quality factor (Q) that depends on the reactance and resistance that constitute its impedance. A quality factor represents the ratio of inductive reactance to the resistance of an inductor at a given frequency. The quality factor of an inductor is calculated by applying the following formula:

$$Q = \frac{X_L}{R_L}$$

where

Q = quality factor

X_L = inductive reactance (in Ω)

R_L = inductive resistance (in Ω)

Example: What is the quality factor of an inductor in a series resonant circuit that has inductive reactance of 714 Ω and inductive resistance of 200 Ω?

$$Q = \frac{X_L}{R_L}$$

$$Q = \frac{714\ \Omega}{200\ \Omega}$$

$Q = \mathbf{3.57}$

Typical quality factors range from 10 to several hundred. The quality factor in the example is below the expected range of values because of the high resistance in the circuit. A series resonant circuit with a high quality factor (above 100) has low resistance and high current. A series resonant circuit with a medium quality factor (between 10 and 100) has medium resistance and current. A low quality factor (below 10) in a series resonant circuit indicates high resistance and low current. **See Figure 19-5.**

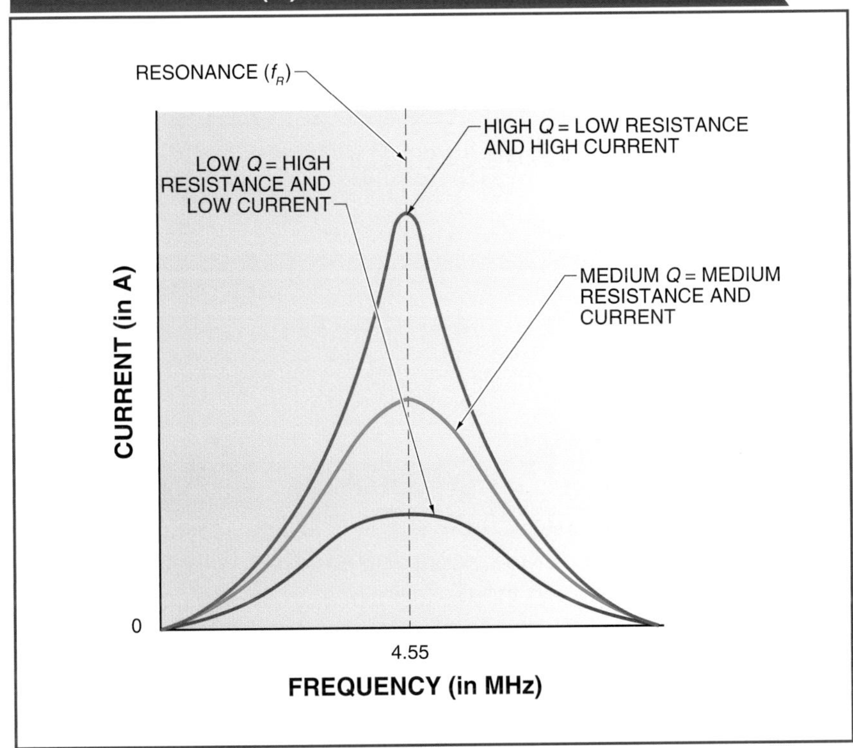

Figure 19-5. A series resonant circuit with a low quality factor has high resistance and low current, and with a high quality factor has low resistance and high current.

The quality factor of an inductor is different at each frequency because the inductive reactance of the inductor changes with frequency. Quality factor is calculated at the resonant frequency.

The quality factor of an inductor is greater at high resonant frequency since inductive reactance is greater than at low resonant frequency. Resistance is constant because it is not affected by frequency. Resonant frequency is used for many purposes, such as in frequency filters that decrease the strength of unwanted frequencies while amplifying desired frequencies.

Voltage drop across an inductor at resonant frequency is calculated because levels may exceed normal line current and voltage. These currents and voltages may be hundreds of times normal or expected values. In turn, underrated components will be destroyed due to insulation breakdown caused by the high currents and voltages. The quality factor of an inductor can be used to calculate the voltage drop across an inductor at resonant frequency, and is calculated by applying the following formula:

$$V_L = V_A \times Q$$

where

V_L = voltage drop across inductor (in V)

V_A = source voltage (in V)

Q = quality factor

Example: What is the voltage drop across an inductor with a quality factor of 3.57, located in a 100 V circuit?

$V_L = V_A \times Q$

$V_L = 100\text{ V} \times 3.57$

$V_L = \mathbf{357\text{ V}}$

Nondestructive Testing

Resonance is used to test the integrity of certain types of equipment and materials without destruction of the objects under test. A sonic probe transmits resonance (oscillations) from an electronic test instrument, such as an ultrasonic tester, into the test object. The resonances of the test object are picked up by a receiver probe and evaluated in the test instrument. The instrument not only defines elastic constants, density, hardness, and dimensions, but also detects and classifies imperfections such as cracks, cavities, porosities, chips, and missing teeth on gearwheels. When imperfections are detected, they can also be defined in dimension and location. Imperfections, depending on location, size, and mode of the vibration, cause a shift or split in some resonant frequencies. This allows the test instrument to collect data about the location and size of an imperfection and to separate each imperfection by changes in density or dimensions of the test object. Objects that are typically tested in this manner include items such as railroad rails, railroad tie-plates, steel axles, steel wheels, and welds used on structural I-beams. Resonance can also be used to test items made from stone, ceramic, and powdered metal.

Bandwidth in Series Resonant Circuits

Bandwidth is the range of frequencies that the circuit passes without a significant reduction in the signal magnitude. The quality factor affects the shape of current sine waves in a series resonant circuit. A change in the shape of the sine waves affects the bandwidth of the circuit. A *half-power point* is a frequency where the circuit power is half of the resonant power. Signals reduced to half power or lower typically do not interfere with the signals the circuit is designed to pass. Half power in a series resonant circuit occurs when the resistive current and voltage have been reduced to 0.707 of their maximum values at resonance. **See Figure 19-6.**

Terms

Bandwidth is the range of frequencies that the circuit passes without a significant reduction in the signal magnitude.

A **half-power point** is a frequency where the circuit power is half of the resonant power.

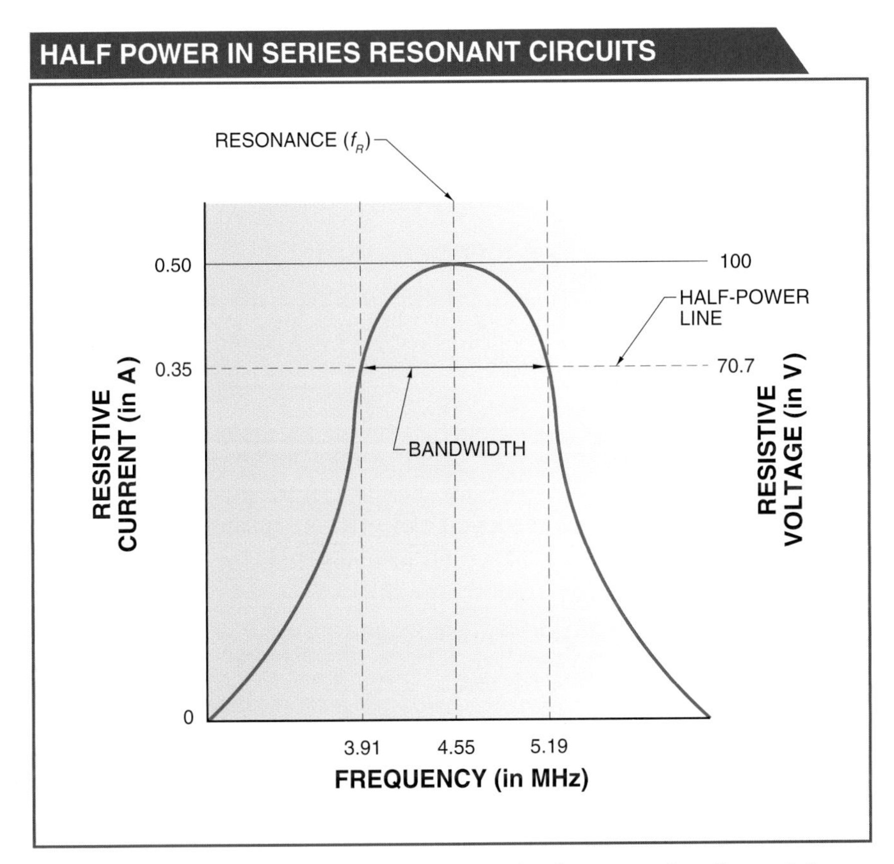

Figure 19-6. Half power in a series resonant circuit occurs when the resistive current and voltage have been reduced to 0.707 of their maximum values at resonance.

In any resonant circuit, the bandwidth includes all frequencies between the half-power points of the power curve. A circuit with a high quality factor has a narrower bandwidth than a circuit with a low quality factor. **See Figure 19-7.**

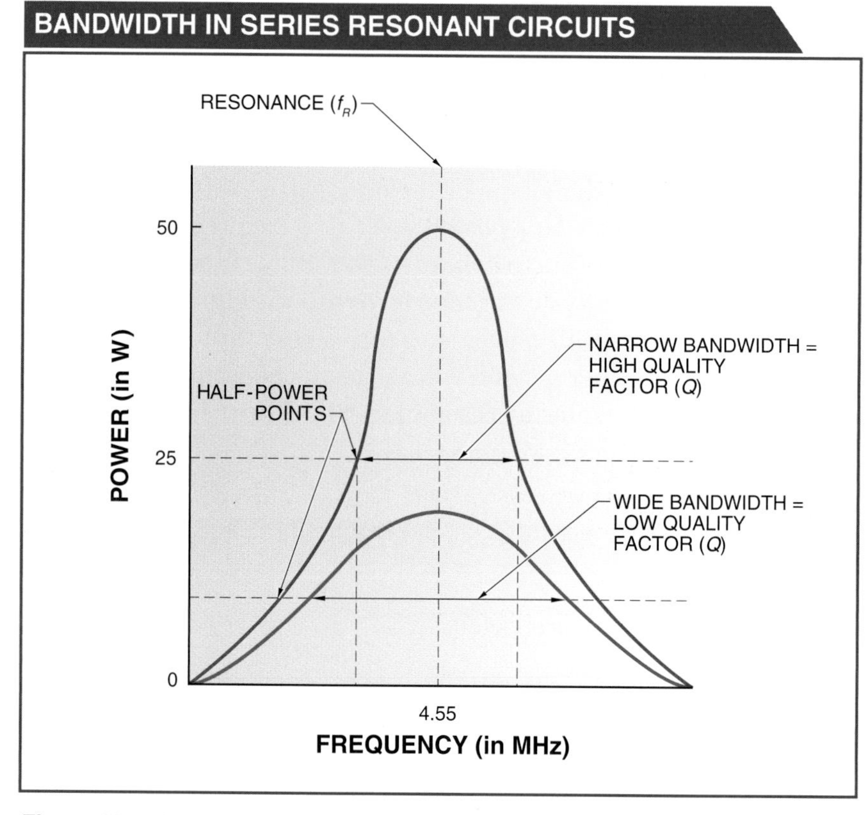

Figure 19-7. A circuit with a high quality factor (Q) has a narrower bandwidth than a circuit with a low quality factor.

Bandwidth in a series resonant circuit determines which signal frequencies are accepted and which are rejected. Bandwidth is calculated by applying the following formula:

$$B_W = \frac{f_R}{Q}$$

where

B_W = bandwidth (in Hz)

f_R = resonant frequency (in Hz)

Q = quality factor

Example: What is the bandwidth in a series resonant circuit with a frequency of 4.55 MHz and a quality factor of 3.57?

$$B_W = \frac{f_R}{Q}$$

$$B_W = \frac{4.55 \text{ MHz}}{3.57}$$

$$B_W = \mathbf{1.275\ MHz}$$

A series resonant circuit is affected by an increase or decrease in parameters such as inductance, capacitance, and resistance. **See Figure 19-8.** When inductance, capacitance, or resistance changes, other parameters such as capacitive reactance, inductive reactance, total current, apparent power, true power, voltage, angle theta, and power factor are affected. For example, an increase in inductance in a series resonant circuit causes an increase in inductive reactance but a decrease in most other circuit parameters.

EFFECTS OF PARAMETER CHANGES IN SERIES RESONANT CIRCUITS

	Inductance (*L*)		Capacitance (*C*)		Resistance (*R*)	
	Increase	Decrease	Increase	Decrease	Increase	Decrease
X_L	↑	↓	–	–	–	–
X_C	–	–	↓	↑	–	–
Z	↑	↑	↑	↑	↑	↓
I_T	↓	↓	↓	↓	↓	↑
P_{APP}	↓	↓	↓	↓	↓	↑
P_{TRUE}	↓	↓	↓	↓	↓	↑
V_L	↓	↓	↓	↓	↓	↑
V_C	↓	↓	↓	↓	↓	↑
V_R	↓	↓	↓	↓	–	–
θ	↑	↓	↑	↓	–	–
PF	↓	↓	↓	↓	–	–

Figure 19-8. A series resonant circuit is affected by an increase or decrease in parameters such as inductance, capacitance, and resistance.

PARALLEL RESONANT CIRCUITS

A *parallel resonant circuit* is a circuit in which the circuit elements are connected in parallel and the inductive and capacitive reactance are equal (as in a series resonant circuit). Major characteristics of a parallel resonant circuit include:

Terms

Parallel resonant circuit is a circuit in which the circuit elements are connected in parallel and the inductive and capacitive reactance are equal (as in a series resonant circuit).

- Maximum impedance
- Minimum allowable current in circuit
- High inductive and capacitive currents
- Angle theta equal to 0°

Voltage is used as the reference point in a parallel resonant circuit because the voltage across all parallel components is the same. Voltage is not measured in a parallel resonant circuit because it is typically known and is common across all parallel components. Parameters that are typically measured in a parallel resonant circuit include frequency, impedance, current, power, quality factor, and bandwidth. These parameters are used to understand the operation of the circuit and to ensure a safe design. Typical applications for parallel resonant circuits include oscillators and filters.

Frequency in Parallel Resonant Circuits

The resonant frequency in a parallel resonant circuit is the same as in a series resonant circuit. Resonance occurs when inductive reactance equals capacitive reactance. A *tank circuit (tuned circuit)* is an inductive-capacitive circuit that can store electric energy over a band of frequencies that are continuously distributed on a resonant single frequency. Parallel resonant circuits are sometimes referred to as tank circuits because of their ability to store energy. **See Figure 19-9.** For example, when the switch is in position 1, the capacitor charges (stores energy) to a peak value (V_p) of 100 V. When the switch is in position 2, the capacitor discharges through the inductor.

TANK CIRCUITS (PARALLEL RESONANT CIRCUITS)

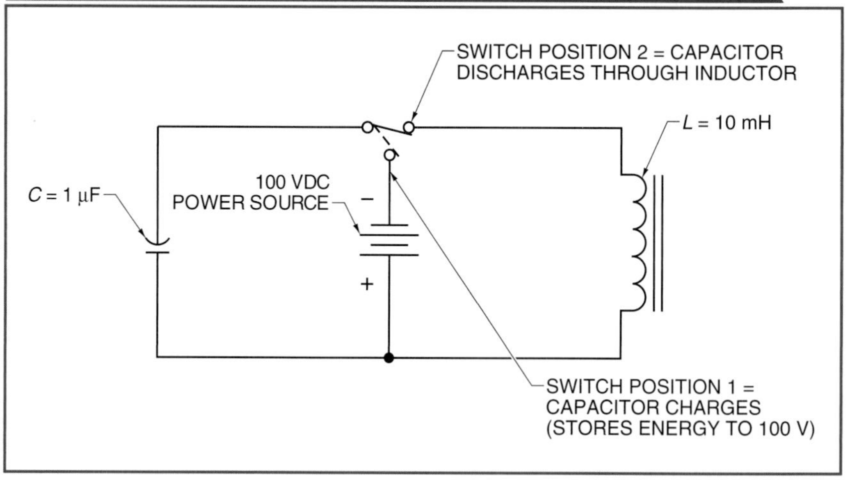

Figure 19-9. Parallel resonant circuits are sometimes referred to as tank circuits because of their ability to store energy.

When the capacitor in a parallel resonant circuit charges, the magnetic field expands out from the inductor due to the changing current flow through it. When the capacitor discharges, the magnetic field around the inductor collapses, causing current to flow in the opposite direction. Once the magnetic field is collapsed, the cycle is repeated, with the capacitor discharging through the inductor. Sine waves are generated by these actions. In this theoretical tank circuit, sine waves at resonant frequency are continuous for an infinite amount of time since no power is consumed. A *damped sine wave* is a sine wave that becomes reduced due to a resistance consuming power. When a resistance is added to a tank circuit, the sine wave oscillations become damped and eventually cease. **See Figure 19-10.**

Terms

A **damped sine wave** is a sine wave that becomes reduced due to a resistance consuming power.

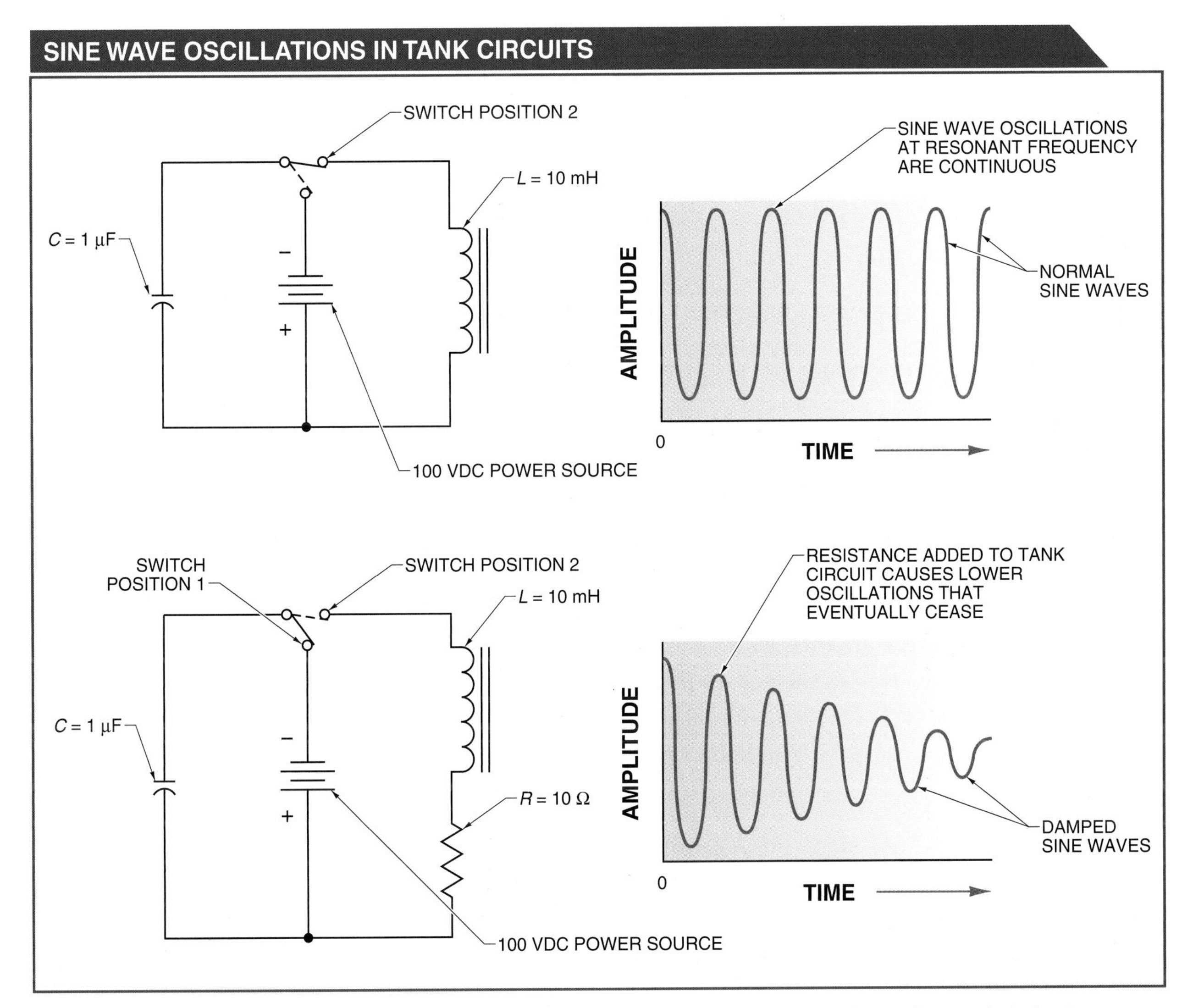

Figure 19-10. When a resistor is added to a tank circuit, sine wave oscillations cease after a short period, due to energy consumed by the resistor.

Resonant frequency in a parallel resonant circuit is calculated by applying the following formula:

$$f_R = \frac{1}{2\pi\sqrt{L \times C}}$$

where

f_R = resonant frequency (in Hz)
2π = 6.28 (in radians per cycle)
L = inductance (in H)
C = capacitance (in F)

Example: What is the resonant frequency in a parallel resonant circuit with inductance of 10 mH and capacitance of 1 μF?

$$f_R = \frac{1}{2\pi\sqrt{L \times C}}$$

$$f_R = \frac{1}{6.28\sqrt{(10 \times 10^{-3})\text{H} \times (1 \times 10^{-6})\text{F}}}$$

$$f_R = \frac{1}{6.28\sqrt{(0.0100)\text{H} \times (0.000001)\text{F}}}$$

$$f_R = \frac{1}{6.28\sqrt{0.00000001}}$$

$$f_R = \frac{1}{0.000628}$$

$$f_R = \mathbf{1592\ Hz}$$

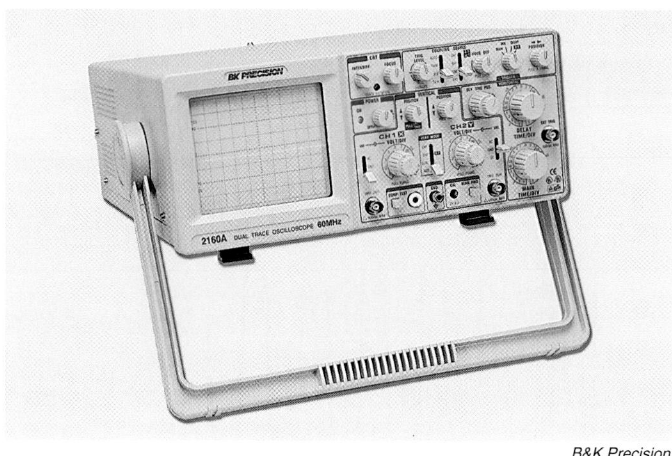

B&K Precision

Oscilloscopes are test instruments that use resonant circuits to test electronic components.

Impedance in Parallel Resonant Circuits

Impedance in a parallel resonant circuit is calculated as part of a parallel resonant circuit analysis. Impedance of each branch is calculated in order to determine the current through each branch. Angle theta in a parallel resonant circuit is calculated by applying the following formula:

$$\theta_C = \tan^{-1}\left(\frac{-X_C}{R_C}\right)$$

where

θ_C = capacitive impedance angle theta (in °)
X_C = capacitive reactance (in Ω)
R_C = capacitive resistance (in Ω)

Example: What is the impedance capacitive angle theta in a parallel resonant circuit with a capacitive reactance of 100 Ω and resistance of 0 Ω?

$$\theta_C = \tan^{-1}\left(\frac{-X_C}{R_C}\right)$$

$$\theta_C = \tan^{-1}\left(\frac{-100\ \Omega}{0\ \Omega}\right)$$

$\theta_C = \tan^{-1} - \infty$

$\theta_C = \mathbf{-90°}$

Capacitive impedance in a parallel resonant circuit is calculated by applying the following formula:

$Z_C = X_C$

where

Z_C = capacitive impedance (in Ω)

X_C = capacitive reactance (in Ω)

Example: What is the capacitive impedance in a parallel resonant circuit with capacitive reactance of 100 Ω?

$Z_C = X_C$

Z_C = **100 Ω at –90°**

Inductive impedance angle theta in a parallel resonant circuit is calculated by applying the following formula:

$$\theta_L = \tan^{-1}\left(\frac{X_L}{R_L}\right)$$

where

θ_L = inductive impedance angle theta (in °)

X_L = inductive reactance (in Ω)

R_L = inductive resistance (in Ω)

Example: What is the inductive impedance angle theta in a parallel resonant circuit with inductive reactance of 100 Ω and inductive resistance of 10 Ω?

$$\theta_L = \tan^{-1}\left(\frac{X_L}{R_L}\right)$$

$$\theta_L = \tan^{-1}\left(\frac{100\ \Omega}{10\ \Omega}\right)$$

$\theta_L = \tan^{-1} 10$

$\theta_L = \mathbf{84.3°}$

Inductive impedance in a parallel resonant circuit is calculated by applying the following formula:

$$Z_L = \sqrt{R_L{}^2 + X_L{}^2}$$

where

Z_L = inductive impedance (in Ω)
R_L = inductive resistance (in Ω)
X_L = inductive reactance (in Ω)

Example: What is the inductive impedance in a parallel resonant circuit with inductive resistance of 10 Ω and inductive reactance of 100 Ω?

$$Z_L = \sqrt{R_L{}^2 + X_L{}^2}$$

$$Z_L = \sqrt{(10\ \Omega)^2 + (100\ \Omega)^2}$$

$$Z_L = \sqrt{100\ \Omega + 10{,}000\ \Omega}$$

$$Z_L = \sqrt{10{,}100\ \Omega}$$

Z_L = **100.5 Ω at 84.3°**

Current in Parallel Resonant Circuits

Current in the individual branches of a parallel resonant circuit is calculated to determine the total line current, phase, and circuit impedance. In a parallel resonant circuit, inductive reactance equals capacitive reactance. The level of total circuit current is limited by the reactance of the two branches. Branch current in a parallel resonant circuit is calculated by applying the following procedure:

1. Calculate current through each capacitive branch. Current through a capacitive branch is calculated by applying the following formula:

   $$I_C = \frac{V_A}{Z_C}$$

   where

   I_C = capacitive branch current (in A)
   V_A = source voltage (in V)
   Z_C = capacitive branch impedance (in Ω)

2. Calculate current through each inductive branch. Current through an inductive branch is calculated by applying the following formula:

   $$I_L = \frac{V_A}{Z_L}$$

where

I_L = inductive branch current (in A)

V_A = source voltage (in V)

Z_L = inductive branch impedance (in Ω)

Example: What are the branch currents for a 100 V parallel resonant circuit with capacitive branch impedance of 100 Ω at –90° and inductive branch impedance of 100.5 Ω at 84.3°?

1. Calculate current through capacitive branch.

$$I_C = \frac{V_A}{Z_C}$$

$$I_C = \frac{100\text{ V}}{100\ \Omega\text{ at } -90^\circ}$$

I_C = **1 A at 90°**

2. Calculate current through inductive branch.

$$I_L = \frac{V_A}{Z_L}$$

$$I_L = \frac{100\text{ V}}{100.5\ \Omega\text{ at } 84.3^\circ}$$

I_L = **0.995 A at –84.3°**

Total line current is equal to the sum of the current through each branch. Total line current is calculated by applying the following procedure:

1. Calculate vertical components of each current. The vertical component is calculated by applying the following formula:

   $I_V = I \times \sin\theta$

   where

   I_V = vertical component of current (in A)

   I = branch current (in A)

   θ = angle theta (in °)

2. Calculate horizontal components of each current. The horizontal component is calculated by applying the following formula:

   $I_H = I \times \cos\theta$

   where

   I_H = horizontal component of current (in A)

   I = branch current (in A)

   θ = angle theta (in °)

3. Calculate total line current. Total line current is calculated by applying the following formula:

$$I_T = \sqrt{(I_{HL} + I_{HC})^2 + (I_{VL} + I_{VC})^2}$$

where

I_T = total line current (in A)

I_{HL} = horizontal component of inductive branch current (in A)

I_{HC} = horizontal component of capacitive branch current (in A)

I_{VL} = vertical component of inductive branch current (in A)

I_{VC} = vertical component of capacitive branch current (in A)

4. Calculate angle theta. Angle theta is calculated by applying the following formula:

$$\theta = \tan^{-1}\left(\frac{I_V}{I_H}\right)$$

where

θ = angle theta (in °)

I_V = vertical component of current (in A)

I_H = horizontal component of current (in A)

Example: What is the total line current and angle theta in a parallel resonant circuit with inductive branch current of 0.995 A at –84.3° and capacitive branch current of 1 A at 90°?

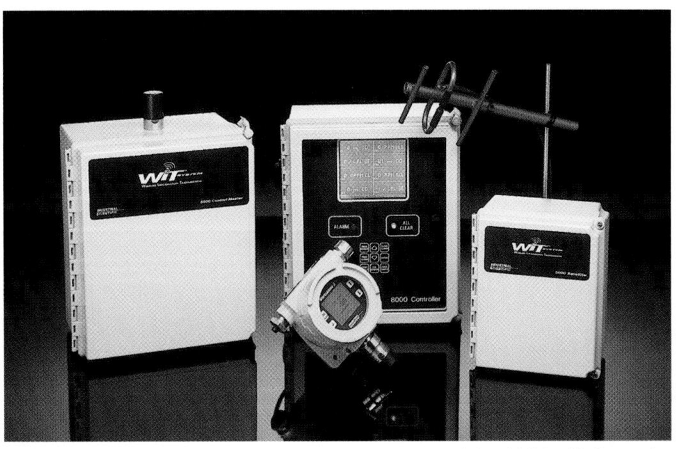

Industrial Scientific Corporation

Resonance is present in applications such as industrial-use wireless transmitters and controllers.

Note: The capacitive branch does not have a horizontal component.

1. Calculate vertical components of each current.

$I_{VL} = I_L \times \sin\theta$

$I_{VL} = 0.995\text{ A} \times \sin -84.3°$

$I_{VL} = \mathbf{-0.990\ A}$

$I_{VC} = I_C \times \sin\theta$

$I_{VC} = 1.0\text{ A} \times \sin 90°$

$I_{VC} = \mathbf{1.0\ A}$

2. Calculate horizontal components of each current.

$I_{HL} = I_L \times \cos\theta$

$I_{HL} = 0.995\text{ A} \times \cos -84.3°$

$I_{HL} = \mathbf{0.099\ A}$

$I_{HC} = I_C \times \cos\theta$

$I_{HC} = 1.0\text{ A} \times \cos 90°$

$I_{HC} = \mathbf{0\ A}$

3. Calculate total line current.

$$I_T = \sqrt{(I_{HL} + I_{HC})^2 + (I_{VL} + I_{VC})^2}$$

$$I_T = \sqrt{(0.099 \text{ A} + 0 \text{ A})^2 + (-0.990 \text{ A} + 1.0 \text{ A})^2}$$

$$I_T = \sqrt{(0.099 \text{ A})^2 + (0.01 \text{ A})^2}$$

$$I_T = \sqrt{0.0098 \text{ A} + 0.0001 \text{ A}}$$

$$I_T = \sqrt{0.0099 \text{ A}}$$

$$I_T = \mathbf{0.0995 \text{ A}}$$

4. Calculate angle theta.

$$\theta = \tan^{-1}\left(\frac{I_V}{I_H}\right)$$

$$\theta = \tan^{-1}\left(\frac{0.01 \text{ A}}{0.099 \text{ A}}\right)$$

$$\theta = \mathbf{5.8°}$$

Total circuit impedance in a parallel resonant circuit can be calculated when the values of source voltage and total circuit current are known. Impedance is calculated in parallel resonant circuits when working with applications such as audio amplifiers, speaker systems, and receivers. Total circuit impedance in a parallel resonant circuit is calculated by applying the following formula:

$$Z_T = \frac{V_A}{I_T}$$

where

Z_T = total circuit impedance (in Ω)

V_A = source voltage (in V)

I_T = total circuit current (in A)

Example: What is the total circuit impedance in a parallel resonant circuit with source voltage of 100 V and total circuit current 0.0995 A at 5.8°?

$$Z_T = \frac{V_A}{I_T}$$

$$Z_T = \frac{100 \text{ V}}{0.0995 \text{ A at } 5.8°}$$

$$Z_T = \mathbf{1005 \ \Omega \text{ at } -5.8°}$$

Power in Parallel Resonant Circuits

When total circuit current and the source voltage are known, apparent power, true power, and power factor can be calculated. Power in parallel resonant circuits is measured to calculate consumed power and the circuit power factor. Apparent power in a parallel resonant circuit is calculated by applying the following formula:

$P_A = I_T \times V_A$

where

P_A = apparent power (in VA)

I_T = total circuit current (in A)

V_A = source voltage (in V)

Power factor in a parallel resonant circuit is equal to the cosine of angle theta. Power factor is calculated using the cosine of angle theta because the value of angle theta is known. Power factor in a parallel resonant circuit is calculated by applying the following formula:

$PF = \cos\theta \times 100$

where

PF = power factor (in %)

θ = angle theta (in °)

True power in a parallel resonant circuit is calculated by applying the following formula:

$P_T = P_A \times PF$

where

P_T = true power (in W)

P_A = apparent power (in VA)

PF = power factor (in %)

Tech Tip

The power factor of an electric motor is lowest when the motor is not loaded and highest when the motor is loaded from 50% to the full-load rating.

Example: What are the apparent power, power factor, and true power in a parallel resonant circuit with total circuit current of 0.0995 A at 5.8° and a source voltage of 100 V?

Calculate apparent power.

$P_A = I_T \times V_A$

$P_A = 0.0995\text{ A} \times 100\text{ V}$

$P_A = \mathbf{9.95\text{ VA}}$

Calculate power factor.

$PF = \cos\theta \times 100$

$PF = \cos 5.8° \times 100$

$PF = \mathbf{99.49\%}$

Calculate true power.

$P_T = P_A \times PF$

$P_T = 9.95\text{ VA} \times 0.9949$

$P_T = \mathbf{9.9\text{ W}}$

Quality Factor in Parallel Resonant Circuits

The quality factor (Q) of a parallel resonant circuit is the ratio of the tank current to the total current. This approach takes into consideration resistance in the capacitive branch as well as the inductive branch. Quality factor in a parallel resonant circuit is calculated by applying the following formula:

$$Q = \frac{I_{TANK}}{I_T}$$

where

Q = quality factor

I_{TANK} = tank current (in A)

I_T = total current (in A)

Example: What is the quality factor of a parallel resonant circuit with tank current of 1 A and total current of 0.0995 A?

$$Q = \frac{I_{TANK}}{I_T}$$

$$Q = \frac{1\text{ A}}{0.0995\text{ A}}$$

$$Q = \mathbf{10}$$

The circulating tank current is typically the current flow in the capacitive branch of a parallel resonant circuit. This is because the current in the capacitive branch is normally larger at resonance due to the inherent resistance of the inductive branch. In effect, the line current from the source is used to compensate for the power consumed by the resistance in the tank circuit.

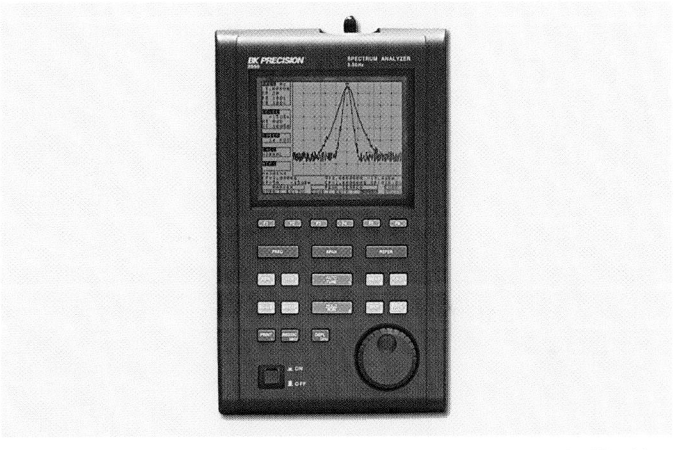

B&K Precision

Spectrum analyzers are used for testing the frequency of wireless signals and other high-frequency applications.

The quality factor of a parallel resonant circuit can also be calculated if the values of inductive reactance and tank circuit resistance are known. When the values of inductive reactance and tank circuit resistance are known, the quality factor of a parallel resonant circuit is calculated by applying the following formula:

$$Q = \frac{X_L}{R_{TANK}}$$

where

Q = quality factor

X_L = inductive reactance (in Ω)

R_{TANK} = tank circuit resistance (in Ω)

Tech Tip

The formula for the quality factor of a nonresonant parallel circuit is the reciprocal of the formula for the quality factor of a nonresonant series circuit.

Example: What is the quality factor of a parallel resonant circuit with inductive reactance of 100 Ω and tank circuit resistance of 10 Ω?

$$Q = \frac{X_L}{R_{TANK}}$$

$$Q = \frac{100\ \Omega}{10\ \Omega}$$

$$Q = \mathbf{10}$$

The quality factor in a parallel resonant circuit can be calculated by using different calculations related to the circuit impedance. The quality factor in a parallel resonant circuit can be calculated in the same manner as that of a series resonant circuit. A parallel resonant circuit appears to be a series circuit within the circulating tank, with the capacitor and inductor alternating as the source and then the load.

Bandwidth in Parallel Resonant Circuits

Bandwidth in a parallel resonant circuit is typically calculated to determine the frequencies over which the circuit operates. Bandwidth is calculated by applying the following formula:

$$B_W = \frac{f_R}{Q}$$

where

B_W = bandwidth (in Hz)

f_R = resonant frequency (in Hz)

Q = quality factor

Example: What is the bandwidth in a series resonant circuit with a frequency of 1600 Hz and a quality factor of 10?

$$B_W = \frac{f_R}{Q}$$

$$B_W = \frac{1600\ \text{Hz}}{10}$$

$$B_W = \mathbf{160\ Hz}$$

Once the bandwidth and the quality factor of a parallel resonant circuit are known, the upper and lower limits of the bandwidth can be calculated. The bandwidth lower limit is calculated by applying the following formula:

$$f_1 = f_R - \left(\frac{B_W}{2}\right)$$

where

f_1 = bandwidth frequency lower limit (in Hz)

f_R = resonant frequency (in Hz)

B_W = bandwidth (in Hz)

The bandwidth upper limit is calculated by applying the following formula:

$$f_2 = f_R + \left(\frac{B_W}{2}\right)$$

where

f_2 = bandwidth frequency upper limit (in Hz)

f_R = resonant frequency (in Hz)

B_W = bandwidth (in Hz)

Example: What are the lower and upper limits of the bandwidth of a 1600 Hz parallel resonant circuit with a bandwidth of 160 Hz?

Calculate bandwidth lower limit.

$$f_1 = f_R - \left(\frac{B_W}{2}\right)$$

$$f_1 = 1600\text{ Hz} - \left(\frac{160\text{ Hz}}{2}\right)$$

$$f_1 = 1600\text{ Hz} - 80\text{ Hz}$$

$$f_1 = \mathbf{1520\ Hz}$$

Calculate bandwidth upper limit.

$$f_2 = f_R + \left(\frac{B_W}{2}\right)$$

$$f_2 = 1600\text{ Hz} + \left(\frac{160\text{ Hz}}{2}\right)$$

$$f_2 = 1600\text{ Hz} + 80\text{ Hz}$$

$$f_2 = \mathbf{1680\ Hz}$$

Lower and upper bandwidth limits are the half-power points. The bandwidth is between the half-power points that occur at 0.707 times the maximum impedance and at the minimum current divided by 0.707. In a parallel resonant circuit, there is minimum current flow and maximum impedance at the resonant frequency. **See Figure 19-11.**

Unlike a series resonant circuit, the resonant point (where inductive reactance equals inductive capacitance) does not correspond exactly to the point where angle theta equals 0° or to when the current is at minimum. Each of the three frequency points is slightly different.

This is due to the higher impedance of the resistance included in the inductive branch. From a practical approach, the minimum current point is used when tuning a tank circuit using an ammeter. The circuit is tuned to resonance at the point of minimum line current.

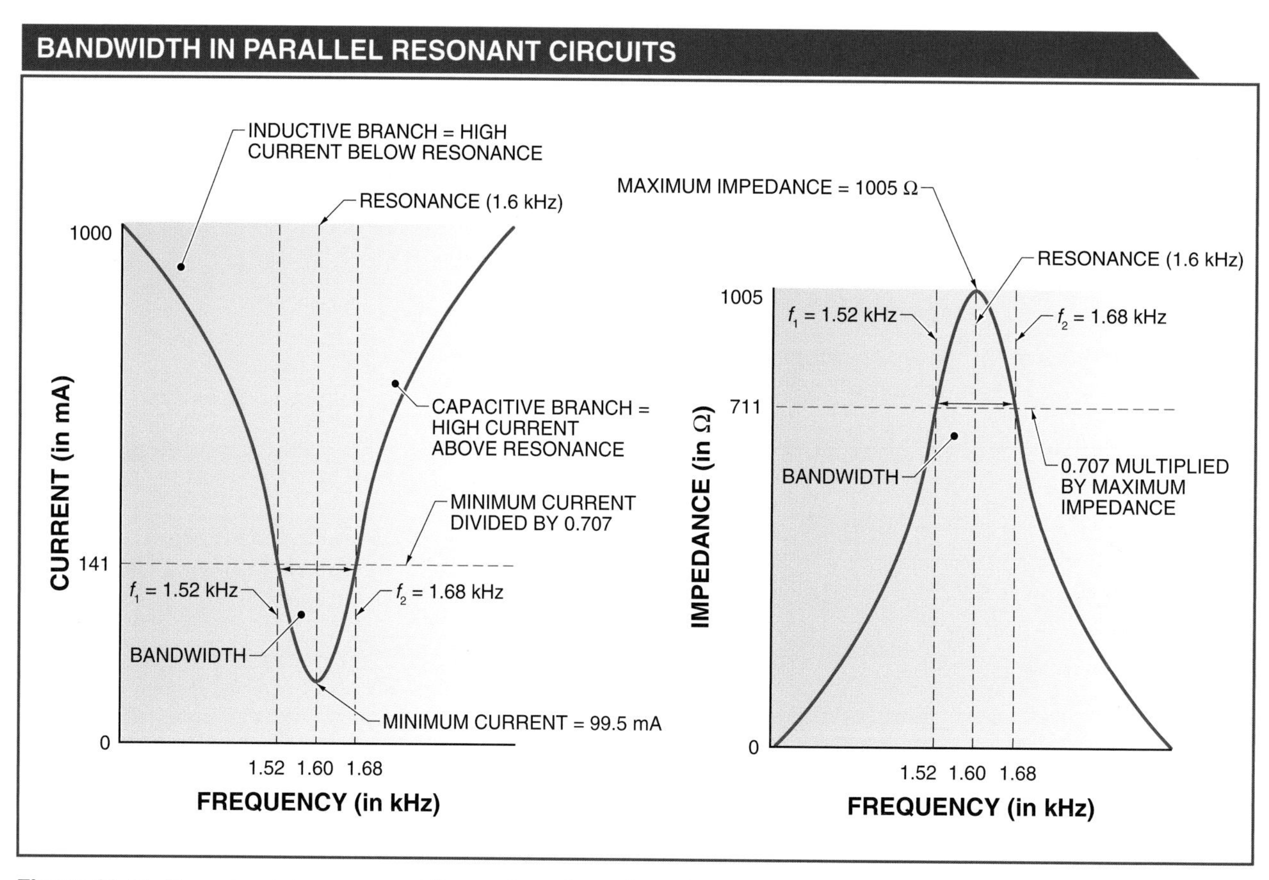

Figure 19-11. There is minimum current flow and maximum impedance in a parallel resonant circuit at resonant frequency.

The main difference between series and parallel resonant circuits is in their ability to pass AC. A series resonant circuit passes AC at the resonant frequency and attenuates current flow at nonresonant frequencies. A parallel resonant circuit attenuates AC at the resonant frequency and passes current at nonresonant frequencies.

Properties affected when either a series or parallel resonant circuit is tuned above or below resonance include reactance, impedance, and angle theta. **See Figure 19-12.** At resonance, both the series and parallel inductive, resistive, and capacitive circuits are resistive, so angle theta is equal to 0°.

RESONANT CIRCUIT PROPERTIES				
Property	**Below Resonance**		**Above Resonance**	
	Parallel	**Series**	**Parallel**	**Series**
Reactance	I_L Greater	X_C Greater	I_C Greater	X_L Greater
Impedance	Low	High	Low	High
Angle theta (θ)	Lagging current	Leading current	Leading current	Lagging current
Resulting circuit	Inductive	Capacitive	Capacitive	Inductive

Figure 19-12. Properties affected when either a series or parallel resonant circuit is tuned above or below resonance include reactance, impedance, and angle theta.

FREQUENCY FILTERS

A *frequency filter* is a circuit that passes certain frequencies and reduces signal levels at other frequencies. Frequency filters are combinations of inductors, capacitors, and resistors that make it possible to select or reject bands of frequencies. The characteristics of different combinations of these components make it possible to select broadcast from one radio or television station while excluding others. Federal Communications Commission (FCC) regulations limit AM stations to a bandwidth of 10 kHz, while FM stations are limited to a bandwidth of 150 kHz, and television stations are limited to a bandwidth of 6 MHz. Different bandwidths are used to transfer broadcast information and to separate the different broadcast stations.

The half-power point is used to separate the different station transmission frequencies. For example, the signal strength of an AM station broadcasting at 1 MHz must be at the half-power point (± 5 kHz) of the resonant frequency. If an AM station has a power output of 10 kW, then the power radiated must be at 5 kW at the half-power points, since these are the upper and lower limits of the bandwidth. Frequency range for the half-power points of an FM station broadcasting at 100 MHz is 99.85 MHz to 100.15 MHz. A television station broadcasting at 60 MHz has a bandwidth range of 57 MHz to 63 MHz.

Frequency filters are used for power applications as well as for communications applications. A *harmonic* is a frequency that is an integer (whole number) multiple (second, third, fourth, fifth, etc.) of the fundamental frequency. For example, certain equipment using variable-speed motors create harmonics of 60 Hz. Harmonics can cause excessive current and voltage peaks on the power line and are generally undesirable.

Terms

A **frequency filter** is a circuit that passes certain frequencies and reduces signal levels at other frequencies.

A **harmonic** is a frequency that is an integer (whole number) multiple (second, third, fourth, fifth, etc.) of the fundamental frequency.

Terms

Attenuation is a reduction in the strength of a frequency signal.

Frequency filters are used to eliminate harmonics. Power lines also pick up interference (noise) caused by lightning strikes and machine-made radiation from devices such as high-power transmitters or radiation devices. This type of interference can cause problems with other types of devices that may be on the line. Frequency filters are used to remove these undesirable signals.

Attenuation is a reduction in the strength of a frequency signal. Frequency filters are used for attenuation of undesirable frequency signals and are typically rated in decibels (dB). For example, a reduction of 3 dB represents the half-power point of the frequency signal. Frequency filters are designed to pass or reject a selected band of frequencies. It is not uncommon for a frequency filter to reduce an interfering signal by 60 dB. Frequency filter types are low-pass, high-pass, band-pass, and band-reject. **See Figure 19-13.**

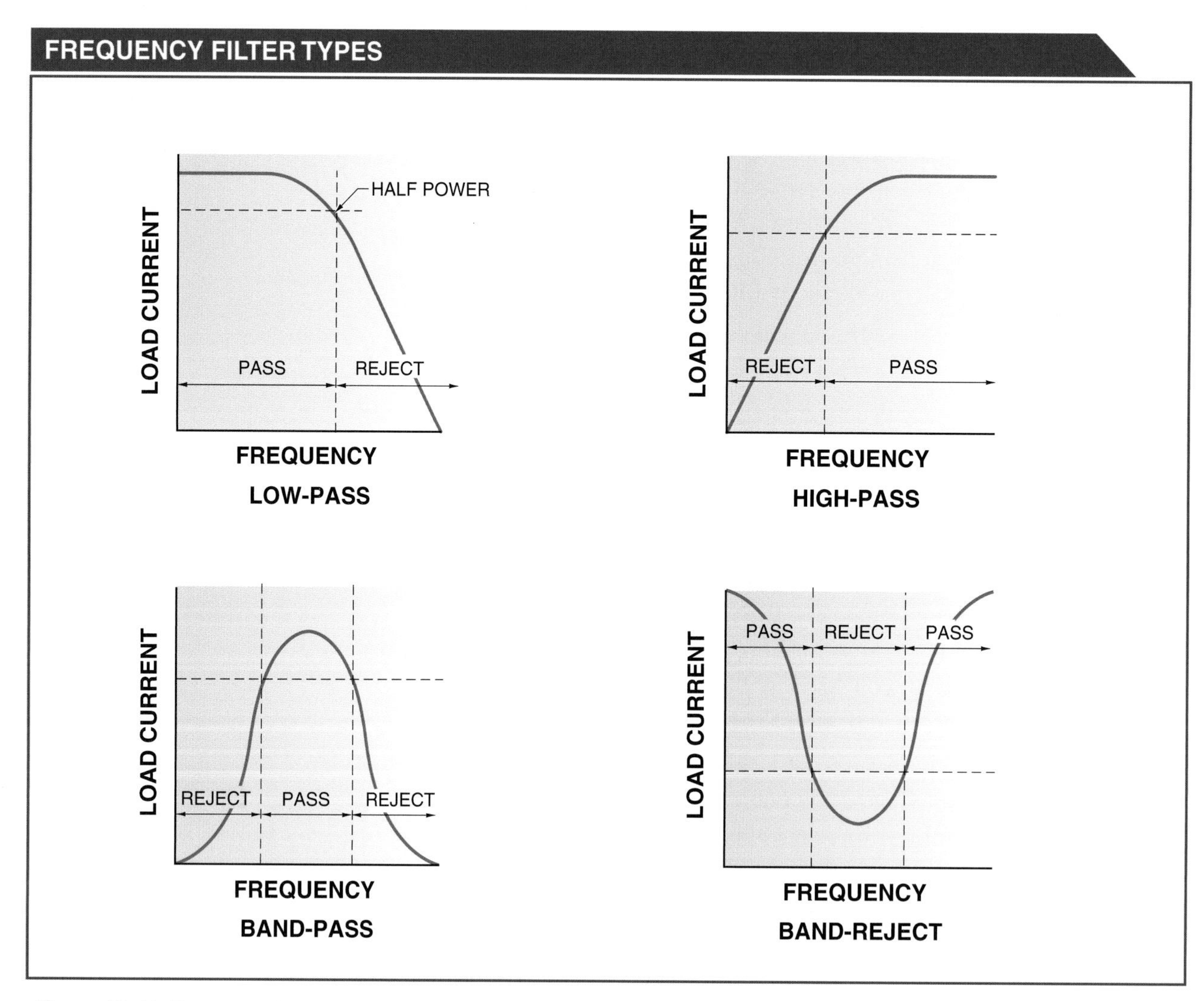

Figure 19-13. Frequency filter types are low-pass, high-pass, band-pass, and band-reject.

Low-Pass Frequency Filters

A *low-pass frequency filter* is a filter that allows all frequencies below a selected frequency to be applied to a load and blocks all frequencies above that point. **See Figure 19-14.** The simplest of these filters consist of an inductor connected in series with a load. Low-pass filters are typically used in DC power and in audio applications.

At 0 Hz, an inductor has zero reactance and the total source voltage is across the load. Under this condition there is maximum current flow through the load. As frequency increases, reactance increases in direct proportion. Some of the source voltage is dropped across the inductor, reducing current flow through the load.

Terms

A **low-pass frequency filter** is a filter that allows all frequencies below a selected frequency to be applied to a load and blocks all frequencies above that point.

Bode Plots...

Bode plots were created by Bell Labs scientist Hendrik Wade Bode in the 1930s. This type of plot is a simple but accurate method for graphing the magnitude and phase of a circuit over frequency. Usually frequency is plotted on a log scale, while magnitude is shown in decibels (dB) and phase in degrees. The circuit magnitude in decibels is 20 times the log (base 10) of the voltage output divided by the voltage input. The relationship between voltage output and voltage input, or the transfer function, is usually expressed as a mathematical formula that includes circuit and frequency elements. For example, a simple LR low-pass filter can be expressed as follows:

$$\frac{V_{OUT}}{V_{IN}} = \frac{\frac{R}{L}}{s + \frac{R}{L}}$$

where
V_{OUT} = output voltage
V_{IN} = input voltage
R = resistance (in Ω)
L = inductance (in H)
$s = j\omega$

The letter "j" indicates 90°, similar to inductive reactance, and ω is equal to $2\pi f$, where f is frequency in hertz (Hz). Substituting for s, the formula becomes the following:

$$\frac{V_{OUT}}{V_{IN}} = \frac{\frac{R}{L}}{2\pi f \text{ at } 90° + \frac{R}{L}}$$

To find the magnitude over frequency, the Pythagorean theorem is used and vector addition is performed on the denominator. To find the magnitude in decibels, the log of the equation is multiplied by 20.

$$\text{magnitude } (f) \text{ in dB} = 20 \times \log \frac{\frac{R}{L}}{\sqrt{(2\pi f)^2 + \left(\frac{R}{L}\right)^2}}$$

...Bode Plots

Plotting the result over frequency generates a magnitude Bode Plot.

Example:

What is the magnitude equation for an LR low-pass filter with an inductance of 0.01 H and a resistance of 100 Ω?

1. The formula for the equation is the following:

$$\frac{V_{OUT}}{V_{IN}} = \frac{\frac{R}{L}}{s + \frac{R}{L}}$$

2. Substituting for *L*, *R*, and *s*, the equation becomes the following:

$$\frac{V_{OUT}}{V_{IN}} = \frac{\frac{100}{0.01}}{2\pi f \text{ at } 90° + \frac{100}{0.01}}$$

3. The magnitude in decibels is the following:

$$\text{magnitude } (f) \text{ in dB} = 20 \times log\left(\frac{10{,}000}{\sqrt{(2\pi f)^2 + 10{,}000^2}}\right)$$

4. Plotting this equation over frequency generates the following graph:

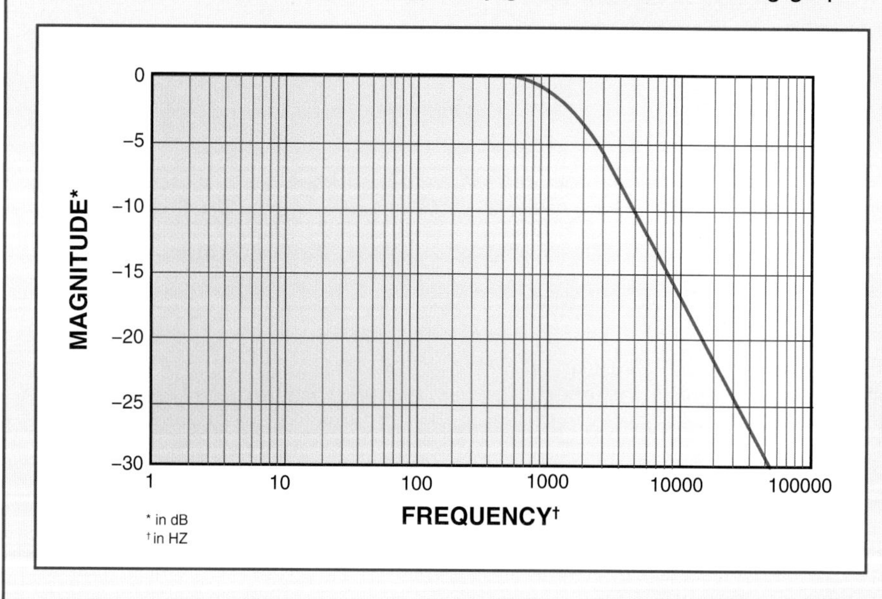

Terms

A **cutoff frequency** is the half-power frequency point.

Note: The current sine wave seen in graphic displays can be replaced by a resistive voltage sine wave. Half-power points for either current or voltage sine waves are equal to 0.707 of the maximum values.

A *cutoff frequency* is the half-power frequency point. A filter response diagram can be used to show signal level versus frequency.

A filter response diagram shows which frequencies are allowed to pass as well as which frequencies are rejected. The value of an inductor can be calculated if the values of the load and the cutoff frequency are known. To obtain these characteristics for the filter, the inductive reactance should equal the resistance of the load at the cutoff frequency. The filter pass band is the range of frequencies where the signal is greater than the half-power points. Inductance in a resonant circuit with a low-pass filter is calculated by applying the following formula:

$$L = \frac{X_L}{2\pi \times f}$$

where
L = inductance (in H)
X_L = inductive reactance (in Ω)
2π = 6.28 (in radians per cycle)
f = frequency (in Hz)

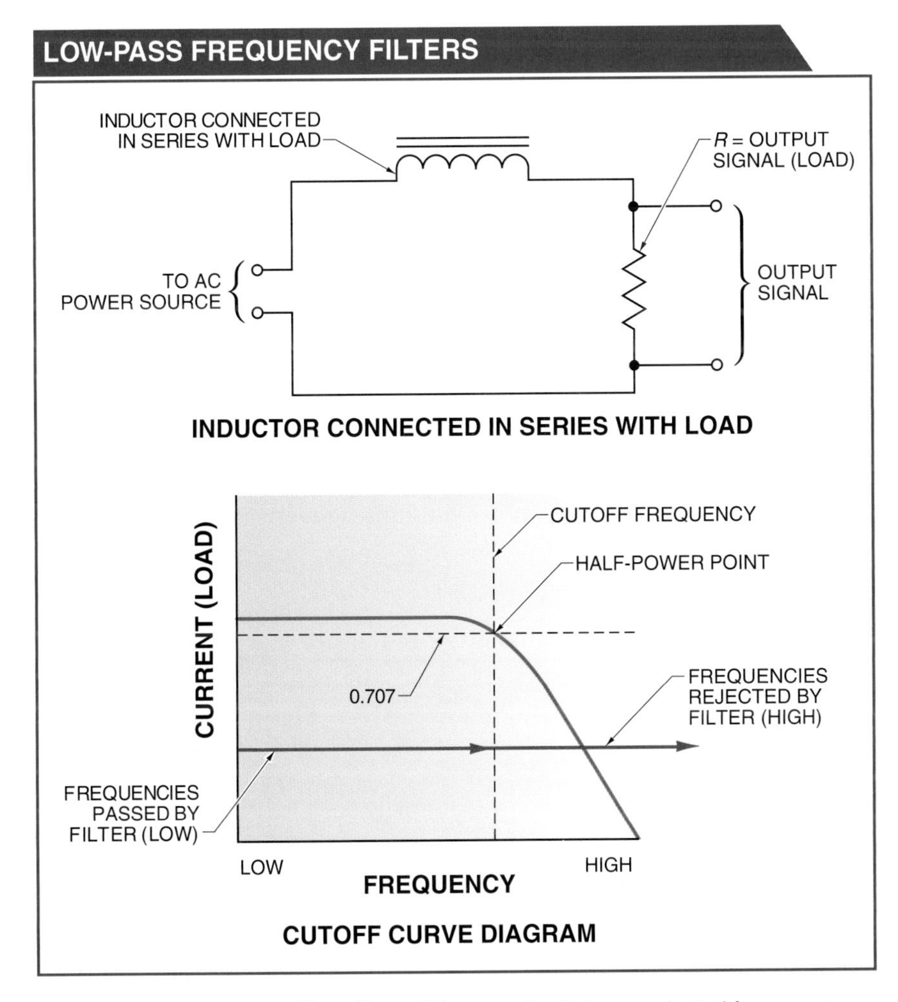

Figure 19-14. A low-pass filter allows all frequencies below a selected frequency band to be applied to a load and blocks all frequencies above that point.

Example: What is the inductance of a resonant circuit with a low-pass filter that has a load of 72 Ω and a cutoff frequency of 1 MHz?

$$L = \frac{X_L}{2\pi \times f}$$

$$L = \frac{72\ \Omega}{6.28 \times 10^6\ \text{Hz}}$$

$$L = \frac{72\ \Omega}{6.28 \times 1{,}000{,}000\ \text{Hz}}$$

$$L = \frac{72\ \Omega}{6{,}280{,}000\ \text{Hz}}$$

L = **1.15 μH**

Some low-pass filters are designed to use a capacitor. While two resistances are connected in series, a capacitor is connected in parallel with the load resistance. **See Figure 19-15.** A capacitor connected in parallel reduces the current through the load at higher frequencies. At low frequencies, a capacitor has high reactance, and most of the current passes through the load. As the frequency is increased, the capacitive reactance decreases so that more current flows through the capacitor and less current flows through the resistance. For all practical purposes, when the capacitive reactance is 1/10 the value of the load, all current passes through the capacitive branch.

A current-limiting resistor is typically connected in series with the circuit to limit excessive current as capacitive reactance approaches zero. Excessive current can damage or destroy components and devices in the circuit. The pass band is below the frequency where the capacitive reactance equals the load resistance. Typical low-pass filter applications include smoothing the output of a DC power supply. Power supplies are used in most consumer electronics to transform 120 VAC line voltage into the lower DC voltages used by these types of products.

When the value of the capacitive reactance is known along with the frequencies to be passed to the load, the value of capacitance can be calculated by applying the following formula:

$$C = \frac{1}{2\pi \times f \times X_C}$$

where

C = capacitance (in F)

2π = 6.28 (in radians per cycle)

f = frequency (in Hz)

X_C = capacitive reactance (in Ω)

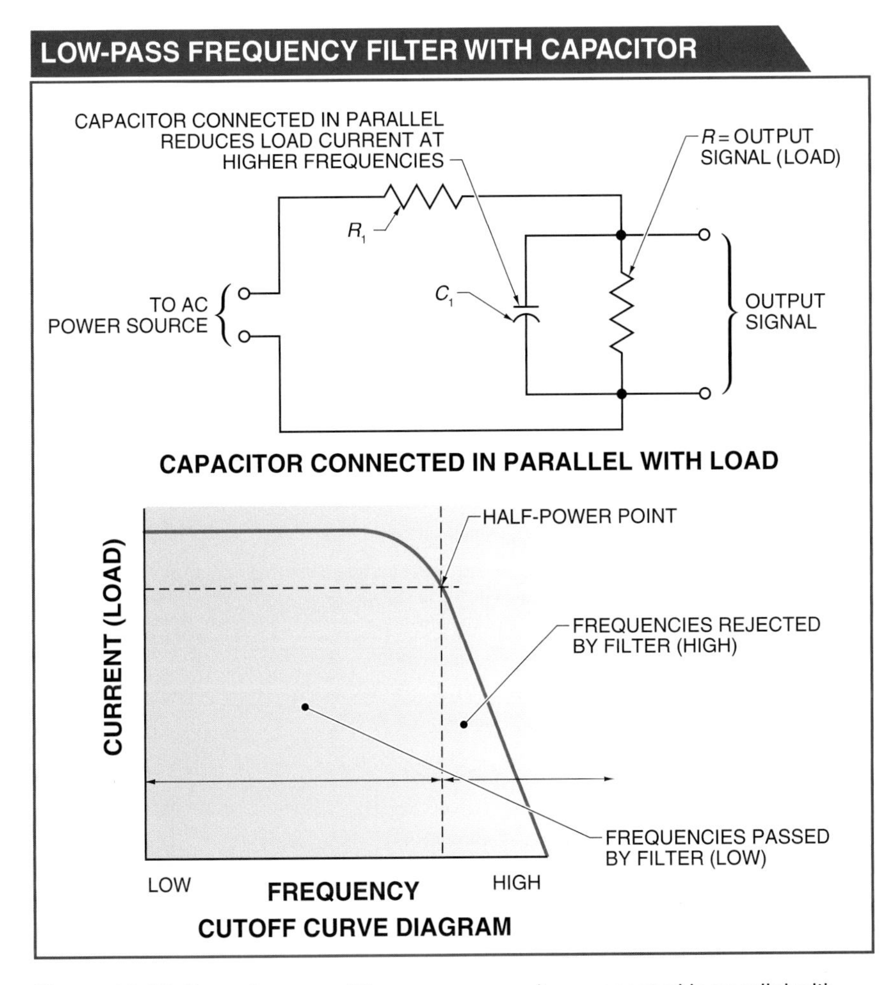

Figure 19-15. Some low-pass filters use a capacitor connected in parallel with the load resistance to reduce the current through the load at higher frequencies.

Bypass Capacitors

Bypass capacitors are used in circuits that have pulsating DC. When pulsating DC flows through a resistance, the voltage across it varies, which can have undesirable effects on a circuit. A bypass capacitor is placed in parallel with the resistance, which causes the voltage variations to equalize across the resistance.

Example: What is the capacitance of a resonant circuit with a low-pass filter that has a frequency of 1 MHz and capacitive reactance of 72 Ω?

$$C = \frac{1}{2\pi \times f \times X_C}$$

$$C = \frac{1}{6.28 \times f \times X_C}$$

$$C = \frac{1}{6.28 \times 10^6 \text{ Hz} \times 72\ \Omega}$$

$$C = \frac{1}{6.28 \times 1{,}000{,}000\ \text{Hz} \times 72\ \Omega}$$

$$C = \frac{1}{452{,}160{,}000}$$

C = **0.0000000022 F (2.2 nF)**

Low-pass filters that include inductors and capacitors result in a sharper cutoff curve. **See Figure 19-16.** These characteristics result in a circuit designed such that resonance occurs at or near the cutoff frequency. A slight peak in the current flow through the load indicates resonance. As frequency moves above the resonant point, current is reduced much more rapidly than when only an inductor or capacitor is used to implement a low-pass filter.

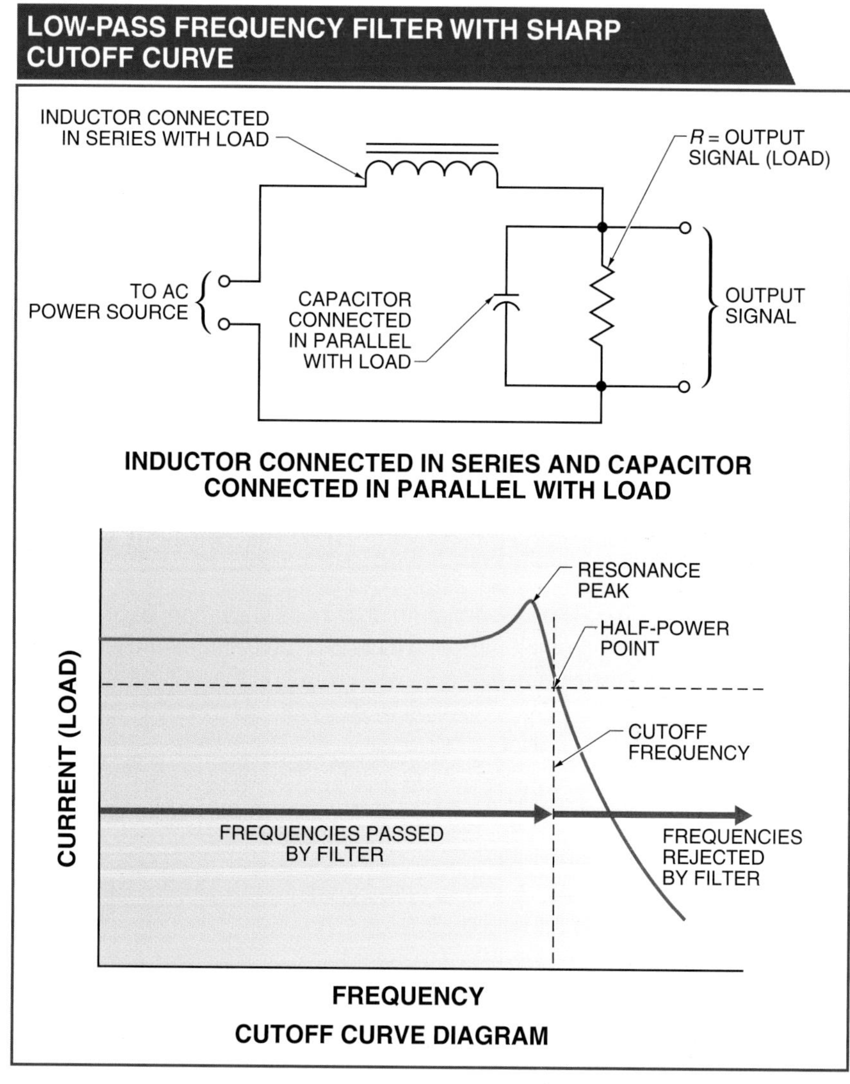

Figure 19-16. Low-pass filters designed with inductors and capacitors result in a sharp cutoff curve.

High-Pass Frequency Filters

A *high-pass frequency filter* is a filter that allows all frequencies above a selected frequency to be applied to a load and blocks all frequencies below that point. **See Figure 19-17.** Like low-pass filters, combinations of inductive-resistive-capacitive circuits are designed for this purpose. The simplest of these filters is a capacitor in series with a load or an inductor in parallel with a load. High-pass filters are typically used in applications related to communications, such as in radio and television broadcast, and in residential appliances to eliminate the effects of 60 Hz noise.

Terms

A **high-pass frequency filter** is a filter that allows all frequencies above a selected frequency to be applied to a load and blocks all frequencies below that point.

HIGH-PASS FREQUENCY FILTERS

CAPACITOR CONNECTED IN SERIES WITH LOAD

INDUCTOR CONNECTED IN PARALLEL WITH LOAD

CUTOFF CURVE DIAGRAM

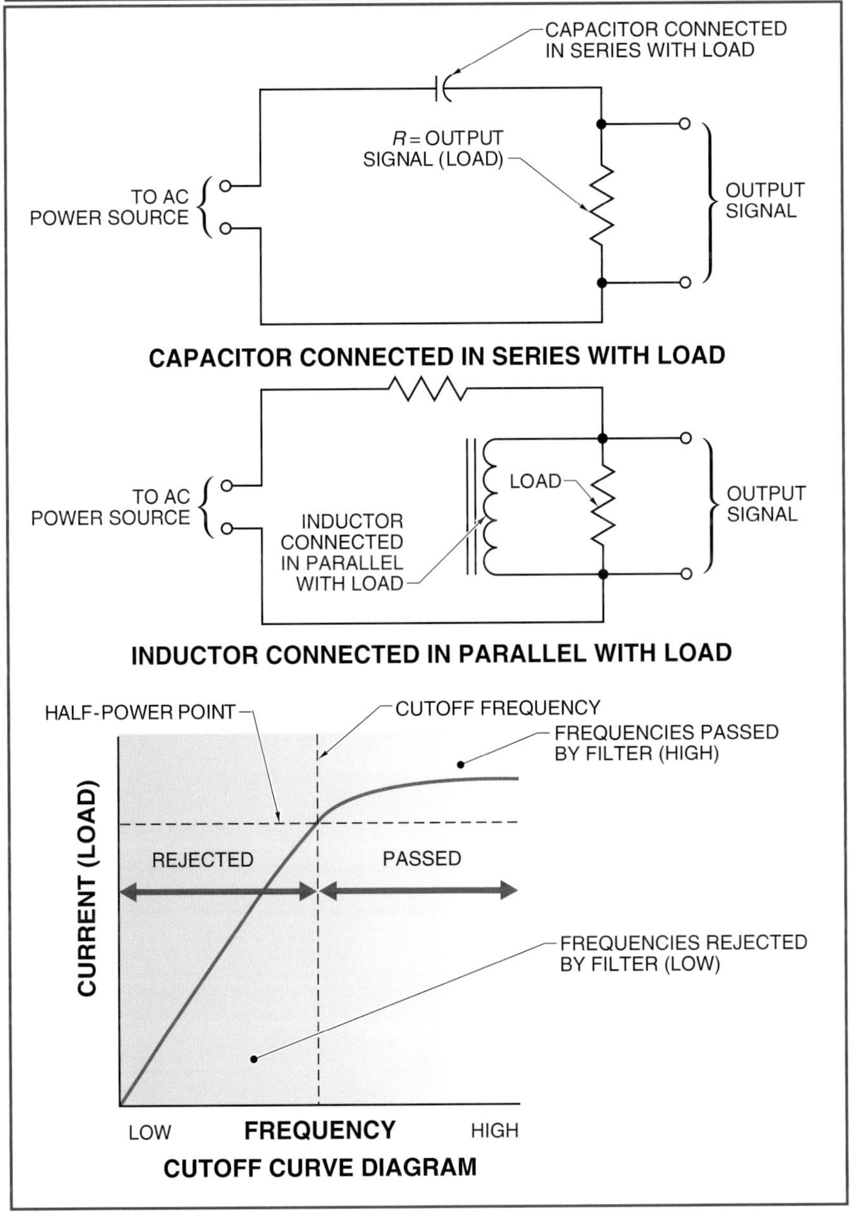

Figure 19-17. A high-pass filter allows all frequencies above a selected frequency to be applied to a load and blocks all frequencies below that point.

Terms

Reject frequencies are frequencies that are rejected because of amplitude or size.

Reject frequencies are frequencies that are rejected because of amplitude or size. At reject frequencies, the capacitor presents a high reactance to the source, limiting the current flow through the load. As frequency increases, reactance decreases until capacitive reactance equals load resistance and reaches the high-pass band. The inductor presents a low reactance at the reject frequencies, which shunts the current around the load resistor. As frequency increases, inductive reactance also increases, until it is equal to the value of the load. A sharp cutoff curve for a high-pass filter can be obtained using a resonant circuit. **See Figure 19-18.** At this point, the high-pass band of frequencies flows through the load, delivering half power or more.

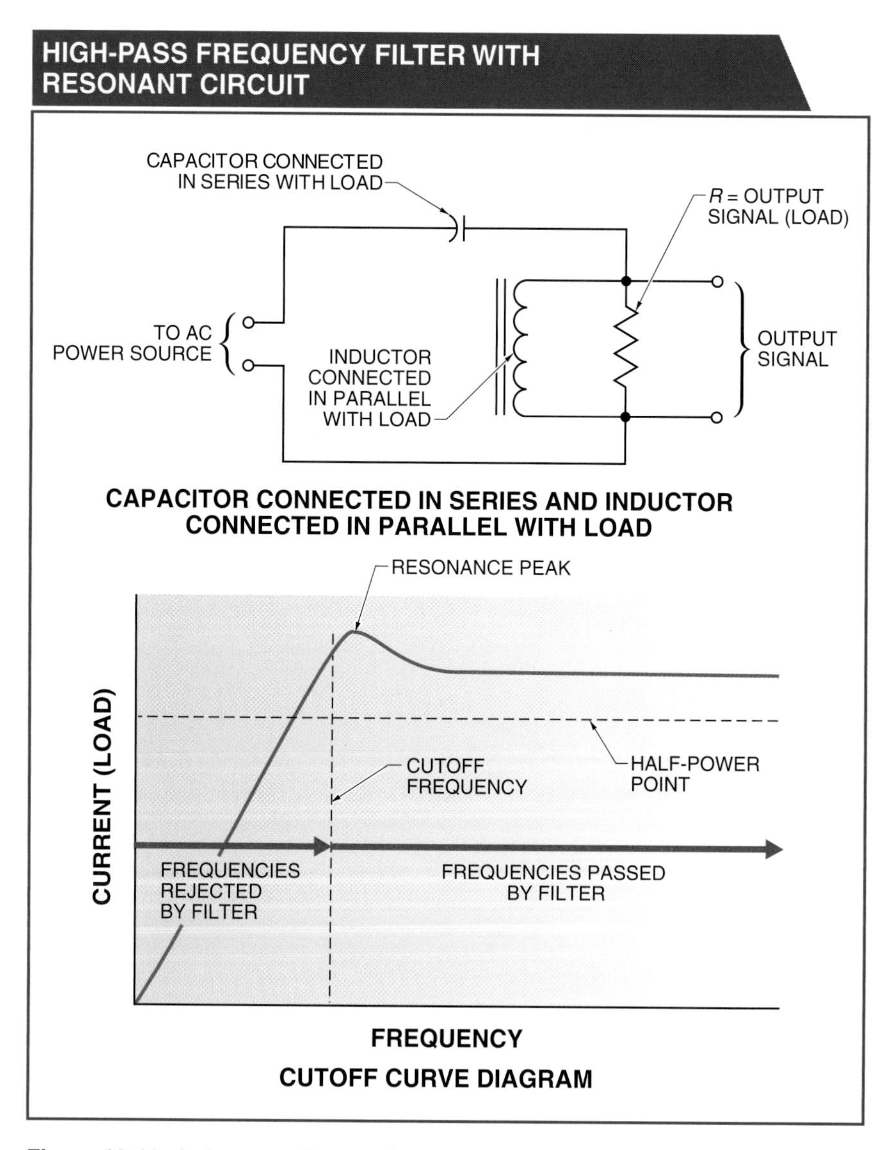

Figure 19-18. A sharp cutoff curve for a high-pass filter can be obtained using a resonant circuit.

If the impedance of the load and the pass band are known, the values of the capacitor and the inductor can be calculated by using the same equations as for the low-pass filter. As with a low-pass filter, the inductive or capacitive reactance is equal to the load at the cutoff frequency.

A sharp cutoff curve for a high-pass filter is the mirror image of a sharp cutoff curve for a low-pass filter. Because of resonance, the peak is part of the curve, but the slope is greater, with a sharper cutoff.

Terms

A **band-pass frequency filter** is a filter that passes signals between selected frequencies.

A **crossover network** is a system of band-pass filters used to route the appropriate signal to the appropriate loudspeaker.

Band-Pass Frequency Filters

A *band-pass frequency filter* is a filter that passes signals between selected frequencies. Resonance can be used to allow one band of frequencies to be delivered to a load. The simplest of these circuits are a series resonant circuit and a parallel resonant circuit. Band-pass filters are typically used in communications applications for the purpose of tuning frequency bands.

Fluke Corporation

Resonance is used to filter selected frequencies in applications such as radio broadcast antennas.

In a series resonant circuit, capacitance has high reactance below resonant frequency, and inductance has low reactance below the resonant frequency. Above resonant frequency, the capacitance has a low reactance, and inductance has high reactance. At resonant frequency, the inductive reactance equals capacitive reactance and maximum current passes through the load resistor. By using a band-pass filter, the band of frequencies passed to the load includes those between the half-power points of the curve. **See Figure 19-19.**

The same band-pass circuit can be obtained using a parallel resonant circuit in parallel with the load. Below resonant frequency, the inductive reactance is low and current bypasses the load. Above resonant frequency, the capacitive reactance is low and shunts the current around the load. At resonant frequency, the impedance of the parallel circuit is at its maximum value, causing the maximum current flow to the load. Resistance limits current flow in the circuit when either the capacitive reactance or the inductive reactance is equal to zero. Band-pass filters are used in the reproduction of audio frequencies in stereo speakers. Typical stereo speaker systems have three loudspeakers including small (tweeter), midsize, and large (woofer). A *crossover network* is a system of band-pass filters used to route the appropriate signal to the appropriate loudspeaker.

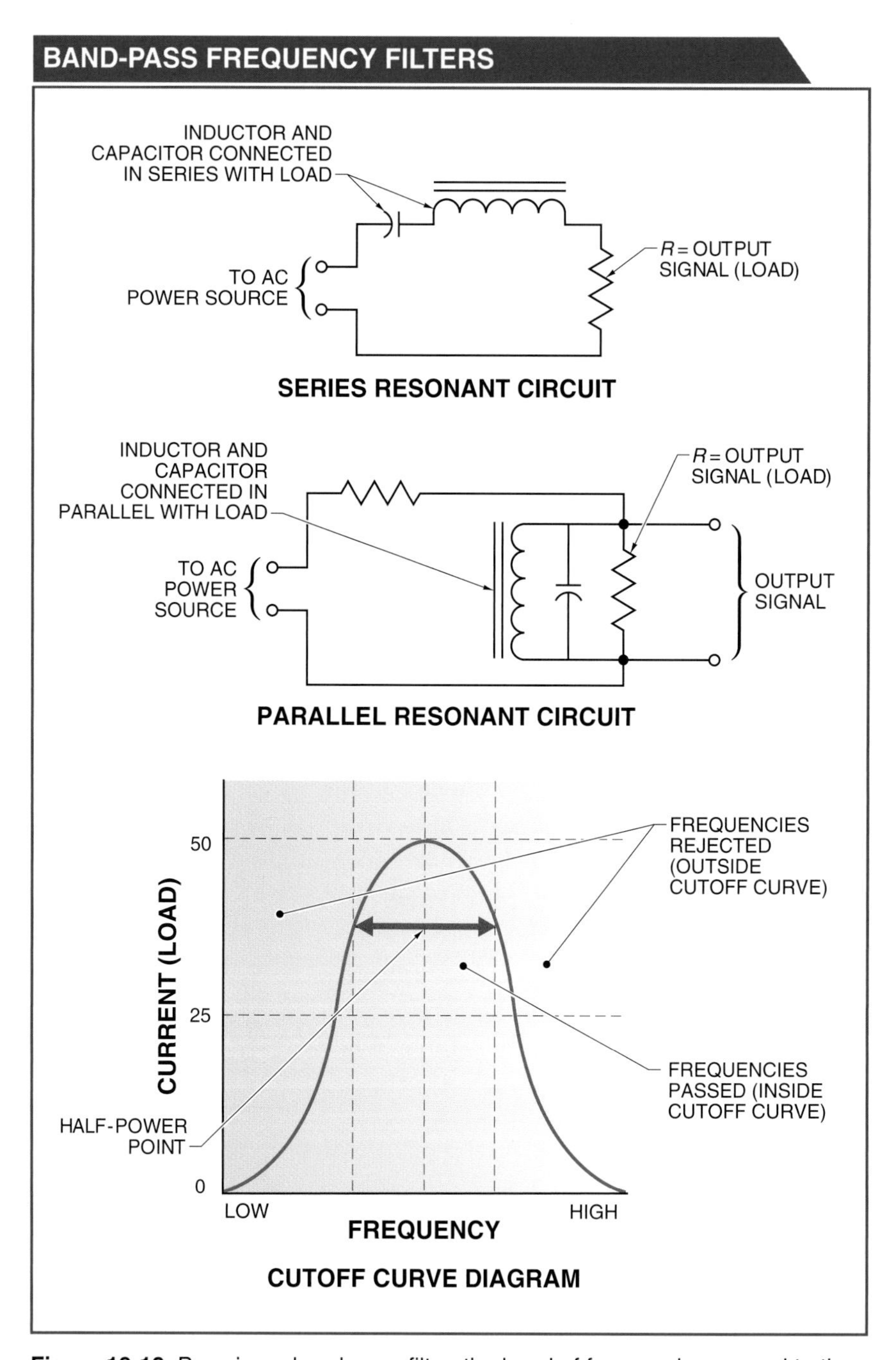

Figure 19-19. By using a band-pass filter, the band of frequencies passed to the load includes those between the half-power points of the curve.

Band-Reject Frequency Filters

A *band-reject frequency filter* is a filter that reduces signals between selected frequencies. By using a band-reject filter, selected frequency bands can be prevented from reaching the load. **See Figure 19-20.** Filtering of selected frequency bands is accomplished by using a

Terms

A **band-reject frequency filter** is a filter that reduceses signals between selected frequencies.

series/parallel resonant circuit designed with a rejection band. Band-reject filters are typically used in applications to remove a certain frequency, such as 60 Hz.

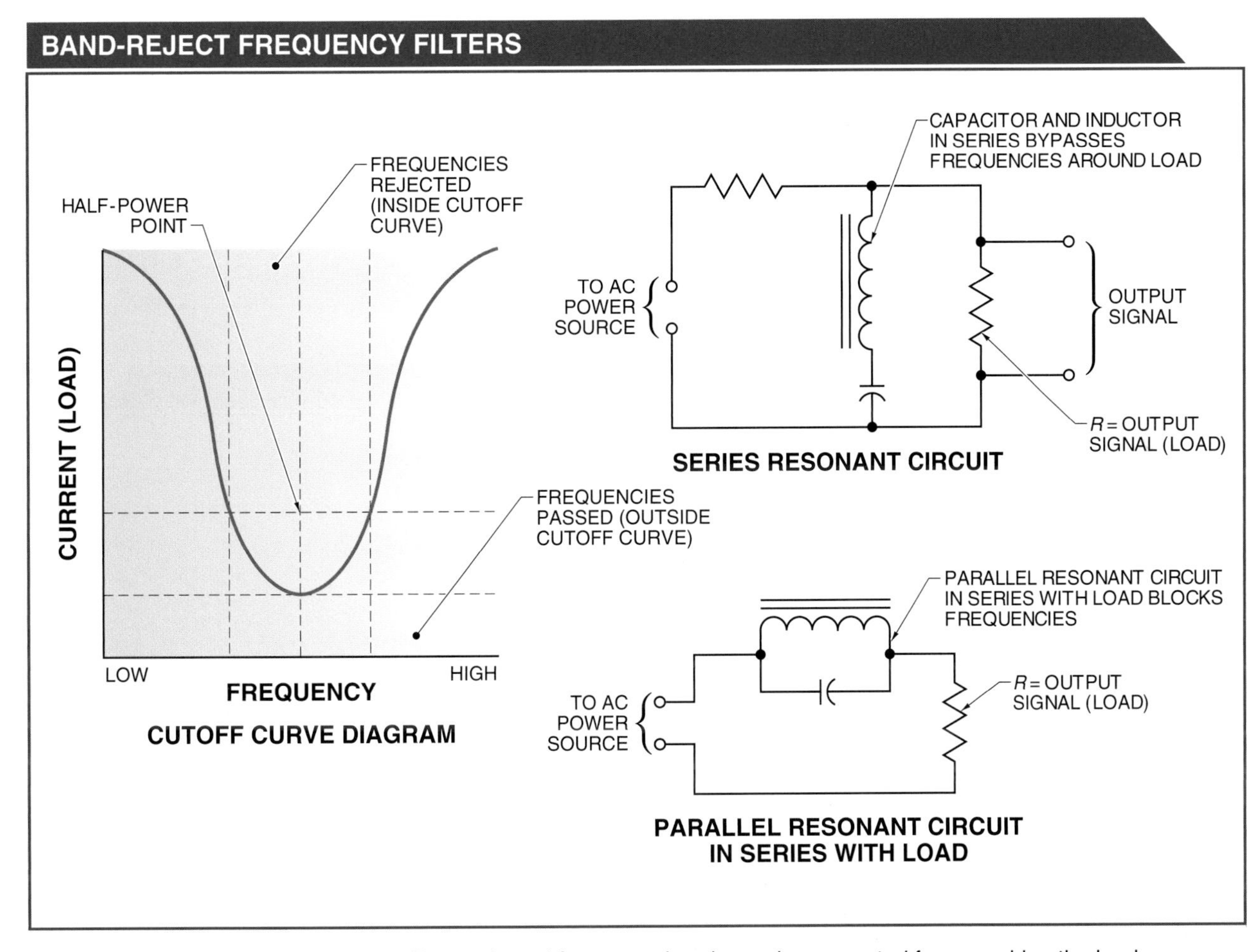

Figure 19-20. By using a band-reject filter, selected frequency bands can be prevented from reaching the load.

When a series resonant circuit is placed in shunt with the load resistor, the series resonant circuit, at its resonant frequency, will short most of the current around the load resistor. Since this arrangement acts as a short circuit, a resistor is used to limit current flow from the source to a safe level.

Current flow from the source can also be limited by placing a parallel resonant circuit in series with the load. At resonance, placing a parallel resonant circuit in series with the load presents very high impedance to the source, limiting current flow through the load. Maximum current flow in this circuit is limited by the value of the load and the value of the source voltage.

Terms

A **frequency filter** network is a complex design of frequency filters arranged to improve filter performance.

FREQUENCY FILTER NETWORKS

A *frequency filter network* is a complex design of frequency filters arranged to improve filter performance. Simple frequency filters, such as low-pass, high-pass, band-pass, and band-reject, are useful, but in many applications they are not able to remove or pass frequencies in a sufficient manner. For this reason, more complex filter networks are often required. Complex filter networks are often referred to as half-sections, because they represent only a part of the filter. Complex filter networks are also referred to as "pi" (π) networks and "T" networks, because of the shape of the circuits. **See Figure 19-21.** It is not uncommon to group as many T networks or pi networks together as necessary to achieve a desired result. Filter networks are used in applications that require greater filter performance than can be achieved with simple resonant circuits.

PI AND T NETWORKS . . .

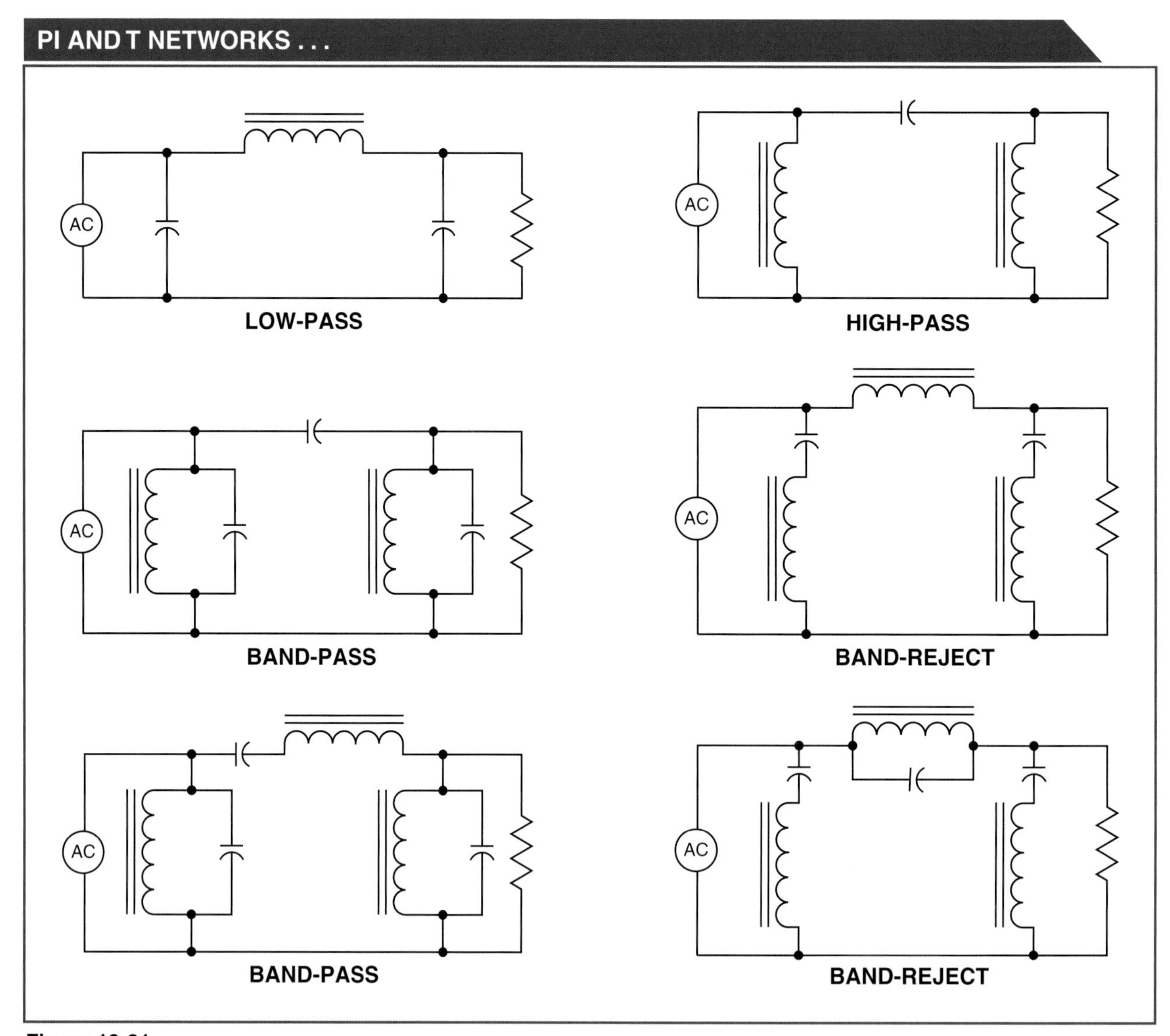

Figure 19-21...

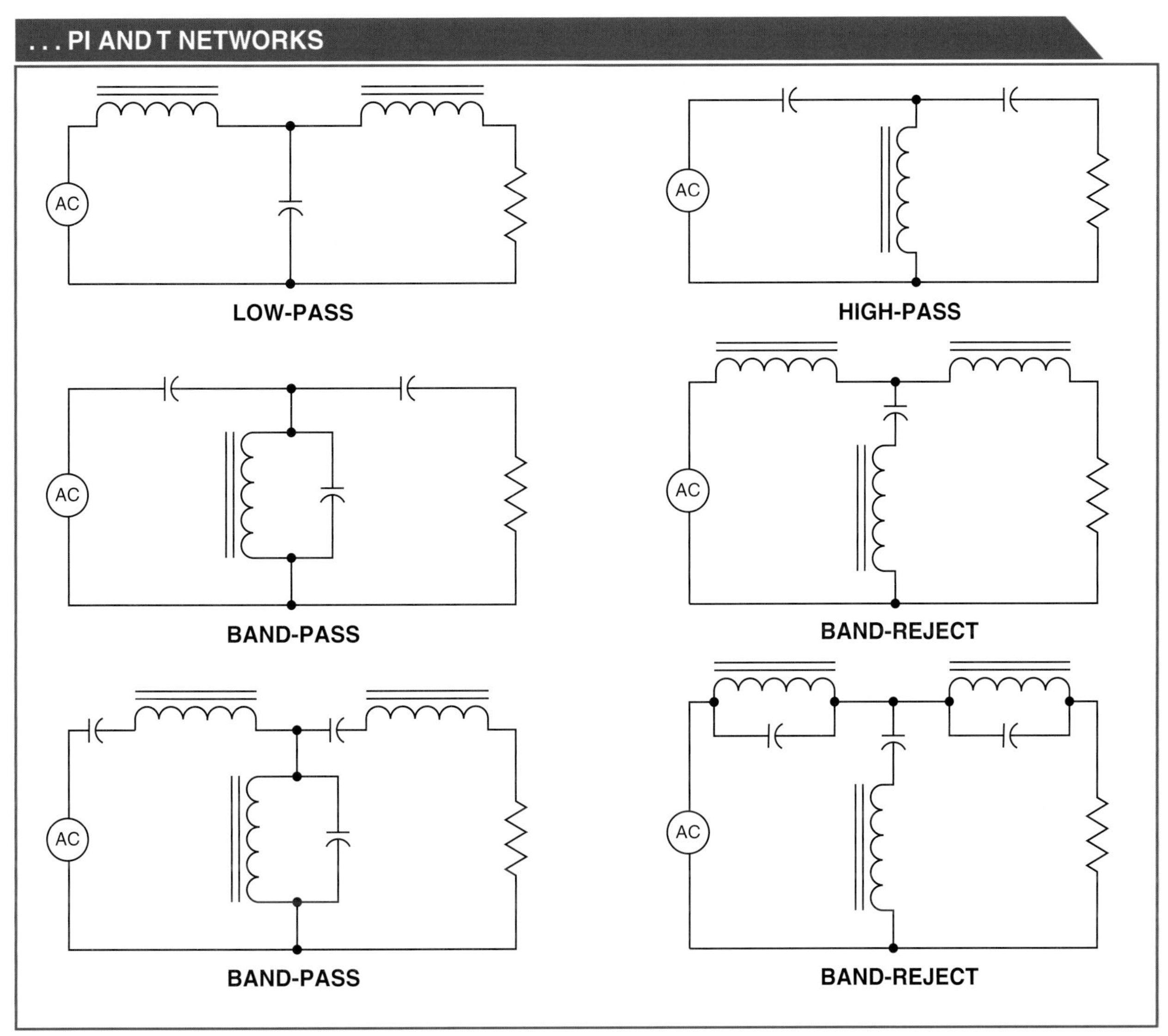

...Figure 19-21. Frequency filter networks are also referred to as pi (π) networks and T networks, because of the shape of the circuits.

SUMMARY

Resonance occurs when the inductive reactance is equal to the capacitive reactance. It is one of the most useful concepts in electricity. Resonance allows separation of one radio station or television broadcast channel from all others. It is through resonance that modern communications are possible.

Both series and parallel circuits can exhibit resonance. Major characteristics of a series resonant circuit are minimum impedance and a corresponding maximum current at 0° θ. Both capacitive and inductive voltages can be greater than the source voltage.

Major characteristics of a parallel resonant circuit include maximum impedance and a minimum line current with angle theta at 0°. A parallel resonant circuit can be used to create a tank circuit. Currents greater than the line current are possible in a tank circuit. In the parallel resonant circuit, where the resistance in the inductive branch is considered, there is a slight difference in frequency between resonance, minimum line current, and angle theta at 0°.

Another important concept is the quality factor (Q) of a resonant circuit. Quality factor is equal to the inductive reactance divided by the resistance in both series and parallel resonant circuits. The quality factor is selected to adjust the bandwidth of the filter response curve.

The primary application of inductive-resistive-capacitive combination circuits is to provide filtering of electrical signals. The most common filter types are low-pass, high-pass, band-pass, and band-reject types. The most common arrangements of filter networks have T and pi (π) shapes.

APPLICATION: FREQUENCY FILTERS...

Note: This application requires the use of a scientific calculator.

Background: Frequency filters are circuits that attenuate, or reduce, signals based on their frequency. Frequency filters have applications in audio systems, communication systems, radars, sonars, amplifiers, power supplies, and oscillators.

Some frequency filters use passive components, such as inductors, capacitors, and resistors, while other frequency filters use active components, such as operational amplifiers. The difference is that passive frequency filters do not require external power, while active frequency filters require a voltage to operate.

Frequency filters are classified by their transfer function. A transfer function is a comparison of the voltage output with the voltage input and is usually measured in decibels (dB). To find the transfer function in decibels, apply the following formula:

$$TF = 20 \times log\left(\frac{V_{out}}{V_{in}}\right)$$

where

TF = transfer function (in dB)

20 = constant

log = constant

V_{out} = output voltage (in V)

V_{in} = input voltage (in V)

When the output voltage is equal to the input voltage, the transfer function is 0 dB since log (1) is zero. When the output voltage is half of the input voltage, the transfer function is -6 dB. The filter pass band, which is the range of frequencies that can pass a filter unaffected, is between 0 dB and –3 dB (70.7%).

...APPLICATION: FREQUENCY FILTERS...

Frequency filters also affect the signal phase. Based on frequency, the phase relationship between the input signal and output signal is between 0° and 90°. The phase shift at –3 dB is 45°.

For low-pass and high-pass filters, RC and RL circuits provide the same filter characteristics. It is also possible to combine or cascade low-pass and high-pass filters to create a band-pass filter. A band-pass filter passes frequencies from a lower frequency to a higher frequency. **See Figure 1.**

FREQUENCY FILTER CHARACTERISTICS				
Frequency Filter	**Passive Circuit**	**Gain**	**Phase**	**Cutoff Frequency**
RC Low-Pass	R, C	$V_{out} = V_{in} \frac{X_c}{\sqrt{R^2 + X_c^{\ 2}}}$	$\varphi = \tan^{-1}(-2\pi fRC)$	$f_c = \frac{1}{2\pi RC}$
RL Low-Pass	L, R	$V_{out} = V_{in} \frac{R_c}{\sqrt{R^2 + X_L^{\ 2}}}$	$\varphi = \tan^{-1}\left(-\frac{2\pi fL}{R}\right)$	$f_c = \frac{R}{2\pi L}$
RC High-Pass	C, R	$V_{out} = V_{in} \frac{R}{\sqrt{R^2 + X_c^{\ 2}}}$	$\varphi = \tan^{-1}\left(\frac{1}{2\pi fRC}\right)$	$f_c = \frac{1}{2\pi RC}$
RL High-Pass	R, L	$V_{out} = V_{in} \frac{X_L}{\sqrt{R^2 + X_L^{\ 2}}}$	$\varphi = \tan^{-1}\left(\frac{1}{2\pi fL}\right)$	$f_c = \frac{R}{2\pi L}$

Figure 1. For low-pass and high-pass filters, RC and RL circuits provide the same filter characteristics.

Key Points: frequency filters; transfer function; impedance

Problem: The design task is to create a crossover network for a three-way speaker. A three-way speaker has three speakers: a woofer, a midrange speaker, and a tweeter. Each speaker type is optimized for a specific frequency range. The woofer is designed for low-frequency signals and works well with signals ranging from 30 Hz to 3 kHz. The midrange speaker is designed for frequencies ranging from 200 Hz to 4 kHz. The tweeter is designed for high-frequency signals between 2 kHz and 30 kHz. **See Figure 2.**

...APPLICATION: FREQUENCY FILTERS...

SPEAKER PROPERTIES		
Speaker Type	Frequency Range	Impedance
Woofer	30 Hz to 3 kHz	8 Ω
Midrange	200 Hz to 4 kHz	8 Ω
Tweeter	2 kHz to 30 kHz	8 Ω

Figure 2. Speaker properties are different for the woofer, midrange speaker, and tweeter.

A crossover network is a set of frequency filters that limits the signal frequencies sent to a specific speaker. For example, the crossover network would attenuate the signal level of frequencies above 3 kHz when connected to the woofer since the woofer is not designed to operate above 3 kHz. For the midrange speaker, the crossover network would reduce the low- and high-frequency signals. Finally, only high-frequency signals are passed to the tweeter.

Solution: The crossover network for the three-way speaker includes a filter for each speaker. The filter must not only pass frequencies supported by the speaker but also limit the amount of common frequencies sent to more than one speaker. First, each filter's pass band must be selected. Since there is overlap, there is some flexibility in selecting the pass bands. For this filter design, the following pass band and filter types are selected as design inputs.

Each filter uses a speaker impedance of 8 Ω. Therefore, each filter output is a resistive element of 8 Ω. The midrange speaker requires a pass-band filter. Cascading a high-pass filter with f_c = 500 Hz and a low-pass filter with f_c = 4 kHz provides the frequencies for the midrange speaker. **See Figure 3.**

SPEAKER FILTERS		
Speaker Type	Filter	Cutoff Frequency
Woofer	Low-pass	1 kHz
Midrange	Pass band	500 Hz to 4 kHz
Tweeter	High-pass	3 kHz

Figure 3. The crossover network for the three-way speaker includes a filter for each speaker.

The pass-band filter of the midrange speaker cascades a low-pass filter with f_c = 4 kHz and a high-pass filter with f_c = 500 Hz. Both filters use RC networks. The high-pass filter (R) uses a speaker impedance of 8 Ω. Therefore, capacitance (C) is calculated by applying the following formula:

$$C = \frac{1}{2\pi \times R \times f_c}$$

$$C = \frac{1}{6.28 \times 8\ \Omega \times 500\ \text{Hz}}$$

$$C = 40\ \mu\text{F}$$

...APPLICATION: FREQUENCY FILTERS...

The formula for the transfer function of the high-pass filter based on capacitance is the following:

$$V_{out} = V_{in} \frac{R}{\sqrt{R^2 + X_c^2}}$$

The formula for the transfer function in decibels is the following:

$$\frac{V_{out}}{V_{in}} \text{dB} = 20 \times \log\left(\frac{8\ \Omega}{\sqrt{64\ \Omega^2 + \left(\frac{1}{2\pi f \times 40\ \mu\text{F}}\right)^2}}\right)$$

The low-pass filter (R) is not constrained to the speaker impedance. To simplify the equations, R for the low-pass filter is also 8 Ω.

$$C = \frac{1}{2\pi \times R \times f_c}$$

$$C = \frac{1}{6.28 \times 8\ \Omega \times 4\ \text{kHz}}$$

$$C = 5\ \mu\text{F}$$

The formula for the transfer function of the low-pass filter based on frequency is the following:

$$V_{out} = V_{in} \frac{X_C}{\sqrt{R^2 + X_C^2}}$$

The formula for the transfer function in decibels is the following:

$$\frac{V_{out}}{V_{in}} \text{dB} = 20 \times \log\left(\frac{\frac{1}{2\pi f \times 5\ \mu\text{F}}}{\sqrt{64\ \Omega^2 + \left(\frac{1}{2\pi f \times 5\ \mu\text{F}}\right)^2}}\right)$$

The decibel sum of the low- and high-pass filters creates the band-pass-filter transfer function. **See Figure 4.**

Based on the speaker impedance and frequency range, a passive filter is selected that passes low frequencies for the woofer, high frequencies for the tweeter, and combined low- and high-pass filters for the midrange speaker.

...APPLICATION: FREQUENCY FILTERS

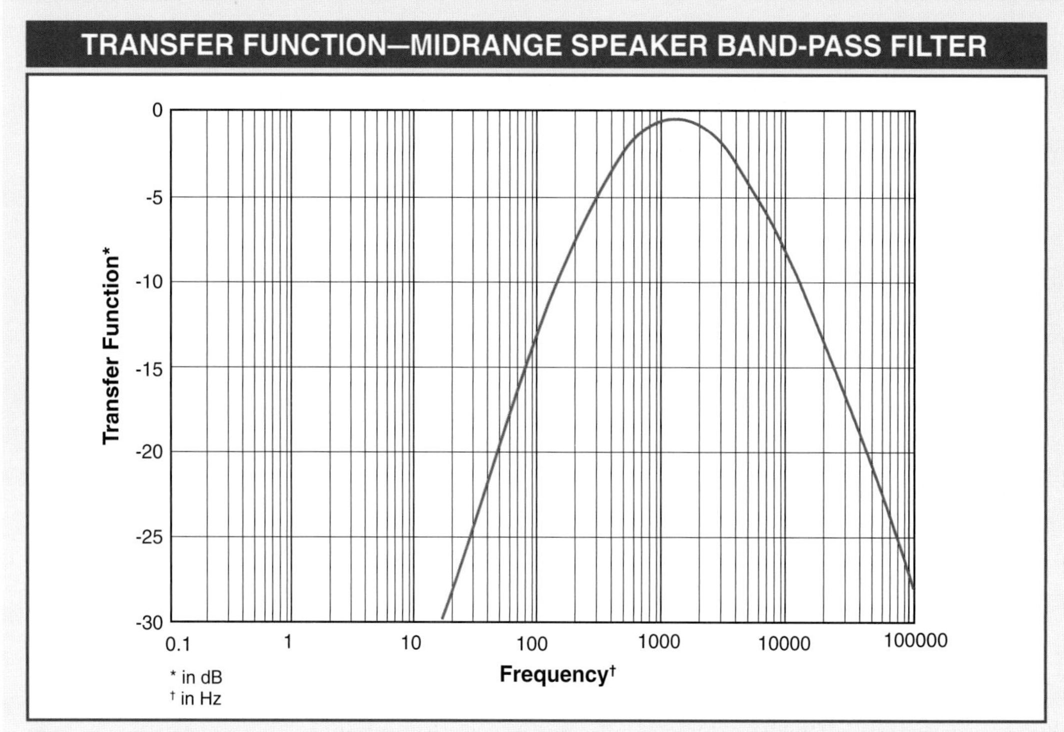

Figure 4. Frequency for the midrange speaker reaches its maximum value around 1000 Hz.

Chapter Review

1. What is a half-power point?
2. Why are complex filter networks sometimes referred to as "pi" and "T" networks?
3. What is the quality factor (Q) of a parallel resonant circuit?
4. What is a crossover network?
5. List three major characteristics of a parallel resonant circuit.
6. What is cutoff frequency?
7. What are reject frequencies?
8. What is the main difference between series and parallel resonant circuits?
9. Explain the composition of a frequency filter.
10. What does quality factor represent in a series resonant circuit?
11. What is a tank circuit?
12. What is a band-pass frequency filter?
13. What is a high-pass frequency filter?
14. What occurs with sine waves when resistance is added to a tank circuit?
15. What is resonance?
16. What properties are affected when a resonant circuit is tuned above or below resonance?
17. Is resistance and current high or low in a series resonant circuit with a low quality factor (below 10)?
18. Why is the quality factor of an inductor different at each frequency in a series resonant circuit?
19. How does the bandwidth of a circuit with a high quality factor compare with that of a circuit with a low quality factor?
20. Is resistance and current high or low in a series resonant circuit with a medium quality factor (between 10 and 100)?
21. What applications typically use high-pass frequency filters?
22. What is a harmonic?
23. Why are inductive and capacitive reactance values not required to calculate impedance at resonance?
24. List three applications in which impedance is calculated in parallel resonant circuits.
25. Is resistance and current high or low in a series resonant circuit with a high quality factor (above 100)?

20 Three-Phase AC

OBJECTIVES

- Explain three-phase power and power generation.
- List the test instruments used to analyze three-phase AC systems.
- Explain power factor and power factor correction.
- Describe three-phase wiring systems.
- Analyze three-phase reactive circuits.

Commercially produced alternating current (AC) is almost exclusively three-phase power. Three-phase AC power distribution has multiple advantages over single-phase AC power distribution, such as reductions in the size, weight, and cost of a power system. Three-phase loads are more constant, and three-phase generators and motors are less complex. A comparison of three-phase and single-phase AC power distribution systems reveals differences in power generation, frequency, generated and induced voltage, power measurement methods, power factor, power factor correction, delta and wye connections, and reactive circuits.

THREE-PHASE POWER

Three-phase power is a common form of electrical power. Three-phase power systems use three individual conductors to transmit power. The voltage of each conductor is 120° out of phase with that of the other conductors. Single-phase power has only one conductor transmitting power.

Phase is often represented by the Greek letter phi (ϕ). For example, "three-phase" is typically expressed in formulas and calculations as "3ϕ". The letters A, B, and C are typically used to designate the individual phases. Individual phases may be referenced as Aϕ, Bϕ, and Cϕ. A *generator (alternator)* is a machine that converts mechanical energy into electrical energy. Generators are used to produce three-phase (3ϕ) power. Advantages of using 3ϕ power rather than single-phase (1ϕ) power include the following:

- The amount of copper (conductors) used to deliver a given amount of power is reduced.
- A more constant load is provided to the generator.
- Three-phase motors are less complex than 1ϕ motors and have constant torque (rotating force).
- A smoother DC output is created from 3ϕ power, using less filtering than with 1ϕ power.

For 3ϕ and 1ϕ power systems to deliver the same amount of power to their respective loads, a 1ϕ power system requires about 1.15 times the copper required for a 3ϕ power system. With a resistive load, current and voltage are in phase with each other. For 1ϕ power, current and voltage produce no power when changing polarity. With 3ϕ power, at least two of the phases are providing power at all times, causing a generator to have a more constant load.

Three-phase motors are less complex than 1ϕ motors. Most AC motors start only if the magnetic field rotates. With 3ϕ motors, a rotating magnetic field is created by the rotation of the field windings, with a separate phase on each pole. With a 1ϕ motor, the rotating magnetic field must be created using two sets of poles with currents that are out of phase with each other. Single-phase motors usually require more components to operate. Because of this, 1ϕ motors also require more maintenance than 3ϕ motors.

A smooth DC signal is achieved using 3ϕ power. A *rectifier* is a device that converts AC voltage to DC voltage by allowing the AC voltage and current to flow in only one direction. A rectifier is typically used to eliminate either the negative or the positive portion of the AC waveform. A *full-wave rectifier* is a device that converts both positive and negative portions of an AC waveform into positive values.

Terms

A **generator (alternator)** is a machine that converts mechanical energy into electrical energy.

A **rectifier** is a device that converts AC voltage to DC voltage by allowing the AC voltage and current to flow in only one direction.

A **full-wave rectifier** is a device that converts both positive and negative portions of an AC waveform into positive values.

A full-wave rectifier is also used to convert AC voltage into DC voltage. The output of a 1ϕ full-wave rectifier falls to 0 V at each alternation. A 3ϕ full-wave rectifier maintains a more consistent high level of DC voltage at all times. A 1ϕ rectifier falls to 0 V each cycle, while a 3ϕ rectifier never falls below 86.6% of peak value. **See Figure 20-1.**

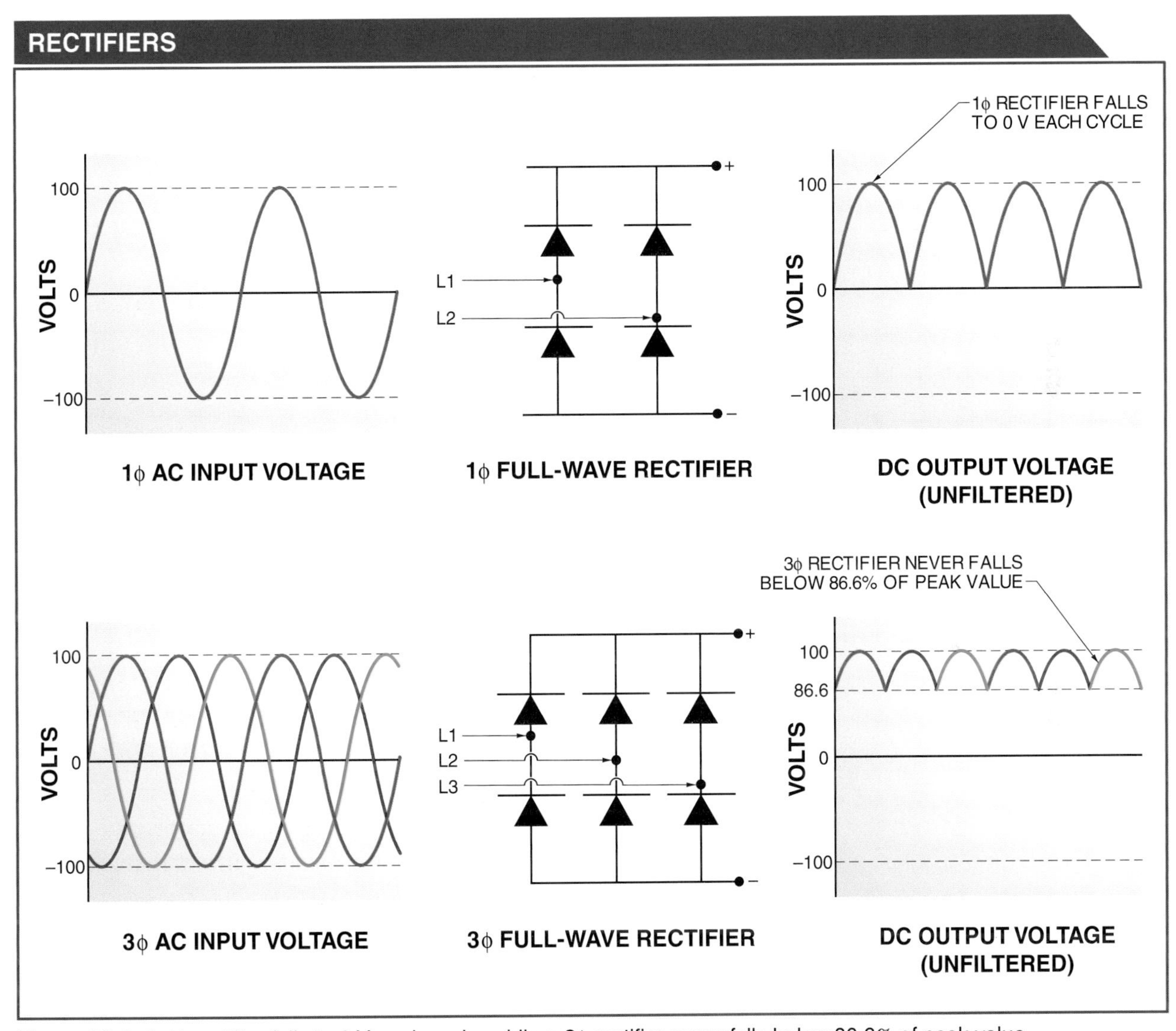

Figure 20-1. A 1ϕ rectifier falls to 0 V each cycle, while a 3ϕ rectifier never falls below 86.6% of peak value.

THREE-PHASE POWER GENERATION

Three-phase power is produced by an AC generator. AC generators are rated in kVA, rather than kW as with DC generators, because current and voltage can be out of phase in an AC system.

Terms

An **armature** is the rotating part of a DC motor.

Slip rings are metallic rings that are connected to the ends of the armature and are used to connect the induced voltage to the brushes.

A **rotor** is the rotating part of an AC motor.

A **brush** is the sliding contact that rides against the slip rings and is used to connect the armature to the external circuit.

A **stator** is a part of a generator that remains in a stationary position.

When the voltage is out of phase in an AC system, the generator must deliver more current than is required for the power generated. The amount of current a generator can deliver is limited due to its own internal resistance.

An *armature* is the rotating part of a DC motor. The two basic types of generators are the revolving-armature generator and the revolving-field generator. A revolving-armature generator has an armature that rotates through a stationary magnetic field. A revolving-field generator has a rotating magnetic field and a stationary armature. An armature is generally wrapped around a laminated iron core. *Slip rings* are metallic rings that are connected to the ends of the armature and are used to connect the induced voltage to the brushes. A *rotor* is the rotating part of an AC motor. A *brush* is the sliding contact that rides against the slip rings and is used to connect the armature to the external circuit.

Typically, AC generators have their stator windings as the armature, which eliminates the need for brushes and slip rings to remove the high voltage and current output. A *stator* is a part of a generator that remains in a stationary position. Using stator windings allows greater power to be removed from the AC generator through hardwiring, while reducing the downtime for maintenance required with the use of brushes and slip rings (as in DC generators). AC stator windings are wound in slots in the cylindrical core and help provide a better path for the magnetic lines of force. **See Figure 20-2.**

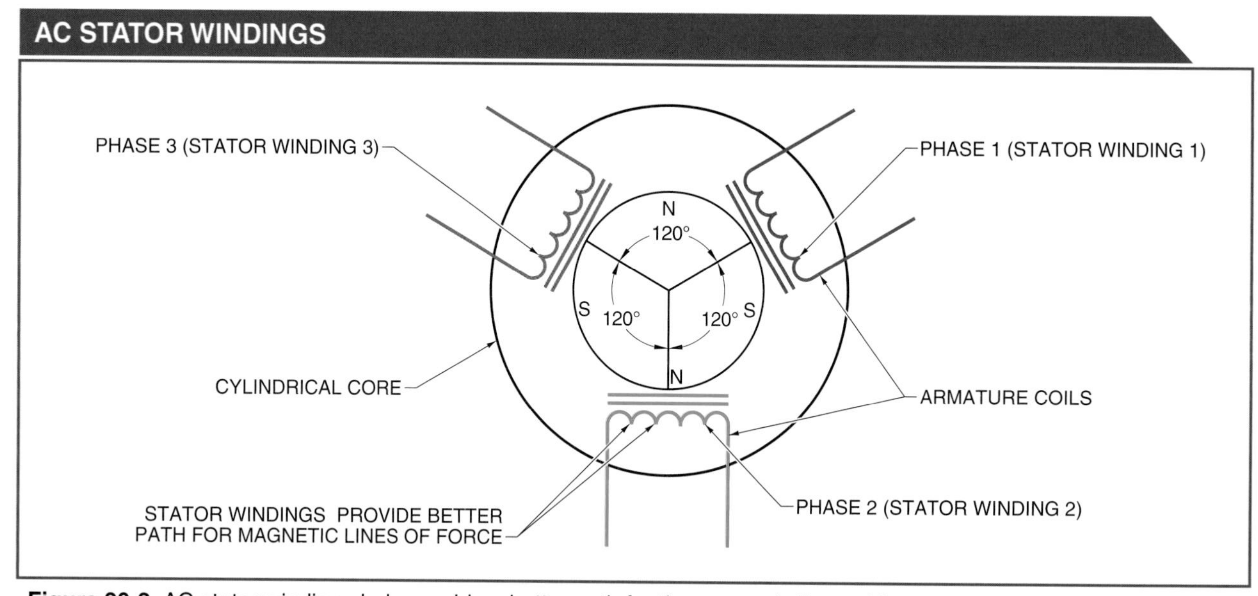

Figure 20-2. AC stator windings help provide a better path for the magnetic lines of force.

Excitation current is the DC used to produce electromagnetism in the fields of a DC motor or generator or in the rotor of an AC generator. An AC generator cannot produce output voltage until the rotor has been excited. Excitation current is sometimes supplied by a second generator and full-wave rectifier mounted on a common shaft with the armature windings. **See Figure 20-3.** Excitation current is also sometimes supplied through brushes and slip rings. An internal exciter is often used to reduce the maintenance requirements of brushes and slip rings. A second armature is placed on the main shaft along with a rectifier system to supply DC exciter current to the field of a 3ϕ generator. Excitation current causes the iron pole pieces to act as eletromagnets. Varying current to the field windings of the exciter causes the induced voltage in the armature of the exciter to increase. Increased voltage causes a larger current to the field windings of a 3ϕ generator. Increased current increases the magnetic field, and thus the output voltage of the 3ϕ system.

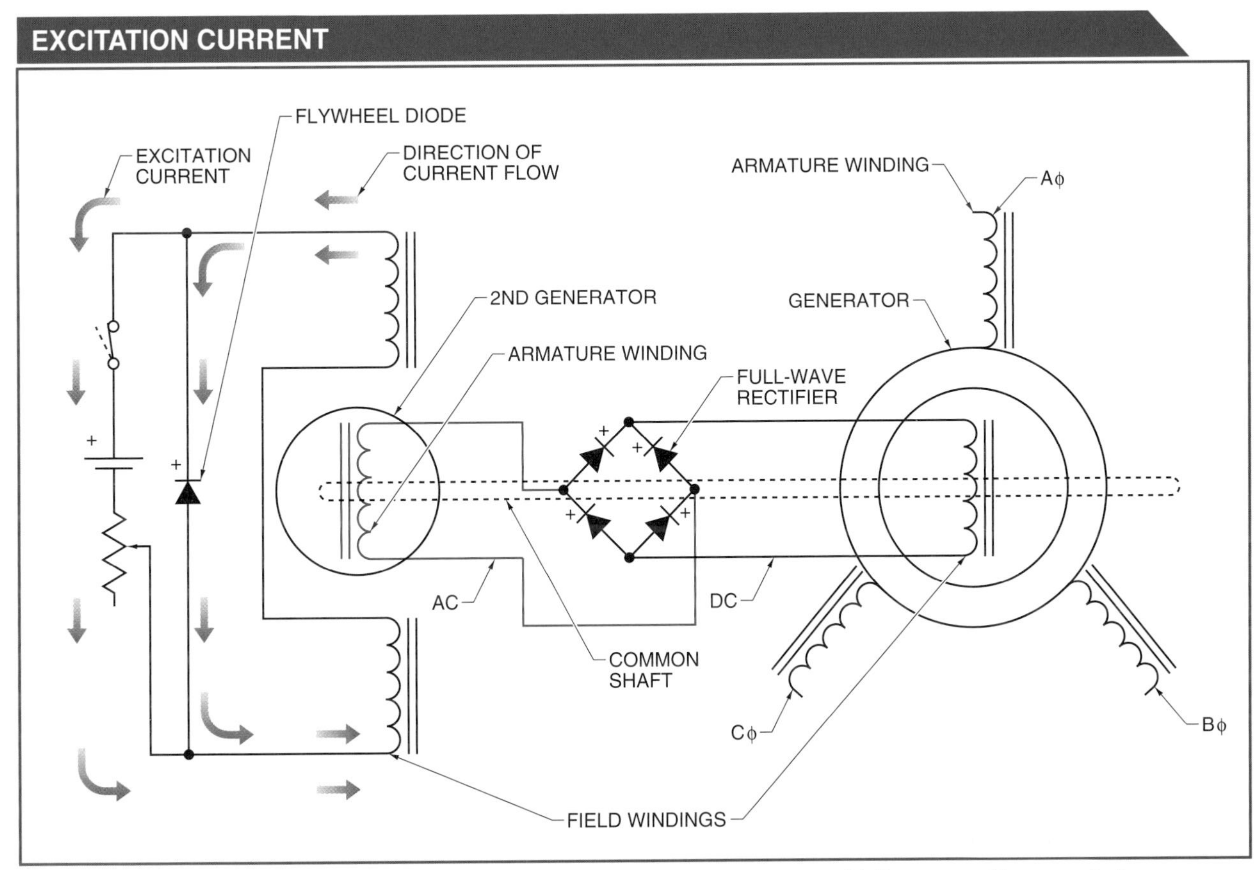

Figure 20-3. Excitation current is sometimes supplied by a second generator and full-wave rectifier mounted on a common shaft with the armature windings.

Terms

A **flywheel diode** is a diode that is used in a generator to suppress the inductive voltage that occurs when the exciter switch is opened.

A **swamping resistor** is a resistor that is used to suppress the high inductive kick of an exciter circuit.

A *flywheel diode* is a diode that is used in a generator to suppress the inductive voltage that occurs when the exciter switch is opened. The inductive voltage can be high enough to cause damage to the windings and associated components.

When the exciter switch is closed, the flywheel diode acts like high resistance because it is reverse biased. When the switch is open, induced voltage is in such a direction as to keep current flow in the same direction. Current flow in the same direction causes the diode to be forward biased. The diode presents low impedance so that the high inductive voltage is suppressed by the current flow through the diode.

A *swamping resistor* is a resistor that is used to suppress the high inductive kick of an exciter circuit. When a swamping resistor is used for this purpose, the switching is such that resistance is across the inductor when power is applied and/or when the switching removes it from the circuit. When power is removed from the swamping resistor circuit, the switching puts resistance across the field windings before the power is removed. A system that uses a swamping resistor is more complicated and costly than one that uses a single diode for the same purpose. A swamping resistor is used instead of a single diode when the bandwidth of a tuned circuit must be increased.

Tech Tip

One method of widening the band-pass range of a tuned circuit is to use a swamping resistor. A swamping resistor connected in parallel with a tuned circuit will result in a much wider pass band.

For example, with a 3ϕ generator, armature windings are spaced mechanically 120° apart and are fixed so that they do not move. The armature windings are electrically isolated from each other. As the rotor field rotates, a sine wave voltage is induced in each winding in sequence. Each sine wave is displaced from each other by 120° (electrically).

With the generated voltage sine waves, the sum of the instantaneous voltages of the three phases is 0 V. The sum is always equal to 0 V, regardless of the degree of generation. The degrees of the three points are relative to phase 1. For example, when phase 1 is at 0°, phase 3 is at 120° and phase 2 is at 240°.

With 10 V induced at 0°, phase 1 has 0 V, phase 3 has 8.66 V induced, and phase 2 has −8.66 V. The sum of the voltages of the three phases is equal to 0 V. At 150°, phases 1 and 2 have 5.0 V and phase 3 is at −10.0 V. The sum of the voltages of the three phases is still equal to 0 V. At 260°, phase 2 is at 6.43 V, phase 3 at 3.42 V, and phase 1 is at −9.85 V. Again, the sum of the voltages of the three phases is equal to 0 V. Two of the phases are always equal to the magnitude of the third phase.

Understanding three-phase power generation requires knowledge of frequency, generated and induced voltage, power measurement, power factor, and power factor correction within the generator.

Frequency

Frequency is the number of cycles per second (cps) in an AC sine wave. A pole is a magnetic north or south polarity. Frequency of generated voltage depends upon the number of poles and the speed of rotation. The faster the rotor turns, the higher the frequency that is generated. The greater number of poles, the greater the frequency of the generated voltage. If the number of poles is increased, the speed of rotation can be reduced to obtain the same frequency. For example, a two-pole generator needs to be rotated at twice the speed to obtain the same frequency as a four-pole generator. Frequency is calculated by applying the following formula:

$$f = P \times \left(\frac{S}{120} \right)$$

where

f = frequency (in Hz)

P = number of poles

S = speed (in rpm)

120 = constant

To convert to pairs of poles, the number of poles is divided by 2, because a magnetic field has a north and south pole. The speed of rotation is divided by 60 to convert from minutes to seconds, to conform to the definition of Hz (in cycles per sec).

Example: What is the frequency of a four-pole generator when its rotor has rotation of 1800 rpm?

$$f = P \times \left(\frac{S}{120} \right)$$

$$f = 4 \times \left(\frac{1800 \text{ rpm}}{120} \right)$$

$$f = \frac{7200}{120}$$

$$f = \mathbf{60\ Hz}$$

Generated and Induced Voltage

Generated voltage is a voltage produced in a closed path or circuit by the relative motion of the circuit or its parts with respect to magnetic lines of force. *Induced voltage* is a voltage produced around a closed path or circuit by a change in magnetic lines of force linking that path. The difference between generated voltage and induced voltage is that with generated voltage the speed of the relative motion of the circuit or its parts is varied to cause a change in voltage. With

Terms

Frequency is the number of cycles per second (cps) in an AC sine wave.

Generated voltage is a voltage produced in a closed path or circuit by the relative motion of the circuit or its parts with respect to magnetic lines of force.

Induced voltage is a voltage produced around a closed path or circuit by a change in magnetic lines of force linking that path.

Terms

The **pitch of a coil** is how tightly the windings of a coil are wound.

A **wattmeter** is a test instrument that is used to measure true power in a circuit.

induced voltage, the magnetic field is varied to cause a change in voltage. With generated voltage, the faster a conductor cuts a magnetic field, the greater the generated voltage. Voltage can be increased by increasing the number of magnetic lines of force, and decreased by reducing the number of magnetic lines of force, for the same speed of the rotating conductor. If a conductor is arranged in a loop of several turns, then voltage is increased as a multiple of the number of turns. The angle at which the conductor cuts the field also determines the level of induced voltage. Minimum induced voltage occurs when the conductor is moving parallel with the field. Maximum induced voltage occurs when the conductor cuts the field at 90°. Rotating conductors generate sine waves.

Taking these factors in consideration, the generated voltage in each phase of the armature of a 3ϕ generator is calculated by applying the following formula:

$$V_\phi = 2.22 \times \Phi \times N \times (f \times 10^{-8}) \times K$$

where

V_ϕ = generated voltage (in V)

2.22 = constant

Φ = number of magnetic lines of force (in Mx)

N = number of conductors (in series per phase)

f = frequency (in Hz)

K = machine constant

The machine constant (K) for any given machine (generator) is used to compensate for factors such as non-uniform conductors in the windings. None of the conductors cuts the magnetic field at the same time or at the same angle. The *pitch of a coil* is how tightly the windings of a coil are wound. Coils may have a different pitch. Depending on the pitch of the coil, the induced voltage may be greater or lesser.

Example: What is the generated voltage of a 3ϕ, 60 Hz generator with a K of 0.90, 100 series conductors per phase, and a field winding that produces 5 x 106 magnetic lines of force?

$$V_\phi = 2.22 \times \Phi \times N \times (f \times 10^{-8}) \times K$$

$$V_\phi = 2.22 \times (5 \times 10^6) \times 100 \times (60 \text{ Hz} \times 0.00000001) \times 0.90$$

$$V_\phi = 2.22 \times 5{,}000{,}000 \times 100 \times 0.0000006 \times 0.90$$

$$V_\phi = \mathbf{599.4\ V}$$

Power Measurement Methods

Power measurement methods are performed with test instruments such as wattmeters, clamp-on ammeters, and power quality meters. A *wattmeter* is a test instrument that is used to measure true power in a circuit. A wattmeter is provided with a scale that is usually graduated in watts (W), kilowatts (kW), or megawatts (MW). If the scale is graduated in kW or MW, the test instrument is usually designated as a kilowattmeter or megawattmeter. A wattmeter is used to analyze circuits by measuring the true power in a 3ϕ system. Depending on the system connections, multiple wattmeters can be used to measure true power in a 3ϕ system. **See Figure 20-4.**

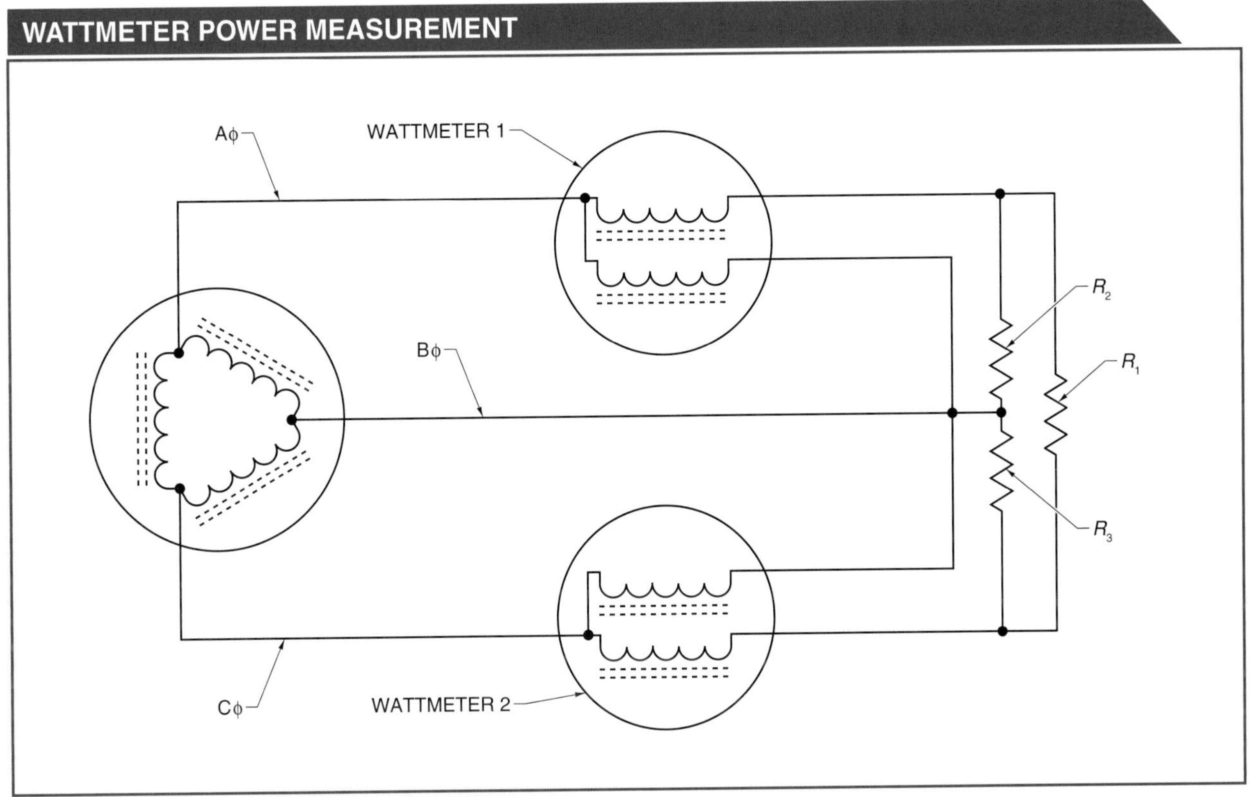

Figure 20-4. Depending on system connections, multiple wattmeters can be used to measure power in a 3ϕ system.

A wattmeter has two coils. One coil is a low-resistance movable current coil that is connected in series with the load. The other coil is a high-resistance fixed potential coil that is connected across the load. Current flow through the two coils sets up a magnetic field that causes an indicator (pointer) attached to the low-resistance movable current coil to indicate the wattage of a circuit. For a given value of current and voltage, maximum deflection is obtained when current and voltage are in phase with each other. When current and voltage are 90° out of phase, no force exists between the two coils, and the wattmeter indicator is at 0 W. When three wattmeters are used to analyze a circuit, the total true power of the 3ϕ system is equal to the sum of the three separate readings.

Terms

A **clamp-on ammeter** is a meter that measures current in a circuit by measuring the strength of the magnetic field around a single conductor.

A **power quality meter** is a test instrument that measures, displays, and records voltage, current, and power, as well as power problems such as sags, swells, transients, and harmonics.

When analyzing circuits with two separate wattmeters, the low-resistance movable current coils are connected in series with the loads on two phases. The high-resistance fixed potential coils of the wattmeter are connected to the common third phase.

When using wattmeters for analysis, true power equals the algebraic sum of the two separate wattmeter readings. If one of the wattmeters has a low reading, its high-resistance fixed potential coil connections are reversed so the pointer indicates a high value. The high value is then subtracted from the reading of the other wattmeter. Most wattmeters have a toggle switch to reverse the coils.

When the power factor of the first load is less than 50% (with balanced loads), the true power equals the difference between the two loads. If the power factor is exactly 50%, one wattmeter shows 0 W, while the second wattmeter indicates the total true power of the three phases. If the power factor is greater than 50%, both wattmeters indicate a high reading. The values of both wattmeters are added together to obtain the total true power of the system.

In addition to measuring voltage, current, and resistance, ammeters can take continuity and frequency measurements.

A *clamp-on ammeter* is a meter that measures current in a circuit by measuring the strength of the magnetic field around a single conductor. To measure the apparent power of a balanced load, the line current and the line voltage must be measured separately. **See Figure 20-5.** Typically, a clamp-on ammeter is used to obtain line current measurements, and a DMM set to measure voltage is used to obtain voltage measurements. The use of a clamp-on ammeter eliminates the need to turn the power OFF to open the line to insert an ammeter in series. Since voltage is equal in each phase of a 3ϕ system, typically only one line voltage needs to be measured, but all three voltages can be measured if required.

A *power quality meter* is a test instrument that measures, displays, and records voltage, current, and power, as well as power problems such as sags, swells, transients, and harmonics. A power quality meter can be used to obtain more valuable troubleshooting data than other types of test instruments, such as wattmeters and clamp-on ammeters, because a power quality meter can capture and record measurement values over a long period, such as minutes, hours, or days. Power quality meters are designed for troubleshooting power quality problems that occur over time in a distribution circuit, by analyzing problems such as phase sequence, voltage sags, voltage swells, and overloaded circuits.

APPARENT POWER MEASUREMENTS

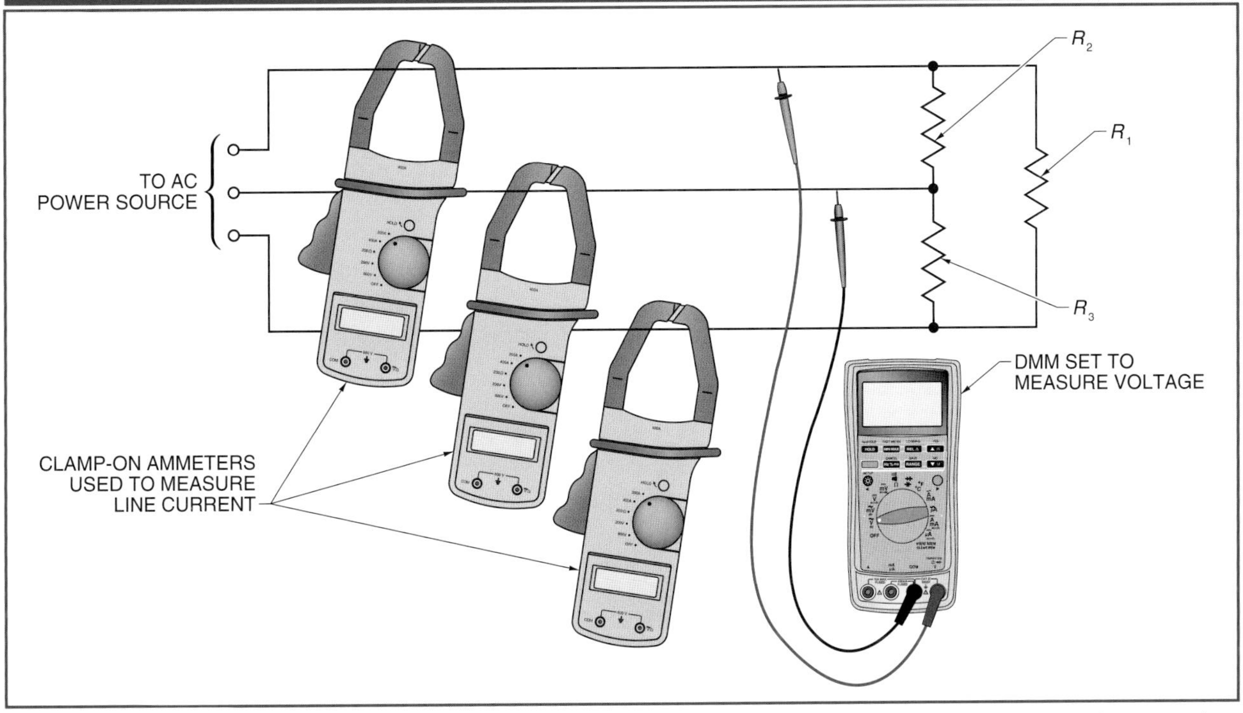

Figure 20-5. To measure the apparent power of a balanced load, line current and line voltage must be measured separately.

Power Factor

Power factor (PF) is the ratio of true power used in an AC circuit to apparent power delivered to the circuit. A power triangle can be used to show the relationship of true power, apparent power, reactive power, and power factor in a circuit. **See Figure 20-6.** A power triangle uses basic trigonometric principles to show the mathematical relationships between the different types of power in a system. In practice, a power quality meter can show the relationship between the different types of power and the power factor by displaying circuit measurements. A power quality meter can display true power (W or kW), apparent power (VA or kVA), reactive power (VAR or kVAR), and power factor (PF) in a circuit.

The power factor indicates the electrical efficiency of the circuit. Power factor is lagging for an inductive load, leading for a capacitive load, and in phase for a resistive load. As circuit reactance increases, angle theta (θ) also increases. Likewise, as circuit reactance decreases, angle theta also decreases. Angle theta can be used to calculate the power factor because the cosine of angle theta is equal to the circuit power factor. As in a 1ϕ system, the power factor of a 3ϕ system is calculated by applying the following formula:

Terms

Power factor (PF) is the ratio of true power used in an AC circuit to apparent power delivered to the circuit.

$$PF = \frac{P_T}{P_A} \times 100$$

where

PF = power factor (in %)

P_T = true power (in W)

P_A = apparent power (in VA)

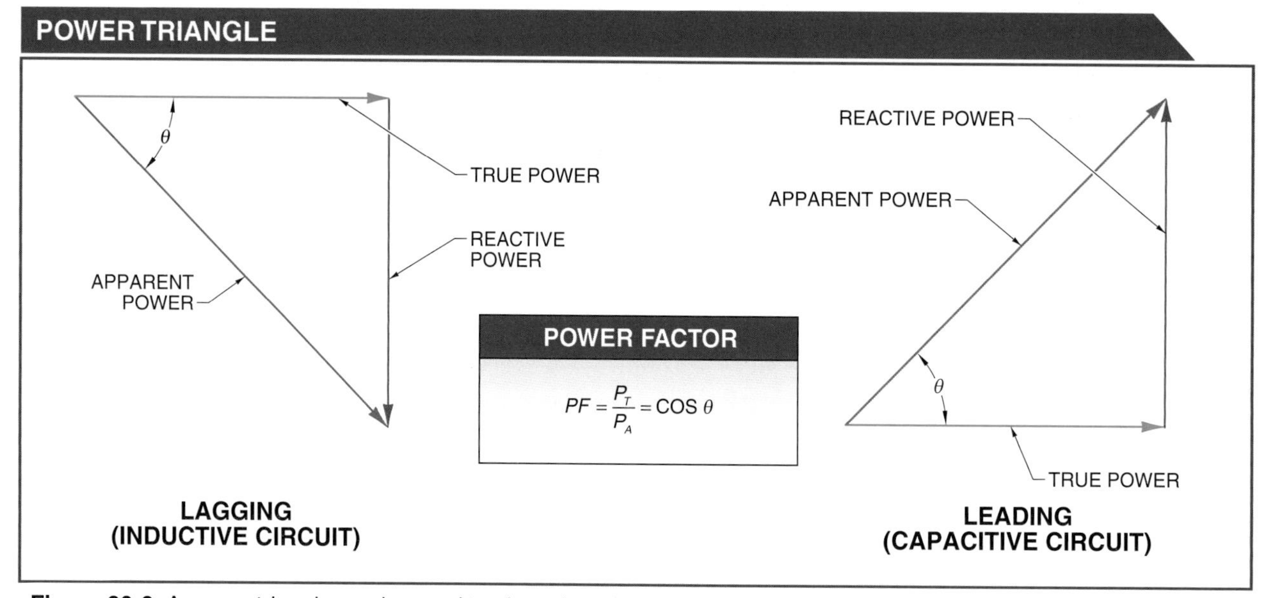

Figure 20-6. A power triangle can be used to show the relationships between true power, apparent power, reactive power, and power factor in a circuit.

Example: What is the power factor in a circuit that has true power of 1000 W and apparent power of 2000 VA?

$$PF = \frac{P_T}{P_A} \times 100$$

$$PF = \frac{1000\text{ W}}{2000\text{ VA}} \times 100$$

PF = **50%**

Power factor is equal to the cosine of angle theta of the power triangle. In a purely resistive circuit, cosine of angle theta is equal to 1 (unity). If the loads are purely reactive, no true power is consumed, and cosine of angle theta is equal to 0. If the circuit is both reactive and resistive, then cosine of angle theta varies between 0 and 1. It is desirable to keep power factor as close to 100% as possible because a low power factor reduces power distribution efficiency.

Appliances that produce heat or provide incandescent light are examples of resistive loads. The power consumed by resistive loads is measured in watts (W). **See Figure 20-7.** Examples of resistive loads include toasters, coffee makers, steam irons, dishwashers, lamps, and water heaters.

Loads can be both inductive and resistive. For example, a water pump can be both inductive and resistive. Without a load, the motor is mostly inductive. Under full load, the motor is mostly resistive. Under partial load, the motor is between inductive and resistive. For most motors to operate, an inductive field is necessary to cause the rotor to turn. Other examples of inductive loads include transformers, ballasts of gaseous tube lighting, and induction furnaces. These devices require a magnetic field to produce heat, light, or change voltage levels. A lagging power factor is often caused by a large number of inductive devices connected in the circuit. In an inductive circuit, measuring circuit voltage and current with a DMM is not useful in determining true power. When the voltage measurement is multiplied by the current measurement, true power cannot be determined because the calculation does not account for the phase shift between voltage and current. When measuring power in an inductive circuit, a power quality meter must be used. A power quality meter can be used to measure reactive power and true power in a circuit.

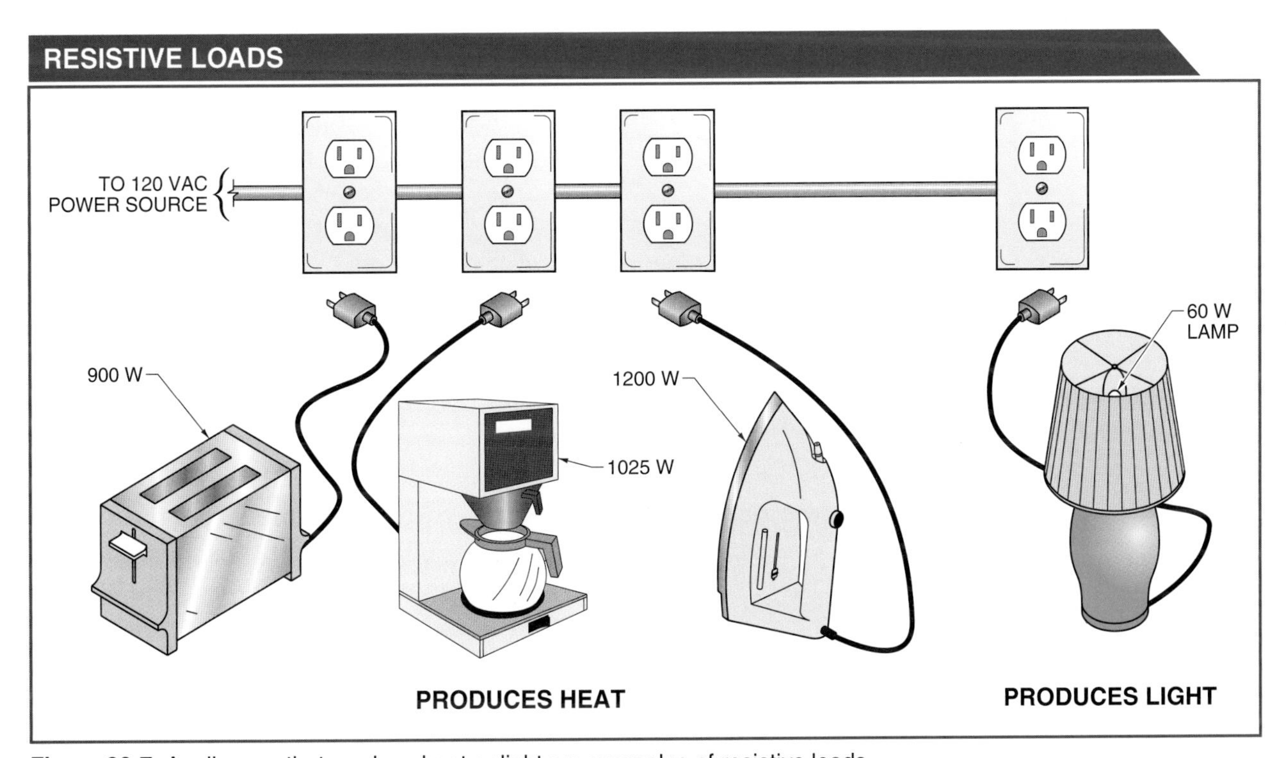

Figure 20-7. Appliances that produce heat or light are examples of resistive loads.

Terms

Power factor correction is the process of correcting power factor to be between 90% (0.90) and 97% (0.97).

Power Factor Correction

Power factor correction is the process of correcting power factor to be between 90% (0.90) and 97% (0.97). A power factor of less than 100% increases the overall cost of a power distribution system. The lower the power factor, the higher the operating cost. Power factors of less than 90% (0.90) are generally corrected. Anytime the circuit power factor drops to less than 90%, the circuit has a poor power factor. A poor power factor can result in heat damage to insulation and other circuit components, can reduce the amount of useful power available, and requires increased conductor and equipment sizes. The lower the power factor requirements, the higher the current required to supply the loads. **See Figure 20-8.** For example, a 35 kW circuit load requires a 35 kVA transformer if the circuit power factor is 100%. However, if the circuit power factor is only 70%, a 50 kVA transformer is required. Using higher current than necessary to perform electrical work increases the cost of the system.

POWER FACTOR REQUIREMENTS

Load Required Power*	Load Power Factor†	Required Transformer Size‡	Circuit Current§	AWG Number Conductor Size
35	100	35.00	76.08	3
35	90	38.88	84.52	2
35	85	41.17	89.50	2
35	70	50.00	108.69	0
35	65	53.84	117.04	0
35	60	58.33	126.80	00
35	50	70.00	152.17	000

WIRE SIZE INCREASING

* in kW
† in %
‡ in kVA
§ in A

Figure 20-8. The lower the power factor requirements, the higher the current required to supply the loads.

Electric utilities (power suppliers) measure and record the kW and kVA used by an end user (customer). The kVA measurement includes the reactive power value that kW meters do not record. The kW and kVA measurements are used to gauge the power factor from an end user. For example, if a power factor is corrected from 70% to 90% from a line load of 142 kVA, then the line load is reduced to 111 kVA. The amount of kVA required to deliver 100 kW of useful power is reduced by 31 kVA. Typically, no attempt is made to correct the power factor to 100%. Rather, a range of about 90% to 97% is attempted, due to the changing loads on most lines. The changing loads on most lines require different power factors at different times. Electric utilities often compensate for a poor power factor by increasing the cost of power to the end user.

A common method for correcting the power factor of a circuit is to connect capacitors across the load. **See Figure 20-9.** A lagging poor power factor is caused by inductive loads and is normally corrected by adding capacitors across the load. Capacitors have a leading power factor, which is used to cancel the lagging power factor of an inductive load.

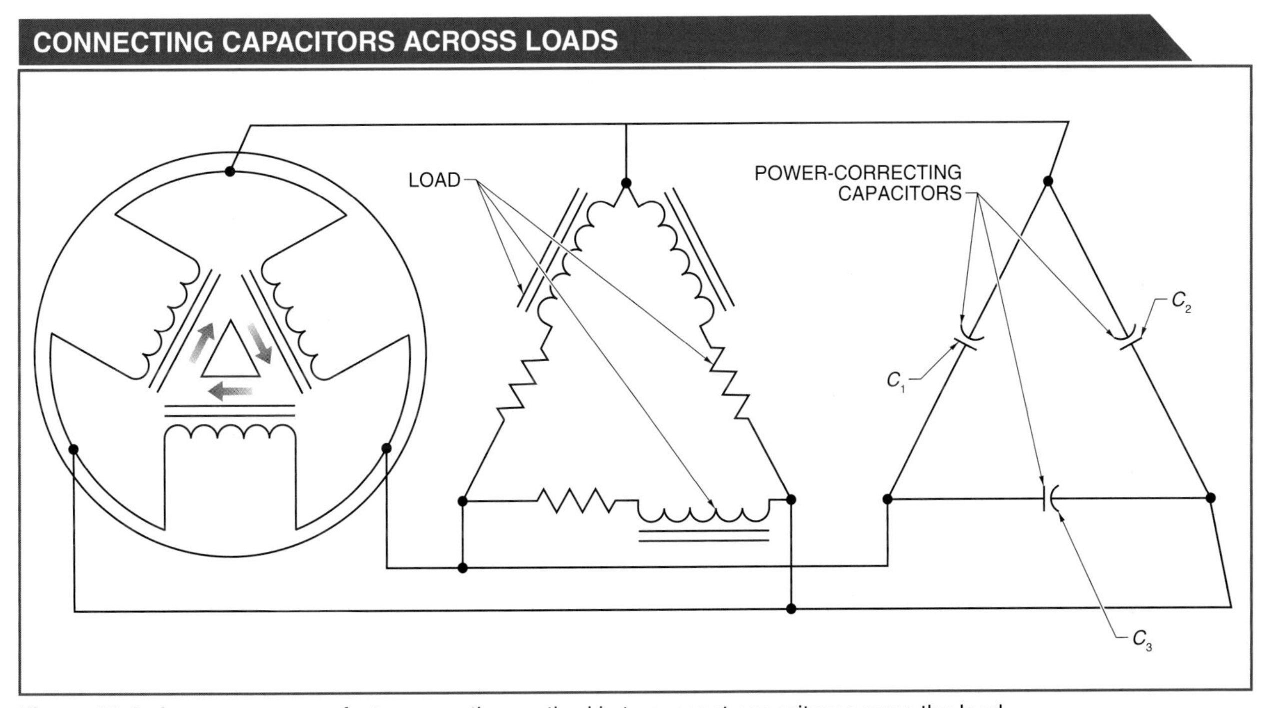

Figure 20-9. A common power factor correction method is to connect capacitors across the load.

Power factor analysis is typically used to determine if power factor correction is necessary. *Horsepower (HP)* is the unit of power equal to 746 W or 550 lb-ft per second (33,000 lb-ft per minute). Horsepower value can be required when performing power factor analysis. Power factor analysis is performed by applying the following procedure:

1. Calculate apparent power for a 3ϕ circuit. Apparent power is calculated by applying the following formula:

Terms

Horsepower (HP) is the unit of power equal to 746 W or 550 lb-ft per second (33,000 lb-ft per minute).

$$P_A = \frac{V_T \times I_T \times \sqrt{3}}{1000}$$

where

P_A = apparent power (in kVA)

V_T = total circuit voltage (in V)

I_T = total circuit current (in A)

$\sqrt{3} = 1.73$ (constant)

1000 = kVA conversion factor (constant)

2. Calculate true power for a 3ϕ circuit. True power for a 3ϕ circuit is calculated by applying the following formula when motor output HP is known:

$$P_T = \frac{HP \times 746}{1000}$$

where

P_T = true power (in kW)

HP = motor output (in HP)

746 = horsepower conversion factor (in W)

1000 = kW conversion factor (constant)

3. Calculate power factor. Power factor is calculated by applying the following formula:

$$PF = \frac{P_T}{P_A} \times 100$$

where

PF = power factor (in %)

P_T = true power (in kW)

P_A = apparent power (in kVA)

4. Calculate angle theta. Angle theta is calculated by applying the following formula:

$\theta_1 = \cos^{-1} PF$

where

θ_1 = angle theta (in °)

$\cos^{-1}$ = inverse cosine function

PF = power factor (in decimal)

5. Calculate the reactive power of the circuit. The reactive power of the circuit is calculated by applying the following formula:

$P_{VAR} = \tan \theta_1 \times P_T$

where

P_{VAR} = reactive power (in kVAR)

θ_1 = angle theta (in °)

P_T = true power (in kW)

Tech Tip

Heat is a major cause of electrical system failure. All electrical equipment, including receptacles, switches, circuit breakers, panelboards, and wire nuts, has a listed temperature rating. Heat resistance is only as high as the lowest temperature rating of any electrical system component.

6. Correct power factor to required value. Correcting power factor to the required value is calculated by applying the following formula:
   $\theta_2 = \cos^{-1} PF$
   where
   θ_2 = corrected angle theta (in °)
   $\cos^{-1}$ = inverse cosine function
   PF = desired power factor (in decimal)

7. Calculate capacitor kilovoltamperes reactive (CkVAR). CkVAR is calculated by applying the following formula:
   $CkVAR = (\tan\theta_1 - \tan\theta_2) \times P_T$
   where
   $CkVAR$ = capacitor kilovoltamperes reactive (in CkVAR)
   θ_1 = angle theta (in °)
   θ_2 = corrected angle theta (in °)
   P_T = true power (in kW)

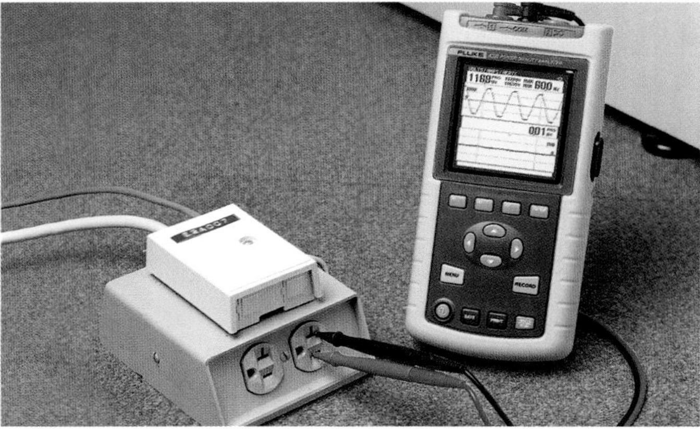

Fluke Corporation

Power quality meters can be used to troubleshoot power quality problems in distribution circuits.

Note: The total CkVAR determines the size of capacitor required to correct the power factor.

8. Calculate corrected apparent power. Corrected apparent power is calculated by applying the following formula:

$$P_A = \sqrt{P_T^{\,2} + \left(P_{VAR} - CkVAR\right)^2}$$

   where
   P_A = apparent power (in kVA)
   P_T = true power (in kW)
   P_{VAR} = reactive power (in kVAR)
   $CkVAR$ = capacitor kilovoltamperes reactive (in CkVAR)

9. Calculate corrected line current. Corrected line current is calculated by applying the following formula:

$$I_{LINE} = \frac{P_A}{\left(V_T \times \sqrt{3}\right)}$$

   where
   I_{LINE} = corrected line current (in A)
   P_A = apparent power (in kVA)
   V_T = total voltage (in kV)
   $\sqrt{3}$ = constant

Example: What is the power factor analysis of a 600 V, 35 A circuit that has a 25 HP motor and needs to be operating with a power factor of 95% (0.95)? **See Figure 20-10.**

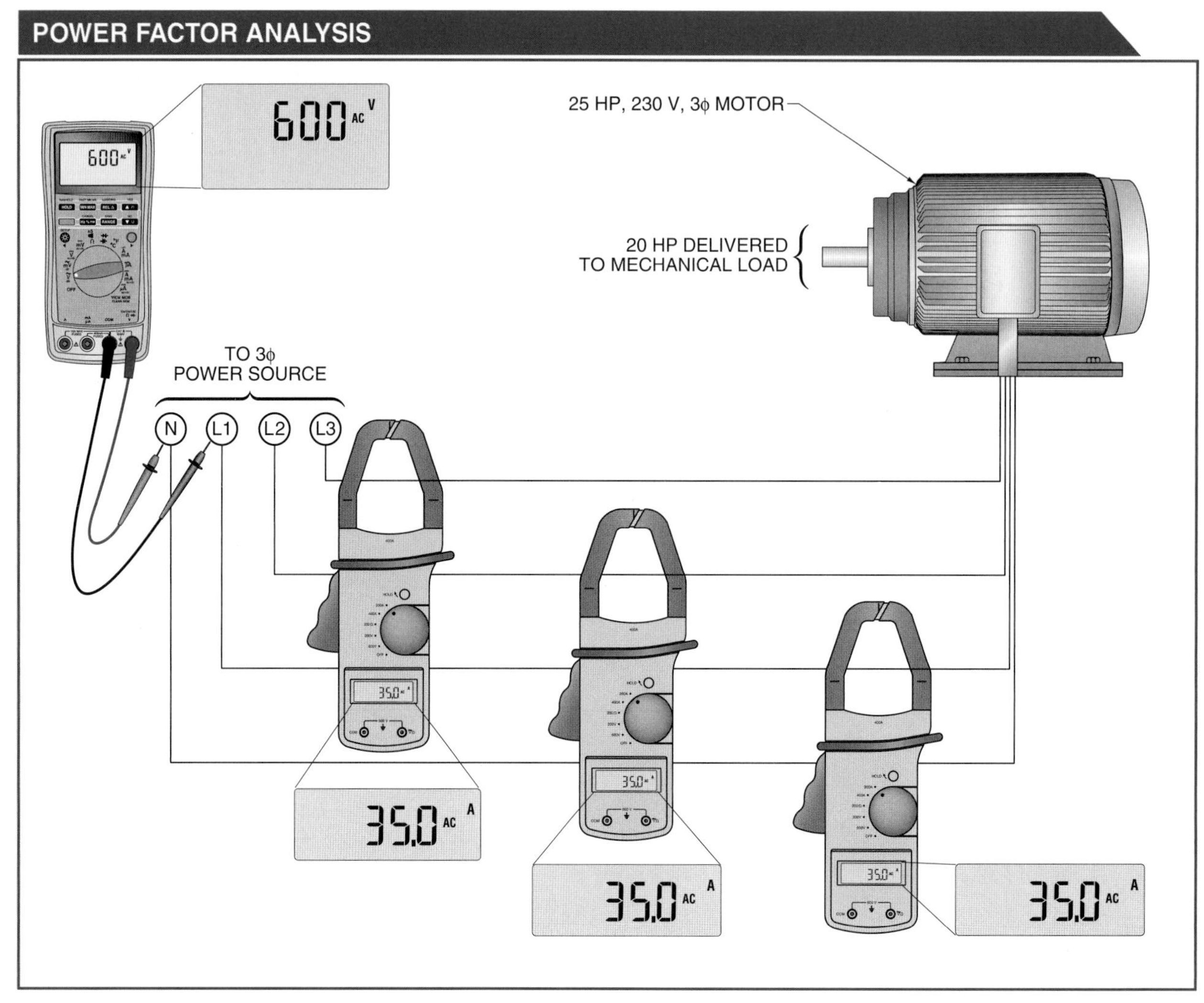

Figure 20-10. A 25 HP, 230 V, 3φ motor is loaded so that it delivers 20 HP to its mechanical load.

1. Calculate apparent power.

$$P_A = \frac{V_T \times I_T \times \sqrt{3}}{1000}$$

$$P_A = \frac{600\text{ V} \times 35\text{ A} \times \sqrt{3}}{1000}$$

$$P_A = \frac{600\text{ V} \times 35\text{ A} \times 1.73}{1000}$$

$$P_A = \frac{36{,}330}{1000}$$

$$P_A = 36.3\text{ kVA}$$

2. Calculate true power for a 3ϕ circuit.

$$P_T = \frac{HP \times 746}{1000}$$

$$P_T = \frac{20 \times 746}{1000}$$

$$P_T = \frac{14{,}920}{1000}$$

$$P_T = 14.92 \text{ kW}$$

3. Calculate power factor.

$$PF = \frac{P_T}{P_A}$$

$$PF = \frac{14.92 \text{ kW}}{36.3 \text{ kVA}}$$

$$PF = 0.41\ (41\%)$$

4. Calculate angle theta.

$$\theta_1 = \cos^{-1} PF$$

$$\theta_1 = \cos^{-1} 0.41$$

$$\theta_1 = 65.7°$$

5. Calculate the reactive power of the circuit.

$$P_{VAR} = \tan \theta_1 \times P_T$$

$$P_{VAR} = \tan 65.7° \times 14.92 \text{ kW}$$

$$P_{VAR} = 2.2 \times 14.92 \text{ kW}$$

$$P_{VAR} = 32.8 \text{ kVAR}$$

6. Correct power factor to required value.

$$\theta_2 = \cos^{-1} PF$$

$$\theta_2 = \cos^{-1} 0.95$$

$$\theta_2 = 18.2°$$

7. Calculate *CkVAR*.

$$CkVAR = (\tan \theta_1 - \tan \theta_2) \times P_T$$

$$CkVAR = (\tan 65.7° - \tan 18.2°) \times 14.92 \text{ kW}$$

$$CkVAR = (2.21 - 0.329) \times 14.92 \text{ kW}$$

$$CkVAR = 1.89 \times 14.92 \text{ kW}$$

$$CkVAR = 28.2 \text{ CkVAR}$$

Since this is a balanced load, a 3ϕ capacitor rated as close to 28 CkVAR as possible is needed to correct the power factor to 95%. If three 1ϕ capacitors are used, one-third of the calculated CkVAR is placed across each phase of the loads. The value is about 9 CkVAR to 10 CkVAR for each phase. Power factor correction capacitors are available in standard sizes. Therefore, a standard value is selected that meets the requirement for a correction to 95%. Care should be taken not to overcompensate to the point that the lagging current is replaced by a leading current, because this could also cause changes in power factor. A power triangle can be used to show the difference between the original power factor and the corrected power factor. **See Figure 20-11.**

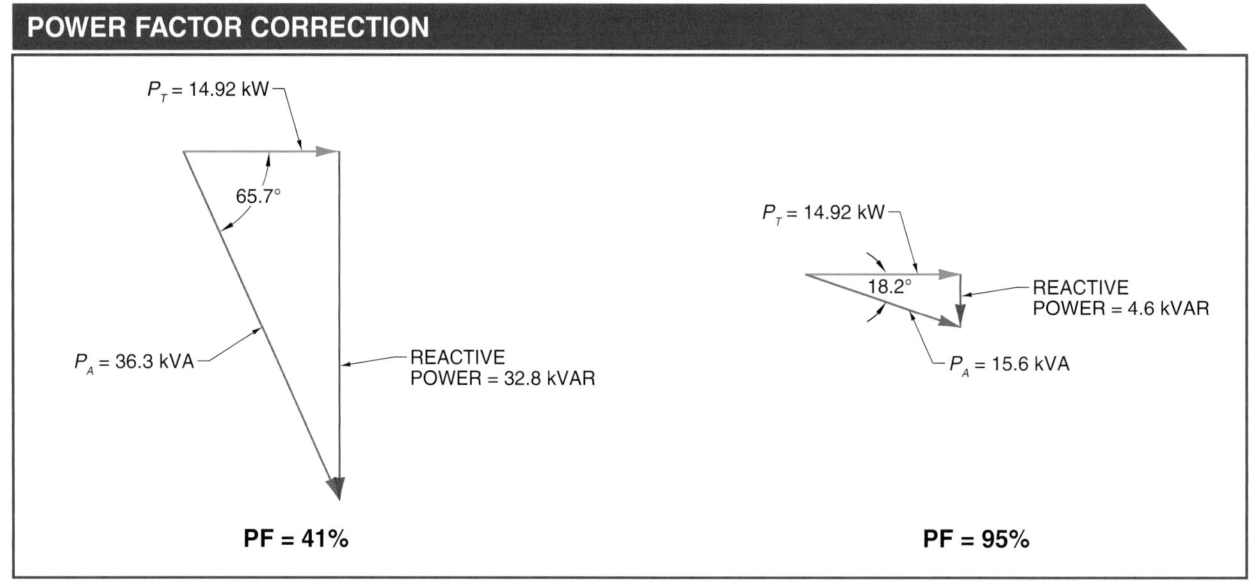

Figure 20-11. A power triangle can be used to show the difference between the original power factor and the corrected power factor.

8. Calculate corrected apparent power.

$$P_A = \sqrt{P_T^2 + (kVAR - CkVAR)^2}$$

$$P_A = \sqrt{14.92^2 + (32.8 - 28.2)^2}$$

$$P_A = \sqrt{222.61 + 21.16}$$

$$P_A = \sqrt{243.77}$$

$$P_A = 15.6 \text{ kVA}$$

9. Calculate corrected line current.

$$I_{LINE} = \frac{kVA}{V \times \sqrt{3}}$$

$$I_{LINE} = \frac{15.6}{0.600 \times 1.73}$$

$$I_{LINE} = \frac{15.6}{1.038}$$

$$I_{LINE} = \mathbf{15.0\ A}$$

Transient Voltages

Power quality meters are typically used to monitor transient voltages. The time, size, and duration of transient voltages can be displayed at a later time when using power quality meters. Transient voltage measurements must be taken any time equipment prematurely fails and a transient voltage is suspected. Transient voltage measurements are taken to verify that surge protection devices are working. A transient voltage measurement is different from a standard voltage measurement in that a standard voltage measurement displays the root-mean-square (RMS) voltage value of a circuit, and a transient voltage measurement displays the peak voltage value of a circuit. For example, a standard voltmeter that displays 115 VAC is displaying the RMS voltage measurement of the circuit. The peak voltage of a circuit with an RMS value of 115 VAC is about 163 V (RMS voltage multiplied by 1.414 equals peak voltage).

Line current falls from 35 A to 15 A, which represents a 20 A reduction in line current. There is 35 A of current flowing into the original load. The placement of the capacitor(s) compensates for the high line current only upstream from the capacitor. The load, which is downstream, still has the original kVA loading of 36.3 kVA. This is because the load has not changed. Only the input to the load with a power factor correction capacitor has changed.

In addition to the size of the capacitor needed to correct the power factor, other factors considered for selection of the device to be used and where to install it with respect to the load include:

- Type of load. The type of load includes the number of phases (1ϕ or 3ϕ) and the amount of HP.
- Constancy of loads. The constancy of load is the amount of time that the load is applied. *Jogging* is the frequent starting and stopping of a motor for short periods. *Plugging* is a method of motor braking in which the motor connections are reversed so that the motor develops a countertorque that acts as a braking force. Jogging and plugging affect the constancy of load because these procedures require additional power.
- Capacity. The capacity of the load applied to transformers, conductors, and related equipment must be considered, as well as future load additions or reductions.
- Economics. Economic considerations include how the electric utility (power company) compensates for a poor power factor. Typically, circuits that are overloaded must be corrected.

Terms

Jogging is the frequent starting and stopping of a motor for short periods.

Plugging is a method of motor braking in which the motor connections are reversed so that the motor develops a countertorque that acts as a braking force.

Terms

A **power-correcting capacitor** is a capacitor used to improve a facility's power factor by improving voltage levels, increasing system capacity, and reducing line losses.

A *power-correcting capacitor* is a capacitor used to improve a facility's power factor by improving voltage levels, increasing system capacity, and reducing line losses. Facilities with many inductive loads tend to have poor power factors because the voltage is leading the current. It is a common practice to install capacitors to improve power factor, because capacitors cause the current to lead the voltage. Power-correcting capacitors must be placed before the electric motor drive in the AC line supplying power to the drive, not between the drive and motor. **See Figure 20-12.**

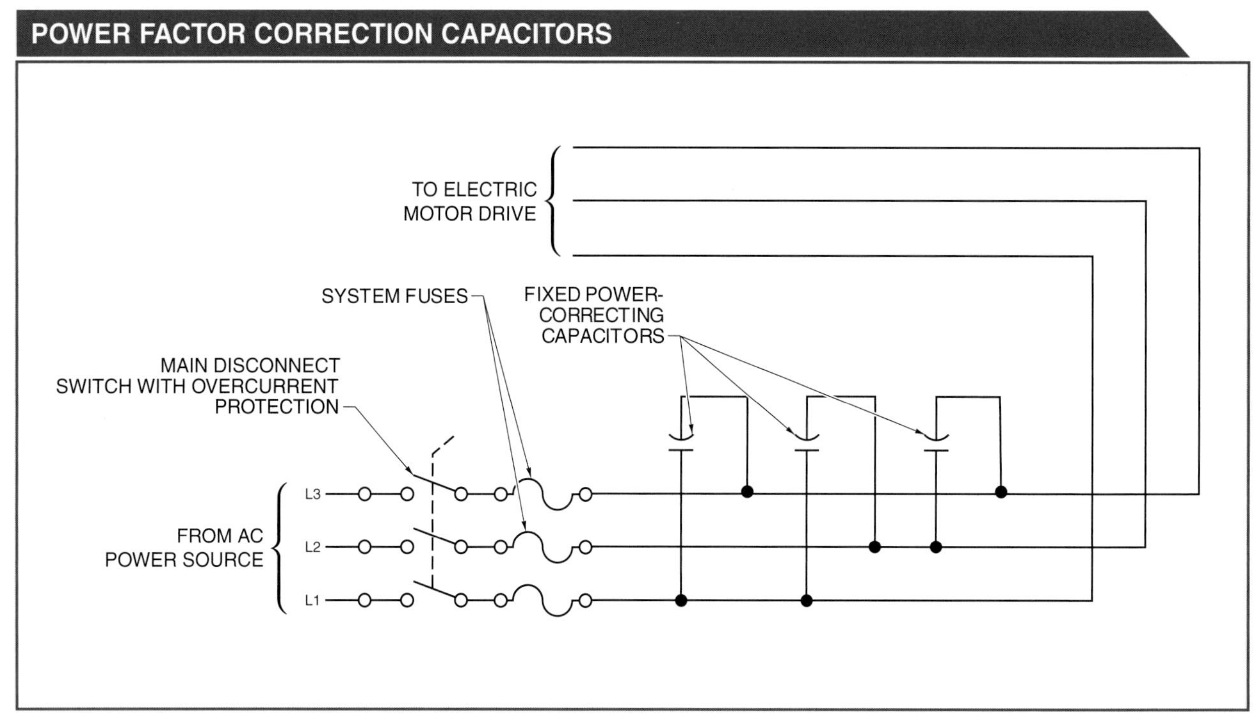

Figure 20-12. Power factor correction capacitors must be placed before the electric motor drive in the AC line supplying power to the drive, not between the drive and motor.

For example, if a capacitor is placed at the motor, the overloads need to be replaced according to the NEC® and/or the manufacturer's specifications. The higher-current-rated overloads should remain in the circuit if capacitors are placed before the overloads. Capacitors must be disconnected from the line when power is removed from the motor to eliminate the possibility of having a leading power factor, which may cause problems on electrical utility lines. Problems on electrical utility lines can generate excessive heat and destroy a motor. Maximum benefit is achieved by having a power-correcting capacitor connected directly to the machine for which it is designed.

When a motor is disconnected, the power-correcting capacitor can be removed from the system with a fused safety switch. Anytime a motor is jogged, plugged, or reversed as part of its operation, the power-correcting capacitor should be connected between the fuses and the motor starter or further upstream between the motor starter and the meter.

Connecting power-correcting capacitors further upstream between the motor starter and the meter typically has the advantage of a lower cost per CkVAR installed. Large power-correcting capacitors with a CkVAR value of 100 or higher can be used to compensate for many different combined loads. A large capacitor requires only one installation, rather than a separate installation at each machine. When using a large capacitor, the NEC® requires a main disconnect switch with overcurrent protection for each ungrounded conductor and each capacitor bank.

The NEC® requires that each piece of equipment in a three-phase system have accessible disconnect switches with overcurrent protection.

Installation of capacitors on the main power switchgear (bus) permits maximum utilization of system ampacity. Adding capacitors to or removing capacitors from the line can compensate for changes in reactive power. Automatic switching systems are available for use in applications that have varied loads where rapid switching prohibits manual adjustments.

THREE-PHASE WIRING SYSTEMS

The principles of a 3ϕ wiring system are the same as for a 1ϕ wiring system, except that in a 3ϕ system there are three equally spaced armature windings 120° out of phase with each other. **See Figure 20-13.** For example, the output of a 3ϕ AC generator results in three output voltages 120° out of phase with each other.

A 3ϕ generator has six leads coming from the armature windings. When the six leads are brought out from the generator, they are connected so that only three leads appear for connection to the load. The manner in which the leads are connected to the load determines the electrical characteristics of the generator output. Armature windings can be connected in a delta connection or a wye connection.

THREE-PHASE AC GENERATORS

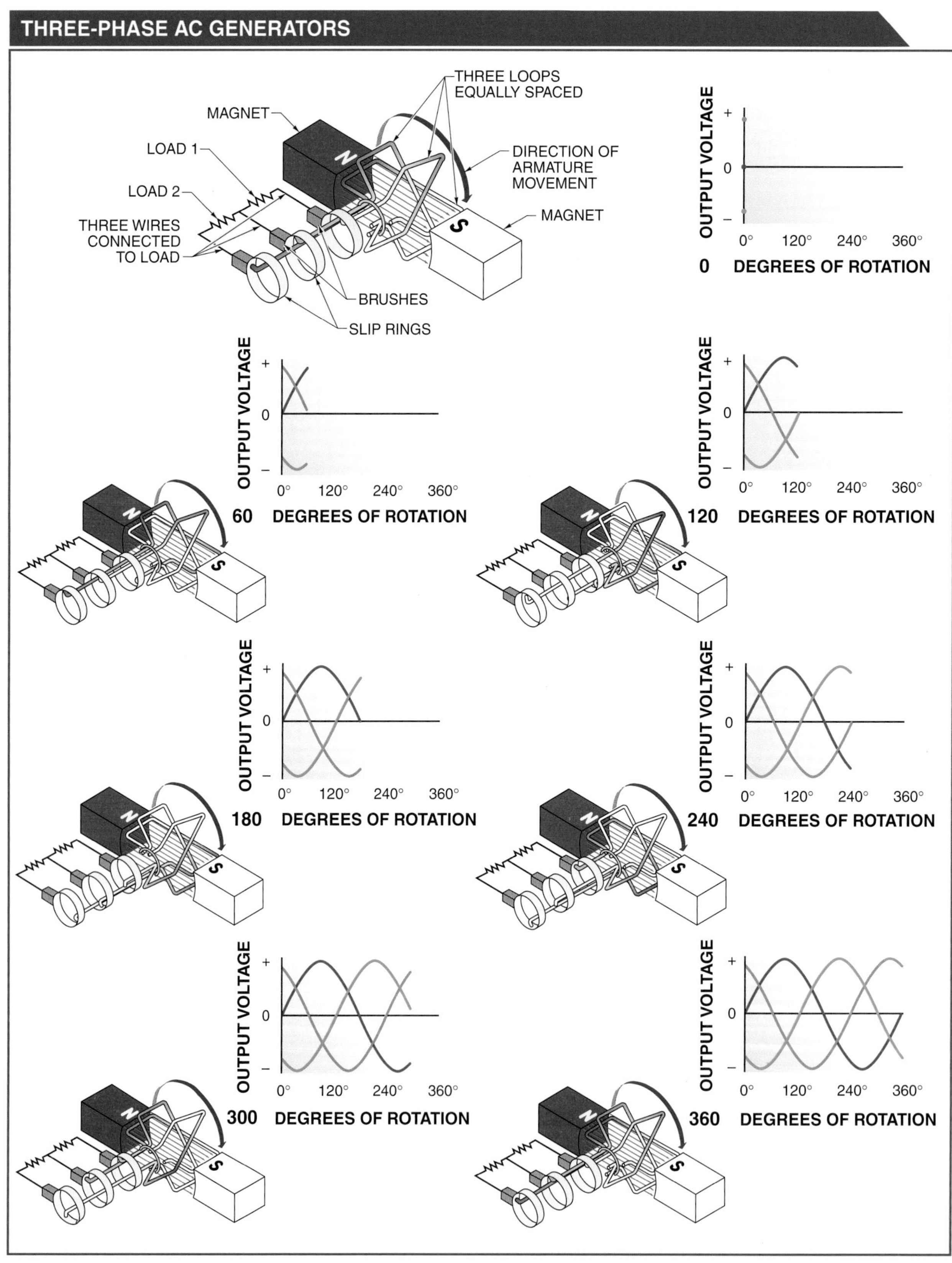

Figure 20-13. A 3ϕ AC generator has three equally spaced armature windings 120° out of phase with each other.

Delta Connections

A *delta connection* is a connection that has each coil (winding) connected end-to-end to form a closed loop. **See Figure 20-14.** The leads are connected to the nodes (connection points) where two windings are joined. A delta connection is often represented by the Greek letter delta (Δ).

Terms

A **delta connection** is a connection that has each coil (winding) connected end-to-end to form a closed loop.

In a delta connection, the start connection point of one winding is connected to the finish connection point of the adjacent winding. These connection points are brought out of the 3ϕ generator that supplies voltage to the resistive load that is also delta connected.

When the generator windings are properly delta connected, there is no current flow within the delta loop when no voltage is applied to the generator. If any one of the connections is reversed where an end-to-end or start-to-start connection is made, a short-circuit current flows, even when no load is applied. A short-circuit condition can damage windings.

To minimize the possibility of a short circuit, the delta circuit should be tested before the final connection is made. Either a DMM set to measure voltage or a fuse should be connected across the open connection. If the fuse blows, or a voltage is indicated on the display of the DMM, then the delta connection has been improperly made.

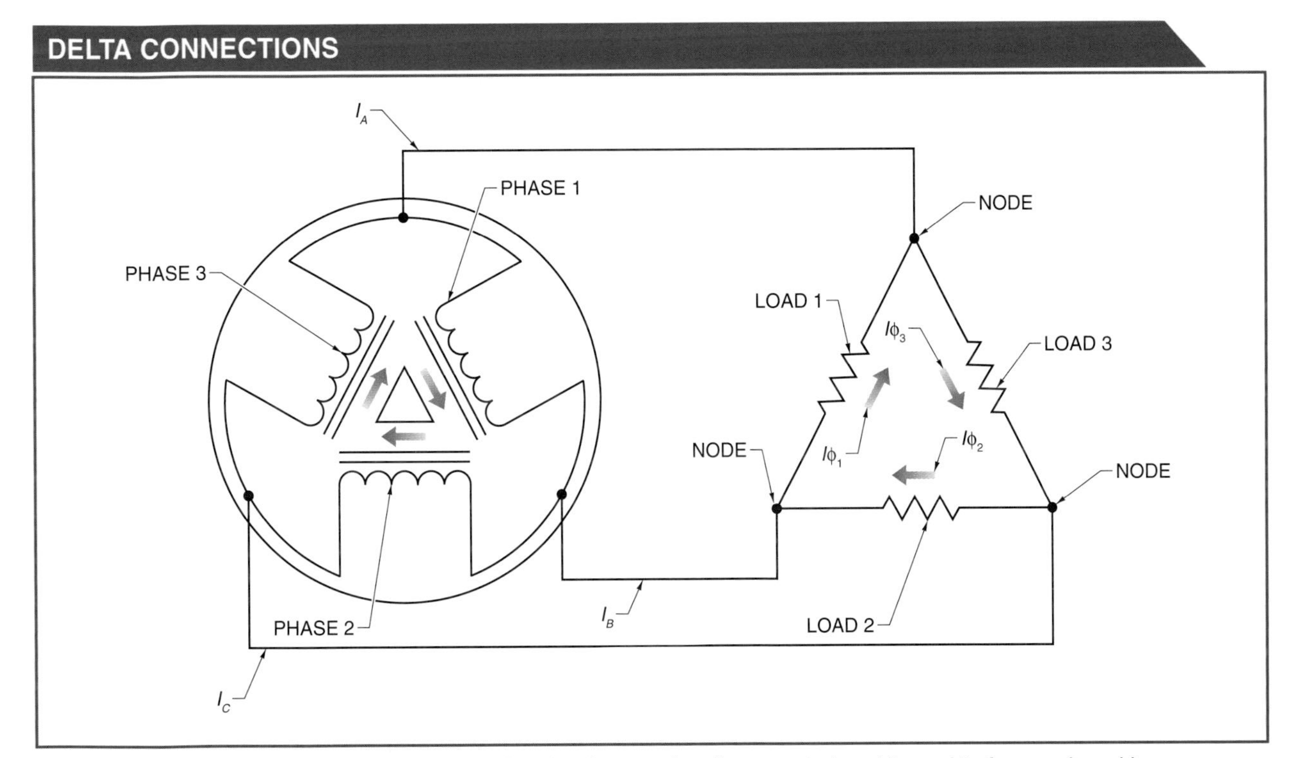

Figure 20-14. A delta connection is a connection that has each coil connected end-to-end to form a closed loop.

Kirchhoff's voltage law (KVL) is the basic law used to analyze series circuits in electronics applications.

Note: When a generator is delta connected, the phase voltage of the winding is equal to the line voltage, which results in the following conditions:

$V_{\phi 1} = V_{AB}$

$V_{\phi 2} = V_{BC}$

$V_{\phi 3} = V_{AC}$

or

$V_{PHASE} = V_{LINE}$

According to Kirchhoff's voltage law, the algebraic sum of the voltages around a closed loop must equal zero. For example, the 3ϕ currents in the windings of the generator, $I_{\phi 1}$, $I_{\phi 2}$, and $I_{\phi 3}$, are indicated by arrows that show the direction of current flow through the windings. In an AC system, the current in each winding periodically changes in direction of flow. As with voltages, the 3ϕ currents are 120° out of phase with each other. In this system, the 3ϕ load is balanced within each branch. Therefore, current in each phase is equal to current in each other phase.

Line current (I_A) on the Aϕ is supplied by $I_{\phi 1}$ and $I_{\phi 3}$. Line current (I_B) on the Bϕ is supplied by $I_{\phi 1}$ and $I_{\phi 2}$, and line current (I_C) on the Cϕ is supplied by $I_{\phi 2}$ and $I_{\phi 3}$.

Line currents adhere to Kirchhoff's current law. Therefore, line currents I_A, I_B, and I_C are equal to the vector sum of the currents supplied by the windings of the generator. Current flow into the line is considered to be positive, and current flow away from the line is considered to be negative. Vector diagrams can be used when performing calculations between I_A, I_B, and I_C of a delta connection. **See Figure 20-15.**

Because the load is balanced and purely resistive, phase currents are in phase with phase voltages and equal to load currents. Current flow through the loads can be calculated using Ohm's law, by applying the following formula:

$$I_1 = \frac{V_{AB}}{R_1}$$

where

I_1 = line 1 current (in A)

V_{AB} = line A-B voltage (in V)

R_1 = line 1 resistance (in Ω)

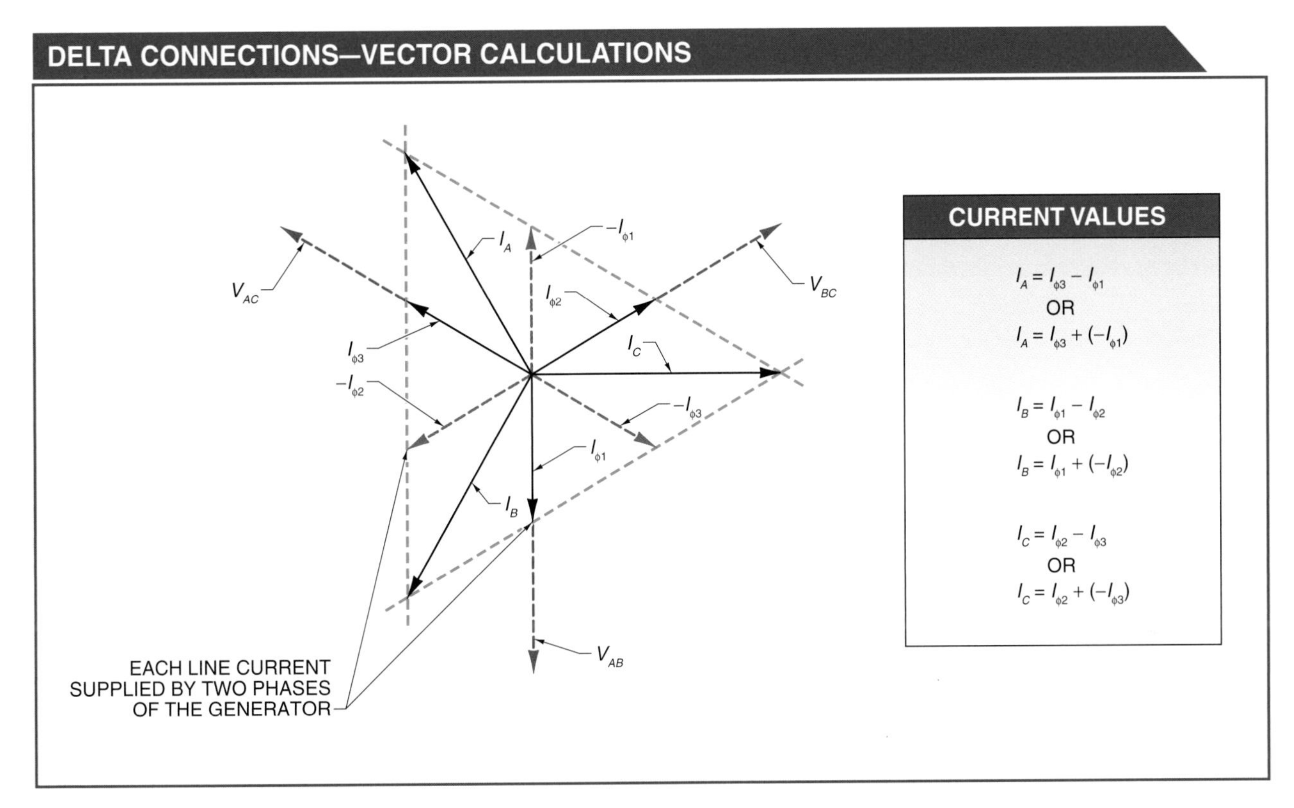

Figure 20-15. Vector diagrams can be used when performing calculations between I_A, I_B, and I_C of a delta connection.

Example: What is the current for line 1 if its resistance is at 10 Ω and line A-B is 10 V?

$$I_1 = \frac{V_{AB}}{R_1}$$

$$I_1 = \frac{10\text{ V}}{10\ \Omega}$$

$$I_1 = \mathbf{1\ A}$$

In this example, all line voltages are 10 V and the loads are 10 Ω, so the current flow through all the loads is the same, and $I_1 = I_2 = I_3$. In a 3ϕ system, each line current must equal the two other 1 A phase currents that are 120° out of phase with each other.

The graphical vector method may be used to verify the current readings. For example, the two currents supplied by the windings of the generator are added together. This results in a current of 1.73 A that is 30° out of phase with the two other phase currents. **See Figure 20-16.** Solutions for the other two line currents yield the same result, with each line current out of phase with the other line currents by 120°.

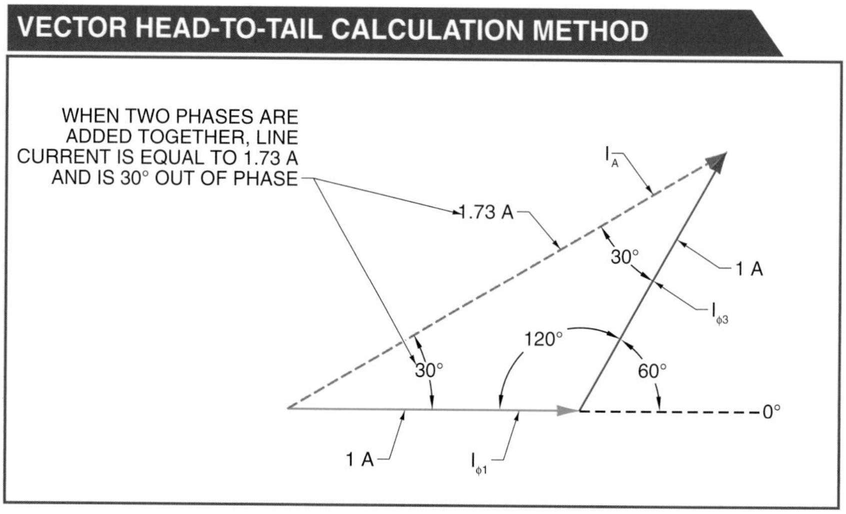

VECTOR HEAD-TO-TAIL CALCULATION METHOD

WHEN TWO PHASES ARE ADDED TOGETHER, LINE CURRENT IS EQUAL TO 1.73 A AND IS 30° OUT OF PHASE
1.73 A
I_A
30°
1 A
$I_{\phi 3}$
120°
30°
60°
0°
1 A
$I_{\phi 1}$

Figure 20-16. A vector solution of the phase currents shows that 1 A flowing in each phase produces 1.73 A of line current.

Anytime the magnitudes are equal and 120° apart, the resultant vector can be obtained by multiplying one vector by 1.73 ($\sqrt{3}$). This solution is comparable to that for two-phase power, where one side of the 90° angle is multiplied by 1.414 ($\sqrt{2}$). These values are equal to the values of the square roots of 2 and 3 because they are derived from the solution for the hypotenuse of a triangle with two sides of equal length.

A delta-connected generator has the following characteristics:

- Generated voltages and load voltages have equal values and are located 120° apart.
- Line currents have equal values and are located 120° apart.
- Line currents are 30° out of phase with line voltages when the power factor is at unity.
- Line current equals the individual phase current or individual load current multiplied by 1.73.

Vector Addition

When adding vectors, consideration must be given to the magnitude and direction of the vector displacements. Because addition of vectors is geometric rather than algebraic, it is possible for the magnitude of a vector sum to be less than the magnitude of either of the individual vector displacements.

Wye Connections

A *wye connection* is a connection that has one end of each coil connected together and the other end of each coil left open for external connections. In a wye connection, three leads (one from each winding) are connected together, while the other three leads are brought out for connecting to the load. A wye connection resembles the letter Y, and is sometimes referred to as a star connection. **See Figure 20-17.** A wye connection requires that one end of each winding be connected to a common point. The common point may or may not be brought out of the generator case. When the common-point connection is used, it is referred to as the neutral line. A neutral line is not required for a properly balanced 3ϕ load. However, if 1ϕ power is used with unbalanced loads, a neutral line is required. With a wye-connected generator, the current on any phase is the same as on any other phase, since the phase winding and the load are in series with each other. A three-phase wye connection can be either a 120/208 V 4-wire connection or a 277/480 V 4-wire connection.

Terms

A **wye connection** is a connection that has one end of each coil connected together and the other end of each coil left open for external connections.

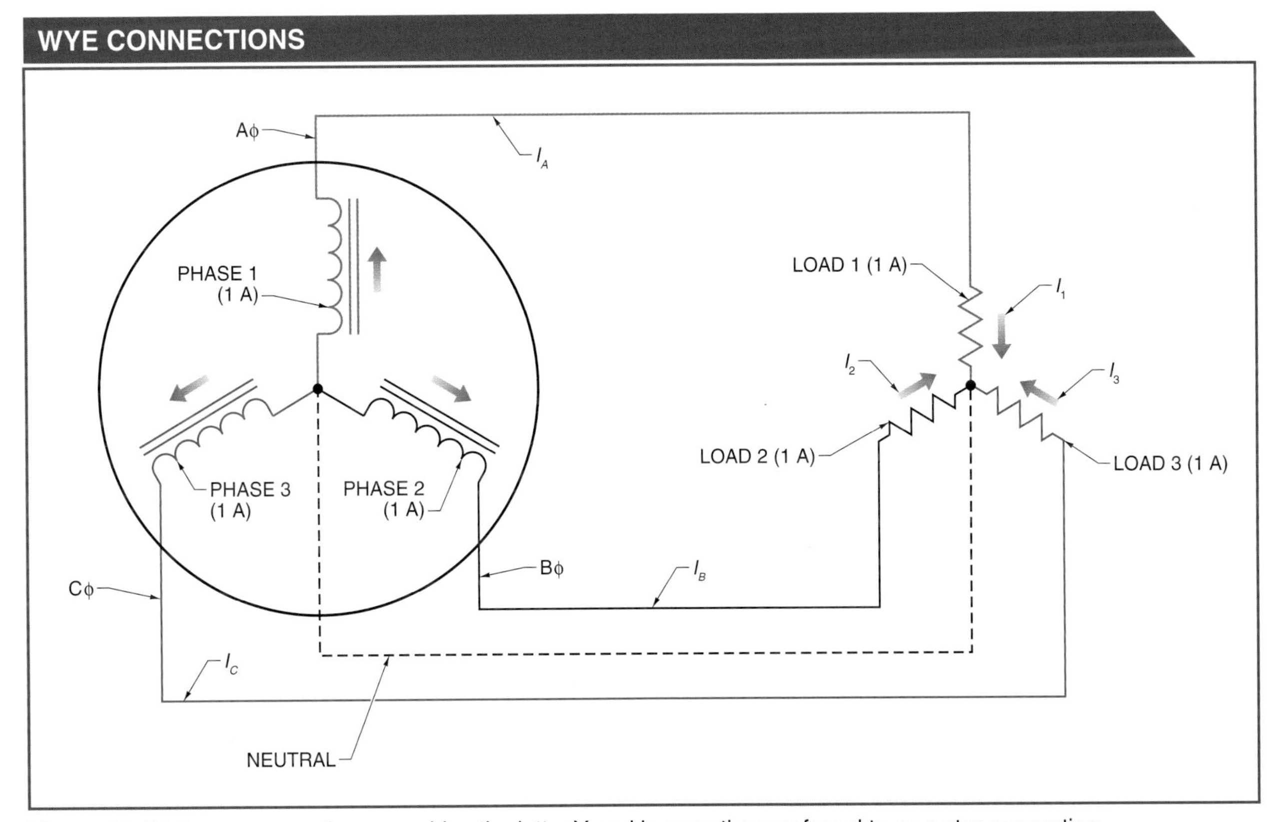

Figure 20-17. A wye connection resembles the letter Y and is sometimes referred to as a star connection.

Current in one phase is out of phase with the currents in the other two phases by 120°. As long as the load is balanced, no potential exists between the neutral points of the generator and the load. If the loads become unbalanced, a potential develops between the two neutral points, with a resulting change in the potential across the branches (legs) of the wye load. To prevent a change in load voltages, the neutral points are normally connected in most systems. The neutral line can carry current unbalance. As established by Kirchhoff's current law, the sum of the currents at the neutral points must equal 0 A.

Tech Tip

It is best to use a meter with an automatic shutoff function to save batteries. This function automatically shuts OFF the meter if the meter is not used in a set amount of time.

Note: When a generator is wye-connected, the voltage of two generator windings is across the load, resulting in the following conditions:

$V_{AB} = V_{\phi 1} - V_{\phi 2}$

$V_{BC} = V_{\phi 2} - V_{\phi 3}$

$V_{AC} = V_{\phi 1} - V_{\phi 3}$

where

V_{AB} = line AB voltage (in V)

$V_{\phi 1}$ = phase 1 voltage (in V)

$V_{\phi 2}$ = phase 2 voltage (in V)

V_{BC} = line BC voltage (in V)

$V_{\phi 3}$ = phase 3 voltage (in V)

V_{AC} = line AC voltage (in V)

or

$$V_{LINE} = V_{PHASE} \times \sqrt{3}$$

where

V_{LINE} = line voltage (in V)

V_{PHASE} = phase voltage (in V)

$\sqrt{3} = 1.73$ (constant)

Example: What is the line voltage in a wye-connected generator that has three lines at 10 V that are 120° out of phase?

$V_{LINE} = V_{PHASE} \times \sqrt{3}$

$V_{LINE} = 10\text{ V} \times 1.73$

$V_{LINE} = \mathbf{17.3\text{ V}}$

When combining AC voltages, the direction of current flow for the positive maximum value as well as the value of the phase voltage must be known. **See Figure 20-18.** Different results are achieved if the phases are not properly connected.

A wye-connected generator has the following characteristics:

- Line voltages have equal values and are located 120° apart.
- Line voltages equal the phase voltage multiplied by $\sqrt{3}$ (1.73).
- In a properly balanced system, line currents are out of phase with line voltages by 30°.
- In a properly balanced system, line currents are out of phase with each other by 120°.

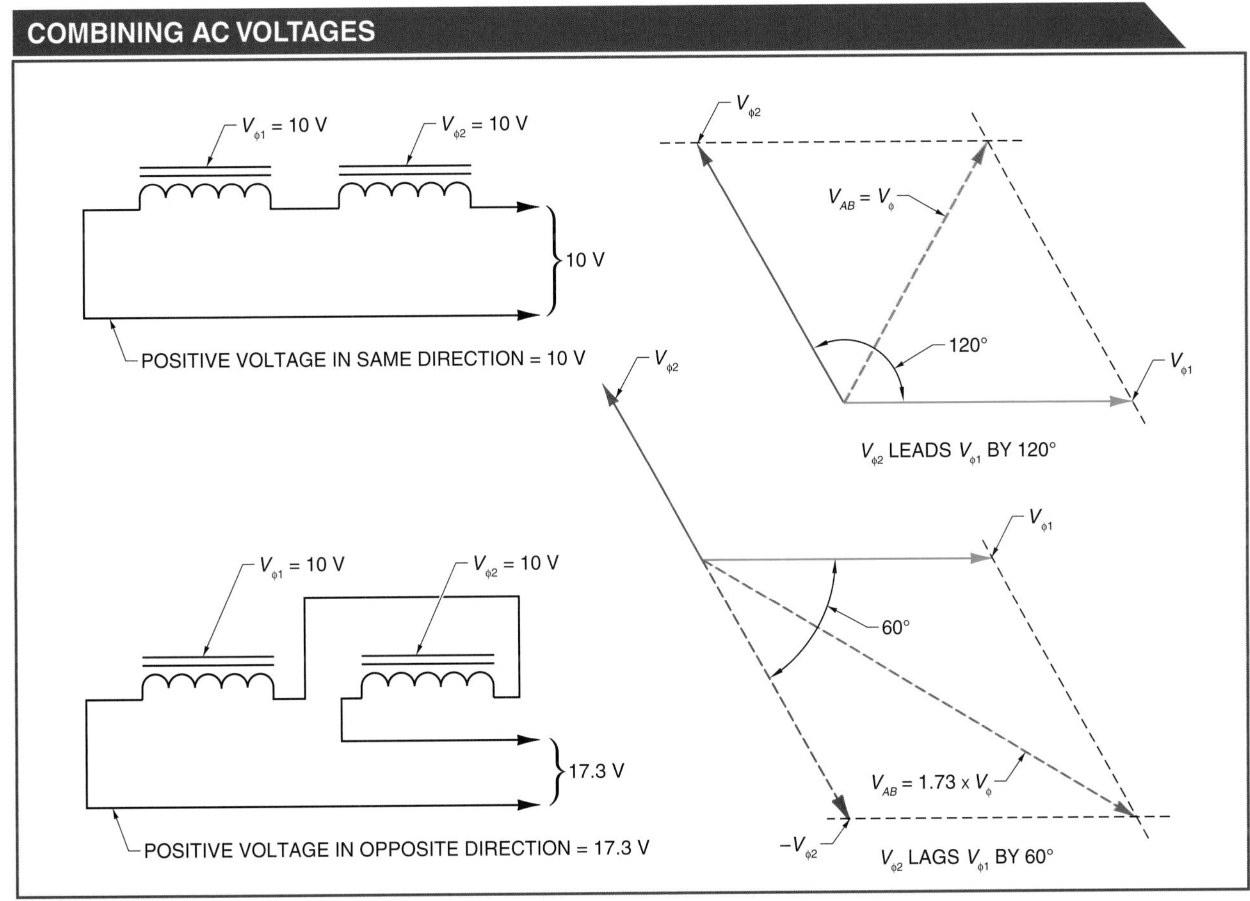

Figure 20-18. When combining AC voltages, the direction of current flow for the positive maximum value as well as the value of the phase voltage must be known.

REACTIVE CIRCUITS

A *reactive circuit* is a circuit that is either capacitive or inductive and has a power factor of less than 1. Typical reactive circuit applications are circuits that have motors, HVAC systems, machinery, and appliances as loads. In a capacitive reactive circuit, current leads the voltage. Current in each capacitive branch is calculated by applying the following formula:

Terms

A **reactive circuit** is a circuit that is either capacitive or inductive and has a power factor of less than 1.

$$I_{CL} = \frac{V_{LOAD}}{X_{CL}}$$

where

I_{CL} = capacitive branch current (in A)
V_{LOAD} = load voltage (in V)
X_{CL} = branch capacitive reactance (in Ω)

Terms

Continuity tester is a test instrument that tests for a complete path for current to flow.

Continuity Testers

A *continuity tester* is a test instrument that tests for a complete path for current to flow. A continuity test is commonly used to test components such as switches, fuses, electrical connections, and individual conductors. The test instrument emits an audible response (beeps) when there is a complete path. Indication of a complete path can be used to determine whether a component is in an open or a closed condition.

Example: What is the capacitive current in a branch circuit with an applied voltage of 10 V and branch capacitive reactance of 10 Ω?

$$I_{CL} = \frac{V_{LOAD}}{X_{CL}}$$

$$I_{C1} = \frac{V_{AB}}{X_{C1}}$$

$$I_{C1} = \frac{10 \text{ V}}{10 \text{ }\Omega}$$

$$I_{C1} = \mathbf{1\ A}$$

In a pure capacitive circuit, current leads phase voltage by 90°. With a balanced load, current in the phase windings reflects the load. Current in the phase windings is equal to the load current and also leads the source voltage by 90°. Current in capacitive branch 1 is equal to current in capacitive branches 2 and 3.

In a pure inductive circuit, current lags phase voltage by 90°. When a delta-connected generator is connected to a delta-connected balanced inductive load, inductors with a reactance of 10 Ω are used to replace the capacitors. Balanced current lags phase voltages by 90°. **See Figure 20-19.**

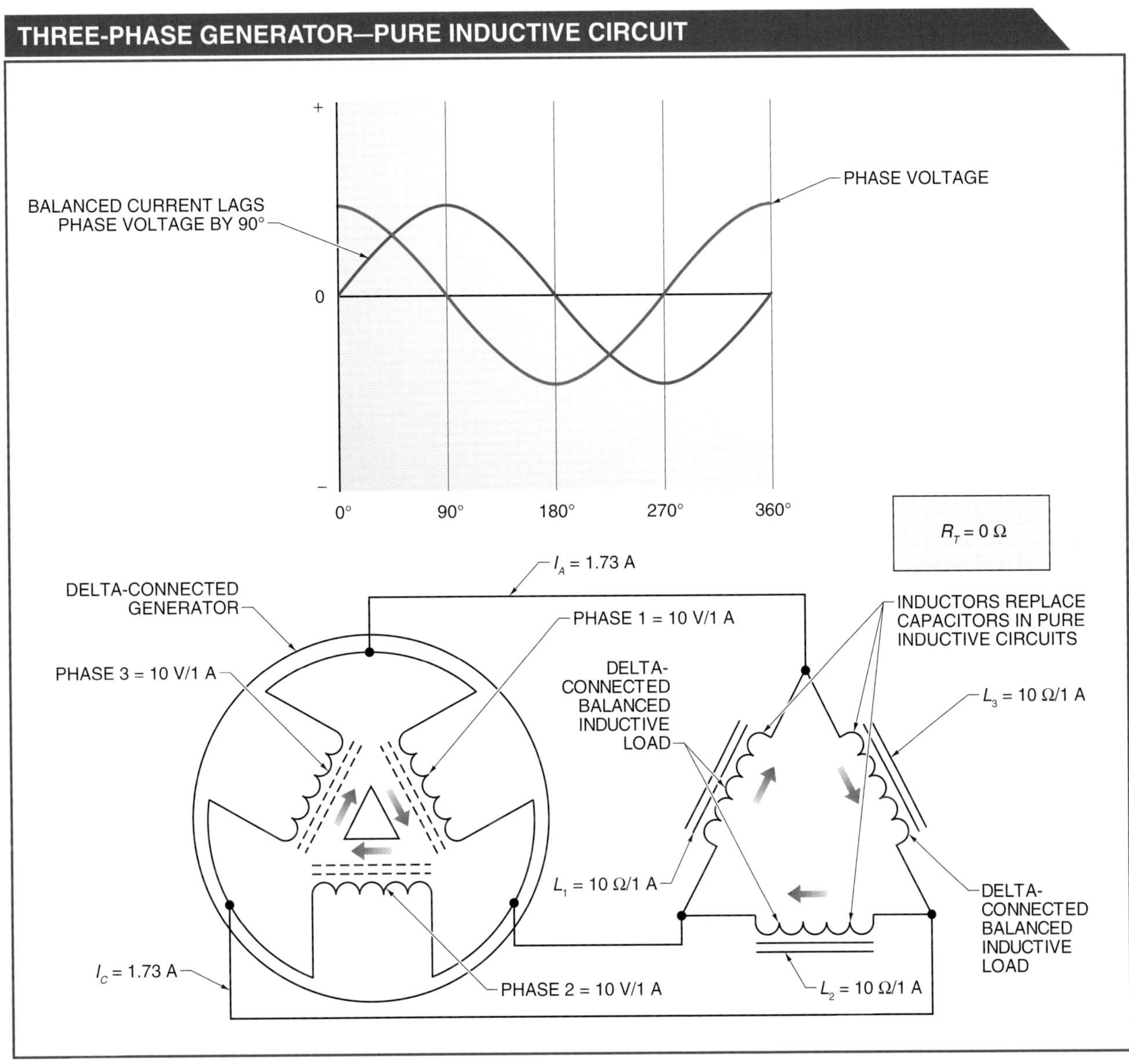

Figure 20-19. When a delta-connected generator is connected to a delta-connected balanced inductive load, inductors with a reactance of 10 Ω are used to replace the capacitors.

In real-world reactive circuit applications, reactive circuits are never purely reactive because there is always some amount of resistance present. However, if the value of the reactive element is ten times the value of the resistive element, then the circuit is usually analyzed as purely reactive.

Most circuits are not purely inductive. Rather, most practical circuits are inductive-resistive, so that current lags the source voltage between 0° and 90°. **See Figure 20-20.** When resistance equals inductive reactance in each phase, current lags source voltage by 45°.

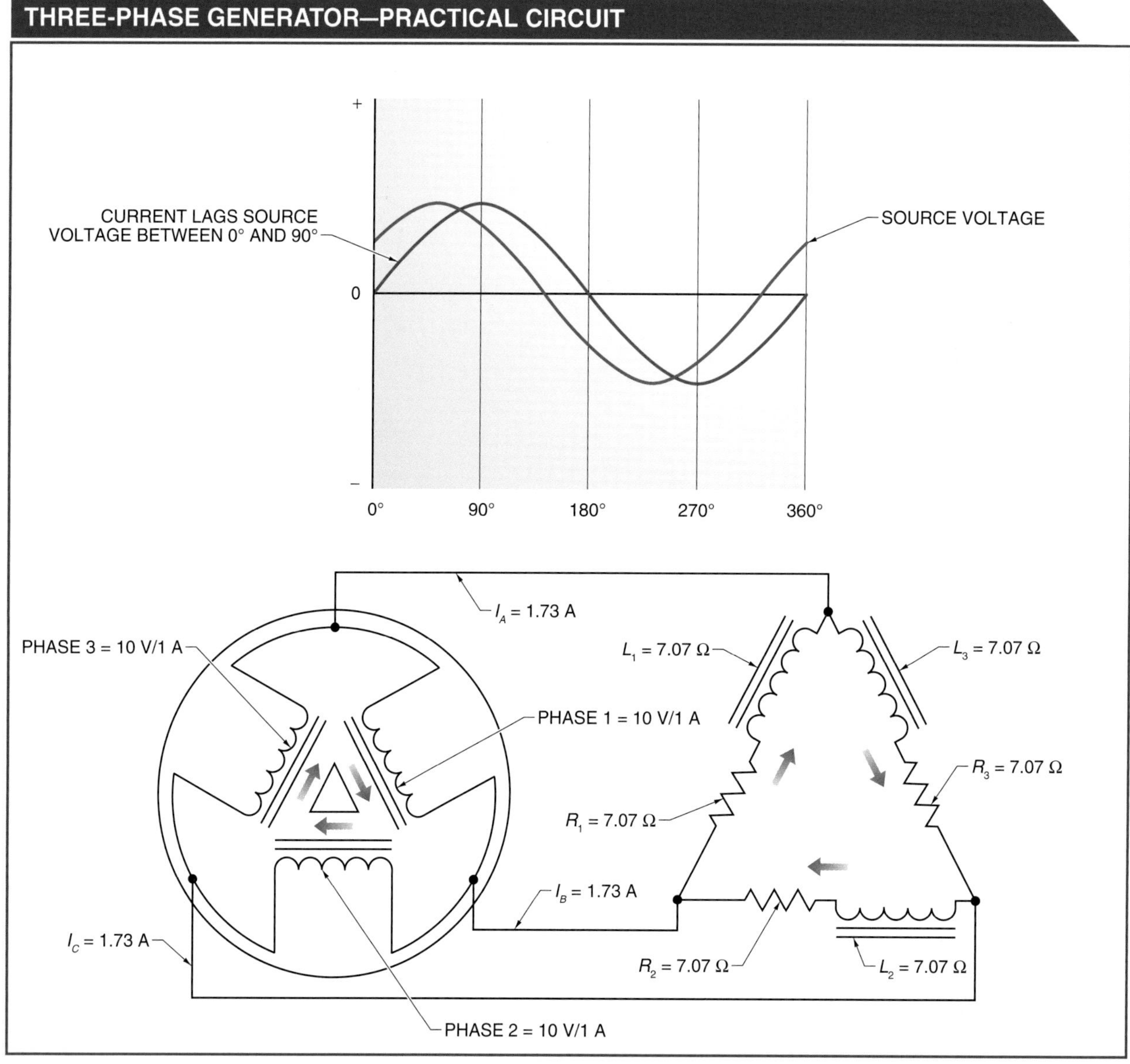

Figure 20-20. Most practical circuits are inductive-resistive circuits, which means current lags the source voltage between 0° and 90°.

Branch impedance can be calculated by using the $\sqrt{2}$ (1.41) multiplier since the values are the same and 90° out of phase. Each of the branches of the delta connection has the same values. Therefore, their impedance is also 10 Ω at 45°. Impedance has both a magnitude and an angle, which makes it a phasor quantity. Analysis of a reactive branch circuit includes determining the values of impedance, angle theta, and current. Reactive branch circuit analysis is performed by applying the following procedure:

1. Calculate reactive branch circuit impedance. Reactive branch circuit impedance is calculated by applying the following formula:

   $Z_B = \sqrt{2} \times R$

   where

   Z_B = branch impedance (in Ω)

   R_B = branch resistance (in Ω)

2. Calculate angle theta. Angle theta is calculated by applying the following formula:

   $$\theta = \tan^{-1}\left(\frac{X_{LB}}{R_B}\right)$$

   where

   θ = angle theta (in °)

   X_{LB} = branch inductive reactance (in Ω)

   R_B = branch resistance (in Ω)

3. Calculate branch current. Branch current is calculated by applying the following formula:

   $$I_B = \frac{V_{LOAD}}{Z_B}$$

   where

   I_B = branch current (in A)

   V_{LOAD} = load voltage (in V)

   Z_B = branch impedance (in Ω)

Example: What is the analysis of a reactive branch circuit that has resistance and inductive reactance of 7.07 Ω and a source voltage of 10 V?

1. Calculate reactive branch circuit impedance.

   $Z_B = \sqrt{2} \times R_B$

   $Z_B = 1.414 \times 7.07\ \Omega$

   $Z_B = 10\ \Omega$

2. Calculate angle theta.

   $$\theta = \tan^{-1}\left(\frac{X_{LB}}{R_B}\right)$$

   $$\theta = \tan^{-1}\left(\frac{7.07\ \Omega}{7.07\ \Omega}\right)$$

   $\theta = \tan^{-1} 1.000$

   $\theta = 45°$

3. Calculate branch current.

$$I_B = \frac{V_{LOAD}}{Z_B}$$

$$I_B = \frac{10\ \text{V}}{10\ \Omega\ \text{at}\ 45°}$$

$$\mathbf{I_B = 1\ A\ at\ -45°}$$

Current through I_B lags V_{LOAD} by 45°. Similarly, the currents in the other two branches are also 1 A and lag their respective voltages by 45°.

Analysis of 3ϕ circuits becomes more complex if the loads are not balanced and if there are various types of loads in each branch.

SUMMARY

Three-phase power has significant advantages over 1ϕ power. Accordingly, most commercial power is 3ϕ power. The advantages of 3ϕ systems over 1ϕ systems include the following:

- The amount of copper (conductors) used to deliver the same amount of power is reduced.
- A more constant load is provided to the generator.
- Three-phase motors are less complex than 1ϕ motors and have constant torque (rotating force).
- A smoother DC output is created from 3ϕ power, using less filtering than with 1ϕ power.

For a delta-connected system, the phase voltage equals the line voltage, and the line current is 1.73 ($\sqrt{3}$) times the phase current. In a wye-connected system, the line voltage is equal to 1.73 ($\sqrt{3}$) times the phase voltage, and the line current equals the phase current.

In a 3ϕ system, the phase voltages are separated by 120°. Depending on the type of load, current can lead, lag, or be in phase with voltage. Most power systems are inductive-resistive with a lagging current due to motors and other inductive devices. Electrical utilities penalize end users that have a poor power factor. Mathematical methods of measuring the necessary parameters are used to calculate the power factor of a circuit. Power factor correction is achieved by connecting capacitors across the loads.

APPLICATION: CALCULATING THREE-PHASE, 208-Y/120-V PHASE CURRENT...

Note: A scientific calculator is required for understanding and solving this application.

Background: Information technology (IT) data centers have seen a significant increase in power requirements due to the increase in the number of servers, mass storage devices, backup systems, networking equipment, and other related computer equipment. To meet this demand, an increasing number of data centers are being powered using three-phase (3ϕ), 208-Y/120-V power. High-power applications require less copper to supply the same amount of power when using 3ϕ distribution.

Typical power available in North America is 120 V. When 3ϕ, 120-V power is used in a wye configuration, each line is 120° out of phase and the line-to-line voltage is 208 V. **See Figure 1.**

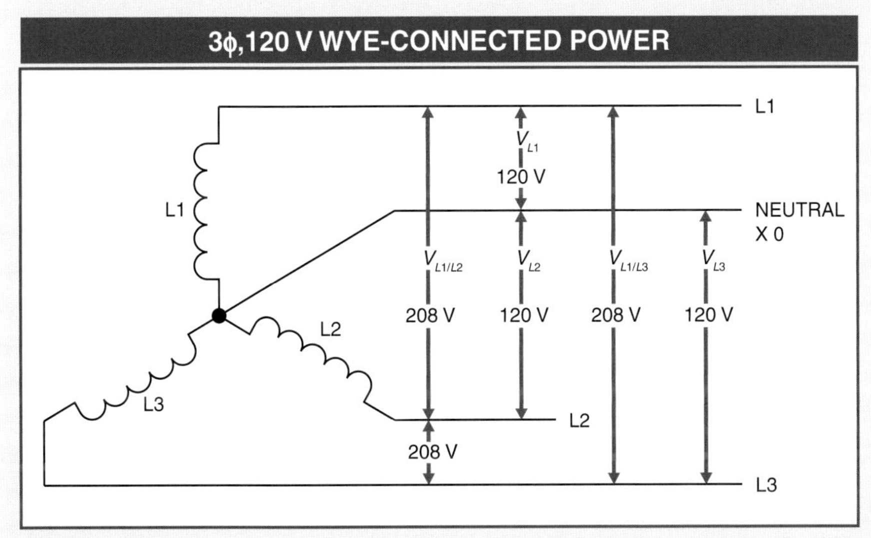

Figure 1. When 3ϕ, 120-V power is used in a wye configuration, each line is 120° out of phase and the line-to-line voltage is 208 V.

The circuits are provided by lines 1 and 2 (L1/L2), lines 2 and 3 (L2/L3), and lines 3 and 1 (L3/L1). The line-to-line voltage is found by adding the two phases together using vector sums. The line-to-line voltage is equal to the following:

$$V_{L1/L2} = V_{L1} \times \sqrt{3}$$
$$V_{L1/L2} = 120\text{ V} \times \sqrt{3}$$
$$V_{L1/L2} = 208\text{ V}$$

IT equipment is connected to the power supplied by L1/L2, L2/L3, or L3/L1. It is important to understand the increase in current on each phase as equipment is added. The calculation of current used on each phase does not involve simple addition. For example, if 10 A is added to L1/L2 and 10 A is also added to L2/L3, the current on L2 is 17.3 A, not 20 A. This is due to the difference in phase between the lines. The amount of current in each line is calculated using the law of cosines, which takes into account the phase relationship. The general equation that expresses the law of cosines is the following:

$$c = \sqrt{a^2 + b^2 - 2 \times a \times b \times \cos\theta}$$

...APPLICATION: CALCULATING THREE-PHASE, 208-Y/120-V PHASE CURRENT...

For 3ϕ power in which each phase is 120° out of phase, the equation becomes the following:

$$c = \sqrt{a^2 + b^2 - 2 \times a \times b \times \cos(120)}$$
$$c = \sqrt{a^2 + b^2 - 2 \times a \times b \times -0.5}$$
$$c = \sqrt{a^2 + b^2 + a \times b}$$

To find the current in each phase, the following formulas are used:

$$I_{L1} = \sqrt{(I_{L1/L2})^2 + (I_{L3/L1})^2 + I_{L1/L2} \times I_{L3/L1}}$$
$$I_{L2} = \sqrt{(I_{L2/L3})^2 + (I_{L1/L2})^2 + I_{L2/L3} \times I_{L1/L2}}$$
$$I_{L3} = \sqrt{(I_{L3/L1})^2 + (I_{L2/L3})^2 + I_{L3/L1} \times I_{L2/L3}}$$

where

I_{L1} = current on line 1 (L1)
I_{L2} = current on line 2 (L2)
I_{L3} = current on line 3 (L3)

Using the earlier example, 10 A on L1/L2 and 10 A on L2/L3, the current on L2 is the following:

$$I_{L2} = \sqrt{(I_{L2/L3})^2 + (I_{L1/L2})^2 + I_{L2/L3} \times I_{L1/L2}}$$
$$I_{L2} = \sqrt{(10\text{ A})^2 + (10\text{ A})^2 + 10\text{ A} \times 10\text{ A}}$$
$$I_{L2} = \sqrt{300\text{ A}^2}$$
$$I_{L2} = 17.3\text{ A}$$

Key Points: Three-phase wye power

Problem: An IT data center needs to power three 5 kW loads. The current requirements of 1ϕ, 120-V wye power must be compared to 3ϕ, 208-V/120-V wye power distribution.

Solution: For the 1ϕ, 120-V power distribution, all three 5 kW loads are connected in parallel. The current required by the load is found by applying the following formula:

$$I = \frac{P}{V}$$

where

I = current (in A)
P = power (in W)
V = voltage (in V)

...APPLICATION: CALCULATING THREE-PHASE, 208-Y/120-V PHASE CURRENT

Total power is 15 kW. Therefore, total current for the 1ϕ circuit is the following:

$$I = \frac{P}{V}$$

$$I = \frac{15 \text{ kW}}{120 \text{ V}}$$

$$I = 125 \text{ A}$$

For 3ϕ, 208-V/120-V wye power, the 5 kW loads are added to L1/L2, L2/L3, and L3/L1. The current required for each load is as follows:

$$I = \frac{P}{V}$$

$$I = \frac{5 \text{ kW}}{208 \text{ V}}$$

$$I = 24 \text{ A}$$

Since a common load is added to each phase, only one calculation is necessary. The current for L1 is the following:

$$I_{L1} = I_{L2} = I_{L3} = \sqrt{\left(I_{L1/L2}\right)^2 + \left(I_{L3/L1}\right)^2 + I_{L1/L2} \times I_{L3/L1}}$$

$$I_{L1} = I_{L2} = I_{L3} = \sqrt{(24 \text{ A})^2 + (24 \text{ A})^2 + 24 \text{ A} \times 24 \text{ A}}$$

$$I_{L1} = I_{L2} = I_{L3} = \sqrt{1728 \text{ A}^2}$$

$$I_{L1} = I_{L2} = I_{L3} = 41.6 \text{ A}$$

Using 1ϕ power in this application requires three 1 AWG conductors (hot, neutral, and ground), while the 3ϕ power requires four 8 AWG conductors (hot, hot, neutral, and ground). The wiring gauges are for insulated conductors according to NEC® table 310.15(B)(16) with a conductor temperature rating of 75°C. The 3ϕ power requires less total copper, reducing overall conductor weight and cost.

Chapter Review

1. What are the characteristics of a properly delta-connected generator?
2. How does current act in a pure inductive circuit?
3. Explain why a power quality meter is more useful than other types of meters when gathering information.
4. How does current act in a pure capacitive circuit?
5. What level of power factor is considered poor?
6. What occurs with a generator when voltage is out of phase in an AC system?
7. What is power factor correction?
8. Why is it desirable to keep power factor as close to 100% as possible?
9. What occurs to the frequency of generated voltage when the number of poles is increased?
10. List three advantages of using 3ϕ power over 1ϕ power.
11. Why are capacitors added to loads with lagging power factor?
12. What is a reactive circuit?
13. With a wye-connected load, when would a neutral line be required?
14. Describe excitation current.
15. What is the difference in the copper requirement between 1ϕ and 3ϕ systems?
16. Explain the difference between jogging and plugging.
17. What is the main purpose of a wattmeter?
18. Describe the difference between generated and induced voltage.
19. What occurs with a flywheel diode when the exciter switch is closed?
20. What is a rectifier?
21. List five examples of typical resistive loads found in residential applications.
22. Explain the difference between minimum and maximum induced voltage.
23. Explain how power factor reacts in an inductive, a capacitive, and a resistive load.
24. What is a swamping resistor?
25. How many phases supply power at all times in a 3ϕ circuit?

21

Transformers

OBJECTIVES

- Describe transformer theory and ratings.
- Describe how loads affect transformers.
- List different types of transformer ratios.
- List and describe the types of transformer losses.
- Explain transformer efficiency.
- Describe the different transformer classifications.
- Explain the differences between single-phase and three-phase transformer connections.

Transformers are a critical component used in power distribution systems. Transformers operate using electromagnetism and are generally used to change one voltage level to another. Voltage is either increased or decreased based on need and the type of transformer. Because transformers do not create power, whenever voltage is increased, current is decreased, and whenever voltage is decreased, current is increased.

Transformer parameters include transformer ratings, loads, losses, efficiency, and classification. Transformers are identified by their classification and connection method.

Terms

A **transformer** is an electrical device that uses electromagnetism to change voltage from one level to another or to isolate one voltage from another.

Mutual induction is the ability of an inductor in one circuit to induce a voltage in another circuit.

A **wall transformer** is a large, box-shaped plug used with many electronic devices that consists of a transformer and prongs for connection to an electrical receptacle.

A **primary winding** is the winding (coil) of a transformer that draws power from the source.

A **secondary winding** is the winding (coil) of a transformer that delivers energy at the transformed, or changed, voltage to the load.

A **transformer schematic symbol** is a symbol used to identify transformers in drawings and prints.

A **step-up transformer** is a transformer that has a secondary voltage greater than the primary voltage.

A **step-down transformer** is a transformer that has a secondary voltage less than the primary voltage.

An **isolation transformer** is a transformer that has equal primary and secondary voltages and is used to electrically isolate the source from the load.

TRANSFORMERS

A *transformer* is an electrical device that uses electromagnetism to change voltage from one level to another or to isolate one voltage from another.

Transformers do not create power. Transformers use mutual induction to change available power to a desired level of voltage and current required by a load. *Mutual induction* is the ability of an inductor in one circuit to induce a voltage in another circuit. Transformers are used extensively in power distribution, control circuits, and electronic systems. For example, a *wall transformer* is a large, box-shaped plug used with many electronic devices that consists of a transformer and prongs for connection to an electrical receptacle.

Most transformers have two or more windings. A *primary winding* is the winding (coil) of a transformer that draws power from the source. A *secondary winding* is the winding (coil) of a transformer that delivers energy at the transformed, or changed, voltage to the load. A *transformer schematic symbol* is a symbol used to identify transformers in drawings and prints. **See Appendix.** In a transformer schematic symbol, the primary winding connection is indicated by an "H" and the secondary winding connection is indicated by an "X". Transformers may have a primary winding and a secondary winding wound around an iron core. **See Figure 21-1.**

A *step-up transformer* is a transformer that has a secondary voltage greater than the primary voltage. In this configuration the source is connected to the winding with the fewest turns and the load is connected to the winding with the most turns. A *step-down transformer* is a transformer that has a secondary voltage less than the primary voltage. In this case the source is connected to the winding with the most turns and the load is connected to the winding with the fewest turns. An *isolation transformer* is a transformer that has equal primary and secondary voltages and is used to electrically isolate the source from the load. This transformer has an equal number of turns on both the primary and secondary windings.

Early Transformers

Pure iron was used for the cores of early transformers. Iron worked well as an electrical conductor, but eddy-current losses were great. In the early 1900s, a new process began to be used, adding silicon to the iron. Silicon greatly reduced magnetic losses, making current transformers closer to an ideal transformer.

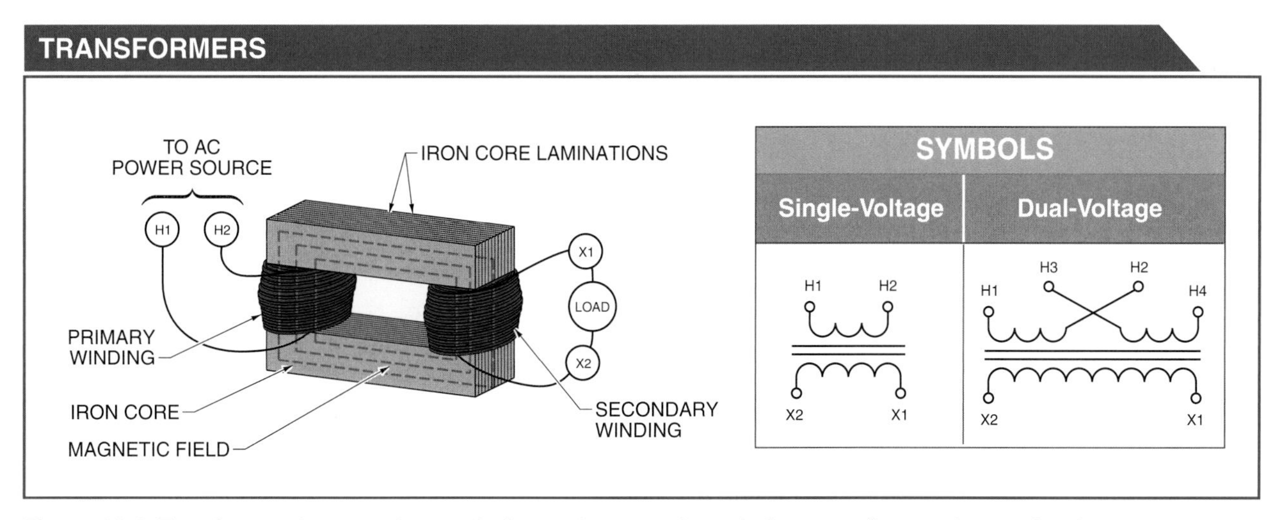

Figure 21-1. Transformers have a primary winding and a secondary winding, usually wound around an iron core.

The *turns ratio* is the ratio of the number of turns in the primary winding to the number of turns in the secondary winding. If twice as many turns are on the secondary winding as on the primary winding, twice the amount of primary voltage is induced on the secondary winding. The turns ratio of primary winding to secondary winding is 1:2, making the transformer a step-up transformer. If only half as many turns are on the secondary winding, then only half of the voltage is induced on the secondary winding. The turns ratio of primary to secondary windings is 2:1, making the transformer a step-down transformer. **See Figure 21-2.**

Terms

The **turns ratio** is the ratio of the number of turns in the primary winding to the number of turns in the secondary winding.

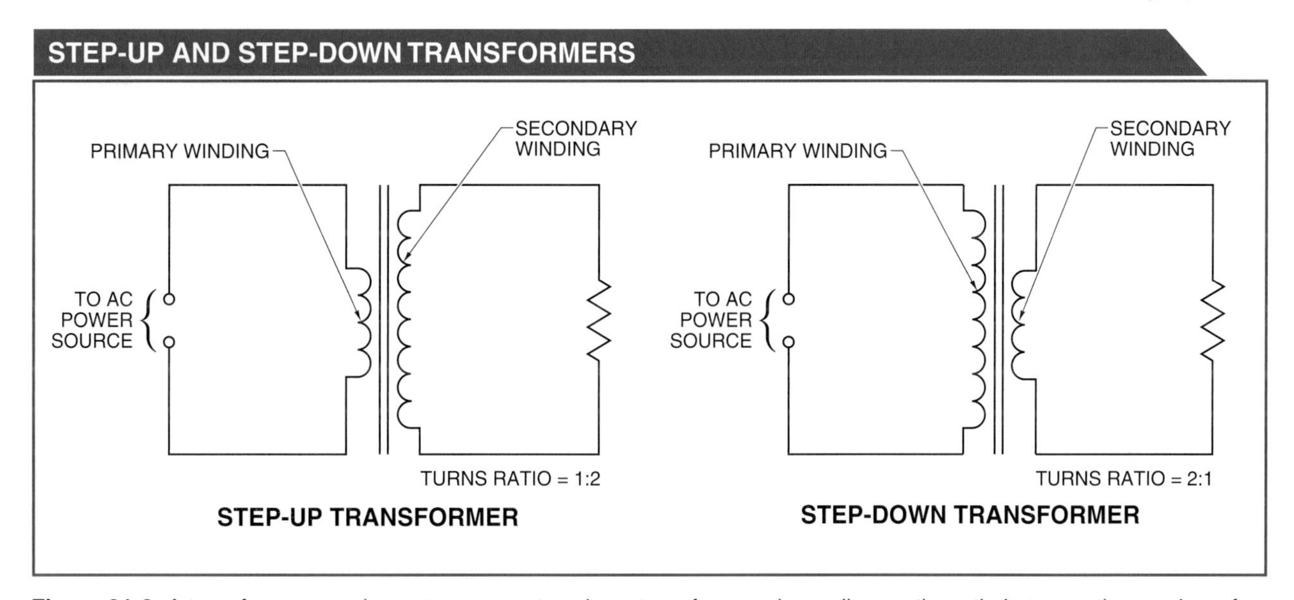

Figure 21-2. A transformer may be a step-up or step-down transformer, depending on the ratio between the number of conductor turns in the primary winding and secondary winding of the transformer.

In a step-up transformer, a ratio of 1:2 doubles the voltage. A ratio of 1:2 may seem to be a gain in voltage without any sacrifice. However, the amount of power transferred in a transformer is equal on both the primary and the secondary windings, excluding small losses within the transformer.

Tech Tip

Magnetic lines of force of opposite polarity cancel each other when they occupy the same space.

Because power is equal to voltage times current ($P = V \times I$) and power is always equal on both sides of a transformer, voltage cannot change without causing a change in current. For example, when voltage is stepped down from 240 V to 120 V (2:1 ratio), current increases from 1 A to 2 A, keeping the power equal on each side of the transformer. By contrast, when voltage is stepped up from 120 V to 240 V (1:2 ratio), the current is reduced from 2 A to 1 A to maintain power balance. Voltage and current may be changed, but power is constant.

One advantage of increasing voltage and reducing current is that power may be transmitted through smaller-gauge conductors, thereby reducing the cost of power lines. For this reason, the voltages generated from the utility (power company) are stepped up to higher levels for distribution across large distances, and then stepped back down to meet consumer requirements. Although both the voltage and current can be stepped up or down, when used with transformers the terms "step-up" and "step-down" always apply to voltage.

TRANSFORMER THEORY

Terms

Flux linkage is the number of the flux lines (magnetic lines of force) linking the primary and secondary transformer windings.

Magnetizing current (exciting current) is the current through a primary winding core when no load is on the secondary winding.

Transformer theory is based on relationships between flux linkage, mutual inductance, and magnetizing current. *Flux linkage* is the number of the flux lines (magnetic lines of force) linking the primary and secondary transformer windings. In order to transfer electrical energy from the primary windings to the secondary windings, a transformer must have good flux linkage and high mutual inductance. *Magnetizing current (exciting current)* is the current through a primary winding core when no load is on the secondary winding. Application of transformer theory includes a proper understanding of coefficient of coupling, voltage-current phase relationships, and the secondary winding polarity.

Coefficient of Coupling

Terms

The **coefficient of coupling (mutual inductance)** of a transformer is a measure of the efficiency with which power is transferred from the primary winding to the secondary winding.

The *coefficient of coupling (mutual inductance)* of a transformer is a measure of the efficiency with which power is transferred from the primary winding to the secondary winding. **See Figure 21-3.** If the power transfer is perfect, the coefficient of coupling is 1. If there is no power transfer, the coefficient of coupling is 0. The coefficient of coupling depends on the transformer design. The most important factor of the transformer design is the position of each winding with respect to the other. If the windings are wound over one another and each magnetic line of force (flux) from the primary winding cuts a turn in the secondary winding, the coefficient of coupling is close to a value of 1. If any flux is lost, the coefficient of coupling value is less than 1. The coefficient of coupling of typical transformers ranges from 0.950 to 0.999, depending on the transformer design and purpose.

A *tight coefficient of coupling* is a coefficient of coupling close to 1. A *loose coefficient of coupling* is a coefficient of coupling that is much less than 1. *Critical coupling* is the point separating a tight coefficient of coupling from a loose coefficient of coupling.

Terms

A **tight coefficient of coupling** is a coefficient of coupling close to 1.

A **loose coefficient of coupling** is a coefficient of coupling that is much less than 1.

Critical coupling is the point separating a tight coefficient of coupling from a loose coefficient of coupling.

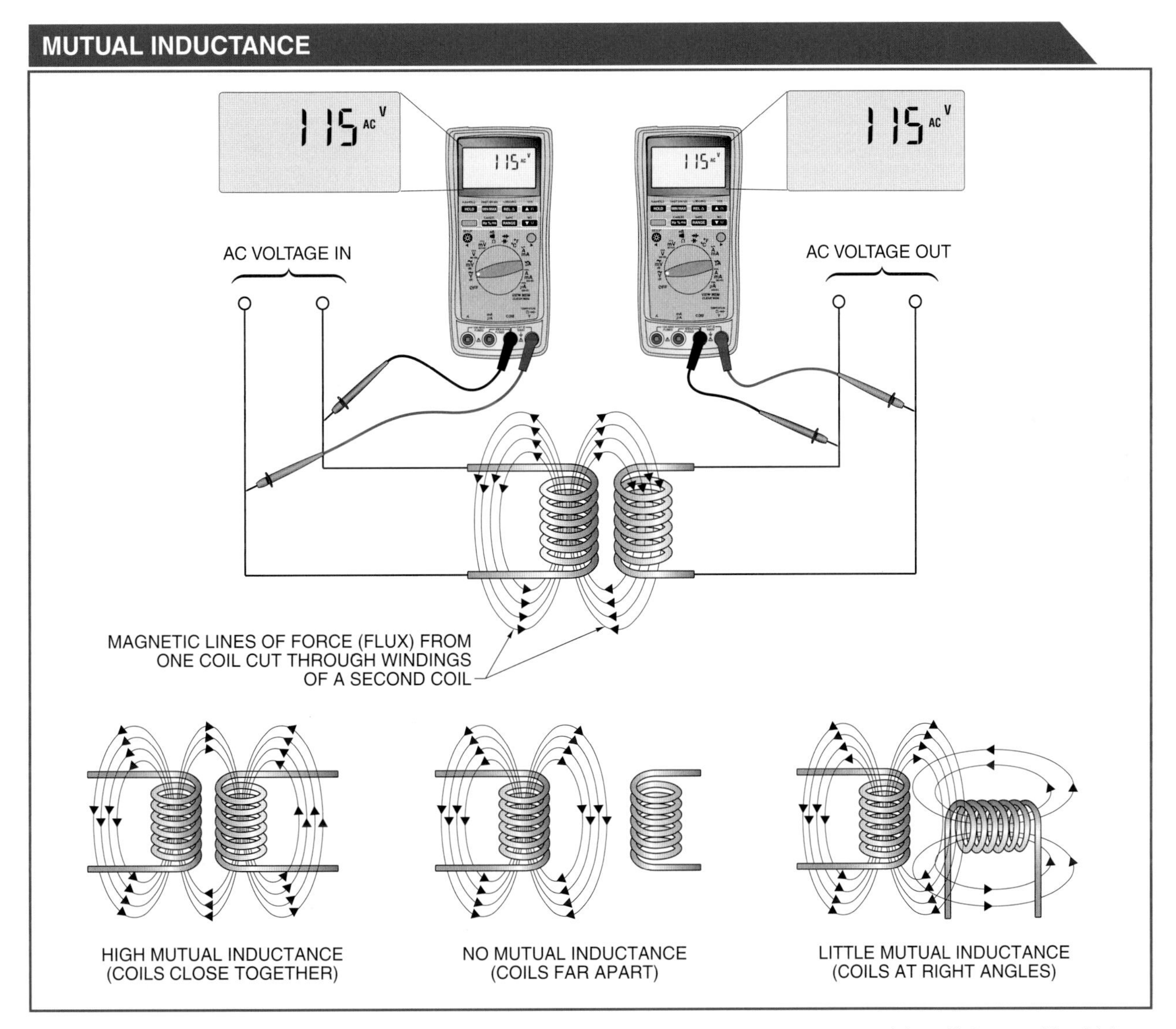

Figure 21-3. The coefficient of coupling (mutual inductance) of a transformer is a measure of the efficiency with which power is transferred from the primary winding to the secondary winding.

Voltage-Current Phase Relationships

Voltage-current phase relationships in a transformer are used to understand transformer theory and transformer operation. Voltage-current phase relationships between the primary and secondary windings are sometimes represented with vector diagrams. **See Figure 21-4.**

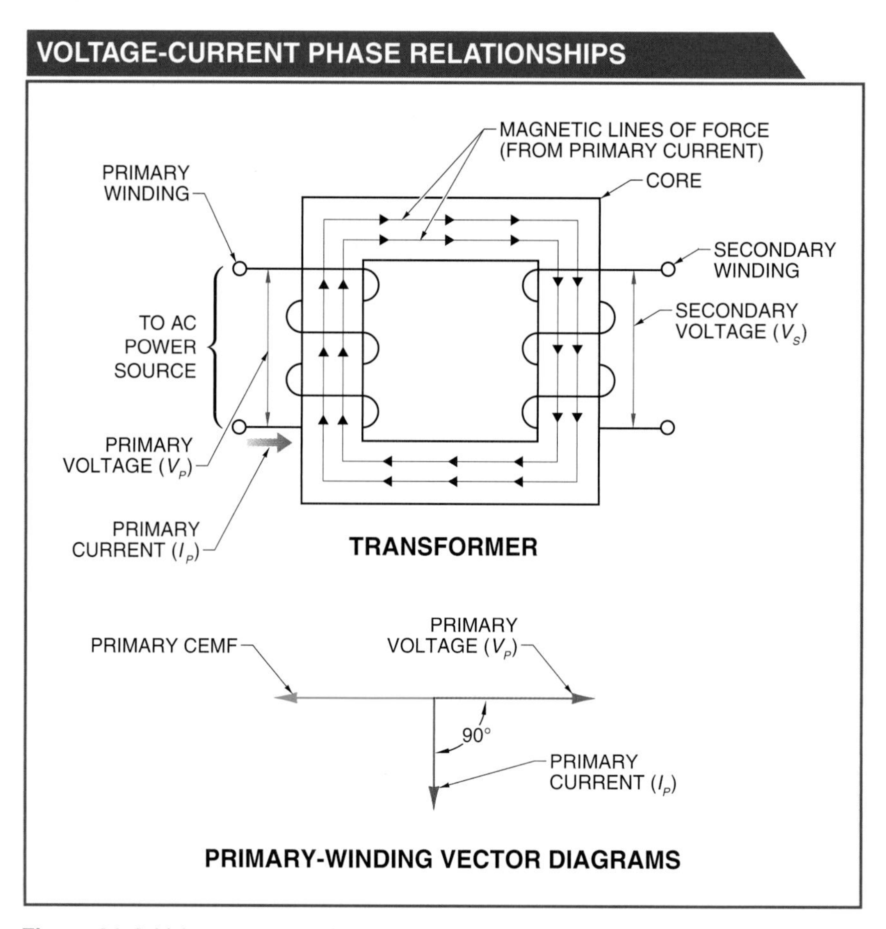

Figure 21-4. Voltage-current phase relationships between the primary and secondary windings are sometimes represented with vector diagrams.

With power applied to the primary winding, current flow is established that is limited by the value of inductive reactance. Because inductance is large in power transformers, inductive reactance is also large, and the primary current is small. In a pure inductive circuit, current in the primary circuit lags source voltage by 90°. **See Figure 21-5.** When the primary circuit lags the source voltage by 90°, the cosine of angle theta (cos θ) is equal to 0°, and true power consumed is equal to 0 W.

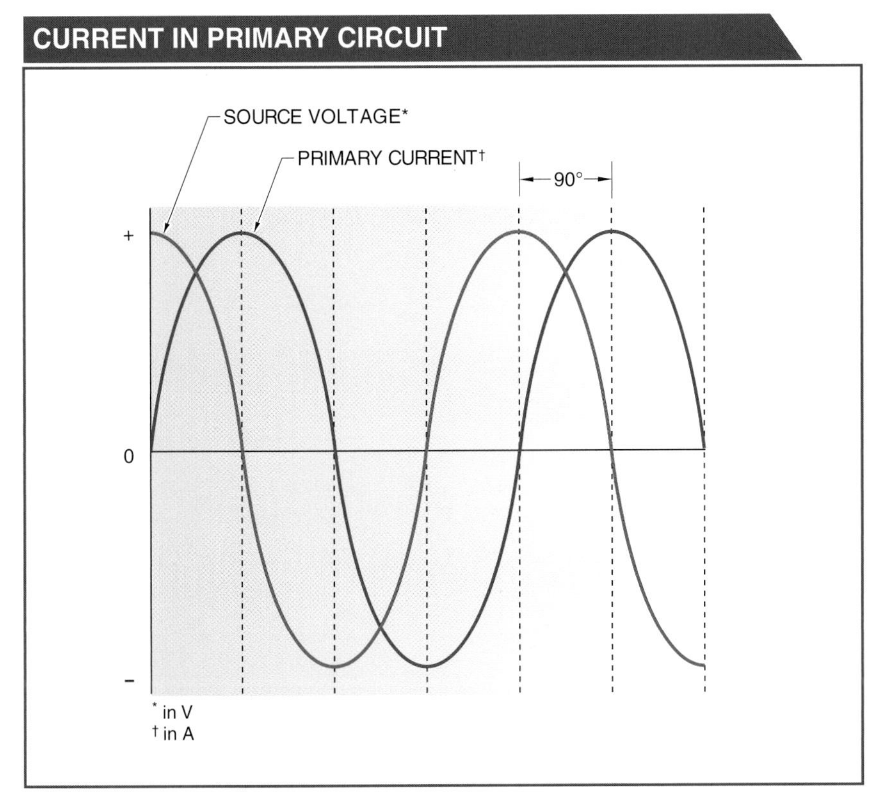

Figure 21-5. In a pure inductive circuit, current in the primary circuit lags source voltage by 90°.

The left-hand rule is the relationship between the current in a conductor and the magnetic field existing around the conductor. The left-hand rule states that if the fingers of the left hand are wrapped in the direction of the current flow in a coil, the left thumb points to the magnetic north pole of the coil. **See Figure 21-6.**

For example, the current flow in the primary winding causes a counter-electromotive force (CEMF) that opposes the change of current through the winding. The CEMF is 180° out of phase with the source voltage across the primary winding, thereby limiting the current flow.

The voltage induced into the secondary winding, due to the constantly changing flux field caused by the primary winding, is 180° out of phase with the primary voltage. The leads of windings are marked with an "H" or an "X". An "H" indicates high-voltage leads, and an "X" indicates low-voltage leads. Windings are also marked with the number 1 on one end and the number 2 on the opposite end. The number 1 is the starting point of each winding, and the number 2 is the ending point of each winding. Marking windings allows proper connection of the windings.

Current flows when a load is placed across the secondary winding of a transformer. Current flow is always in a direction that opposes the magnetic lines of force created by the primary winding, in accordance with Lenz's law. Lenz's law states that an induced voltage or current opposes the motion that caused it and is the basic principle underlying all induction theory.

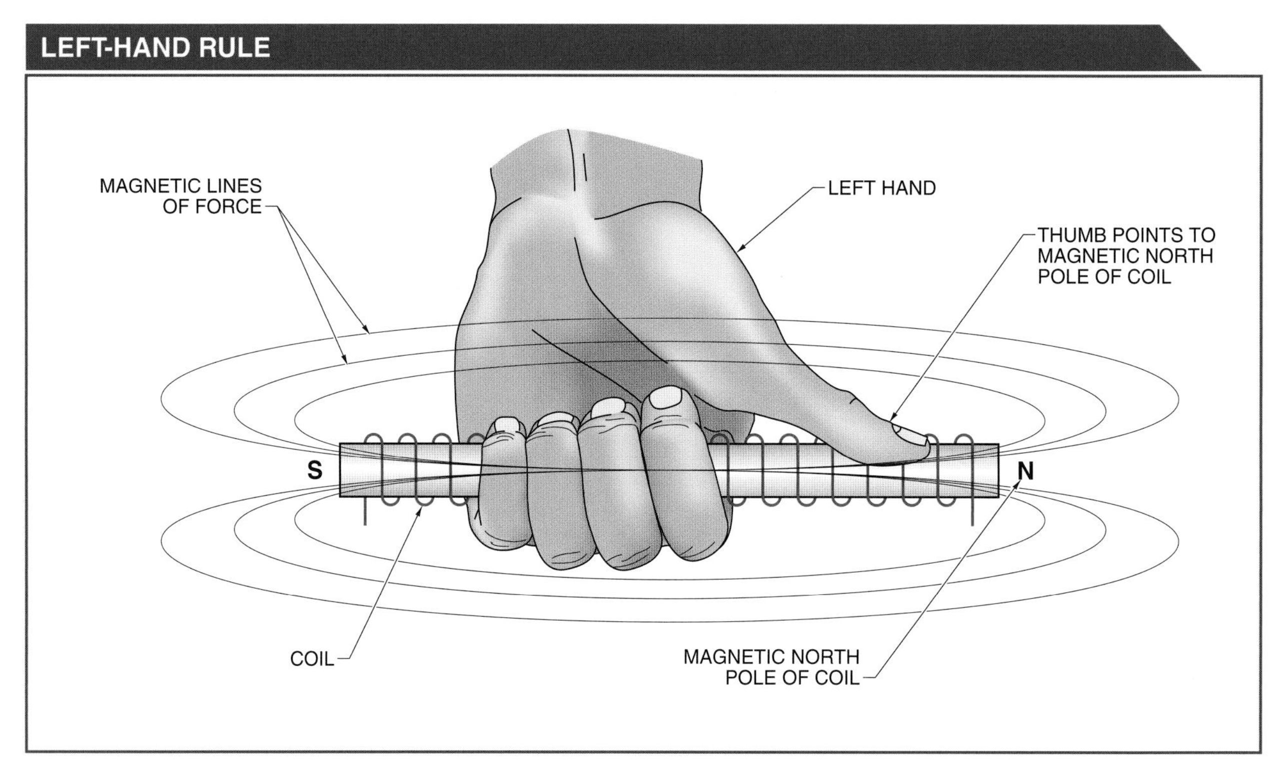

Figure 21-6. The left-hand rule states that if the fingers of the left hand are wrapped in the direction of the current flow in a coil, the left thumb points to the magnetic north pole of the coil.

The secondary winding of a transformer is simply an inductor in which voltage has been induced by the changing magnetic lines of force created in the primary circuit. Secondary-winding voltage is opposite that of the primary-winding voltage and its CEMF is in phase with the primary-winding voltage. This combination forms a simple series circuit consisting of induced voltage, the secondary windings, and the load. Effects of the load are ignored in order to isolate the relationships of the secondary winding circuit's voltage and current. Thus, the circuit is mostly inductive, and the secondary-winding current lags the secondary-winding voltage by 90°. **See Figure 21-7.**

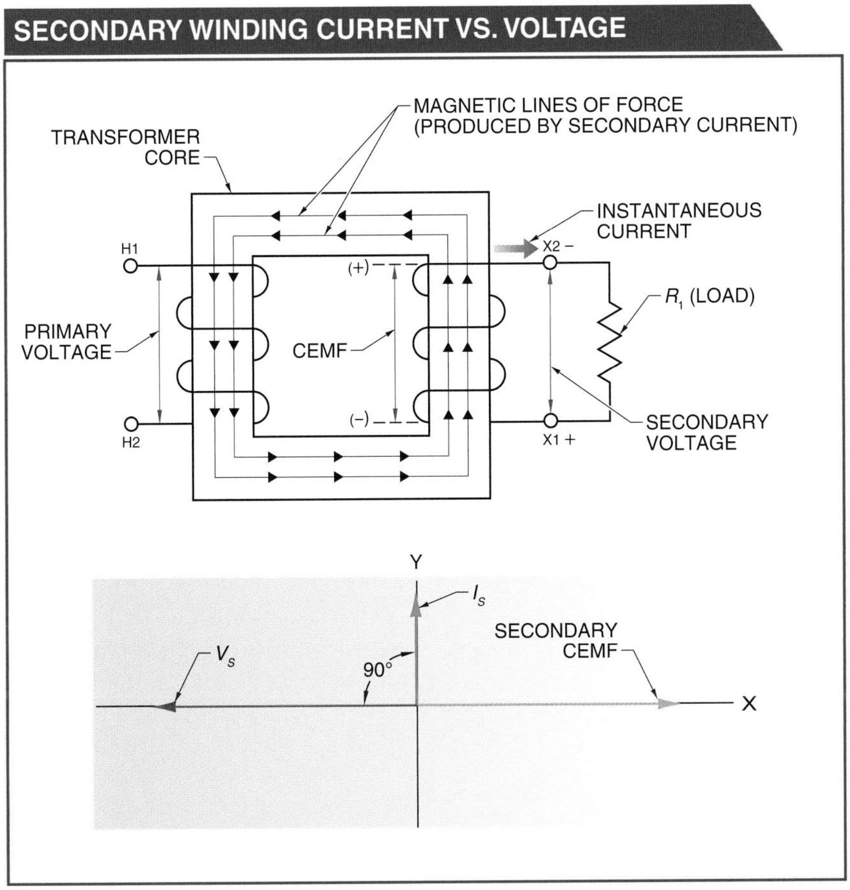

SECONDARY WINDING CURRENT VS. VOLTAGE

Figure 21-7. Transformer secondary winding current lags transformer secondary winding voltage by 90°.

Secondary-winding current sets up a CEMF that is 180° out of phase with the secondary-winding voltage. The CEMF tends to cause current flow in the secondary winding and cause magnetic lines of force in a counterclockwise direction.

The magnetic lines of force created by the secondary-winding current cancel some of the primary magnetic lines of force, so that the total number of magnetic lines of force in the core is reduced. Reduction of the magnetic lines of force in the core tends to reduce the voltage across the primary winding. Since the primary winding is across the source voltage, Kirchhoff's voltage law (the algebraic sum of all voltages around a closed loop circuit is equal to 0 V) must be satisfied, and the voltage remains the same. Due to the reduced magnetic lines of force, the inductive reactance of the primary circuit has been reduced. To maintain balance in the circuit, primary circuit current increases to maintain the original number of magnetic lines of force in the circuit.

Secondary-Winding Polarity

Secondary-winding polarity depends on the manner in which the secondary winding is wound in respect to the primary winding. Secondary windings of transformers can be wound so that the primary and secondary terminals opposite each other can be either out of phase or in phase with each other. **See Figure 21-8.** The phase relationship between the primary and secondary terminals is used in residential service where two 120 V out-of-phase services are used to create a 240 V service for large appliances such as air conditioners.

To determine secondary-winding polarity, apply the left-hand rule to the secondary winding of the out-of-phase transformer. By applying the left-hand rule, current flow is from bottom to top of the secondary winding, which is now the source voltage and corresponds to the internal current flow of a source voltage from the positive terminal to the negative terminal. An excessive number of electrons are available at the top of the secondary winding, making it the negative terminal. The current through the load is from the negative terminal of the source to the positive terminal.

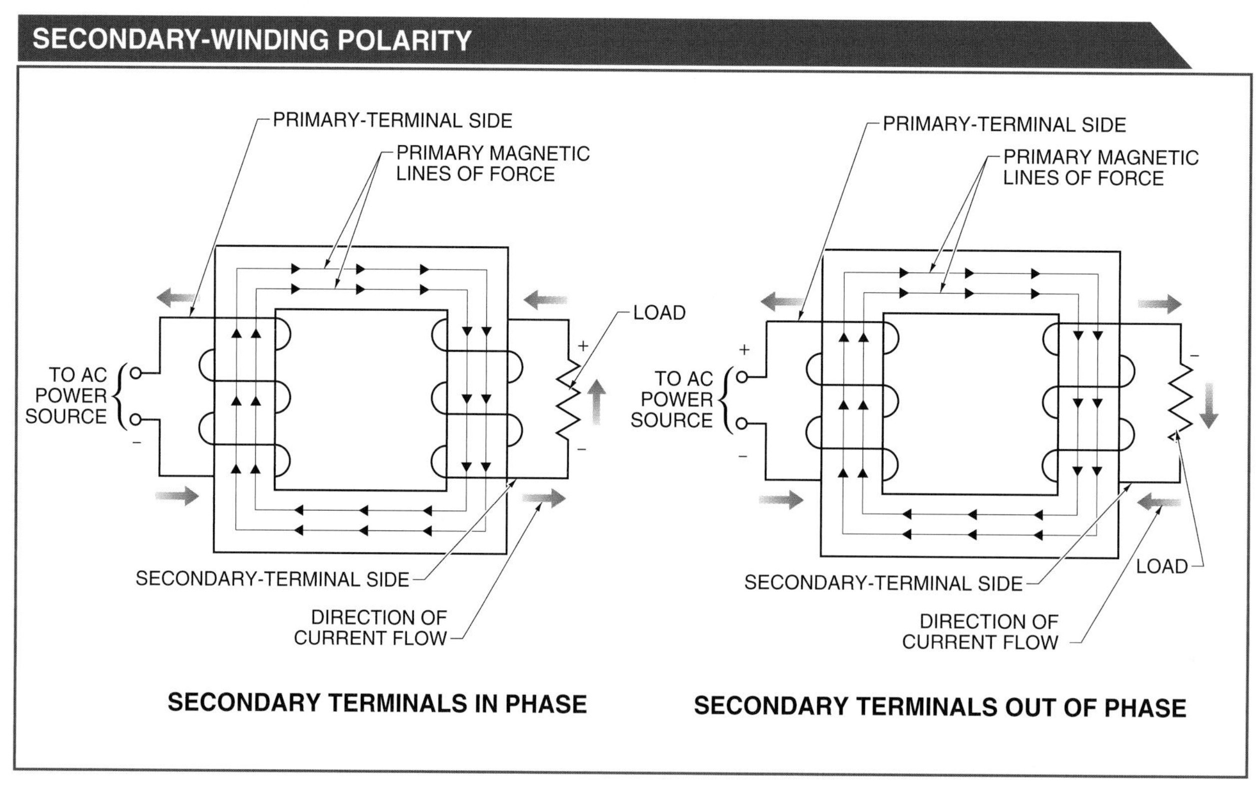

Figure 21-8. Secondary windings of transformers can be wound so that the primary and secondary terminal sides opposite each other can be either out of phase or in phase with each other.

To determine secondary-winding polarity for the in-phase transformer, the primary winding is wound in the same direction as the out-of-phase transformer. The secondary winding is wound in the opposite direction. Current flow within the voltage source is also from positive to negative and the top of the secondary winding is now positive. The current flow through the load remains from the negative terminal of the source to the positive terminal.

In both examples, current flow in the primary windings is in the same direction. This means that the rules of interaction between the secondary and primary windings are the same, and change in the load of the secondary winding is in such a direction that it opposes the magnetic field that caused it. Increased power use in the secondary winding is reflected as increased power use in the primary winding. Different schematic symbols are used to distinguish in-phase, out-of-phase, air-core, and iron-core transformers. **See Appendix.**

TRANSFORMER RATINGS

As with AC generators, transformers are rated in kilovoltamperes (kVA) for a given frequency. When working with transformers, the voltage and current ratings of the device must be known. By knowing the values of voltage and current, the power and kVA ratings can be determined.

Tech Tip

For every 50°F (10°C) rise in temperature above a transformer's rated limit, the life of the transformer is reduced by about 50%.

Voltage Ratings

Transformer original equipment manufacturers (OEMs) always state the voltage rating of the primary and secondary windings. Operating a transformer above its primary voltage rating usually results in the transformer overheating. Stress on the insulation is also increased on both the primary and secondary windings, and insulation breakdown may occur under this condition. Transformers can be operated below their primary voltage rating, but the secondary voltage will be lower than the secondary voltage rating. For example, when a transformer primary voltage rated for 240 V that has a ratio of 2:1 is used with 120 V, the secondary output voltage is only 60 V instead of the 120 V when the primary voltage is 120 V.

Rated voltages of the secondary windings of a transformer are specified for the full-load condition with the rated primary voltage applied. When the full load is removed from the secondary winding, the voltage on each winding is usually 5% to 10% higher than its rated voltage.

Current Ratings

Current ratings are usually specified for the secondary windings only. As long as the secondary-winding current is not exceeded, the current capacity of the primary winding is not exceeded. An overload in the secondary winding circuit will cause the secondary-winding voltage to fall below its rated value. Voltage readings below the rated value are an indication that the circuit is using excessive current. This condition increases heating loss in the secondary winding that can cause deterioration and eventual failure of the transformer.

AC/DC Adapters

AC/DC adapters accept AC input voltage directly from a wall outlet, and output DC voltage. The primary configurations are wall mount, where the adapter unit plugs directly into a wall and the DC lead then goes to the DC device, and desk mount, where the adapter unit is in-line between the AC plug and the DC output. Many DC components use this type of adapter or power supply for applications such as household electronics, computer printers, low-voltage DC boards, cellular and cordless telephone battery chargers, and cordless power-tool battery chargers.

Power Ratings

Terms

True power rating is the amount of power a transformer can deliver to a resistive load.

Transformer OEMs typically specify transformer power ratings in watts (W). The power rating of a transformer can be determined by multiplying the rated voltage and current of the secondary winding. If there is more than one secondary winding, the sums of the kVA calculations are added together. For example, the current in each secondary winding must not exceed its individual rating even if the total power rating for all the windings is not exceeded. Each winding has a maximum current that must not be exceeded. The *true power rating* of a transformer is the amount of power a transformer can deliver to a resistive load. The true power rating cannot exceed the kVA rating of the transformer.

Kilovoltampere Ratings

Most transformers are rated in kilovoltamperes (kVA). The kVA rating is applicable to any inductive, resistive, or capacitive load. As with the power rating, the kVA rating given for the transformer equals the sum of the ratings for each secondary winding. Current ratings for individual secondary windings cannot safely be exceeded. A transformer can be loaded to its full kVA rating while delivering only a fraction of its power rating when the load is inductive or capacitive. A transformer can be damaged, however, when subjected to an overload. To prevent damaging a transformer due to an overload, voltage greater than the rated voltage of the primary winding should never be applied. In addition, current greater than the rated current should never be drawn from any secondary winding.

TRANSFORMER LOADS

A transformer is both a load and a source for voltage. A transformer is a load to the source voltage that is supplying its primary winding. It is also the source voltage to the load connected to its secondary winding. Transformer loads may be large, small, inductive, resistive, capacitive, or a combination inductive-resistive-capacitive circuit. The transformer load action under any of these conditions can be analyzed through the use of vector diagrams.

When primary voltage is equal to source voltage, and secondary voltage is connected across a large resistive load, little current is drawn from the transformer circuit. The transformer circuits under this condition are almost completely inductive, and transformer voltages and currents are 180° out of phase with each other. Currents lag their respective voltages by 90°.

Slowly decreasing the amount of resistance causes the resistive load current to increase. Decreasing resistance results in an equal reduction in the angle theta in the secondary and primary circuit. With greater resistive current being drawn, the secondary and primary currents approach 0° with their respective voltages. **See Figure 21-9.**

Angle theta must exist when the transformer is operated at its full-load kVA rating. The inductive component must be present for the magnetic transfer of power to take place. It is assumed that maximum transfer occurs after the value of resistance is one-tenth or less than the inductive reactance in the circuit, thereby making the load current mostly resistive. This value is assumed because of the effect of circuit resistance on circuit impedance.

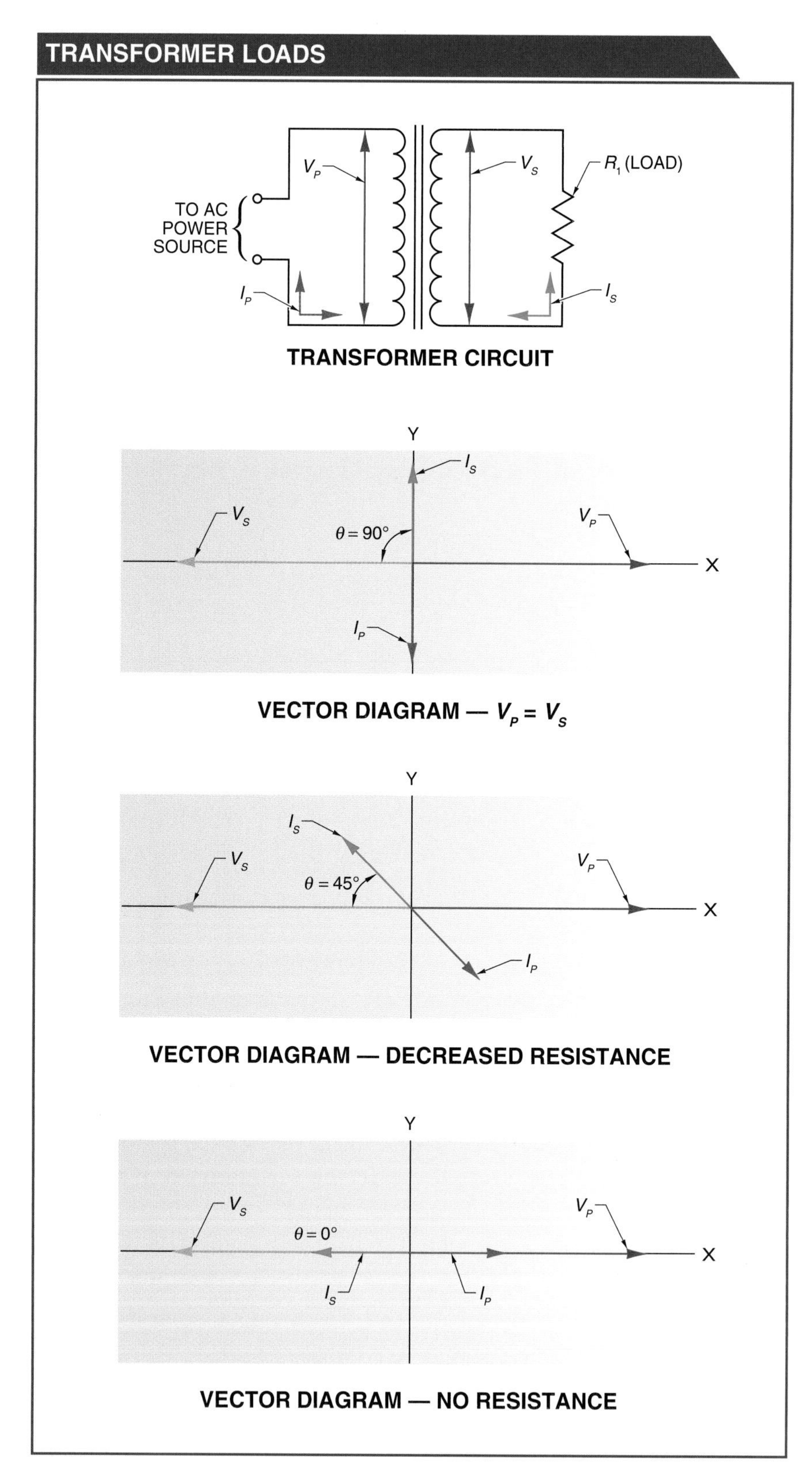

Figure 21-9. Transformer circuits are almost totally inductive when a large resistive load is connected to the secondary. Decreasing the resistance, thereby increasing the resistive load, reduces the angle theta towards 0°.

Tech Tip

The addition of a reactor or a choke to a circuit can create a series- or parallel-resonant circuit, especially where power factor correction capacitors are used.

Transformers may be connected to inductive-resistive-capacitive (L-R-C) loads. This type of circuit is different from a resistive circuit because the circuit operation is based on the frequency applied by the power source. At resonance, the circuit is purely resistive; above resonance, the circuit is inductive-resistive; below resonance, the circuit is resistive-capacitive. **See Figure 21-10.**

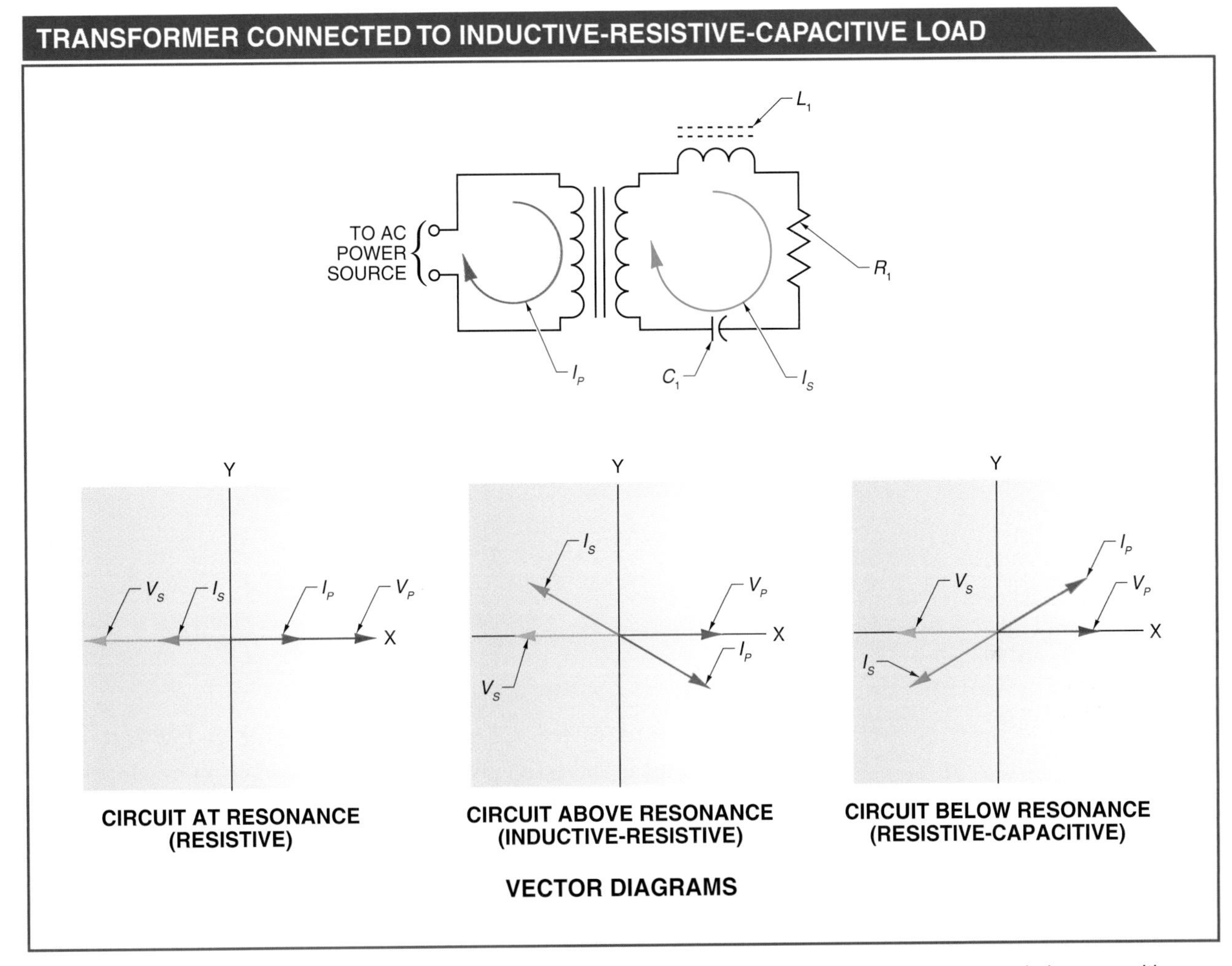

Figure 21-10. An L-R-C circuit is resistive at resonance, inductive-resistive above resonance, and resistive-capacitive below resonance.

The *reflected load* of a transformer is the primary impedance of the transformer that is directly proportional to the secondary load. When secondary current lags secondary voltage, the secondary load is inductive. However, when the secondary load is inductive, primary current leads primary voltage by the same angle as the secondary current lags by. This indicates that the reflected load is capacitive.

Terms

The **reflected load** of a transformer is the primary impedance of the transformer that is directly proportional to the secondary load.

When secondary current leads secondary voltage, the secondary load is capacitive and current in the primary winding lags the primary voltage. This indicates that the reflected load is inductive. **See Figure 21-11.**

TRANSFORMER REFLECTED LOADS

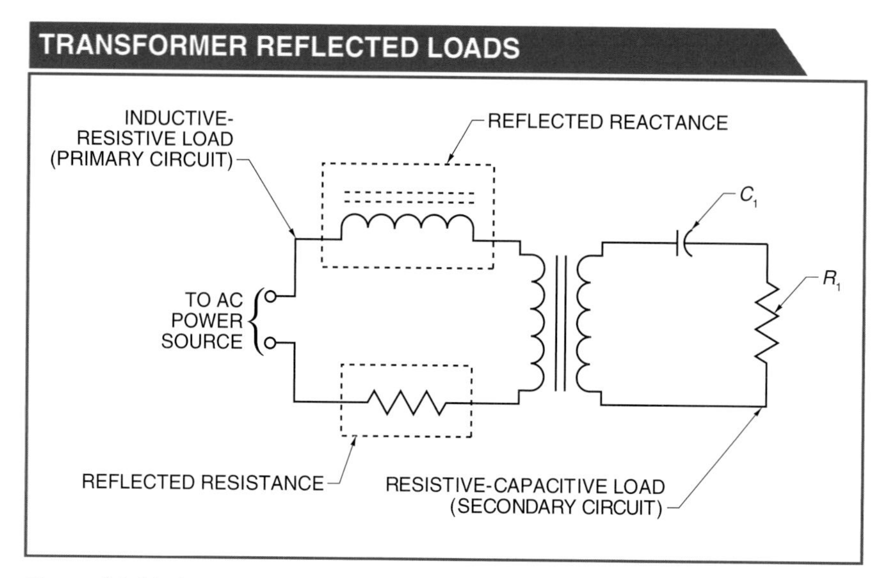

Figure 21-11. A secondary-circuit resistive-capacitive load is reflected into the primary circuit as a series inductive-resistive load.

The difference between a parallel inductive-resistive-capacitive circuit and a series inductive-resistive-capacitive circuit is that a parallel circuit acts in a manner opposite the series circuit. Below resonance a parallel circuit is inductive, while above resonance a parallel circuit is capacitive. At resonance, parallel and series circuits are identical, with circuit impedance equal to the value of the resistance.

TRANSFORMER TURNS RATIO

Terms

The **turns ratio** or **turns-to-turns** ratio is the ratio of the number of turns in the primary winding to the number of turns in the secondary winding.

The *turns ratio* or *turns-to-turns ratio* is the ratio of the number of turns in the primary winding to the number of turns in the secondary winding. The turns ratio is written with two numbers, such as 2:1 or 2 to 1. The first number represents the relative number of turns in the primary winding and the second number represents the relative number of turns in the secondary winding. The turns ratio of a transformer affects the voltage ratio, volts per turn, current ratio, and impedance parameters such as impedance ratio, impedance matching, and reflected impedance of transformers. The turns ratio of a transformer is calculated by applying the following formula:

$$\frac{N_P}{N_S}$$

where

N_P = number of turns in primary winding

N_S = number of turns in secondary winding

Example: What is the turns ratio of a transformer with 500 turns in the primary winding and 1000 turns in the secondary winding?

$$\frac{N_P}{N_S}$$

$$\frac{500}{1000}$$

1:2

Transformer Testing

All electrical tests performed on transformers must be performed carefully to ensure that there is no risk from a potentially powered circuit. Most tests are performed with the power removed from the transformer. Anyone performing a test must stand clear of all parts of the circuits. All terminals must be treated as if there may be a shock hazard. The test leads must be disconnected from the transformer before disconnecting the leads from the test meter. The ground connection must be the first connection made and the last connection removed. Only authorized procedures should be performed. Proper personal protective equipment (PPE) must be worn during all testing procedures.

Voltage Ratio

The *voltage ratio* is the ratio between the primary-winding voltage and the secondary-winding voltage. The voltage ratio is used to indicate the change in voltage between the primary and secondary windings. Current flow through the primary winding of a transformer constantly changes at a rate that causes variations in the sine waves. The changing current in the primary winding sets up a changing magnetic field that is magnetically coupled to the secondary winding. As with a generator, the rate of change of the magnetic field and the number of conductors cut by the magnetic field, the greater the voltage induced. Since the angular velocity of a sine wave is calculated by frequency, the secondary voltage is calculated by the number of turns in the secondary winding.

Terms

The **voltage ratio** is the ratio between the primary-winding voltage and the secondary-winding voltage.

The CEMF induced in the primary windings is not equal to the voltage induced in the secondary windings. Since the CEMF is opposite the source voltage, an equation can be established to express the turns ratio to the voltage ratio of any transformer. The turns ratio to voltage ratio of a transformer is calculated by applying the following formula:

$$\frac{N_P}{N_S} = \frac{V_P}{V_S}$$

where

N_P = number of turns in primary winding

N_S = number of turns in secondary winding

V_P = primary-winding voltage (in V)

V_S = secondary-winding voltage (in V)

Example: What is the turns ratio of a transformer that has primary voltage of 600 V and secondary voltage of 120 V?

$$\frac{N_P}{N_S} = \frac{V_P}{V_S}$$

$$\frac{N_P}{N_S} = \frac{600 \text{ V}}{120 \text{ V}}$$

$$\frac{N_P}{N_S} = \frac{5}{1}$$

$$\frac{N_P}{N_S} = \mathbf{5:1}$$

The solution indicates that the step-down transformer has five turns in the primary winding per each turn in the secondary winding. If the transformer has voltage of 120 V across the primary winding and voltage of 600 V across the secondary winding, it is a step-up transformer, and the turns ratio is equal to 1:5. In this format, primary winding turns are typically given first.

Volts per Turn. The turns ratio can be used to explain the related concept of volts per turn. *Volts per turn (V/turn)* is the voltage dropped across each turn of a winding or the voltage induced into each turn of the secondary winding. Each transformer has a design value for the volts per turn. For example, if a transformer primary winding has 120 turns with a power source of 120 V, then it has 1 V per turn. The secondary winding has the same design value. For example, if the secondary winding has 24 turns, the secondary voltage is 24 V. Therefore, a transformer with a turns ratio of 120:24, V/turn of 1, and a primary voltage of 120 V has a secondary voltage of 24 V.

Terms

Volts per turn (V/turn) is the voltage dropped across each turn of a winding or the voltage induced into each turn of the secondary winding.

Current Ratio

The *current ratio* is the ratio of the primary-winding current to the secondary-winding current. The current ratio is used to calculate the current of the primary winding. The current ratio of a transformer is the inverse of the voltage ratio. Thus, in a step-up transformer the current is stepped down, and in a step-down transformer the current is stepped up. Because a transformer does not create power, the secondary power must equal the primary power.

Since the current ratio is the inverse of the voltage ratio, the current ratio is calculated by applying the following formula:

$$\frac{I_S}{I_P} = \frac{N_P}{N_S}$$

where

I_S = secondary-winding current (in A)

I_P = primary-winding current (in A)

N_P = number of turns in primary winding

N_S = number of turns in secondary winding

Terms

The **current ratio** is the ratio of the primary-winding current to the secondary-winding current.

Tech Tip

Most motor control circuits are powered from step-down transformers that reduce the voltage to the control circuit. A step-down transformer reduces the voltage to the control circuit to a level of 24 V or 12 V, as needed.

Example: What is the primary-winding current for a secondary-winding current of 30 A in a transformer with a 5:1 turns ratio?

$$\frac{I_S}{I_P} = \frac{N_P}{N_S}$$

$$\frac{30\text{ A}}{I_P} = \frac{5}{1}$$

$$\frac{I_P}{30\text{ A}} = \frac{1}{5}$$

$$I_P = \frac{30\text{ A}}{5}$$

I_P = **6 A**

Impedance Ratio

The *impedance ratio* is the ratio of the impedance of the primary winding to the impedance of the secondary winding. The impedance ratio is equal to the square of the turns ratio. The amount of primary current in a transformer is determined by the amount of secondary current. When large impedance is connected to the secondary winding, a small secondary current flows. Large secondary impedance is reflected as small primary current, indicating large primary impedance. Small impedance connected to the secondary winding results in large secondary current that causes large primary current. Large secondary and primary current indicates small primary impedance.

Terms

The **impedance ratio** is the ratio of the impedance of the primary winding to the impedance of the secondary winding.

The impedance of the secondary circuit controls the impedance of the primary circuit of a transformer. The impedance ratio of a transformer is calculated by applying the following formula:

$$\frac{Z_P}{Z_S} = \frac{{N_P}^2}{{N_S}^2}$$

where

Z_P = primary impedance (in Ω)

Z_S = secondary impedance (in Ω)

N_P = number of turns in primary winding

N_S = number of turns in secondary winding

Example: What is the primary impedance of a transformer with a secondary impedance of 100 Ω and a turns ratio of 100:1?

$$\frac{Z_P}{Z_S} = \frac{{N_P}^2}{{N_S}^2}$$

$$Z_P = 100\ \Omega \times \left(\frac{100^2}{1^2}\right)$$

$$Z_P = \frac{1{,}000{,}000\ \Omega}{1}$$

Z_P = **1,000,000 Ω**

Terms

Impedance matching (load matching) is the process of adjustment of the load-circuit impedance to produce the desired energy transfer from the power source to the load.

Tech Tip

A zigzag connection is a transformer wiring method where the windings are divided over several legs of the transformer core.

Impedance Matching. *Impedance matching (load matching)* is the process of adjustment of the load-circuit impedance to produce the desired energy transfer from the power source to the load. In order to obtain maximum transfer of energy from a power source, the load's impedance must equal the impedance of the power source. Transformers are useful devices in matching the impedances so that the maximum power transfer occurs. In electrical and electronic circuits, it is often necessary to match low-impedance power sources to high-impedance loads, or to match high-impedance power sources to low-impedance loads. In either of these cases, the desired effect can be achieved by coupling the load to the power source with a transformer with a turns ratio selected for this purpose. Impedance can be matched by applying the following formula:

$$\frac{{N_P}^2}{{N_S}^2} = \frac{Z_P}{Z_S}$$

where

N_P = number of turns in primary winding

N_S = number of turns in secondary winding

Z_P = primary impedance (in Ω)

Z_S = secondary impedance (in Ω)

Example: What transformer turns ratio must be used to match a 300 Ω receiver antenna to a 75 Ω coaxial cable?

$$\frac{N_P^2}{N_S^2} = \frac{Z_P}{Z_S}$$

$$\frac{N_P^2}{N_S^2} = \frac{300\ \Omega}{75\ \Omega}$$

$$\frac{N_P^2}{N_S^2} = \frac{4}{1}$$

$$\frac{N_P}{N_S} = \frac{\sqrt{4}}{\sqrt{1}}$$

$$\frac{N_P}{N_S} = \mathbf{2{:}1}$$

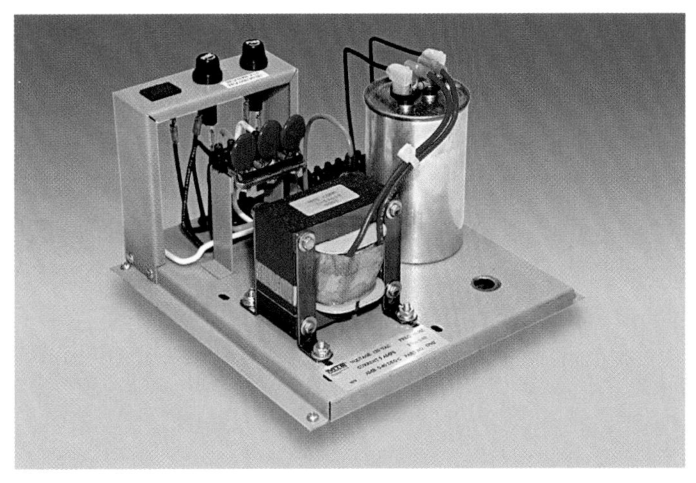

MTE Corporation

Reactors are used as harmonic filters in electronic power supplies.

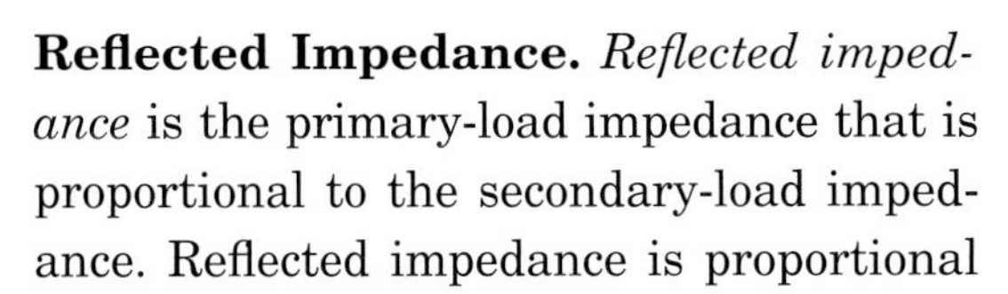

Reflected Impedance. *Reflected impedance* is the primary-load impedance that is proportional to the secondary-load impedance. Reflected impedance is proportional to the square of the turns ratio. Reflected impedance is calculated by using the secondary impedance and the transformer turns ratio. Reflected impedance is calculated by applying the following procedure:

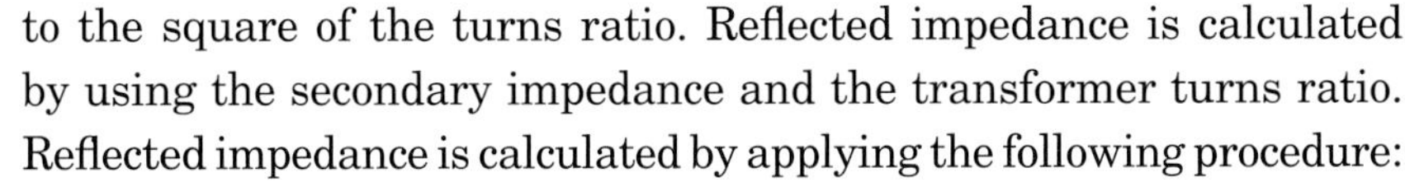

Terms

Reflected impedance is the primary-load impedance that is proportional to the secondary-load impedance.

1. Calculate reflected reactance. Reflected reactance is calculated by applying the following formula:

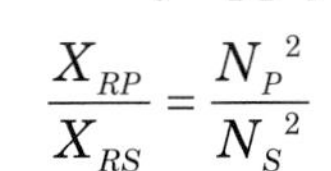

$$\frac{X_{RP}}{X_{RS}} = \frac{N_P^2}{N_S^2}$$

where

X_{RP} = primary reactance (in Ω)

X_{RS} = secondary reactance (in Ω)

N_P = number of turns in primary winding

N_S = number of turns in secondary winding

2. Calculate reflected resistance. Reflected resistance is calculated by applying the following formula:

$$\frac{R_P}{R_S} = \frac{N_P^2}{N_S^2}$$

where

R_P = primary resistance (in Ω)

R_S = secondary resistance (in Ω)

N_P = number of turns in primary winding

N_S = number of turns in secondary winding

3. Calculate reflected impedance. Reflected impedance is calculated by applying the following formula (Pythagorean theorem):

$$Z_P = R_P^{\,2} + X_{RP}^{\,2}$$

where

Z_P = primary reflected impedance (in Ω)

R_P = primary reflected resistance (in Ω)

X_{RP} = primary reflected reactance (in Ω)

Example: What is the reflected impedance of a transformer that has a turns ratio of 4:1, secondary inductance of 300 Ω, and secondary resistance of 400 Ω?

1. Calculate reflected reactance.

$$\frac{X_{RP}}{X_{RS}} = \frac{N_P^{\,2}}{N_S^{\,2}}$$

$$X_{RP} = 300\ \Omega \times \left(\frac{4^2}{1^2}\right)$$

$$X_{RP} = 300\ \Omega \times \left(\frac{16}{1}\right)$$

$$X_{RP} = \frac{4800\ \Omega}{1}$$

$$X_{RP} = 4800\ \Omega$$

2. Calculate reflected resistance.

$$\frac{R_P}{R_S} = \frac{N_P^{\,2}}{N_S^{\,2}}$$

$$R_P = 400\ \Omega \times \left(\frac{4^2}{1^2}\right)$$

$$R_P = 400\ \Omega \times \left(\frac{16}{1}\right)$$

$$R_P = \frac{6400\ \Omega}{1}$$

$$R_P = 6400\ \Omega$$

3. Calculate reflected impedance.

$$Z_P = R_P^{\,2} + X_{RP}^{\,2}$$

$$Z_P = \sqrt{6400\ \Omega^2 + 4800\ \Omega^2}$$

$$Z_P = \sqrt{(40.96\ \Omega \times 10^6) + (23.04\ \Omega \times 10^6)}$$

$$Z_P = \sqrt{64\ \Omega \times 10^6}$$

$Z_P = \sqrt{64,000,000\ \Omega}$

$Z_P = \mathbf{8000\ \Omega}$

Note: The reflected capacitive reactance in the transformer cancels some of the primary winding inductive reactance. The primary circuit remains inductive. Even with perfect coupling, the reflected capacitive reactance can never be greater than the inductive reactance of the primary windings.

Leakage Inductance

Leakage inductance causes voltage to drop across a transformer winding and can negatively affect efficiency. The best way to reduce leakage inductance is to use a long, narrow window in the core of a concentric winding or to use interleaved windings.

TRANSFORMER LOSSES

While ideal transformers do not have losses, real transformers, like other electromagnetic devices, have power losses. The power output from a transformer is always slightly less than the power input to a transformer. These power losses end up as heat that must be removed from the transformer. The four main types of loss are resistive loss, eddy current loss, hysteresis loss, and flux loss.

Resistive Loss

Resistive loss (I^2R loss) is power loss in a transformer caused by the resistance of the copper conductors used to make the windings. Since higher frequencies cause the electrons to travel toward the outer circumference of the conductor (skin effect), harmonics on the line have the effect of reducing the conductor size and increasing resistive loss. These losses are the same as the power losses in any conductor and are calculated by applying the following formula:

$P = I^2R$

where

P = power (in W)

I = current (in A)

R = resistance (in Ω)

Terms

Resistive loss (I^2R loss) is power loss in a transformer caused by the resistance of the copper conductors used to make the windings.

Example: What is the resistive loss in a transformer primary winding that is wound with copper conductors that carry 15 A and have a resistance of 0.1588 Ω?

$P = I^2R$

$P = (15\text{ A})^2 \times 0.1588\ \Omega$

$P = 225\text{ A} \times 0.1588\ \Omega$

$P =$ **35.7 W**

The transformer primary winding consumes 35.7 W of power that is lost as heat. If the transformer is not cooled properly, this heat loss increases the temperature of the transformer and the conductors. This increased temperature causes an increase in the conductor resistance and in the voltage dropped across the conductor. This heat loss varies with the current and is always present in the primary winding when it is energized. The secondary winding has very little heat loss when there is no load applied to it. Current flow through the windings also generates heat loss. Resistive losses are the greatest losses in any transformer.

Eddy Current Loss

Eddy current loss is power loss in a transformer or motor due to currents induced in the metal field structure from the changing magnetic field. Any conductor that is in a moving magnetic field has a voltage and current induced in it. Induced currents in an iron core are undesirable because such currents use power and produce heat. The resultant heat is unavailable for useful electrical output and tends to heat the primary and the secondary winding, increasing the resistance and the power loss in both coils. The iron core offers low reluctance to the magnetic lines of force for mutual induction. The magnetic lines of force induce current at right angles to the flux. This means that current is induced through the core. This current causes heating in the core. Heat produced by eddy currents increases as the square of the frequency. For example, the third harmonic (180 Hz) has nine times the heating effect of the fundamental (60 Hz) frequency.

Terms

Eddy current loss is power loss in a transformer or motor due to currents induced in the metal field structure from the changing magnetic field.

Constructing the core of thin sheets of 29-gauge alloy (0.014″) iron laminated together can minimize this loss. The thin sheet-iron layers shorten the current path and minimize the eddy currents. Each thin sheet of iron is coated with a varnish that isolates each layer from the others and keeps the currents in the sheets of iron to a minimum. The thin sheets of iron are manufactured from silicon-iron or nickel-iron alloys that can be magnetized more readily than pure iron. The use of alloys in the core materials also improves the longevity of the core.

Hysteresis Loss

Hysteresis is the property of ferromagnetic materials where the magnetic induction of a winding lags the magnetic field that is charging the winding. *Hysteresis loss* is loss caused by magnetism that remains (lags) in a material after the magnetizing force has been removed. Magnetic domains are small sections of a magnetic material that act together when subjected to an applied magnetic field. Magnetic domains have magnetic properties and move in the iron when influenced by a magnetic field. When the iron is subject to a magnetic field in one polarity, the magnetic domains will be forced into alignment with the field. Power is consumed by the changing of polarity, which occurs twice per cycle. This realignment reduces the efficiency of the transformer. The movement of the molecules produces friction in the iron, which in turn produces heat. The presence of harmonics may cause the current to reverse direction more often, resulting in greater hysteresis loss.

Terms

Hysteresis is the property of ferromagnetic materials where the magnetic induction of a winding lags the magnetic field that is charging the winding.

Hysteresis loss is loss caused by magnetism that remains (lags) in a material after the magnetizing force has been removed.

Flux Loss

Flux loss is a power loss that occurs in a transformer when some of the magnetic lines of force from the primary winding do not travel through the core to the secondary winding. There are two main reasons for the magnetic lines of force to travel through air instead of through the core. The first reason is that the iron core can become saturated with magnetic lines of force so that the core cannot accept any more magnetic lines of force. The magnetic lines of force then travel through the air and are not cut by the secondary winding. The second reason is that there is more operating space in the air than there is in the core for the magnetic lines of force to travel. The ratio of the reluctance of the air and the core in the unsaturated region is typically about 10,000:1, meaning that for every 10,000 magnetic lines of force through the core, there is 1 magnetic line of force through the air. Flux loss is generally minimal in a well-designed transformer.

Terms

Flux loss is a power loss that occurs in a transformer when some of the magnetic lines of force from the primary winding do not travel through the core to the secondary winding.

TRANSFORMER EFFICIENCY

Transformer efficiency is the ratio of a transformer's output power to its input power. Transformer efficiency is used to measure the effects of the transformer losses, and is usually expressed as a percentage. Transformer efficiency is calculated by applying the following formula:

$$Eff = \frac{P_{OUT}}{P_{IN}}$$

Terms

Transformer efficiency is the ratio of a transformer's output power to its input power.

where

Eff = transformer efficiency (in %)

P_{OUT} = transformer output power (in W)

P_{IN} = transformer input power (in W)

Example: What is the efficiency of a transformer that has output power of 1500 W and input power of 1525 W?

$$Eff = \frac{P_{OUT}}{P_{IN}}$$

$$Eff = \frac{1500\text{ W}}{1525\text{ W}}$$

Eff = **98.36%**

Power transformers typically have efficiencies that range from 97% to 99%. Input power must equal the power delivered to the load plus the resistive, eddy current, hysteresis, and flux losses. Input power is always greater than output power.

Turns Ratio Testing

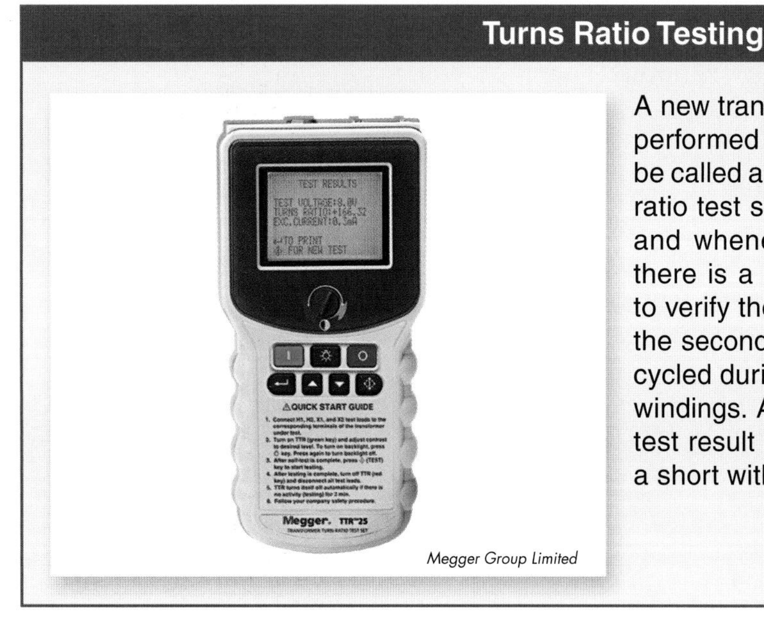

Megger Group Limited

A new transformer should have a turns ratio test performed when it is installed. This test may also be called a turns-to-turns ratio (TTR) test. A turns ratio test should be repeated on a regular basis and whenever there is reason to believe that there is a problem. The turns ratio test is used to verify the turns ratio between the primary and the secondary. A load tap changer must also be cycled during the turns ratio test to check all the windings. A change from the original turns ratio test result indicates there is a problem, such as a short within the windings.

TRANSFORMER CLASSIFICATIONS

Transformers are classified by the application for which the transformer is designed. The frequency at which the transformer operates determines its classification. The most common transformer classifications include power transformers, instrument transformers, audio transformers, and radio frequency transformers.

Power Transformers

A *power transformer* is a transformer that is used to raise or lower voltage as required to serve transmission or distribution circuits. Power transformers must have good flux linkage and high mutual inductance in order to efficiently transfer electrical energy from the primary windings to the secondary windings. At power line frequencies of 50 Hz and 60 Hz, this requires that the transformers have iron cores as indicated by the two lines between the windings on the schematic diagram. When the iron core is removed from a power transformer, the inductive reactance decreases greatly and the primary-winding current rises appreciably even without a load on the secondary winding. Therefore, it is desirable to have high inductance, high inductive reactance, and low magnetizing current to establish the flux linkage in power transformers.

Power transformers are typically used in changing voltages for transmission over long distances and for conversion back to voltage levels required by end users. Power transformers are usually designed to operate at 50 Hz or 60 Hz, which are the standard frequencies used for power distribution by the majority of countries throughout the world. The United States and Canada use a power distribution frequency of 60 Hz. Power transformers with high primary voltage ratings of 300 kV and higher are used for transmission and distribution purposes to take advantage of lower currents with reduced line losses. Small power transformers used in electrical and electronic components serve the same purpose as large power transformers. Small power transformers convert 120 VAC to the levels of voltage and current required by the component they serve.

Power transformers should be loaded to close to 90% of their rated power. Loading power transformers to 90% of their rated power brings the voltage and current close to being in phase and the power factor close to unity. This practice helps prevent excessive voltage and current spikes in the circuit caused by high inductive reactance. When power transformers are operated at close to 90% of their load, they typically have an efficiency of about 95%. Power transformer classifications include tapped-winding, multiple-winding, isolation, autotransformer, buck-boost, and core-type transformers.

Tapped-Winding Transformers. A *tapped-winding transformer* is a transformer with a single tapped winding that is used to convert voltages in residential areas. With a tapped-winding transformer, the distribution voltage to the primary winding is 7.2 kV, and the distribution voltage to the secondary winding is 240 V. **See Figure 21-12.** The secondary winding is center-tapped so that there are two 120 V circuits with a common conductor that is normally grounded.

Terms

A **power transformer** is a transformer that is used to raise or lower voltage as required to serve transmission or distribution circuits.

Terms

A **tapped-winding transformer** is a transformer with a single tapped winding that is used to convert voltages in residential areas.

TAPPED-WINDING TRANSFORMERS

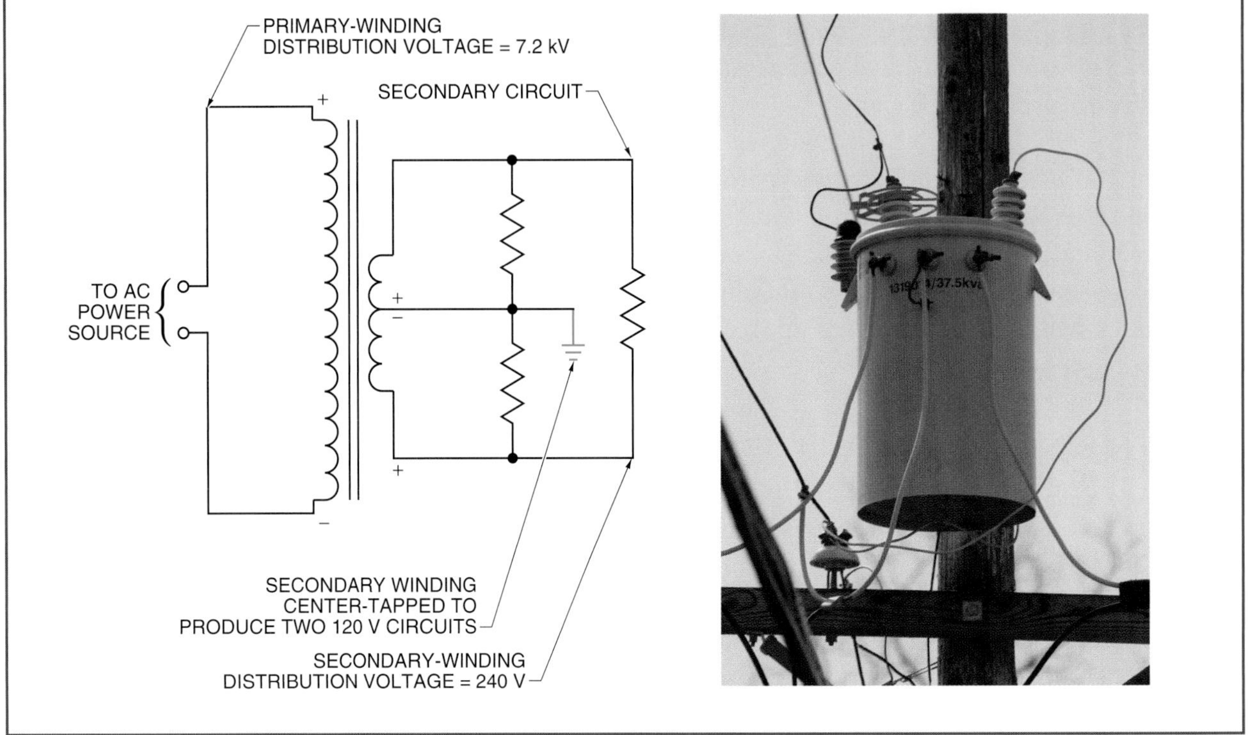

PRIMARY-WINDING DISTRIBUTION VOLTAGE = 7.2 kV
SECONDARY CIRCUIT
TO AC POWER SOURCE
SECONDARY WINDING CENTER-TAPPED TO PRODUCE TWO 120 V CIRCUITS
SECONDARY-WINDING DISTRIBUTION VOLTAGE = 240 V

Figure 21-12. A typical residential transformer converts high transmission voltage (7.2 kV) to a lower secondary voltage of 240 V, with the secondary winding tapped to provide two 120 V circuits.

In order to step down the 7.2 kV distribution voltage to the 240 V utilization voltage, a 30:1 turns ratio is used on a tapped-winding transformer. A three-conductor system is used for the secondary circuit that provides power to three separate loads. Two of the loads are supplied with 120 V, which is the voltage utilized by many small home appliances and lighting systems. The third load is supplied with 240 V, the voltage that is utilized by larger home appliances such as electric ranges, air conditioners, water heaters, and clothes dryers.

Terms

A **multiple-winding transformer** is a transformer that has more than two windings on its core.

Multiple-Winding Transformers. A *multiple-winding transformer* is a transformer that has more than two windings on its core. Multiple-winding transformers are used when two or more utilization voltages are required by the component. Most multiple-winding transformers have multiple secondary windings, each with a different voltage to meet the requirements of multiple loads. **See Figure 21-13.** Although multiple secondary windings are more common, multiple-winding transformers may also have multiple primary windings to accommodate different values of source voltage.

MULTIPLE-WINDING TRANSFORMERS

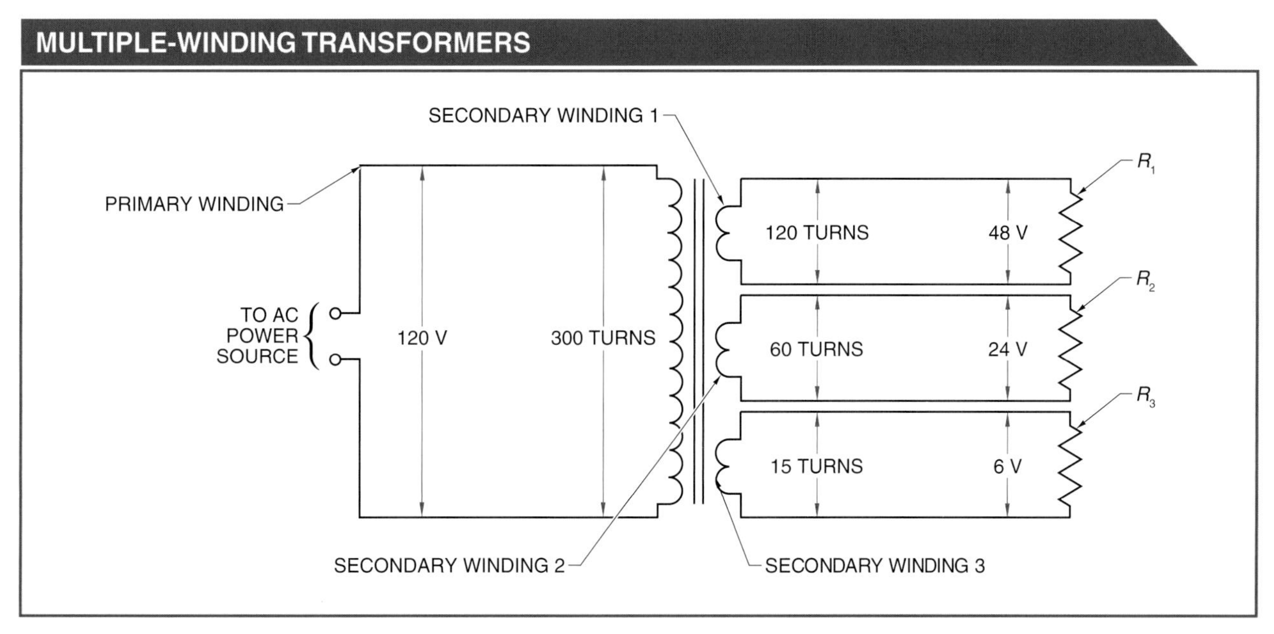

Figure 21-13. Most multiple-winding transformers have multiple secondary windings, each with a different voltage to meet the requirements of multiple loads.

Isolation Transformers. An *isolation transformer* is a transformer that has complete electrical separation between the primary and the secondary windings. Isolation transformers are used to isolate primary and secondary voltages. Devices designed without transformers and/or polarized plugs can present an electrical shock hazard to ground. **See Figure 21-14.** Dangerous conditions that could result in electrical shock or arc blast hazards can be present when working on electrical components without transformers and/or polarized plugs, such as video monitors. For example, when working on a video monitor, the white lead from the chassis of the video monitor is connected to the ungrounded lead of the power source. This condition places the 120 V of the power source on the chassis of the video monitor while the black lead of the video monitor is connected to the ground of the power source. The video monitor operates under this condition as it would if the plug in the receptacle with the chassis were at ground potential, causing a potential shock hazard. This condition is not allowed, per the NEC®.

Unlike a video monitor, an oscilloscope uses a polarized plug and has a transformer. The grounding (green) lead of the polarized plug is attached to the chassis of the oscilloscope. A potential of 120 V exists between the chassis of the oscilloscope and the chassis of the video monitor. An electric shock hazard exists if the hands are placed between the two components. If an attempt is made to attach the ground lead of the oscilloscope to the video monitor, the power line will be shorted.

Terms

An **isolation transformer** is a transformer that has complete electrical separation between the primary and the secondary windings.

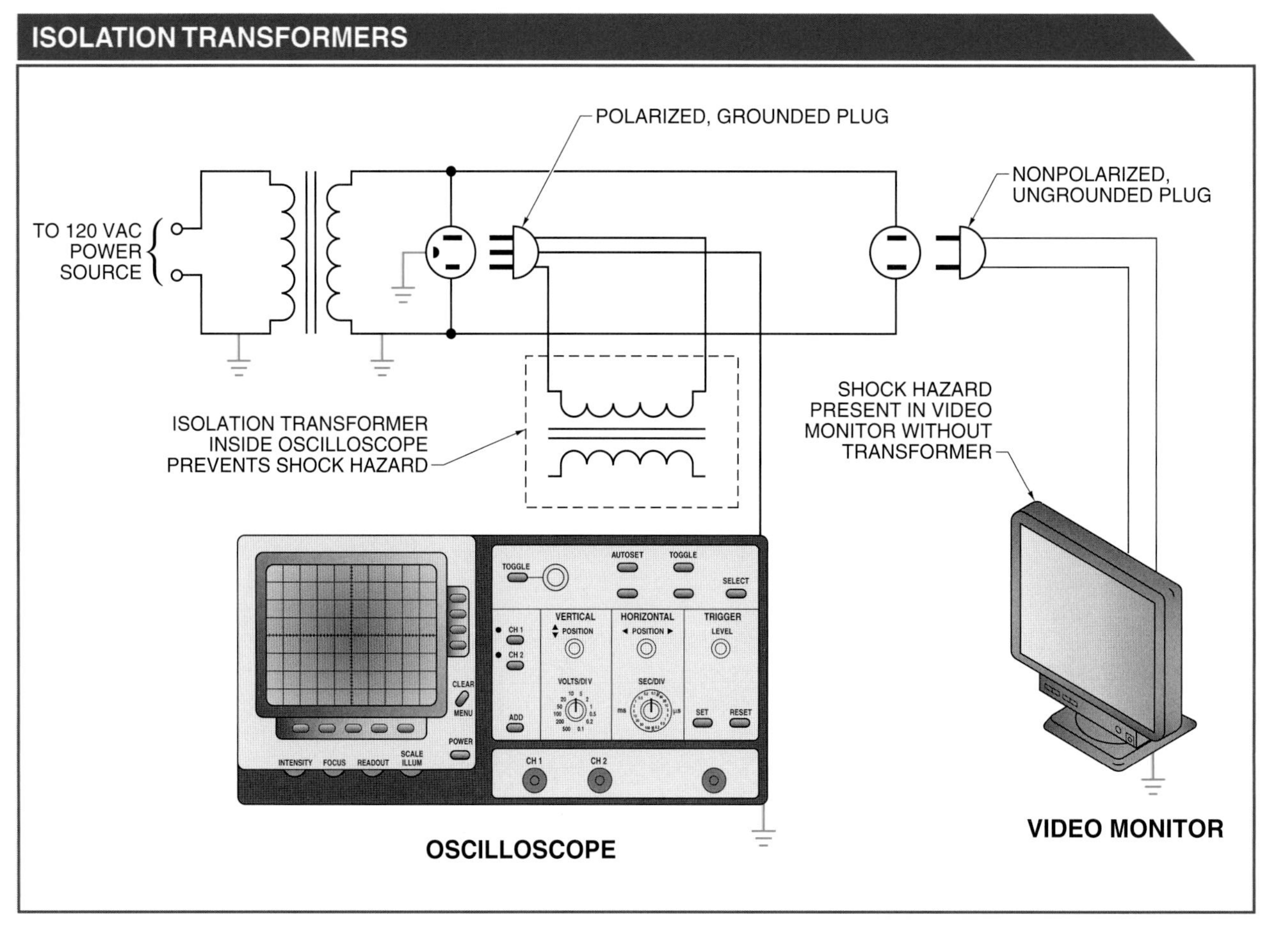

Figure 21-14. Devices designed without transformers and/or polarized plugs can present an electrical shock hazard to ground.

To overcome this problem, an isolation transformer can be used between the power line and the video monitor. An isolation transformer has a 1:1 turns ratio, so the primary voltage is equal to the secondary voltage. With the isolation transformer in place, it does not matter from which direction the monitor is plugged into the receptacle. Either side of the secondary winding of the isolation transformer can be brought to ground without causing a hazard.

Terms

An **autotransformer** is a transformer in which the primary and secondary circuits have portions of their two windings in common.

Autotransformers. An *autotransformer* is a transformer in which the primary and secondary circuits have portions of their two windings in common. An autotransformer is a self-induced transformer. If a voltage is impressed across the connected windings, current may be taken from the transformer between the point where the two windings are connected and either of the two ends of the winding. An autotransformer is a single-winding tapped transformer that can be used to step up or step down voltage. **See Figure 21-15.**

AUTOTRANSFORMERS

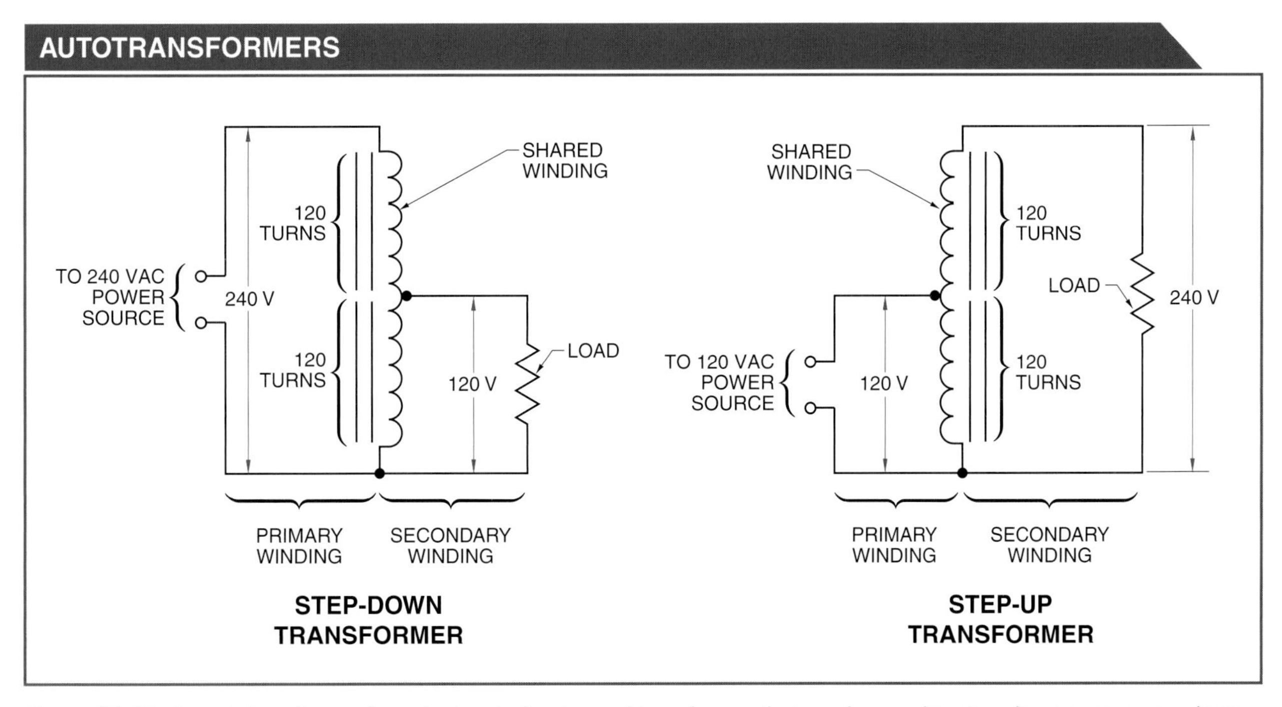

Figure 21-15. An autotransformer is a single-winding tapped transformer that can be used to step down or step up voltage.

When an autotransformer is used to step up voltage, the primary winding is part of the secondary winding. When an autotransformer is used to step down voltage, the secondary winding acts as part of the primary winding. The same equations used to calculate unknown parameters such as voltage, number of turns, and current in other types of power transformers can also be used with autotransformers. The primary disadvantage of using autotransformers is that the autotransformer does not provide electrical isolation, as is the case when only magnetic coupling is used between the primary and secondary windings.

Autotransformers have certain advantages over other types of power transformers due to the primary- and secondary-winding currents being 180° out of phase. These currents cancel each other, which reduces the amount of current flow through the transformer windings. As with all transformers, the amount of power consumed by the secondary load must be supplied by the primary voltage source.

Buck-Boost Transformers. A *buck-boost transformer* is a small transformer that is designed to lower or raise line voltage. Transformers with dual primary and secondary windings are often used to lower or raise line voltage. When a transformer is used to lower the line voltage, it is said to "buck" the line voltage. When the transformer is used to raise the line voltage, it is said to "boost" the line voltage.

Terms

A **buck-boost transformer** is a small transformer that is designed to lower or raise line voltage.

Voltage across any winding of a transformer has polarity. Because of polarity, the primary and secondary voltages can be made to add to or subtract from the line voltage. In order to reduce or increase source voltage, buck-boost transformers are connected in series as autotransformers. **See Figure 21-16.**

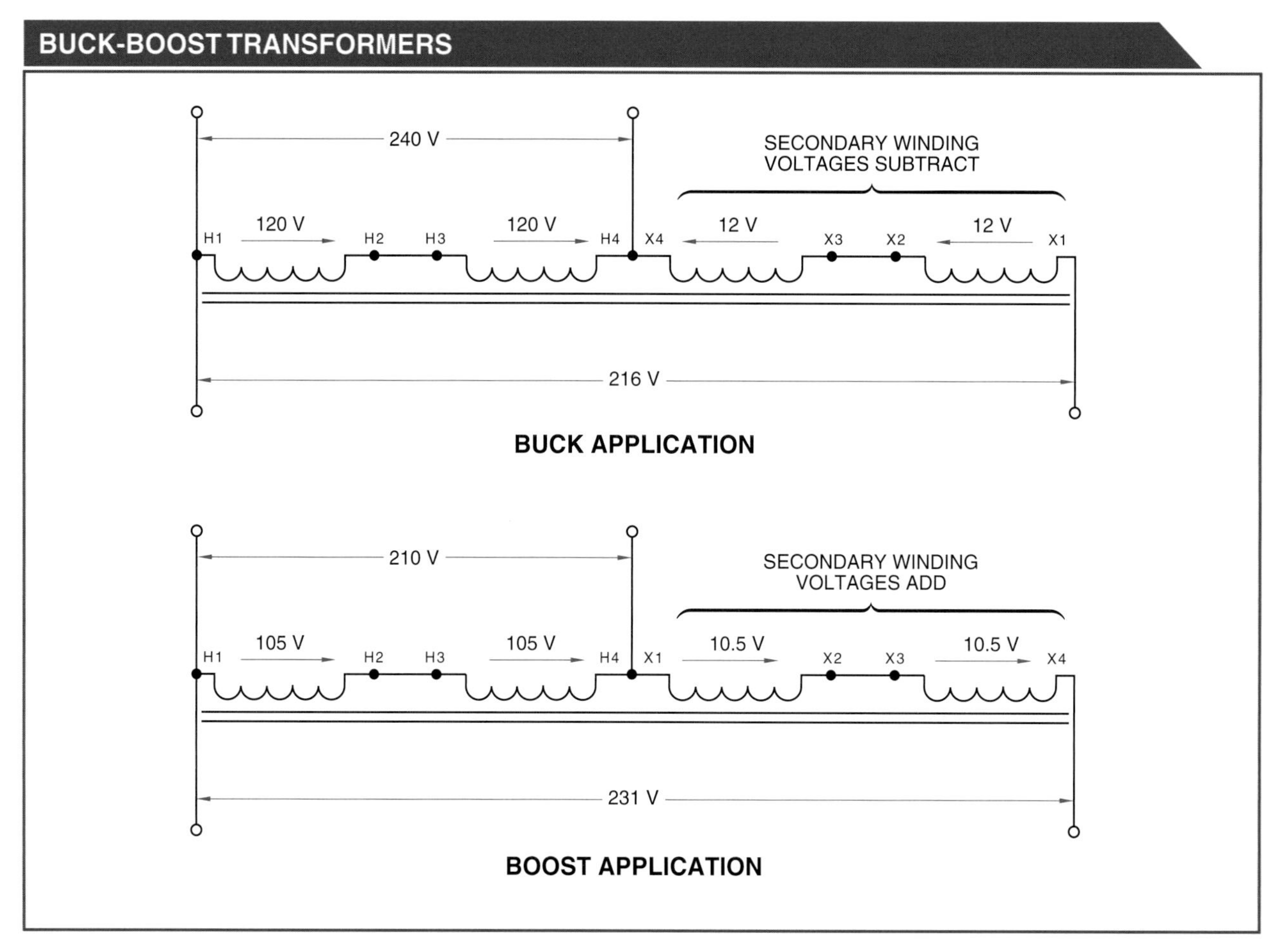

Figure 21-16. Buck-boost transformers are connected in series as autotransformers to reduce or increase source voltage.

Terms

A **core-type transformer** is a transformer that has windings placed around each leg of the core material.

Core-Type Transformers. Cores are constructed of legs and yokes. The vertical legs support the coils, and upper and lower yokes connect the legs. A *core-type transformer* is a transformer that has windings placed around each leg of the core material. **See Figure 21-17.** A thick layer of insulating material is wrapped around the legs to prevent electrical contact between the wire of the coil and the iron of the leg. Three-phase (3ϕ) transformers typically use a three-leg design. Single-phase (1ϕ) transformers typically use a two-leg design.

Core-type transformers are generally less expensive than other types of transformers because less iron is used on the core and the sheet-steel transformer enclosure is smaller. When transforming 3ϕ power, either a single 3ϕ core-type transformer or three individual 1ϕ core-type transformers can be used. An advantage to using a single 3ϕ core-type unit is that it is more economical and more efficient than using three individual 1ϕ units. A single 3ϕ core-type transformer requires less copper for the same load and is smaller, thereby requiring less space. Three-phase transformers in single sheet-metal enclosures are also easier to install, because they are prewired.

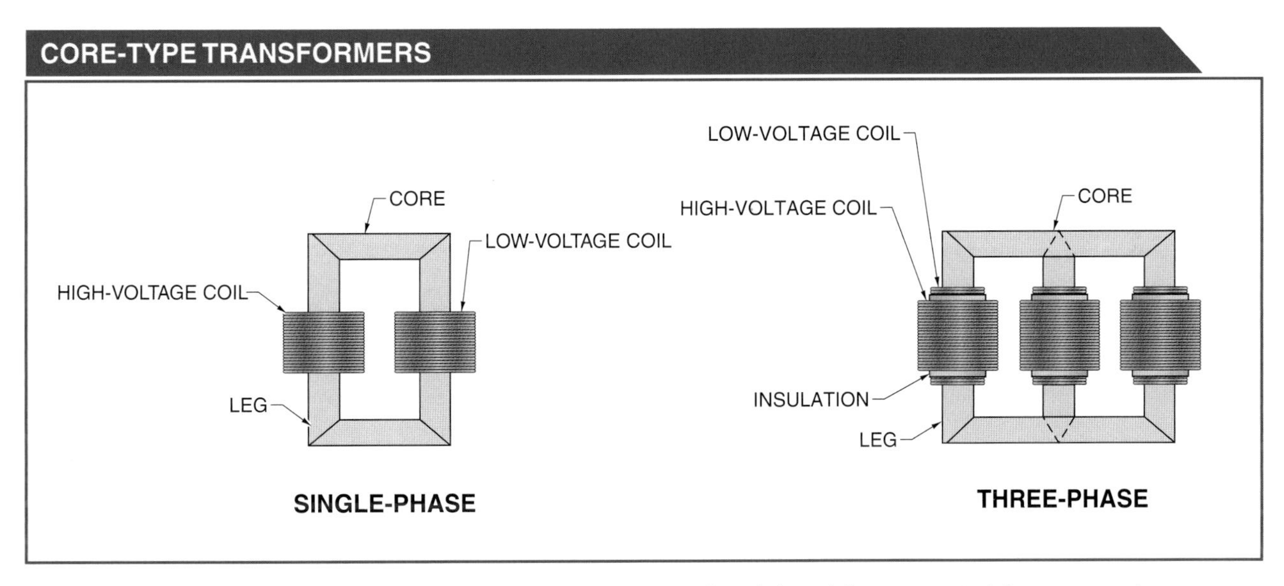

Figure 21-17. A core-type transformer has windings placed around each leg of the core material.

The primary disadvantage of 3ϕ core-type transformers is that if one phase fails, the entire unit must be taken off-line and repaired or replaced. This is due to the use of a common core for the three phases. When one phase fails, the core area of the defective phase quickly becomes magnetically saturated due to the lack of an opposing magnetic field. This large magnetic field escapes from the core, inducing high eddy currents into the sheet-metal enclosure. The eddy currents cause severe heating that can exceed the temperature ratings of the insulation material used on the windings, further damaging the transformer. The heat generated by the eddy currents can be sufficient to cause a fire. This type of situation can be prevented if three separate 1ϕ transformers are used, since each phase has its own dedicated core. The failed unit can be removed while the other two phases continue delivering power to the load.

Terms

An **instrument transformer** is a transformer that steps down the voltage or current of a circuit to a low value that can be effectively and safely used for the operation of instruments such as ammeters, voltmeters, wattmeters, relays used for various protective purposes, telemetering used for indications in remote areas, and dispatching energy.

A **potential transformer** is a precision two-winding transformer that is used to step down high voltage to allow safe voltage measurement.

A **current transformer** is a transformer that is used to isolate an ammeter to prevent the hazards caused by connecting to high-voltage lines.

An **audio transformer** is used to transfer complex signals containing energy at a large number of frequencies from one circuit to another.

Instrument Transformers

An *instrument transformer* is a transformer that steps down the voltage or current of a circuit to a low value that can be effectively and safely used for the operation of instruments such as ammeters, voltmeters, wattmeters, relays used for various protective purposes, telemetering used for indications in remote areas, and dispatching energy. Although high-voltage and high-current instruments are available for certain applications, it is unusual to directly measure high voltages and high currents in power applications. Unless a high-voltage circuit is grounded at the instrument, a lethal voltage-to-ground potential can exist between the instrument and ground. An electrocution hazard is caused when there is contact between the ground and the instrument. In addition, the accuracy of instruments is reduced when they are operated in the strong electrostatic field of a high-voltage system.

To overcome instrument transformer problems, low-voltage instruments are used to take measurements. The two most common types of instrument transformers are potential transformers and current transformers.

A *potential transformer* is a precision two-winding transformer that is used to step down high voltage to allow safe voltage measurement. A potential transformer operates on the same principle as a power transformer. The main difference is that the capacity of a potential transformer is relatively small compared to that of a power transformer. Potential transformers are typically used to measure high voltage but are also found in protection relays and devices. A *current transformer* is a transformer that is used to isolate an ammeter to prevent the hazards caused by connecting to high-voltage lines. In addition to insulating the meter from high voltage, current transformers step down the current by a known and accurate ratio for measurement.

Audio Transformers

An *audio transformer* is used to transfer complex signals containing energy at a large number of frequencies from one circuit to another. Audio transformers are required to respond uniformly to signal voltages over a large frequency range, such as 10 Hz to 100,000 Hz, and must be designed so that most of the magnetic lines of force threading through one winding also pass through the other. Audio transformers are used in audio applications for impedance matching. While the frequency of a power transformer is either 50 Hz or 60 Hz (depending on the country where the power transformer is used), the frequency of an audio transformer ranges between 20 Hz and 20 kHz, which are generally considered the limits of human hearing. **See Figure 21-18.**

Tech Tip

Items such as thermometers and bushings are often used as accessories to transformers. However, transformers themselves can also be used as external accessories when used to match the impedance of two pieces of electronic equipment, such as a microphone and an amplifier.

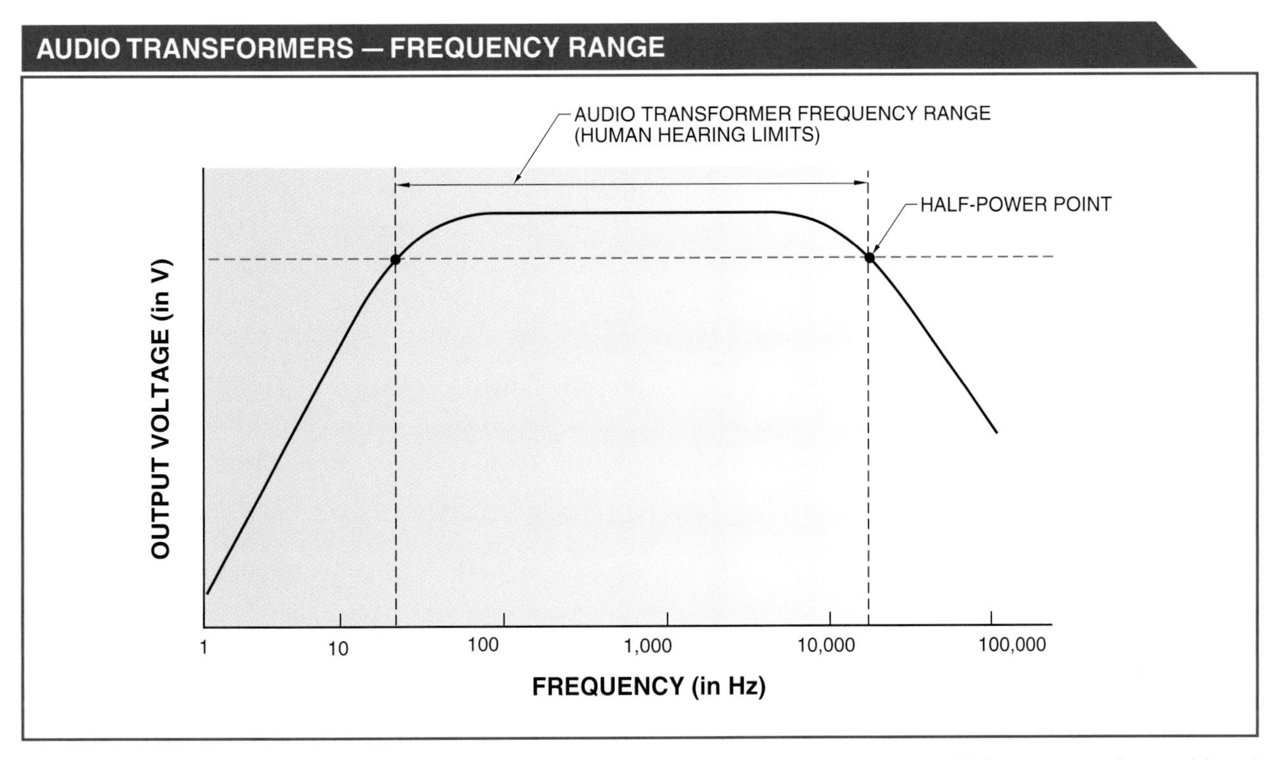

Figure 21-18. The frequency of an audio transformer ranges between 20 Hz and 20 kHz, which are generally considered the limits of human hearing.

As frequency in an audio transformer increases above 60 Hz, eddy current and hysteresis losses increase and the distribution capacitance between windings becomes apparent. For any given frequency, these losses are constant, but since the audio frequency is always changing, the losses are different for different frequencies. Thus, an ideal audio transformer does not exist. For example, most audio transformers are designed to pass frequency bands of 30 Hz to 15 kHz. When only voice applications are required, inexpensive transformers with pass bands that range from 150 Hz to 3 kHz are sufficient. When designing an audio amplifier, special compensation circuits such as bass and treble control are incorporated into the design.

Radio Frequency Transformers

A *radio frequency (RF) transformer* is used to transfer energy in narrow frequency bands from one circuit to another. The difference between an RF transformer and an audio transformer is that an audio transformer is designed to work over a large frequency range, while a radio frequency transformer is designed to operate in a very narrow frequency range. Radio frequency transformers and audio transformers both have a high level of mutual inductance, gained by using a

Terms

A **radio frequency (RF) transformer** is used to transfer energy in narrow frequency bands from one circuit to another.

large number of turns and an iron core. At radio frequencies above 20 kHz, hysteresis and eddy current losses become prohibitive in an iron core. For that reason, a graphite-compound or powdered-iron core is used at lower radio frequencies. At higher radio frequencies, graphite-compound or powdered-iron cores have prohibitive losses, and only air cores are used. Radio frequency transformers have several different types of symbols. **See Appendix.** Radio frequency transformers can have either fixed or variable values for inductance and coefficient of coupling.

Air-core transformers have a maximum possible coefficient of coupling of about 65%. A coefficient of coupling at this level makes the ratio equations for turns, voltage, and current invalid for RF transformers. These ratios become invalid because of the loss of flux linkage. Radio frequency transformers are typically used to couple circuits rather than to change the level of voltages or currents.

Terms

A **balun** is a special radio frequency transformer used to match impedance between a balanced 2-wire transmission line to an unbalanced transmission line or component.

A *balun* is a special radio frequency transformer used to match impedance between a balanced 2-wire transmission line to an unbalanced transmission line or component. A common application for a balun is to connect the balanced twisted-pair cables used in telecommunications systems to unbalanced coaxial cables. There is often lower performance with baluns, as a signal cannot run as far on twisted-pair cables as it can on coaxial cables.

Because the inductive reactance of relatively few turns is high at radio frequencies, high mutual reactance exists between the primary and secondary windings of an RF transformer. High mutual reactance makes RF transformers effective as tuned circuits when placing a capacitor across the windings. At resonance, these transformers provide high voltage gain and power transfer. Radio frequency transformers are typically used with a signal receiver to connect an antenna to the input of an RF amplifier. The primary winding of the transformer is not tuned. The secondary winding of the transformer is tuned with a variable capacitor across the secondary windings. Tuning the secondary winding allows matching of the impedance of the antenna and the input of the RF amplifier. The tuned secondary winding is a filter that presents high impedance to one frequency, which is sent to the input of the amplifier. All other frequencies are rejected.

Tech Tip

The most common type of transformer has a delta-connected primary winding and a wye-connected secondary winding.

TRANSFORMER CONNECTIONS

Transformer windings, like other electrical devices, can be connected in various configurations. Polarity must be observed when making transformer connections. Separate 1ϕ and 3ϕ wiring schemes are used when making transformer connections.

Single-Phase Transformer Connections

Single-phase transformers can be connected in many different configurations to provide the necessary voltage and current. For example, a 1ϕ power transformer is used with dual primary and dual secondary windings. When a properly rated 1ϕ power transformer is selected, the method in which the windings are connected is dependent on the value of source voltage and the requirements of the load. Transformers can be utilized below their voltage and current ratings for any given winding but they cannot be operated above these ratings.

Dual windings can be connected either in series or in parallel with each other. When dual windings are connected in series with the proper polarity, the voltage values add together, and the current rating of the windings is that of the lowest-rated winding. When windings are connected in parallel, the voltage ratings must be equal, and the current values add together.

For example, when a 1ϕ transformer has line voltage of 480 V with load requirements of 120 V and 30 A, the primary windings are rated at 240 V, and the dual windings must be connected in series with each other so that there are 240 V across each winding, for a total of 480 V. **See Figure 21-19.** To deliver 120 V and 30 A to the load, the secondary windings must be connected in parallel. Each secondary winding can deliver 20 A at 120 V to the load. When connected in parallel, the secondary windings can deliver up to 40 A.

Note: The polarity of the windings should be observed and noted because if the output is wired out of phase, twice as much voltage is generated on the secondary winding.

SINGLE-PHASE TRANSFORMER CONNECTIONS

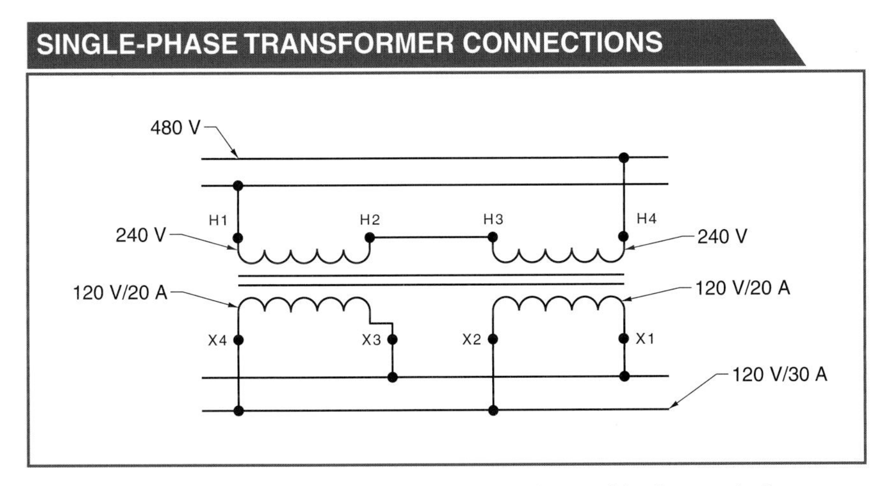

Figure 21-19. With a 1ϕ transformer connection, the 240 V primary windings are connected in series to accommodate the 480 V line, and the 120 V/20 A secondary windings are connected in parallel to provide 30 A to the load.

Three-Phase Transformer Connections

When there are dual windings, each phase of a 3ϕ transformer can have its windings connected in series or in parallel. The procedure for connecting a 3ϕ transformer is the same as the procedure for connecting a 1ϕ transformer. The phase windings of each phase are connected in either a wye (Y) or a delta (Δ) configuration, in the same manner as that for a 3ϕ generator. Three-phase transformers can be connected in delta-delta, delta-wye, wye-wye, and wye-delta configurations. **See Figure 21-20.** One primary and one secondary lead of the same polarity are marked with a white polarity mark. It is important that the polarity is observed when making these connections because improper connections can generate undesirable voltages. **See Appendix.**

When three-phase windings are connected in a delta configuration, the line voltage is equal to the phase voltage. The line voltage of a 3ϕ transformer connected in a delta configuration is calculated by applying the following formula:

$$V\phi = V_{LINE}$$

where

$V\phi$ = phase voltage (in V)

V_{LINE} = line voltage (in V)

THREE-PHASE TRANSFORMER CONNECTIONS

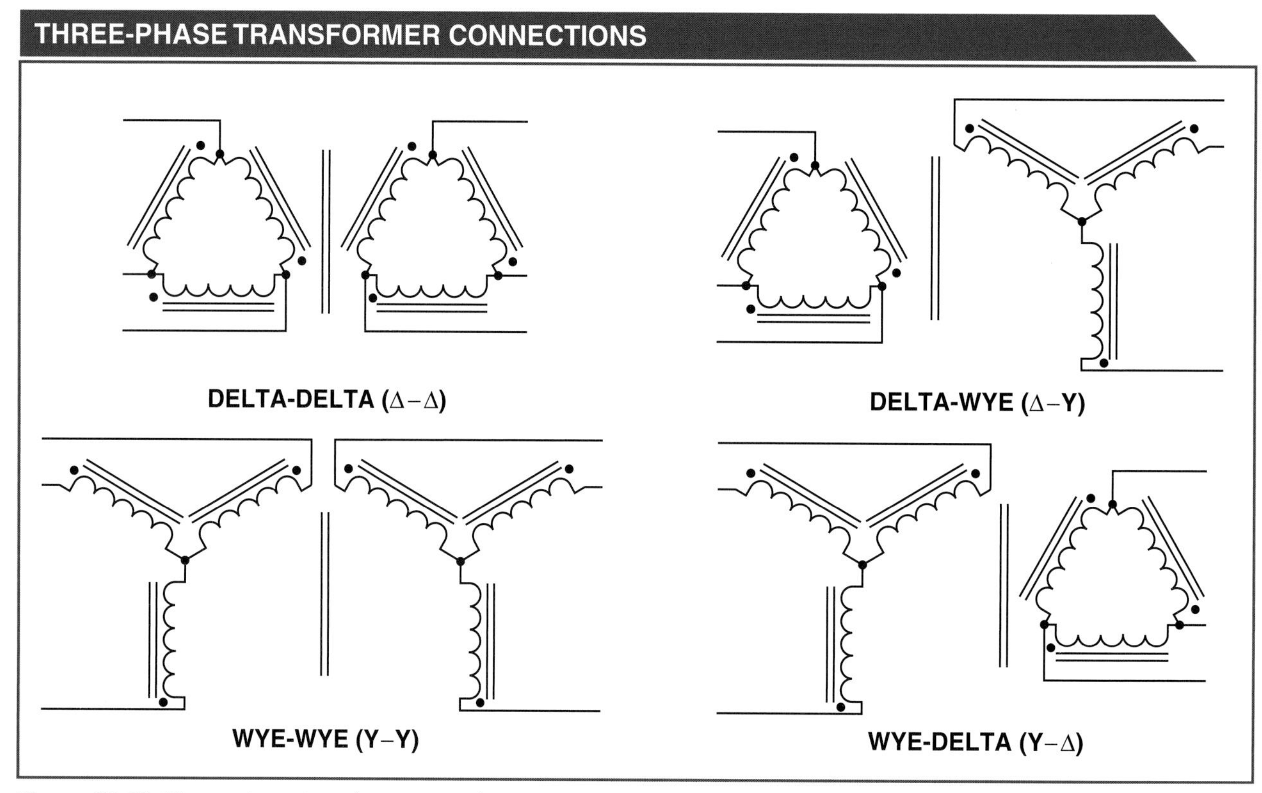

Figure 21-20. Three-phase transformers can be connected in delta-delta, delta-wye, wye-wye, and wye-delta configurations.

Example: What is the phase voltage of a delta-connected 3ϕ transformer that has line voltage of 230 V?

$V\phi = V_{LINE}$

$V\phi = \mathbf{230\ V}$

For delta-connected transformers with a balanced load, the line current is calculated by applying the following formula:

$I_{LINE} = 1.73 \times I\phi$

where

I_{LINE} = line current (in A)

1.73 = constant

$I\phi$ = phase current (in A)

Example: What is the line current of a delta-connected 3ϕ transformer with phase current of 15 A?

$I_{LINE} = 1.73 \times I\phi$

$I_{LINE} = 1.73 \times 15\ A$

$I_{LINE} = \mathbf{25.9\ A}$

When the 3ϕ windings are connected in a wye configuration, the value of the line voltage is calculated by applying the following formula:

$V_{LINE} = 1.73 \times V\phi$

where

V_{LINE} = line voltage (in V)

1.73 = constant

$V\phi$ = phase voltage (in V)

Example: What is the line voltage of a wye-connected, 3ϕ transformer with phase voltage of 240 V?

$V_{LINE} = 1.73 \times V\phi$

$V_{LINE} = 1.73 \times 240\ V$

$V_{LINE} = \mathbf{415.2\ V}$

When working with wye-connected transformers, the transformer line current is equal to phase current, because the line and phase windings are connected in series with each other. Phase current in a wye-connected transformer is calculated by applying the following formula:

$I\phi = I_{LINE}$

where

$I\phi$ = phase current (in A)

I_{LINE} = line current (in A)

Tech Tip

To overcome the tuning problem of radio frequency amplification being limited to three stages, American electrical engineer and inventor Edwin H. Armstrong invented the superheterodyne circuit in 1912. The superheterodyne circuit amplified signals that were previously too weak to be considered useful into stronger signals. This type of circuit is the basic circuit used in about 98% of all modern radio and television receivers.

Example: What is the phase current of a wye-connected 3ϕ transformer that has line current of 15 A?

$I_\phi = I_{LINE}$

$I_\phi = \mathbf{15\ A}$

These equations can be applied to systems that use three separate 1ϕ transformers with dual primary and secondary windings. Primary windings are rated at 600 V and secondary windings are rated at 120 V. As with a 1ϕ transformer, the method in which 3ϕ transformers are connected is determined by the amount of source voltage and voltage requirements of the load. **See Figure 21-21.**

For example, 120 V load requirements are met by center-tapping one of the secondary phases. There is 120 V present between Bϕ and neutral and between Cϕ and neutral. Only one phase can be used to obtain this voltage. A short circuit will develop between phases if another phase has its center tap connected to ground.

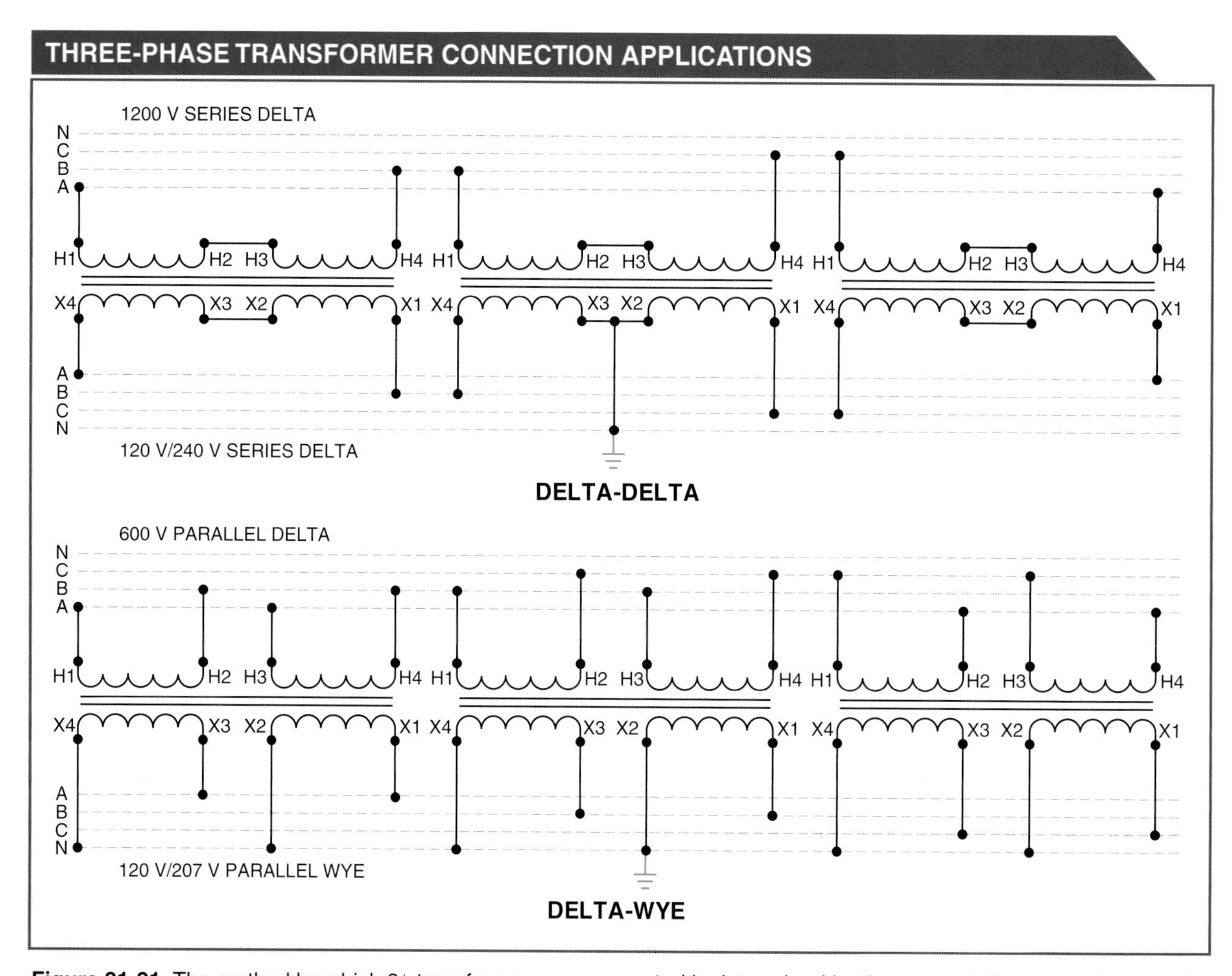

Figure 21-21. The method by which 3ϕ transformers are connected is determined by the amount of source voltage and the voltage requirements of the load.

There are 240 V present between phases A-B, B-C, and C-A. It is difficult to balance the current load with delta secondary windings where the load requires two values of voltages. The main advantage of delta connections is that if one phase is lost, 3ϕ power is still available to the load, although with a reduced kVA rating to all three phases. Phase current equals line current and the 1.73 multiplier advantage has been lost.

If primary windings are connected in parallel delta, so that the 600 V line voltage is impressed across each primary winding and the secondary windings are connected in a parallel wye configuration, the connection provides 120 V to the neutral for each phase. Between phases, the voltage is equal to 207 V (120 V × 1.73). The advantage of having the secondary winding in a parallel-connected wye configuration is that 120 V can be obtained from all phases. This makes it easier to balance the load on the transformer. The biggest disadvantage of a parallel-connected wye configuration is that 3ϕ power is lost if any phase fails.

SUMMARY

Transformers use mutual inductance to transfer energy from one circuit to another. The source voltage is connected to the primary winding of a transformer and the load is connected to the secondary winding. Transformer primary and secondary windings may be connected in different configurations to produce proper secondary output voltage and current.

When the coefficient of coupling has a high value, such as with an iron-core transformer, the ratio of primary-winding voltage to secondary-winding voltage is directly proportional to the turns ratio of the transformer. The current ratio is inversely related to the turns ratio. This causes the primary-winding VA rating to be equal to the secondary-winding VA rating. The impedance ratio is directly proportional to the square of the turns ratio.

Transformers are classified by their frequencies and applications. Power transformers are usually rated in kVA at 50 Hz to 60 Hz. Voltage and current rating values are also specified for the windings. Voltage and current rating values should never be exceeded.

Other transformer types include tapped-winding, multiple-winding, isolation, autotransformer, buck-boost, core-type, instrument, audio, and radio frequency transformers. These classifications are based on the application that the transformer is designed to perform.

APPLICATION: TRANSFORMER IMPEDANCE MATCHING...

Background: Transformers are used to step up or step down voltage and are a key element of AC power distribution. In AC power distribution systems, voltage is normally stepped up for long-distance transmission and stepped down for local distribution. This improves distribution efficiency and reduces cable losses.

Transformers can also be used for impedance matching. Impedances must match to maximize the power transfer between the source voltage and load. **See Figure 1.**

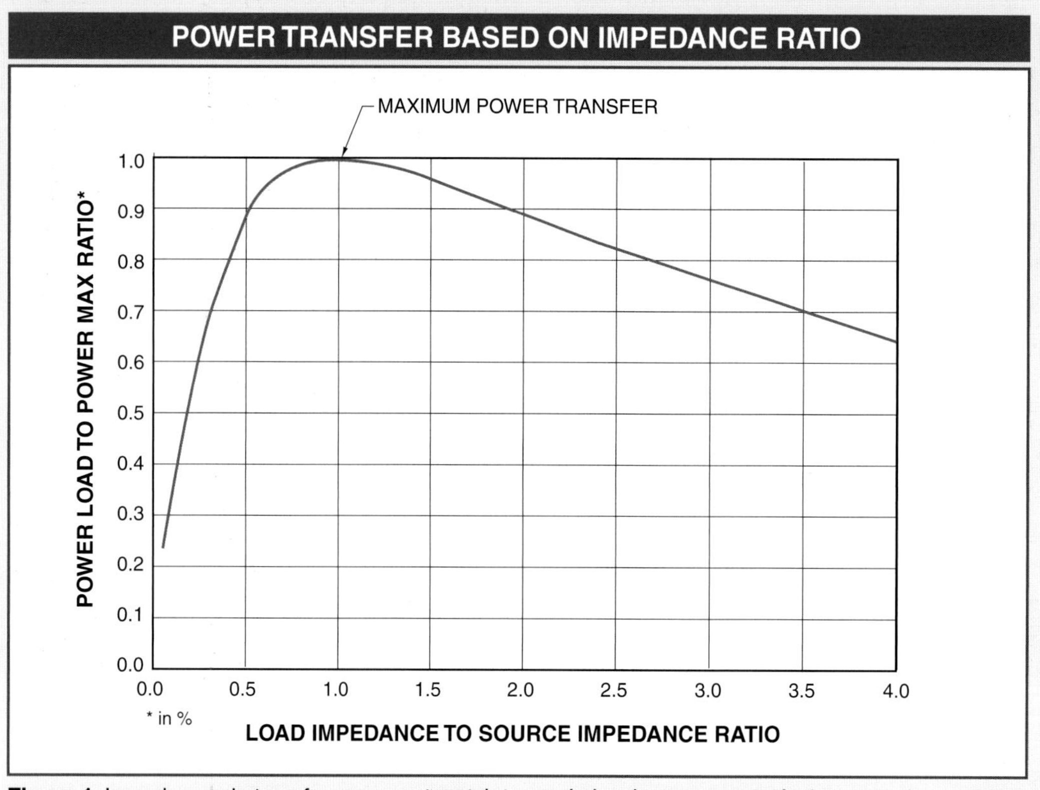

Figure 1. Impedances in transformers must match to maximize the power transfer between the source voltage and load.

Maximum power transfer occurs where the impedances match (ratio = 1). The power transfer ratio is worst with a high-impedance source and low-impedance load. When the impedances are different, a transformer can be used to match the impedances by selecting the correct number of turns between the primary and secondary winding.

The transformer voltage-ratio equation relates the turns ratio and voltages on the primary and secondary windings and is calculated by applying the following formula:

$$\frac{N_P}{N_S} = \frac{V_P}{V_S}$$

where

N_p = number of turns (primary winding)
N_s = number of turns (secondary winding)
V_p = primary winding voltage (in V)
V_s = secondary winding voltage (in V)

...APPLICATION: TRANSFORMER IMPEDANCE MATCHING...

The transformer current-ratio equation relates the turns ratio and currents on the primary and secondary windings. This equation is inversely proportional to the voltage ratio equation because a transformer does not create power. An increase voltage is proportional to a decrease in current and is calculated by applying the following formula:

$$\frac{N_P}{N_S} = \frac{I_S}{I_P}$$

where
Z_s = secondary winding impedance (in Ω)
Z_p = primary winding impedance (in Ω)
N_p = number of turns (primary winding)
N_s = number of turns (secondary winding)

From the voltage and current ratio equations, impedances can be related to the turns ratios by applying the following formulas:

$$\left(\frac{N_P}{N_S}\right)^2 = \frac{V_P}{V_S} \times \frac{I_S}{I_P}$$

$$Z_S\left(\frac{N_P}{N_S}\right)^2 = Z_P$$

$$\frac{Z_P}{Z_S} = \left(\frac{N_P}{N_S}\right)^2$$

$$\sqrt{\frac{Z_P}{Z_S}} = \frac{N_P}{N_S}$$

where
Z_s = secondary winding impedance (in Ω)
Z_p = primary winding impedance (in Ω)
N_p = number of turns (primary winding)
N_s = number of turns (secondary winding)

Key Points: transformer impedance matching; turns ratio

Problem: A homeowner requires a set of speakers for a 100 W/channel home stereo system. The stereo has an optimum output impedance to operate with 16 Ω speakers. The new speakers are 4 Ω. How does the homeowner match the impedance of the stereo system with the new 4 Ω speakers?

Solution: Since there is a higher impedance source (home stereo = 16 Ω) driving a lower impedance load (new speakers = 4 Ω), power transfer is only 64%. A transformer can be used to match the impedance between the stereo and the new speakers. The turns ratio is found by applying the formula:

...APPLICATION: TRANSFORMER IMPEDANCE MATCHING

$$\sqrt{\frac{Z_P}{Z_S}} = \frac{N_P}{N_S}$$

$$\sqrt{\frac{16}{4}} = \frac{N_P}{N_S}$$

$$\sqrt{4} = \frac{N_P}{N_S}$$

The transformer turns ratio is 2:1 and must support the maximum power of 100 W. An impedance matching transformer is required for each speaker. The turns ratio in this application is common and is found when matching a 300 Ω antenna with a 75 Ω video coaxial cable, although this matching transformer is not designed for 100 W.

Chapter Review...

1. What is mutual induction?
2. Why are core-type transformers typically less expensive than other types of transformers?
3. What is the phase voltage of a delta-connected 3ϕ transformer that has line voltage of 208 V?
4. What is the line voltage of a wye-connected, 3ϕ transformer with phase voltage of 208 V?
5. Why are voltages from the utility typically stepped-up for transmission over long distances?
6. What telecommunications applications are baluns commonly used for?
7. What is the phase current of a wye-connected 3ϕ transformer that has line current of 21 A?
8. What do the numbers in a turns ratio represent?
9. What is the primary-winding current for a secondary-winding current of 20 A in a transformer with a 3:1 turns ratio?
10. Briefly explain the difference between a transformer primary winding and secondary winding.
11. Briefly describe the coefficient of coupling of a transformer.
12. What is the line current of a delta-connected 3ϕ transformer with phase current of 15 A?
13. How is transformer efficiency used with power transformer applications?

...Chapter Review

14. What is the left-hand rule as applied to transformers?
15. What is a turns ratio in relation to transformers?
16. What is the transformer efficiency of a transformer that has output power of 600 W and input power of 650 W? Explain why this efficiency is considered good or bad.
17. Explain the difference between a tight and loose coefficient of coupling.
18. What is the resistive loss in a transformer primary winding that is wound with copper conductors that carry 20 A and have a resistance of 0.2764 Ω?
19. What is the reflected load of a transformer?
20. Briefly explain the difference between a step-up transformer and a step-down transformer.
21. What is the turns ratio of a transformer with 1000 turns in the primary winding and 3000 turns in the secondary winding?
22. What is impedance matching and why is it used?
23. Explain I^2R loss in a transformer.
24. Briefly explain Lenz's law.
25. What is exciting current?

22

AC Motors

OBJECTIVES

- List the different types of AC motors.
- Describe how motor rotation, motor speed, and rotor types affect three-phase induction motors.
- Explain the differences between split-phase, capacitor, and shaded-pole motors.

All motors convert electrical energy into mechanical energy. Because alternating current (AC) is the primary source of electrical energy, it follows that most motors are designed to run on AC. AC motors are typically less expensive than DC motors. Whereas DC motors require rectifiers when using AC power, rectifiers are not required for AC motors. Additionally, the most commonly used AC motors do not require brushes and a commutator. The same terms used with DC motors, such as the stator, rotor, armature, and field windings, apply to AC motors. AC motors are manufactured in different sizes and shapes with a variety of voltage and power ratings and may be either polyphase or single phase.

THREE-PHASE INDUCTION MOTORS

Three-phase (3ϕ) induction motors are typically used to turn multiple-horsepower loads. Because there is always current flow to the motor, the motor is capable of turning its rated load at a constant speed. Three-phase motors are less complex than single-phase (1ϕ) motors and are more economical to manufacture. Preventive and corrective maintenance in 3ϕ motors is generally less than that required by 1ϕ motors because there are fewer parts in 3ϕ motors.

In order for the armature of any motor to turn, it must be in the rotating magnetic field caused by the stator windings. The rotating magnetic field for a 3ϕ power source can be shown at different rotation points. **See Figure 22-1.** The different points are located at 60° intervals of the 3ϕ power sine waves. At each 60° interval, one of the phases is at its maximum current of 10 A, while the other two phases supply 5 A each.

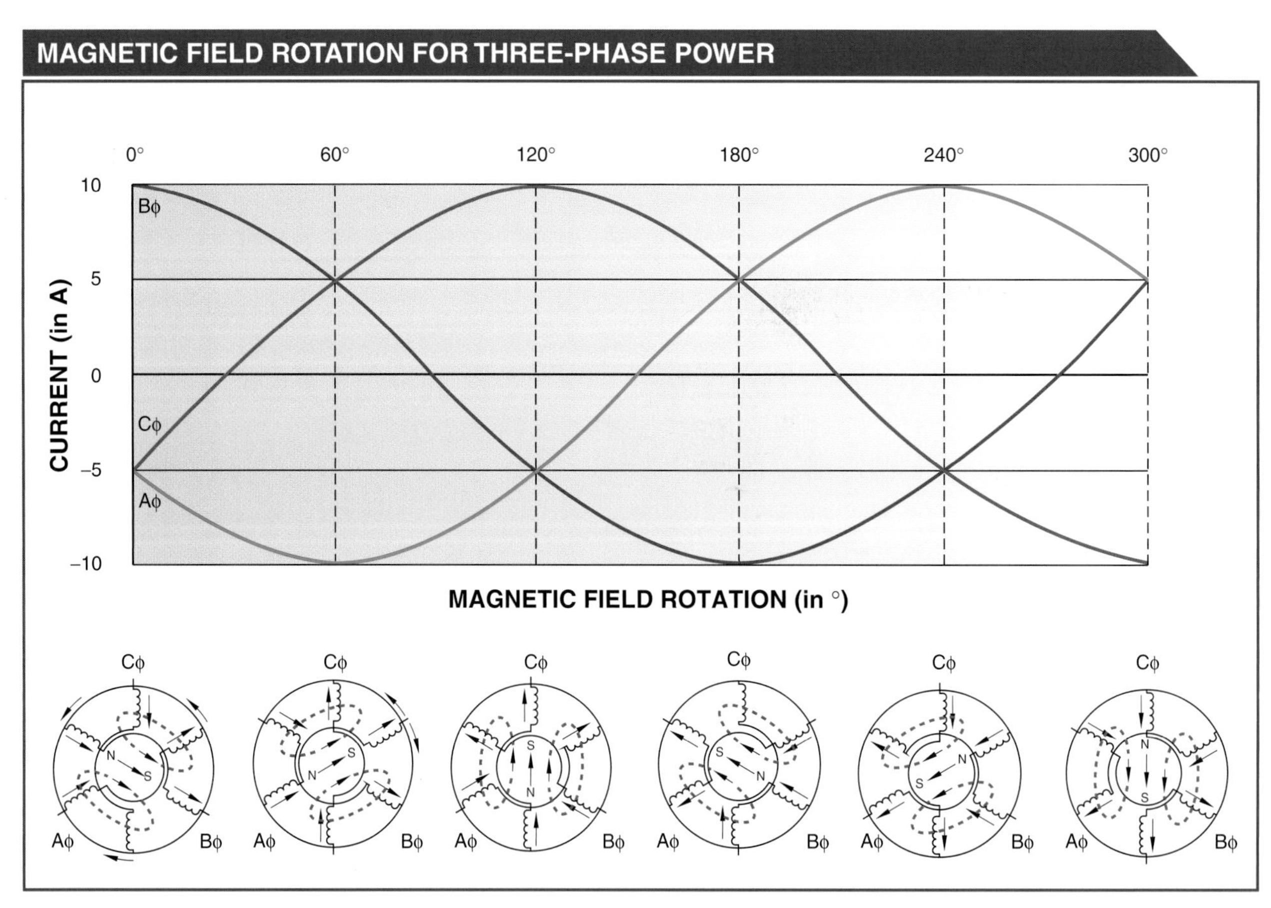

Figure 22-1. The rotating magnetic field for a three-phase (3ϕ) power source produces a rotating magnetic field used to rotate the rotor.

For example, at 0°, Bϕ is at maximum current of 10 A, while Aϕ is at 5 A going more negative, and Cϕ is at 5 A going toward positive. At 60°, the magnetic field rotates counterclockwise by 60°. The magnetic field is aligned with the Aϕ stator windings, which have –10 A of current flow through them. The strongest armature field is aligned with the stator winding that has 10 A. This armature field is aided by the smaller fields created by the stator winding that has 5 A of current flow through it. Rotation of the field continues, returning to 0°, with cycles repeating as long as 3ϕ power is applied. Three-phase induction motors have the ability to switch motor rotation, vary motor speed, and use one of two different types of rotors.

Motor Rotation

A 3ϕ induction motor can have its direction of rotation reversed by interchanging the connections of any two phases. For example, line 1 is connected to Aϕ, line 2 is connected to Bϕ, and line 3 is connected to Cϕ.

When line current reaches the maximum values in the sequence of lines 1, 2, and 3, the rotation becomes Aϕ, Bϕ, and Cϕ. When lines 1 and 2 are reversed, the sequence of rotation becomes Bϕ, Aϕ, and Cϕ. **See Figure 22-2.** Regardless of the connections and the direction of rotation of the running winding, changing any two connections causes the rotation of the magnetic field to reverse.

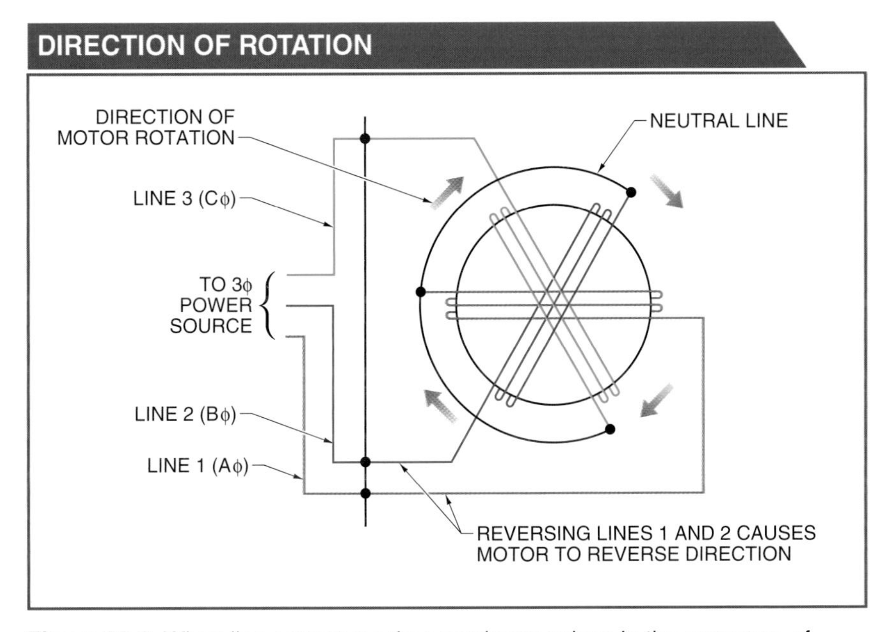

Figure 22-2. When line current reaches maximum values in the sequence of lines 1, 2, and 3, the sequence of rotation is Aϕ, Bϕ, Cϕ. When lines 1 and 2 are reversed, the sequence of rotation becomes Bϕ, Aϕ, Cϕ.

Motor Speed

Synchronous speed is the theoretical speed of a motor based on line frequency and the number of poles of the motor. For example, with line frequency of 60 Hz applied, the field rotates at 60 revolutions per second, or 3600 (60 × 60) revolutions per minute (rpm). For a 3ϕ, two-pole motor, the armature rotates at the synchronous speed. However, if the number of poles is doubled, producing a four-pole field, the speed of rotation is cut by half to 1800 rpm. Therefore, the synchronous speed of a 3ϕ motor is directly proportional to its frequency and inversely proportional to its number of poles. The synchronous speed of a 3ϕ motor is calculated by applying the following formula:

$$\Omega_S = 120 \times \left(\frac{f}{P}\right)$$

where

Ω_S = synchronous speed (in rpm)

120 = constant

f = source voltage frequency (in Hz)

P = number of motor poles

Terms

Synchronous speed is the theoretical speed of a motor based on line frequency and the number of poles of the motor.

Tech Tip

Failure to derate or provide auxiliary cooling to motors operating at low speeds results in premature failure of the motors.

Example: What is the synchronous speed of a 4-pole, 60 Hz, 3ϕ motor?

$$\Omega_S = 120 \times \left(\frac{f}{P}\right)$$

$$\Omega_S = 120 \times \left(\frac{60\text{ Hz}}{4}\right)$$

$$\Omega_S = \frac{7200}{4}$$

Ω_S = **1800 rpm**

If the armature were to rotate exactly at the synchronous speed of the field, no current would be induced into the running windings, and there would be no relative motion between the changing magnetic field and the conductors. The armature actually rotates at a slightly slower speed in order for the stator field to cut the running windings and induce a current.

The speed of the stator's rotating magnetic field is constant, provided the line frequency is constant, and is independent of the mechanical load placed on the motor. *Slip* is the difference between the synchronous speed and the actual speed of a motor. As armature speed is reduced with an increased load, slip increases. As slip increases, so does torque (turning force). Increased torque is caused by the increase in current flow in the armature. Under a constant load, a 3ϕ motor maintains a constant speed.

Terms

Slip is the difference between the synchronous speed and the actual speed of a motor.

Rotor Types

The two types of rotors used on AC motors are a cage rotor and a wound rotor. **See Figure 22-3.** A cage rotor consists of a series of parallel copper or aluminum bars placed in slots on the rotor core. The copper or aluminum bars are connected together at each end by shorting bars of the same material as that from which the bars are made. The copper and aluminum bars are good conductors of electricity and carry relatively high current at low voltage. The ability of copper and aluminum bars to carry high current at low voltage makes it unnecessary to insulate the bars from the core, because current takes the path of least resistance. A cage rotor is sometimes referred to as a "squirrel-cage rotor" or "squirrel-cage motor" due to its similarity to a device used to exercise squirrels and other small mammals.

A wound rotor has windings similar to 3ϕ stator windings. Wound-rotor windings are typically wye connected, with the free end of each winding connected to a slip ring. A wye connection allows the rotor windings to be connected through brushes to an external, wye-connected, variable resistance.

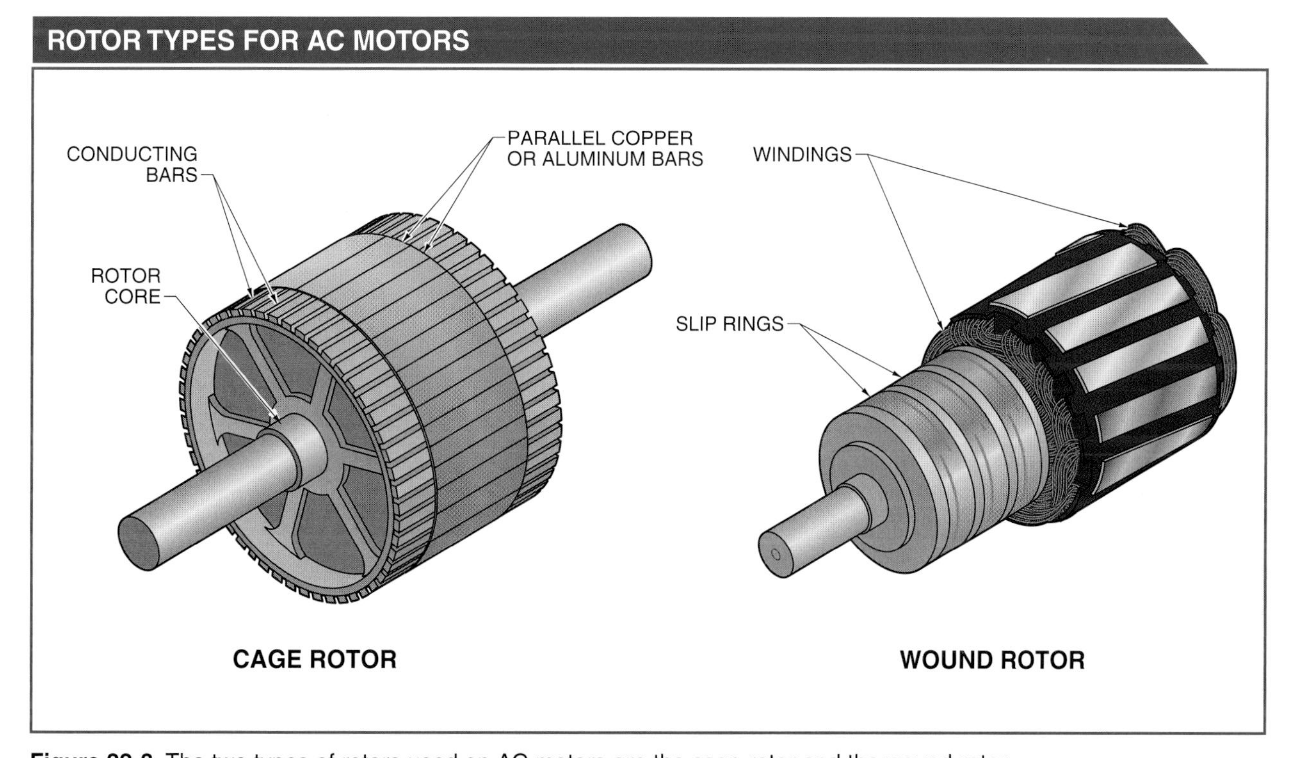

Figure 22-3. The two types of rotors used on AC motors are the cage rotor and the wound rotor.

A variable resistance is connected in series with the rotor windings to limit inrush current at motor startup. Because motor inrush currents may be 8 to 10 times higher than normal operating current, the inrush current must be limited. Also, by limiting the inrush current, the motor has greater startup torque at lower startup current. Increasing rotor resistance increases rotor torque. As the motor increases speed, the resistance is removed from the circuit and a short circuit is placed across the slip rings. At this point, the wound rotor is similar in operation to a cage rotor.

Advantages of a wound-rotor induction motor over a cage motor include the following:

- Moderate starting current and high starting torque
- Smooth acceleration under heavy loads
- Reduced heating of rotor during starting
- Adjustable speed control

Disadvantages of a wound-rotor induction motor over a cage motor include the following:

- Higher initial cost
- Higher maintenance cost due to brushes and slip rings

SYNCHRONOUS MOTORS

When supplied with the proper voltage, an AC generator will run as a synchronous motor, with practically no change in construction or rating. For example, two AC generators operating in parallel and with the driving force removed from one of the generators will become a generator and a synchronous motor and continue to run as such. When running as a synchronous motor, the speed is the same as that of the generator that is supplying power to it.

A *synchronous motor* is a three-phase motor that runs at synchronous speed. The stator windings on a synchronous motor are identical to stator windings on a 3ϕ induction motor. In order to achieve synchronous speed, the rotor windings are supplied with a DC power source. The DC power source causes the rotor windings to operate as an electromagnet. **See Figure 22-4.**

When the magnetic field approaching the armature pole is of opposite polarity, the tendency is for the armature to rotate in a counterclockwise direction. When the field approaching the armature pole is of the same polarity, the tendency is for the armature to rotate in a clockwise direction. Because the starting cycle completes quickly, the magnetic field is not capable of overcoming the inertia of the armature and is frozen in the magnetic field.

Terms

A **synchronous motor** is a three-phase motor that runs at synchronous speed.

Tech Tip

Depending on the design, electric motors can typically be mounted in any position or at any angle, but some motors, such as drip-proof motors, are manufactured for a specific mounting position.

SYNCHRONOUS MOTORS

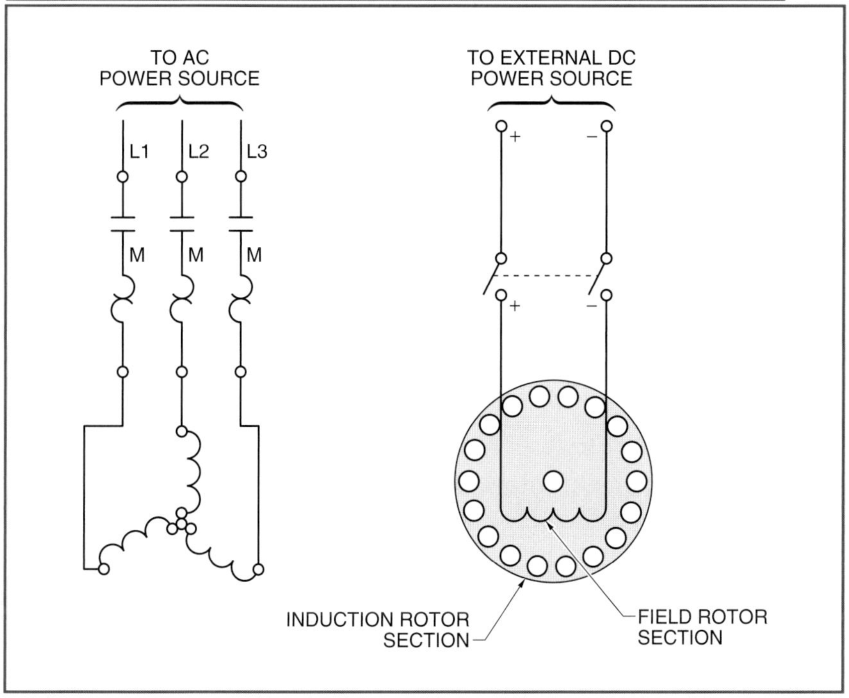

Figure 22-4. A synchronous motor is started as an induction motor, with DC power supplied to the rotor to bring the motor up to synchronous speed.

Synchronous motors are not self-starting. A small induction motor can be used to bring the rotor up to synchronous speed. This is a very common approach. Another method of starting a synchronous motor is to use a DC motor coupled to the shaft to achieve synchronous speed. At approximately 95% of synchronous speed, the DC motor is switched to operate as a DC generator to provide the DC excitation for the rotor.

Another starting method is accomplished by placing a cage rotor winding on the shaft of a synchronous motor, which makes the rotor self-starting since it is a small induction motor. When starting a small induction motor, the DC magnetizing field is removed and the stator voltage is reduced during the start mode. When the rotor is close to synchronous speed, the rotor is excited by a DC generator mounted on the shaft of the synchronous motor. Concurrently, line voltage is increased, and field rheostats are adjusted for minimum line current.

Synchronous motors can have either a leading or lagging power factor, which makes them ideal for use in power factor correction. Operating synchronous motors with a leading power factor to perform useful work eliminates the need to install power factor correction capacitors.

SINGLE-PHASE INDUCTION MOTORS

There are many types of single-phase motors. Single-phase induction motors are used for applications such as analog clocks, timers on ranges, washing machines, clothes dryers, small fans, air conditioners, furnaces, sump pumps, and computer equipment.

A 1ϕ induction motor can be compared to a transformer. The primary winding of the transformer is stationary and is comparable to the stator winding of the motor. The secondary winding of the transformer is movable and is comparable to the rotor winding of the motor. **See Figure 22-5.**

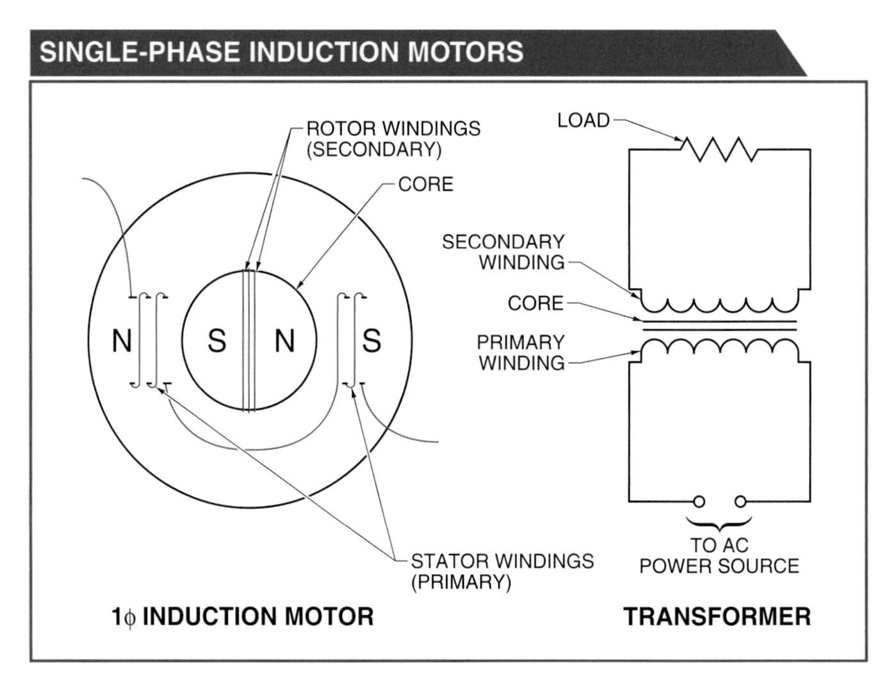

Figure 22-5. A single-phase (1ϕ) AC induction motor can be compared to a transformer with the stator winding as the primary and the rotor winding as the secondary.

The stator winding of the motor is wound on two opposite poles. The rotor winding of the motor is wound as a closed loop on a movable rotor. Magnetic polarity is based on the direction of current flow. Lenz's law states that the direction of current flow produces a magnetic pole that is opposite the polarity of the pole that produced the current. As the current polarity through the stator winding changes, its magnetic polarity also changes. The current in the rotor winding now changes direction to produce the opposite magnetic pole.

Since unlike poles attract each other, the rotor of a 1ϕ induction motor is always attracted to the stator pole closest to it. This causes the rotor to be locked in position and not rotate. A 1ϕ induction motor lacks a rotating magnetic field to cause the rotor to rotate (as in a 3ϕ motor). If the rotor shaft is given a turn, the rotor increases speed to the rated speed in the direction it is turned. That is, it rotates at synchronous speed minus the slip. As with a 3ϕ motor, if the motor were to turn at exactly synchronous speed, the magnetic field of the stator would no longer cut the rotor winding and no current would be induced.

Several methods can be used to provide the necessary starting torque for 1ϕ motors. These methods are also used to identify the motor type. Single-phase motor types include split-phase motors, capacitor motors, and shaded-pole motors.

Split-Phase Motors

A *split-phase motor* is a single-phase AC motor that includes a running winding (main winding) and a starting winding (auxiliary winding). Split-phase motors are AC motors of fractional horsepower, usually 1/20 HP to 1/3 HP. Split-phase motors are commonly used to operate washing machines, oil burners, small pumps, and blowers.

A split-phase motor has a rotating part (rotor), a stationary part consisting of the running winding and starting winding (stator), and a centrifugal switch that is located inside the motor to disconnect the starting winding at approximately 60% to 80% of full-load speed. **See Figure 22-6.** A *centrifugal switch* is a switch that opens to disconnect the starting winding when the rotor reaches a preset speed and reconnects the starting winding when the speed falls below the preset value.

The starting winding has fewer turns and smaller conductors than the running winding. Thus, the starting winding has a lower inductance and a higher resistance than the running winding. The starting winding and running winding are connected in parallel with each other. Because these windings are connected in parallel, voltage is used as the reference point for current flow through both sets of windings.

In any inductive circuit, current lags voltage. Current through the starting winding (starting current) is smaller than the current through the running winding (running current). Running current is more inductive and lags the source voltage by a greater amount than in the starting windings. The difference in inductance to resistance (L/R) ratios between the windings creates a phase shift of less than 30° between the two windings, which creates a moving magnetic field that starts the motor.

Tech Tip

Line voltage applied to the load terminals of an electric motor drive for an AC motor severely damages the drive.

Terms

A **split-phase motor** is a single-phase AC motor that includes a running winding (main winding) and a starting winding (auxiliary winding).

A **centrifugal switch** is a switch that opens to disconnect the starting winding when the rotor reaches a preset speed and reconnects the starting winding when the speed falls below the preset value.

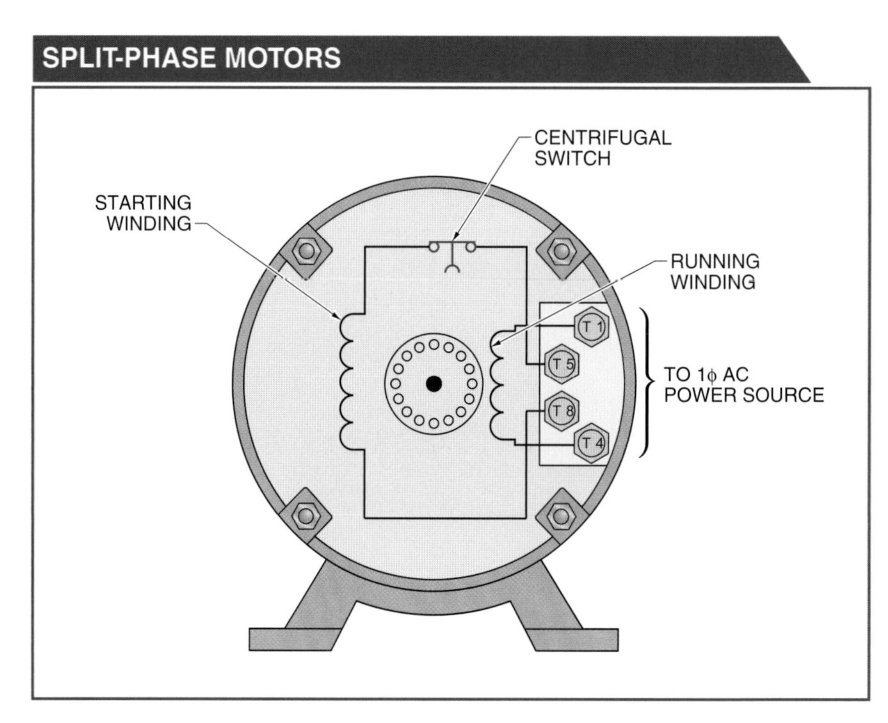

Figure 22-6. A split-phase motor includes a running winding, starting winding, and a centrifugal switch that removes the starting winding after the motor has reached 60% to 80% of the full-load speed.

Due to the physical displacement and the out-of-phase currents, the starting and running windings set up a rotating magnetic field at synchronous speed. The rotating magnetic field cuts across the rotor conductors and induces a voltage in them. Rotor current initially lags the induced voltage by 90° due to the high reactance of the rotor. Interaction of the rotor's field and the stator's field causes the rotor to accelerate in the direction in which the stator field is rotating. The starting torque is in the range of 150% to 200% of the running torque. Starting current is in the range of six to eight times that of the running current.

When a rotor rotates at about 75% of synchronous speed, the centrifugal switch opens, disconnecting the starting windings from the circuit. If the centrifugal switch fails to open, the starting windings will overheat, and in time, the motor will be destroyed. The rotating magnetic field is maintained by the interaction of the rotor and stator's magnetomotive force.

The rotor voltage causes a rotor current with an angle theta whose tangent is the ratio of the rotor's reactance to the rotor's resistance (L/R). The angle is small because the slip angle is small. However, the angle is sufficient to keep the rotor turning in the rotating magnetic field.

The direction of rotation of the rotor on a split-phase motor can be changed. To reverse the rotation of the rotor, the leads on either the running winding or starting winding must be interchanged. Usually, the leads on the starting winding are interchanged. Changing the leads on either the starting or running winding results in the same change in rotation. Normally, motors are connected to run in a counterclockwise direction when the shaft is viewed from the shaft end of the motor.

Split-phase motors are also designed to operate with two separate voltages. To operate with two separate voltages, split-phase motors have two sets of running windings and a single starting winding designed to take the higher voltage for a short period of time. Running windings are connected in series to operate on higher voltage, and are connected in parallel to operate on lower voltage.

AC Motor Maintenance

To perform AC motor maintenance as safely as possible, the information given on the motor nameplate should be checked to ensure that the proper voltage and current are being used before putting a motor into operation. When replacing a motor, it is important to ensure that the motor enclosure meets the proper specifications. The frame of a motor should be grounded, especially if the motor will be used in a damp location. If a motor shaft does not rotate after the switch has been turned ON, the motor should be turned OFF immediately to prevent the windings from becoming seriously overheated. Keeping the air openings on the motor frame clear at all times helps to prevent the motor from becoming overheated. When oiling a motor, oil should be applied to the bearings only. Excessive oil on a motor damages the winding insulation and causes the motor to accumulate dirt and dust.

Capacitor Motors

A *capacitor motor* is a single-phase AC motor that includes a capacitor in addition to the running and starting windings. Capacitor motor sizes range from ⅛ HP to 10 HP. Capacitor motors are used to operate refrigerators, compressors, washing machines, and air conditioners. The construction of a capacitor motor is similar to that of a split-phase motor, except that in a capacitor motor a capacitor is connected in series with the starting winding. The addition of a capacitor in the starting winding gives a capacitor motor more torque than a split-phase motor, because phase angle between the run and start fields is increased. The three types of capacitor motors are capacitor-start, capacitor-run, and capacitor-start-and-run motors.

Terms

A **capacitor motor** is a single-phase AC motor that includes a capacitor in addition to the running and starting windings.

A capacitor-start motor operates similarly to a split-phase motor, in that it uses a centrifugal switch that opens at approximately 60% to 80% of full-load speed. **See Figure 22-7.** In a capacitor-start motor, the starting winding and the capacitor are removed when the centrifugal switch opens. The capacitor used in the starting winding gives a capacitor-start motor high starting torque. Capacitance values in a capacitor-start motor range from 40 μF to 1200 μF.

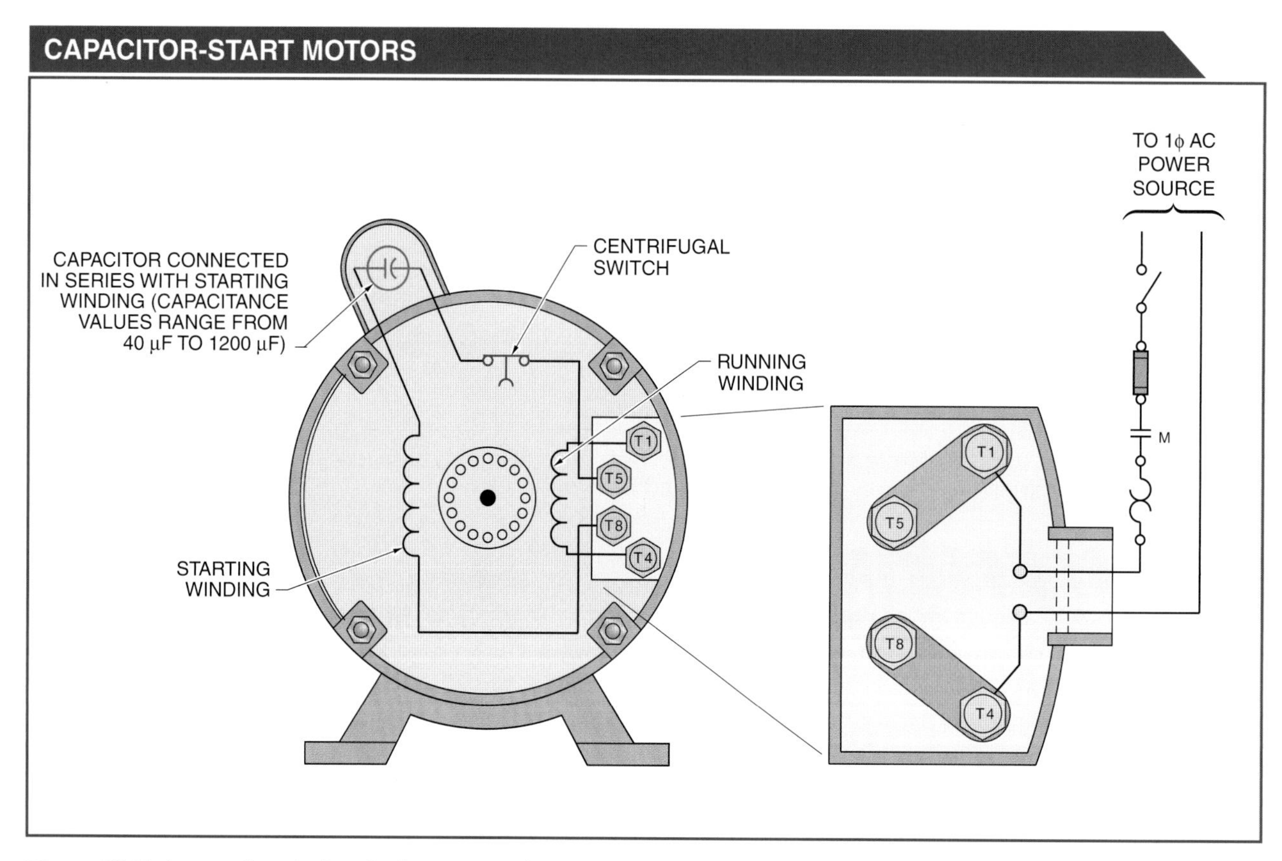

Figure 22-7. A capacitor-start motor has a capacitor connected in series with the starting winding, which gives the motor a high starting torque.

A centrifugal switch is not required in a capacitor-run motor because a capacitor-run motor has the starting winding and capacitor connected in series at all times. A low-capacitance capacitor is used in a capacitor-run motor because a low-capacitance capacitor remains in the circuit at full-load speed. A low-capacitance capacitor gives a capacitor-run motor medium starting torque and higher running torque than a capacitor-start motor. **See Figure 22-8.** Capacitance values in a capacitor-run motor range from 2 μF to 60 μF.

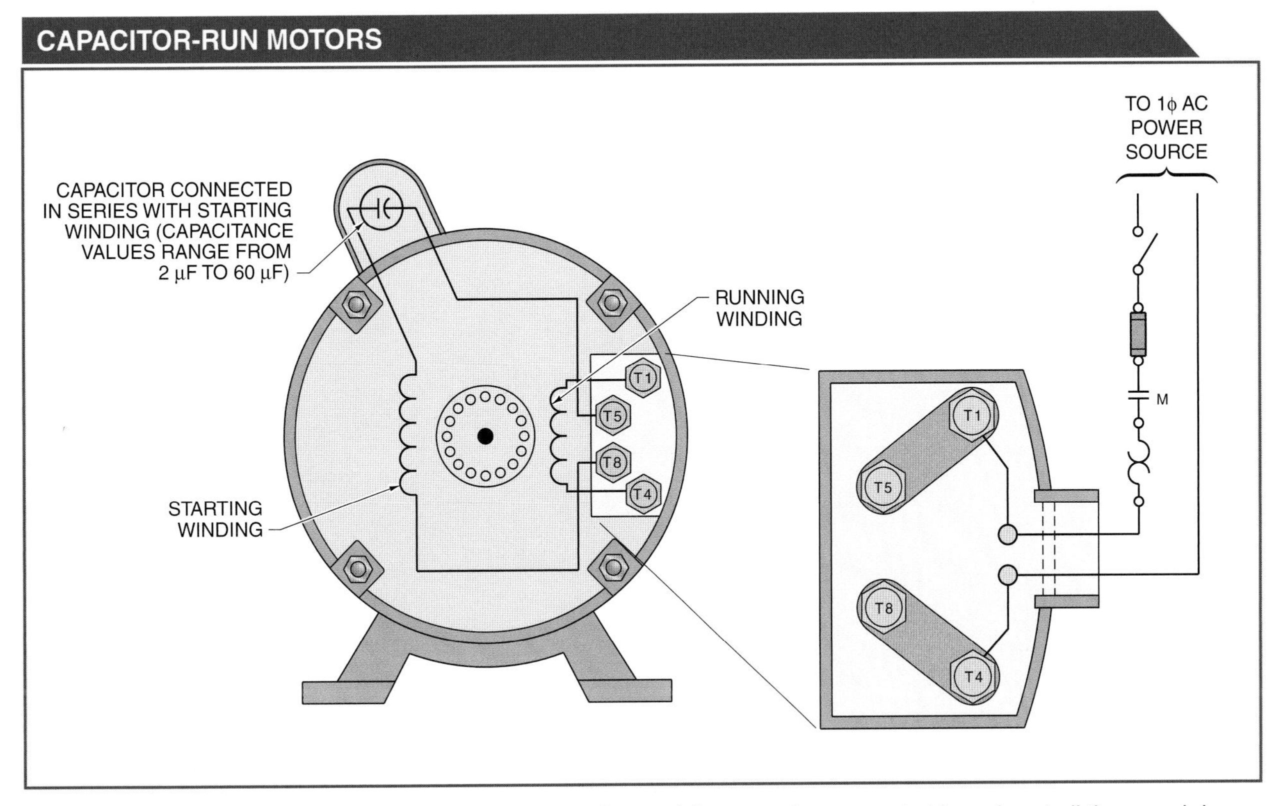

Figure 22-8. A capacitor-run motor has the starting winding and the capacitor connected in series at all times and does not require a centrifugal switch.

A capacitor-start-and-run motor (dual-capacitor motor) uses two capacitors. A capacitor-start-and-run motor starts with a capacitor connected in series with the starting winding and runs with a capacitor of a different value connected in series with the starting winding. **See Figure 22-9.**

A capacitor-start-and-run motor has the same starting torque as a capacitor-start motor. A capacitor-start-and-run motor has more running torque than a capacitor-start motor or capacitor-run motor because the capacitance is better matched for starting and running.

Baldor Electric Co.

A capacitor-start motor has a capacitor in the starting winding, which gives the motor a high starting torque.

In a typical capacitor-start-and-run motor, one capacitor is used for starting the motor and the other capacitor remains in circuit while the motor is running. A high-capacitance capacitor is used for starting, and a low-capacitance capacitor is used for running. Capacitor-start-and-run motors are commonly used to run compressors.

CAPACITOR START-AND-RUN MOTORS

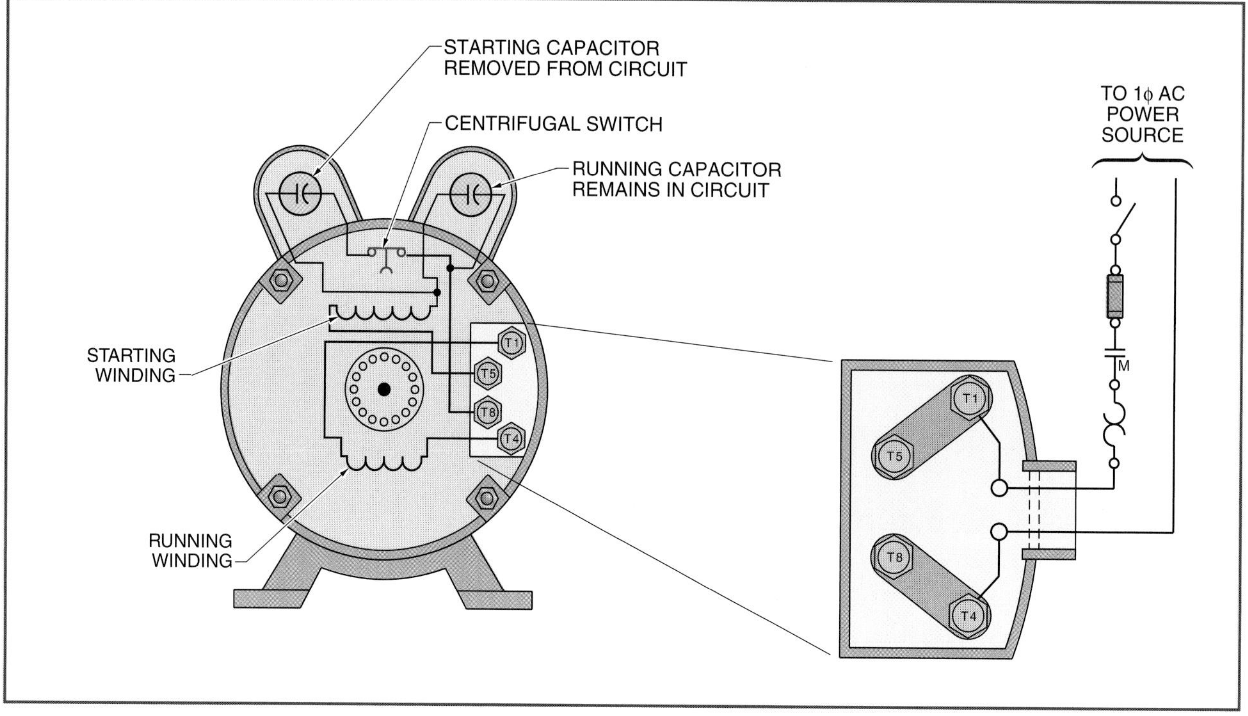

STARTING CAPACITOR REMOVED FROM CIRCUIT
CENTRIFUGAL SWITCH
RUNNING CAPACITOR REMAINS IN CIRCUIT
TO 1ϕ AC POWER SOURCE
STARTING WINDING
RUNNING WINDING
T1
T5
T8
T4
M

Figure 22-9. In a capacitor-start-and-run motor, the starting capacitor is removed when the motor reaches full-load speed, but the running capacitor remains in the circuit.

Shaded-Pole Motors

Terms

A **shaded-pole motor** is a single-phase AC motor that uses a shaded stator pole for starting.

A **shaded pole** is a single pole, normally made of a single turn of heavy-gauge copper wire, that is used for starting a single-phase motor and is added to a slot cut into the motor pole.

A **salient pole** is a pole that consists of a separate radial projection having its own iron pole and field coil.

A *shaded-pole motor* is a single-phase AC motor that uses a shaded stator pole for starting. A *shaded pole* is a single pole, normally made of a single turn of heavy-gauge copper wire, that is used for starting a single-phase motor and is added to a slot cut into the motor pole. Shaded-pole motors are usually very small machines, manufactured in sizes of $\frac{1}{20}$ HP or less. The starting torque is very small. Shaded-pole motors are used in such light-duty applications as analog clocks and small cooling fans located in computers and home entertainment systems. Shaded-pole motor operation is similar to split-phase motor operation. Simplicity of manufacture, high reliability, and low cost make shaded-pole motors ideal for light-duty applications.

Shaded-pole motors have a salient-pole stator and a cage rotor. A *salient pole* is a pole that consists of a separate radial projection having its own iron pole and field coil. The iron poles resemble those of a DC motor except the poles are of laminate (layer-type) construction. In addition, there is a slot in each iron pole to allow a short-circuited copper wire (shaded pole) to be mounted. **See Figure 22-10.** Shaded-pole motors are typically designed with either two salient poles (two-pole motors) or four salient poles (four-pole motors).

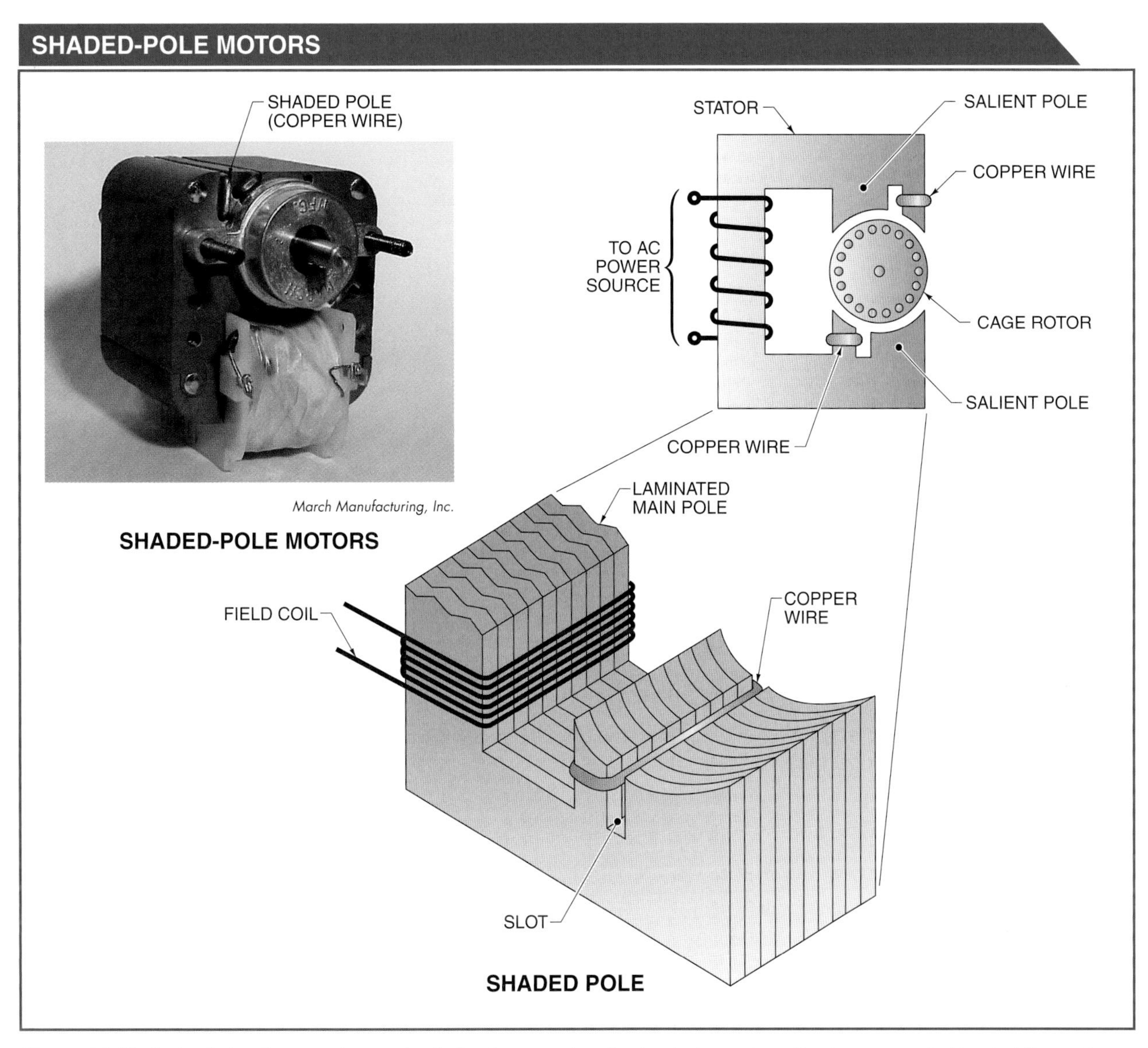

Figure 22-10. A shaded-pole motor has a shaded pole on each salient pole to produce the torque needed to start the motor.

The shaded pole opposes the magnetic field produced by the stator, which creates the single-phase rotating magnetic field. **See Figure 22-11.** Shading causes the magnetic field at the pole area to be positioned approximately 90° from the magnetic field of the main stator pole. As the field winding current progresses through the sine wave, it creates a magnetic field that cuts the shaded pole, causing a current flow. The resulting magnetic field opposes the magnetic field of the running winding, causing the combined magnetic fields to move toward the shaded pole. The offset magnetic field causes the rotor to rotate from the main pole toward the shaded pole. The rotor rotation determines the starting direction of a shaded-pole motor.

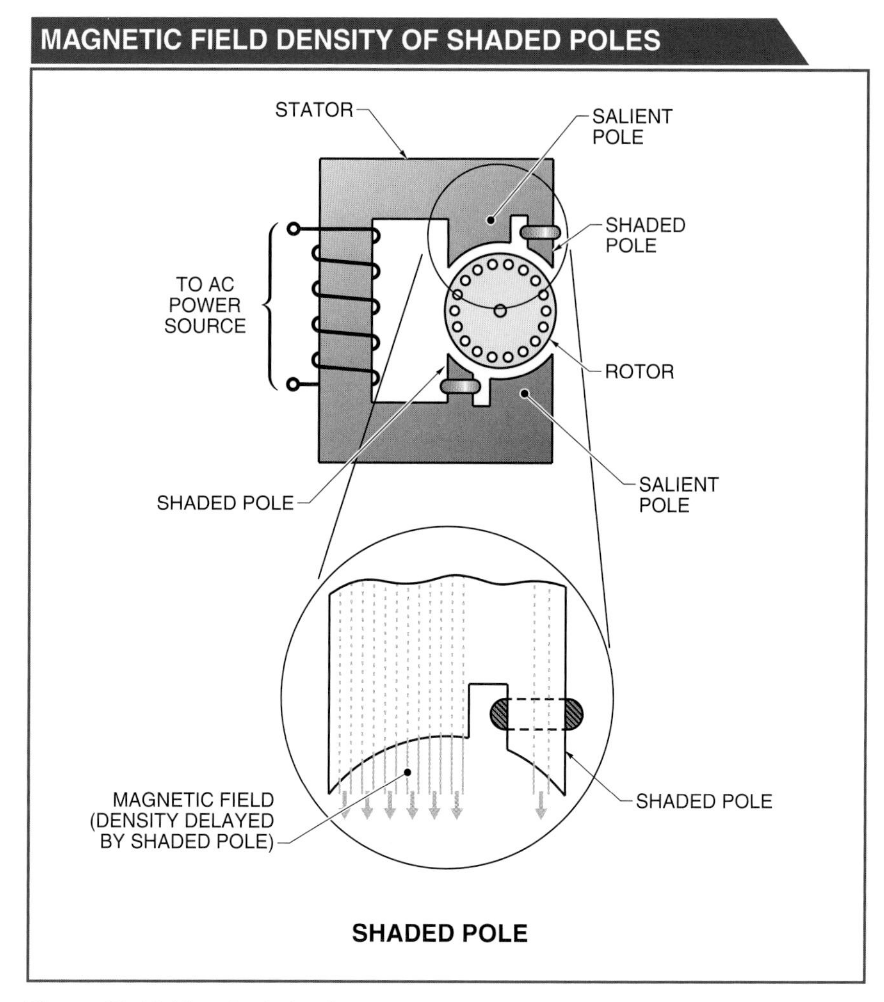

Figure 22-11. The shaded pole opposes the magnetic field produced by the stator, which creates the single-phase rotating magnetic field.

REPULSION MOTORS

Terms

A **repulsion motor** is a motor with the rotor connected to the power supply through brushes that ride on a commutator.

A **hard neutral position** is a position where brushes are aligned directly with stator poles.

A *repulsion motor* is a motor with the rotor connected to the power supply through brushes that ride on a commutator. About 5% of 3ϕ motors used in industry are repulsion motors. A repulsion motor is similar to a DC motor in that it has a commutator and brushes. Unlike a DC motor, however, where the brushes are connected to the electrical power source, the brushes on a repulsion motor are shorted together. The location of the brushes determines the direction of current flow in the armature windings. **See Figure 22-12.**

A *hard neutral position* is a position where brushes are aligned directly with stator poles. Current is induced in the armature windings by the changing stator field and produces like poles between them. As with an induction motor, a rotating magnetic field is not present and the rotor does not rotate from the hard neutral position.

REPULSION MOTORS

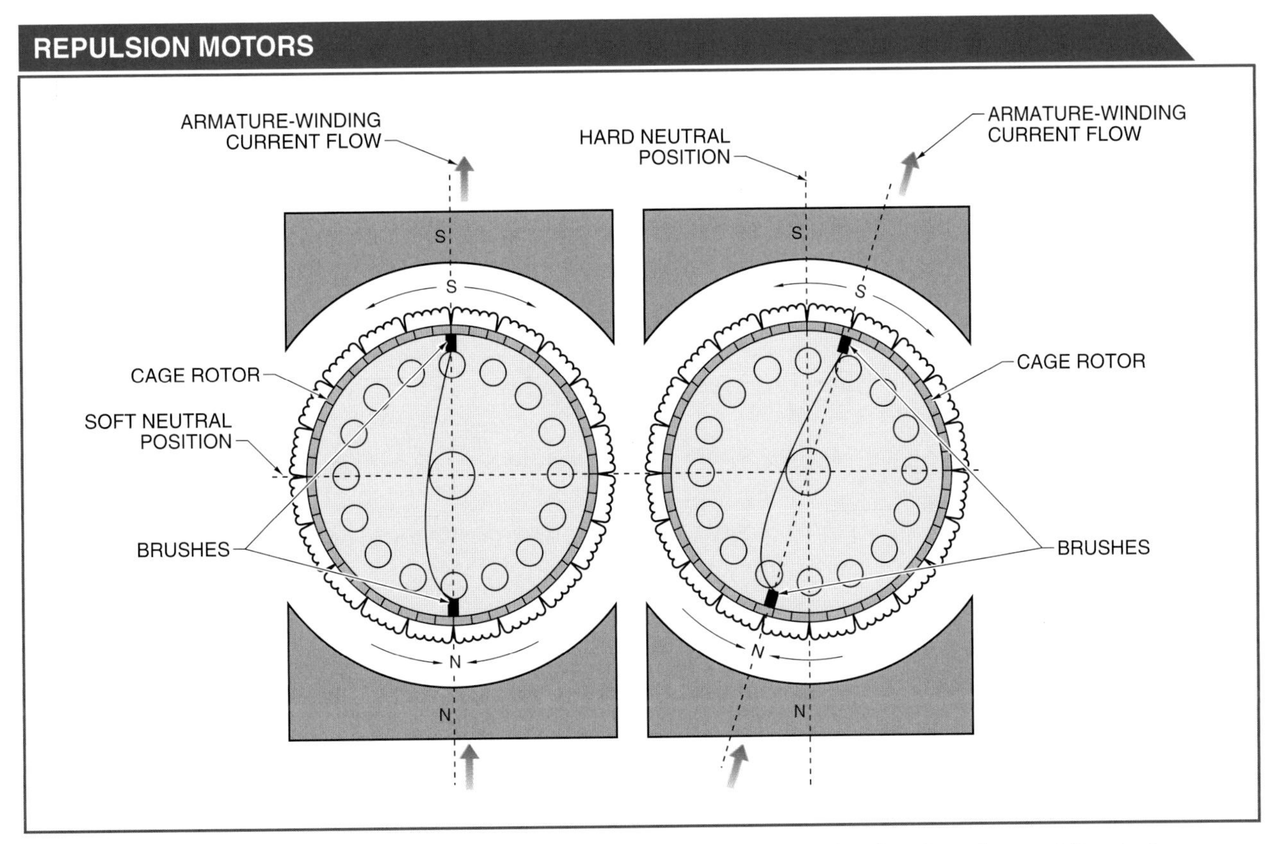

Figure 22-12. In a repulsion motor, the location of the brushes determines the direction of current flow in the armature windings.

If the brushes shift about 17° from the hard neutral position, unlike poles of the stator and rotor repel each other, and the rotor rotates due to the repulsion between the magnetic fields. The rotor accelerates and, when it reaches about 75% of its synchronous speed, a centrifugal device lifts the brushes off the commutator, shorting the commutator rings. A cage rotor embedded in the armature keeps the motor running as an induction motor. The cage rotor rotates in the direction of the brush shift. If the brush is shifted counterclockwise from hard neutral position, the rotor also rotates counterclockwise.

A *soft neutral position* is a position where the brushes are aligned 90° from the stator poles. A soft neutral position exists at 90° from the hard neutral position. As with the hard neutral position, if the brushes are set in the soft neutral position, the rotor does not rotate. The motor will usually start from the soft neutral position by shifting the brush. However, starting torque is greatly reduced since the repelling magnetic fields are further apart. When the motor starts from soft neutral position, the rotor rotates in the opposite direction of the brush shift. Repulsion motors should always be started from the hard neutral position.

Terms

A **soft neutral position** is a position where the brushes are aligned 90° from the stator poles.

Although a repulsion motor has a high starting torque, it does not readily increase speed when started under a load. Due to the use of a commutator, brushes, and centrifugal mechanism to short-circuit the commutator and lift the brushes, repulsion motors are more expensive and require greater maintenance than other types of motors. Because of the higher cost and greater amount of maintenance for repulsion motors, capacitor-start motors are typically chosen over repulsion motors when high starting torque is required.

AC SERIES MOTORS

AC series motors are similar to DC series motors and will operate on either type of current. The direction of rotation of a DC series motor is independent of the polarity of the power source. For a DC series motor to operate on AC, certain modifications are required, including the following:

- The field poles, yoke, and armature are laminated to reduce eddy currents produced by AC.
- High-permeability silicon steel is used for the laminated parts to reduce hysteresis loss caused by AC.
- Shallow pole pieces with few conductor turns, low flux density, and a narrow air gap are used to minimize the reactance of field windings. For large motors, a low frequency (25 Hz) is used to further reduce the reactance of field windings.

A *universal motor* is an AC series motor that has brushes and a wound armature. Universal motors have construction similar to that of DC motors. *Armature reactance* is inductance of the armature. Armature reactance is higher for an AC series motor than for a DC series motor due to the large number of conductors, and is reduced by the use of a compensation winding. A *compensation winding* is an inductor used to reduce armature reactance effects.

Terms

A **universal motor** is an AC series motor that has brushes and a wound armature.

Armature reactance is inductance of the armature.

A **compensation winding** is an inductor used to reduce armature reactance effects.

To reduce armature reactance, a compensation winding may be connected in series with the running winding (conductive method) or may be short-circuited away from the running winding (inductive method). **See Figure 22-13.** AC motors are connected with the conductive method. The inductive method is used with DC motors only. The axis of the compensating winding is displaced from the main magnetic field by 90°.

Unless a load is maintained on the motor, the motor speed will increase to a dangerously high speed. Motors can reach speeds that are high enough to cause solder and segments of the armature to break away and destroy the motor. To prevent excessive motor speed, the load usually is directly coupled to the shaft of the motor.

Tech Tip

Electric motors powered and controlled by AC drives should be inverter rated.

Fractional-horsepower AC series motors can operate on AC at any power frequency, or on DC. Fractional-horsepower motors are small and do not have compensating windings. Fractional-horsepower AC series motors are used for applications such as fans and cordless power tools. Fractional-horsepower motor outputs are ¼ HP, ⅓ HP, ½ HP, ¾ HP, 1 HP, 1½ HP, and 2 HP, and operate at speeds ranging from 3000 rpm to 15,000 rpm.

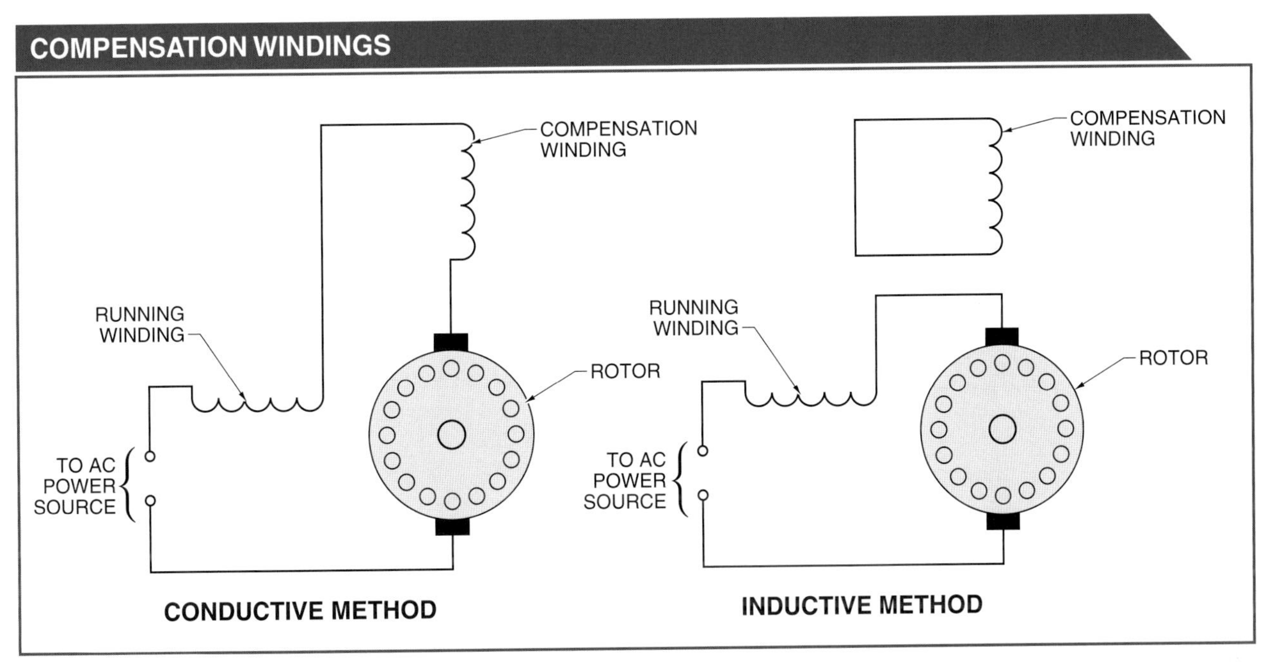

Figure 22-13. In an AC motor, a compensation winding may be connected in series with the running winding (conductive method) or short-circuited away from the running winding (inductive method).

SUMMARY

AC motors are used to convert electrical energy into mechanical energy for many different applications, ranging from handheld appliances to large industrial applications. AC motors range in size from fractional horsepower to hundreds of horsepower.

Three-phase induction motors are typically used when multiple AC phases are available. When provided with 3ϕ power, a 3ϕ induction motor turns its rated load at a constant speed. Three-phase motors are less complicated than 1ϕ motors, making them less expensive to manufacture and maintain, due to the use of fewer components.

AC motors may use different types of rotors, such as wound rotors or cage rotors. Wound-rotor motors have modest starting current but high starting torque, have smooth acceleration under heavy loads, and have adjustable speed. Cage-rotor motors are less expensive to purchase and have lower maintenance costs because no brushes or slip rings are required.

Three-phase synchronous motors are used when loads require a constant speed. These motors have a high efficiency when converting electrical energy into mechanical power. Three-phase synchronous motors have DC power supplied to the rotor windings and may have leading or lagging power factor, which makes them ideal for use in power factor correction applications.

When multiple AC phases are not available, 1ϕ induction motors are used. A 1ϕ induction motor is similar to a transformer, where the stator is comparable to the primary winding and the rotor is comparable to the secondary winding. The stator is wound with two poles on opposite sides of the rotor. The rotor winding is wound as a closed loop. Single-phase motors need to create a rotating magnetic field for starting.

AC series motors, which are also known as universal motors, can be used with either AC or DC systems. AC series motors require different designs than DC series motors because of losses that occur in AC systems, such as eddy current loss, hysteresis loss, and armature reactance.

APPLICATION: POWER FACTOR CORRECTION...

Background: When a load is not purely resistive, current and voltage are out of phase with each other. With inductive loads, such as AC motors, the current lags the voltage. In this case, the apparent power being supplied to the load is greater than the working power consumed by the load. The lower the power factor, the lower the system efficiency, and the higher the overall costs. Industry power factor examples include 45% to 60% for saw mills; 60% to 70% for machine tool/ stamping equipment; 65% to 75% for plating, textile, and chemical plants as well as breweries; and 70% to 80% for hospitals, granaries, and foundries.

Power factor is the percentage ratio of working power to apparent power. The power factor formula is the following:

$$\frac{PF = P_T}{P_A = \cos\theta}$$

where

PF = power factor (between 0 and 1)
P_T = true power (in kW)
P_A = apparent power (in kVA)
$\cos\theta$ = power triangle θ

For example, a 100 kW motor with a power factor of 75% requires 133.33 kVA, while a similar system with a 95% power factor only requires 105.26 kVA for the same amount of work.

The power triangle shows the relationship between true power, apparent power, and reactive power. **See Figure 1.** Reactive power is the power needed to sustain the inductive magnetic field, which does not perform any useful work. Only true power performs useful work.

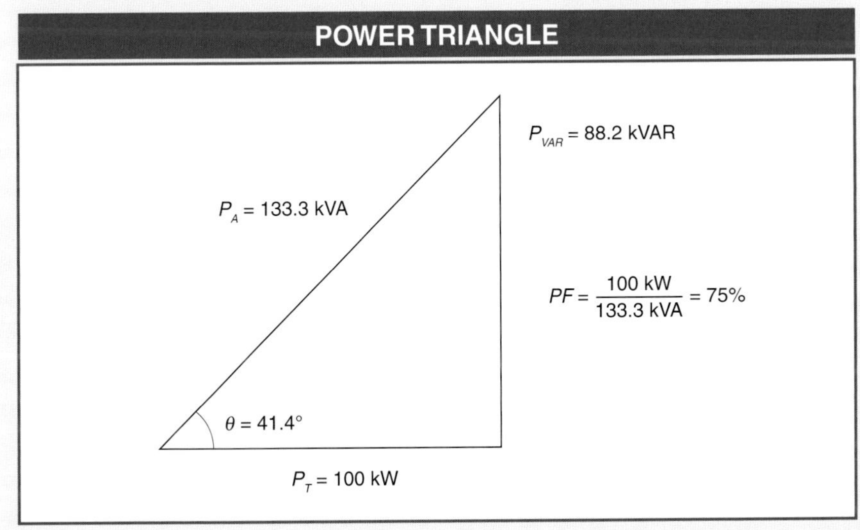

Figure 1. The power triangle shows the relationship between reactive power, true power, and apparent power.

...APPLICATION: POWER FACTOR CORRECTION

A parallel capacitor is used to correct an inductive load with a poor power factor. The size of the capacitor is selected for the desired power factor correction. Usually, a power factor of 95% is desired. To find the required size of the power correction capacitor, the following formula is used:

$C_{RP} = (\tan \theta_1 - \tan \theta_2) \times P_T$

where

C_{RP} = capacitor reactive power (in CkVAR)

θ_1 = uncorrected power factor (in °)

θ_2 = corrected power factor (in °)

P_T = true power (in kW)

Key Points: power factor

Problem: A 100-HP AC motor is being used in an industrial application. The uncorrected power factor is 60%. What size capacitor is required for a corrected power factor of 95%?

Solution: The value of the true power is needed. To find the system true power, the relationship between horsepower and watts is needed. One horsepower is equal to 746 W. Therefore, the system true power is the following:

$$P_T = \frac{\text{HP} \times 746\text{ W}}{\text{HP}}$$

$$P_T = \frac{100\text{ HP} \times 746\text{ W}}{\text{HP}}$$

$$P_T = 74.6\text{ kW}$$

The power factor angle θ is found using the power factor formula as follows:

$$\frac{PF = kW}{kVA = \cos\theta}$$

$$\theta = \cos^{-1}(PF)$$

$$\theta_1 = \cos^{-1}(60\%)$$

$$\theta_1 = 53.13^\circ$$

$$\theta_2 = \cos^{-1}(95\%)$$

$$\theta_2 = 18.19^\circ$$

The required size of the power correction capacitor is the following:

$C_{RP} = (\tan \theta_1 - \tan \theta_2) \times P_T$

$C_{RP} = (\tan 53.13^\circ - \tan 18.19^\circ) \times 74.6\text{ kW}$

$C_{RP} = 75\text{ kVAR}$

Power correction capacitors are available in standard sizes and are sized for the individual motor loads, motor frame type, revolutions per minute, and horsepower requirement. Many suppliers have tables that indicate the capacitor-reactive-power rating needed for a power factor of 95%.

Chapter Review

1. Describe a shaded pole.
2. List three advantages of a wound-rotor induction motor over a cage motor.
3. Describe a split-phase motor.
4. List the modifications that are required to allow a DC motor to operate on AC power.
5. List six common applications in which single-phase induction motors are used.
6. List two common applications for shaded-pole motors.
7. Describe a capacitor motor.
8. What is slip?
9. List three common applications for split-phase motors.
10. List the two types of rotors used on AC motors.
11. List three common applications in which capacitor motors are installed.
12. What is synchronous speed?
13. Describe a repulsion motor.
14. Why are capacitor-start motors typically chosen over repulsion motors when high starting torque is required?
15. List two disadvantages of wound-rotor induction motors compared to cage motors.
16. What is the most common application in which capacitor-start-and-run motors are used?
17. Describe the difference between a hard neutral and soft neutral position in a repulsion motor.
18. What causes torque to increase in a three-phase motor?
19. Why are synchronous motors ideal for power factor correction?
20. Why is it unnecessary to insulate the bars from the core on a cage rotor?

Appendix

ANNEALED SOLID COPPER WIRE CHARACTERISTICS						
AWG Gauge	Diameter (mils)	Cross-Sectional Area		Resistance*		Weight†
		Circular mils	Square inches	25°C	65°C	
0000	460.0	212,000.0	0.166	0.0500	0.0577	641.0
000	410.0	168,000.0	0.132	0.0630	0.0727	508.0
00	365.0	133,000.0	0.105	0.0795	0.0917	403.0
0	325.0	106,000.0	0.0829	0.100	0.116	319.0
1	289.0	83,700.0	0.0657	0.126	0.146	253.0
2	258.0	66,400.0	0.0521	0.159	0.184	201.0
3	229.0	52,600.0	0.0413	0.201	0.232	159.0
4	204.0	41,700.0	0.0328	0.253	0.292	126.0
5	182.0	33,100.0	0.0260	0.319	0.369	100.0
6	162.0	26,300.0	0.0206	0.403	0.465	79.5
7	144.0	20,800.0	0.0164	0.508	0.586	63.0
8	128.0	16,500.0	0.0130	0.641	0.739	50.0
9	114.0	13,100.0	0.0103	0.808	0.932	39.6
10	102.0	10,400.0	0.00815	1.02	1.18	31.4
11	91.0	8230.0	0.00647	1.28	1.48	24.9
12	81.0	6530.0	0.00513	1.62	1.87	19.8
13	72.0	5180.0	0.00407	2.04	2.36	15.7
14	64.0	4110.0	0.00323	2.58	2.97	12.4
15	57.0	3260.0	0.00256	3.25	3.75	9.86
16	51.0	2580.0	0.00203	4.09	4.73	7.82
17	45.0	2050.0	0.00161	5.16	5.96	6.20
18	40.0	1620.0	0.00128	6.51	7.51	4.92
19	36.0	1290.0	0.00101	8.21	9.48	3.90
20	32.0	1020.0	0.000802	10.4	11.9	3.09
21	28.5	810.0	0.000636	13.1	15.1	2.45
22	25.3	642.0	0.000505	16.5	19.0	1.94
23	22.6	509.0	0.000400	20.8	24.0	1.54
24	20.1	404.0	0.000317	26.2	30.2	1.22
25	17.9	320.0	0.000252	33.0	38.1	0.970
26	15.9	254.0	0.000200	41.6	48.0	0.769
27	14.2	202.0	0.000158	52.5	60.6	0.610
28	12.6	160.0	0.000126	66.2	76.4	0.484
29	11.3	127.0	0.0000995	83.4	96.3	0.384
30	10.0	101.0	0.0000789	105.0	121.0	0.304
31	8.9	79.7	0.0000626	133.0	153.0	0.241
32	8.0	63.2	0.0000496	167.0	193.0	0.191
33	7.1	50.1	0.0000394	211.0	243.0	0.152
34	6.3	39.8	0.0000312	266.0	307.0	0.120
35	5.6	31.5	0.0000248	335.0	387.0	0.0954
36	5.0	25.0	0.0000196	423.0	488.0	0.0757
37	4.5	19.8	0.0000156	533.0	616.0	0.0600
38	4.0	15.7	0.0000123	673.0	776.0	0.0476
39	3.5	12.5	0.0000098	848.0	979.0	0.0377
40	3.1	9.9	0.0000078	1230.0	1420.0	0.0299

* in Ω/1000′
† in lb/1000′

PERMEABILITY OF MAGNETIC SUBSTANCES

Type of Substance	Maximum Permeability*	Saturation Flux Density B Gausses
Cobalt	170	3000
Iron-Cobalt Alloy (Co 34%)	13000	8000
Heusler Alloy (Cu 60%, Mn 24%, Al 26%)	200	2000
Iron, purest commercial annealed	6000-8000	6000
Nickel	400-1000	1000-3000
Permalloy (Ni 45%, FE 21.5%)	>80000	5000
Pewminvar (Ni 45%, FE 30%, Co 25%)	2000	4
Silicon steel (Si 4%)	5000-10000	6000-8000
Steel, cast	1500	7000
Steel, open-hearth	3000-7000	6000
Cold Rolled Steel	2000	21,000
Copper	0.999991	–
Vacuum	1	–
Air	1.0000004	–

*in A at ambient temperatures below 30°C

MAGNETIZING FORCE (H) AND PERMEABILITY (μ) OF IRON AND STEEL

	Soft Iron or Annealed Steel		Soft Cast Steel		Cast Iron	
Φ/cm^2	$H = B/\mu$	μ	$H = B/\mu$	μ	$H = B/\mu$	μ
3100	0.49	6260	3.25	955	8.38	370
4650	0.74	6260	4.77	978	16.03	290
6200	0.99	6260	6.36	978	31.00	300
7750	1.26	6130	7.99	970	62.00	125
9300	1.64	5640	10.111	920	103.33	90
10850	2.43	4470	13.07	830	–	–
12400	3.96	3130	17.34	715	–	–
14000	7.60	1840	26.32	535	–	–
15500	19.14	810	46.97	330	–	–
17100	63.33	270	90.00	190	–	–

COIL INSULATION TYPE					
Gauge	Number of Turns per Linear Inch				Feet/Ohm
	Enamel	Single Silk Covered	DSC*/SCC†	Double Cotton Covered	
10	9.6	–	9.3	8.9	1001.0
12	10.7	–	11.5	10.9	629.6
14	15.0	–	14.2	13.8	396.0
16	18.9	18.9	17.9	16.4	249.0
18	23.6	23.6	22.0	19.8	156.5
20	29.4	29.4	27.0	23.8	98.5
22	37.0	36.5	34.1	30.0	61.95
24	46.3	45.3	41.5	35.6	38.96
26	58.0	55.6	50.2	41.8	24.50
28	72.7	68.6	60.2	48.5	15.41
30	90.5	83.3	71.5	55.5	9.691
32	113.0	1201.0	83.6	62.6	6.095
34	143.0	120.0	97.0	70.0	3.833
36	175.0	143.0	111.0	77.0	2.411
38	224.0	166.0	126.0	83.6	1.516
40	282.0	181.0	140.0	89.7	0.9534

* DSC = Double Silk Covered
† SCC = Single Cotton Covered

OHM'S LAW

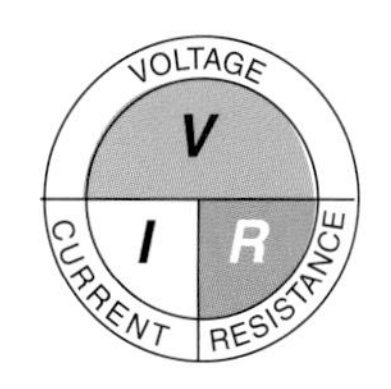

V = VOLTAGE (IN V)
I = CURRENT (IN A)
R = RESISTANCE (IN Ω)

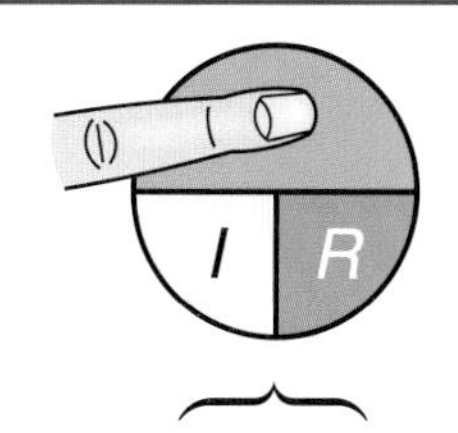

$V = I \times R$

VOLTAGE = CURRENT × RESISTANCE

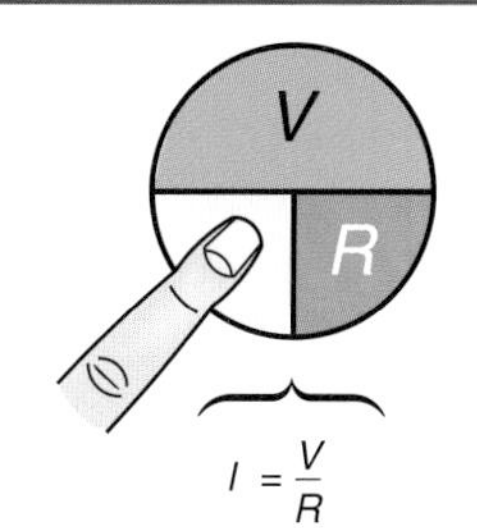

$I = \frac{V}{R}$

$\text{CURRENT} = \frac{\text{VOLTAGE}}{\text{RESISTANCE}}$

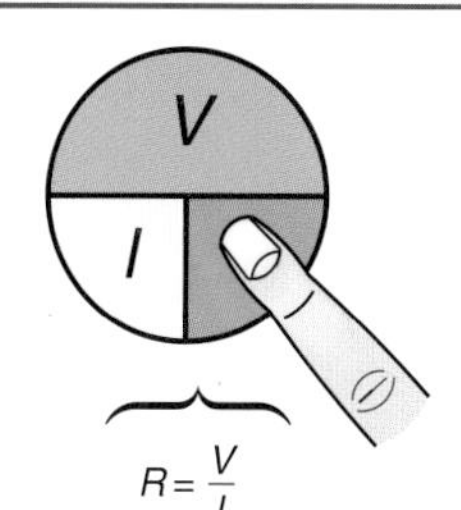

$R = \frac{V}{I}$

$\text{RESISTANCE} = \frac{\text{VOLTAGE}}{\text{CURRENT}}$

POWER FORMULA

P = POWER (IN W)
V = VOLTAGE (IN V)
I = CURRENT (IN A)

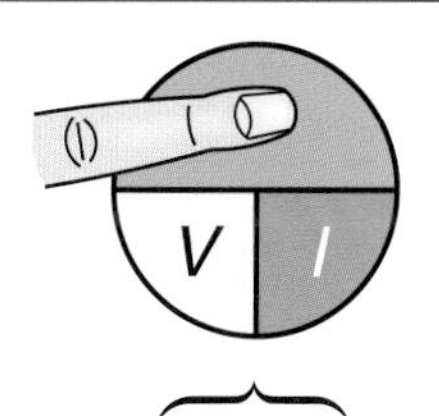

$P = V \times I$

POWER = VOLTAGE × CURRENT

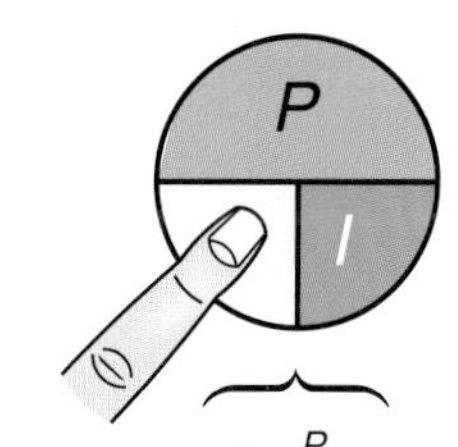

$V = \frac{P}{I}$

$\text{VOLTAGE} = \frac{\text{POWER}}{\text{CURRENT}}$

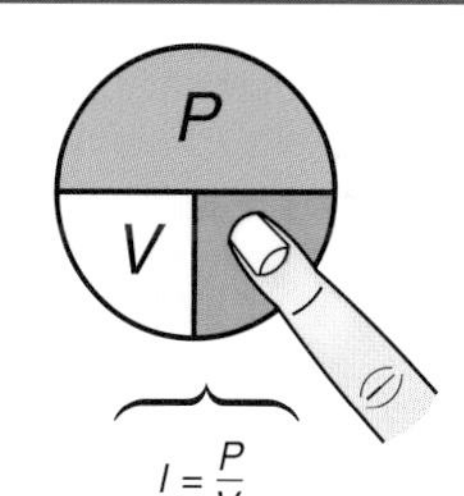

$I = \frac{P}{V}$

$\text{CURRENT} = \frac{\text{POWER}}{\text{VOLTAGE}}$

VOLTAGE, CURRENT, AND IMPEDANCE RELATIONSHIP

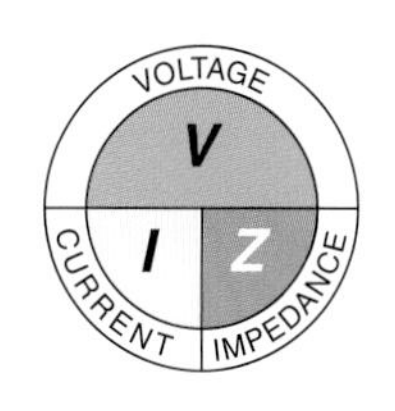

V = VOLTAGE (IN V)
I = CURRENT (IN A)
Z = IMPEDANCE (IN Ω)

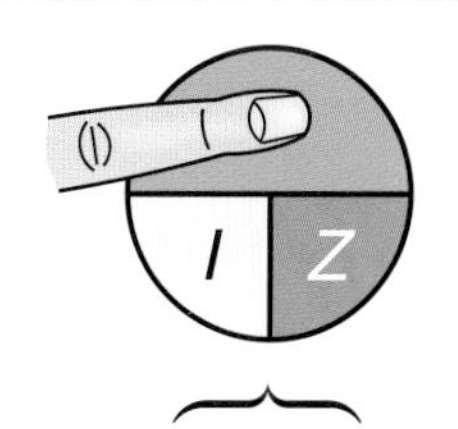

$V = I \times Z$

VOLTAGE = CURRENT × IMPEDANCE

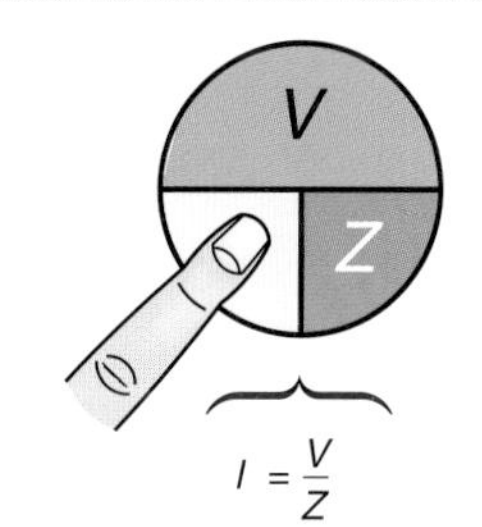

$I = \frac{V}{Z}$

$\text{CURRENT} = \frac{\text{VOLTAGE}}{\text{IMPEDANCE}}$

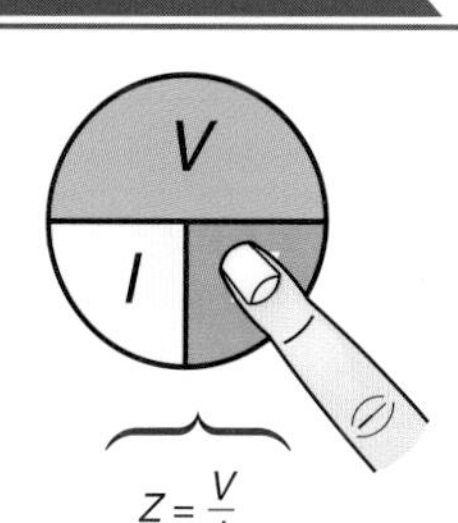

$Z = \frac{V}{I}$

$\text{IMPEDANCE} = \frac{\text{VOLTAGE}}{\text{CURRENT}}$

PARALLEL CIRCUIT CALCULATIONS

RESISTANCE

$$R_T = \frac{R_1 \times R_2}{R_1 + R_2}$$

where

R_T = total resistance (in Ω)
R_1 = resistance 1 in (in Ω)
R_2 = resistance 2 in (in Ω)

VOLTAGE

$V_T = V_1 = V_2 = \ldots$

where

V_T = total applied voltage (in V)
V_1 = voltage drop across load 1 in (in V)
V_2 = voltage drop across load 2 in (in V) V)

CURRENT

$I_T = I_1 + I_2 + I_3 + \ldots$

where

I_T = total circuit current (in A)
I_1 = current through load 1 (in A)
I_2 = current through load 2 (in A)
I_3 = current through load 3 (in A)

SERIES CIRCUIT CALCULATIONS

RESISTANCE

$R_T = R_1 + R_2 + R_3 + \ldots$

where

R_T = total resistance (in Ω)
R_1 = resistance 1 in (in Ω)
R_2 = resistance 2 in (in Ω)
R_3 = resistance 3 in (in Ω)

VOLTAGE

$V_T = V_1 + V_2 + V_3 + \ldots$

where

V_T = total applied voltage (in V)
V_1 = voltage drop across load 1 (in V)
V_2 = voltage drop across load 2 (in V)
V_3 = voltage drop across load 3 (in V)

CURRENT

$I_T = I_1 = I_2 = I_3 = \ldots$

where

I_T = total circuit current (in A)
I_1 = current through load 1 (in A)
I_2 = current through load 2 (in A)
I_3 = current through load 3 (in A)

POWER FORMULAS — 1ϕ, 3ϕ

Phase	To Find	Use Formula	Example		
			Given	Find	Solution
1ϕ	I	$I = \frac{VA}{V}$	32,000 VA, 240 V	I	$I = \frac{VA}{V}$ $I = \frac{32{,}000\ VA}{240\ V}$ $I =$ **133 A**
1ϕ	VA	$VA = I \times V$	100 A, 240 V	AV	$VA = I \times V$ $VA = 100\ A \times 240\ V$ $VA =$ **24,000 VA**
1ϕ	V	$V = \frac{VA}{I}$	42,000 VA, 350 A	V	$V = \frac{VA}{I}$ $V = \frac{42{,}000\ VA}{350\ A}$ $V =$ **120 V**
3ϕ	I	$I = \frac{VA}{V \times \sqrt{3}}$	72,000 VA, 208 V	I	$I = \frac{VA}{V \times \sqrt{3}}$ $I = \frac{72{,}000\ VA}{360\ V}$ $I =$ **200 A**
3ϕ	VA	$VA = I \times V \times \sqrt{3}$	2 A, 240 V	VA	$VA = I \times V \times \sqrt{3}$ $VA = 2 \times 416$ $VA =$ **831 VA**

MOLDED INDUCTOR COLOR CODES

Color	Significant Figure	Multiplier	Tolerance*
Black	0	1	1
Brown	1	10	2
Red	2	100	3
Orange	3	1000	–
Yellow	4	–	–
Green	5	–	–
Blue	6	–	–
Violet	7	–	–
Gray	8	–	–
White	9	–	–
None†	–	–	20
Silver	–	–	10
Gold	Decimal Point	–	5

* in %

†Body color only

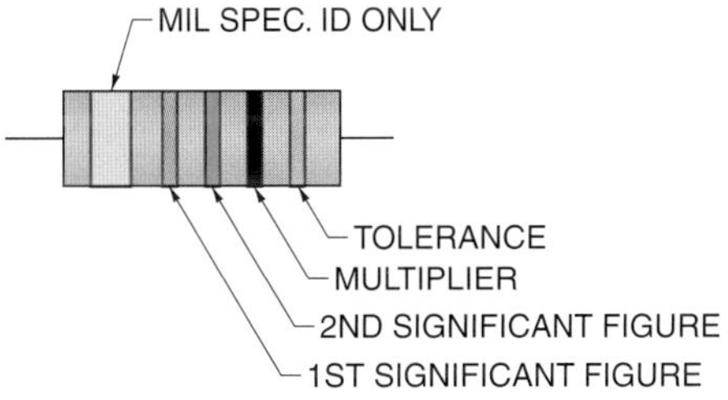

MIL SPEC. ID ONLY
TOLERANCE
MULTIPLIER
2ND SIGNIFICANT FIGURE
1ST SIGNIFICANT FIGURE

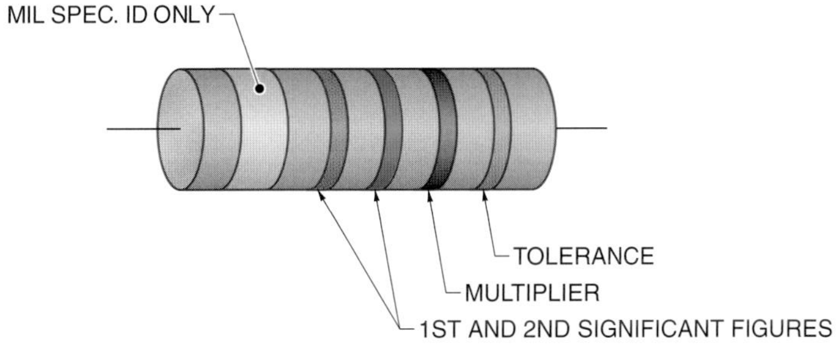

MIL SPEC. ID ONLY
TOLERANCE
MULTIPLIER
1ST AND 2ND SIGNIFICANT FIGURES

TUBULAR CAPACITOR IDENTIFICATION CODES

Color	Capacitance*			Tolerance†	Voltage Rating‡	
	1st Digit	2nd Digit	Multiplier		1st Digit	2nd Digit
Black	0	0	1	20	0	0
Brown	1	1	10	–	1	1
Red	2	2	100	–	2	2
Orange	3	3	1000	30	3	3
Yellow	4	4	10,000	40	4	4
Green	5	5	100,000	5	5	5
Blue	6	6	1,000,000	–	6	6
Violet	7	7	–	–	7	7
Grey	8	8	–	–	8	8
White	9	9	–	10	9	9

* in pF
† in ± %
‡ in V

CAPACITANCE = 47 × 10,000 = 0.47 μF ±20%

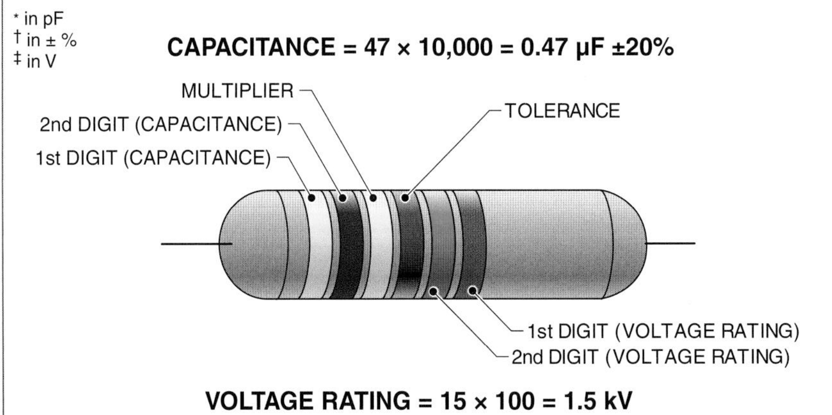

VOLTAGE RATING = 15 × 100 = 1.5 kV

FILM-TYPE CAPACITOR IDENTIFICATION CODES

Number	Multiplier		Tolerance	
			<10*	>10†
0	1	B	0.1	–
1	10	C	0.25	–
2	100	D	0.5	–
3	1000	F	1.0	1
4	10,000	G	2.0	2
5	100,000	H	–	3
6	–	J	–	5
7	–	K	–	10
8	0.01	M	–	20
9	0.1	–	–	–

* in pF
† in %

2ND DIGIT
1ST DIGIT
MULTIPLIER
TOLERANCE
561K

DIPPED TANTALUM CAPACITOR IDENTIFICATION CODES

Color	Capacitance		Tolerance*	Voltage Rating†
	Number	Multiplier		
Black	0	–	–	4
Brown	1	–	–	6
Red	2	–	–	10
Orange	3	–	–	15
Yellow	4	10,000	–	20
Green	5	100,000	–	25
Blue	6	1,000,000	–	35
Violet	7	10,000,000	–	50
Grey	8	–	–	–
White	9	–	–	3
Gold	–	–	5	–
Silver	–	–	10	–
Body Color	–	–	20	–

* in %
† in V

CERAMIC CAPACITOR IDENTIFICATION CODES

Color	Capacitance		Tolerance*		Temperature Coefficient†
	Number	Multiplier	>10 pF	<10 pF	
Black	0	1	20	2.0	0
Brown	1	10	1	–	–30
Red	2	100	2	–	–80
Orange	3	1000	–	–	–150
Yellow	4	10,000	–	–	–220
Green	5	–	5	0.5	–330
Blue	6	–	–	–	–470
Violet	7	–	–	–	–750
Grey	8	0.01	–	0.25	+30
White	9	0.10	10	1.0	+500

* in %
† in PPM/C

RADIAL LEAD TYPE CAPACITOR

5-DOT

Color Codes
A–Temperature Coefficient
B–1st digit (in pf)
C–2nd digit (in pf)
D–Multiplier
E–Tolerance
F–Voltage 100 V Brown 250 V Red 400 V Yellow

AXIAL LEAD TYPE CAPACITOR

3-DOT

DOT-SYSTEM CAPACITOR IDENTIFICATION CODES

Type (6-Dot Only)	Color	Capacitance			Tolerance*	Voltage Rating† (5-Dot Only)	Characteristic or Class (6-Dot Only)
		1st Digit	2nd Digit	Multiplier			
JAN, Mica	**Black**	0	0	1	–	–	Applies to the temperature coefficient or method of testing
	Brown	1	1	10	1	100	
	Red	2	2	100	2	200	
	Orange	3	3	1000	3	300	
	Yellow	4	4	10,000	4	400	
	Green	5	5	100,000	5	500	
	Blue	6	6	1,000,000	6	600	
	Violet	7	7	10,000,000	7	700	
	Grey	8	8	100,000,000	8	800	
EIA, Mica	**White**	9	9	1,000,000,000	9	900	
	Gold	–	–	0.1	–	1000	
Paper	**Silver**	–	–	0.01	10	2000	
	Body Color	–	–	–	20	–	

* in ± %
† in V

MULTIPLIER
2ND DIGIT
1ST DIGIT
TOLERANCE
NO COLOR
VOLTAGE

5-DOT MICA CAPACITOR IDENTIFICATION CODE

TOLERANCE
NO COLOR
VOLTAGE
MULTIPLIER
2ND DIGIT
1ST DIGIT

5-DOT MOLDED PAPER CAPACITOR IDENTIFICATION CODE

2ND DIGIT
1ST DIGIT
TYPE
MULTIPLIER
TOLERANCE
CHARACTERISTIC OR CLASS

6-DOT CAPACITOR IDENTIFICATION CODE

MULTIPLIER
2ND DIGIT
1ST DIGIT

3-DOT IDENTIFICATION CODE

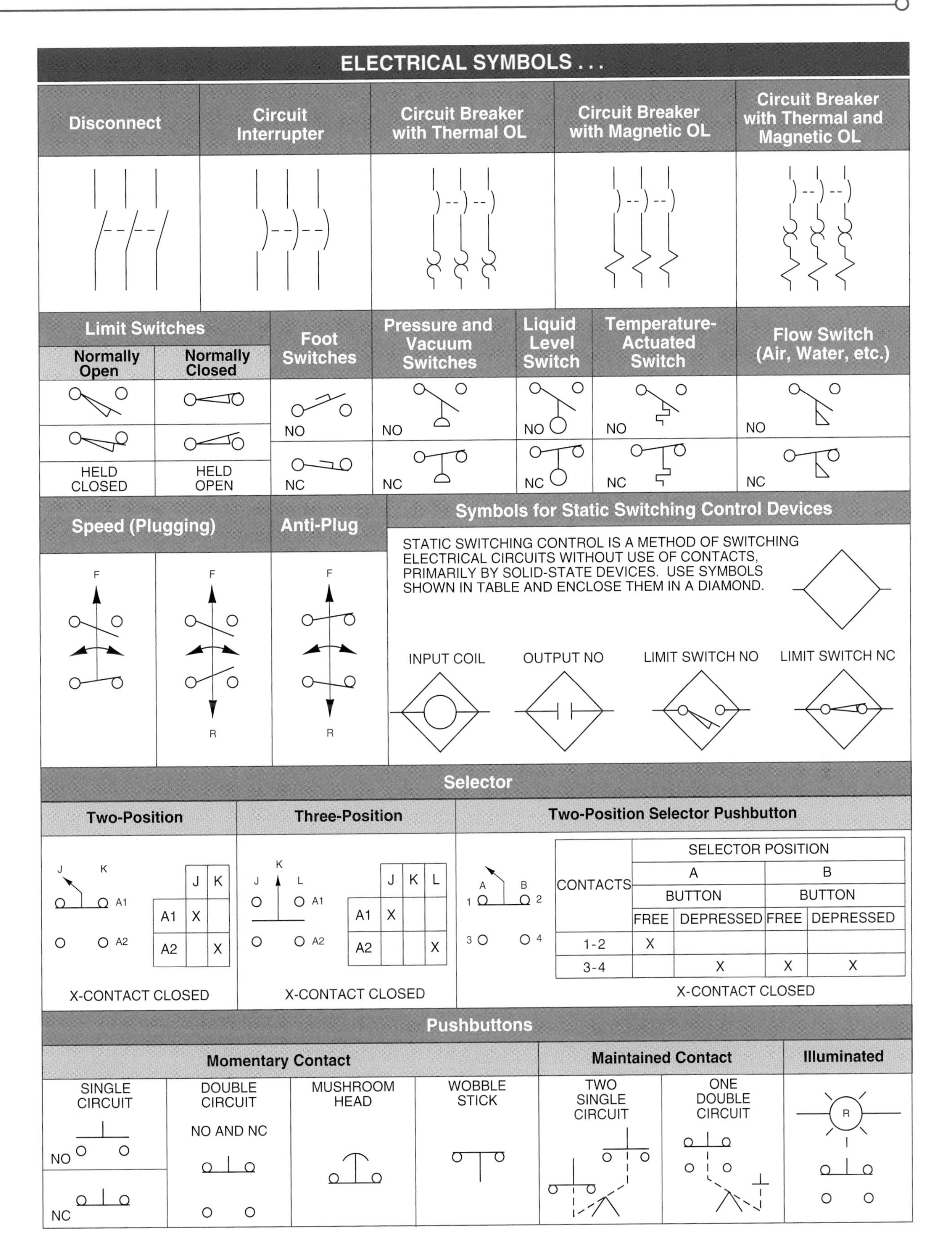
ELECTRICAL SYMBOLS . . .
Disconnect
Circuit Interrupter
Circuit Breaker with Thermal OL
Circuit Breaker with Magnetic OL
Circuit Breaker with Thermal and Magnetic OL
Limit Switches
Normally Open
Normally Closed
HELD CLOSED
HELD OPEN
Foot Switches
NO
NC
Pressure and Vacuum Switches
NO
NC
Liquid Level Switch
NO
NC
Temperature-Actuated Switch
NO
NC
Flow Switch (Air, Water, etc.)
NO
NC
Speed (Plugging)
F
F
R
Anti-Plug
F
R
Symbols for Static Switching Control Devices
STATIC SWITCHING CONTROL IS A METHOD OF SWITCHING ELECTRICAL CIRCUITS WITHOUT USE OF CONTACTS, PRIMARILY BY SOLID-STATE DEVICES. USE SYMBOLS SHOWN IN TABLE AND ENCLOSE THEM IN A DIAMOND.
INPUT COIL
OUTPUT NO
LIMIT SWITCH NO
LIMIT SWITCH NC
Selector
Two-Position
J
K
A1
A2
J K
A1 X
A2 X
X-CONTACT CLOSED
Three-Position
K
J
L
A1
A2
J K L
A1 X
A2 X
X-CONTACT CLOSED
Two-Position Selector Pushbutton
A
B
1
2
3
4
CONTACTS
SELECTOR POSITION
A
B
BUTTON
BUTTON
FREE DEPRESSED FREE DEPRESSED
1-2 X
3-4 X X X
X-CONTACT CLOSED
Pushbuttons
Momentary Contact
SINGLE CIRCUIT
NO
NC
DOUBLE CIRCUIT
NO AND NC
MUSHROOM HEAD
WOBBLE STICK
Maintained Contact
TWO SINGLE CIRCUIT
ONE DOUBLE CIRCUIT
Illuminated
R

. . . ELECTRICAL SYMBOLS . . .

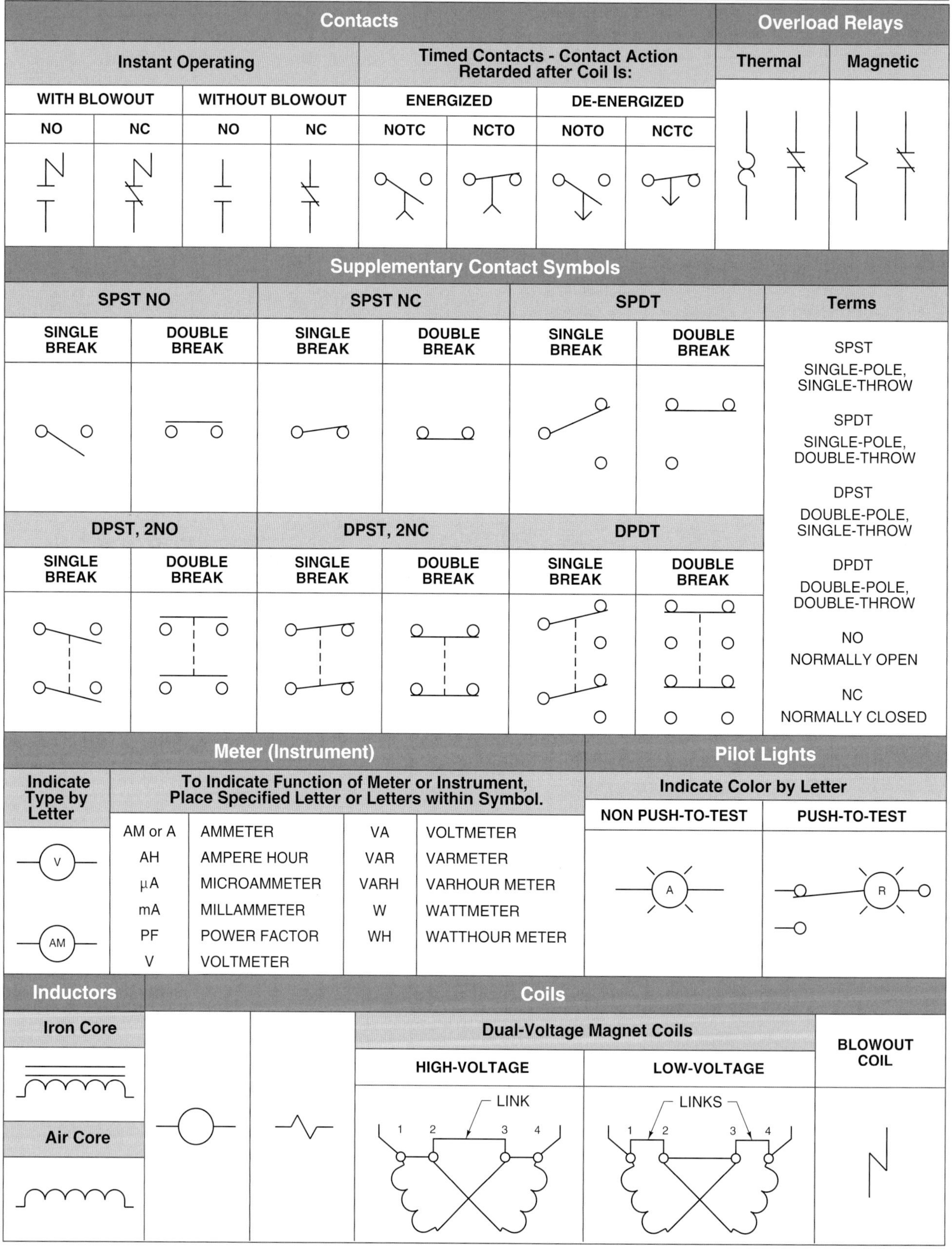

Contacts

Instant Operating				Timed Contacts - Contact Action Retarded after Coil Is:			
WITH BLOWOUT		WITHOUT BLOWOUT		ENERGIZED		DE-ENERGIZED	
NO	NC	NO	NC	NOTC	NCTO	NOTO	NCTC

Overload Relays

Thermal	Magnetic

Supplementary Contact Symbols

SPST NO		SPST NC		SPDT	
SINGLE BREAK	DOUBLE BREAK	SINGLE BREAK	DOUBLE BREAK	SINGLE BREAK	DOUBLE BREAK

DPST, 2NO		DPST, 2NC		DPDT	
SINGLE BREAK	DOUBLE BREAK	SINGLE BREAK	DOUBLE BREAK	SINGLE BREAK	DOUBLE BREAK

Terms

SPST
SINGLE-POLE, SINGLE-THROW

SPDT
SINGLE-POLE, DOUBLE-THROW

DPST
DOUBLE-POLE, SINGLE-THROW

DPDT
DOUBLE-POLE, DOUBLE-THROW

NO
NORMALLY OPEN

NC
NORMALLY CLOSED

Meter (Instrument)

Indicate Type by Letter: V, AM

To Indicate Function of Meter or Instrument, Place Specified Letter or Letters within Symbol.

AM or A	AMMETER	VA	VOLTMETER
AH	AMPERE HOUR	VAR	VARMETER
μA	MICROAMMETER	VARH	VARHOUR METER
mA	MILLAMMETER	W	WATTMETER
PF	POWER FACTOR	WH	WATTHOUR METER
V	VOLTMETER		

Pilot Lights

Indicate Color by Letter

NON PUSH-TO-TEST	PUSH-TO-TEST
A	R

Inductors

Iron Core

Air Core

Coils

		Dual-Voltage Magnet Coils		BLOWOUT COIL
		HIGH-VOLTAGE	LOW-VOLTAGE	
		LINK 1 2 3 4	LINKS 1 2 3 4	

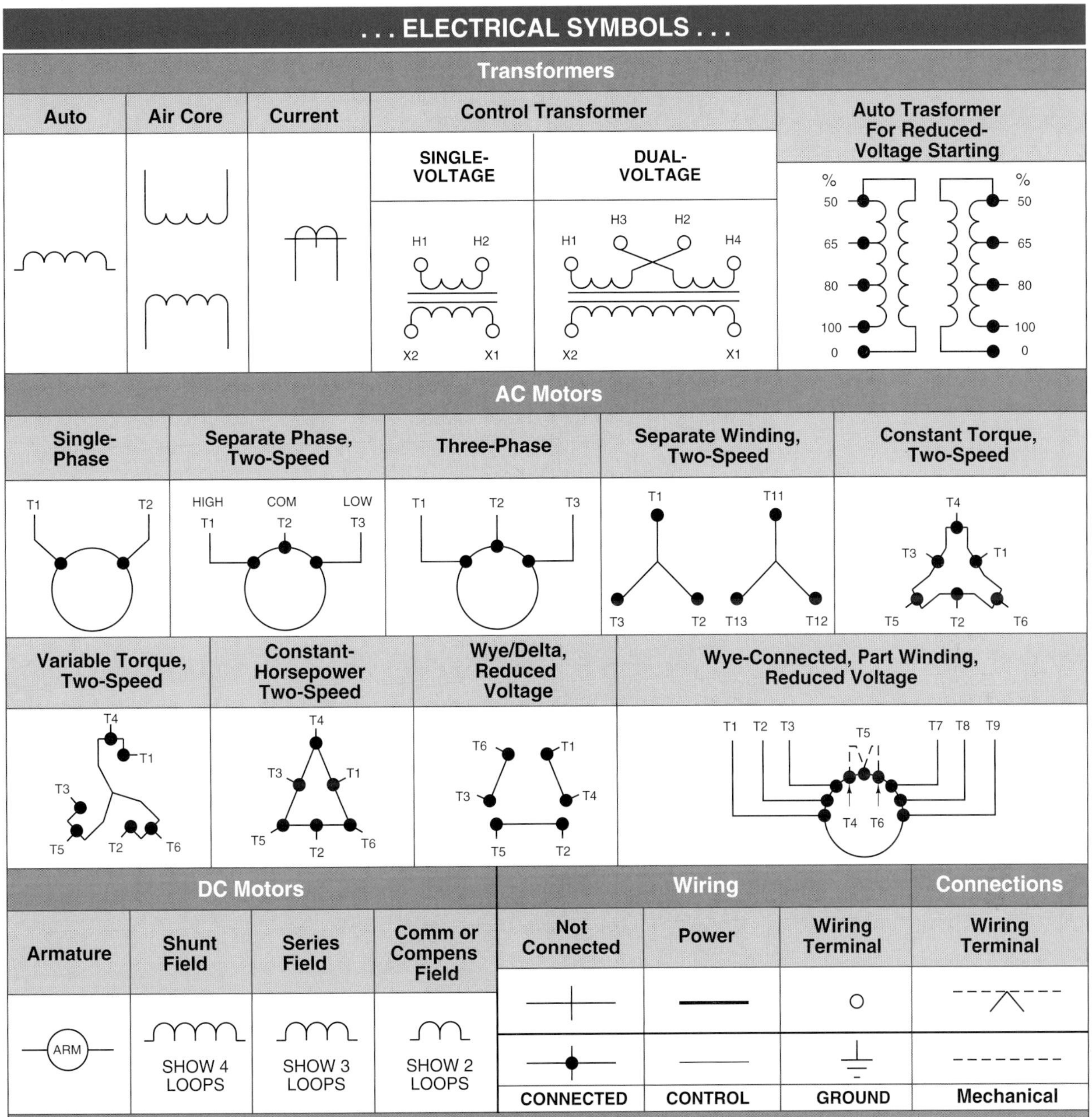

. . . ELECTRICAL SYMBOLS . . .

Transformers

Auto	Air Core	Current	Control Transformer		Auto Trasformer For Reduced-Voltage Starting
			SINGLE-VOLTAGE	DUAL-VOLTAGE	
			H1 H2 X2 X1	H1 H3 H2 H4 X2 X1	% 50 65 80 100 0 / % 50 65 80 100 0

AC Motors

Single-Phase	Separate Phase, Two-Speed	Three-Phase	Separate Winding, Two-Speed	Constant Torque, Two-Speed
T1 T2	HIGH COM LOW T1 T2 T3	T1 T2 T3	T1 T11 T3 T2 T13 T12	T4 T3 T1 T5 T2 T6

Variable Torque, Two-Speed	Constant-Horsepower Two-Speed	Wye/Delta, Reduced Voltage	Wye-Connected, Part Winding, Reduced Voltage
T4 T1 T3 T5 T2 T6	T4 T3 T1 T5 T2 T6	T6 T1 T3 T4 T5 T2	T1 T2 T3 T5 T7 T8 T9 T4 T6

DC Motors / Wiring / Connections

Armature	Shunt Field	Series Field	Comm or Compens Field	Not Connected	Power	Wiring Terminal	Wiring Terminal
ARM	SHOW 4 LOOPS	SHOW 3 LOOPS	SHOW 2 LOOPS	CONNECTED	CONTROL	GROUND	Mechanical

Control and Power Connections – 600 V or Less Across-the-Line Starters

		1ϕ	2ϕ, 4-WIRE	3ϕ
LINE MARKINGS		L1, L2	L1, L3 PHASE 1 L2, L4 PHASE 2	L1, L2, L3
GROUND WHEN USED		L1 IS ALWAYS UNGROUNDED	–	L2
MOTOR RUNNING OVERCURRENT UNITS IN	1 ELEMENT	L1	–	–
	2 ELEMENT	–	L1, L4	–
	3 ELEMENT	–	–	L1, L2, L3
CONTROL CIRCUIT CONNECTED TO		L1, L2	L1, L3	L1, L2
FOR REVERSING INTERCHANGE LINES		–	L1, L3	L1, L3

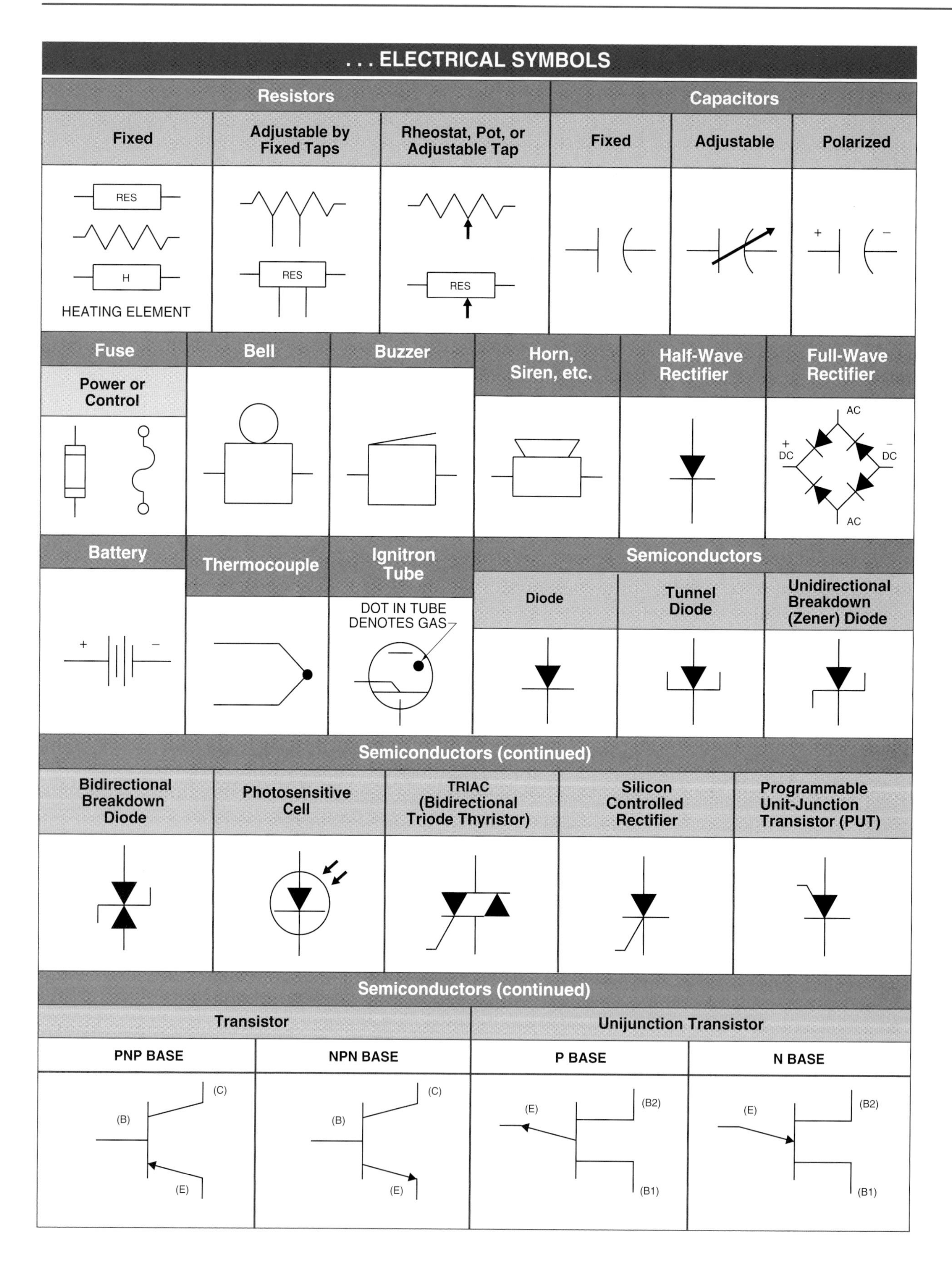

. . . ELECTRICAL SYMBOLS
Resistors
Fixed
Adjustable by Fixed Taps
Rheostat, Pot, or Adjustable Tap
Capacitors
Fixed
Adjustable
Polarized
RES
H
HEATING ELEMENT
RES
RES
+
−
Fuse
Power or Control
Bell
Buzzer
Horn, Siren, etc.
Half-Wave Rectifier
Full-Wave Rectifier
AC
+
DC
−
DC
AC
Battery
Thermocouple
Ignitron Tube
Semiconductors
Diode
Tunnel Diode
Unidirectional Breakdown (Zener) Diode
+
−
DOT IN TUBE DENOTES GAS
Semiconductors (continued)
Bidirectional Breakdown Diode
Photosensitive Cell
TRIAC (Bidirectional Triode Thyristor)
Silicon Controlled Rectifier
Programmable Unit-Junction Transistor (PUT)
Semiconductors (continued)
Transistor
Unijunction Transistor
PNP BASE
NPN BASE
P BASE
N BASE
(C)
(B)
(E)
(C)
(B)
(E)
(E)
(B2)
(B1)
(E)
(B2)
(B1)

ELECTRICAL/ELECTRONIC ABBREVIATIONS/ACRONYMS . . .

Abbr/ Acronym	Meaning	Abbr/ Acronym	Meaning
A	Ammeter; Ampere; Anode; Armature	DPST	Double-Pole, Single-Throw
AC	Alternating Current	DS	Drum Switch
AC/DC	Alternating Current; Direct Current	DT	Double-Throw
A/D	Analog to Digital	DVM	Digital Voltmeter
AF	Audio Frequency	EMF	Electromotive Force
AFC	Automatic Frequency Control	F	Fahrenheit; Fast; Forward; Fuse; Farad
Ag	Silver	FET	Field-Effect Transistor
ALM	Alarm	FF	Flip-Flop
AM	Ammeter; Amplitude Modulation	FLC	Full-Load Current
AM/FM	Amplitude Modulation; Frequency Modulation	FLS	Flow Switch
ARM.	Armature	FLT	Full-Load Torque
Au	Gold	FM	Frequency Modulation
AU	Automatic	FREQ	Frequency
AVC	Automatic Volume Control	FS	Float Switch
AWG	American Wire Gauge	FTS	Foot Switch
BAT.	Battery (electric)	FU	Fuse
BCD	Binary-Coded Decimal	FWD	Forward
BJT	Bipolar Junction Transistor	G	Gate; Giga; Green; Conductance
BK	Black	GEN	Generator
BL	Blue	GRD	Ground
BR	Brake Relay; Brown	GY	Gray
C	Celsius; Capacitance; Capacitor, Coulomb	H	Henry; High Side of Transformer; Magnetic Flux
CAP.	Capacitor	HF	High Frequency
CB	Circuit Breaker; Citizen's Band	HP	Horsepower
CC	Common-Collector; Configuration	Hz	Hertz
CCW	Counterclockwise	I	Current
CE	Common-Emitter Configuration	IC	Integrated Circuit
CEMF	Counter-Electromotive Force	INT	Intermediate; Interrupt
CKT	Circuit	IOL	Instantaneous Overload
cmil	Circular Mil	IR	Infrared
CONT	Continuous; Control	ITB	Inverse Time Breaker
CPS	Cycles Per Second	ITCB	Instantaneous Trip Circuit Breaker
CPU	Central Processing Unit	J	Joule
CR	Control Relay	JB	Junction Box
CRM	Control Relay Master	JFET	Junction Field-Effect Transistor
CT	Current Transformer	K	Kilo; Cathode
CW	Clockwise	kWh	kilowatt-hour
D	Diameter; Diode; Down	L	Line; Load; Coil; Inductance
D/A	Digital to Analog	LB-FT	Pounds Per Foot
DB	Dynamic Braking Contactor; Relay	LB-IN.	Pounds Per Inch
DC	Direct Current	LC	Inductance-Capacitance
DIO	Diode	LCD	Liquid Crystal Display
DISC.	Disconnect Switch	LCR	Inductance-Capacitance-Resistance
DMM	Digital Multimeter	LED	Light-Emitting Diode
DP	Double-Pole	LRC	Locked Rotor Current
DPDT	Double-Pole, Double-Throw	LS	Limit Switch

... ELECTRICAL/ELECTRONIC ABBREVIATIONS/ACRONYMS			
Abbr/ Acronym	**Meaning**	**Abbr/ Acronym**	**Meaning**
LT	Lamp	RF	Radio Frequency
M	Motor; Motor Starter; Motor Starter Contacts	RH	Rheostat
MAX.	Maximum	rms	Root-Mean-Square
MB	Magnetic Brake	ROM	Read-Only Memory
MCS	Motor Circuit Switch	rpm	Revolutions Per Minute
MEM	Memory	RPS	Revolutions Per Second
MED	Medium	S	Series; Slow; South; Switch; Second; Siemen
MIN	Minimum	SCR	Silicon-Controlled Rectifier
MMF	Magnetomotive Force	SEC	Secondary
MN	Manual	SF	Service Factor
MOS	Metal-Oxide Semiconductor	1 PH; 1ϕ	Single-Phase
MOSFET	Metal-Oxide Semiconductor Field-Effect Transistor	SOC	Socket
MTR	Motor	SOL	Solenoid
N; NEG	North; Negative; Number of Turns	SP	Single-Pole
NC	Normally Closed	SPDT	SIngle-Pole, Double-Throw
NEUT	Neutral	SPST	Single-Pole, Single-Throw
NO	Normally Open	SS	Selector Switch
NPN	Negative-Positive-Negative	SSW	Safety Switch
NTDF	Nontime-Delay Fuse	SW	Switch
O	Orange	T	Tera; Terminal; Torque; Transformer
OCPD	Overcurrent Protection Device	TB	Terminal Board
OHM	Ohmmeter	3 PH; 3 ϕ	Three-Phase
OL	Overload Relay	TD	Time-Delay
OZ/IN.	Ounces Per Inch	TDF	Time-Delay Fuse
P	Peak; Positive; Power; Power Consumed	TEMP	Temperature
PB	Pushbutton	THS	Thermostat Switch
PCB	Printed Circuit Board	TR	Time-Delay Relay
PH	Phase	TTL	Transistor-Transistor Logic
PLS	Plugging Switch	U	Up
PNP	Positive-Negative-Positive	UCL	Unclamp
POS	Positive	UHF	Ultrahigh Frequency
POT.	Potentiometer	UJT	Unijunction Transistor
P-P	Peak-to-Peak	UV	Ultraviolet; Undervoltage
PRI	Primary Switch	V	Violet; Volt
PS	Pressure Switch	VA	Voltampere
PSI	Pounds Per Square Inch	VAC	Volts Alternating Current
PUT	Pull-Up Torque	VDC	Volts Direct Current
Q	Transistor; Quality Factor	VHF	Very High Frequency
R	Radius; Red; Resistance; Reverse	VLF	Very Low Frequency
RAM	Random-Access Memory	VOM	Volt-Ohm-Milliammeter
RC	Resistance-Capacitance	W	Watt; White
RCL	Resistance-Inductance-Capacitance	w/	With
REC	Rectifier	X	Low Side of Transformer
RES	Resistor	Y	Yellow
REV	Reverse	Z	Impedance

TRIG FUNCTION VALUES

Angle (Degrees)	Sine	Cosine	Tangent
0	0.0000	1.0000	0.0000
1	0.0175	0.9998	0.0175
2	0.0349	0.9994	0.0349
3	0.0523	0.9986	0.0524
4	0.0698	0.9976	0.0699
5	0.0872	0.9962	0.0875
6	0.1045	0.9945	0.1051
7	0.1219	0.9925	0.1228
8	0.1392	0.9903	0.1405
9	0.1564	0.9877	0.1584
10	0.1736	0.9848	0.1763
11	0.1908	0.9816	0.1944
12	0.2079	0.9781	0.2126
13	0.2250	0.9744	0.2309
14	0.2419	0.9703	0.2493
15	0.2588	0.9659	0.2679
16	0.2756	0.9613	0.2867
17	0.2924	0.9563	0.3057
18	0.3090	0.9511	0.3249
19	0.3256	0.9455	0.3443
20	0.3420	0.9397	0.3640
21	0.3584	0.9336	0.3839
22	0.3746	0.9272	0.4040
23	0.3907	0.9205	0.4245
24	0.4067	0.9135	0.4452
25	0.4226	0.9063	0.4663
26	0.4384	0.8988	0.4877
27	0.4540	0.8910	0.5095
28	0.4695	0.8829	0.5317
29	0.4848	0.8746	0.5543
30	0.5000	0.8660	0.5774
31	0.5150	0.8572	0.6009
32	0.5299	0.8480	0.6249
33	0.5446	0.8387	0.6494
34	0.5592	0.8290	0.6745
35	0.5736	0.8192	0.7002
36	0.5878	0.8090	0.7265
37	0.6018	0.7986	0.7536
38	0.6157	0.7880	0.7813
39	0.6293	0.7771	0.8098
40	0.6428	0.7660	0.8391
41	0.6561	0.7547	0.8693
42	0.6691	0.7431	0.9004
43	0.6820	0.7314	0.9325
44	0.6947	0.7193	0.9657

TRIG FUNCTION VALUES . . .

Angle (Degrees)	Sine	Cosine	Tangent
45	0.7071	0.7071	1.0000
46	0.7193	0.6947	1.0355
47	0.7314	0.6820	1.0724
48	0.7431	0.6691	1.1106
49	0.7547	0.6561	1.1504
50	0.7660	0.6428	1.1918
51	0.7771	0.6293	1.2349
52	0.7880	0.6157	1.2799
53	0.7986	0.6018	1.3270
54	0.8090	0.5878	1.3764
55	0.8192	0.5736	1.4281
56	0.8290	0.5592	1.4826
57	0.8387	0.5446	1.5399
58	0.8480	0.5299	1.6003
59	0.8572	0.5150	1.6643
60	0.8660	0.5000	1.7321
61	0.8746	0.4848	1.8040
62	0.8829	0.4695	1.8807
63	0.8910	0.4540	1.9626
64	0.8988	0.4384	2.0503
65	0.9063	0.4226	2.1445
66	0.9135	0.4067	2.2460
67	0.9205	0.3907	2.3559
68	0.9272	0.3746	2.4751
69	0.9336	0.3584	2.6051
70	0.9397	0.3420	2.7475
71	0.9455	0.3256	2.9042
72	0.9511	0.3090	3.0777
73	0.9563	0.2924	3.2709
74	0.9613	0.2756	3.4874
75	0.9659	0.2588	3.7321
76	0.9703	0.2419	4.0108
77	0.9744	0.2250	4.3315
78	0.9781	0.2079	4.7046
79	0.9816	0.1908	5.1446
80	0.9848	0.1736	5.6713
81	0.9877	0.1564	6.3138
82	0.9903	0.1392	7.1154
83	0.9925	0.1219	8.1443
84	0.9945	0.1045	9.5144
85	0.9962	0.0872	11.4301
86	0.9976	0.0698	14.3007
87	0.9986	0.0523	19.0811
88	0.9994	0.0349	28.6363
89	0.9998	0.0175	57.2900

. . . TRIG FUNCTION VALUES

Angle (Degrees)	Sine	Cosine	Tangent
90	1.0000	0.0000	∞
91	0.9998	–0.0175	–57.2900
92	0.9994	–0.0349	–28.6363
93	0.9986	–0.0523	–19.0811
94	0.9976	–0.0698	–14.3007
95	0.9962	–0.0872	–11.4301
96	0.9945	–0.1045	–9.5144
97	0.9925	–0.1219	–8.1443
98	0.9903	–0.1392	–7.1154
99	0.9877	–0.1564	–6.3138
100	0.9848	–0.1736	–5.6713
101	0.9816	–0.1908	–5.1446
102	0.9781	–0.2079	–4.7046
103	0.9744	–0.2250	–4.3315
104	0.9703	–0.2419	–4.0108
105	0.9659	–0.2588	–3.7321
106	0.9613	–0.2756	–3.4874
107	0.9563	–0.2924	–3.2709
108	0.9511	–0.3090	–3.0777
109	0.9455	–0.3256	–2.9042
110	0.9397	–0.3420	–2.7475
111	0.9336	–0.3584	–2.6051
112	0.9272	–0.3746	–2.4751
113	0.9205	–0.3907	–2.3559
114	0.9135	–0.4067	–2.2460
115	0.9063	–0.4226	–2.1445
116	0.8988	–0.4384	–2.0503
117	0.8910	–0.4540	–1.9626
118	0.8829	–0.4695	–1.8807
119	0.8746	–0.4848	–1.8040
120	0.8660	–0.5000	–1.7321
121	0.8572	–0.5150	–1.6643
122	0.8480	–0.5299	–1.6003
123	0.8387	–0.5446	–1.5399
124	0.8290	–0.5592	–1.4826
125	0.8192	–0.5736	–1.4281
126	0.8090	–0.5878	–1.3764
127	0.7986	–0.6018	–1.3270
128	0.7880	–0.6157	–1.2799
129	0.7771	–0.6293	–1.2349
130	0.7660	–0.6428	–1.1918
131	0.7547	–0.6561	–1.1504
132	0.7431	–0.6691	–1.1106
133	0.7314	–0.6820	–1.0724
134	0.7193	–0.6947	–1.0355

TRIG FUNCTION VALUES . . .

Angle (Degrees)	Sine	Cosine	Tangent
135	0.7071	–0.7071	–1.0000
136	0.6947	–0.7193	–0.9657
137	0.6820	–0.7314	–0.9325
138	0.6691	–0.7431	–0.9004
139	0.6561	–0.7547	–0.8693
140	0.6428	–0.7660	–0.8391
141	0.6293	–0.7771	–0.8098
142	0.6157	–0.7880	–0.7813
143	0.6018	–0.7986	–0.7536
144	0.5878	–0.8090	–0.7265
145	0.5736	–0.8192	–0.7002
146	0.5592	–0.8290	–0.6745
147	0.5446	–0.8387	–0.6494
148	0.5299	–0.8480	–0.6249
149	0.5150	–0.8572	–0.6009
150	0.5000	–0.8660	–0.5774
151	0.4848	–0.8746	–0.5543
152	0.4695	–0.8829	–0.5317
153	0.4540	–0.8910	–0.5095
154	0.4384	–0.8988	–0.4877
155	0.4226	–0.9063	–0.4663
156	0.4067	–0.9135	–0.4452
157	0.3907	–0.9205	–0.4245
158	0.3746	–0.9272	–0.4040
159	0.3584	–0.9336	–0.3839
160	0.3420	–0.9397	–0.3640
161	0.3256	–0.9455	–0.3443
162	0.3090	–0.9511	–0.3249
163	0.2924	–0.9563	–0.3057
164	0.2756	–0.9613	–0.2867
165	0.2588	–0.9659	–0.2679
166	0.2419	–0.9703	–0.2493
167	0.2250	–0.9744	–0.2309
168	0.2079	–0.9781	–0.2126
169	0.1908	–0.9816	–0.1944
170	0.1736	–0.9848	–0.1763
171	0.1564	–0.9877	–0.1584
172	0.1392	–0.9903	–0.1405
173	0.1219	–0.9925	–0.1228
174	0.1045	–0.9945	–0.1051
175	0.0872	–0.9962	–0.0875
176	0.0698	–0.9976	–0.0699
177	0.0523	–0.9986	–0.0524
178	0.0349	–0.9994	–0.0349
179	0.0175	–0.9998	–0.0175

. . . TRIG FUNCTION VALUES			
Angle (Degrees)	Sine	Cosine	Tangent
180	0.0000	−1.0000	0.0000
181	−0.0175	−0.9998	0.0175
182	−0.0349	−0.9994	0.0349
183	−0.0523	−0.9986	0.0524
184	−0.0698	−0.9976	0.0699
185	−0.0872	−0.9962	0.0875
186	−0.1045	−0.9945	0.1051
187	−0.1219	−0.9925	0.1228
188	−0.1392	−0.9903	0.1405
189	−0.1564	−0.9877	0.1584
190	−0.1736	−0.9848	0.1763
191	−0.1908	−0.9816	0.1944
192	−0.2079	−0.9781	0.2126
193	−0.2250	−0.9744	0.2309
194	−0.2419	−0.9703	0.2493
195	−0.2588	−0.9659	0.2679
196	−0.2756	−0.9613	0.2867
197	−0.2924	−0.9563	0.3057
198	−0.3090	−0.9511	0.3249
199	−0.3256	−0.9455	0.3443
200	−0.3420	−0.9397	0.3640
201	−0.3584	−0.9336	0.3839
202	−0.3746	−0.9272	0.4040
203	−0.3907	−0.9205	0.4245
204	−0.4067	−0.9135	0.4452
205	−0.4226	−0.9063	0.4663
206	−0.4384	−0.8988	0.4877
207	−0.4540	−0.8910	0.5095
208	−0.4695	−0.8829	0.5317
209	−0.4848	−0.8746	0.5543
210	−0.5000	−0.8660	0.5774
211	−0.5150	−0.8572	0.6009
212	−0.5299	−0.8480	0.6249
213	−0.5446	−0.8387	0.6494
214	−0.5592	−0.8290	0.6745
215	−0.5736	−0.8192	0.7002
216	−0.5878	−0.8090	0.7265
217	−0.6018	−0.7986	0.7536
218	−0.6157	−0.7880	0.7813
219	−0.6293	−0.7771	0.8098
220	−0.6428	−0.7660	0.8391
221	−0.6561	−0.7547	0.8693
222	−0.6691	−0.7431	0.9004
223	−0.6820	−0.7314	0.9325
224	−0.6947	−0.7193	0.9657

TRIG FUNCTION VALUES . . .			
Angle (Degrees)	Sine	Cosine	Tangent
225	−0.7071	−0.7071	1.0000
226	−0.7193	−0.6947	1.0355
227	−0.7314	−0.6820	1.0724
228	−0.7431	−0.6691	1.1106
229	−0.7547	−0.6561	1.1504
230	−0.7660	−0.6428	1.1918
231	−0.7771	−0.6293	1.2349
232	−0.7880	−0.6157	1.2799
233	−0.7986	−0.6018	1.3270
234	−0.8090	−0.5878	1.3764
235	−0.8192	−0.5736	1.4281
236	−0.8290	−0.5592	1.4826
237	−0.8387	−0.5446	1.5399
238	−0.8480	−0.5299	1.6003
239	−0.8572	−0.5150	1.6643
240	−0.8660	−0.5000	1.7321
241	−0.8746	−0.4848	1.8040
242	−0.8829	−0.4695	1.8807
243	−0.8910	−0.4540	1.9626
244	−0.8988	−0.4384	2.0503
245	−0.9063	−0.4226	2.1445
246	−0.9135	−0.4067	2.2460
247	−0.9205	−0.3907	2.3559
248	−0.9272	−0.3746	2.4751
249	−0.9336	−0.3584	2.6051
250	−0.9397	−0.3420	2.7475
251	−0.9455	−0.3256	2.9042
252	−0.9511	−0.3090	3.0777
253	−0.9563	−0.2924	3.2709
254	−0.9613	−0.2756	3.4874
255	−0.9659	−0.2588	3.7321
256	−0.9703	−0.2419	4.0108
257	−0.9744	−0.2250	4.3315
258	−0.9781	−0.2079	4.7046
259	−0.9816	−0.1908	5.1446
260	−0.9848	−0.1736	5.6713
261	−0.9877	−0.1564	6.3138
262	−0.9903	−0.1392	7.1154
263	−0.9925	−0.1219	8.1443
264	−0.9945	−0.1045	9.5144
265	−0.9962	−0.0872	11.4301
266	−0.9976	−0.0698	14.3007
267	−0.9986	−0.0523	19.0811
268	−0.9994	−0.0349	28.6363
269	−0.9998	−0.0175	57.2900

. . . TRIG FUNCTION VALUES

Angle (Degrees)	Sine	Cosine	Tangent
270	−1.0000	0.0000	∞
271	−0.9998	0.0175	−57.2900
272	−0.9994	0.0349	−28.6363
273	−0.9986	0.0523	−19.0811
274	−0.9976	0.0698	−14.3007
275	−0.9962	0.0872	−11.4301
276	−0.9945	0.1045	−9.5144
277	−0.9925	0.1219	−8.1443
278	−0.9903	0.1392	−7.1154
279	−0.9877	0.1564	−6.3138
280	−0.9848	0.1736	−5.6713
281	−0.9816	0.1908	−5.1446
282	−0.9781	0.2079	−4.7046
283	−0.9744	0.2250	−4.3315
284	−0.9703	0.2419	−4.0108
285	−0.9659	0.2588	−3.7321
286	−0.9613	0.2756	−3.4874
287	−0.9563	0.2924	−3.2709
288	−0.9511	0.3090	−3.0777
289	−0.9455	0.3256	−2.9042
290	−0.9397	0.3420	−2.7475
291	−0.9336	0.3584	−2.6051
292	−0.9272	0.3746	−2.4751
293	−0.9205	0.3907	−2.3559
294	−0.9135	0.4067	−2.2460
295	−0.9063	0.4226	−2.1445
296	−0.8988	0.4384	−2.0503
297	−0.8910	0.4540	−1.9626
298	−0.8829	0.4695	−1.8807
299	−0.8746	0.4848	−1.8040
300	−0.8660	0.5000	−1.7321
301	−0.8572	0.5150	−1.6643
302	−0.8480	0.5299	−1.6003
303	−0.8387	0.5446	−1.5399
304	−0.8290	0.5592	−1.4826
305	−0.8192	0.5736	−1.4281
306	−0.8090	0.5878	−1.3764
307	−0.7986	0.6018	−1.3270
308	−0.7880	0.6157	−1.2799
309	−0.7771	0.6293	−1.2349
310	−0.7660	0.6428	−1.1918
311	−0.7547	0.6561	−1.1504
312	−0.7431	0.6691	−1.1106
313	−0.7314	0.6820	−1.0724
314	−0.7193	0.6947	−1.0355

TRIG FUNCTION VALUES

Angle (Degrees)	Sine	Cosine	Tangent
315	−0.7071	0.7071	−1.0000
316	−0.6947	0.7193	−0.9657
317	−0.6820	0.7314	−0.9325
318	−0.6691	0.7431	−0.9004
319	−0.6561	0.7547	−0.8693
320	−0.6428	0.7660	−0.8391
321	−0.6293	0.7771	−0.8098
322	−0.6157	0.7880	−0.7813
323	−0.6018	0.7986	−0.7536
324	−0.5878	0.8090	−0.7265
325	−0.5736	0.8192	−0.7002
326	−0.5592	0.8290	−0.6745
327	−0.5446	0.8387	−0.6494
328	−0.5299	0.8480	−0.6249
329	−0.5150	0.8572	−0.6009
330	−0.5000	0.8660	−0.5774
331	−0.4848	0.8746	−0.5543
332	−0.4695	0.8829	−0.5317
333	−0.4540	0.8910	−0.5095
334	−0.4384	0.8988	−0.4877
335	−0.4226	0.9063	−0.4663
336	−0.4067	0.9135	−0.4452
337	−0.3907	0.9205	−0.4245
338	−0.3746	0.9272	−0.4040
339	−0.3584	0.9336	−0.3839
340	−0.3420	0.9397	−0.3640
341	−0.3256	0.9455	−0.3443
342	−0.3090	0.9511	−0.3249
343	−0.2924	0.9563	−0.3057
344	−0.2756	0.9613	−0.2867
345	−0.2588	0.9659	−0.2679
346	−0.2419	0.9703	−0.2493
347	−0.2250	0.9744	−0.2309
348	−0.2079	0.9781	−0.2126
349	−0.1908	0.9816	−0.1944
350	−0.1736	0.9848	−0.1763
351	−0.1564	0.9877	−0.1584
352	−0.1392	0.9903	−0.1405
353	−0.1219	0.9925	−0.1228
354	−0.1045	0.9945	−0.1051
355	−0.0872	0.9962	−0.0875
356	−0.0698	0.9976	−0.0699
357	−0.0523	0.9986	−0.0524
358	−0.0349	0.9994	−0.0349
359	−0.0175	0.9998	−0.0175

POWERS OF 10

1×10^4	=	10,000	=	$10 \times 10 \times 10 \times 10$		Read ten to the fourth power
1×10^3	=	1000	=	$10 \times 10 \times 10$		Read ten to the third power or ten cubed
1×10^2	=	100	=	10×10		Read ten to the second power or ten squared
1×10^1	=	10	=	10		Read ten to the first power
1×10^0	=	1	=	1		Read ten to the zero power
1×10^{-1}	=	0.1	=	$^1/_{10}$		Read ten to the minus first power
1×10^{-2}	=	0.01	=	$1/(10 \times 10)$ or $^1/_{100}$		Read ten to the minus second power
1×10^{-3}	=	0.001	=	$1/(10 \times 10 \times 10)$ or $^1/_{1000}$		Read ten to the minus third power
1×10^{-4}	=	0.0001	=	$1/(10 \times 10 \times 10 \times 10)$ or $^1/_{10,000}$		Read ten to the minus fourth power

UNITS OF ENERGY

Energy	Btu	ft lb	J	kcal	kWh
British thermal unit	1	777.9	1.056	0.252	2.930×10^{-4}
Foot-pound	1.285×10^{-3}	1	1.356	3.240×10^{-7}	3.766×10^{-7}
Joule	9.481×10^{-4}	0.7376	1	2.390×10^{-7}	2.778×10^{-7}
Kilocalorie	3.968	3.086	4.184	1	1.163×10^{-3}
Kilowatt-hour	3.413	2.655×10^6	3.6×10^6	860.2	1

UNITS OF POWER

Power	W	ft lbs	HP	kW
Watt	1	0.7376	0.341×10^{-3}	0.001
Foot-pound/sec	1.356	1	0.818×10^{-3}	1.356×10^{-3}
Horsepower	745.7	550	1	0.7457
Kilowatt	1000	736.6	1.341	1

VOLTAGE CONVERSIONS

To Convert	To	Multiply By
rms	Average	0.9
rms	Peak	1.414
Average	rms	1.111
Average	Peak	1.567
Peak	rms	0.707
Peak	Average	0.637
Peak	Peak-to-peak	2

PERIODIC TABLE OF THE ELEMENTS

PERIODIC TABLE OF THE ELEMENTS

Key: 92 U Uranium 238.029 -21-9-2

Atomic Number
Symbol
Name
Atomic Weight
Electron Configuration

	I	II											III	IV	V	VI	VII	0
	1 H Hydrogen 1.0079 1																	2 He Helium 4.00260 2
	3 Li Lithium 6.941 2-1	4 Be Beryllium 9.01218 2-2											5 B Boron 10.81 2-3	6 C Carbon 12.011 2-4	7 N Nitrogen 14.0067 2-5	8 O Oxygen 15.9994 2-6	9 F Fluorine 18.99840 2-7	10 Ne Neon 20.179 2-8
	11 Na Sodium 22.98977 2-8-1	12 Mg Magnesium 24.305 2-8-2											13 Al Aluminum 26.98154 2-8-3	14 Si Silicon 28.086 2-8-4	15 P Phosphorus 30.97376 2-8-5	16 S Sulfur 32.06 2-8-6	17 Cl Clorine 35.453 2-8-7	18 Ar Argon 39.948 2-8-8
2	19 K Potassium 39.098 -8-8-1	20 Ca Calcium 40.08 -8-8-2	21 Sc Scandium 44.9559 -8-9-2	22 Ti Titanium 47.90 -8-10-2	23 V Vanadium 50.9414 -8-11-2	24 Cr Chromium 51.996 -8-13-1	25 Mn Manganese 54.9380 -8-13-2	26 Fe Iron 55.847 -8-14-2	27 Co Cobalt 58.9332 -8-15-2	28 Ni Nickel 58.70 -8-16-2	29 Cu Copper 63.546 -8-18-1	30 Zn Zinc 65.38 -8-18-2	31 Ga Gallium 69.72 -8-18-3	32 Ge Germanium 72.59 -8-18-4	33 As Arsenic 74.9216 -8-18-5	34 Se Selenium 78.96 -8-18-6	35 Br Bromine 79.904 -8-18-7	36 Kr Krypton 83.80 -8-18-8
2 8	37 Rb Rubidium 85.4678 -18-8-1	38 Sr Strontium 87.62 -18-8-2	39 Y Yttrium 88.9059 -18-9-2	40 Zr Zirconium 91.22 -18-10-2	41 Nb Niobium 92.9064 -18-12-1	42 Mo Molybdenum 95.94 -18-13-1	43 Tc Technetium 97 -18-13-2	44 Ru Ruthenium 101.07 -18-15-1	45 Rh Rhodium 102.9055 -18-16-1	46 Pd Palladium 106.4 -18-18-0	47 Ag Silver 107.868 -18-18-1	48 Cd Cadmium 112.40 -18-18-2	49 In Indium 114.82 -18-18-3	50 Sn Tin 118.69 -18-18-4	51 Sb Antimony 121.75 -18-18-5	52 Te Tellurium 127.60 -18-18-6	53 I Iodine 126.9045 -18-18-7	54 Xe Xenon 131.30 -18-18-8
2 8 18	55 Cs Cesium 132.9054 -18-8-1	56 Ba Barium 137.34 -18-8-2	57–71	72 Hf Hafnium 178.49 -32-10-2	73 Ta Tantalum 180.9479 -32-11-2	74 W Tungsten 183.85 -32-12-2	75 Re Rhenium 186.207 -32-13-2	76 Os Osmium 190.2 -32-14-2	77 Ir Iridium 192.22 -32-15-2	78 Pt Platinum 195.09 -32-17-1	79 Au Gold 196.9665 -32-18-1	80 Hg Mercury 200.59 -32-18-2	81 Tl Thallium 204.37 -32-18-3	82 Pb Lead 207.2 -32-18-4	83 Bi Bismuth 208.9804 -32-18-5	84 Po Polonium 209 -32-18-6	85 At Astatine 210 -32-18-7	86 Rn Radon 222 -32-18-8
2 8 18 32	87 Fr Francium 223 -18-8-1	88 Ra Radium 226.0254 -18-8-2	89–103	104 Rf Rutherfordium 261 -32-10-2	105 Db Dubnium 262 -32-11-2	106 Sg Seaborgium 263 -32-12-2	107 Bh Bohrium 262 -32-13-2	108 Hs Hassium 265 -32-14-2	109 Mt Meitnerium 266 -32-15-2	110 Ds Darmstadtium 269 -32-17-1	111 Uuu (Temporary name) 272 -32-18-1	112 Uub (Temporary name) 277 -32-18-1	113 Uut (Temporary name)	114 Uuq (Temporary name) 289	115 Uup (Temporary name)	116 Uuh (Temporary name) 289		

57 La Lanthanum 138.9055 -18-9-2	58 Ce Cerium 140.12 -19-9-2	59 Pr Praseodymium 140.9077 -21-8-2	60 Nd Neodymium 144.24 -22-8-2	61 Pm Promethium 145 -23-8-2	62 Sm Samarium 150.4 -24-8-2	63 Eu Europium 151.96 -25-8-2	64 Gd Gadolinium 157.25 -25-9-2	65 Tb Terbium 158.9254 -26-9-2	66 Dy Dysprosium 162.50 -28-8-2	67 Ho Holmium 164.9304 -29-8-2	68 Er Erbium 167.26 -30-8-2	69 Tm Thulium 168.9342 -31-8-2	70 Yb Ytterbium 173.04 -32-8-2	71 Lu Lutetium 174.97 -32-9-2
89 Ac Actinium 227 -18-9-2	90 Th Thorium 323.0381 -18-10-2	91 Pa Protactinium 231.0359 -20-9-2	92 U Uranium 238.029 -21-9-2	93 Np Neptunium 237.0482 -22-9-2	94 Pu Plutonium 244 -24-8-2	95 Am Americium 243 -25-8-2	96 Cm Curium 247 -25-9-2	97 Bk Berkelium 247 -27-8-2	98 Cf Californium 251 -28-8-2	99 Es Einsteinium 254 -29-8-2	100 Fm Fermium 257 -30-8-2	101 Md Mendelevium 258 -31-8-2	102 No Nobelium 255 -32-8-2	103 Lr Lawrencium 260 -32-9-2

AC/DC FORMULAS

To Find	DC	AC		
		1ϕ, 115 or 220 V	1ϕ, 208, 230, or 240 V	3ϕ – All Voltages
I, HP known	$\frac{HP \times 746}{E \times Eff}$	$\frac{HP \times 746}{E \times Eff \times PF}$	$\frac{HP \times 746}{E \times Eff \times PF}$	$\frac{HP \times 746}{1.73 \times E \times Eff \times PF}$
I, kW known	$\frac{kW \times 1000}{E}$	$\frac{kW \times 1000}{E \times PF}$	$\frac{kW \times 1000}{E \times PF}$	$\frac{kW \times 1000}{1.73 \times E \times PF}$
I, kVA known		$\frac{kVA \times 1000}{E}$	$\frac{kVA \times 1000}{E}$	$\frac{kVA \times 1000}{1.73 \times E}$
kW	$\frac{I \times E}{1000}$	$\frac{I \times E \times PF}{1000}$	$\frac{I \times E \times PF}{1000}$	$\frac{I \times E \times 1.73 \times PF}{1000}$
kVA		$\frac{I \times E}{1000}$	$\frac{I \times E}{1000}$	$\frac{I \times E \times 1.73}{1000}$
HP (output)	$\frac{I \times E \times Eff}{746}$	$\frac{I \times E \times Eff \times PF}{746}$	$\frac{I \times E \times Eff \times PF}{746}$	$\frac{I \times E \times 1.73 \times Eff \times PF}{746}$

Eff = efficiency

HORSEPOWER FORMULAS

To Find	Use Formula	Example		
		Given	Find	Solution
HP	$HP = \frac{I \times E \times Eff}{746}$	240 V, 20 A, 85% Eff	HP	$HP = \frac{I \times E \times Eff}{746}$ $HP = \frac{20\text{ A} \times 240\text{ V} \times 85\%}{746}$ $HP =$ **5.5**
I	$I = \frac{HP \times 746}{E \times Eff \times PF}$	10 HP, 240 V, 90% Eff, 88% PF	I	$I = \frac{HP \times 746}{E \times Eff \times PF}$ $I = \frac{10\text{ HP} \times 746}{240\text{ V} \times 90\% \times 88\%}$ $I =$ **39 A**

VOLTAGE DROP FORMULAS—1ϕ, 3ϕ

Phase	To Find	Use Formula	Example		
			Given	Find	Solution
1ϕ	VD	$VD = \frac{2 \times R \times L \times I}{1000}$	240 V, 40 A, 60 L, 0.764 R	VD	$VD = \frac{2 \times R \times L \times I}{1000}$ $VD = \frac{2 \times 0.764 \times 60 \times 40}{1000}$ $VD =$ **3.67 V**
3ϕ	VD	$VD = \frac{2 \times R \times L \times I}{1000} \times 0.866$	1208 V, 110 A, 75 L, 0.194 R, 0.866 multiplier	VD	$VD = \frac{2 \times R \times L \times I}{1000} \times 0.866$ $VD = \frac{2 \times 0.194 \times 75 \times 110}{1000} \times 0.866$ $VD =$ **2.77 V**

$^{*}\frac{\sqrt{3}}{2} = .866$

HORSEPOWER-TO-TORQUE CONVERSION

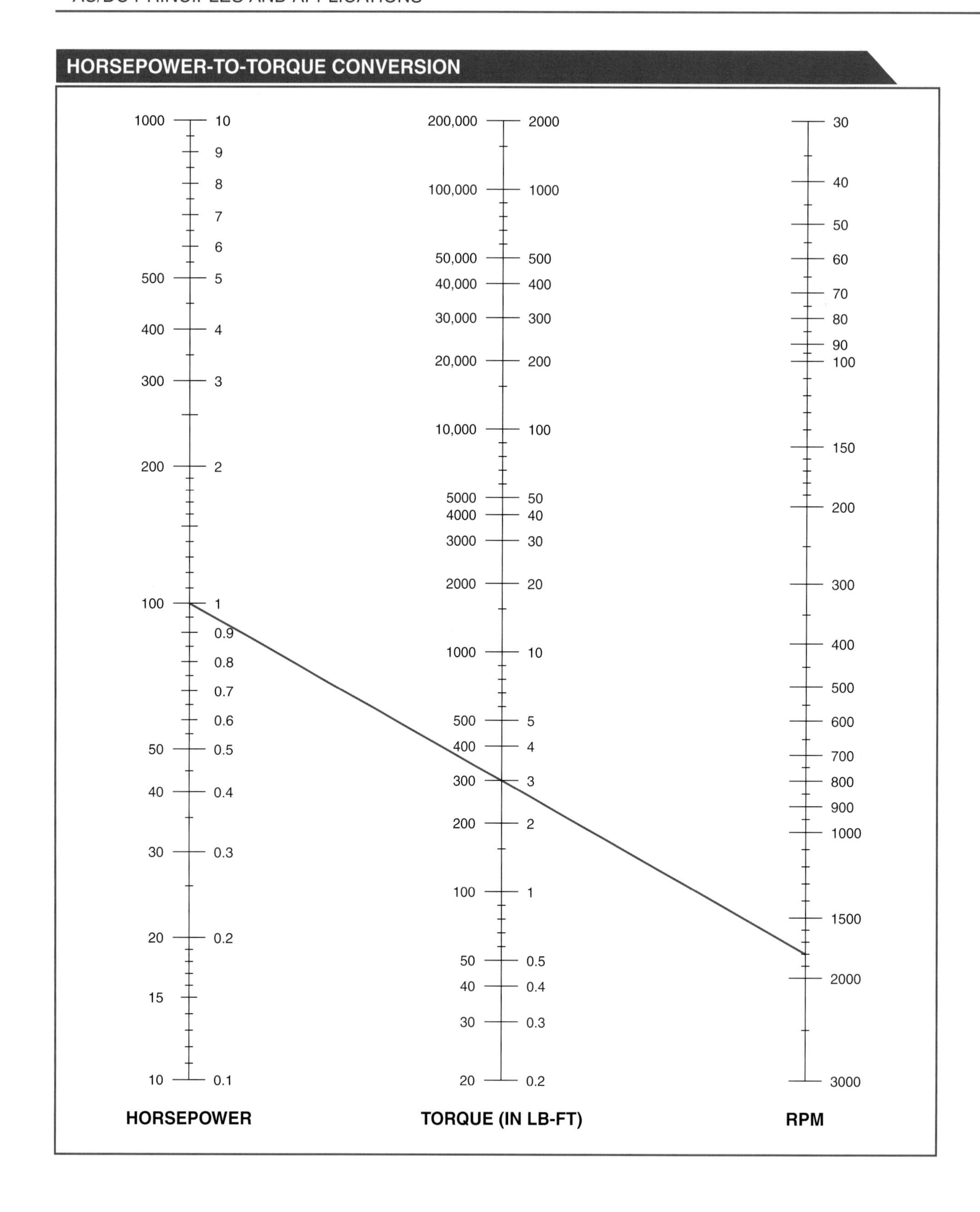

ENGLISH TO METRIC EQUIVALENTS

Category	Measure	Unit	Metric Equivalent
LENGTH		mile	1.609 km
		rod	5.029 m
		yard	0.9144 m
		foot	30.48 cm
		inch	2.54 cm
AREA ($A = l \times w$)		square mile	2.590 k^2
		acre	0.405 hectacre, 4047 m^2
		square rod	25.293 m^2
		square yard	0.836 m^2
		square foot	0.093 m^2
		square inch	6.452 cm^2
VOLUME ($V = l \times w \times t$)		cubic yard	0.765 m^3
		cubic foot	0.028 m^3
		cubic inch	16.387 cm^3
CAPACITY — WATER, FUEL, ETC.	U.S. liquid measure	gallon	3.785 l
		quart	0.946 l
		pint	0.473 l
		gill	118.294 ml
		fluidounce	29.573 ml
		fluidram	3.697 ml
		minim	0.061610 ml
VEGETABLES, GRAIN, ETC.	U.S. dry measure	bushel	35.239 l
		peck	8.810
		quart	1.101 l
		pint	0.551 l
DRUGS	British Imperial liquid and dry measure	bushel	0.036 m^3
		peck	0.0091 m^3
		gallon	4.546 l
		quart	1.136 l
		pint	568.26 cm^3
		gill	142.066 cm^3
		fluidounce	28.412 cm^3
		fluidram	3.5516 cm^3
		minim	0.059194 cm^3
MASS AND WEIGHT — COAL, GRAIN, ETC.	avoirdupois	short ton	0.907 t
		long ton	1.016 t
		pound	0.454 kg
		ounce	28.350 g
		dram	1.772 g
		grain	0.0648 g
GOLD, SILVER, ETC.	troy	pound	0.373 kg
		ounce	31.103 g
		pennyweight	1.555 g
		grain	0.0648 g
DRUGS	apothecaries'	pound	0.373 kg
		ounce	31.103 g
		dram	3.888 g
		scruple	1.296 g
		grain	0.0648 g

METRIC TO ENGLISH EQUIVALENTS

	Unit	English Equivalent
LENGTH	kilometer	0.62 mi
	hectometer	109.36 yd
	dekameter	32.81′
	meter	39.37″
	decimeter	3.94″
	centimeter	0.39″
	millimeter	0.039″
AREA A = $l \times w$	square kilometer	0.3861 sq mi
	hectare	2.47 A
	acre	119.60 sq yd
	square centimeter	0.155 sq in.
VOLUME V = $l \times w \times t$	cubic centimeter	0.061 cu in.
	cubic decimeter	61.023 cu in.
	cubic meter	1.307 cu yd

CAPACITY (WATER, FUEL, ETC.; VEGETABLES, GRAIN, ETC.; DRUGS)		Cubic	Dry	Liquid
	kiloliter	1.31 cu yd		
	hectoliter	3.53 cu f.	2.84 bu	
	dekaliter	0.35 cu f.	1.14 pk	2.64 gal.
	liter	61.02 cu in.	0.908 qt	1.057 qt
	cubic decimeter	61.02 cu in.	0.908 qt	1.057 qt
	deciliter	6.1 cu in.	0.18 pt	0.21 pt
	centiliter	0.61 cu in.		338 fl oz
	milliliter	0.061 cu in.		0.27 fl dr

MASS AND WEIGHT (COAL, GRAIN, ETC.; GOLD, SILVER, ETC.; DRUGS)	Unit	English Equivalent
	metric ton	1.102 t
	kilogram	2.2046 lb
	hectogram	3.527 oz
	dekagram	0.353 oz
	gram	0.035 oz
	decigram	1.543 gr
	centigram	0.154 gr
	milligram	0.015 gr

Glossary

A

abscissa: The x-value of a trigonometric function.

acute angle: An angle that is fewer than 90°.

adjustable resistor: A resistor whose resistance value can be changed by moving one or more contacting elements.

admittance: The ability of a circuit containing both resistance and reactance to conduct current.

alternating current (AC): Current that reverses its direction of flow at regular intervals.

alternation: Half of a cycle.

alternator: *See* generator.

aluminum electrolytic capacitor: A capacitor composed of aluminum that contains an electrolyte.

ampacity: The maximum amount of current a conductor can carry continuously without exceeding its temperature rating.

ampere (A): One coulomb of electrons passing a given point in one second.

analog multimeter: An electromechanical device that indicates readings by the mechanical motion of a pointer.

angle theta: The phase difference between the source voltage and resistive voltage.

anode: A positive lead of a diode.

apparent power (P_A): The product of the voltage and current in a circuit, calculated without considering the phase shift that may be present between the voltage and current in the circuit.

arc-rated face shield: Any eye and face protection device that covers the entire face with a plastic shield and is used for protection from flying objects.

armature: The current-carryingg part of a DC generator.

armature reactance: The inductance of an armature.

atom: The smallest particle that an element can be reduced to and still maintain the properties of that element.

attenuation: A reduction in the strength of a frequency signal.

audio transformer: A transformer that is used to transfer complex signals containing energy at a large number of frequencies from one circuit to another.

autoranging DMM: A DMM that automatically adjusts to a higher range setting if the range is not high enough.

autotransformer: A transformer in which the primary and secondary circuits have portions of their two windings in common.

average value: In an AC voltage or current in a resistive AC circuit, the mathematical mean of all instantaneous values in a sine wave.

B

balanced resistance bridge: A Wheatstone bridge with the resistances adjusted so that there is zero potential across the bridge.

ballast resistor: A resistor whose resistance increases and decreases rapidly with changes in current.

balun: A special radio frequency transformer used to match impedance between a balanced 2-wire transmission line to an unbalanced transmission line or component.

band-pass frequency filter: A filter that passes signals between selected frequencies.

band-reject frequency filter: A filter that reduces signals between selected frequencies.

bandwidth: The range of frequencies that a circuit passes without a significant reduction in the signal magnitude.

battery: A DC voltage source that converts chemical energy to electrical energy.

bias tee: A three-port device that combines and separates radio frequency (RF) signals and direct current (DC) voltage.

bleeder resistor: A resistor that is used to discharge a capacitor.

break: A place on a switch contact that opens or closes a circuit.

brush: The sliding contact that rides against the commutator and is used to connect the armature to the external output circuit.

buck-boost transformer: A small transformer that is designed to lower or raise line voltage.

C

capacitance (C): The ability of a component or circuit to store energy in the form of an electrical charge.

capacitive reactance: The opposition to current flow by a capacitor (capacitance).

capacitor: An electrical device designed to store electrical energy by means of an electrostatic field.

capacitor motor: A single-phase AC motor that includes a capacitor in addition to the running and starting windings.

carbon composition resistor: A resistor that is constructed using carbon graphite mixed with clay.

cathode: A negative lead of a diode.

cell: A device that produces electricity at a fixed voltage and current level.

centrifugal switch: A switch that opens to disconnect the starting winding when the rotor reaches a preset speed and reconnects the starting winding when the speed falls below the preset value.

ceramic capacitor: A capacitor that uses ceramic-coated cylindrical tubes or flat metal disks as the capacitor and the dielectric.

cermet: A mixture of fine particles of glass or ceramic and powdered metals such as silver, platinum, or gold.

circuit breaker: An overcurrent protective device with a mechanical mechanism that manually or automatically opens a circuit when a short circuit or overload occurs.

circular mil (cmil): An area of a circle with a diameter of 1 mil and is the standard unit of wire cross-sectional area used by the AWG and the English wire tables.

clamp-on ammeter: A meter that measures current in a circuit by measuring the strength of the magnetic field around a single conductor.

closed circuit: An electrical circuit in which electrons flow uninterrupted from the negative terminal of the voltage source, through the load, and back to the positive terminal of the voltage source.

coefficient of coupling: **1.** The percent of magnetic lines of force created by one coil that cut the windings of the second coil. **2.** The effect of one coil inducing a voltage into another coil. Also known as mutual inductance.

coil: A circular wound wire (winding) consisting of insulated conductors arranged to produce lines of magnetic flux.

commutator: The part of an armature that connects each armature coil to the brushes by using copper bars (commutator segments) that are insulated from each other with pieces of mica.

compact fluorescent light: A fluorescent light that fits into an existing light bulb socket.

compensation winding: An inductor used to reduce armature reactance effects.

compound: A combination of the atoms of two or more elements.

conductance (G): The ability of voltage to produce electron flow through resistance.

conductor: A material that has low electrical resistance and permits electrons to move through it easily.

confined space: A space large enough for an employee to physically enter and perform assigned work but has limited or restricted means for entry and exit and is not designed for continuous employee occupancy.

contact: The conducting part of a switch that operates with another conducting part to make or break a circuit.

contactor: A control device that uses a small input current to energize or de-energize the load connected to it.

continuity: The presence of a complete path for current flow.

continuity checker: An instrument that indicates an open or closed circuit in a circuit in which all power is off.

continuity tester: A test instrument that tests for a complete path for current to flow.

conventional current flow: Current flow from positive to negative.

converter: A device that changes AC power to DC power and DC power to AC power.

core-type transformer: A transformer that has windings placed around each leg of the core material.

cosine of a right triangle (*cos*): A representation of the ratio of the length of the side adjacent to an acute angle to the length of the hypotenuse.

coulomb (C): The quantity of electric charge that passes any cross-section of a conductor in one second when the current is maintained at 1 A; a total of 6.241×10^{18} electrons.

counter-electromotive force (cemf): The property of a conductor to oppose any change in current.

critical coupling: The point separating a tight coefficient of coupling from a loose coefficient of coupling.

crossover network: A system of band-pass filters used to route the appropriate signal to the appropriate loudspeaker.

current (I): The flow of electrons through an electrical circuit.

current divider: A parallel circuit that divides current proportionally between branches.

current ratio: The ratio of the primary-winding current to the secondary-winding current.

current transformer: A transformer that is used to isolate an ammeter to prevent the hazards caused by connecting to high-voltage lines.

cutoff frequency: The half-power frequency point.

cycle: One complete positive and negative alternation of a waveform.

D

damped sine wave: A sine wave that becomes reduced due to resistance in consuming power.

delta connection: A connection in which each coil (winding) is connected end-to-end to form a closed loop.

diamagnetic material: A material only slightly repelled by a magnetic field.

diamagnetic substance: A substance with permeability less than 1 and that is repelled by either pole of a magnet.

dielectric: A nonconductor of DC current.

dielectric breakdown: The breakdown of insulation between the plates of a capacitor.

dielectric constant (k): The ratio of the capacitance of a capacitor with a given dielectric to the capacitance of a capacitor with air (vacuum) as the dielectric.

dielectric strength: The ability of a dielectric material to withstand voltage applied across it during a breakdown.

digital multimeter: An electronic device that indicates readings as numerical values.

diode: An electronic device that allows current to flow in only one direction.

dipped tantalum capacitor: A small capacitor that has tantalum dielectric and resembles the head of a match in shape.

direct current (DC): Current that flows in only one direction.

direct current (DC) generator: A generator that operates on the principle that when a conductor coil is rotated in a magnetic field, voltage is induced in the coil.

displacement current: Current that exists in addition to normal current.

double-break contacts: Contacts that break an electrical circuit in two places.

dyne: A force that produces an acceleration of one centimeter per second per second on 1 gram of mass.

E

eddy current: Unwanted current induced in the metal structure of a device due to the rate of change in the induced magnetic field.

eddy current loss: Power loss in a transformer or motor due to currents induced in the metal field structure from the changing magnetic field.

effective value: In an AC voltage or current, the value of a sine wave that produces the same amount of heat in a pure resistive circuit as DC of the same value.

efficiency: The ratio of the power output to the power input of a system.

electrical circuit: An assemblage of conductors and electrical devices through which electrons flow.

electrical shock: A condition that results any time a body becomes part of an electrical circuit.

electricity: The movement of electrons from atom to atom.

electric motor: A rotating device that converts electrical power into a rotating, mechanical force.

electrification: The process of charging an object.

electrolyte: A nonmetallic electric conductor in which current is carried by the movement of ions.

electromagnetic induction: The production of a potential difference across a conductor when it is exposed to a varying magnetic field.

electromagnet: A coil that has a core material made of soft iron.

electromagnetism: The magnetism produced when electricity passes through a conductor.

electromotive force (emf): Electrical pressure applied to a circuit.

electron: A negatively charged particle in an atom.

electron current flow: Current flow from negative to positive.

electrostatic discharge (ESD): The movement of electrons from a source to an object.

electrostatic precipitor: A device that uses electricity to remove particles from flue gases.

element: A substance that cannot be chemically broken down and contains atoms of only one variety.

end bell: *See* end frame.

end frame: The end part of a frame. Also known as an end bell.

energy: The ability to do work and is measured in the same units as work.

even harmonic: An even multiple of fundamental frequency (second, fourth, sixth, etc.).

exciting current: *See* magnetizing current.

F

farad (F): A unit equal to the capacitance of the accumulated charge of one coulomb for each volt applied.

ferromagnetic substance: A substance that has permeability greater than 1 as well as greater than a paramagnetic substance.

field winding: A magnet used to produce the magnetic field in a generator.

filter: A selective network of resistances, capacitances, and inductances that provides little opposition to certain frequencies or DC while blocking or attenuating other frequencies.

filter capacitor: A capacitor utilized with inductors and/or resistors for controlling harmonics problems in the power source, such as reducing voltage distortion due to large rectifier loads.

fixed resistor: A resistor that has only one resistance value.

flux density: The measure of the magnetic lines of force per unit area taken at right angles to the direction of flux.

flux linkage: The number of the flux lines (magnetic lines of force) linking the primary and secondary transformer windings.

flux loss: A power loss that occurs in a transformer when some of the magnetic lines of force from the primary winding do not travel through the core to the secondary winding.

flywheel diode: A diode that is used in a generator to suppress the inductive voltage that occurs when the exciter switch is opened.

force (F): Any cause that changes the position, motion, direction, or shape of an object.

forward breakover voltage: The voltage required to switch an SCR into a conductive state from a nonconductive state.

four-way switch: A doublepole, doublethrow (DPDT) switch that changes the electrical connections inside the switch from straight to diagonal.

frequency: The number of cycles per second (cps) in an AC sine wave.

frequency filter: A circuit that passes certain frequencies and reduces signal levels at other frequencies.

frequency filter network: A complex design of frequency filters arranged to improve filter performance.

friction: The resistance to motion that occurs when two surfaces move against each other.

full-wave rectifier: A device that converts both positive and negative portions of an AC waveform into positive values.

fuse: An overcurrent protective device with a fusible link that melts and opens a circuit when an overload condition or short occurs.

G

gas: Matter that has no definite volume or definite shape.

gauss: A unit of flux density equal to one magnetic flux line per square centimeter.

generated electricity: Electricity produced either by pressure (piezoelectricity), light (photocell), heat (thermocouple), chemical action (battery), or magnetism (generator).

generated voltage: A voltage produced in a closed path or circuit by the relative motion of the circuit or its parts with respect to magnetic lines of force.

generator: A machine that converts mechanical energy into electrical energy by means of electromagnetic induction. Also known as an alternator.

gilbert (Gb): The unit of measure in the cgs system for mmf.

goggles: An eye protection device with a flexible frame that is secured on the face with an elastic headband.

ground: A low-resistance conducting connection between electrical and electronic circuits, equipment, and the earth.

grounding: The connection of all exposed non-current-carrying metal parts to earth.

H

half-power point: A frequency in which the circuit power is half of the resonant power.

hard neutral position: A position in which brushes are aligned directly with stator poles.

harmonic: A frequency that is an integer (whole number) multiple (second, third, fourth, fifth, etc.) of the fundamental frequency.

heat sink: A device that conducts and dissipates heat away from an electrical component.

henry (H): The unit of inductance equal to the amount of voltage induced per the rate of current change.

hertz: The international unit of frequency equal to one cycle per second.

high induced voltage: An energy release caused by the coil in an inductive circuit.

high-pass frequency filter: A filter that allows all frequencies above a selected frequency to be applied to a load and blocks all frequencies below that point.

horsepower (HP): The unit of power equal to 746 W or 550 lb-ft per second (33,000 lb-ft per minute).

hypotenuse: The side of a right triangle that is opposite the right angle.

hysteresis: The property of ferromagnetic materials in which the magnetic induction of a coil (winding) lags the magnetic field that is charging the coil.

hysteresis loss: Loss caused by magnetism that remains (lags) in a material after the magnetizing force has been removed.

I

I^2R loss: *See* resistive loss.

ideal inductive circuit: A circuit that has an ideal inductor connected to a battery through a switch.

ignitron tube: A single-anode pool tube in which an igniter is used to initiate the cathode spot before each conducting period.

impedance (I): The total opposition of any combination of resistance, inductive reactance, and capacitive reactance offered to the flow of alternating current.

impedance matching: The process of adjusting load-circuit impedance to produce the desired energy transfer from a power source to a load. Also known as load matching.

impedance ratio: The ratio of the impedance of a primary winding to the impedance of a secondary winding.

incandescence: The emission of light energy from a substance when the substance is heated to a high temperature.

incandescent lamp: A device that produces light from the flow of current through a tungsten filament inside a sealed glass bulb.

induced voltage: A voltage produced around a closed path or circuit by a change in magnetic lines of force linking that path.

inductance (L): The property of a device or circuit that causes it to oppose a change in current due to energy stored in a magnetic field.

inductive kick: The energy release in a coil anytime the inductive current is abruptly stopped or reversed.

inductive reactance: The opposition of an inductor to alternating current.

instrument transformer: A transformer that steps down the voltage or current of a circuit to a low value that can be effectively and safely used for the operation of instruments such as ammeters, voltmeters, wattmeters, and relays used for various protective purposes, telemetering used for indications in remote areas, and dispatching energy.

insulator: A material that has a very high resistance to the flow of electrons.

intensity of magnetizing force: A measure of magnetic strength per unit length.

interrupting rating: The maximum amount of current that a fuse can safely interrupt without rupturing or arcing over.

ion: An electrically charged atom.

ionization: The separation of atoms and molecules into particles that have electrical charges.

isolation transformer: **1.** A transformer that has equal primary and secondary voltages and is used to electrically isolate the source from the load. **2.** A transformer that has complete electrical separation between the primary and the secondary windings.

J

jogging: The frequent starting and stopping of a motor for short periods.

joule (J): The unit for energy measurement.

K

kinetic energy: The energy of motion.

Kirchhoff's current law (KCL): A scientific law used in the evaluation of electrical circuits that states that the algebraic sum of all currents entering a branch point (junction) in a circuit must equal zero.

Kirchhoff's voltage law (KVL): A scientific law used in the evaluation of electrical circuits that states that the algebraic sum of the voltages around any closed-loop circuit must equal zero.

L

law of electric charges: A law that states that opposite charges attract and like charges repel.

leakage current: The small amount of current that flows through insulation.

leather protector: Gloves worn over rubber insulating gloves to prevent penetration of the rubber insulating gloves and to provide added protection against electrical shock.

length: The measurement of linear units; standard units of length measurement are the inch (in.), foot (ft), yard (yd), and mile (mi) in the English system and the millimeter (mm), centimeter (cm), meter (m), and kilometer (km) in the metric system.

Lenz's law: The basic principle used to determine the direction of an induced voltage or current.

light-emitting diode (LED): A semiconductor device that emits light energy when current flows through it.

lightning: A transient, high-current electrostatic discharge that occurs in the atmosphere.

lightning stroke: An initial electrostatic discharge and the return discharge.

liquid: Matter that has definite volume but not a definite shape.

load: A device that converts electrical energy to motion, light, heat, or sound.

load matching: *See* impedance matching.

lockout: The process of removing the source of power and installing a lock that prevents the power from being turned ON.

loose coefficient of coupling: A coefficient of coupling that is much less than 1.

low-pass frequency filter: A filter that allows all frequencies below a selected frequency to be applied to a load and blocks all frequencies above that point.

LRC circuit: *See* series inductive-resistive-capacitive circuit.

M

magnet: A substance that attracts iron and produces a magnetic field.

magnetic circuit: The path or paths taken by the magnetic lines of force leaving the north pole of a magnet and returning to the south pole of the magnet.

magnetic field: An invisible field produced by a current-carrying conductor, a permanent magnet, or the earth that develops a north and south polarity.

magnetic flux (ϕ): A measure of magnetic induction represented by magnetic lines of force.

magnetic lines of force: The invisible lines of force that make up a magnetic field.

magnetism: The invisible force exerted by a magnet.

magnetizing current: The current through a primary winding core when no load is on the secondary winding. Also known as exciting current.

magnetomotive force (mmf): The current through a coil multiplied by the number of turns on the coil.

mass: The measurement of matter contained in an object.

matter: Anything that has mass and occupies space.

maxwell (Mx): The centimeter-gram-second (cgs) unit of flux and is defined as one line of flux.

mechanical power: The rate of doing work through mechanical means.

mechanical resistance: Any force that tends to hinder the movement of an object.

megohmmeter: A device that detects insulation deterioration by measuring high resistance values under high-voltage conditions.

metal film resistor: A resistor that consists of a metal film of a selected metal alloy or carbon deposited in a vacuum on a small ceramic cylinder to which leads are attached.

meter loading: Inaccurate readings due to a meter's resistance being in parallel with the component under test.

mica capacitor: A capacitor that is composed of metal foil plates separated by layers of dielectric material.

mil: One thousandth of an inch (0.001″).

molded inductor: An inductor that is encapsulated in plastic or other insulating materials with a conductor at each end for connecting or soldering into a circuit.

molecule: The smallest particle that a compound can be reduced to and still possess the chemical properties of the original compound.

motor: A rotating device that converts electrical energy into mechanical energy.

multilayer coil: A coil that has more than one layer of turns and is wound on an air-core form.

multimeter: A test tool used to measure two or more electrical values.

multiple-winding transformer: A transformer that has more than two windings on its core.

mutual inductance: The effect of one coil inducing a voltage into another coil. Also known as coefficient of coupling.

mutual induction: The ability of an inductor in one circuit to induce a voltage in another circuit.

N

National Electrical Code® (NEC®): A standard on practices for the design and installation of electrical products.

National Fire Protection Association (NFPA): A national organization that provides guidance in assessing the hazards of the products of combustion.

negative static charge: The accumulation of excessive electrons on a body.

negative temperature coefficient: The result of a decrease in the resistance of a material with an increase in temperature.

neutron: A neutral particle with a mass approximately the same as a proton and exists in the nucleus of an atom.

nodal equation: A network equation based on Kirchhoff's current law.

node: A junction point in an electrical circuit where connections are made between different circuit paths.

noninductive coil inductor: An inductor that has coils wound in such a manner that the inductance created by one coil turn is canceled by an adjacent coil turn.

normally closed (NC) contacts: Contacts that are closed before being activated.

normally open (NO) contacts: Contacts that are open before being activated.

north pole: The pole of a magnet that points to the geographic north pole of the Earth.

Norton's theorem: A method of circuit analysis in which a complex circuit is reduced to a circuit with one current source and a parallel resistance.

O

odd harmonic: An odd multiple of fundamental frequency (third, fifth, seventh, etc.).

ohm: The amount of resistance in a circuit when 1 ampere (A) flows with 1 volt (V) applied.

ohmmeter: An instrument used to measure the continuity and resistance of a circuit or component.

Ohm's law: The relationship between voltage (*V*), current (*I*), and resistance (*R*) in an electrical circuit.

oil capacitor: A capacitor that uses oil or oil-impregnated paper as a dielectric.

open circuit: An electrical circuit in which the flow of electrons is interrupted.

optocoupler: A device that normally consists of an LED as the input stage and photodiode or phototransistor as the output stage of a fiber-optic system.

ordinate: The y-value of a trigonometric function.

overcurrent: A condition that exists in an electrical circuit when the normal load current is exceeded.

overcurrent protective device (OCPD): A fuse or circuit breaker used to provide overcurrent protection in a circuit.

overload: A condition that occurs when circuit current rises above the current level at which a load and/or circuit is designed to operate.

overload relay: An overload device that responds to electrical loads and operates at a preset value.

P

padder capacitor: A small adjustable capacitor that is connected in series in the tuning circuit of a radio.

pancake coil: A short coil with only a few conductor turns along its length.

paper capacitor: A capacitor that has thin sheets of metallic foil, which act as plates, separated by waxed paper, which acts as a dielectric.

parallax error: The inaccuracy created by the difference in apparent position of an analog meter pointer when viewed from different angles not directly perpendicular to the pointer and scale.

parallel circuit: A circuit that has more than one current path.

parallel resonant circuit: A circuit in which the circuit elements are connected in parallel and the inductive and capacitive reactance are equal (as in a series resonant circuit).

parallelogram: A four-sided plane figure with opposite sides parallel and equal.

parallelogram calculation method: A graphical method of calculation that uses parallelogram values to perform vector calculations.

paramagnetic material: A partially magnetic material.

paramagnetic substance: A substance with a permeability greater than 1.

peak-to-peak value: The value measured from the maximum positive alternation to the maximum negative alternation of a sine wave.

peak value: The maximum instantaneous value of either the positive or negative alternation of the sine wave of an AC voltage or current in a resistive circuit.

peltier effect: The absorption or emission of heat at the junctions of two dissimilar metals when electrons flow through the junctions.

peltier heat: The heat either emitted or absorbed at the junction of two dissimilar metals.

period: The time required to produce one complete cycle of a waveform.

permanent magnet: A magnet that retains its magnetism after the magnetizing force has been removed.

permeability (μ)**:** The ability of a material to carry magnetic lines of force.

permeance: The property of a core material that allows the passage of magnetic lines of force.

personal protective equipment (PPE): Clothing and/or equipment worn by a technician to reduce the possibility of injury in a work area.

phase (ϕ)**:** The time relationship of a sine or sound wave to a known time period.

phase shift: The state in which voltage and current in an AC circuit reach their maximum amplitudes and zero levels at different times.

phasor: A vector in which length represents the magnitude of an electrical parameter and direction represents phase angle in electrical degrees.

photocell: *See* photoconductive cell.

photoconductive cell: A transducer that conducts current when energized by light. Also known as a photocell.

photoconductive effect: A change in the electric conductivity of a solid or liquid due to light striking the material.

photodiode: A diode that is switched on and off by a light.

photoelectric effect: The conversion of light energy to electrical energy.

photoelectron: An electron freed by light energy.

photoemissive effect: The release of electrons from a material (normally metal) by radiant energy (light) striking the surface of the material.

phototransistor: A device that combines the effect of a photodiode and the switching capability of a transistor.

photovoltaic cell: A semiconductor device that converts light energy directly to electrical energy.

photovoltaic effect: The production of a voltage potential caused by the absorption of photons across a junction region of a semiconductor.

piezoelectric effect: The electrical polarization of some material when mechanically strained.

piezoelectricity: The generation of voltage by the application of pressure.

piezoelectric transducer: A transducer that operates based on the interaction between the deformation of certain materials and an electric charge.

pigtail: An extended, flexible connection or a braided copper conductor.

pitch (of a coil): How tightly the windings of a coil are wound.

plate: The conductor in a capacitor.

plug fuse: A fuse that has a screw-in base like that of an incandescent lamp.

plugging: A method of motor braking in which the motor connections are reversed so that the motor develops a countertorque that acts as a braking force.

polarity: The positive (+) or negative (–) state of an object.

pole: The number of completely isolated circuits that a relay can switch.

poor power factor: The condition of a circuit in which power factor is less than 90%.

positive static charge: A deficiency of electrons on a body.

positive temperature coefficient: The result of an increase in the resistance of a material and increase in temperature.

potential: The voltage at a point in a circuit with respect to another point in the same circuit.

potential difference: The algebraic difference in potential (voltage) between two different points in a circuit.

potential energy: Stored energy that a body has due to its position, chemical state, or condition.

potential transformer: A precision two-winding transformer that is used to step down high voltage to allow safe voltage measurement.

potentiometer: A low-power three-terminal variable resistor.

power: The rate of doing work or using energy.

power-correcting capacitor: A capacitor used to improve a facility's power factor by improving voltage levels, increasing system capacity, and reducing line losses.

power curve: A graph that shows the changes in circuit power as AC parameters change over time.

power factor (PF): The ratio of the true power used in an AC circuit to apparent power delivered to the circuit.

power factor correction: The process of correcting power factor to be between 90% (0.90) and 97% (0.97).

power formula: The relationship between power, voltage, and current in an electrical circuit.

power loss: The difference between power input and power output.

power quality meter: A test instrument that measures, displays, and records voltage, current, and power as well as power problems such as sags, swells, transients, and harmonics.

power resistor: A resistor with a power rating over 2 W.

power transformer: A transformer that is used to raise or lower voltage as required to serve transmission or distribution circuits.

practical inductive circuit: A circuit that has both resistance and inductance.

primary winding: The winding (coil) of a transformer that draws power from the source.

protective helmet: A hard hat that is used in the workplace to prevent injury from the impact of falling and flying objects and from electrical shock.

proton: A positively charged particle in the nucleus of an atom.

pure capacitive circuit: A theoretical circuit without any resistance.

pure inductive circuit: An AC circuit with an inductor and no resistance.

Pythagorean theorem: A theorem that states that the square of the hypotenuse of a right triangle is equal to the sum of the squares of the other two sides.

Q

qualified person: A person who has knowledge and skills related to the construction and operation of electrical equipment and has received appropriate safety training.

quality factor: The ratio of the inductive reactance to the resistance of an inductor.

R

radio frequency (RF) choke: An inductor designed for high impedance over large frequency ranges.

radio frequency (RF) transformer: A transformer used to transfer energy in narrow frequency bands from one circuit to another.

radius vector: The radius value of a trigonometric function.

reactive circuit: A circuit that is either capacitive or inductive and has a power factor of less than 1.

reactive power: Power supplied to a reactive load.

rectification: The process of changing AC to DC.

rectifier: A device that converts AC voltage to DC voltage by allowing the AC voltage and current to flow in only one direction.

reflected impedance: The primary-load impedance that is proportional to the secondary-load impedance.

reflected load: The primary impedance of a transformer that is directly proportional to the secondary load.

reject frequency: Frequency that is rejected because of amplitude or size.

relative conductance: The ability of a specific conductor to carry electrons as compared to the ability of a copper conductor to carry electrons.

relative motion: The speed at which a conductor cuts a magnetic field.

relative permeability: The ability of a material to conduct magnetic lines of force as compared to air, which has a permeability value of 1.

relative resistance: A comparison of the resistance of a given material to the resistance of copper.

reluctance (R): The property of an electric circuit that opposes magnetic lines of force.

reluctivity: The reluctance per cubic centimeter of a material.

repulsion motor: A motor with the rotor connected to the power supply through brushes that ride on a commutator.

resistance: The opposition to the flow of electrons.

resistance heating element: A conductor that offers enough resistance to produce heat when connected to an electrical power supply.

resistive-capacitive time constant: The amount of time required for a capacitor to charge to 63.2% of the maximum voltage across a resistive-capacitive circuit.

resistive circuit: An electrical circuit that contains only resistance.

resistive loss: Power loss in a transformer caused by the resistance of the copper conductors used to make the windings. Also known as I^2R loss.

resistivity: The resistance of a material of a specific cubic size. Also known as specific resistance.

resistor: An electrical device designed to introduce a specific amount of resistance into a circuit.

resonance: The condition in which inductive reactance (X_L) equals capacitive reactance (X_C) at a given frequency.

resultant vector: The single vector that is calculated by drawing a vector from the tail of the first vector to the head of the last vector.

retentivity: The ability of a core material to retain its magnetism after the magnetizing force is removed.

rheostat: A two-terminal variable resistor.

right angle: An angle that is 90°.

right triangle: A triangle with a right angle.

ripple: A change in DC output voltage level.

ripple voltage: The portion of the output voltage of a power source that is related in frequency to the power source.

root-mean-square voltage value (rms value): The voltage value of a sine wave that produces the same amount of heat in a pure resistive circuit as DC of the same value.

rotary switch: A switch that has one or more poles that can be connected to several positions.

rotor: The rotating part of an AC motor.

rotor plate: The movable plate of a variable capacitor.

rubber insulating gloves: Gloves made of latex rubber and are used to provide maximum insulation from electrical shock.

S

safety glasses: An eye protection device with special impact-resistant glass or plastic lenses, reinforced frames, and side shields.

salient pole: A pole that consists of a separate radial projection having its own iron pole and field coil.

second: A measured duration of time based on cycles of radiation measured with a spectrometer.

secondary winding: The winding (coil) of a transformer that delivers energy at the transformed, or changed, voltage to the load.

Seebeck effect: The production of a voltage potential from a temperature difference between two joined electrical conductors composed of different metals.

self-inductance: The property of a conductor to induce voltage within itself due to changes in current.

semiconductor: A material that exhibits electrical conductivity between that of a conductor (high conductivity) and that of an insulator (low conductivity).

series circuit: A circuit that has only one current path.

series inductive-resistive-capacitive circuit: A circuit in which all inductive, resistive, and capacitive circuit elements are connected such that there is only one current path. Also known as an LRC circuit.

series/parallel circuit: A circuit that contains a combination of both series-connected and parallel-connected components.

shaded pole: A single pole, normally made of a single turn of heavy-gauge copper wire, that is used for starting a single-phase motor and is added to a slot cut into the motor pole.

shaded-pole motor: A single-phase AC motor that uses a shaded stator pole for starting.

shielded inductor: An inductor contained in a shield composed of low-reluctance magnetic material brought to ground potential.

short circuit: Any circuit in which current takes a shortcut around the normal path of current flow.

shunt resistor: A parallel-connected resistor used to increase the range of an ammeter.

silicon-controlled rectifier (SCR): A solid-state rectifier with the ability to rapidly switch currents.

sine of a right triangle (*sin*): A representation of the ratio of the length of the side opposite an acute angle to the length of the hypotenuse.

single-break contacts: Contacts that break an electrical circuit in one place.

single-layer coil: A coil that has only one layer of turns and is wound on an air-core form.

skin effect: The effect that occurs in AC when more current flows near the outer surface (skin) of a conductor and less flows near the center of a conductor at higher frequencies.

slip: The difference between the synchronous speed and the actual speed of a motor.

slip rings: The metallic rings that are connected to the ends of an armature and are used to connect the induced voltage to the brushes.

soft neutral position: A position in which brushes are aligned 90° from stator poles.

solid: Matter that has a definite volume and shape.

sound: Energy that consists of pressure vibrations in the air.

speaker: A device that converts electrical energy into vibrations (sound waves).

specific resistance: *See* resistivity.

split-phase motor: A single-phase AC motor that includes a running winding (main winding) and a starting winding (auxiliary winding).

static charge: The accumulation of excessive electrons on a body.

static electricity: An electrical charge at rest.

stator: The part of a generator that remains in a stationary position.

stator plate: The stationary plate of a variable capacitor.

step-down transformer: A transformer that has a secondary voltage less than the primary voltage.

step-up transformer: A transformer that has a secondary voltage greater than the primary voltage.

stranded conductor: A conductor composed of several strands of solid wire wrapped together to make a single conductor.

sulfidation: The formation of film on a contact surface.

supercapacitor: A capacitor that offers high capacitance in a small volume.

superconductor: A material that exhibits zero resistance and neither opposes electron flow nor consumes power at temperatures close to absolute zero (0° K, −273.15°C, −459.67°F).

superposition theorem: A method of analyzing a circuit with multiple voltage sources by analyzing one voltage source at a time and combining the individual effects to determine circuit parameters.

susceptance: The ability of a reactive circuit to conduct current, measured in siemens (S).

swamping resistor: A resistor that is used to suppress the high inductive kick of an exciter circuit.

switch: A mechanical, electronic, or solid-state electrical device that is used to start, stop, or redirect the flow of electrons in an electrical circuit.

synchronous motor: A three-phase motor that runs at synchronous speed.

synchronous speed: The theoretical speed of a motor based on line frequency and the number of poles of the motor.

T

tagout: The process of placing a danger tag on the source of power, which indicates that the equipment may not be operated until the danger tag is removed.

tangent of a right triangle (*tan*): A representation of the ratio of lengths of the side opposite and adjacent to an acute angle.

tank circuit: An inductive-capacitive circuit that can store electric energy over a band of frequencies that are continuously distributed on a resonant single frequency. Also known as a tuned circuit.

tantalum: A hard metallic material that is resistant to acid.

tapped-winding transformer: A transformer with a single tapped winding that is used to convert voltages in residential areas.

temperature: The measurement of the intensity of heat.

temporary magnet: A magnet that loses its magnetism when the magnetizing force is removed.

thermal conductivity: The higher amount of heat produced the faster the temperature increases in the body being heated.

thermionic emission: The release of electrons from a solid or liquid material based on the heat energy of the material.

thermistor: A device that changes resistance in response to a change in temperature.

thermocouple: A temperature sensor that consists of two dissimilar metals joined at the end where the heat is measured and produces a voltage output at the other end proportional to the measured temperature.

thermoelectricity: The electrical energy produced by the action of heat.

thermopile: An array of thermocouples connected in series, parallel, or series/parallel to provide a higher voltage and/or current output than an individual thermocouple.

Thevenin's theorem: A method of circuit analysis that reduces a complex circuit to one voltage source with a series resistance.

Thevenizing: The process of simplifying a complex circuit into an equivalent circuit containing a single voltage source and resistance connected in series.

three-way switch: A single-pole, double-throw (SPDT) switch.

throw: The number of closed contact positions per pole.

thyristor: A solid-state switching device that switches current ON using a quick pulse of control current.

tight coefficient of coupling: A coefficient of coupling close to 1.

time constant: The time required for the current in an inductive-resistive circuit to reach 63.2% of its maximum value after power is applied to the circuit or to decrease by 63.2% (to 36.8% of maximum power) when power is removed from the circuit.

toroid coil: A doughnut-shaped coil with a closed circular or rectangular core.

transducer: A device that is actuated by power from one system and supplies power in the same or different form to a second system.

transformer: An electrical device that uses electromagnetism to change voltage from one level to another or to isolate one voltage from another.

transformer efficiency: The ratio of a transformer's output power to its input power.

transformer schematic symbol: A symbol used to identify transformers in drawings and prints.

triboelectric effect: The generation of a static charge by friction.

trigonometry: A branch of mathematics in which the relationships between the lengths of the sides of a triangle and the angles of a triangle are used to perform calculations.

trimmer capacitor: A capacitor used with a larger variable capacitor to fine-tune the overall capacitance of the circuit.

true power (P_T): The actual power used in an electrical circuit.

true power rating: The amount of power a transformer can deliver to a resistive load.

tuned circuit: *See* tank circuit.

turns ratio: The ratio of the number of turns in the primary winding to the number of turns in the secondary winding. Also known as turns-to-turns.

turns-to-turns: *See* turns ratio.

two-voltage source T-circuit: A T-shaped circuit with two voltage sources.

two-way switch: A single-pole, single-throw (SPST) switch.

U

unbalanced resistance bridge: A Wheatstone bridge with fixed resistances and the voltage across the bridge proportional to the temperature of the variable resistor.

unit magnetic pole: A force of one dyne between two magnetic poles separated by a distance of 1 cm.

unity power factor: A power factor equal to 1.

universal motor: An AC series motor that has brushes and a wound armature.

utilization voltage: A secondary voltage of transformers applied to rated loads.

V

valence electrons: Electrons in the outermost shell of an atom.

variable capacitor: A capacitor that varies in capacitance value.

variable inductor: An inductor in which the inductance is varied by a core that can be moved into and out of the center of the coil.

variable resistor: A resistor whose resistance value can be changed to any value within its range.

varicap capacitor: A variable solid-state diode that operates under reverse bias.

variometer: An inductor that consists of two coils, one fixed and one movable.

vector: A quantity involving direction and magnitude (scalar value) represented by a straight line with an arrowhead on one end.

vector diagram calculation method: A graphical method of calculation that uses vector values to perform vector calculations.

voltage (V): The amount of electrical pressure in a circuit.

voltage divider: A circuit that is constructed by connecting resistors in series to produce a desired voltage drop across the resistors.

voltage drop: The amount of voltage consumed by a component as current passes through it.

voltage phasor: A vector diagram used to show the magnitude and phase of voltage in a series inductive-resistive AC circuit.

voltage ratio: The ratio between primary-winding voltage and secondary-winding voltage.

volt ampere (VA): The unit of measure for apparent power.

volts per turn (V/turn): The voltage dropped across each turn of a winding or the voltage induced into each turn of a secondary winding.

W

wall transformer: A large box-shaped plug used with many electronic devices that consists of a transformer and prongs for connection to an electrical receptacle.

watt: The equivalent of one joule per second in the metric system.

wattmeter: A test instrument that is used to measure true power in a circuit.

wavelength (λ): The distance covered by one complete cycle of a given frequency of sound as it passes through the air.

weber (Wb): The unit of magnetic lines of force in the meter-kilogram-second (mks) measurement system, or Système International (SI).

weight: A measure of gravity or the force of the Earth's attraction for a body.

Wheatstone bridge: A circuit used to take precise measurements of resistance.

wire wound resistor: A resistor constructed using various types of wire wound on an insulating form.

work (W): The application of a force over a distance.

working voltage: The maximum voltage that can be safely applied to a capacitor.

wye connection: A connection that has one end of each coil connected together and the other end of each coil left open for external connections.

Y

yoke: The center portion of a frame in which field coils are mounted.

Index

**Page numbers in italic refer to figures.*

D

G

H

I

Q

R

S

T

U

V

W

Y